MULTINUCLEAR NMR

MULTINUCLEAR NMR

EDITED BY
JOAN MASON
The Open University
Milton Keynes, Buckinghamshire, England

PLENUM PRESS • NEW YORK AND LONDON

Library of Congress Cataloging in Publication Data

Multinuclear NMR.

Includes bibliographies and index.
1. Nuclear magnetic resonance spectroscopy. I. Mason, Joan, date.
QD96.N8M85 1987 543'.0877 87-12284
ISBN-13: 978-1-4612-8999-9 e-ISBN-13: 978-1-4613-1783-8
DOI: 10.1007/978-1-4613-1783-8

First Printing – August 1987
Second Printing – October 1989

Softcover reprint of the hardcover 1st edition 1987
A Division of Plenum Publishing Corporation
233 Spring Street, New York, N.Y. 10013

CONTRIBUTORS

J. W. Akitt School of Chemistry, University of Leeds, Leeds LS2 9JT, England

Keith R. Dixon Department of Chemistry, University of Victoria, Victoria, British Columbia V8W 2Y2, Canada

R. J. Goodfellow Department of Inorganic Chemistry, University of Bristol, Bristol BS8 1TS, England

Oliver W. Howarth Department of Chemistry, University of Warwick, Coventry CV4 7AL, England

Cynthia J. Jameson Department of Chemistry, University of Illinois, Chicago, Illinois 60680

John D. Kennedy Department of Inorganic and Structural Chemistry, University of Leeds, Leeds LS2 9JT, England

Brian E. Mann Department of Chemistry, University of Sheffield, Sheffield S3 7HF, England

Joan Mason Department of Chemistry, The Open University, Milton Keynes MK7 6AA, England

H. C. E. McFarlane Department of Chemistry, Sir John Cass School of Science and Technology, City of London Polytechnic, London EC3N 2EY, England

W. McFarlane Department of Chemistry, Sir John Cass School of Science and Technology, City of London Polytechnic, London EC3N 2EY, England

Henry W. E. Rattle Biophysics Laboratories, Portsmouth Polytechnic, Portsmouth PO1 2DT, England

Dieter Rehder Institute of Inorganic and Applied Chemistry, University of Hamburg, D-2000 Hamburg 13, Federal Republic of Germany

PREFACE

With the power and range of modern pulse spectrometers the compass of NMR spectroscopy is now very large for a single book—but we have undertaken this. Our book covers the Periodic Table as multinuclear spectrometers do, and introductory chapters are devoted to the essentials of the NMR experiment and its products. Primary products are chemical shifts (including anisotropies), spin–spin coupling constants, and relaxation times; the ultimate product is a knowledge of content and constitution, dynamic as well as static.

Our province is chemical and biochemical rather than physical or technical; only passing reference is made to metallic solids or unstable species, or to practical NMR spectroscopy. Our aim is depth as well as breadth, to explain the fundamental processes, whether of nuclear magnetic shielding, spin–spin coupling, relaxation, or the multiple pulse sequences that have allowed the development of high-resolution studies of solids, multidimensional NMR spectroscopy, techniques for sensitivity enhancement, and so on.

This book therefore combines the functions of advanced textbook and reference book. For reasonably comprehensive coverage in a single volume we have summarized the information in tables and charts, and included all leading references. Special problems are posed by the very well-studied elements, and by the enormous literature of organic and biochemical NMR, from which *in vivo* and medical NMR studies are now taking off. Different strategies have been adopted in the various chapters, including a "multidisciplinary" approach (e.g., coverage of inorganic and physical aspects for hydrogen and organometallic aspects for carbon), and where possible, summarizing principles, as for nitrogen, fluorine, and phosphorus. The different chapters provide references to the relevant biochemical literature, and this is the subject of our final chapters. We owe a particular debt of gratitude to the NMR reference literature listed in Chapter 1, which is constantly cited throughout the text.

The four decades of NMR spectroscopy have seen successive spurts of development following technical breakthroughs. Our book celebrates the exploration by high- (or fairly high-) resolution NMR spectroscopy, during the last decade, of more complex systems, more aspects of the solid state, and more regions of the periodic table. As sensitivity and resolution steadily improve, with higher field working and other advances in instrumentation, the nuclei with lower resonance frequencies or natural abundance, and/or larger quadrupole moments—which constitute so much of the NMR periodic table—are brought into the net.

The next decade will see further application of techniques for sensitivity enhancement, spectral simplification, and better recovery of spectral information content (perhaps by multidimensional NMR and multiple quantum NMR). It will also see the extension of NMR imaging of living systems to other nuclei as well as the proton, and to include spectroscopy. Thus the chemist or other scientist will have an even greater need for handbooks of first resort which can provide a quick résumé of a branch of the subject. We hope our book will fulfill this role for some time to come.

Joan Mason

Milton Keynes, U.K.

FOREWORD

The history of nuclear magnetic resonance spectroscopy is one of the most important, fascinating, intellectually satisfying, and, as if all that were not enough, humanly relevant in all of science. Begun by physicists, soon preempted by chemists, then embraced by biologists and most recently by physicians, NMR is a field of seemingly endless fecundity. It has, deservedly, attracted some of the most ingenious physical chemists of the postwar era.

As early as the 1920s, the existence of nuclear magnetic moments, which the quantum theory allowed one to attribute to the spin of charged nuclei, was recognized by Pauli. In the 1930s the elegant molecular beam experiments of Rabi yielded much detailed information. However, the real beginning of NMR as we know it today was in the work of Purcell, Torrey, and Pound at Harvard and Bloch, Hansen, and Packard at Stanford. In 1946, these groups independently reported that the NMR of protons could be detected in bulk materials such as paraffin and water. In 1952 Bloch and Purcell received the Nobel Prize in physics for these discoveries.

For chemists, the seminal period was the era 1949–1951 (just before I began my own graduate study), when the recognition of the chemical shift, followed by the design and manufacture of high-resolution (40 MHz for ^{1}H) spectrometers, put an awesome new instrumental tool into their hands. For many years, this new tool was mainly of interest to organic chemists, although there were some early applications of proton NMR spectroscopy to inorganic problems, as well as relatively early use of ^{19}F, $^{10}B/^{11}B$, and ^{31}P spectroscopy by inorganic chemists. The advent of practical, routine ^{13}C spectroscopy was largely driven by the needs of organic and biological chemists, but it also opened new doors for inorganic chemists through its applications to metal carbonyls and organometallics.

However, it is the last 10 to 15 years that have seen the real flowering of NMR as a tool for the inorganic chemist. This is the result of two main developments. One is the increasing capacity of commercially available instruments to handle most of the magnetic nuclides in the periodic table in a routine fashion. The other, which is still growing in importance, is the body of techniques that allow solids to be studied at relatively high resolution.

This book, which has been planned, edited, and partially written by one of the leaders in the field of inorganic NMR, should be of tremendous value to all who face the challenge of taking maximum advantage of NMR in their inorganic research. From my examination of the proofs, I would say it is *the* one for taking to that fabled desert island where you can have only one book on any given subject. It contains a succinct but comprehensive summary of the principles, written crisply and with style, as well as "wall-to-wall" coverage of the periodic table. It should become the reference of first and last resort for all users and would-be users of multinuclear NMR.

F. A. Cotton
Department of Chemistry
Texas A&M University
College Station, Texas 77843

CONTENTS

Chapter 4
Spin–Spin Coupling
Cynthia J. Jameson

Chapter 1

INTRODUCTION

JOAN MASON

So as to give a comprehensive account of multinuclear NMR spectroscopy, Chapter 2 describes the essentials of the NMR experiment and the spectroscopic parameters so obtained, Chapters 3 and 4 discuss the principles of nuclear magnetic shielding and spin–spin coupling, respectively, and Chapter 5, relaxation and dynamic processes in general. Chapters 6–21 then cover the Periodic Table from hydrogen to the transition metals and the noble gases, treating egregious elements singly and others in groups. Chapters 22 and 23 afford an introduction to the rapidly growing fields of bioinorganic and biomedical NMR, respectively, with bibliography. Symbols, abbreviations, units and physical constants are given at the end of the book; and nuclear properties of the elements, including NMR receptivities and recommended reference substances, in the Appendix.

In covering the Periodic Table we have tried to summarize voluminous (and burgeoning) literature information in tables and charts, providing as many references as possible, and certainly all key references. Many of the reported ranges of chemical shifts (or coupling constants) are growing apace as new groups of compounds are studied. Rapid growth, as for example in the direct study of many metal nuclei, has necessitated updating during the prolonged gestation of this book, and there is some diversity in effective dates of literature coverage in different chapters.

The organization of the chapters varies, reflecting differences in chemistry and in the NMR history of the nuclei. For long-studied nuclei such as ^{19}F or ^{31}P mainly general principles are discussed; for nitrogen, NMR criteria of structure in ligands and functional groups, and also relaxation processes, given the technical problems (and opportunities) afforded by ^{14}N and ^{15}N; for ^{13}C, organo groups in metal complexes; and for $^{1,2,3}H$, inorganic and physical aspects. In some newer fields more complete coverage is possible.

No specific attention is devoted to practical NMR spectroscopy or instrumentation, which have been ably treated.[1–4] We acknowledge also classics of NMR spectroscopy[5–9] and other general NMR books,[10–17] and salute our colleagues who have aimed at a target similar to our own.[18–20] Other important texts are mentioned in appropriate chapters. We express our gratitude to the invaluable review series,[21–30] and to the NMR Abstracts.[31] Finally a tribute is in order to the achievements and promise of our chosen technique—for example, what other physical method could uncover the "chimpanzee swing" of a water molecule, from oxygen to oxygen, in sodium alumina?[32]

JOAN MASON • Department of Chemistry, The Open University, Milton Keynes MK7 6AA, England.

REFERENCES

1 Martin, M L , Delpuech, J-J , Martin, G J *Practical NMR Spectroscopy*, Heyden London, 1980
2 Shaw, D *Fourier Transform NMR Spectroscopy*, 2nd ed , Elsevier Amsterdam, 1984, 304 pp
3 Brevard, C , Granger, P *Handbook of High Resolution Multinuclear NMR*, Wiley New York, 1981, 229 pp
4 Fukushima, E , Roeder, S B W *Experimental Pulse NMR A Nuts and Bolts Approach*, Addison-Wesley Reading, Massachusetts, 1981, 539 pp
5 Pople, J A , Schneider, W G , Bernstein, H J *High-resolution NMR*, McGraw-Hill New York, 1959, 501 pp
6 Emsley, J W , Feeney, J , Sutcliffe, L H *High Resolution NMR Spectroscopy*, Pergamon Oxford, 1965, Vol 1, 662 pp, Vol 2, 1154 pp
7 Carrington, A , McLachlan, A D *Introduction to Magnetic Resonance*, Harper New York, 1967, reissued 1979, 266 pp
8 Slichter, C P *Principles of Magnetic Resonance*, 2nd ed , Springer-Verlag Berlin, 1978, 397 pp
9 Abragam, A *The Principles of Nuclear Magnetism*, Oxford University Press, corr reprint 1978, 599 pp
10 Lynden-Bell, R M , Harris, R K *NMR Spectroscopy*, Nelson London, 1969, 160 pp
11 McLauchlan, D A *Magnetic Resonance*, Oxford Chemistry Series Oxford, 1972, 105 pp
12 Akitt, J W *NMR and Chemistry*, 2nd ed , Chapman and Hall London, 1983, 263 pp
13 Gunther, H *NMR Spectroscopy*, Wiley Chichester, 1980, 436 pp
14 Becker, E D *High Resolution NMR*, 2nd ed , Academic Press New York, 1980, 354 pp
15 Harris, R K *NMR Spectroscopy A Physicochemical View*, Pitman London, 1983, 250 pp
16 Sanders, J K M , Hunter, B K *Modern NMR Spectroscopy A Guide for Chemists*, O U P Oxford, 1987, 354 pp
17 Kemp, W *NMR in Chemistry a Multinuclear Introduction*, Macmillan London, 1986, 240 pp
18 Harris, R K , Mann, B E , Eds , *NMR and the Periodic Table*, Academic Press London, 1978, 459 pp
19 Lambert, J B , Riddell, F G , Eds , *The Multinuclear Approach to NMR Spectroscopy*, NATO ASI Series, Reidel Dordrecht, 1983, 548 pp
20 Laszlo, P , Ed , *NMR of Newly Accessible Nuclei Chemical and Biochemical Applications*, Academic Press New York, 1983, Vol 1, 298 pp, Vol 2, 436 pp
21 Waugh, J S , Ed , *Adv Magn Reson* , Academic Press New York, 1965–
22 Mooney, E F , Webb, G A , successive Eds , *Ann Rep NMR Spectrosc* Academic Press New York, 1968–
23 Berliner, L J , Reuben, J , Eds , *Biol Magn Reson* , Plenum New York, 1978–
24 Bradbury, J H , Ed , *Bull Magn Reson* , Franklin Institute Press Philadelphia, 1979–
25 C P Poole, Ed , *Magn Reson Rev* , Gordon and Breach, London, 1977–
26 Diehl, P , Fluck, E , Kosfeld, R , Eds , *NMR—Basic Principles and Progress*, Springer-Verlag Berlin, 1969–
27 Kaufman, L , Crooks, L E , Margulis, A R , Eds , *NMR Imaging in Medicine*, Igaku-Shoin New York, 1981–
28 Harris, R K , Abraham, R J , Webb, G A , successive Eds , *Nucl Magn Reson* , *Spec Per Rep Chem Soc London*, 1972–
29 Emsley, J W , Feeney, J , Sutcliffe, L H , Eds , *Progr NMR Spectrosc* , Pergamon, Oxford, 1966–
30 Greenwood, N N , Ebsworth, E A V , Adams, D M , Davidson, G , successive Eds , *Spectrosc Props Inorg Organomet Cpds* , *Spec Per Rep Chem Soc London*, 1968–, Mann, B E "NMR Spectroscopy," p 1
31 NMR (Chemical Aspects), CA Selects, Chemical Abstracts, Columbus, Ohio, Nottingham, U K
32 Kuhns, P L , Richter, L J , Conradi, M S *J Chem Phys* 1982, **76**, 6–9

Chapter 2

THE PARAMETERS OF NMR SPECTROSCOPY

CYNTHIA J. JAMESON and JOAN MASON

1. NUCLEAR PROPERTIES OF THE ELEMENTS AND THE RESONANCE CONDITION

Many nuclei have spin angular momentum, since protons and neutrons have this property, although the nucleons couple to produce nuclei without spin when an even number of each kind is present, as in ^{12}C, ^{16}O, ^{28}Si, ^{32}S, etc. The nuclear angular momentum has an associated magnetic field, comparable to the magnetic field produced by an electric current in a loop. A nuclear spin thus behaves as a *magnetic dipole*, which tends to align with an applied magnetic field and to interact with neighboring dipoles. The *magnetic moment* $\boldsymbol{\mu}$, which determines the potential energy E of the nuclear magnetic dipole in a magnetic field of strength (or flux density) $\mathbf{B}_0$, is defined by

$$E = -\boldsymbol{\mu} \cdot \mathbf{B}_0 \tag{1}$$

Nuclear spin angular momentum, given by the vector I, has total magnitude $\hbar[I(I+1)]^{1/2}$, where I is the *nuclear spin quantum number*, and $I\hbar$ is the maximum observable component of $\mathbf{I}$ in any selected direction (since not all of the angular momentum is along any one direction). The observable components of $\mathbf{I}$ are $M_I h$, where M_I is the magnetic quantum number and takes the values I, $(I-1)$,..., $(-I+1)$, $(-I)$. Nuclei such as ^{12}C with $I=0$ are nonmagnetic and so cannot be observed by NMR. Magnetic nuclei have half-integral values of I (1/2, 3/2, 5/2,...) corresponding to an odd number of nucleons, or, less commonly, integral values of I corresponding to an odd number both of protons and of neutrons (^{2}H, ^{6}Li, ^{10}B, ^{14}N, ^{50}V,...). All nuclei with $I > \frac{1}{2}$ behave as *electric quadrupoles* as well as magnetic dipoles.

The magnetism of a nucleus is usefully described in terms of its *magnetogyric ratio* γ, the proportionality constant which relates the nuclear magnetic moment μ to the nuclear spin angular momentum I:

$$\gamma = \boldsymbol{\mu}/\mathbf{I} = \boldsymbol{\mu}/\hbar[I(I+1)]^{1/2}$$

CYNTHIA J. JAMESON • Department of Chemistry, University of Illinois, Chicago, Illinois 60680. JOAN MASON • Department of Chemistry, The Open University, Milton Keynes MK7 6AA, England.

or

$$\gamma = \mu / I\hbar \tag{2}$$

where μ is the maximum observable component of the vector $\boldsymbol{\mu}$ (and is the value often quoted).[1] For most magnetic nuclei $\boldsymbol{\mu}$ and $\mathbf{I}$ are parallel vectors and γ is positive, but some nuclei (^{15}N, ^{17}O, ^{29}Si) have $\boldsymbol{\mu}$ and $\mathbf{I}$ antiparallel (as for the electron) and γ negative, which can sometimes be disadvantageous.

Nuclear magnetic moments μ are expressed in units of μ_N, the nuclear magneton:

$$\mu = g_N \mu_N I = \gamma \hbar I$$

where

$$\mu_N = e\hbar/2m_p = 5.0508 \times 10^{-27}\ \text{J T}^{-1}$$

and e is the elementary charge, m_p the proton mass, and g_N the *nuclear g factor*, the counterpart of the g factor (with values close to 2) relating an electron's magnetic moment to its spin angular momentum. The magnetic moment of an electron is the Bohr magneton

$$\mu_B = e\hbar/2m_e = 9.2740 \times 10^{-24}\ \text{J T}^{-1}$$

where m_e is the electron mass. Thus the proton, the smallest nucleus, has a magnetic moment which is 658 times smaller than that of the electron. Physical consequences of this disparity are, for example, the use of radiofrequencies for NMR as opposed to microwave frequencies for ESR spectroscopy, and the large perturbations of NMR systems if unpaired electrons are present.

From equations (1) and (2), and the quantization of $\mathbf{I}$ into observable components $M_I\hbar$ along the direction of the magnetic field $\mathbf{B}_0$, we have

$$E = -\gamma \hbar M_I B_0 \tag{3}$$

The $(2I+1)$ values of M_I ranging from $-I$ to $+I$ thus correspond to different nuclear magnetic energy levels separated by the Zeeman splitting $\gamma\hbar B_0$, as shown in Figure 1 for a nucleus with $I = 3/2$. In NMR spectroscopy we induce transitions of the nuclear spin between these levels by the use of electromagnetic radiation, and monitor the absorption. The magnitudes of magnetogyric ratios and practical field strengths B_0 are such as to require radiofrequency (rf) radiation, with frequency (ν) in the range

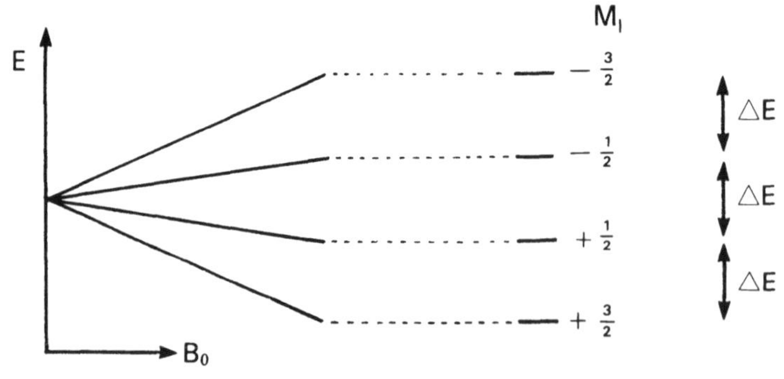

Figure 1. Zeeman splitting for a nucleus with spin $I = 3/2$, $\gamma > 0$.

50–500 MHz or so. The selection rule is $\Delta M_I = \pm 1$, so that for a single bare nucleus all the transitions have the same energy ΔE as in Figure 1:

$$\Delta E = h\nu = \gamma \hbar B_0 \tag{4}$$

Equation (4) is the *resonance condition* which has to be satisfied by the applied frequency in order for absorption to be a maximum. The frequency $\nu_0 = \gamma B_0/2\pi$ is the *Larmor frequency* for that nuclear species in the field B_0.

In NMR as in other forms of spectroscopy net absorption of electromagnetic radiation can occur owing to the greater population of the lower energy levels compared to the upper ones, at thermal equilibrium. But the very small splitting of nuclear energy levels, about 10^{-25} J, makes NMR much less sensitive than optical spectroscopy. The Boltzmann law gives the ratio of populations

$$n_2/n_1 = \exp(h\nu_0/kT) = \exp(\gamma\hbar B_0/kT) \simeq 1 + \gamma\hbar B_0/kT.$$

Thus for ^{13}C in a field of 2.35 T at ambient temperatures the ratio is 1.000004. Luckily measurements at radiofrequencies can be made with very great accuracy.

The signal strength is proportional to the net magnetization, which is given by[2]

$$M = N\gamma^2\hbar^2 I(I+1)\, B_0/3kT$$

for a collection of N nuclear moments. In addition the voltage induced in the detector system increases as the resonance frequency. Depending on the instrumentation, the *sensitivity to NMR detection* increases with increasing γ^3, I^2, and $B_0^{3/2}$. For comparing different nuclei, the *receptivity* R is defined as the product of the natural abundance A of the magnetic isotope (expressed as a percentage) and the sensitivity at constant field, taken as proportional to $\gamma^3 I(I+1)$.[1,2] The values given in the Appendix are those of R^C, relative to that of ^{13}C:

$$R^C = |\gamma^3|\, AI(I+1)/(\gamma_C^3 A_C 3/4) = 3.9472 \times 10^{-28}\, |\gamma^3|\, AI(I+1)/\text{rad}^3\,\text{T}^{-3}\,\text{s}^{-3}\,\%$$

To overcome the intrinsically low sensitivity, NMR signals are accumulated from a series of repeated experiments. Since the signal-to-noise ratio increases only as $\sqrt{n}$, where n is the number of experiments, a factor x in receptivity corresponds to a factor x^2 in time of accumulation. Of peculiar importance is the rate of reestablishment of the Boltzmann distribution (thermal equilibrium) since the repetition rate depends on this. NMR spectroscopy differs from optical spectroscopy in not being able to rely on spontaneous emission of radiation, since the probability of this depends on ν^3 and is extremely low at radiofrequencies. (When the populations of the nuclear magnetic energy levels become equalized, the NMR signal is said to be *saturated*.) Nor can reestablishment of Boltzmann populations be as fast as the molecular collisions which relax vibrational and rotational states. In NMR only local magnetic fields fluctuating at the Larmor frequency can induce radiationless transfers between nuclear magnetic energy levels (*relaxation*).

2. THE NUCLEUS IN A CHEMICAL ENVIRONMENT

The main parameters of NMR spectroscopy, the nuclear magnetic shielding, the indirect spin–spin (J) coupling, and the direct dipolar (D) coupling, are all tensor quantities, since the environment in which the nucleus finds itself is generally not spherically symmetric. Each parameter therefore consists of a 3×3 array of tensor components, in principle. In practice, depending on the symmetry, some of the numbers may be equal and some zero (and in any case only the diagonal elements are important in the analysis of spectra). The NMR parameters are tensors because they are molecular properties arising from the interaction of the nuclear spin vector **I** with another vector, i.e., with $\mathbf{B}_0$ (shielding), with **I** via a nuclear electric quadrupole (quadrupole coupling), or with another nuclear spin $\mathbf{I}'$ (D or J coupling). If all these couplings were evident in the NMR spectrum this would be very complicated indeed. It is greatly simplified in liquid or gas phase measurements (although a wealth of information is then lost) because the disordered motion of a molecule, with rapid rotations interrupted by frequent and random collisions, allows it to sample all possible orientations with equal probability, on the NMR time scale. The *isotropic average* of the tensor is equal to one-third of the trace, which is the sum of the components along the diagonal of the array. The direct dipolar coupling **D** and the quadrupole coupling $e^2\mathbf{q}Q$ are isotropically averaged to zero (their tensors have zero trace) so only the averaged value of the shielding and the (indirect) spin–spin coupling constants contribute to the positions of the NMR spectral lines in fluid phase spectra.

2.1. The Chemical Shift and the Shielding Tensor

Much of the chemical information in NMR spectra arises from chemical shifts due to the local magnetic fields generated at the nucleus by the circulations of the surrounding electrons induced by the applied field. If the resulting magnetic fields oppose the applied field ($\mathbf{B}_0$) the effect is to shield or screen the nucleus from it, and vice versa. The nucleus then experiences an effective field given by

$$\mathbf{B}_{\text{eff}} = \mathbf{B}_0(1 - \boldsymbol{\sigma})$$

where $\boldsymbol{\sigma}$, the nuclear magnetic shielding, is expressed in units of 10^{-6}, or parts per million (ppm). The $\boldsymbol{\sigma}$ tensor elements reflect the symmetry of the electronic environment of the nucleus, depending on the bond type, an averaged value (σ_{av}) being observed in fluid phases.

The resonance condition of equation (4) then becomes

$$\nu_0 = \gamma B_0(1 - \boldsymbol{\sigma})/2\pi$$

The shielding constant $\boldsymbol{\sigma}$ is given by quantum mechanical calculations relative to the bare nucleus, which is not a practical reference. Chemical shifts are commonly measured relative to a standard substance, such as tetramethyl silane (TMS), used for ^{1}H, ^{13}C, and ^{29}Si. The chemical shift is defined by IUPAC convention[(3)] as

$$\delta/\text{ppm} \equiv 10^6(\nu_s - \nu_{\text{ref}})/\nu_{\text{ref}} \tag{5}$$

which can be related to the shielding difference in the limit $\sigma \ll 1$. Then $\delta \simeq \sigma_{\text{ref}} - \sigma_s$, where the subscript s refers to the sample and ref to the reference. The opposite signs of σ (shielding) and δ (shift) can be a source of confusion in the literature. Before 1972 shifts were commonly defined with the opposite sign and more recently there has been a tendency to report shift tensors with either sign, which is unfortunate.

In the *continuous wave* (CW) mode, resonances can be observed with the frequency held constant and the field varied (*field sweep*). Observation of a more shielded nucleus then requires a higher field, giving rise to terms such as "high-field resonance" for higher shielding and vice versa. Usage has now been standardized to refer to the *frequency-sweep* mode, with the field held constant, so that higher shielding should be described in terms of lower frequencies rather than higher fields.

Since the use of reference substances poses practical problems, an alternative which is gaining ground is to refer to an absolute frequency. On the Ξ scale the standard frequency for a given nucleus (X) is the one observed at a field strength (2.35 T) in which TMS protons (under standard conditions) resonate at exactly 100 MHz. (Ξ is the upper case Greek letter xi.) Thus

$$\Xi_X = 100(\gamma_X/\gamma_H)(1 - \sigma_X)/[1 - \sigma_{^1H}(\text{TMS})]$$

The chemical shift is then

$$\delta_X/\text{ppm} = 10^6(\Xi_s - \Xi_r)/\Xi_X$$

and Ξ_X values for the elements are given in the Appendix. For some nuclei (^{195}Pt for example) an arbitrary frequency has been taken as reference because of problems with standard solutions.

This frequency in practice is not as "absolute" as we should like. TMS shifts are temperature dependent, and temperatures are not always accurately reproduced. Further, a bulk susceptibility correction which depends on the shape of the container and its orientation relative to the magnetic field must be made (Section 8.1).

The shielding tensor for a linear molecule has only three nonzero components, of which only two are independent. Clearly, the electronic circulations induced in such a molecule will be different if it is placed parallel or perpendicular to the magnetic field. In this case there are two independent quantities, $\sigma_{\|}$ and $\sigma_{\perp}$, which characterize the nuclear shielding. In a nuclear site with no symmetry at all there are nine different components of $\boldsymbol{\sigma}$,[4-6] but the off-diagonal components have negligible effect on the spectrum (except perhaps for the heaviest nuclei). In fluid phases only the isotropic average is observed:

$$\sigma_{av} = (\sigma_{11} + \sigma_{22} + \sigma_{33})/3$$

but the separate elements can be observed in the spectra of oriented systems (cf. Section 6) or molecular beams.[4,7] Single-crystal studies can give the tensor elements (relative to a standard or to σ_{av}), and the orientation of the principal axes. Polycrystalline samples give characteristic line shapes (as in Figure 2a) from which the tensor elements can be read off or obtained by computer simulation. Figure 2b shows a powder pattern for a nucleus in an axially symmetric location with $\sigma_{11} = \sigma_{22} = \sigma_{\perp}$, and $\sigma_{33} = \sigma_{\|}$, the *anisotropy* being defined as $\Delta\sigma = \sigma_{\|} - \sigma_{\perp}$, or more generally, $\Delta\sigma = \sigma_{33} - (\sigma_{22} + \sigma_{11})/2$, the elements being conventionally taken as $\sigma_{33} \geqslant \sigma_{22} \geqslant \sigma_{11}$. The *asymmetry* is defined as $\eta = (\sigma_{22} - \sigma_{11})/(\sigma_{33} - \sigma_{av})$, which is zero for axially symmetric systems.

Substituent effects in organic molecules are well defined by ^{13}C tensor elements. Thus for the carbonyl group RCOX, where X = H, R, OH, OR etc., σ_{11} and σ_{33} vary rather little, the sensitivity to the substituent X being concentrated in σ_{22}.[8] Similarly, experimental values of tensor elements afford a more stringent test of calculated shieldings than do the isotropic shifts (Chapter 3).

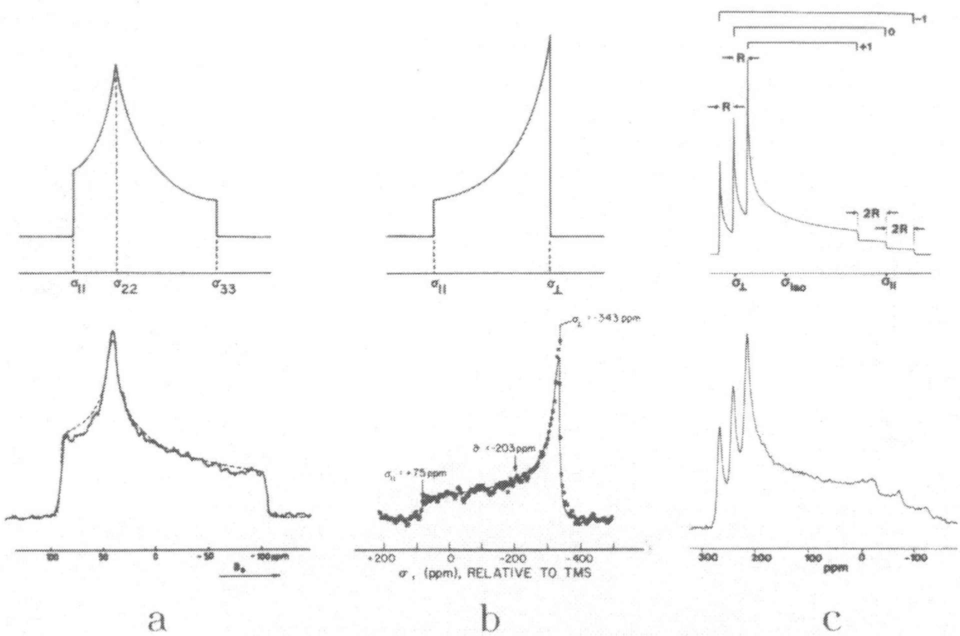

Figure 2. Line shapes for polycrystalline samples showing principal elements of the shielding tensor. (a) Asymmetric (^{31}P in a diethylphosphonate); (b) axially symmetric (^{13}C in [$Mo(CO)_6$]); (c) axially symmetric, with dipolar coupling to one spin $I=1$ (^{13}C in NH_4NCS). Spectra reproduced with permission from *J. Chem. Phys.* 1983, **78**, 5384; *Chem. Phys. Lett.* 1985, **122**, 545; and R. E. Wasylishen.[8a]

2.2. Dipolar (D) Coupling

The presence of neighboring magnetic nuclei alters the local field and therefore the energies of a nucleus. The direct dipolar interaction energy between any two magnetic moments $\boldsymbol{\mu}_1$ and $\boldsymbol{\mu}_2$ separated by a vector $\mathbf{r}$ is

$$E_D = (\boldsymbol{\mu}_1 \cdot \boldsymbol{\mu}_2)/r^3 - 3(\boldsymbol{\mu}_1 \cdot \mathbf{r})(\boldsymbol{\mu}_2 \cdot \mathbf{r})/r^5$$

and for nuclear magnetic dipoles this is

$$E_D = \gamma_1 \gamma_2 \hbar^2 \{(\mathbf{I}_1 \cdot \mathbf{I}_2)/r^3 - 3(\mathbf{I}_1 \cdot \mathbf{r})(\mathbf{I}_2 \cdot \mathbf{r})/r^5\}$$

which is written in abbreviated form as

$$E_D = \gamma_1 \gamma_2 \hbar^2 \mathbf{I}_1 \cdot \mathbf{D} \cdot \mathbf{I}_2$$

where $\mathbf{D}$ is the direct (through space) dipolar coupling tensor. We can see that $\mathbf{D}$ depends only on the relative position coordinates of nuclei 1 and 2.

The effect of the dipolar interaction on the NMR spectrum depends on the angle θ which the internuclear vector makes with the magnetic field $\mathbf{B}_0$. From the $\mathbf{I}_1 \cdot \mathbf{D} \cdot \mathbf{I}_2$ interaction six different kinds of terms arise, of which only two contribute in first order, both having the factor $(3\cos^2\theta - 1)/r^3$. The r^{-3} dependence means that the dipolar coupling is strong at close approach and falls off quite steeply with distance.

The same factor appears in the classical picture, in which the dipolar field B_D at a distance r from a magnet with moment μ_z in the field direction is given by

$$B_D = (\mu_0/4\pi)\,\mu_z(3\cos^2\theta - 1)/r^3 \tag{6}$$

The isotropic average of $3\cos^2\theta - 1$ over all orientations is zero, so the dipolar interactions do not contribute to line positions in a liquid phase NMR spectrum (cf. the fact that the trace of D is zero).

In solids or liquid crystals in which nuclear positions are maintained relative to the applied field, the *direct* interaction of the magnetic dipoles may be observed in the NMR spectrum. A magnetic nucleus X with spin I has $(2I+1)$ orientations in the magnetic field, splitting the resonance of a neighbor Y at a distance r into a $(2I+1)$ multiplet, the splitting depending on $\gamma_X\gamma_Y(3\cos^2\theta - 1)/r^3$. For example, the $+1$, 0, -1 components of ^{14}N split the resonance of ^{13}C in NCS^- ion in Figure 2c, and from the apparent dipolar coupling constant $R = 1295$ Hz, the effective C–N bond length $\langle r^{-3}\rangle^{-1/3} = 1.19 \pm 0.01$ A is obtained.[8a] Because of their high γ, protons have large dipolar coupling constants, and splittings may be 10^2–10^5 Hz. Usually the resonances are greatly broadened by unresolved couplings to neighbors with a range of distances and orientations.

2.3. Indirect Spin–Spin (*J*) Coupling

A coupling between nuclei which does not average to zero in liquids is the indirect spin–spin coupling or J coupling; in fact the total (dipolar) interaction between nuclear magnetic moments is $\gamma_1\gamma_2\hbar^2\mathbf{I}_1\cdot(\mathbf{J}+\mathbf{D})\cdot\mathbf{I}_2$. For oriented molecules the components of $(\mathbf{J}+\mathbf{D})$ are evident in the spectrum. In isotropic phases only the average of $\mathbf{J}$ is observed, and this is nonzero because the interaction is by way of the electrons in the bonds between the nuclei. Whereas $\mathbf{D}$ contains structural information in the form of internuclear distances, which may be intermolecular, J contains bond-related structural information. Thus the observation of a J coupling is proof positive of a chemical bond that is long-lived on the NMR time scale.

Indeed, important information on connectivity within a molecule is carried by the symmetrical multiplet splittings (in high-resolution spectra) due to J coupling of nearby nonequivalent nuclear spins. (The coupling of equivalent spins, with the same chemical shift, splits the energy levels but not the signals, as the transitions have the same energy.) The splittings (J/Hz, where J is the coupling constant) are independent of the applied field, as expected for an intramolecular interaction. Indirect (e.g., double irradiation) methods show that the coupling constant can be positive or negative, but the sign is not normally evident in the NMR spectrum.

There are three main mechanisms for indirect or J coupling, all mediated by the valence electrons. The most important involves the *Fermi contact interaction* of an s electron with the nucleus (with which p, d,... electrons have no contact). This can be envisaged in terms of the Dirac vector model.[9] For simplicity we can consider the HF molecule, since both nuclei have spin 1/2 and positive magnetogyric ratios γ. The bonding pair of electrons *must* have antiparallel spins ($\alpha\beta$) by the Pauli principle, and their motions are correlated such that if the one with α spin is near one nucleus, the one with β spin is likely to be near the other. Within each nucleus the magnetic moments of the nucleus and the electron are more stable when parallel, so that the spins are *antiparallel* (since the electron has a negative γ). The nuclear spins are therefore coupled, having a lower energy when antiparallel than when parallel, in which case the difference hJ is defined as positive (Figure 3). For HF, J has the rather large value of $+530$ Hz, (reflecting the large magnetogyric ratios of ^{1}H and ^{19}F) observed in the molecular beam.

A two-bond coupling, as of ^{1}H and ^{19}F in CH_3F, arises because the bonding electrons on carbon are the more stable when their spins are parallel, as in $H\overset{\alpha\beta}{-}C\overset{\beta\alpha}{-}F$, by Hund's rule of maximum multiplicity. Thus the *parallel* configuration of the ^{1}H and ^{19}F spins is now the more stable, corresponding to a negative sign for $^2J_{HF}$. This model therefore suggests that ^nJ coupling constants are positive for odd values of n and negative for even n, and that ^nJ decreases with increase in n (as the Hund rule is not exclusive). These conclusions have some general validity, particularly for alkanes, but the exceptions show that the reality is more complicated than the model.

The other spin–spin coupling mechanisms involve interactions mediated by p, d,... electrons. The *orbital term* J^o_{AB} arises from the perturbation of the magnetic field due to the electron's orbital motion by the nucleus A, this perturbation being experienced by the nucleus B, and vice versa. The *spin-dipolar term* J^{sd}_{AB} arises from the direct interaction of the magnetic dipole of the nucleus A with that of the orbital electron, which then interacts with nucleus B, and vice versa. The observed coupling constant is then the sum of the *Fermi contact term* J^{Fc}, which usually dominates, and J^o and J^{sd}, which are of course insignificant for hydrogen. These components can be evaluated only by theoretical calculations, which show the importance of nonbonding as well as of σ and π bonding electrons (J^o and J^{sd} are larger when there is multiple bonding).[10]

For all coupling mechanisms the contribution to J_{AB} is proportional to the product $\gamma_A\gamma_B$. Thus the coupling constant for H_2 can be obtained from that of HD, to which it is related by the ratio γ_H/γ_D. For comparisons involving different nuclei it is convenient to use a *reduced coupling constant* K, defined as

$$K_{XY} = 4\pi^2 J_{XY}/h\gamma_X\gamma_Y$$

where J is in Hz and K in N A^{-2} m^{-3}.

Spin–spin coupling in HD produces a doublet in ^{2}D resonance but a triplet in ^{1}H resonance, as the proton interacts with three orientations of the deuteron ($I=1$). All components of the multiplet have near-equal intensity because of the near-equal populations of the nuclear energy levels of the deuteron. Similarly for a nucleus A, coupled to two X spins with $I=1$, the 9 combinations of M_I for the X spins are

	1, 1	1, 0 0, 1	1, −1 −1, 1 0, 0	−1, 0 0, −1	−1, −1
giving $\sum M_I$:	2	1	0	−1	−2
Number of combinations:	1	2	3	2	1

The result is a quintet with intensities in the ratio 1:2:3:2:1.

In general, the coupling of any number of A nuclei to n equivalent I spins gives a multiplet with $(2nI+1)$ lines in the A spectrum. For $I=1/2$ the relative intensities are given by the coefficients of the binomial expansion of $(a+b)^n$ and by the Pascal triangle, and correspondingly for $I>1/2$. The numbers in a row of the triangle give the relative intensities, for n equivalent X spins. Each number in the triangles (for $n>0$) can be obtained as the sum of the $(2nI+1)$ nearest neighbors in the line above, with zero for empty spaces.

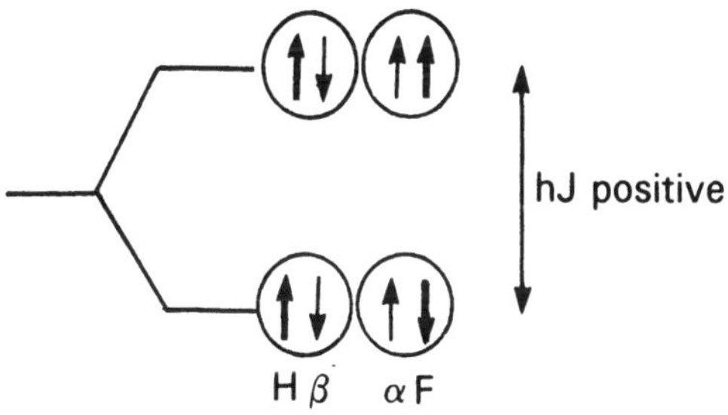

Figure 3. Splitting of ^1H energy level in HF by spin–spin coupling to ^{19}F, mediated by the Fermi contact interaction.

n	$I=1/2$	$I=1$	$I=3/2$
0	1	1	1
1	1 1	1 1 1	1 1 1 1
2	1 2 1	1 2 3 2 1	1 2 3 4 3 2 1
3	1 3 3 1	1 3 6 7 6 3 1	1 3 6 10 12 12 10 6 3 1
4	1 4 6 4 1	1 4 10 16 19 16 10 4 1	1 4 10 20 31 40 44 40 31 20 10 4 1

These simple rules break down if the chemical shift in frequency units between the coupled nuclei becomes comparable with J. The spectrum is then called "second order": there are shifts of intensity from the wings to the center, compared with a first-order spectrum, and further splittings.(11) The calculations of high-resolution spectra in the isotropic phase are a beautiful and straightforward application of quantum mechanics, and have been worked out in detail for systems of spin $\frac{1}{2}$ (12,13) and for quadrupolar nuclei.(14–16) Analytical solutions of this problem are in smaller demand now that higher-field working can restore first-order conditions.

Spin–spin coupling constants are different for members of a chemically equivalent set if *magnetic nonequivalence* is present. In $CF_2=CH_2$, for example, $^3J_{HF}$(cis) is different from $^3J_{HF}$(trans), because of the difference in coupling pathways. The spectra of magnetically nonequivalent systems are complex even in the limit of very high fields (see Figure 4). Computer calculations can be used to determine the chemical shifts and the magnitudes (even the relative signs in some cases) of J by iterative fitting of frequency separations and intensities.(18,19)

2.4. Electric Quadrupole Coupling

All nuclei with $I>1/2$ have an ellipsoidal distribution of charge, as shown in Figure 5, and an electric quadrupole moment eQ; values of Q are given in the Appendix. Q is positive if the nucleus is prolate (lengthened) in the direction of its spin angular momentum, negative if oblate (flattened). The unit is 10^{-28} $m^2=1$ barn (after Fermi's exclamation at the first measurement of a nuclear cross section, that it was as big as a barn). Indeed, a nucleus with $I=n/2$ may have a 2^n moment which is magnetic for n odd, and electric for n even. Thus nuclei with $I\geqslant 3/2$ may have a magnetic octupole and those with $I\geqslant 2$ an electric hexadecapole moment, and so on, although effects of these are rarely observed.

Electrostatic energy is minimized by appropriate alignment of a quadrupole in a field *gradient* (in contrast to dipolar energy, which depends directly on the field). In a molecule, there is an *electric field gradient* (*efg*) at the nucleus because of asymmetry

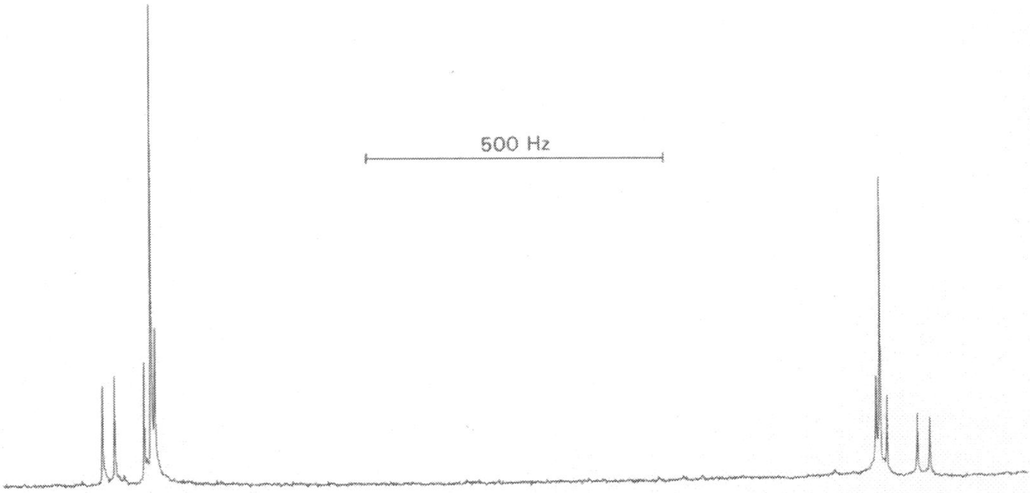

Figure 4. Example of magnetic inequivalence in a chemically equivalent system.[17] The ^{31}P spectrum of *cis*-$[PdCl_2\{PF(OPh)_2\}_2]$ in which the protons are decoupled is that of an $[AX]_2$ spin system, inequivalent because the P–F couplings are unequal. The spectrum is drastically different from the simple 1:2:1 triplet which would have been observed for an A_2X_2 spin system.

in the local charge distribution due to the electrons and other nuclei. The energy of a nuclear quadrupole is quantized according to its orientation in the efg even in the absence of an external magnetic field. Transitions are induced by a radiofrequency, in so-called "pure quadrupole" or *nuclear quadrupole resonance*, NQR (Figure 6). But since transitions of one multipole change the electromagnetic environment of another, the magnetic dipole and electric quadrupole are strongly coupled. When a quadrupolar nucleus is placed in a magnetic field so that the nuclear spin is quantized along the magnetic field direction rather than the efg symmetry axis, the nuclear magnetic energy levels depend on both the efg and the field $\mathbf{B}_0$.

Like $\boldsymbol{\sigma}$, **J**, and **D**, the efg is a tensor, and like **D** it is traceless: the isotropic average of energy terms involving the efg is zero. Thus in the liquid phase the positions of the lines in the NMR spectrum are not affected by the nuclear quadrupole coupling; NMR (and NQR) studies of quadrupolar splittings must be done in the solid state, often at low temperature. In the liquid phase the changes in the local efg with molecular motions induce transitions between the different M_I states, and this is *quadrupolar relaxation* (Section 2.5.4). If very rapid, it causes broadening of the NMR line, which is related to Heisenberg's uncertainty principle. Because of the very

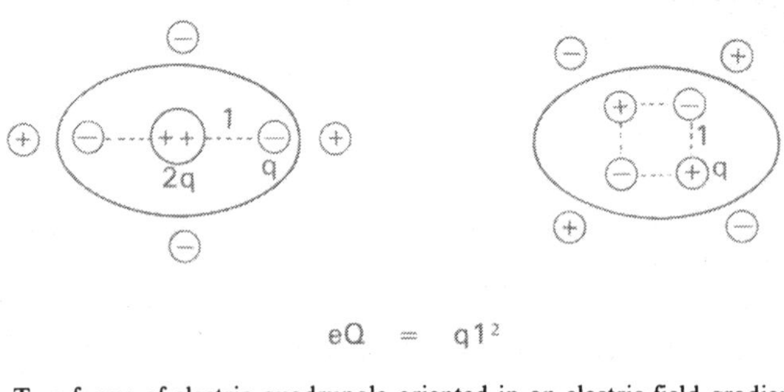

Figure 5. Two forms of electric quadrupole oriented in an electric field gradient. $eQ = ql^2$.

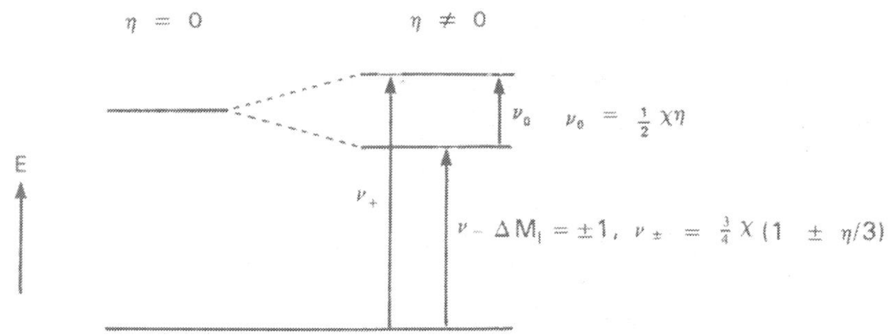

Figure 6. "Pure quadrupole" transitions for a nucleus with $I = 1$ where χ is NQCC.

frequent transitions, the lifetime δt in the excited state is short, and this gives an uncertainty δE in the energy:

$$(\delta E)(\delta t) \geqslant \hbar/2$$

which corresponds to an uncertainty in the frequency ($\delta E = h\,\delta\nu$). The magnitude of the quadrupolar interaction is therefore of great importance to NMR spectroscopy.

The *nuclear quadrupole coupling constant* (NQCC) is a tensor quantity related to the efg tensor. In frequency units

$$\mathrm{NQCC} = e^2 q_{zz} Q/h$$

where q_{zz} is the maximum component of the efg tensor **q** with principal components in the order $|q_{zz}| \geqslant |q_{yy}| \geqslant |q_{xx}|$. The efg tensor is specified by all three components, or else by the largest, q_{zz}, and an asymmetry parameter which lies between 0 and 1, being defined by

$$\eta = (q_{xx} - q_{yy})/q_{zz}.$$

Since the NQCC depends not only on the nuclear quadrupole moment Q, but also on the chemical environment of the nucleus, quadrupole couplings can vary greatly for the same nucleus. Moreover, the NQCC tensor contains useful chemical information on the disposition of bonding and nonbonding electrons which determine the efg. Thus for ^{14}N, with $I = 1$, the NQCC is small (a few kHz) in the symmetrical environment of the ammonium ion $^{14}NH_4^+$, small also in linear groups (as in $R-N\equiv C$), but larger (about 9 MHz) in very asymmetric locations, as in NHF_2.

NQCC values generally range from 10^6 to 10^9 Hz, tending to be larger for heavier nuclei. For small molecules e^2qQ and η can be measured as hyperfine splittings in molecular beam magnetic or electric resonance, or by microwave spectroscopy, since the nuclear spin is coupled to the angular momentum of the rotating molecule. Quadrupole splitting is observed also in Mössbauer spectroscopy. NQR spectroscopy has become a more sensitive technique in recent years.[20] Pulsed double resonance methods can give high-resolution spectra for light nuclei such as ^2H or ^{17}O, even in natural abundance, and even for frozen liquid samples which are not amenable to diffraction methods because of the lack of long-range order. Thus organic and biomolecules can now be studied, as well as heavier nuclei, with NQR usefully complementing NMR spectroscopy.

In solid-state NMR spectroscopy the quadrupolar interaction, if less than 100 kHz, say, can be taken as a perturbation of the Zeeman interaction (the "high-field case").[(2)] Perturbation theory then gives the contribution to the energy (for axial symmetry of the efg, with $\eta = 0$) as follows:

$$\text{in first order, } E^{(1)} = e^2qQh(3\cos^2\theta - 1)[3M_I^2 - I(I+1)]/8I(2I-1);$$

in second order, $E^{(2)}$ is a more complex expression in $(e^2qQ)^2/\gamma$ and M_I^3. Whereas the (zero-order) resonance condition for a quadrupolar nucleus gives a single line at the Larmor frequency ν_0 for $\Delta M_I = \pm 1$, the *first-order quadrupolar interaction* gives a series of $2I$ lines with quadrupolar shifts ν_Q such that (for $\eta = 0$)

$$\nu = \nu_0 + \nu_Q \text{ for the transition } M_I \rightarrow M_I - 1$$

where

$$\nu_Q = (-3e^2qQ/4)(3\cos^2\theta - 1)(M_I - \tfrac{1}{2})/I(2I-1)$$

Thus a spin-1 nucleus (^2H, ^6Li, ^{14}N) gives a doublet with the splitting (for $\eta = 0$)

$$\Delta\nu_Q = (3e^2qQ/4)(3\cos^2\theta - 1) = 3e^2qQ/2 \qquad \text{for} \quad \theta = 0^\circ$$

For half-integral spins ($I = 3/2, 5/2, \ldots$) the frequency of the central line $(+\frac{1}{2} \rightarrow -\frac{1}{2})$ is unchanged to first order, since the expression for ν_Q vanishes if $M_I = \frac{1}{2}$; the other two lines appear as symmetrical satellites (Figure 7). The intensities are in the ratio 3:4:3 for spin 3/2, 5:8:9:8:5 for spin 5/2, and so on.[(2)] The satellites are broadened by molecular motions or crystal imperfections, and for large splittings may disappear out of the observable range.

Second-order quadrupolar interactions may be observed if the quadrupole coupling is larger, relative to the Zeeman splitting. The second-order interaction shifts the entire line $(+\frac{1}{2} \rightarrow -\frac{1}{2})$ but leaves the separation of the satellites unchanged (the components are equally spaced when the energies depend on an odd power of M_I). An interesting example of a second-order quadrupole coupling is that of the ^{27}Al ($I = 5/2$) spectrum of AlN. Because of the high (hexagonal) symmetry the first-order quadrupolar interaction is absent. Only the $(+\frac{1}{2} \rightarrow -\frac{1}{2})$ transitions were observed giving the NQCC as 280 kHz for the aluminum.[(21)]

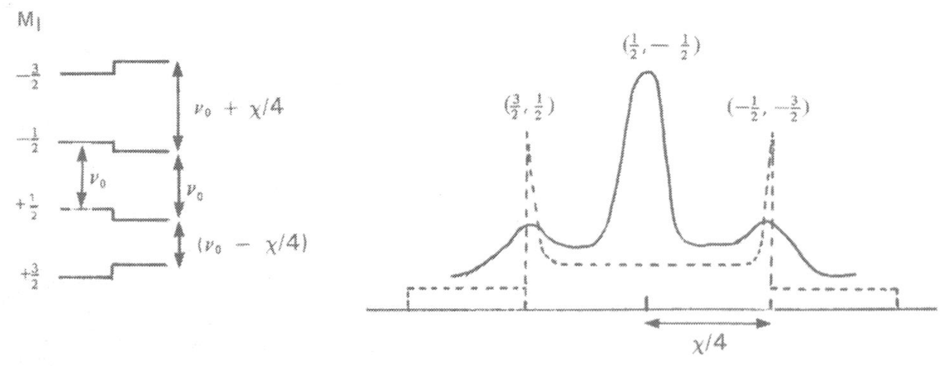

Figure 7. First-order quadrupole coupling for a nucleus with $I = 3/2$, $\gamma > 0$, and axial symmetry. (a) Zeeman splitting with and without quadrupolar shifts; (b) broadened powder spectrum. χ is NQCC.

For some time the measurement of quadrupole splittings in NMR spectra lay effectively in the domain of the physicist, but recent advances are making chemical applications less difficult. Linewidths of a few kilohertz due to dipolar, quadrupolar, or anisotropy broadening may be reduced to 100 Hz or less by the use of higher fields (the second-order interaction is inversely proportional to the field), and by techniques such as high-power irradiation or magic angle spinning, which are discussed in Sections 5 and 6. With the appropriate instrumentation, quadrupole couplings of 1 MHz or less (as for ^{2}H or ^{14}N) are well resolved. Thus a range of e^2qQ and η values has now been recorded for $^{14}NH_4^+$ in different crystals, the variations in efg arising from distortion of the N–H bonds, notably by hydrogen bonding to oxygen.[22]

2.5. Relaxation

The importance of relaxation in NMR spectroscopy may be illustrated by the story of C. J. Gorter, who failed to discover nuclear magnetic resonance in bulk matter in the 1930s and 1940s largely because—owing to the long relaxation time of his specimens[23]—the resonances saturated. Equally hazardous are relaxation times that are too short, as the resonances may then be broadened out of existence by the uncertainty in the energy.

As mentioned in Section 1, spontaneous emission is rare at radiofrequencies. In *spin-lattice relaxation*, nuclei transfer spin energy to the medium or "lattice" by a matching of Larmor frequencies, notably by the rotating fields of nearby magnetic nuclei as the molecules tumble in Brownian motion. These fields can have large local values at any instant, although they average to zero overall. The proton is highly effective because of its large magnetogyric ratio and close approach. Much more effective are unpaired electrons, and paramagnetic materials may only be present in small amounts in NMR experiments. In solids at low temperatures spin-lattice relaxation can be very slow indeed in the absence of paramagnetic impurities.

Relaxation processes follow an exponential law in principle, the rate depending on the excess (n) of excited nuclei compared to the number at thermal equilibrium:

$$dn/dt = -n/T$$

so that the excess at time t after an excitation is given by

$$n_t = n_0 \exp(-t/T)$$

where $n = n_0$ at $t = 0$. T is a time constant characteristic of the system, such that the relaxation rate is T^{-1} T_1, the *spin lattice relaxation time*, gives the rate at which the magnetization M_z in the direction of the applied field B_0 returns to its equilibrium value after excitation, by loss of spin energy to other degrees of freedom. T_1 is thus a *longitudinal relaxation time*, and a corresponding *transverse relaxation time* T_2 is defined by the rate of loss of magnetization in the xy plane.[24]

In nonviscous liquids, T_1 and T_2 are approximately equalized by the molecular motions. They are not necessarily equal, since the radiofrequency (applied in the x direction) which induces the nuclear transitions also makes the nuclei precess in phase in the xy plane, the spectrometer observes the rate of loss of the *phase-coherence*, and this rate may differ from that of the longitudinal relaxation. Thus another name for T_2 is the *spin–spin relaxation time*, because of the contribution from the exchange of spin energy between adjacent antiparallel nuclei, with loss of the phase coherence induced by the excitation, and increase in entropy, but no loss of spin energy to the lattice. This spin exchange is an important mechanism of relaxation and line-broadening in solids, in which the nuclei maintain their relative positions, and T_1 values may be long. In these circumstances $T_2 \ll T_1$. On the other hand, very efficient spin-lattice relaxation (electron–nuclear, J coupling to a quadrupolar nucleus, or chemical exchange) can shorten T_1 relative to T_2.

The natural shape of the NMR signal is Lorentzian, as for damped oscillations, and the linewidth at half-height is given by

$$W_{1/2} = 1/\pi T_2$$

A further contribution to the linewidth and to T_2 is made by the *range* of the static dipolar fields in solids, and of the local fluctuating fields in liquids. These may reinforce or oppose the applied field, and so correspond to a *range* of resonance frequencies. In practice an effective T_2 value called T_2^* is obtained from the observed linewidth, and includes also the contribution from instrumental imperfections, particularly the inhomogeneity of the applied field. (This is a problem for high-resolution NMR spectroscopy, and is reduced by sample spinning.)

2.5.1. Dipole–Dipole Relaxation

Various sources of local fields which can induce relaxation are given in Table 1, the observed relaxation rate $T_{1(\mathrm{obs})}^{-1}$ being the sum of the rates contributed by the mechanisms that are operative.

The major relaxation mode for spin-$\frac{1}{2}$ nuclei, if one or both have large enough γ values, is by direct *dipole–dipole interaction* DD with neighboring spins in random motion (rotations, translations, and collisions) and sometimes in molecular fluxions or chemical exchange. Detailed mathematical description of the motion is complex,(2) particularly when rotations are anisotropic. It is usefully approximated in terms of average values of an angular frequency ω and a *correlation time* τ during which a local field is maintained or "remembered." From the range of frequencies that are present, relaxation occurs at the Larmor frequency $\omega_0 = \gamma B_0$.

The function $2\tau_c/(1+\omega^2\tau_c^2)$ which gives the dependence of T_1 on the fluctuating local fields has a maximum value (relaxation is most efficient) at $\omega\tau_c \simeq 1$. Thus T_1 is shortest for $\tau_c \simeq 10^{-8}$ s, at NMR operating frequencies of ca. 10^8 Hz: it is most favorable, in fact, for large molecules such as lighter polymers, or biomolecules. Spin-$\frac{1}{2}$ nuclei which are difficult to observe because relaxation is too slow (^{183}W for example) may be easier to study in very viscous solution.

For faster molecular motions $T_{1\mathrm{DD}}$ lengthens, and becomes independent of frequency for $\omega\tau_c < 1$ (i.e., $\tau \gg 10^{-8}$ s). This is called the *extreme narrowing* condition, as T_2 then becomes equal to T_1, and the fast motions average the local fields, the range of which contributes to line broadening (this is *motional narrowing*). T_2 depends on $\tau_c/(1+\omega^2\tau_c^2)$ as T_1 does, but the contribution to T_2 from dephasing of the spins requires an additional τ_c dependence that is independent of frequency ω.

For slower molecular motions ($\omega\tau > 1$) T_2 decreases further, then levels off as the dipolar interactions approach those in solids. T_1, however, increases from its minimum value, and becomes dependent on ω^2. Thus high-field working may be disadvantageous for large molecules because of long relaxation times.

Table 1. Mechanisms of Spin-Lattice Relaxation

Mechanism	Rate dependence[a]	Examples	Comments
	For spin-$\frac{1}{2}$ nuclei[b]		
Dipole–dipole (DD)		$S = {}^1H$	
Intramolecular	$n_S\gamma_I^2\gamma_S^2 r_{IS}^{-6}\tau_c$		n_S is the number of S nuclei
Intermolecular	$N_S\gamma_I^2\gamma_S^2/DR_{IS}$ where $D = kT/6V\eta_m$		N_S is concentration R_{IS} is the distance of closest approach $D=(D_I+D_S)/2$ is the mutual translational diffusion coefficient for spherical molecules of volume V η_m = microviscosity $\simeq 0.16\times$ macroviscosity for pure liquids
Electron-nuclear (e)	$N_S\gamma_I^2\mu_{eff}^2 r_{Ie}^{-6}\tau_c$ (rot) $N_S\gamma_I^2\mu_{eff}^2/DR_{IS}$ (transl)	O_2, relaxation reagents	$\tau_c^{-1}=\tau_r^{-1}+\tau_s^{-1}$, where τ_s is the correlation time for the unpaired electron
Spin-rotation (SR)	$IC^2kT\tau_{SR}$ where $C\propto\gamma_I$	P_4	I is the moment of inertia, C the spin-rotation coupling constant (or tensor elements thereof)
Shielding anisotropy (SA)	$\gamma_I^2B_0^2(\Delta\sigma)^2\tau_c$	^{195}Pt in planar complexes	$T_1/T_2=7/6$
J or scalar relaxation (SC)			
1 Coupled to exchanging or fluxional neighbor (S)	$S(S+1)J^2\tau_{ex}/[1+(\tau_{ex}\Delta\omega)^2]$		τ_{ex} is the exchange lifetime
2 Coupled to quadrupolar neighbor (S) with relaxation time $T_{1q}>2\pi J$	$S(S+1)J^2T_q/[1+(\tau_q\Delta\omega)^2]$	$^{13}C^{79}Br$	$\Delta\omega$ is the difference in resonance frequencies of I and S spins
Quadrupolar (Q)	For spin $>\frac{1}{2}$ nuclei $[(NQCC)^2(1+\eta^2/3)(2I+3)/I^2(2I-1)]\tau_c$	All $I>1/2$ nuclei	$NQCC=e^2qQ/h$, η is the asymmetry parameter

[a] For liquids in the extreme narrowing limit ($\omega^2\tau^2 \ll 1$) Note that correlation times τ for different relaxation mechanisms are not necessarily the same All rotational correlation times τ depend on (viscosity/kT)

[b] The factor $I(I+1)=3/4$ has been omitted

2.5.2. Relaxation by Paramagnetic Agents

Most efficient in dipole–dipole relaxation is an unpaired electron, with its large magnetic moment ($\mu_B/\mu_N = 1836$). If close to a nuclear spin (in a free radical or paramagnetic ion, say), the *electron–nuclear interaction* may outweigh the Zeeman effect and ESR replaces NMR spectroscopy. If less close, the effective electron magnetic moment replaces γ_S in the dipole–dipole relaxation rate (see Table 1), and τ_c^{-1} includes a contribution from the relaxation of the electron. There is a contact contribution also if the unpaired electron is close enough to the resonant nucleus.[25] Paramagnetic materials may be present incidentally—dissolved oxygen for example, or transition metal impurities as from a nickel spatula—or they may be deliberately added, for example to promote relaxation when this is too slow (as in unprotonated materials). For nonpolar solvents, a fairly nonspecific unreactive *relaxation reagent* is [Cr(pd)$_3$],* which does not broaden or shift the lines unduly; suitable lanthanide ions may be used in polar solvents. Relaxation reagents which form complexes with the species of interest may be used to probe particular centers in the molecule; the exchange rate τ_M^{-1} may then contribute to τ_c^{-1}.[26]

2.5.3. Other Mechanisms of Relaxation for Spin-$\frac{1}{2}$ Nuclei

Other relaxation mechanisms come into play when dipole–dipole relaxation is less efficient. The *spin-rotation* (SR) *interaction* is important for spin-$\frac{1}{2}$ nuclei in smaller molecules, particularly in the gas phase. The nucleus experiences magnetic fields due to the differential rotation of charge with the molecular frame, and fluctuations of these fields, as with collisions, induce relaxation. This mechanism is distinguished by increase in the rate with increase in temperature and with decrease in viscosity, in contrast to the mechanisms depending on molecular tumbling. This is because the mechanism is more effective with increased population of the higher rotational states. The rate depends on the spin-rotation coupling tensor (C) and its anisotropy (as the square) and appropriate moments of inertia (I), values of which may be available from rotational spectroscopy.

Anisotropy in the shielding of a nucleus affords a mechanism of relaxation with molecular tumbling, and may be observed for ^{13}C in metal carbonyls, transition metals in planar complexes, etc. The *shielding anisotropy* (SA) relaxation rate depends on B_0^2, and may be monitored by measurements at different field strengths. At higher fields it may be large enough to broaden the resonance with loss of spin–spin coupling if τ_c is large, as in solutions of planar complexes of ^{195}Pt, for which $\Delta\sigma$ was measured as ca. 10^3 ppm in the solid state.[27]

Scalar or J coupling of the observed nucleus to a neighboring spin S which is fluctuating rapidly with $\tau^{-1} > J$ washes out the splitting and can induce relaxation.[2] This *scalar relaxation* may arise through chemical exchange of the spin S, with $\tau_{ex}^{-1} > J$. *Scalar relaxation of the second kind* is observed if the S spin is relaxing rapidly because quadrupolar, and has a Larmor frequency close to that of the $I(=\frac{1}{2})$ spin, as in ^{13}C–^{79}Br.

* pd is 2,4-pentanedionate, $(CH_3COCHCOCH_3)^-$.

2.5.4. *Quadrupolar Relaxation*

For the inorganic or organometallic chemist the most important relaxation mechanism may well be the *quadrupolar relaxation* of $I > \frac{1}{2}$ nuclei. This normally dominates over all other mechanisms, so that $T_q(=T_1 \simeq T_2)$ determines the linewidth. As described in Section 2.4, relaxation of the nuclear electric quadrupole with changes in the local electric field gradient (efg) also relaxes the nuclear spin. The consequent line broadening, with loss of information from spin–spin coupling and smaller chemical shifts, has inhibited NMR studies of quadrupolar nuclei.

The rate of quadrupolar relaxation (Table 1) increases as the *square* of the NQCC. Line broadening due to a large value of Q may be offset by a large I value, the *line-broadening factor*[28] being

$$L = f(I)\,Q^2, \qquad \text{where} \quad f(I) = (2I+3)/I^2(2I-1)$$

the function of I decreasing from 5 for $I=1$ to 0.075 for $I=9/2$.

Since the quadrupolar relaxation rate increases also with τ_c, the lines are usually broadened significantly by increase in viscosity, and higher temperatures are then advantageous.

Because of the dependence on the efg, high-resolution spectroscopy is possible for fairly weakly quadrupolar nuclei if the electronic environment is sufficiently symmetric. Examples are ^{2}H, ^{6,7}Li, ^{9}Be, ^{14}N, ^{17}O, or ^{133}Cs. Indeed, quadrupolar relaxation for aqueous $^6Li^+$ is so inefficient that relaxation times of 1000 s have been observed.[29] Asymmetry of the efg is marked for atoms carrying lone pair electrons with *s* character. Thus ^{14}N lines are sharp for NH_4^+ but quite broad ($W_{1/2}$ ca. 200 Hz) for pyridine. The lines for monocovalent bromine or iodine are usually broadened out of existence, but can be observed for aqueous I^- and IO_4^-, for example, because of the high local symmetry (Chapter 17).

3. THE LARMOR PRECESSION AND THE BLOCH EQUATIONS

To understand the potentialities of FT pulsed NMR spectroscopy[30,31] (and to see the full significance of the two relaxation times T_1 and T_2) we need to consider the precession of the nuclear spins. In Figure 8 the nuclear spin angular momentum is shown as a vector at an angle to the z axis such that its projection takes the values

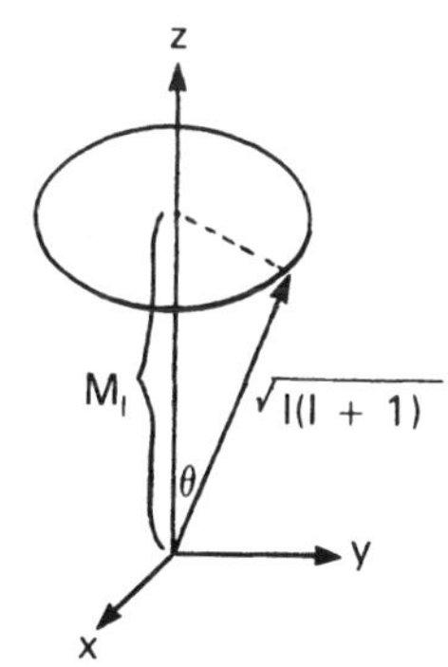

Figure 8. Orientation of a nuclear spin vector.

$M_I = -I, \ldots, +I$. According to the uncertainty principle, if the z component is specified, then the x and y components are undetermined: the vector may be oriented in any direction at the same angle to the z axis, and so lies somewhere on the surface of a cone.

If a magnetic field B_0 is then applied in the z direction the spin vector cannot align with it because of its angular momentum. Like a spinning gyroscope precessing about the vertical (the direction of the gravitational field) the nuclear spin vector precesses about the direction of the magnetic field, its tip tracing a circle in the transverse (xy) plane. The Larmor theorem in classical mechanics gives the effect of a magnetic field (on charged particles) as equivalent to uniform rotation, hence the terms *Larmor precession* for the nuclear spin, and *Larmor frequency*. The angular frequency is given [by equations (3) and (4)] as

$$\omega_0 = 2\pi\nu_0 = -\gamma M_I B_0$$

as shown in Figure 9 for $I = 1$. The sign of M_I gives the sense of the precession, which is clockwise for positive γ.

The transition of the nuclear spin can also be described in these classical terms. The rf field B_1 must exert a *torque* on the spin angular momentum vector in order to change its orientation, and can only do this if applied perpendicular to B_0, and also rotating at the Larmor frequency. B_1 is a *linear* alternating magnetic field, and this is in fact equivalent to two components rotating in opposite senses, with angular velocities $\pm\omega$ (and we can ignore the countercomponent to the Larmor precession).

The vector in Figure 9 represents one nuclear spin. More usefully we can think, as Bloch did,[24] in terms of the bulk magnetic moment (or net magnetization) **M** of the whole ensemble of nuclear spins. This is equal to the sum of the individual moments of the excess (n) in the lower level, say n_0 at equilibrium in the field B_0. The component M_z in the B_0 direction then has an equilibrium value given by

$$M_0 = \gamma\hbar n_0 = \chi_0 B_0$$

where χ_0 is a magnetic susceptibility. At resonance the vector **M** has a steady component M_z in the field direction, and components M_x and M_y rotating in the transverse plane at the Larmor frequency.

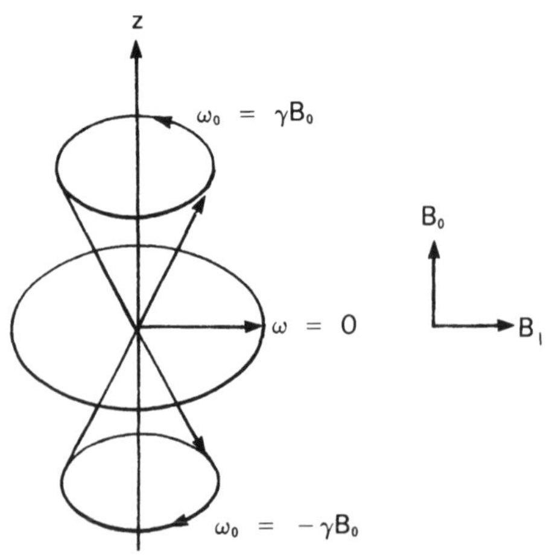

Figure 9. Larmor precession of nuclear spin vectors ($I = 1$) in a magnetic field.

A device which is of great help in understanding the response of spin vectors to pulse experiments is to "stop" the Larmor rotation in the transverse plane. We can do this by choosing axes of coordinates that rotate in the same sense at the same frequency (as we do in our earthly measurements of velocities, etc.). This *rotating frame* has axes x', y', z, where z, the axis of rotation, is unchanged, and the B_1 (probe) field is now fixed in the x' direction. (In multiple resonance experiments a B_2 field may be applied in the y' direction, and so on.)

Bloch defined T_1, the longitudinal relaxation time, by the equation (in the laboratory frame)[24]

$$dM_z/dt = -(M_z - M_0)/T_1$$

and T_2, the transverse relaxation time, by

$$dM_x/dt = -M_x/T_2 \quad \text{and} \quad dM_y/dt = -M_y/T_2$$

the transverse components decaying to zero at thermal equilibrium. The Larmor precession gives a further time dependence:

$$dM_x/dt = \omega_0 M_y \quad \text{and} \quad dM_y/dt = -\omega_0 M_x, \qquad \text{cf. } dM_z/dt = 0$$

where $\omega_0 = \gamma B_0$. Precession in the transverse plane explains the difference between T_1 and T_2 in solids. The angular frequency ($\omega = \gamma B$) changes in response to the *range* of the strong local dipolar fields and their fluctuation with spin–spin relaxation, and also with the range (inhomogeneity) of applied field, so the transverse magnetization decays (T_2 short). The longitudinal magnetization, however, is slow to decay, with minimal molecular motion (T_1 long).

Combination of the equations for the relaxation with those for the precession gives the Bloch equations, describing damped oscillations:

$$\begin{aligned} dM_z/dt &= -(M_z - M_0)/T_1 \qquad \text{(as before)} \\ dM_x/dt &= \omega_0 M_y - M_x/T_2 \\ dM_y/dt &= -\omega_0 M_x - M_y/T_2 \end{aligned}$$

When the B_1 (probe) field is applied (with angular frequency $\omega_1 = \gamma B_1$) the x component induces precession of M_y and M_z about the effective field, and correspondingly for the y component. The phase-sensitive detector monitors M_x and M_y as the u and v mode signals, respectively. In the rotating frame, u is in-phase with B_1 (in the x' direction) and v out-of-phase (in the y' direction), and there is a "fictitious field" ω_0/γ (equivalent to the rotation) which cancels out B_0 at resonance. The *Bloch equations* then become

$$\begin{aligned} dM_z/dt &= -\gamma B_1 v - (M_z - M_0)/T_1, \qquad && \text{where} \quad v = M_y \\ du/dt &= (\omega_0 - \omega_1)\, v - u/T_2, && \text{where} \quad u = M_x \\ dv/dt &= -(\omega_0 - \omega_1)\, u + \gamma B_1 M_z - v/T_2 \end{aligned}$$

with $\omega_1 = \omega_0$ at resonance. Thus the Bloch theory explains the two components, differing in phase by 90°, which make up the NMR signal. The one out of phase with the exciting frequency is the *nuclear induction*, which gives the alternative name nuclear induction spectroscopy. This was used by Bloch, but nowadays is used mainly for memorable acronyms, as for proton-enhanced NIS.[32]

Traditionally the v (out-of-phase) component was monitored as it gives the familiar *absorption signal* which integrates (if not saturated) to give an area which is proportional to the number of resonating nuclei. This line has $W_{1/2} = 1/\pi T_2$, for exponential decay. The u component gives a dispersion line, which can be used to find the center of a broad signal; in this the peak-to-trough distance is $1/\pi T_2$. Spectrometers with quadrature detection monitor both signals (two phase detectors work in quadrature, i.e., 90° out of phase) to increase sensitivity.

The Bloch equations give the line shapes and saturation parameters, and show the dependence of the signal intensity on experimental factors, including $\gamma^3(I+1)$ (at constant field and temperature with $T_1 = T_2$). More generally, the Bloch equations provide a framework for the study of time-dependent processes, including chemical exchange or molecular fluxions, to which NMR spectroscopy has made so great a contribution. The equations are modified to take account of exchange between different resonance frequencies, say, and line shapes computed for different values of the relevant parameters, for comparison with those observed. Because of its simplicity Bloch's semiclassical vector approach is still in use although other formal theories have been developed.

The importance of relaxation in pulsed NMR spectroscopy is discussed in the next section, and these topics are developed to greater depth in Chapter 5.

4. THE FOURIER TRANSFORM TECHNIQUE

NMR spectroscopy now relies on *pulsed Fourier transform* (FT)[30,31] techniques, in which the weak (milliwatt) rf field of the CW experiment, exciting the resonances in turn, is replaced by a powerful (kilowatt) field in the form of a short (microsecond) pulse which contains all the necessary frequencies. The transient response, the *free induction decay* (FID) of all the excited spins, is detected, amplified, digitized (by analog-to-digital converter), and stored in a dedicated computer as a function of *time*. The pulsing is repeated and the FIDs added coherently. The data are then converted by Fourier transformation into the corresponding functions of *frequency*, to give the same spectrum (below saturation) as that obtained by slow sweep in the CW mode.

The saving of time in pulsed NMR as compared with multiple scanning for signal enhancement in the CW mode is large, since the signal-to-noise ratio improves only as the square root of the number of scans, each of which takes minutes. In the FT mode the simultaneous excitation of all the spins means that a transient may be observed literally in a second, depending on the relaxation times. The gain in efficiency depends on the spectral width, which is larger for higher-field working. *Maximum entropy methods* of analyzing FID data are now being developed to maximize the signal-to-noise ratio.[32a]

Relaxation times are peculiarly important in FT NMR work because they determine the rate of pulsing and accumulation; thus fast pulsing is possible for quadrupolar nuclei, and can make up for low receptivity. Conversely, the FT method is peculiarly informative on relaxation rates, because of the signal intensities and frequencies in the FID which are recovered in the Fourier transformed line shape.

The *pulse width* (length, or duration) t_p determines the range of frequencies produced, by the uncertainty principle. The shorter the pulse the wider the range: thus a 10-μs pulse can give a range approaching 10^5 Hz. For very large spectral widths an alternative but related technique is *stochastic excitation*, which uses a noise-modulated rf.[33] The computer then performs a cross-correlation exercise, depending on the manner of the noise-modulation, before the Fourier transform. This method gives added flexibility and may have applications in multinuclear work.

For spin-$\frac{1}{2}$ nuclei or nuclei with weak quadrupolar couplings, the pulse width t_p is much shorter than relaxation times. For nuclei with quadrupole couplings approaching 100 MHz, however, relaxation times are measurable in microseconds. It is then very difficult to monitor the exceedingly short FID, because of interference by aftereffects of the pulse in the material of the probe. These are variously described as *acoustic ringing* or eddy currents, and result in a rolling base line in which it is difficult to distinguish the broad resonance line. Special pulse routines have been devised to minimize these effects by a process of subtraction and improvements in the probe material have been suggested.[33a]

Figure 10 shows effects of a pulse in the *rotating frame* described in the previous section. When the sample (containing magnetically equivalent spin-$\frac{1}{2}$ species) is placed in the B_0 field the steady state magnetization M_0 develops by spin-lattice interactions, and the spin vectors come to rest in the rotating frame with random orientation (Fig. 10b). A B_1 (rf) pulse at the resonance frequency ω_0 supplies a field in the x' direction that rotates the vector M in the $y'z$ plane through a *pulse angle* (or *flip angle*) α_0/rad given by

$$\alpha_0 = \gamma B_1 t_p$$

If the pulse width t_p is chosen so that α_0 is $\pi/2$ rad (90°) the M vector is rotated into the y' axial direction, that of the detector, and gives a maximum signal. A 180° or π pulse inverts the M vector, and no signal is detected.

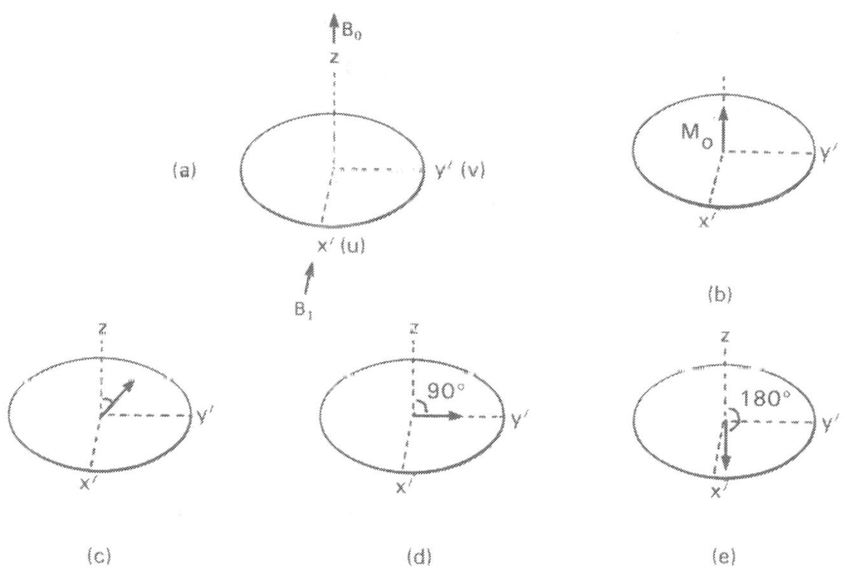

Figure 10. Effects of increasing pulse width on the magnetization vector **M** in the rotating frame. (a) Axis conventions; (b) steady state; (c) pulse angle α; (d) $\pi/2$ pulse; (e) π pulse.

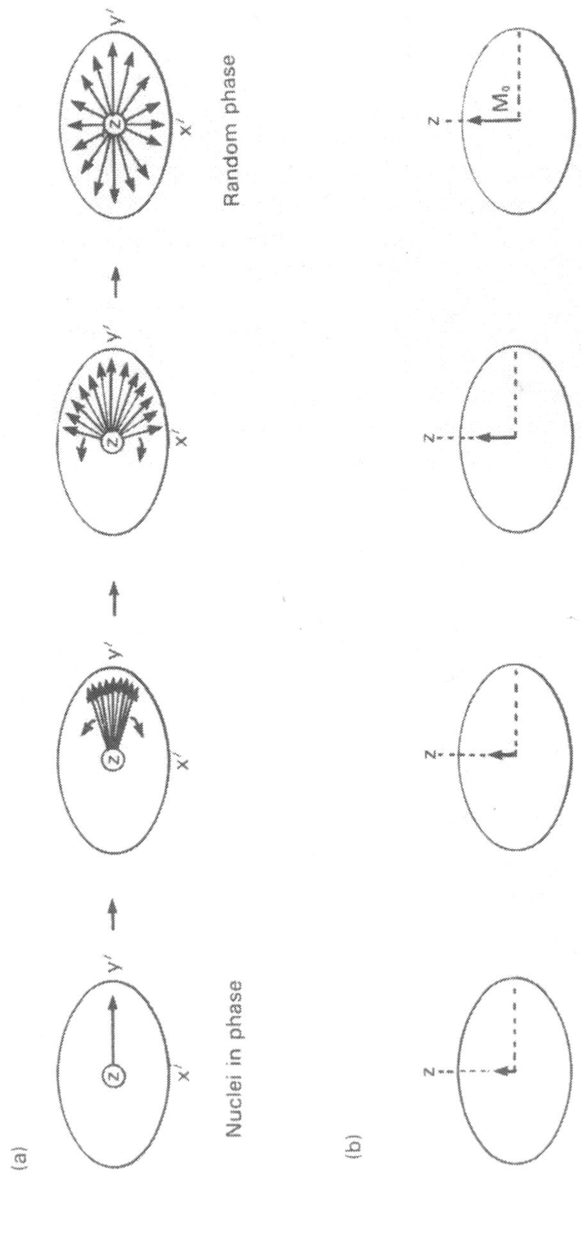

Figure 11. Relaxation after a pulse $0 < \alpha < \pi/2$. (a) The T_2 process, dephasing in the transverse plane; (b) the T_1 process, recovery of the longitudinal magnetization.

After the pulse the magnetization begins its free decay, as shown in Figure 11. Following a 180° pulse the inverted M vector shrinks to zero and then grows to the M_0 value at a rate $1/T_1$, with no signal detected. Other values of the pulse angle, however, induce magnetization in the transverse plane, with the spin vectors bunched together in the M direction. The projection of the magnetization of the z axis returns (undetected) to the M_0 value at the rate $1/T_1$. At the same time, but at the rate $1/T_2^*$, the spin vectors fan out towards random distribution about the z axis. It is this fanning out or *dephasing* of the spins that is monitored as a voltage which is proportional to $\sin \alpha$ and decays exponentially at the rate $1/T_2^*$. This FID signal does not oscillate, since its frequency ν_0 is that of the reference frequency. If, however, the sample contains a second magnetic species which resonates at a frequency that differs from ν_0 by ν_1 Hz, say, then its M vector is precessing at a frequency ν_1, viewed from the rotating frame. The FID now contains the *interference pattern*, or beats, of the two sine waves (due to the two nuclei) each with its own intensity and decay time, since they are in phase immediately after the pulse, and repeatedly go out of and back into phase at intervals $1/\nu_1$. Fourier transformation then gives the line at ν_0 and the second line ν_1 Hz away.

Near-complete relaxation of the spins between pulses needs a time in excess of $4T_1$ (with $T_1 \simeq T_2$). In practice shorter times may be used, a steady state being reached after a few pulses. Maximizing the signal for long relaxation times requires a small pulse angle α_0 or a long pulse interval T_p, the optimal pulse angle being given by

$$\cos \alpha_0 = \exp(-T_p/T_1)$$

α tending to $\pi/2$ for T_p in excess of T_1. Relative intensities of the FT signals may, therefore, be determined by the relative relaxation rates, rather than by the relative concentrations of the different species, as in a CW spectrum. A small signal (as for nonprotonated ^{13}C in organic compounds) may reflect slow relaxation and differential saturation, rather than a low concentration of the resonant nuclei. For *quantitative spectroscopy* it may be necessary to add a relaxation reagent such as $[Cr(pd)_3]$, or perhaps to observe particular resonances selectively under controlled conditions. But high power and saturation in FT NMR, as we have seen, does not broaden and diminish the signal as it does in CW NMR.

In the above discussions we have considered only the simplest FT experiment in which a single pulse (a "read" pulse) is used to obtain a spectrum. Multiple irradiation is used in many ways in FT NMR, the commonest being broad-band irradiation to decouple protons

5. MULTIPULSE AND MULTIPLE RESONANCE TECHNIQUES

In the early work on double resonance the perturbing rf field of frequency ν_2 was applied continuously, and simultaneously with the observing rf field ν_1. The perturbing field has amplitude B_2 (or $\bar{\gamma} B_2$ in frequency units, where $\bar{\gamma} = \gamma/2\pi$). The perturbation of spin states p and q connected by energy difference ν_{pq} is largest when ν_2 coincides with ν_{pq} and falls off with increasing $(\nu_2 - \nu_{pq})$. The rate at which the perturbation drops off depends on the amplitude B_2. A perturbing field B_2 may be considered to have an effective range $(\nu_2 - \nu_{pq})$ of the order $\bar{\gamma} B_2$. Thus, the effect of the irradiation on the observed spectrum depends on whether the frequency γB_2 is of the order of the width of a particular line, the width of an entire multiplet, etc.

In present multiple resonance experiments, irradiation is not necessarily done continuously, but with a single pulse lasting only microseconds, or a train of pulses with some specific delay times between them, or gated (switched on and off at appropriate times) during a single experiment, or switched on at some particular point in the experiment and kept on during the entire observation period. A single experiment usually consists of three periods: preparation, evolution, and detection. A nucleus may be subjected to a number of manipulative pulses during the preparation period, allowed to precess during the evolution period, and subjected to a "read" pulse during the detection period. During the same time periods another nucleus (of the same or a different type) may be subjected to a different set of manipulative pulses, or a saturating continuous radiation, etc. These NMR methods are all based on the ability of an FT spectrometer to sample magnetization immediately (i.e., in times short compared to relaxation times) after a selective perturbation of the spin system.

An experiment is generally repeated a large number of times during which the phases of the various pulses are changed systematically (phase cycling) and the FIDs accumulated. An example of a sequence for a typical multiple resonance experiment is shown in Figure 12. In this particular experiment, the first two pulses applied to the proton excite a spin echo at time 2τ. The π pulse applied to the ^{13}C interchanges the two spin states α_C and β_C. The delay time τ is selected so as to leave the two proton spin states α_H and β_H along the $\pm X$ axes (in the rotating frame) at time 2τ. The third pulse at the proton frequency has a relative phase shift of 90°, and rotates the magnetization vectors about the Y axis into the $\pm Z$ directions. (See Figure 9 of Chapter 5 for the ^1H spin dynamics in the rotating frame.) This corresponds to a population inversion across one proton transition but equilibrium populations across the other. Because the ^1H and ^{13}C share common energy levels, the ^{13}C transitions observed with the $\pi/2$ "read" pulse now have spin population differences appropriate to protons (enhanced by a factor γ_H/γ_C). Maximum enhancement is obtained if τ is set to $[4J(CH)]^{-1}$. The -1:1 doublet of the ^{13}C resonance of a CH group, for example, can

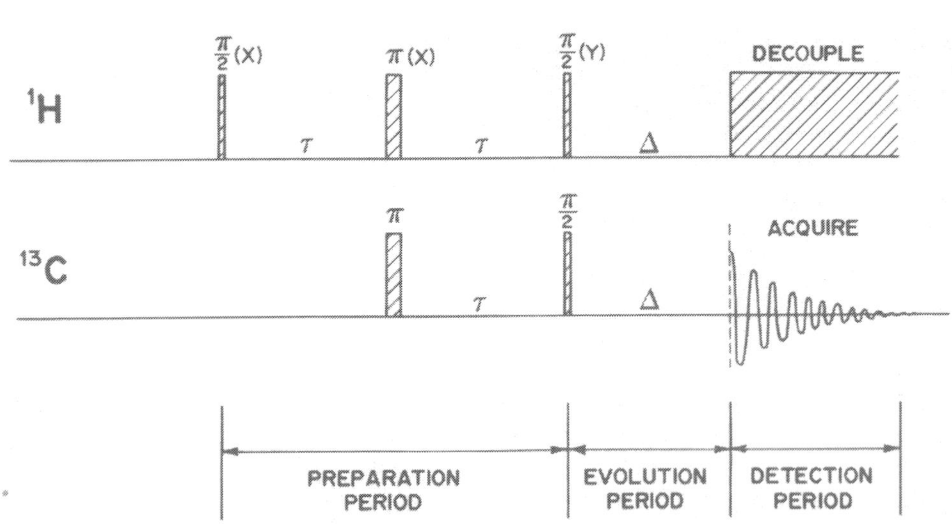

Figure 12. An example of a multipulse experiment, described in the text.

be observed at this stage, which corresponds to an INEPT experiment. If, however, the ^{13}C spins evolve for a time Δ before detection, the ^{13}C relative intensities will vary with Δ. For example, for a CH group the initially opposed ^{13}C spins precess in the rotating frame at rates $J(CH)/2$ and $-J(CH)/2$, leading to signals with the same relative phase at $\Delta = [2J(CH)]^{-1}$. On the other hand for a CH_3 quartet the spins precess at rates $3J(CH)/2$, $J(CH)/2$, $-J(CH)/2$, and $-3J(CH)/2$, leading to signals with the same relative phase at $\Delta = [4J(CH)]^{-1}$ and $3[4J(CH)]^{-1}$.[34,35]

Multiple pulse experiments were carried out in the early days of NMR, but remained largely in the domain of physics until the 1970s. Since then a number have come into use. We now describe some important types.

Spin Echoes. In 1950 Hahn[36] recognized that the decay of transverse magnetization due to the inhomogeneity of the magnetic field is a reversible process, and refocused the dephased magnetization by applying a 180° pulse τ s after an initial pulse had excited the transverse magnetization. Optimum refocusing occurs an equal time τ after the second pulse so that a transverse NMR signal reappears at this time (a *spin echo*). The first cyclic multiple pulse experiment by Carr and Purcell[37] used this sequence to generate spin echoes: $\pi/2$–τ–π–τ–echo–τ–π–τ–echo–···. The first pulse is a preparation pulse; the second pulse refocuses the spins to produce an echo at 2τ. The state at 2τ can now be considered as a new initial state and the process repeated, generating a train of echoes. Those at 2τ, 6τ, 10τ,... are inverted, so that the cycle time of the Carr–Purcell sequence is actually 4τ.

In multiple pulse experiments based on excitation of spin echoes, signal improvement results from regaining losses due to the magnetic field inhomogeneity. Spin echoes are also useful because they are affected by spin–spin coupling in a known way. This is known as *J modulation of spin echoes.* Because the components of a multiplet do not precess at the same rate, after the refocusing pulse the slow and fast components of a doublet, say, continue to diverge. Although the chemical shift and field inhomogeneity are refocused at time 2τ, two echo components occur, and the deviation between the two becomes progressively larger with each echo. With more components, as when there are several multiplets with different J couplings, the echo train is modulated by a beat frequency related to differences in J values. If the J-modulated echo train is Fourier transformed as a function of τ, a *J spectrum* is obtained, showing peaks at frequencies given by sums or differences of all the J values. The J modulation of spin echoes is the basis for many of the applications of multiple resonance techniques to signal enhancement, assigning peaks, and establishing connectivities of atoms in a molecule, as discussed further in Section 5.3. Figure 12 shows one such application, that of the INEPT sequence (Insensitive Nuclei Enhanced by Polarization Transfer). A review of applications of spin echoes to high-resolution work on liquid samples has been given by Freeman and Hill.[38]

Two-Dimensional NMR. There are excellent reviews of two-dimensional (2D) NMR spectroscopy.[39–42] The basis for NMR spectroscopy in two frequency dimensions is spin "memory." Once the nuclear spins in a sample are made to precess, they continue to do so for a time of the order of the spin–spin relaxation time T_2. The behavior of the nuclei during the detection period t_2 can be related to the behavior of the nuclear spins during the evolution period t_1 (when the receiver is inactive) because each spectral component, precessing at its characteristic frequency during t_1, passes on information (usually in the form of phase or amplitude) to the appropriate spectral component during the detection period, as Freeman says, "rather like the individual runners in a relay race passing on their batons to their team-mates running in the next stage of the race."[40] If a series of experiments steps through regular increments in t_1, then the NMR information is in two independent time dimensions $S(t_1, t_2)$. Fourier transformation then gives a spectrum in two frequency dimensions (although a Fourier transformation is not always needed).[41] By imposing various experimental conditions on the nuclei during the evolution period and the detection period, it is possible to separate different NMR parameters such as chemical shifts, J, or D, into the two frequency dimensions. Since the nuclei remember during the period t_2 their past history during t_1, correlations are preserved and appear as cross peaks in the 2D spectrum. The cross peaks can be used to identify nuclei which are related by cross relaxation, chemical exchange, J coupling or dipolar coupling. Two examples of 2D spectra are shown in Chapter 5. The original experiment known as COSY (COrrelated SpectroscopY) is routinely used for proton NMR assignment problems in intermediate to large molecules.

Multiple Quantum NMR. Single quantum transitions in NMR are the allowed $\Delta M_I = \pm 1$ transitions. Multiple quantum transitions do not appear in single-pulse FT experiments, and they can be observed in slow-passage NMR experiments only if sufficiently strong rf fields are applied. The frequencies of these forbidden transitions must therefore be obtained indirectly. The appropriate pulse sequence creates a coherence between nuclear spin energy levels separated by $\Delta M_I \geqslant 2$. A multiple quantum coherence is not directly detectable in the spectrometer because it does not correspond to magnetization, but may be converted into observable transverse nuclear magnetization by a suitable rf pulse, a *mixing pulse* at $t = t_1$. The frequency of a multiple quantum signal can be determined by performing a sequence of experiments at various t_1 values, followed by a 2D Fourier transformation. There are many applications of multiple quantum spectra, in solids,[43,44] and in 2D NMR of liquids.[45,46] We will consider only one here. A unique property of molecules with two carbon-13 spins is the existence of a double quantum energy gap, $\Delta M_I = 2$. This is the basis for a method of discriminating against the strong (ordinary allowed single quantum) signal from molecules with a single carbon-13 nucleus. The carbon-13 spins are excited by a four-pulse sequence which achieves a maximum conversion into double quantum coherence if the time between pulses is set to $1/4J(\mathrm{CC})$. The characteristic phase properties of the double quantum signal are used to distinguish it from all other NMR signals so that only the signals from molecules containing two coupled carbon-13 nuclei will be observed. These doublets normally appear with one line of positive intensity and the other of negative intensity, but can readily be converted into an all-positive-intensity spectrum. In particular, choosing the time between pulses to correspond to $1/4J(\mathrm{CC})$ for directly bonded carbons ensures that only signals from adjacent carbon-13 nuclei will be observed.[46,47] The frequency of the double quantum signal is the sum of the chemical shifts of the two carbon sites, measured relative to the transmitter frequency. This information, plus the magnitudes of C–C coupling constants, can be used for assignment of carbons, and in fact traces out the carbon skeleton of a complex molecule, since every adjacent pair of C-13 nuclei is mapped. In general, an n-quantum spectrum is the equivalent of a spectrum of a mixture of only n-labeled compounds.[48] As well as spectral simplification, multiple quantum experiments can be used for the indirect observation of low-γ nuclei, and as a probe of molecular motions. An excellent review of multiple quantum experiments has appeared.[49]

5.1. Double Resonance Experiments

In a conventional double resonance experiment, designated by A–{X}, one can excite two or more groups of nuclei more or less simultaneously with a main rf B_1^A and a second rf B_2^X. B_2 may excite the same nuclear species (homonuclear double resonance) or a different nuclear species (heteronuclear double resonance). Triple resonance experiments involving a third rf are designated by A – {M, X}, Irradiation with an rf B_2 set at a value corresponding to the energy difference between two spin levels has the effect of equalizing the population of these two levels. Depending on the power of B_2 and the mechanisms whereby these levels are populated and depopulated, the consequences are: changes in intensities of specific parts of the spectrum (increase, decrease, or even inversion), the appearance of additional splittings, or complete collapse of a spin–spin coupled multiplet into a single central peak. The explanation of these double resonance phenomena is given in a detailed early review.[50] Such methods can be used to determine chemical shifts, or the signs and magnitudes of spin–spin coupling constants, to assign peaks and analyze spectra.[51–53] In some applications the second rf B_2 must be powerful enough to affect not just one peak but all components of a spin multiplet equally, or even all the proton nuclei in the molecule simultaneously. Table 2 classifies the possibilities. Here we will give some of the terminology and refer to seminal papers.

INDOR (internuclear double resonance) is a CW technique which has no exact counterpart in FT spectroscopy. The intensity of one of the transitions is plotted out while a second rf ν_2 is slowly swept through a range of values. For a simple AX spin system (see Figure 13), as ν_2 passes through the A transition frequencies, the spins are pumped from the lower to the higher energy levels resulting in changes of the X intensity. These changes may be brought about by spin tickling, in which case only positive intensities will be observed; or spin population transfers, in which case positive and negative intensities will be observed. Heteronuclear INDOR is useful for presentation of the spectra of insensitive nuclei such as ^{57}Fe, ^{103}Rh, ^{107}Ag, or ^{183}W by irradiating ^{1}H, ^{19}F, or ^{31}P. With current multinuclear FT NMR spectrometers and a variety of sensitivity-enhancement techniques, INDOR applications have diminished considerably, although some variations have been proposed.[68,69]

Spin-Tickling. When the strength of the B_2 field becomes comparable with the line width, not only are intensities in the swept spectrum affected because of population changes, but mixing between the original unperturbed states occurs, giving rise to extra peaks.[52,53,55]

Decoupling. Depending on the amplitude of B_2, from the size of J to wideband, various decoupling effects are observed, as listed in Table 2. The main applications of these decoupling schemes are for simplification of spectra and assignment of peaks. Very effective supercycles have been developed for uniform broadband decoupling, and Waugh has given a detailed theoretical treatment.[56]

Selective population transfer (SPT)[59] and *selective population inversion* (SPI)[57] are FT equivalents of INDOR in the sense that the same type of information is obtained, though not yielding spectra that look exactly like INDOR spectra. In SPT,

Table 2. Classification of Conventional Double Resonance Experiments, Listed in Order of Increasing Power of the Second rf

Amplitude of B_2 (Hz)	"Swept" frequencies	Effects on spectrum	Terminology	Reference
$< W_{1/2}$	ν_2	Intensity changes, ν_1 monitored	(CW) INDOR	54
$\simeq W_{1/2}$	ν_1	Additional splittings	Spin tickling	55
(saturation)	ν_2	Intensity changes, ν_1 monitored	(CW) INDOR using spin tickling	54
$< J$		Inversion of some peaks	SPI	57, 58
		Intensities change	SPT	59
$\simeq J$	ν_1	Selective removal of spin coupling	Selective decoupling	60
Width of entire multiplet	ν_1	Collapse of other multiplet into single peak	Coherent spin decoupling	61
		General intensity changes	NOE	62, 63
		Max NOE, spin coupling remains	Gated decoupling	64, 65
		Min NOE, spin coupling removed	Gated decoupling	66
		Same appearance as undecoupled but with reduced J	Off-resonance decoupling	67
Broadband	ν_1	All couplings with nuclear species 2 disappear NOE	Wideband decoupling	67

application of a gated weak irradiating B_2 field ($\gamma B_2 \simeq 0.1$–0.5 Hz) to one of the proton transitions leads to a transfer, though not necessarily an inversion of populations (as opposed to SPI), involving the less sensitive nucleus which is then excited by an intense rf pulse B_1. In the heteronuclear case, the equilibrium population difference between the levels for the A transitions (typically proton) are larger than those for the X transitions (X = ^{13}C, ^{15}N, etc.) by a factor γ_A/γ_X. In our simple example, Figure 13, lines X1 and X2 will thus appear with enhancements $\pm\gamma_A/\gamma_X$. This FT version has the advantage over a spin-tickling method in that the acquisition time may be greatly reduced, as it is characteristic of T_{1H} rather than T_{1X}, which is usually long for spin-$\frac{1}{2}$ X nuclei.

NOE. The nuclear Overhauser effect is the change in the integrated intensity in the signal from one nucleus when all the transitions of a nearby nucleus are saturated.[62,63,70] Saturation is generally brought about by high-power irradiation over a wide range of frequencies, at least the width of the entire multiplet. The NOE arises from cross-relaxation processes. The integrated intensity relative to that in the conventional FT spectrum is $1 + (T_1^{DD}/2T_1)\gamma_S/\gamma_I$ if S (usually proton) transitions are saturated. Thus, only dipolar relaxation contributes to the NOE, and all other mechanisms decrease it. The maximum ratio is $1 + (\gamma_S/2\gamma_I)$; for the $^{13}C-\{^1H\}$ experiment (S = ^{1}H, I = ^{13}C) this number is 2.988. ^{13}C spectra are commonly recorded

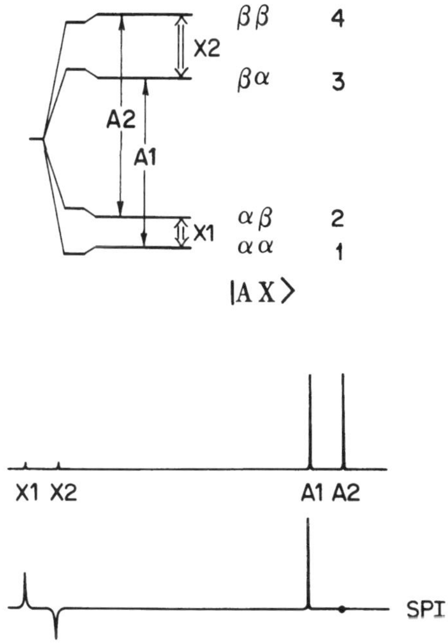

Figure 13. Energy levels and transitions for an AX system. At the bottom of the figure, a population inversion experiment (SPI) is shown. If an rf pulse at exactly the resonance frequency is applied to one of the proton lines (A2) at a power level low enough to affect only A2, then the populations of the energy levels 2 and 4 are inverted. The ^{13}C spins sharing these levels will show an enhanced X1 line and an inverted X2 line.

with simultaneous irradiation of protons with a broadband rf for two purposes: the establishment of the NOE by saturating the protons in the period just prior to ^{13}C excitation, and decoupling during acquisition of the ^{13}C FID. This results in a significant simplification of the ^{13}C spectrum while at the same time improving sensitivity, through the coalescence of multiplets, and the NOE.

For $\gamma_I < 0$ (as in ^{15}N, ^{29}Si, etc.) the NOE may diminish the signal and can be removed by (a) providing a relaxation mechanism that dominates over the dipolar one (e.g., by introducing a relaxation agent[71] or (b) by taking advantage of the fact that NOE builds up slowly in an exponential growth with a time constant of the order of T_{1I}. Thus, if the saturating signal for S had been turned off for a period of the order of at least $3T_{1I}$ prior to the probing pulse B_1, then there would be minimum NOE in the resultant spectrum. This is *gated decoupling.*[66]

An important application of multiple resonance experiments is the determination of relative signs of coupling constants,[51,52,72] as described in Chapter 4 Interpretation of the observed changes in the spectrum under INDOR, SPT, SPI, or spin-tickling is not trivial. In the simplest cases identification of subspectra associated with different nuclear spin states is sufficient for sorting out the relative signs. In more complex spin systems a detailed analysis of the energy level diagram may be necessary. The reasoning involved in the determination of relative signs of coupling constants from an analysis of the energy level diagrams in SPI and SPT is fundamentally the same as that of the analogous spin-tickling experiments used in earlier CW work.[55]. *Double selective population transfer* is a technique for determining relative signs and magnitudes of J between two rare spins.[73]

5.2. Techniques for Signal Enhancement

In all methods, the enhancement depends on a transfer of polarization from a more sensitive nucleus S (usually the proton, but others have also been used) to a less sensitive nucleus I, because S spin population differences are larger than I spin population differences in the same magnetic field. The most common method, Overhauser polarization (NOE), is possible when the relaxation of the low-sensitivity nucleus I is dominated by dipolar interactions with a high-sensitivity nuclear species S.[62,74] *Hartmann–Hahn cross polarization*,[75] as employed in the Pines, Gibby, and Waugh method[76] for solid state NMR, proceeds through cross-relaxation mechanisms that couple the I and S subsystems and permit a transfer of energy when $\gamma_H B_1^H = \gamma_C B_1^C$ (the Hartmann–Hahn condition). In solids this includes intermolecular magnetization transfer. In liquids, this cross-polarization method (*J* cross polarization or JCP) depends on the indirect spin–spin coupling between I and S and is strictly an intramolecular magnetization transfer.[77,78] Adiabatic and quasistochastic *J* cross polarization are modifications of the JCP method in which small deviations from the Hartmann–Hahn condition can be tolerated.[79,80]

The above examples involve net polarization transfer, that is, all the lines in the I spectrum become polarized positively. Differential polarization transfer, on the other hand, results in zero net polarization: the multiplet lines become polarized positively and negatively in such a way that the total polarization of each multiplet vanishes. Proton spin state population changes brought about by proton pulses, as in INEPT,[34,81] result in changes in the population difference for nuclei (^{13}C, ^{15}N, ^{29}Si, etc.) with transitions which share an energy level in common with this proton transition. Because of the progressive and regressive arrangement of these connected transitions, one carbon transition must lose while the other gains an equal amount; no net magnetization is transferred. These techniques are useful for sensitivity enhancement in undecoupled ^{13}C (or other I) spectra. Both SPT[59] and SPI[57] require a relatively precise knowledge of the proton chemical shifts and a resolvable *J* coupling (a long-range *J* will do, so they can be used even for quaternary carbons for which NOE is absent). The application of a selective 180° pulse (in SPI) to one of the proton transitions prior to acquisition of the ^{13}C FID leads to enhancements of $\pm\gamma_H/\gamma_C$ for the carbon-13 transitions connected to this proton, provided the pulse interval is much greater than $T_1(^1\mathrm{H})$.

The zero sum of intensities in a multiplet in these differential polarization transfer experiments means that any decoupling that causes all the lines to coalesce results in exact cancellation of intensities. Several methods have been devised to avoid this problem when decoupling is required. One is *pulse-interrupted precession*,[82] which results in a net polarization transfer, as does the insertion of a properly chosen delay period in the INEPT sequence (as in Figure 12) so that all members of a multiplet appear positively enhanced, i.e., INEPT with refocusing (INEPTR).[35,83] The DEPT (Distortionless Enhancement Polarization Transfer) sequence[88] gives undistorted relative intensities of the multiplet components.

These multiple pulse methods are summarized in Table 3. Examples are the use of SPT for ^{15}N and ^{13}C,[89] and ^{2}D in a liquid crystal solvent;[90] SPI for quaternary carbon;[91] double SPT for ^{29}Si, ^{13}C, and ^{15}N satellites in ^{13}C spectra;[73] INEPT with double quantum spectroscopy for ^{109}Ag and ^{103}Rh;[92] pulse-interrupted precession for ^{15}N;[93] INEPT with delay for ^{29}Si and ^{119}Sn;[94] INEPT using polarization transfer from ^{31}P to M = ^{103}Rh, ^{183}W, and ^{57}Fe,[95] and JCP for ^{15}N in polymers.[96]

Table 3. Multipulse Techniques for Signal Enhancement

Technique	Special requirements	Reference	Intensity[a]	Time[b]
Single $\pi/2$ conventional FT			1	T_{1I}
Overhauser polarization	Dipolar relaxation		$1+0.5\gamma_S/\gamma_I$	c
NOE	dominates, i.e., $T_1 \simeq T_1^{DD}$	62, 74		
Gated double resonance	Precise knowledge of		γ_S/γ_I	T_{1S}
Selective population transfer (SPT)	S chemical shifts	59, 84		
Selective population inversion (SPI)		57		
Double-selective population transfer		73		
Recovery of inhomogeneity broadening			1	$<T_{1I}$
Driven equilibrium FT (DEFT)		85		
Spin echo FT (SEFT)		86		
Differential transfer of polarization	Knowledge of $J(IS)$	87	γ_S/γ_I	T_{1S}
INEPT		81		
Net polarization transfer	Knowledge of $J(IS)$		γ_S/γ_I	T_{1S}
INEPTR		34, 35, 83		
Distortionless enhancement (DEPT)		88		
Hartmann–Hahn polarization				
Pines–Gibby–Waugh (CP)	Solids	8, 76	γ_S/γ_I	T_{1S}
J cross-polarization (JCP)	Special doubly tuned probe	77, 78	γ_S/γ_I	T_{1S}
Adiabatic cross-polarization	Ramped rf fields	79	γ_S/γ_I	d
Quasistochastic JCP		80	$\gamma_S/2\gamma_I$ [e]	T_{1S}

[a] Theoretical maximum normalized intensity.
[b] Experiment repetition rate determined by these times.
[c] Cross-relaxation rates.
[d] Cross-polarization times $\gg 2\,|J_{max}|/J_{min}^2$.
[e] For IS and IS_2 spin systems; $5\gamma_S/8\gamma_I$ for IS_3.

One measure of sensitivity enhancement is the theoretical maximum intensity in a single experiment relative to a conventional single $\pi/2$ FT experiment, as shown in Table 3. In practice, the over-all improvement in sensitivity may be better than this; a large number of replications of the experiment are carried out with a repetition rate determined by T_{1S}, the usually faster proton relaxation rather than the usually longer ^{15}N (or other) T_1. Improvement is smaller for the NOE since the rate at which the experiment can be repeated is determined by cross-relaxation rates which are appreciably slower than proton relaxation rates.

As mentioned above, multipulse techniques which regain the magnetization components dephased as a result of field inhomogeneity are signal-enhancing. Two of these are SEFT and DEFT. *Spin echo FT* (SEFT),[86] uses the Carr–Purcell sequence to produce a series of spin echoes which, considered as back-to-back FIDs, are averaged and transformed to give a spectrum. In *driven equilibrium FT* (DEFT)[85] an initial echo is established and its magnetization is restored to the B_0 direction by a further $\pi/2$ pulse so as to enhance the longitudinal magnetization for the next pulse sequence. These techniques are particularly time-efficient compared to conventional FT when T_1 of the observed nucleus is long, because the optimum delay before the next pulse sequence is short compared to T_1.

The choice of sensitivity enhancement method depends on a variety of factors. The NOE is the most routinely applied since it does not require specialized equipment or specific knowledge of energy levels or coupling constants. However, when $\gamma_1 < 0$ (as for ^{15}N or ^{29}Si) the NOE gives inverted signals, as in Figure 4 of Chapter 5, or loss of intensity (depending on the ratio T_1^{DD}/T_1), and may be better suppressed as described in Section 5.1. Since the direct dipolar relaxation mechanism which varies with the inverse sixth power of the internuclear distance is responsible for the NOE, only insensitive nuclei with directly bonded protons can routinely be assumed to benefit from this. The limitations of the other methods have already been mentioned. The J cross-polarization techniques allow enhancement over a wide range of J values. In the methods requiring a knowledge of J values for setting time delays in the pulse sequence in order to maximize enhancement, only those carbon-13 signals from nuclei with the chosen coupling constants will be enhanced; others with very different magnitudes of coupling will not be enhanced as much or at all. While this causes difficulties, we shall see below that it also provides a technique for distinguishing between different carbons and their connections to each other via coupling to protons.

5.3. Techniques in Aid of Spectral Analysis and Assignment

As the resolution and sensitivity of commercial spectrometers have improved, NMR spectroscopists have been increasingly challenged with spectra of more complex systems. In this section we outline some of the special techniques used in assignment of resonances, determination of coupling networks, differentiation of C, CH, CH_2, and CH_3 groups, etc.[41,42]

One way of facilitating the assignment of peaks is by correlation of chemical shifts of two nuclear species (such as ^1H and ^{13}C) in the same system. In its simplest form *chemical shift correlation spectroscopy* or COSY identifies pairs of proton and carbon resonances that arise from directly bonded atoms.[39] For each proton resonance frequency in the first dimension f_1, the frequency of the coupled carbon may be read off directly from the second dimension f_2. Heteronuclear couplings are absent from both frequency domains although proton–proton coupling structure remains in f_1. Selective proton decoupling can be used to remove some of these homonuclear couplings.[97] Although most applications are to ^{13}C–^1H correlation,[98] this has been applied to ^{31}P [99] and ^{11}B.[100] ^{13}C–^{13}C (directly bonded) correlation maps have also been generated using double quantum transition frequencies, enhanced by INEPT.[101] ^1H–^1H shift correlation maps display cross peaks correlating shifts of groups of spin-coupled protons.[102]

Another approach to spectral assignment is the determination of extended coupling networks. Two-dimensional chemical shift correlation spectroscopy may use a heteronuclear "spy" such as ^{31}P to report on those protons coupled to it.[103] The INADEQUATE (Incredible Natural Abundance Double Quantum Transfer Experiment) experiment allows adjacent carbons to be identified through their couplings.[47] *Relayed coherence transfer spectroscopy*[104] (RELAY) correlates the chemical shifts of carbon-13 nuclei not only with the directly coupled neighbor protons but also with remote protons coupled to the neighboring protons. Adjacent carbon sites in a molecule can be identified through the overlap of coupling networks. *SECSY* (spin echo correlated spectroscopy),[105] which is described further in Chapter 5, is useful in assignments of proton spectra of fairly large molecules; the 2D spectrum discloses the proton resonances connected by spin–spin coupling.

The differentiation of CH_3, CH_2, CH, and quaternary carbons in a complex molecule is particularly useful if the undecoupled ^{13}C spectrum is too crowded for clear identification of individual multiplets. Selective decoupling, or off-resonance decoupling is widely used. If the decoupling frequency is strong but not set precisely at $\nu(H)$, i.e., it is off-resonance by $\Delta\nu$, then the ^{13}C spectrum will have the same multiplicity as in the undecoupled spectrum but with a splitting reduced by the factor $(\Delta\nu/\gamma B_2)$, which can be chosen such that the one-bond couplings are reduced to ca. 5 Hz, making the multiplets more easily recognizable. A multipulse technique known as *J scaling*[106] produces scaled-down multiplets similar to those produced by off-resonance decoupling. Low-level or off-resonance noise decoupling can be used to identify quaternary sites by preferentially broadening the signals of protonated carbons.[107]

Several techniques of spectral editing are based on the modulation of carbon-13 spin echoes by J(CH). A carbon-13 spin echo experiment (Fourier transformation of the carbon-13 spin echo signals) is performed in a fixed evolution time followed by acquisition of the second half of the echo under proton-decoupled conditions. Echo modulation is introduced either by applying a π pulse on the proton system[108] or by gating the proton decoupler[109]; the components of the multiplets either reinforce or cancel when the proton decoupler is switched on under a proper choice of the evolution period. A subspectrum containing only CH and CH_3 carbons or containing only CH_2 and quaternary carbons can then be obtained.[108] Subspectra can also be produced by taking sums or differences of spectra from two or more experiments. This is a simple procedure in FT spectrometers since FIDs can be added and subtracted quite easily before Fourier transformation. For example, by summing over the results from experiments with different spin echo delay times, it is possible to eliminate all carbon signals except $C+CH_2$.[110] Still other techniques of determining ^{13}C multiplicity are based on a proper choice of the delay time in INEPT or flip angle in DEPT; ^{13}C resonances can be either removed or their phases changed depending on the number of protons.[35,85] For routine use the DEPT sequence is preferred. At a flip angle of 135° CH_2 signals will appear negative while CH and CH_3 signals remain positive.

Chain-tracing techniques for large molecules may use combinations of 2D J spectra, 2D NOE, and exchange correlation maps. A 2D INADEQUATE spectrum identifies sections of the carbon backbone of a molecule since the presence of a pair of double quantum peaks in the spectrum normally indicates a CC bond. An example of the power of 2D methods in spectral assignment and determination of structure is in the application to biopolymers.[111,112]

6. ORIENTED SYSTEMS

In a single crystal, the orientations of the internuclear vectors with respect to the magnetic field are stationary and not randomly directed. In a polycrystalline or powder sample the orientations of these vectors are stationary and randomly distributed. The same is true for molecules trapped in a clathrate or in a glassy matrix. On the other hand, molecules partly oriented by a liquid crystal or a strong electric field generally move freely in all other directions around the axis of orientation. NMR spectra in these oriented systems differ from those in the isotropic phase because of this incomplete averaging of interactions over all orientations. The full NMR Hamiltonian is needed to describe the spin system, and the quantities which are tensors (shielding, indirect spin–spin coupling, quadrupole coupling, and direct dipolar coupling) then appear in their full glory.

Because of the $1/r^3$ factor in the direct dipolar interaction, the contributions to the signal of a resonant spin due to dipolar interactions with the other nuclear spins can be very large (as from the other proton in the same H_2O molecule) or much smaller (as from the protons in neighboring H_2O molecules). It is these contributions due to a large number of neighboring spins at various distances over the entire sample which cannot be resolved, resulting in broad envelopes which can mask other information such as anisotropy of the shielding, and the indirect spin–spin coupling. Thus, the appearance of spectra of oriented systems will depend on the relative magnitudes of the components of these tensors. For example, the components of the proton shielding tensor are generally small and differences between them are small,[113] whereas the components of the ^{19}F tensor of a single molecule can have a range which exceeds the isotropic chemical shift range for ^{19}F nuclei.[7] Thus, ^{19}F NMR spectra of oriented molecules are generally dominated by the anisotropic chemical shift and proton spectra by dipolar coupling. Quadrupolar nuclei, on the other hand, generally show clearly visible quadrupolar splittings unless the electronic environment is spherically symmetric or nearly so.

6.1. High-Resolution NMR Techniques for Solids

Solid state NMR, especially of spin-1/2 nuclei, has developed greatly mainly thanks to Waugh and others, who succeeded in designing pulse techniques which drastically improved the resolution of solid state spectra and in addition, significantly enhanced the sensitivity for nuclei with low natural abundance.

6.1.1. Magic Angle Spinning, MAS

There are two ways of removing dipolar broadening. One is by manipulation of the sample by magic angle spinning and the other is by manipulation of the spins by applying a train of pulses. With magic angle spinning (*MAS*), widths as narrow as a few hertz have been obtained, although this is not always possible. Rapid spinning of a sample about an axis inclined at an angle β with respect to the magnetic field B_0 introduces a factor $(3\cos^2\beta - 1)/2$ into the time averaged NMR Hamiltonian representing the dipolar interaction, the anisotropic shift interaction, the anisotropic spin–spin coupling, and the first-order electric quadrupole interactions. Thus, if β is chosen as the "magic" angle which makes $3\cos^2\beta - 1$ vanish, all these interactions vanish from the spectrum under sufficiently fast rotation, revealing fine structure which was previously obscured.[114,115] The first spin multiplet to be resolved in the NMR spectrum of a solid was that of ^{19}F in $KAsF_6$.[115b] The shielding tensor components can be recovered from the spinning sideband intensities at lower spinning frequencies, either directly[116a] or by a moment analysis.[116b]

6.1.2. Multipulse Line-Narrowing Experiments

An alternative technique for the removal of dipolar broadening is by introducing a time dependence into the spin system by applying a train of rf pulses while the sample remains stationary. Pulse repetition rates can easily be made greater than the static dipolar line width so in effect line narrowing can be achieved with pulse spacings of a few microseconds equivalent to rotation speeds in excess of 100 kHz. The first line narrowing pulse sequence to achieve high resolution in solids was the four-pulse sequence of Waugh, Huber, and Haeberlen (WAHUHA-4), and several others are currently used.[117] This method is different from MAS in that (a) the pulse train affects interacting spins only if they have the same Larmor frequency (only homonuclear broadening is removed); (b) rotation of the spins does not affect the anisotropic chemical shift interaction (since this is first order in the spin). Thus the components of shielding can be obtained.

6.1.3. *High-Power Dipolar Decoupling*

Dipolar broadening by unlike spins is removed by strong irradiation of these spins at their resonance frequency.[118,119] This is analogous to proton decoupling in liquids: the indirect spin–spin coupling of the order of a few hundred hertz also disappears, but the power has to be set much higher so as to wipe out ^{13}C–^{1}H dipolar coupling of the order of 20 kHz.

6.1.4. *Cross-Polarization (CP)*

Introduced by Pines, Gibby, and Waugh,[8,76] this experiment (described in Section 5.2) exploits abundant spins I (e.g., protons) with large magnetogyric ratio γ_I as a reservoir of polarization for the rare spins S (e.g., ^{13}C) to enhance the NMR signal of S. This is done by repeatedly transferring polarization from I to S by application of two strong rf fields B_1^I and B_1^S at the I and S resonance frequencies such that the Hartmann–Hahn condition $\gamma_I B_1^I = \gamma_S B_1^S$ is satisfied. It is in fact the dipolar interaction which provides the I–S contact (cross polarization). Following this, the S spins are directly detected while the I spins are simultaneously decoupled by the same irradiating field B_1^I.

For liquid-like high-resolution ^{13}C spectra in solids, the combination of dipolar decoupling, cross polarization, and magic angle spinning may be used.[120,121] Variations on this basic experiment have been reviewed,[122–124] and methods are suitable also for observation of other dilute spins such as ^{15}N, ^{29}Si, ^{119}Sn.[125] Although most of the applications are to spin $\frac{1}{2}$, high-resolution spectra of quadrupolar nuclei in solids are also being obtained.[126] A growth area is the application of solid state NMR techniques to biological systems.[127]

6.2. Experimental Determination of Tensor Components of σ, J, D, and q

All of these quantities can (in principle) be obtained in the gas phase or in molecular beams with high-resolution microwave spectroscopy and molecular beam magnetic or electric resonance. Such data[128] provide isolated molecule values (usually selected as to vibrational and rotational state) with which to compare liquid crystal and solid state values. These techniques reveal the desired tensor components by their relationship to the principal inertial axes of the molecule.

In partially oriented systems (in liquid crystal solution or in strong electric fields) only linear combinations of tensor elements, such as $\sigma_{\|} - \sigma_{\perp}$ or $J_{aa} - (J_{bb} + J_{cc})/2$, may be obtained. The spin Hamiltonian for partially oriented molecules, spectral analysis, determination of NMR constants, and results from many applications have been described in detail.[129] Typically, the molecule is dissolved in a nematic mesophase which is aligned by the magnetic field, so that measurements made are of those components of the chemical shift, spin–spin coupling, and dipolar and quadrupolar coupling along the field direction. Both solute and solvent molecules are in rapid random motion in all the other directions and this averages intermolecular dipole–dipole interaction to nearly zero. NMR spectra with sharp lines are obtained which are more complex than isotropic phase spectra. The relationship between the observed quantities and the desired tensor components involves the ordering matrix elements S_{ab} introduced by Saupe.[130] This matrix depends on the symmetry of the interacting pair of nuclei for **J** or **D** or the symmetry of the nuclear site for $\boldsymbol{\sigma}$ or **q**. The partially averaged component of the chemical shift tensor along the director of the liquid crystal can be obtained and from this a linear combination of shielding components. For example, for the nuclear site symmetry of F in CH_3F, the S matrix has only one independent element, which is called S_{33} (for the threefold axis), and the F nucleus in CH_3F has an axially symmetric shielding tensor, so that $\Delta\delta = \delta_{\text{nem}} - \delta_{\text{av}} = (2/3)\, S_{33}(\sigma_{\parallel} - \sigma_{\perp})$. In general, $\Delta\delta = \delta_{\text{nem}} - \delta_{\text{av}} = (2/3) \sum_{a,b} S_{ab}\sigma_{ab}$. Similarly, linear combinations of spin–spin coupling components may be obtained. For example, for 1,1-difluoroethylene, with the b axis along the C=C bond and a axis perpendicular to the molecular plane,

$$\Delta J(\text{FF}) = (2/3)\, S_{aa}[J_{aa} - (J_{bb} + J_{cc})/2] + (1/3)(S_{bb} - S_{cc})(J_{bb} - J_{cc})$$

With six independent measurements of the ordering parameter and ΔJ (by varying concentration, for example), it is possible to determine separately the linear combinations $J_{aa} - (J_{bb} + J_{cc})/2$ and $(J_{bb} - J_{cc})$.

Because the intermolecular direct dipolar interaction is averaged out in a liquid crystal solvent, it is possible to observe the dipolar coupling constants characteristic of a vibrating molecule, and from these dipolar coupling constants, vibrationally averaged internuclear distances. For quadrupolar nuclei, splittings proportional to the nuclear quadrupole coupling constants are also observed. A summary of measurements in liquid crystal solutions of anisotropies in spin–spin coupling and shielding and quadrupole couplings is given by Emsley and Lindon.[129] There are extensive reviews of NMR of molecules oriented in the nematic phase,[131] in thermotropic liquid crystals,[132] and in the lyotropic phase.[133] Analyses of NMR spectra of molecules partially oriented by a strong electric field have been given.[4,134,135]

Before the advent of multipulse techniques the determination of the components of $\boldsymbol{\sigma}$, **J**, **q**, and **D** in solids could be carried out only for favorable systems (with very large shielding anisotropies or quadrupole coupling constants) by complete line shape analysis or second moment analysis of the wide line NMR spectrum obtained. Now NMR spectrometers are commercially available for high-resolution spectroscopy in solids. Depending on the nature of the system and of the information desired, some combination of the previously mentioned techniques is used. We give only a few examples here because the field is still taking fast strides.

To obtain the components of the shielding tensor in a single crystal or powder sample a line-narrowing pulse sequence such as WAHUHA-4 or its more recent versions may be used.[117] Many ^{19}F shielding components have been determined by this method.[7,113] The shielding tensor for a dilute spin such as ^{13}C may be obtained by cross polarization with high-power proton decoupling.[136] Zilm, Grant and co-workers have determined ^{13}C shielding components in small molecules in an argon matrix at 2–4 K.[137–139] A ^{31}P nucleus in a solid such as $(MeO)_3P$ can also be treated as a dilute spin, and many ^{31}P shielding components have been obtained by this method.[140,141]

With removal of dipolar broadening, smaller quadrupolar couplings (less than 1 MHz) can now be well resolved in solids, yielding values of q_{zz} as well as η.[22,142] Quadrupolar information can be obtained from spectra of neighboring nuclei as well. A well-known example of splitting due to a quadrupolar neighbor is observed for ^{13}C attached to ^{14}N which is in an asymmetric environment. A broadened asymmetric doublet is then obtained in the ^{13}C solid state spectrum with MAS because the axis of quantization of the ^{14}N spin is tilted away from $\mathbf{B}_0$. From an analysis of the line shape, values of e^2qQ and η for the nitrogen can be measured.[143a]

High-resolution dipolar spectra can also be obtained for solids. If there are large chemical shifts between magnetically dilute spins, as in the carbons in glycine ($NH_3^+CH_2COO^-$), CP with proton decoupling splits each ^{13}C line into a triplet because of the dipolar interaction with the ^{14}N nucleus. From this high-resolution dipolar spectrum it is possible to determine the internuclear distances and do temperature-dependent motion studies.[143b] Cross polarization can be extended to 3-spin systems such as ^1H, ^{13}C, ^{15}N.[122] In *double cross polarization*, magnetization is transferred from the protons to either ^{13}C or ^{15}N, then the proton channel is turned off to allow C–N spin contact. ^{13}C or ^{15}N spectra containing ^{13}C ^{15}N coupling (and the turnover of C–N bonds in biosystems) can then be observed with good sensitivity.[123]

The arsenal of techniques used for solutions is being applied to solids or liquid crystal solutions as well. For example, multiple quantum spectroscopy and two-dimensional NMR have been combined with the techniques described above.[144] Density matrix theory[42] and the individual spin operator formalism[145] have been used to understand the mechanisms of multiple pulse sequences and the design of new ones. Since FT spectrometers offer many of the techniques mentioned in this Section 6 as standard features, the chemist can utilize an array of multiple resonance multiple pulse techniques without benefit of theory, using Refs. 41–42 as guides.

7. THE NMR TIME SCALE

It is generally true for any kind of spectroscopy that one can obtain rather direct information about the dynamic behavior of molecules if the characteristic frequencies of the motion are comparable with frequency separations in the spectrum. Since frequency separations in NMR usually range from a few hertz to a few kilohertz in liquids or solutions even in superconducting magnets, the types of motions which are studied by NMR in solution have characteristic rates which are "slow" compared with other spectroscopy techniques. However, there are factors which extend the "characteristic" times of motion which can be studied by NMR, as we shall see below.

In liquids or solutions, the separations between peaks are determined by isotropic chemical shifts $(\sigma_A - \sigma_B)$ and isotropic spin–spin coupling constants, J. These are a few hertz to a few kilohertz depending on the nucleus and the magnetic field used. Some types of processes which have been studied extensively by band shape analysis in solution are intermolecular chemical exchange, rearrangements such as scrambling and bridge-terminal exchanges, internal rotation (conformational changes), and fluxional rearrangements.[146–150]

In the simplest uncoupled two-site exchange system, $A \underset{}{\overset{k_A}{\rightleftarrows}} B$, the appearance of the spectrum is fairly simple. The data needed to calculate the band shape are k_A, p_A, $\delta\nu = \nu_A - \nu_B$, T_{2A}, and T_{2B}. The rate constant k_A and the fractional population p_A vary with temperature. At slow exchange rates, $k_A \ll (\nu_A - \nu_B)$, two sharp signals are observed, separated by $\nu_A - \nu_B$, with intensities related to the fractional populations p_A and $(1 - p_A)$, and line widths dependent on effective T_{2A} and T_{2B}. At fast exchange rates, $k_A \gg (\nu_A - \nu_B)$, a single peak centered at the population-weighted mean of ν_A and ν_B is observed. At intermediate rates of exchange, greatly broadened or coalesced spectra are observed. Multisite exchange spectra are more complicated, but the basic changes in appearance are similar: sharp separated peaks broaden, then coalesce, with the emergence of fewer sharp peaks centered at population-averaged positions. The ambiguous "fast," "slow" in the above discussion has of course a different meaning for each system. For methyl groups undergoing enantiomerization, we take $\delta\nu = 20$ Hz, $J = 7$ Hz, $T_2 = 0.5$ s [146]; "fast" in the NMR time scale means $k \sim 10^5\ s^{-1}$; "slow" means $k \sim 10^{-1}\ s^{-1}$. The intermediate $k = 10$–$10^3\ s^{-1}$ show greatly broadened or coalesced peaks. One to $10^4\ s^{-1}$ is a typical range of rates of processes which can be studied in solution.

By the use of double resonance, one can measure slower rates of exchange than is possible with bandshape analysis using a single frequency. These double irradiation methods depend on the fact that an exchange process provides a mechanism by which the magnetization of one nucleus can affect the magnetization of the other nucleus. In the *spin saturation transfer* method[151] one of the sites is strongly irradiated; saturation is transferred to all sites directly connected to it by the exchange. The intensity of the signals for the latter will have a characteristic time dependence, from which the rate constant for the exchange can be derived. In the *selective inversion* experiment[152,153] a "soft" selective pulse is used to invert the signals of one site (e.g., one conformer). Reestablishment of the equilibrium value for these signals and transient effects noted in the magnetization of the second site (e.g., the other conformer) are related to T_{1A}, T_{1B} and the lifetimes before interchange. Lifetimes are of course inversely related to rates of exchange for first-order kinetics. A quantitative comparison of these double resonance methods with line shape analysis shows that they are equally reliable.[154]

Multipulse techniques have been used to identify exchange networks (i.e., routes of exchange in multisite exchange systems) with direct measurements of rate constants in solution, as in two-dimensional exchange spectroscopy[155,156] (discussed in Chapter 5, Section 8), "accordion" spectroscopy (involving a third Fourier transformation, thus 3D spectroscopy),[157,158] and time-dependent NOE.[159]

The frequencies of the motions which can be studied in the solid phase depend on the field and the origin of the peak separations. In solids the separation between the peaks are determined by the anisotropy of the shielding tensor, anisotropy of the quadrupole coupling, or if neither is significant, by the magnitude of dipolar broadening. For nuclei with small shielding anisotropies (such as protons), slower motions (with frequencies 1–20 kHz) can be studied. Examples are the tetrahedral motion of the protons in ice[160] or the slow jumps of protons in a single crystal of $CaSO_4 \cdot 2H_2O$ between 90 and 300 K.[161] For other spin-$\frac{1}{2}$ nuclei with larger anisotropies, processes with rates up to several hundred kHz can be studied if superconducting magnets are used. For example, jump frequencies of 88–1800 kHz have been measured for P_4 in solid white phosphorus at 104–126 K.[160] With ^{2}D still faster reorientations can be studied since the quadrupole splittings can reach 2 MHz. For example, D_2O molecules undergo 180 deg rotational jumps (flips) about their twofold axes at the rate of 10^4–$10^6\ s^{-1}$ from 170 to 190 K in $LiSO_4 \cdot D_2O$.[162]

It is also possible to distinguish between different types of motion, such as internal rotation, molecular rotation about a single fixed axis, tetrahedral reorientation, etc. The changes in line shape as a function of temperature in a solid provide information about the nature of the motions. In a completely rigid lattice, the various $(1 - 3\cos^2\theta_{ij})$ and r_{ij} which appear in the direct dipole–dipole interactions have fixed values. The observed spectrum is broad because of the contributions from all near and far neighbors of the observed spin. If reorientation is fast enough, the averages of $(1 - 3\cos^2\theta_{ij})$ and r_{ij} determine the shape of the observed spectrum. The average of $(1 - 3\cos^2\theta)$ over a sphere is zero. Thus, there are no linewidth contributions due to dipolar interactions of the observed nucleus with any collection of nuclei rotating as a unit in such a way that all orientations relative to the observed nucleus are taken with equal probability. The same is true for averaging over the vertices of a regular polyhedron as for rapid reorientation of an NH_4^+ ion such that the H nuclei exchange positions. The dipolar broadening produced by interaction of the observed spin with a rapidly reorienting group can be calculated for a given type of motion. Thus, the proton NMR line shape in $[Co(NH_3)_6]Cl_3$ at 100 K was matched by a theoretical spectrum for a rotating triangle model, the NH_3 reorienting rapidly about the Co–N bonds, whereas the line shape at 300 K was matched by a rotating ion model, the entire $Co(NH_3)_6^{3+}$ ion reorienting isotropically in the lattice. This beautiful piece of work was done in the early days of NMR.[163] Recent interpretation of the deuteron resonance spectra and Co NMR spectra of the single crystal supports this early work and provides more detailed information about the uniaxial reorientations and isotropic motions which take place in the temperature range 93–404 K.[164]

NMR relaxation studies provide information about dynamic behavior of liquids, complementing that obtained from other "fast" methods such as ir, Raman, and picosecond spectroscopy. This is due to the fact that the reorientation of the molecules in a liquid does not occur with a well-defined angular frequency, and Fourier analysis of this motion gives appreciable intensities at low frequencies as well as much higher ones. Quantitative information about molecular reorientation in liquids can be obtained from all the relaxation mechanisms of dipole–dipole and quadrupole coupling, as well as anisotropic shielding and spin-rotation interaction. By selecting the experimental conditions (nucleus, temperature, magnetic field) it may be possible to isolate particular mechanisms. From the spin-rotation interaction mechanism, one can generally obtain the correlation time of rotational angular momentum. Thus a ^{13}C or ^{15}N relaxation time of $\geqslant 10$ s implies a correlation time which typically is less than 0.1 ps![160] In the gas phase spin-rotation relaxation times provide cross-sections for rotational energy transfer and molecular reorientation.

Intramolecular relaxation by quadrupole coupling offers the possibility of studying anisotropic rotational motions in liquids, especially when both the quadrupole coupling tensor and the diffusion tensor have axial symmetry.[165,166] For planar molecules such as benzene, the anisotropy of the motion (the rotation about the sixfold axis being about 3.5 times as fast as rotation about axes in the molecular plane) can be studied through a wide range of temperatures by NMR because the relaxation through anisotropic shielding on the one hand, and dipolar and quadrupolar relaxation on the other hand, can be determined separately.

All these types of NMR studies, from measurements of second moments in wideline solid spectra, to complete bandshape analysis and measurements of relaxation times in liquid solutions and gases, allow determination of the details of molecular dynamics which have characteristic times ranging from hundreds of seconds to picoseconds.[160]

8. PHYSICAL EFFECTS ON THE NMR PARAMETERS

8.1. Medium and Temperature Effects

Medium effects include that of bulk susceptibility, which depends on the geometry of the sample and the composition of the solution. For a perfectly spherical sample, the nuclear magnetic shielding contribution σ_b of the bulk susceptibility is zero. For other shapes a susceptibility correction is needed because of polarization of the sample near the liquid surface. For an infinitely long cylinder, oriented with the cylindrical axis perpendicular to the direction of the magnetic field as in an electromagnet, the frequency shift due to bulk susceptibility of a pure solvent (as for a molecule observed in infinite dilution) is related to the molecular magnetic susceptibility χ, or the volume susceptibility χ_V of the solvent by

$$\sigma_b = +(2\pi/3)\,\chi_V = +(2\pi/3)\,\chi NC \times 10^3$$

where N is Avogadro's number, C (mol/liter) is the concentration of the neat liquid solvent, which is a function of temperature since a liquid usually expands with increasing temperature. For an infinitely long cylinder, oriented parallel to the field, as in a superconducting magnet

$$\sigma_b = -(4\pi/3)\,\chi_V$$

For diamagnetic substances χ and χ_V are negative, so that for example, for 85% phosphoric acid at room temperature, σ_b is -0.37 ppm in the conventional sample configuration in an iron magnet but $+0.73$ ppm in superconducting magnet, a difference of 1.1 ppm.[167] It has been found that a long axis/diameter ratio of 5 for a cylindrical sample is sufficiently large for the infinitely long cylinder formulas to hold to an accuracy of 5% in σ_b. This correction is of course the same for all observed nuclei in the molecule. When an external reference standard is used in conventional cylindrical sample tubes, the corrected chemical shift is then given by

$$\delta = [\nu(\text{sample}) - \nu(\text{ref})]/\nu(\text{ref}) - k[\chi_V(\text{sample}) - \chi_V(\text{ref})]$$

where $k = 2\pi/3$ for B_0 perpendicular and $k = -4\pi/3$ for B_0 parallel to the sample axis. The correct chemical shift is measured directly if the sample cell has concentric spherical chambers.

χ_V can be measured by any of several convenient methods. For spectrometers with the sample axis perpendicular to the magnetic field, the methods are based on the splitting of the signal when the reference liquid is placed in the annular region of a nonspinning coaxial sample tube. The splitting depends on the χ_V of the sample in the inside tube.[168–171] For an NMR spectrometer with a superconducting magnet, a different method has to be used.[172,173]

Further medium contributions to nuclear magnetic shielding arise from the van der Waals interactions,[174] electric field effects (if electric moments, polar groups, or net charges are present),[175] and magnetic anisotropy effects (if solute or solvent molecules have preferred orientations and anisotropic magnetic susceptibility).[176] For example, planar conjugated molecules have such a pronounced magnetic anisotropy effect on other molecules that this solvent shift has been given a specific name: ASIS (aromatic solvent induced shifts).[177] These interactions are binary (solute + one other) as well as many-body (solute + many others), so the concentration dependence of the shift is not always linear. Extrapolation of the resonance frequency to infinite dilution in a solution excludes only the effects of interactions between molecules of the same kind; the other effects still persist. Thus, even at infinite dilution, the resonance frequency depends on the composition or density of the solvent and is temperature dependent, because of the temperature dependence of the density (and therefore the concentration), and of the van der Waals (etc.) interactions between the solute and the solvent.

When an external homonuclear reference is used the reference resonance frequency depends on the geometrical arrangement, so that if the reference material is contained in a capillary tube this should be coaxial with the NMR tube. The reference resonance frequency also depends on the temperature, so that any variable temperature studies on the sample necessitate corrections due to the temperature dependence of the reference itself. The latter can only be obtained from observations relative to a truly temperature-independent standard such as an isolated xenon atom.[178] When an internal homonuclear reference is used, the same temperature dependence corrections have to be made, and in addition the reference frequency will be concentration dependent owing to the van der Waals and other interactions with the components of the sample.[179] Of course, in this case there will be no bulk susceptibility corrections to the chemical shift, since all observed molecules, including the reference, are subject to the same σ_b.

Referencing shifts in the presence of a paramagnetic substance illustrates the problems associated with referencing chemical shifts in general. If an internal reference is used, as noted above, then the measured quantity is the differential effect of the paramagnetic agent upon the molecule of interest and the reference. On the other hand, use of an external reference may involve a very large contribution to the shift due to the bulk magnetic susceptibility of a sample containing a paramagnetic agent. Thus, in both cases a correction must be made before the desired chemical shift difference $\Delta\delta = \delta(\text{with } p) - \delta(\text{without } p)$ can be reported[180]:

$$\text{True paramagnetic shift} = [\delta_p^{\text{ext}} - \delta_0^{\text{ext}}] - [\chi_V(\text{with } p) - \chi_V(\text{without } p)]$$

or

$$= [\delta_p^{\text{int}} - \delta_0^{\text{int}}] + \delta_r$$

where $k = 2\pi/3$ or $-4\pi/3$, as discussed above, and $\chi_V < 0$ for diamagnetic, or $\chi_V > 0$ for paramagnetic substances. δ_r is the contribution of the paramagnetic agent to the shielding of the reference molecule (and can be positive or negative). The solvent effects on the $m \cdot p$ complex may be different from those on m itself; this difference is usually neglected. The nature of solvent interactions with diamagnetic substances is described in Chapter 3 Section 5.1. The nature of the paramagnetic interactions which contribute to δ_r and the paramagnetic shift is described in Section 8.3 and in Chapter 3 Section 6.

These medium and temperature effects of a nonspecific nature have to be taken into account in variable temperature studies and especially in comparative studies involving small changes in chemical shifts or coupling constants. Magnitudes of medium effects for a variety of nuclear species are tabulated in several reviews, both for chemical shifts[179,181,182] and for coupling constants.[183,184] Except for bulk susceptibility, medium effects are usually related to the range of chemical shifts for the nucleus, and are proportionately large for heavy nuclei.

8.2. Isotope Effects

Isotope effects are generally small. There are primary isotope effects: the nitrogen chemical shift between $^{15}NH_3$ and $CH_3{}^{15}NO_2$ is slightly different from that between $^{14}NH_3$ and $CH_3{}^{14}NO_2$, even if all other conditions are kept the same. The ratio of the ^{13}C–H to the corresponding ^{13}C–D spin–spin coupling constant is not exactly the same for various molecular isotopomers and is not exactly equal to the ratio of γ_H/γ_D. There are also secondary isotope effects which involve comparisons between the same nuclear species (shifts or couplings) in different isotopomers. Magnitudes of isotope effects on chemical shifts depend on the range of chemical shifts of the observed nucleus, being less than 0.1 ppm for proton shifts, and up to 5.2 ppm per D substitution for ^{59}Co; they are tabulated in review articles.[181,185,186]

Larger isotope effects (equilibrium isotope effects) arise because chemical or conformational equilibrium constants are themselves mass dependent, leading to different equilibrium populations of conformers or exchange-related species. These are a consequence of the well-known kinetic isotope effect, and can lead to sizable differences in average chemical shifts or average coupling constants.[187] For example, an ion dissolved in D_2O can have a sizable shift (up to 3 ppm for F^- ion, up to 13 ppm for I^- ion) compared to the corresponding H_2O solution because of isotope effects on the hydrogen-bonding equilibria, liquid structure, etc.[188]

8.3. Effects of Paramagnetic Substances

NMR spectroscopy of paramagnetic species is a whole field in itself.[189–192] Nuclei in paramagnetic complexes interact with the unpaired electron spin, depending on the probability of the unpaired electron spin being found in the vicinity of the nucleus of interest. If the nucleus is part of the paramagnetic molecular species, or has nonspecific or weakly complexing intermolecular interactions with it, there will be observable effects on the chemical shifts, coupling constants, and line widths; there are large effects even in the gas phase, where the interacting pair have fairly short-lived encounters.[193] No spectrum may be observable if the lines are unduly broadened or shifted.

The chemical shift arising from interaction with the unpaired spin is a sum of two terms: contact and dipolar (sometimes called pseudocontact). The *Fermi contact interaction* is the interaction between the nuclear spin and a net electron spin density at the nucleus, giving rise to a_N, the electron hyperfine interaction constant at the nucleus N, as observed in ESR experiments. The net electron spin density at the nucleus may arise directly when the unpaired electron wave function involves an s orbital centered on the nucleus, and indirectly when the closed s shells centered on the nucleus are spin-polarized by the unpaired electron in a d or p orbital.

When the electron spin relaxation time T_{1e} is sufficiently short that, for a_N in joules,

$$T_{1e}^{-1} \gg a_N/\hbar$$

then the nucleus sees only a time-averaged local magnetic field proportional to the mean value $\langle S \rangle_{B0}$ of the electron spin component along the applied magnetic field B_0. This is nonvanishing because the electron spin states have significantly different populations in a magnetic field. The effective field due to this contact interaction is

$$(\Delta B)_{\text{con}} = -(a_N/g_N\mu_N)\langle S \rangle_{B0}$$

The mean value $\langle S \rangle_{B0}$ can be expressed as a function of B_0 and the temperature, to give a ratio

$$(\Delta B/B_0)_{\text{con}} = -(g_e\mu_B/g_N\mu_N)\, a_N S(S+1)/3kT$$

This assumes that the magnetic moment of the paramagnetic species contains no orbital contribution except that produced by spin–orbit coupling and that there is only one thermally accessible paramagnetic state with spin S. These conditions are not strictly true for some transition metal complexes.

The *dipolar shift* arises from the direct magnetic interaction of the electron spin and the nuclear spin. For molecules with tetragonal symmetry, the dipolar shift is[189]

$$(\Delta B/B_0)_{\text{dip}} = -(3\cos^2\theta - 1)\, r^{-3}\mu_B^2 S(S+1)\, F(g)/3kT$$

where r is the separation between the nucleus N and the unpaired electrons, and θ is the angle between the vector **r** and the symmetry axis of the molecule. $F(g)$ is a function depending on the g tensor components which reduces to $(g_{\parallel}^2 - g_{\perp}^2)/3$ for axial symmetry. Small contact shifts, large dipolar shifts, and axial symmetry are all desirable properties of shift reagents in most applications.

Small traces of paramagnetic species not only produce large shifts but also dramatically increase the rate of nuclear spin relaxation. The short electron spin relaxation time of many paramagnetic ions causes the local fields to fluctuate rapidly and induce fast transitions between the nuclear spin states, leading to line broadening. The line width then yields information on structure and bonding and also kinetic information.[194] A discussion of contributions of paramagnetic interactions to T_1 and T_2 is in Chapter 5. At high fields, the severe broadening caused by chemical exchange with lanthanide shift reagents can seriously impair resolution. Spin-echo methods can remove broadened peaks selectively and controllably without distorting the remaining signals.[195]

Paramagnetic species are added to a sample for several reasons. *Shift reagents* are used to increase the shifts between nuclear sites so as to reduce proton second-order NMR spectra to easily recognizable first-order spectra. Examples are lanthanide complexes, which coordinate to basic (e.g., ketonic) sites, and produce large shifts without substantial line broadening. Structural information can be obtained if the dipolar shifts are dominant over the contact shifts. From the relative shifts of different nuclei in various sites in the molecule, the relative r^{-3} values of different nuclei from the paramagnetic center can be deduced. *Nonspecific relaxation agents* such as [Cr(pd)$_3$] provide an alternative dominant relaxation mechanism for nuclei with long T_1 (unprotonated ^{13}C, ^{15}N, etc.) so as to shorten acquisition times. Paramagnetic species with a chiral center may be used to separate the NMR spectra of enantiomers.

There are some unwanted effects of paramagnetic additives. They may wash out spin–spin coupling, either by too fast relaxation (for small J), or by promoting decoupling by fast exchange. The magnitudes of coupling constants may be altered, inhibiting their use for structure determination. Furthermore, no paramagnetic relaxation reagent is truly "shiftless," although the bulk susceptibility correction can often be neglected in practice; the presence of 0.1 *M* [$Cr(pd)_3$] in nitromethane produces an ^{15}N shielding of 2.75 ppm (Chapter 12). [$Gd(dpm)_3$] induces significant increases in shielding by interaction with basic nitrogen sites.[196] No paramagnetic relaxation reagent is truly unreactive; thus the "aromatic" proton in the 2,4-pentanedionate ligand in [$Cr(pd)_3$] and other related complexes is susceptible to electrophilic substitution.[197] Finally, Overhauser effects of electron–nuclear cross relaxation (secondary changes in the populations of the nuclear spin states) may alter the relative intensities in a spectrum so that it does not reflect the relative numbers of nuclei quantitatively.

REFERENCES

1. (a) Fuller, G. H. *J. Phys. Chem. Ref. Data*, 1976, **5**, 835–864. (b) Lederer, C. M.; Shirley, V. S. *Table of Isotopes*, 7th ed., Wiley: New York, 1978; Appendix VII.
2. Abragam, A. "The Principles of Nuclear Magnetism," corrected reprint, Oxford, 1978.
3. *Pure Appl. Chem.* 1972, **29**, 627; 1976, **45**, 217–219.
4. Buckingham, A. D.; Pople, J. A. *Trans. Faraday Soc.* 1963, **59**, 2421–2430.
5. Buckingham, A. D.; Malm, S. M. *Mol. Phys.* 1971, **22**, 1127–1130.
6. Raynes, W. T. *NMR Spec. Per. Rep. C.S.* 1978, **7**, 1–25.
7. Mehring, M. *NMR Basic Princ. Prog.* 1976, **11**, 1–243.
8. Pines, A.; Gibby, M. G.; Waugh, J. S. *Chem. Phys. Lett.* 1972, **15**, 373–376.
8a. Wasylishen, R. E., personal communication; Gleeson, J. W.; Vaughan, R. W. *J. Chem. Phys.* 1983, **78**, 5384–5392; Klose, G.; Trahms, L.; Möps, A. *Chem. Phys. Lett.* 1985, **122**, 545–549.
9. Lynden-Bell, R. M.; Harris, R. K. *NMR Spectroscopy*, Nelson: London, 1969.
10. Venanzi, T. J. *J. Chem. Educ.* 1982, **59**, 144.
11. Abraham, R. J. *The Analysis of High Resolution NMR Spectra*, Elsevier: Amsterdam, 1971.
12. Emsley, J. W.; Feeney, J.; Sutcliffe, L. H. *High Resolution NMR Spectroscopy*, Pergamon: Oxford, 1965; Volumes I and II.
13. Hoffman, R. A.; Forsen, S.; Gestblom, B. *NMR Basic Princ. and Prog.* 1971, **5**, 1–165.
14. Siddall, T. H.; Flurry, R. L. Jr. *J. Magn. Reson.* 1980, **39**, 487–497.
15. Siddall, T. H.; Flurry, R. L. Jr. *J. Magn. Reson.* 1981, **43**, 357–372.
16. Siddall, T. H. *J. Phys. Chem.* 1982, **86**, 91–96.
17. Crocker, C. Ph.D. thesis, University of Bristol, 1978.
18. Haigh, C. W. *Annu. Rep. NMR Spectrosc.* 1971, **4**, 311–362.
19. Diehl, P.; Kellerhals, H.; Lustig, E. *NMR Basic Princ. and Prog.* 1972, **6**, 1–96.
20. Edmonds, D. T. *Int. Rev. Phys. Chem.* 1982, **2**, 103–126.
21. Sears, R. E. J. *Phys. Rev. Ser. B* 1980, **22**, 1135–1140.
22. Wolff, E. K.; Griffin, R. G.; Watson, C. J. *Chem. Phys.* 1977, **66**, 5433–5438.
23. Gorter, C. J. *Physica* 1936, **3**, 995–998; Gorter, C. J.; Broer, L. F. J. *Physica* 1942, **9**, 591–596.
24. Bloch, F. *Phys. Rev.* 1946, **70**, 460–474; 1956, **102**, 104–135.
25. Pople, J. A.; Schneider, W. G.; Bernstein, H. J. *High-Resolution NMR*; McGraw-Hill: New York, 1959.
26. Martin, M. L.; Delpuech, J. J.; Martin, G. J. *Practical NMR Spectroscopy*; Heyden: London, 1980, 492 pp.
27. Doddrell, D. M.; Barron, P. F.; Clegg, D. E.; Bowie, C. *J. Chem. Soc., Chem. Commun.* 1982, 575–576.
28. Akitt, J. W. *J. Magn. Reson.* 1970, **3**, 411–414.
29. Wehrli, F. W. *J. Magn. Reson.* 1978, **30**, 193–209.
30. Shaw, D. *Fourier Transform NMR Spectroscopy*, 2nd Ed.; Elsevier: Amsterdam, 1984, 304 pp.
31. Farrar, T. C.; Becker, E. D. *Pulse and Fourier Transform NMR*, Academic Press; New York, 1973, 115 pp.; Fukushima, E.; Roeder, S. B. *Experimental Pulse NMR: A Nuts and Bolts Approach*, Addison-Wesley: Reading, 1981, 539 pp.
32. Gibby, M.; Pines, A.; Waugh, J. S. *J. Am. Chem. Soc.* 1972, **94**, 6231–6232.

32a. Laue, E. D.; Skilling, J.; Staunton, J.; Sibisi, S.; Brereton, R. G. *J. Magn. Reson.* 1985, **62**, 437–452.
33. Ernst, R. R. *J. Magn. Reson.* 1970, **3**, 10–27; Bluemich, B.; Kaiser, R. *J. Magn. Reson.* 1983, **54**, 486–501.
33a. Belton, P. S.; Cox, I. J.; Harris, R. K. *J. Chem. Soc., Faraday Trans. 2* 1985, **81**, 63–75; Morris, G. A.; Toohey, M. J. *J. Magn. Reson.* 1985, **63**, 629–633; Canet, D.; Brondeau, J.; Marchal, J. P.; Robin–Lherbier, B. *Org. Magn. Reson.* 1982, **20**, 51–53; Lambert, J. B.; Riddell, F. G. Eds.; *The Multinuclear Approach to NMR Spectroscopy*, NATO ASI Series, Reidel; Dordrecht, 1983, p. 22 (Brévard, C); p. 310 (Drakenberg, T.; Forsén, S.; Ellis, P. D.).
34. Morris, G. A. *J. Am. Chem. Soc.* 1980, **102**, 428–429.
35. Doddrell, D. M.; Pegg, D. T. *J. Am. Chem. Soc.* 1980, **102**, 6388–6390.
36. Hahn, E. L. *Phys. Rev.* 1950, **80**, 580–584.
37. Carr, H. Y.; Purcell, E. M. *Phys. Rev.* 1954, **94**, 630–638.
38. Freeman, R.; Hill, H. D. W. In *Dynamic NMR Spectroscopy*, Jackman, L. M.; Cotton, F. A., Ed.; Academic Press: New York, 1975; pp. 131–162.
39. Freeman, R.; Morris, G. A. *Bull. Magn. Reson.* 1979, **1**, 5–26.
40. Freeman, R. *Proc. R. Soc. London, Ser. A* 1980, **373**, 149–178.
41. Morris, G. A. *Magn. Reson. Chem.* 1986, **24**, 371–403; *Top.* ^{13}C *NMR Spectrosc.* 1984, **4**, 179–196; Benn, R.; Günther, J. *Angew. Chem. Int. Ed. Engl.* 1983, **22**, 350–380; Turner, C. J. *Prog. NMR Spectrosc.* 1985, **16**, 311–358; Sanders, J. K. M.; Hunter, B. K. *Modern NMR Spectroscopy*, Oxford, 1987.
42. Bax, A. *Two-Dimensional NMR Spectroscopy in Liquids*, Reidel: Hingham, 1982; *Top.* ^{13}C *NMR Spectrosc.* 1984, **4**, 197–238; Ernst, R. R.; Bodenhausen, G.; Wokaun, A. *Principles of NMR in One and Two Dimensions*; Clarendon Press: Oxford, 1986; Derome, A. E. *Modern NMR Techniques for Chemistry Research*, Pergamon: Oxford, 1987.
43. Drobny, G.; Pines, A.; Sinton, S.; Weitekamp, D. P.; Wemmer, D. Faraday Symp. C. S. (Pulsed NMR in Solids), 1978, **13**, 49–55.
44. Drobny, G.; Pines, A.; Sinton, S.; Warren, W. S.; Weitekamp, D. P. *Phil. Trans. R. Soc. London, Ser. A* 1981, **299**, 585–592; Munowitz, M.; Pines, A. *Science* 1986, **233**, 525–531.
45. Turner, D. L. *J. Magn. Reson.* 1982, **49**, 175–178.
46. Bax, A.; Freeman, R.; Frenkiel, T. A.; Levitt, M. H. *J. Magn. Reson.* 1981, **43**, 478–483.
47. Bax, A.; Freeman, R.; Frenkiel, T. A. *J. Am. Chem. Soc.* 1981, **103**, 2102–2104.
48. Warren, W. S.; Pines, A. *J. Am. Chem. Soc.* 1981, **103**, 1613–1617.
49. Bodenhausen, G. *Prog. NMR Spectrosc.* 1981, **14**, 137–173.
50. Hoffman, R. A.; Forsen, S. *Prog. NMR Spectrosc.* 1966, **1**, 15–204.
51. McFarlane, W. *Annu. Rep. NMR Spectrosc.* 1972, **5a**, 353–393; *Annu. Rep. NMR Spectrosc.* 1979, **9**, 319–414.
52. McFarlane, W.; Rycroft, D. S. *Annu. Rep. NMR Spectrosc.* 1985, **16**, 293–364.
53. McFarlane, W. and others, annual chapters. *NMR Spec. Per. Rep. C. S.* 1977–1986, **6–15**.
54. Baker, E. B. *J. Chem. Phys.* 1962, **37**, 911–912.
55. Freeman, R.; Anderson, W. A. *J. Chem. Phys.* 1962, **37**, 2053–2073.
56. Levitt, M. H.; Freeman, R.; Frenkiel, T. *J. Magn. Reson.* 1982, **50**, 157–160; Waugh, J. S. *J. Magn. Reson.* 1982, **49**, 517–521.
57. Pachler, K. G. R.; Wessels, P. L. *J. Magn. Reson.* 1973, **12**, 337–339.
58. Chalmers, A. A.; Pachler, K. G. R.; Wessels, P. L. *J. Magn. Reson.* 1974, **15**, 415–419.
59. Sorensen, S.; Hansen, R. S.; Jakobsen, H. J. *J. Magn. Reson.* 1974, **14**, 243–245.
60. Anderson, W. A. *Phys. Rev.* 1956, **102**, 151–167.
61. Bloch, F. *Phys. Rev.* 1954, **93**, 944.
62. Solomon, I.; Bloembergen, N. *J. Chem. Phys.* 1956, **25**, 261–266.
63. Kaiser, R. *J. Chem. Phys.* 1963, **39**, 2435–2442.
64. Feeney, J.; Shaw, D.; Pauwells, P. J. S. *J. Chem. Soc., Chem. Commun.* 1970, 554–555.
65. Freeman, R.; Hill, H. D. W. *J. Magn. Reson.* 1971, **5**, 278–280.
66. Freeman, R.; Hill, H. D. W.; Kaptein, R. *J. Magn. Reson.* 1972, **7**, 327–329.
67. Ernst, R. R. *J. Chem. Phys.* 1966, **45**, 3845–3861.
68. DeMilou, M. E.; Kowalewski, V. J. *J. Magn. Reson.* 1982, **46**, 54–60.
69. Kessler, H.; Krack, G.; Zimmermann, G. *J. Magn. Reson.* 1981, **44**, 208–211.
70. Noggle, J. H.; Schirmer, R. E. *The Nuclear Overhauser Effect: Chemical Applications*, Academic Press: New York, 1971, 259 pp.
71. Freeman, R.; Pachler, K. G. R.; LaMar, G. D. *J. Chem. Phys.* 1971, **55**, 4586–4593.
72. McFarlane, W. *Annu. Rep. NMR Spectrosc.* 1968, **1**, 135–163.

73 Sorensen, O W , Bildsoe, H , Jakobsen, H J *J Magn Reson* 1981, **45**, 325–336
74 Kuhlmann, K F , Grant, D M *J Am Chem Soc* 1968, **90**, 7355–7357
75 Hartmann, S R , Hahn, E L *Phys Rev* 1962, **128**, 2042–2053
76 Pines, A , Gibby, M G , Waugh, J S *J Chem Phys* 1972, **56**, 1776–1777, *J Chem Phys* 1973, **59**, 569–590
77 Bertrand, R D , Moniz, W B , Garroway, A N , Chingas, G C *J Am Chem Soc* 1978, **100**, 5227–5229
78 Bertrand, R D , Moniz, W B , Garroway, A N , Chingas, G C *J Magn Reson* 1978, **32**, 465–467
79 Chingas, G C , Garroway, A N , Moniz, W B , Bertrand, R D *J Am Chem Soc* 1980, **102**, 2526–2528
80 Doddrell, D M , Pegg, D T , Bendall, M R *J Magn Reson* 1982, **49**, 181–196
81 Morris, G A , Freeman, R *J Am Chem Soc* 1979, **101**, 760–762
82 Maudsley, A A , Muller, L , Ernst, R R *J Magn Reson* 1977, **28**, 463–469
83 Burum, D T , Ernst, R R *J Magn Reson* 1980, **39**, 163–168, Bolton, P H *J Magn Reson* 1980, **41**, 287–292
84 Jakobsen, H J , Linde, S A , Sorensen, S *J Magn Reson* 1974, **15**, 385–388
85 Becker, E D , Ferretti, J A , Farrar, T C *J Am Chem Soc* 1969, **91**, 7784–7785
86 Allerhand, A , Cochran, D W *J Am Chem Soc* 1970, **92**, 4482–4484
87 Maudsley, A A , Ernst, R R *Chem Phys Lett* 1977, **50**, 368–372
88 Doddrell, D M , Pegg, D T , Bendall, M R *J Magn Reson* 1982, **48**, 323–327
89 Martineau, H , Trierweiler, M , Martin, M L *Org Magn Reson* 1981, **17**, 182–185, Sarkar, S K , Bax, A *J Magn Reson* 1985, **62**, 109–112
90 Voigt, J , Jakobsen, J P *J Magn Reson* 1981, **45**, 510–513
91 Pachler, K G R , Wessels, P L *J Magn Reson* 1977, **28**, 53–61
92 Brevard, C , Van Stein, G C , Van Koten, G *J Am Chem Soc* 1981, **103**, 6746–6748
93 Staedeli, W , Bigler, P , Von Philipsborn, W *Org Magn Reson* 1981, **16**, 170–172
94 Doddrell, D M , Pegg, D T , Brooks, W , Bendall, M R *J Am Chem Soc* 1981, **103**, 727–728
95 Brevard, C , Schimpf, R *J Magn Reson* 1982, **47**, 528–534
96 Holmes, B S , Chingas, G C , Moniz, W B , Ferguson, R C *Macromol* 1981, **14**, 1785–1787
97 Bolton, P H , Bodenhausen, G *J Magn Reson* 1981, **43**, 339–342
98 Morris, G A , Hall, L D *J Am Chem Soc* 1981, **103**, 4703–4711
99 Van Divender, J M , Hutton, W C *J Magn Reson* 1982, **48**, 272–279
100 Finster, D C , Hutton, W C , Grimes, R N *J Am Chem Soc* 1980, **102**, 400–401
101 Sorensen, O W , Freeman, R , Frenkiel, T , Mareci, T H , Schuck, R *J Magn Reson* 1982, **46**, 180–184
102 Bax, A , Freeman, R , Morris, G A *J Magn Reson* 1981, **42**, 164–168
103 Bolton, P H , Bodenhausen, G *J Am Chem Soc* 1979, **101**, 1080–1084
104 Bolton, P H *J Magn Reson* 1982, **48**, 336–340, Bax, A *J Magn Reson* 1983, **53**, 149–153, Bax, A , Drobny, G *J Magn Reson* 1985, **61**, 306–320
105 Nagayama, K , Wuthrich, K , Ernst, R R *Biochem Biophys Res Commun* 1979, **90**, 305–311
106 Freeman, R , Morris, G A *J Magn Reson* 1978, **29**, 173–176, Morris, G A , Nayler, G L , Shaka, A J , Keeler, J , Freeman, R *J Magn Reson* 1984, **58**, 155–160, 161–166
107 Opella, S J , Cross, T A *J Magn Reson* 1980, **37**, 171–175
108 Bendall, M R , Doddrell, D M , Pegg, D T *J Am Chem Soc* 1981, **103**, 4603–4605
109 Brown, D W , Nakashima, T T , Rabenstein, D L *J Magn Reson* 1981, **45**, 302–314
110 Bolton, P H *J Magn Reson* 1981, **45**, 418–421
111 Nagayama, K , Wuthrich, K *Eur J Biochem* 1981, **114**, 365–374
112 Wagner, G , Kumar, A , Wuthrich, K *Eur J Biochem* 1981, **114**, 375–384, Wuthrich, K *NMR of Proteins and Nucleic Acids*, Wiley New York, 1986, 432 pp
113 Haeberlen, U *Adv Magn Reson* 1976, Suppl 1, 1–186
114 Andrew, E R *Prog NMR Spectrosc* 1971, **8**, 1–39
115 (a) Andrew, E R *Phil Trans R Soc London Ser A* 1981, **299**, 505–520, *Int Rev Phys Chem* 1981, **1**, 195–224 (b) Andrew, E R , Farnell, L F , Gledhill, T D *Phys Rev Lett* 1967, **19**, 6–7
116 (a) Herzfeld, J , Berger, A E *J Chem Phys* 1980, **73**, 6021–6030 (b) Maricq, M M , Waugh, J S *J Chem Phys* 1979, **70**, 3300–3316
117 (a) Waugh, J S , Huber, L M , Haeberlen, U *Phys Rev Lett* 1968, **20**, 180–182 (b) Rhim, W K , Burum, D P , Elleman, D D *J Chem Phys* 1979, **71**, 3139–3141
118 Sarles, L R , Cotts, R M *Phys Rev* 1958, **111**, 853–859

119 Mehring, M , Pines, A , Rhim, W -K , Waugh, J S *J Chem Phys* 1971, **54**, 3239–3240
120 Schaefer, J , Stejskal, E O *J Am Chem Soc* 1976, **98**, 1030–1032
121 Earl, W L , VanderHart, D L *J Magn Reson* 1982, **48**, 35–54
122 *Phil Trans R Soc London, Ser A* 1981, **299**, (especially pages 593–663)
123 Wasylishen, R E , Fyfe, C A *Annu Rep NMR Spectrosc* 1982, **12**, 1–80
124 Fyfe, C A *Solid State NMR for Chemists*, CFC Press Guelph, Ontario, Canada, 1983
125 Lippmaa, E T , Alla, M A , Pehk, T J , Engelhardt, G *J Am Chem Soc* 1978, **100**, 1929–1931
126 (a) Oldfield, E , Schramm, S , Meadows, M D , Smith, K A , Kinsey, R A , Ackerman, J *J Am Chem Soc* 1982, **104**, 919–920, Ganapathy, S , Schramm, S , Oldfield, E *J Chem Phys* 1982, **77**, 4360–4365, Meadows, M D , Smith, K A , Kinsey, R A , Rothgeb, T M , Scarjune, R P , Oldfield, E *Proc Natl Acad Sci USA* 1982, **79**, 1351–1355, Schramm, S , Oldfield, E *J Am Chem Soc* 1984, **106**, 2502–2506, Oldfield, E , Timken, H K C , Montez, B , Ramachandran, R *Nature* 1985, **318**, 163–165, (b) Lippmaa, E , Samoson, A , Magi, M *J Am Chem Soc* 1986, **108**, 1730–1735
127 Opella, S J *Annu Rev Phys Chem* 1982, **33**, 533–562
128 Landolt–Bornstein, *Zahlenwerte und Funktionen aus Naturwissenschaften und Technik*, New Series, Group II, Springer-Verlag Berlin, 1974, Vol 6, pp 2–383 to 2–434
129 Emsley, J W , Lindon, J C *NMR Spectroscopy Using Liquid Crystal Solvents*, Pergamon Oxford, 1975, 367 pp , Khetrapal, C L , Kunwar, A C *Adv Liq Cryst* 1983, **6**, 173–242
130 Saupe, A *Z Naturforsch A* 1964, **19**, 161–171
131 Daborn, G T , Ferguson, W I , Handy, N C *Chem Phys* 1980, **50**, 255–263
132 Khetrapal, C L , Kunwar, A C *Adv Magn Reson* 1977, **9**, 302–422
133 Khetrapal, C L , Kunwar, A C , Tracey, A S , Diehl, P *NMR Basic Princ Prog* 1975, **9**
134 Buckingham, A D , McLauchlan, K A *Prog NMR Spectrosc* 1967, **2**, 63–109
135 Haeberlen, U , Waugh, J S *Phys Rev* 1968, **175**, 453–467
136 Veeman, W S *Phil Trans R Soc London, Ser A* 1981, **299**, 629–641, *Prog NMR Spectrosc* 1984, **16**, 193–235
137 Zilm, K W , Beeler, A J , Grant, D M , Michl, J , Chou, T -C , Allred, E L *J Am Chem Soc* 1981, **103**, 2119–2120
138 Zilm, K W , Conlin, R T , Grant, D M , Michl, J *J Am Chem Soc* 1980, **102**, 6672–6676, 1978, **100**, 8038–8039
139 Facelli, J C , Orendt, A M , Beeler, A J , Solum, M S , Depke, G , Malsch, K D , Downing, J W , Murthy, P S , Grant, D M , Michl, J *J Am Chem Soc* 1985, **107**, 5649–6754, Beeler, A J , Orendt, A M , Grant, D M , Cutts, P W , Michl, J , Zilm, K W , Downing, J W , Facelli, J C , Schindler, M S , Kutzelnigg, W *J Am Chem Soc* 1984, **106**, 7672–7676, Strub, H , Beeler, A J , Grant, D M , Michl, J Cutts, P W , Zilm, K W *J Am Chem Soc* 1983, **105**, 3333–3334
140 Dutasta, J P , Robert, J B , Wiesenfeld, L *ACS Symp Ser* 1981, **171**, 581–584
141 Robert, J B , Wiesenfeld, L *Mol Phys* 1981, **44**, 319–327
142 Bodenhausen, G , Stark, R E , Ruben, D J , Griffin, R G *Chem Phys Lett* 1979, **67**, 424–427
143 (a) Opella, S J , Hexem, J G , Frey, M H , Cross, T A *Phil Trans R Soc London, Ser A* 1981, **299**, 665–683, Hexem, J G , Frey, M H , Opella, S J *J Am Chem Soc* 1981, **103**, 224–226, (b) Griffin, R G , Bodenhausen, G , Haberkorn, R A , Huang, T H , Munowitz, M , Osredkar, R , Ruben, D J , Stark, R E , van Willigen, H *Phil Trans R Soc London, Ser A* 1981, **299**, 547–563
144 Drobny, G P *Annu Rev Phys Chem* 1985, **36**, 451–489
145 Sørensen, O W , Ernst R R *J Magn Reson* 1983, **51**, 477–489, Sørensen, O W , Eich, G , Levitt, M H , Bodenhausen, G , Ernst, R R *Prog NMR Spectrosc* 1983, **16**, 163–192
146 Sandstrom, J *Dynamic NMR Spectroscopy*, Academic Press New York, 1982, 226 pp
147 Steigel, A In *NMR Basic Princ Prog* 1978, **15**, 1–53
148 Mann, B E Annu *Rep NMR Spectrosc* 1982, **12**, 263–286, Martin, M L , Sun, X Y , Martin, G J *Annu Rep NMR Spectrosc* 1985, **16**, 187–364
149 *Dynamic NMR Spectroscopy*, Jackman, L M , Cotton, F A , Ed , Academic Press New York, 1975, 660 pp
150 Oki, M , Ed , *Applications of Dynamic NMR Spectroscopy to Organic Chemistry*, VCH Verlag Weinheim, 1985, 423 pp
151 Forsen, S , Hoffmann, R A *J Chem Phys* 1964, **40**, 1189–1196, 1963, **39**, 2892–2901
152 Alger, J R , Prestegard, J H *J Magn Reson* 1977, **27**, 137–141
153 Campbell, I D , Dobson, C M , Ratcliffe, R G , Williams, R J P *J Magn Reson* 1978, **29**, 397–417
154 Baine, P , Gerig, J T , Stock, A D *Org Magn Reson* 1981, **17**, 41–45

155. Jeener, J.; Meier, B. H.; Bachmann, P.; Ernst, R. R. *J. Chem. Phys.* 1979, **71**, 4546–4553.
156. Macura, S.; Huang, Y.; Suter, D.; Ernst, R. R. *J. Magn. Reson.* 1981, **43**, 259–281.
157. Bodenhausen, G.; Ernst, R. R. *J. Am. Chem. Soc.* 1982, **104**, 1304–1309.
158. Bodenhausen G.; Ernst, R. R. *J. Magn. Reson.* 1981, **45**, 367–373.
159. Dobson, C. M.; Olejniczak, E. T.; Poulsen, F. M.; Ratcliffe, R. G. *J. Magn. Reson.* 1982, **48**, 97–110.
160. Spiess, H. W. *NMR Basic Princ. Prog.* 1978, **15**, 55–214.
161. McKnett, C. L.; Dybowski, C. R.; Vaughan, R. W. *J. Chem. Phys.* 1975, **63**, 4578–4581.
162. Berglund, B.; Tegenfeldt, J. *J. Magn. Reson.* 1977, **27**, 315–323.
163. Waugh, J. S. *Ann. N. Y. Acad. Sci.* 1958, **70**, 900–926.
164. Reynhardt, E. C. *J. Magn. Reson.* 1977, **28**, 441–458.
165. Huntress, W. T. *Adv. Magn. Reson.* 1970, **4**, 1–37.
166. Jonas, J. *Adv. Magn. Reson.* 1973, **6**, 73–139.
167. Batley, M.; Redmond, J. W. *J. Magn. Reson.* 1982, **49**, 172–174.
168. Reilly, C. A.; McConnell, H. M.; Meisenheimer, R. G. *Phys. Rev. A* 1955, **98**, 264–265.
169. Zimmerman, J. R.; Foster, M. R. *J. Phys. Chem.* 1957, **61**, 282–289.
170. Li, N. C.; Scruggs, R. L.; Becker, E. D. *J. Am. Chem. Soc.* 1962, **84**, 4650–4654.
171. Douglass, D. C.; Fratiello, A. *J. Chem. Phys.* 1963, **39**, 3161–3162.
172. Lagodzinskaya, G. V.; Klimenko, I. Y. *J. Magn. Reson.* 1982, **49**, 1–7.
173. Kovesdi, I. *J. Magn. Reson.* 1981, **43**, 1–7.
174. Raynes, W. T.; Buckingham, A. D.; Bernstein, H. J. *J. Chem. Phys.* 1962, **36**, 3481–3488.
175. Buckingham, A. D. *Can. J. Chem.* 1960, **38**, 300–307.
176. Buckingham, A. D.; Schaefer, T.; Schneider, W. G. *J. Chem. Phys.* 1960, **32**, 1227–1233.
177. Laszlo, P. *Prog. NMR Spectrosc.* 1968, **3**, 231–402.
178. Jameson, C. J.; Jameson, A. K.; Cohen, S. M. *J. Magn. Reson.* 1975, **19**, 385–392.
179. Rummens, F. H. A. *NMR Basic Princ. Prog.* 1975, **10**, 1–118.
180. Qiu, Z. W.; Grant, D. M.; Pugmire, R. J. *J. Am. Chem. Soc.* 1982, **104**, 2747–2753.
181. Jameson, C. J. *Bull. Magn. Reson.* 1980, **3**, 3–28.
182. Jameson, C. J.; Jameson, A. K.; Burrell, P. M. *J. Chem. Phys.* 1980, **73**, 6013–6020.
183. Barfield, M.; Johnston, M. D. Jr. *Chem. Rev.* 1973, **73**, 53–73.
184. Emsley, J. W.; Phillips, L.; Wray, V. *Prog. NMR Spectrosc.* 1975, **10**, 83–756.
185. Batiz-Hernandez, H.; Bernheim, R. A. *Prog. NMR Spectrosc.* 1967, **3**, 63–85; Hansen, P. E. *Annu. Rep. NMR Spectrosc.* 1983, **15**, 105–233; Forsyth, D. A. *Isotopes in Organic Chemistry*, 1984, **6**.
186. Jameson, C. J. *NMR Spec. Per. Rep. R.S.C.* 1980, **10**, 1–17; Jameson, C. J.; Osten, H. J. *Annu. Rep. NMR Spectrosc.* 1986, **17**, 1–78.
187. Anet, F. A. L.; Bagus, V. J.; Hewett, A. P. W.; Saunders, M. *J. Am. Chem. Soc.* 1980, **102**, 3945–3946.
188. Lindman, B.; Forsén, S. *NMR Basic Princ. Prog.* 1976, **12**, 1–365.
189. Webb, G. A. *Annu. Rep. NMR Spectrosc.* 1975, **6A**, 1–145.
190. Orrell, K. G. *Annu. Rep. NMR Spectrosc.* 1979, **9**, 1–124.
191. Reuben, J. *Prog. NMR Spectrosc.* 1973, **9**, 1–70.
192. LaMar, G. N.; Horrocks, W. DeW., Jr.; Holm, R. H., Ed.; *NMR of Paramagnetic Molecules*, Academic Press: New York, 1973.
193. Jameson, C. J.; Jameson, A. K.; Cohen, S. M. *Mol. Phys.* 1975, **29**, 1919–1927; *J. Chem. Phys.* 1976, **65**, 3397–3400; Johnson, C. S.; Waugh, J. S. *J. Chem. Phys.* 1961, **35**, 2020–2024; Lipsicas, M.; Siegel, M.; Piott, J.; Higgins, T.; Gerber, J. A. *J. Magn. Reson.* 1976, **22**, 303–314.
194. Swift, T. J. pp. 53–177 in ref. 192.
195. Bulsing, J. M.; Sanders, J. K. M.; Hall, L. D. *J. Chem. Soc., Chem. Commun.* 1981, 1201–1203.
196. Witanowski, M.; Stefaniak, L.; Kamienski, B.; Biernet, S.; Webb, G. A. *J. Magn. Reson.* 1981, **43**, 456–462.
197. Brown, A. J.; Howarth, O. W.; Moore, P. *J. Am. Chem. Soc.* 1978, **100**, 713–718.

Chapter 3

THE CHEMICAL SHIFT

CYNTHIA J. JAMESON AND JOAN MASON

1. NUCLEAR MAGNETIC SHIELDING AND THE CHEMICAL SHIFT

1.1. The Absolute Shielding Tensor

The chemical shift (as defined by IUPAC) in terms of measured frequencies

$$\delta = (\nu - \nu_{\text{ref}})/\nu_{\text{ref}} \tag{1}$$

is related to a fundamental molecular property, the nuclear magnetic shielding, $\boldsymbol{\sigma}$. For a molecule held in a fixed orientation in a uniform magnetic field **B**, the field at a selected point in the vicinity of the molecule is different from **B** because of the small field arising from the circulations of the electrons induced by the magnetic field*:

$$\mathbf{B}_{\text{local}} = (1 - \boldsymbol{\sigma})\mathbf{B} \tag{2}$$

The term "magnetic shielding" implies that any magnetic dipole of very small magnitude, such as that of a nucleus placed at that selected point, would be shielded from the full effect of **B** by the influence of the induced electronic motions. For free atoms σ is always positive because this circulation generates a shielding field which opposes the applied field. In a molecule the presence of other nuclei hinders this circulation to an extent which depends on the electronic distribution and may even lead to a negative σ.

The components of the magnetic shielding at a nucleus in the molecule are defined as follows:

$$(\mathbf{B}_{\text{local}})_\alpha = (\mathbf{1} - \boldsymbol{\sigma})_{\alpha\beta}\, \mathbf{B}_\beta, \qquad \alpha, \beta = x, y, z \tag{3}$$

That is, the induced field in the α direction due to the currents induced in the molecular electrons by a uniform field in the β direction is given by $\sigma_{\alpha\beta} B_\beta$. In general $\sigma_{\alpha\beta} \neq \sigma_{\beta\alpha}$ so that nine numbers (components) specify the nuclear magnetic shielding. However, the symmetry of the molecular electronic distribution at the nuclear site renders some of these components equivalent or zero. By observing the NMR spectra of

* Equation (2) applies strictly to diamagnetic samples. In paramagnetic samples there are additional local magnetic fields generated by the paramagnetic center. These are discussed in Section 5 of this chapter.

CYNTHIA J. JAMESON • Department of Chemistry, University of Illinois, Chicago, Illinois 60680. JOAN MASON • Department of Chemistry, The Open University, Milton Keynes MK7 6AA, England.

oriented or partially oriented molecules in a single crystal, a polycrystalline solid, a liquid crystal, or a solid argon matrix, it is possible to find three principal elements δ_{11}, δ_{22}, and δ_{33}, which can be related to the absolute shielding principal components σ_{11}, σ_{22}, and σ_{33} with a knowledge of the absolute shielding of the reference. For systems with more than three independent shielding tensor elements, some information is lost, for the observed frequencies depend on some linear combination of σ_{ii} and $(\sigma_{ij})^2$.

In fluid media the shielding tensor is averaged over all orientations of the molecule during its tumbling motion. If the fluid is isotropic, all orientations are equally probable and the average value of the shielding tensor is its isotropic part, a single quantity $\sigma = (\sigma_{11} + \sigma_{22} + \sigma_{33})/3$ for each nucleus. Under such conditions the experimentally observed quantity, the chemical shift δ, is related to σ as follows:

$$\delta = (\nu - \nu_{\text{ref}})/\nu_{\text{ref}} = (\sigma_{\text{ref}} - \sigma)/(1 - \sigma_{\text{ref}}) \tag{4}$$

Only when σ_{ref} is negligible compared to unity can the chemical shift be written as

$$\delta \simeq \sigma_{\text{ref}} - \sigma$$

This is not valid for very accurate comparisons, or for heavy nuclei for which σ_{ref} may be of the order of 10^{-3} or 10^{-2}. The choice of a highly shielded species as a reference so as to avoid minus signs in chemical shifts may therefore be undesirable.

1.2. The Diamagnetic and Paramagnetic Contributions to Shielding

Nuclear magnetic shielding was discovered by physicists occupied with accurate measurements of nuclear magnetic moments. Lamb proposed a classical model in 1941 for the field induced within an atom by its own electrons in an external magnetic field.[1] This induces a diamagnetic current density in the electrons, which produces an induced field at the nucleus which is proportional to the applied field and opposite in sign. The proportionality constant or nuclear magnetic shielding was estimated by Lamb to be $0.319 \times 10^{-4} Z^{4/3}$ for an atom of atomic number Z. Just as Van Vleck[2] added a paramagnetic contribution (χ^p) to Langevin's diamagnetic term (χ^d) for the magnetic susceptibility, so did Ramsey add a paramagnetic contribution for the nuclear magnetic shielding in molecules,[3] $\sigma = \sigma^d + \sigma^p$:

$$\boldsymbol{\sigma}^d = (\mu_0/4\pi)(e^2/2m)\langle 0| \sum_k r_{kN}^{-3}(\mathbf{r}_k \cdot \mathbf{r}_{kN}\mathbf{1} - \mathbf{r}_k\mathbf{r}_{kN})\,|0\rangle \tag{5}$$

$$\boldsymbol{\sigma}^p = -(\mu_0/4\pi)(e^2/2m^2) \sum_{n \neq 0} ({}^1E_n - E_0)^{-1} \left\{ \begin{array}{c} \langle 0| \sum_k r_{kN}^{-3} \boldsymbol{l}_{kN}\,|n\rangle\langle n| \sum_k \boldsymbol{l}_k\,|0\rangle \\ + \\ \langle 0| \sum_k \boldsymbol{l}_k\,|n\rangle\langle n| \sum_k r_{kN}^{-3}\boldsymbol{l}_{kN}\,|0\rangle \end{array} \right\} \tag{6}$$

where $\mathbf{r}_k$ is the position vector for electron k and $\boldsymbol{l}_k$ the angular momentum operator with respect to the chosen origin (gauge origin), whereas $\mathbf{r}_{kN}$ and $\boldsymbol{l}_{kN}$ are with respect to the observed nucleus N. Equations (5) and (6) as written provide all the components σ_{zz}, σ_{xy}, etc., by use of appropriate components of $\mathbf{r}$ and $\boldsymbol{l}$.

The paramagnetic term arises from the mixing of certain excited states with the ground electronic state in the presence of a magnetic field. Because of the nature of the integrals in equation (6), the only excited states which can contribute to the paramagnetic term are those which are connected to the ground state by magnetic-dipole-allowed transitions. These transitions correspond to electron excitations between atomic or molecular orbital energy levels which have nonzero matrix elements of the x, y, and z components of orbital angular momentum, such as $p_x \to p_y$, $d_{xy} \to d_{x^2-y^2}$. Also, because the x, y, z components of the angular-momentum operators in equation (6) have the same symmetry as rotation operations about x, y, and z axes, these nonzero integrals connect orbitals such that electronic charge appears to rotate in going from one orbital to another (such as $p_x \to p_y$). This is the basis for language such as "charge rotation" used in discussions of the paramagnetic shielding term, referring to orbital angular momentum unquenched by the magnetic field.

Consider the electron density currents that are set up in the molecule due to the presence of the magnetic field. In an atom these currents are circular and in a direction such as to produce a field opposing the applied field (diamagnetic). In a molecule the currents are modified by the presence of the other nuclei as shown by maps of current densities for diatomic molecules.[4] The contour lines connecting points of equal current density, with arrows indicating the direction, provide the basis for the simple picture of a shielding component along a given axis being produced by "circulation" of electrons in the plane perpendicular to the axis and containing the nucleus. The magnetic dipole of the nucleus interacts with these currents from its vantage point (hence the r^{-3} distance dependence).

For a free atom, the natural choice of gauge origin is the nucleus in question, and the shielding then becomes equal to σ^d in Ramsey's formula, which simplifies to

$$\sigma = (\mu_0/4\pi)(e^2/3m)\langle 0|\sum_k r_{kN}^{-1}|0\rangle \qquad \text{for a free atom.} \tag{7}$$

Values of σ^d have been calculated[5] for neutral free atoms for $Z = 2$ up to 86 from nonrelativistic ground state atomic functions. Relativistic calculations show that these values are too small for neutral atoms with high atomic number or for highly charged positive ions.[6]

For molecules, there are several possible choices of origin. Since the measured physical observable cannot depend on this choice, moving the gauge origin means adding equal canceling contributions to the diamagnetic and paramagnetic parts. In this book, gauge origin is implied to be at the observed nucleus unless otherwise specified. The advantage of this particular choice is described below.

1.3. The Relationship between σ^p and the Nuclear Spin-Rotation Constant

With the gauge origin at the nucleus in question, σ^p in Ramsey's expression is related to another molecular property, the nuclear spin-rotation constant.[3,7] The nuclear spin-rotation constant arises from the coupling of the magnetic moment of a nucleus with the magnetic field generated by the molecular rotation at that nucleus. Ramsey[3] and Flygare[7] have shown that

$$\sigma^p_{gg} = (m_p/2mg_N)\,C_{gg}/G^{(e)}_{gg} - (\mu_0/4\pi)(e^2/2m)\sum_{N'} Z_{N'}[R^2_{NN'} - (R^2_{NN'})_{gg}]/R^3_{NN'} \tag{8}$$

C_{gg} and σ^p_{gg} are the diagonal components of the spin-rotational tensor and the paramagnetic shielding tensor in the principal axis system of the molecular moment of inertia. $G_{gg}^{(e)}$ are the rotational constants at the equilibrium configuration. The σ_{gg}^{p} can be related to the components along the principal axes of the shielding tensor by a rotational transformation using the known molecular geometry. The second term in equation (8) is the nuclear contribution which depends only on the coordinates and atomic number of all the other nuclei N' in the molecule. Since σ^d can be calculated from the ground state wave function of the molecule, and in fact can be estimated to within 0.1 to a few ppm from the known free neutral atom σ,[8] the absolute shielding tensor $\boldsymbol{\sigma}$ for a nucleus in a molecule can be determined from the nuclear spin-rotation constants.

1.4. Molecular Symmetry and Nuclear Magnetic Shielding

Nuclei are *isochronous* (have identical shielding) if they are equivalent (that is, interchangable by C_n, a proper rotation about an n-fold symmetry axis)[9] or enantiotopic (interchangable by S_n, an improper rotation) and in an achiral environment.[10] Nuclei that are *diastereotopic* are not interchangable by any symmetry operation and are therefore intrinsically anisochronous.

Nuclear site symmetry depends on molecular symmetry and determines the number of nonvanishing independent components of the nuclear magnetic shielding tensor. The higher the nuclear site symmetry the fewer the independent components. For example, a nucleus in a linear molecule has only three nonzero components, of which only two are independent, which we can identify as $\sigma_{\parallel}$ and $\sigma_{\perp}$. On the other hand, ^{1}H in H_2O has five nonzero components, of which five are independent, which we can identify as σ_{11}, σ_{22}, σ_{33}, σ_{12}, σ_{21}, if we designate by 1 and 2 the principal axes in the molecular plane, and 3 the axis perpendicular to it. In the same molecule, ^{17}O has three nonzero independent components, σ_{11}, σ_{22}, and σ_{33}. The numbers of nonvanishing and independent components of the nuclear magnetic shielding tensor have been described for systems of various symmetry.[11,12]

We have already noted that for a spherically symmetric electronic distribution, as in an atom, $\sigma^p = 0$. The symmetry of the electronic distribution around the nucleus determines the paramagnetic contribution σ^p. The molecular wave function may change sign under some symmetry operations, but an integral which corresponds to a physical observable such as nuclear shielding cannot change sign, and can only satisfy the symmetry conditions by being zero. Thus, the component of σ^p along the internuclear axis, $\sigma_{\parallel}^p$, is identically zero for all nuclei in a linear molecule because symmetry restrictions on integrals contributing to it in equation (6) cause all terms in $\sigma_{\parallel}^p$ to vanish.[3] The entire paramagnetic contribution to nuclear shielding in linear molecules therefore comes from $\sigma_{\perp}^p$. Symmetry restrictions were applied to the determination of the shielding of the central atom in an octahedral complex by Griffith and Orgel,[13] and extended to square planar symmetry, and to the shielding of a hydride ligand by Buckingham and Stephens.[14]

When a symmetry element of the electronic distribution is destroyed by substitution, the zeros dictated by this symmetry element in the terms of equation (6) no longer occur, and this alone can lead to marked changes in σ^p. $\sigma_{\parallel}^p$ is symmetry quenched (zero) in $HC{\equiv}CH$ but not in $H_2C{=}CH_2$. The effect of off-axis protons not even directly bonded to the atom of interest, as for $\overset{*}{C}$ in $CH_2{=}\overset{*}{C}{=}CH_2$ or $CH_2{=}\overset{*}{C}{=}O$ can be seen in $\sigma_{\parallel}^p$ values of the order of -260 and -160 ppm, compared with zero in $O{=}C{=}\overset{*}{C}{=}C{=}O$ or $O{=}\overset{*}{C}{=}O$.[15] Examples of higher shielding following higher symmetry are, for carbon, $\sigma(CO_2) > \sigma(CO)$, and $\sigma(CO_2) > \sigma(COS)$, for nitrogen or oxygen $\sigma(NO_2^+)$, linear $> \sigma(NO_2^-)$, bent. When a nucleus is more shielded in a less symmetrical molecule, as, in fluorine resonance, $\sigma(ClF) > \sigma(F_2)$, the higher shielding may again be traced to symmetry: in this example, fluorine in ClF is shielded by the $\pi^* \rightarrow \sigma^*$ circulation, which is symmetry-forbidden in F_2.[16,17]

1.5. Absolute Shielding Scales

Since the applied magnetic field itself cannot be measured independently with high precision, accurate shielding measurements can only be carried out by simultaneous measurement of two frequencies in the same physical sample in the field. The proton shielding in a spherical sample of pure liquid H_2O at 34.7°C = 25.790 ± 0.014 ppm[18] is established by two experiments. The first one was a simultaneous measurement of the frequencies of an electronic transition and a nuclear magnetic transition in atomic hydrogen.[18b] The second one was a simultaneous measurement of NMR frequencies of protons in a spherical sample of liquid water at 34.7°C and in a spherical bulb filled with atomic hydrogen in the same magnetic field.[18a] The first experiment establishes the g factor of the bare proton. With these results, the second experiment allows the determination of the nuclear magnetic shielding of protons in liquid water.

There are several methods of establishing an absolute shielding scale.

Method 1: This involves simultaneous NMR measurements in the free atoms and in a convenient liquid reference or free molecule, as in the experiment described above for protons.[18]

Method 2: An alternative is the simultaneous precise measurement of the signals of ^2D in liquid D_2O and the nucleus N of interest in a well-defined sample.[19] From the ratio of frequencies, and the previously measured ratio of resonance frequencies of ^2D in liquid D_2O to ^1H in liquid H_2O[20] and ^1H in H atoms to ^1H in liquid H_2O,[18] combined with the results of an atomic beam magnetic resonance measurement or an optical pumping experiment on the atomic vapor of the nucleus in question, the shielding difference, $[\sigma_N(\text{sample}) - \sigma_N(\text{free atom})]/[1 - \sigma_N(\text{free atom})]$ is obtained. Since $\sigma_N(\text{free atom})$ can be calculated with good accuracy,[5,6] the absolute shielding, $\sigma_N(\text{sample})$, can be derived. The absolute shielding scales for Li, Na, K, Rb, Cs, Zn, Cd, Ga, Hg, and Pb were determined in this way.

Method 3: Another method of establishing an absolute shielding scale is by using the relationship [equation (8)] between σ^p and the spin-rotation constant measured in a molecular beam magnetic or electric resonance experiment or by high-resolution microwave spectroscopy.[21] Combined with an accurate calculation of σ^d in the molecule, the absolute shielding of a suitable primary reference molecule can be obtained. For ^{13}C and ^{17}O the primary reference molecule is CO; for ^{31}P it is PH_3. The spin-rotation constant may also be obtained from the ratio of the relaxation times of two nuclei in the same molecule in the gas phase when both are wholly relaxed by the spin-rotation mechanism. In SeF_6, for example

$$T_1(^{77}Se)/T_1(^{19}F) = [C_{av}^2 + \tfrac{4}{45}(C_{\parallel} - C_{\perp})^2](^{19}F)/C_{av}^2(^{77}Se)$$

in which the spin-rotation tensor C is known for the fluorine (see Table 2 in Chap. 16).

Table 1. Absolute Shielding Scales Established by Methods 1 and 2 in the Text (Continued in Table 2)[a]

Nucleus	Primary reference molecule[b]	Method	Reference	Secondary reference liquid	Method	Reference
^{1}H	H atom	H maser	18	H_2O 307.9 K	1	18
	$\sigma = 17.733$	theory	18	$\sigma = 25.790(14)$		
^{6,7}Li	Li atom	ABMR	25	Li^+ ∞ D_2O, 297 K	2	25, 26
	$\sigma = 101.45$	theory		$\sigma = 90.0(8)$ for ^{6}Li		
				$\sigma = 90.4(7)$ for ^{7}Li		
^{23}Na	Na atom	ABMR	25	Na^+ ∞ D_2O, 297 K	2	25, 26
	$\sigma = 628.90$	theory		$\sigma = \sigma(\text{atom}) - 60.5(10)$		
	637.1	rel				
	Na_2, 448 K	OP	27		1	27
	$\sigma = \sigma(\text{atom}) + 29(16)$					
^{39,41}K	K atom	ABMR	25	K^+ ∞ H_2O, 298 K	2	25, 28
	$\sigma = 1329.36$	theory		$\sigma = \sigma(\text{atom}) - 105.1(10)$		
	1376	rel		for ^{39}K		
	K_2, 381 K	OP	29			
	$\sigma = \sigma(\text{atom}) + 99(85)$					
^{85,87}Rb	Rb atom	OP	30	Rb^+ ∞ D_2O, 297 K	2	30, 31
	$\sigma = 3366.80$	theory		$\sigma = \sigma(\text{atom}) - 211(2)$ ^{85}Rb		
	3760	rel		$\sigma = \sigma(\text{atom}) - 211.6(12)$ ^{87}Rb		
	Rb_2, 345 K	OP	29		1	29
	$\sigma = \sigma(\text{atom}) + 120(17)$					
^{133}Cs	Cs atom	OP	30	Cs^+ ∞ D_2O, 297 K	2	26
	$\sigma = 5780$	theory		$\sigma = \sigma(\text{atom}) - 344.0(15)$		
	7280	rel				
	Cs_2, 323 K	OP	27			
	$\sigma = \sigma(\text{atom}) + 221(12)$					
^{67}Zn	Zn atom	OP	32	Zn^{2+} ∞ D_2O, 303 K	2	33
	$\sigma = 2521.67$	theory		$\sigma = \sigma(\text{atom}) - 690(10)$		
	2710	rel				
^{69,71}Ga	Ga atom	ABMR	34	Ga^{3+} ∞ D_2O, 299 K	2	35
	$\sigma = 2638.57$	theory		$\sigma = \sigma(\text{atom}) - 800(45)$		
	2855	rel				
^{111}Cd	Cd atom	OP	36	Cd^{2+} ∞ D_2O, 306 K	2	37
	$\sigma = 4813$	theory		$\sigma = \sigma(\text{atom}) - 1106(4)$		
	5740	rel				
^{199}Hg	Hg atom	OP	38	Hg^{2+} ∞ D_2O, 300 K	2	39
	$\sigma = 9729$	theory		$\sigma = \sigma(\text{atom}) - 2432(5)$		
	16 100	rel				
^{207}Pb	Pb atom	OP	40	Pb^{2+} ∞ D_2O, 302 K	2	41
	$\sigma = 10\,060$	theory		$\sigma = \sigma(\text{atom}) - 17\,810(60)$		
	17 000	rel				
^{35}Cl	Cl atom					
	$\sigma = 1142.64$	theory		Cl^- ∞ H_2O		42
	1175	rel		$\sigma = \sigma(\text{atom}) - 170(12)$		

[a] All numbers are absolute shielding, in ppm, in spherical samples (or corrected to a spherical sample). ∞ stands for infinite dilution in H_2O or D_2O. OP, optical pumping; ABMR, atomic beam magnetic resonance; theory, nonrelativistic calculation[5]; rel, estimate based on relativistic calculations for rare gas atoms by Kolb *et al.*[6]

[b] The free atoms or molecules.

Table 2. Absolute Shielding Scales Based on Measured Spin-Rotation Constants (Method 3 in the Text) from Molecular Beam Magnetic/Electric Resonance or High-Resolution Microwave Spectroscopy

Nucleus	Primary reference molecule	Reference	Secondary reference neat liquid, spherical	Reference	Other Reference[a]
^{13}C	CO	43	TMS, 298 K	44	12, 45, 46
	$\sigma = 3.20(27)^b$		$\sigma = 185.4^b$		
			(see T dependence)	47	
^{15}N	NH_3	48	CH_3NO_2, 300 K	24	24, 49, 46
	$\sigma = 264.54(5)$		$\sigma = -135.0$		
			(see T dependence)		
^{17}O	CO	50, 51	H_2O, 298 K	51	22
	$\sigma = -42.3(17.2)$		based on $\sigma(CO) = -42.3$		
			$\sigma = 307.9$		
^{19}F	HF	52, 53	$CFCl_3$, 303 K	53	24, 53, 54, 55
	$\sigma = 410(6)$		$\sigma = 189.9$		
			(see T dependence)	24	
^{31}P	PH_3	8, 56	H_3PO_4 (85% soln)	22	
	$\sigma = 597$		$\sigma = 356$		
^{33}S	OCS	57, 58	CS_2	58	
	$\sigma = 843(12)$		$\sigma = 581$		

[a] The last two columns are references to reviews of free molecule values and the shielding scale, and references to shielding based on spin-rotation constants in addition to those cited in the reviews

[b] ^{13}C chemical shift data have been converted to absolute shielding using these numbers More recent (1987) values differ by about 1 ppm

The ^{77}Se spin-rotation constant can thus be obtained, establishing the ^{77}Se shielding scale on the basis of the ^{19}F scale, and similarly for ^{125}Te.

Method 4: For a linear molecule $\sigma_{\|}^p = 0$, so $\sigma_{\|} = \sigma_{\|}^d$. Since $\sigma_{\perp} = \sigma_{\|} - \Delta\sigma$, the absolute shielding tensor is completely specified by a measured anisotropy $\Delta\sigma$ and a theoretically calculated diamagnetic term $\sigma_{\|}^d$ for the nucleus in question.

Tables 1 and 2 give a list of absolute shielding scales which are in use, *cf.* earlier summaries.[17,22] These have been updated regularly.[23] In principle, once the frequency of a secondary liquid reference has been measured relative to a primary isolated molecule reference, the absolute shielding for the liquid reference is determined; and all other chemical shifts relative to this secondary liquid reference can be converted to the absolute scale. Problems associated with establishing absolute shielding scales have been discussed.[24]

1.6. Experimental Methods for Determining the Shielding Anisotropy

While chemical shifts are useful indicators of the electronic environment around the nucleus, the values measured in isotropic media provide only a fraction of the information available. Experimental techniques have provided a means of obtaining the shielding tensor elements σ_{11}, σ_{22}, σ_{33} which are evolving as a new physical index of the chemical bond. As a shielding component along a given axis is basically determined by circulation of electrons in the plane perpendicular to the axis and containing the nucleus, individual components from molecule to molecule can be compared in a more meaningful way than the isotropic average chemical shifts. Methods of determination of the anisotropy ($\Delta\sigma = \sigma_{\|} - \sigma_{\perp}$ in axially symmetric nuclear sites) or the individual components of the shielding tensor, require samples in which there is some preferred orientation.

The incomplete averaging of the shielding tensor in solid samples is accompanied by the incomplete averaging of direct dipolar coupling interactions. A variety of techniques have been developed to reduce the direct dipolar broadening.(59) Multiple pulse experiments, notably the WAHUHA pulse cycle,(60) and its later versions have been used in powders and in single crystals to reduce dipolar and quadrupolar interactions. This was applied extensively to ^{19}F and ^{1}H shielding anisotropies.(61) For nuclei with low natural abundance of spins (^{13}C, ^{17}O, etc.) the homonuclear dipolar broadening is nil because the distances between like rare spins are large. In order to obtain a high-resolution spectrum in the solid with improved signal-to-noise ratio, the dilute spins are first polarized by spin transfer from the abundant spins (such as ^{1}H). Then, decoupling of the abundant spins during the observation of the rare spins represses the heteronuclear dipolar broadening. This cross-polarization technique was devised by Pines, Gibby, and Waugh,(62) and has been extensively applied to determination of shielding tensors of ^{13}C nuclei. Abundant spins such as ^{31}P can be studied by this CP technique if chemically dilute as in $O{=}PR_3$.(63) In matrix-isolation methods(64) the molecules of interest are randomly oriented in a spinless matrix, such as argon, a technique which is especially useful for molecules which do not form single crystals. The isotropic argon matrix minimizes anisotropic intermolecular contributions to the shielding, which in the single crystal or powder may lead to changes in both the orientation and magnitude of the tensor elements. The relation of the dipolar tensor to the molecular frame can be used to orient the shielding tensor. The low temperature at which the experiment is carried out (2–20 K) is sensitivity-enhancing, and the inert matrix allows observation of ions and reactive species. Small molecules trapped in clathrate lattices(65) give the same type of NMR spectra as powders or matrix-isolated molecules, as shown for example in Chap. 2 Fig. 2. The same static powder pattern spectrum envelope is traced out by the spinning sidebands at slow MAS, so that the chemical shift tensor elements can be extracted from the sideband intensities either directly or by a moment analysis.(65a)

Preferred orientation can be achieved by dissolving the molecule in a liquid crystal,(66) or by imposing an external electric field on a polar molecule.(11) This useful method for studying anisotropies suffers from several limitations: The order parameters which characterize the molecular orientation with respect to the magnetic field are intimately involved with the shielding components in the measured chemical shift anisotropy: $\Delta\delta = \sum_{a,b} S_{ab}(\sigma_{ab} + \sigma_{ba})/3$, where S_{ab} are functions of the angles between the molecular axes and the magnetic field direction, if the liquid crystal is completely aligned. These order parameters therefore have to be known accurately for the individual components of the shielding to be deduced. In addition, vibrational averaging of the intramolecular dipolar interactions is necessary. More important is that the liquid crystal itself imposes an additional anisotropy not originally present in the free molecule; e.g., anisotropies of $+6.6$ to -19.4 ppm are induced in spherical ions such as Cs^+ and F^- ions by liquid crystalline media.(67)

As noted in equation (8), the σ^p components can be obtained directly from the moments of inertia, the internuclear distances, and the spin-rotation constants measured in a molecular beam experiment. These can be combined with accurate theoretical σ^d components. Advantages of this method are that absolute shielding components are obtained, and are not subject to medium effects. Spin-rotational constants published prior to 1974 are summarized in Landolt–Börnstein tables.(21) Some absolute shielding components derived from spin-rotation constants are given in Table 2.

Shielding anisotropies can be obtained from measured relaxation times when the chemical shift anisotropy relaxation mechanism or the spin-rotation mechanism is dominant. The correlation time has to be estimated or otherwise derived. There are large uncertainties associated with this method.

2. THEORETICAL DESCRIPTION

2.1. Computational Schemes

In Ramsey's shielding theory [equations (5) and (6)] calculations are invariant to the choice of gauge origin. In practice, accurate calculation of the paramagnetic term is difficult and the equal and oppositely signed terms in the diamagnetic and paramagnetic parts upon change of origin are not calculated equally accurately and therefore do not exactly cancel. Calculation methods have therefore been devised to use local gauge origins instead of a common gauge origin, such as a different gauge origin for each atomic orbital[68] or for each molecular orbital.[69] While these methods may be criticized as violating current conservation laws, they nevertheless appear to give results which are closer to experiment than the standard (common origin) calculations using the same numbers of basis orbitals.

If Ramsey's equations are used directly, as in equations (5) and (6), then the method involves a sum over the molecular excited states (SOS) which are mixed into the ground state unperturbed wave function by the presence of an external magnetic field and a nuclear spin.[70] In the coupled Hartree-Fock (CHF) method,[71] however, the perturbed wave functions are treated directly rather than as expansions in the unperturbed states; the simultaneous equations are coupled together by the electron-repulsion interaction terms. Another approach, finite perturbation theory (FPT), is to add a small (finite) magnetic field term to the energy, then solve for the wave functions.[72] The derivatives of the calculated energies with respect to the magnetic field are evaluated and extrapolated to the limit of zero field. In the "equations-of-motion method" one well-known approximate solution, the random phase approximation (RPA)[73] is equivalent to the CHF approach. These four methods (SOS, CHF, FPT, and RPA) are equivalent and should lead to the same level of accuracy. Examples of results of very good *ab initio* calculations are given in Table 3. A method of calculation of σ at the configuration interaction (CI) level (to include electron correlation effects on shielding) has been developed[74a] but not yet widely applied. Correlation effects on the σ^d contribution for nuclei in molecules are known to be small.[74b] A promising start has been made on the application of the $X\alpha$-scattered wave method to shielding calculations, which will be useful for larger molecules when faster analytic integration techniques become available.[75]

Additional approximations which give less accurate results but which can be applied to large molecular systems, or to discussions of general empirical trends, are described in Section 2.3. Appleman and Dailey,[27] Lipscomb,[4,71] and Schastnev and Cheremisin[76] provide a guide to methods and results. Reviews of nuclear magnetic shielding calculations are updated annually.[23]

Table 3. Selected Calculations of Shielding Compared with Experimental Values[a]

Nucleus		Reference	$\sigma_{\parallel}$	$\sigma_{\perp}$	σ_{av}	$\sigma_{\parallel} - \sigma_{\perp}$
^{1}H	CH_4	77	38.36	27.90	31.39[30.611(24)][12]	
	NH_3	77	41.59	25.95 26.69	31.41[32.43][48]	
	HF	77	44.06	19.74	27.85[28.51(20)][12]	24.32[24][78]
	PH_3	77	33.74	25.30 27.16	28.73[28.3][56,8]	
	HCl	77	45.45	22.13	29.90[31.5][78,8]	23.32[33][78,8]
^{13}C	CH_4	77			195.8[197.35(34)][12]	0
	C_2H_4	77	84.3[56.4][64]	177.9[156.4][64] −81.1[−53.6][64]	60.4[66.79(35)][12]	
	C_2H_2	77	279.4[275.4][64]	39.0[35.4][64]	119.1[121.35][12]	240.4[240][64]
^{15}N	NH_3	77	239.2	279.6	266.1[264.54][48]	−40.4[−40][48]
	N_2	73	339.0	−335.5	−110.6[−61.6][24]	674.5[603(28)][79]
^{19}F	HF	77	481.6	381.7	415.0[410(6)][53]	99.9[108][78]
	F_2	69	488.6	−503.0	−172.5[−232.6][53] [−219][49,8]	991.6[1050(50)][80]
^{31}P	PH_3	77	560.6	597.0	584.9[597][56,8]	−36.4[−55(5)][81]
^{35}Cl	HCl	77	1148.6	853.0	951.5[952][78,8]	295.6[292][78,8]

Nucleus		Paramagnetic term at nucleus N
^{1}H	CH_4	$\sigma^{p}_{av} = -55.80[-56.45]$[77]
	NH_3	$\sigma^{p}_{\parallel \mathrm{to\,N-H\,bond}} = -15.53[-4.78]$[77,48], $\sigma^{p}_{\perp} = -93.81[-104.84]$ $\sigma^{p}_{\perp'} = -85.56[-84.18]$
	HF	$\sigma^{p}_{\perp} = -120.96[-119]$[53]
	HCl	$\sigma^{p}_{\perp} = -167.99[-166]$[78,8]
^{15}N	NH_3	$\sigma^{p}_{\parallel} = -116.8[-118.27]$[48], $\sigma^{p}_{\perp} = -74.1[-75.96]$
	N_2	$\sigma^{p}_{av} = -464.6[-483(20)]$[49]
^{19}F	HF	$\sigma^{p}_{av} = -67.2[-62.8(3)]$[53]
^{31}P	PH_3	$\sigma^{p}_{\parallel} = -419.5[-421]$[56,8], $\sigma^{p}_{\perp} = -384.4[-370]$
^{35}Cl	HCl	$\sigma^{p}_{\perp} = -293.8[-300]$[78,8]

[a] Experimental values are given in square brackets. Vibrational corrections have not been included. Other calculations are reviewed in Ref. 23, 1987, Vol. 16.

2.2. Relativistic Effects

Shielding (and coupling constant) calculations based on Ramsey's theory using the nonrelativistic hyperfine Hamiltonian [equations (5) and (6)] require large corrections for heavy nuclei. For a single-electron atom with a nuclear charge of 80, the 1*s* electron's average velocity is roughly 58% of the velocity of light and the electron mass is roughly 20% larger than the rest mass. Thus, the relativistic average radius of a 1*s* electron in Hg is expected to be about 20% smaller than the nonrelativistic one. This *relativistic contraction* of the *s* and *p* shells has a large effect on the diamagnetic shielding of a heavy nucleus. There is an accompanying *expansion of the d and f* atomic orbitals due to the greater screening of the nuclear charge by the contracted *s* and *p* orbitals, and a *spin–orbit interaction* leading to $p_{1/2}$ and $p_{3/2}$, $d_{3/2}$ and $d_{5/2}$, etc. subshells. All three relativistic effects affect magnitudes of nuclear shielding. For example, for the rare gases He, Ne, Ar, Kr, and Xe, σ^d values are, respectively, 59.95, 558.6, 1276, 3598, and 7040 ppm,[(6)] to be compared with the nonrelativistic values of 59.90,

552.3, 1237.6, 3245.6, and 5642.32 ppm, respectively.[5] Relativistic corrections to σ^d(atom) increase dramatically with atomic number, e.g., 1400 ppm for $Z = 54$. These relativistic values were calculated using a theory (RRPA) which reduces to the CHF technique in the nonrelativistic limit. We have used the above results to determine the relativistic free atom values shown in Table 1. It is important to keep relativistic effects in mind when the absolute shielding of nuclei of high atomic number are considered. On the other hand, relativistic corrections are largest for the 1*s* core, which is least disturbed when a heavy atom forms a molecule, so that relativistic effects on heavy atom chemical shifts may be considerably less than those on absolute shielding.

No relativistic calculations of nuclear shielding have been carried out for molecules, except for proton shielding in H_2^+,[82] although the theory has been formulated.[83] The relativistic counterpart of the magnetic hyperfine integrals which appear in the paramagnetic term in Ramsey's theory [equation (6)] are plotted in Figure 1 for some of the heavy atom valence *p* shells, and in Figure 2 for the transition element *d* shells. These atomic values reflect the three relativistic effects cited above and shielding of heavy nuclei in molecules is affected proportionately.

A part of the relativistic corrections due to a heavy neighboring atom can be approximated by third-order perturbation theory with nonrelativistic functions. The contribution to shielding comes from the cross term involving the spin–orbit coupling, **I** · **S** interaction, and the external field-orbit interaction.[84] In nonrelativistic terms, the shielding mechanism is that the external field induces an orbital angular momentum on the electrons of the heavy atom (Br or I), this produces a polarization in the electron spin by spin–orbit coupling, and the spin polarization is transferred to the resonant nucleus by a Fermi contact and by a nuclear spin-electron spin dipolar

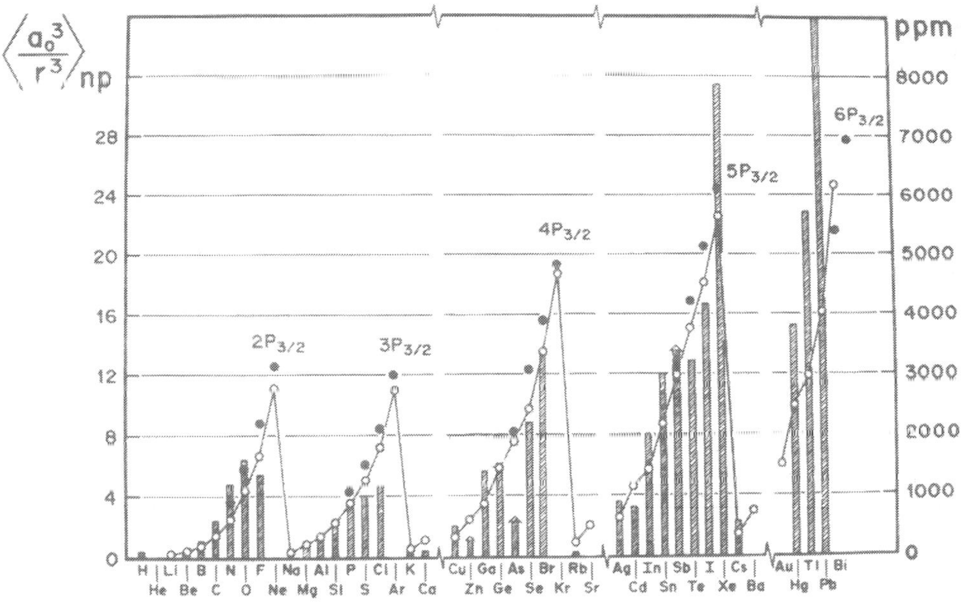

Figure 1. Ranges of chemical shifts of nuclei of main group and post-transition elements. The $\langle r^{-3} \rangle_{np}$ values are shown as dimensionless ratios $\langle a_0^3/r^3 \rangle_{np}$, where a_0 is the atomic unit of length ($a_0 = 0.529$ Å). These were calculated by the method of Barnes and Smith[86] from spin–orbit splittings in atomic spectra[100] without relativistic correction (open circles). Values calculated from relativistic Hartree–Fock–Slater atomic wave functions[101] are included for comparison (filled circles).

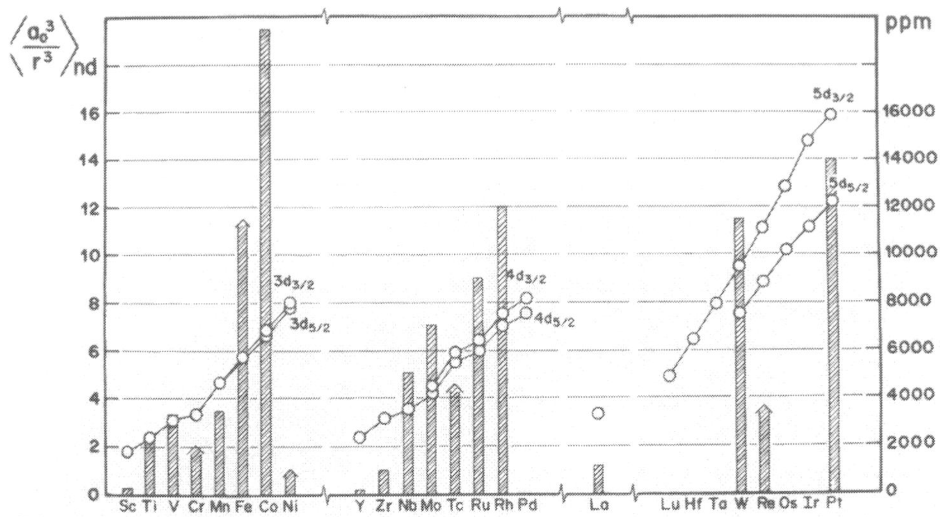

Figure 2. Ranges of chemical shifts of nuclei of transition elements. The $\langle r^{-3} \rangle_{nd}$ are calculated relativistic values.[101] Other experimental and theoretical values are available.[103]

interaction. This shielding contribution depends on excitation energies as $(\Delta E)^{-2}$ and on the magnitude of the spin–orbit coupling parameter, ζ, in the heavy atom. The latter depends on atomic number roughly as[85] $\zeta_p = 0.450 Z^{2.23}/n^{*3}$ cm^{-1} for valence p shells, n^* being the effective principal quantum number. Hence ζ has a periodic behavior with Z. In the first and second transition series ζ_{3d} and ζ_{4d} behave in a similar fashion but are smaller owing to the larger orbital quantum number. Typically F, Cl, Br, I have spin–orbit coupling parameters equal to 272, 587, 2460, and 5060 cm^{-1},[86] whereas for the atoms of the transition series ζ_{3d} are 45–1100 cm^{-1},[87] and for rare earths ζ_{4f} are 750–3400 cm^{-1}.[88] This theory provides a semiquantitative explanation of nonadditive effects in systematic substitution with halogens.[84]

2.3. Approximate Calculations and Models

One of the most useful approximations of nuclear shielding is that for the diamagnetic term by Flygare *et al.*[8] With the origin at the nucleus N, σ^d can be written as a free atom contribution plus the contribution of the electronic point charges centered at the other nuclei N':

$$\sigma_{xx}^d = \sigma^d(\text{free neutral atom}) + (\mu_0/4\pi)(e^2/2m) \sum_{N'} Z_{N'}(y_{N'}^2 + z_{N'}^2)/R_{NN'}^3 + \text{correction terms} \tag{9}$$

The correction terms allow for the displacement of the "point" charges from the centers of the other N' nuclei and for their spatial extension. They need to be included in the individual components, and can be summed up for a molecule from the tabulated atomic dipole contributions.[8] The corrections nearly cancel in the sum over the components and can be ignored in calculations of the isotropic average σ. Thus,

$$\sigma_{av}^d = \sigma(\text{free atom}) + 9.393 \sum_{N'} Z_{N'}/R_{NN'}$$

for σ in ppm and R in Å.

The success of this approximation, which allows the calculation of $\boldsymbol{\sigma}^d$ for molecules without knowledge of the wave functions, continues to be demonstrated as more accurate calculations of $\boldsymbol{\sigma}^d$ (including electron correlation) become available.

Approximate models have been devised to minimize the errors arising from imperfect cancellation of large positive and negative terms which arise in σ^d and σ^p. One approach is to neglect both σ^d and σ^p contributions of electrons on atoms which are distant from the nucleus in question by (a) including only orbitals on the atom in question (atom-in-a-molecule approach), or (b) including orbitals on the nearest neighbors as well (atom-plus-ligand approach). The other atoms are then treated as if their contributions to σ^d and σ^p cancel, or, their contributions are approximated by a neighbor magnetic anisotropic term as described below.

The "atom-in-a-molecule" approach is a one-center approximation to the shielding tensor.[13,89,90] The calculation is restricted to terms centered on the nucleus in question, which is taken to be the origin. The expression for the shielding tensor components is then unsuitable for accurate prediction of individual shieldings, but is simple enough to be used for discussion of general empirical trends:

$$\boldsymbol{\sigma}^p \simeq (-\mu_0/4\pi)(4\mu_B^2/\Delta E)[\langle r^{-3}\rangle_{np}\,\mathbf{P}_i + \langle r^{-3}\rangle_{nd}\,\mathbf{D}_i] \tag{10}$$

Here $\mathbf{P}_i$ and $\mathbf{D}_i$ represent the degree of imbalance of valence electrons in the p and d orbitals on atom A.[89] $\mathbf{P}_i$ may be written in terms of orbital populations or in terms of qualitative constants such as ionicity of the bond, pi bond order, degree of s character; $\langle r^{-3}\rangle_{np}$ and $\langle r^{-3}\rangle_{nd}$ are values of these averages for the free atom. (Figure 1 shows the variation of $\langle r^{-3}\rangle_{np}$ with atomic number.[86,89]) Karplus and Das[91] expressed the shielding tensor in independent electron molecular orbital terms, such that the sum over excited states becomes expressible in terms of orbital energy differences. Their expression, if reduced to the atom-in-molecule model, leads to $\mathbf{P}_i$ written in p orbital populations: For example, for σ^p_{xx}, the xx component of $\mathbf{P}_i$ is

$$(P_i)_{xx} = p_{yy} + p_{zz} - p_{zz}\,p_{yy} + p_{yz}\,p_{zy} \tag{11}$$

Jameson and Gutowsky[89] used the independent electron molecular orbital and the valence bond approaches, including d orbitals explicitly. Their equations were applied extensively to ^{31}P shielding by Letcher and Van Wazer.[92]

Equivalent expressions by Karplus and Pople[93,94] for main-group elements and by Griffith and Orgel[13] for transition metals are

$$\boldsymbol{\sigma}^p \simeq -(\mu_0/4\pi)\,4\mu_B^2(\Delta E)^{-1}\langle r^{-3}\rangle_{np}\sum \mathbf{Q} \tag{12}$$

$$\boldsymbol{\sigma}^p \simeq -(\mu_0/4\pi)\,4\mu_B^2(\Delta E)^{-1}\langle r^{-3}\rangle_{(n-1)d}\langle 0|\,\mathbf{L}^2\,|0\rangle \tag{13}$$

In equation (12) the xx component of $\mathbf{Q}$ is

$$\sum_B Q_{AB}^{(xx)} = \tfrac{1}{2}\sum_B [p_{yy}(2\delta_{AB} - p_{zz}) + p_{zz}(2\delta_{AB} - p_{yy}) + 2p_{yz}\,p_{zy}] \tag{14}$$

for all B bonded to the nucleus in question, with the orbital populations p_{yy}, etc. in each AB bond. As in equation (6), equations (10), (12), and (13) as written provide all the components xx, yy, zz, by use of appropriate components of $\mathbf{P}$, $\mathbf{Q}$, and $\mathbf{L}^2$. The isotropic average value is one-third of the sum of the xx, yy, and zz components.

Going beyond the one-center approximation, the "atom-plus-ligand" model gives the total shielding of the nucleus N in terms of the contributions from orbitals on N and on its directly bonded neighbors (ligands L), and neglecting all other (longer-range) terms[14,17]:

$$\sigma^N = \sigma^d(\text{on } N \text{ and } L) + \sigma^p(\text{on } N \text{ and } L) + \sigma(\text{long-range}) \qquad (15)$$

Explicit equations for the atom plus ligand terms were given by Buckingham and Stephens[14] to explain the large proton shifts in transition metal hydrides.

To include the effects of distant parts of the molecule, long-range terms may be added on. One contribution arises from the anisotropy of the magnetic susceptibility $\Delta\chi$ of a neighboring group at distance R[68,95]

$$\sigma_{\text{magn}}(\text{long-range}) = \Delta\chi(1 - 3\cos^2\theta)/3R^3 \qquad (16)$$

θ is the angle between the vector R and the symmetry axis of the group. When conjugated ring systems are involved, the long-range contributions to shielding have been estimated using the ring current model.[96] Another long-range contribution is that due to electric fields E generated by polar bonds or groups,[97]

$$\sigma_{\text{el}}(\text{long-range}) = k \times 10^{-12}E_z + 10^{-18}E^2 \qquad (17)$$

where k is proportional to the dipole moment of the polar group and E_z depends on the geometry of the polar group relative to the observed nucleus. For nuclei other than protons the coefficient of the E^2 term is larger than 10^{-18} and not necessarily negligible.

An intermediate approach is one which preserves the sum-over-states form of the Karplus and Das independent-electron model.[91] This is the analog of the Pople and Santry model for spin–spin coupling, and considers specific singlet excited states obtained by one-electron promotions. Thus the paramagnetic term in the individual components of the ^{13}C shielding tensor in organic carbonyl compounds can be associated with the mixing of these excitations with the ground state[98]: σ^p_{33} (in the sp^2 plane and perpendicular to the CO bond) with $\sigma \rightarrow \pi^*$ and $\pi \rightarrow \sigma^*$ excitations, σ^p_{22} (parallel to the CO bond) with $n \rightarrow \pi^*$, and σ^p_{11} (perpendicular to the sp^2 plane) with $\sigma \rightarrow \sigma^*$. The observed ^{13}C shielding is largest for the applied field perpendicular to the sp^2 plane, first, because the diamagnetic shielding is largest for this direction, and second, because the paramagnetic antishielding is generally smaller since the $\sigma \rightarrow \sigma^*$ excitations have higher energies than $n \rightarrow \pi^*$ and $\sigma \rightarrow \pi^*$. It was by considering individual excitations of this type (e.g., $\pi^* \rightarrow \sigma^*$, which corresponds to electron circulation in opposite senses on the two atoms) that Cornwell accounted for the unusual positive paramagnetic term for ^{19}F in ClF.[16] This approach is valuable in discussions of the shielding of transition metal nuclei, which is strongly dependent on d–d excitations, and also of the shielding of ligating atoms, which is influenced by the ligand field splitting.[99]

3. PATTERNS OF CHEMICAL SHIFTS

3.1. Chemical Shift Ranges of the Nuclei

The magnitude of the chemical shift varies considerably with the probe nucleus, and may be used as a measure of the "shielding sensitivity." This may be estimated from the shifts between analogous compounds, say from the chloride to iodide. A more general though less precise comparison is afforded by the overall range of chemical shifts for a given nucleus.

The periodic variation with atomic number of the ranges of chemical shifts observed for the nuclei of main-group and post-transition elements was pointed out in 1964 by Jameson and Gutowsky.[89] The tendency of the ranges to increase across a given row of the Periodic Table, and down a given group, was compared with the similar variation of the radial term $\langle r^{-3} \rangle_{np}$ for the valence p electrons in the free atom, since the paramagnetic shielding term σ^p has this radial dependence [equations (10)–(13)].[89] Figure 1 shows the periodicity of the ranges as now known and the corresponding periodicity of $\langle r^{-3} \rangle_{np}$ values. The relatively large shift ranges for the first row elements compared with heavier elements can be linked with well-known "first row anomalies" such as the readiness to form $p\pi$ bonds (C=O, C=C, N=N, etc.) and relatively high electronegativity.

Figure 2 shows the chemical shift ranges of transition elements, and relativistic values calculated for $\langle r^{-3} \rangle_{(n-1)d}$,[101] consistent with those in Figure 1 (cf. equation 13). The increase in $\langle r^{-3} \rangle_d$ for the $5d$ series reflects lanthanoid and relativistic contraction of the $5d$ subshell.[102] Figure 2 shows the expected increase in shift ranges across the transition series, by comparison with the change in $\langle r^{-3} \rangle_d$, but the vertical trends are less consistent.

3.2. Scaling of Chemical Shifts

The chemical shift ranges reveal broad patterns of periodicity despite wide differences in structure. A more precise comparison of chemical shifts is that for compounds with corresponding structure. These also appear to be scaled by ratios of $\langle r^{-3} \rangle$. For example, a plot of the ^{125}Te vs. the ^{77}Se chemical shifts for the analogous compounds gives a straight line with a slope of 1.8 which can be compared with 1.56 for the ratio of $\langle r^{-3} \rangle_{np}$ for the free atoms. Corresponding figures for ^{207}Pb and ^{119}Sn are 3.0 and 1.84, and for ^{199}Hg and ^{111}Cd, 4.6 and 2.2. For ^{51}V compared with ^{41}Ti the scale factor is 1.33 for the shift ranges and 1.28 for $\langle r^{-3} \rangle_d$, and for ^{95}Mo and ^{93}Nb the figures are 1.22 and 1.25, respectively. In the comparisons of elements in the same group the shifts for the heavier element are larger than the increase in the free atom radial term would suggest. Several properties of heavier elements may increase the shielding range. They have relatively large ranges of coordination number, bond strength, and of atomic or ionic charge, and an increased d-orbital participation, as evidenced by the necessity to include polarization functions in *ab initio* calculations. Relativistic effects also are important, these increasing approximately as Z^4.

3.3. General Factors in the Shielding of Main-Group and Transition Metal Nuclei

The approximate models of Section 2.3 form a useful basis for the discussion of broad empirical patterns. For these, variations in the local diamagnetic term and in nonlocal contributions are usually neglected. The paramagnetic (deshielding) terms are greater the closer the $p, d, f, \ldots$ electrons to the nucleus (the larger is $\langle r^{-3} \rangle$), the greater the imbalance of charge in the valence shell, and the smaller the appropriate excitation energy. The approximate equations (10)–(13) all contain $\langle r^{-3} \rangle$ and $(\Delta E)^{-1}$ factors and a measure of the imbalance of electron distribution on the atom in question. This last factor is expressed by the P_i and D_i terms for np and $(n-1)d$ electrons, respectively, in equation (10), by the Q terms in equation (12), and by the angular momentum terms in equation (13). The relatively small range for fluorine is related to the small number of bonding orbitals,[89] the factor P having a maximum value of 1 compared with 2 for nitrogen and 3/2 for carbon; the greater range now observed for chlorine than for phosphorus supports this conclusion.

Excitations which contribute to the paramagnetic term must be magnetic dipole allowed. Thus the ΔE used in the approximate theory usually corresponds to the lowest energy transition of this type (in terms of one-electron excitations, $n \rightarrow \pi^*$, $n \rightarrow \sigma^*$, $\pi \rightarrow \sigma^*$, $\sigma \rightarrow \pi^*$, etc.) or else some effective average over a few of them, the most important involving orbitals centered on the nucleus of interest. Such transitions have, e.g., $p_x \rightarrow p_y$, $d_{xy} \rightarrow d_{x^2-y^2}$, or corresponding $f \rightarrow f$ components which involve a circulation of charge.[94] For transition metal nuclei $d\pi \rightarrow d\sigma^*$ circulations largely determine the metal shielding, e.g., $d_{xy} \rightarrow d_{x^2-y^2}$ in the d^6 octahedral case,[13] $d_{xy} \rightarrow d_{x^2-y^2}$ and $d_{yz,xz} \rightarrow d_{x^2-y^2}$ in d^8 square planar complexes,[14] and similarly for formally d^0 or d^{10} configurations, because of ligand-to-metal or metal-to-ligand π-delocalization, or metal s, p, d orbital mixing. Thus it is not surprising that the relatively small ligand field splitting in the first transition series dramatically increases the ranges for these nuclei, as shown in Figure 2. The divergence of trends down the transition metal groups from the $\langle r^{-3} \rangle$ pattern is due in part to the increase in ligand field splitting from the first to the second or third transition series. As to adjacent elements, nitrogen and carbon shift ranges in comparable organic or inorganic compounds show scaling by the $\langle r^{-3} \rangle_{2p}$ ratio, but the nitrogen range extends to lower shielding in compounds with low-lying $n_N \rightarrow \pi^*$ states (since a lone pair of electrons is uncommon for carbon), showing the influence of the ΔE term.[121]

3.4. Dependence of Nuclear Shielding on Charge Density, Oxidation State, and Substituent Electronegativity

Deshielding with decrease in atomic electron density is a well-known pattern for all nuclei. It may result (e.g.) from a more negative ionic charge, or from increase in oxidation number or in the electronegativity of substituents. The effect can be related to increase in the magnitude of the paramagnetic term [equations (10)–(13)], via an increase in $\langle r^{-3} \rangle_{p,d}$, at least in part. For example, the radial terms for C^-, C, and C^+ are estimated by Slater's rules as 1.0, 1.4, and 1.9 a.u., respectively, a near-doubling from the anion to the cation. Indeed, carbon is strongly deshielded in carbonium ions. Similarly, $\langle r^{-3} \rangle_{(n-1)d}$ values for various $(n-1)\,d^m ns^{1,2}$ configurations show an increase with increase in oxidation number.[87] As expected, "hard" atoms show little effect of the radial term (as in ^{17}O or ^{19}F shielding). Proton shieldings are not normally dominated by the local paramagnetic term, but the shielding due to the diamagnetic term also increases with charge density.

As the charts in later chapters demonstrate, the shielding of all elements in different oxidation states shows distinct ranges, tending to decrease with increase in oxidation number: thus xenon shielding decreases as Xe(0) > Xe(II) > Xe(IV) $\gtrsim$ Xe(VI). Xenon chemistry is rather simple, however, and for other elements differences in bond types complicate the comparisons. In general, transition metals show high shielding in zero or negative oxidation states, as in d^6 (octahedral) configurations in midseries or d^{10} for group VIII metals. Similarly they show low shielding when highly oxidized, as in d^0 oxo-anions (vanadate, molybdate, permanganate).

Effects of atomic charge on the radial term may be reinforced by parallel changes in the energy term. Thus, in aromatic compounds decrease in charge density also reduces the shielding by reducing $\Delta E(\sigma \rightarrow \pi^*)$.[104] Alternatively, effects of changes in charge may be overwhelmed by concomitant changes in the energy factor. From NH_3 to NH_4^+, nitrogen is deshielded by 25 ppm with the increase in positive charge, consistent with contraction of the nitrogen orbitals and increase in the radial term. From pyridine to pyridinium ion, however, nitrogen is shielded by 100 ppm, as for the protonation of nitrogen in other compounds in which it is part of a π system. The increase in the ΔE term with removal of low-lying $n_N \rightarrow \pi^*$ states outweighs the effect of the delocalized cationic charge. Similarly, unusually low shielding is observed for Mo in quadruply bonded compounds (because of low-lying δ^* and π^* orbitals) compared with Mo(VI) in molybdates.[105a]

Other counterexamples show the influence of the other factors. Whereas xenon shielding decreases with increase in coordination number, phosphorus shielding decreases as $PCl_6^- > PCl_5 > PCl_4^+ > PCl_3$,[106] also $PO_4^{3-} > PO_3^{3-}$ and $PF_5 > PF_3$; the higher shielding correlates with higher symmetry (smaller P_i, D_i factors). Among the main-group elements, higher shielding is usually observed for higher local symmetry and coordination numbers (with comparable ligands), i.e., octahedral > tetrahedral > planar. For transition metals, however, such as rhodium and platinum, the shielding may decrease from square planar to octahedral symmetry, with decrease in the ligand field splitting.

The general increase in shielding accompanying an increase in negative charge has led to many correlations between chemical shifts and calculated charge densities. For example, there are many δ/q relationships in carbon shielding,[107–115] where δ is a chemical shift and q an atomic charge density from the Mulliken population analysis. ^{13}C shifts for aromatic ring carbons,[109,110] *para* carbons in substituted benzenes, etc., show a correlation with net $(\sigma + \pi)$ charge which is better than with the π charge alone, but the proportionality coefficients vary greatly, depending on the method used to estimate q.[108,114] Ethylenic and carbonyl carbon show similar correlations, but an opposite correlation is found for tetrahedral carbon in hydrocarbons. It is suggested that π-enrichment correlates with shielding and σ-enrichment with deshielding.[114] Thus, ^{17}O shifts in carbonyls but not ethers correlate with charge densities, and in substituted nitrobenzenes the ^{17}O shifts correlate with charge densities but ^{14}N shifts show the opposite trend.[116] A related and complex field is that of structure-reactivity (linear free energy) relationships of organic derivatives.[108,111] The substituent parameters of Hammett, Yukawa, Swain, Taft, Dewar, and others, which reflect σ- and π-inductive effects of substituents, are found empirically to correlate with chemical shifts of a range of nuclei. Indeed, the transmission of electronic effects is often inferred from such correlations, or degrees of electron transfer in donor–acceptor complexes.[111,112]

Greater clarity will follow as more measurements of shielding tensors are made. The archetypal δ/q_π relation, Spiesecke and Schneider's near-linear plot for aromatic species,[109] has been illuminated by the cryogenic study of $C_7H_7^+$, C_6H_6, and $C_5H_5^-$ by Grant and co-workers.[117] Linear correlation with q_π is observed only for the in-plane tensor elements which arise from $(\sigma \leftrightarrow \pi)$ circulations. The out-of-plane component arising from in-plane $(\sigma \leftrightarrow \sigma^*)$ circulations is more positive, as for alkane carbon, and its lack of correlation with q_π is responsible for the slight departure from linearity of the isotropic average in the original liquid phase measurement.[109] But the plurality of the factors dependent on charge remains.

Correlations of chemical shifts with substituent electronegativity are well known in proton, carbon,[118] and fluorine[119] resonance. As with many "electronegativity" correlations, counterexamples demonstrate a deeper periodic dependence on the substituent X.[120,17] In the binary hydrides XH_n there is a trend to lower proton shielding across the second row ($C \sim N > O > F$), but this reverses in subsequent rows. The trend to higher proton shielding down groups VI and VII (e.g., $F < Cl < Br < I$) with decrease in electronegativity, reverses in groups III–V. Particularly in group IV the trends zigzag, reflecting the contractions that follow the filling of the d and f shells. Carbon or fluorine shifts with electronegative substituents[118] are similarly periodic.[17,54] Carbon in XMe_n compounds is deshielded with increase in electronegativity of the atom X across the first two rows, with a reversal thereafter. As in the XH_n plot, the trend down the groups swivels round, and there is a pronounced zigzag down group IV ($C \ll Si \sim Ge < Sn > Pb$).

3.5. Correlations with Electronic Excitation and Ionization Energies

In 1952, Ramsey[3] noted the deshielding of nuclei in molecules having low-lying magnetically active excited states. In these molecules, a particular ΔE may dominate the paramagnetic term in shielding, leading to a correlation of δ with $(\Delta E)^{-1}$ or the wavelength λ of that particular (spin-allowed) absorption band, if observable. For main-group elements well-known δ/λ correlations involve $n \rightarrow \pi^*$ states, as for azabenzene, nitroso, or diazene compounds in nitrogen resonance[121] and nitroso or carbonyl groups in oxygen resonance.[122,123] In these compounds the $n \rightarrow \pi^*$ absorption band at longer wavelengths is readily distinguished from higher-energy excitations of bonding electrons; $n \rightarrow \pi^*$ bands have an $s \rightarrow p$ electric dipole allowed component (increasing the intensity) when the lone pair has some s character, as in many nitrogen groups. For azabenzenes such as pyridine the large $n_N \rightarrow \pi^*$ contribution explains the increase in shielding with protonation or perfluorination,[124] which stabilize the $(\sigma + n_N)$ orbitals relative to the π.

In general, however, magnetically allowed transitions are electric dipole forbidden and give weak bands in the electronic spectrum. These may be obscured by stronger allowed bands such as $\pi \rightarrow \pi^*$ in aromatic systems, that are magnetically inactive since the charge displacement is linear. In these, δ/λ correlations are usually unsuccessful since the magnetically active $\pi \rightarrow \sigma^*$ and $\sigma \rightarrow \pi^*$ (and $\sigma \rightarrow \sigma^*$) excitations are submerged in the vacuum uv spectrum. Further, the optical spectrum arises from excitations in the chromophore as a whole, whereas the local orbitals are the most important to the shielding.

In saturated systems, the relevant $n \rightarrow \sigma^*$ or $\sigma \rightarrow \sigma^*$ energies are larger and the forbidden $\sigma \rightarrow \sigma^*$ bands difficult to identify. Correlations may, however, be observed with the ionization energies of lone pair homos (variations in the lumos being smaller) as in the δ/I_i plots for ^{17}O in aliphatic alcohols and ethers.[125]

Chemical shifts may correlate with excitations that are not magnetically active (or not highly so) if these have energies moving in parallel with those that are. This was demonstrated for d^0 tetroxo species from $[VO_4]^{3-}$ to $[OsO_4]$, in which the ^{17}O shifts correlate with the lowest-energy optical excitation (effectively $n_O \rightarrow d\pi^*$). This has very little magnetic character, but shares a homo or lumo with the active excitations ($n_O \rightarrow \sigma^*$, $\sigma \rightarrow \sigma^*$, and $\pi \rightarrow \sigma^*$); and similarly for polyoxometalates and related compounds.[122]

Transition metal complexes may yield good $\delta/\lambda(d-d)$ relationships, which are discussed in terms of substituent effects in Section 3.6.1. Again, these magnetically active bands are relatively weak, and may be overlaid by stronger (magnetically inactive) metal-to-ligand or ligand-to-metal charge transfer bands. The shielding components show better correlations with the relevant excitation energies than does the isotropic average since the individual components can sometimes be associated with specific single electron excitations whereas the average includes all of them. Shielding anisotropies have been correlated with the anisotropies of optical spectra (difference in absorptivity for parallel and perpendicular plane polarized light) for various symmetries of octahedral complexes with mixed ligands.[126] In d^{10} or $d^{10}s^2$ configurations of late transition or post-transition elements, *s*, *p*, *d* mixing may allow relatively low-energy paramagnetic ($d-d$ or $p-p$) circulations.[127] In Pt(0) alkyne complexes $[Pt(RC{=}CR')(PPh_3)_2]$ the ^{195}Pt shielding correlates with a charge transfer ($d_{xz,x^2-y^2} \rightarrow \delta^*$) band energy, with *p*, *d* mixing in the homo, the lumo being formed from *d* and alkyne π^* orbitals.[128]

An interesting spin-off from the deshielding associated with a small homo–lumo gap is that it may correlate with catalytic or other reactivity, as observed for organometallic (sandwich) compounds in ^{59}Co[128a] or ^{95}Mo[105b] resonance.

3.6. Substituent Effects

A striking feature of Figure 2 is the large range spanned by the ^{59}Co shifts. This is not simply explained by the NMR accessibility of this nucleus, nor by the large values of the "electron imbalance" factor D_i (which takes values up to 12 for d^6 octahedral or d^4 tetrahedral complexes) since most of the range is spanned by fully octahedral complexes. There are evidently large effects of the ligands.

3.6.1. Chemical Shifts and the Spectrochemical Series of the Ligands

Ligand field splittings Δ are approximate excitation energies for the shielding of transition metal nuclei, and this is often observed to decrease with increase in $\lambda(d-d)$, particularly when Δ is small. The spectrochemical series of the ligands (increase in Δ) largely depends on the ligating atom, and is given by optical spectroscopy as $I < Br < Cl < S < F < O < N < C$, and also $Se < S < O$, but $As \sim P > N$, since synergic ($\sigma + \pi$) bonding increases Δ.[129,130] Thus the highest metal shielding is found in complexes of strong ligands such as CO, PF_3, H^-, and alkyl$^-$; and particularly for low metal shieldings, parallelisms are evident in the periodicity of substituent effects in optical[129] and NMR spectroscopy.[99]

Correlation of chemical shifts with the spectrochemical series of the ligands has been called "inverse halogen dependence"[131] although parallel dependencies are observed for ligators from other groups (VI, V, etc.). Increase in metal shielding as $I < Br < Cl$, following the spectrochemical series, is the inverse of the "normal halogen dependence" commonly observed for main group elements and many transition metal complexes.[131–133] The position of fluorine may be anomalous because of additional factors such as $p\pi$ bonding, or the extreme electronegativity.

The spectrochemical ordering (inverse ligand dependence) is observed for transition metals in d^0 configurations, also for d^{10} configurations of Cu and Ag, and the alkali metals. Thus (d^0) niobium is highly shielded in $[NbF_6]^-$ (compared to the other hexahalides), but (d^6) platinum is strongly deshielded in $[PtF_6]^{2-}$. The correlation with low-lying excited states is clearly shown by d^0 tetroxo ions. ^{95}Mo in molybdate $[MoO_4]^{2-}$ is increasingly strongly deshielded, and the low-energy absorption moves to longer wavelengths in the visible spectrum, as O is replaced by S, then Se. The spectrochemical dependence of the metal shielding in the tetroxo ions, and also in formally d^0 tetrahedral or octahedral halides, may be understood in terms of $(\sigma + \pi) \rightarrow \pi^*$ charge circulation since inductive destabilization of the bonding MOs and conjugative stabilization of the unoccupied MOs both increase down the group of the ligand. This explanation that inverse ligand dependence is due to $(\Delta E)^{-1}$ is supported by the corresponding (inverse) ligand dependence observed for nitrogen and oxygen shielding. In nitrosyl compounds XN=O the nitrogen shielding decreases with X in the order F > Cl > Br > OR > SR, showing a chemical shift correlation with $\lambda(n \rightarrow \pi^*)$.(134) A similar ordering is observed for N in cyclophosphazenes $(X_2PN)_n$, and also for the oxygen shielding in acyl halides RCOX.(135)

Chemical shifts of a transition metal, ^{55}Mn,(136a) and post-transition metal (^{63}Cu, ^{109}Ag, ^{67}Zn, and ^{113}Cd)(136b) nuclei have been calculated by an *ab initio* FPT method. In d^6 complexes $[Mn(CO)_5L]$ the ^{55}Mn shielding decreases as the ligand L becomes more π-donating, as CN < H < CH_3 < Cl, and also as L becomes a harder ligand, reducing the interaction with the soft Lewis acid $Mn(CO)_5^+$. These decreases in shielding are as expected, on an independent-electron model, for raising of the $d\pi$ homo and lowering of the $d\sigma^*$ lumo. In formally d^0 complexes, metal nucleus deshielding arising from $L \rightarrow d(M)$ donation has been mentioned. In formally d^{10} complexes of Cu, Ag, Zn and Cd *ab initio* calculations(136b) show that deshielding can arise from $d\pi$ back-donation or from $L \rightarrow M$ donation. For Cu the shielding decreases with increase in π-acceptor ability of the ligand (Cl < NH_3 < CN), i.e., with increase in the hole in the d shell. For Zn and Cd, however, the shielding decreases with decrease in ligand electronegativity (e.g., from H_2O to Cl to CH_3) with increase in valence p-electron density on the metal. For the Ag complexes the p and d contributions are competitive.

3.6.2. *Nephelauxetic Effects: The Radial Term*

Systematic divergences from δ/λ correlations are well known,(131–133) particularly the decrease in slope from first to second to third row ligands, as documented for Co(III)(137a) or for Pt(II).(137b) These were ascribed at least in part to decrease in the radial factor $\langle r^{-3} \rangle_d$ with increase in covalence or polarizability of the ligand, appropriate orbital coefficients being inserted into the expression for the paramagnetic term [equations (10)–(13)].

Decrease in $\langle r^{-3} \rangle$ can be related to the nephelauxetic (cloud expanding) effect of the ligand.(129) This is quantified by optical spectroscopic measurements as a ratio (β) of values of an interelectron repulsion parameter (the Racah B parameter) in the complex compared to the gaseous ion. The effect is a property of the ligating group rather than the atom, but there is a tendency for nephelauxetic ability to decrease across the row (C > N > O > F), with increase in electronegativity. This periodicity therefore reinforces that of the Δ term, so effects of the increase in $\langle r^{-3} \rangle$ are subsumed in δ/λ correlations for ligators from the same row. This helps to explain the large shielding range for Co(III) complexes, also the difficulty of extrapolating the δ/λ plots to a common origin. Here again there are parallelisms in the substituent effects in optical and NMR spectroscopy, and the quantification of nephelauxetic effects(129) is relevant to both.(99)

The explicit incorporation of the nephelauxetic ratio β into the Ramsey expression for σ^p [(equation (6)] has been demonstrated[138] for orthoaxial complexes of Co(III) which include ligators from different rows of the periodic table. Angular momentum integrals were calculated for intermediate ligand fields from optical spectral parameters. The cobalt shifts are linearly related to the optical parameter $(h_1/\beta)^{-1}$, where h_1 is the energy of the first spin-allowed *d–d* band, so that $\langle r^{-3}\rangle/\beta$ can be factored out. It seems, however, that Rh(III) shifts are better described by an optical parameter containing β^3 rather than β.[138]

Down the group of the ligating atom the radial term decreases and the nephelauxetic effect tends to oppose the effects of the energy term for ligands in main groups VII and VI, though not in groups V or IV. Whereas ^{11}B shielding in R_2BX increases as X = Cl < Br < I, it decreases as X = C > Si > Sn > Pb. Increased shielding in the nephelauxetic sequence (F <) Cl < Br < I of the ligating atom ("normal halogen dependence") is commonly observed for nuclei of main groups III and IV, and for transition elements with partly filled or (formally) filled *d* shells.[131] Corresponding sequences are commonly observed for ligating atoms from groups VI and V also, the metal shielding increasing in the order N < P < As < Sb, and O < S < Se < Te, as the ligand becomes more polarizable, or "soft."[129,130]

It is clear that either "inverse" or "normal" ligand dependence may result from the opposed effects of the ligand on the ΔE and $\langle r^{-3}\rangle$ factors. Inverse or spectrochemical dependence is found when low-lying excited states dominate the paramagnetic term, as for a metal such as Mo at the lower end of its shielding range, or for the N in nitrosyls XN=O. Otherwise the dependence may be mixed or "normal," i.e., nephelauxetic, as for metals (such as Mo) at the higher end of the shielding range. Another source of "normal" ligand dependence is the relativistic contribution to the shielding, as this increases down the group of the ligand.

3.6.3. Relativistic Effects

Relativistic contributions to the shielding were described in Section 2.3, and have been calculated for halogen ligands. In the hydrogen halides the contribution to the proton shielding increases from 0.2 ppm in HF to 13.6 in HI,[84] and in the methyl halides the contribution to the carbon shielding is reported as 0.8, 6.2, and 14.6 ppm, respectively, for CH_3Cl, CH_3Br, and CH_3I. The contributions are not linearly additive, the values for CH_2I_2, CHI_3, and CI_4 being calculated as 46.6, 123.7, and 307.3 ppm, respectively.[84] The effects are thus rather large for the heavier halogens, exaggerating apparent nephelauxetic effects.

3.6.4. Additivity of Substituent Effects

Additivity of substituent shift contributions is quite common for main-group and transition element nuclei. Shoolery's parameters[118] give proton shifts in methane derivatives for a wide range of substituents (alkyl, alkenyl, alkynyl, $CONH_2$, NR_2, SR, CN, COR, I, Br, OR, Cl, NO_2) and corresponding parameters give ^{13}C shifts in substituted alkanes, alkenes, and benzenes.[115] Grant and Paul's parameters give carbon shifts in alkanes,[139] and similar rules give nitrogen shifts in amines. Both in normal and in inverse dependence the shift per halogen ligand for a given nucleus is fairly constant from Cl to Br, or from Br to I.[131] Linear additivity may fail if there is interaction between substituents. Pairwise additivity is often found, and has been rationalized in terms of mutual perturbations.[140]

Deviations from additivity are evident in the U-shaped or sagging curves which are commonly observed for a central atom with progressive substitution of hydrogen, alkyl, or aryl groups by halogen, NR_2, OR, etc., notably for ^{27}Al, ^{29}Si, ^{119}Sn, and ^{31}P, in tetrahedral or pyramidal molecules.[92,141] The greatest deviations are found when there are strong π interactions, as in aminoboranes or halostannanes. Lower shielding for the partially substituted species may be explained by lower local electronic symmetry, by splitting of energy levels that are degenerate in the unsubstituted molecules, and (or) by the additional excitations that become magnetically allowed with the lowering of the symmetry.[142,143]

Deviations from additivity sometimes give hill-shaped curves. In planar molecules such as BX_3, higher ^{11}B shielding with mixed substituents is attributable to raising of the π lumo by a π-donor substituent.[142] For octahedral complexes of Co(III) with NH_3, NO_2^-, or CN^- ligands, higher shielding in the mixed species is related to an increase in ligand field splitting as t_{2g} and e_g degeneracies are removed.[144] With heavy substituents such as iodine nonadditive relativistic effects may be involved.

3.6.5. *Stereochemical Effects*

NMR spectroscopy is a powerful instrument for structure determination because of the sensitivity of the chemical shift to differences in geometry, despite our lack of understanding of many consistent effects. One such is the angle dependence of the γ-carbon shielding in alkanes,[139] which reliably predicts the microstructure of polymers.[145] Shielding in *syn* and *anti* compounds can differ substantially: the ^{31}P shielding is ca. 100 ppm lower in the *syn* than in the *anti* 7-phosphano norbornene system, for example.[146]

Many series of metal complexes which show a general additivity of substituent effects show small and systematic differences for geometric isomers. In Co(III) complexes with second row or Cl^- ligands the cobalt shielding increases and the linewidth decreases from the *trans* to the *cis* and from the *mer* to the *fac* isomer, with decrease in the local electric field gradient.[147,148] These correlations may vary,[149] depending on the *trans* effects of the ligands.

3.6.6. *Coordination Shifts*

The chemical shifts of ligands in metal complexes are usefully discussed in terms of the coordination shift, defined as the difference in shielding from the free to the coordinated ligand. Characteristic patterns are often observed which are of value in spectral assignment, and carry information on the nature of the coordinate bond. Coordination shifts are relatively small (e.g., a few tens of ppm for ^{13}C in the carbonyl ligand) unless there is particularly strong bonding to the metal. If a ligand does not exist in the free state, an alkyl derivative can be used as a base line: thus linear nitrosyls show fairly small ^{15}N shifts relative to the free ligand NO^+, but strongly bent nitrosyls (formally NO^-) show strong nitrogen deshielding, as in nitrosoalkanes RNO, which are bent at nitrogen. The range of coordination shifts for a given ligand may be stretched by the presence of low-lying excited states, as small changes in a small ΔE bulk large.

Coordination shifts may be of either sign. Deshielding is observed for ^{13}C and ^{17}O in CO (except for the heaviest metals, and some d^{10} complexes), ^{31}P in phosphines, phosphites, PF_3, etc., ^{15}N in amines, and terminal nitrogen in N_2, azide, or thiocyanate ($-SCN$). Increased shielding is usually observed for H^-, ligating ^{13}C in alkyl, aryl, η-arene or -alkene, and CN^-, and for ligating ^{15}N in NH_3, NCO^-, NCS^-, and azide. For some ligands, including NO^+, $C^{15}N^-$, and ligating nitrogen in N_2, the free ligand resonates well inside the range for coordinated ligands.

The high shielding often observed for hydride ligands was shown to be due to *d–d* circulations on the metal giving positive $\sigma^p_\perp$ tensor elements for ^1H, $\sigma^p_\parallel$ being small[(14)]; indeed, low shielding is observed for d^0 hydrides of early transition metals. Such neighbor effects fall off rather sharply, as R^{-3}, and so are important mainly for protons. Increased shielding for a ligating atom may be associated with the stabilization of the free ligand *n* orbital on complexation, when low-lying excited states do not compensate, as for N in NH_3, N_3^-, NCS^-, or pyridine.

Deshielding may occur when coordination involves (π) back-bonding, as observed for ^{13}C and ^{17}O in CO and ^{31}P in phosphines. The ^{13}C and ^{17}O tensor measurements in carbonyl complexes[(151)] show that the parallel component does not vary greatly from the free ligand: the main variation is in $\sigma_\perp$, which arises from $\sigma \leftrightarrow \pi$ excitations. The σ bonding to the metal, with (π) back-bonding, tends to stabilize the σ manifold relative to the π, favoring $\pi \rightarrow d\sigma^*$ circulations. This more than compensates for a smaller reduction in the $\sum Q$ (electron imbalance) term of equation (12).[(150)] The strong ^{13}CO deshielding early in the transition series correlates with high back-donation,[(150a)] which falls off as positive charge on the metal increases in the d^6 series. The deshielding correlates with other indicators of back-bonding, such as increase in the CO bond length and decrease in the stretching force constant [and increased $\pi(M-C)$ bonding]. The deshielding is enhanced by a negative charge on the complex (and reduced by a positive charge), by *trans*-ligands, which are better charge donors than carbonyls (polyhapto arenes for example),[(152)] and with filling of the *d* subshell for a given metal (e.g., from d^6 to d^8 to d^{10} complexes of cobalt).

The shielding tensors[(152b)] of ring carbons in [Cr(arene)$(CO)_3$] complexes, similarly, illustrate the importance of the $(\Delta E)^{-1}$ factor. The change on coordination is small for $\sigma_\perp$, which is dominated by $\sigma \rightarrow \sigma^*$ circulations in the CC and CH bonds, but the tangential component σ_t increases somewhat, and there is a sizeable increase in the radial component σ_r which arises from $\sigma \leftrightarrow \pi$ (mainly $\sigma \rightarrow \pi^*$) excitations in CC and MC bonds. Thus the coordination shift may be related to the increase in splitting of the arene π, π^* orbitals on complexation, as well as to decrease in $\sum Q$ [equation (12)] with change in the carbon hybridization.[(105,150)] The back-bonding, which is less efficient than for CO (in which the π^* orbital is lower-lying) further destabilizes the lumos, increasing the shielding, which therefore decreases from the *bis*-arene to the arene-metal tricarbonyl. The N_2 ligand shows deshieldings in cyclopentadienyl complexes of Ti and Zr particularly when bridging two metal ions. (Interstitial atoms in metal clusters often suffer large coordination shifts, and can be considered to be multiply coordinated.)

Coordination shifts show strong periodicities, as shown in Figure 3 for ^{13}C in CO. The vertical trend, here of increased shielding of the ligating atom down the group of the neighbor, is usually the same whether the neighbor is a main group element or a transition metal. This radial term dependence is reinforced by relativistic effects of heavy atom neighbors, as in the high ^{13}C shielding of carbonyls of Ir or Pt. The horizontal trends often differ. In main-group chemistry the shielding of the ligating

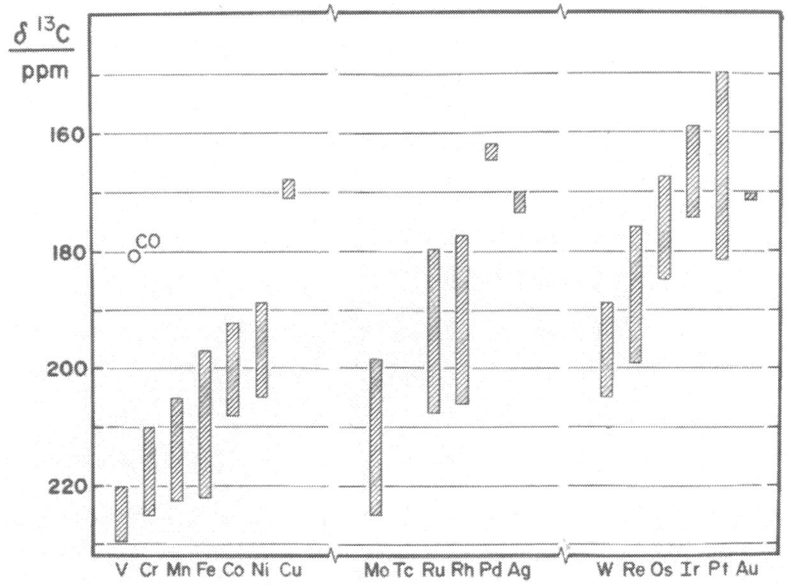

Figure 3. Ranges of ^{13}C shifts for the carbonyl group in mononuclear complexes (without polyhapto coligands).

atom tends to decrease across the row of the neighbor, effects of the radial term being larger than effects of the energy term [as $(\Delta E)^{-1}$ tends to increase across the row]. With coordination to a transition metal, however, the horizontal trend for many ligands is for the shielding to increase across the transition metal series, as well as down the group. These trends appear in comparisons of corresponding complexes; indeed, they appear in the perpendicular shielding tensor element ($\sigma_{\perp}$) for ^{13}C and ^{17}O, and also in $\sigma_{\parallel}$ for ^{17}O, in Cr, Mo, and W carbonyls.(151) They appear also in a plot of the overall ranges as in Figure 3. The ranges may be affected by chemical considerations, such as the carbonyls of Cu, Ag, Au being limited to $+1$ oxidation state, compared with -1, $+1$, $+3$ for Co. The π-cyclopentadienyl ligand shows maximal ^{13}C shielding for the d^6 Fe, Ru, Os triad, which may reflect stronger binding. This horizontal pattern, of increase in ligand shieldings across the transition series, appears to run parallel to the increase in ligand field splitting for a given d electron configuration,(150) and with the tendency for the splitting to increase with filling of the d shell, as from d^6 octahedral to d^8, square planar, as well as with decreased back-bonding. The dominance of "spectrochemical" over "nephelauxetic" influences across the transition series is not surprising, given the ligand field splittings, and the smaller increase in electronegativity across the row compared with main group elements. Interestingly, *de*crease in shielding across the transition series is observed for the oxo ligand in (formally) d^0 complexes(122) and fluorides(120) of a given d-electron configuration. The deshielding parallels the decrease in energy of the ligand-to-metal charge transfer bands, since the major paramagnetic circulation is of ligand $p\pi$ electrons in lumos which become more stable across the series. The ^{19}F shielding increases sharply from d^0 to d^6 complexes as the lumos are filled.

An example of sizable coordination shifts for nonligating atoms is afforded by ^{19}F

in CF_2 or CF_3 ligands in d^6 octahedral complexes (as compared with compounds of main-group metals), and explained by paramagnetic circulations of n_F electrons in $d\sigma^*$ orbitals.[153]

4. CORRELATIONS OF CHEMICAL SHIFTS WITH OTHER MOLECULAR PROPERTIES

4.1. Nuclear Quadrupole Coupling Constants

The nuclear quadrupole coupling constants (NQCC), e^2qQ (Chapter 2, Section 2.4) monitors the electric field gradient (efg) at the nuclear site, the largest component being given by[154]

$$q_{zz} = \langle\psi|\,(3\cos^2\theta - 1)/r^3\,|\psi\rangle \tag{18}$$

The efg differs from the shielding in being a property of the electronic ground state only. Both, however, have effectively the same dependence on the ground state charge distribution if the shielding is approximately described as in equations (10)–(13). A linear δ/NQCC relation may therefore be observed if excitation energies are comparable, as for chloromethanes.[155] Shielding and NQCC tensors have been compared for single crystals of complexes of quadrupolar nuclei such as ^{59}Co, ^{55}Mn, and ^{187}Re.[156] Mössbauer measurements of quadrupole splittings have been compared with NMR data for ^{119}Sn,[157] and this will become a possibility for ^{57}Fe. Chemical shifts may correlate with the quotient NQCC/ΔE if an effective excitation energy is identifiable, as in nitroso or nitrosyl compounds $R^{14}NO$ or $X^{14}NO$ which have a low-lying $n_N \rightarrow \pi^*$ state.[158]

NQCCs and NMR shifts together find applications to questions of σ and π bonding[159] and of additivity of substituent effects, as in NMR/NQR comparisons of σ-donor vs. π-acceptor properties made in ^{59}Co[160] and in ^{55}Mn[161] resonance. A useful concept developed in NMR and Mössbauer[162] studies is the additivity of partial field gradients (pfg) for individual ligands. Cobaloxime data could be fitted by cooperative variation of the pfg parameters of the glyoximato and axial ligands, the ^{59}Co shielding decreasing with increase in the ratio of the equatorial to the axial pfg.

4.2. Van Vleck Paramagnetism, and the Electronic g Tensor

Van Vleck's theory[7] gives magnetic susceptibility as an algebraic sum of paramagnetic and diamagnetic (χ^p and χ^d) contributions which are described as in equations (6) and (7) with the omission of the r^{-3} terms. Thus, approximate relations between shielding and magnetic susceptibility such as $\sigma^d \simeq -2\langle r^{-3}\rangle\,\chi^d$ and $\sigma^p \simeq -2\langle r^{-3}\rangle\,\chi^p$ have been used.[68,163] Likewise, the contribution to the shielding by a neighboring group with anisotropic susceptibility $\Delta\chi$, at a distance R from the observed nucleus has been expressed in equation (16).

The correlation of Van Vleck or temperature-independent paramagnetism with low shielding is well established, as in the MnO_4^- ion,[122,164] and related to low-lying excited states. The converse may not hold: molecules containing strongly deshielded nuclei are usually diamagnetic because the χ^p contributions are outweighed by the diamagnetism of the rest of the molecule.

The electronic g tensor of ESR spectroscopy has a similar form to that of the shielding tensor [equation (6) or (13)]. The orbital contribution is given by

$$\mathbf{g} = -\frac{\mu_0}{4\pi} 2\zeta \sum_{n \neq 0} (E_n - E_0)^{-1} \langle 0 | \mathbf{L} | n \rangle \langle n | \mathbf{L} | 0 \rangle \tag{19}$$

where ζ is the spin–orbit coupling constant. Species which are isostructural but differ in that one has an unpaired electron may therefore show δ/g correlations. The $^{51}V(V)$ shielding in diamagnetic molybdenum and tungsten heteropolyanions containing a single vanadium has been found to decrease in parallel with the g factor for reduced forms containing V(IV).[(165)]

4.3. Spin–Spin Coupling Constants and Relaxation Times

Some series of compounds show δ/J correlations. Fluorine shielding decreases with increase in the absolute value of $^1J(XF)$ for compounds of boron, carbon, or xenon, and this correlation can suggest relative signs of coupling constants.[(166)] Geminal $|^2J(FF)|$ values decrease with decrease in the averaged fluorine shielding in fluoroalkenes $CF_2{=}CX_2$, and similarly for perfluorovinyl derivatives of metals and semimetals.[(167)] There are several ways in which δ and J values may be connected, since the expressions for σ^p and the Fermi contact term both contain factors which correlate with atomic charge and excitation energies which may vary in parallel. Overall, absolute values of ^1J or ^2J tend to increase, and shielding to decrease, with increase in electronegativity of substituents, but anisotropies and the presence of lone pairs are complicating factors.

Chemical shifts may correlate with quadrupolar linewidths (if ΔE terms are similar, or vary consistently), since σ^p is related to the NQCC, and the linewidth to $(NQCC)^2$. Examples have been reported in chlorine[(155,168)] and in manganese[(161)] resonance.

Relaxation by the spin-rotation mechanism can lead to δ/T_1 correlations, as observed in ^{119}Sn resonance for methyltin halides,[(169)] since σ^p is directly related to the spin-rotation tensor, and the spin-rotation relaxation rate to its square. With the use of higher magnetic fields, shielding anisotropy relaxation is more often observed: $\Delta\sigma/T_1$, T_2 relationships are reported for Pt(IV) complexes with ^{195}Pt anisotropies of 10^3 ppm.[(170)]

4.4. Bond Properties

Some useful correlations of chemical shifts with structural properties have been observed, the number increasing with the development of solid state studies. Thus in polyoxoanions, the oxygen shielding increases with increase in M–O bond length and with decrease in bond order.[(171a)] Silicon shielding in silicates increases with decrease in Si–O bond length, with increase in Si–O–Si bond angle, and with increase in anion condensation, from isolated tetrahedra to chains, layers, and framework structures (zeolites, etc.)[(171b)] Corresponding relationships are found for ^{31}P in polyphosphates,[(171c)] ^{27}Al in aluminosilicates[(171d)] and so on, and the shifts reflect the Si:Al ratio.

In organometallic chemistry deshielding of the metal nucleus (and reduction in spin–spin coupling constants) is often observed with bulky substituents, and may be related to bond-lengthening and weakening, and (or) to bond angle distortions. Chelate rings also show regularities depending on ring size.[171d]

"For a tightly defined series of compounds it is possible to obtain linear relationships between ^{13}C chemical shifts and almost any other experimental property."[172] Such correlations with bond lengths, bond energies, stretching frequencies or force constants, and the like, have chemical interest nonetheless. ^{13}C shifts in organomercury compounds vary with the C–Hg bond energies, the effective excitation energy being $\sigma \rightarrow \sigma^*$ in the C–Hg bond.[173] In closely related transition metal carbonyls the carbon shielding decreases with decrease in CO force constant[174]: π back bonding increases the M–C bond order and decreases the C≡O bond order, deshielding being associated with M=C and C=O double bond character. Deshielding of a transition metal nucleus may correlate with reactivity, as in organometallic compounds of cobalt[128a] or molybdenum.[105b]

5. SHIFTS IN PARAMAGNETIC SYSTEMS

NMR spectroscopy of paramagnetic systems[175] is a large and growing field. The line broadening varies greatly, and may be offset by the spectral expansion: up to 730 ppm is reported for the range of proton shifts in vanadocene derivatives.[176] Changes in nuclear shielding due to interaction with unpaired electron spin density are called paramagnetic shifts (or in metals or other conducting solids, Knight shifts).[177] These shifts are observed for a nucleus in a paramagnetic molecule, generally in a ligand attached to a metal ion or in a stable radical such as a nitroxide.[177a]

The portion of the isotropic nuclear resonance shift arising from the Fermi contact hyperfine interaction is called the *contact* contribution (as discussed in Chapter 2 Section 8.3). Contact shifts of a nucleus in a molecule with electron spin S and an axially symmetric g tensor are given by

$$\left(\frac{\Delta B}{B_0}\right)_{\text{con}} = \frac{-a_N \mu_B S(S+1)}{g_N \mu_N 3kT} \frac{g_{\parallel} + 2g_{\perp}}{3} \tag{20}$$

a_N is the hyperfine interaction constant which depends on the net electron spin density at the observed nucleus. The other two hyperfine terms arising from the electron spin-nuclear spin interaction together give rise to the *pseudocontact* contribution to the paramagnetic shift. These are due to contributions from (a) unpaired electrons in orbitals centered at the metal ion and from (b) unpaired electrons in orbitals belonging to the ligand nucleus which is being observed. If the ligand-centered contribution is negligible, the pseudocontact shift is dipolar in form. Dipolar contributions are observed only in complexes which are magnetically anisotropic (having unequal magnetic susceptibility tensor components). When the magnetic susceptibility of the molecule is axially symmetric,

$$\left(\frac{\Delta B}{B_0}\right)^{\text{M}}_{\text{dip}} = -\frac{1}{3r^3}(3\cos^2\theta - 1)(\chi_{\parallel} - \chi_{\perp}) \tag{21}$$

If only one thermally populated multiplet with spin S is considered, and zero-field splitting is ignored, equation (21) simplifies to the form

$$\left(\frac{\Delta B}{B_0}\right)^{\mathrm{M}}_{\mathrm{dip}} = -\frac{1}{3r^3}(3\cos^2\theta - 1)(g_{\parallel}^2 - g_{\perp}^2)\,\mu_B^2\,\frac{S(S+1)}{3kT} \tag{22}$$

The dipolar shift due to unpaired electrons in a ligand orbital depends on the admixture of metal d and ligand p orbitals resulting in the delocalization of the unpaired electrons, and has to be added to equation (22). This second contribution gives a dipolar shift even for molecules with an isotropic g tensor. For the free electron spin the g tensor is isotropic and equal to $g_e = 2.0023$. This is the case for many organic free radicals as well. A g tensor with an isotropic average which differs from g_e can come from the coupling between electron spin and electron orbital angular momentum. However, it is not always true that the electron orbital angular momentum contribution is accounted for entirely by the g tensor components. A general theoretical description of paramagnetic shifts has been given.[178]

In some cases, one observes a nucleus in a diamagnetic molecule which forms a weak complex with a paramagnetic molecule. When the rate of exchange between the free and complexed ligand is slow on the NMR time scale, separate signals are detected. Usually, however, there is rapid exchange between the diamagnetic and paramagnetic sites giving rise to a single averaged signal such that the observed isotropic shift is directly proportional to the fraction of ligand molecules coordinated to the paramagnetic molecule. A careful analysis of the linewidths of the exchange averaged signals can give valuable information on the kinetics of ligand exchanges.

Paramagnetic shift reagents are used to induce large shifts so as to simplify spectra. They are also used to determine distances. Methods have been developed for separating the contact and dipolar terms from the experimentally observed shifts by using a variety of lanthanides. The geometric factor in the dipolar term gives the desired distances as for determination of conformations in biomolecules with a lanthanide shift reagent attached to oxygen.

When the dipolar contribution is absent, or separable, it is possible to measure the contact shift. This is proportional to the hyperfine interaction constant a_N, and so to the electron spin density at the nucleus. The direction of the contact shift gives the sign of a_N and the electron spin density; this is normally indeterminate in an ESR experiment, since it appears as a splitting. NMR is capable of higher sensitivity and gives simpler spectra than ESR, although the NMR theory is more complicated. The two methods are complementary, since NMR requires the frequency of electron spin relaxation or exchange to be greater than a_N (thus collapsing the doublet due to coupling with the unequally populated electron spin states). The linewidth also is proportional to a_N, so if this is large the NMR line is unobservable.

The sign of the contact shift usually alternates along a chain if the spin-delocalization is mainly via π orbitals; σ delocalization gives like signs, and sharper attenuation with distance from the spin source, Simple considerations of electron spin correlation, such as the Pauli principle and Hund's rule, give overall guidance, but the reality is often complex (e.g., Ref. 176).

Delocalization of d-electron spin from a transition metal has been used to study structure, equilibria, and reactions of complexes, particularly ligand exchange and solvation. Other applications are to organometallics, such as metallocenes, metallobiomolecules such as porphyrins (haemoglobins, cytochromes), iron–sulfur clusters, and so on. Organic radicals such as nitroxides or aromatic radical ions are useful spin probes also. The field of paramagnetic NMR spectroscopy has been reviewed biennially.[179]

6. EFFECTS OF INTERMOLECULAR INTERACTIONS AND INTRAMOLECULAR DYNAMICS

Because molecules studied by NMR are not rigid isolated bodies, there will always be some effects due to the surrounding molecules and also due to small and large amplitude motions within the molecule.[180] It may be necessary to correct for these effects. Alternatively, they can provide subtle information about the environment of a molecule as well as the dynamics of the intramolecular motional averaging that takes place in the NMR time scale.

6.1. Medium Effects

Medium effects on the molecule of interest, the homonuclear reference molecule, and on the lock substance, may need to be taken into account in an NMR experiment. These effects have been studied in the dilute gas phase, in which the shielding can be written in the form of a virial expansion in powers of density, ρ:

$$\sigma(T, \rho) = \sigma_0(T) + \sigma_1(T)\rho + \sigma_2(T)\,\rho^2 + \cdots \tag{23}$$

$\sigma_0(T)$, the independent molecule value of the shielding, is a function of temperature due to averaging over intramolecular motions such as rotation, vibration, pseudorotation and internal rotation. The terms dependent on the density characterize the intermolecular effects, with $\sigma_1(T)$ being a measure of the change in shielding due to binary collisions, $\sigma_2(T)$ due to ternary collisions, etc. In a mixture of gases there will be a term for each constituent, $\sigma_1(A-A)\,\rho_A + \sigma_1(A-B)\,\rho_B + \sigma_1(A-C)\,\rho_C + \cdots$. In the low-density limit, at a given temperature the resonance frequency increases linearly with increasing gas density; that is, σ_1 appears to be uniformly negative, intermolecular interactions leading to deshielding. The additional terms, $\sigma_2(T)\,\rho^2 + \cdots$, due to many-body effects, become important at higher densities and are usually opposite in sign to $\sigma_1(T)\rho$. Thus, at higher densities the change of resonance frequencies with increasing density tends to be somewhat less.

$\sigma_1(T)$ values have been measured for ^{1}H, ^{11}B, ^{13}C, ^{15}N, ^{19}F, ^{31}P, ^{129}Xe in a large number of interacting pairs of molecules.[180,181] The magnitude of $\sigma_1(T)$ depends to a large extent on the range of chemical shifts of the observed nucleus. For example, for a density 50 times the density of an ideal gas at standard conditions, $\sigma_1(T)\rho$ is -0.005 to -0.4 ppm for ^{1}H, -0.10 to -0.50 ppm for ^{13}C, -0.1 to -2.5 ppm for ^{15}N, -0.3 to -3.5 ppm for ^{19}F, -0.1 to -13 ppm for ^{31}P, and -15 and -40 ppm for ^{129}Xe. It depends also on the specific pair of interacting molecules, being larger for more polarizable pairs.

Contributions to the shielding of a nucleus due to the presence of a neighboring molecule are made by effects of bulk susceptibility, van der Waals interactions, electric fields, and magnetic anisotropy[182]:

$$\sigma_1 = \sigma_{1b} + \sigma_{1w} + \sigma_{1E} + \sigma_{1a}$$

The bulk susceptibility contribution is the same for all nuclei in the sample and depends only on the molar magnetic susceptibility such that its contribution to $\sigma_1(\mathrm{A-B})\,\rho_\mathrm{B}$, for example, will depend on χ of molecule B: σ_{1b} is $(2\pi/3)\chi$ or $-(4\pi/3)\chi$ for cylindrical gas samples perpendicular or parallel to the magnetic field. For protons and ^{13}C this correction may be a large part of the medium shift. The van der Waals contribution, σ_{1w}, results from both dispersive (long-range) and repulsive (short-range) interactions, and depends on polarizabilities and on the intermolecular potential. There are electric field contributions, σ_{1E}, if the observed molecule and/or its collision partner have electrical (dipole or quadrupole) moments or local polar groups. Magnetic anisotropy effects, σ_{1a}, are present if the collision partner has a non-isotropic magnetic susceptibility and if the interacting pair of molecules have preferred mutual orientations. This is the major contribution in the aromatic-solvent-induced shifts (ASIS). This binary collision model is quantitatively successful in explaining general trends, especially for ^1H shifts in systems dominated by bulk susceptibility and electric field contributions. It is less successful for other nuclei, especially for systems dominated by van der Waals contributions.

The same contributions (bulk susceptibility, electrical, and magnetic anisotropy) can be used to explain medium effects in condensed phases.[97] Gas-to-neat liquid shifts range from a few tenths to 200 ppm.[24] Since the bulk susceptibility contribution to medium effects for a sample in a superconducting magnet is opposite in sign to the intermolecular contributions, the liquid signal can appear on either side of the gas signal. The intermolecular effects generally cause the nucleus to be less shielded in the liquid than the gas except when there is strong association. Typical magnitudes of gas-to-solution shifts observed in an iron magnet for a nonpolar solute in various non-polar solvents are -0.225 to 0.781 ppm for ^1H in CH_4, 5.3 to 12.1 ppm for ^{13}C in CH_4, 3.04 to 7.90 ppm for ^{19}F in CF_4, 21.1 to 93.0 ppm for ^{31}P in P_4,[180] and 148 to 335 ppm for Xe in various organic liquids. The reaction field model and the equivalent solvaton model[181,183] have been very useful in the systematic classification of some solvent effects on NMR chemical shifts, and have the advantage that only bulk properties such as refractive index are necessary in the various models. They are most successful in correlating one solute with various solvents of roughly comparable dielectric constants, less successful in interpreting larger bodies of data especially for non-polar solvents.

An intermolecular effect on nuclear shielding which is subtle and expected to be small, is that which is due to the differing interactions between chiral solute and solvent molecules. Corresponding nuclei of an enantiomeric pair of molecules in an optically active solvent will have different average shielding. A chiral solute molecule R_A and its enantiomer S_A form diastereomeric pairs $R_A \cdots R_B$ and $S_A \cdots R_B$ with chiral solvent molecule R_B. Because stereospecific interactions between such solute and solvent molecules at short range lead to correlated intermolecular geometry, corresponding nuclei in S_A and R_B in the diastereomeric pairs will experience different shielding via the usual mechanisms operating in intermolecular effects on shielding (e.g., magnetic anisotropy, electrical, and dispersive effects),[182] leading in many instances to experimentally significant magnetic shielding nonequivalence. NMR spectroscopy can therefore be used for the direct determination of optical purity and of absolute configurations. Observed chiral shifts are of the order of 0.02 ppm for ^1H,[184] 0.04 ppm for ^{19}F,[185] and 0.16–0.7 ppm for ^{15}N.[186] Chiral solvents and their use in the NMR method of absolute configuration determination have been reviewed.[187]

6.2. Rovibrational Averaging and Isotope Effects

As a molecule rotates, the atoms tend to move away from the center of mass; this is known as centrifugal distortion. The vibrations of a molecule are anharmonic, so that a vibrationally excited molecule has longer bonds on average than a molecule fixed at its equilibrium configuration. The thermal average of the bond length, as observed on the NMR time scale, therefore increases with increasing temperature with consequent changes in nuclear magnetic shielding, even in the absence of intermolecular effects.[180]

This intrinsic temperature dependence of the shielding σ in a molecule can be obtained by observing the resonance frequencies in samples of known density and effectively extrapolating to zero density as for ^1H in HCl and HBr, ^{13}C in CO, CO_2, ^{15}N in NNO, N_2, NH_3, ^{19}F in many molecules, and ^{31}P in PH_3, PF_3, POF_3, and PF_5. In all cases (except for ^{15}N in NH_3 and ^{31}P in PH_3) the shielding decreases (the resonance frequency increases) with increasing temperature. The magnitudes of these shifts, for a 50 degree temperature change, are 0.01–0.02 ppm for ^1H, 0.06–0.45 ppm for ^{15}N, and from 0.1 to 1.9 ppm for ^{19}F. Theoretical calculations predict a decrease in shielding with increasing bond length in diatomic molecules (with the exception of Li in LiH). Temperature-dependent chemical shifts in the isolated molecule may be obtained from the rovibrational average over these shielding functions.[188,189] Conversely, the observed temperature dependence can give empirical derivatives of shielding with respect to bond extension.[180]

NMR isotope shifts arise from the same rovibrational averaging. Because a lighter atom undergoes greater amplitude motions than a heavier one, substitution of a heavy isotope alters the shielding of the nucleus replaced (primary isotope effect) and of other nuclei in the molecule (a secondary isotope effect).

$$\begin{aligned}\text{Secondary isotope shift} = {}^n\Delta X({}^{M/m}\mathrm{Y}) &= [\nu_\mathrm{X}({}^M\mathrm{Y}) - \nu_\mathrm{X}({}^m\mathrm{Y})]/\nu(\mathrm{ref}) \\ &= \delta_\mathrm{X}({}^M\mathrm{Y}) - \delta_\mathrm{X}({}^m\mathrm{Y}) \\ &\simeq \sigma_\mathrm{X}({}^m\mathrm{Y}) - \sigma_\mathrm{X}({}^M\mathrm{Y}) \qquad (24)\end{aligned}$$

Here the X nucleus is observed in the isotopomers with ^{M}Y or ^{m}Y (mass $M > m$) n bonds away from X. The primary isotope effect is more difficult to assess experimentally. An absolute measure requires independent accurate knowledge of the ratio of g factors of the bare nuclei; thus only relative primary isotope shifts are reported.[190]

Since isotope shifts are a consequence of differences in rovibrational averaging (in the NMR time scale) of heavy and light isotopomers, they can be interpreted in the same theoretical framework as the intrinsic temperature dependence.[191] The theory by Jameson and Osten[191] successfully explains the following general empirical trends.[192]

(1) Substitution with a heavy isotope shifts the NMR signal of a nearby nucleus toward lower frequencies (greater shielding); ${}^n\Delta\mathrm{X}({}^{M/m}\mathrm{Y}) < 0$.

(2) The isotope shift decreases with increase in distance of the isotopic substitution from the observed nucleus.[193]

(3) The magnitude of the shift reflects the overall range of chemical shifts observed for the resonant nucleus. For example, while ^1H shifts due to a single deuterium substitution are 0.012–0.040 ppm, ^{31}P shifts are 0.85–1.30 and ^{59}Co shifts are about 5 ppm.

(4) The isotope shift is largest where the fractional change in mass upon substitution is greatest. Shifts due to substitution of ^{13}C for ^{12}C or ^{18}O for ^{16}O are proportionately smaller than ^2D for ^1H.

(5) The isotope shift is approximately proportional to the number of atoms isotopically substituted. This leads to a series of equally spaced isotopomer resonances.

Isotope effects induced by ^{2}D, ^{13}C, and ^{18}O in ^{13}C and ^{31}P spectra can yield mechanistic and structural information. The isotope shifts(194) depend on the same factors (such as bond type) that influence absolute shielding and on the location of the bond in question. The isotope shift due to ^{18}O substitution at a peripheral location is generally larger than at a skeletal position, as expected. For the same reason, ^{13}C isotope shifts due to $^{13}C/^{12}C$ substitution in small ring systems can be of either sign.(195)

Because of isotope effects on the chemical equilibrium constant, a phenomenon which is well understood, there are additional effects on the NMR spectrum by isotopic substitution. Since the equilibrium constant for a conformational equilibrium is different for different isotopically related molecules, the relative concentrations of conformers vary.(196) These are *equilibrium isotope effects* and quite apart from the intrinsic isotope effects we have discussed above. The two types are no longer distinct when the barrier separating the two species is very low.(197)

There are also solvent isotope effects which are not completely understood(198,199) for alkali, halide, and other nuclei in H_2O vs. D_2O aqueous solutions.(198,41) The difference between the ^{207}Pb ion shieldings in H_2O and D_2O is 31 ppm for 0.04 mole/mole solution. Observed shifts are generally to lower frequency in the heavier solvent, although opposite shifts have been observed as well, e.g., for ^{67}Zn ions.(200) If the hydrated ion in solution, $M(D_2O)_n$ or $M(H_2O)_n$, is considered as an isolated entity, then there should be an intrinsic isotope effect on shielding no different from that on ^{13}C in CD_4 versus CH_4. In addition, there will be effects due to the different liquid structures of the solvents (hydrogen bonding in D_2O being significantly different from that in H_2O). The deuterium isotope effect on ^{17}O in H_2O is -3.08 ppm in the liquid and -4.04 ppm in the gas (lower ^{17}O frequency in D_2O).(201)

6.3. Dynamic Processes: Fluxional, Conformational, and Exchange Equilibria

NMR has long been used to study specific interactions such as hydrogen bonding and complex formation, and fluxional or conformational equilibria. When the exchange of nuclear sites in these inter- and intramolecular rearrangements occurs at a fast rate in the NMR time scale (see Section 7 of Chapter 2), then only an average spectrum is observed, with averaged values of nuclear shielding. For example, $\sigma_{av} \simeq p\sigma_A + (1-p)\,\sigma_B$ for a two-site system with average fractional populations p and $(1-p)$ in the A and B sites. These populations are generally temperature dependent because A and B sites correspond to different energy minima. The temperature dependence of the equilibrium constant of association or conformational change or rearrangement may overwhelm the temperature dependence due to medium effects or rovibrational averaging. It is nonetheless important to keep in mind the foregoing discussions of these effects in interpreting temperature and concentration studies in such systems. We have already seen that even in the absence of site-exchange equilibria there are sizable concentration and temperature dependences of resonance frequencies. A partial correction for these has been attempted for acceptor–donor molecular complex studies in solution.(202) To eliminate medium effects, some studies of fluxional, conformational, and association equilibria have been carried out in the gas phase.(203)

REFERENCES

1. Lamb, W. *Phys. Rev.* 1941, **60**, 817–819.
2. Van Vleck, J. H. *Electric and Magnetic Susceptibilities*, Oxford University Press: London, 1932; p. 275.
3. Ramsey, N. F. *Phys. Rev.* 1950, **77**, 567; **78**, 699–703; 1951, **83**, 540–541; 1952, **86**, 243–246.
4. Lipscomb, W. N. *International Review of Science, Phys. Chem. Ser. One*, Buckingham, A. D., Ed.; Butterworths: London, 1972; Vol. 1, pp. 167–196.
5. Malli, G.; Froese, C. *Int. J. Quantum Chem., Suppl.*, 1967, **1**, 95–98.
6. Kolb, D.; Johnson, W. R.; Shorer, P. *Phys. Rev. A* 1982, **26**, 19–31.
7. Flygare, W. H. *J. Chem. Phys.* 1964, **41**, 793–800; *Chem. Rev.* 1974, **74**, 653–687.
8. Flygare, W. H.; Goodisman, J. *J. Chem. Phys.* 1968, **49**, 3122–3125; Gierke, T. D.; Tigelaar, H. L.; Flygare, W. H. *J. Am. Chem. Soc.* 1972, **94**, 330–338; Gierke, T. D.; Flygare, W. H. *J. Am. Chem. Soc.* 1972, **94**, 7277–7283.
9. Abragam, A. *The Principles of Nuclear Magnetism*, Oxford University Press: Oxford, 1961; p. 480 (also 1978 corrected reprint).
10. Mislow, K.; Raban, M. *Top. Stereochem.* 1967, **1**, 1–38.
11. Buckingham, A. D.; Malm, S. M. *Mol. Phys.* 1971, **22**, 1127–1130; Buckingham, A. D.; Pople, J. A. *Trans. Faraday Soc.* 1963, **59**, 2421–2430.
12. Raynes, W. T. *NMR, Spec. Per. Rep. C. S.* 1978, **7**, 1–25.
13. Griffith, J. S. *The Theory of Transition Metal Ions*, Cambridge University Press: Cambridge, 1961; Section 12.5; Griffith, J. S.; Orgel, L. E. *Trans. Faraday. Soc.* 1957, **53**, 601–606.
14. Buckingham, A. D.; Stephens, P. J. *J. Chem. Soc. A* 1964, 2747–2759, 4583–4587.
15. Beeler, A. J.; Orendt, A. M.; Grant, D. M.; Cutts, P. W.; Michl, J.; Zilm, K. W.; Downing, J. W.; Facelli, J. C.; Schindler, M. S.; Kutzelnigg, W. *J. Am. Chem. Soc.* 1984, **106**, 7672–7676.
16. Cornwell, C. D. *J. Chem. Phys.* 1966, **44**, 874–880; Davis, R. E.; Muenter, J. S. *J. Chem. Phys.* 1967, **57**, 2836–2888.
17. Mason, J. *Adv. Inorg. Chem. Radiochem.* 1976, **18**, 197–228; 1979, **22**, 199–237.
18. (a) Phillips, W. D.; Cooke, W. E.; Kleppner, D. *Phys. Rev. Lett.* 1975, **35**, 1619–1622; (b) Myint, T.; Kleppner, D.; Ramsey, N. F.; Robinson, H. G. *Phys. Rev. Lett.* 1966, **17**, 405–407; Winkler, P. F.; Kleppner, D.; Myint, T.; Walther, F. G. *Phys. Rev. A* 1972, **5**, 83–114.
19. Sahm, W.; Schwenk, A. *Z. Naturforsch. A.* 1974, **29**, 1763–1766.
20. Smaller, B. *Phys. Rev.* 1951, **83**, 812–820.
21. Landolt-Börnstein, *Zahlenwerte und Funktionen aus Naturwissenschaften und Technik*, Neue Serie, Group II; Springer-Verlag: Berlin, 1974; Vol. 6, pp. 2-383–2-434.
22. Appleman, B. R.; Dailey, B. P. *Adv. Magn. Reson.* 1974, **7**, 231–320.
23. NMR Spec. *Per. Rep. C.S. (or R.S.C.)* 1972–. The first chapter of every volume is an annual review of nuclear magnetic shielding, especially theoretical and physical aspects.
24. Jameson, C. J.; Jameson, A. K.; Burrell, P. M. *J. Chem. Phys.* 1980, **73**, 6013–6020; Jameson, C. J.; Jameson, A. K.; Oppusunggu, D.; Wille, S.; Burrell, P. M.; Mason, J. *J. Chem. Phys.* 1981, **74**, 81–88.
25. Beckman, A.; Boklen, K. D.; Elke, D. *Z. Phys.* 1974, **270**, 173–186.
26. Lutz, O. *Z. Naturforsch. A* 1968, **23**, 1202–1209.
27. Huber, R.; König, F.; Weber, H. G. *Z. Phys. A* 1977, **281**, 25–33.
28. Sahm, W.; Schwenk, A. *Z. Naturforsch. A* 1974, **29**, 1754–1762.
29. Huber, R.; Knapp, M.; König, F.; Reinhard, H.; Weber, H. G. *Z. Phys. A* 1980, **296**, 95–99.
30. White, C. W.; Hughes, W. M.; Hayne, G. S.; Robinson, H. G. *Phys. Rev.* 1968, **174**, 23–32; 1973, **7**, 1178–1182.
31. Lutz, O. *Phys. Lett. A* 1967, **25**, 440–441; Lutz, O.; Nolle, A. *Z. Naturforsch. A* 1972, **27**, 1577–1581.
32. Spence, P. W.; McDermott, M. N. *Phys. Lett. A* 1967, **24**, 430–431.
33. Epperlein, B. W.; Kruger, H.; Lutz, O.; Schwenk, A. *Phys. Lett. A* 1973, **45**, 255–256.
34. Fowler, T. R. UCRL-18331, 1968, cited in Ref. 35.
35. Kodweiss, J.; Lutz, O.; Messner, W.; Mohn, K. R.; Nolle, A.; Stutz, B.; Zepf, D. *J. Magn. Reson.* 1981, **43**, 495–498.
36. Leduc, M.; Brossel, J.; Lehmann, J. C. *C. R. Acad. Sci. Paris Ser. B* 1966, **263**, 740–743.
37. Kruger, H.; Lutz, O.; Schwenk, A.; Stricker, G. *Z. Phys. A* 1974, **266**, 233–237.
38. Cagnac, B. *Ann. Phys.* (Paris) 1961, **6**, 467–526.
39. Kruger, H.; Lutz, O.; Nolle, A.; Schwenk, A. *Z. Phys. A* 1975, **273**, 325–330.
40. Gibbs, H. M.; White, C. W. *Phys. Rev.* 1969, **188**, 180–187.

41. Lutz, O.; Stricker, G. *Phys. Lett. A* 1971, **35**, 397–398.
42. Gauss, W.; Gunther, S.; Haase, A. R.; Kerber, M.; Kessler, D.; Kronenbitter, J.; Kruger, H.; Lutz, O.; Nolle, A.; Schrade, P.; Schule, M.; Siegloch, G. E. *Z. Naturforsch A* 1978, **33**, 934–939.
43. Neumann, D. B.; Moskowitz, J. W. *J. Chem. Phys.* 1969, **50**, 2216–2236; Ozier, I.; Crapo, L. M.; Ramsey, N. F. *J. Chem. Phys.* 1968, **49**, 2314–2321.
44. Mason, J. J. *Chem. Soc., Perkin Trans. 2* 1976, 1671–1674.
45. Jameson, C. J. *NMR Spec. Per. Rep. R.S.C.* 1981, **11**, 1–17.
46. Flygare, W. H.; Weiss, V. W. *J. Chem. Phys.* 1966, **45**, 2785–2792; Garvey, R. M.; DeLucia, F. C. *J. Mol. Spectrosc.* 1974, **50**, 38–44.
47. Morin, F. G.; Solum, M. S.; Withers, J. D.; Grant, D. M.; Dalling, D. K. *J. Magn. Reson.* 1982, **48**, 138–142.
48. Kukolich, S. G. *J. Am. Chem. Soc.* 1975, **97**, 5704–5707.
49. Reinartz, J. M. L. J.; Meerts, W. L.; Dymanus, A. *Chem. Phys.* 1978, **31**, 19–29; Casleton, K. H.; Kukolich, S. G. *J. Chem. Phys.* 1975, **62**, 2696–2699; Chan, S. I.; Baker, M. R.; Ramsey, N. F. *Phys. Rev. A* 1964, **136**, 1224–1228; Baker, M. R.; Anderson, C. H.; Ramsey, N. F. *Phys. Rev. A* 1964, **133**, 1533–1536.
50. Frerking, M. A.; Langer, W. D. *J. Chem. Phys.* 1981, **74**, 6990; Garber, A. R.; Ellis, P. D.; Seidman, K.; Schade, K. *J. Magn. Reson.* 1979, **34**, 1–30.
51. Wasylishen, R. E.; Mooibroek, S.; Macdonald, J. B. *J. Chem. Phys.* 1984, **81**, 1057–1059.
52. Baker, M. R.; Nelson, H. M.; Leavitt, J. A.; Ramsey, N. F. *Phys. Rev.* 1961, **121**, 807–815.
53. Hinderman, D. K.; Cornwell, C. D. *J. Chem. Phys.* 1968, **48**, 4148–4154.
54. Mason, J. *J. Chem. Soc., Dalton Trans.* 1975, 1426–1431.
55. Fabricant, B.; Muenter, J. S. *J. Chem. Phys.* 1977, **66**, 5274–5277; Ozier, I.; Yi, P. N.; Ramsey, N. F. *J. Chem. Phys.* 1977, **66**, 143–146; Casleton, K. H.; Gierke, T. D.; Wang, J. H. S.; Kukolich, S. G. *J. Chem. Phys.* 1976, **64**, 471–476; Muenter, J. S.; Klemperer, W. *J. Chem. Phys.* 1970, **52**, 6033–6037; Davis, R. E.; Muenter, J. S. *J. Chem. Phys.* 1967, **57**, 2836–2838; Weiss, V. M.; Todd, H. D.; Lo, M. K.; Gutowsky, H. S.; Flygare, W. H. *J. Chem. Phys.* 1967, **47**, 4021–4029.
56. Davies, P. B.; Neumann, R. M.; Wofsy, S. C.; Klemperer, W. *J. Chem. Phys.* 1971, **55**, 3564–3568.
57. Reinartz, J. M. L. J.; Dymanus, A. *Chem. Phys. Lett.* 1974, **24**, 346–351.
58. Wasylishen, R. E.; Connor, C.; Friedrich, J. O. *Can. J. Chem.* 1984, **62**, 981–985.
59. Fyfe, C. A. in Solid State NMR for Chemists, C.F.C. Press: Guelph, Ontario, Canada, 1983.
60. Waugh, J. S.; Huber, L. M.; Haeberlen, U. *Phys. Rev. Lett.* 1968, **20**, 180–182.
61. (a) Mehring, M. *NMR Basic Princ. Prog.* 1976, **11**, 1–243; (b) Haeberlen, U. *Adv. Magn. Reson.* 1976, Suppl. 1, 1–186.
62. Pines, A.; Gibby, M. G.; Waugh, J. S. *J. Chem. Phys.* 1972, **56**, 1776–1777.
63. Robert, J. B.; Wiesenfeld, L. *Mol. Phys.* 1981, **44**, 319–327.
64. Zilm, K. W.; Conlin, R. T.; Grant, D. M.; Michl, J. *J. Am. Chem. Soc.* 1978, **100**, 8038–8039; Zilm, K. W.; Beeler, A. J.; Grant, D. M.; Michl, J.; Chou, T.-C.; Allred, E. L. *J. Am. Chem. Soc.* 1981, **103**, 2119–2120; Zilm, K. W.; Grant, D. M. *J. Am. Chem. Soc.* 1981, **103**, 2913–2922; Zilm, K. W.; Conlin, R. T.; Grant, D. M.; Michl, J. *J. Am. Chem. Soc.* 1980, **102**, 6672–6676.
65. Garg, S. K.; Ripmeester, J. A.; Davidson, D. W. *J. Magn. Reson.* 1980, **39**, 317–323; *J. Magn. Reson.* 1979, **35**, 145–156; *J. Magn. Reson.* 1979, **36**, 325–331.
65a. Lippmaa, E.; Alla, M.; Tuherm, T. *Proc. 19th Congress Ampere*, Heidelberg, 1976, 113–118; Stejskal, E. O.; Schaefer, J.; McKay, R. A. *J. Magn. Reson.* 1977, **25**, 569; Herzfeld, J.; Berger, A. E. *J. Chem. Phys.* 1980, **73**, 6021–6030; Maricq, M. M.; Waugh, J. S. *J. Chem. Phys.* 1979, **70**, 3300–3316.
66. Saupe, A. *Z. Naturforsch. A* 1964, **19**, 161–171; Saupe, A.; Englert, G. *Phys. Rev. Lett.* 1963, **11**, 462–464; Diehl, P.; Khetrapal, C. L. *NMR Basic Princ. Prog.* 1969, **1**, 1–95; Khetrapal, C. L.; Kunwar, A. C. *Adv. Liq. Cryst.* 1983, **6**, 173–242.
67. Fujiwara, F. Y.; Reeves, L. W. *Mol. Phys.* 1978, **36**, 1897–1900; Persson, N.-O.; Lindblom, G. *J. Phys. Chem.* 1979, **83**, 3015–3019; Soderman, O.; Khan, A.; Lindblom, G. *J. Magn. Reson.* 1979, **36**, 141–146.
68. Pople, J. A. *J. Chem. Phys.* 1962, **37**, 53–59; Ditchfield, R. *J. Chem. Phys.* 1972, **36**, 5688–5691.
69. Schindler, M.; Kutzelnigg, W. *J. Chem. Phys.* 1982, **76**, 1919–1933; Schindler, M.; Kutzelnigg, W. *J. Am. Chem. Soc.* 1983, **105**, 1360–1370; *Mol. Phys.* 1983, **48**, 781–798.
70. Barfield, M.; Grant, D. M. *J. Chem. Phys.* 1977, **67**, 3322–3329.
71. Lipscomb, W. N. *Adv. Magn. Reson.* 1966, **2**, 137–176.
72. Pople, J. A.; McIver, J. W. Jr.; Ostlund, N. S. *J. Chem. Phys.* 1968, **49**, 2960–2964; Ditchfield, R.; Miller, D. P.; Pople, J. A. *J. Chem. Phys.* 1970, **53**, 613–619; Rohlfing, C. M.; Allen, L. C.; Ditchfield, R. *Chem. Phys.* 1984, **87**, 9; Chesnut, D. B.; Foley, C. K. *J. Chem. Phys.* 1986, **84**, 852–861.

73. Rowe, D. J. *Rev. Mod. Phys.* 1968, **40**, 153–166; Lamanna, U. T.; Guidoti, C.; Arrighini, G. P. *J. Chem. Phys.* 1977, **67**, 604–610; Hansen, A. E.; Bouman, T. D. *J. Chem. Phys.* 1985, **82**, 5035–5047.
74. (a) Daborn, G. T.; Handy, N. C. *Chem. Phys. Lett.* 1981, **81**, 201–208; Daborn, G. T.; Ferguson, W. I.; Handy, N. C. *Chem. Phys.* 1980, **50**, 255–263. (b) Amos, R. D. *Mol. Phys.* 1978, **35**, 1765–1775; 1980, **39**, 1–14; Pyper, N. C.; Edwards, P. P. *J. Am. Chem. Soc.* 1986, **108**, 78–81.
75. Freier, D. G.; Fenske, R. F.; You, X.-Z. *J. Chem. Phys.* 1985, **83**, 3526–3537.
76. Schastnev, P. V.; Cheremisin, A. A. *Zh. Strukt. Khim.* 1982, **23**, 139–170 (English transl. p. 440–472).
77. Höller, R.; Lischka, H. *Mol. Phys.* 1980, **41**, 1017–1040; 1041–1050.
78. Kaiser, W. *J. Chem. Phys.* 1970, **53**, 1686–1703; de Leeuw, F. H.; Dymanus, A. *J. Mol. Spectrosc.* 1973, **48**, 427.
79. Ishol, L. M.; Scott, T. A. *J. Magn. Reson.* 1977, **27**, 23–28.
80. O'Reilley, D. E.; Peterson, E. M.; El Saffar, Z. M.; Scheie, C. E. *Chem. Phys. Lett.* 1971, **8**, 470–472.
81. Zumbulyadis, N.; Dailey, B. P. *Mol. Phys.* 1974, **27**, 633–640.
82. Hegstrom, R. A. *Phys. Rev. A* 1979, **19**, 17–30.
83. Pyper, N. C. *Chem. Phys. Lett.* 1983, **96**, 211–217; 204–210; Pyykkö, P. *Chem. Phys.* 1983, **74**, 1–7.
84. Morishima, I.; Endo, K.; Yonezawa, T. *J. Chem. Phys.* 1973, **59**, 3356–3364; Volodicheva, M. I.; Rebane, T. K. *Theor. Exp. Chem. (English Transl.)* 1978, **14**, 447–455; Cheremisin, A. A.; Schastnev, P. V. *J. Magn. Reson.* 1980, **40**, 459–468.
85. Fano, U.; Martin, W. C. *Top. Modern Phys.* 1971, 147–152.
86. Barnes, R. G.; Smith, W. V. *Phys. Rev.* 1954, **93**, 95–98.
87. Dunn, T. M. *Trans. Faraday Soc.* 1961, **57**, 1441–1444.
88. Watson, R. E.; Freeman, A. J. *Hyperfine Interactions*, Freeman, A. J.; Frankel, R. B., Ed.; Academic Press: New York, 1967; pp. 53–94.
89. Jameson, C. J.; Gutowsky, H. S. *J. Chem. Phys.* 1964, **40**, 1714–1724.
90. Schneider, W. G.; Buckingham, A. D. *Discuss. Faraday Soc.* 1962, **34**, 147–155.
91. Karplus, M.; Das, T. P. *J. Chem. Phys.* 1961, **34**, 1683–1692.
92. Letcher, J. H.; Van Wazer, J. R. *Top. Phosphorus Chem.* 1967, **5**, 75–167.
93. Karplus, M.; Pople, J. A. *J. Chem. Phys.* 1963, **38**, 2803–2807; Kolker, H. J.; Karplus, M. *J. Chem. Phys.* 1963, **39**, 2011–2016.
94. Pople, J. A. *Mol. Phys.* 1964, **7**, 301–306.
95. McConnell, H. M. *J. Chem. Phys.* 1957, **27**, 226–234.
96. Haigh, C. W.; Mallion, R. B. *Prog. NMR Spectrosc.* 1979, **13**, 303–344; Mallion, R. B. *J. Chem. Phys.* 1981, **75**, 793–797.
97. Buckingham, A. D.; Schaefer, T.; Schneider, W. G. *J. Chem. Phys.* 1960, **32**, 1227–1233.
98. Kempf, J.; Spiess, H. W.; Haeberlen, U.; Zimmerman, H. *Chem. Phys.* 1974, **4**, 269–276.
99. Mason, J. *Chem. Rev.*, in press.
100. Moore, C. E. "Atomic Energy Levels," National Bureau of Standards Circular No. 467, Vols. 1–3, revised 1971.
101. Carlson, T. A.; Lu, C. C.; Tucker, T. C.; Nestor, C. W.; Malik, F. B. "Eigenvalues, Radial Expectation Values, and Potentials for Free Atoms from $Z=2$ to 126 as calculated from Relativistic Hartree–Fock–Slater Atomic Wave Functions," Oak Ridge National Laboratory, 1970, pp. 1–29.
102. Pyykkö, P.; Desclaux, J.-P. *Acc. Chem. Res.* 1979, **12**, 276–281; Pitzer, K. S. *Acc. Chem. Res.* 1979, **12**, 271–276.
103. Büttgenbach, S. *Hyperfine Structure in 4d- and 5d-Shell Atoms*, Springer-Verlag: Berlin, 1982.
104. Tokuhiro, T.; Fraenkel, G. *J. Am. Chem. Soc.* 1969, **91**, 5005–5013.
105. (a) Grieves, R. A.; Mason, J. *Polyhedron*, 1986, **5**, 415–421, and work to be published; (b) Green, J. C., Grieves, R. A.; Mason, J. *J. Chem. Soc., Dalton Trans.* 1986, 1313–1316.
106. Andrew, E. R. *Phil. Trans. R. Soc. London Ser. A* 1981, **299**, 505–520.
107. Farnum, D. G. *Adv. Phys. Org. Chem.* 1975, **11**, 123–175; Nelson, G. L.; Williams, E. A. *Prog. Phys. Org. Chem.* 1976, **12**, 229–342; Martin, G. J.; Martin, M. L.; Odiot, S. *Org. Magn. Reson.* 1975, **7**, 2–17.
108. Ewing, D. F. In *Correlation Analysis in Chemistry: Recent Advances*, Chapman, N. B.; Shorter, J., Eds.; Plenum: New York, 1978, pp. 357–396; *Org. Magn. Reson.* 1979, **12**, 499–523.
109. Spiesecke, H.; Schneider, W. G. *Tetrahedron Lett.* 1961, 468–472.
110. Olah, G. A.; Mateescu, G. D. *J. Am. Chem. Soc.* 1970, **92**, 1430–1432.
111. Memory, J. D.; Wilson, N. K. *NMR of Aromatic Compounds*, Wiley: New York, 1982.
112. Griffith, R. C.; Grant, D. M.; Roberts, J. C. *J. Org. Chem.* 1975, **40**, 3726–3729.
113. Hehre, W. J.; Taft, R. W.; Topsom, R. D. *Prog. Phys. Org. Chem.* 1976, **12**, 159–187.
114. Fliszar, S.; Cardinal, G.; Beraldin, M. T. *J. Am. Chem. Soc.* 1982, **104**, 5287–5292.

115. Stothers, J. B. ^{13}C *NMR Spectroscopy*, Academic Press: New York, 1972; Levy, G. C.; Lichter, R. L.; Nelson, G. L. ^{13}C *NMR Spectroscopy*, Wiley: New York, 1980.
116. Craik, D. J.; Levy, G. C.; Brownlee, R. T. C. *J. Org. Chem.* 1983, **48**, 1601–1606.
117. Strub, H.; Beeler, A. J.; Grant, D. M.; Michl, J.; Cutts, P. W.; Zilm, K. W. *J. Am. Chem. Soc.* 1983, **105**, 3333–3334.
118. Pople, J. A.; Schneider, W. G.; Bernstein, H. J. *High-Resolution NMR*, McGraw-Hill: New York, 1959.
119. Gutowsky, H. S.; Hoffman, C. J. *J. Chem. Phys.* 1951, **19**, 1259–1267.
120. Mason, J. *J. Chem. Soc., Dalton, Trans.* 1975, 1422–1426 and refs. therein.
121. Mason, J. *Chem. Rev.* 1981, **81**, 205–227.
122. Figgis, B. N.; Kidd, R. G.; Nyholm, R. S. *Proc. R. Soc. London, Ser. A* 1962, **269**, 469–480; Jung, W.-S.; Tomiyasu, H.; Fukutomi, H. *Bull. Chem. Soc. Jpn.* 1985, **58**, 938–942, and refs. therein.
123. De Jeu, W. H. *Mol. Phys.* 1970, **18**, 31–37; St. Amour, T. E.; Burgar, M. I.; Valentine, B.; Fiat, D. *J. Am. Chem. Soc.* 1981, **103**, 1128–1136.
124. Mason, J. *J. Chem. Soc., Faraday Trans. 2* 1982, **78**, 1539–1549.
125. Sugawara, R.; Kawada, Y.; Katoh, M.; Iwamura, H. *Bull. Chem. Soc. Jpn.* 1979, **52**, 3391–3396.
126. Fujiwara, S.; Yajima, F.; Yamasaki, A. *J. Magn. Reson.* 1969, **1**, 203–210; Juranic, N.; Celap, M. B.; Vucelic, D.; Malinar, M. H.; Radivojsa, P. N. *J. Magn. Reson.* 1979, **35**, 319–327.
127. Orgel, L. E. *J. Chem. Soc.* 1958, 4186–4190; *Mol. Phys.* 1958, **1**, 322–325.
128. Koie, Y.; Shinoda, S.; Saito, Y. *J. Chem. Soc., Dalton Trans.* 1981, 1082–1088.
128a. Bönnemann, H.; Brijoux, W.; Brinkmann, R.; Meurers, W.; Mynott, R. *J. Organomet. Chem.* 1984, **272**, 231.
129. Jørgensen, C. K. *Absorption Spectra and Chemical Bonding*, Pergamon: Oxford, 1962; Jørgensen, C. K. *Modern Aspects of Ligand Field Theory*, North-Holland: Amsterdam, 1971.
130. Phillips, C. S. G.; Williams, R. J. P. *Inorganic Chemistry*, Oxford University Press: Oxford, 1965.
131. Kidd, R. G. *Annu. Rep. NMR Spectrosc.* 1980, **10A**, 1–79; Kidd, R. G.; Goodfellow, R. J. in *NMR and the Periodic Table*, Harris, R. K.; Mann, B. E., Eds. Academic Press: London, 1978; chap. 8, 195–278.
132. Dechter, J. J. *Prog. Inorg. Chem.* 1985, **33**, 393–507; Rehder, D. *Magn. Reson. Rev.* 1984, **9**, 125–237.
133. Laszlo, P. (chapter 9); Mann, B. E. (chapter 11) in *NMR of Newly Accessible Nuclei*, Laszlo, P., Ed.; Academic Press: New York, 1983, vol. 2.
134. Andersson, L.-O.; Mason, J.; van Bronswijk, W. *J. Chem. Soc. A* 1970, 296–299.
135. Christ, H. A.; Diehl, P.; Schneider, R.-H., Dahn, H. *Helv. Chim. Acta* 1961, **44**, 865–880.
136. (a) Kanda, K.; Nakatsuji, H.; Yonezawa, T. *J. Am. Chem. Soc.* 1984, **106**, 5888–5892. (b) Nakatsuji, H.; Kanda, K.; Endo, K.; Yonezawa, T. *J. Am. Chem. Soc.* 1984, **106**, 4653–4660.
137. (a) Martin, R. L.; White, A. H. *Nature* 1969, **223**, 394–396; Weiss, R.; Verkade, J. G. *Inorg. Chem.* 1979, **18**, 529–530. (b) Goggin, P. L.; Goodfellow, R. J.; Haddock, S. R.; Taylor, B. F.; Marshall, I. R. H. *J. Chem. Soc. Dalton Trans.* 1976, 459–467; Pidcock, A.; Richards, R. E.; Venanzi, L. M. *J. Chem. Soc. A* 1968, 1970.
138. Bramley, R.; Brorson, M.; Sargeson, A. M.; Schäffer, C. E. *J. Am. Chem. Soc.* 1985, **107**, 2780–2787.
139. Grant, D. M.; Paul, E. G. *J. Am. Chem. Soc.* 1964, **86**, 2984–2990.
140. Vladimiroff, T.; Malinowski, E. R. *J. Chem. Phys.*, 1967, **46**, 1830–1841; Malinowski, E. R. *J. Am. Chem. Soc.* 1969, **91**, 4701.
141. Schraml, J.; Bellama, J. M. *Determination of Organic Structures by Physical Methods*, Nachod, F. C.; Zuckerman, J. J.; Randall, E. W., Ed.; Academic Press: New York, 1976; Vol. 6, pp. 203–269; Grimmer, A. R. *Z. Anorg. Allg. Chem.* 1973, **400**, 105–114.
142. Ionov, S. P.; Lyubimov, V. S. *Russ. J. Phys. Chem.* 1980, **54**, 5–12.
143. Lyubimov, V. S.; Ionov, S. P. *Russ. J. Phys. Chem.* 1972, **46**, 486–489; 1755–1758.
144. Ionov, S. P.; Lyubimov, V. S. *Russ. J. Phys. Chem.* 1971, **45**, 1407–1409.
145. Ando, I.; Asakura, T. *Annu. Rep. NMR Spectrosc.* 1980, **10A**, 81–132.
146. Quin, L. D.; Mesch, K. A. *J. Chem. Soc., Chem. Commun.* 1980, 959.
147. Au-Yeung, S. C. F.; Eaton, D. R. *J. Magn. Reson.* 1983, **52**, 351–365.
148. Akitt, J. W.; McDonald, W. S. *J. Magn. Reson.* 1984, **58**, 401–412.
149. Yajima, F.; Koike, Y.; Yamasaki, A.; Fujiwara, S. *Bull. Chem. Soc. Jpn.* 1974, **47**, 1442–1446.
150. Brown, D. A.; Chester, J. P.; Fitzpatrick, N. J. *Inorg. Chem.* 1982, **21**, 2111–2112; 1977, **16**, 2497–2500.
150a. Beach, N. A.; Gray, H. B. *J. Am. Chem. Soc.* 1968, **90**, 5713–5721.
151. Gleeson, J. W.; Vaughan, W. R. *J. Chem. Phys.* 1983, **78**, 5385–5392; Oldfield, E.; Keniry, M. A.; Shinada, S.; Schramm, S.; Brown, T. L.; Gutowsky, H. S. *J. Chem. Soc., Chem. Commun.* 1985, 791–793.

152. (a) Carroll, D. G.; McGlynn, S. P. *Inorg. Chem.* 1968, **7**, 1285–1290; (b) Maricq, M. M.; Waugh, J. S.; Fletcher, J. L.; McGlinchey, M. J. *J. Am. Chem. Soc.* 1978, **100**, 6902–6904.
153. Pitcher, E.; Buckingham, A. D.; Stone, F. G. A. *J. Chem. Phys.* 1962, **36**, 124–129.
154. Semin, G. K.; Rabushkina, T. A.; Yakobson, G. G. *Nuclear Quadrupole Resonance in Chemistry*, Khimiya: Leningrad 1972; English translation, Halsted-Wiley: Chichester, 1975; Lucken, E. A. C. *Nuclear Quadrupole Coupling Constants*, Academic Press: New York, 1969.
155. Saito, Y. *Can. J. Chem.* 1965, **43**, 2530–2534.
156. Slater, J. L.; Pupp, M.; Sheline, R. K. *J. Chem. Phys.* 1972, **57**, 2105–2108 and refs. therein.
157. Onaka, S.; Sasaki, Y.; Sano, H. *Bull. Chem. Soc. Jpn.* 1971, **44**, 726–730; Wuyts, L. F.; Van der Kelen, G. P. *J. Mol. Struct.* 1974, **23**, 73–79.
158. Mason, J. *J. Chem. Soc., Faraday Trans. 2* 1976, **72**, 2064–2068.
159. Cotton, F. A.; Harris, C. B. *Proc. Natl. Acad. Sci. U.S.A.*, 1966, **56**, 12–15; Lucken, E. A. C. *Fortschr. Chem. Forsch.* 1972, **30**, 156–171.
160. Brown, T. L. *Acc. Chem. Res.* 1974, **7**, 408–415.
161. Bancroft, G. M.; Clark, H. C.; Kidd, R. G.; Rake, A. T.; Spinney, G. C. *Inorg. Chem.* 1973, **12**, 728–731.
162. Parish, R. V. *Prog. Inorg. Chem.* 1972, **15**, 101–200; Bancroft, G. M.; Platt, R. H. *Adv. Inorg. Chem. Radiochem.* 1972, **15**, 59–258.
163. Pople, J. A. *Discuss. Faraday Soc.* 1962, **34**, 7–14; *J. Chem. Phys.* 1962, **37**, 60–66.
164. Carrington, A. *Mol. Phys.* 1960, **3**, 271–275; Jørgensen, C. K. *Prog. Inorg. Chem.* 1970, **12**, 101–158.
165. Kazanskii, L. P.; Spitsyn, V. I. *Dokl. Akad. Nauk. S.S.S.R.* 1975, **223**, 381–384 (721–724 in English).
166. Gillespie, R. J.; Schrobilgen, G. J. *Inorg. Chem.* 1974, **13**, 765–770; 1455–1459; 2370–2374.
167. Reuben, J.; Shvo, Y.; Demiel, A. *J. Am. Chem. Soc.* 1965, **87**, 3995–3996.
168. Johnson, K. J.; Hunt, J. P.; Dodgen, H. W. *J. Chem. Phys.* 1969, **51**, 4493–4496.
169. Lassigne, C. R.; Wells, E. J. *Can. J. Chem.* 1977, **55**, 927.
170. Doddrell, D. M.; Barron, P. F.; Clegg, D. E.; Bowie, C. *J. Chem. Soc., Chem. Commun.* 1982, 575–576.
171. (a) Klemperer, W. G. *Angew. Chem. Intl. Ed.* 1978, **17**, 246–254. (b) Grimmer, A.-R.; Radeglia, R. *Chem. Phys. Lett.* 1984, **106**, 262–265; Grimmer, A.-R.; Von Lampe, F.; Lippmaa, E. *Monatsh. Chem.* 1984, **115**, 561–564; Klinowski, J. *Prog. NMR Spectrosc.* 1984, **16**, 237–309. (c) Grimmer, A.-R. *Z. Chem.* 1984, **24**, 194–195; Lippmaa, E.; Samoson, A.; Mägi, M. *J. Am. Chem. Soc.* 1986, **108**, 1730–1735; e.g., Garrou, P. *Chem. Rev.* 1981, **81**, 229–266.
172. Mann, B. E.; Taylor, B. F. *^{13}C NMR Data for Organometallic Compounds*, Academic Press: London, 1981.
173. Brown, A. J.; Howarth, O. W.; Moore, P. *J. Chem. Soc., Dalton Trans.* 1976, 1589–1592.
174. Mann, B. E. *J. Chem. Soc., Dalton Trans.* 1973, 2012–2015; Braterman, P. S.; Milne, D. W.; Randall, E. W.; Rosenberg, E. *J. Chem. Soc., Dalton Trans.* 1973, 1027–1031.
175. LaMar, G. N.; Horrocks, W. DeW.; Holm, R. H., Eds.; *NMR of Paramagnetic Molecules*, Academic Press: New York, 1973.
176. Kohler, F. H.; Hofmann, P.; Prossdorf, W. *J. Am. Chem. Soc.* 1981, **103**, 6359–6367.
177. Knight, W. D. *Solid State Phys.* 1956, **2**, 93–136.
177a. Grigor'ev, I. A.; Volodarsky, L. B., Gogolev, A. Z.; Sagdeev, R. Z. *Chem. Phys. Lett.* 1985, **122**, 46–50.
178. Kurland, R. J.; McGarvey, B. R. *J. Magn. Reson.* 1970, **2**, 286; McGarvey, B. R. *J. Chem. Phys.* 1970, **53**, 86–91.
179. Orrell, K. G. *NMR Spec. Per. Rep. R.S.C.* 1985, **4**, 318–349; also **6**, **8**, **10**, **12**.
180. Jameson, C. J. *Bull. Magn. Reson.* 1980, **3**, 3–28.
181. Rummens, F. H. A. *NMR Basic Princ. Prog.* 1975, **10**, 1–118.
182. Raynes, W. T., Buckingham, A. D.; Bernstein, H. J. *J. Chem. Phys.* 1962, **36**, 3481–3488.
183. Homer, J. *Appl. Spectrosc. Rev.* 1975, **9**, 1–132; Ando, I.; Webb, G. A. *Org. Magn. Reson.* 1981, **15**, 111–130; Jallali-Heravi, M.; Lamphun, B. N.; Webb, G. A.; Ando, I.; Kondo, M.; Watanabe, S. *Org. Magn. Reson.* 1980, **14**, 92–97; 116–118; Jallali-Heravi, M.; Webb, G. A. *Org. Magn. Reson.* 1980, **13**, 116–118; Orville-Thomas, W. J.; Ratajczak, H.; Rao, C. N. R., Eds.; *Top. Mol. Interactions*, Elsevier: Amsterdam, 1985.
184. Pirkle, W. H.; Beare, S. D. *J. Am. Chem. Soc.* 1969, **91**, 5150–5155.
185. Pirkle, W. H. *J. Am. Chem. Soc.* 1966, **88**, 1837.
186. Dyllick-Brenzinger, R.; Roberts, J. D. *J. Am. Chem. Soc.* 1980, **102**, 1166–1167.
187. Pirkle, W. H.; Hoover, D. J. *Topics in Stereochemistry*, Allinger, N. L., Eliel, E. L., Ed.; Interscience: New York, 1982; Vol. 13, pp. 263–331.
188. Ditchfield, R. *Chem. Phys.* 1981, **63**, 185–202.

189. Fowler, P. W.; Raynes, W. T. *Mol. Phys.* 1981, **43**, 65–82.
190. Altman, L. J.; Laungani, D.; Gunnarsson, G.; Wennerstrom, H.; Forsén, S. *J. Am. Chem. Soc.* 1978, **100**, 8264–8266.
191. Jameson, C. J. *J. Chem. Phys.* 1977, **66**, 4983–4988; Jameson, C. J.; Osten, H. J. *J. Chem. Phys.* 1984, **81**, 4293–4299; 4300–4305; *Ann. Rep. NMR Spectrosc.* 1986, **17**, 1–78.
192. Batiz-Hernandez, H.; Bernheim, R. A. *Prog. NMR Spectrosc.* 1967, **3**, 63–85.
193. Aydin, R.; Günther, H. *J. Am. Chem. Soc.* 1981, **103**, 1301–1303; Jurlina, J. L.; Stothers, J. B. *J. Am. Chem. Soc.* 1982, **104**, 4677–4678.
194. Jameson, C. J. In *Nuclear Magnetic Resonance*, Webb, G. A., Ed.; Royal Society of Chemistry: London, 1980; Vol. 10, pp. 1–17.
195. Hansen, P. E. *Annu Rep. NMR Spectrosc.* 1983, **15**, 105–233; Forsyth, D. A. *Isotopes in Organic Chemistry*, 1984, **6**.
196. Saunders, M.; Kates, M. R. *J. Am. Chem. Soc.* 1977, **99**, 8071–8072; Saunders, M.; Jaffe, M. H.; Vogel, P. *J. Am. Chem. Soc.* 1971, **93**, 2558–2559; Saunders, M.; Vogel, P. *J. Am. Chem. Soc.* 1971, **93**, 2559–2561, 2561–2562; Saunders, M.; Telkowski, L.; Kates, M. R. *J. Am. Chem. Soc.* 1977, **99**, 8072–8073; Anet, F. A. L.; Basus, V. J.; Hewett, A. P. W.; Saunders, M. *J. Am. Chem. Soc.* 1980, **102**, 3945–3946; Kuroda, Y.; Fujiwara, Y. *J. Phys. Chem.* 1982, **86**, 4913–4918; Lycka, A.; Hansen, P. E. *Org. Magn. Reson.* 1985, **23**, 973–6; 1984, **22**, 569–572; Mason, J.; Mingos, D. M. P.; Sherman, D.; Wardle, R. W. M. *J. Chem. Soc., Chem. Commun.* 1984, 1223.
197. Anet, F. A. L.; Dekmezian, A. H. *J. Am. Chem. Soc.* 1979, **101**, 5449–5451.
198. Lindman, B.; Forsen, S. *NMR Basic Princ. Prog.* 1976, **12**, 1–365.
199. Deverell, C.; Schaumburg, K.; Bernstein, H. J. *J. Chem. Phys.* 1968, **49**, 1276–1283.
200. Epperlein, B. W.; Kruger, H.; Lutz, O.; Schwenk, A. *Z. Naturforsch. A* 1974, **29**, 660–661.
201. Lutz, O.; Oehler, H. *Z. Naturforsch. A* 1977, **32**, 131–133; Raynes, W. T. *Mol. Phys.* 1983, **49**, 443–447.
202. Stamm, H.; Lamberty, W.; Stafe, J. *J. Am. Chem. Soc.* 1980, **102**, 1529–1532.
203. Harris, R. K.; Spragg, R. A. *J. Chem. Soc., Chem. Commun.* 1967, 362–363; Carter, R. E.; Drakenberg, T.; Bergman, N. A. *J. Am. Chem. Soc.* 1975, **97**, 6990–6996; Lumbroso-Bader, N.; Coupry, C.; Baron, D.; Clague, D. H.; Govil, G. *J. Magn. Reson.* 1975, **17**, 386–392; Pachler, K. G. R.; Wessels, P. L. *J. Mol. Struct.* 1980, **68**, 145–159; Ross, B. D.; True, N. S.; Decker, D. L. *J. Phys. Chem.* 1983, **87**, 89–94; Ross, B. D.; True, N. S. *J. Am. Chem. Soc.* 1983, **105**, 1382–1383.

Chapter 4

SPIN–SPIN COUPLING

CYNTHIA J. JAMESON

In most routine applications of NMR only the magnitude of the isotropic average of the coupling constant appears, and this merely as spacings in symmetric multiplet patterns. However, the additional information which comes from the sign of the isotropic average coupling constant and the anisotropy of the coupling tensor is of even greater use as a diagnostic parameter and as an index of the chemical bond. We consider the general empirical patterns in coupling constants which have been very useful in structural applications, taking a global (i.e., entire periodic table) view and giving many examples of each general trend. Because the dependences on structural factors and bond parameters are more easily sorted out in one-bond coupling constants, we have used these to illustrate the dependence on atomic *s* densities, coordination number, oxidation state, formal hybridization, and electronegativity of substituents on the coupled nuclei, factors which affect the magnitude of coupling constants over any number of bonds. In addition, general trends involving geometric factors (e.g., bond angles) and stereospecificity of one-bond, two-bond, and three-bond couplings are discussed. The theoretical interpretation follows, with a comparative overview of the different computational schemes. Approximate models are used to explain the empirical trends where appropriate. Since a large volume of data has become available for the heavier nuclei, for which relativistic effects cannot be neglected, we quote results of relativistic calculations as well.

1. GENERAL CONSIDERATIONS

The interaction of nuclear spins is composed of a direct dipolar coupling and an indirect interaction by way of the electrons. The interaction energy is

$$E_{NN'} = h\mathbf{I}_N \cdot (\mathbf{D} + \mathbf{J}) \cdot \mathbf{I}_{N'} = hJ_S \mathbf{I}_N \mathbf{I}_{N'} + h\mathbf{I}_N \cdot (\mathbf{D} + \mathbf{J}_T) \cdot \mathbf{I}_{N'} \tag{1}$$

where J_S is the scalar part and J_T the tensor part of **J**. The direct dipolar coupling **D** which we discussed in Chapter 2 (Section 2.2) is symmetric and traceless. In an oriented system, both **D** and **J** contribute to the observed spectrum. Since **D** may be calculated from internuclear distances, the components of the indirect nuclear coupling constant **J** may be deduced by difference. A positive J as conventionally defined in equation (1) results from an interaction which minimizes the energy $E_{NN'}$ when the two nuclear spins are antiparallel.

CYNTHIA J. JAMESON • Department of Chemistry, University of Illinois, Chicago, Illinois 60680.

1.1. Mechanisms of Spin–Spin Coupling

The indirect spin–spin coupling arises by means of the several mechanisms whereby one nucleus perturbs the electrons surrounding it and these electrons produce a magnetic field at the other nucleus. We have already described the Fermi-contact, orbital, and spin-dipolar mechanisms of J coupling in Chapter 2. Ramsey formulated the theory of indirect spin–spin coupling in which the electron-coupled nuclear spin–nuclear spin interaction arises from the magnetic interactions between the nuclei and the molecular electrons. These magnetic interactions, the terms of which are given in Figure 1,[1] will contribute to the total energy of the molecule. r is the electron–

	$\mathscr{H}_{1b}$	$\mathscr{H}_2$	$\mathscr{H}_3$
$\mathscr{H}_{1b}$	orbital mechanism $J_S^{(1b)}+J_T^{(1b)}$	zero	zero
$\mathscr{H}_2$	zero	spin-dipolar mechanism $J_S^{(2)}+J_T^{(2)}$	spin-dipolar contact mechanism $J_T^{(4)}$
$\mathscr{H}_3$	zero	spin-dipolar contact mechanism $J_T^{(4)}$	Fermi contact mechanism $J_S^{(3)}$

$$\mathscr{H}=\mathscr{H}_{1a}+\mathscr{H}_{1b}+\mathscr{H}_2+\mathscr{H}_3 \tag{2}$$

$$\mathscr{H}_{1a}=e\hbar\mu_B(\mu_0/4\pi)^2\sum_{NN'}\gamma_N\gamma_{N'}\sum_k r_{kN}^{-3}r_{kN'}^{-3} \times[(\mathbf{I}_N\cdot\mathbf{I}_{N'})(\mathbf{r}_{kN}\cdot\mathbf{r}_{kN'})-(\mathbf{I}_N\cdot\mathbf{r}_{kN'})(\mathbf{I}_{N'}\cdot\mathbf{r}_{kN})] \tag{3}$$

$$\mathscr{H}_{1b}=(2\mu_B\hbar/i)(\mu_0/4\pi)\sum_N\gamma_N\mathbf{I}_N\cdot\sum_k r_{kN}^{-3}(\mathbf{r}_{kN}\times\mathbf{\nabla}_k) \tag{4}$$

$$\mathscr{H}_2=2\mu_B\hbar(\mu_0/4\pi)\sum_N\gamma_N\sum_k[3(\mathbf{S}_k\cdot\mathbf{r}_{kN})(\mathbf{I}_N\cdot\mathbf{r}_{kN})\,r_{kN}^{-5}-(\mathbf{S}_k\cdot\mathbf{I}_N)\,r_{kN}^{-3}] \tag{5}$$

$$\mathscr{H}_3=(16\pi\mu_B\hbar/3)(\mu_0/4\pi)\sum_N\gamma_N\sum_k\delta(r_{kN})\,\mathbf{S}_k\cdot\mathbf{I}_N \tag{6}$$

$$\mathbf{J}=\mathbf{J}^{(1a)}+\mathbf{J}^{(1b)}+\mathbf{J}^{(2)}+\mathbf{J}^{(3)}+\mathbf{J}^{(4)} \tag{7}$$

$$\mathbf{J}^{(1a)}=(8m/h)\,\mu_B^2(\mu_0/4\pi)^2\,\gamma_N\gamma_{N'}\langle 0|\sum_k(\mathbf{r}_{kN}\cdot\mathbf{r}_{kN'})\,r_{kN}^{-3}r_{kN'}^{-3}\,|0\rangle \tag{8}$$

$$\mathbf{J}^{(1b)}=(-2/h)(2\mu_B\hbar)^2(\mu_0/4\pi)^2\,\gamma_N\gamma_{N'}\sum_n({}^1E_n-E_0)^{-1}\,\langle 0|\sum_k r_{kN}^{-3}(\mathbf{r}_{kN}\times\mathbf{\nabla}_k)\,|n\rangle\langle n|\sum_j r_{jN'}^{-3}(\mathbf{r}_{jN'}\times\mathbf{\nabla}_j)\,|0\rangle \tag{9}$$

$$\mathbf{J}^{(2)}=(-2/h)(2\mu_B\hbar)^2(\mu_0/4\pi)^2\,\gamma_N\gamma_{N'}\sum_n({}^3E_n-E_0)^{-1}\,\langle 0|\sum_k 3r_{kN}^{-3}(\mathbf{S}_k\cdot\mathbf{r}_{kN})\,\mathbf{r}_{kN}-r_{kN}^{-3}\mathbf{S}_k\,|n\rangle \times\langle n|\sum_j 3r_{jN'}^{-3}(\mathbf{S}_j\cdot\mathbf{r}_{jN'})\,\mathbf{r}_{jN'}-r_{jN'}^{-3}\mathbf{S}_j\,|0\rangle \tag{10}$$

$$\mathbf{J}^{(3)}=(-2/h)(16\pi\mu_B\hbar/3)^2(\mu_0/4\pi)^2\,\gamma_N\gamma_{N'}\sum_n({}^3E_n-E_0)^{-1} \times\langle 0|\sum_k\delta(r_{kN})\,\mathbf{S}_k\,|n\rangle\langle n|\sum_j\delta(r_{jN'})\,\mathbf{S}_j\,|0\rangle \tag{11}$$

$$\mathbf{J}^{(4)}=(-4/h)(2\mu_B\hbar)(16\pi\mu_B\hbar/3)(\mu_0/4\pi)^2\,\gamma_N\gamma_{N'}\sum_n({}^3E_n-E_0)^{-1} \times\langle 0|\sum_k\delta(r_{kN})\,\mathbf{S}_k\,|n\rangle\langle n|\sum_j r_{jN'}^{-3}(\mathbf{S}_j\cdot\mathbf{r}_{jN'})\,\mathbf{r}_{jN'}-r_{jN'}^{-3}\mathbf{S}_j\,|0\rangle \tag{12}$$

Figure 1. Mechanisms for indirect coupling of nuclear spins arising from second-order perturbation theory according to the formulation of Ramsey.[1] Contributions to the isotropic part J_S and the anisotropic part J_T of equation (1) are shown here. There is an additional term which arises from first-order perturbation, which is considered part of the orbital mechanism, given by $\mathscr{H}_{(1a)}$ in equation (3).

nuclear distance, $\delta(r_{kN})$ is a Dirac delta function which picks out the value at $r_{kN}=0$ in any integration over the coordinates of electron k. The other symbols conform to standard usage and have been defined in Chapter 2. All of these types of interactions can be derived in a natural way from the relativistic Dirac equation for an electron in an external field if the magnetic vector potential of a nucleus is taken to be that of a point dipole.[2] In term (1b) the magnetic moment of the nucleus interacts with the magnetic field produced by orbital motion of the surrounding electrons. The spin dipolar term (2) involves the interaction of the electron spin magnetic dipole with the nuclear spin magnetic dipole. The Fermi contact term (3) arises from the interaction of the nuclear spin magnetic dipole with the electron spin which has a finite density at the nucleus.

All of the interactions in the Hamiltonian given in Figure 1 are small compared with the electrostatic terms in the molecular Hamiltonian. Therefore, perturbation theory may be applied. The terms which are bilinear in the coupled nuclear spins can be selected from the perturbation energy and equated to $h\mathbf{I}_N \cdot \mathbf{J}_{NN'} \cdot \mathbf{I}_{N'}$. The term $\mathscr{H}_{(1a)}$ is already bilinear in the nuclear angular momentum operator; that is, it already is a sum of terms of the form $h\mathbf{I}_N \cdot \mathbf{J}^{(1a)}_{NN'} \cdot \mathbf{I}_{N'}$. The other terms are linear in the nuclear angular momentum. Thus, they contribute to the electron coupled nuclear spin–nuclear spin interaction in second-order perturbation theory. The sum of all the terms will give the total value of the coupling constant. Ramsey has shown that some of the second-order terms vanish as indicated in Figure 1. Only the contact term $J^{(3)}$ is purely isotropic, only the cross term $J^{(4)}$ is purely anisotropic; the others have isotropic and anisotropic parts. Since some of the contributions to the coupling tensor are in general nonsymmetric ($J^{(1)}$ and $J^{(2)}$), the coupling tensor $\mathbf{J}$ itself is in general nonsymmetric. The symmetry properties of the molecule determine the number of independent components of $\mathbf{J}$; the relevant symmetry is that appropriate to the two coupled nuclei. Buckingham *et al.*[3,4] have tabulated the number of independent components of the coupling tensor for some important group symmetries. In those molecules (e.g., linear molecules) where the internuclear axis coincides with a threefold or higher symmetry axis of the molecule, the off-diagonal components of the tensor vanish and two of the diagonal components become equal. Then the coupling tensor has only two independent components, and they may be taken as $J_{\parallel}$ and $J_{\perp}$. We will consider the anisotropy of the coupling tensor in the next section.

For different isotopes of the same element, the coupling constants differ because of different values of γ_N; however, the electronic environments are only slightly affected by the different nuclear properties. It is therefore convenient to write the interaction energy in terms of the nuclear moments instead of the nuclear spin angular momentum

$$E_{NN'} = h\mathbf{I}_N \cdot \mathbf{J}_{NN'} \cdot \mathbf{I}_{N'} = \boldsymbol{\mu}_N \cdot \mathbf{K}_{NN'} \cdot \boldsymbol{\mu}_{N'} \tag{13}$$

so that the reduced coupling constant $\mathbf{K}_{NN'}$ defined by Pople and Santry[5] is related to $\mathbf{J}_{NN'}$ by

$$\mathbf{K}_{NN'} = 4\pi^2 \mathbf{J}_{NN'}/(h\gamma_N \gamma_{N'}) \tag{14}$$

and is of the order $1 \times 10^{20}\ \mathrm{cm}^{-3}$ (in cgs units) or $1 \times 10^{19}\ \mathrm{N\,A^{-2}\,m^{-3}}$ (in SI units), whereas J is in Hz. While the coupling constant J is the directly measured quantity for the practicing NMR spectroscopist, comparisons of coupling constant magnitudes and especially signs are best done with the reduced coupling constant. Thus, couplings involving nitrogen can be compared with one another when the data include couplings

$^1K(HX)$

H +42.9	B +20.8	C +41.3	N +50.3	O +48.5	F +46.9
	Al (+96.1)	Si +84.9	P +37.8	S (+28.5)	Cl +32
		Ge +232	As ±45 (+36)	Se +28.4	Br ±19 (−69)
		Sn +430	Sb (+5)	Te ±15.5 (−98)	I −33 (−223)
		Pb +938	Bi (−572)		

a

$^1K(CX)$

H +41.3	B +4.85	C +49	N +14.7	O	F −57
Li +12.8		Si +83.3	P −11	S (+26)	Cl ±78.5
		Ge	As	Se −107	Br ±55
		Sn +300	Sb ±622	Te −169	I < 96
	Tl +1096	Pb +396	Bi		

b

$^1K(P^{III}X)$ / $^1K(P^{IV,V,VI}X)$

H +37.8 / +112	B +28.5 / +64.2	C −11 / +76	N −120 / +54.6	O −23.2 / +182	F −314 / −154
		Si −43.7 /	P − / ∓	S / (+4)	Cl ∓252 /
		Ge	As	Se −221 / −752	Br ∓288 /
		Sn −460 / −119	Sb	Te / −1113	I
		Pb −131.5 /	Bi		

c

$^1K(FX)$

H +46.9	B −23.1	C −91	N ∓195	O ∓196	F (+200)
Li +48		Si −79	P −314 −154	S ∓289	Cl +970 ±304
Na		Ge ∓455	As ∓438	Se ∓645	Br ±558
K ±60	In −865	Sn −546	Sb ∓687	Te −1030	I ±930 ±1210
Rb +210	Tl −2040	Pb	Bi ∓2103		
Cs +400					

d

Figure 2. Observed values of (a) $^1K(HX)$, (b) $^1K(CX)$, (c) $^1K(PX)$, and (d) $^1K(FX)$ displayed according to the position of X in the Periodic Table. Calculated values from Ref. 13 are shown in parentheses. Units are 10^{19} N A^{-2} m^{-3}. (a) Data for H_2, BH_4^-, CH_4, NH_3, OH_2, FH, AlH_3, SiH_4, PH_3, SH_2, ClH, GeH_4, AsH_3, SeH_2, BrH, SnH_4, SbH_3, TeH_3, IH, TlH, $PbHMe_3$, BiH_3. (b) Data for CH_4, BMe_3, CMe_4, $MeNH_2$, MeOH, MeF, $(MeLi)_4$, $SiMe_4$, PMe_3, MeSH, $CHCl_3$, Me_2Se, CD_3Br. $SnMe_4$, Ph_3Sb, Me_2Te, CH_3I, $TlMe_3$, $PbMe_4$. (c) Data for PH_3, PH_4^+, Me_2PBX_2, $(MeO)_3PBH_3$, Me_3P, $Me_3P{=}O$, $(Me_2N)_3P$, $(Me_2N)_3P{=}O$, $(MeO)_3P{=}O$, $Me_3P{=}O$, PF_3, PF_6^-, R_3PAlX_3, $P(SiH_3)_3$, $Me_3P{=}S$, PCl_3, Me_2PSeMe, $Me_3P{=}Se$, PBr_3, $(Me_3Sn)_3P$, $(Me_3Sn)_3PM(CO)_5$, $Me_3P{=}Te$, Me_3PbPH_2. (d) Data for HF, HBF_2, CF_4, NF_3, OF_2, F_2, LiF, SiF_4, PF_3 and PF_6^-, SF_6, ClF and ClF_6^+, GeF_4, AsF_6^-, SeF_6, BrF_6^+, KF, InF, R_3SnF, SbF_6^-, TeF_6^-, IF_7 and IF_6^+, RbF, TlF, BiF_6^-, and CsF.

with either ^{14}N or ^{15}N. The reduced coupling constants are also important in the interpretation of multiple resonance experiments which yield relative signs of coupling constants. It is the signs of the reduced coupling constants K which are actually compared in these experiments, not the signs of J.[6]

1.2. Anisotropy of the Spin–Spin Coupling

This has been reviewed by Emsley and Lindon.[7] The experimental determination of the coupling anisotropy in oriented molecules (in liquid crystalline solvents or in the solid state) is complicated by the direct dipolar contributions to the observed coupling. The direct dipolar contribution **D** can in principle be separated out exactly from the measured components of the coupling tensor since it depends only on nuclear coordinates. However, the molecular geometry is not always completely known (the molecular geometry obtained with high accuracy from microwave spectra in the gas phase may not be the same as that of the molecule in a liquid crystal). In addition, experimental measurements can usually yield only linear combinations of K tensor components so that only one useful piece of information can be obtained, the difference $K_{cc} - (K_{aa} + K_{bb})/2$, where c is the axis of highest symmetry of the molecule. This provides all the needed information if there are only two independent components as in the example below, but falls short of the required number in the general case.

Indirect reduced spin coupling tensor K

$$\begin{pmatrix} a & b & c \\ K_\perp = K - \frac{1}{3}\Delta K & 0 & 0 \\ 0 & K_\perp = K - \frac{1}{3}\Delta K & 0 \\ 0 & 0 & K_\parallel = K + \frac{2}{3}\Delta K \end{pmatrix}$$

Isotropic part $= K$
Anisotropy $= \Delta K = K_\parallel - K_\perp$

Direct reduced spin coupling tensor D

$$\begin{pmatrix} a & b & c \\ 1/R^3 & 0 & 0 \\ 0 & 1/R^3 & 0 \\ 0 & 0 & -2/R^3 \end{pmatrix}$$

Isotropic part $= 0$
Anisotropy $= \Delta D = D_\parallel - D_\perp = -3/R^3$

Results of theoretical calculations indicate that, for molecules containing light atoms such as H, C, and F, the anisotropy in the electron coupled spin–spin interaction, $K_\parallel - K_\perp$ is less than 1% of the anisotropy of the direct dipole–dipole interaction $D_\parallel - D_\perp$.[3] This overwhelmingly large direct dipolar contribution can lead to large uncertainties in the $K_\parallel - K_\perp$, especially in nonrigid polyatomic molecules. Some of the early results for anisotropies of spin–spin coupling in molecules oriented in liquid crystals were subject to additional errors in determination of the order parameter. More recent results appear to be reasonably consistent with theoretical estimates.[8] Examples are $^1K(\mathrm{Hg-C})$ in $HgMe_2$, which is characterized by $K_\parallel - K_\perp = 599$ and the isotropic coupling $K = 1264$[9]; $^1K(\mathrm{CdC})$ in $CdMe_2$ with $K_\parallel - K_\perp = 279$, $K = 797$[10]; $^1K(\mathrm{PSe})$ in $Se{=}PMe_3$ with $K_\parallel - K_\perp = -630 \pm 165$, $K = -650$,[11] all in units of $10^{19}\ \mathrm{N\ A^{-2}\ m^{-3}}$.

Since the Fermi contact mechanism leads to a purely isotropic contribution to the coupling constant, large anisotropies are usually taken to indicate that noncontact mechanisms are making significant contributions to the isotropic average as well.

1.3. Methods of Determining Signs of Coupling Constants

The results of a single or multiple resonance experiment in a freely tumbling system are invariant with respect to a reversal of the signs of all of the coupling constants when relaxation rates are neglected. Therefore, the absolute signs of the coupling constants in a molecule can only be obtained if the sign of one of them is known on an absolute basis. The one whose sign is considered known on an absolute basis is generally $^1K(\mathrm{CH})$, although any one-bond coupling to a proton will do.

McFarlane has come to the conclusion that all $^1K(X-H)$ are positive.[12] However, relativistic calculations[13] indicate that coupling of H to heavy (fifth or sixth row) atoms with lone pairs will be negative.

An absolute sign determination of a coupling constant is a difficult experiment and cannot be carried out for each molecule; however, there is very extensive relative sign information primarily due to McFarlane and co-workers. These provide, by comparisons of trends for ^1K, ^2K, and ^3K for various pairs of nuclei, a huge interlocking network of relative signs which is firmly anchored at several points by absolute sign experiments. Thus, whenever relative signs of coupling constants are determined, they can usually be translated to absolute signs.

1.3.1. Absolute Sign Determination

In a molecular beam resonance experiment, the sign and magnitude of the coupling constant can be obtained. For example, the signs of $^1K(\mathrm{HD})$ and $^1K(\mathrm{HF})$ were obtained in this way.[14,15] In condensed phases, absolute signs of coupling constants can be obtained if the direct dipolar coupling is not quite averaged out to zero such that the observed spectrum depends on $(\mathbf{D}+\mathbf{K})$. The conditions of the experiment are such that the intermolecular dipole–dipole coupling is still averaged to zero but the intramolecular coupling is not. Since the reduced direct dipolar coupling tensor $\mathbf{D}$ is dependent only on nuclear positions, its absolute sign is known and its magnitude can be obtained by calculation in a molecule with a known geometry. By comparison of the coupling constant $|\mathbf{D}+\mathbf{K}|$ observed under these conditions with that observed in the freely tumbling molecule, $|K|$, the absolute sign of K can be known. The classic first experiment was carried out by Buckingham and McLauchlan, in which the absolute sign of $^3K(\mathrm{HCCH})$ was determined in *p*-nitrotoluene molecules partially oriented in a liquid by the application of a strong electric field.[16] Since the relative signs of $^3K(\mathrm{HCCH})$ and $^1K(\mathrm{CH})$ in this molecule (and in many others) had previously been determined, this experiment established that the absolute sign of $^1K(\mathrm{CH})$ is positive. Absolute sign determinations have been carried out in CH_3CN, benzene, hexafluorobenzene, and fluorobenzene oriented in liquid crystalline media.[17] All results are consistent with $^1K(\mathrm{CH})$ being absolutely positive. $^1K(\mathrm{PH})$ has been shown to be positive from ^{1}H NMR of partially oriented PH_3 in a nematic solvent.[18]

It has been mentioned that the appearance of a single or multiple resonance high-resolution spectrum is invariant to a change in sign of all spin coupling constants when relaxation effects are neglected. However, when relaxation effects are taken into account, the appearance of the NMR spectrum does indeed depend on the absolute sign of the coupling constant provided that cross terms between different relaxation mechanisms are important. It has been shown theoretically[19,20] that in a system relaxed by dipole–dipole interaction and by an anisotropy in the chemical shift, the relaxation rate of different lines in a multiplet will depend on the absolute sign of the coupling constant. This was applied experimentally to the determination of the absolute sign of $^2K(\mathrm{HF})$ in $CHFCl_2$.[21]

Although signs of coupling constants can in principle also be obtained by noting the variation (increase or decrease) of its magnitude with an increase of some property such as dielectric constant of the solvent, this is not a reliable method of determining signs. The solvent effects on coupling constants are not yet understood to the point of being able to unequivocally state that an increase in the solvent dielectric constant will always result in an algebraic increase of all K.[22]

1.3.2. Methods of Determining Relative Signs

The established methods involve (a) use of second-order features of single resonance spectra, (b) double resonance methods, and (c) triple resonance methods. All of these methods depend on the fact that the topological structure of the nuclear spin energy level diagram depends on the relative signs of coupling constants. Therefore, the problem of determination of relative signs is related to the construction of energy level diagrams and assignment of individual spectral lines.

The use of single resonance experiments for sign determination depends on the presence of second-order features which in favorable cases incidentally lead to relative sign information. This has been widely used in magnetically inequivalent systems such as AA'XX'. While the early relative sign data were obtained in this fashion, multiple resonance techniques are the preferred direct method.

Determination of relative signs of reduced coupling constants by selective double resonance experiments of various kinds has been reviewed.[6,23,24] The sign determination depends on identification of the various subspectra associated with different nuclear spin states. For example, in a double resonance experiment on a spin system $A_pM_qX_r$, with the $A-\{X\}$ experiment (as discussed in Chapter 2, this denotes observation of the A nucleus while irradiating the X nucleus), in selective decoupling and spin tickling experiments the affected peaks belong to the same subspectrum as the irradiated one. In off-resonance decoupling experiments the degree of off-resonance is different for X peaks associated with the different M spin orientations. Thus, it is determined which A and X transitions are associated with a particular spin state of M. This gives the relative signs of $K(\mathrm{AM})$ and $K(\mathrm{MX})$. It is usually possible to obtain relative signs of several pairs of couplings from a single off-resonance decoupled spectrum.

Triple resonance experiments for the determination of relative signs usually involve the same interpretation as in the normal double resonance experiment, with the third irradiating field being used to lift degeneracies so as to make sign determination possible.[25] For example, in $A_pM_qX_r$ and related spin systems with $K(\mathrm{AX})=0$, it is not possible to perturb the A spectrum by irradiating at the resonant frequency of X. This makes it impossible to use $A-\{X\}$ experiments to determine the relative signs of $K(\mathrm{AM})$ and $K(\mathrm{MX})$. However, the degeneracy of the transitions in the A spectrum associated with different spin states of X can be lifted by weak (spin-tickling) irradiation of one of the nondegenerate M transitions. The results can then be interpreted in the same way as an experiment in a system with nonvanishing $K(\mathrm{AX})$. A recent example is the determination of the relative signs of $^3K(\mathrm{POCH})$ and $^1K(\mathrm{SeP})$ in $(\mathrm{MeO})_3\mathrm{PSe}$ [$\mathrm{A}={}^1\mathrm{H}$, $\mathrm{M}={}^{31}\mathrm{P}$, $\mathrm{X}={}^{77}\mathrm{Se}$].[6]

Details of the interpretation of double resonance spectra in the determination of relative signs of coupling constants are found in several references.[26,27] With polarization transfer techniques described in Chapter 2, it is possible to determine relative signs of coupling constants involving less sensitive nuclei such as $^{15}\mathrm{N}$.[28]

2. EMPIRICAL PATTERNS OF COUPLING CONSTANTS

The application of coupling constants to the elucidation of structure of inorganic and organometallic compounds depends on being able to predict values of coupling constants in new systems based on general patterns of the dependence of the signs and magnitudes of coupling constants on various aspects of structure. In this section we will describe some important generalizations which have been useful and temper their wide application with examples of exceptions where known.

2.1. Signs and General Magnitudes of $^nK(XY)$

The signs and magnitudes of coupling constants are determined by structural factors to a significant extent, and we will consider them in the three sections following this one. However, there are global, Periodic-Table-wide trends that should be noted and are treated in this section.

2.1.1. Dependence of $^nK(XY)$ on Number of Bonds Separating X and Y, Multiple Paths, and Spatial Proximity

It is generally observed that for a given pair of nuclei $^1K(XY) \gg {}^3K(XY) > {}^2K(XY) > {}^4K(XY)$ and that longer-range couplings are two or more orders of magnitude smaller than one-bond couplings. This does not mean, however, that these constants have mutually exclusive ranges. Depending on the structure, it is indeed possible for some three-bond couplings to be smaller than two-bond couplings, and in some special cases a one-bond coupling may be smaller than expected, leading to some confusion in assignments. Nevertheless, this inequality is sufficiently widespread, for a large variety of nuclei, to be useful in making spectral assignments. It is also generally found that long-range couplings ^nJ decrease overall with increasing n, although the magnitudes for odd and even n may be reversed, as in $^nJ > {}^{n-1}J$ due to structural factors.[29]

In cyclic and polycyclic systems there are multiple bond paths connecting a pair of nuclei. It is commonly assumed, and data appear to support the contention, that the net coupling constant which is observed is an algebraic sum of the couplings through alternative bond paths. The presence of more than one path should always be taken into consideration in assignments and interpretation of couplings in cyclic systems. Where contributing couplings are opposite in sign, the observed coupling constant may be unusually small, even of exceptional sign. This is the obvious explanation for the very unusual negative "$^1J(CC)$" observed for the coupling between the bridgehead carbons in bicyclobutane derivatives.[30,31] The three paths involve one $^1J(CC)$ and two two-bond paths. Since $^2J(CCC)$ is usually negative, the one-bond coupling itself may be unexceptional (positive). Unusually large values of $^4J(SnC)$, 7.0 and 15.8 Hz, respectively, in 1-Me_3Sn-adamantane and 1-Me_3Sn-cubane, can be attributed to two and six pathways of favorable geometry.[32] $^3J(PP)$ in polycyclic trivalent phosphorus systems and trialkyl phosphites are known to be larger than in their open chain analogs.[33,34]

Multiple paths are introduced in metal complexes when bidentate ligands are involved. What would normally be a two-bond coupling involving the metal nucleus—for example, the PtC coupling constant in diamino-ethane platinum complexes[35,36]—becomes a combination of $(^2J + {}^3J)$ when a five-membered ring is formed. Since ^2J and ^3J are commonly opposite in sign, the observed "two-bond" coupling with the metal nucleus may be quite small compared to that involving open chain ligands. For example, the PP couplings can be attributed to the PMP (where M is the metal) and PCCP or PCCCP paths in Pt(II) chelates with ligands containing a $P(CH_2)_n P$ ($n = 2, 3$) moiety.[37] Not surprisingly, compared to the PP coupling in the

Table 1. Absolute Signs of One-Bond Couplings (^1K)[a]

	I	II	M[b]	III	IV	V	VI	VII	0
I	+		+	+	+	+	+	+/−	
II								?	
M	+		+	?	+	+	+/−	+/−	
III	+			+	+	+		−	
IV	+		+	+	+/−	+/−	−	−	
V	+		+	+	+/−	+/−	+/−	−	
VI	+		+/−		−	+/−	+	−	?
VII	+/−	?	+/−	−	−	−	−	+/−	?
0							?	?	

[a] References and notes for this table are given in Table 2.
[b] M stands for elements in the transition series.

free ligand, a decrease in magnitude is observed upon formation of a five-membered chelate ring (opposite sign $^2J + {}^3J$) and an increase upon formation of a six-membered ring (same sign $^2J + {}^4J$).

"Through-space" coupling is a term used to denote a coupling contribution by direct electronic interaction in systems where the coupled nuclei are in close spatial proximity. This appears to be important in cases of sizable long-range couplings (over four or more bonds) involving fluorine. The "through-space" mechanism has been reviewed(38) and many examples are given of ^nJ for $n = 4$–8 in annual reviews of spin–spin couplings.(39) The magnitude of ^nJ should be related to the internuclear separation if the mechanism is as described.(40)

2.1.2. *Signs of One-Bond Couplings*

Table 1 shows the signs of one-bond couplings ^1K between pairs of nuclei of various groups in the Periodic Table. Except for VI–VI, which is based on a single sign determination, these signs are well established, based on a large number of systems involving different members of the groups as well as different structural environments. Where both signs are observed, the change in sign usually occurs with a change in some structural aspect, such as presence of lone pairs, independent of position in the group. However, there are some experimental examples of systematic sign changes upon descending in a column of the Periodic Table. (References are given in Table 2.) In IV–IV coupling the sign of ^1K is positive but at least one case of negative ^1K(PbC) has been reported. In IV–V coupling ^1K(XN) for X = C, Si, Sn, Pb goes as +, +, +/−, −. In V–VI coupling ^1K(P=O) is positive, whereas ^1K(P=Se, Te) are negative. In VII–VII coupling ^1K(ClF) is positive while ^1K(II) is negative. There are also predictions by theory of a general sign change for couplings involving certain groups of the Periodic Table (VI and VII).(13) These trends are shown in Figure 2. The theoretical bases for these trends are discussed in Section 4.4 in terms of spin-polarization(209) or by the Pople–Santry model.(5)

2.1.3. *Trends in Magnitudes of the One-Bond Coupling ^1K down a Family of the Periodic Table and along a Row*

There is a wide variation in magnitudes of one-bond coupling constants, some part of which is due to the wide range of magnetogyric ratios. We eliminate this by

Table 2. References and Notes for Table 1, Arranged According to the Families of Pairs of Nuclei for Which the Sign of the One-Bond Coupling Has Been Determined

I–I	The only experimental value available is that for HD.[14] The sign was determined by a molecular beam resonance experiment.
I–M	Positive for PtH[44,45] and RhH.[46] Magnitudes only for WH,[47] OsH,[48,49] ScH,[41] MoH,[42] TcH,[43] and PtLi.[49a]
I–III	Positive for BH.[50,51] Magnitudes only for AlH and TlH.[52]
I–IV	Positive for CH,[53] CLi,[54] SiH,[55–57] SnH,[58] PbH.[58] Magnitude for GeH.[59]
I–V	Positive for NH in pyramidal,[60,61] and trigonal[62] nitrogen and, for PH in P(III),[18,63–66] $P(IV)^+$,[67] and P(V).[65,66,68] Magnitude only for AsH[69,70] and for linear ≡NH.[71]
I–VI	Positive for OH[12] and SeH.[57,72,73] Magnitude only for TeH.[74]
I–VII	Positive for HF, LiF, RbF, CsF, LiCl, LiBr. Negative for KCl, HI, NaI. These are all molecular beam results.[75] Magnitudes only for HCl and HBr.[76]
II–VII	Magnitudes only for BeF.[77,78]
M–M	It is difficult to characterize the coupling pathway in many of the binuclear complexes as exclusively one-bond or two-bond, so the range of values is wide: 90–9000 Hz for ^1J(PtPt).[47] Sign has been determined only for ^1K(PtPt) (positive).[47] For copper metal ^1K(CuCu) is positive.[79,80] Magnitudes of RhRh,[81] PtRh,[82] PtAg,[82a] HgHg,[82b] and HgW[83] have been reported.
M–III	Magnitude of ZrB in $Zr(BH_4)_4$.[84]
M–IV	Positive for PtC,[85] CdC,[86,87] HgC,[88,89] PtSi,[45] WSn.[90] Magnitudes of C coupling with V, Fe, Co, Y, Mo, Rh, and W have been reported in reviews,[47,77,81,91] and C coupling with Os[92] also of Si coupling with Pt and Hg,[93] and of Sn with Pt,[94] Os, Rh,[94a] and Ru,[95] Pb with Pt.[95a]
M–V	Positive for RhP,[46,96] CdP,[47] WP,[97–99] PtP,[44,45,100] and HgP.[101] Magnitudes of MP couplings recently reviewed[41,102,103] include M = Sc, V,[104,105] Co,[77] Cu,[77] Mo, Rh, Ag, Cd, W, Pt, and Hg.[106] Magnitudes of N couplings with Fe, Co, Rh, Cd, W, Pt, Mo, and Hg have also been reported.[40,41,103,107,107a]
M–VI	Positive for PtSe,[108,109] PtTe,[109] and RhTe[47] with one exception, a negative ^1K(PtSe) for $[PtI_5(SeMe_2)]^-$.[109] Negative for HgSe in $HgX_2[P(Se)Bu_3]_2$.[110] Magnitudes of couplings with O of V, Cr, Mn, Mo, and Re have been reviewed.[74,111] Magnitudes reported for RuO,[112] TcO,[113] and CdSe[114] couplings also.
M–VII	Positive for RhF,[115] only magnitudes reported for others; however, in a few cases the two ^1K(WF) coupling constants in the same molecule are found to be of opposite relative signs and comparable magnitudes.[116,117] Magnitudes of coupling with F for Sc, Ti, V, Nb, Rh, W, Pt, and U have been reviewed.[47,77]
III–III	Positive for BB,[118] only magnitudes available for AlB[52] and TlTl.[119]
III–IV	Positive for BC,[90,118] BSi,[51,120] BSn,[51,90,121] BPb,[121] TlC,[87,122] and TlSn.[123] Magnitudes only for AlC.[52,124]
III–V	Positive for BN,[124a] BP[51,125] for P(III) to P(V). Magnitudes of AlN, GaN,[125a] AlP.[52]
III–VII	Negative for BF[50,126] and TlF, InF, TlCl, TlBr, TlI[75]; however, can become small positive in BF_4^-.[51,126] Magnitudes of AlCl, AlBr, and GaCl.[77]
IV–IV	Positive for CC,[127–129] SiC,[56,130,131] SnC,[131–134] PbC,[131,134–136] SnSi,[90,137] SnSn,[121,138] PbSn,[121] and PbPb[139]; however, becomes negative for SnC[140] and SnSn[141] in the presence of a lone pair on one of the Sn atoms as in trimethyl stannyl lithium[140] and $Li[Sn_A(Sn_BMe_3)_3] \cdot$ 3thf.[141] Negative sign also found for PbC in $Et_3PbC{\equiv}CH$.[142]
IV–V	Positive signs have been found for CN in all types of N bonding situations: pyramidal, trigonal, linear,[143–149] with three known exceptions of small negative values changing to positive upon protonation.[148,150,151] Positive for SiN,[152,153] +/− for SnN,[90,123] negative for PbN.[121] Negative for CP (3-coord),[15,34,66,154–156] and for aryl carbon,[157–160] positive for CP (4,5,6-coord),[34,66,154,160,161] negative for SiP (3-coord),[137,162] negative for SnP (3,4-coord),[137,163] negative for PbP (3-coord).[121] Only a magnitude for CSb.[164]
IV–VI	Negative for CSe[57,73,165,166] in Se(II), Se(IV), and Se(VI) compounds, CTe,[57,73,165] SiSe,[57] SiTe,[57] SnSe,[24,90,133,167] SnTe,[90,133] and PbSe.[121]
IV–VII	Negative for CF,[21,168] SiF,[55,169,170] and SnF.[171] Magnitudes for CBr,[172] SnCl, SnBr, SnI, and PbCl.[76]

Table 2. (*Continued*)

V–V	Negative for NN in NNO,[173] and in a substituted pyrazole,[28] magnitudes for all others.[107,174] Negative for tricoordinate P, +/− observed for 4,5-coordinate P, for NP,[152,175–180] as well as PP.[36,63,181–187] Even for P(V)–P(V), both negative and positive couplings have been observed.[188]
V–VI	Negative for PO,[189] PSe[190–193] for P(III) as well as P(V). Positive P=O in $Me_3P=O$, negative for P=Se and P=Te in $Me_3P=Se$, Te.[194]
V–VII	Negative for PF for tricoordinate,[156,170,195] 4-,[196,197] 5-,[161] and 6-[198] coordinate P, and for BiF.[199] Magnitudes of NF,[200] PCl,[76] PBr,[76] and AsF.[201]
VI–VI	Only one sign determination: Positive ^1K(TeSe) in $CH_3TeSeCH_3$.[57] Magnitudes for SeSe and TeTe.[202,203]
VI–VII	Negative for TeF.[204] Magnitudes for OF, SeF, OCl.[74]
VII–VII	Negative for II in an excited electronic state of molecular iodine.[205] Positive for ClF.[75] Magnitudes only for BrF and IF.[76]
0–VI,VII	Magnitudes only for XeN,[205a] XeO[206] and XeF.[207,208]

comparing reduced coupling constants. There will still be the dependence on structural factors to be discussed in Section 2.2; however, underlying these are factors which become apparent when we compare ^1K magnitudes for analogous compounds. This has been done for nuclei in group IV. The ratio $^1K(\text{SiX})/^1K(\text{CX}) \simeq 1.9$[210] for several analogous compounds with X=H, B, C, Si, Sn, and $^1K(\text{SnX})/^1K(\text{SiX}) \simeq 4.2$[210] for the analogous Sn compounds. For coupling with the same set of X nuclei, separate studies using analogous Sn and C compounds gave $^1K(\text{SnX})/^1K(\text{CX}) \simeq 9.0$[90] and an analogous set of Pb and Sn compounds gave $^1K(\text{PbX})/^1K(\text{SnX}) \simeq 1.6$.[121] However, for X = N, P, and Se, where lone pairs allow a greater influence of structural factors, the ratios $K(\text{SnX})/K(\text{CX})$ [90] and $K(\text{PbX})/K(\text{SnX})$ [121] are not consistent. Analogous hydrides of group III have also been compared: ^1K(XH) in XH_4^- (X = B, Al, Ga) are 20.8, 54.25, 178 (in units of 10^{19} N A^{-2} m^{-3}).[211] The ratios Ga/Al and Al/B for the coupling constants are 3.28 and 2.61, which may be compared with the ratios of electron densities at the nuclei of the atoms, which are 3.29 and 1.72, respectively. This seems to indicate that outside of structural effects, some factor in ^1K(XY) is intrinsic to the identity of X and Y. The ratios of the electron densities at the nuclei for ground state atoms are Si/C = 1.41, Sn/Si = 4.61, and Pb/Sn = 3.37.[13] From equations (8)–(12) we see that all the contributions to the coupling constant depend on the electronic distribution at the coupled nuclei or in their immediate vicinity. Although the magnitude of the coupling constant is modulated by the molecular structure, the overriding dependence on the identity of the coupled nuclei can be seen in terms of the densities $|\Psi_{ns}(0)|^2$ and the averages $\langle r^{-3} \rangle$. Figure 3 shows the variation of the calculated neutral atom ground state electron densities at the nucleus with atomic number. The atomic averages of r^{-3} (shown in Figure 1 of Chapter 3) are known to vary with atomic number in a similar manner.[215] Underlying the variations in magnitude and sign due to structural factors is a periodicity in magnitudes of ^1K(XF), which mimics the variation of atomic X densities at the nucleus [see filled circles indicating ^1K(XF) in Figure 3]. We minimize the effects of structure by comparing binary fluorides. Figure 4 illustrates the variation of ^1K(XF) in these binary fluorides with the calculated electron density at the nucleus in the X atom. A linear relation between ^1K(XF) and the square of the atomic number of X within a family had previously been found.[216,217]

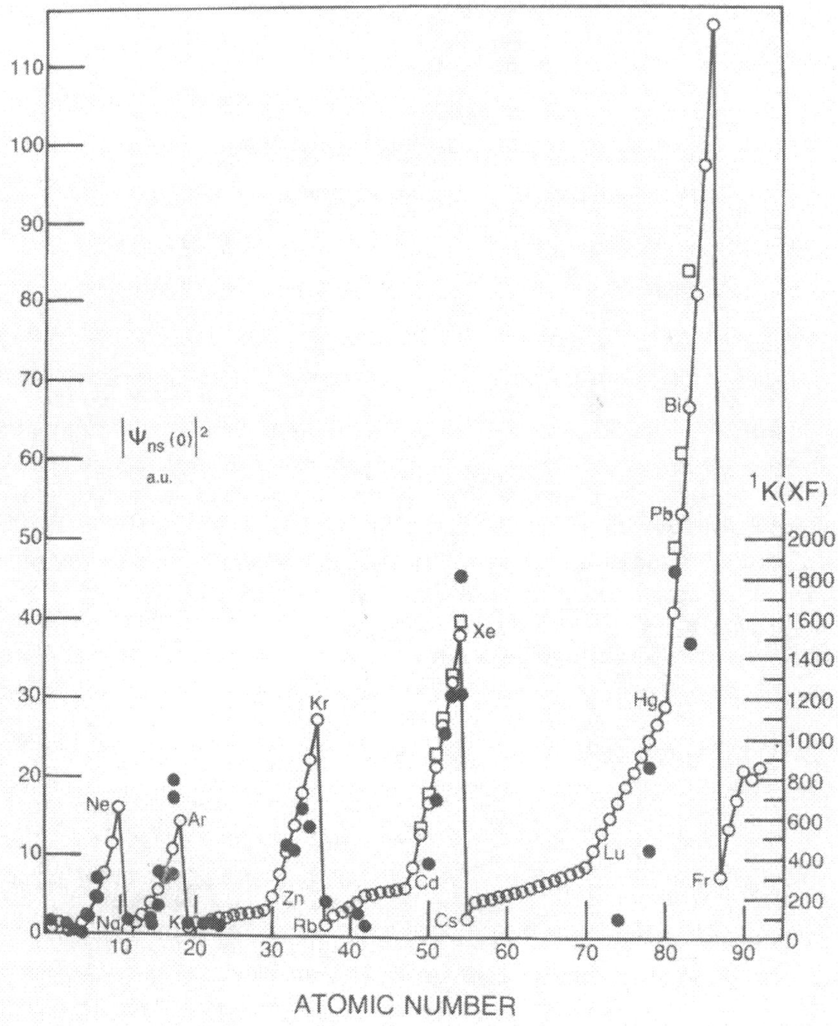

Figure 3. Electron densities at the nucleus (left scale) for ground state atoms[102] corrected for relativistic effects with the Breit factors given by Pyykkö *et al.*[212] and $^1K(XF)$ in binary fluorides[24,213,214] (right scale) as a function of atomic number of X. More accurate relativistic values for selected atoms have been calculated by Pyykkö.[13] These are shown only for the heavier nuclei (□) where the difference between the more accurate calculations and the use of the Breit factors become significant. However, the periodicity of the more accurate relativistic values is preserved. Nonrelativistic values published elsewhere[102] are largely incorrect for the heavier nuclei (atomic number > 40); e.g., it should be 39.644 a.u. instead of 26.688 for Xe, 60.752 instead of 20.53 for Pb, 83.79 instead of 24.86 for Bi!

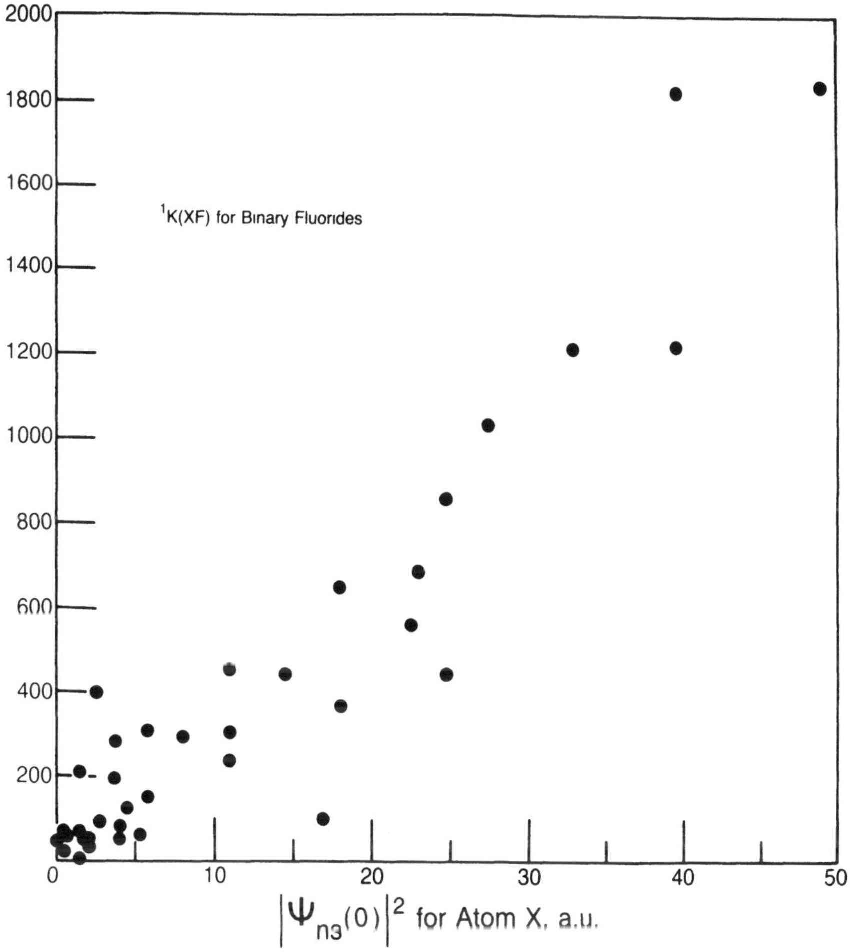

Figure 4. Variation of $^1K(XF)$ in binary fluorides XF_n (24,213,214) (units: 10^{19} N A^{-2} m^{-3}) with the relativistic electron density at the nucleus in atomic X.(13)

2.2. Structural Factors Affecting ^1K

2.2.1. The Lone Pair

The presence of one or more lone pairs on one or both of the coupled nuclei seems to be associated with a negative one-bond coupling constant except in coupling to H. Thus, we find in Tables 1 and 2 that, excepting coupling to H, most $^1K(XF)$ are negative; $^1K(OX)$, $^1K(SeX)$, $^1K(TeX)$ are negative except for TeSe in $CH_3TeSeCH_3$ (57) and PtSe and PtTe, but even in this case, one $^1K(PtSe)$ was found to be negative(109); the only negative $^1K(SnC)$ and $^1K(SnSn)$ reported are those in which the Sn has a lone pair.(140,141) PP coupling between tricoordinate phosphorus atoms with lone pairs are negative. PP couplings involving P(V), which has no lone pair, are positive or at least considerably less negative than in either P(V)–P(III) or P(III)–P(III); $^1K(PC)$ are negative when the phosphorus has a lone pair, positive when this lone pair is used in bonding, as in transition metal complexes or in adducts such as X_3B-PY_3 (see references in Table 2). $^1K(SeC)$ is negative in all cases for Se with 0–3 lone pairs but substantially less so when Se has no lone pair.[166,218] Nitrogen does not behave in a clear cut way as P in this respect. ^1K(CN) is usually positive, and ^1K(SnN) is + (small) or − (large) in $Me_3SnNXPh$[121,180] and K(SiN) is + (small) for $(H_3Si)_3N$.[152] Nevertheless, the importance of the N lone pair is indicated when it is removed by protonation or by oxide formation, upon which the ^1K(NC) changes from − to + or becomes larger positive,[148,219] i.e., an algebraic increase in ^1K(NC), just like the algebraic increase in ^1K(OH) by 33% upon protonation of H_2O to H_3O^+,[220] and ^1J(AsH) upon protonation of AsH_3 to AsH_4^+ increases from 92.1[69] to 555 Hz.[70] Just as dramatic is the algebraic increase in ^1J(PH) upon removal of the lone pairs on phosphorus: in going from PH_2^- to PH_3 to PH_4^+, ^1J(PH) changes from 138–140 to 182–195 to 545.7–548.3 Hz.[221] The removal of the lone pair by complexation results in an algebraic increase in the coupling. For example, ^1K(SeP) (which are negative) increase algebraically (i.e., the magnitudes are reduced from that in the free ligand) upon coordination of $R_3P{=}Se$ with SO_2, Cd(II), or HgX_2 [222,223]; ^1K(PC) and ^1K(SnP), which are negative in the free phosphine ligands, increase algebraically [it has been shown that ^1K(PC) actually undergoes a change in sign from − to + upon complexation with a metal].[102,163] With the partial removal of the lone pair ^1K(SnH) changes from (+)25 in SnH_3^- [224] to (+)430 in SnH_4 [225]; complete removal of the lone pair increases ^1K(SnH) further to (+)660 in SnH_3^+ [226] (all in units of 10^{19} N A^{-2} m^{-3}). The same dramatic increase in ^1K(SnH) is observed in going from $SnR_nH_{3-n}^-$ to SnR_nH_{4-n}.[227] All these examples indicate that the presence of one or more lone pairs on one or both of the coupled nuclei is associated with a negative contribution to the one-bond coupling constant. We explain this empirical generalization in Section 4.4 in terms of the Pople–Santry model.

2.2.2. Coordination Number, Oxidation State, and Formal Hybridization

The reduction of the magnitude of a coupling constant (in the free ligand) by coordination is commonly used as a diagnostic tool for complexation. Although this is consistent with an algebraic increase in ^1K with apparent loss of a lone pair, as discussed in Section 2.2.1, a more general decrease in the magnitude of ^1K (positive or negative) upon increase in coordination seems to be observed. For example, P=Se coupling constants (−724 to −1046 Hz) are much greater in magnitude than P−Se− couplings (−100 to −391 Hz).[193,228] SiF coupling in SiF_4, SiF_5^-, SiF_6^{2-} are, respectively, 170, 140, 108 Hz.[229] ^1K(Ag or Hg, P) decreases with increasing n in phosphine complexes $[AgP_n]X$, $[HgP_n]X_2$ ($n=1$–4).[102,106,230,231] For example, in $[(EtO)_3P]_n Ag^+X^-$, $^1J(^{107}AgP)$ varies with counterion but decreases in a monotonic fashion with increasing coordination n as follows: 794.1, 535.3, 402.2, 341.8 Hz, respectively for $n=1, 2, 3, 4$, with SCN^- counterion (1059.0, 689.4, 454.1, 340.8 Hz with NO_3^- counterion).[230] $^1J(^{199}HgP)$ varies with coordination number too and the change is even more dramatic: from about 3000 to about 10,000 Hz in going from $n=3$ to 1 in $[Ph_3P]_n HgX_2$, where $X=OCOCH_3$ or $OCOCF_3$.[231]

Bridging atoms have a higher coordination number than terminal ones. It is commonly observed that the magnitudes of ^1K to bridging atoms are smaller than those to terminal atoms. For example ^1J(AlC) = 19 and 110 Hz, respectively, for bridging and terminal methyls in Me_6Al_2.[124] ^1J(PtC) = 357 and 1937 Hz, respectively, for bridge and terminal isocyanide ligands[232]; ^1J(PtH) = 540 and 810 Hz for bridging hydrides compared to 1138 to 1184 Hz for terminal ones.[233]

An apparent exception to this general decrease in magnitude of ^1K with increasing coordination number is the increase in ^1K upon additional coordination with more electronegative atoms. As we shall see in Section 2.2.3, there is a rather pronounced increase in coupling constant with increasing electronegativity of substituents at the coupled atoms. Thus, when an increase in coordination number involves addition of highly electronegative ligands, we are no longer seeing just the effect of increasing coordination number but also the effect of electronegative substituents, the latter being more pronounced. Thus, ^1K(PbC) increases (rather than decreases) with additional coordination at Pb with donor solvent molecules and with acetate ions.[234] A similar apparent exception is the increase in ^1K(SnC) with increasing coordination at Sn brought about by addition of halogen, OMe and other electronegative ligands: The ranges of ^1J(SnC) values are 300–340 Hz for tetraalkyltin (SnR_4), 330–390 (SnR_3X), 370–480 (SnR_2X_2), 450–480 (SnR_3X_2), 900–970 (SnR_2X_4), with 0, 1, 2, 2, and 4 electronegative substituents, respectively.[235]

Oxidation State. It has been noted that the oxidation state of the metal influences the magnitude of the metal–phosphorus coupling constant in phosphine complexes or the metal–carbon coupling in carbonyls. For example, ^1J(Rh(III)C) < ^1J(Rh(I)C),[236] ^1J(Rh(III)P) : ^1J(Rh(I)P) = 0.50–0.73 : 1,[46] in octahedral versus planar complexes with carbonyl or phosphine ligands. It has also been noted that ^1J(Pt(IV)P) ≤ ^1J(Pt(II)P) ≤ ^1J(Pt(0)P) with values ranging from 1086 to 2074 Hz for Pt(IV), 1686 to 5859 for Pt(II), 2837 to 9150 for Pt(0).[102] Because there are usually other structural factors besides oxidation state of the metal, the range of values of ^1J may overlap and the dependence on oxidation state not be immediately obvious. However, when ^1J(PtP) of complexes of different oxidation states of Pt are plotted against PtP bond lengths from x-ray data, the values fall into separate curves, for a given bond length, the Pt(0)P coupling is larger than the Pt(II)P coupling, despite the widely differing structures (tetrahedral, planar, trigonal pyramidal) for the Pt(0) complexes included.[237]

Formal Hybridization. The dependence of the one-bond coupling on the lone pair, coordination number, and oxidation state is often described in terms of the qualitatively useful construct: "hybridization." It is assumed that one can classify nuclear environments in terms of idealized equivalent hybrid orbital language (meaningful only when employing a minimum basis valence orbital description), e.g., for the second row of the Periodic Table, sp^3 for tetrahedral or pyramidal coordination, sp^2 for trigonal planar, and sp for linear. While not quantitatively justifiable, this does provide a useful taxonomy. A very good example of this is the classification of ^1J(CC) values according to the formal hybridization of the two carbons, as in Figure 1 in the review of carbon couplings by Wray and Hansen.[238] In Pt(II) complexes, ^1J(PtC) is 457–698 for alkyl groups, 750–850 for sp^2 hybridized C, and 990–2240 for carbonyls.[81] ^1J(WC) in the carbene and carbyne complexes of W with the general structures $(CO)_5W{=}CRR'$ and $X(CO)_4W{\equiv}CR$, X = halogen, can be used to distinguish the two types, ^1J(WC) = 87.9–102.5 Hz to the carbene carbons and 168.5–197.5 Hz to the carbyne carbon atoms.[239] ^1J(WC) is 80, 120, and 210 Hz in one and the same complex for alkyl, alkylidene, and alkylidyne ligands, respectively.[240] It is also helpful in classifying ^1K(NH) and ^1K(NC) values since it is generally found that these nitrogen couplings are in the order pyramidal < trigonal < linear,[107] i.e., formally sp^3, sp^2, sp. When the coordination number exceeds 4, d orbitals are generally included in the qualitative description, and classifications by formal hybridization such as d^2sp^3 (octahedral) dsp^3 (trigonal bipyramidal), or dsp^2 (square planar) generally places $^1J(MP)$ or $^1J(MC)$ into nonoverlapping ranges.[81,102] We have already seen that ^1J(Rh(III), C or P) in octahedral complexes < ^1J(Rh(I), C or P) in planar complexes.

In summary we note that an increase in coordination number of X (in going from linear to trigonal, tetrahedral, square planar, octahedral; or terminal to bridging; or free ligand to coordinated ligand), leads to decrease in the magnitude of $^1K(XY)$ for the general reason that there is an increase in the number of bonds over which the *s* orbital is distributed. On the other hand, the removal of the lone pair on X accompanying an increase in coordination number of X [as in P(III) → P(V)] causes an algebraic increase in $^1K(XY)$ since lone pairs are generally associated with a negative contribution to ^1K.

2.2.3. *Electronegativity of Substituents*

The magnitude of ^1K increases with increasing electronegativity of substituents regardless of its sign. This first came to attention in the case of $^1K(CH)$ in substituted methanes but has been found to be a very general phenomenon. This is an important factor in any coupling; the magnitude of $^1K(XY)$ has been found to change by a factor of up to 2 or 3 with changes in substituents on X or Y. We shall see in Section 2.3.2 that this is the case also for ^2K. Regardless of the choice of electronegativity measure (Allred-Rochow, Huggins, Sanderson, Pauling) a linear plot is generally found for ^1K versus electronegativity of substituents on one of the coupled nuclei. Linear relationships between ^1K and electronegativity have been found in many cases: WP,[241,242] SeP,[243] MoP,[244] PP,[245] upon changing substituents on phosphorus. This is especially dramatic in $K(PP)$.[24,34,187] In the case of $^1K(MP)$, changing the electronegativity of other ligands on the metal appears to have less effect than changing the electronegativity of substituents on P.[102] As we shall see in Section 4, this is also described in terms of "*s* character." Maciel *et al.* have shown that the *s* character calculated by the INDO method increases with increasing electronegativity of substituents on the carbon, and leads to larger CH coupling constants.[246]

2.2.4. *Stereospecificity of* ^1K*: Sensitivity to Geometrical Arrangement of the Atoms Around the Coupled Nuclei*

2.2.4a. Axial vs. Equatorial. In a trigonal bipyramidal molecule, it is found that $|^1K(MX_{eq})| > |^1K(MX_{ax})|$. Many examples of $^1J(PF)_{eq} > {}^1J(PF)_{ax}$ in five-coordinate phosphorus compounds have been reported; the ratio $^1J(PF)_{eq}/{}^1J(PF)_{ax}$ in the same compound is 1.2.[213,214] PC couplings are more variable in magnitude: $^1J(PC_{eq})$ are 169–448 Hz and $^1J(PC_{ax})$ are 0–115 Hz in an extensive series of phosphoranes.[247,248] A similar dependence of $^1J(RhP)$ and $^1J(PtP)$ has been observed in the trigonal bipyramidal complexes, for example, $^1J(RhP_{eq}) = 146$ Hz, $^1J(RhP_{ax}) = 116$ Hz in one case, 206 and 143 Hz, respectively, in another case,[102] while $^1J(PtP_{eq})$ and $^1J(PtP_{ax})$ were 2730 and 2125 Hz, respectively.[249] In a square pyramidal ion TeF_5^-, $^1J(^{125}TeF_{eq}) = 1375.7$ and $^1J(^{125}TeF_{ax}) = 2883.3$ Hz,[250] the reverse order from the trigonal bipyramidal case. The difference in equatorial/axial couplings in trigonal bipyramidal and square pyramidal molecules is only superficially a stereospecific effect since it is commonly found that bond lengths to axial substituents differ significantly (are longer in trigonal bipyramidal and shorter in square pyramidal) from those to equatorial ones. The stereospecificity in the substituted octahedral case is clear: In Me_3NAsF_5 $^1J(AsF) = 1048$ to the equatorial F and 840 Hz to the axial one,[251] analogous to that for the phosphorus compound Me_3NPF_5, in which the values of $^1J(PF)$ are 848 and 747, respectively.[251]

2.2.4b. Cis vs. Trans, the Trans Influence. "*Trans* influence" is a term used to describe an indirect effect on a bond that a ligand *trans* to it exerts. It has been defined as the "tendency of a ligand to weaken the bond *trans* to itself."[252] A review of the *trans* influence has been published.[253] This manifests itself as an alteration of the bond length and therefore has an effect on the magnitude of the one-bond coupling. The magnitude of $^1K(XY)$ is smaller if the XY bond is *trans* to a ligand which has a stronger *trans* influence. The empirical order of ligands according to their ability to affect the bond *trans* to them is $MePh_2Si > Ph > Me \gg R_3P > (PhO)_3P$, $CN > Et_3As > NO_2 >$ amine, $\underline{N}CO$, $\underline{N}CS > Cl$, Br, $I > \underline{O}NO_2 > F$.[254] Thus, $^1K(MP)$ *trans* to $Cl > {}^1K(MP)$ *trans* to PR_3 with the ratio being 1.3–1.5 : 1 for $^1J(PtP)$, the range of values being 3200–3600 Hz vs. 2200–2400 Hz, respectively.[102] $^1K(MF)$ *trans* to $F > {}^1K(MF)$ *trans* to OH, Cl, OMe, or O [for example, in $PtF_5(OH)^{2-}$ there are two $^1J(PtF)$ values: $^1J(PtF)$ *trans* to $F = 1936$ Hz, whereas $^1J(PtF)$ *trans* to $OH = 1049$ Hz[255]]. In square planar platinum complexes, $|{}^1J(PtN)|$ *trans* to $Cl >$ $|{}^1J(PtN)|$ *trans* to amine, 351 and 290 Hz, respectively, and $|{}^1J(PtN)|$ *trans* to $NO_2 >$ $|{}^1J(PtN)|$ *trans* to PR_3, 592 and 453 Hz, respectively.[102] There has been an extensive study of the ligand *trans* effect on $^1J(PtN)$.[256] $|{}^1J(Pt^{125}Te)|$ *trans* to $Cl > |{}^1J(PtTe)|$ *trans* to Te, 900 and 544 Hz, respectively, in *cis* and *trans* $PtCl_2[TeR_2]_2$.[257]

2.2.4c. Dihedral Angle Dependence. As we shall see in Section 3, a large temperature dependence of a coupling constant is observed when some internal rotation is taking place and there is a dihedral angle dependence of the coupling constant. However, in this section we will limit our examples to cases where the coupling is observed in a more or less rigid configuration so that the dependence of the one-bond coupling on a dihedral angle can be ascertained by observing NMR spectra of the isomers. We will consider two cases:

(a) The dihedral angle of interest is that between the bond connecting the coupled nuclei and a lone pair or a bond to an electronegative substituent. In a geometrical arrangement such as shown in Figure 5, $|{}^1K(XY)|$ appears to be larger when a lone pair on Z has a large dihedral angle with the X–Y bond. For example, $^1J(NF) = 203$ and 190 Hz in *cis* and *trans* FN=NF respectively.[200] $^1J(CH)$ in oxaziridines are consistently larger when the CH bond is *cis* to the lone pair on the nitrogen (Z isomer) than when it is *trans* (E isomer).[150] The effect of a bond to an electronegative substituent such as O^- or OMe is opposite to that of the lone pair; that is, $|{}^1K(XY)|$ is larger when the Z–OMe bond is *cis* to the XY bond.

(b) The dihedral angle of interest is that between lone pairs on the coupled nuclei. $|{}^1K(XY)|$ appears to be smaller when the lone pair on X and the lone pair on Y are *trans* to each other. For example, the ^{15}NN coupling is 17 and 21 Hz, respectively, for *trans* and *cis* isomers of $Ph-N=N-CR_3$.[275] In general the NN couplings in $X-N=N-Y$ are smaller for *trans* than for *cis*.[107] $^1J(PP)$ for the *trans* MePPMe moieties (in which the lone pairs are *trans* to each other) are smaller than in the *cis* MePPMe moieties in alkyl substituted cyclomonocarbaphosphanes, e.g., 219.8(*trans*) versus 297.5(*cis*) Hz.[277] These general observations are summarized schematically in Figure 5.

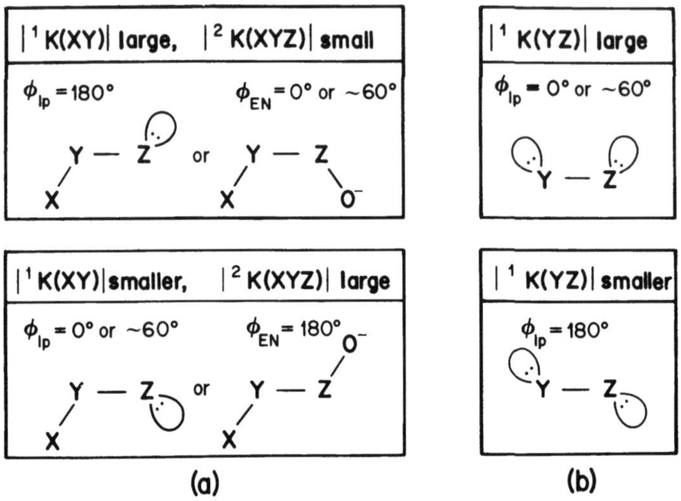

Figure 5. (a) Dependence of $|^1K|$ and $|^2K|$ on the dihedral angle ϕ_{lp} between the bond of interest and a lone pair, or the angle ϕ_{EN} with an electronegative substituent. Examples: XYZ = CNC,[258] HCP,[34] CCN,[107,259–262] CCP,[157,263] HCN,[107,150,264] CNP,[265–267] HCTe,[268] CCSe,[270] FCN,[271] FNN,[200] CNN,[272] PCP,[273] PNP.[274] (b) Dependence of $|^1K|$ on the dihedral angle between the lone pairs. Examples: YZ = NN,[107,275] PP.[185,276–278]

2.2.4d. Diastereotopic Nuclei. Geometric isomers have different one-bond coupling constants. For example, $|^1J(NC)|$ is consistently larger for the E isomer compared to the Z isomer of an oxime.[107,260] $^1J(PP)$ in the two diastereomers of 1,2-dimethyl-1,2-diphenyl-biphosphine are −215 and −234 Hz.[186] In diastereomeric phosphorinanes one-bond couplings to the P nucleus are larger for the diastereomer with the coupled nucleus in the equatorial position. For example, $^1J(PN_{eq})$ and $^1J(PN_{ax})$ are 53 and 42, respectively,[279] and $^1J(PC_{eq})$, $^1J(PC_{ax})$ are 94.9 and 75.0 Hz, respectively.[280] $[MeBu^tP(Se)]_2$ Se exists in diastereomeric forms and two values are observed for both the P−Se− and the P=Se coupling constants, respectively, −346.2, −760.4 for one form and −390.6, −751.1 for the other form.[228]

2.3. Structural Factors Affecting the Sign and Magnitude of ^2K

Signs of two-bond couplings, unlike one-bond couplings, are not always characteristic of the coupled nuclei. The sign usually depends on some or all of the same factors affecting the magnitudes.

The magnitudes of ^2K are in a gross way related to atomic densities at the nuclei. For example, in analogous complexes, for X = Al, Ga, In, $^2K(XOP)$ increases down the group, 1.0, 1.42, and 2.87×10^{19} N A^{-2} m^{-3}, respectively.[281] In addition, both sign and magnitude depend on oxidation state, coordination numbers, formal hybridization, and electronegativity of substituents on the coupled atoms, just as one-bond couplings do. In fact, ^2K plotted vs. ^1K shows this relationship: for example, plots of $^1K(SnSn)$ vs. $^2K(SnSnH)$,[141] of $^1K(SnC)$ vs. $^2K(SnCH)$,[140] of $^1K(TlC)$ vs. $^2K(TlCH)$,[282] and of $^1K(PC)$ vs. $^2K(PCH)$,[34] indicate a very good correlation between ^1K and ^2K. This correlation is so good that if one of ^1K or ^2K changes sign in a series of compounds, but the other does not, it is possible to obtain the sign of the first from the plot if only the magnitudes of ^1K and ^2K for the compound are known.

The ratio $^{2}K(XCH)/^{1}K(XC)$ changes systematically with the position of X in the Periodic Table.[87]

The sign of $^{1}K(XY)$ relative to $^{2}K(XMY)$ is dependent on the XMY bond angle, but it is usual for $^{1}K(XY)$ and $^{2}K(XMY)$ to be opposite in sign for most XMY bond angles.

2.3.1. *Oxidation States, Coordination Numbers, and Formal Hybridization of Coupled Nuclei*

The dependence of $^{2}K(XMY)$ on the oxidation states, coordination numbers, and formal hybridization of X and Y is basically the same as in $^{1}K(XY)$. For example, it is found that for a formally sp^3 hybridization on the intervening carbon the magnitude of $|^{2}K(CCC)|$ goes in the order $(sp^3)(sp^3) < (sp^3)(sp^2) < (sp^3)(sp)$ according to the formal hybridization of the coupled C atoms.[258,283] While there is an algebraic increase in $^{1}K(PP)$ in going from P(III) to P(V), $^{2}K(PNP)$ has large positive values, sensitive to relative orientation of lone pairs in three-coordinate P, decreases greatly to a still positive value on increase of coordination numbers of the coupled P atoms.[245,284] This opposite algebraic behavior has to do with the opposite signs of ^{2}K and ^{1}K. Similarly, $^{2}J(SeCH)$ is negative when Se has zero lone pairs and is positive and increasing when Se goes from 1 to 3 lone pairs.[166] ^{2}K is smaller for coupling to a bridging atom compared to coupling to a terminal one: for example $^{2}J(PtPtC) = 180$ and 0 Hz for terminal and bridging carbons.[232] Like ^{1}K, ^{2}K values also decrease when a coordination compound is formed, and this decrease can be used as a diagnostic tool for complexation.

2.3.2. *Substituents on Coupled Atoms*

Like $|^{1}K|$, $|^{2}K|$ increases with increasing electronegativity of substituents regardless of sign. Many examples are available for $^{2}K(PMP)$,[241] as well as for other pairs of nuclei.[39,103] $^{2}K(XMY)$ changes by as much as a factor of 2 or 3 upon substitution at either X or Y or both with electronegative groups or atoms.

2.3.3. *Dependence of $^{2}K(XMY)$ on the Coordination Number of M, Position of M in a Family*

The coordination number of M also determines the XMY bond angle (see Section 2.3.4), so these two factors are not always mutually exclusive. ^{2}K increases algebraically with decreasing coordination number of M. For example, in bis(trimethylsilyl)aminobis(trimethylsilylimino) phosphorane, $R_2N-P(=NR)_2$ ($R = SiMe_3$), the coupling of the phosphorus to a silicon $^{2}J(PNSi)$ through the three-coordinate N is 1.2 Hz and through the two-coordinate N is 16.7 Hz.[285] $^{2}J(^{17}OCH)$ changes from 5–10 Hz to 38 Hz when the carbon goes from tetrahedral to trigonal coordination[286] while $^{2}K(NCH)$ goes from negative to positive.[287]

When comparing analogous compounds, $^{2}K(XMY)$ increases algebraically as M descends in a family. For example, $^{2}J(HgCC) = 29.2$ Hz, $^{2}J(HgSiC) = 92.3$ Hz in $Hg(EMe_3)_2$.[288] $^{2}J(HXH)$ shows a regular algebraic increase in XH_4 in going from C to Pb.[58] $^{2}J(PtXPt) = 199$ and 380 Hz with X = Cl and I, respectively, for the dinuclear complexes *trans* $[Pt(II)X_2PBu_3]_2$.[289] The average $^{2}J(SnXSn)$ over various SnXSn moieties in several cyclic heterotristannanes is 198, 234, and 247 Hz for X = S,

Se, Te, respectively.[290] ^2J(PXP in $[(CF_3)_2P]_2X$ is 108.3 and 234.2 Hz for X = O, S, respectively.[291] [There are exceptions; for example, ^2J(FXF) is 351 Hz in $Kr_2F_3^+$,[292] 308 Hz in $Xe_2F_3^+$.[293]] ^2J(PMP) also increases algebraically as M descends in the triad Cr, Mo, W,[102] for analogous *cis* disubstituted complexes and for *trans* complexes. ^2J(PMC) for M = Cr, Mo, W in analogous $(CO)_5MPR_3$ complexes show a similar algebraic increase in going from M = Cr to Mo; however, Mo and W values are comparable, e.g., −14.4, 10.0, 7.5 for $^2J(PMC)_{cis}$ and −8.5, 20.1, 18.6 for $^2J(PMC)_{trans}$.[294]

2.3.4. *Dependence of $^2K(XMY)$ on the XMY Bond Angle*

That ^2K(XMY) has a dependence on the XMY bond angle was brought to our attention by theoretical calculations on the CH_2 fragment.[295] That ^2J(HCH) increases algebraically with increasing HCH angle in both H_3C- and $H_2C=$ fragments is indeed found experimentally, and semiempirical calculations can reproduce these trends.[296] There are extensive data on a variety of nuclei to support this idea that ^2K(XMY) depends on the XMY bond angle. However, they are not easily drawn to one overall generalization, like the dihedral angle dependence in ^3K for example, because other factors which influence ^2K(XMY), such as coordination number or formal hybridization of M, have consequences in the XMY bond angle. The cases we will consider are those in which the ^2K(XMY) involve the same intervening M atom, as to minimize such other factors. In the octahedral and square planar complex there are two angles to consider, 90° (*cis*) and 180° (*trans*) and in the trigonal bipyramidal complex there are 90° (axial–equatorial), 120° (equatorial–equatorial), and 180° (axial–axial). The general rule that emerges is that there is a very pronounced difference in magnitude, |*trans*| > |*cis*| and sometimes even in sign so that algebraically, $^2K(XMY)_{trans} > {}^2K(XMY)_{cis}$. This is so general that it serves as the basis for identification of octahedral and square planar isomers. The most extensive data are for ^2K(PMP)[102]; however, the same general rule has been found to hold in PMC,[294,297] PMN,[256,298–300] PMSn,[301–304] PMH,[249,305,306] PMHg,[146] PMF,[307] for M = any transition series and also SnFeC.[308] In the latter case, the data are consistent only with $^2K(SnFeC)_{trans}$ (positive small) > $^2K(SnFeC)_{cis}$ (negative large), which is the same algebraic inequality as in all the other ^2K(XMY) cases mentioned above. Some care should be taken in applying this algebraic inequality to structural determination when the relative signs of the *cis* and *trans* couplings may not be the same. Except in coupling involving fluorine, which we shall see below, there are no clear exceptions, although there are reports of $|^2J(PMH)_{cis}| > |^2J(PMH)_{trans}|$ in which the signs of the couplings were not determined: In square pyramidal complexes of Cr, Mo, W, and Mn the magnitudes of ^2J(PMH) were 54–78 Hz for *cis* couplings and 21–28 Hz for *trans*[309]; in octahedral $HM(PF_3)(CO)_4$ complexes magnitudes of *cis* PMH couplings were 70.3 and 34, respectively, for Mn and Re but *trans* couplings were very small (0–2).[310]

Couplings involving fluorine do not present as consistent a picture. In many cases the ^2K does increase algebraically with bond angle. For example, in SF_3CF_3, a trigonal bipyramidal structure with an equatorial CF_3 group, $^2J(F_{eq}SC)(120°) > {}^2J(F_{ax}SC)(90°)$, 19.8 and 11.8 Hz, respectively.[311] In the octahedral F-bridged binuclear fluoroanions MNF_{11}^-, where M, N = Nb, Ta, $^2J(FF)_{trans} > {}^2J(FF)_{cis}$.[312] However, ^2J(HF) and ^2J(FF) couplings in octahedral systems do not always follow these trends. In some cases, they increase in magnitude as the FMF angle decreases:

cis couplings (90°) are large (40–150 Hz, positive where the sign has been determined) whereas *trans* couplings (180°) are very small (0–8 Hz). Examples are ^2J(HPF) and ^2J(FPF),(198,313) ^2J(FAsF) and $^2J(FSbF_3^+)$.(314,315)

The generalization of an "algebraic increase of K(XMY) with increasing XMY bond angle" should be used with great caution when comparing couplings in different compounds because other structural factors come into play. However, it may be helpful in differentiating between possible structures. For example, the FF couplings in $Kr_2F_3^+$ and $Xe_2F_3^+$ are much larger than those observed for other inert gas fluoride ions such as XeF_3^+, XeF_5^+ and $XeOF_3^+$ (351 and 308 Hz versus 174, 176, and 103 Hz). In $Kr_2F_3^+$ and in $Xe_2F_3^+$ the FMF bond angle is ~180° since the structure is essentially that of two linear FXF moieties sharing a common F,(292,293) whereas the bond angles in the other cations are ~90°.

2.3.5. *Stereospecificity in $^2K(XMY)$*

It is well known that the magnitudes and signs of the coupling constants ^2K(PX) and ^3K(PX) depend mainly on the orientation of the P lone pair. For example both ^2K(PNC)(267) and ^3K(PNCH)(316) have large positive values when the dihedral angle ϕ (between the NC or the CH bond and the P lone pair direction) is small (*cis* or syn periplanar). As ϕ approaches 180° (*trans* or antiperiplanar) these two couplings become small and may even become negative for ^3K(PNCH),(316) ^2K(PNC).(267) The dependences on the dihedral angle involving the lone pair and that involving a bond to an electronegative atom such as O or Cl are opposite. These are shown schematically in Figure 5, in which references for several examples are given.

2.4. Structural Factors Affecting the Sign and Magnitude of ^3K

The three-bond coupling depends on some of the same factors as one- and two-bond coupling: coordination number, formal hybridization, electronegativity of substituents at the coupled atoms. However, these effects are not clearly understood; because there are now four atoms involved in the coupling path, it is not very easy to isolate individual factors. In addition, three-bond couplings exhibit a dependence on the dihedral angle, which has been found empirically in many cases to be of the form called the Karplus equation, after the original theoretical calculation of the dependence of ^3K(HH) on the geometry of the HCCH fragment(317):

$$^3K = C \cos 2\phi + B \cos \phi + A \tag{15}$$

This equation serves as a good qualitative guide. However, since coupling constants depend on the other factors mentioned previously, in addition to the dihedral angle, one can expect to use the equation to determine approximate conformations only if the examples on which the empirical values of A, B, and C have been based are similar to those to which the equation will be applied. It is especially important to take into account the electronegativity of substituents at the coupled nuclei since this has a pronounced effect.(318) The other cautionary comment that needs to be made is that there are other contributions, such as that of the pi system (if any), which may not be negligible, especially when the dihedral dependence gives a low value ($\phi \sim 90°$), and that of "through-space" coupling when ϕ is close to zero. Although the Karplus equation was originally applied extensively to ^3K(HCCH), it is continuing to provide a way of categorizing the three-bond coupling ranges observed in a wide variety of coupling frameworks such as PtCCC, PWNN, PCPSe, etc. Many examples continue to be reviewed annually.[39] Used with caution, the observed ^3J and a Karplus equation make a reasonable conformational probe.

In addition to the dihedral angle the three-bond coupling depends on the bond angles between any two adjacent bonds out of the three. This becomes apparent when the dihedral angle dependence is averaged out by free rotation of one end relative to the other. For example, the *cis* and *trans* ^3J(PAuCH) couplings in a square planar Au III complex are $+8.5$ and -8.9 Hz, respectively, for PAuC bond angles of 90° and 180°.[319] *Cis* and *trans* ^3J(PMCF) values in such complexes, as well as in square planar Pt(II) complexes, are all positive but fall into distinct ranges for *cis* and *trans* coupling paths.[319]

3. EFFECTS OF INTERMOLECULAR INTERACTIONS AND INTRAMOLECULAR DYNAMICS ON SPIN–SPIN COUPLING

These effects are important to consider because no molecule is observed in isolation or at its rigid equilibrium conformation. What is observed in NMR is always an average, the average over several species. The species may be molecules in different vibrational and rotational levels (as in the dilute gas phase), or different rotameric species, or different clusters of observed molecule with solvent molecule(s).

Spin–spin splittings have been used to derive mechanistic conclusions in systems undergoing intramolecular rearrangement or intermolecular exchange, by band shape analyses of the temperature-dependent NMR spectra.[320] A review of multisite exchange in coupled systems[91] gives several examples in which the analysis of the coupling patterns is used to derive the mechanism for exchange or at least eliminate several others while supporting one or two mechanisms. In this section we will consider only the effects of such intra- and intermolecular exchange on the *average* coupling constant derived from a spectrum in the fast exchange limit.

3.1. Averaging via Rotameric Equilibria and Intramolecular Rearrangement

A molecule capable of undergoing rearrangement (such as PF_5), or internal rotation such as CH_3CF_3, may be considered as having several minimum energy wells in a multidimensional potential energy surface. The observed coupling constant depends on the time average (for a single molecule) over the values of the coupling constant as the molecule goes from well to well via lowest-energy pathways. We have already seen the dependence of coupling constants on spatial configuration. The pathways on this energy surface are not known, but the configurations corresponding to the wells can usually be determined. Thus, we tend to think in terms of averaging over populations (fractions of total number of molecules in the different wells) with the assumption that the molecule spends relatively little time between wells. In the simplest case of two wells (or sites) A and B, the observed coupling in the limit of rapid site exchange (inversion, pseudorotation, or internal rotation) is given by

$$\langle J(XY)\rangle = pJ(XY)_A + (1-p)\,J(XY)_B \tag{16}$$

where the mole fractions in the two sites are p and $(1-p)$, $J(XY)_A$ and $J(XY)_B$ are the coupling constants characteristic of the configuration in either well A or B. In some cases, it is possible to go down to temperatures sufficiently low that the site exchange is slow compared to the NMR time scale (as defined in Chapter 2), in which case we can observe both sites A and B and measure $J(XY)_A$ and $J(XY)_B$ directly. It is sometimes possible to relate the relative populations to the free energy difference between wells.

The earliest experimental evidence for a dependence of a one-bond coupling, $^1J(PP)$, upon rotation about the P–P bond was based on just the indirect information obtained from time-averaged coupling constants.[185,186] McFarlane and McFarlane observed a difference of 19 Hz in the coupling for the two diastereomers of 1,2-dimethyl-1,2-diphenyl biphosphine.[186] They pointed out that owing to differences in steric interaction the rotamer populations of the two diastereomers will be different. Thus, any difference in $^1J(PP)$ between different rotamers will show up as a difference between the two diastereomers. Similar indirect evidence was provided in the form of a temperature-dependent $^1J(PP)$ in F_2P-PF_2, which changed from -230.3 Hz at -101°C to -228.5 Hz at -1°C.[185] The dependence of one-bond couplings on the dihedral angles between lone pairs in this type of molecule has been observed in many cases since then. In fact, McFarlane's argument is now turned around to be used as a means of assigning structures to diastereomers. For example, the meso and dl forms of $(CF_3)_2P-P(CF_3)_2$ were assigned on the basis of the temperature dependence of $^1J(PP)$, the meso $^1J(PP)$ becoming more negative and the dl less negative with increasing temperature.[321]

Bridging groups usually undergo relatively large displacements about one or more equilibrium position. At room temperature the average over these excursions may give a single coupling constant, leading to the false impression that the bridging group is symmetrically located. For example, at 75°C the bridging hydrogen in diplatinum complexes such as $[Pt(\mu-H)-(SiOEt_3)(PMe_3Bu^t_2]_2$ shows a $^1J(PtH)$ of 681 Hz.[322] However, at -60°C two couplings (465 and 894 Hz) are observed, indicating that the μ-hydrido ligand is unsymmetrically bonded to the two Pt atoms. When an untypical coupling constant is observed, the possibility of averaging due to intramolecular motion should be considered. The large temperature dependence of $^2J(PSP)$ in F_2PSPF_2 ($+274$ Hz at 31°C to $+393$ Hz at -120° C)[185] is untypical. Vibrational averaging around one equilibrium configuration cannot account for such a large temperature dependence. Thus, it must be presumed that some fluxion around the S atom may be taking place.

3.2. Isotope Effects

The effects of intramolecular dynamics (ordinary rotational and vibrational averaging in addition to changes in conformation or rearrangements) are manifested in yet another way. Because such averaging is mass dependent, there are isotope effects on coupling constants. There is a well-known mass effect on the breaking of a bond in a reaction; the rate of this step is less for the heavier atom. In a chemical equilibrium—for example, one in which hydrogen bonds are formed and broken, or one in which different conformations are undergoing internal rotation—there are mass effects on the magnitudes of equilibrium constants. Thus, the relative populations of different exchange sites may be different for the heavy isotopomer compared to the light, leading to a different average coupling constant. There are many examples of these effects of isotopic substitution on equilibrium constants, as in conformational equilibria or hydrogen bonding equilibria. In addition, there is an intrinsic isotope effect: for a molecule not involved in a chemical equilibrium, there may still be a mass effect on the coupling constant because the rotational and vibrational averaging is different for different isotopomers. In a diatomic molecule the rovibrationally averaged bond length is slightly shorter for the heavier isotopomer. Since the coupling constant has a different value at each internuclear separation, the rovibrationally averaged coupling constant may be sufficiently different for two isotopomers for an isotope effect to be observed. By considering the anharmonicity of the HF molecule potential function, a small (about 1%) isotope effect between the HF and DF reduced coupling constant was predicted.[323] While isotope effects on nuclear shielding have been widely observed, there are very few reports of isotope effects on coupling constants.

The primary isotope effect, the difference in the coupling constants from observations on different isotopes of the coupled nuclei, may be expressed as follows:

$$\Delta_p J(\mathrm{X}^{2/1}\mathrm{H}) = |J(\mathrm{XD})(\gamma_\mathrm{H}/\gamma_\mathrm{D})| - |J(\mathrm{XH})| \tag{17}$$

where $\gamma_\mathrm{H}/\gamma_\mathrm{D} = 6.514\,398\,04(120)$,[324] for deuterium substitution, or better still, in terms of the reduced coupling constant,

$$\Delta_p K(\mathrm{X}^{2/1}\mathrm{H}) = |K(\mathrm{XD})| - |K(\mathrm{XH})| \tag{18}$$

The primary isotope effect (as defined above) is usually positive regardless of the sign of the reduced coupling constant; that is, the magnitude of the reduced coupling for the heavy isotope is greater than that for the light one. $\Delta_p\,{}^1K$ for one-bond couplings have been reported for PH in $(CH_3O)_2P(O)H^*$,[325] $PO_2H_2^*$,[326] and other three- and four-coordinate phosphorus compounds,[327] and in PH_3^* [328]; for SeH in H_2^*Se.[57,329] $\Delta_p\,{}^2K$ for two-bond couplings have also been reported for FH in CH^*F_3,[330] HD in SiH_3D, $SiHD_3$,[331] and HH in benzylmethylsulfoxide,[332] for example. These primary isotope effects are all negative with the exception of H_2^*Se,[329] PH_3^*,[328] and other three-coordinate phosphorus.[327] Their magnitudes are small, the largest being 2.75×10^{19} N A^{-2} m^{-3} over 6% of the reduced PH coupling in PH_3.[328]

The secondary isotope effect, the change in the coupling constant due to isotopic substitution of neighboring nuclei, C,

$$\Delta_s J(\mathrm{AB})[C] = |J(\mathrm{AB})_{\mathrm{heavy}}| - |J(\mathrm{AB})_{\mathrm{light}}| \tag{19}$$

has usually been found to be negative. The magnitude of the coupling decreases monotonically with increasing deuterium substitution at neighboring nuclei, and is roughly additive. Examples of $\Delta_s^1 K$ have been reported for ${}^1K$(BH) in BH_nD_{4-n} with various cations,[333,334] ${}^1K$(PH) and ${}^1K$(PD) in PH_nD_{3-n},[328] ${}^1K$(PH) and ${}^1K$(PD) in $PO_2H_nD_{2-n}$,[326] ${}^1K$(SiH) in SiH_2H^*I,[58] ${}^1K$(SiF) in H^*SiF_3,[335] ${}^1K$(SeH) and ${}^1K$(SeD) in $H_nD_{2-n}Se$.[329] The magnitudes are small, the largest being about 1.5% (per D) of the SeH coupling in H_2Se. Magnitudes and signs of isotope effects on spin–spin coupling have been interpreted by Jameson and Osten.[336]

3.3. Chemical Exchange and Medium Effects

Intermolecular exchange involving a separation of the coupled atoms will appear as zero coupling if the exchange is fast on the NMR time scale, but when the intermolecular exchange involves no separation of the coupled atoms an average coupling constant will be observed. From a practical point of view one should expect the observed coupling constant to vary with the chemical environment to the extent that the environment affects the factors (coordination number, etc.) that have been discussed in Section 2. By their nature, strong specific interactions by a solvent molecule with one of the coupled atoms can result in a large change in the coupling constant (as much as 50% or more) whereas nonspecific (van der Waals) interactions between the observed molecule and a solvent molecule may result in a small change (less than 5%) in the coupling constant. The degree of interactions with the environment is a continuum, so there is no clear dividing line between what is usually called "solvent effects" and the effects of intermolecular chemical exchange.

Coupling constants involving Tl have very large intermolecular effects. For example, ^1J(TlC) in $Tl(OAc)_2$-norbornyl derivatives have much higher values in pyridine solution (6321–6655 Hz) than in chloroform solution (5471–5764 Hz).[337,338] ^1J(TlC) in $[Me_2Tl][NO_3]$ is 2503 Hz in D_2O, 2918 Hz in DMSO, and 3054 Hz in pyridine, whereas ^2J(TlCH) differs by about 100 Hz in these solvents.[339] ^1J(PbC) in alkyl lead compounds varies by a factor of 1.3–1.8 with a change in solvent.[234] The coordination number of Pb in these solutions is thought to change from 4 to 5–8 by complex formation with the solvent.[234] At the other end of the scale are the modest solvent shifts: ^2J(HCF) in CHF_3 molecule changes up to 1% in 21 different solvents.[340] ^1J(SnC) in $Me_3SnC{\equiv}C{-}C{\equiv}CSnMe_3$, is 372 Hz in $CDCl_3$ and 391 Hz in C_6D_6.[341] ^1J(PH) in NaH_2PO_2 varies with its molal concentration in acetic acid: $J = 537.2 - 2.45$ m.[342]

There is a gas-to-liquid shift in the BF coupling constant in BF_3.[343] The density dependence of this coupling in the gas phase and the temperature dependence in the pure liquid (liquid density varies with temperature) are both measures of the effects of van der Waals interactions on the BF coupling constant. The effects of solvent on coupling constants have been reviewed by Barfield and Johnston.[22]

4. THEORETICAL DESCRIPTION

We have seen (in Section 1.1) Ramsey's formulation for the different terms contributing to the coupling constant. In this section we will consider the computational schemes used in the application of equations (8)–(12) to specific systems, the relative importance of the Fermi contact, spin dipolar, and orbital terms to the isotropic coupling constant, relativistic effects, and approximate calculations and models. Where possible we will note the theoretical basis for the general trends enumerated in the previous sections.

4.1. Computational Schemes

Excellent reviews of the methods of calculations of coupling constants, as well as results for specific systems published in 1969–1981, have been given by Kowalewski.[344,345] Results of calculations are also given in the annual reviews of coupling constants,[346,347] and theoretical calculations prior to 1969 were reviewed by Barfield and Grant[348] and Murrell.[349] Because the second-order perturbation equations (8)–(12) as given by Ramsey require knowledge of all the excited states of the molecule, including the continuum, various approaches have been developed using truncated sets of excited state wave functions of varying degrees of sophistication. We write the coupling constant as $J=(\mu_0/3\pi)^2\,\mu_B^2 h\gamma_N\gamma_{N'}K'$, the reduced coupling as $K=(4/9)\,\mu_0^2\mu_B^2 K'$. The working equations for some of the entries in the figure are shown below (only the Fermi contact term is given; the other terms can be written in analogous form).

McConnell[354]:

$$K'=s_N^2(0)\,s_{N'}^2(0)\,P_{s_Ns_{N'}}^2(^3\Delta E)^{-1} \tag{20}$$

where the bond order is given by

$$P_{s_Ns_{N'}}=2\sum_i^{\text{occ}} c_{iN}c_{iN'}$$

$s_N^2(0)=\langle s_N|\,\delta(r_{kN})\,|s_N\rangle$, c_{iN} is the coefficient of the valence s orbital centered on N in the ith occupied molecular orbital, and $^3\Delta E$ is some average excitation energy, chosen *ad hoc*.

Pople and Santry[5]:

$$K'=s_N^2(0)\,s_{N'}^2(0)\,\pi_{NN'} \tag{21}$$

where the mutual polarizability of the orbitals on N and N' is given by

$$\pi_{NN'}=-4\sum_i^{\text{occ}}\sum_a^{\text{unocc}} c_{iN}c_{aN}c_{iN'}c_{aN'}/(\varepsilon_a-\varepsilon_i)$$

where ε_a and ε_i are molecular orbital energies, i are the occupied and a the virtual molecular orbitals.

Loeve and Salem[363]:

$$K'=-4\sum_i^{\text{occ}}\sum_a^{\text{unocc}}(^3\Delta E_{i\to a})^{-1}\langle i|\,\delta(r_{kN})\,|a\rangle\langle a|\,\delta(r_{kN'})\,|i\rangle \tag{22}$$

where $^3\Delta E_{i\to a}$ is the excitation energy corresponding to the excitation of an electron from the orbital i to virtual orbital a, the resulting electron configuration being a triplet.

Pople, McIver, and Ostlund[351,361,362] and Blizzard and Santry[358–360]:

$$K'=s_N^2(0)\,s_{N'}^2(0)\cdot 2P_{s_Ns_{N'}}^{(1)} \tag{23}$$

where the first-order density matrix is given by

$$P_{\lambda\nu}^{(1)}=\sum_i^{\text{occ}}(c_{i\lambda}^{(0)}c_{i\nu}^{(1)}+c_{i\lambda}^{(1)}c_{i\nu}^{(0)})$$

$c_{i\lambda}^{(0)}$ is the coefficient of the λth atomic orbital in the ith molecular orbital obtained by an SCF calculation. $c_{i\nu}^{(1)}$ is the first-order correction to $c_{i\nu}^{(0)}$ due to the Fermi contact interaction.

Ditchfield and Snyder[364,367]: The working equations are the same as in the above method except that nonempirical calculations are used, in particular two-center integrals are evaluated.

Barfield and Karplus[369]:

$$K'=\phi_N^2(0)\,\phi_{N'}^2(0)\cdot 2P_{NN'}^{\text{PD}}(^3\Delta E)^{-1} \tag{24}$$

where $\phi_N(0)$ is the electron density at nucleus N, $P^{PD}_{NN'}$ is the Penney–Dirac bond order between atoms N and N', which is obtained by summing over valence bond structures.

Ditchfield and Murrell[356,357,364]:

$$K' = -4 \sum_n ({}^3\Delta E_n)^{-1} \sum_i^{\text{occ}} \sum_a^{\text{unocc}} \sum_j^{\text{occ}} \sum_b^{\text{unocc}} d^n_{i\to a} d^n_{j\to b} \langle i|\, \delta(r_{kN})\, |a\rangle \langle b|\, \delta(r_{kN'})\, |j\rangle \qquad (25)$$

where $d^n_{i\to a}$ is the coefficient of the singly excited $(i \to a)$ configuration in the nth triplet state wave function, obtained by diagonalizing the Hamiltonian matrix in the basis of configurations, and ${}^3\Delta E_n$ is the nth eigenvalue.

Kowalewski[373–375]:

$$K' = 8 \sum_\mu c^{(1)}_{\mu N} \langle \mu |\, \sum_k \delta(r_{kN'})\, S_{zk}\, |0\rangle \qquad (26)$$

where $c^{(1)}_{\mu N}$ is the coefficient of the excited triplet configuration μ in the first-order correction to the wave function due to the Fermi contact interaction at nucleus N. The sum is over all the singly and doubly excited triplet configurations that can be constructed from the SCF MOs; $|0\rangle$ is the singlet wave function obtained by configuration interaction.

The working equations are elegantly simple; however, the computational effort is determined largely by the size of the basis functions used to describe the unperturbed wave function. There are some very good recent calculations on the coupling constants in HD,[376] CH_4, NH_3, H_2O, HF[377–379] and the vicinal HCCH coupling.[380] The results on CH_4, NH_3, H_2O, and HF are compared with experiment in Table 3.

The recent theoretical calculation of the dependence of the three-bond HH coupling on the dihedral angle in the HCCH fragment[380] compares very favorably with the empirical equation of the Karplus form which had been least-squares fitted to data from rigid systems of known geometry. In ${}^3J(\text{HH}) = A + B\cos\phi + C\cos 2\phi$, theory gives A, B, $C = 6.9$, -1.1, and 6.1 to be compared with the empirical values, A, B, $C = 7$, -1, and 5,[381] respectively. In cases such as these, in which the level of accuracy of the theoretical calculation becomes comparable with the precision of the experiment, the comparison of theoretical calculations with experiment must be made with caution. The theoretical value must include the entire isotropic coupling, not just

Table 3. Theoretical Contributions of Various Mechanisms to the Reduced One- and Two-Bond Coupling Constants in CH_4, NH_3, H_2O, and HF[a]

	Contact[b]	Orbital[c]	Spin dipolar[c]	Total	Expt.
CH	37.7	0.53	−0.13	38.1	41
NH	43.9	2.42	−0.23	46.1	50
OH	40.2	7.49	0.43	48.1	48
FH	31	17.42	−0.858	47.6	46.9
HCH	−1.03	0.225	0.03	−0.77	−1.03
HNH	−1.06	0.391	0.07	−0.60	−0.87
HOH	−1.04	0.591	0.12	−0.33	−0.60

[a] Units: 10^{19} N A^{-2} m^{-3} or 10^{20} cm^{-3}. Theoretical values not corrected for vibration, experimental values not extrapolated to zero pressure limit.
[b] Reference 378.
[c] Reference 377.

the contact term, and be averaged over vibrational motions of the molecule.[374,382–384] The experimental value must be that corresponding to an isolated molecule (gas phase, zero-pressure limit) not undergoing fluxion or internal rotation.

4.2. Relative Importance of the Fermi Contact, Spin-Dipolar, and Orbital Terms

Because of early interest in couplings involving protons, the Fermi contact term has received the most attention. From the equations (9)–(12) we can see under what general conditions the noncontact terms can become important. If only one-center integrals over the atomic orbitals are included, $J^{(1a)}$ vanishes; $J^{(1b)}$ and $J^{(2)}$ differ from zero only if non-*s* orbitals centered on both coupled nuclei are contained in the basis set. Therefore, for couplings involving a proton the noncontact terms are expected to be small. The authors who developed the computational schemes shown above considered all the coupling mechanisms. The analogous working equations for the noncontact terms involve $\langle r_N^{-3} \rangle$ instead of the electron density at the nucleus, and integrals of the orbital angular momentum operator instead of the Dirac delta function $\delta(r_{kN})$, but otherwise the equations for the orbital and spin-dipolar terms are analogous to equations (20)–(26), with the sums being over singlet instead of triplet states for the orbital part. By examining the bond orders involved in the expressions obtained for the noncontact terms when the average energy approximation is used, Pople and Santry were able to show that $J^{(1b)}$ will be nonvanishing for one-bond couplings only if there exists a multiple bond between the coupled nuclei. To the same approximation the spin-dipolar contribution also vanishes for one-bond couplings except under these conditions. What this means is that the noncontact contributions calculated by more sophisticated methods will be important in systems with multiple bonds and very likely be considerably smaller than the contact term in all others. To what extent this is true can be seen in Table 3. Relativistic calculations by Pyykkö and Wiesenfeld reveal that the relativistic term corresponding to the nonrelativistic Fermi contact term almost invariably dominates in $^1K(\mathrm{XH})$, $^1K(\mathrm{XC})$, $^1K(\mathrm{XP})$, and $^1K(\mathrm{XX'})$, where X = C, Si, Ge, Sn, Pb, Tl, O, S, Se, Te. The only exceptions found are the coupling constants between two group VI or group VII atoms such as Se–Se or I–I.[13]

Some calculations indicate that in a few cases, the orbital and spin-dipolar terms may be important and may even dominate. For example, coupled Hartree–Fock calculations of $^1J(\mathrm{CC})$ in benzenes by Lazzeretti *et al.* indicate that the orbital term may not be neglected (up to 20% of the observed coupling), whereas the spin-dipolar terms may be small (about 5%).[386] The same level of calculation of $^1J(\mathrm{CN})$ by Schulman and Venanzi in cyanides, nitriles, and pyridine[387] indicates that in acetonitrile and pyridine, the calculated Fermi contact term is smaller than the orbital and spin-dipolar terms. In the molecules they considered, there are instances in which each of the three coupling mechanisms makes the largest contribution: spin dipolar dominates in HCN and in CH_3CN, the orbital term dominates in CN^- ion, and the Fermi contact term dominates in couplings involving pyramidal nitrogen. In further work, Lee and Schulman show that the orbital term, although small in most cases, can account for significant percentages of the experimental couplings in several instances [up to 67% for $^2J(\mathrm{CCH})$ in ethylene].[388,389]

Although the orbital and spin-dipolar contributions may be relatively small compared to the contact contributions for coupling between elements from H to N, this may no longer be the case for coupling between heavier nuclei. The counterpart of $s_N^2(0)$ in the noncontact terms is $\langle r_N^{-3} \rangle_{np}$. This, like $s_N^2(0)$, varies in a periodic fashion, giving a plot much like Figure 3. (In fact, see Figure 2 of Chapter 3.) On the other hand, the heavier atoms have less tendency to form multiple bonds, which may offset the larger values of $\langle r_N^{-3} \rangle_{np}$. Estimates of the dipolar and orbital contributions to ^1K(PSn) and ^1K(PPb) for certain formal hybridizations on P, Sn, and Pb show that these estimated noncontact terms are under 10% of typical experimental ^1K(PSn) and ^1K(PPb) values.[102] In transition metal complexes which could have substantial π back-bonding between the coupled metal and ligand atoms, the dipolar and orbital contributions may have to be considered. Upper limits for the spin-dipolar term for one-bond couplings ^1K(XC) and ^1K(XF) in the absence of multiple bonds have been calculated for elements X = Li to Bi and displayed in Periodic Table form.[209]

4.3. Relativistic Effects

The form of the magnetic dipole hyperfine Hamiltonian given in equations (3)–(6) is correct for a point nucleus at the nonrelativistic limit. Relativistic effects are important to consider in any molecular electronic properties of the heavier elements. These effects are particularly significant for properties such as the indirect spin–spin coupling, which depend on electronic wave functions near nuclei where the electrons move fast. Here we consider only the general consequences of relativistic effects. Relativistic effects on the electronic structure of atoms and molecules consist of a contraction of s and p shells, the spin–orbit splitting of the non-s shells and the relativistic self-consistent field expansion of d and f shells.[390,391] The contraction of s and p shells leads to larger hyperfine interactions for atoms and molecules than are given by equation (6). The corrections due to this effect can be large for heavy atoms. A correction of equation (6) by a multiplicative factor $B(n, Z)$ depending on the principal quantum number n and the nuclear charge Z was suggested by Breit.[392] For the 6s shell of lead, $B(6, 82)$ is 2.592, for the 5s of Sn, $B(5, 50) = 1.348$.[212] This factor is roughly proportional to Z^2. For couplings involving the heavier nuclei the relativistic effects can be very important. Since the ratio of Pb/C values from the nonrelativistic values is 7.4, use of an uncorrected equation (6) would yield roughly this ratio of ^1K(PbH)/^1K(CH). Thus, the 23-fold change in ^1K(XH) in the XH_4 molecules in going from X = C to Pb can only be reproduced by including relativistic correction factors.[393,394] An additional problem in calculation of coupling constants (and indeed other magnetic properties) for molecules containing heavy atoms is that the spin–orbit splitting of the non-s shells in relativistic atoms means that Russell–Saunders coupling, which is assumed in Ramsey's formulation of the theory, no longer holds for such systems. Pyykkö has derived a relativistic analog to Ramsey's theory, using a relativistic hyperfine Hamiltonian and j–j or intermediate coupling.[395] Equations (4), (5), and (6) are replaced by a single hyperfine Hamiltonian, and equation (3) in relativistic theory is replaced by a second-order contribution. In order to use this theory, one needs relativistic molecular wave functions for the ground state and excited states. However, even approximate relativistic molecular calculations yield reasonable results provided the hyperfine integrals used are taken from good relativistic atomic wave functions.[396] The Pople–Santry level of approximation in this relativistic theory of coupling constants gives reasonable estimates of the ^1K(HgC) and ^1K(CdC) isotropic average and anisotropy $(K_{\parallel} - K_{\perp})$.[9] Results are shown in Table 4. Relativistic effects increase both K and ΔK by a factor of about 2–4, although the ratio $\Delta K/K$ is changed by a smaller factor. These results are found to be general,[13] and

Table 4. Importance of Relativistic Corrections for Coupling Involving Heavy Nuclei[a]

	Nonrelativistic[b,c]	Relativistic[b,c] Without vibrational correction	Relativistic[b,c] With vibrational correction	Experiment[d]	Reference
K(HgC) in $HgMe_2$	591.9	1561.6	1261	1290	385
ΔK(HgC) in $HgMe_2$	141.2	499.6		710	385
K(CdC) in $CdMe_2$	322.3	431.1	798	797.7	10
ΔK(CdC) in $CdMe_2$	77.7	115.2		279.0	10
K(HgC) in $MeHgNO_3$		2990		(nem)3150 ± 20 (lyo)3333 ± 2	9
ΔK(HgC) in $MeHgNO_3$		777		(nem)1080 ± 185 (lyo)600 ± 90	9

[a] K is the isotropic part of the coupling constant and $\Delta K = K_{\parallel} - K_{\perp}$
[b] Reference 9 (in SI Units of 10^{19} N A^{-2} m^{-3} or cgs Units of 10^{20} cm^{-3}).
[c] Calculations are based on single hybridization description of bonding: *sp* for Hg in $HgMe_2$ and pure *s* for $HgNO_3$, *sp* for Cd in $CdMe_2$.
[d] Experiments were carried out on partially oriented molecules in the nematic phase, except for $MeHgNO_3$, in which values measured in the nematic and lyotropic phases have been reported.

since relativistic effects increase roughly as Z^2, the ratio $\Delta K/K$ for analogous couplings is expected to increase in descending a group.

Although all terms of Ramsey's nonrelativistic theory are combined into only one equation in the relativistic theory of coupling constants, it is nevertheless possible to separate a "contact" part as the contribution involving the *s* atomic orbitals on both nuclei. In this, the relativistic *ns* spin densities $|\psi(0)|^2 = -(137.036/2\pi)\, v_{-1}$, where v_{-1} are radial hyperfine integrals calculated by Pyykkö[(13)] using Desclaux' program.[(396)] These relativistic *s* densities are given by Pyykkö for selected nuclei. Pyykkö's calculations show that the terms involving these *s* densities invariably dominate one-bond couplings. The only exceptions found are the couplings between group VI atoms or between group VII atoms, such as Se–Se or I–I.[(13)]

Pyykkö's relativistic calculations for ^1K(XH), ^1K(XC), and ^1K(XX), X = C, Si, Ge, Sn, Pb, O, S, Se, Te and for ^1K(XP), X = C, O, S, Se, Te, and for ^1K(XTl), X = H, C, Cl, Br, I for representative molecules, reproduce the observed empirical trends such as the following: For systems without lone pairs, having local symmetry which is tetrahedral, trigonal planar, or linear symmetric, ^1K(XH) and ^1K(XC) both increase with increasing atomic number of X, especially so when relativistic effects are included. For systems with lone pairs, having the local symmetry C_{3v}, C_{2v}, C_s, or $C_{\infty v}$, the one-bond reduced couplings algebraically diminish with increasing atomic number, making them large and negative for the heaviest nuclei of the group. Instances of experimentally verified sign changes as Z increases in a group are noted in Table 2: the one-bond reduced coupling constant for HF is positive, for HI it is negative, for P=O it is positive, and for P=Se and P=Te it is negative.

4.4. Approximate Calculations and Models

One might say that an experimental result can be considered "understood" only if it can be reproduced exactly by a nonempirical calculation. The HD molecule provides a good case: at all stages of theoretical sophistication, from Ramsey's initial calculation using an average energy approximation up to that of Kowalewski *et al.*[374] [using a basis set of 40 functions and full configuration interaction (all singly and doubly excited singlets and triplets) with vibrational averaging], calculations give the same correct sign and the correct order of magnitude, the latest one in complete agreement with experiment to within experimental precision.[374] Kowalewski's excellent review has shown that some computational schemes for coupling constants are capable of reproducing the experimental coupling in a few small molecules provided that the basis set of functions is sufficiently large and correlation of electrons is included.[344] On the other hand, even a molecule as small as HF has eluded quantitative agreement with experiment—theories at various stages of sophistication gave the wrong sign and wrong magnitude.[344] HF is probably an extreme case, so one need not be unduly pessimistic.

The general trends observed in Section 2 need some explanation. Some of the systems involve a large number of electrons so that only semiempirical level of calculations can be cited in support. There are also approximate models which may be useful if not overinterpreted. We will consider some of them.

For positive $^1K(XY)$ it is actually possible to do comparative studies of coupling constants using only localized orbitals and formal hybridization.[397] In the simplest description using McConnell's equation (20), the Fermi contact term of the coupling is proportional to $P^2_{s_X s_Y}/^3\Delta E$. If we further simplify to localized bonds, the bond order $P^2_{s_X s_Y}$ will be proportional to the s characters of X and Y in the XY bond. If we assume equivalent hybrids, the carbon s character in CH bonds in H_3C-CH_3, $H_2C=CH_2$, and $HC\equiv CH$ will be roughly $\frac{1}{4}(sp^3)$, $\frac{1}{3}(sp^2)$ and $\frac{1}{2}(sp)$, so that comparative values of $^1J(CH)$ will be

$$^1J\,(\text{in } H_3CCH_3) : {}^1J\,(\text{in } H_2C=CH_2) = 124.9/156.4 \simeq (1/4)/(1/3)$$

$$^1J\,(\text{in } H_3CCH_3) : {}^1J\,(\text{in } HC\equiv CH) = 124.9/248.7 \simeq (1/4)/(1/2)$$

General trends discussed in Section 2.2.2 with respect to coordination number can be explained in this localized picture. The formal hybridization inferred from a linear, trigonal planar, and tetrahedral arrangement are sp, sp^2, sp^3, respectively, as in $Hg(PMe_3)_n^{2+}$ with $n=2, 3,$ and 4.[398] Here the ^{199}HgP couplings are 5505:3580:2255 Hz, respectively, which roughly equals 1/2:1/3:1/4. One cannot expect quantitative interpretations with such a simple description, but it does allow the dependence of the one-bond coupling on the coordination number of either coupled atom to be understood. This connection between the magnitude of the coupling constant and the so-called "s character" first came to attention in CH coupling.[397] Indeed, a plot of experimental $^1J(CC)$ versus $(\%s)(\%s)$ calculated from INDO functions gives a good straight line,[246,260] a satisfying result anticipated earlier.[399]

The trend noted in Sections 2.2.5 and 2.3.2 that the coupling constant increased in magnitude, regardless of sign, upon substitution of electronegative atoms at the coupled nuclei, can also be understood in terms of s character. The model of Walsh[400] rediscovered by Bent[401] states that "If group X attached to carbon is replaced by a more electronegative group Y, then the carbon valency towards Y has more p character than it had towards X." Thus, a CH bond will have increasing s character in going from CH_4 to CH_3Y, CH_2Y_2, and CHY_3. This model was applied to the substituted methanes CHXYZ, to explain the nearly additive changes in $^1J(CH)$ with various substituents.[402] Theoretical calculations by Maciel *et al.*[246] at the FPT INDO level, on a large number of compounds—methanes, ethylenes, acetylenes, substituted benzenes—do indeed show a systematic increase in the s bond order between

C and H upon substitution at the carbon with a wide variety of electronegative substituents. The computed ${}^1J(CH)$ is found to be substantially more sensitive to substituent variation than the corresponding P^2_{ss} values. Thus, one can calculate large substituent effects on ${}^1J(CH)$ without giving such large variations in carbon *s* character.[246] This means that one may not take too literally a calculated *s* character from an observed coupling constant. After all, the direct proportionality between the two is a literal one only in the localized bond average energy approximation. Nevertheless, this explains why MP coupling constants tend to give linear plots against various electronegativity numbers for electronegative substituents at the phosphorus, as was seen in Section 2.2.5. The use of formal hybridization can also explain the general trend that the magnitude of ${}^1K$ is greater for equatorial coupling than for axial coupling in a trigonal bipyramidal arrangement, which we noted in Section 2.2.4. In the equatorial bonds the central atom can be considered to have a formal hybridization of sp^2, as opposed to the axial bonds, which may be formally described with a *dp* hybrid orbital. This is consistent with the observation that equatorial bonds are generally shorter than axial ones in trigonal bipyramidal molecules.

In independent molecular orbital theory the coupling arises from a one-electron *delocalization* term, which depends on the overlap between the localized orbitals. In the average energy approximation, this spin delocalization can only give rise to positive coupling constants. What is missing is a *spin polarization* mechanism, a way by which an electron transmits its spin information to another electron in a different orbital. This is that part of the coupling which arises from two-electron exchange integrals between the localized orbitals; it is not necessarily zero even if two orbitals are strictly orthogonal (zero overlap). Two-electron exchange integrals have a dependence on the angle between the symmetry axes of the two orbitals. For example, the two CH bonding orbitals in the CH_2 fragment have a negative exchange integral whose magnitude depends on the bond angle. For a HC–CH fragment, the most important exchange integral determining the spin correlation between electrons in different bonds is that between vicinal orbitals, and that has a dihedral angle dependence, minimum for 90° and slightly larger for *trans* than for *cis*. Murrell has shown how the different exchange integrals in the CH_2, the HC–CH and the HC–CH–CH fragments combine to give a geminal, vicinal, and a meta coupling which changes in sign in the same way as the experimental values: i.e., ${}^2J(HH)$ is $-$ for CH_4, $+$ for ethylene, ${}^3J(HCCH)$ is $+$ large and ${}^4J(HH)$ is $+$ small for ortho and meta coupling in benzene; ${}^3J(HCCH)$ is -0.54 for axial–axial, $+0.43$ for axial–equatorial, and $+1.80$ for equatorial–equatorial coupling in cyclohexane.[349] Thus, by including two-electron exchange integrals to introduce spin polarization, the molecular orbital method can give rise to the correct signs of coupling constants over two or more bonds. The mutual polarizability of the *s* orbitals $\pi_{NN'}$ in equation (21) can be positive or negative so that unless the average energy approximation is used, the Pople–Santry equation can formally result in either sign of coupling constants.

A simple model which accounts for the signs of couplings over two, three, or more bonds relative to one-bond coupling is the Dirac vector model, used very early by McConnell[403] and nicely explained by Lynden-Bell and Harris.[404] With this model, it is easy to understand why the ${}^2K(HCH)$ is opposite in sign to ${}^1K(CH)$ and ${}^3K(HH)$ in ethane.

With this pictorial model, there is no *a priori* reason for the Fermi contact interaction to lead to a positive ${}^1K(XY)$ at all times. A single electron in an orbital will have a positive Fermi contact interaction with the nucleus provided the orbital has a finite value at the origin. (The sign is defined conventionally as positive if this tends to make the nuclear and electron spins antiparallel.) However, except for hydrogen, an atom has several electrons. Thus, while an electron in non-s orbital may not have a direct Fermi contact interaction, it has an exchange interaction with the same-spin electron but not with the opposite-spin electron in a closed s shell. This like-spin-only exchange interaction leads to different densities at the nucleus of the α and β electrons of the closed s shell, therefore leads to either a parallel or antiparallel nuclear spin alignment. This model was used by Jameson and Gutowsky to explain the change in sign of one-bond couplings in going across the Periodic Table.[209] It successfully predicted most of the signs in Table 1 and trends in the signs of two-bond couplings as well.[405,406] However, it fails completely for P–P coupling between like P(III) groupings (as in H_2P-PH_2),[407] and indeed for any negative one-bond coupling across a covalent bond between two identical atoms or groups. Any model based on an average energy approximation is bound to fail for some systems.

In general, the negative contribution of the lone pair, discussed in Section 2.2.1, is still puzzling, although semiempirical calculations can give rise to negative PP coupling in H_2P-PH_2 when the lone pairs are gauche.[408,409] The negative lone pair effect is a more general phenomenon which is looking for an explanation. It is generally supposed that lone pairs tend to have more s character,[401] which leaves more p character in the bond, in which case the spin polarization model discussed above would lead to a negative contact interaction. For a more satisfying explanation, we must abandon the average energy approximation.

Models which do not use an average energy approximation can still be fairly simple. Sometimes use of just one or two virtual (i.e., unoccupied) molecular orbitals can lead to meaningful analyses. The simplest of these models was that of Pople and Santry.[5] In this model, the Fermi contact term in the reduced coupling $K_{NN'} =$ constants $\times\, s_N^2(0)\, s_{N'}^2(0)\, \pi_{NN'}$. For each occupied MO ($i$) which includes s orbitals on both nuclei N and N', there will be a contribution to $\pi_{NN'}$ by each virtual MO (a) which also includes s orbitals on both nuclei: $\pi_{NN'} = -4c_{iN}c_{aN}c_{iN'}c_{aN'}(\varepsilon_a - \varepsilon_i)^{-1}$. For a system without lone pairs, and with high symmetry, say octahedral or tetrahedral, only the totally symmetric MOs contain s orbitals on both the coupled central and ligand atoms; the positive contact contribution to ^1K arises from a single $a_1 \rightarrow a_1^*$ excitation. Here a_1 would be a bonding molecular orbital involving the coupled nuclei, which in the simplest case would be a combination such as $\phi_X + \lambda\phi_H$ including some s character on each of the coupled atoms X and H. On the other hand, the a_1^* virtual molecular orbital would be a combination such as $\phi_H - \lambda\phi_X$. If ϕ_X has 25% s character as in CH_4, then π_{XH} is approximately $-4(\frac{1}{2})(-\lambda/2)(\lambda)(1)/(\varepsilon_{a_1}^* - \varepsilon_{a_1}) > 0$. Thus, the contact contribution would be positive. For atoms with no lone pairs all the s character is isolated in bonding MOs of the proper symmetry (a_1 usually). On the other hand, for systems with lone pairs, some of the s character is in the lone pair orbitals. For example, in passing from XMe_4 to XMe_3^- a part of the t_2 molecular orbital becomes a lone pair, having some s character on X. In a linear XH molecule, the nonbonding molecular orbital 2σ is a lone pair. Because the latter nonbonding MO is antibonding with respect to the coupled X–C or X–H nuclei, it will give a negative contribution to the ^1K coupling constant, as seen in Figure 6 for HI. With increasing atomic number along a group with lone pairs, the ns_X AO participation decreases in the bonding $1a_1$ or 1σ MO and increases in the nonbonding 2σ (or other) MO. The negative lone pair contribution then begins to dominate, since the orbital energy separation between the 2σ and the virtual MO is

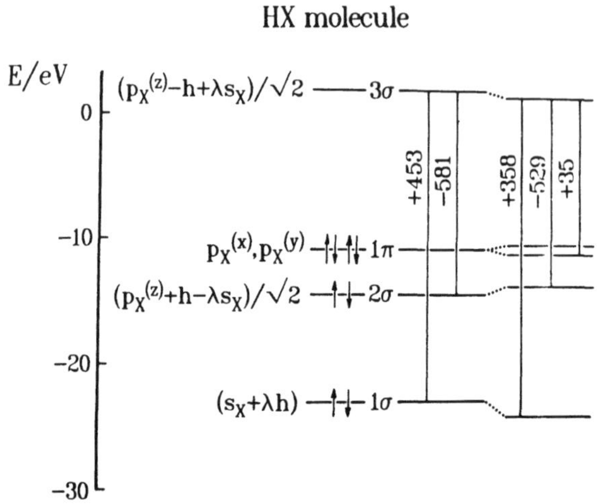

Figure 6. Molecular orbital energies for the HX molecule and the contributions to ${}^1K(\mathrm{HX})$. The numerical results are from a figure by Pyykkö and Wiesenfeld for HI.[13] The oversimplified molecular orbitals are shown merely to illustrate the signs of the contributions of the AOs to each MO which can lead to a net negative ${}^1K$ when there are lone pairs on one of the coupled atoms. In Pople and Santry's model,[5] ${}^1K(\mathrm{HX}) \simeq \phi_{\mathrm{H}}(0)^2\,\phi_{\mathrm{X}}(0)^2 \;\sum\sum -4c_{i\mathrm{H}}c_{a\mathrm{H}}c_{i\mathrm{X}}c_{a\mathrm{X}}/(\varepsilon_a-\varepsilon_i)$. For $i=1\sigma$, $a=3\sigma$ we have $-4(\lambda)(-1)(1)(\lambda)/(\varepsilon_{3\sigma}-\varepsilon_{1\sigma})>0$. On the other hand, for $i=2\sigma$, $a=3\sigma$ we have $-4(1)(-1)(-\lambda)(\lambda)/(\varepsilon_{3\sigma}-\varepsilon_{2\sigma})<0$. Of course, in a realistic calculation the magnitudes of the $(1\sigma\rightarrow 3\sigma)$ and the $(2\sigma\rightarrow 3\sigma)$ terms need not always be in the inverse order of the orbital energy separations so that positive reduced couplings for atoms having lone pairs can also be observed. The 1σ MO becomes more stabilized and the 2σ destabilized with increasing atomic number of X, especially so when relativistic effects are included,[13] as shown on the right-hand side.

less than that between the 1σ or $1a_1$ MO and the virtual MO. Pyykkö has shown that relativistic effects enhance these tendencies.[13]

Spin polarization is also the mechanism used in the models for pi contributions to long-range coupling constants. Hyperfine constants are used here, as in the Jameson–Gutowsky model, to provide an empirical measure of the magnitudes and signs of the spin polarization. For pi radicals both + and − hyperfine interactions are possible with different nuclei in the same molecule: negative for protons attached to the pi-electron-bearing carbon, positive for protons on a methyl attached to the pi electron-bearing carbon. McConnell has shown that it is possible to interpret proton hyperfine coupling in organic pi radicals by using an empirical parameter as a measure of the sigma-pi spin polarization.[410] Using this, he developed a model for pi contributions to long-range NMR coupling constants.[411] A nucleus interacts with an electron in a σ orbital with some s character, the electron in the σ orbital spin-polarizes an electron in a π orbital, the electron in the π system is delocalized all over the molecule and has a finite probability of being found in a π orbital which by another σ–π spin polarization communicates the spin information to the electron in the σ orbital with some s character and Fermi contact interaction with the other nucleus. Karplus derived analogous equations which do not use the average energy approximation.[412] Modifications to the Karplus equation for pi contribution to nuclear coupling have been given by Ditchfield and Murrell.[413] Being based on McConnell's model, all these equations give a pi contribution to the NN' coupling which is written in terms of the ESR hyperfine constants $a_N a_{N'}$. Thus, large pi contributions might be expected for nuclei with large ESR hyperfine constants. Notable examples of long-range couplings which may have sizable pi contributions are the ${}^5J({}^{119}\mathrm{Sn}\,{}^{119}\mathrm{Sn})$ value of 196.7 to 202.1 Hz observed for coupling through the cumulative bonds of a tin allene compound,[414] and the $^{n}J(PP)$, $n = 5$–9 in $CH_2P(O)(OEt)_2$-substituted benzenes and naphthalenes, ranging from 1.2 to 9.0 Hz for five bonds and 2.6 to 4.0 for nine bonds.[301] A new method of evaluating π (and σ) electron contributions to coupling constants[415] has been applied to CC, CH, and HH couplings in ethylene, acetylene, and benzene,[415] and butadiene, hexatriene, octatetraene, and butatriene.[416] The calculated π contributions show the empirically known characteristics: sign alternations of π contributions to K depending on the number of bonds separating the coupled nuclei, and no dependence of the absolute values of π contribution to K either on the zigzag path or on the number of bonds.

REFERENCES

1. Ramsey, N. F. *Phys. Rev.* 1953, **91**, 303–307.
2. Blinder, S. M. In *Advances in Quantum Chemistry*, Lowdin, P. O., Ed.; Academic Press: New York, 1965; Vol. 2, pp. 47–91.
3. Buckingham, A. D.; Love, I. *J. Magn. Reson.* 1970, **2**, 338–351.
4. Buckingham, A. D.; Pyykkö, P.; Robert, J. B.; Wiesenfeld, L. *Mol. Phys.* 1982, **46**, 177–182.
5. Pople, J. A.; Santry, D. P. *Mol. Phys.* 1964, **8**, 1–18.
6. McFarlane, W.; Rycroft, D. S. *Ann. Rep. NMR Spectrosc.* 1979, **9**, 319–414.
7. Emsley, J. W.; Lindon, J. C. *NMR Spectroscopy Using Liquid Crystal Solvents*, Pergamon: Oxford, 1975; Chap. 7.
8. Diehl, P.; Jokisaari, J.; Amrein, J.; Väänänen, T.; Pyykkö, P. *J. Magn. Reson.* 1982, **48**, 495–502.
9. Jokisaari, J.; Raisanen, K.; Kuonanoja, J.; Pyykkö, P.; Lajunen, L. *Mol. Phys.* 1980, **39**, 715–723.
10. Jokisaari, J.; Raisanen, K.; Lajunen, L.; Passoja, A.; Pyykkö, P. *J. Magn. Reson.* 1978, **31**, 121–132.
11. Cogne, A.; Grand, A.; Laugier, J.; Robert, J. B.; Wiesenfeld, L. *J. Am. Chem. Soc.* 1980, **102**, 2238–2242.
12. Jenkins, W. B.; McFarlane, W. *J. Chem. Soc., Chem. Commun.* 1977, 922–923.
13. Pyykkö, P.; Wiesenfeld, L. *Mol. Phys.* 1981, **43**, 557–580.
14. Code, R. F.; Ramsey, N. F. *Phys. Rev. A* 1971, **4**, 1945–1959.
15. Albrand, J. P.; Gagnaire, D. *J. Chem. Soc., Chem. Commun.* 1970, 874–875.
16. Buckingham, A. D.; McLauchlan, K. A. *Proc. Chem. Soc., London* 1963, 144.
17. Buckingham, A. D.; McLauchlan, K. A. *Prog. NMR Spectrosc.* 1967, **2**, 63–109.
18. Zumbulyadis, N.; Dailey, B. P. *Mol. Phys.* 1974, **27**, 633–640.
19. Shimizu, H. *J. Chem. Phys.* 1964, **40**, 3357–3364.
20. Anderson, J. M. *Mol. Phys.* 1964, **8**, 505–511.
21. Mackor, E. L.; MacLean, C. *J. Chem. Phys.* 1966, **44**, 64–69
22. Barfield, M.; Johnston, Jr., M. D. *Chem. Rev.* 1973, **73**, 53–73.
23. McFarlane, W. *Ann. Rep. NMR Spectrosc.* 1972, **5A**, 353–393.
24. McFarlane, W. *Q. Rev. Chem. Soc.* 1969, **23**, 187–203.
25. Cohen, A. D.; Freeman, R.; McLauchlan, K. A.; Whiffen, D. H. *Mol. Phys.* 1963, **7**, 45–56
26. McFarlane, W. *Ann. Rep. NMR Spectrosc.* 1968, **1**, 135–163.
27. Hoffman, R. A.; Forsen, S. *Prog. NMR Spectrosc.* 1966, **1**, 15–204.
28. Berkhoudt, T.; Jakobsen, H. J. *J. Magn. Reson.* 1982, **50**, 323–327.
29. Hansen, P. E.; Berg, A.; Jakobsen, H. J.; Manzara, A. P.; Michl, J. *Org. Magn. Reson.* 1977, **10**, 179–187.
30. Finkelmeier, H.; Luttke, W. *J. Am. Chem. Soc.* 1978, **100**, 6261–6262.
31. Pomerantz, M.; Fink, R.; Gray, G. A. *J. Am. Chem.* 1976, **98**, 291–292.
32. Della, E. W.; Patney, H. K. *Aust. J. Chem.* 1979, **32**, 2243–2248.
33. McFarlane, W.; Nash, J. A. *J. Chem. Soc., Chem. Commun.* 1969, 127–128.
34. Mavel, G. *Annu. Rep. NMR Spectrosc.* 1973, **5B**, 1–441.
35. Erickson, L. E.; Sarneski, J. E.; Reilley, C. N. *Inorg. Chem.* 1975, **14**, 3007–3017.
36. Schultz, C. W.; Rudolph, R. W. *J. Am. Chem. Soc.* 1971, **93**, 1898–1903.
37. Tau, K. D.; Uriarte, R.; Mazanec, T. J.; Meek, D. W. *J. Am. Chem. Soc.* 1979, **101**, 6614–6619.
38. Hilton, J.; Sutcliffe, L. H. *Prog. NMR Spectrosc.* 1975, **10**, 27–39.

39 Ewing, D F *NMR Spec Per Rep RSC* chapter on Applications of Spin–spin Coupling, 1980 –, especially volumes 9–11
40 Buckingham, A D, Cordle, J E *J Chem Soc, Faraday Trans* **2**, 1974, **70**, 994–1004
41 Mancini, M, Bougeard, P, Burns, R C, Mlekus, M, Sayer, B G, Thompson, J I A, McGlinchey, M J *Inorg Chem* 1984, **23**, 1022
42 Masters, A F, Bossard, G E, Georfe, T A, Brownlee, R T C, O'Connor, M J, Wedd, A G *Inorg Chem* 1983, **22**, 908
43 Franklin, K J, Lock, C J L, Sayer, B G, Schrobilgen, G J *J Am Chem Soc* 1982, **104**, 5303
44 McFarlane, W *J Chem Soc, Chem Commun* 1967, 772–773
45 Anderson, D W W, Ebsworth, E A V, Rankin, D W H *J Chem Soc, Dalton Trans* 1973, 2370–2373
46 Hyde, E M, Kennedy, J D, Shaw, B L, McFarlane, W *J Chem Soc, Dalton Trans* 1977, 1571–1576
47 Kidd, R G, Goodfellow, R J In *NMR and the Periodic Table*, Harris, R K, Mann, B E, Ed, Academic Press London, 1978, Chapter 8
48 Mann, B E, Masters, C, Shaw, B L *J Chem Soc, Chem Commun* 1970, 1041
49 Vancea, L, Graham, W A G *J Organomet Chem* 1977, **134**, 219–227
49a Sebald, A, Wrackmeyer, B, Theocaris, C R, Jones, W *J Chem Soc Dalton Trans* 1984, 747
50 Whipple, E B, Brown, T H, Farrar, T C, Coyle, T D *J Chem Phys* 1965, **43**, 1841–1842
51 Fusstetter, H, Noth, H, Wrackmeyer, B, McFarlane, W *Chem Ber* 1977, **110**, 3172–3182
52 Hinton, J F, Briggs, R W In *NMR and the Periodic Table*, Harris, R K, Mann, B E, Ed, Academic Press London, 1978, Chap 9
53 Buckingham, A D, McLauchlan, K A *Proc Chem Soc, London* 1963, 144
54 McFarlane, W, Rycroft, D S *J Organomet Chem* 1974, **64**, 303–305
55 Danyluk, S L *J Am Chem Soc* 1964, **86**, 4504–4505
56 McFarlane, W *J Chem Soc A* 1967, 1275–1276
57 Pfisterer, G, Dreeskamp, H *Ber Bunsenges Phys Chem* 1969, **73**, 654–661
58 Schumann, C, Dreeskamp, H *J Magn Reson* 1970, **3**, 204–217
59 Dreeskamp, H *Z Naturforsch A* 1964, **19**, 139–142
60 Johannesen, R B *J Chem Phys* 1968, **48**, 1414–1415
61 McFarlane, W, Dean, R R *J Chem Soc A* 1968, 1535–1538
62 Chuck, R J, Gillies, D G, Randall, E W *Mol Phys* 1969, **16**, 121–130
63 Lynden-Bell, R M *Trans Faraday Soc* 1961, **57**, 888–892
64 Manatt, S L, Juvinall, G L, Wagner, R I, Elleman, D D *J Am Chem Soc* 1966, **88**, 2689–2697
65 Albrand, J P, Gagnaire, D, Robert, J P *Bull Soc Chim, Fr* 1968, 479–482
66 McFarlane, W *Proc R Soc London Ser A* 1968, **306**, 185–199
67 Dreeskamp, H, Elser, H, Schumann, C *Ber Bunsenges Phys Chem* 1966, **70**, 751–755
68 McFarlane, W *J Chem Soc A* 1967, 1148–1150
69 Burnett, L J, Zeltmann, A H *J Chem Phys* 1972, **56**, 4695–4701
70 Balimann, G, Pregosin, P S *J Magn Reson* 1977, **26**, 283–289
71 Olah, G A, Kiovsky, T E *J Am Chem Soc* 1968, **90**, 4666–4672
72 McFarlane, W *J Chem Soc, Chem Commun* 1967, 963–964
73 Dreeskamp, H, Pfisterer, G *Mol Phys* 1968, **14**, 295–297
74 Rodger, C, Sheppard, N, McFarlane, C, McFarlane, W In *NMR and the Periodic Table*, Harris, R K, Mann, B E, Ed, Academic Press London, 1978, Chapter 12
75 Landolt-Bornstein Zahlenwerte und Funktionen aus Naturwissenschaften und Technik, Neue Serie, Group II, Springer-Verlag Berlin, 1974, Vol 6, pp 2-398–2-401
76 Lindman, B, Forsen, S In *NMR Basic Principles and Progress*, Diehl, P, Fluck, E, Kosfeld, R, Ed, Springer-Verlag Berlin, 1976, Vol 12, pp 1–365
77 Wehrli, F W *Annu Rep NMR Spectrosc* 1979, **9**, 123–219
78 Feeney, J, Haque, R, Reeves, L W, Yue, C P *Can J Chem* 1968, **46**, 1389–1398
79 Andrew, E R, Hinshaw, W S *Phys Lett A* 1973, **43**, 113–114
80 Andrew, E R, Carolan, J L, Randall, P J *Phys Lett* 1971, **37a**, 125–126
81 Pregosin, P S *Annu Rep NMR Spectrosc* 1981, **11A**, 227–271
82 Fumagalli, A, Martinengo, S, Chini, P, Albinati, A, Bruckner, S, Heaton, B T *J Chem Soc, Chem Commun* 1978, 195–196
82a van der Ploeg, A F M H, van Koten, G, Brevard, C *Inorg Chem* 1982, **21**, 2878
82b Gillespie, R J, Granger, P, Morgan, K R, Schrobilgen, G J *Inorg Chem* 1984, **23**, 887

83 Albright, M J, Oliver, J P *J Organomet Chem* 1979, **172**, 99–107
84 Sayer, B G, Thompson, J I A, Nguyen, H, Birchall, T, Eaton, D R, McGlinchey, M J *Inorg Chem* 1981, **20**, 3748–3750
85 Iwayanagi, T, Saito, Y *Chem Lett* 1976, 1193–1196
86 Dreeskamp, H, Hildenbrand, K *Z Naturforsch A* 1968, **23**, 940–941
87 Hildenbrand, K, Dreeskamp, H *Z Phys Chem (Frankfurt)* 1970, **69**, 171–182
88 McLauchlan, K A, Wiffen, D H, Reeves, L W *Mol Phys* 1966, **10**, 131–140
89 Anet, F A L, Sudmeier, J L *J Magn Reson* 1969, **1**, 124–138
90 Kennedy, J D, McFarlane, W, Pyne, G S, Wrackmeyer, B *J Chem Soc, Dalton Trans* 1975, 386–390
91 Mann, B E, Taylor, B F *^{13}C NMR Data for Organometallic Compounds*, Academic Press London, 1981, 336 pp
92 Kordize, A A, Kizas, O A, Astakhova, N M, Petrovskii, P V, Grishin, Y K *J Chem Soc, Chem Commun* 1981, 853–855
93 Williams, E A, Cargioli, J D *Annu Rep NMR Spectrosc* 1979, **9**, 221–318
94 Starzewski, K A O, Pregosin, P S *Adv Chem Ser* 1982, **196**, 23–41
94a Farrugia, L J, James, B R, Lassigne, C R, Wells, E F, *Can J Chem* 1982, **60**, 1304, Kretschmer, M, Pregosin, P S, Ruegger, H *J Organomet Chem* 1983, **241**, 87, Kretschmer, M, Pregosin, P S, Garralda, M *J Organomet Chem* 1983, **244**, 175
95 Farrugia, L J, James, B R, Lassigne, C R, Wells, E J *Inorg Chim Acta* 1981, **53**, L261–263
95a Carr, S, Colton, S, Dakternieks, D *J Organomet Chem* 1982, **240**, 143
96 Goggin, P L, Goodfellow, R J, Knight, J R, Norton, M G, Taylor, B F *J Chem Soc, Dalton Trans* 1973, 2220–2226
97 McFarlane, W, Rycroft, D S *J Chem Soc, Chem Commun* 1973, 336
98 McFarlane, H C E, McFarlane, W, Rycroft, D S *J Chem Soc, Dalton Trans* 1976, 1616–1622
99 Colquhoun, I J, Grim, S O, McFarlane, W, Mitchell, J D, Smith, P H *Inorg Chem* 1981, **20**, 2516–2521
100 McFarlane, W *J Chem Soc A* 1967, 1922–1923
101 McFarlane, W, Rycroft, D S *J Chem Soc, Faraday Trans 2* 1974, **70**, 377–385
102 Pregosin, P S, Kunz, R W *NMR Basic Princ Prog* 1979, **16**, 1–156
103 Pachler, K G R, Chalmers, A A *NMR Spec Per Rep C S* 1979, **8**, 63–122
104 Rehder, D *Transition Met Chem (Weinheim, Ger)* 1977, **2**, 41–45
105 Rehder, D *J Magn Reson* 1977, **25**, 177–181
106 Alyea, E C, Dias, S A, Goel, R G, Ogini, W O, Pilon, P, Meek, D W *Inorg Chem* 1978, **17**, 1697–1700
107 Martin, G J, Martin, M L, Gouesnard, J P *NMR Basic Princ Prog* 1981, **18**, 1–382
107a Minelli, M, Hubbard, J L, Christensen, K A, Enemark, J H *Inorg Chem* 1983, **22**, 2652
108 McFarlane, W *J Chem Soc, Chem Commun* 1968, 755–756
109 Goggin, P L, Goodfellow, R J, Haddock, S R *J Chem Soc, Chem Commun* 1975, 176–177
110 Colquhoun, I J, McFarlane, W *J Chem Soc, Dalton Trans* 1981, 658–660
111 Kintzinger, J P *NMR Basic Princ Prog* 1981, **17**, 1 64
112 Brevard, C, Granger, P *J Chem Phys* 1981, **75**, 4175–4176
113 Buckingham, M J, Hawkes, G E, Thornback, J R *Inorg Chim Acta* 1981, **56**, L41–L42
114 Carson, G K, Dean, P A W *Inorg Chim Acta* 1982, **66**, 37–39
115 Colquhoun, I J, McFarlane, W *J Magn Reson* 1982, **46**, 525–528
116 McFarlane, W, Noble, A M, Winfield, J M *J Chem Soc, A* 1971, 948–953
117 Chambers, O R, Rycroft, D S, Sharp, D W A, Winfield, J M *Inorg Nucl Chem Lett* 1976, **12**, 559–561
118 Zosulin, A J, Jakobsen, H J, Moore, T F, Garber, A R, Odom, J D *J Magn Reson* 1980, **41**, 458–466
119 Schneider, W G, Buckingham, A D *Discuss Faraday Soc* 1962, **34**, 147–155
120 Biffar, W, Noth, H *Angew Chem Int Ed Engl* 1980, **19**, 58–59
121 Kennedy, J D, McFarlane, W, Wrackmeyer, B *Inorg Chem* 1976, **15**, 1299–1302
122 Burke, P J, Mathews, R W, Gillies, D G *J Organomet Chem* 1976, **118**, 129–134
123 Harris, R K, Kennedy, J D, McFarlane, W In *NMR and the Periodic Table*, Harris, R K, Mann, B E, Ed, Academic Press London, 1978, Chap 10
124 Yamamoto, O, Yanagisawa, M *Bull Chem Soc Jpn* 1978, **51**, 1231–1232
124a Wrackmeyer, B *J Magn Reson* 1983, **54**, 174

125. McFarlane, H. C. E.; McFarlane, W.; Rycroft, D. S. *J. Chem. Soc., Faraday Trans. 2* 1972, **68**, 1300–1305.
125a. Tarasov, V. P.; Privalov, V. I.; Kirakosyan, G. A.; Buslaev, Y. A. *Z. Naturforsch. B*, 1983, **38**, 1173.
126. Gillespie, R. J.; Hartman, J. S.; Parekh, M. *Can. J. Chem.* 1968, **46**, 1601–1610.
127. Lynden-Bell, R. M.; Sheppard, N. *Proc. R. Soc. London Ser. A* 1962, **269**, 385–403.
128. McLauchlan, K. A. *J. Chem. Soc., Chem. Commun.* 1965, 105–106.
129. Grant, D. M. *J. Am. Chem. Soc.* 1967, **89**, 2228–2229.
130. Dean, R. R.; McFarlane, W. *Mol. Phys.* 1967, **12**, 289–291.
131. Dreeskamp, H.; Stegmeier, G. *Z. Naturforsch. A* 1967, **22**, 1458–1464.
132. McFarlane, W. *J. Chem. Soc., A* 1967, 528–530.
133. Kennedy, J. D.; McFarlane, W. *J. Organomet. Chem.* 1975, **94**, 7–14.
134. Wrackmeyer, B. *J. Organomet. Chem.* 1979, **166**, 353–363.
135. Clark, R. J. H.; Davies, A. G.; Puddephatt, R. J.; McFarlane, W. *J. Am. Chem. Soc.* 1969, **91**, 1334–1339.
136. McFarlane, W. *Mol. Phys.* 1967, **13**, 587–588.
137. Elser, H.; Dreeskamp, H. *Ber. Bunsenges. Phys. Chem.* 1969, **73**, 619–624.
138. McFarlane, W. *J. Chem. Soc., A* 1968, 1630–1634.
139. Kennedy, J. D.; McFarlane, W. *J. Organomet. Chem.* 1974, **80**, C47–C49.
140. Kennedy, J. D.; McFarlane, W. *J. Chem. Soc., Chem. Commun.* 1974, 983–984.
141. Kennedy, J. D.; McFarlane, W. *J. Chem. Soc., Dalton Trans.* 1976, 1219–1223.
142. Wrackmeyer, B. *J. Magn. Reson.* 1981, **42**, 287–297.
143. McFarlane, W. *Mol. Phys.* 1966, **10**, 603–604.
144. McFarlane, W. *J. Chem. Soc. A* 1967, 1660–1662.
145. Paolillo, L.; Becker, E. D. *J. Magn. Reson.* 1970, **3**, 200–203.
146. Yeh, H. J. C.; Ziffer, H.; Jerina, D. M.; Boyd, D. R. *J. Am. Chem. Soc.* 1973, **95**, 2741–2743.
147. Koole, N. J.; Krol, D.; deBie, M. J. A. *J. Magn. Reson.* 1976, **21**, 499–502.
148. Bundgaard, T.; Jakobsen, H. J. *Tetrahedron Lett.* 1976, 1621–1624.
149. Friesen, K. J.; Wasylishen, R. E. *J. Magn. Reson.* 1982, **48**, 152–154.
150. Jennings, W. .B.; Boyd, D. R.; Watson, C. G.; Becker, E. D.; Bradley, R. B.; Jerina, D. M. *J. Am. Chem. Soc.* 1972, **94**, 8501–8504.
151. Christl, M. *Org. Magn. Reson.* 1975, **7**, 349–350.
152. Anderson, D. W. W.; Bentham, J. E.; Rankin, D. W. H. *J. Chem. Soc., Dalton Trans.* 1973, 1215–1217.
153. Ebsworth, E. A. V.; Edward, J. M.; Rankin, D. W. H. *J. Chem. Soc., Dalton Trans.* 1976, 1673–1677.
154. McFarlane, W. *J. Chem. Soc., Chem. Commun.* 1967, 58.
155. Albrand, J. P.; Gagnaire, D. *J. Chem. Soc., Chem. Commun.* 1970, 874–875.
156. Dreeskamp, H.; Schumann, C.; Stelzer, O. *J. Chem. Soc., Chem. Commun.* 1970, 619.
157. Sorensen, S.; Hansen, R. S.; Jakobsen, H. J. *J. Am. Chem. Soc.* 1972, **94**, 5900–5902.
158. Bundgaard, T.; Jakobsen, H. J. *Acta Chem. Scand.* 1972, **26**, 2548–2549.
159. Jakobsen, H. J.; Bundgaard, T.; Hansen, R. S. *Mol. Phys.* 1972, **23**, 197–201.
160. Lequan, R.-M.; Pouet, M.-J.; Simonnin, M.-P. *J. Chem. Soc., Chem. Commun.* 1974, 475.
161. Dreeskamp, H.; Schumann, C.; Schmutzler, R. *J. Chem. Soc., Chem. Commun.* 1970, 671–672.
162. Crosbie, K. D.; Sheldrick, G. M. *Mol. Phys.* 1971, **20**, 317–320.
163. McFarlane, W.; Rycroft, D. S. *J. Chem. Soc., Dalton Trans.* 1974, 1977–1979.
164. Wilkie, C. A. *J. Am. Chem. Soc.* 1978, **100**, 2288–2291.
165. McFarlane, W. *Mol. Phys.* 1967, **12**, 243–248.
166. McFarlane, W.; Rycroft, D. S.; Turner, C. J. *Bull. Soc. Chim. Belg.* 1977, **86**, 457–463.
167. Kennedy, J. D.; McFarlane, W. *J. Chem. Soc., Dalton Trans.* 1973, 2134–2139.
168. Tiers, G. V. D. *J. Am. Chem. Soc.* 1962, **84**, 3972–3973.
169. Johannesen, R. B. *J. Chem. Phys.* 1967, **47**, 955–960.
170. Johannesen, R. B. *J. Chem. Phys.* 1967, **47**, 3088–3089.
171. McFarlane, W.; Wood, R. J. *J. Chem. Soc., Chem. Commun.* 1969, 262.
172. Hayashi, S.; Hayamizu, K.; Yamamoto, O. *J. Magn. Reson.* 1980, **37**, 17–29.
173. Bhattacharyya, P. K.; Dailey. B. P. *J. Chem. Phys.* 1973, **59**, 5820–5823.
174. Schultheiss, H.; Fluck, E. *Z. Naturforsch. B* 1977, **32**, 257–264.
175. Cowley, A. H.; Schweiger, J. R.; Manatt, S. L. *J. Chem. Soc., Chem. Commun.* 1970, 1491–1492.
176. Schweiger, J. R.; Cowley, A. H.; Cohen, E. A.; Kroon, P. A.; Manatt, S. L. *J. Am. Chem. Soc.* 1974, **96**, 7122–7123.
177. Arnold, D. E. J.; Rankin, D. W. H. *J. Chem. Soc., Dalton Trans.* 1975, 889–894.
178. Arnold, D. E. J.; Rankin, D. W. H. *J. Chem. Soc., Dalton Trans.* 1976, 1130–1134.

179. Gray, G. A.; Albright, T. A. *J. Am. Chem. Soc.* 1976, **98**, 3857–3861.
180. McFarlane, W.; Wrackmeyer, B. *J. Chem. Soc., Dalton Trans.* 1976, 2351–2355.
181. Harris, R. K.; Hayter, R. G. *Can. J. Chem.* 1964, **42**, 2282–2291.
182. Finer, E. G.; Harris, R. K. *Mol. Phys.* 1967, **13**, 65–75.
183. Harris, R. K.; Finer, E. G. *J. Chem. Soc., Chem. Commun.* 1968, 110–112.
184. Harris, R. K.; Woplin, J. R.; Stec, W. *J. Chem. Soc., Chem. Commun.* 1970, 1391–1392.
185. Rudolph, R. W.; Newmark, R. A. *J. Am. Chem. Soc.* 1970, **92**, 1195–1199.
186. McFarlane, H. C. E.; McFarlane, W. *J. Chem. Soc., Chem. Commun.* 1971, 1589–1590.
187. McFarlane, H. C. E.; McFarlane, W.; Nash, J. A. *J. Chem. Soc. Dalton Trans.* 1980, 240–244.
188. McFarlane, H. C. E. and McFarlane, W. *J. Chem. Soc., Chem. Commun.* 1975, 582–583.
189. McFarlane, H. C. E.; McFarlane, W. *J. Chem. Soc., Chem. Commun.* 1978, 531–532.
190. McFarlane, W.; Nash, J. A. *J. Chem. Soc., Chem. Commun.* 1969, 913.
191. McFarlane, W.; Rycroft, D. S. *J. Chem. Soc., Chem. Commun.* 1972, 902–903.
192. McFarlane, W.; Rycroft, D. S. *J. Chem. Soc., Dalton Trans.* 1973, 2162–2166.
193. Anderson, D. W. W.; Ebsworth, E. A. V.; Meikle, G. D.; Rankin, D. W. H. *Mol. Phys.* 1973, **25**, 381–385.
194. Cogne, A. Ph.D. thesis, 1976, Grenoble, France (cited in ref. 13).
195. Manatt, S. L.; Elleman, D. D.; Cowley, A. H.; Burg, A. B. *J. Am. Chem. Soc.* 1967, **89**, 4544–4545.
196. Dean, R. R.; McFarlane, W. *J. Chem. Soc., Chem. Commun.* 1967, 840–841.
197. VanderHart, D. L.; Gutowsky, H. S.; Farrar, T. C. *J. Chem. Phys.* 1969, **50**, 1058–1065.
198. McFarlane, W.; Nixon, J. F.; Swain, J. R. *Mol. Phys.* 1970, **19**, 141–143.
199. Fukushima,. E. *J. Chem. Phys* 1971, **55**, 2463–2466; Morgan, K., Sayer, B. G.; Schrobilgen, G. J. *J. Magn. Reson.* 1983, **52**, 139–142.
200. Noggle, J. H.; Baldeschwieler, J. D.; Colburn, C. B. *J. Chem. Phys.* 1962, **37**, 182–189.
201. Harris, R. K. In *NMR and the Periodic Table*, Harris, R. K.; Mann, B. E., Ed.; Academic Press: London, 1978; Chap. 11.
202. Lassigne, C. R.; Wells, E. J. *J. Chem. Soc., Chem. Commun.* 1978, 956–957.
203. Schrobilgen, G. J.; Burns, R. C.; Granger, P. *J. Chem. Soc., Chem. Commun.* 1978, 957–960.
204. Fraser, G. W.; Peacock, R. D.; McFarlane, W. *Mol. Phys.* 1969, **17**, 291–296.
205. Bunker, P. R.; Hanes, G. R. *Chem. Phys. Lett.* 1974, **28**, 377–379.
205a. Sawyer, J. F.; Schrobilgen, G. J.; Sutherland, S. J. *J. Chem. Soc., Chem. Commun.* 1982, 210; *Inorg. Chem.* 1982, **21**, 4064.
206. Shamir, J.; Selig, H.; Samuel, D.; Reuben, J. *J. Am. Chem. Soc.* 1965, **87**, 2359–2360
207. Schrobilgen, G. J.; Holloway, J. H.; Granger, P.; Brevard, C. *Inorg. Chem.* 1978, **17**, 980–987.
208. Schrobilgen, G. J. *NMR and the Periodic Table*, Harris, R. K.; Mann, B. E., Ed.; Academic Press: London, 1978; Chap. 14.
209. Jameson, C. J.; Gutowsky, H. S. *J. Chem. Phys.* 1969, **51**, 2790–2803.
210. Wrackmeyer, B.; Biffar, W. *Z. Naturforsch. B* 1979, **34**, 1270–1274.
211. Tarasov, V. P.; Bakum, S. I. *J. Magn. Reson.* 1975, **18**, 64–68.
212. Pyykkö, P.; Pajanne, E.; Inokuti, M. *Int. J. Quantum Chem.* 1973, **7**, 785–806
213. Emsley, J. W.; Phillips, L.; Wray, V. *Prog. NMR Spectrosc.* 1975, **10**, 83–756.
214. Wray, V. *Annu. Rep. NMR Spectrosc.* 1980, **10B**, 1–507.
215. Jameson, C. J.; Gutowsky, H. S. *J. Chem. Phys.* 1964, **40**, 1714–1724.
216. Inglefield, P. T.; Reeves, L. W. *J. Chem. Phys.* 1964, **40**, 2425–2426.
217. Reeves, L. W. *J. Chem. Phys.* 1964, **40**, 2423–2424.
218. McFarlane, W.; Rycroft, D. S. *J. Chem. Soc., Chem. Commun.* 1973, 10–11.
219. Levy, G. C.; Lichter, R. L. *Nitrogen-15 Nuclear Magnetic Resonance Spectroscopy*, John Wiley: New York, 1979.
220. Mateescu, G. D.; Benedikt, G. M. *J. Am. Chem. Soc.* 1979, **101**, 3959–3960.
221. Lazzeretti, P.; Rossi, E.; Taddei, F.; Zanasi, R. *J. Chem. Phys.* 1982, **77**, 408–414.
222. Dean, P. A. W.; Hughes, M. K. *Can. J. Chem.* 1980, **58**, 180–190.
223. Grim, S. O.; Walton, E. D.; Satek, L. C. *Inorg. Chim. Acta* 1978, **27**, L115–L116.
224. Birchall, T.; Pereira, A. *J. Chem. Soc., Chem. Commun.* 1972, 1150–1151.
225. Flitcroft, N.; Kaesz, H. D. *J. Am. Chem. Soc.* 1963, **85**, 1377–1380.
226. Webster, J. R.; Jolly, W. L. *Inorg. Chem.* 1971, **10**, 877–879.
227. Birchall, T.; Pereira, A. R. *J. Chem. Soc., Dalton Trans.* 1975, 1087–1092.
228. Colquhoun, I. J.; McFarlane, H. C. E.; McFarlane, W.; Nash, J. A.; Keat, R.; Rycroft, D. S.; Thompson, D. G. *Org. Magn. Reson.* 1979, **12**, 473–475.
229. Marsmann, H. *NMR Basic Princ. Prog.* 1981, **17**, 65–235.

230 Colquhoun, I J, McFarlane, W *J Chem Soc, Chem Commun* 1980, 145–147
231 Allman, T, Goel, R G, Pilon, P *Can J Chem* 1979, **57**, 91–97
232 Green, M, Howard, J A K, Murray, M, Spencer, J L, Stone, F G A *J Chem Soc, Dalton Trans* 1977, 1509–1514
233 Brown, M P, Puddephatt, R J, Rashidi, M, Seddon, K R *J Chem Soc, Dalton Trans* 1978, 516–522
234 Cox, R H *J Magn Reson* 1979, **33**, 61–70, De Vos, D, Van Beelen, D C, Wolters, J *J Organomet Chem* 1979, **172**, 303–308
235 Mitchell, T N *J Organomet Chem* 1973, **59**, 189–196
236 Bresler, L S, Buzina, N A, Varshavsky, Yu S, Kiseleva, N V, Cherkasova, T G *J Organomet Chem* 1979, **171**, 229–235
237 March, F C, Mason, R, Scollary, G R *Aust J Chem* 1977, **30**, 2407–2415
238 Wray, V, Hansen, P E *Annu Rep NMR Spectrosc* 1981, **11A**, 99–181
239 Kohler, F H, Kalder, H J, Fischer, E O *J Organomet Chem* 1976, **113**, 11–22
240 Clark, D N, Schrock, R R *J Am Chem Soc* 1978, **100**, 6774–6776
241 Verkade, J G *Coord Chem Rev* 1972, **9**, 1–106
242 Schumann, H, Kroth, H J *Z Naturforsch B* 1977, **32**, 768–770
243 Kroshefsky, R D, Weiss, R, Verkade, J G *Inorg Chem* 1979, **18**, 469–472
244 Milbrath, D S, Verkade, J G, Clark, R J *Inorg Nucl Chem Lett* 1976, **12**, 921–924
245 Finer, E G, Harris, R K *Prog NMR Spectrosc* 1971, **6**, 61–118
246 Maciel, G E, McIver, J W, Ostlund, N S, Pople, J A *J Am Chem Soc* 1970, **92**, 1–11
247 Cavell, R G, Gibson, J A, The, K I *J Am Chem Soc* 1977, **99**, 7841–7847
248 The, K I, Cavell, R G *Inorg Chem* 1977, **16**, 2887–2894
249 English, A D, Meakin, P, Jesson, J P *J Am Chem Soc* 1976, **98**, 422–436
250 Morris, R J, Moss, K C *J Fluorine Chem* 1979, **13**, 551–559
251 Lunazzi, L, Brownstein, S *J Magn Reson* 1969, **1**, 119–123
252 Pidcock, A, Richards, R E, Venanzi, L M *J Chem Soc A* 1966, 1707–1710
253 Appleton, T G, Clark, H C, Manzer, L E *Coord Chem Rev* 1973, **10**, 335–422
254 Allen, F H, Sze, S N *J Chem Soc, A* 1971, 2054–2056
255 Nefedov, V I, Yarzhemsky, V G, Tarasov, V P *Chem Phys* 1976, **18**, 417–430
256 Motschi, H, Pregosin, P S, Venanzi, L M *Helv Chim Acta* 1979, **62**, 667–677
257 Gysling, H J, Zumbulyadis, N, Robertson, J A *J Organomet Chem* 1981, **209**, C41–C44
258 Hansen, P E *Annu Rep NMR Spectrosc* 1981, **11A**, 65–98
259 Buchanan, G W, Dawson, B A *Can J Chem* 1978, **56**, 2200–2204
260 Wasylishen, R E *Annu Rep NMR Spectrosc* 1977, **7**, 246–292
261 Axenrod, T, Mangiaracina, P, Watnick, C M, Wieder, M J, Bulusu, S *Org Magn Reson* 1980, **13**, 197–199
262 Buchanan, G W, Dawson, B A *Can J Chem* 1977, **55**, 1437–1439
263 Sorensen, S, Hansen, R S, Jakobsen, H J *J Am Chem Soc* 1973, **95**, 5080–5081
264 Sorensen, O W, Scheibye, S, Lawesson, S O *Org Magn Reson* 1981, **16**, 322–324
265 Hargis, J H, Worley, S D, Jennings, W B, Tolley, M S *J Am Chem Soc* 1977, **99**, 8090–8091
266 Bulloch, G, Keat, R, Rycroft, D S *J Chem Soc, Dalton Trans* 1978, 764–769
267 Simonnin, M P, Lequan, R M, Wehrli, F W *J Chem Soc, Chem Commun* 1972, 1204–1205
268 Dewan, J C, Jennings, W B, Silver, J, Tolley, M S *Org Magn Reson* 1978, **11**, 449–452
270 Reich, H J, Trend, J E *J Chem Soc, Chem Commun* 1976, 310–311
271 Jakobsen, H J, Brey, W S *J Chem Soc, Chem Commun* 1979, 478–479
272 Schwotzer, W, Leuenberger, C, Hoesch, L, Dreiding, A S, Philipsborn, W V *Org Magn Reson* 1977, **9**, 382–384
273 Colquhoun, I J, Grim, S O, McFarlane, W, Mitchell, J D *J Magn Reson* 1981, **42**, 186–192
274 Keat, R, Monojlovic-Muir, L, Muir, K W, Rycroft, D S *J Chem Soc, Dalton Trans* 1981, 2192–2198
275 Porter, N A, Dubay, G R, Green, J G *J Am Chem Soc* 1978, **100**, 920–925
276 Albrand, J P, Robert, J B *J Chem Soc, Chem Commun* 1974, 644–645
277 Baudler M, Tolls, E, Clef, E, Kloth, B, Koch, D *Z Anorg Allg Chem* 1977, **435**, 21–32
278 Albrand, J P, Gagnaire, D, Robert, J B *J Am Chem Soc* 1973, **95**, 6498–6500
279 Stec, W J, Zielinski, W S *Tetrahedron Lett* 1980, **21**, 1361–1364
280 MacDonell, G D, Berlin, K D, Baker, J R, Ealick, S E, van der Helm, D, Marsi, K L *J Am Chem Soc* 1978, **100**, 4535–4540

281 Rodehuser, L , Rubini, P R , Delpuech, J S *Inorg Chem* 1977, **16**, 2837–2840
282 Hoad, C S , Matthews, R W , Thakur, M M , Gillies, D G *J Organomet Chem* 1977, **124**, C31–C36
283 Hansen, P E , Poulsen, D K , Berg, A *Org Magn Reson* 1976, **8**, 632–637
284 Colquhoun, I J , McFarlane, W *J Chem Soc*, *Dalton Trans* 1977, 1674–1679
285 Niecke, E , Flick, W *Angew Chem Int Ed Engl* 1974, **13**, 134–135
286 Earl, W L , Niederberger, W *J Magn Reson* 1977, **27**, 351–353
287 Randall, E W , Gillies, D G *Prog NMR Spectrosc* 1971, **6**, 119–174
288 Muller, H , Rosch, L , Erb, W , Zeisberg, R *J Organomet Chem* 1977, **140**, C17–C20
289 Kiffen, A A , Masters, C , Visser, J P *J Chem Soc*, *Dalton Trans* 1975, 1311–1315
290 Blecher, A , Mathiasch, B , Mitchell, T N *J Organomet Chem* 1980, **184**, 175–180
291 Kang, D K , Servis, K L , Burg, A B *Org Magn Reson* 1971, **3**, 101–112
292 Gillespie, R J , Schrobilgen, G J *J Chem Soc*, *Chem Commun* 1974, 90–92
293 Gillespie, R J , Netzer, A , Schrobilgen, G J *Inorg Chem* 1974, **13**, 1455–1459
294 Guns, M F , Claeys, E G , Van der Kelen, G P *J Mol Struct* 1979, **53**, 45–53
295 Karplus, M , Anderson, D H *J Chem Phys* 1959, **30**, 6–10
296 Maciel, G E , McIver, J W , Ostlund, N S , Pople, J A *J Am Chem Soc* 1970, **92**, 4151–4157
297 Hamner, E R , Kemmitt, R D W , Smith, M A R *J Chem Soc*, *Dalton Trans* 1977, 261–265
298 Carty, A J , Jacobson, S E *J Chem Soc*, *Chem Commun* 1975, 175–176
299 Carty, A J *Inorg Chem* 1976, **15**, 1956–1959
300 Meij, R , Stufkens, D J , Vrieze, K , Roosendaal, E , Schenk, H *J Organomet Chem* 1978, **155**, 323–336
301 Ernst, L *J Chem Soc*, *Chem Commun* 1977, 375–376
302 Butler, G , Eaborn, C , Pidcock, A *J Organomet Chem* 1979, **181**, 47–59
303 Pregosin, P S , Sze, S N *Helv Chim Acta* 1978, **61**, 1848–1855
304 Eaborn, C , Pidcock, A , Steele, B R *J Chem Soc*, *Dalton Trans* 1976, 767–776
305 Dingle, T W , Dixon, K R *Inorg Chem* 1974, **13**, 846–851
306 Eaborn, C , Pidcock, A , Steele, B R *J Chem Soc*, *Dalton Trans* 1975, 809–813
307 Cairns, M A , Dixon, K R , McFarland, J J *J Chem Soc*, *Dalton Trans* 1975, 1159–1164
308 Pomeroy, R K , Vancea, L , Calhoun, H P , Graham, W A G *Inorg Chem* 1977, **16**, 1508–1514
309 Flood, T C , Rosenberg, E , Sarhangi, A *J Am Chem Soc*, 1977, **99**, 4334–4340
310 Miles, W J , Clark, R J *J Organomet Chem* 1977, **131**, 93–104
311 Braun, R W , Cowley, A H , Cushner, M C , Lagow, R J *Inorg Chem* 1978, **17**, 1679–1681
312 Brownstein, S *Inorg Chem* 1973, **12**, 584–589
313 Nixon, J F , Swain, J R *J Chem Soc*, *A* 1970, 2075–2080
314 Bacon, J , Dean, P A W , Gillespie, R J *Can J Chem* 1970, **48**, 3413–3424
315 Brownstein, S *Can J Chem* 1969, **47**, 605–609
316 Cowley, A H , Dewar, M J S , Jackson, W R , Jennings, W B *J Am Chem Soc* 1970, **92**, 1085–1086
317 Karplus, M *J Chem Phys* 1959, **30**, 11–15
318 Haasnoot, C A G , DeLeeuw, F A A M , Altona, C *Tetrahedron* 1980, **36**, 2783–2792
319 Kennedy, J D , McFarlane, W , Puddephatt, R J *J Chem Soc*, *Dalton Trans* 1976, 745–748
320 Steigel, A *NMR Basic Princ Prog* 1978, **15**, 1–53
321 Albrand, J P , Robert, J B , Goldwhite, H *Tetrahedron Lett* 1976, 949–952
322 Ciriano, M , Green, M , Howard, J A K , Proud, J , Spencer, J L , Stone, F G A , Tsipis, C A *J Chem Soc*, *Dalton Trans* 1978, 801–808
323 Murrell, J N , Turpin, M A , Ditchfield, R *Mol Phys* 1970, **18**, 271–274
324 Smaller, B , Yasaitis, E , Anderson, H L *Phys Rev* 1951, **81**, 896–897
325 McFarlane, W , Rycroft, D S *Mol Phys* 1972, **24**, 893–895
326 Diehl, P , Leipert, Th *Helv Chim Acta* 1964, **47**, 545–557
327 Borisenko, A A , Sergeyev, N M , Ustynyuk, Y A *Mol Phys* 1971, **22**, 715–719
328 Jameson, A K , Jameson, C J *J Magn Reson* 1978, **32**, 455–457
329 Jakobsen, H J , Zozulin, A J , Ellis, P D , Odom, J D *J Magn Reson* 1980, **38**, 219–227
330 Banwell, C N , Murrell, J N , Turpin, M A *J Chem Soc*, *Chem Commun* 1968, 1466–1468
331 Ebsworth, E A V , Turner, J J *J Chem Phys* 1962, **36**, 2628–2634
332 Fraser, R R , Petit, M A , Miskow, M *J Am Chem Soc* 1972, **94**, 3253–3254
333 James, B D , Smith, B E , Newman, R H *J Chem Soc*, *Chem Commun* 1974, 294–295, Smith, B E , James, B D , Peachey, R M *Inorg Chem* 1977, **16**, 2057–2062

334. Oxton, I. A.; McInnes, A. G.; Walter, J. A. *Can. J. Chem.* 1979, **57**, 503–505.
335. Murray, M. *J. Magn. Reson.* 1973, **9**, 326–328.
336. Jameson, C. J.; Osten, H. J. *J. Am. Chem. Soc.* 1986, **108**, 2497–2503.
337. Barron, P. F.; Doddrell, D.; Kitching, W. *J. Organomet. Chem.* 1977, **132**, 351–358.
338. Morishima, I.; Inubushi, T.; Uemura, S.; Miyoshi, H. *J. Am. Chem. Soc.* 1978, **100**, 354–356.
339. Schramm, C.; Zink, J. I. *J. Magn. Reson.* 1977, **26**, 513–521.
340. Schaefer, T.; Hutton, H. M.; Salman, S. R. *Can. J. Chem.* 1979, **57**, 1877–1880.
341. Holzl, F.; Wrackmeyer, B. *J. Organomet. Chem.* 1979, **179**, 397–401.
342. Clark, J. H.; Emsley, J.; Middleton, T. B. *J. Chem. Soc., Dalton Trans.* 1979, 1693–1696.
343. Jameson, A. K.; Moyer, J. W.; Jameson, C. J. *J. Chem. Phys.* 1978, **68**, 2873–2877.
344. Kowalewski, J. *Prog. NMR Spectrosc.* 1977, **11**, 1–78.
345. Kowalewski, J. *Annu. Rep. NMR Spectrosc.* 1982, **12**, 82–176.
346. Pachler, K. G. R. *NMR Spec. Per. Rep. R.S.C.* 1981, **10**, 56–73.
347. Pachler, K. G. R. *NMR Spec. Per. Rep. R.S.C.* 1980, **9**, 49–67.
348. Barfield, M.; Grant, D. M. *Adv. Magn. Reson.* 1965, **1**, 149–194.
349. Murrell, J. N. *Prog. NMR Spectrosc.* 1971, **6**, 1–60.
350. Nakatsuji, H. *J. Chem. Phys.* 1974, **61**, 3728–3736.
351. Pople, J. A.; McIver, J. W.; Ostlund, N. S. *J. Chem. Phys.* 1968, **49**, 2960–2964.
352. McWeeny, R. *Phys. Rev.* 1962, **126**, 1028–1034.
353. Hiroike, E. *J. Phys. Soc. Jpn.* 1967, **22**, 379–392.
354. McConnell, H. M. *J. Chem. Phys.* 1956, **24**, 460–467.
355. Denis, A.; Malrieu, J. P. *Mol. Phys.* 1972, **23**, 581–600.
356. Ditchfield, R.; Murrell, J. N. *Mol. Phys.* 1968, **14**, 481–486.
357. Ditchfield, R. *Mol. Phys.* 1969, **17**, 33–39.
358. Blizzard, A. C.; Santry, D. P. *J. Chem. Soc., Chem. Commun.* 1970, 87–88.
359. Blizzard, A. C.; Santry, D. P. *J. Chem. Phys.* 1971, **55**, 950–963.
360. Blizzard, A. C.; Santry, D. P. *J. Chem. Phys.* 1973, **58**, 4714.
361. Pople, J. A.; McIver, J. W.; Ostlund, N. S. *Chem. Phys. Lett.* 1967, **1**, 465–466.
362. Pople, J. A.; McIver, J. W.; Ostlund, N. S. *J. Chem. Phys.* 1968, **49**, 2965–2970.
363. Loeve, P.; Salem, L. *J. Chem. Phys.* 1965, **43**, 3402–3403.
364. Ditchfield, R.; Ostlund, N. S.; Murrell, J. N.; Turpin, M. A. *Mol. Phys.* 1970, **18**, 433–440.
365. Barbier, C.; Gagnaire, D.; Berthier, G.; Levy, B. *J. Magn. Reson.* 1971, **5**, 11–22.
366. Barbier, C.; Levy, B.; Millie, P. *Chem. Phys. Lett.* 1972, **17**, 122–126.
367. Ditchfield, R.; Snyder, L. C. *J. Chem. Phys.* 1972, **56**, 5823–5829.
368. Maestro, M.; Moccia, R.; Zandomeneghi, M. *Chem. Phys. Lett.* 1974, **28**, 373–376.
369. Barfield, M.; Karplus, M. *J. Am. Chem. Soc.* 1969, **91**, 1–10.
370. Barfield, M. *J. Chem. Phys.* 1967, **46**, 811–812.
371. Barfield, M. *J. Chem. Phys.* 1968, **48**, 4458–4462.
372. Armour, E. A. G.; Stone, A. J. *Proc. R. Soc. London, Ser. A* 1967, **302**, 25–34.
373. Kowalewski, J.; Vestin, R.; Roos, B. *Chem. Phys. Lett.* 1971, **12**, 25–29.
374. Kowalewski, J.; Roos, B.; Siegbahn, P.; Vestin, R. *Chem. Phys.* 1974, **3**, 70–77; 1975, **9**, 29–39.
375. Laaksonen, A.; Saunders, V. R. *Chem. Phys. Lett.* 1983, **95**, 375.
376. Rayez-Meaume, M. T.; Hoarau, J. *Chem. Phys.* 1978, **27**, 73–78.
377. Guest, M. F.; Saunders, V. R.; Overill, R. E. *Mol. Phys.* 1978, **35**, 427–443.
378. Kowalewski, J.; Laaksonen, A.; Roos, B.; Siegbahn, P. *J. Chem. Phys.* 1979, **71**, 2896–2902.
379. Fukui, H.; Sanyoshi, A.; Miura, K. *J. Chem. Phys.* 1978, **69**, 943–944.
380. Laaksonen, A.; Kowalewski, J.; Siegbahn, P. *Chem. Phys. Lett.* 1980, **69**, 109–112.
381. Bothner-By, A. A. *Adv. Magn. Reson.* 1965, **1**, 195–316.
382. Raynes, W. T.; Riley, J. P. *Mol. Phys.* 1974, **27**, 337–343.
383. Emmanuel, R. V. *Mol. Phys.* 1970, **19**, 399–403.
384. Schulman, J. M.; Kaufman, D. N. *J. Chem. Phys.* 1972, **57**, 2328–2332.
385. Jokisaari, J.; Diehl, P. *Mol. Phys.* 1979, **37**, 981–983.
386. Lazzeretti, P.; Taddei, F.; Zanasi, R. *J. Am. Chem. Soc.* 1976, **98**, 7989–7993.
387. Schulman, J. M.; Venanzi, T. *J. Am. Chem. Soc.* 1976, **98**, 4701–4705.
388. Lee, W. S.; Schulman, J. M. *J. Am. Chem. Soc.* 1979, **101**, 3182–3184.
389. Lee, W. S.; Schulman, J. M. *J. Chem. Phys.* 1979, **70**, 1530–1537.
390. Pyykkö, P. *Adv. Quantum Chem.* 1978, **11**, 353–409.
391. Pyykkö, P.; Desclaux, J. P. *Acc. Chem. Res.* 1979, **12**, 276–281.

392. Breit, G. *Phys. Rev.* 1930, **35**, 1447–1451.
393. Pyykkö, P.; Jokisaari, J. *J. Chem. Phys.* 1975, **10**, 293–301.
394. Dalling, D. K.; Gutowsky, H. S. *J. Chem. Phys.* 1971, **55**, 4959–4966.
395. Pyykkö, P. *Chem. Phys.* 1977, **22**, 289–296.
396. Desclaux, J. P. *Comput. Phys. Commun.* 1975, **9**, 31–45.
397. Muller, N.; Pritchard, D. E. *J. Chem. Phys.* 1959, **31**, 1471–1476.
398. Schmidbaur, H.; Rathlein, K. H. *Chem. Ber.* 1973, **106**, 2491–2507.
399. Frei, K.; Bernstein, H. J. *J. Chem. Phys.* 1963, **38**, 1216–1226.
400. Walsh, A. D. *Discuss. Faraday Soc.* 1947, **2**, 18–25.
401. Bent, H. A. *Chem. Rev.* 1961, **61**, 275–311.
402. Juan, C. S.; Gutowsky, H. S. *J. Chem. Phys.* 1962, **37**, 2198–2208.
403. McConnell, H. M. *J. Chem. Phys.* 1955, **23**, 2454.
404. Lynden-Bell, R. M.; Harris, R. K. *Nuclear Magnetic Resonance Spectroscopy*, Appleton-Century-Crofts: New York, 1969; pp. 102–103.
405. Jameson, C. J. *J. Am. Chem. Soc.* 1969, **91**, 6232–6234.
406. Jameson, C. J.; Damasco, M. C. *Mol. Phys.* 1970, **18**, 491–504.
407. Emmanuel, R. V. *J. Chem. Phys.* 1970, **53**, 856–857.
408. Gray, G. A.; Albright, T. A. *J. Am. Chem. Soc.* 1977, **99**, 3243–3250.
409. Albrand, J. P.; Faucher, H.; Gagnaire, D.; Robert, J. B. *Chem. Phys. Lett.* 1976, **38**, 521–523.
410. McConnell, H. M. *J. Chem. Phys.* 1956, **24**, 764–766.
411. McConnell, H. M. *J. Mol. Spectrosc.* 1957, **1**, 11–17.
412. Karplus, M. *J. Chem. Phys.* 1960, **33**, 1842–1849.
413. Ditchfield, R.; Murrell, J. N. *Mol. Phys.* 1968, **15**, 533–539.
414. Wrackmeyer, B. *Z. Naturforsch. B* 1978, **33**, 385–389.
415. Fukui, H.; Tsuji, T.; Miura, K. *J. Am. Chem. Soc.* 1981, **103**, 3652–3653.
416. Fukui, H.; Miura, K.; Ohta, K.; Tsuji, T. *J. Chem. Phys.* 1982, **76**, 5169–5170.

Chapter 5

RELAXATION AND RELATED TIME-DEPENDENT PROCESSES

OLIVER HOWARTH

1. IMPORTANCE

The study of time-dependent processes adds a new dimension to NMR. It can reveal the courses and rates of chemical reactions, both directly and via the effect of such reactions on line shape. More subtly, it can yield a wealth of detail concerning the rates of energy transfer between spin quantum states, and also of precessional dephasing within one state.

A knowledge of these rates is important for many reasons. Perhaps the most important for the inorganic NMR spectroscopist is that they determine many of the conditions necessary for obtaining spectra. For example, the spin-lattice relaxation rate, $1/T_1$, represents the mean rate of transfer of energy between a spin and its surroundings. Even in the liquid phase at room temperature this rate can vary over many orders of magnitude. For the nucleus ^{33}S it may be $10^4\ s^{-1}$ or more, whereas for ^{29}Si and for many lighter metal nuclei it will frequently be $10^{-2}\ s^{-1}$ or less, and for ^{129}Xe gas it can be $10^{-4}\ s^{-1}$. As the optimum pulse repetition rate in Fourier-transform NMR is close to $1/T_1$, it is clear that one must usually know the latter at least approximately before attempting any experiment.

An accurate knowledge of relaxation parameters, together with their temperature and magnetic field variation where relevant, can also yield many details of the underlying molecular motions upon which these parameters depend. The parameters are first factorized into contributions from distinct physical interactions, and from there it is often possible to estimate rates of overall molecular motion and even rates and average extents of internal motions. In some cases the relaxation data can also reveal chemical shift anisotropies, and couplings that would otherwise be hidden. When selective relaxation data are measurable for a set of neighboring protons, these can yield bond lengths directly. Indeed, a single selective irradiation experiment can often demonstrate the proximity of two nuclei which are not scalar-coupled to each other.

An important extension of such experiments is to perturb the relaxation processes by introducing an electron spin, in the form of a shift or relaxation agent. This may

OLIVER HOWARTH • Department of Chemistry, University of Warwick, Coventry CV4 7AL, England.

serve simply to shorten T_1 or to simplify a spectrum, but a more sophisticated use is to obtain the average distances of atoms from the paramagnetic center.

One further application of T_1 measurements is of growing importance. NMR spin-imaging methods can be adapted to detect the spatial variation of T_1 and T_2, typically that of protons, throughout a body. In this way one can detect objects such as tumors which otherwise merge with their surroundings.

This chapter attempts to outline the principles and practice of relaxation and dynamic measurements. Chapter 2 contains introductory material, and later chapters contain more detailed examples.

2. DEFINITIONS

The most important relaxation parameters are T_1, the spin-lattice relaxation time, T_2, the spin–spin relaxation time, and the nuclear Overhauser enhancement, or NOE, where present. A second spin-lattice relaxation time, $T_{1\rho}$, may also be measured in the rotating frame via spin-locking experiments (Section 2.4), and T_1 itself may be subdivided into separate contributions as indicated above. Also the NOE depends upon which other nuclei are irradiated.

2.1. Macroscopic Definition of T_1 and T_2: Bloch Equations

Both T_1 and T_2 were defined in the earliest days of NMR in terms of an analysis based on the classical mechanics of forced, damped oscillators.[1] This explained NMR absorption and dispersion and line shapes in terms of the motion of the macroscopic magnetic dipole which arises from the combined dipoles of the relevant nuclei. Bloch successfully accounted for these three fundamental characteristics of NMR spectroscopy by assuming that the magnetic dipole, in addition to its stimulated motions, has an inherent tendency to decay exponentially. The decay in the magnetic field direction, conventionally z, is towards the static magnetization M_z, and has a time constant T_1 for decay to $1/e$ of its original value. The corresponding decay in any direction normal to the field is towards a zero magnetization component, and has the time constant T_2, which is not necessarily the same as T_1.

The equations that Bloch constructed using these definitions are solved in many texts[2,3] (cf. also p. 19ff.). Their most important prediction is that the nuclear magnetic resonance measured under the usual absorption-mode conditions in the presence of a radiofrequency field $2B_1 \cos \omega t$ will be proportional to

$$\omega_0 T_2 B_1 / [1 + (\omega - \omega_0)^2 T_2^2 + \gamma^2 B_1^2 T_1 T_2] \quad (1)$$

where ω_0 is the Larmor frequency in rad s^{-1} and γ is the magnetogyric ratio. Thus the linewidth at half height is $1/\pi T_2$ Hz at low B_1, and T_1 may in principle be measured once T_2 and B_1 are known. For the derivative-mode spectra obtainable on some older CW spectrometers the corresponding peak-to-peak separation is $1/\sqrt{3}\pi T_2$, plus any contribution from modulation.

Equation (1) also explains why a knowledge of T_1 and T_2 determines the optimum setting of B_1 in CW NMR. Related considerations apply to Fourier-transform NMR.

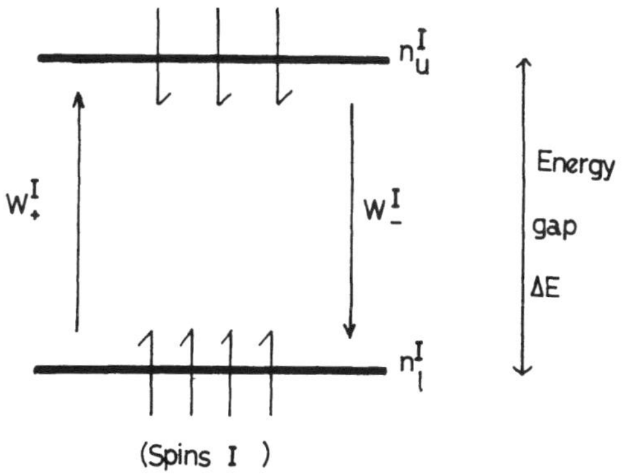

Figure 1. Transition probabilities and populations for a single $I = 1/2$ spin.

2.2. Microscopic Interpretation

M_z, the parallel component of bulk nuclear magnetization, is determined by the quantized nuclear spin orientations and populations, whereas the normal components M_x and M_y may be thought of as arising from spins precessing with nonrandom phase. Thus T_1 relaxation is determined only by transitions which change the nuclear spin state and hence energy, whereas T_2 will also be affected by isoenergetic processes affecting only the phase angle of the nuclear wave function, such as chemical exchange or sample flow. Indeed T_2 is often limited by the field inhomogeneity over the sample volume, because spins precessing at slightly different Larmor frequencies soon change in relative phase so as to lose their overall coherence. T_2 may also be affected by overmodulation of B_0 in older spectrometers.[4] But these last contributions to T_2 are so variable and so trivial that authors often refer to a T_2 value which includes them as T_2^*. The macroscopic definition of T_1 may readily be related to microscopic transition rates for a two-level system.[5] In Figure 1, n_u^I and n_l^I are the spin populations of the upper and lower levels, respectively, and W_+^I and W_-^I the transition probabilities for gain or loss of spin energy. Now

$$W_+^I/W_-^I = \exp(-\gamma\hbar B_0/kT) = n_u^I/n_l^I \qquad (2)$$

The first equality arises because W_+^I is less than W_-^I in proportion to the slightly lower availability of lattice phonons with excess energy $\gamma_I\hbar B_0$, which are necessary for the W_+^I transition. The second equality arises because each spin population is dynamically constant, so that

$$W_+^I n_l^I = W_-^I n_u^I \qquad (3)$$

But the nuclear magnetization is proportional to $(n_l^I - n_u^I)$. A simple kinetic argument shows that after perturbation from equilibrium this magnetization decays exponentially to its Boltzmann value, with decay constant

$$1/T_1^I = W_+^I + W_-^I \qquad (4)$$

2.2.1. The Two-Spin System: Cross Relaxation

When two or more physically interacting spins are present in a molecule, such as with a $^{13}C-{}^1H$ fragment, or with physically close protons in an organic moiety, then

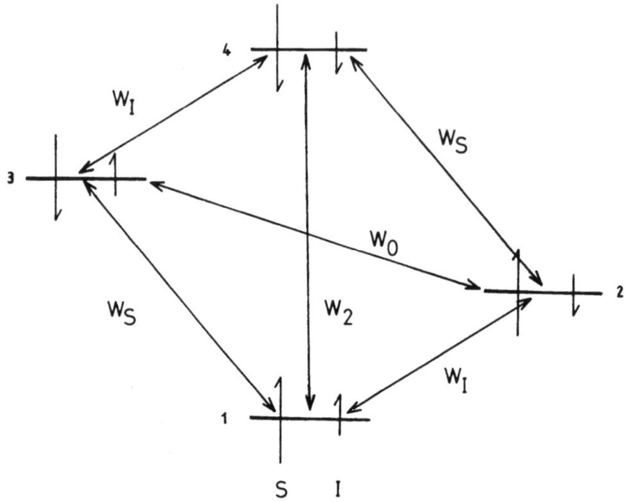

Figure 2. Averaged transition probabilities for two $I=1/2$ spins, I and S.

the simple two-level considerations of the previous section are not valid, and indeed the relaxation of one or more spins may not even be exponential. This may be illustrated by an analysis[6] of a system containing two interacting spins I and S, which has combined energy states as pictured in Figure 2. For the moment we ignore the small differences between the probabilities of absorption and emission within a given transition, so that our present analysis is only valid for recovery from some imagined very large perturbation to the populations. (A fuller analysis shows populations reverting at the same rate to their equilibrium populations rather than to zero.)

We can immediately write down four equations, one of which is not independent, to describe the four rates of population change.

$$dn_1/dt = (n_2 - n_1)\,W_I + (n_3 - n_1)\,W_S + (n_4 - n_1)\,W_2 \tag{5}$$

$$dn_2/dt = (n_1 - n_2)\,W_I + (n_3 - n_2)\,W_0 + (n_4 - n_2)\,W_S \tag{6}$$

$$dn_3/dt = (n_1 - n_3)\,W_S + (n_2 - n_3)\,W_0 + (n_4 - n_3)\,W_I \tag{7}$$

$$dn_4/dt = (n_1 - n_4)\,W_2 + (n_2 - n_4)\,W_S + (n_3 - n_4)\,W_I \tag{8}$$

We may now consider three distinct cases.

2.2.1a. I Observed While S is Irradiated Continuously. Here $n_1 = n_3$ and $n_2 = n_4$ because of the S irradiation. We wish to inspect the time evolution of the population difference $n_1 - n_2$, or equivalently $n_3 - n_4$. Now in this case, using the above equalities,

$$d(n_1 - n_2)/dt = -2(n_1 - n_2)\,W_I - (n_1 - n_2)\,W_2 - (n_1 - n_2)\,W_0 \tag{9}$$

This may be compared with the simpler case in the previous section, where only the $2W_I = W_+ + W_-$ term was present. In the present case the relation is once again exponential, but now

$$1/T_1^I = 2W_I + W_2 + W_0 \tag{10}$$

This is the normal expression for $1/T_1$ used in, for example, $^{13}C\{^1H\}$ NMR.

2.2.1b. I Alone Is Inverted and Observed. In this case we may note that provided all perturbations on I are independent of the state of *S*, then $n_1 - n_3 = n_2 - n_4$ throughout. This will be the case for all except very highly selective rf pulses, and has the consequence that the terms in W_S disappear from the expression for $d(n_1 - n_2)/dt$. However, an extra term does appear in $W_0 - W_S$, so that, on gathering terms in the difference of equations (5) and (6), we obtain

$$d(n_1 - n_2)/dt = -(n_1 - n_2)(2W_I + W_2 + W_0) - (n_1 - n_3)(W_2 - W_0) \qquad (11)$$

The extra term means that the relaxation is no longer exponential. However, the perturbation of I, for example by a 180° pulse, will not normally depend on the spin state of S, and in this case $n_1 = n_3$ and $n_2 = n_4$ initially. Thus the *initial* relaxation rate of I is the same as in the previous case, where S was continuously irradiated, with

$$1/T_1^I(\text{initial}) = 2W_I + W_2 + W_0 \qquad (12)$$

although the relaxation is subsequently slowed by replenishment of n_3 from n_2.

One example of this experiment is the selective inversion of a single resonance multiplet in a proton NMR spectrum. Although the recovery of this resonance is not in general exponential, the initial recovery rate may be related to W_I, W_0, and W_S, and hence to internuclear distances (see Section 4.1).

2.2.1c. I and S Are Both Subjected to a 180° Pulse. We also assume here for simplicity that I and S have essentially the same magnetogyric ratio. Once again the W_S terms in the difference of equations (5) and (6) disappear, giving

$$d(n_1 - n_2)/dt = -2(n_1 - n_2)\, W_I + (n_4 - n_1)\, W_2 + (n_3 - n_2)\, W_0 \qquad (13)$$

But now our initial condition is that $n_2 = n_3$. Hence the term in W_0 disappears and $n_4 - n_1 = -2(n_1 - n_2)$, so that the initial relaxation rate is given by

$$1/T^{I(S)}_{1(\text{initial})} = 2W_I + 2W_2 \qquad (14)$$

In the case of dipolar relaxation (see below) and of relatively rapid motion, it can be shown that $W_0 : W_2 : W_I - 2:12:3$, which means that the initial relaxation rate after full homonuclear spin inversion is 3/2 times that following selective inversion. The irradiation of the S spins has the effect of driving the relaxation of the I spins. If the experimental ratio is less than 3/2 then the presence of competing relaxation mechanisms is indicated. Therefore the "3/2 test" is a useful check before any attempt to convert ^1H relaxation data into structural information.[7]

2.3. Nuclear Overhauser Enhancement

If the relaxation of spin I is significantly affected by the motions of a different spin, S, then in general any deviation from Boltzmann equilibrium of the S spins will also render the I spin populations non-Boltzmann.[8] The resulting change in the intensity of resonance I is called a nuclear Overhauser enhancement (NOE) when spins I and S both belong to nuclei. The nuclear Overhauser enhancement factor is conventionally η, so that $1 + \eta_{IS}$ is defined as the intensity of the I resonance when the S spin populations are equalized, divided by the intensity of the I resonance when the

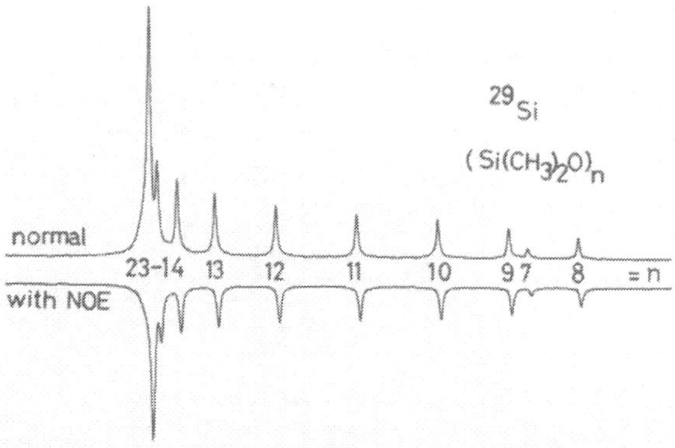

Figure 3. $^{29}Si\{^1H\}$ NOE of siloxane rings. The NOE is negative because γ_{Si} is negative. However, it is reduced, especially for the larger rings, because of a competing relaxation mechanism. Sample kindly loaned by Professor R. K. Harris.

S spin populations are at Boltzmann equilibrium. Figure 3 gives an example of the NOE effect.

η, like T_1, depends on the nature of the interaction between spins I and S, the possible presence of different types of interaction, and the time dependence of all these. It also invariably depends upon γ_S/γ_I, as the following argument illustrates.

2.3.1. A Simple Example

Let us suppose for simplicity that in the example outlined in Figure 2, W^I_+ and W^I_- are in fact totally dominated by the motions of spin *S*, in the specific sense that all W^I_+ transitions are the result of W^S_+ transitions, and all W^I_- of W^S_-. We may first check that equation (2) still holds, provided the *S* spins are at Boltzmann equilibrium. The joint (double quantum) transition probability $W^I_+ W^S_+$ is less than $W^I_- W^S_-$ by the fraction $\exp[-(\gamma_I+\gamma_S)\hbar B_0/kT]$, for the same reason as in the previous section. But the W^S_+ transitions that on occasion give rise to $W^I_+ W^S_+$ are also more probable than W^S_- transitions, because the S spin populations differ according to $n^S_l/n^S_u = \exp(\gamma_S \hbar B_0/kT)$. Hence W^I_+/W^I_- under these circumstances becomes equal to the product of the above two exponentials, i.e., $\exp(-\gamma_I \hbar B_0/kT)$ as anticipated.

However, if n^S_l is made equal to n^S_u, for example by irradiation of the S resonance before and possibly during observation of the I resonance, then the second exponential becomes equal to unity. Hence W^I_+/W^I_- rises to $\exp[-(\gamma_I+\gamma_S)\hbar B_0/kT]$, and so, using the relation $e^x \approx 1+x$ when x is small, we find that the fractional population difference $(W^I_- - W^I_+)/W^I_+$ now equals $(\gamma_I+\gamma_S)\hbar B_0/kT$, where beforehand it was just $\gamma_I \hbar B_0/kT$. Hence the intensity increase $1+\eta_{IS} = 1+\gamma_S/\gamma_I$. This expression of course only holds for the physically simple case described above. The general case is described below.

In general, I transitions will be affected in more complex ways by S transitions. Dipolar coupling alone will give an enhancement similar to the above, but reduced by some concomitant transitions such as $W^I_+ W^S_-$, as described below. If the interspin vector tumbles slowly, there will be a further reduction of NOE, and for some spin combinations, notably ^1H–^{19}F, the NOE may change sign as the molecular rotation slows. The case of the I–S interaction being predominantly scalar has so far only been identified when S is an electron spin, as in A. W. Overhauser's original experiments on metals.[9] It may turn out to be important for, e.g., ^{31}P – (spin 1/2 metal) couplings, for which commercial heteronuclear decoupling equipment is becoming available. Interestingly, only $W^{\mathrm{I}}_{+}W^{\mathrm{S}}_{-}$ and $W^{\mathrm{I}}_{-}W^{\mathrm{S}}_{+}$ transitions can occur via scalar coupling, so that in this case $\eta_{\mathrm{IS}} = -\gamma_{\mathrm{S}}/\gamma_{\mathrm{I}}$, with the opposite sign to that arising from dipolar coupling.

Nuclear Overhauser enhancements are not just an experimental curiosity. They are independent relaxation parameters which can be used to investigate molecular motions in detail. Also, the technique of measuring specific ^{1}H ^{1}H Overhauser enhancements by NOE difference spectroscopy detects spatial rather than chemical proximity and hence adds a powerful new tool to conformational and structural analysis.

2.3.2. *The More General Case*

We may now borrow the general two-spin expression for T_1 in the presence of S irradiation, equation (10), and extend it to include the small differences between absorptive and emissive transition probabilities.[(10)] If

$$i = \gamma_{\mathrm{I}}\hbar B_0/2kT \quad \text{and} \quad s = \gamma_{\mathrm{S}}\hbar B_0/2kT \tag{15}$$

then, for either I transition in Figure 3 we can write

$$\frac{n_1}{n_2} = \frac{n_3}{n_4} = e^{2i}\left(\frac{2W_{\mathrm{I}} + W_2 e^{s} + W_0 e^{-s}}{2W_{\mathrm{I}} + W_2 e^{-s} + W_0 e^{s}}\right) \tag{16}$$

Now $e^{\pm s} = 1 \pm s$ to a very good approximation, so that the term in parentheses becomes

$$\frac{s(W_2 - W_0) + 2W_{\mathrm{I}} + W_2 + W_0}{2W_{\mathrm{I}} + W_2 + W_0 + s(W_0 - W_2)} = 1 + 2s\frac{W_2 - W_0}{2W_{\mathrm{I}} + W_2 + W_0} = e^{2s\xi} \tag{17}$$

when we remember that $(a+x)/(a-x) = 1 + 2x/a$ when x is small, and write ξ for $(W_2 - W_0)/(2W_{\mathrm{I}} + W_2 + W_0)$. Therefore the population difference $n_2 - n_1$ equals $2i + 2s\xi$, whereas previously it was simply $2i$, and this means a nuclear Overhauser enhancement of

$$1 + \xi s/i = 1 + \xi\gamma_{\mathrm{S}}/\gamma_{\mathrm{I}} = 1 + \eta_{\mathrm{IS}} \tag{18}$$

In the simple case discussed previously, $W_0 = W_{\mathrm{I}} = 0$, making $\xi = +1$. This has no physical counterpart, although pure scalar coupling renders $W_2 = W_{\mathrm{I}} = 0$ and hence $\xi = -1$. In the more complex dipolar case, where $W_0{:}W_2{:}W_{\mathrm{I}} = 2{:}12{:}3$, one obtains $\xi = +1/2$. If other relaxation mechanisms are significant, these will increase W_{I} only, and hence reduce ξ towards zero and suppress any NOE.

2.4. Relaxation in the Rotating Frame: $T_{1\rho}$

If an equilibrium spin magnetization is subjected to a 90° pulse immediately followed by continuous rf irradiation having a 90° rf phase shift relative to the pulse, it will precess in exact alignment with the appropriate rotating magnetic component of the rf, and is then said to be spin-locked. Its decay after a given interval may readily be measured, and hence one obtains a relaxation time in the rotating frame, $T_{1\rho}$. In most cases $T_{1\rho} = T_2$, but even so the measurement of $T_{1\rho}$ can be preferable to measuring a linewidth, because the linewidth may be inflated by the overlap of separate resonances. More precisely, one may obtain detailed expressions for $T_{1\rho}$ by replacing the frequency-independent term τ_c in the expression for T_2 by $\tau_c/(1+\Omega^2\tau_c^2)$, where Ω is the relatively low Larmor precession frequency about the rf locking field.

3. MICROSCOPIC THEORY

We must now investigate how such transition probabilities as W_+^{I} and $W_+^{\mathrm{I}}\, W_-^{\mathrm{S}}$ are derived from first principles.(2,10–12) The underlying law is expressed in the (time-dependent) wave equation $i\hbar\,\partial\psi/\partial t = H(\psi)$, which is the quantum-mechanical version of Newton's second law.(13) Let us once again consider a simple two-level spin system, described by the spin wave function

$$\psi = C_l\psi_l \exp(-i\varepsilon_l t/\hbar) + C_u\psi_u \exp(-i\varepsilon_u t/\hbar) \tag{19}$$

C_u and C_l will in general be time-dependent, and subject to the constraint that $|C_u|^2+|C_l|^2=1$. When (19) is substituted into the wave equation then, in the presence of the original Hamiltonian H_0 plus a smaller interaction $Vf(t)$, the standard projection technique gives the equation

$$i\hbar\,\partial C_u/\partial t = C_l V_{ul} f(t)\exp[i(\varepsilon_u-\varepsilon_l)\,t/\hbar] \tag{20}$$

together with its counterpart for $\partial C_l/\partial t$, and their complex conjugates. Here $V_{ul} = \int \psi_u^* V\psi_l\, d\tau = \int \psi_l^* V^*\psi_u\, d\tau$. $Vf(t)$ also produces small time-dependent fluctuations in E_u and E_l, which affect T_2 but not T_1, and are ignored for the present. The time dependence of the small interaction is taken to be contained entirely within $f(t)$.

3.1. Spin-Lattice Relaxation

Experimentally, one typically observes the rate of lattice-induced energy absorption or loss

$$d\,|C_u|^2/dt = C_u^*\, dC_u/dt + C_u\, dC_u^*/dt = W_+ \text{ (per spin)} \tag{21}$$

For simplicity, we consider this as an absorption rate to be observed subject to the condition that C_u is very small, and hence $C_l=1$. (This implies some very efficient spin-polarization mechanism at work. A more general case, although algebraically more complex, is not intrinsically different.) Under these conditions equation (19) may readily be integrated to give

$$C_u = (V_{ul}/i\hbar)\int_0^t f(t')\exp(i\omega_{ul}t')\,dt' \tag{22}$$

where ω_{ul} is defined as $(E_u-E_l)/\hbar$. Note that in this equation t' is variable and t a fixed interval of time. Hence

$$\frac{d\,|C_u|^2}{dt} = (|V_{ul}|^2/\hbar^2)\left\{\int_0^t f(t')\,f(t)\exp[i\omega_{ul}(t-t')]\,dt' + \int_0^t f(t')\,f(t)\exp[-i\omega_{ul}(t-t')]\,dt'\right\} \tag{23}$$

Now if we let $t' = t+\tau$ (24)

we obtain

$$\frac{d\,|C_u|^2}{dt} = (|V_{ul}|^2/\hbar^2)\left[\int_{-t}^{0} f(t+\tau)\,f(t)\exp(-i\omega_{ul}\tau)\,d\tau + \int_{-t}^{0} f(t+\tau)\,f(t)\exp(i\omega_{ul}\tau)\,d\tau\right] \tag{25}$$

A change of variable from τ to $-\tau$ cannot affect the value of the first definite integral, which therefore becomes

$$\int_{0}^{t} f(t-\tau)\,f(t)\exp(i\omega_{ul}\tau)\,d\tau \tag{26}$$

So far this derivation has not specified $f(t)$. We now suppose $f(t)$ to be the time-dependence of some physical interaction (such as a scalar coupling), and hence to fluctuate in a random way. As a result, $f(t-\tau)\,f(t)$ will on average be the same as $f(t)\,f(t+\tau)$, because this average, $\overline{f(t)\,f(t+\tau)}$, is simply the extent to which the value of $f(t)$ correlates on average with its value τ seconds earlier. This average property is clearly independent of the moment of measurement t. $\overline{f(t)\,f(t+\tau)}$ is called the autocorrelation function of $f(t)$, and is conventionally written as $G(\tau)$ to emphasize its independence of t. Because $G(\tau)$ could equally well be expressed as $\overline{f(t-\tau)\,f(t)}$ the integral (25) may be combined (after averaging) with the second integral in (24) to give

$$\int_{-t}^{t} \overline{f(t+\tau)\,f(t)}\exp(i\omega_{ul}\tau)\,d\tau \tag{27}$$

Furthermore, (27) will become altogether independent of t once t exceeds the value beyond which $f(2t)$ no longer correlates significantly with $f(t)$. Thus the integral limits in (27) can be altered to $\pm\infty$, so that (27) simply becomes the expression for the Fourier transform of $G(\tau)$ evaluated at the frequency ω_{ul}. Equation (26) is normally described as $J(\omega_{ul})$, the spectral density of $f(t)$ at ω_{ul}, i.e.,

$$J(\omega_{ul}) = \int_{\infty}^{\infty} G(\tau)\exp(i\omega_{ul}\tau)\,d\tau \tag{28}$$

Finally, (6) becomes

$$1/T_1 = W_{l} + W = 2\,|V_{ul}|^2\,J(\omega_{ul})/\hbar^2 \tag{29}$$

3.2. Spin–Spin Relaxation

Spin transitions which give rise to spin-lattice relaxation by reducing the spin magnetization parallel to the field will also tend to reduce the magnetization perpendicular to the field. In a simple two-level system each spin jump is bound to alter the contribution to the parallel magnetization, but has only a 50% chance of altering that to the perpendicular magnetization. Therefore $1/T_2$ from this source alone equals $1/2T_1$. This relationship becomes more complex in systems with more levels.

In addition, $1/T_2$ is increased by time-dependent fluctuations in $E_u - E_l$, in close analogy to the line broadening produced by an exchange process in the fast-exchange limit. For many forms of interaction V in the limit of rapid tumbling motions, this second contribution to $1/T_2$ equals the first, so that overall $T_2 = T_1$. However, this is not the case, for example, when chemical shift anisotropy dominates the relaxation, and $T_2 = \frac{6}{7}T_1$ even in the rapid motion limit. More importantly, $T_2 \ll T_1$ when the molecular tumbling is slow, because the second contribution to $1/T_2$ increases linearly with the reciprocal rate of tumbling, just as it would with a reciprocal rate of chemical exchange. This is why NMR is not very generally applicable to slowly tumbling macromolecules, such as many *in vivo* biochemicals. The linewidths become too large.

A fuller introduction to the calculation of T_2 is given by Carrington and McLachlan.[14]

3.2.1. *Linewidths in Solids*

A rigid solid represents the slow-exchange limit in the analogy between T_2 and exchange-broadening, and in general any resonance will be broadened or split by dipolar interactions and chemical shift anisotropy (CSA). Nuclei of spin $>1/2$ will also be affected by interactions of their electric quadrupoles with local electric field gradients.

In theory, both the dipolar and the anisotropy broadening may be removed by spinning the sample at a rate greater than the broadenings and couplings at an angle of $\cos^{-1}(1/\sqrt{3})$ to B_0. This is because both dipolar coupling and CSA varies with angle θ to B_0 as $(3\cos^2\theta - 1)$. In practice, only CSA and the dipolar coupling of nuclei which are distant or with low γ may be removed in this way, because of limitations on achievable spinning speeds. The remaining dipolar coupling may, however, be removed by high-power rf decoupling, using a complex pulse sequence. Both these methods effectively convert the static part of $1/T_2$ into a contribution sensitive to lattice motions at the spinning frequency (or its pulse equivalent). In a rigid solid, such motions are small and so the resonance can be narrowed dramatically. However, even in a highly viscous fluid they are not small, and little narrowing results.

3.3. Dependence of Spectral Density upon Frequency

$G(\tau)$ is often, but not necessarily, a simple function decaying exponentially to zero on either side of $\tau = 0$ with a time constant τ_c, the "correlation time" for the motion under consideration. For example, random jumps of otherwise fixed interactions will give $G(\tau)$ the same first-order decay as occurs with random radioactive fission of nuclei. Random angular diffusion will achieve the same. In the exponential case, for any frequency ω,

$$J(\omega) = \int_{-\infty}^{\infty} \exp(-|\tau|/\tau_c)\exp(i\omega\tau)\,d\tau = \frac{2\tau_c}{1+\omega^2\tau_c^2} \tag{30}$$

This implies the well-known dependence of T_1 upon τ_c.[15] When $\tau_c \ll \omega$ then $T_1 \propto \tau_c$, but for slow motions where $\tau_c \gg \omega$, one obtains $T_1 \propto 1/\tau_c$. Minimum T_1 occurs when $\omega = 1/\tau_c$. This is illustrated in Figure 4. Note that in equations such as (29), ω is always in *radians*/s.

3.3.1. *More Complex Motions*

Many molecular motions are more complex than simple angular jumps or isotropic diffusion. Although this means that $G(\tau)$ will often not be a simple exponen-

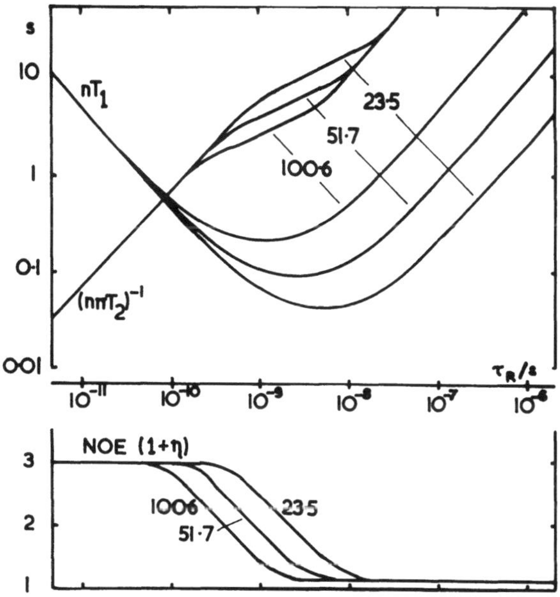

Figure 4. T_1, T_2, and NOE for ^{13}C nuclei relaxed solely via the dipoles of n directly bonded ^1H atoms, plotted as a function of the isotropic rotational correlation time τ_R. In many real molecules the T_1 minima are partially flattened and hence raised, because of internal motions of limited extent which are superimposed upon the overall rotation. Other protonated spin-1/2 nuclei give qualitatively similar curves in most cases. The numbers by the curves are the ^{13}C frequency in MHz.

tial, it may, nevertheless, often be calculated, leading to experimentally testable predictions of relaxation parameters. One common example is the anisotropic angular diffusion of a nonspherical molecule, for which detailed calculations exist.[16,17] An extreme organometallic illustration of anisotropic rotation is the methylmercury(II) cation.[18] This has a 120:1 ratio of angular diffusion rates perpendicular and parallel to the C–Hg bond. The three equivalent protons also show correlated motions (see Section 3.5), and the combined result is highly nonstandard relaxation behavior.[5] The decoupled ^{13}C spectrum shows strikingly nonexponential relaxation (Figure 5), and the proton-coupled spectrum yields markedly different initial T_1 and NOE for the inner and outer lines of the CH_3 quartet. Similar effects have been found in ^1H and ^{19}F spectra.[19,20] A useful rule in such cases is that the initial decoupled relaxation rate equals that which would have pertained in the absence of cross-relaxation.

Another tractable case is that of relatively rapid internal rotations or librations superimposed on a slower overall molecular tumbling. The presence of such motions is revealed by unexpected NOE and T_2's, and by a nonstandard variation of T_1 with spectrometer frequency. In this case also the angular correlation function does not decay exponentially. Instead, it decays at a rate determined by the internal motion to some fraction of its original value. This fraction quantitatively represents the mean angular constraint on the internal motion, in that it is the angular correlation that remains when the internal motion alone has completed its randomizing work. $G(\tau)$ then continues to decay more slowly to zero at a rate (here assumed to be much slower) determined by the overall molecular tumbling.

Analyses of this type reveal plausible and consistent internal motions for such diverse cases as proline ring-puckering,[21] floppy peptides,[22] proteins[23] including the mobile binding of chloride ion,[24] nucleic acids,[25] and even "solid" rubber.[26]

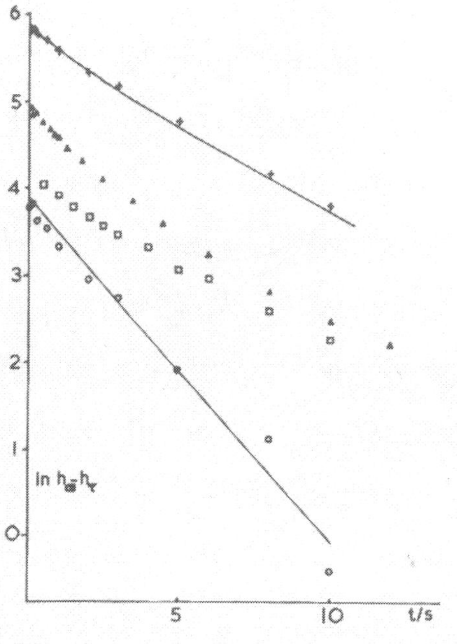

Figure 5. Nonexponential ^{13}C relaxation in CH_3HgCl and $CH_3HgO_2CH_3$. (Δ $^{13}CH_3HgO_2CCH_3$, decoupled; + and ○, inner and outer quartet lines, respectively, of coupled $^{13}CH_3HgO_2CCH_3$; □, $^{13}CH_3HgCl$ decoupled. Vertical position is arbitrary. Solid lines: theoretical fit for $D_{\|}/D_{\perp} = 121$.) From Ref. 18.

They can also measure very low-energy barriers to the rotation of methyl groups,[27] and they may find application in the study of highly fluxional inorganic compounds where the internal exchanges make a measurable contribution to relaxation.

3.4. The Static Part V

V will usually be a sum of terms in real molecules, rather than the single potential or operator treated above. Section 4 deals with the breakdown of V into specific parts. One notes in general that V may involve mutual spin flips of several kinds, so that weighted sums of terms such as $J(\omega_{ul}^{\mathrm{I}} + \omega_{ul}^{\mathrm{S}})$ will appear in the more general version of equation (9). In addition, some interactions change sign more than once for every full rotation, so that τ_c may become the angular correlation time for the interaction rather than for overall molecular tumbling.

3.5. More Complete Treatments

The preceding derivation of an expression for $1/T_1$, equation (29), was for a simple case with two eigenstates and only one interaction V. A more general case will involve further simultaneous equations, and relaxation times that depend on more than one integral of type V_{ul}. Such cases can often be made more tractable by replacing expression (19) with a density matrix, whose elements are the averaged products $\overline{C_m^* C_n}$. In this representation the wave equation converts to an equation involving this matrix. Its solution was first studied by Kubo and Tomita,[28] and then developed by Redfield[29] and many others. Their approach is essential in complex cases of relaxation, for example involving correlated motions of several spins, because in such cases the spin system may no longer be adequately described in terms of simple eigenstate populations $C_m^* C_m$. This may be seen by considering the two isochronous protons, e.g., in the methylene group of ethanol. Because the separate nuclei are precessing at the same rate, the magnetic moment components normal to B_0 of one nucleus will not average to zero as viewed from the second nucleus. Instead, the spins will be forced to precess with a mutual phase relationship defined by the quantum laws of angular momentum, much as in *ortho*- and *para*-hydrogen. Thus any adequate description of their NMR must include off-diagonal components of the density matrix, which are equivalent to measures of M_x and M_y.

4. SPECIFIC MECHANISMS

The general method for obtaining the relaxation contribution of a specific interaction is to reexpress the interaction in terms of quantum-mechanical operators, and from this to calculate the sizes and ratios of such variables as W_0, W_1, and W_2 in Figure 2. We may take scalar coupling as a simple example. A coupling constant J between two nuclei I and S is physically equivalent to the direct overlap of two bar magnets, arising from (vectorial) angular momenta **I** and **S**. The energy of interaction is simply $J\mathbf{I}\cdot\mathbf{S}$. This scalar product may be expressed in terms of the Cartesian vector components as

$$J(I_x S_x + I_y S_y + I_z S_z) \tag{31}$$

and upon conversion to quantum operators this becomes

$$J(\tfrac{1}{2}I^+ S^- + \tfrac{1}{2}I^- S^+ + I_z S_z) \tag{32}$$

where $I^+ = I_x + I_y$, etc. $I^+ S^-$ and $I^- S^+$ permit transitions W_0 only, which explains the comments made in Section 2.3.2. The more complex dipolar case is clearly explained by Slichter.[30]

4.1. Dipole–Dipole Relaxation

The perturbation of one spin by the nearby magnetic dipole of another is the commonest, and in some cases the only significant relaxation mechanism for spin-1/2 nuclei. Quantitative calculation of this contribution to relaxation is easier when the two spins lie at a fixed distance, and so this case is treated first.

4.1.1. Intramolecular Dipolar Relaxation

If spin I is relaxed solely by the dipolar interaction of a different type of spin S, tumbling with a correlation time τ_c at a fixed distance r, then $1/T_1 = W_0 + 2W_1 + W_2$ [essentially equation (10)] and $1/T_2 = 1/2T_1 + 6a/(1 + \omega_s^2 \tau_c^2) + 4a$. A full calculation based on the principles of the previous sections gives

$$W_0 = 2a/[1 + (\omega_\mathrm{I} - \omega_\mathrm{S})^2 \tau_c^2] \tag{33}$$

$$W_\mathrm{I} = 3a/(1 + \omega_\mathrm{I}^2 \tau_c^2) \tag{34}$$

$$W_2 = 12a/[1 + (\omega_\mathrm{I} + \omega_\mathrm{S})^2 \tau_c^2] \tag{35}$$

Here $a = \mu_0^2 \gamma_I^2 \gamma_S^2 \hbar^2 \tau_c S(S+1)/240\pi^2 r^6$ and ω_I, ω_S are in rad s^{-1}. The W_0 term represents mutually antiparallel spin transitions, the W_2 term mutually parallel ones, and the W_I term represents transitions of spin I parallel to B_0 which are induced by motions of spin S normal to B_0.

When spin S is of the same type as I, then a different formula applies:

$$1/T_{1dd}(\text{intra}) = \frac{\mu_0^2 \gamma_I^4 \hbar^2 I(I+1)}{40\pi^2 r^6} \left(\frac{\tau_c}{1+\omega_I^2 \tau_c^2} + \frac{4\tau_c}{1+4\omega_I^2 \tau_c^2} \right) \tag{36}$$

In a typical heteronuclear example, such as I = ^{13}C and S = ^{1}H, the r^{-6} term ensures that only directly bound protons, if present, relax the ^{13}C nucleus significantly. It is normal in this case to set $r = 0.109$ nm, although this assumption has been disputed. One should note that the r^{-6} term is really $(\overline{r^{-3}})^2$, where the average is over vibrational motions. This average will typically increase r^{-6} by 5%–10% over the equilibrium r_e^{-6} value, and is thus experimentally significant.

Although directly bound protons dominate dipolar relaxation, the effect of more distant protons, including those from solvent, is important in the relaxation of unprotonated atoms, and may be exploited to investigate structures and the solvent-accessibility of large molecules via differential relaxation or NOE.

4.1.2. Intermolecular Dipolar Relaxation

The calculation of $1/T_{1dd}$(inter) depends on the model chosen for mutual motion. In a simple model which ignores orientational and steric limitations beyond those of freely diffusing hard spheres, one obtains in a limit of rapid motions

$$1/T_{1dd}(\text{inter}) = \frac{\mu_0^2 N_S \gamma_I^2 \gamma_S^2 \hbar^2 S(S+1)}{90\pi D_{IS} a} \tag{37}$$

where N_S is the concentration of spins S per unit volume, a the distance of closest approach, and D_{IS} the mutual translational self-diffusion constant.

4.1.3. Nuclear Overhauser Enhancement: Specific Examples

The theory of the NOE was developed in Section 2.3, and fuller details may be found in Noggle and Schirmer.[(31)] We now note some important practical cases. When I = ^{13}C and S = ^{1}H then T_1 will usually be entirely dipolar in origin if S is directly bound to I. In this case η will vary from +1.98 at short τ_c to +0.15 at very long τ_c. When both I and S = ^{1}H, then η = +0.5 for short τ_c in the absence of dipolar interactions from other protons which have not been saturated. This latter condition is not usually fulfilled, so that specific ^{1}H–^{1}H η's are usually +0.01 to +0.30, and can be even less when the ^{1}H–^{1}H separation exceeds (typically) 0.5 nm. Only highly isolated protons or very rapidly spinning isolated methyl groups show NOEs from irradiation of other protons beyond this distance. Figure 6 shows an inorganic example of NOE difference spectroscopy.

The positive sign of η in the proton–proton case is convenient, because secondary Overhauser enhancements can occur between spin I and further spins, once I has had its spin populations perturbed. These "knock-on" enhancements may be readily detected by their opposite sign, which arises because the I_z polarization increases when S_z is

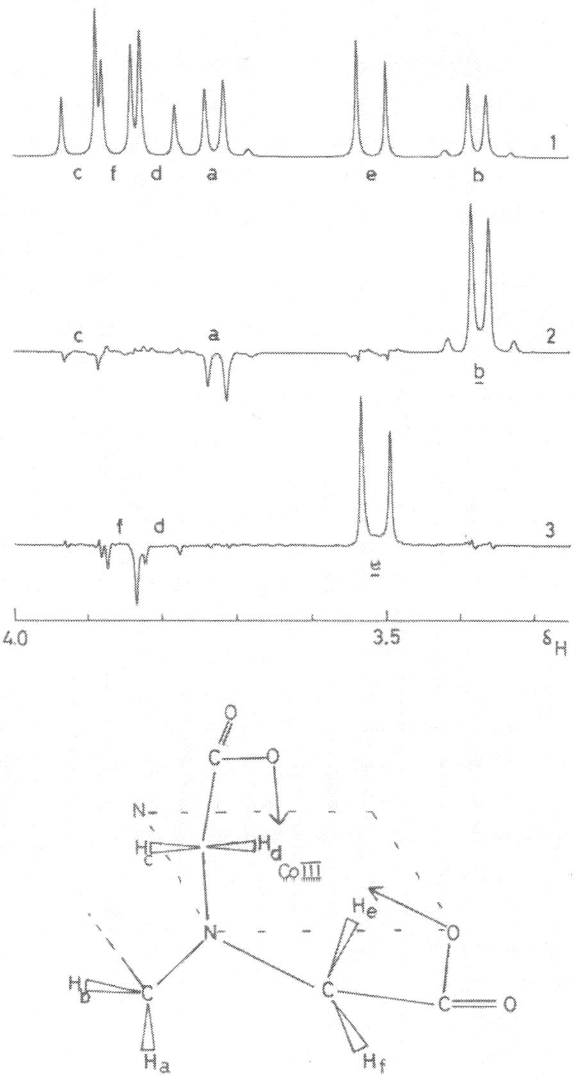

Figure 6. ^{1}H spectrum (1) and NOE difference spectra (2 and 3) of $[Co^{III}(edta)]^-$. The difference spectra reveal nonbonded contacts, which permit a full assignment of (1). In (2), protons *b* are irradiated, with resulting NOE at *a* and *c*. In (3), irradiation of *e* effects *d* and *f*. The structural diagram shows only half of the ligand for clarity: the ion has C_2 symmetry.

decreased. If necessary, they can be removed by a truncated NOE experiment,[32] where the irradiation time of spin S is limited to ca. T_1^I and then a long relaxation delay is enforced.

Unfortunately, when τ_c is very long, then $\eta(^1H-^1H) = -1$. This means that in molecules of moderate complexity one may encounter the intermediate case of $\eta = 0$. It occurs with high-field spectrometers at $\tau_c = 1\text{–}2 \times 10^{-10}$ s, which is common for molecules of M.W. 500–1000 in light solvents. In this case some expected internuclear enhancements may not appear, although at lower temperatures they appear with opposite sign.

A similar problem reduces the usefulness of proton-decoupled ^{19}F spectra of specifically fluorinated macromolecules. In this case η can be -1, giving zero ^{19}F intensity in the ordinary spectrum upon irradiation of ^{1}H. The same problem of $1+\eta=0$ occurs even at short τ_c with nuclei such as ^{15}N and ^{29}Si which have negative magnetogyric ratios. Here it may be obviated by suitable gating or by paramagnetic doping.

NOEs do not have to be intramolecular in origin. Irradiation of solvent can reveal exposed carbon atoms on a biomolecule, and ligand–biomolecule interactions may be similarly studied.

4.2. Scalar Interactions

In principle spin–spin coupling to another nucleus, suitably modulated by, for example, exchange, could also produce intramolecular relaxation in a similar fashion to dipolar coupling. However, in practice the couplings are less by two or more orders of magnitude, so that their effect is negligible. This normally unobserved process is called scalar relaxation of the first kind. Abragam[(11)] gives a rare example.

Scalar relaxation of the second kind occurs when the S nucleus relaxes rapidly, typically because of its quadrupole. This will have a marked effect on the T_2 of nucleus I, by tending to collapse the $\mathrm{I}\cdot\mathrm{S}$ scalar coupling in a manner analogous to the effects of exchange. More precisely

$$1/T_2^{\mathrm{I}}(\text{scalar}) = 1/2T_1^{\mathrm{I}}(\text{scalar}) + (4/3)\,\pi^2 J^2 S(S+1)\,T_1^{\mathrm{S}} \tag{38}$$

where J is the coupling constant and T_1^{S} the spin-lattice relaxation time for nucleus S. Here

$$1/T_1^{\mathrm{I}}(\text{scalar}) = (8/3)\,\pi^2 J^2 S(S+1)\,T_2^{\mathrm{S}}/[1+(\omega_{\mathrm{I}}-\omega_{\mathrm{S}})^2(T_2^{\mathrm{S}})^2] \tag{39}$$

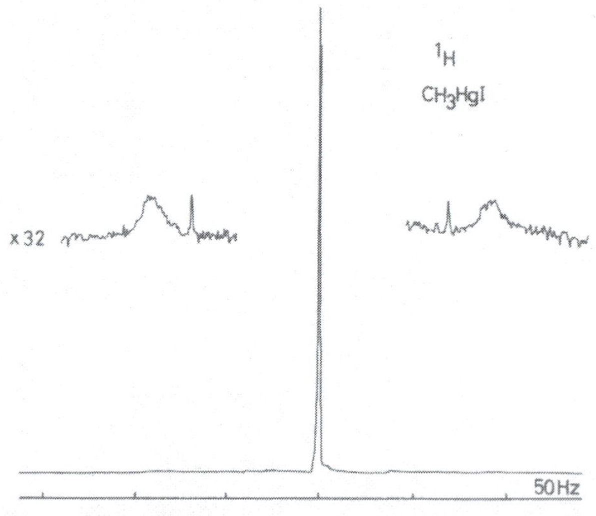

Figure 7. ^{1}H NMR spectrum of CH_3HgI, (9.4 T, 295 K, $CDCl_3$). The ^{199}Hg (16.9% abundant) sidebands are broadened so that they are no higher than ^{13}C (1.1% abundant) sidebands. The broadening (in Hz) depends approximately on B_0^{-2}, which shows that it is not primarily due to exchange (no field dependence in this limit) or to SA broadening (depends on B_0^2). Instead, it is a rare example of relaxation by scalar coupling of the second kind, which depends on B_0^{-2} provided T_1^{S} dominates.

This contribution to T_1 (scalar) is simply that of the coupling perturbation $J\mathbf{I}\cdot\mathbf{S}$ modulated by the spin relaxation of S. As $1/T_2^S$ will rarely exceed 10^6 s, it follows that for normal spectrometers ω_I must be reasonably close to ω_S for the mechanism to be significant. One of the few cases for which it has been established,[33] and for which exchange has been excluded as a possibility, is the relaxation of ^{199}Hg by ^{127}I in CH_3HgI, illustrated in Figure 7. Here the $^1J_{HgI}$ must be very large indeed, as will be the relaxation rate of the quadrupolar nucleus ^{127}I. Also the resonance frequency difference between ^{127}I and ^{199}Hg using typical spectrometers is only about 2 MHz, which is not totally out of proportion to the ^{127}I relaxation rate. Indeed, $T_2^{127I}/[1+(\omega_{Hg}-\omega_I)^2(T_2^{127I})^2]$ may be as large as 10^{-9} in such a case, which for $J=10^5$ Hz would make $1/T_1^{199Hg}\simeq 200\ s^{-1}$. Similar broadenings, but less by an order of magnitude, are observed for CH_3HgBr.[34] T_2 (scalar) broadenings are much more common, and if T_1^S is known then J may be at least approximately extracted despite the absence of splitting. However, it should be noted that equations (38) and (39) are not strictly accurate when J is very large, or when the relaxation perturbation of spins is comparable in energy to its Zeeman splitting.

4.3. Shielding Anisotropy

The nuclear screening and hence chemical shift of a nucleus in a molecule will usually be anisotropic, so that when the molecule tumbles the nucleus will experience a fluctuating magnetic field. The resulting expressions for T_1 and T_2(SA) are analogous to those for dipolar relaxation, except that the spin system is only two-level. It may, however, be complicated by the possibility of asymmetry in the screening tensor.

In the simple case of a symmetric tensor ($\sigma_{zz}\neq\sigma_{xx}=\sigma_{yy}$)

$$1/T_1(\mathrm{SA})=(1/15)\,\gamma_I^2B_0^2(\Delta\sigma)^2[2\tau_c/(1+\omega_I^2\tau_c^2)] \qquad (40)$$

and

$$1/T_2(\mathrm{SA})=(1/90)\,\gamma_I^2B_0^2(\Delta\sigma)^2\{[6\tau_c/(1+\omega_I^2\tau_c^2)]+8\tau_c\} \qquad (41)$$

Here $\Delta\sigma=\sigma_{zz}-\sigma_{xx}$. Once again, the expression for $1/T_2$ is half that for $1/T_1$ plus a frequency-independent term arising from spin precession phase jumps induced by the fluctuating field. However, in the motional narrowing limit ($\omega_I\tau_c\ll 1$) one now finds that $T_1(\mathrm{SA})/T_2(\mathrm{SA})=7/6$. This relationship has yet to find clear experimental verification.

$\Delta\sigma$ is of the order of 200 at ^{13}C for a carbonyl double bond, but may easily rise to 5000 ppm and more, for example around linear ^{199}Hg or square-planar ^{195}Pt. Thus these resonances will often be noticeably broadened in a high-field spectrometer ($B_0\simeq 9$ T), with $1/T_1$ commonly $50\ s^{-1}$. The presence of SA relaxation is unambiguously proved by its dependence upon the square of B_0.

When the nuclear screening is completely anisotropic, or even symmetric but with principal axes different from those for molecular tumbling (e.g., a *meta* or *ortho* phenyl carbon), then the appropriate equations are more complex. They may be found in a paper by Lynden-Bell.[35]

4.4. Spin-Rotation Interactions

When a molecule rotates, its electrons do not quite "catch up" with the constantly changing nuclear positions, which gives the molecule a small magnetic moment proportional to its angular momentum even in the absence of an external magnetic field. Molecular collisions will change the direction of this moment after an average interval $\tau(\mathrm{SR})$, and hence provide a further spin-relaxation mechanism which is unique in becoming increasingly effective as temperature increases. Quantitatively, for a spherical top

$$1/T_1(\mathrm{SR}) = 2IkT\overline{C^2}\tau(\mathrm{SR})/3\hbar^2 \tag{42}$$

where I is the molecular moment of inertia, and $\overline{C^2}$ is the average of the square of the spin-rotation tensor diagonal elements. In a liquid $\tau(\mathrm{SR}) \simeq I/6kT\tau_c$. $\overline{C^2}$ may sometimes be available either from calculations or from absolute chemical shift measurements.[36] It is of the order of $(2\ \mathrm{kHz})^2$ for ^{19}F but considerably less for ^1H. $T_1(\mathrm{SR})$ is usually dominant for small molecules in mobile liquids, and even more so in the gaseous state, where $\tau(\mathrm{SR})$ is lengthened by the reduced collision rate to such an extent that even ^1H resonances are typically 100 Hz in width.

4.5. Electric Quadrupole Interactions

All nuclei with $I > 1/2$ must be nonspherical, according to physical theory, and hence must possess nuclear electric quadrupole moments. These moments cause nuclear quadrupole splittings in the solid state, when the nucleus is aligned by electric field gradients (efg's) at the nucleus. Such splittings range from 10^5 Hz (^2H) to 10^7 Hz and more, unless the efg is near zero. They may be investigated by gas-phase microwave spectroscopy and by nuclear quadrupole resonance; alternatively they may cause complex splittings in solid-state NMR spectra. The smaller splittings may be partially but not completely removed by magic-angle spinning, so that, for example, the CP-MAS spectrum of glycine shows the α-carbon broadened and split by residual interaction with ^{14}N. The sign of the nuclear quadrupole moment may be obtained by experiments at spin temperatures of ca. 0.01 K obtained via NOE pumping. Mobile but oriented molecules show reduced splittings.

Full molecular rotation averages nuclear quadrupole splittings to zero, but as with dipolar coupling the quadrupole/efg interaction becomes a relaxation mechanism, sufficiently powerful to dominate T_1 for all but the lightest and most mobile nuclei. Indeed, heavy nuclei with $I > 1/2$, such as ^{81}Br, ^{127}I, and ^{91}Zr, give resonances almost undetectably broad when asymmetrically bound. In general, for an isotropically tumbling molecule

$$\frac{1}{T_1(Q)} = \frac{3\chi^2}{200}\left(\frac{2I+3}{I^2(2I-1)}\right)\left(1+\frac{\eta^2}{3}\right)\left(\frac{\tau_c}{1+\omega^2\tau_c^2}+\frac{4\tau_c}{1+4\omega^2\tau_c^2}\right) \tag{43}$$

and

$$\frac{1}{T_2(Q)} = \frac{\chi^2}{400}\left(\frac{2I+3}{I^2(2I-1)}\right)\left(1+\frac{\eta^2}{3}\right)\left(9\tau_c+\frac{15\tau_c}{1+\omega^2\tau_c^2}+\frac{6\tau_c}{1+4\omega^2\tau_c^2}\right) \tag{44}$$

where $\chi = e^2q_{zz}Q/\hbar$, $\eta = (\chi_{yy} - \chi_{xx})/\chi_{zz}$, eq_{ii} are the principal values of the efg at the nucleus, and eQ is the nuclear electric quadrupole moment. η is called the asymmetry parameter; it is not related to the NOEF. The increase of $1/T_2(Q)$ with increasing τ_c means that even ^{2}H resonances become uninformatively broad in macromolecules. It may, however, be exploited, for example in ^{1}H spectra of protons coupled to ^{11}B, in that on lowering the temperature the ^{11}B relaxes so rapidly that its coupling to ^{1}H no longer affects the latter's spectrum. This is sometimes called relaxation decoupling. At the other extreme, where $\tau_c \ll 1/\omega$, $T_1(Q) = T_2(Q)$. Equations extended to include anisotropic diffusion are given by Lehn and Kintzinger.[37]

Even in the case of isotropic diffusion, equations (43) and (44) are in error by up to 10% when $I > 1$ and at the same time $\tau_c \not< \omega^{-1}$. In these cases the relaxation becomes nonexponential; for example, when $I = 3/2$ then the z magnetization decays as if the two τ_c terms in equation (43) governed separate exponentials.

Most spectra of nuclei with $I > 1$ become unmanageably broad before τ_c approaches ω^{-1}. However, the above exception is relevant to the investigation of biomolecules via the kinetically brief binding of a small fraction of, e.g., Cl^- ions. Not only may the line shape become a superimposition of broader and narrower components from the one spin, but also a small frequency shift between the components is at least theoretically possible, which would give the overall resonance an appearance of being poorly phased.[38]

4.5.1. *Quadrupolar Relaxation of Ions in Solution*

Quadrupolar nuclei symmetrically coordinated in ions, such as $[^{27}Al(OH_2)_6]^{3+}$ or $[^{35}ClO_4]^-$, have no permanent efg to relax them. However, because quadrupolar relaxation is so efficient, they can be markedly affected even by the fluctuating efg's that continually arise from outer-sphere motions. One simplified equation that has been proposed[39] for $T_1(Q)$ or $T_2(Q)$ in the limit of short τ_c is

$$\frac{1}{T(Q)} = \frac{8\pi}{5}\left(\frac{\mu_B eQ}{\hbar}\right)^2\left[\frac{2I+3}{I^2(2I-1)}\right]\left(\frac{\mu^2 c_s}{r_0^5}\tau_s + \sum_j \frac{z_j^2}{3a_j^3}c_j\tau_j\right) \tag{45}$$

In this equation the term in μ describes the effect of these solvent electric dipoles on I, and the term in z describes the interaction with ions j; c_s, τ_s, and c_j, τ_j are the solvent and j-ion concentrations and correlation times, respectively. a_j is the mean distance of closest approach, and $\mu_B = P(1-\gamma_\infty)$; $P = (2\varepsilon + 3)/5\varepsilon$, ε being the dielectric constant of the solvent. γ_∞ is the Sternheimer antishielding factor[40] described below. Equation (45) is only approximate, and certainly requires further terms when, for example, hydrogen bonding to the ion is present.

4.5.2. *Calculation of Field Gradients*

In theory, efg's should be calculable. In practice, the approximate wave functions that must necessarily be employed for all but light atoms are least accurate near the nucleus, so that calculations are at present only useful (i.e., within 20%) for first row atoms. Even these are only possible for isolated molecules, although it is known that intermolecular interactions, particularly involving lone pairs, can affect eq by up to 50%.

Equation (45) above represents one attempt at calculating intermolecular contributions to χ. In crystals, the term corresponding to z_j depends on orientation, as an ion of charge z_je at distance r from I and angle θ to B_0 contributes $(3\cos^2\theta - 1)/r^3$ to the efg at I in that direction. Such contributions do not decrease rapidly with distance, so that calculations involving crystals present difficulties similar to those found in calculating Madelung constants. The electrons around I contribute similarly if I is unsymmetrically coordinated, and approximate expressions have been developed by Townes and Dailey[41,42] relating *p*- and *d*-orbital occupancies to χ.

Perhaps the most serious limitation on both these approximate calculations is that the influence of both outer bound and coordinated charges must be scaled by a factor γ_∞ in order to allow for the amplifying effect of inner electron motions on the externally applied field gradient. γ_∞ is called the Sternheimer antishielding factor,[40] and lies typically between two and ten. Yet a further efg arises from the polarization of the inner orbitals by the electric quadrupole of I itself, but this term may readily be allowed for in any calculation.

Because of all the above difficulties, accentuated by the fact that $1/T_2(Q)$ depends on the square of χ, only qualitative insights are available concerning most quadrupole-dominated linewidths.[41] Some useful but approximate generalizations are as follows:

(i) The principal component of the efg usually lies approximately parallel to the axis of the lone pair, when present, and the efg is generally reduced when the lone pair becomes involved in bonding.

(ii) Linear molecules generally show smaller efg than ones bent at I, for a similar reason.

(iii) Tetrahedral or octahedral coordination of a spherical ion removes the efg to first order, giving very narrow lines. The *fac*-isomer of octahedrally coordinated $M_IX_3Y_3$ also has a very small efg, although now there may be a substantial electric (dipole) field at I. In contrast, square planar coordination will usually give a large efg despite the absence of electric field at I. These principles are occasionally useful in distinguishing mixed-ligand isomers. Even slight departures from symmetrical coordination can produce substantial line-broadening (as in $V(O_2)_4^{3-}$, linewidth 300 Hz, compared with VO_4^{3-}, linewidth 2 Hz; in the former species the peroxo-ligand is unsymmetrically coordinated, although the nearest four oxygen atoms form a tetrahedron around the vanadium).

(iv) Townes and Dailey's approach predicts that each component of the efg will be proportional to r^{-3} averaged over the relevant wave functions. The same r^{-3} term appears in the local term approximation to σ_p (Chapter 3). Hence some proportionality of δ^2 to quadrupolar-dominated linewidth may be anticipated provided that δ is determined more by variation in orbital size than by variations of excitation energy. Experimental verification comes from ^{35}Cl NMR studies[43] and from ^{55}Mn NMR of $R_3Sn\text{-}Mn(CO)_5$ compounds.[44] However, it does not seem to pertain for ^{17}O resonances of oxo-species, where terminal oxo ligands commonly give much sharper resonances than bridging ones.[34]

4.5.3. Quadrupolar Nuclei in the Solid State

Almost all quadrupole couplings are much too large to be averaged by rotating the sample rapidly at a suitable angle. However, the $M_I = 1/2 \leftrightarrow M_I = -1/2$ transition of quadrupolar nuclei with half-integral I is only affected to second order by quadrupole splittings, and hence may be observed in such experiments at realistic spinning rates. The resonance is nevertheless usually quite broad, and as mentioned above the optimum spinning angle is not precisely $\cos^{-1}(1/\sqrt{3})$.

5. METHODS OF MEASUREMENT

5.1. T_1 Measurements

T_1 is measurable in principle by any method which detects the effects on the nuclear magnetization of any perturbation of the Boltzmann populations. It may be measured either statically or dynamically. The original static method was progressive saturation, which exploits equation (1) by increasing B_1 until its effects on the resonance are substantial. The method requires an independent measure of the rf field of $2B_1$, which is more conveniently available in the method's Fourier-transform approximation, the $(90° - t)_n$-observe sequence. Here one obtains T_1 from the dependence upon t of the observed reduction in intensity between $n = 1$ and $n = \text{large}$ using the equation

$$M_z(\infty) = M_0(1 - \exp(-t/T_1)) \tag{46}$$

Accuracy of the 90° pulse is not highly critical, and can be corrected via a theoretical factor.[45]

The commonest dynamic method, illustrated in Figure 8, is the (180°-t-90°-observe-long delay)$_n$ sequence, in which n need not be large and "long delay" means at least five times the longest T_1 to be measured. If this is pared down to $3T_1$ an error of about 5% will result, and if it is as short at T_1 itself the error will be nearer 50%. A modification exists which is more suitable for spectrometers with unstable gain, or for less stable samples.[45] Various pulse sequences have been suggested to reduce the long delay, but many of them are only useful for rather long T_1's, and are also less tolerant of imperfect pulse lengths than the simple sequence. This is particularly important in higher-field spectrometers, where it may be difficult or impossible uniformly to stimulate all resonances, say of ^{13}C, in a molecule, because of limitations on pulse power. Some of the deleterious effects of pulse imperfections and of necessarily long experiments may be reduced by cycling the interpulse delays t, and by phase cycling. The commonest precaution, however, is to avoid assuming that the initial supposedly 180° inversion is complete, and to rely on the gradient of the plot of $\ln(h_t/h_\infty)$ against t.

For a quadrupolar nucleus in a rapidly tumbling molecule, T_1 may often be directly deduced from the linewidth. Even so, it may be advisable to check with a direct measurement made as above, provided that such a measurement is not vitiated by interference from magnetization components normal to the field which have not

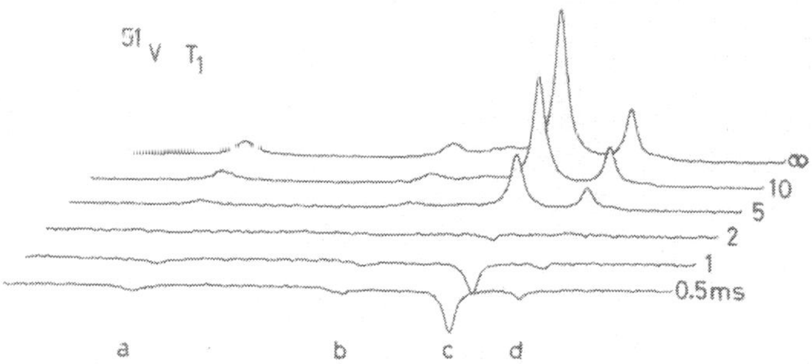

Figure 8. Inversion-recovery ^{51}V spectra of vanadate species, with interpulse delays from 0.5 to 10 ms. a = monomer, b = dimer, c = cyclic tetramer, d = cyclic pentamer. Note that T_1 is slightly less for the cyclic pentamer, but that because of faster exchange the linewidth is slightly greater for the cyclic tetramer.

decayed within time t, and which add a spurious component of apparently random phase to each observed signal. Such effects may be reduced either by improving the accuracy of the 180° pulse, or, in the case of narrower resonances, by applying a substantial pulsed field gradient during the interval t, or by suitable phase cycling.

T_1 may also be extracted as the time constant for recovery from such perturbations as nuclear Overhauser enhancement, saturation, or even simply entering B_0. As specific NOEs may be measured between chosen pairs of protons, and as T_1 depends on whether or not other coupled spins are inverted (Section 2.2.1.), it is possible to factor the intramolecular T_1(dipolar) into contributions from separate neighbors and hence to measure proton–proton distances. Such possibilities have been explored for organic molecules,(46) but although they could be most useful for organometallics, they require the absence of interproton chemical exchange.

5.2. Nuclear Overhauser Enhancement Measurements

NOE measurements are notoriously difficult to repeat reproducibly. Even greater care must be taken than with T_1 measurements to ensure constant sample conditions and deoxygenation. Furthermore, the long relaxation delay must be at least $9T_1$ in order to eliminate cross-correlation effects.(47) The measurement itself is nominally simple; one compares spectra with and without decoupling at the second nucleus. This decoupling must be sufficient to ensure the latter's complete saturation. In the case of $^{13}C\{^1H\}$ NOEs the decoupling remains on before and during collection of the enhanced FID, is switched off for a relaxation period, and is then switched on just before collecting the unenhanced free induction decay. One may note that this latter irradiation does not affect the FID because NOE only affects M_z magnetization, whereas the FID arises from M_x and M_y magnetization. However, the prepulse irradiation may heat the sample, thus introducing errors. Such errors are normally reduced by computer-controlled alternation of measurements with and without NOE. They may be further reduced in spectrometers with computer control of decoupler power by switching to lower irradiation power during the prepulse period, because NOE buildup requires considerably lower power than decoupling.

For the measurement of homonuclear NOE it is normal to switch off the decoupler just before measurement, in order to reduce artifacts. Small NOE's on sensitive nuclei such as ^1H are conveniently identified by computer subtraction of the enhanced and unenhanced spectra, i.e., by NOE difference spectroscopy. As this subtraction is rarely perfect, it is wise to check for the presence of an overall peak integral to confirm that an apparent residual resonance is not merely a subtraction artifact. The percentage NOE of a peak is readily calculated by comparison of its area with the (negative) area of the large peak which will always arise from saturation at the position of irradiation. Irradiation sufficient for such spectroscopy can be specific within a typical range of ± 40 Hz.

In some cases it is possible to obtain even greater sensitivity enhancements by using other polarization transfer techniques, some of which are described at the end of the next section.

5.3. T_2 Measurement and Other Uses of Spin Echoes

Although T_2 may often be readily obtained from the linewidth $1/\pi T_2$, there are also frequent cases where this linewidth is obscured by overlapping or, in the case of long T_2, by B_0 inhomogeneity. In these cases T_2 is most commonly obtained via a spin-echo technique. The two main methods, Carr–Purcell A and B,[48] are described in many texts.[2,49] They rely on the refocusing of M_{xy} magnetization spatially dephased by field inhomogeneities, this refocusing being achieved at time $2t$ by a 180° pulse at time t. If the refocusing is perfect, M_{xy} will then have decayed by $\exp(-2t/T_2)$, plus a contribution from spatial diffusion if present, whereas in the absence of the 180° pulse it would have decayed by $\exp(-2t/T_2^*)$. Method A outlines the exponential decay by stepwise variation of t, whereas method B uses a single, fairly short t and a succession of 180° pulses at times $(2n+1)\,t$, where n is an integer. T_2 is then obtained from the envelope of the successive spin-echoes at times $2nt$. It is a quicker experiment than method A, but relies more stringently upon accurately achieving a 180° pulse throughout the entire sample volume. Several variations have been suggested[49] to reduce the effects of pulse imperfections. $T_{1\rho}$ measurements may be regarded as an extreme case of the Carr–Purcell B technique.

One limitation on the spin-echo measurement of T_2 is that if the 180° pulse inverts the resonances of spins coupled to each other, then this is equivalent to symmetrical switching of the Larmor precession frequencies of each multiplet component. Thus although each component, say of a doublet, will refocus adequately, it will do so with a phase error of $\pm tJ$, where J is the doublet separation. The echo train will thus be modulated according to J.

Fourier transformation of the echo permits one to inspect the time evolution of such individual resonances. Indeed, their evolution may be treated as a new type of FID, and therefore Fourier-transformed in a second dimension. This is the basis of J-resolved two-dimensional spectroscopy, which is described in a more general context in Section 8.

One might have hoped that heteronuclear decoupling would have been free from the above difficulties, so that, say, proton-decoupled ^{31}P T_2's might be obtained as easily as proton-free ^{31}P T_2's. But unfortunately two problems can occur. One is simply that unless the heteronuclear decoupling is very efficient there will be a residual linewidth on the observed nucleus which will also appear as a contribution to T_2. The other is that under some decoupling conditions, including most broadband decoupling, the two nuclear magnetizations can exchange energy via cross-polarization.[49] As a practical example, this limits meaningful $^{13}C\{^1H\}$ T_2 measurements to resonances where $T_2 < 0.1$ s.

5.3.1. Cross-Polarization and Its Uses

This exchange of energy via cross-polarization may be simply envisaged in the following way. Consider two spins I and S, whose magnetizations are each separately being rotated by radiofrequency fields B_1^I and B_1^S. If I and S have very different magnetogyric ratios, then each spin magnetization will only sense the net z-magnetization of the other spin. Their rotating frames are different, and each magnetization sees the other executing linear sinusoidal oscillations. However, if the "Hartmann–Hahn" condition,[50]

$$\gamma_I B_1^I = \gamma_S B_1^S \tag{47}$$

is fulfilled, the two oscillations will have the same frequency. Thus if the I and S spins are coupled, either by J-coupling as above, or more strongly by unaveraged dipolar coupling in the solid state, then they will exchange energy and hence polarization in the standard manner of coupled oscillators.

This may be turned to advantage in solid-state NMR because achievement of the Hartmann–Hahn condition may be timed so as to transfer polarization repetitively from a convenient spin, such as ^{1}H, to an inconvenient or dilute spin such as ^{13}C. This transfer may be repeated many times within the T_1 of the ^{13}C, to give it a large polarization. This application of cross-polarization is usually combined with a pulse sequence that provides the very high-power proton decoupling that is usually necessary to obtain reasonably narrow resonances from solid samples containing protons.

5.3.2. INEPT

A related technique which uses a different kind of one-shot cross-polarization, and is achievable on standard instruments fitted with a pulse programmer, is called Insensitive Nuclei Enhanced by Polarization Transfer, or INEPT.[51] The method involves carrying out what is initially a Carr–Purcell A sequence on a spin S, having a large γ and a relatively short relaxation time (typically ^{1}H or ^{31}P) and also having a spin–spin coupling J with spin I, which has a small γ and a relatively long relaxation time. However, the 180°(S) pulse is accompanied by a 180°(I) pulse, which deliberately introduces the phase modulations which are unavoidable when homonuclear coupling is present. If the t-interval is set to equal $1/2J$, in the case of I and S both being single spin-1/2 nuclei, then the phases of the two components of the S resonance after time t will be $\pm 90°$ from those in the absence of the I inversion. If at this moment a 90° pulse is applied in a direction normal both to z and to the original 90° pulse, the one S component will revert to the z direction and the other to the $-z$ direction. Thus one component of the S doublet has been selectively inverted (barring T_2 losses). This sequence is outlined in Figure 9.

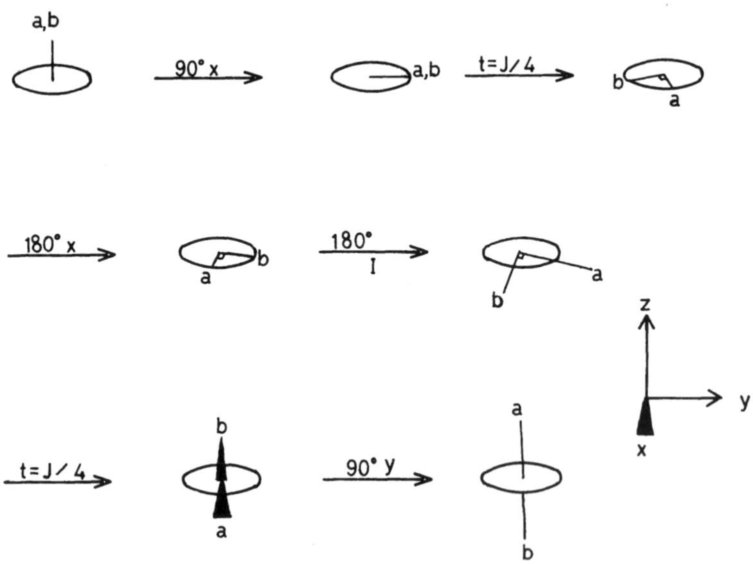

Figure 9. S-spin dynamics for the INEPT experiment. a and b are the two components of the S resonance resulting from coupling J to $I = 1/2$. The second 180° pulse inverts the I resonance. All the figures are viewed in the S rotating frame.

At this point consider the population of the four spin quantum states S(α) I(α), S(α) I(β), S(β) I(α), and S(β) I(β). Normally these have populations determined by Boltzmann's law, as described in Section 2.3. If $\gamma_S \gg \gamma_I$, then the population ratio S(α) I(α)/S(α) I(β) will be relatively small, compared with, say, S(α) I(α)/S(β) I(α). But if the populations of S(α) I(α) and S(β) I(α) are exchanged via INEPT, then the ratio S(α) I(α)/S(α) I(β) will become almost as large as S(α) I(α)/S(β) I(α), in a negative sense, and hence that I resonance will be simultaneously inverted and enhanced approximately in the ratio γ_S/γ_I. The other I resonance will be enhanced but not inverted. As a further bonus, this polarization may be reestablished as soon as the S spins have relaxed, it being unnecessary to wait for relaxation of the I spins.

The INEPT method has already proved of value for sensitivity enhancement in $^{15}N\{^1H\}$ studies, and in the study of nuclei such as ^{57}Fe and ^{183}W coupled to ^{31}P.[52] It may also be exploited in $^{13}C\{^1H\}$ studies because a suitable choice of t can invert CH_3 and CH resonances while leaving CH_2 and unprotonated C resonances upright. Variations such as DEPT are also available which overcome some of the artifacts inherent in INEPT.

5.3.3. *Practical Problems Presented by Very Short T_2 Values*

The measurement of very short FIDs presents serious problems of spectrometer design. The pulse length and the computer dwell time must each be sufficiently short so that the signal is captured with adequate digital resolution before it decays. High pulse repetition rates will usually be necessary to obtain sufficient signal. But an even bigger problem, especially at lower frequencies, is an electronic artifact normally described as "acoustic ringing." Ringing occurs because the pulse causes brief vibrations in the metal of the probe at the rf frequency, which masquerade as signals. It commonly lasts for about 100 μs, that is, about as long as the FID from a line 10 kHz wide. It may be removed from longer FIDs simply be zeroing the first few points of the FID; using a long preaccumulation delay time achieves the same purpose at the price of phase distortions. Spin-echo methods are also possible if the 180° pulse is short enough and T_2 long enough.

Acoustic ringing is unfortunately an incoherent resonant phenomenon, so that like noise, it cannot generally be removed by subtraction of blanks. The baseline roll which it causes severely limits wide-line NMR experiments on less-sensitive nuclei. It can be suppressed over a limited bandwidth by various techniques which cycle pulse phase and power, and may also be reduced at some frequencies by the use of a plastic rather than a metal probe can.

6. LINE BROADENING DUE TO CHEMICAL EXCHANGE

A single resonating nucleus which jumps frequently, and within a short time interval, between two or more sites will produce a nonsinusoidal contribution to the FID which is a collage of random-length components whose frequencies alternate between those for the two sites. The total xy magnetization will therefore decay by interference within a time closely related to the reciprocal of the weighted frequency difference of the components, and the line shape will be correspondingly broadened. In the case of faster exchange, the components of the collage are sufficiently short for the dephasing to average towards zero, and one obtains a single, only slightly broadened, line at the weighted mean of the original frequencies.

The effects of exchange processes upon NMR spectra are possibly even more important in inorganic NMR than for organic compounds. Lability is commoner and more versatile and also the chemical shift range is often much greater, so that at very high fields one may detect residual broadening even from kinetic processes whose rates (e.g., $3\times10^{8}\,s^{-1}$) approach diffusion-control. Thus T_2 for some solvated ions of heavier nuclei may possibly be determined by outer-sphere association rates.

This aspect of NMR is sometimes called dynamic NMR, perhaps sweepingly in view of the plethora of other dynamic effects. Several useful books are available. The text by Sandström[27] is particularly oriented towards practical problems, and that by Kaplan and Fraenkel[53] presents the full theory. The earlier books edited by La Mar, Horrocks, and Holm[54] and Jackman and Cotton[55] are rich in examples from organometallic and coordination chemistry.

Simple kinetic systems may be quantitatively analyzed using formulas derived by adding exchange terms to the Bloch equations. For example, the line shape $v(\omega)$ in the case of exchange at rate $k_{\mathrm{A\ or\ B}}$ between two sites with populations P_{A} and P_{B} at resonance frequencies ω_{A} and ω_{B} (in rad s^{-1}) and with natural widths $1/T_{2\mathrm{A\ or\ B}}$ is given by[27,56]

$$v(\omega)\propto\left[P(1+\tau)\left(\frac{P_{\mathrm{B}}}{T_{2\mathrm{A}}}+\frac{P_{\mathrm{A}}}{T_{2\mathrm{B}}}\right)+QR\right]\Big/(P^2+R^2) \tag{48}$$

where

$$P=\tau\left(\frac{1}{T_{2\mathrm{A}}\cdot T_{2\mathrm{B}}}-(\Delta\omega)^2+\frac{(\delta\omega)^2}{4}\right)+\frac{P_{\mathrm{A}}}{T_{2\mathrm{A}}}+\frac{P_{\mathrm{B}}}{T_{2\mathrm{B}}} \tag{49}$$

$$Q=\tau(\Delta\omega-\delta\omega(P_{\mathrm{A}}-P_{\mathrm{B}})/2) \tag{50}$$

$$R=\Delta\omega\left[1+\tau\left(\frac{1}{T_{2\mathrm{A}}}+\frac{1}{T_{2\mathrm{B}}}\right)\right]+\frac{\tau\,\delta\omega}{2}\left(\frac{1}{T_{2\mathrm{B}}}-\frac{1}{T_{2\mathrm{A}}}\right)+\frac{\delta\omega}{2}(P_{\mathrm{A}}-P_{\mathrm{B}}) \tag{51}$$

$$\tau=\frac{P_{\mathrm{A}}}{k_{\mathrm{A}}}=\frac{P_{\mathrm{B}}}{k_{\mathrm{B}}} \tag{52}$$

$$\delta\omega=\omega_{\mathrm{A}}-\omega_{\mathrm{B}} \tag{53}$$

$$\Delta\omega=\frac{1}{2}(\omega_{\mathrm{A}}-\omega_{\mathrm{B}})-\omega \tag{54}$$

In the case where $P_{\mathrm{A}}=P_{\mathrm{B}}$, and hence $k_{\mathrm{A}}=k_{\mathrm{B}}$, the merging bandshape is flat-topped when $k=\delta\omega/2\sqrt{2}$, provided that $2\,\delta\omega\gg T_2^{-1}$.

In more complex systems there will be several distinct rate constants awaiting extraction, and the resulting bandshapes as one varies temperature over the interesting range will be sufficiently complex to require total fitting by least-squares analysis. When second-order spin–spin splitting is also present, the theory requires density matrices and the programming can be very demanding. But the presence of transitions with varied frequency separations yields information which may not be obtainable from first-order spectra.

Kinetic exchange processes can contribute to T_1 as well as to T_2 and $T_{1\rho}$ provided they are fast enough. Thus T_1 measurements have been used to measure free energy barriers for methyl group rotation, and associated enthalpies of activation.[27] More sophisticated analyses have also yielded kinetic information on the rate and extent of puckering motions in proline rings, and on ion binding to macromolecules (see Section 3.3.1), and such methods have potential for the investigation of fluxional processes which are far too fast for standard treatment.

At the other kinetic extreme, reactions with $k < 10^3$ may be investigated by flow[57] or stopped-flow NMR ($k < 1$).[58] In this manner one may extend the temperature range of Arrhenius plots so as to reduce their all too prevalent errors.[59] Observed line shapes in flow FT NMR can become complex.

6.1. T_1 in the Presence of Chemical Exchange

When two species are in rapid chemical exchange then the apparent T_1 of each merged peak is simply the weighted average of the T_1 values of the constituent resonances. This fact is commonly exploited in the study of ligand binding to macromolecules or to paramagnetic ions, because a large excess of ligand increases sensitivity and avoids the difficulties associated with the measurement of very short T_1 values and of very broad resonances. Care must be taken, however, to ensure that the exchange really is rapid, and that the T_1 in the absence of the broadening agent is accurately known. Also the detailed interpretation may need to allow for more than one type of ligand interaction. If the ligand visits many parts of a macromolecule, then it may be used as the perturbing agent. For example, even residual solvent protons may be irradiated so that NOE transfer from them reveals unprotonated carbon atoms, such as carboxylate groups, exposed on the exterior of a protein.

A second important case arises when the chemical exchange is sufficiently slow to permit resolution of separate peaks, but still takes place largely within the time interval T_1. In this case one can irradiate one resonance and measure the fractional reduction in intensity at the other, as the spin saturation transfers. Hoffmann and Forsén[60] have shown that in the case of a reaction

$$\mathrm{A} \underset{1/\tau_A}{\overset{1/\tau_B}{\rightleftharpoons}} \mathrm{B},$$

where B is saturated, the fractional reduction is to $\tau_A/(\tau_A + T_{1A})$. T_{1A} may either be measured directly or extracted by measuring the fractional intensity as a function of time. This varies as

$$\left[\left(1 - \frac{\tau_{1A}}{T_{1A}}\right)\exp\left(-\frac{t}{\tau_{1A}}\right) + \frac{\tau_{1A}}{T_{1A}}\right] \tag{55}$$

where

$$\frac{1}{\tau_{1A}} = \frac{1}{T_{1A}} + \frac{1}{\tau_A} \tag{56}$$

In this way one can measure τ for moderately slow-exchange reactions.[61] More complex chemical systems require multiple saturation experiments. The technique may also be used qualitatively with spin-1/2 nuclei to reveal the course of such reactions as metal-hydride catalyzed hydrogenation[62] and organophosphate metabolism *in vivo*.[63]

7. PARAMAGNETIC INTERACTIONS

The very large magnetogyric ratio of the electron ensures that it dominates T_1 and T_2 in almost all samples containing unpaired electrons. Exceptions to this arise only when the paramagnetic center remains sterically remote, as for some resonances of metalloproteins. In more normal cases, such as metal ions in solution, one can extract many details concerning solvation, kinetics, and binding sites by studying paramagnetically induced relaxation and shifts. Such studies are particularly appropriate with lighter $I > 1/2$ nuclei because the linewidths of such resonances are less liable to misinterpretation.

7.1. Kinetics

Swift and Connick[64,65] first developed a generally applicable equation for the line shape $v(\omega)$ of resonance A of mole fraction P_A in the presence of a small fraction P_B of paramagnetic species B:

$$v_A(\omega) = \frac{\omega_1 M_0(T_{2A}^{-1} + T_{2P}^{-1})}{(T_{2A}^{-1} + T_{2P}^{-1})^2 + (\Delta\omega_A + \Delta\omega_P)^2} \tag{57}$$

where

$$T_{2P}^{-1} = \frac{1}{\tau_A}\left[\frac{T_{2B}^{-2} + (T_{2B}\tau_B)^{-1} + (\omega_{0(B)} - \omega)^2}{(T_{2B}^{-1} + \tau_B^{-1})^2 + (\omega_{0(B)} - \omega)^2}\right] \tag{58}$$

and

$$\Delta\omega_P = \frac{(\omega_{0(B)} - \omega)/\tau_A\tau_B}{(T_{2B}^{-1} + \tau_B^{-1})^2 + (\omega_{0(B)} - \omega)^2} \tag{59}$$

$$\Delta\omega_A = \omega_{0(A)} + \omega \tag{60}$$

In these equations, ω_1 is the Larmor frequency, M_0 the static magnetization, $\omega_{0(B)}$ the resonance position of the paramagnetically bound species, and T_{2B} its T_2. The times τ_A and τ_B are mean residence times sites A and B, assuming first-order kinetics. The derivation of equations (57)–(60) is essentially from equations (48)–(54).

Figure 10 shows a typical plot of log T_2 against temperature assuming an Arrhenius temperature dependence of reaction rate, for a case where $\omega_{0(B)} - \omega$ is large compared with T_{2B}^{-1}. A simple explanation of the shape of the curve is that $1/T_2$ reaches a maximum when the exchange rate between the A and B sites becomes comparable to their frequency separation.

Swift and Connick's equation is not easy to apply to experimental linewidths in the common case where $\omega_{0(B)}$ and T_{2B} are not directly measurable, and fitting errors can become especially large when in addition only a portion of the curve corresponding to that in the upper part of Figure 10 is delineated.[66] However, further information is obtainable from the measurement of chemical shift, as in the lower part of Figure 10, which is also described by the same equation. Shift and width data should be simultaneously fitted in order to obtain a more accurate measure of the unknown parameters. There is as yet no agreed relative weighting for these two data items, which means that the same data can be manipulated to give variable results. However, $\omega_{0(B)}$ can in principle be obtained both through curve fitting and also directly, and where this has been done there is good agreement provided the Swift–Connick equations are modified as below.

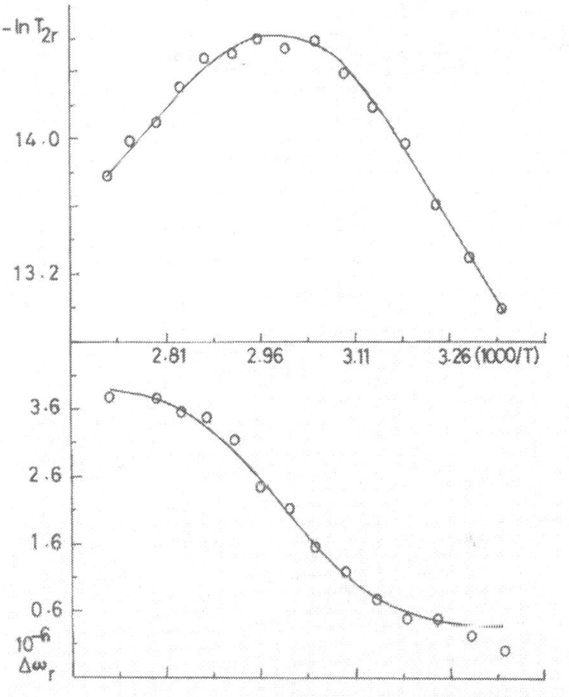

Figure 10. Temperature variation of the linewidths (upper) and shifts (lower) for the ^{17}O resonance of dimethylsulfoxide containing $[Co^{II}(dmso)_6]^{2+}$, fitted to the Swift–Connick equations. The substantial paramagnetic broadening of the directly bound ^{17}O atom, together with its significant natural linewidth, obviates the need for an outer-sphere correction term. Spectra kindly provided by Drs. P. Moore and C. H. McAteer.

Several workers, including the originators, have commented on the approximations inherent in the original equations. One minor source of error is that the electron T_1 is assumed to be negligibly short. Johnson and Grant[(67)] discuss the limits of this approximation, and show that it only brings serious error for such species as spin-labels. But more important errors arise in the estimation of T_{2A} and in the assumption that $\omega_{0(D)}$ is independent of temperature. T_{2A} is normally taken to be the solvent linewidth in the absence of paramagnetic species, or preferably that in the presence of a chemically similar diamagnetic species. But in either of these cases errors will arise because even in the absence of exchange some broadening will arise from outer-sphere association between, or even from the statistical proximity of the ligand and the paramagnetic center. This will not only reduce T_{2A} but will also have a different activation enthalpy from the main inner-sphere process. Both these parameters must be obtained in any complete fit to experimental data, and indeed such data offer one of the few available methods for investigating outer-sphere interactions. Outer-sphere broadening is less significant when quadrupolar nuclei are observed.

7.1.1. *Scalar Shifts*

The second main source of error arises because $\omega_{0(B)}$, the resonance position of the bound species, is not independent of temperature. One simple assumption to allow for this is that of a Curie-law dependence, as described below, i.e., $\omega_{0(B)} \propto 1/T$.

However, better fits are obtained when a term in $1/T^2$ is also added. The alternative practice of adding a temperature-independent term has been criticized.[66]

Ideally, one would like to be able to calculate $\omega_{0(\mathrm{B})} - \omega_{0(\mathrm{A})}$ from first principles. Bloembergen[68] was the first to attempt this for exchanging species, and McConnell and Chesnut[69] for inert species. They started by noting that the scalar paramagnetic shift arises because an electron-nuclear scalar coupling exists in the bound species, and that this is of sufficient magnitude to ensure unequal population of the two components of the NMR spectrum. Therefore, although rapid exchange will largely average the position of the two $m_s = \pm 1/2$ components, the resulting resonance position will be biased towards the lower-energy component. The resulting scalar shift, in units of $10^{-6}\,\delta$, will be

$$+2\pi g_{\mathrm{eff}} \mu_{\mathrm{B}} a_{\mathrm{N}} S(S+1)/3h^2 \gamma_N kT \tag{61}$$

where g_{eff} is the effective electronic g value, μ_{B} the electronic Bohr magneton, and S its spin quantum number. a_{N} is the scalar coupling constant in hertz and k is Boltzmann's constant.

The above expression was derived assuming a simple two-level electron Zeeman system. Deviations from the Curie law which it represents may arise from such complexities[70] as other couplings, unquenched orbital angular momentum, electron relaxation, and zero-field splittings. If the paramagnetic center is anisotropically coordinated then two further corrections become necessary. A minor correction is that g_{eff} is replaced by $(g_{\parallel} + 2g_{\perp})/3$. A major correction arises because the asymmetry of g ensures that dipolar coupling between the nucleus under study and the electronic magnetic moment no longer averages to zero. Thus a pseudocontact term must be added[70] to expression (61). If θ is the angle between the electron–nucleus vector (of length r) and the axis of symmetry of g, then the new term is

$$\frac{\mu_{\mathrm{B}}^2 S(S+1)}{9kTr^3}(1 - 3\cos^2\theta)(g_{\parallel}^2 - g_{\perp}^2) \tag{62}$$

in the same units of $10^{-6}\,\delta$. Little study has been made of such pseudocontact interactions in the case where the entire coordination shell is highly labile, but in better-defined cases it has been much exploited in the mapping of molecular conformations in solution.

Contact and possibly pseudocontact shifts of a few parts per million can arise via weak interactions which are commonly ignored, such as those between stable organic free radicals and halocarbons. [Cr(acac)$_3$] is commonly used as a relaxation agent because of its relative freedom from such interactions, but care must be taken even here. Care must also be taken to use internal standards in the experimental determination of paramagnetic shifts, in order to avoid the need for bulk susceptibility corrections.

7.2. Paramagnetic Contributions to T_1 and T_2

The paramagnetic contribution of an electron spin S to the T_1 of a nucleus I bound at a distance r is given by

$$1/T_1 = \frac{2\gamma_{\mathrm{I}}^2 g_{\mathrm{eff}}^2 \mu_{\mathrm{B}}^2 S(S+1) f(\tau)}{15r^6} + \frac{2S(S+1)\,a_{\mathrm{N}}^2}{3\hbar^2}\left[\frac{\tau_{e2}}{1 + (\omega_{\mathrm{S}} - \omega_{\mathrm{I}})^2 \tau_{e2}^2}\right] \tag{63}$$

In the first term, which arises from dipolar coupling, $f(\tau)$ is slightly modified from the nuclear–nuclear case, being given by

$$f(\tau)=\frac{\tau_{c2}}{1+(\omega_S-\omega_I)^2\tau_{c2}^2}+\frac{3\tau_{c1}}{1+\omega_I^2\tau_{c1}^2}+\frac{6\tau_{c2}}{1+(\omega_S+\omega_I)^2\tau_{c2}^2} \tag{64}$$

where

$$1/\tau_{c1}=1/T_{1e}(\text{electron})+1/\tau_c(\text{rotation})+1/\tau_e(\text{electron exchange}) + 1/\tau(\text{chemical exchange}) \tag{65}$$

and

$$1/\tau_{c2}=1/T_{2e}(\text{electron})+\text{same three minor terms} \tag{66}$$

S, g_{eff}, μ_B, a_N, and r are as above. In the second, scalar term, T_{2e} normally dominates τ_{2e}, and the term is only dominant if a large a_N occurs to a relatively distant nucleus. When this is not the case, and T_{1e} dominates, and τ_c is short, then equation (63) reduces to

$$1/T_1=4S(S+1)\,\gamma_I^2\, g_{\text{eff}}^2\mu_B^2\, T_{1e}/3r^6 \tag{67}$$

The corresponding full equations for $1/T_2$ are

$$1/T_2=\frac{\gamma_I^2 g_{\text{eff}}^2\mu_B^2 S(S+1)\,f'(\tau)}{15r^6}+\frac{S(S+1)\,a_N^2}{3\hbar^2}\left[\tau_{e1}+\frac{\tau_{e2}}{1+(\omega_S-\omega_I)^2\tau_{e2}^2}\right] \tag{68}$$

where

$$f'(\tau)=4\tau_{c1}+f(\tau)+\frac{6\tau_{c2}}{1+\omega_s^2\tau_{c2}^2} \tag{69}$$

The above relationships are exploited in the use of shift reagents. These reagents, mainly stable complexes of Pr(III), Eu(III), and Dy(III) with a vacant coordination site, have electronic configurations which give them exceptionally short T_{1e}. Hence $f'(\tau)$ becomes quite small, and good resolution is maintained despite large pseudocontact shifts. This advantage is somewhat lost at high magnetic fields, where chemical exchange terms as in Section 7.1 become significant.

Gd(III) and some of its complexes provide a counterpart to shift reagents. They are used as relaxation agents, because T_{1e} is relatively long, and S is substantial. Also the g tensor is almost isotropic, so that pseudocontact shifts are small. Their value lies particularly in the r^{-6} dependence of $1/T_2$, which means that nuclei near to the Gd binding site are relaxed highly preferentially. When the binding is kinetically labile then T_2 may be conveniently controlled by using mixtures of Gd(III) and diamagnetic La(III) in varying proportion.

8. TWO-DIMENSIONAL NMR

Two-dimensional NMR is now generally available on larger commercial spectrometers with hard disks, and it offers elegant alternatives[71] for studying both chemical exchange and some relaxation processes, as well as other problems. The essence of two-dimensional NMR is to trace the effects of some interaction or perturbation of the spins under study as a function of "before" and "after" frequencies. Chemical exchange provides a simple and elegant example.[72] Consider the following multiple pulse experiment on two uncoupled spins A and B with Larmor frequencies ν_A and ν_B. Both magnetizations are first turned through 90° about the x axis in the rotating frame by a single pulse. They then precess at their separate frequencies in the xy plane. After a suitable interval they are subjected to a second 90° pulse about the x axis.

For an instant their magnetization is forced to lie in the xz plane, but then it continues to precess about the z axis at $\Delta\nu_A$ or $\Delta\nu_B$, the residual rotating frame frequencies. A period of typically 0.05 s now ensues during which two things occur. The magnetization decays via T_2 (and to a much smaller extent via T_1) onto the z axis.

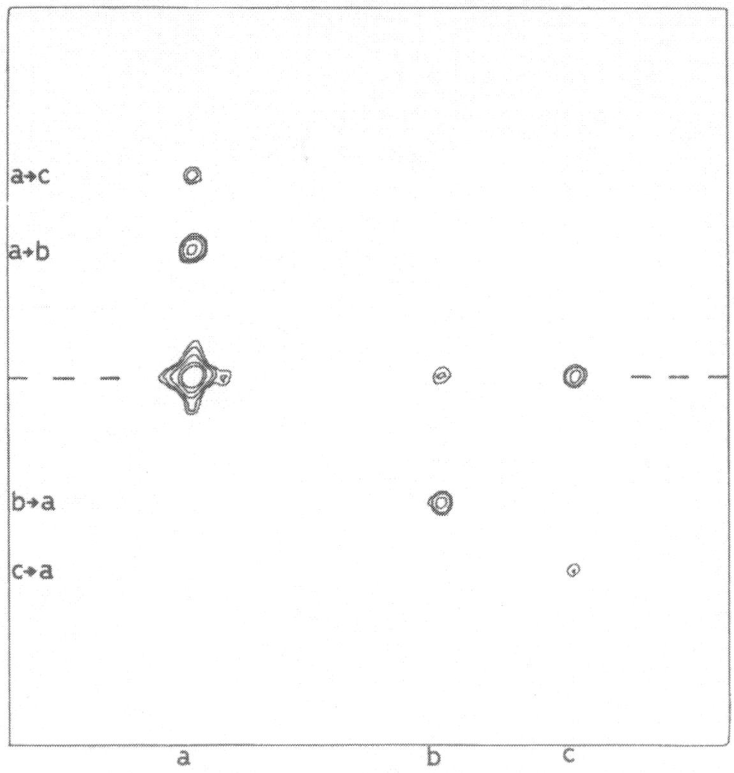

Figure 11. ^{51}V two-dimensional exchange experiment on vanadate equilibria. The unexchanged peaks, viewed from above, lie along the horizontal line indicated by dashes. Peaks lying off this line, on a halved vertical scale, arise from spins which have exchanged environments during the 0.02-s period allowed for exchange. a = monomer, b = dimer, c = cyclic tetramer ($V_4O_{12}^{4-}$). Monomer–dimer exchange is prominent, and as expected the less abundant dimer has a higher probability of becoming monomer than of remaining unchanged. Total time ca. 1/2 h.

Also chemical exchange occurs. The experiment is completed at the end of this period by a third 90° pulse immediately followed by data collection. The final frequency of precession may be measured directly as the NMR signal, and the initial frequency of precession may also be measured, indirectly, because the restarting angle in the xz plane following the second pulse equals the angle reached in the original xy plane after time t. Experimentally, the final signal is sinusoidally modulated in amplitude at the initial frequency, and hence this frequency may be extracted by repeating the experiment for many t values and then performing a second Fourier transformation of the normally transformed signals using the successive values of t as the timebase. The initial and final frequencies are then plotted on a square graph.

Of course, if no exchange takes place, then the resulting two-dimensional spectrum only has peaks along its diagonal, and reduces to no more than an expensive and time-consuming way of doubly measuring the ordinary spectrum. However, if some spins change $A \leftrightarrow B$ during this interval, these will then have different "before" and "after" frequencies, and so will appear as off-diagonal peaks in the two-frequency plot. The value of this method is that when the exchange processes are complex, the two-dimensional spectrum will often clearly reveal the predominant exchange pathways via the larger off-diagonal peaks, and exclude other possibilities. It is thus preferable in many respects to line-shape analysis. Figure 11 is an example of such an experiment.

8.1. Shift-Correlation Experiments

Chemical exchange is not the only way in which two spins may reveal a mutual connection not obvious in their 1D spectra. The basic method given above, for remembering an initial Larmor frequency by converting it into the modulation of a magnetization, may in principle be extended. The kinetic exchange may be replaced by any process capable of conveying magnetization between different spins. If these processes only involve the two spins in question, they are called single-quantum processes. If they involve more spins, including perhaps some not directly observed, then they are termed multiple-quantum processes.

8.1.1. Homonuclear Correlated 2D NMR Spectroscopy (COSY, SECSY)

The object of this family of experiments is to correlate the chemical shifts of mutually coupled spins. Thus they replace tedious multiple decoupling experiments, and have the additional advantage of high specificity and an absence of Bloch–Siegert frequency distortions.

The most useful experiment of all is the homonuclear COSY (two-dimensional homonuclear COrrelated SpectroscopY) experiment.[73] This yields a square matrix of data. The diagonal, projected onto either frequency axis, is the normal one dimensional spectrum, and off-diagonal multiplets appear with the x coordinate of resonance A and the y coordinate of resonance B or vice versa, when A and B are mutually spin-coupled. A two-dimensional spectrum of this type is most conveniently presented as a contour diagram, as viewed from above, even though such maps seem at first strange to one-dimensional spectroscopists.

In the COSY experiment the mutual transfer of polarization between A and B is accomplished by an INEPT-type method, simplified by omission of the two 180° pulses. One component of a spin multiplet (say of A) is inverted relative to the other after a time t which ideally equals $1/2J_{AB}$, and this alters the B magnetization as described in Section 5.3.2. Thus the B magnetization becomes modulated by the Larmor frequency of A, and vice versa.

This is very simply achieved in practice either by the repeated sequence (90°–t–90°–FID) or (90°–t–45°–FID), where t is a variable delay whose range is chosen to suit the probable range of J values in the system. The 45° option has slightly reduced sensitivity, but greater clarity near the diagonal. In a typical case the COSY experiment will detect J couplings appreciably smaller than the linewidth. It is thus most useful for tracing ^{11}B–^{11}B connectivities in boranes, as well as for spin-1/2 nuclei such as ^{1}H, ^{31}P and ^{29}Si, and some examples are given in the relevant chapters of this book. It has even been employed to trace ^{183}W couplings in polytungstate complexes.

A further advantage of COSY is its relative insensitivity to spectrometer error. More elaborate two-dimensional experiments may well be degraded by poor pulse settings, variations of pulse amplitude over the sample volume or the frequency range, and phase errors. Some of the errors that do arise in COSY can be concealed by symmetrization, a computer process that enforces exact symmetry about the two-dimensional diagonal by cutting down anomalously high peaks. This advantage is lost in the closely related SECSY experiment, where the data is not symmetrically displayed. The gain in SECSY is the use of less data points. COSY and SECSY yield J values directly at high resolution, and they can reveal the relative signs of coupling constants in appropriate cases.

A comparable range of heteronuclear shift correlation experiments is also available, with and without multiplet decoupling. Again, the spins are correlated by polarization transfer. A more elaborate homonuclear shift correlation experiment, INADEQUATE, may be performed to detect ^{13}C–^{13}C connectivities[74] and hence identify carbon frameworks. The elaborations involve double-quantum processes to eliminate the otherwise dominant signals from ^{13}C bound to ^{12}C. These and other methods are fully discussed by Sanders[74] and by Ernst *et al.*[75,76]

8.1.2. Two-Dimensional NOE Spectroscopy (NOESY)

The NOESY experiment[77] gives data similar to the above SECSY or COSY experiment, but now the off-diagonal peaks reveal dipolar-coupled resonances, such as might be studied one-dimensionally via NOE difference spectroscopy. The sequence in its simplest form is

$$90°–t–90°–\text{dipolar mixing delay}–90°–\text{measure} \qquad (70)$$

During the dipolar mixing delay, where this delay $\simeq T_1$, the relatively weak dipole–dipole interactions have time to produce essentially the same transfer of magnetization as chemical exchange, albeit less efficiently. The second and third 90° pulses serve to turn the initial rotating magnetization into an amplitude modulation of the final one. Inspection of the above sequence reveals that the first two pulses simultaneously perform a SECSY or COSY experiment, so that off-diagonal peaks can arise from both scalar and dipolar coupling. However, the scalar peaks may be markedly reduced in intensity if the mixing delay is randomly varied, and the resulting FID's added so as to permit phase cancellation.

Many other variations are also possible in the above experiments. One usually introduces phase-cycling to eliminate artifacts, and one may adjust the delay patterns so as to turn the diagonal into an x axis with the cross-peaks altered only in their y

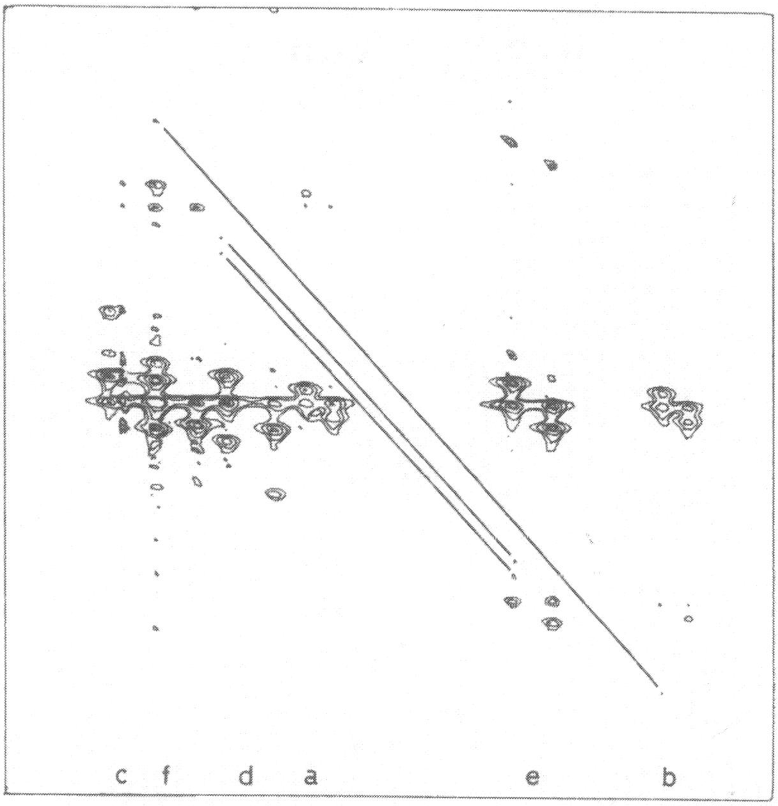

Figure 12. ^{1}H two-dimensional NOE experiment on $[Co^{III}(edta)]^-$, as an alternative to the experiments shown in Figure 6. The format is as in Figure 11, but the spectra are complicated by the additional appearance of all *J*-coupled peaks. The important nonbonded interactions *bc* and *de* are marked. The experiment also confirms the absence of interactions such as *af*, which could not be checked via NOE difference experiments because of crowding. Total time ca. 2 h.

coordinates. Figure 12 is an example of such a NOESY spectrum. It comes from the same $[Co^{III}(edta)]^-$ species as in Figure 6, but with the advantage that now one can penetrate the most crowded region of the spectrum, and thus be sure that no other pairwise dipolar interactions have been missed. There is, however, one disadvantage. The cross-peaks are quite weak, and are not always reliably distinguishable from noise.

8.2. *J*-Resolved Two-Dimensional Spectroscopy

This is perhaps the best known two-dimensional experiment. Its purpose is to turn all spin multiplets through 90°, and thus to disentangle their overlaps.[71] The experiment has obvious value for crowded coupled spectra such as those of silicate oligomers.

The actual experiment is deceptively similar to SECSY, with the only important difference being that the second 90° pulse is replaced by a 180° one. This latter pulse *does not mix* the A and X spins, however. It merely inverts each, so that each component of each multiplet simply becomes its partner. Thus where one A spin was initially precessing faster than average due to the extra field from an "up" X spin, it is bound to finish precessing slower than average under the influence of the same spin, not "down." In the homonuclear case A and X are necessarily inverted together, whereas in the heteronuclear case separate pulses are required, usually on parallel coils.

In this experiment all motion after the first pulse occurs in the *xy* plane, and the second frequency is detected as a phase modulation of the resulting spin-echo (Section 5.3).

Suitable scaling produces the desired spectrum, which also has the advantage that the J axis is not affected by inhomogeneity in B_0, so that very accurate coupling constants may be obtained. Once again, however, there is a drawback. Markedly second-order multiplets give unsatisfactory spectra with unwanted extra peaks, and even first-order spectra have nonstandard peak intensity ratios. The method is also prone to artifacts.

REFERENCES

1. Bloch, F. *Phys. Rev.* 1946, **70**, 460–474.
2. Pople, J. A.; Schneider, W. G.; Bernstein, H. J. *High-Resolution Nuclear Magnetic Resonance*; McGraw-Hill: New York, 1959; pp. 31–40.
3. Lynden-Bell, R. M.; Harris, R. K. *Nuclear Magnetic Resonance Spectroscopy*; Nelson: London, pp. 122–132.
4. Howarth, O. W.; Richards, R. E. *Prog. Nucl. Magn. Reson. Spectrosc.* 1966, **1**, 1–14.
5. Slichter, C. P. *Principles of Magnetic Resonance*; Harper and Row: New York, Evanston, London, 1963.
6. Campbell, I. D.; Freeman, R. *J. Magn. Reson.* 1973, **11**, 143–162.
7. Hall, L. D.; Hill, H. D. W. *J. Am. Chem. Soc.* 1976, **98**, 1269–1270.
8. Solomon, I.; Bloembergen, N. *J. Chem. Phys.* 1956, **25**, 261–266.
9. Overhauser, A. W. *Phys. Rev.* 1953, **92**, 411–415.
10. Carrington, A.; McLachlan, A. D. *Introduction to Magnetic Resonance*; Harper and Row: New York, 1979, pp. 229–233.
11. Abragam, A. *The Principles of Nuclear Magnetism*; Oxford, 1961; pp. 333–338.
12. Poole, C. P. Jr.; Farach, H. C. *Relaxation in Magnetic Resonance*; Academic Press: New York and London, 1971.
13. Howarth, O. W. *Theory of Spectroscopy*; Nelson: London, 1973; Chap. 4.
14. Reference 10; pp. 254–257.
15. Reference 2; p. 205.
16. Woessner, D. E. *J. Chem. Phys.* 1962, **36**, 1–4.
17. Woessner, D. E. *J. Chem. Phys.* 1962, **37**, 647–654.
18. Brown, A. J.; Howarth, O. W.; Moore, P.; Bain, A. D. *J. Magn. Reson.* 1977, **28**, 317–319.
19. Mayne, C. L.; Grant, D. M.; Alderman, D. W. *J. Chem. Phys.* 1976, **65**, 1684–1695.
20. Van Putte, K. *J. Magn. Reson.* 1971, **5**, 367–375.
21. London, R. E. *J. Am. Chem. Soc.* 1978, **100**, 2678–2685.
22. Howarth, O. W. *J. Chem. Soc., Faraday Trans. 2*, 1979, **75**, 863–873.
23. Howarth, O. W.; Lian, L. Y. *J. Chem. Soc., Perkin Trans. 2*, 1982, 263–267.
24. Forsén, S.; Lindman, B. *Chem. Br.* 1978, **14**, 29–35.
25. Keepers, J. W.; Janies, T. L. *J. Am. Chem. Soc.* 1981, **104**, 929–939.
26. Howarth, O. W. *J. Chem. Soc., Faraday Trans. 2*, 1980, **76**, 1219–1223.
27. Sandström, J. *Dynamic NMR Spectroscopy*; Academic Press: New York and London, 1982; pp. 59–61
28. Kubo, R.; Tomita, K. *J. Phys. Soc. Jpn* 1954, **9**, 888–919.
29. Redfield, A. G. *IBM J. Res. Dev.* 1957, **1**, 19.
30. Reference 5, p. 47.
31. Noggle, J. H.; Schirmer, R. E. *The Nuclear Overhauser Effect*; Academic: New York, 1971.
32. Poulsen, F. M.; Hoch, J. C.; Dobson, C. M. *Biochemistry* 1980, **19**, 2597–2607.

33. Ham, N. S.; Jeffery, E. A.; Mole, T.; Stuart, S. N. *J. Chem. Soc. Chem. Commun.* 1967, 254–255.
34. Howarth, O. W., unpublished observations.
35. Lynden-Bell, R. M. *Mol. Phys.* 1975, **29**, 301–303.
36. Gierke, T. D.; Flygare, W. H. *J. Am. Chem. Soc.* 1972, **94**, 7277–7283.
37. Lehn, J. M.; Kintzinger, J. P. *Nitrogen NMR*; Witanowski, M.; Webb, G. A., Ed.; Plenum: London, 1973; p. 108.
38. Halle, B.; Wennerstrom, H. *J. Magn. Reson.* 1981, **44**, 89–100.
39. Hertz, H. G. *Z. Electrochem.* 1961, **65**, 20–36.
40. Lucken, E. A. C. *Nuclear Quadrupole Coupling Constants*; Academic: London, 1969; Chap. 5.
41. Townes, C. H.; Dailey, B. P. *J. Chem. Phys.* 1949, **17**, 782–796.
42. Reference 40, Chap. 7.
43. Saito, Y. *Can. J. Chem.* 1965, **43**, 2530–2534.
44. Bancroft, G. M.; Clark, H. C.; Kidd, R. G.; Rake, A. T.; Spinney, H. G. *Inorg. Chem.* 1973, **12**, 728–731.
45. Freeman, R.; Hill, H. D. W. *J. Chem. Phys.* 1971, **54**, 3367–3377.
46. Grant, C. W. M.; Hall, L. D.; Preston, C. M. *J. Am. Chem. Soc.* 1973, **95**, 7742–7747.
47. Harris, R. K.; Newman, R. H. *J. Magn. Reson.* 1976, **24**, 449–456.
48. Carr, H. Y.; Purcell, E. M. *Phys. Rev.* 1954, **94**, 630–638.
49. Freeman, R.; Hill, H. D. W. *Dynamic Nuclear Magnetic Resonance Spectroscopy*; Jackman, L. M.; Cotton, F. A., Ed.; Academic: New York, 1975, Chap. 5.
50. Hartmann, S. R.; Hahn, E. L. *Phys. Rev.* 1962, **128**, 2042–2053.
51. Morris, G. A.; Freeman, R. *J. Am. Chem. Soc.* 1979, **101**, 760–762.
52. Brevard, C.; Van Koten, G.; Van Stein, G. *J. Am. Chem. Soc.* 1981, **103**, 6746–6748.
53. Kaplan, J. I.; Fraenkel, G. *NMR of Chemically Exchanging Systems*; Academic Press: New York and London, 1980.
54. *NMR of Paramagnetic Molecules*, La Mar, G. N.; Horrocks, W. de W. Jr.; Holm, R. H., Ed.; Academic: New York, 1973.
55. *Dynamic Nuclear Magnetic Resonance Spectroscopy*, Jackman, L. M.; Cotton, F. A., Ed.; Academic Press: New York and London, 1975.
56. Swift, T. J. as Ref. 54, Chap. 2.
57. Fyfe, C. A.; Cocivera, M.; Damji, S. W. H. *Acc. Chem. Res.* 1978, **11**, 277–282.
58. Couch, D. A.; Howarth, O. W.; Moore, P. *J. Phys. E.* 1975, **8**, 831–833.
59. Brown, A. J.; Howarth, O. W.; Moore, P.; Parr, W. J. E. *J. Chem. Soc., Dalton Trans.* 1979, 1776–1781.
60. Forsén, S.; Hoffmann, R. A. *J. Chem. Phys.* 1964, **40**, 1189–1196.
61. Mann, B. E. *J. Magn. Reson.* 1977, **25**, 91–94.
62. Howarth, O. W.; McAteer, C. H.; Moore, P.; Morris, G. E. *J. Chem. Soc. Chem. Commun.* 1981, 506–507.
63. Nunnally, R. A.; Hollis, D. P. *Biochemistry* 1979, **18**, 3642–3646.
64. Swift, T. J.; Connick, R. E. *J. Chem. Phys.* 1962, **37**, 307–319.
65. Swift, T. J. in Ref. 54, pp. 65–66.
66. Newman, K. E.; Meyer, F. K.; Merbach, A. E. *J. Am. Chem. Soc.* 1979, **101**, 1470–1476.
67. Johnston, E. R.; Grant, D. M. *J. Magn. Reson.* 1982, **47**, 282–291.
68. Bloembergen, N. *J. Chem. Phys.* 1957, **27**, 595–596.
69. McConnell, H. M.; Chesnut, D. B. *J. Chem. Phys.* 1958, **28**, 107–117.
70. Jesson, J. P. in Ref. 55, Chap. 1.
71. Aue, W. P.; Bartholdi, E.; Ernst, R. R. *J. Chem. Phys.* 1976, **64**, 2229–2246.
72. Meier, B. H.; Ernst, R. R. *J. Am. Chem. Soc.* 1979, **101**, 6441–6442.
73. Nagayama, K.; Wütrich, K.; Ernst, R. R. *Biochem. Biophys. Res. Commun.* 1979, **90**, 305–311.
74. *Modern NMR Spectroscopy*, Sanders, J. K. M.; Hunter B. K.; Oxford: 1987
75. *Principles of Nuclear Magnetic Resonance in One and Two Dimensions*, Ernst, R. R.; Bodenhausen, G.; Wokaun, A.; Oxford: 1986.
76. Bax, A.; Freeman, R.; Frenkiel, T. A. *J. Am. Chem. Soc.* 1981, **103**, 2102–2104.
77. Jeener, J.; Meier, B. H.; Bachmann, P.; Ernst, R. R. *J. Chem. Phys.* 1979, **71**, 4546–4553.

Chapter 6

HYDROGEN AND ITS ISOTOPES

Hydrogen, Deuterium, and Tritium

J. W. AKITT

1. INTRODUCTION

Hydrogen, known affectionately yet confusingly as "the proton" to NMR spectroscopists, can be regarded as the mother nucleus of high-resolution study. Its place and contribution to organic chemistry are well documented and its use is extensive, both because of its early emergence in the commercial spectrometer world and because it is the most sensitive, and receptive, of the natural nuclei. There is a very large body of data dealing with the chemical shifts of protons attached to carbon, nitrogen, or oxygen. This will only be discussed here in the form of a summary chart since the space available is insufficient for more. The shift range is small, an 8 ppm span covering the majority of the resonances encountered, and the shifts are determined principally by anisotropy of neighboring groups and atoms, or the electric fields of polar groups. The shift range is extended to low field for hydrogen bonded protons, conventionally thought to be due to even greater electric field effects, and to high field when bonding to the heavier metals is involved because of the effects of the electronic orbitals on these elements. The full chemical shift range for all types of proton is some 70 ppm and the region of interest to organic chemists then is only a fraction of this. Because the proton has spin 1/2, it shows spin–spin coupling to nearby nuclei, including other hydrogen atoms. Coupling to other nuclear species is dealt with principally in the chapters dealing with those nuclei, though we will mention some specialized coupling topics here. Neither will interproton coupling be discussed, though this is a subject which has proved a source of so much structural and conformational information. The study of dynamic processes will, however, receive a brief though inevitably inadequate mention.

Deuterium NMR is now finding wide use, and we are able, proportionally, to provide a reasonably full survey. It is, of course, difficult to dissociate deuterium from proton NMR; it is used extensively as a lock substance, in solvents—particularly where accurate relaxation measurements are required—and as a label. These latter topics do not, however, form a part of ^2H NMR itself.

Tritium NMR is not yet widespread, and has found very little use in inorganic chemistry. Sufficient comment is included to allow a reader to find easy access to the literature for this nucleus.

J. W. AKITT • School of Chemistry, University of Leeds, Leeds LS2 9JT, England.

The nuclear properties of these three nuclei are summarized in the main table of the Appendix. Deuterium is the only one of the three to possess a quadrupole moment. This is comparable in magnitude to that of ^{133}Cs, and so is quite small, though the quadrupole coupling constants of deuterium are such as to make quadrupole relaxation its principal mechanism of relaxation.[1] The resulting linewidths, however, should not be large, at least in small molecules. The small magnetic moment of deuterium also means that substitution of hydrogen will cause marked changes in the relaxation behavior of nearby nuclei, and this can be used to distinguish different contributions to proton relaxation. The quadrupole moment gives rise to a quadrupole splitting of the resonance in oriented systems which is due to local effects, the small magnetic moment eliminating the long-distance spin interactions.

Tritium is a spin-1/2 nucleus and is more sensitive than the proton. It is radioactive and suitable precautions are required when working with it. Fortunately, its high sensitivity, coupled with Fourier transform techniques, means that quite low degrees of enrichment can be used.

Deuterium and tritium NMR have been reviewed,[1,2] but proton NMR has long passed the days of comprehensive reviewing. Many articles will be found dealing with specialized topics and the reader's attention is drawn to three review series.[3–5] Reference to other specialized articles will be found below.

2. EXPERIMENTAL TECHNIQUES

The experimental aspects of proton spectroscopy have recently been described in great detail[6] and require no further comment here. The receptivity of deuterium is rather low, and while it can be observed in natural abundance, it is normal to use it in enriched forms which are fortunately widely available. Its sensitivity is then satisfactory, indeed average when compared with the great majority of nuclei, as is of course required for the standard lock nucleus. Its relaxation times are not so long as those of the proton, though the difference may not be great, but this helps in the rapid accumulation of data. The techniques of tritium NMR parallel those of the proton, the main problem being the containment of radioactivity. This has been found not to be an insuperable problem though it introduces its own unique problems such as that of standardization. Tritiated TMS is, for instance, too volatile for safe handling and other, safer standards have been sought.[7]

3. HYDROGEN OR THE PROTON, OR PROTIUM

The chemical shift range of the proton is summarized in the chart of Figure 1. Three regions may be distinguished: hydrogen bonded protons to low field, protons in organic compounds from about 8 to 0 ppm, and transition metal hydride protons to high field together with other main-group hydrides. Since the organic region is so crowded on this scale, an expanded chart is given in Figure 2 for this region alone. Chemical shift theories have been discussed in an earlier chapter, but one can summarize the data on these two figures by noting that the low field shifts arise because of electric field effects, and the high field shifts of the hydrides by the anisotropy of the metal orbitals. The smaller variations throughout can be ascribed to weaker and long-distance effects of similar type. We will deal briefly with a series of topics moving from

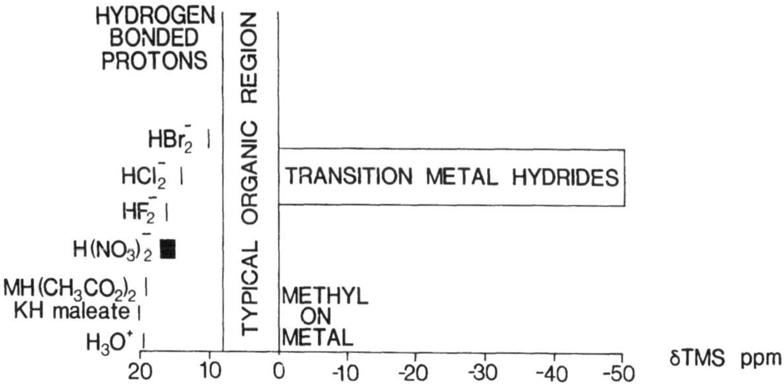

Figure 1. Chemical shift chart for hydrogen, covering the full range of compounds studied. Reference is TMS.

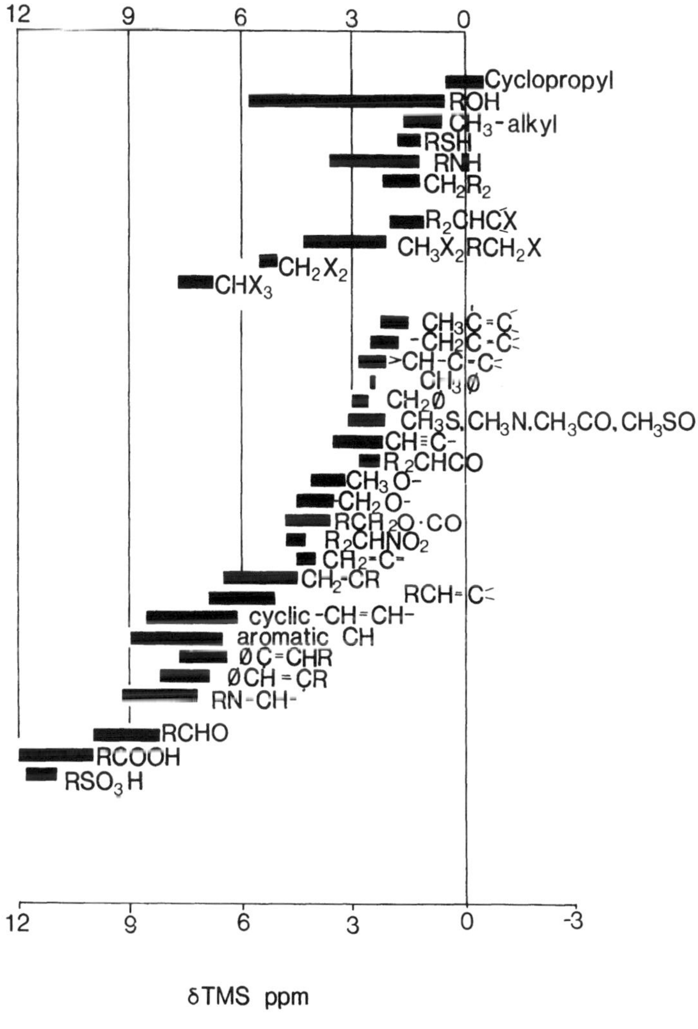

Figure 2. Chemical shift chart for hydrogen in organic compounds. X = halogen, R = organo substituent, ϕ = a phenyl group.

protons with the lowest field shifts to those at highest field and conclude with a mention of dynamic NMR. The discussion will of course be biased towards inorganic and physical chemistry.

3.1. Strong or Weak Hydrogen Bonds

Emsley has reviewed the field of very strong hydrogen bonding, with the declared purpose of convincing the reader that strong and weak bonds are quantifiably different.[8] Several criteria may be used to distinguish the two, principally crystallography and vibrational spectroscopy. The strong hydrogen bond is one in which the hydrogen nucleus is situated roughly equidistant between the two atoms it bridges, which are separated by less than their van der Waals radii. The bond energy is also higher than normal. The chemical shift of the proton involved in the bond is strongly to low field, some typical values being illustrated on Figure 1. The proton in normal, asymmetric hydrogen bonds is also shifted to low field, though not so strongly, and this is often ascribed to an electric field effect, the approach of the negatively charged center to the HX group causing a drift of electrons away from the hydrogen, along the H−X bond. Electric field calculations also provide quite reasonable figures for the shifts of strong hydrogen bonds,[9] even though the proton seems in such cases to be embedded in a bridging electron cloud and the concept of electron drift seems hardly appropriate. There appear to be strong paramagnetic effects about these nuclei.

Weak hydrogen bonds have been much studied and feature in many NMR texts. The shifts are an average due to exchange between bonded and nonbonded molecules and so can be varied by changing such factors as concentration or temperature. Examination of such features allows information about the interaction to be extracted.[10] Water is a particularly important example of hydrogen bonding since this interaction determines its unique structure, and has been studied using all available techniques,[11,12] including NMR. The shift is strongly temperature dependent, changing by some 1 ppm between 0 and 100°C, and can be explained either on the basis of a mixture of bonded and nonbonded molecules (broken or complete hydrogen bonds) or of distorted or bent bonds. The latter model can explain the chemical shifts observed if polarization effects are taken into account.[13–15] A recent article[16] which deals with the changes in the chemical shift of water present in small amounts in a variety of solvent systems illustrates both the properties of the weak hydrogen bond and the way information on the species present may be obtained—in this case monomer H_2O, S · HOH, and S · HOH · S with S = a donor solvent.

3.2. Ionic Solvation

When an inorganic salt is dissolved in certain solvents it is found to dissociate into anions and cations. This phenomenon of ionization has been studied intensively and NMR spectroscopy is a relative newcomer to this field, though one that has supplied new perspectives. The early work (e.g., conductivity measurements) tended to be applicable only to rather dilute solutions, whereas the sensitivity problems of NMR required much more concentrated solutions to be used. The results were thus not strictly comparable, though it is remarkable how well the concepts of the early work have been found to be applicable to the NMR results. With modern instrumentation, of course, we can approach more nearly the low concentrations, though little NMR work of this type has yet appeared. Two different types of investigation may be made using proton spectroscopy depending upon whether the rates of solvent exchange between solvation sphere and bulk solvent are slow or fast on the NMR time scale. With slow exchange, the cationic solvation complex may be observed and its solvation number determined, ion pairing may be observed, or a variety of species may be detected in a mixed solvent system. If exchange is fast then the chemical shift is an ensemble average of all the environments in a solution and has to be studied as a function of some variable such as temperature or concentration for analysis. The results have in both cases provided strong evidence that cationic and anionic solvation are in fact different, while the slow exchange results remind us that solvation spheres have structure and that continuum models are not appropriate in more concentrated solutions.

Systems exhibiting slow exchange are invariably those containing small, highly charged cations, and in order to obtain resolved resonances, it may be necessary to work at low temperatures (ca. $-50°C$), and this means that the freezing point of the solutions must be depressed in some way. For aqueous solutions this is achieved by using high salt concentrations (e.g., $AlCl_3$ at $3\,M$ in water and $-47°C$[17]) or by adding an inert "antifreeze" solvent such as acetone or acetone/Freon. Solvation numbers of six are obtained for Mg^{2+}, the group III metals, Zn^{2+}, Ni^{2+}, and certain lanthanoid cations in the solvents water, DMF, alcohols, and organophosphates, though not all possible combinations of cation and solvent have been studied. For Be^{2+} the solvation number is invariably near four. A selection of references is given to this work, which was carried out principally in the decade prior to the mid 1970s following Schuster and Fratiello's early discovery.[17–27] Where the solvation number obtained is lower than six or four then it is concluded that ion pair formation with the anion is occurring.[25,28–31] These total solvation numbers are retained in mixed solvents and it is then possible to study competitive solvation, though recourse may be necessary to data from other nuclides (^{13}C, ^{17}O or the cation resonance) for a full interpretation. One particularly illustrative example has been published for Al^{3+} in moist methyl cyanide,[32] and Figure 3 shows the sort of information which can be

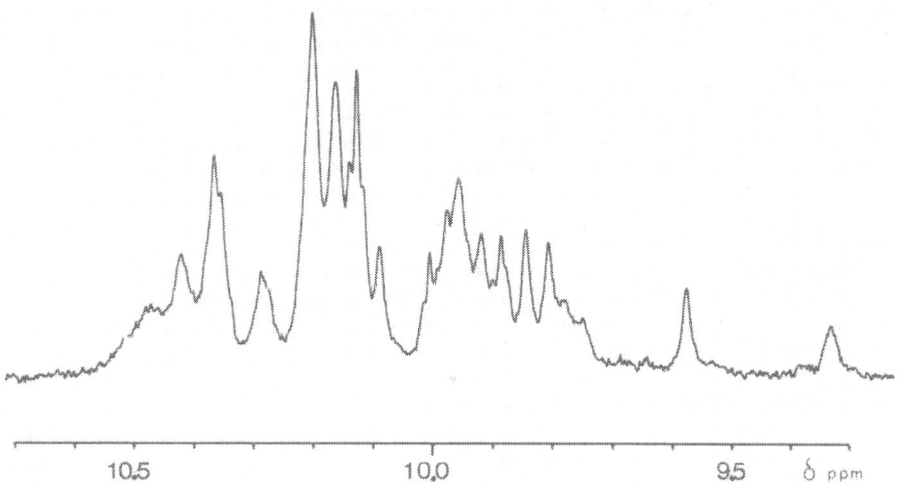

Figure 3. 400-MHz proton spectrum at $-20°C$ of the bound OH in a solution of $AlCl_3$ in moist ethanol. Complexes are formed which contain MeOH and H_2O (seven possible) or MeOH, H_2O and Cl^- in various proportions (15 possible) and which are likely to have differentiable geometrical isomers. Some 18 resonances are observed.

obtained at very high magnetic fields for the system $AlCl_3/EtOH/H_2O$. It will be clear to the reader that, since these spectra are sensitive to temperature changes, kinetic data can be obtained for the exchange of solvent molecules on the cations, and indeed where the solvent is an alcohol it is possible to test whether the rates of hydroxyl proton and whole molecule exchange are different.

Systems where exchange is fast give only a single resonance for the solvent, and information has then to be obtained by observing the shift or linewidth changes which take place when the concentration or temperature is varied. The environment close to paramagnetic cations is much probed in this way and information as to solvent exchange or ion pairing can readily be obtained.[32a] In the case of solutions of diamagnetic salts in water it is found that both the water chemical shift and the rate at which this changes with temperature are altered by the solute. Malinowski has pointed out[33–36] that the change in temperature sensitivity appears to be related to the number of water molecules trapped in the cationic first hydration sphere and which therefore have properties different from those molecules in the bulk liquid. Where the hydration number is known, the correspondence seems good[37–38] and gives some confidence that the values around 4 obtained for the alkali metals by Malinowski are reasonably accurate. Combination of these results with the known shift of bound water allows the anionic effect to be derived, though the results contain a large error due to the poor accuracy with which the temperature may be known in NMR spectrometers. A significant increase in the accuracy of such temperature measurements is needed before further progress can be made in this field.[39,40] Proton spectroscopy has also been used to estimate the degree of dissociation of strong acids. References to the early work are given in Ref. 41 and measurements have been reported quite recently for hydrobromic acid.[42]

3.3. Chemical Shifts of Adducts

The determination of the proton chemical shifts which occur in both donor and acceptor molecules when these form an adduct was a popular field of endeavor in the 1960s. The direction of the shift changes was that to be expected from the direction of electron drift, and NMR offered the prospect of an easy method of determining adduct bond energies. It became clear, however, that the changes observed were quite small, bore little relation to the bond formation constants, and in some cases were in the opposite direction to that required by the simple theory.[43–45] With hindsight it is clear that the shifts are caused by such factors as the effect of the anisotropy of one component of the adduct on the other, or of the electric field of the polar adduct bond. Molecular geometry will therefore be the most important variable in determining the shifts. Where different adducts give distinguishably different shifts it is nevertheless possible to determine relative adduct stabilities in a mixed system, and with a high degree of accuracy.

3.4. Hydrogen on Carbon

This vast subject can receive only a mention here, and that mention must be biased towards inorganic chemistry. The spectra of organic ligands in inorganic compounds are in general what would be predicted from organic NMR spectroscopy, though there are some important differences. Thus large anisotropic ring current effects may be encountered in metallic complexes or metallo-porphyrins, and methyl protons of methyl groups attached to metal atoms often have unusually high field shifts, high field of TMS. The arrangement of the ligands around a metal will lead to differentiation of the ligand signals according to the symmetry of the whole complex. If other nuclei in the complex are NMR active this will also modify the ligand signals, the well-known 1:3.92:1 multiplicity of ligands spin coupled to platinum being a case in point. Coupling to phosphorus proves useful in complexes containing two or more phosphine ligands since *cis* and *trans* ligands can be differentiated due to the effects of the P–P coupling manifested in the proton multiplets of the ligands.[46]

We conclude by recommending a recent and well-illustrated example[47] which indicates how information may be obtained today with available sophisticated techniques. The unsymmetrical porphyrin *N*-methylprotoporphyrin IX (dimethyl ester) can have the *N*-methyl group at one of four distinguishable positions. Identification of the four isomers involved measuring the relative rates of deuteriation of the meso protons followed by a series of NOE difference spectroscopy experiments to identify the various ring methyl protons. Anomolously large chemical shifts of the *N*-methyl (to −4.6 ppm) and meso (to +10.0 to +10.5 ppm) protons are observed due to the anisotropy of the large porphyrin ring.

We give few references in this section and the various specialist reviews should be consulted by the interested reader,[3–5,48] which include an extensive review of the data for metal carbonyl complexes.[49]

3.5. Hydride Protons

These are defined as protons bonded directly to metallic or metalloid elements such as transition metals or boron. In the former case their chemical shifts are strongly to high field of TMS, and this is believed to arise because of anisotropy of the electronic orbitals around the metal.[50] The actual value of the shift in a given complex depends upon the metal complexed and the nature of the other ligands, though it does not seem to be possible at this time to systematize the measurements. If there is more than one hydride hydrogen present in a given complex, then the ligand arrangement may be such that they are chemically inequivalent and so chemically shifted. In such a situation, the hydride hydrogens may show mutual spin–spin coupling. The chemical shifts range from about −3 ppm [e.g., $HRe(PPh_3)_3(CO)_2$] to −50 ppm [$HIrCl_2(Phosphine)_2$],[51] though there are a few complexes with hydride shifts to low field of this range. The hydride protons may also couple to other magnetically active nuclei such as ^{31}P and the magnitude of this coupling in phosphine complexes is indicative of the stereochemistry, being 80–160 Hz to P *trans* to H and only 10–40 Hz if *cis*.[52] This latter reference tabulates data for complexes of hydrides of Ru, Rh, and Ir. The hydride hydrogen in metallic cluster complex hydrides can be present in terminal positions, as bridge hydrogen or coordinated interstitially by the metal atoms and at the center of the cluster. The terminal hydrides come generally at 0 to −15 ppm with the bridging hydrogens to higher field at −15 to −23 ppm, though there are some exceptions.[53] Hydride hydrogen atoms within polyhedra often appear to high field, though in certain cases such as $HRu_6(CO)_{18}^-$ or $HCo_6(CO)_{15}^-$ the shifts are equally strongly to low field (16.4 and 23.2 ppm, respectively). The reason for these differences is not known. The suggestion that the low field protons were associated with a carbonyl group and so were formyl protons is now discounted, and it has been recently suggested that in solution, the central hydride atom in these complexes may

Table 1. Hydride Proton Chemical Shifts of Some Selected Mononuclear Transition Metal Complexes[a]

				Shift δ/ppm			Solvent[b]
Manganese complexes(49)							
$HMn(P(CF_3)_3)(CO)_4$		−6.7 (H, P *cis*), −8.5 (H, P *trans*)					
$HMn(P(CF_3)_2F)(CO)_4$		−6.1 (H, P *cis*), −9.1 (H, P *trans*)					
$HMn(CO)_5$		−7.73					
Cobalt complexes(49)							
$HCo(PF_3)_3(CO)$		−11.4					Et_2O
$HCo(PBu_3)_3(CO)$		−13.9					C_6H_6
$HCo(PPh_3)_3(CO)$		−13.0					THF
		−12.3					C_6H_6
$HCo(PBu_3)_2(CO)_2$		−11.15					C_6H_{12}
		−10.96					THF
		−10.83					DMF
		−10.62					C_6H_6
		−10.55					Pyridine
$HCo(PPh_3)_2(CO)_2$		−10.35					C_6H_6
Ruthenium complexes(52)							
$HRu(NO)(PPh_2\mathbf{R})_3$	**R**	Ph−	Me−	Pr^i−	C_6H_{11}−		CH_2Cl_2
	δ	−6.6	−5.0	−7.2	−7.4		
$HRu\mathbf{X}(PPh_3)_3$	**X**	$MeCO_2$	Cl	Br	NO		
	δ	−19.89	−17.75	−17.11	−6.6		
$HRu\mathbf{X}(HPPh_2)_4$	**X**	SCN	Cl	Br	I	$SnCl_3$	CH_2Cl_2
	δ	−16.23	−17.7	−16.49	−14.08	−10.15	
trans $HRuX(Et_2PCH_2$-$CH_2PEt_2)_2$	**X**		Cl	Br	I		$CHCl_3$
	δ		−22.3	−21.5	−29.8		
$HRu\mathbf{X}(CO)(PPhPr^n_2)_3$	**X**		Cl	Br	I		$CHCl_3$
	δ		−7.5	−7.8	−8.5		
$HRuCl(CO)(PPhPr^n_2)_2\mathbf{L}$	**L**	$PPhPr^n_2$	$PPhMe_2$	PEt_3	$P(OEt)_3$	$AsPhMe_2$	C_6H_6
	δ	−7.5	−6.8	−6.9	−6.2	−8.46	
$HRuCl(CO)_2(PPh_3)_2$	$\delta = -4.5$						
$H_2Ru(CO)(PPh_2Me)_3$	Two signals, singlet at -8.02δ doublet at -6.8δ						THF
Rhodium complexes(52)							
$HRh(PPh_3)_3$		−8.9					
$HRh(PPh_3)_4$		−8.55					
$HRh(CO)_n(PPh_3)_{4-n}$		−9.2 ($n=1$), −9.1 ($n=2$)					
$HRhCl_2(PBu^t_2R)_2$		−31.4 (R = Me)[c]					
$HRhCl_2(PPh_3)_2 \cdot \frac{1}{2}CH_2Cl_2$		−16.1					
$HRhBr(PPh_3)_2SiCl_3$		−13.4					
$H_2RhCl(PPh_3)_2CDCl_3$		−18.2, −10.15					
$HRh(NH_3)_4(H_2O)^{2+}$		−22					
Rhenium complexes(49)							
$HRe(PPh_3)_3(CO)_2$		−3.1					$CHCl_3$
$HRe(PPh_3)_2(CO)_3$		−5.35					$CHCl_3$
$HRe(PPh_3)(CO)_4$		−4.33					
$HRe(PEt_3)_2(CO)_3$		−6.53					
$HRe(PEt_3)(CO)_4$		−5.66					C_6H_6
$HRe(Ph_2PCH_2CH_2PPh_2)(CO)$		−5.2					
$HRe(Ph_2PCH_2PPh_2)(CO)_3$		−3.23					

Table 1. (*Continued*)

		Shift δ/ppm				Solvent[b]
Osmium complexes[49]						
$HOs(Et_2PCH_2CH_2PEt_2)(CO)^+$		-10.4				$CHCl_3$
$H_2Os(PPh_3)_2(CO)_2$		-7.65				THF
$H_2Os(PPh_3)(CO)_3$		-8.00				THF
$H_2Os(CO)_4$		-8.84				THF
$HOs(CO)_4^-$		-10.24				THF
Iridium complexes[52]						
$HIrCl_2(PBu^t_2Me)_2$		-50.5 (record high field shift?)				C_6H_6
$HIr(CO)(PPh_3)_3$		-10.3				C_6H_6
$HIr(CO)_2(PPh_3)_2$		-10.97				CH_2Cl_2
HIr(pyridine-2-thiol)$_2$-(PPh$_3$)$_2$		-21.5				C_6H_6
$HIrCl(PPh(OEt)_2)_4$[c]		*trans* -18.4, *cis* -10.8				$CHCl_3$
HIrX$_2$(P(O-*p*tolyl)$_3$)$_3$	X	Cl$-$	Br$-$	I$-$		C_6H_6
	δ	-16.45	-15.95	-14.70		
$HIrCl_2L_3$	L	P(O-*p*tolyl)$_3$	P(*p*MeOC$_6$H$_4$)$_3$	PPhPr$^n_2$	$AsPh_3$	C_6H_6
	δ	-16.45	-18.7	-21.5	-19.9	
$HIrCl_2(PBu^t_2Me)_2L$	L	CO	$P(OMe)_3$	MeNC	Pyridine	C_6H_6
	δ	-8.36	-11.55	-11.75	-23.8	
$HIrCl_2(CO)(PBu^t_2Et)_2$		-16.24				C_6H_6
$H_2IrCl(PPhMe_2)_3$		-22.0, -10.1				C_6H_6
$H_2IrCl(PPhMe_2)_2(CO)$		-18.36, -7.58				C_6H_6

[a] The data were extracted from the tables of Refs. 49 and 52. It was noted when using the latter compilation that a few δ values have been inserted as if they were τ values, a source of confusion which it is to be hoped will be eliminated in the future by the universal adoption of the δ scale.
[b] Only the hydrogen form of the solvent is noted, though the deuterated form was used in some cases—i.e., it is assumed that any solvent isotope effect is negligible. [c]$CDCl_3$ was used for the first, methylene chloride for the remainder.
[c] Little change with R or different halide.

be able to move away from the central position using some undefined exchange pathway.[54]

Some selected values of hydride chemical shifts are collected in Table 1. These have been arranged to attempt to indicate some trends with change in constitution across a group of related molecules. It is clear, where the evidence is available, that the hydride shifts are very solvent dependent. [A range of 0.7 ppm is reported for $HCo(CO)(PPh_3)_3$ and of 0.6 ppm for $HCo(CO)_2(PBu_3)_2$, for example.[49]] Changing a ligand will alter the interaction with the solvent, and this could account in part for the trends observed with halide substitution and also for these being in opposite directions in $HRuX(PPh_3)_3$ and $HRuX(CO)(PPhPr^n_2)_3$, for instance. Apart from this, the data have been selected simply to provide examples covering the whole range of chemical shifts observed. An example is illustrated in Figure 4.

In agreement with Buckingham and Stephens' theory,[50] hydrogen atoms attached to nontransition elements resonate at lower fields. Thus in the boron hydrides, the terminal boron hydrogens appear within the normal range for organic compounds and the bridging hydrogens are only a little to high field of TMS.[55] The shift of the latter contrasts strongly with that of hydrogen-bonded hydrogen.

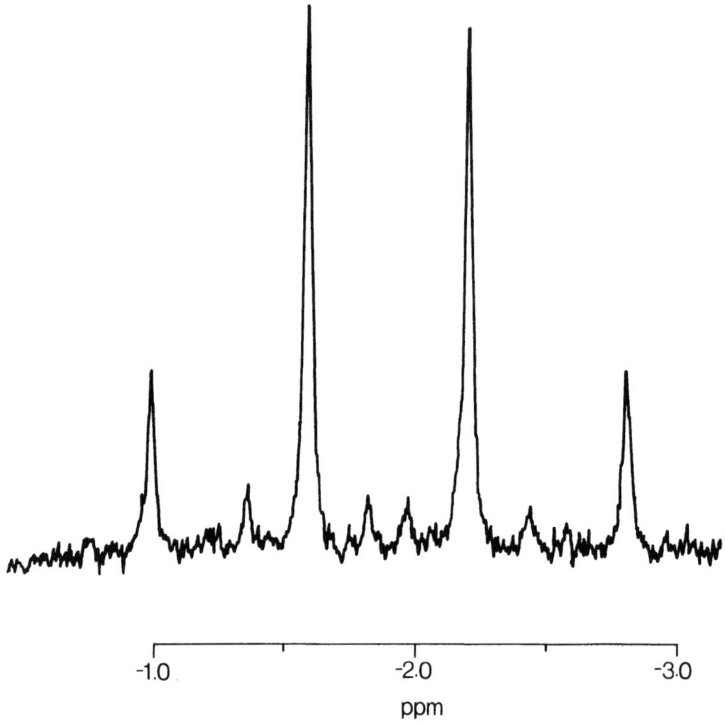

Figure 4. 60-MHz proton spectrum of the hydride proton region of the complex $WH_6(PMe_2Ph)_3$ taken in benzene at 30°C. The spectrum is a quartet due to coupling to three phosphorus atoms [J(P–H) = 36 Hz] and satellite lines can be seen constituting the intense components of the two quartets due to the molecules containing ^{183}W [natural abundance 14.4%, J(W–H) = 28 Hz]. The chemical shift is to high field of TMS at −1.91 ppm. An interesting feature is that there is no way six protons can be equivalent in such a complex so that the simple form of the spectrum must mean that a dynamic process is taking place and scrambling the protons. It is found that no spectral changes occur as the sample is cooled to the lowest attainable temperatures so that the scrambling process is facile. (Example kindly supplied by Professor B. L. Shaw.)

3.6. Dynamic Processes

NMR can be used to follow the rates of a reaction quite simply by taking spectra at intervals as the reaction proceeds, and in this way is used like any other analytical technique. The reaction must take much longer than the time needed to obtain an individual spectrum, though the advent of FT spectroscopy has shortened the time scale considerably and stopped-flow experiments are now feasible.[6] Where NMR has made an unusually important contribution, however, is in the way it has allowed the study of dynamic behavior in systems at equilibrium (see Chapter 2). If a given proton may exist in more than one chemically different environment in such a system, then it will in general have different chemical shifts in these environments. If it moves between these environments at a rate which is comparable with the frequency separation between the chemically shifted lines, then its resonance frequencies are no longer precisely defined and the form of the spectrum changes dramatically. It is possible to compute line shapes and so to estimate the rates of exchange of environments. The form of the spectra changes markedly with temperature, as the rate changes, and various temperature–rate plots then allow thermodynamic parameters to be derived. The techniques for the study of DNMR (the initials seem now to be the accepted label, and one that excludes the use of DMR to describe ^{2}H NMR) have recently been described in detail, and the ways in which the now well-known errors associated with the technique may be minimized have been discussed.[6,58] The ideal appears to be to use the signals of more than one nucleus (e.g., ^{1}H, ^{13}C) and spectrometers operating at different frequencies. Several coalescence points can then be identified and the rate constants obtained without recourse to computation. Like most ideals, this one will not always be attainable. Dynamic processes are common with organometallic compounds, which may exhibit fluxional behavior of a ligand or interchange of ligands between complex molecules, and there is an extensive literature.[56,57]

4. DEUTERIUM

Two reviews of deuterium NMR have been published,[1,2] the second being particularly comprehensive. The nucleus is used principally in two ways: either as an adjunct to proton spectroscopy to help solve certain problems or to investigate oriented materials. There is also an important body of solid state work.

4.1. Deuterium NMR in Isotropic Liquids

It is possible to observe deuterium at natural abundance, though large numbers of transients are necessary.[59] Because proton–deuterium coupling constants are small (1/6.51 of the ^{1}H–^{1}H value), the resonances appear as singlets, though the resolution can be improved by proton broad band irradiation. This means that despite the lower chemical shift dispersion of deuterium, the spectra are simplified and chemical shifts can be easily extracted which may not be available in the second-order proton spectrum of the same compound.[60] If coupling is visible it may not relate to a single interaction and be deceptively simple.[2] Where a strict value of a coupling can be extracted, then the corresponding ^{1}H–^{1}H value can be calculated and so coupling between isochronous protons can be measured by deuterium substitution.[61] This may be better measured from the proton spectrum. However, for an accurate determination, it is necessary that the deuterium quadrupolar relaxation time should not be too short since proton spectral collapse occurs as $\pi T_{1D} J(H-D)$ is reduced to unity.[62] Some values of coupling constants to ^{1}H or ^{13}C are tabulated.[63,64]

Substitution of hydrogen by deuterium in a complex molecule will cause changes in the chemical shift of nearby protons or carbon. Typical close range effects are of the order of 0.01 ppm for ^{1}H or 0.2–1.33 ppm for ^{13}C. Such isotope shifts are measured in the spectrum of the affected nucleus, a good example being that of the carbon spectrum of a mixture of CH_3CN and CD_3CN.[64,65] It is worth noting that the intensities to be expected for coupling to three deuterons is 1:3:6:7:6:3:1 and not the 1:2:3:4:3:2:1 ratio quoted, which is appropriate to two nuclei with $I = 3/2$. A recent application based on this effect notes that in molecules with coordinated cyclopentadiene, it is possible to distinguish between pentahapto and exchanging monohapto forms by measuring the ^{13}C chemical shift change which occurs on deuteration of the cyclopentadiene.[66] There appear to be no isotope effects on the potassium or boron resonances,[67,68] though an appreciable solvent effect exists for the ^{55}Mn resonance of MnO_4^-, of -0.76 ppm between H_2O and D_2O.[69] There is also a small isotope effect on the coupling to various nuclei, ΔJ, which is defined as $J(H-X) - 6.51J(D-X)$. ΔJ

varies from -0.6 to 2.2 Hz when $X = {}^{13}C$, though the errors are large,[64] or from -4.7 to 5.0 Hz when $X = {}^{31}P$.[70]

Relaxation times of deuterium are not particularly short and fall typically in the range 5.04 s in toluene-d_3 to 0.05 s in larger molecules such as 2-benzamido-4,5-d_2-norborneol. There is nevertheless a maximum molecular size, above which high-resolution deuterium NMR will not be possible owing to the increasing effectiveness of the quadrupole relaxation mechanism. The relaxation behavior is considered in detail in Ref. 1.

The relaxation time of deuterons in a molecule, when coupled with knowledge of the nuclear quadrupole coupling constant (which may be assumed or experimental) and information gathered from other nuclides in the molecule, can often give detailed information about the molecular motions. Thus the methyl group behavior (internal motion) and whole molecular motion can be investigated in deuterated Me_2Hg[71] or Me_4Sn.[72] A similar approach has been used to test the hydrodynamic relationship:

$$\tau_c = V\eta/kT$$

for a group of approximately spherical molecules.[73] The rather rigid (and therefore smooth) spheres adamantane-d_{16} or $Et_4N^+ - d_2$ have much shorter correlation times than predicted by this relationship, whereas the larger and irregularly shaped $Bu_4N^+ - d_2$ or Al(acac-d)$_3$ have correlation times close to those which would be predicted, and appear to "mesh" with the solvent.

In other studies, the correlation time of $Co(en)_3^{3+}$ is measured as 30–35 ps but increases to about 160 ps in the presence of PO_4^{3-} due to the formation of ion pairs.[74] ^2H measurements on mesitylene-d_6 or CD_3SiF_3 support the proposition that that their motion follows the *M* diffusion model, i.e., one where the energy of internal motion is unchanged during collisions.[75] ^1H and ^2H relaxation measurements on liquid NH_3 and ND_3 support a small step diffusion model. It was noted that the quadrupole coupling constant of the deuterium is smaller in the liquid than in the gas, an observation consistent with the presence of hydrogen bonding in the liquid.[76] The interaction of LiCl or $LaCl_3$ with HMPTA has also been investigated by ^2H relaxation measurements.[77]

Reaction mechanisms may be followed using deuterium, where the spectra may be much simpler than the corresponding proton spectra. Thus both butane isomerization and deuteron scrambling occur during thermal decomposition of (η^5-C_5H_5) Fe(CO) (PPh_3)(1,1-d_2-n-butyl),[78] and the thermolysis of $DCCo_3(CO)_9$ or $CH_3CCo_3(CO)_9$ to give complexes containing larger alkyl groups can be followed.[79] The stereospecificity of hydrogen addition to rings of organometallic complexes can be followed by ^2H NMR.[80] The spectra in Figure 5 show the addition of D to $C_7H_7Mo(CO)_3$ to have been only slightly stereospecific, and that the deuterium migrates around the ring after heating for one hour. The products of the reaction

$$Mo_2(CH_2CD_3)_2(NMe_2)_4 + CO_2 \rightarrow Mo_2(O_2CNMe_2)_4 + CD_2CH_2 + CH_2DCD_3$$

were analyzed by ^2H NMR.[81] Methyl group redistribution on $TiMe_4 - Al_2Me_6 - Et_2O$ is easily followed by ^2H NMR, whereas solvent peaks interfere in proton spectroscopy—an inversion of an old theme.[82] Examples of monitoring exchange processes have been reported[83,84] and the use of the technique in the fields of organic and biochemistry has been described.[2,85–87]

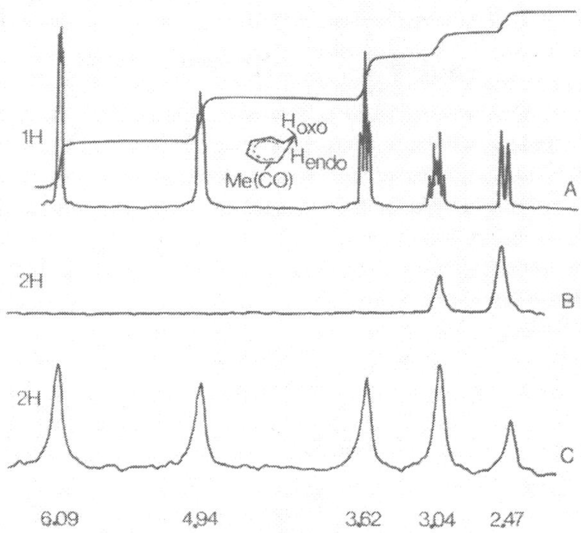

Figure 5. (A) 270-MHz proton spectrum of $C_7H_8Mo(CO)_3$ in $CDCl_3$. (B) 41.4-MHz deuterium spectrum of $C_7H_7DMo(CO)_3$ in $CHCl_3$ displayed to the same scale. The methylene resonances are at highest field. (C) as (B) after heating for 1 h in octane. (B) and (C) were obtained with broad band proton decoupling. (Reprinted with permission from Faller, *Inorg. Chem. 1980,* **19**, 2859.)

4.2. Deuterium NMR of Liquid Crystalline Phases

The deuterium NMR of anisotropic systems has assumed considerable importance, as may be judged from the fact that one-third of Ref. 2 is devoted to this topic while the subject is reviewed regularly in the Specialist Periodical Reports of the Royal Society of Chemistry dealing with NMR. The properties of oriented fluids, of which thermotropic or lyotropic liquid crystals are important examples, and of biological membranes are currently of great interest[88] and their investigation by ^2H NMR is complemented by studies of cations (see Chapter 7). The systems can, of course, be studied by proton NMR, but problems arise due to the signals from background resonances and interpretation is complex. If hydrogen is substituted by deuterium then the interaction becomes principally quadrupolar, there is no background interference, and specific deuteration can provide very fine details about the behavior of the oriented molecules.

If a ^2H nucleus resides in a molecule whose motion is restrained so as to be anisotropic, being for instance partially aligned along a magnetic or electric field, then the magnitude of the interaction of the quadrupole moment of the nucleus with the electric field gradient at that point in the molecule will depend on the spin quantum state of the nucleus, and for $I = 1$, a doublet splitting results whose magnitude depends on the quadrupole coupling constant of the ^2H and the order parameter S, which indicates the extent to which the motion of a given C–D bond is restricted.[89–91] Thus local differences within a molecule can be probed, to inform us about the detailed internal motions, or changes throughout the bulk of a sample may be studied.

When a sample is placed in a magnetic field it may take some appreciable time before the ordering has reached equilibrium, and this can be followed as changes occur in the spectra.[93] The compounds studied typically consist of a more or less rigid or immobile group and a long chain aliphatic tail or tails. Studies of deuterated chains, completely deuterated[94-96] or specifically deuterated at each position separately[97,98] indicate that the degree of ordering falls off steadily as one looks at CD_2 groups further away from the rigid part of the molecule. However, it has now been shown in thermotropic liquid crystalline mesophases that this is incorrect and the decrease in order parameter is irregular, with a particularly noticeable increase in going from carbon 4 to carbon 5.[99-101] Phase changes in micellar systems are also conveniently studied by ^{2}H NMR.[102] The effect of the liquid crystalline environment on dissolved, small deuterated molecules has also been much studied,[103-106] and values for quadrupole coupling constants extracted. Somewhat surprisingly, it has been found that small symmetrical molecules also exhibit quadrupole splitting and this is ascribed to small distortions of the solutes due to the anisotropic environment.[107-113] The last paper[113] also contains an example of the pattern obtained for spin–spin coupling to four deuterons in an isotropic liquid, i.e., a 1:4:10:16:19:16:10:4:1 nonet. The effects of counter cations in certain systems have also been studied.[114-116] The method is much used in biochemistry, and for instance, the orientation of peptide groups may be studied,[117] or the orientation of water absorbed onto biopolymers.[118-121] Specifically deuterated lipids have been used, though in some cases the quadrupolar doublets may all be resolved with a fully deuterated species.[122-124]

4.3. Deuterium NMR in Solids and Heterogeneous Systems

It is not possible to give these subjects more than a brief mention here. Many references will be found listed in two of the Specialist Periodical Reports of the Royal Society of Chemistry: that dealing with NMR contains chapters dedicated to these areas, and the more inorganic applications are dealt with by Mann in the series dealing with the Spectroscopic Properties of Inorganic and Organometallic Compounds. We also note some recent discussions of deuterium spin echoes in solids.[125-128]

5. TRITIUM

Only a few laboratories are equipped to work with tritium, because of the radioactive hazard. This is minimized by using quite low isotopic concentrations of the element, and the high sensitivity means that spectra may be obtained in reasonable times: 0.038% ^{3}H in water, for instance, will give a spectrum in one scan[129] though accumulation times of the order of an hour are quoted.[130] Much of the chemistry studied using tritium can be classified as organic, and has been concerned with the best means of standardization[7,130] or isotope effects. Thus small differences exist between ^{3}H and ^{1}H shieldings in organic molecules[130] and the ^{3}H shieldings themselves may depend upon the nature of the hydrogenic isotope in other parts of the molecule.[131] Several reviews of tritium NMR are available[1,2,132] and one is promised for the future in *Progress in NMR Spectroscopy*.

It is obviously of importance for these studies to have a variety of means available for the incorporation of tritium into organic molecules. This is done using catalytic systems and is where the subject takes on a particular interest to the organometallic chemist. Indeed, the high sensitivity of ^{3}H NMR makes it an ideal tool for the study of surface catalysis, an area of research which has already been reviewed,[133] or for the mechanistic study of metal catalyzed reactions.[134] Tritiation can be carried out with tritium gas over Raney nickel for organosilanes[7,135] and tritiated water will react with aromatic compounds in the presence of heated zeolites, much poorer incorporation being obtained with aliphatics.[136] Other catalysts are investigated[137] and Elvidges group have worked extensively in this area.[138–141] ^{3}H is also being used as a probe into biological interactions[142–144] and the isotopomer distribution in labeled borohydride has been investigated.[145]

REFERENCES

1. Brevard, C.; Kintzinger, J. P. "*Deuterium and Tritium*," in *NMR and the Periodic Table*, Harris, R. K., and Mann, B. E., Eds.; Academic: New York, 1978.
2. Mantsch, H. H.; Saito, H.; Smith, I. C. P. *Progr. NMR Spectrosc.* 1977, **11**, 211; E. A. Evans *et al.*, John Wiley, Chichester and New York, 1985.
3. *Nuclear Magnetic Resonance*; Specialist Periodical Reports of the Royal Society of Chemistry: London, Vols. 1–11 and onward.
4. *Annual Reports in NMR Spectroscopy*, Academic: New York, Vols. 1–11 and onward.
5. *Progress in NMR Spectroscopy*, Emsley, J. W., Feeney, J. and Sutcliffe, L. H., Eds.; Pergamon: New York, Vols. 1–13 and onward (first volume 1966).
6. Martin, M. L.; Delpuech, J.-J.; Martin, G. J. *Practical NMR Spectroscopy*, Heyden & Son: London, 1980.
7. Long, M. A., Garnett, J. L.; Lukey, C. A. *Org. Magn. Reson.* 1979, **12**, 551.
8. Emsley, J. *Chem. Soc. Revs.* 1980, **9**, 91.
9. Akitt, J. W. *J. Chem. Soc. Faraday I* 1977, **73**, 1622.
10. Emsley, J. W.; Feeney, J.; Sutcliffe, L. H. *High Resolution Nuclear Magnetic Resonance Spectroscopy*; Pergamon: New York, 1965, Vol. 1, p. 534.
11. Eisenberg, D.; Kauzmann, W. *The Structure and Properties of Water*, Oxford University Press: London, 1969.
12. *Water: A Comprehensive Treatise*, Franks, F., Ed.; in 7 Vols., Plenum: New York, 1972–1982.
13. Pople, J. A. *Proc. R. Soc. London* 1951, **A205**, 163.
14. Hindman, J. C. *J. Chem. Phys.* 1966, **44**, 4582.
15. Akitt, J. W. *J. Chem. Soc. Dalton Trans.* 1974, 175.
16. Coetzee, J. F.; Hussam, A. *J. Soln. Chem.* 1982, **11**, 395.
17. Schuster, R. E.; Fratiello, A. *J. Chem. Phys.* 1967, **47**, 1554.
18. Fratiello, A.; Miller, D. P.; Schuster, R. E. *Mol. Phys.* 1967, **12**, 111.
19. Fratiello, A.; Schuster, R. E. *J. Phys. Chem.* 1967, **71**, 1948.
20. Movius, W. G.; Matwiyoff, N. A. *Inorg. Chem.* 1967, **6**, 847.
21. Fratiello, A.; Lee, R. E.; Nishida, V. M.; Schuster, R. E. 1968, **48**, 3705; Matwiyoff, N. A.; Darley P. E.; Movius W. G. *Inorg. Chem.* 1968, **7**, 2173.
22. Matwiyoff, N. A.; Taube, H. *J. Am. Chem. Soc.* 1968, **90**, 2796.
23. Wawro, R. G.; Swift, T. J. *J. Am. Chem. Soc.* 1968, **90**, 2792.
24. Matwiyoff, N. A.; Darley, P. E.; Movius, W. G. *Inorg. Chem.* 1968, **7**, 2173.
25. Fratiello, A.; Lee, R. E.; Schuster, R. E. *Mol. Phys.* 1970, **18**, 191.
26. Fratiello, A.; Kubo, V.; Peak, S.; Sanchez, B.; Schuster, R. E. *Inorg. Chem.* 1971, **10**, 2552.
27. Delpuech, J.-J.; Peguy, A.; Khaddar, M. R. *J. Magn. Reson.* 1972, **6**, 325.
28. Fratiello, A.; Kubo, V.; Vidulich, G. A. *Inorg. Chem.* 1973, **12**, 2066.
29. Stockton, G. W.; Martin, J. S. *Can. J. Chem.* 1974, **52**, 744.
30. Delpuech, J.-J.; Peguy, A.; Khaddar, M. R.; Rubini, P. *J. Chem. Soc. Chem. Commun.* 1974, 154.
31. Richardson, D.; Alger, T. D. *J. Phys. Chem.* 1975, **79**, 1733.
32. Ruben, Y.; Reuben, J. *J. Phys. Chem.* 1976, **80**, 2394.
32a. Orrell, K. G.; Ref. 4, 1979, Vol. **9**, 1.
33. Malinowski, E. R.; Knapp, P. S.; Feuer, B. *J. Chem. Phys.* 1966, **45**, 4274.
34. Knapp, P. S.; Waite, R. O.; Malinowski, E. R. *J. Chem. Phys.* 1968, **49**, 5459.
35. Malinowski, E. R.; Knapp, P. S. *J. Chem. Phys.* 1968, **48**, 4989.
36. Deverell, C. *Progr. NMR Spectrosc.* 1969, **4**, 235.
37. Akitt, J. W.; Duncan, R. H. *J. Chem. Soc. Faraday Trans. I* 1976, **72**, 2132.
38. Akitt, J. W. *Faraday Disc. Chem. Soc.* 1978, **64**, 102, 125–135.
39. Akitt, J. W. *J. Chem. Soc. Dalton Trans.* 1973, 42.

40. Akitt, J. W. *J. Chem. Soc. Faraday Trans. I* 1982, **78**, 607.
41. Akitt, J. W. *J. Chem. Soc. Dalton Trans.* 1973, 49.
42. Soffer, N.; Marcus, Y.; Shamir, J. *J. Chem. Soc. Faraday Trans. I* 1980, **76**, 2347.
43. Satchell, D. P. N.; Satchell, R. S. *Chem. Revs.* 1969, **69**, 251.
44. Akitt, J. W. *Ann. Rev. NMR Spectrosc.* 1972, **5A**, 510.
45. Hartman, J. S.; Miller, J. M. *Adv. Inorg. Radiochem.* 1978, **21**, 147.
46. Harris, R. K. *Can. J. Chem.* 1964, **42**, 2275.
47. Kunze, K. L.; Ortiz de Montellano, P. R. *J. Am. Chem. Soc.* 1981, **103**, 4225.
48. *Spectroscopic Properties of Inorganic and Organometallic Compounds*; Specialist Periodical Reports of the Royal Society of Chemistry: London, Vols. 1–13 and onward.
49. Hickmoll, P. W.; Cais, M.; Modiano, A., Ref. 4, 1977, **6C**, 1–635.
50. Buckingham, A. D.; Stephens, P. J. *J. Chem. Soc.* 1964, 2747, 4583.
51. Masters, C.; Shaw, B. L.; Stainbank, R. E. *J. Chem. Soc. Chem. Commun.* 1971, 209.
52. Geoffroy, G. L.; Lehman, J. R. *Adv. Inorg. Radiochem.* 1977, **20**, 190.
53. Humphries, A. P.; Kaesz, H. D. *Progr. Inorg. Chem.* 1979, **25**, 145.
54. Hart, D. W.; Teller, R. G.; Wei, C.-Y.; Bau, R.; Longoni, G.; Camanella, S.; Chini, P.; Koetzle, T. F. *Angew. Chem. Internatl. Edn.* 1979, **18**, 80.
55. Farrar, T. C.; Johannensen, R. B.; Coyle, T. D. *J. Chem. Phys.* 1968, **49**, 281.
56. *Dynamic Nuclear Magnetic Resonance Spectroscopy*, Jackman, L. M.; Cotton, F. A., Eds.; Academic: New York, 1975.
57. *Advances in Organometallic Chemistry*, Stone, F. G. A.; West, R., Eds.; Academic: New York, 1977, Vol. 16.
58. Sandstrom, J. *Dynamic NMR Spectroscopy*; Academic: New York, 1982.
59. Bielmann, J.-F.; d'Orchymont, H.; Schmitt, J.-L. *J. Am. Chem. Soc.* 1979, **101**, 3283.
60. Briggs, J. M.; Farnell, L. F.; Randall, E. W. *J. Chem. Soc. Chem. Commun.* 1973, 70.
61. Abraham, R. J.; Bernstein, H. J. *Can. J. Chem.* 1961, **39**, 216.
62. Moriarty, R. M.; Kim, J. P.; Druck, S. J.; Lustig, E. *Tetrahedron* 1969, **25**, 1261.
63. Bernstein, H. J.; Sheppard, N. *J. Chem. Phys.* 1962, **37**, 3012.
64. Colli, H. N.; Gold, V.; Pearson, J. E. *J. Chem. Soc. Chem. Commun.* 1973, 408.
65. Gutowsky, H. S.; Karplus, M.; Grant, D. M. *J. Chem. Phys.* 1959, **31**, 1278.
66. Faller, J. W.; Murray, H. S.; Sanders, M. *J. Am. Chem. Soc.* 1980, **102**, 2306.
67. Sahm, W.; Schwenk, A. *Z. Naturforsch.* 1974, **29A**, 1754.
68. Epperlein, B. W.; Lutz, O.; Schwenk, A. *Z. Naturforsch.* 1975, **30A**, 955.
69. Lutz, O.; Steinkilberg, W. *Z. Naturforsch.* 1974, **29A**, 1467.
70. Borisenko, A. A.; Sergeyev, N. M.; Ustinyuk, Y. A. *Mol. Phys.* 1971, **22**, 715.
71. Lassigne, C. R.; Wells, E. J. *Can. J. Chem.* 1977, **55**, 1303.
72. Lassigne, C. R.; Wells, E. J. *J. Magn. Reson.* 1977, **26**, 55.
73. McGaughy, T. W.; Wu, S. Y. C.; Fung, B. M. *J. Chem. Phys.* 1978, **69**, 473.
74. Craighead, K. C.; Jones, P.; Bryant, R. G. *J. Chem. Phys.* 1975, **63**, 1856.
75. Bull, T. E. *J. Chem. Phys.* 1979, **70**, 571.
76. Hamer, H.; Lang, E.; Ludemann, H.-D. *Ber. Buns. Phys. Chem.* 1979, **83**, 1262.
77. Kessler, Y. M.; Mishustin, A. I.; Lyalina, R. B.; Podkovyrin, A. I.; Bratishko, R. K. *Z. Fiz. Khim.* 1977, **51**, 2960.
78. Reger, D. L.; Culbertson, E. C. *J. Am. Chem. Soc.* 1976, **98**, 2789.
79. Seyferth, D.; Nivert Rudie, C.; Merola, J. S. *J. Organomet. Chem.* 1978, **162**, 89.
80. Faller, J. W. *Inorg. Chem.* 1980, **19**, 2859.
81. Chisholm, M. H.; Haitko, D. A. *J. Am. Chem. Soc.* 1980, **101**, 6784.
82. Bresler, L. S.; Khachaturov, A. S.; Podburnyi, I. Y. *J. Organometal. Chem.* 1974, **64**, 335.
83. Copenhafer, W. C. ; Rieger, P. H. *Inorg. Chem.* 1977, **16**, 2431.
84. Behrendt, S.; Borsdorf, R.; Kleinpeter, E.; Grundel, D.; Hartschmann, A. *Z. Chem.* 1976, **16**, 405.
85. Lindman, B. *J. Magn. Reson.* 1978, **32**, 39.
86. Casadevall, E.; Metzger, P. *Tetrahedron Lett.* 1970, **48**, 4199.
87. Moore, W. R.; Mogolesko, P. D.; Traficante, D. D. *J. Am. Chem. Soc.* 1972, **94**, 4753.
88. Khetrapal, C. L.; Kunwar, A. C.; Tracey, A. C.; Diehl, P. *NMR Basic Principles and Progress*, Diehl, P., Fluck, E., and Kosfeld, R., Eds.; Springer-Verlag: Berlin, 1975, Vol. 9.
89. Rowell, J. C.; Phillips, W. D.; Melby, L. R.; Panar, M. *J. Chem. Phys.* 1965, **43**, 3442.
90. Diehl, P.; Khetrapal, C. L. *NMR: Basic Principles and Progress* 1969, **1**, 1.
91. Emsley, J. W.; Lindon, J. C. *NMR Spectroscopy using Liquid Crystal Solvents*; Pergamon: New York, 1975.

92. Khetrapal, C. L.; Kunwar, A. C. *Adv. Magn. Reson.* 1977, **9**, 302–470.
93. Lawson, K. D.; Flautt, T. J. *J. Am. Chem. Soc.* 1967, **89**, 5489.
94. Emsley, J. W.; Lindon, J. C.; Luckhurst, G. R. *Mol. Phys.* 1975, **30**, 1913.
95. Deloche, B.; Charvolin, J.; Liebert, L.; Strzelecki, J. *J. Phys. (Paris)* 1975, **36C-1**, 21.
96. Charvolin, J.; Manneville, P.; Deloche, B. *Chem. Phys. Lett.* 1973, **23**, 345.
97. Seelig, J.; Niederberger, W. *Biochem.* 1974, **13**, 1585.
98. Niederberger, W.; Seelig, J. *Ber. Buns. Phys. Chem.* 1974, **78**, 947.
99. Boden, N.; Bushby, R. J.; Clark, L. D. *Chem. Phys. Lett.* 1979, **64**, 519.
100. Boden, N.; Bushby, R. J.; Clark, L. D. *Mol. Phys.* 1979, **38**, 1683.
101. Boden, N.; Clark, L. D.; Bushby, R. J.; Emsley, J. W.; Luckhurst, G. R.; Stockley, C. P. *Mol. Phys.* 1981, **42**, 565.
102. Boden, N.; Jackson, P. H.; McMullen, K. *Chem. Phys. Lett.* 1979, **65**, 476.
103. Diehl, P.; Khetrapal, C. L. *Can. J. Chem.* 1969, **47**, 1411.
104. Diehl, P.; Khetrapal, C. L. *J. Magn. Reson.* 1969, **1**, 524.
105. Caspary, W. J.; Millet, F.; Reichbach, M.; Dailey, B. P. *J. Chem. Phys.* 1969, **51**, 623.
106. Gill, D.; Klein, M. P.; Kotowycz, G. *J. Am. Chem. Soc.* 1968, **90**, 6870.
107. Ader, R.; Loewenstein, A. *J. Am. Chem. Soc.* 1974, **96**, 5336.
110. Reeves, L. W.; Tracey, A. S. *J. Am. Chem. Soc.* 1974, **96**, 365.
111. Boden, N.; Folland, R. *Chem. Phys. Lett.* 1975, **32**, 127.
112. Bailey, D.; Buckingham, A. D.; Fujiwara, F.; Reeves, L. W. *J. Magn. Reson.* 1975, **18**, 344.
113. Chen, D. M.; Glickson, J. D. *J. Magn. Reson.* 1977, **28**, 9.
114. Radley, K.; Reeves, L. W.; Tracey, A. S. *J. Phys. Chem.* 1976, **80**, 124.
115. Persson, N.-O.; Lindman, B.; *J. Phys. Chem.* 1975, **79**, 1410.
116. Abdolall, K.; Bunell, E. E.; Valic, M. I. *Chem. Phys. Lipids* 1977, **20**, 115.
117. Chapman, G. E.; Campbell, I. D.; McLaughlin, K. A. *Nature* 1970, **225**, 639.
118. Berendsen, H. J. C. *J. Chem. Phys.* 1962, **36**, 3297.
119. Migchelsen, C.; Berendsen, H. J. C. *Proc. XIVth Colloque. Ampere, Ljubljana*, 1967, 761.
120. Migchelsen, C.; Berendsen, H. J. C. *J. Chem Phys.* 1973, **59**, 296.
121. Dehl, R. E. *J. Chem. Phys.* 1968, **48**, 831.
122. Stockton, G. W.; Smith, I. C. P. *Chem. Phys. Lipids* 1976, **17**, 251.
123. Smith, I. C. P.; Stockton, G. W.; Tulloch, A. P.; Polnaszek, C. F.; Johnson, K. G. *J. Colloid Interface. Sci.* 1977, **58**, 439.
124. Boden, N.; Kahol, P. K. *Mol. Phys.* 1980, **40**, 1117.
125. Boden, N.; Clark, L. D.; Hanlon, S. M.; Mortimer, M. *Faraday Diss. Chem. Soc., Faraday Symp.* **13**, 1979, 109.
126. Boden, N. Kahol, P. K. *Chem. Phys. Lett.* 1979, **68**, 158.
127. Boden, N.; Hanlon, S. M.; Levine, Y. K.; Mortimer, M. *Mol. Phys.* 1978, **36**, 519.
128. Boden, N.; Hanlon, S. M.; Levine, Y. K.; Mortimer, M. *Chem. Phys. Lett.* 1978, **57**, 151.
129. Bloxidge, A.; Elvidge, J. A.; Jones, J. R.; Evans, E. A. *Org. Magn. Reson.* 1971, **3**, 127.
130. Bloxidge, A.; Elvidge, J. A.; Jones, J. R.; Mane, R. B.; Saljoughian, M. *Org. Magn. Reson.* 1979, **12**, 574.
131. Angelini, G.; Speranza, M.; Segre, A. L.; Altman, L. J. *J. Org. Chem.* 1980, **45**, 3291.
132. Elvidge, J. A.; Jones, J. R.; Chambers, V. M. A.; Evans, E. A. *Isotopes in Organic Chemistry*, Bruncel, E., and Lee, C. C., Eds.; Elsevier: New York, 1978, Vol. 4, Chap. 1.
133. Garnett, J. L.; Long, M. A.; Odell, A. L. *Chem. Aust.* 1980, **47**, 215.
134. Garnett, J. L.; Long, M. A.; Luke, C. A. *J. Chem. Soc. Chem. Commun.* 1979, 634.
135. Long, M. A.; Garnett, J. L.; Lukey, C. A.; Williams, P. G. *Aust. J. Chem.* 1980, **33**, 1393.
136. Long, M. A.; Garnett, J. L.; Williams, P. G.; Mole, T. *J. Am. Chem. Soc.* 1980, **103**, 1571.
137. Long, M. A.; Garnett, J. L.; West, J. C. *Tetrahedron Lett.* 1978, 4171.
138. Buncel, E.; Elvidge, J. A.; Jones, J. R.; Walkin, K. T. *J. Chem. Res. (S)* 1980, 272.
139. Buncel, E.; Norris, A. R.; Elvidge, J. A.; Jones, J. R.; Walkin, K. T. *J. Chem. Res. (S)* 1980, 326.
140. Al-Rawi, J. M. A.; Elvidge, J. A.; Jones, J. R.; Mane, R. B.; Saieed, M. *J. Chem. Res. (S)* 1980, 298.
141. Evans, E. A.; Kitcher, J. P.; Warrell, D. C.; Elvidge, J. A.; Jones, J. R.; Lenk, R. *J. Labelled Compd. Radiopharm.* 1979, **16**, 697.
142. Libor, S.; Bloxidge, J. P.; Elvidge, J. A.; Jones, J. R.; Woods, L. F. J.; Wiseman, A. *Biochem. Soc. Trans.* 1980, **8**, 99.
143. Woods, L. F. J.; Wiseman, A.; Libor, S.; Jones, J. R.; Elvidge, J. A. *Biochem. Soc. Trans.* 1980, **8**, 98.
144. Aberhart, D. J.; Tann, C. H. *J. Am. Chem. Soc.* 1980, **102**, 6377.
145. Altman, L. J.; Thomas, L. *Analyt. Chem.* 1980, **52**, 992.

Chapter 7

THE ALKALI AND ALKALINE EARTH METALS

Lithium, Sodium, Potassium, Rubidium, Cesium, Beryllium, Magnesium, Calcium, Strontium, and Barium

J. W. AKITT •

1. INTRODUCTION TO GROUPS I AND II

The first two groups of the Periodic Table have in common that their members exist principally in solution as cations, and their study is thus that of solute–solvent interactions and of ion–ion interactions. The elements Li, Be, and Mg have in addition a covalent chemistry and their covalently bonded compounds are also studied. From the point of view of NMR spectroscopy, all these elements have nuclei with quadrupole moments, though ^{6}Li and ^{133}Cs have two of the smallest moments known. In the case of the group I ions, the ion–solvent interactions are relatively weak, so that we think of the ions as existing in spherical *s* states with low field gradients. We would thus expect relaxation times to be a maximum in these solutions and modified by the interaction with the solvent. Weak interaction means fast exchange and so only singlets will be observed giving data for the ensemble average of all the possible states experienced by the ions. Where strong interactions occur, these are immediately detected by the metal NMR. In the case of the smaller group II ions, a definite solvation sheath is formed which can be detected by proton NMR for Be or Mg only, though it probably persists for the heavier ions, only being differentiated from the group I ions by the time scale of the solvent exchange.

Some nuclear properties of these elements are outlined in Table 1. The relaxation time is determined by three principal factors in a given chemical environment (Chapter 5, Section 4.5), the spin quantum number, *I*, the quadrupole moment, *Q*, and the Sternheimer quadrupolar antishielding factor, γ_∞, which terms appear in the various equations for quadrupole relaxation in the form

$$\frac{2I+3}{I^2(2I-1)}Q^2(1+\gamma_\infty)^2 = \text{width factor in Table 1} \qquad (1)$$

The width factors given in the table are normalized to that for $^{7}Li = 1$ and in principle

J. W. AKITT • School of Chemistry, University of Leeds, Leeds LS2 9JT, England.

Table 1. Some Nuclear Properties of the Alkali Metal and Alkaline Earth Elements

Nucleus	Quadrupole moment, Q (10^{-28} m^2)	$1+\gamma_{\text{ion}}$	Width factor WF ^{7}Li = 1	T_1 (dil aquo) (s)	T_1(WF)^c	D^p/WF
^{6}Li	6.4×10^{-4}	0.74	2.0×10^{-3}	830(D_2O)	Other mechanisms intervene	—
^{7}Li	-3.7×10^{-2}	0.74	1	37(D_2O)	37	2.7×10^{-1}
^{23}Na	1.0×10^{-1}	5.1	343	0.0625	21	2.7×10^{-4}
^{39}K	5.5×10^{-2}	18.3	1359	0.057	77	5.2×10^{-7}
^{41}K	6.0×10^{-2}	—	—	—	—	—
^{85}Rb	2.7×10^{-1}	48.2	54200	0.0024	130	2.5×10^{-7}
^{87}Rb	1.3×10^{-1}	—	—	—	—	1.6×10^{-6}
^{133}Cs	-3×10^{-3}	111(95.2)	15(11)	13.3	199(146)	4.6×10^{-3}
^{9}Be	5.2×10^{-2}	0.81	3.6(no γ_∞)	8	29	8.4×10^{-3}
^{25}Mg	2.2×10^{-1}	4.32	284	0.22	62	1.4×10^{-6}
^{43}Ca	-5×10^{-2}	13.1	59	1.33	79	2.2×10^{-7}
^{87}Sr	3.6×10^{-1a}	41	16200	0.0042	68	1.7×10^{-8}
^{135}Ba	1.8×10^{-1}	123(76.2)	660000(250000)	0.0006	396(152)	ca. 10^{-9}
^{137}Ba	2.8×10^{-1}	—	1610000(618000)	—	—	ca. 10^{-9}

[a] A newer, and smaller value of Q is quoted in the Appendix located on pp. 625–629. Adoption of this value gives T_1(WF) as 13 which is clearly not compatible with the values for ^{43}Ca or ^{135}Ba and the old value is preferred here.

allow us to obtain some idea of the relative relaxation behavior of this group of nuclei. γ_∞ is not a nuclear property but is a property of the cation. In the free ion, the electronic orbitals are distorted by any external electric fields to create additional fields which magnify the field gradient at the nucleus for all the cations except Li^+ and Be^{2+}. Clearly this will have a marked influence upon the nuclear relaxation in these ions. It should be stressed that the Sternheimer factors are calculated for and relate to the free ion only. If covalent bonds are formed to the ion then the effect will be much reduced. Thus a charged, bonded ligand should have little effect on relaxation times via its electric field, whereas the same ligand may influence relaxation times markedly if no bond is formed and γ_∞ is substantial. The bonded ligand will, however, alter the electric symmetry at the nucleus and increase the relaxation rate for this reason. It is thus essential to know when it is appropriate to use the Sternheimer factors in calculations involving these nuclei, and conceivable that it may not always be clear when to do so. γ_∞ can be large, and now that situations involving relatively slow exchange are being encountered with solutions of group I elements, we may well find that this consideration may no longer be ignored.

In order to illustrate these comments, at least on an order of magnitude basis, we have included in Table 1 representative relaxation times in infinitely dilute aqueous solution and the product of this with the width factors. If all aqueous environments were the same, then these products should all be equal. In fact, the precision with which the infinite dilution T_1's are known varies somewhat and we should also remember both that γ_∞ is a calculated quantity and that Q is derived from calculations of field gradients in molecules whose quadrupole coupling constants have been measured, so that exact equality is unlikely. The products T_1(WF) for the alkali metals are reasonably constant, but increase from ^7Li to ^{133}Cs. ^6Li is anomalous because its relaxation is determined by several effects and is not dominated by the quadrupolar mechanism. In the case of ^{133}Cs, the usually accepted value of γ_∞ may be high. A recent, smaller value[1] is given in parentheses and this gives a reduced spread in T_1(WF). The numbers obtained with the group II cations are similar in magnitude, which is perhaps surprising since we expect much slower water exchange to occur, and indeed know that Be^{2+} forms a long-lived hydration complex.[2,3] In this case we ignore γ_∞, though this in fact makes little difference since $1+\gamma_\infty$ is close to unity. The Sternheimer effect does, however, seem necessary to explain the behavior of Mg^{2+},[4] though exchange on this cation should also be relatively slow.[5] We are thus led to suggest that it may indeed be difficult to distinguish covalency and electrostatic effects in quadrupolar relaxation, the Sternheimer effect having the consequence that the two are governed by related physical constants. We should note that in this second part of Table 1 we have included a recently amended value of Q for ^{43}Ca,[4] and present a recent, lower value of γ_∞ for Ba^{2+}.[1]

Several of the elements represented in this chapter are of very great importance in the field of biochemistry, and Na^+, K^+, Mg^{2+}, and Ca^{2+} are intimately involved in the chemistry of life. The study of only one of these, ^{23}Na, is easy for the NMR spectroscopist, though the remaining three are now accessible without too much difficulty where high-field, FT equipment is available. Their study has thus become of wide interest during the last few years, greatly reinforcing the earlier interest based on the "simple" solution chemistry of the salts of these cations. As a direct result of this interest, there are now many reviews available devoted wholly or partly to the NMR of the elements of groups I and II, some of them recent.[4,6–16] We will therefore in this chapter provide a summary review of the state of the field up to approximately 1978 followed by a discussion of the work which has appeared since then up to early 1983. Aqueous, ionic solutions have received little attention in this later period, though mixed and nonaqueous solvents have proved of interest. The bulk of the new work, however, deals with interaction between cations and molecules of increasing complexity, from crown ethers through to biopolymers or liquid crystal systems, and there is much to report, with the most popular nuclei being ^{23}Na and ^{133}Cs.

In addition, the use of ^{6}Li has allowed some most interesting investigations to be carried out, particularly into the covalent compounds of this element.

2. EXPERIMENTAL TECHNIQUES

It is not possible to lay down any universal rules since receptivities vary widely and relaxation times may be very long, especially for ^{6}Li, or extremely short. Where the relaxation times are not too long, data are easy to obtain in the FT mode and the disadvantages of shorter relaxation times and broad lines can often be overcome to a great extent since short acquisition times and rapid pulse repetition rates are permissible, especially if some line-shape distortion can be tolerated.[4] For maximum use to be made of such techniques access to memory sizes less than 1K must be possible.

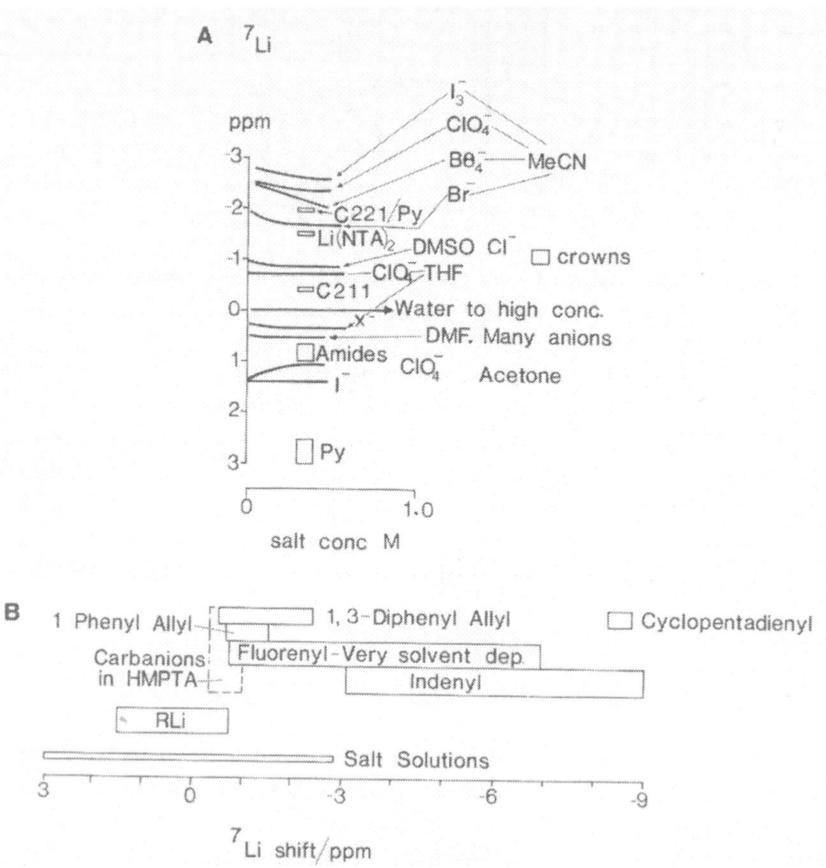

Figure 1. Chart of lithium chemical shifts, (A) salt solutions and (B) organic compounds.[91,107,110,183,184,194,200–203]

Obviously, if narrow lines are encountered then the technique must be modified to take account of an appropriate relaxation time. The only difficult alkali metal nuclei are ^{6}Li and ^{39}K on account of their poor receptivities. This problem is easily overcome with the former since highly enriched ^{6}Li is available—one of the more acceptable types of fallout from hydrogen bomb manufacture! With ^{39}K, acceptable spectra can be obtained in about 40 min using a conventional iron magnet(17); 2000 scans into 8K with a spectral width of 3 kHz are reported for 4.2 MHz operating frequency. Scanning could in fact have been more rapid than this, and of course a superconducting magnet system will be even faster.(120,230) In the case of the alkaline earth nuclei, only ^{9}Be can be regarded as easy and the remainder have low receptivities. ^{87}Sr and ^{135}Ba have been observed,(18,19) but the difficulties involved, which are compounded by the width of the lines, will probably limit the studies of these nuclei. ^{25}Mg and ^{43}Ca have even lower receptivities, but because of their biological importance they have received, and will receive, a great deal of attention. Isotopic enrichment is possible at a price and is quite widely used. Though this is a

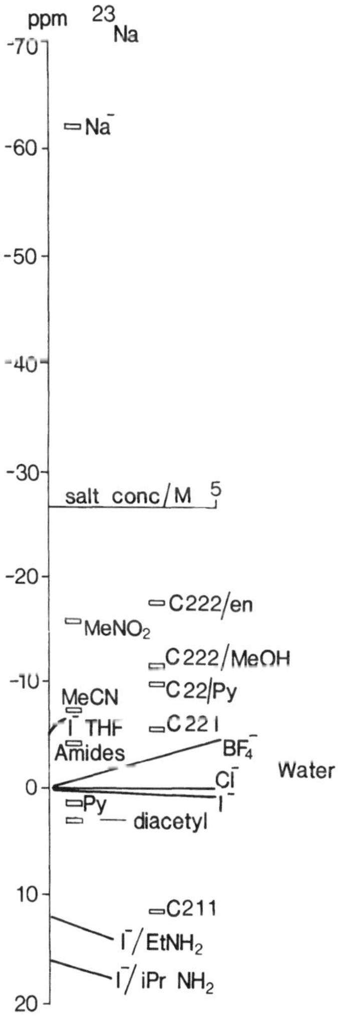

Figure 2. Chart of sodium chemical shifts.(45,71,83,90,104,107,109,110,112,115–117,204–208)

costly option, it is worth noting that specially extracted, rare biological materials may be even more expensive, and enrichment then becomes the normal option. With the use of special spectrometer techniques it is then possible to obtain satisfactory spectra with about 100 transients from 1 mM ^{43}Ca (61% enriched) or 5 mM ^{25}Mg (90% enriched).[4] For comparison the limit for ^{87}Sr at natural abundance (7.02%) may be as high as 100 mM.

NMR Parameters

We will deal with the group I and group II nuclei together, which will allow a more ready comparison between them. The topics covered will be (1) aqueous solutions of the simple salts, where we will also review the various theories about quadrupole relaxation and chemical shifts; (2) mixed and nonaqueous solutions; (3) complexes with various types of ligands, including liquid crystals; and (4) metal organic compounds. Table 1 allows a comparison to be made between the relaxation behavior of the nuclei, the infinite dilution values in water being near the longest of the relaxation times recorded for each. The chemical shift ranges of various salt solutions are illustrated in Figures 1–8 which show chemical shift charts of the group I and II elements. The shift scales have been drawn in the approximate ratios of the number of electrons for the free ions, viz., 2:10:18:36:54, though the scale for cesium has had to be split to include the unusually large shift range of its complexes. In order to accommodate the fact that in many salt solutions the shift is a function of concentration, the charts have been made a mix of shift/concentration plots and bar graphs,

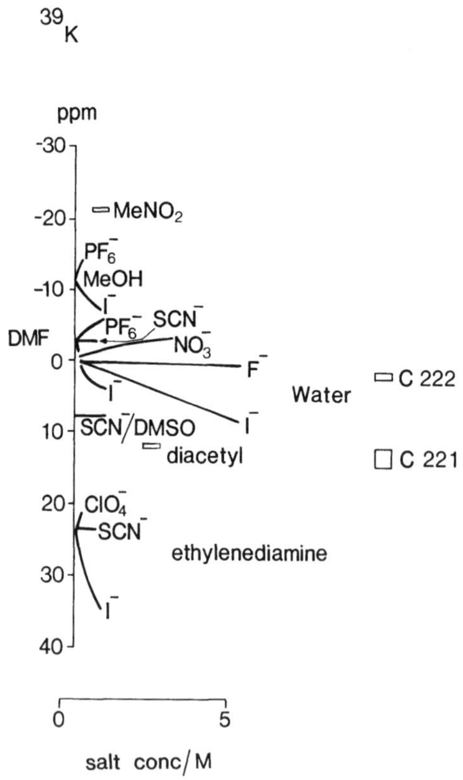

Figure 3. Chart of potassium chemical shifts.[67,85,109,120,204,209]

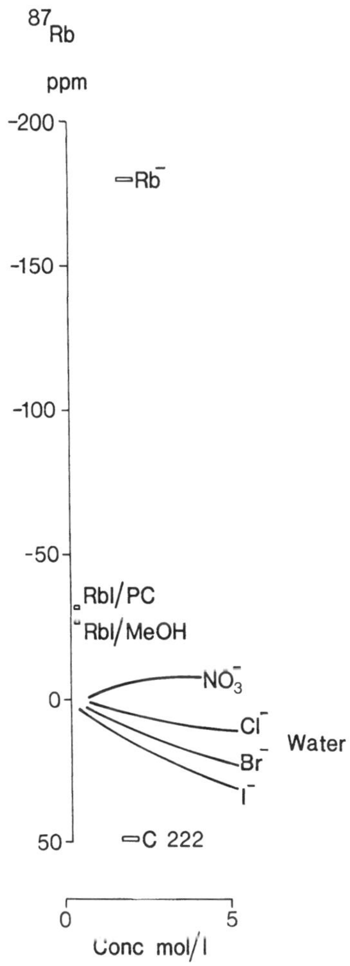

Figure 4. Chart of rubidium chemical shifts.[110,203,204,210,225]

the latter being appropriate where covalent compounds are formed or where definite complexes have been characterized. The plots are necessarily on a small scale and therefore cannot be accurate. The data have also been selected to give a clear, uncluttered diagram and so are not complete. References are, however, given to the relevant literature. The solvent and counteranion are indicated on the plots where appropriate.

3. AQUEOUS SOLUTIONS OF SIMPLE SALTS

The subject has an extensive literature and there are several excellent reviews available.[4,6–9] The metallic ions were an obvious target for study since they had been investigated by so many other physical techniques and solution theory was well developed, if not well understood. All the systems investigated throughout this early period underwent fast exchange on the NMR time scale and only singlet resonances were observed. The studies were thus concerned with how the chemical shifts (if any) or relaxation times varied with changes in solution composition. Relaxation times varied with viscosity, and there is still implicit discussion of whether changes in viscosity should be allowed for when comparing relaxation times of a nucleus in dif-

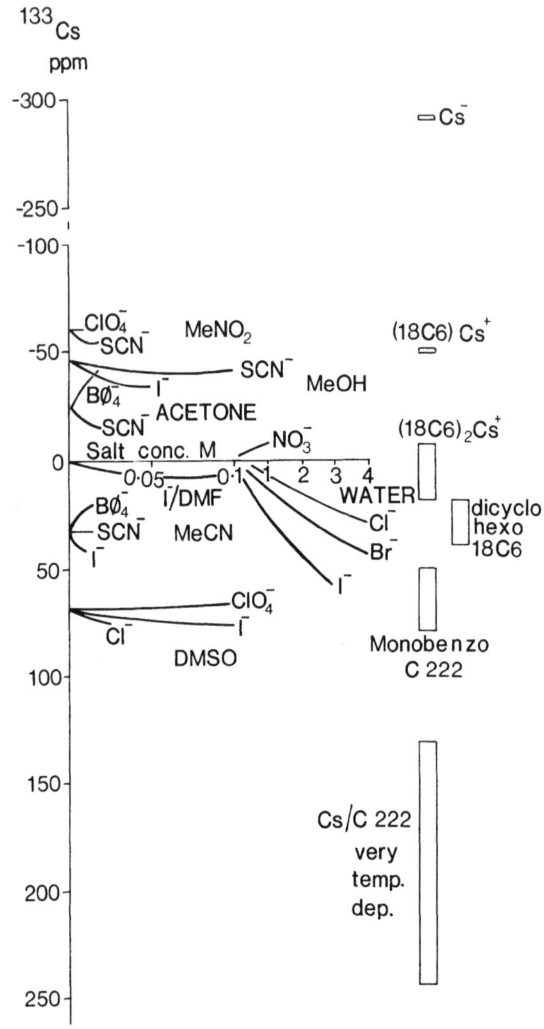

Figure 5. Chart of cesium chemical shifts. The figures for the cryptate and crown complexes depend strongly upon solvent, solution composition, and temperature.[84,110,121,122,204,211,212,226] Note that the concentration scale is split into two ranges.

ferent solutions. Chemical shifts were affected by bulk magnetic susceptibility differences between sample and standard, since where exchange is rapid, the standard must of necessity be external. The relaxation times depended upon temperature (which has to be controlled), upon concentration (partly via the viscosity change) and upon the nature of the anion. Extrapolation to low concentration allows an infinite dilution value to be obtained, which should be independant of anion type. The chemical shifts vary smoothly with increase in concentration and the magnitude and direction of the change depends upon the anion: I^- gives a strong downfield shift, Cl^- a smaller downfield shift while the oxo anions cause upfield shifts, ClO_4^- and BF_4^- being particularly effective. Some typical results are shown in Figures 2–5, 9 and 10. Clearly, the measurements are detecting the results of competitive cation–water and cation–anion (and indeed cation–cation) interactions and considerable effort has been expended on attempts to account for the observations. While all the nuclei show relaxation effects, chemical shifts are not detected for the strongly hydrated cations Li^+, Mg^{2+}, or Be^{2+}

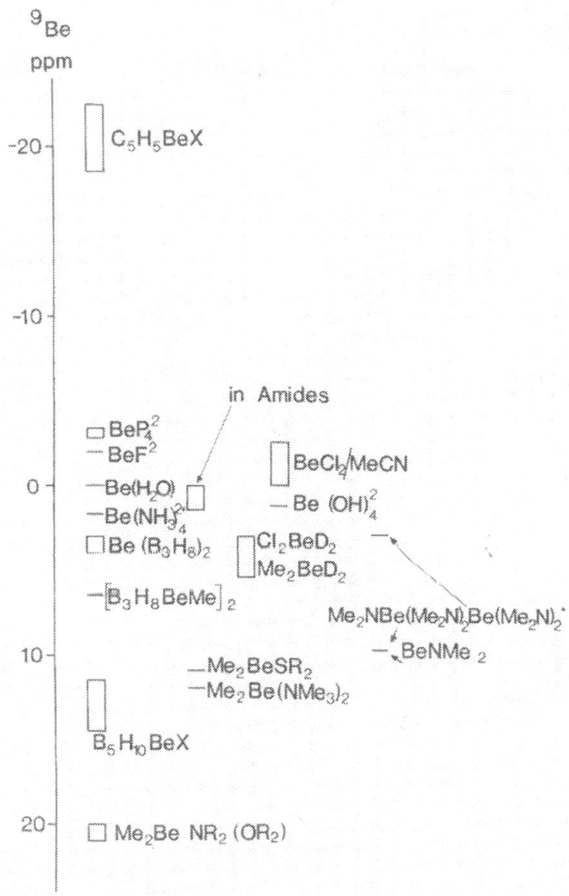

Figure 6. Chart of beryllium chemical shifts. The values recorded are very solvent dependent, even though most of these species may be considered to be entirely covalent. D represents a donor molecule and P a phosphate or HMPTA.[7,86,107,196–198]

(not even during hydrolysis of the latter[199]), where the strong solvation sheath inhibits the access of counterions and for which nuclei the overall chemical shift ranges are in any case small.[20,21] Relaxation of these two group II cations may be entirely viscosity controlled,[20] though Hertz' group believe that anions cause distortion of the first sphere at high concentrations.[237]

3.1. Nuclear Relaxation

3.1.1. Infinite Dilution Values

Some representative infinite dilution relaxation values are set out in Table 1. Theoretical attempts to explain the variation with ion type were developed for the alkali metals and later extended to group II. It is usual to ascribe the relaxation of a quadrupolar nucleus entirely to quadrupole interactions, though the small Q of ^{6}Li and ^{133}Cs, and the small γ_{∞} of Li^{+} and Be^{2+}, suggest that this might not be true for these nuclei in the relatively symmetrical environment of aqueous solution. The

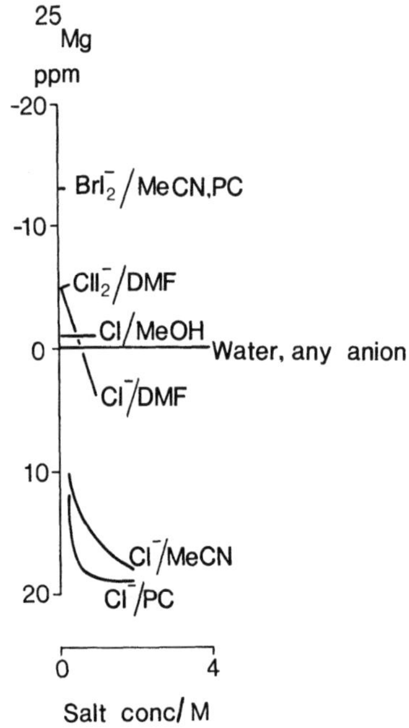

Figure 7. Chart of magnesium chemical shifts. The shifts cannot be measured to very small concentrations because of the low receptivity of ^{25}Mg, and this makes for very real difficulties in estimating infinite dilution values. There is for instance an initial very rapid down field shift as $MgCl_2$ is added to MeCN or propylene carbonate (PC).(20,21,82,109)

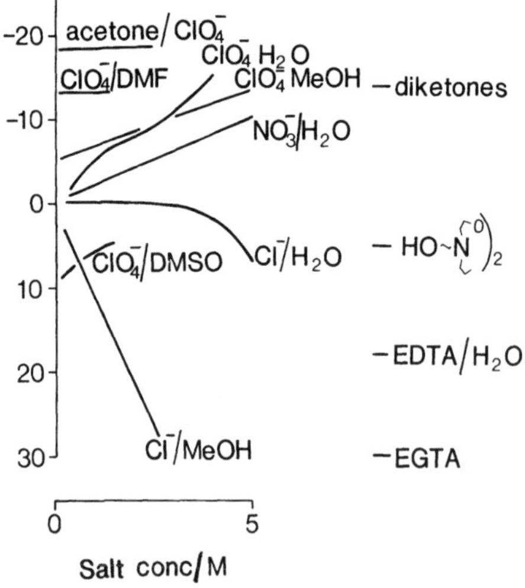

Figure 8. Chart of calcium chemical shifts. As for Mg, infinite dilution shifts are often difficult to estimate; see for example the plots for $CaCl_2$ or $Ca(ClO_4)_2$ in methanol.(15,21,109,198)

various relaxation contributions can be separated by measuring T_1 in normal and heavy water with strong proton irradiation in the first.[22] Thus there is no quadrupole contribution for $^6Li^+$ which has an NOE of 2.61 in H_2O. In D_2O at 80°C, its T_1 is 1040 s, the longest so far recorded.[23] A spin-rotation contribution was also detected. On the other hand, ^{133}Cs relaxation is entirely quadrupolar,[24] probably because its γ_∞ is so much larger than that of Li^+. In the case of ^9Be, there is a pronounced maximum in T_1 at 40° and an NOE at the lower temperatures so that D D, quadrupole, and SR effects are all present in different proportions depending on temperature. It should be possible to extract a correlation time for rotation of $Be(H_2O)_4^{2+}$, but lack of knowledge of the length of the Be–H vector prevents this.[25] An H–D isotope effect also exists for ^7Li relaxation in concentrated, supercooled aqueous LiCl and LiI. A common motion determines the relaxation of the three nuclear species, ^1H, ^2H, ^7Li, and is the Brownian motion in the system. Relaxation is due to both the DD and quadrupolar mechanisms in this case also.[26,27] However, the values of Q and γ_∞ for the remaining nuclei indicate that the relaxation mechanism should be entirely quadrupolar.

Two theories have been developed to attempt to explain the different quadrupolar relaxation rates of these cations. The electrostatic model due to Valiev[28] and Hertz[29,30] is currently favored and considers the relaxation to be caused by the electric field gradients due to the dipole moments of the solvent molecules modulated by their rotational and translational motion. The fully random distribution (FRD) version of the model leads in the extreme narrowing limit to

$$\frac{1}{T_1}=\frac{1}{T_2}=\frac{24\pi^3(2I+3)}{5I^2(2I-1)}\left(\frac{eQ}{h}\right)^2[\mu(1+\gamma_\infty)P]^2\frac{C_s\tau_s}{r_0^5} \tag{2}$$

where μ is the solvent dipole moment, P is a polarization factor of assumed value, C_s is solvent concentration, τ_s is solvent correlation time and is estimated from measurements of the proton relaxation time, and r_0 is the ion–H_2O distance. If the ion has a strongly bound first hydration sphere the effects of the second sphere are also taken into account in the fully oriented solvation version (FOS) in which terms are added to the equation to include the orientation and symmetry of the second sphere water molecules [see equation (45) in Chapter 5]. A fuller description will be found in Ref. 8. The model describes the observed behavior well, though it suffers the disadvantage of several adjustable parameters and has been criticized on this account.

The alternative to this theory is the electronic distortion model developed by Deverell and Richards,[16,31] which is also, unlike the electrostatic theory, concerned with chemical shift effects. Collisions between an ion and other molecules or ions lead to distortions in the outer orbitals of the ion which create electric field gradients at the nuclei and also affect the paramagnetic screening term. The effect may be described mathematically by using the overlap integrals between the interacting species, a concept developed to explain the screening in alkali halide crystals. The overlap terms are related to the paramagnetic screening term, and Deverell obtained

$$\frac{1}{T_1}=\frac{1}{T_2}=\frac{3\pi^2}{1000}\frac{2I+3}{I^2(2I-1)}\left(\frac{e^2Q}{h}\right)^2\left(\frac{\sigma_p\Delta E}{\alpha^2}\right)^2\tau_c \tag{3*}$$

* There is a mistake in the versions of equation (3) quoted in Refs. 7 and 8, which both miss the second square superscript.

where ΔE is a mean excitation energy, α is the fine structure constant, and σ_p is the paramagnetic shielding term. The significance of τ_c seems to be disputed.[8] In the infinite dilution situation, there are no ions to collide and the only possible interaction is with water molecules. It then becomes difficult to calculate relaxation rates, and for this reason the theory seems to have fallen out of favor. It is supported by plots of $(\text{linewidth})^{1/2}$ against shift, and we shall meet a recent application of this idea below. On the other hand, the electrostatic theory has received support from ^{131}Xe relaxation measurements in a variety of solvents where the rate of relaxation is enhanced in polar solvents.[235] In fact the two theories seem to this author to have a common factor in that both predict relaxation rate to depend upon orbital distortion, either directly or via the antishielding. If this view is correct, it suggests that it may be difficult to distinguish the two. We should note the way in which the large observed range of relaxation rates is reduced once the nuclear properties I and Q and the Sternheimer effect are taken into account, from $60000\times$ to just $5\times$ in Table 1. It is of interest that Hertz' group have recently suggested a rapprochement with the electronic viewpoint.[27] Two novel attacks on this problem are also of interest.[228,231] Very recently it has been found that in very dilute solutions, relaxation rate is proportional to the square root of concentration.[242]

Viscosity and Relaxation. The rate of relaxation of an ion depends upon the rate of motion of the electric field gradient at its nucleus created by the motion of solvent and other solute molecules near to it. The calculation of the overall nature of such fluctuations on a tensor quantity is not straightforward and so other criteria for rate of motion have to be sought. One obvious indication of internal motion is the solution viscosity, though there is no direct relation between this bulk property and the rate of motion near an ion. However, rates of relaxation commonly vary linearly with viscosity changes, and it is customary when making comparisons of relaxation and chemical shift to support the electronic theory, to normalize the former to constant viscosity, since the shifts are independent of viscosity.

The viscosity of a real solution (only aqueous solutions are considered here) depends on the concentration of the solute, following the well-known Jones–Dole relationship[243]

$$\eta/\eta_0 = 1 + AC^{1/2} + BC \tag{4}$$

the viscosity being influenced by ion–ion (A coefficients) and ion–solvent (B coefficients) effects. The B coefficients vary widely, and the derived single-ion values even have different signs for different types of ion, corresponding to the local effects. The ionic relaxation rates vary linearly with viscosity (more exactly, with η/T) as they should from the Debye hydrodynamic relationship (see Chapter 6, p. 182 and Ref. 73), though if the plots are reproduced to $\eta/T = 0$, then a finite relaxation rate is predicted, which is not in accord with Debye.[32a] The relaxation rate follows a law $1/T_1 = (A[H_2O] + BC)\eta/T + D$, where A, B, D are constants unrelated to those in equation (4). The significance of this equation is not understood.

On the other hand the proponents of the electronic theory ignore viscosity and use instead an estimate of τ_s in equation (2) made from proton relaxation times in the solvent, though this method still uses a bulk property to estimate a local rate of motion. An acceptable value can be given to τ_s, but an adjustable counterion symmetry parameter is still needed to obtain reasonable predicted relaxation times.

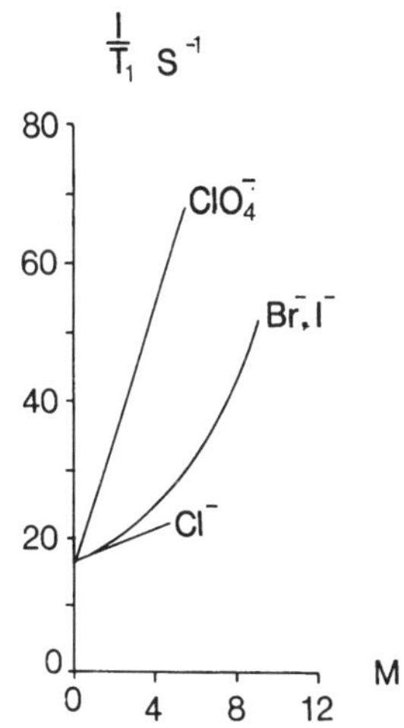

Figure 9. ^{23}Na relaxation rates as a function of concentration in aqueous NaCl, NaBr, NaI, and $NaClO_4$ at 25°C. The results for NaBr and NaI differ slightly but cannot be separated on this diagram.[220]

3.1.2. Actual Solutions

Some relaxation measurements made as a function of concentration are shown in Figures 9 and 10. Rates of relaxation generally increase with concentration, though there are some exceptions. The halides do not differ greatly in their effects but the molecular anions ClO_4^- and BF_4^- are appreciably better at producing relaxation, especially at the lower concentrations. These results have been interpreted in terms of the electrostatic theory, and a complex model evolved,[30] a feature of which is that the field gradient is quenched owing to the existence of an anionic cloud around the cation rather than just a single counterion. The electronic theory is again difficult to test, and suffers the disadvantage that the relaxation behavior and chemical shifts do not follow the same sequences.

A comprehensive set of references have already been given.[8] Recent work includes a multinuclear investigation of ion–solvent encounter configurations of ClO_4^-

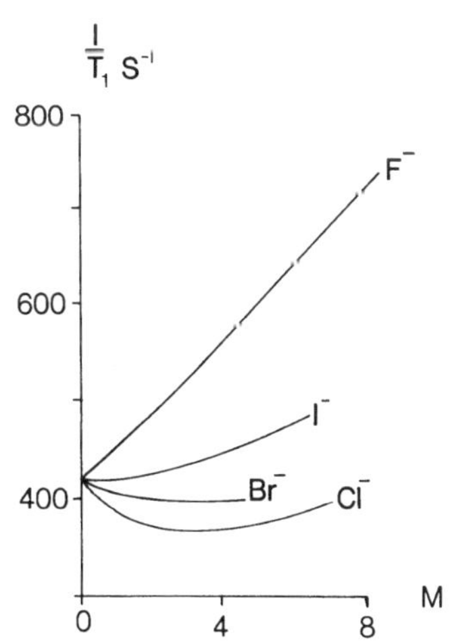

Figure 10. Relaxation rates of ^{85}Rb as a function of concentration in aqueous RbF, RbCl, RbBr, and RbI.[220]

or BF_4^- with Li^+ or Na^+. Solvent separated ion encounter pairs are formed with various configurations.[32] ^7Li relaxation in various formate solutions has been studied,[33] and a brief report of a rapid method for obtaining the relaxation times of the alkali metals has been given.[34] Measurements are reported on NaN_3 and RbN_3 solutions and inflections occur when the rates for ^{23}Na or ^2H in the former are plotted against concentration, at H_2O/NaN_3 ratos of 24:1 and 12:1.[35] Similar studies suggest that the hydration number of Li^+ is 4 at low temperatures and increases to 6 at high temperatures.[36] It is claimed that the activation energy of relaxation of ^{23}Na in $NaClO_4$ and of ^7Li in LiCl is different above and below 15–20°C.[37] ^7Li[38] and ^{23}Na[39] relaxation have been studied in the presence of paramagnetic cations. The alkali metal relaxes by interaction with the electron dipole, and it is shown in each case that the two cations are separated by one molecule of water, i.e., $Li^+(H_2O)_6Ni^{2+}$ and $Na^+(H_2O)_6Mn^{2+}$. More specifically, for the encounter pair with $Niaq^{2+}$, it is believed that the Li–Ni distance is sufficiently short that the lithium must take up a position in the dimple between three water molecules of the nickel hydration complex and be at the distance of closest approach.

T_1 has been measured in a time-consuming experiment for the strongly hydrated cation $^{25}Mg^{2+}$ [40] or for ^{25}Mg, ^{43}Ca, ^{87}Sr, and ^{137}Ba in aqueous MX_2 salt solutions.[237] The infinite dilution value for Mg is similar in all solutions and is 0.21 s, but T_1 is dependent on concentration. There is estimated to be a high degree of symmetry quenching of the electric field gradient and the increase of relaxation rate with increase in concentration is ascribed to distortion of the hydration sphere by collision with anions.

3.1.3. Relaxation of Nuclei with $I > 1$ Outside the Extreme Narrowing Limit

The extreme narrowing region is that regime of molecular motion where the correlation time is so short that it can be described by a frequency which is much higher than that being used for the NMR experiment being undertaken. The condition is normally written $\omega\tau_c \ll 1$. The equations used so far in this chapter assume this condition to be met and predict simple exponential decay of any nonequilibrium magnetization, with $T_1 = T_2$. If the nucleus in question becomes attached to a large, slowly moving molecule, such as is encountered in biological situations, then the condition of extreme narrowing may no longer be met, $\omega\tau_c \geqslant 1$, and $T_1 \neq T_2$. In addition, for a quadrupolar nucleus, the magnetic field direction and electric field gradient direction take particular orientations for times appreciable relative to a Larmor period, and several energy states exist for which the transition probabilities are no longer equal. Because the motion is isotropic, the transition frequencies remain essentially the same and we do not see the quadrupole splittings which occur in the solid state or in liquid crystals. The nuclear population nevertheless behaves as if it were made up of several different groups with different interactions. Thus for $I = 1$ there is still only one decay rate; for $I = 3/2$ or 2 there are two; for $I = 5/2$ or 3 there are three; for $I = 7/2$ or 4 there are four, etc.[41] The form of the relaxation behavior can be obtained analytically only for the case of $I = 3/2$, and numerical methods have to be used for the more complex cases.[42–44,229] In many cases, of course, the linewidths would be too large for such signals to be observed, but if exchange is possible between the bound ion and the ion in a simple solvation state then the behavior of the former will be manifested in the narrow resonance of the free ion.[4,6] Clearly, analysis of such situations as they arise will be helpful in attempting to understand ion binding effects, though as yet this aspect of NMR spectroscopy is at an early stage of development. Dispersion/absorption plots (DISPA) allow the effect to be detected and correlation times to be estimated.[233]

3.2. Chemical Shifts

The infinite dilution values depend only on the nature of the solvent and will be considered in the next section. What concerns us here is how, and why, the shifts in the simple salt solutions vary with concentration. Some experimental results are given in Figures 1–8. For those ions where effects are observed (all but Li^+, Be^{2+}, and Mg^{2+}; neither have any yet been seen for Ba^{2+} [19]) their magnitude and direction depend upon the nature of the counterion. The halides cause downfield shifts increasing in the order $Cl^- < Br^- < I^-$, whereas the oxo anions cause upfield shifts, though NO_3^-, ClO_4^-, SO_4^{2-} can cause shifts in either direction, depending on the cation studied. Interestingly, other cations also influence the shift of a cocation, the larger ones giving the greatest paramagnetic effect. The detailed results will be found in Refs. 4, 8, and 9. BF_4^- also has a marked upfield influence.[45] There is an interesting asymmetry between the chemical shifts and relaxation effects of the anions. In almost all cases, the anions cause an increase in the relaxation rate of cations but the shift effects may be strongly opposed. The downfield effect of the halides and the fact that this increases with size led to early rejection of the possibility that the ionic electric field or the strength of interaction with the cation might cause the shifts. Instead the evidence pointed to direct ion–ion contacts with the shifts caused by overlap between their outer electrons. A theory was available for the shifts of alkali halide crystals[46–48] and was developed by Richards' group at Oxford[16,31,49,50] for the shifts in solution. Thus the shift is made up of overlap contributions from water molecules, the anions, and other cations and the measured quantity is the average of all possibilities in the molecular ensemble. Overlap requires direct ion–ion contact and the picture that the theory produces of these solutions is of anions continually displacing water molecules to form short-lived contact ion pairs.

The model is supported also by the fact that the strongly hydrated ions (including Al^{3+}) show negligible chemical shifts. The high field shifts in the presence of the oxo ions have to be explained by assuming that they have smaller overlap than does water. This is quite reasonable since formally these ions have a substantial positive charge at their centers which must reduce the electron density at the periphery. Qualitatively, then, the shifts seem to be adequately explained, though it is not easy to produce satisfactory numbers by calculation. This is due partly to the fact that we do not know how the various contact terms vary with ion type, but also because the contact interaction itself is not fully understood. Recent measurements of solid salt chemical shifts, for instance, do not agree with theory.[51] There is a small isotope shift for metal cations on going from H_2O to D_2O solution, and this is also explicable in terms of an overlap effect.[52] The overlap theory finds less favor when applied to relaxation effects, though it links the two via equation (3), one problem being the upfield shifts and increased relaxation obtained with the polyatomic anions. One possible cause of the difference may be the strong multipole electric moments of these ions, which must mean that the detailed motions in their solutions are not comparable with those in the alkali halide solutions.[37,45]

These spectroscopic results lead us to make some general observations regarding the phenomenon and nomenclature of ion pairing. Classically, thermodynamic ion pairing arises in dilute solutions when oppositely charged ions approach one another more closely than average, and approach sufficiently closely to affect the total level of electrostatic interactions. The ions may be in contact but need not be, though as the theory assumes a solvent continuum without structure the term "solvent separated ion pair" is without significance in this context. If we allow the solvent to be a structured, molecular solvent, it is easy to show that at the low concentrations used for most physical solution measurements there is a large number of water molecules between ions, for instance more than 32 at 0.001 *M*. NMR spectroscopy only exhibits measurable effects at concentrations somewhat higher than this (quite apart from limitations of sensitivity), and at a typical lower limit of 0.05 *M* there are at most only 8 molecules between ions, and measurements are made to concentrations where the average ion–ion distances are *less* than the diameter of one water molecule (3.0 *M* for $NaBF_4$). It seems generally agreed that the NMR shifts are symptomatic of ion–ion contacts between ions of both opposite or the same sign of charge. Are these ion pairs? Clearly, the minimum distance between colliding ions is the sum of their radii. If this distance is not appreciably different from the average ion–ion distance then there will be no thermodynamic consequences. In fact the average ion–ion separation is a somewhat misleading concept here since there will be a range of distances, so that an ion–ion collision represents only an extreme of this range in the more concentrated solutions and can be accommodated in the normal radial distribution description. Ions in collision are then not paired in any thermodynamic sense and are detected because the NMR parameters are sensitive only to very short-range interactions: shifts in particular distinguish between collisional and close, noncollisional associations. We therefore need further criteria to define ion pairing in concentrated solution, and we suggest that the time scale of association is the critical factor. Thus if the collision lasts for a time which is long compared with the general rate of motion within the radial distribution around a given ion, then an ion pair is formed. In the limit of very long time of association this becomes a new species. The definition given also of course embraces second sphere ion pairs, if long lived. Thus the concept of ion pairing is useful at all concentrations, but we should distinguish the use of the term to denote a change in the interactional energies in solution from its use to provide a simple physical picture of the way ions may be arranged such as "contact ion pairs" or "solvent separated ion pairs."

Contact interactions are also responsible for alkali metal chemical shifts which occur in the presence of many paramagnetic transition metal cations,[8] and lately interest in this topic has been revived for both relaxation[38,39] and chemical shift studies. Two papers report effective means of distinguishing the signals from intra- and extracellular ^{23}Na by finding shift reagents which will interact with the latter but remain outside the cell sufficiently long for measurements to be made. The measurements are of biological interest because a high intracellular Na^+ concentration is associated with a high growth rate and so with cancer development. Dy-tripolyphosphate gives upfield shifts of 17 ppm in a blood cell suspension or 5.6 ppm with skeletal muscle at 5 mM of reagent concentration and without serious line broadening.[53] Gd-EDTA causes relaxation of Na and Li nuclei and $Fe(CN)_6^{3-}$ or $Pr(EDTA)^-$ have negligible effect. Two reagents were effective, $Dy(dipicolinate)_3^{3-}$ and $Dy(nitrilotriacetate)_2^{3-}$, the latter being the most soluble and a solution 50 mM in this reagent shifts 2 mM Na^+ some 5.5 ppm up field. It is suggested that it is used by equilibrating cells with the reagent and then deactivating the extracellular reagent by adding $Lu(NO_3)_3$, which causes the less effective Dy^{3+} and $DyNTA^0$ to form.[54] A survey of shift reagents has also been carried out in an attempt to find an axially symmetrical one which would give entirely dipolar shifts in aqueous solution. Tb, Dy, or Ho complexed with the cyclic ligand $\text{-}(CH_2N(COO^-)CH_2)_4\text{-}$ seem most active and give shifts of the order of 2 ppm with Na^+.[55]

4. MIXED AND NONAQUEOUS SOLUTIONS OF SIMPLE SALTS

4.1. Nuclear Relaxation

There are extensive reports dealing with the relaxation of the alkali metal ions in a variety of solvents[56–70] or mixed solvents.[62,63,68,72–77] These data are interpreted on the basis of the electrostatic model,[8] though it is probably fair to say that a full analysis of the implications of the measurements is still to be carried out. The relaxation rates increase with concentration, much as in water, but the increase is in general more marked because of the lower solvating ability of these solvents and the greater likelihood of ion pairing. Studies of changes with composition in mixed solvents thus have to use values extrapolated to infinite dilution at each solvent composition to avoid introducing the complication of a variable counterion effect. Such studies provide usually highly curved plots of relaxation rate against proportion of one solvent component, which may have a marked maximum and which correlates with average solvent correlation time or with the viscosity of the mixed solvents.[220]

Recent work with pure solvents shows that relaxation rates may be correlated with donor number,[60,68]* and this supports the electronic model, as does some work by Laszlo.[71] Hertz's work on ^7Li relaxation in formic acid is of particular interest.[69] The quadrupole moment of this nucleus is sufficiently low that dipole–dipole effects can be detected. The relaxation rate is thus different in the four isotopic forms of the acid; DCOOD, HCOOD, DCOOH, or HCOOH, increasing in that order. Once the dipolar contribution has been isolated, it becomes possible to calculate the Li–H distances and obtain the conformation of the solvent molecule relative to the ion. As would be predicted, the carbonyl oxygen is directed towards the ion.

More current effort, however, seems to be devoted to investigating preferential solvation in mixed systems via relaxation measurements. ^{23}Na and ^{87}Rb measurements in mixtures of H_2O or D_2O with methanol, formamide, or DMSO show that cations and anions exhibit different changes in relaxation behavior with composition in the water-rich region. Na^+ and Rb^+ are preferentially hydrated in water–methanol mixtures, but Na^+ prefers DMSO. In pure water, the isotope effect is related to the different viscosities of the solvents.[79] In an extension of this work with ^{23}Na relaxation, it is shown that water–formamide and water–*N*-methyl formamide deviate only slightly from nonpreferential solvation whereas in water–di-*N*-methyl formamide the sodium cation is preferentially solvated by the major component in the solvent mixture.[78] ^7Li relaxation has also been studied in binary mixed solvents over a range of compositions. All the curves show a maximum at intermediate compositions and assumption of four coordination allows the free energy of transfer between solvents to be calculated.[80] Nuclear relaxation of the alkali metal nuclei in $Li/MeNH_2$ or Na/NH_3 solutions has been reviewed. Dramatic changes occur across the metal–nonmetal transition.[81] The linewidth of ^{23}Na in $NaCl/AlCl_3$ melts indicates that the free cation is present.[82] Molten nitrates have also been examined.[215–217]

* "Donor number" is defined in Ref. 227.

4.2. Chemical Shifts

The chemical shift of the cations is in general very sensitive to the nature of the solvent and to the coanion and salt concentration. Further, the interpretation of the measurements is not complicated by the necessity to know any details about molecular motion or solvent structure, which makes the discussions of the relaxation phenomena so difficult. As a result the determination of cation shifts has interested many researchers and has an extensive literature.[8] Some infinite dilution shifts are given for four of the alkali metal cations and two alkaline earths dissolved in eight common solvents in Table 2.[15,83–87,105,107] The accuracy of the figures is not uniform and is particularly open to doubt for Mg^{2+} and Ca^{2+}, where they have had to be extracted from published plots, and where the counterion effects appear to be large. These shifts correlate with donor numbers to a greater or lesser extent, though it is clear that the order of shifts is not the same for each solvent as one looks across the table. It is instructive to plot these shifts versus the cation type, and this has been done in Figure 11, making the following assumptions: that the individual chemical shift scales shall be in the proportion of the number of electrons in the free ions (2:10:18:54) and that there is no reason why the chemical shifts of the aqueous solutions should all appear on the same horizontal line. The individual scales have thus been moved vertically to attempt to give a reasonably smooth series of plots. This has the conceptually satisfying result that the points of standardization for the M^+ are all similar and differ from the octahedrally coordinated M^{2+}. Li^+ as always is a little anomalous, and it is helpful to move it slightly in the opposite direction. The plot then shows that the overall behavior of the solvents chosen can be quite different. MeCN and DMSO run roughly parallel and cause upfield shifts for Li^+ but downfield shifts for Cs^+, whereas DMF, acetone, and propylene carbonate also form a distinct group where the shifts tend to be mostly high field. Methanol, on the other hand, behaves similarly to water. Within these groups the shifts remain in the order of donor number, the higher numbers giving lowest field shifts, but there are also other factors at work of which it may be suggested that solvent structure and ion size are important.

The correlation with donating power suggests that the shifts arise through the

Table 2. Infinite Dilution Chemical Shifts (ppm) for the Cations $^{7}Li^+$, $^{23}Na^+$, $^{39}K^+$, ^{87}Rb, $^{133}Cs^+$, $^{25}Mg^{2+}$, and $^{43}Ca^{2+}$ in a Variety of Solvents[a]

Solvent	Li	Na	K	Rb	Cs	Mg	Ca
$MeNO_2$	−0.36	−15.6	−21.1		−59.8		
MeCN	−2.80	−7	−0.41		+32.0	−13	
DMSO	−1.01	−0.11	+7.77		+68.0	−3	+10
Propylene carbonate	−0.61	−9.4	−11.48	−25	−35.2	−13	−22
MeOH	−0.54	−3.8	−10.05	−21	−45.2	−1	0±2
DMF	+0.45	−5.0	−2.77		−0.5	−5	−13
Acetone	+1.34	−8.4	−10.48		−26.8	−5	−18
Pyridine	+2.54	+1.35	+0.82		−31		

[a] References were infinitely dilute solutions except for ^{7}Li (4.0 M aqueous $LiClO_4$) and ^{23}Na (3.0 M aqueous $NaClO_4$) and all measurements were corrected for the different volume magnetic susceptibilities of the solvents. The figures are taken from Refs 15, 83–85, and 87. It should be emphasized that the infinite dilution shifts for ^{25}Mg and ^{43}Ca are difficult to obtain (see Figures 7 and 8) and the compiler has been bolder than the original workers in presenting the estimates,[15,87] which may well have to be substantially revised in the future. The Rb data are very recent.[225]

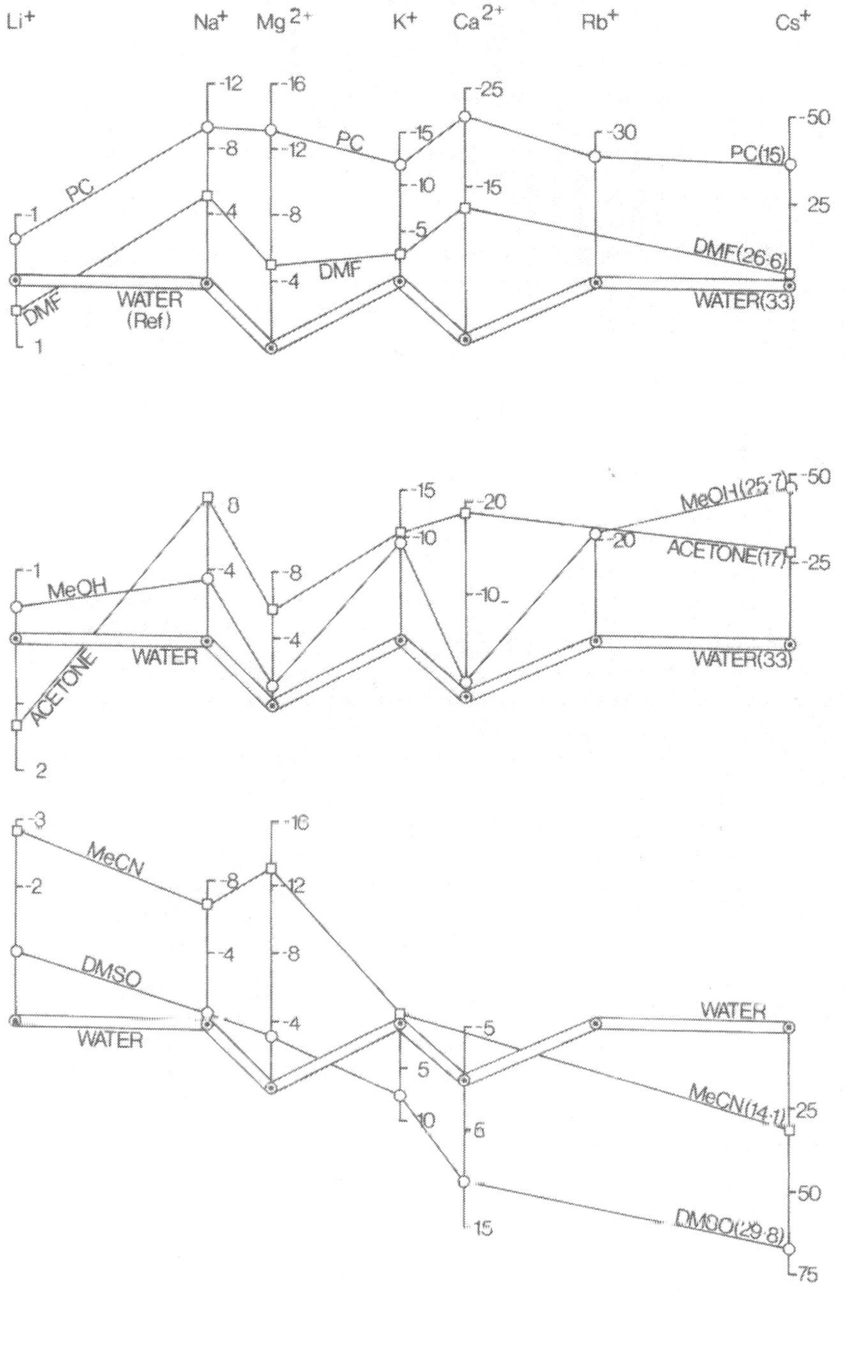

Figure 11. Plot of the infinite dilution shifts of Table 2 across the range of cations considered here. The shift axes are scaled in the proportion of the number of electrons in the free ions, and those for Mg^{2+} and Ca^{2+} have been displaced in the low-field direction to take some account of the fact that the coordination of these cations differs from that of the alkali metals. (See text for details.)

paramagnetic screening mechanism,[83–85,88–90] though this is not expected to work well for lithium, where paramagnetic screening should not be dominant,[83,91] and this may explain the crossings over of many of the lines between Li^+ and Na^+ in the figure. Changes of shift which occur as salt concentration is increased resemble qualitatively those seen in aqueous solutions and may be up or down field. The plots are often highly curved and are believed to reflect ion pairing interactions.

Cation shifts in mixed solvents, in the absence of counterion effects, should reflect preferential solvation by the solvent components. Care is required as in the relaxation studies because changes in solvent composition may also cause changes in the ion pairing present, and measurements have to be extrapolated to infinite dilution, and it may also be wise to use more than one counterion. The deductions made tend to be qualitative, and for instance Popov's group have found for the sodium cation that solvation ability increases in the order $MeNO_2 \ll MeCN < \text{Pyridine} < \text{TMU} \approx \text{DMSO} \simeq \text{HMPT}$.[92,93] A more systematic approach uses the formalism developed by Covington,[94–97] where the shifts are determined via the equilibrium populations of the possible solvates. The assumption has to be made that the chemical shifts of the solvates vary in proportion to the numbers of constituent molecules of one type in each solvated ion, though it is possible to handle changes in coordination number or equilibrium constants across a series of solvates. The method has proved very useful, though it has been shown that the assumption of regular chemical shift changes with solvation status may not be correct.[98]

There is an appreciable amount of recent shift data to report for several of these cations. In the case of Li^+, association of Li^+ with SCN^- has been monitored, principally using ^{15}N NMR, but ^{7}Li also suffers detectable shifts as the anion is added in DMF or dimethylcarbonate solvents. The species formed in the two solvents are different but the ^{7}Li shift changes are similar in both at about 0.3 ppm.[99] Solutions containing $LiAlH_4$ have been examined using both ^{27}Al and ^{7}Li. The chemical shifts seen with the latter indicate that the cation is six-coordinate in glycol dimethylethers where it has low-field shifts between 0.87 and 1.13 ppm. It is tetracoordinate in ether or THF and the shift is then -0.3 or -0.2 ppm upfield, respectively.[100] The breadths of the two metal resonances change with concentration, but not in parallel.[101] Data are also reported for related systems,[102] and quantum mechanical calculations of ^{7}Li shifts in the solvents H_2O, NH_3, and MeOH have been undertaken.[103] Laszlo has published four papers reporting ^{23}Na shifts in mixed solvent systems. It was demonstrated that this cation is probably tetracoordinate in mixed ether/alcohol solvents.[104] The ^{23}Na shifts differ by 20.3 ppm in dilute solutions of $NaClO_4$ in THF or diethylene triamine. The same shift range is observed in mixtures but is confined to the THF-rich region; once the amine is in excess the shift is constant. The data are interpreted in terms of an equilibrium between $Na(THF)_4^+$, $Na(dien)(THF)^+$, and $Na(dien)_2^+$.[105] This work has been extended to mixtures of THF with monodentate amines and reported in a very useful paper which contains an extensive bibliography.[71,106] The ^{23}Na shifts are analyzed in terms of successive displacement equilibria involving solvated cations of general formula $Na(THF)_n(amine)_{4-n}^+$. The linewidths were also measured as a function of composition and, provided these were normalized for viscosity changes, were found to fit well to Deverell's equation. These results thus support the electronic theory of relaxation. The viscosity correction seems to be less controversial in this system since the viscosity is a function of solvent composition and is influenced little by the dissolved salt. The relaxation–composition plots are humped, as has been noted in many other systems, but here this is explained at least partly by viscosity and so interaction between the solvent components, which form the matrix within which to measure the cation parameters. Solvation by amides has also been investigated in detail and the effects of varied substitution at the nitrogen studied.(107) The changes in the series formamide, its N−Me, N−di−Me, N−Et, and N−di−Et derivatives are quite small and are in opposite directions for Na and Li. The changes correlate with the calculated electron densities on the oxygen: a reduced density shields Li (and Be) but deshields Na. Shift changes indicate that there is little preferential solvation of Na^+ by the components of MeOH/DMSO mixtures.(241) ^{133}Cs shifts in H_2O/DMF or D_2O/DMF mixtures have been noted to move outside the extrema of the shifts in the pure solvents, which is further evidence that caution should be exercised in interpreting such systems. It is suggested that in this case the isotope effect may be a better measure of competitive solvation.(98) The nucleus ^{87}Rb has, however, been found to give rather broad lines, and spectroscopy in nonaqueous solutions is not easy. It is predicted that this nucleus will give little information.(225)

^9Be chemical shifts have been measured in methyl cyanide or amide solvents. Several species can be observed in the former at high fields and low temperatures.(86) Be is affected in the same way as is lithium by change in N-substitution of the amides.(107) Unrecorded chemical shifts are seen when small quantities of water are added to methanolic magnesium salt solutions indicating that Mg^{2+} is solvated preferentially by water.(108) The shifts in other nonaqueous solvents are sensitive to concentration and anion,(87) and adding iodine to, e.g., $MgCl_2$ solutions to form $MgCl_2I_4$ solutions causes marked shift changes. ^{43}Ca shifts are also reported for calcium salts in a variety of solvents, and increasing salt concentrations in methanol cause marked shifts to high or low field, respectively, for $Ca(NO_3)_2$ or $CaCl_2$. Large shifts are also recorded in the presence of EDTA or the crown ether 18C6.(15)

5. COMPLEXES BETWEEN THE CATIONS AND VARIOUS TYPES OF LIGANDS

The systems discussed up to the present exhibit complexing interactions of the solvation type. The species formed are weakly bound and exist for very short periods of time and are difficult to characterize. In this section we shall deal with species which give separate signals or which have averaged signals whose behavior can be interpreted in terms of the presence of defined species rather than radial distributions.

5.1. Complexes with Low-Molecular-Weight Compounds

Such complexes have been reviewed recently(7–9) and a selection are given in the charts (Figures 1–8) with references. Complexes with aminocarboxylic acids have been characterized for Li and Ca,(238) and recently for a series of diketones,(109) and interaction with many systems of biological interest seems well established.(8,218,219) Interaction with the crown ethers and cryptands has been of particular interest(7) and has also permitted the exciting observation of the resonances of the anions Li^-, Na^-, Rb^-, and Cs^-,(110,111) though recent work has shown that the presence of the cryptand is not necessary if a suitable solvent is chosen such as HMPTA.(222) Recent work in this area is reported.(244) The cation chemical shifts in the presence of these agents may have a definite single value if the ion is protected fully by the molecules complexing it, or they may depend upon solvent and composition if the complex is weaker or if the complex includes solvent ligands. The kinetics of these complexes has thus been much studied using linewidth or relaxation studies, or where two resonances exist, line-shape analysis as a function of temperature.[112] Recent work in this field involves ^{7}Li,[113,114] ^{23}Na,[114–119] ^{39}K,[120] ^{87}Rb,[225] and ^{133}Cs.[114,116,121–123] For example, complexes between Li^+ and some crown ethers vary inversely in stability with the Gutmann donor number of the solvent and both 1:1 and 1:2 sandwich complexes are formed with 12C4 in $MeNO_2$.[113] ^{23}Na chemical shifts vary considerably in the presence of different crown or cryptand ligands.[115–117] Spiro *bis* crown ethers will complex 1 or 2 sodium cations, the first going in at near the diffusion controlled rate and the second more weakly.[118] Ion pairing and ion pair dimerization of sodium ethylacetoacetate have also been studied with the aid of crown ethers.[119] Separate bound and free ^{39}K signals have been seen in the presence of C222 and C221 if K^+ was in excess. With C211, a singlet was observed and its chemical shift varied over 24 ppm and was used to obtain formation constants in the various solvents.[120] Exclusive and inclusive complexes between Cs^+ and various cryptands have been further investigated,[122] and ^{133}Cs relaxation has been used to discover the effects of crown complexation.[123] A review of these and related ligands will be of interest, though not an NMR article.[124] ^{7}Li has been used to study complexes formed with polydentate amide ethers in nonaqueous solvents[125] and the ligand $ArO(CH_2O)_4Ar$ wraps itself around Na^+ (Ar is *p* substituted $C_6H_4NHCOCH_3$).[126,136]

5.2. Complexes with Synthetic Polymeric Ligands

This area has recently been reviewed in detail for ^{25}Mg and ^{43}Ca.[4] A recent report shows that both cadmium NMR and magnesium NMR may be used to study competition between these ions for polystyrene sulfonate.[127] The remainder of the work in this area uses almost exclusively the ^{23}Na resonance. Thus, a 130 ppm high-field shift of this resonance has been observed in the presence of a perfluorosulfate ion exchanger with reduced water content.[128,129] The cations in these systems may complex strongly or may exist in an "atmospheric" sheath around the polyelectrolyte.[130] This interaction has been studied with both poly(acrylic) or poly(methacrylic) acids,[131–133] and as would be expected its study is complicated by the presence of multiexponential behavior. The presence of polyelectrolyte enhances the relaxation rate of alkali metal ions in the presence of paramagnetic Mn^{2+}.[133] Complexation by poly ethers or amides has also been investigated, and highly temperature dependent shifts are observed for Na^+ of 8–12 ppm downfield. Two resonances are seen in certain circumstances.[134,135]

5.3. Complexes with Biopolymers

This area also has been covered recently.[4,6] Several techniques, including ^{7}Li NMR, have been used to study sarcoplasmic reticular calcium ion transport adenosine triphosphatase. The lithium relaxation rate is high in the presence of the Gd^{3+} form of the enzyme because the two ions are close together. Adding Ca^{2+} reduces the lithium relaxation rate as Gd^{3+} is displaced. Ion–ion distances are estimated.[137] Comparison of the relaxation rates of ^{7}Li with ^{6}Li on the Mn^{2+} form of pyruvate kinase allows the molecular correlation time to be estimated at 3700 ps so that the Mn–Li distance must be 570 ± 20 pm.[138] ^{23}Na NMR in the presence of DNA favors the theory that the

Na^+ condenses around the DNA,[139] and linewidth changes in Na^+/serotin bimaleate/lasalocid mixtures permit the interaction between the latter two components to be studied.[140] The sodium cations in skeletal muscle are associated mostly with macromolecular charged sites, and relaxation measurements show T_1 to be a single exponential whereas T_2 is a complex double exponential.[141] ^{23}Na NMR also shows that there are three types of sodium ion in bacterial pellets.[142,224] Laszlo's group have published three papers dealing with the self-assembly of 5′-guanosine monophosphate in the presence of cations.[17,143,144] Multiexponential relaxation is discussed first and it is demonstrated how to deconvolute the two components. The ^{23}Na signal broadens and shifts above a critical concentration as hexamers and octadecamers form. Deconvolution of the two signals allows a correlation time to be extracted which is not related to the viscosity of the solution and is interpreted as being the sodium ion residence time of about 30 ns. The roles of the cations Na^+, K^+, and Rb^+ are discussed in a following paper,[17] and it is shown that when the first two are present together, the K^+ is sandwiched between two 5′GMP tetramers with the Na^+ around the outside to reduce the repulsion between the anions, the structure being $(GMP_4K\,GMP_4)\,Na_4^{11-}$. Extra K^+ does not bind but condenses around the assembly. ^{23}Na and ^{39}K behave quite differently in carragenate gels, the former giving a narrow line and the latter a powder pattern, a difference which is reflected in the rheology of the systems.[145] An intriguing paper deals with the relaxation of ^{39}K in whole body newborn mice.[146] The samples have to be allowed to run around between the determination of each point of the decay curves and the results are somewhat scattered, though the claimed oscillatory decay does not seem to have been satisfactorily established. Earlier work from this group has been criticized,[223] though the experimental difficulties are considerable.

^{25}Mg and ^{43}Ca (generally enriched) have also received recent attention. The ^{25}Mg linewidth is a function of pH in the presence of peptides and there is increased binding above pH 7.[147] A dissociation constant for Mg on peptides containing γ-carboxyglutamic acid is found in the same way and it is suggested that Mg^{2+} may also condense on the peptide surface.[148] ^{43}Ca, on the other hand, shows no significant line broadening but a complexing shift of 14 ppm downfield.[148] ^{25}Mg and ^{43}Ca NMR are compared with other techniques for examining how prothrombin fragments interact with cations and were found to give different results.[149] It is of course believed that the bound cation resonances are not observed in these experiments and it is of interest to report that signals have now been obtained from both ^{25}Mg and ^{43}Ca bound to proteins.[150] Enrichment, Fourier transform techniques, high magnetic field, and a special probe with a solenoid detector coil were all applied to the problem, and an observation time of 12 h and two million transients were required.

5.4. Cations in Liquid Crystals

The use of ^{23}Na NMR for the study of these systems has been discussed briefly.[6] If a cation is present in an oriented system and is complexed for part of the time to, say, a head group of the crystalline component, then it experiences a situation where the transition energies contain a contribution from the interaction of the nuclear quadrupole with the slowly moving electric field gradient caused by the complexing interaction. The magnitude of this contribution depends upon the orientation of the nuclear quadrupole relative to the field gradient direction so that the nuclear transitions now have different energy and so frequency. The resonance is split into a multiplet (e.g., a triplet if $I=3/2$) whose splitting is governed by the magnitude of the quadrupole coupling constant and the degree of ordering. The spectra normally observed contain a reduced splitting which arises because of exchange between free and complexed cation. Clearly the alkali metal resonance offers the opportunity for investigating order and phase (order–disorder) transitions in such systems and for probing the head group surfaces which may be formed,[151–155] though the conclusions available yet seem to be tentative. Recent work with a variety of alkali metals is reported. The ^{23}Na relaxation times are single valued in aqueous solutions of sodium fatty acids if these are transparent micellar solutions or microemulsions but become double exponentials in turbid emulsions where the particle diameter is greater than 30–40 nm.[156,157] In sodium laurate solutions the quadrupole splitting of Na and of deuterium in the first few CD_2 groups of the chain are correlated.[158] Structural changes in sodium laurate/H_2O which occur following concentration and temperature changes have been examined.[159] The intrinsic chemical shifts of bound ^{23}Na in ampiphilic systems are not sensitive to chain length but to the nature of the end group in the sequence $OSO_3 > ArSO_3 > RSO_3 >$ carboxyl.[160] An attempt has been made to measure ^{133}Cs and ^{23}Na chemical shift anisotropies in Cs decylsulfate/Na_2SO_4/H_2O systems.[161,162] Quadrupole splitting of ^6Li, ^7Li, and ^{23}Na in the related Li/Na system show that the two cations interact quite differently with the head group since the ^{23}Na splitting decreases with increasing temperature whereas the ^7Li splitting increases.[163] A two-dimensional liquid model is suggested to be appropriate in melts of Na *n*-butyrate or Na *iso*-valerate to explain high ionic mobility coupled with a large quadrupole coupling constant.[164] The dynamics of these systems are also discussed.[165] In lyotropic liquid crystals, it is suggested that distortion, perhaps of the hydration water near the ion, due to the thinness of the water layer at the lipid/water interface is the principal cause of the electrical field gradients and so the quadrupole splitting.[166–168] The sodium, however, plays a passive role in the hexadecylpyridinium/hexadecyl trimethylammonium system.[169]

Lithium behaves differently from sodium or cesium also in the monooctanoin Li/H_2O system.[170] Quadrupole splittings for all the alkali metal cations are reported in decylammonium mesophases.[171] A recent report finds ^{23}Na quadrupole splittings in a nonaqueous liquid crystal in which the sodium is complexed by the cryptand C222, and this is believed to indicate that the splittings arise because of ordering effects rather than electrostatic interactions.[172]

6. GROUP I AND II METAL ORGANIC COMPOUNDS

Three of the elements considered in this chapter have a covalent chemistry: lithium, beryllium, and magnesium. Covalent compounds of the first two have been studied and those of the more difficult magnesium have just recently received attention.[245–247] Covalent bonding means, of course, that in such cases spin–spin coupling is possible, and we will encounter in this section the sole examples of coupling in the two groups. In addition the phenomenon of ion pairing between alkali metal cations and organic anions has been much studied. ESR signals can be obtained from the radical anions and often show hyperfine interaction with the alkali metal, indicating the existence of contact ion pairs.[173–177] Some recent papers deal with ENDOR double resonance studies involving the alkali metals in such compounds in solution[178–180] or in frozen solution.[181] A purely NMR example will be found below.

6.1. Lithium Organic Compounds

The early measurements made with the metal nucleus in such compounds used, naturally, the ^{7}Li isotope, which despite its appreciable quadrupole moment gives quite well-resolved spectra. Much attention was given to the alkyl lithium compounds since the method proved ideal for studying the degree of aggregation through ^{13}C enrichment and the coupling patterns then observed. This area of study has been fully reviewed quite recently.(7,8) The lithium chemical shifts observed in these compounds are believed to be determined entirely by neighbor anisotropy effects.(182–189) The results of this period are summarized, shifts in Figure 1 and coupling constants in Table 3. Only a few papers have appeared in the last four years (to 1982), but these seem to herald further advances in this field, to be achieved by replacing ^{7}Li by ^{6}Li. The relaxation times of this latter nuclide in simple organolithium compounds are much longer than those of ^{7}Li and contributions from the quadrupolar and dipole–dipole mechanisms can be separated by measuring the NOE. Indeed ^{6}Li behaves very much like a spin-1/2 nucleus in these systems, as the results in Table 4 show.(22,190) The way this approach may be used is demonstrated for $^{12}C_2H_5^{13}CH_2{}^6Li$ and its ^{7}Li form by Fraenkel's group.(191,192) Five species are detected by both ^{13}C and ^{7}Li NMR and the coupling patterns support their degrees of aggregation to be 6, 8, and three forms with 9. Exchange occurs between aggregates and much more rapid exchange within aggregates. ^{7}Li–^{13}C coupling has also been observed in some halo-organic lithium compounds.(193) Where coupling to the metal does not exist, as in the ion pairs, there is probably little to be gained by enriching in ^{6}Li, as some recent excellent spectra testify.(194) The solution structure of the peralkylcyclohexadienyl-lithium ion pairs depends upon the solvent and in THF, glymes, or HMPT exist as

Table 3. Coupling Constants to ^{6}Li, ^{7}Li, and ^{9}Be

		Coupling constant	Reference
Of ^{7}Li to ^{13}C in	$(MeLi)_4$	14.5–15 Hz	187, 188
in	$(Bu^nLi)_4$	$^{1}J = 14$ Hz	186
in	$(Bu^tLi)_4$	$^{1}J = 10–14$ Hz	186
in	Br_3CLi	43,40 Hz[a]	193
in	$^{13}CH_3Br_2{}^{13}CLi$	$^{1}J = 45$ Hz	193
Of ^{6}Li to ^{13}C in	$(Pr^nLi)_6$	$^{1}J = 3.35$ Hz[b]	192
	$(Pr^nLi)_8$	$^{1}J = 2.48$ Hz[b]	192
	$(Pr^nLi)_9$	$^{1}J = 2.22$ Hz[b]	192
Of ^{6}Li to ^{7}Li	alkyl Li	<0.3 Hz	213
Of ^{7}Li to ^{31}P in	$[(Me_2N)_2PON(Me)PO(NMe_2)_2]_2Li^+$	2 Hz	232
in	$Ph_2PLi_2PPh_2$	45 Hz	234
Of ^{9}Be to ^{19}F in	BeF_4^-	33, 34 Hz	22, 195
	BeF_3OH^-	38 Hz	22
	$BeF_2(OH)_2^-$	28 Hz	196
Of ^{9}Be to ^{1}H on B in	$\eta5–C_5H_5BeBH_4$	10.2 Hz	198
Of ^{9}Be to ^{11}B in	$\eta5–C_5H_5BeBH_4$	3.6 Hz	198
Of ^{9}Be to ^{31}P in	$[(NMe_2)_3PO]_4Be^{2+}$	6.1 Hz	213
	$(RR'R''PO)_4Be^{2+}$	4–6 Hz	197
	$[(Me_2N)_2PON(Me)PO(NMe_2)_2]_2Be^{2+}$	2.6 Hz	232

[a] Nature of species uncertain.
[b] These are in fact averages of ^{1}J and longer-range couplings.

Table 4. Relaxation in Alkyl Lithium Compounds[a]

	T_1 (s)				NOE – ^{1}H irrad	
Compound	^{13}C of C_1	^{6}Li	^{6}Li$_{DD}$	^{7}Li	^{6}Li	^{13}C
$(MeLi)_4$	8 9	72	350	1 44	0 7	1 73
$(BuLi)_6$	1 9	19 8	57	0 083	1 19	full
$(PhLi)_2$	—	12 8	74	0 047	0 59	full

[a] Reference 190 The ^{6}Li dipole–dipole relaxation time is calculated The maximum theoretical NOE for ^{6}Li when irradiating protons is 3 4 The literature values of degree of association have been used in compiling the table

AnLiAn$^-$ $\cdots$ Li$^+$ with one lithium sandwiched between two cyclohexadiene rings and the other loosely associated with this triple anion. The normal monomeric ion pair may coexist and is favored at low concentrations. The spectra are shown in Figure 12. Such triple ions with enolate anions are also formed if cryptand C211 is introduced.[(221)]

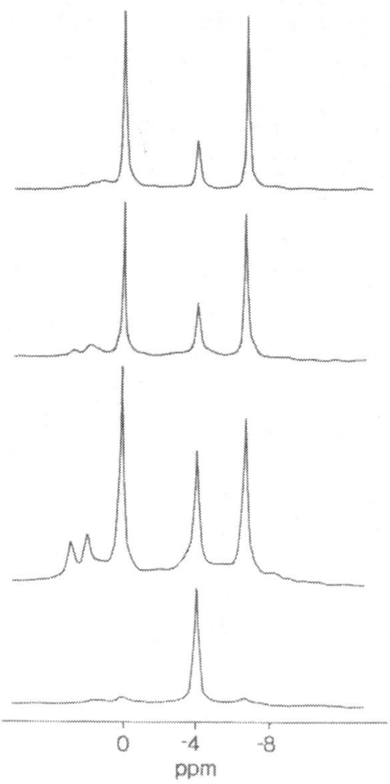

Figure 12. ^{7}Li spectra at 23 319 MHz of a peralkylcyclohexadienyllithium compound in cyclopentane/THF mixtures at −40°C The concentration of THF was varied from 4 4 *M* (uppermost spectrum) through 2 8 *M*, 1 4 *M*, to 0 73 *M* (lowest spectrum), while the concentration of compound was kept at 0 64 *M* The two major peaks are due to AnLiAn$^-$ and Li$^+$ and the central one to the monomonomeric ion pair The small peaks to low field are due to MeLi (1.86 ppm) and lithium acetaldehyde enolate (2.76 ppm).[(194)] (Reprinted with permission from Fraenkel and Hallden-abberton, *J. Am. Chem. Soc.*, **103**, 5657, copyright (1981) American Chemical Society.)

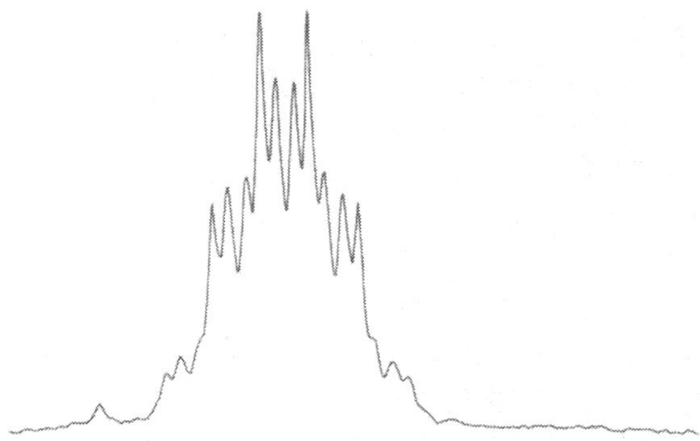

Figure 13. The ^{9}Be spectrum at 14 06 MHz of (η^5-C_5H_5) $BeBH_4$ in C_6F_6 solution Coupling is observed to the boron and hydrogen of the BH_4 group.[198] (Reprinted with permission from Gaines, Coleson, and Hillenbrand, *J. Magn. Reson.*, 1981, **44**, 84.)

6.2. Beryllium Covalent Compounds

This area has also been well reviewed[7,8] and there is very little new to report. The chemical shift data are summarized in Figure 6 and the coupling data are in Table 3.[22,195,197] One recent article deals with the ^{9}Be NMR of the berylloboranes and presents some excellent spectra,[198] one being shown in Figure 13. The chemical shifts support neighbor anisotropy effects as being important, and for instance the high field shifts of the cyclopentadienyl compounds probably arise from ring currents.[198,214] The coordination number also has its effect and four-coordinate beryllium is high field of the three coordinate compounds. Few linewidths are recorded, though these tend to be quite small, and indeed this is a prerequisite for the observation of coupling.

REFERENCES

1 Schmidt, P C, Sen, K D, Das, T P, Weiss, A *Phys Rev B* 1980, **22**, 4167
2 Fiat, D N, Connick, R E *J Chem Phys* 1963, **39**, 1349
3 Fratiello, A, Lee, R E, Nishida, V M, Schuster, R E *J Chem Phys* 1968, **48**, 3705
4 Forsen, S, Lindman, B *Ann Rep NMR Spectrosc* 1981, **11A**, 183
5 Akitt, J W *J Chem Soc Faraday Trans I* 1982, **78**, 289, and references quoted therein
6 Laszlo, P *Angew Chem Int Edn* 1978, **17**, 254
7 Wehrli, F W *Ann Rep NMR Spectrosc* 1979, **9**, 126
8 Lindman, B, Forsen, S *NMR and the Periodic Table*, Harris, R K, and Mann, B E, Eds, Academic New York, 1978, Chap 6, "The Alkali Metals"
9 Lindman, B, Forsen, S *NMR and the Periodic Table*, Harris, R K, and Mann, B E, Eds, Academic New York, 1978, Chap 7, "The Alkaline Earth Metals"
10 Popov, A I *Proc Symp Spectrosc Electrochem Charact Solute Species in non Aqueous Solvents* 1976, **2**, 47, 271
11 Lindman, B *J Magn Reson* 1978, **32**, 39
12 Civan, M M, Shporer, M *Biol Magn Reson* 1978, **1**, 1
13 Van Geet, A L *Progr Analyt Chem* 1973, **6**, 155
14 Granger, P *J Biophys Med Nucl* 1981, **5**, 137
15 Farmer, R M, Popov, A I *Inorg Nucl Chem Lett* 1981, **17**, 51
16 Deverell, C *Progr NMR Spectrosc* 1969, **4**, 278
17 Detellier, C, Laszlo, P *J Am Chem Soc* 1980, **102**, 1135
18 Banck, J, Schwenk, Z *Z Phys* 1973, **265**, 165

19. Lutz, O.; Oehler, H. *Z. Phys. (A)* 1978, **288**, 11.
20. Simeral, L.; Maciel, G. E. *J. Phys. Chem.* 1975, **80**, 552.
21. Lutz, O.; Schwenk, A.; Uhl, A. *Z. Naturforsch.* 1975, **30A**, 1122.
22. Wehrli, F. W. *J. Magn. Reson.* 1978, **30**, 193.
23. Wehrli, F. W. *J. Magn. Reson.* 1976, **23**, 527.
24. Wehrli, F. W. *J. Magn. Reson.* 1977, **25**, 575.
25. Wehrli, F. W. *J. Magn. Reson.* 1976, **23**, 181.
26. Harmon, J. F.; Sutter, E. J. *J. Phys. Chem.* 1978, **82**, 1938.
27. Weingartner, H. *J. Magn. Reson.* 1980, **41**, 74.
28. Valiev, K. A. *J. Struct. Chem.* 1964, **5**, 477; 1962, **3**, 630; 1966, **7**, 470 (with Zaripov, M. M.).
29. Hertz, H. G. *Ber. Bunsenges. Phys. Chem.* 1973, **77**, 531, 688.
30. Hertz, H. G.; Holtz, M.; Keller, G.; Versmold, H.; Yoon, C. *Ber. Bunsenges. Phys. Chem.* 1974, **78**, 493.
31. Deverell, C. *Mol. Phys.* 1969, **16**, 491.
32. Contreras, M.; Hertz, H. G. *Faraday Diss. Chem. Soc.* 1978, **64**, 33.
32a. Hall, C.; Richards, R. E.; Schultz, G. N.; Sharp, R. R. *Mol. Phys.* 1969, **16**, 529.
33. Mazitov, R. K.; Itkina, L. S.; Petrova, E. V.; Nadzhanyan, K. A. *Z. Neorg. Khim.* 1980, **25**, 1200.
34. Mazitov, R. K. *Magn. Reson. Relat. Phenom., Proc. 20th Congr. Ampere*, 1978, p. 474.
35. Chizik, V. I.; Mikhailov, V. I.; Ermolaeva, T. T.; Karpenko, G. V.; Maiorova, T. N. *Zh. Strukt. Khim.* 1981, **22**, 178(631).
36. Chizik, V. I.; Mikhailov, V. I.; Ermolaeva, T. T. *Magn. Reson. Relat. Phenom., Proc. 20th Congr. Ampere* 1978, p. 467.
37. Berger, C. L.; Emons, H.-H.; Pohl, D. *Z. Phys. Chem. (Leipzig)* 1977, **258**, 632.
38. Hirata, F.; Friedman, H. L.; Holtz, M.; Hertz, H. G. *J. Chem. Phys.* 1980, **73**, 6031.
39. Reimarsson, P. *J. Magn. Reson.* 1980, **38**, 245.
40. Holtz, M.; Gunther, S.; Lutz, O.; Nolle, A.; Schrade, P.-G. *Z. Naturforsch.* 1979, **34A**, 944.
41. Hubbard, P. S. *J. Chem. Phys.* 1970, **53**, 985.
42. Rubinstein, M.; Baram, A.; Luz, Z. *Mol. Phys.* 1971, **20**, 67.
43. Reuben, J.; Luz, Z. *J. Phys. Chem.* 1976, **80**, 1357.
44. Bull, T.; Forsen, S.; Turner, D. *J. Chem. Phys.* 1979, **70**, 3106.
45. Akitt, J. W. *J. Chem. Soc. Faraday Trans. I* 1975, **71**, 1557.
46. Kondo, J.; Yamashita, J. *J. Phys. Chem. Solids* 1959, **10**, 245.
47. Ikenberry, D.; Das, T. P. *J. Chem. Pys.* 1965, **43**, 2199.
48. Ikenberry, D.; Das, T. P. *J. Chem. Phys.* 1966, **45**, 1361.
49. Halliday, J. D.; Richards, R. E.; Sharp, R. R. *Proc. R. Soc. London* 1969, **313A**, 45.
50. Hall, C.; Richards, R. E.; Sharp, R. R. *Proc. R. Soc. London* 1974, **337A**, 297.
51. Gauss, W.; Guenther, S.; Haase, A. R.; Kerber, M.; Kessler, D.; Kronenbitter, J.; Krueger, H.; Lutz, O.; Nolle, A. *Z. Naturforsch.* 1978, **33A**, 934.
52. Halliday, J.; Hill, H. D. W.; Richards, R. E. *J. Chem. Soc. Chem. Commun.* 1969, 219.
53. Gupta, J. K.; Gupta, P. *J. Magn. Reson.* 1982, **47**, 344.
54. Pike, M. M.; Springer, C. S. *J. Magn. Reson.* 1982, **46**, 348.
55. Bryden, C. C.; Reilly, C. N.; Desreux, J. F. *Analyt. Chem.* 1981, **53**, 1418.
56. Geiger, A.; Hertz, H. G. *Adv. Mol. Relaxation Proc.* 1976, **9**, 293.
57. Melendres, C. A.; Hertz, H. G. *J. Chem. Phys.* 1974, **61**, 4156.
58. Mishustin, A. I.; Sidorova, R. A. *Vestn. Mosk. Univ., Fiz. Asron.* 1970, **11**, 528.
59. Buslaev, Y. A.; Evsikov, V. V.; Buslaeva, M. N.; Mazitov, R. V.; Plakhotnik, V. N. *Proc. Acad. Sci. USSR, Phys. Chem.* 1973, **213**, 1120.
60. Kessler, Y. M.; Mishustin, A. I.; Podkovyrin, A. I. *J. Soln. Chem.* 1977, **6**, 111.
61. Shchori, E.; Jagur-Grodinski, J.; Luz, Z.; Shporer, M. *J. Am. Chem. Soc.* 1971, **93**, 7133.
62. Green, R. D.; Martin, J. S. *Can. J. Chem.* 1972, **50**, 3935.
63. Van Geet, A. L. *J. Am. Chem. Soc.* 1972, **94**, 5583.
64. Grotens, A. M.; Smid, J.; DeBoer, E. *J. Chem. Soc. Chem. Commun.* 1971, 759.
65. Canters, G. W. *J. Am. Chem. Soc.* 1972, **94**, 5230.
66. Detellier, C.; Laszlo, P. *Bull. Soc. Chim. Belg.* 1975, **84**, 1087.
67. Sahm, W.; Schwenk, A. *Z. Naturforsch.* 1974, **29A**, 1754.
68. Mishustin, A. I.; Kessler, Y. M. *J. Soln. Chem.* 1975, **4**, 779.
69. Hertz, H. G.; Weingartner, H.; Bode, B. M. *Ber. Bunseges. Phys. Chem.* 1975, **79**, 1190.
70. Weingartner, H.; Hertz, H. G. *Ber. Bunseges. Phys. Chem.* 1977, **81**, 1204.

71 Delville, A, Detellier, C, Gerstmans, A, Laszlo, P *J Magn Reson* 1981, **42**, 14
72 Craig, R A, Richards, R E *Trans Faraday Soc* 1963, **59**, 1972
73 Richards, R E, Yorke, B A *Mol Phys* 1963, **6**, 289
74 Neggia, P, Holtz, H, Hertz, H G *J Chim Phys* 1974, **71**, 56
75 Detellier, C, Laszlo, P *Helv Chim Acta* 1976, **59**, 1346
76 Goldsmith, M, Hor, D, Damadian, R *J Chem Phys* 1976, **65**, 1708
77 Holtz, M, Weingartner, H, Hertz, H G *J Chem Soc Faraday Trans I* 1977, **73**, 71
78 Holtz, M, Rau, C K *J Chem Soc Faraday Trans I* 1982, **78**, 1899
79 Holtz, M *J Chem Soc Faraday Trans I* 1978, **74**, 644
80 Mishustin, A I, Podkoyyin, A L, Kessler, Y M *Dokl Akad Nauk SSSR (Phys Chem)* 1979, **245**, 1420
81 Edwards, P P *J Phys Chem* 1980, **84**, 1215
82 Anders, U, Plambeck, J A *J Inorg Nucl Chem* 1978, **40**, 387
83 Popov, A I *Pure Appl Chem* 1975, **41**, 275
84 DeWitte, W J, Schoening, R C, Popov, A I *Inorg Nucl Chem Lett* 1976, **12**, 251
85 Shih, J S, Popov, A I *Inorg Nucl Chem Lett* 1977, **13**, 105
86 Wehrli, F W, Wehrli, S L *J Magn Reson* 1982, **47**, 151
87 Heubel, P H, Popov, A I *J Soln Chem* 1979, **8**, 283
88 Bloor, E G, Kidd, R G *Canad J Chem* 1968, **46**, 3425
89 Erlich, R H, Popov, A I *J Am Chem Soc* 1971, **93**, 5620
90 Herlem, M, Popov, A I *J Am Chem Soc* 1972, **94**, 1431
91 Cahen, Y M, Handy, P R, Roach, E T, Popov, A I *J Phys Chem* 1975, **79**, 80
92 Erlich, R H, Greenberg, M S, Popov, A I *Spectrochim Acta* 1973, **29A**, 543
93 Greenberg, M S, Popov, A I *Spectrochim Acta* 1975, **31A**, 697
94 Covington, A K, Newman, K E, Lilley, T H *J Chem Soc Faraday Trans I* 1973, **69**, 973
95 Covington, A K, Newman, K E, Lilley, T H, Porthouse, G A *J Chem Soc Faraday Trans I* 1973, **69**, 963
96 Covington, A K, Lantzke, I R, Thain, J M *J Chem Soc Faraday Trans I* 1974, **70**, 1869
97 Covington, A K, Thain, J M *J Chem Soc Faraday Trans I* 1974, **70**, 1879
98 Gustavsson, H, Ericsson, T, Lindman, B *Inorg Nucl Chem Lett* 1978, **14**, 37
99 Vaes, J, Chabanel, M, Martin, M L *J Phys Chem* 1978, **82**, 2420
100 Noth, H, Rurlaender, R, Wolfgardt, P *Z Naturforsch* 1981, **36B**, 31
101 Noth, H *Z Naturforsch* 1980, **35B**, 119
102 Horne, D A *J Am Chem Soc* 1980, **102**, 6011
103 Ribas-Prado, F, Giessner-Prettre, C, Daudey, J-P, Pullman, A, Hinton, J F, Yuong, G, Harpool, D *J Magn Reson* 1980, **37**, 431
104 Detellier, C, Laszlo, P *Helv Chim Acta* 1976, **59**, 1333
105 Detellier, C, Gerstmans, A, Laszlo, P *Inorg Nucl Chem Lett* 1979, **15**, 93
106 Delville, A, Detellier, C, Gerstmans, A, Laszlo, P *J Am Chem Soc* 1980, **102**, 6558
107 Rode, B M, Pontani, T, Heckmann, G *J Chem Soc Faraday Trans I* 1978, **74**, 71
108 Bouhoutsos-Brown, E, Bryant, R G *J Inorg Nucl Chem* 1981, **43**, 3213
109 Kraft, H G, Peringer, P, Rode, B M *Inorg Chem Acta* 1981, **48**, 135
110 Dye, J L, Andrews, C W, Ceraso, J M *J Phys Chem* 1975, **79**, 3076
111 Ceraso, J M, Dye, J L *J Am Chem Soc* 1973, **95**, 4432
112 Ceraso, J M, Smith, P B, Landers, J S, Dye, J L *J Phys Chem* 1977, **81**, 760
113 Smetana, A J, Popov, A I *J Soln Chem* 1980, **9**, 183
114 Hourdakis, A, Popov, A I *J Soln Chem* 1977, **6**, 299
115 Lin, J D, Popov, A I *J Am Chem Soc* 1981, **103**, 3773
116 Shamsipur, M, Popov, A I *J Am Chem Soc* 1979, **101**, 4051
117 Grandjean, J, Laszlo, P, Offermann, W, Rinaldi, P L *J Am Chem Soc* 1981, **103**, 1380
118 Bouquant, J, Delville, A, Grandjean, J, Laszlo, P *J Am Chem Soc* 1982, **104**, 686
119 Cornelius, A, Laszlo, P, Cambillau, C *J Chem Res (S)* 1978, 462
120 Shih, J-S, Popov, A I *Inorg Chem* 1980, **19**, 1689
121 Mei, E, Liu, L, Dye, J L, Popov, A I *J Soln Chem* 1977, **6**, 771
122 Kauffmann, E, Dye, J L, Lehn, J-M, Popov, A I *J Am Chem Soc* 1980, **102**, 2274
123 Wehrli, F W *J Magn Reson* 1977, **25**, 575
124 Vogtle, F, Weber, E *Angew Chem Internatl Edn* 1979, **18**, 753
125 Olsher, U, Elgavish, G A, Jagur-Grodzinski, J *J Am Chem Soc* 1980, **102**, 3338

126. Grandjean, J.; Laszlo, P.; Vogtle, F.; Sieger, H. *Angew. Chem. Internatl. Edn.* 1978, **17**, 856.
127. Linse, P.; Gustavsson, H.; Lindman, B.; Drakenberg, T. *J. Magn. Reson.* 1981, **45**, 133.
128. Komoroski, R. A.; Mauritz, K. A. *J. Am. Chem. Soc.* 1978, **100**, 7487.
129. Komoroski, R. A. *Adv. Chem. Ser.* 1980, **187**, 155.
130. Gustavsson, H.; Lindman, B.; Siegel, G. *Magn. Reson. Relat. Phenom., Proc. 20th Congr. Ampere* 1978, 468.
131. Meurer, B.; Spegt, P.; Weill, G. *Chem. Phys. Lett.* 1978, **60**, 55.
132. Gustavsson, H.; Lindman, B.; Bull, T. *J. Am. Chem. Soc.* 1978, **100**, 4655.
133. Westra, S. W. T.; Leyte, J. C. *Ber. Bunseges. Phys. Chem.* 1978, **83**, 678.
134. Ricard, A. *Eur. Polymer J.* 1979, **15**, 1.
135. Ptitsyna, N. V.; Shupik, A. N.; Kazanskii, K. S.; Arkhipovich, G. N. *Dokl. Akad. Nauk. SSSR (Phys. Chem.)* 1979, **247**, 1412.
136. Laszlo, P. *Nachr. Chem. Tech. Lab.* 1979, **27**, 710.
137. Stephens, E. M.; Grisham, C. M. *Biochem.* 1979, **18**, 4876.
138. Raushel, F. M.; Villafranca, J. J. *J. Am. Chem. Soc.* 1980, **102**, 6618.
139. Anderson, C. F.; Record, M. T.; Hart, P. A. *Biophys. Chem.* 1978, **7**, 301.
140. Grandjean, J.; Laszlo, P. *Angew. Chem. Internatl. Edn.* 1979, **18**, 153.
141. Chang, D. C.; Woessner, D. E. *J. Magn. Reson.* 1978, **30**, 185.
142. Melamud, R.; Risk, M.; Gilboa, H. *Biochim. Biophys. Acta.* 1981, **678**, 311.
143. Delville, A.; Detellier, C.; Laszlo, P. *J. Magn. Reson.* 1979, **34**, 301.
144. Borzo, M.; Detellier, C.; Laszlo, P.; Paris, A. *J. Am. Chem. Soc.* 1980, **102**, 1124.
145. Morris, V. J.; Belton, P. S. *J. Chem. Soc. Commun.* 1980, 983.
146. Cope, F. W.; Damadian, R. *Physiol. Chem. Phys.* 1979, **11**, 143.
147. Robertson, P.; Koehler, K. A.; Hiskey, R. G. *Biochem. Biophys. Res. Commun.* 1979, **86**, 265.
148. Robertson, P.; Hiskey, R. G.; Koehler, K. A. *J. Biol. Chem.* 1978, **253**, 5880.
149. Marsh, H. C.; Robertson, P.; Scott, M. E.; Koehler, K. A.; Hiskey, R. G. *J. Biol. Chem.* 1979, **254**, 10268.
150. Andersson, T.; Drakenberg, T.; Forsen, S.; Thulin, E.; Sward, M. *J. Am. Chem. Soc.* 1982, **104**, 576.
151. Wennerstrom, H.; Lindblom, G.; Lindman, B. *Chem. Scr.* 1974, **6**, 97.
152. Chen, D. M.; Radley, K.; Reeves, L. W. *J. Am. Chem. Soc.* 1974, **96**, 5251.
153. Lindblom, G.; Persson, N-O.; Arvidson, G. *Adv. Chem. Ser.* 1975, **152**, 121.
154. Lindblom, G.; Lindman, B.; Tiddy, G. J. T. *Acta. Chem. Scand.* 1975, **29A**, 876.
155. Rosenholm, J. B.; Lindman, B. *J. Colloid Interface Sci.* 1975, **57**, 362.
156. Monoi, H.; Uedaira, H. *Biophys. J.* 1979, **25**, 535.
157. Monoi, H.; Uedaira, H. *J. Magn. Reson.* 1980, **38**, 119.
158. Abdolall, K.; MacKay, A. L.; Bloom, M. *J. Magn. Reson.* 1978, **29**, 309.
159. Nishikawa, H. *Kyoto-Juritsu Ika Daigaku Zasshi* 1979, **88**, 1219.
160. Gustavsson, H.; Lindman, B. *J. Am. Chem. Soc.* 1978, **100**, 4647.
161. Fujiwara, F. Y.; Reeves, L. W. *Mol. Phys.* 1978, **36**, 1897.
162. Persson, N.-O.; Lindblom, G. *J. Phys. Chem.* 1979, **83**, 3015.
163. Chen, D. M. *Mol. Cryst. Liquid Cryst.* 1978, **47**, 201.
164. Bonekamp, J.; Eguchi, T.; Jonas, J. *Chem. Phys. Lett.* 1980, **25**, 360.
165. Kanazawa, Y.; Kamei, H. *Bull. Chem. Soc. Jpn* 1979, **52**, 895.
166. Tiddy, G. J. T.; Lindblom, G.; Lindman, B. *J. Chem. Soc. Faraday Trans. I* 1978, **74**, 1290.
167. Lindblom, G.; Lindman, B.; Tiddy, G. J. T. *J. Am. Chem. Soc.* 1978, **100**, 2299.
168. Loewenstein, A.; Brenman, M.; Schwarzmann, R. *J. Phys. Chem.* 1978, **82**, 1744.
169. Reeves, L. W.; Tracey, A. S.; Tracey, M. M. *Can. J. Chem.* 1979, **57**, 747.
170. Persson, N.-O.; Lindblom, G. *Chem. Scr.* 1979, **13**, 193.
171. Gauss, W.; Kronenbitter, J.; Lutz, O.; Nolle, A.; Siegloch, G. E. *Z. Naturforsch.* 1978, **33A**, 105.
172. Loewenstein, A.; Brenman, M. *J. Phys. Chem.* 1980, **84**, 340.
173. Sharp, J. H.; Symons, M. C. R. *Ions and Ion Pairs in Organic Reactions*, Szwarc, M., Ed.; Wiley-Interscience: New York, 1972, Vol. 1, Chap. 5.
174. Canters, G. W.; DeBoer, E. *Mol. Phys.* 1973, **26**, 1185.
175. DeBoer, E.; Hendriks, B. M. P. *Pure Appl. Chem.* 1974, **40**, 259.
176. Canters, G. W.; DeBoer, E. *Mol. Phys.* 1974, **27**, 665.
177. Takeshita, T.; Hirota, N. *J. Chem. Phys.* 1973, **58**, 3745.
178. Meise-Gresch, K.; Muller-Warmuth, W. *Ber. Bunseges. Phys. Chem.* 1979, **83**, 586.
179. Lubitz, W.; Plato, M.; Moebius, K.; Biehl, R. *J. Phys. Chem.* 1979, **83**, 3402.

180. Plato, M.; Lubitz, W.; Moebius, K. *J. Phys. Chem.* 1981, **85**, 1202.
181. Van Willigen, H.; Mulks, C. F. *J. Chem. Phys.* 1981, **75**, 2135.
182. Lewis, H. L.; Brown, T. L. *J. Am. Chem. Soc.* 1970, **92**, 4664.
183. Scherr, P. A.; Hogan, R. J.; Oliver, J. P. *J. Am. Chem. Soc.* 1974, **96**, 6055.
184. Cox, R. H.; Terry, H. W. *J. Magn. Reson.* 1974, **14**, 317.
185. Hogan, R. J.; Scherr, P. A.; Weibel, A. T.; Oliver, J. P. *J. Organomet. Chem.* 1975, **85**, 265.
186. McKeever, L. D.; Waack, R. *J. Chem. Soc. Chem. Commun.* 1969, 750.
187. McKeever, L. D.; Waack, R.; Doran, M. A.; Baker, E. B. *J. Am. Chem. Soc.* 1968, **90**, 3244.
188. McKeever, L. D.; Waack, R.; Doran, M. A., Baker, E. B. *J. Am. Chem. Soc.* 1969, **91**, 1057.
189. Ladd, J. A.; Parker, J. *J. Chem. Soc. Dalton Trans.* 1972, 930.
190. Wehrli, F. W. *Organic Magn. Reson.* 1978, **11**, 106.
191. Fraenkel, G.; Fraenkel, A. M.; Geckle, M. J.; Schloss, F. *J. Am. Chem. Soc.* 1979, **101**, 4745.
192. Fraenkel, G.; Henrichs, M.; Hewitt, J. M.; Biing Ming Su; Geckle, M. J. *J. Am. Chem. Soc.* 1980, **102**, 3345.
193. Siegel, H.; Hiltbrunner, K.; Seebach, D. *Angew. Chem. Internatl. Edn.* 1979, **18**, 785.
194. Fraenkel, G.; Hallden-abberton, M. P. *J. Am. Chem. Soc.* 1981, **103**, 5657.
195. Kotz, J. C.; Schaeffer, R.; Clouse, A. *Inorg. Chem.* 1967, **6**, 620.
196. Kovar, R. A.; Morgan, G. L. *J. Am. Chem. Soc.* 1970, **92**, 5067.
197. Delpuech, J. J.; Peguy, A.; Rubini, P.; Steinmetz, J. *Nouv. J. Chim.* 1976, **1**, 133.
198. Gainess, D. F.; Coleson, K. M.; Hillenbrand, D. F. *J. Magn. Reson.* 1981, **44**, 84.
199. Akitt, J. W.; Duncan, R. H. *J. Chem. Soc. Faraday Trans. I* 1980, **76**, 2212.
200. Akitt, J. W.; Downs, A. J. *Chem. Soc. Spec. Publ. The Alkali Metals*, No 22, 1967, 199.
201. Maciel, G. E.; Huncock, J. A.; Lafferty, L. F.; Mueller, P. A.; Musker, W. K. *Inorg. Chem.* 1966, **5**, 554.
202. Akitt, J. W.; Parekh, M. *J. Chem. Soc.* 1968, 2195.
203. Cahen, Y. M.; Dye, J. L.; Popov, A. I. *J. Phys. Chem.* 1975, **79**, 1289.
204. Deverall, C.; Richards, R. E. *Mol. Phys.* 1966, **10**, 551.
205. Templeman, G. J.; Van Geet, A. L. *J. Am. Chem. Soc.* 1972, **94**, 5578.
206. Relich, R. H.; Roach, E.; Popov, A. I. *J. Am. Chem. Soc.* 1970, **92**, 4989.
207. Haynes, D. H.; Pressman, B. C.; Kowalsky, A. *Biochemistry* 1971, **10**, 852.
208. Kintzinger, J. P.; Lehn, J. M. *J. Am. Chem. Soc.* 1974, **96**, 3313.
209. Bloor, E. G.; Kidd, R. G. *Can. J. Chem.* 1972, **50**, 3926.
210. Lutz, O.; Nolle, A. *Z. Naturforsch.* 1972, **27A**, 1577.
211. Mei, E.; Popov, A. I.; Dye, J. L. *J. Am. Chem. Soc.* 1977, **99**, 6532.
212. Lutz, O. *Z. Naturforsch.* 1967, **22A**, 286.
213. Delpuech, J.-J.; Peguy, A.; Khaddar, M. R. *J. Magn. Reson.* 1972, **6**, 325.
214. Drew, D. A.; Morgan, G. L. *Inorg. Chem.* 1977, **16**, 1704.
215. Wolney Filho, W.; Havill, R. L.; Titman, J. M. *J. Magn. Reson.* 1982, **49**, 296.
216. Harold-Smith, D. *J. Chem. Phys.* 1973, **59**, 4771.
217. Harold-Smith, D. *J. Chem. Phys.* 1974, **60**, 1405.
218. Magnusson, J. A.; Bothner-By, A. A. *Magnetic Resonance in Biological Research*, Franconi, A., Ed.; Gordon and Breach: London, 1969.
219. Bryant, R. G. *J. Magn. Reson.* 1972, **6**, 159.
220. Hertz, H. G.; Holz, M.; Keller, G.; Versmold, H.; Yoon, C. *Ber. Bunseges. Phys. Chem.* 1974, **78**, 493.
221. Cambillau, C.; Ourevitch, M. *J. Chem. Soc. Chem. Commun.* 1981, 996.
222. Edwards, P. P.; Guy, S. C.; Holton, D. M.; McFarlane, W. *J. Chem. Soc. Chem. Commun.* 1981, 1185.
223. Civan, M. M.; McDonald, G. C.; Pring, M.; Shporer, M. *Biophys. J.* 1976, **16**, 1385.
224. Lyon, R. C.; Magnuson, N. S.; Magnuson, J. A. *Extreme Environments*, Heinrich, M. R., Ed.; Academic Press: New York, 1976, p. 305.
225. Khazaeli, S.; Dye, J. L.; Popov, A. I. *Spect. Acta.* 1983, **39A**, 19.
226. DeWitte, W. J.; Liu, L.; Mei, E.; Dye, J. L.; Popov, A. I. *J. Soln. Chem.* 1977, **6**, 1977.
227. Gutmann, V. *Coordination Chemistry in Nonaqueous Solvents*; Springer: Berlin, 1968.
228. Engstrom, S.; Jonsson, B.; Jonsson, B. *J. Magn. Reson.* 1982, **50**, 1.
229. Werbelow, L. G.; Ikenberry, D. A.; Pouzard, G. *J. Magn. Reson.* 1983, **51**, 409.
230. Neurohr, K. J.; Drakenberg, T.; Forsen, S.; Lilja, H. *J. Magn. Reson.* 1983, **51**, 460.
231. Hynes, J. T.; Wolynes, P. G. *J. Chem. Phys.* 1981, **75**, 395.
232. Rubini, P. R.; Rodehüser, L.; Delpuech, J.-J. *Inorg. Chem.* 1979, **18**, 2962.
233. Marshall, A. G.; Wang, T.-C. L.; Cottrell, C. E.; Werbelow, L. G. *J. Am. Chem. Soc.* 1982, **104**, 7665.

234. Colquhoun, I. J.; McFarlane, H. C. E.; McFarlane, W. *J. Chem. Soc. Chem. Commun.* 1982, 220.
235. Stengle, T. R.; Reo, N. V.; Williamson, K. L. *J. Phys. Chem.* 1984, **88**, 3225.
236. Contreras, M.; Hertz, H. G. *J. Soln. Chem.* 1978, **7**, 99.
237. Helm, L.; Hertz, H. G. *Z. Phys. Chem. N. F.* 1981, **127**, 23.
238. Drakenberg, T. *Acta Chem. Scand.* 1982, **36A**, 79.
239. Cambillan, C.; Ourevitch, M. *J. Chem. Soc. Chem. Commun.* 1981, 996.
240. Holton, D. M.; Edwards, P. P.; Johnson, D. C.; Page, C. J.; McFarlane, W.; Wood, B. *J. Chem. Soc. Chem. Commun.* 1984, 740.
241. Baltzer, L.; Bergman, N.-A.; Drakenberg, T. *Acta. Chem. Scand.* 1981, **35A**, 759.
242. Sacco, A.; Holz, M.; Hertz, H. G. *J. Magn. Reson.*, in press.
243. Jones, G.; Dole, M. *J. Am. Chem. Soc.* 1979, **51**, 2950.
244. Ellaboudy, A.; Tinkham, M. L.; van Eck, B.; Dye, J. L.; Smith, P. B. *J. Phys. Chem.* 1984, **88**, 3852; Philips, R. C.; Khazaeli, S.; Dye, J. L. *J. Phys. Chem.* 1985, **89**, 606.
245. Benn, R.; Lehmkuhl, K.; Mehler, K.; Rufinska, A. *Angew. Chem. Internatl. Edn.* 1984, **23**, 534.
246. Benn, R.; Rufinska, A. *Angew. Chem. Internatl. Edn.* 1986, **25**, 861.
247. Lehmkuhl, K.; Mehler, K.; Benn, R.; Rufinska, A.; Kriger, C. *Chem. Ber.* 1986, **119**, 1054.

Chapter 8

BORON

JOHN D. KENNEDY

1. NUCLEAR PROPERTIES AND GENERAL CONSIDERATIONS

Although boron-11 is a quadrupolar nucleus, line-broadening problems are generally not severe; chemical shift and a certain amount of coupling constant data are therefore readily available by straightforward continuous-wave or Fourier-transform NMR spectroscopy. The nucleus is of high NMR receptivity and of high chemical interest, and, in common with the proton, phosphorus-31, and fluorine-19, has been well studied since the early days of NMR.(1) The progress of boron NMR spectroscopy has therefore largely developed synergically with the dramatic advancements in inorganic, organic, and polyhedral boron chemistry that have occurred over the last three decades, and the literature both old and new is extensive.(2–13)

The nuclear properties of the naturally occurring isotopes ^{10}B and ^{11}B are given in Table 1. Both primary(14) and secondary(15) isotope effects are significant (up to ca. 0.1 ppm in examples measured so far), but are small enough to be neglected for most purposes in inorganic and organometallic chemistry. The more abundant ^{11}B is generally the preferred nucleide for study from most points of view. Although the greater NMR line broadening expected from the greater quadrupole moment of ^{10}B is compensated for by its higher spin factor,(15) this advantage is generally countered by its lower resonance frequency.(16) Advantages in using ^{10}B spectroscopy include isotopic labeling experiments, relaxation studies in which the differences in quadrupolar coupling may be of use, and the study of internuclear spin–spin coupling in certain cases, for example, in a chemically symmetrical $\{B_2\}$ system. In this last the predominant isotopomer observable in the ^{11}B spectrum will be the $\{^{11}B^{11}B\}$ one, no coupling $J(^{11}B-^{11}B)$ will be observed, and the resonance line would generally obscure any lines arising from the $\{^{11}B-^{10}B\}$ isotopomer. The latter would, however, be the predominant species observed in the spectrum of the magnetically more dilute ^{10}B nucleide, and permit any coupling structure arising from $J(^{11}B-^{10}B)$ to be observed.

There are generally no problems of sensitivity in ^{11}B observation and it has been advocated as a reliable (and of course nondestructive) quantitative analytical method for 10-mg samples.(17) The typical relaxation rates of a few ms^{-1} permit the rapid accumulation of successive transients in FT NMR work, and reports of NMR spectroscopy on solutions made up from single crystals used for x-ray diffraction analysis(18) imply satisfactory detection in these cases of microgram quantities of ^{11}B in

JOHN D. KENNEDY • Department of Inorganic and Structural Chemistry, University of Leeds, Leeds LS2 9JT, England.

Table 1. Summary of the Properties of the Boron Nuclei[a]

Nucleus	^{10}B	^{11}B
Receptivity, R^H	0.0039	0.133
Natural abundance, %	19.6	80.4
Ξ/MHz[b]	10.743 657	32.083 971
Spin, I	3	3/2
Quadrupole moment, $Q/10^{-28}$ m^2	0.074	0.0355
$(2I+3)/I^2(2I-1)$	0.2	1.33
Width factor (^{27}Al = 1)[a,c]	0.154	0.236
Relative intensity[d]	0.025	0.563
Dispersion factor[e]	70	136

[a] Compare other group III nuclei, Table 3 in Chapter 9.
[b] $[BF_3(OEt_2)]$, 15% in $CDCl_3$ at +21°C.
[c] Width factor = $Q^2(2I+3)/I^2(2I-1)$ and is the nuclear contribution to quadrupole relaxation.
[d] Given by (receptivity/width factor).
[e] Given by (Ξ/width factor).

particular chemical sites. A problem that may arise at extreme dilutions is the appearance of dominant broad signals from any boron-containing glasses or ceramics used in spectrometer probe construction, or in the sample tubes, but these effects may often be easily removable by baseline-subtraction procedures, and suitable spin-echo pulse sequences may also be of use (Chapter 2, Section 2.5). These become less satisfactory as the linewidth of the sample signal increases (see Chapter 9, Section 2.1). Ideally non-boron-containing materials such as silica glass should be used throughout, but this is not always practicable.

The currently accepted chemical shift standard is $[BF_3(OEt_2)]$, although $[B(OMe)_3]$ is also used in the older literature to a significant extent, and occasionally others such as $[BH_4]^-$, $[BCl_3]$, etc. Also much of the older literature (pre-1976) uses the "high-field positive" sign convention, which may lead to additional confusion. For external capillary references $\delta(^{11}B)$ for $[B(OMe)_3]$ is generally taken to be at ca. +18.1 ppm to high frequency ("low field") of that for $[BF_3(OEt_2)]$,[5,8] but corrected Ξ values for these as neat liquids are not available. $\Xi(^{11}B)$ and $\Xi(^{10}B)$ for $[BF_3(OEt)_2]$, 15% v/v in $CDCl_3$ at +21°C, are 32,083,971(1) and 10,743,657(1) Hz, respectively, relative to internal 1% TMS, and $\Xi(^{11}B)$ for $[B(OMe)_3]$ under the same conditions is 32,084,566 Hz [it may be noted at this point that $\gamma(^{11}B)/\gamma(^{10}B)$ = 2.986317 and not "precisely" 3.0 as is sometimes[8] stated]. There is significant variation in Ξ with solution conditions, for example $\Xi(^{11}B)$ for $[BF_3(OEt_2)]$, 15% v/v in C_6D_6, is 32,083,988 Hz versus internal TMS, i.e., a change of +0.53 ppm. Presumably neat liquid samples often used as external standards will also have different chemical shifts from these. For example, a $[BF_3(OEt_2)]$ capillary used in the author's laboratory typically gives Ξ values of ca. 32,083,958 when immersed in $CDCl_3$ solution. Factors such as this and changes with temperature probably account for the variations of up to ca. 2 ppm sometimes seen between results reported from different laboratories, but these discrepancies are often not important for the gross comparisons which characterize much of the usage of boron NMR spectroscopy, and comparative work performed within a particular laboratory will generally be internally self-consistent. However, most modern multielement NMR spectrometers permit the ready use of frequency standards for most experiments, and the adoption of a particular frequency, preferably that of 32,083,971 Hz given above for $[BF_3(OEt_2)]$ in $CDCl_3$, as reference standard, is recommended.

^{11}B may exhibit relaxation times of up to several seconds in tetrahedral environments such as in $[BH_4]^-$ in the liquid phase. Long relaxation times also sometimes occur in ostensibly asymmetric environments when the efg at the nucleus is fortuitously close to zero.(19,20) These long relaxation times are exceptional, however, and in general a significant asymmetry of the efg at the nucleus prevails and moderately efficient quadrupolar relaxation occurs (Chapter 5, Sec. 4.5).(21) Relaxation times $T_2(^{11}B)$ vary typically from >ca. 10^{-1} s for reasonably symmetric environments in smaller molecules in mobile solutions to $<10^{-3}$ s for bulkier species such as polyhedral metallaboranes with several triarylphosphine ligands attached to the metal centers. Natural linewidths therefore typically vary from a few hertz to several hundred hertz, although occasionally some sharper lines occur.

For these reasons resolution problems may easily arise in ^{11}B spectra of mixtures of smaller boron-containing species, or in spectra of pure samples of higher polyboron compounds. Line narrowing,(22–24) increases in temperature and/or decreases in solution viscosities,(24,25) "partial relaxation" utilizing the π delay–$\pi/2$ pulse sequence,(18,26–29) selective ^1H–$\{^{11}B\}$ spectroscopy,(29–32) boron–boron correlation spectroscopy,(33) etc., all have some specific or general uses in tackling these problems, either for chemical shift dispersion or coupling constant resolution, or both. Dramatic effects arise from the improvements in dispersion afforded by the increasingly high

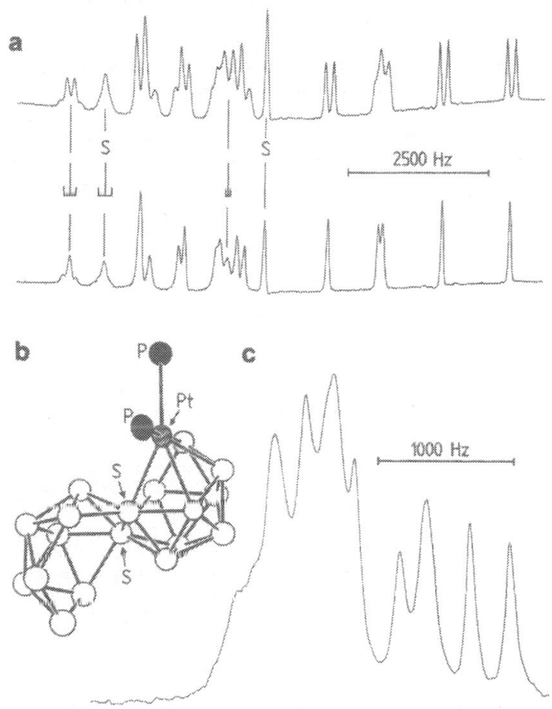

Figure 1. ^{11}B FT NMR spectra of $[(PMe_2Ph)_2Pt\text{-}syn\text{-}B_{18}H_{20}]$ at 32 MHz (bottom trace) and at 128 MHz (top two traces; some resolution enhancement). The bottom two traces were recorded with $\{^1H$(broad-band noise)$\}$ decoupling. The 32-MHz spectrum is relatively uninformative, whereas the 128-MHz spectra permit the identification of (a) the 18 ^{11}B shifts, (b) the B atoms which do not have directly bound *exo*-terminal H atoms (singlet peaks S), and (c) the fine structure arising from couplings $^1J(^{195}Pt-^{11}B)$.

fields now used on commercially available superconducting magnet instruments (e.g., Ref. 34 compared to Ref. 35; see also Figure 1).

Boron chemistry (together with the concomitant use of NMR spectroscopy) falls into two categories: firstly that of compounds with trigonal and tetrahedral boron coordination environments involving straightforward two-electron two-center bonds, and secondly that of the polyhedral boron-containing compounds which generally involve multicenter bonding and often a delocalized σ framework. These categories are conveniently dealt with separately.

2. TRIGONAL AND TETRAHEDRAL COMPOUNDS

The electronic structures of these compounds are generally typified by straightforward σ frameworks of localized two-electron two-center bonds with occasional additional π bonding; geometry and constitution then readily specify familiar concepts of electronic structure which permit empirical correlations with NMR data to be made.

The trigonal and tetrahedral species are nicely dealt with in the monograph "Nuclear Magnetic Resonance Spectroscopy of Boron Compounds,"[(5)] and it is recommended that the reader consult this treatise for an exhaustive listing and a much more detailed and thorough critical discussion than is possible here. The monograph deals with published work to early 1977, but there have been few fundamental advances since then and so the treatment is reasonably contemporary. An exception to this generalization is a very recent review on the analysis of organoboron compounds; this has a particular emphasis on the use of NMR, and now constitutes a primary information resource in this area.[(13)]

2.1. Chemical Shifts

2.1.1. General Considerations

Ranges of ^{11}B chemical shift for selected bonding environments are given in Figures 2 and 3, respectively, for tetrahedral and trigonal boron bonding geometries. As with all charts of this type, the ranges are of course a function of the particular compounds that happen to have been studied, and future chemistry will almost certainly change these ranges and their centroids.[(36)] In each case therefore the position of the simplest methyl derivative (X) is also shown. Shifts and substituent effects are very well documented for a very large variety of bonding environments,[(5)] so that $\delta(^{11}\mathrm{B})$ values have a great utility as very specific structural diagnostics.

The total ^{11}B shielding range for nonparamagnetic species covers some 200 ppm. It can be seen that the shielding ranges of the four-coordinate sp^3 species (δ ca. +20 to −128 ppm; Figure 2) are generally several tens of ppm to higher field of the nearest equivalent three-coordinate species (δ +92 to −8 ppm, Figure 3) which generally have sp^2 trigonal boron bonding environments. This difference has considerable use as a diagnostic in nonpolyhedral boron chemistry.[(5)] These gross differences may be rationalized by a simplistic *p*-orbital occupancy and electronic imbalance approach (Chapter 3, Section 2.3) and how this affects the "excited state" contribution σ^p to the nuclear shielding expressions (Chapter 3, Section 1–3): the term P_i of equation (10) in Chapter 3 will be a maximum for sp^2 occupancy so the trigonal compounds will have

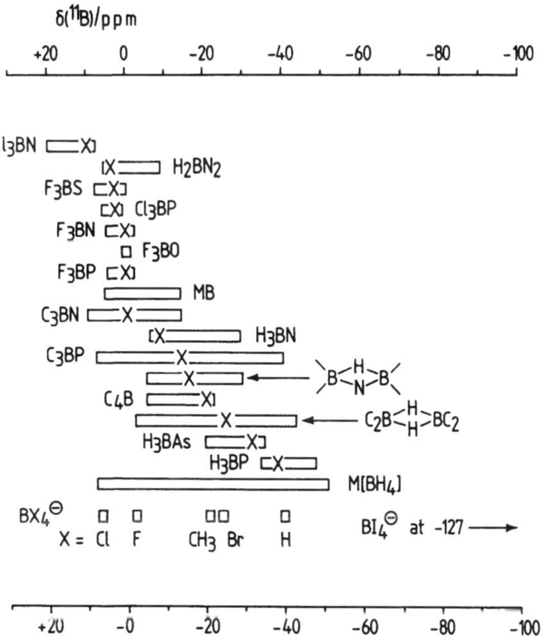

Figure 2. Chemical shift values and ranges for ^{11}B in four-coordinate boron compounds [BWXYZ]. Where appropriate, "X" indicates the shift of the simplest methyl derivative. See Ref. 5 for an extensive survey, and Ref. 8 for more extensive charts.

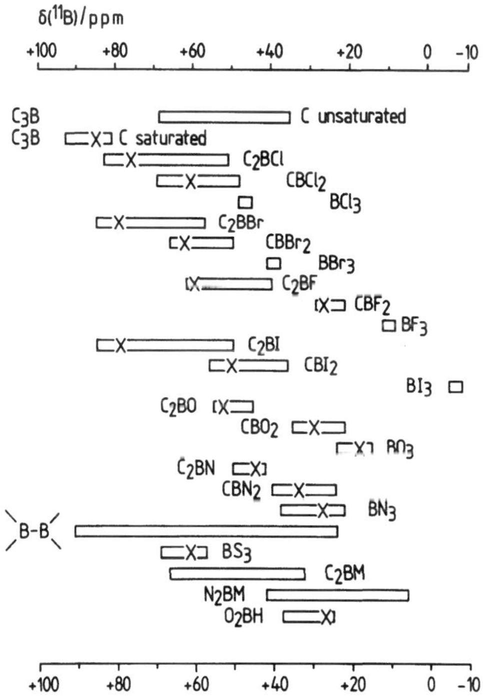

Figure 3. Chemical shift values and ranges for ^{11}B in three-coordinate compounds [BXYZ]. Where appropriate, "X" indicates the shift of the simplest methyl derivative. See Ref. 5 for an extensive survey, and Ref. 8 for more extensive charts. Note that in donor solvents many species will be four-coordinate.

lowest nuclear shielding. Conversely P_i will be a minimum for sp^3 occupancy and so the four-coordinate species will exhibit highest shieldings.

Theoretical treatments (which so far have not been extended much beyond CNDO-type levels)[37–41] also show that calculated boron 2*p* density matrix elements exhibit the best correlations with experimentally determined shieldings.[41] This applies to both trigonal and tetrahedral species. These treatments also indicate that in general there is little reliable correlation with calculated *total* electron density,[42] which tends to confirm the expected dominance of variations in the excited state term σ^p (Chapter 3, Section 2 and 3) in these compounds.

2.1.2. Tetrahedral Species

For tetrahedral sp^3-type species (Figure 2) substituent effects within the gross shielding range are qualitatively similar to those which typify most of the covalent *p*-block chemistry (see Chapter 3, Section 3, and Chapters 10–15). For example, replacement of the hydrogens by saturated organic groups can result in an ^{11}B deshielding of up to ca. 18 ppm per group, and halogen substituents behave as if they have a deshielding effect, roughly proportional to their electronegativities, together with a shielding effect[40,43] which increases with their atomic weight. Similar considerations apply to other electronegative substituents (e.g., OR, SR, SeR, NR_2, etc.) which have formally nonbonding electron pairs in their valence shell, although the first-row substituents induce somewhat higher shielding than expected on this simple basis. For species $[BX_4]^-$ the ^{11}B shielding is in the order $Cl < F \sim OR < Ar < R < Br < H < I$. Mixed species $[BX_{4-n}Y_n]$ generally have chemical shifts which are approximately linearly intermediate between $[BX_4]$ and $[BY_4]$ unless X or Y have lone-pair valence electrons, in which case a plot of $\sigma(^{11}B)$ versus *n* often exhibits the "sagging dependence" also well established, for example, in group IV chemistry [Chapters 3 (Section 3.6.4), 10, and 11]. Pairwise additivity parameters are often useful for representing this behavior.[44]

As expected there are strong parallels in the central nucleus shielding patterns of the tetrahedral boron compounds $[BWXYZ]^{n-}$ and the very extensively investigated isoelectronic tetrahedral carbon analogs $[CWXYZ]^{(n-1)-}$. A plot of $\delta(^{13}C)$ versus $\delta(^{11}B)$ for these isoelectronic four-coordinate species has a gradient $\delta(^{11}B){:}\delta(^{13}C)$ of ca. 0.4:1 in the general case,[44] although more limited series of compounds have gradients that differ from this.[45–47] The intercept in these plots is dependent on the two shift reference standards chosen, and has limited fundamental significance in the absence of accurate absolute shielding standards. The parallels with tetrahedral nitrogen chemistry have also been explored to some extent in amine–borane adducts,[45,49] but $\delta(^{11}B)$ and $\delta(^{14}N)$ values for identical bonding environments—i.e., $[BWXYZ]^{n-}$ versus $[NWXYZ]^{(n-2)-}$—have not been compared.

2.1.3. Trigonal Species

The trigonal sp^2-type boron species (Figure 3) are also subject to similar substituent effects, although with unsaturated and lone-pair substituents positive shielding contributions are often more pronounced than for the tetrahedral compounds. To account for this, π interaction with the otherwise vacant *p*-type orbital on trigonal boron is often invoked,[5,48] and, conversely, higher-field chemical shifts have sometimes been invoked to "prove" increased π interaction. Presumably the shielding effects derive at least in part from ground-state $p\pi$ donation, from the lone pair of the π cloud of the substituent, although it is not always apparent how these may be related at a less empirical level to the shielding expressions discussed in Chapter 3. At the simplest levels additional p-orbital occupancy would increase the electron density on the boron center and also reduce the p-orbital imbalance, of which both will be expected to increase the nuclear shielding (Chapter 3).

Some calculations have been carried out which suggest linear correlations of experimental $\delta(^{11}B)$ values with calculated π-electron density at boron.[41] Factors that would be expected to decrease this π bonding, such as angle deformation resulting from ring-size or steric crowding, generally decrease the shielding.[5] This latter contrasts to the effect of distortions in the σ framework, which in the trigonal compounds tends to increase the shielding somewhat (e.g., boraadamantane at ca. 3.5 ppm to high field of the trialkylboranes): these factors often interfere with additivity approaches to describe the shielding variation in the trigonal species.

Again the overall shielding trends are paralleled in carbon chemistry, and a plot of $\delta(^{13}C)$ for classical carbenium ions $[CXYZ]^+$ versus $\delta(^{11}B)$ for the corresponding trigonal boron compounds [BXYZ] has a gradient $\delta(^{11}B)$:$\delta(^{13}C)$ of ca. 0.38:1 for the species examined.[45,46,49,50] The deviations from this correlation are somewhat greater than for the tetrahedral species, particularly for lone pair and for unsaturated substituents. Presumably the greater effective nuclear charge of the carbon center has a proportionately larger effect on the $p\pi$ donation to the central atom.

2.2. Coupling Constants

The observation of ^{11}B coupling constants $^nJ(^{11}B-X)$ in spectra of both ^{11}B and the other nucleus X is often hindered or excluded by broadening induced by the relatively rapid ^{11}B relaxation (Sections 1 above and 3.4 below). This applies particularly to smaller couplings which include most couplings over more than one bond. Often coupling constant resolution may be effected by the use of line-narrowing techniques,[22-24] etc., or improved by the use of less viscous solvents and/or higher temperatures[24,25] provided that these do not otherwise unacceptably change the chemistry of the system being studied.

Representative couplings for trigonal and tetrahedral compounds are given in Tables 2 and 3, respectively. Most data that are available are for one-bond coupling constants. For these a dominance of the Fermi contact mechanism (Chapter 2, Section 2.2.3, and Chapter 4, Section 4) would require a dependence on hybridization at boron,[42,135] and where reasonably direct comparisons are possible {e.g., $^1J(^{13}C-^{11}B)$ in $[BMe_4]^-$ of ca. 25 Hz and in $[BMe_3]$ of ca. 45 Hz, both presumably positive}[51] there is a qualitative agreement with the Fermi contact mean excitation energy approximation (Chapter 4, Section 4.1) for this. It can also be seen from Tables 2 and 3 that there is also a very marked dependence upon the electronegativities of the other substituents on the atoms of the two coupled nuclei.[42,135] For $^1J(^{11}B-^1H)$, $^1J(^{13}C-^{11}B)$, etc., this dependence can be generally rationalized at a simplistic level in terms of the Fermi contact mean excitation energy approximation (Chapter 4, Section 4) and on the basis of "isovalent rehybridization"[198] and/or changes in effective nuclear charges[199] (see also Chapter 4). The few quantum mechanical calculations that have been carried out (e.g., Refs. 42 and 135) are generally taken to justify this qualitative treatment, and, using these assumptions, the magnitudes of coupling constants are often usefully diagnostic of particular B–X environments.[5]

Table 2. Selected Coupling Constants $^1J(^{11}B\text{–}X)$ for Trigonally Coordinated Boron Compounds[a]

Other nucleus X	Compound	$^1J(^{11}B\text{–}X)$ (Hz)
^1H	F_2BH	−211
^1H	Cl_2BH	196
^1H	$(MeO)_2BH$	141
^1H	$(Me_2N)_2BH$	126
^1H	$(HMe_2CMe_2C)_2BH$	114
^1H	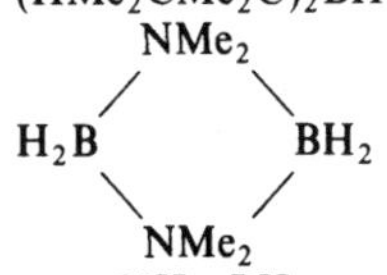	114
^1H	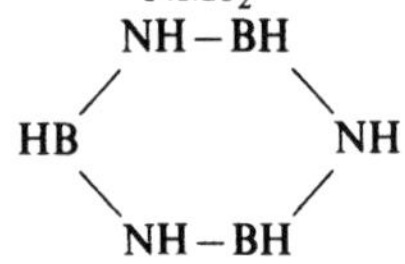	137
^1H	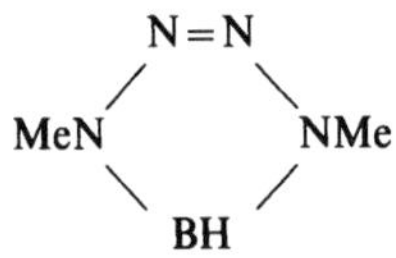	164
^{13}C	BMe_3	47
^{13}C	$MeB(OMe)_3$	76
^{13}C	$(HC{\equiv}C)_2BNEt_2$	132
^{15}N	$B(NHMe)_3$	−45
^{19}F	BrIBF	189
^{19}F	Me_2BF	119
^{19}F	MeBF(OMe)	85
^{19}F	$MeBF_2$	76
^{19}F	Cl_2BF	73
^{19}F	$ClBF_2$	33
^{19}F	BF_3	14.5
^{19}F	$(Et_2N)\,BF_2$	15
^{19}F	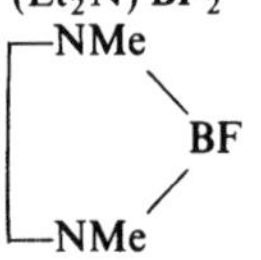	30.5
^{31}P	$Me_2P{-}B(NMe_2)_2$	50
^{29}Si	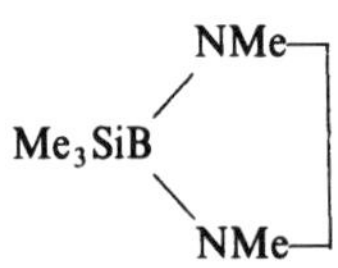	−97
^{119}Sn	$[Me_3SnBCl(NMe_2)]$	−1007
^{119}Sn	$[(Me_3Sn)_2BNMe_2]$	−657
^{119}Sn	$Me_3SnBH_3^-$	−554
^{207}Pb	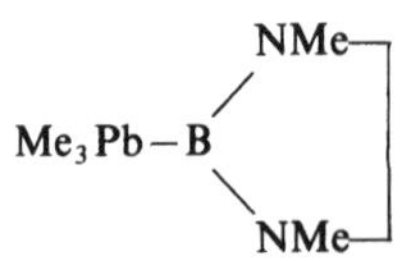	+1330

[a] For comprehensive listings see Ref. 5.

Table 3. Selected Coupling Constants $^1J(^{11}B–X)$ for Tetrahedrally Coordinated Boron Compounds[a]

Other nucleus X	Compound	$^1J(^{11}B–X)$ (Hz)
^1H	$[HCl_2BWH_2Cp_2]$	164
^1H	$HCl_2B \cdot PH_3$	154
^1H	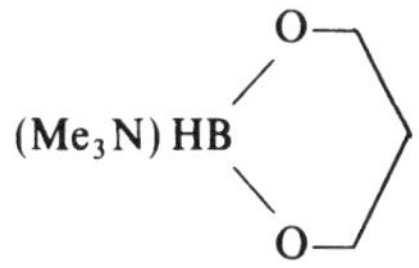	152
^1H	$HF_2B \cdot NH_3$	140
^1H	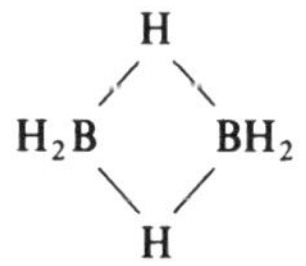	135 (terminal), 45 (bridge)
^1H	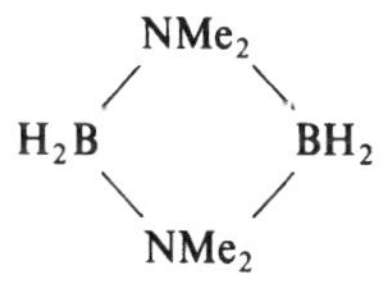	114
^1H	$H_3B \cdot CO$	105
^1H	$H_3B \cdot PH_3$	104
^1H	$(H_2B \cdot NH_2)_3$	101
^1H	$H_3B \cdot AsMe_3$	100
^1H	$H_3B \cdot P(OMe)_3$	97
^1H	$H_3B \cdot NH_3$	94
^1H	$[H_3B(CN)]$	92
^1H	$[Al(BH_4)_3]$	87
^1H	$H_2IB \cdot PPh_3$	83
^1H	BH_4^-	80[b]
^{13}C	$[BPh_4]^-$	49(^3J 2.2 Hz, ^2J ca. zero)
^{13}C	$H_3B \cdot CO$	30
^{13}C	$[BMe_4]^-$	22
^{15}N	$F_3B \cdot NMe_3$	ca. 15
^{15}N	$Br_3B \cdot NMe_3$	13
^{19}F	$HF_2B \cdot NH_3$	70
^{19}F	$Et_2FB \cdot NEt_3$	67
^{19}F	$F_3B \cdot CH_2 \cdot PMe_3$	53
^{19}F	$F_3B \cdot PMe_3$	50
^{19}F	$(F_2B \cdot NEt_2)_2$	43
^{19}F	$F_3B \cdot SMe_2$	24
^{19}F	$F_3B \cdot NEt_3$	16
^{19}F	$F_3B \cdot O \cdot NEt_3$	8
^{19}F	BF_4^-	1–5[b]

Table continued

Table 3. (*Continued*)

Other nucleus X	Compound	$^1J(^{11}B\text{–}X)$ (Hz)
^{31}P	$F_3B \cdot P\overset{/}{\underset{\backslash}{-}}$	165–217
^{31}P	$Cl_3B \cdot P\overset{/}{\underset{\backslash}{-}}$	140–166
^{31}P	$Br_3B \cdot P\overset{/}{\underset{\backslash}{-}}$	125–165
^{31}P	$I_3B \cdot P\overset{/}{\underset{\backslash}{-}}$	108–134
^{31}P	$H_3B \cdot P(OMe)_3$	97
^{31}P	$H_3B \cdot PEt_3$	64
^{31}P	$[(H_3B)_2PH_2]^-$	61
^{31}P	$H_2IB \cdot PH_3$	60
^{31}P	$H_3B \cdot PF_3$	39
^{31}P	$[B(PEt_2)_4]^-$	23–32
^{31}P	$H_3B \cdot PH_3$	27
^{31}P	$H_3B \cdot PF_2I$	<15
^{103}Rh	$[Cl_3B \cdot Rh(CO)Cl(PPh_3)_2]$	150

[a] For exhaustive listings see Ref. 5.
[b] Depends on solution conditions.

These generalizations, however, do not appear to hold for couplings $^1J(^{31}P-{}^{11}B)$ which exhibit a range of variation of somewhat more than an order of magnitude (13–217 Hz),(5) and for these there at present appears to be no satisfactory theoretical or empirical rationalization.(5) Although, as with polyhedral species,(52,53) an increase of substituent electronegativity often increases the magnitude of the coupling $^1J(^{31}P-{}^{11}B)$ (e.g., $[BH_3(PH_3)]$ 27 Hz, $[BBr_3(PH_3)]$ 127 Hz, and $[BH_3\{P(OMe)_3\}]$ 97 Hz),(54) this is far from the general case: the bulk of the available data are all for tetrahedrally coordinated phosphorus, and so a breakdown of the mean excitation energy approximation(55–57) on the basis of "closed" versus "open" available valence *s*-shell configurations(200) (Chapter 4, Section 4) cannot be invoked. The sign of $^1J(^{31}P-{}^{11}B)$ is generally positive, not only for quaternary phosphorus species, which constitute the majority of investigated compounds, but also for the isolated ternary phosphorus species (e.g., $[Me_2P-B(NMeCH_2)_2]$ at +44 Hz)(57) that have been measured. This behavior appears to contrast with corresponding couplings $^1J(^{13}C-{}^{31}P)$ in which sign inversion occurs upon phosphorus quaternization (Chapter 4, Section 4.4), and the indications are that the "sign inversion" in couplings $^1J(^{11}B-X)$ as X is varied (Chapter 4, Section 2.1.3) may generally occur somewhat further to the right in the Periodic Table than for $^1J(^{13}C-X)$. The (reduced) coupling involving ^{15}N is also positive in the unquaternized nitrogen compound $[B(NHMe)_3]$ $[^1J(^{15}N-{}^{11}B)\ -45\ \text{Hz}]$,(58) but couplings to ^{15}N generally have a much smaller magnitude than this(59) and it is thought that sign inversion may easily occur.(5)

Couplings to fluorine $^1J(^{19}F-^{11}B)$ are well studied,[60–62] and those of larger magnitude are generally of negative sign. Of particular interest is the coupling in the anion $[BF_4]^-$, which is of very small magnitude, and varies dramatically with solution conditions. For this and other small couplings it may well be that a sign change to positive can occur. As with other couplings to ^{19}F (Chapter 16, Section 3), it is thought that mechanisms other than the Fermi contact one may be very significant for $^1J(^{19}F-^{11}B)$.[5]

Couplings to metal nuclei are of interest to the inorganic chemist, but few data are yet available. Some comparative work has been carried out for couplings to silicon, tin, and lead,[55–57] but there are only isolated data for other metals, such as beryllium,[159] aluminum,[63] scandium,[188] yttrium,[64] zirconium,[160] and rhodium.[65] Some of these latter involve multicenter bonding, and should perhaps therefore be regarded more as polyhedral fragment species (Section 3 below).

2.3. Relaxation Studies

Although well over 3000 trigonal and tetrahedral boron compounds have been examined by ^{11}B NMR spectroscopy, liquid-state relaxation behavior appears to be essentially unexamined, even though the potential usefulness of work in this area was demonstrated some two decades ago (for example, Ref. 66). Even linewidths, which are readily obtained by straightforward spectroscopy and can vary substantially with compound and solvent, have only been reported for a handful of species. It would be useful if these could be reported routinely. There would seem to be much scope for systematic study within this general area, which can in principle yield much information about electronic environment, geometry, molecular motion, states of aggregation, etc. (see Chapter 4).

3. POLYHEDRAL BORON-CONTAINING SPECIES

NMR is a crucial tool in polyhedral boron chemistry[67]; the literature is extensive and detailed; excellent comprehensive compilations of the data published to mid-1981 are in Refs. 3, 6, and 7, but unfortunately no recent critical and comparative comprehensive account is available. Citations given should themselves be considered as sources of further references.

3.1. General Considerations

It is important to recognize that the structural types encountered in polyhedral boron chemistry are not necessarily unique to boron, but represent points on continua of geometric and electronic behavior between condensed metal systems and the more open mononuclear systems of classical inorganic chemistry.[68] For example, many "π-bonded" organometallic systems (see Chapter 10) have strong analogies to boron clusters,[69] and there are corresponding gross parallels in NMR behavior.[70] Thus the *nido* 16-cluster-electron pentagonal pyramidal motif (structure **1**) exemplified by

1

nido-$[B_6H_{10}]$ is mirrored by that of $[C_6Me_6]^{2+}$ and also is common to a variety of mono- and bimetallic metallaborane and organometallic systems (structures **2–4**), of which one is the familiar half-metallocene cluster **4**. All these are generally characterized by lower ^{11}B or ^{13}C nuclear shielding for the basal cluster atoms, and a higher shielding for the apical atoms (see also Section 3.2.3 and Chapter 10).(70)

2

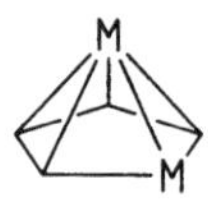

3

4

It should be pointed out that the polyhedral boranes are generally "electron deficient" only in the sense that they have fewer bonding electrons than *atomic* valence-shell orbitals of the component atoms. The *molecular* bonding orbitals are generally all occupied, and indeed a number of polyhedral borane species, again exemplified by B_6H_{10}, are *basic* towards Lewis acids such as metal centers(68) and the proton.(67) Only one neutral binary borane, the unstable $[B_8H_{12}]$, is in fact believed to be electron deficient in the sense that it will form complexes with Lewis bases in the same manner as vacant-orbital species such as $[BF_3]$.(71) Solvent effects, in the absence of specific chemical reaction, on ^{11}B and ^1H NMR parameters of polyhedral species are therefore generally expected to be of the same order of magnitude as those found for ^{13}C and ^1H in straightforward organic molecules (see, for example, Table 4), although electric field effects in strongly dipolar solvents such as CH_2Cl_2 may occasionally produce larger effects if polarizable features, such as those associated with metal centers, are present. It may therefore be misleading to invoke them, for example, to account for discrepancies between calculated and observed values for nuclear screening,(72) or for other ostensibly puzzling effects.

Table 4. Solvent Effects on Selected *nido*-$B_{10}H_{14}$ Shieldings

	$\delta(^{11}B)^a$				
Solvent	1, 3	6, 9	5, 7, 8, 10	2, 4	$\delta(^1H)$(bridge)
$CDCl_3$	+12.1	+10.2	+0.55	−36.1	−2.14
CD_2Cl_2	+11.8	+10.3	+0.4	−36.1	−2.08
CS_2 + 10% CD_2Cl_2	+12 3	+10.05	+0.7	−35.8	−2.13
C_6H_{12} + 10% CD_2Cl_2	+12.5	+9.9	+0.6	−35.8	−2.24
C_6D_6	+12.3	+10.2	+0.1	−35.8	−2.82
$CD_3C_6D_5$	+12.3	+10.2	0.0	−36.0	−2.81
CH_3COOH + 10% CD_2Cl_2	+11.4	+10.7	−0.4	−36.0	−1.82
$BF_3(OEt)_2$ + 10% CD_2Cl_2	+11.4	+10.7	—	−36.35	−1.87
$B(OMe)_3$ + 10% CD_2Cl_2	+12.4	+10.2	+0.4	−36.2	−2.12
THF + 10% CD_2Cl_2	+11.4	+10.4	−0.4	−35.9	−1.75
CD_3CN	+10.9	+10.9	−0.7	−35.7	−1.90
$(CD_3)_2CO$	+11.1	+11.1	−0.7	−35.9	−1.53
SMe_2 + 10% CD_2Cl_2	+11.3	+10.5	−0.3	−35.4	−1.88
$MeNO_2$ + 10% CD_2Cl_2	+11.2	+10.7	−0.3	−36 2	−1.81
CD_3OD	+12.0	+10.4	−1.1	−36.2	—

[a] Chemical shifts $\delta(^{11}B)$ (ppm) ±0 2 relative to Ξ 32 083 971 = zero, measured using 1% TMS [Ξ(^1H) 100] as the internal secondary standard (J D Kennedy and B E Mann, unpublished observations, 1981) (for more recent work in this area see Ref 148)

3.2. Boron Chemical Shifts

3.2.1. General Considerations

In the chain and ring compounds and mononuclear complexes which typify the bulk of classical organic and inorganic chemistry, a particular combination of geometry and substituents is usually sufficient effectively to define a specific gross electronic structure in terms of familiar concepts such as π-bonding, localized two-electron two-center σ-bonding, etc. This enables useful correlations to be made between readily recognizable structural features and observed NMR parameters, even though the precise relationship between the electronic structure and the observed parameter may sometimes be far from clear. In polyhedral cluster species, however, ostensibly mutually similar polyhedral fragments can often accommodate a variety of electronic structures (which themselves are little understood), and there is at present therefore little reliably unambiguous direct general correlation between NMR features and gross molecular structure.

This ability to accommodate different electronic structures is exemplified by the open ten-vertex C_{2v} cluster structure **5**, which is the geometry found both for *nido*

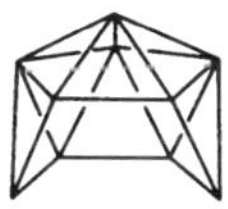

5

$[B_{10}H_{14}]$ and for the *arachno* anion $[B_{10}H_{14}]^{2-}$. In these the ordering of the four $\delta(^{11}B)$ assignments for the *nido* species is the *inverse* of that for the *arachno* species (see Figures 9 and 10 below). When the resonances are assignable this is of course diagnostic of *nido* versus *arachno* ten-vertex behavior.(73,74) However, because the $\delta(^{11}B)$ shielding ranges are very similar for both species, it follows that boron shielding patterns in the absence of any other evidence would not distinguish between asymmetrically substituted *nido* and *arachno* ten-vertex clusters, or indeed between these and a variety of other systems that would contain ten boron atoms. Structures **6** and **7** show valence-bond descriptions of the $\{B_{10}\}$ bonding in these two species; although these are obviously different it is not clear how they may be simply related to nuclear shielding. For example, at the simplest level it is not apparent how the presence of a three-center rather than a two-center bond to a particular atom (obviously important in this area of chemistry) affects its chemical shift.(8)

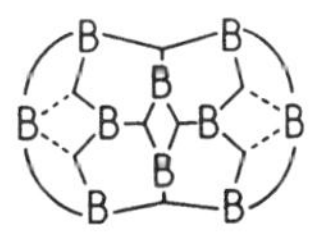

6

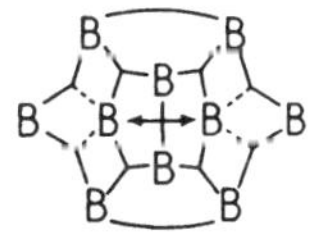

7

Since initial considerations by Lipscomb and co-workers,(3,75–77) who concluded that there was little correlation between $\delta(^{11}B)$ and total electron density on a polyhedral boron atom, and therefore that σ^p contributions (Chapter 3, Section 1.2) were important, relatively few calculations related to nuclear shieldings in the polyhedral boranes have been attempted (see, for example, Refs. 78–80). Recent attempts include one at the CNDO/S level[42] and one using a semiempirical approach based on gauge-invariant atomic orbitals and parametrization of the atomic β values.[72] The former includes results for $[B_2H_6]$, $[B_4H_{10}]$, $[B_5H_9]$, $[B_5H_{11}]$, and $[B_{10}H_{14}]$ as well as some borane anions and some dicarbaboranes and shows some correlation between $\sigma(^{11}B)$ and total p electron density on the atom (compare Section 2.1.1 above).[42] The latter study includes the same boranes, and the results could be taken to indicate, perhaps not surprisingly, that multicenter contributions to the $\sigma(^{11}B)$ shielding expression are important.[72]

In the discussion of shielding at an empirical level, the observation made by Williams nearly two decades ago, that "the exceptions outnumber the rules," still often applies, as does his feeling of relief upon the confirmation by x-ray studies of a structure previously deduced by NMR spectroscopy.[4] Because of these exceptions, discussions at present must necessarily be couched in relatively nebulous terms of trends and tendencies, rather than in terms of definitive ranges associated with particular structural features. For example, some of the shielding and coupling behavior of smaller clusters, and of some of the more open *arachno*, *hypho*, and *klado* aspects of larger clusters, in which polyhedral character is becoming indistinct, will constitute exceptions to generalizations made on the more condensed and/or larger boranes, and will often have intermediate mononuclear/polyhedral character. Workers in a particular area typically develop an essentially intuitive working appreciation of shielding patterns associated with particular classes of compounds that they may be interested in, but this is difficult to quantify. Occasionally particular cluster units, such as the quasiaromatic pentagonal pyramid,[81,82] can be identified as an obvious electronic as well as geometric fragment of a larger cluster, which may then be used in attempts to build up pictures of the total cluster bonding pattern.[83]

3.2.2. Assignments and Boron–Boron Correlation Spectroscopy

A major problem in the development of the use and interpretation of boron chemical shifts in polyhedral species is that of assignment. Although the assignments for many of the principal simple basic boranes and carboranes have been established,[6,7,11,67] much smaller proportions of other heteroborane structural classes have been assigned. In simple cases this has been done on the basis of symmetry and relative intensity, but for more ambiguous cases painstaking isotopic labeling and substituent chemistry has had to be carried out.

Methods of assignment which avoid the tedious chemistry rely on the presence of intracluster nearest-neighbor couplings $^1J(^{11}B-^{11}B)$ of up to a few tens of hertz (see Section 3.3.3 below). An identification of mutually coupled pairs of boron nuclei will assign nearest neighbors. If this is done for all the boron nuclei, then the entire spectrum can in principle be assigned because it is generally possible to pinpoint, from other considerations, the assignment of particular resonances. Of prime importance also is that the connectivity information also enables the establishment of gross structural geometry in the absence of diffraction information.

Early experiments in this area utilized $^{11}B-\{^{11}B\}$ homonuclear CW INDOR spectroscopy (Chapter 2) at 80.2 MHz,[84] applied to some derivatives of $[B_{10}H_{14}]$ and *ortho*-$[C_2B_{10}H_{12}]$,[85–87] and to some somewhat more demanding 18-vertex species.[88] The experiments were very successful for the compounds examined, but unfortunately the work has not since been extended further. However, apart from the assignments, a number of important features, such as an effective absence of coupling

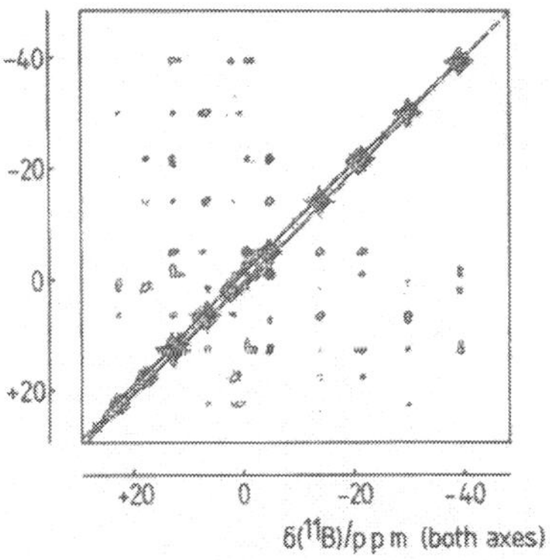

Figure 4. 128 MHz ^{11}B–^{11}B COSY contour plot (protons decoupled) for $[(PMe_2Ph)_2Pt\text{-}syn\text{-}B_{18}H_{20}]$. Taken together with the singlet and satellite information of Figure 1, sufficient nearest-neighbor interboron correlations are observed reasonably to assign all 18 resonance positions to specific structural sites (work by O. W. Howarth and J. D. Kennedy).

between certain pairs of hydrogen-bridged boron atoms, and between boron atoms bound by the long *nido*-decaboranyl B(5)–B(10)-type linkages, were established.

Current commercially available FT instrumentation permits these correlations to be examined by homonuclear two-dimensional correlation spectroscopy using routine applications of straightforward Jeener-type multiple pulse sequences ("COSY" spectroscopy; see Chapter 5). It is likely that the most important developments in the application of NMR to polyhedral boron chemistry will lie in this area, particularly when allied with ^1H–^{11}B and ^1H–^1H correlation work (see Section 3.5 below and Refs. 9 and 104). Although a number of initial studies on boron compounds have been reported,[89–91] and preliminary considerations reviewed,[92] most applications published at time of writing have been illustrative rather than analytical. Genuine research-tool usage is, however, now starting to emerge,[83,93–97,104,189,190,196] for example for the previously unassigned *nido*-$[B_9H_{12}]^-$ anion,[93] for the neutral nine-vertex *nido* species $[(PMe_2Ph)_3H_2ReB_8H_{11}]$,[189] for 12-vertex $B_{12}H_{16}$,[190] for the 17-vertex metallaborane cluster compound $[(PMe_2Ph)PtB_{16}H_{18}(PMe_2Ph)]$,[83] and for the examination of spin–spin coupling constants in the *nido*-$[Me_2TlB_{10}H_{12}]^-$ anion.[95] Figure 4 illustrates a 128-MHz ^{11}B–^{11}B COSY two-dimensional contour plot for the 19-vertex platinaborane $[(PMe_2Ph)_2Pt\text{-}\eta^4\text{-}syn\text{-}B_{18}H_{20}]$.[33] The 18 separate ^{11}B resonance positions may be distinguished in the straightforward ^{11}B-{^1H(broad band noise)} spectrum (Figure 1 above), and sufficient off-diagonal correlation peaks are apparent to enable complete assignment.

It should be emphasized that the observation of a correlation using these techniques depends critically upon the existence of a finite internuclear spin–spin coupling and upon the magnitude of this relative to the relaxation rates of the boron nuclei involved. If the ^{11}B relaxation rate excessively exceeds the coupling constant, or if the coupling constant is effectively zero (even though a strong bond may exist), then no correlation will be seen. This last commonly happens, for example, in two-electron three-center B–H–B bridging bonds and also in long interboron distances such as those in the B(5)–B(10) linkages in *nido*-decaboranyl clusters, although exceptions to both these generalizations are already recognized.[33,96,97,196] Conversely, if longer-range interboron spin–spin couplings happen to be large, then correlations can in

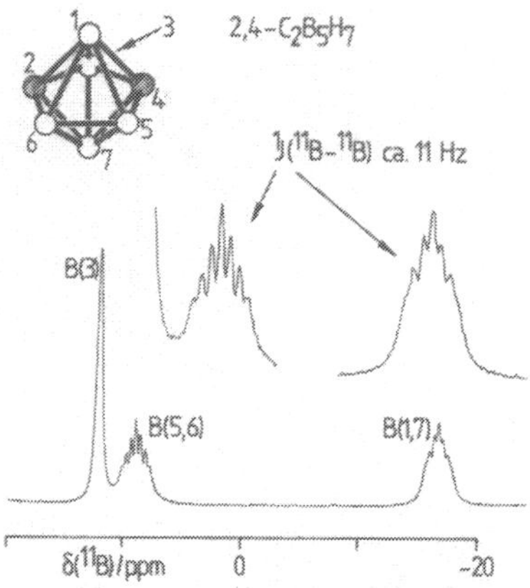

Figure 5. 28 MHz ^{11}B spectrum (protons decoupled) for *closo*-2,4-$C_2B_5H_7$. Coupling $^1J(^{11}B–^{11}B)$ is readily apparent between $^{11}B(1,7)$ and $^{11}B(5,6)$, but not between $^{11}B(3)$ and the other boron nuclei, although the somewhat broader resonance for $^{11}B(1,7)$ implies a small coupling $^1J[^{11}B(3)–^{11}B(1,7)]$ (Data from Ref. 89). Note that boron atoms flanked by carbon atoms in polyhedral species often exhibit small intracluster couplings to other atoms, and that the coupling $^1J(^{13}C–^{11}B)$ between the flanked and flanking atoms is conversely sometimes quite marked (see, for example, Refs. 97a and 97b).

principle be observable between nonadjacent boron centers, particularly so if the relaxation times are long. Some of these considerations are exemplified in Figures 5–8.

3.2.3. Nuclear Shielding Patterns

3.2.3a. General Shielding Patterns. The chemical shifts of representative binary boron hydride species are given in Figures 9–11; schematic structural representations showing numbering schemes are given in Figures 12–14. In these species $\delta(^{11}B)$ varies between $+20$ and -55 ppm. If heteroboranes, particularly metal-containing species, are included, then this range is extended, principally downfield to $+150$ ppm or so. From this it follows that extreme low-field resonances are diagnostic of electronic and/or structural features not found in the binary boron hydride species themselves. The overall range of about $+150$ to -60 ppm is generally exceeded only by paramagnetic species, and the reader is referred to Refs. 98 and 99 for a more

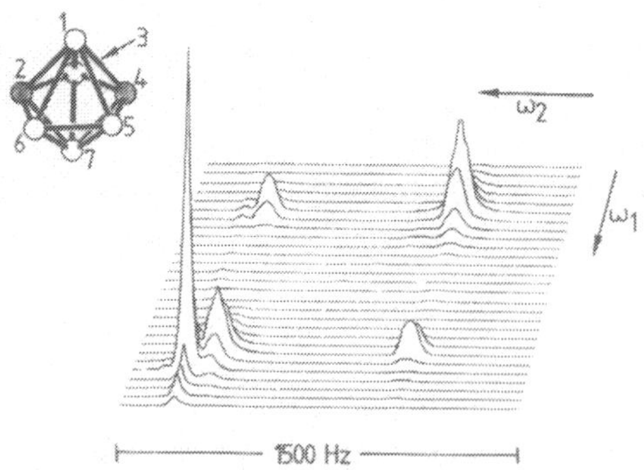

Figure 6. 28-MHz $^{11}B–^{11}B$ COSY stacked plot (protons decoupled) for *closo*-2,4-$C_2B_5H_7$ (compare Figure 5). Strong correlations are observed between the B(1,7) and B(5,6) sites, but little if any between B(3) and the other sites (Data from Ref. 89).

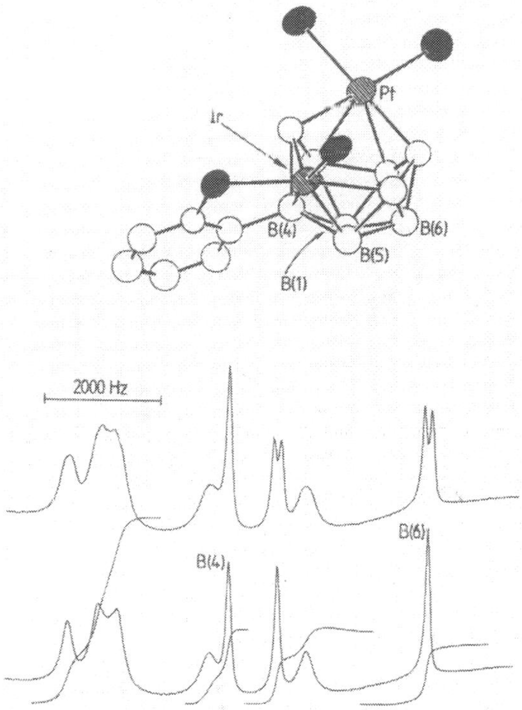

Figure 7. 128-MHz ^{11}B FT NMR spectra of *nido*-7,9-[$(PMe_3)_2Pt(PPh_3)(Ph_2PC_6H_4)HIrB_9H_{10}$-4]: the lower trace is with proton decoupling, the upper one without. There is considerable variation in the broadness of the resonances (see Figure 8).

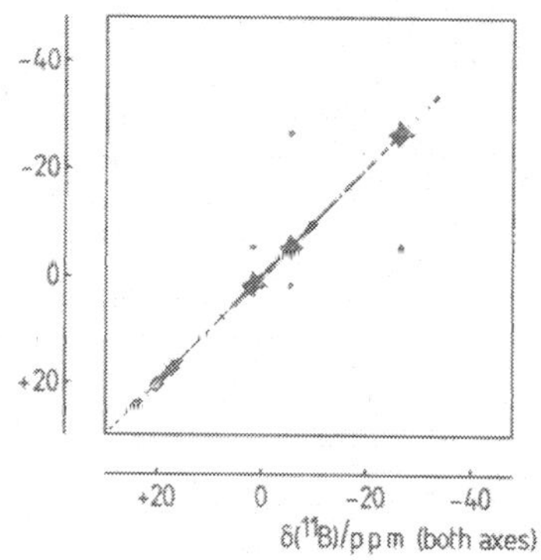

Figure 8. 128-MHz ^{11}B–^{11}B COSY contour plot (protons decoupled) for the *nido* eleven-vertex dimetallaborane of Figure 7. Correlations are observed only among the three sharp resonances of Figure 7, which are assigned to B(4), B(6), and one of B(1) or B(5) (J. Bould, O. W. Howarth, *et al.*, unpublished observations).

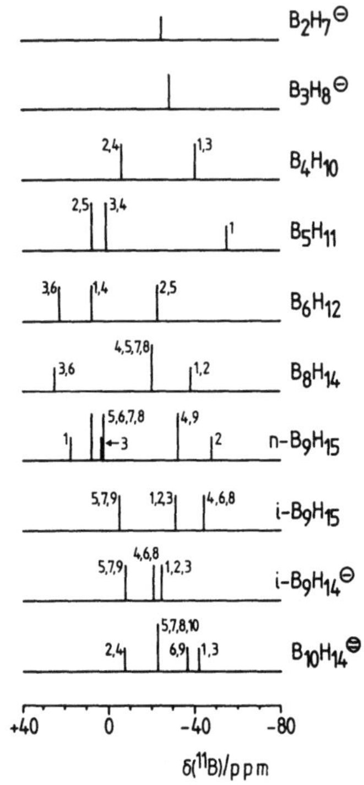

Figure 9. Representation of the chemical shifts and relative intensities in the ^{11}B NMR spectra of selected basic binary *arachno* borane fragments. For numbering and structures see Figure 12. $B_3H_8^-$ and $B_9H_{14}^-$ are fluxional in solution and therefore have simpler spectra than implied by their static molecular symmetries.

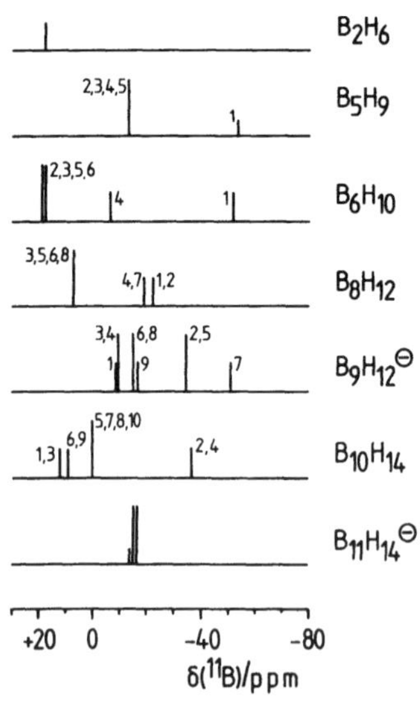

Figure 10. Representation of the chemical shifts and relative intensities in the ^{11}B NMR spectra of selected basic binary *nido* borane fragments. For numbering and structures see Figure 13. Note that at ambient temperatures B_6H_{10} is fluxional with respect to basal bridging H atom exchange among the five available sites and exhibits a single mean resonance position for B(2), B(3), B(4), B(5), and B(6) (Ref. 67).

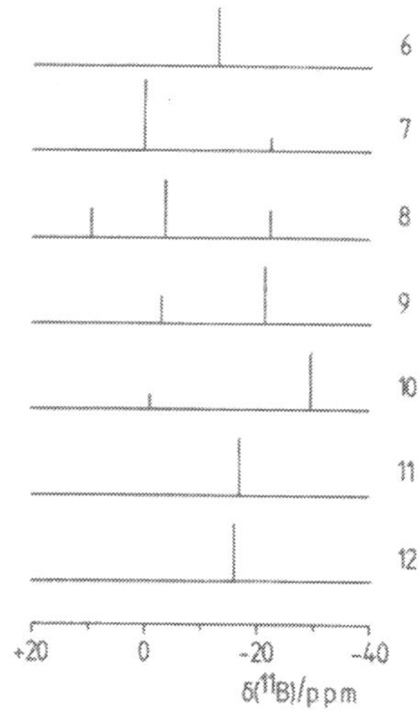

Figure 11. Representation of the chemical shifts and relative intensities in the ^{11}B NMR spectra of the *closo* borane anions $B_nH_n^{2-}$, for which the basic geometries are given in Figure 14. The eight-vertex $B_8H_8^{2-}$ does not exhibit the 4:4 intensity pattern expected from the dodecahedral solid-state symmetry, and solution equilibrations with C_{2v} forms (*inter alia*) have been postulated to account for this. $B_{11}H_{11}^{2-}$ is fluxional with respect to interchange of the {BH} units among all the available structural sites.

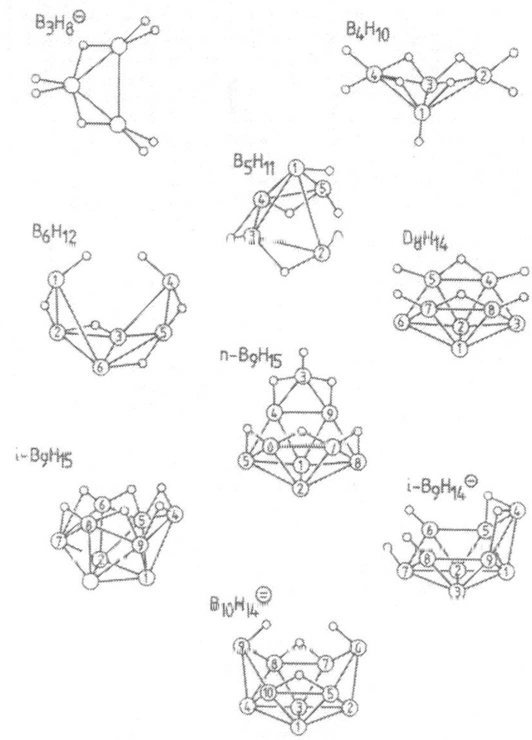

Figure 12. Representations of the molecular structures of the *arachno* borane fragments, of which the spectra are in Figure 9. $B_3H_8^-$ is fluxional with respect to effective exchange of all three B atoms among all three sites, *and* with respect to exchange of all the H atoms among all eight sites (Ref. 181). In the solid state the *endo*-terminal H atom on B(1) in B_5H_{11} shows bonding interaction with B(2), and there may be similar interaction between the *endo*-terminal H atoms of B_8H_{14} and its B(3), B(6) positions. $B_9H_{14}^-$ is fluxional in solution with respect to exchange of the five bridging/*endo*-terminal H atoms among the six available sites and therefore exhibits average threefold symmetry in the [11]B spectrum.

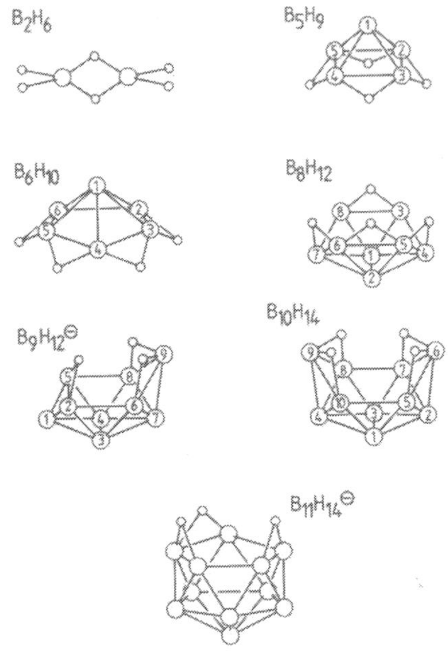

Figure 13. Representations of the molecular structures of the *nido* borane fragments of which the spectra are in Figure 10. B_6H_{10} and $B_{11}H_{14}^-$ are each fluxional with respect to intramolecular nondissociative exchange of the bridging H atoms among the five available open-face sites, and therefore exhibit average fivefold symmetry in their ^{11}B spectra at ambient temperatures. $B_9H_{12}^-$ is fluxional with respect to exchange of the three bridging H atoms among the three sites available, but this does not affect the gross appearance of the ^{11}B spectrum.

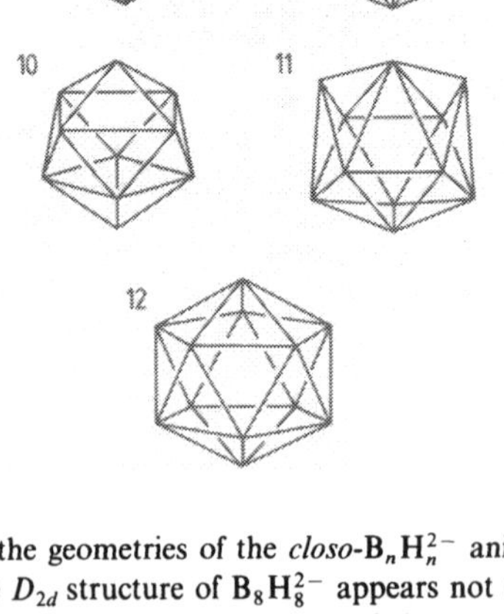

Figure 14. Representations of the geometries of the *closo*-$B_nH_n^{2-}$ anions, where $n = 6$–12. The ^{11}B spectra are in Figure 11. The solid-state D_{2d} structure of $B_8H_8^{2-}$ appears not to be retained in solution; the precise nature of the various $B_8H_8^{2-}$ solution species, and also their rates of mutual equilibration, depend on solution conditions such as pH, etc. (for details see Ref. 11, and other references cited therein). $B_{11}H_{11}^{2-}$ is highly fluxional (Ref. 182).

thorough discussion of these. As summarized in Sections 3.2.1 and 3.2.2 above, a general correlation of chemical shift ranges with particular structural features is not at present feasible, and much use is made of gross spectral similarities in attempts to assess compound type; this together with ^{1}H shielding behavior (see Section 3.5 below) often gives valuable information about structural classes.

"Basal" boron atoms, i.e., those in open faces of *nido* and *arachno* clusters, and also a number of apical boron atoms, generally have chemical shifts between $\delta(^{11}B)$ $+25$ and -25 ppm or so, and it is usual for polyhedral boron-containing compounds to have a number of resonances in this intermediate region. Those for the *closo* anions $[B_nH_n]^{2-}$ are all in this area, for example. Marked shifts away from this central region usually derive from a particular structural feature and sometimes are diagnostic. For example, particular apical (i.e., non-open-face) resonances of normal cluster connectivity five occur at higher field; these include the unique boron B(1) on $[B_6H_{10}]$, the (2,4) resonance in $[B_{10}H_{14}]$, and the (1,3) resonance in $[B_{10}H_{14}]^{2-}$ (Figures 9 and 10). These appear to be associated with delocalized bonding modes in the pentagonal pyramidal fragment (e.g., structure **1**) which approximate to an aromatic sextet (e.g., valence bond structures **8–10**), so that the apical atoms on the axis of circulation are shielded, and those at the periphery are correspondingly less shielded (see Refs. 51 and 82). Note that this is a function of geometry *plus* electronic structure because many five-connected apical resonances do *not* occur at high field. A collection of empirical pointers known informally as "Stan's Rules" help to locate the centers of higher shielding,[100] and are often of use provided heteroatoms do not have a dominant effect.[201]

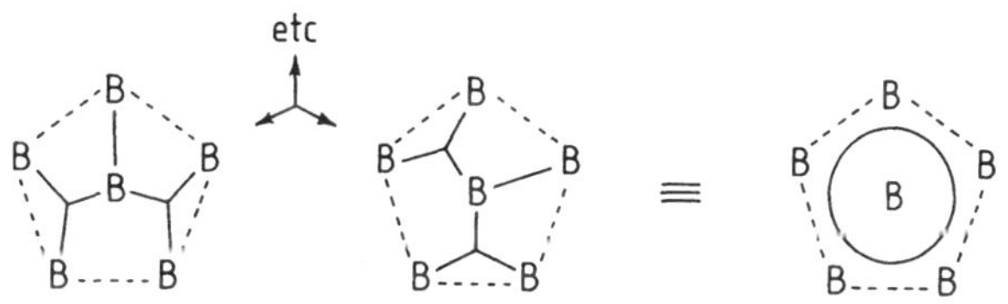

8 **9** **10**

An apical high-field resonance at δ ca. -53 ppm is also apparent for the four-connected B(1) atom in square pyramidal *nido*-$[B_5H_9]$ (schematic geometry **11**) which has similar bonding to $[B_6H_{10}]$. Again, however, the high shielding is not universal for the four-connected apical geometry: in $[B_{10}H_{10}]^{2-}$, (geometry **12**), for example the resonance is at δ $+4$ ppm, whereas in the neutral heteroboranes $[SB_9H_9]$[19] and $[(PMe_2Ph)_2NiB_9H_7Cl_2]$[101] it occurs at $+74.5$ and $+88.6$ ppm, respectively (geometry **13**). In these last two it is of interest that there is such a marked deshielding at cluster positions antipodal to the heteroatom, which in each case is believed to supply four, rather than the two electrons formally associated with a {BH} unit, to the notional "cluster electron count."*

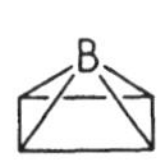

11

12

13

* Rules for predicting shielding changes in *closo* boranes and *closo* heteroboranes have very recently been reported: Teixidor, F.; Viñas, C.; Rudolph, R. W. *Inorg. Chem.* 1986, **25**, 3339–3345.

From the 140-ppm range of these four-connected vertices it is apparent that, in contrast to the effect of coordination number on $\delta(^{11}B)$ in mononuclear species (Section 2.1.1 above), there is no *dominant* effect of cluster connectivity on $\delta(^{11}B)$ in polyhedral species when different compounds are compared. Within individual compounds, however, it is often found that lower cluster connectivities are associated with lower or intermediate shielding, and higher connectivities with intermediate or higher shieldings.(4) When lower-connectivity cluster boron atoms are adjacent to cluster metal centers, they often have very low nuclear shielding,(102) for example those of cluster connectivity four in the *isocloso* metallaboranes such as $[(PMe_3)_2HIrB_9H_9]$ (δ ca. +94 ppm),(103) $[(C_5Me_5)RhB_{10}H_{10}]$ (δ ca. +118 ppm),(104) and $[(C_6Me_6)RuB_{10}H_{10}]$ (δ ca. +112 ppm).(125)

Thus the resulting ^{11}B 3:3:3 resonance pattern, with one group of three resonances at very low field, and the 2:4:2:2 resonance pattern, with two resonances at very low field, are, respectively, diagnostic of these two metallaborane structural types **14** and **15** (which do not have counterparts in binary boron hydride chemistry).(105) Low shielding also often occurs with boron atoms adjacent to two or more linked metal centers, e.g., δ + 60 ppm for $\{Pt_2B\}$ in $[(PMe_2Ph)_2Pt_2B_{12}H_{18}]$,(101) δ + 106 ppm for $\{Fe_4B\}$ in $[(CO)_{12}HFe_4BH_2]$,(106) etc., which may be associated with lower excitation energies introduced by the metal center in $\{M_nB\}$ multicenter bonding. Occasionally also lower shieldings have been ascribed to boron centers which are believed to have some vacant-orbital character, e.g., B(3)(6) in $[B_8H_{14}]$ and the corresponding B(7) position in $[B_9H_{13}L]$ (compare vacant-orbital trigonal sp^2 species in Section 2.1).

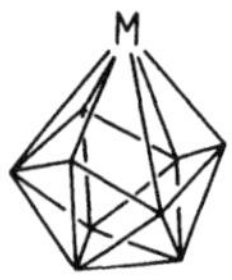

14

15

3.2.3b. Substituent Effects. Two broad categories of substituent effects on an *n*-vertex borane may be defined: that in which an *exo*-terminal hydrogen atom is replaced by the substituent atom or group, and that in which one or more *endo*-terminal or bridging hydrogen atoms are replaced by the substituent.

(i) *Exo*-bound substituents. Straightforward replacement of *exo*-terminal hydrogen atoms by σ-bound substituents produces α-shielding effects similar to those found for mononuclear boranes (see Section 2.1). Thus replacement of H by F typically produces chemical shift changes $\Delta\delta(^{11}B)$ of +10 to +20 ppm (though few examples are known), by Cl of ca. 0 to +25 ppm, by OR of +10 to +30 ppm, by alkyl of +5 to +20 ppm, by Br of −10 to +15 ppm, by I of −5 to −25 ppm, and by σ-bound borane of −5 to +10 ppm.

These ranges show that there is variation of substituent effect with the character of the substituted α-boron atom. There are generally also noticeable β- and γ-effects on the other cluster boron atoms. For example, small β-shift effects of either sign (depending on the cluster type, and on whether or not the substituent exhibits non-σ electronic interaction with the cluster),(85,107,108) are usually associated with the deshielding α effects. In addition, differential β effects of up to several ppm in the same cluster are encountered when the α atom is a transition metal center with mixed ligands that have different *trans* effects.(109,110) γ effects are often less marked;(108)

exceptions to this arise in smaller clusters where the γ atom is antipodal to the substituent site.(111) In the larger polyhedra based on icosahedral fragments the antipodal atom is δ to the substituent site, and the corresponding *antipodal shifts* are significant and have been investigated in some detail for a variety of systems,(112–115) particularly the 12-vertex dicarbadodecaboranes.(114)

Generally an increase in the electron-withdrawing power of the substituent results in an increase in the shielding of the nucleus in the antipodal "*para*" position. Thus halogen substituents can produce ^{11}B antipodal shifts of up to 12 ppm or so; ^{13}C and ^{31}P chemical shifts of appropriate carba- and phosphaboranes also exhibit analogous changes.(112) There are parallels between the antipodal effect and the mesomeric *para* effect in aromatic ^{13}C NMR.(113)

These *exo* substituent effects are generally approximately additive,(86,111b,116–118) at least for two or three substituents. Often the cumulative result of the α, β, γ, etc. influences results in "conservation of cluster chemical shift," and in this context it is interesting that the perhaloanions, for example $[B_9X_9]^{2-}$, have quite similar shifts to their $[B_nH_n]^{2-}$ analogs.(119) In these perhalo species, however (Table 5), it is apparent that the electronic structures now deviate significantly from those of their binary borane analogs. For example, in contrast to (unknown) neutral $[B_9H_9]$, neutral $[B_9Br_9]$ is stable. It has a very low value of $\delta(^{11}B)$(120) which may arise from

Table 5. Boron Chemical Shifts of Perhaloborane Species[a]

Compound	$\delta(^{11}B)$ (ppm)		
B_4Cl_4	+85.0 (4B)		
$B_6Cl_6^{2-}$	−17.4 (6B)		
$B_6Br_6^{2-}$	−18.5 (6B)		
$B_6I_6^{2-}$	−27.5 (6B)		
$B_6H_6^{2-}$	−13 (6B)		
$B_9Cl_9^{2-}$	+1.5 (3B)	−5.5 (6B)	
$B_9Br_9^{2-}$	−0.8 (3B)	−8.2 (6B)	
$B_9I_9^{2-}$	−3.1 (3B)	16.0 (6B)	
$B_9H_9^{2-}$	−3.4 (3B)	−21.5 (6B)	
B_9Br_9	+60.4 (9B)	[presumably fluxional]	
$1-(C_5H_5)NiB_9H_9^-$	+78.6 (1B),	+29.0 (4B),	−1.3 (4B)
$1-(C_5H_5)NiB_9Cl_9^-$	+54.3 (1B),	+38.5 (4B),	+18.4 (4B)
$B_{10}Cl_{10}^{2}$	−10.6 (8B)	−5.3 (2B)	
$B_{10}Br_{10}^{2-}$	−15.4 (8B)	−3.8 (2B)	
$B_{10}H_{10}^{2-}$	−28.5 (8B)	+4.0 (2B)	
$B_{12}Cl_{12}^{2-}$	−12.9 (12B)		
$B_{12}Br_{12}^{2-}$	−12.6 (12B)		
$B_{12}I_{12}^{2-}$	−6.4 (12B)		
$B_{12}H_{12}^{2-}$	−16.0 (12B)		

[a] Relative intensities given in parentheses.

lower HOMO–LUMO excitations, the origins of the LUMO in the neutral compound presumably approximating to the HOMO in the anionic species.

(ii) *Endo*-bound substituents. *Endo*-protons (i.e., those in *endo*-terminal or bridging positions) are bound by electrons which are often more intimately concerned with the intracluster bonding scheme, and replacement of these by substituents can sometimes have markedly greater effects on the entire cluster bonding. For example, although replacement of a bridging hydrogen atom in $[B_5H_9]$ by a transition metal center generally results in an α-deshielding of only 2 or 3 ppm or so at B(2) B(3), the corresponding β-deshielding effects on the apical B(1) position are relatively large at up to 10 ppm. Replacement by the vacant-orbital center $\{BMe_2\}$ induces an even larger β-deshielding effect (ca. 20 ppm), again much larger than the α-effect.[121] These effects may result from a disruption of the quasiaromatic six-electron system (Section 3.2.3.1, near structures **8–10**).

A common *endo* substitution mode involves the replacement of two or more *endo* hydrogen atoms by a corresponding number of bonds to a multivalent substituent. Often this may be accompanied by the (synergic) involvement of additional substituent valence-shell orbitals and/or electrons in cluster bonding, in which case changes in the borane cluster shielding give a measure of the degree of involvement of the substituent in the entire cluster bonding.[69] When this additional involvement is extensive, it is often useful to regard the substitution in terms of the replacement of a *boron* center in the next higher $(n+1)$-vertex borane by the substituent, rather than in terms of a replacement of *hydrogen* atoms in the *n*-vertex borane. Thus the NMR characteristics of a variety of *nido*-type 10-vertex metallaboranes of basic geometry **16** bear little resemblance to starting *arachno* substrates of basic geometry **17**, and are best considered as *nido* decaboranyl analogs in which a boron center in $[B_{10}H_{14}]$ (schematic geometry **5** and **18**) has been replaced by a transition-metal center. Thus derivatives $[MB_9H_{13}]$, where $M = \{WH_2(PR_3)_3\}$, $\{ReH(PR_3)_3\}$, $\{Os(PR_3)_3\}$, and $\{IrH(PR_3)_2\}$, all show marked ^{11}B NMR similarities to *nido*-$[B_{10}H_{14}]$, thereby confirming the electronic structural type.[122] On the other hand, species such as $[(PR_3)_2PtB_9H_{11}(PR_3)]$, also of basic geometry **16**, show marked NMR similarities to *arachno*-$[B_{10}H_{14}]^{2-}$, and also to *arachno*-$[B_9H_{14}]^-$, typifying it as an *arachno*-type metallaborane, and also showing that a much smaller modification of the *arachno* nine-vertex starting substrate structure has occurred.[73,123]

16

17

18

Conversely, for example, the *nido*-metallaboranes $[(C_5Me_5)RhB_8H_{10}(PMe_2Ph)]$,[196] $[(CO)(PMe_3)_2IrB_8H_{11}]$,[124] and $[(PMe_2Ph)_3H_2ReB_8H_{11}]$[122,125,189] (schematic geometry **19**) appear at first sight to have few shielding parallels with either the ostensibly isoelectronic and isostructural binary borane anion nido-$[B_9H_{12}]^-$ (structure **20**) or with any known eight-vertex eight-boron species such as $[B_8H_{12}]$ or $[B_8H_{14}]$, perhaps therefore indicating markedly modified electronic structure.[124]

19

20

In this general area the ^{11}B shielding behavior of carboranes versus their isoelectronic borane analogs merits comparison, but again few generally applicable diagnostics arise. Comparison of *closo*-$[B_{12}H_{12}]^{2-}$ shieldings [$\delta(^{11}B)$ ca. −16 ppm] with isoelectronic and isostructural 1,2-$[C_2B_{10}H_{10}]$ and 1,7-$[C_2B_{10}H_{10}]$(112) (structures **21** and **22**) reveals little perturbation of the basic pattern by the more electronegative carbon centers. The major change is an antipodal shielding effect of about 10 ppm (compare $[B_{10}H_{10}]^{2-}$ analogs, structures **12** and **13** above, and see also Ref. 126).

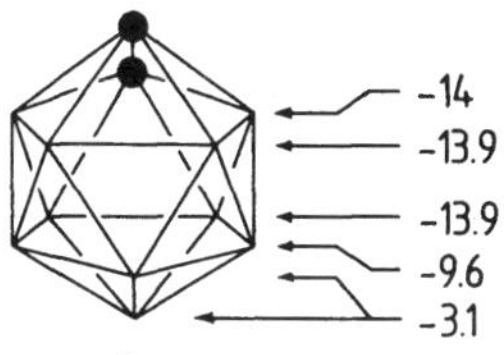

21

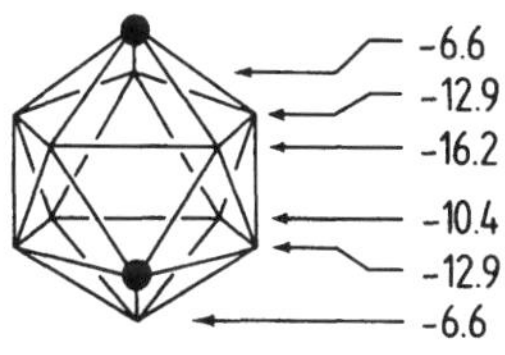

22

By contrast, comparison of ten-vertex *nido*-5,7-$[C_2B_8H_{12}]$(127) with isoelectronic *nido*-$[B_{10}H_{14}]$ indicates more substantial differences: the $^{11}B(2)$ and (4) resonances are now differentially shielded by some 17 ppm, and the shielding of other nuclei changed by up to 20 ppm or so, so that the basic *nido*-decaboranyl pattern is obscured somewhat. This may be related to the changing incidence of open-face bridging hydrogen atoms,(96a) and similar effects occur in 5,6-$[C_2B_8H_{12}]$.(128) On the other hand, in the successive replacement of boron by carbon in the isoelectronic series of *nido* compounds $[B_6H_{10}]$,(129) $[CB_5H_9]$,(130) $[C_2B_4H_8]$,(131) $[C_3B_5H_7]$,(132) and $[C_4B_2H_6]$,(130,133) the simpler basic *nido*-hexaboranyl shielding pattern is readily seen to be retained, even when as many as four of the six boron atoms have been replaced (Figure 15).

3.3. Coupling Constants

3.3.1. General Considerations

A popular factorization of boron bonding in the *closo* polyhedral boranes involves principal contributions from *sp* hybrid orbitals radially directed towards the

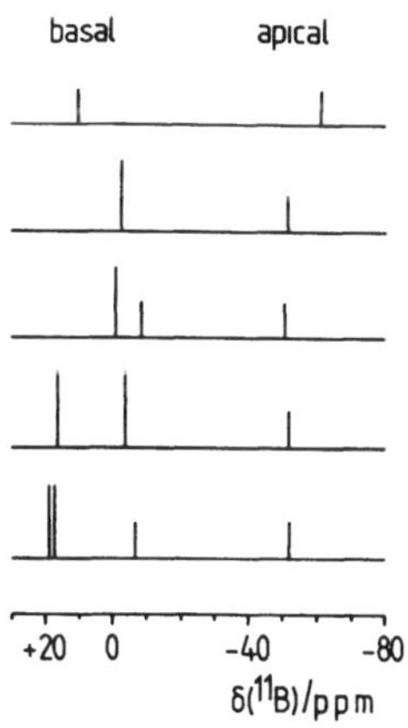

Figure 15. Representation of the chemical shifts and relative intensities in the ^{11}B NMR spectra of the *nido* six-vertex cluster compounds (A) $C_4B_2H_6$, (B) $Me_2C_3B_3H_5$ (the nearest equivalent to $C_3B_3H_7$ for which data are available), (C) $C_2B_4H_8$, (D) CB_5H_9, and (E) B_6H_{10} (data from Refs. 129–133).

center of the cluster and towards the *exo*-terminal bonds, together with surface bonding involving the remaining p^2 system. A dominance of the Fermi contact mechanism (Chapter 4) is then in accord with the observed larger couplings between polyhedral ^{11}B and *exo*-bound nuclei, and much smaller couplings to *endo*-bound nuclei. As clusters become progressively more open, these differences often become less distinct as some open-face structural features may then involve boron atoms with bonding approximating more towards a more balanced tetrahedral sp^3 hybridisation, and larger couplings to the nuclei of *endo*-bound atoms will result. It must be emphasized that for *endo* cluster bonding involving atomic orbitals with low *s* character, smaller or even zero coupling constant magnitudes may be expected, even though a strong bond may exist (see Figures 5 and 6 above). Conversely, the absence of an observable coupling does not necessarily imply the lack of a bond or the lack of electron density between the two atoms. The observation of boron coupling constants (particularly the smaller ones) is often hindered by the quadrupolar broadening effects of the boron nuclei; this applies both to boron spectra and to the spectra of the other nucleides.

3.3.2. Boron–Proton Coupling Constants

One-bond couplings to *exo*-terminal protons $^1J(^{11}B-{}^1H)(exo)$ are typically in the range 120–170 Hz, and presumably have positive sign. To a crude approximation, the magnitude is consistent with *sp* hybridization of an *exo*-bond at boron on the basis of sp^3 hybridization for $[BH_4]^-$ which has $^1J(^{11}B-{}^1H)$ *ca.* 80 Hz. These *exo* couplings are generally resolved in a polyhedral boron hydride ^{11}B spectrum, which therefore commonly appears as a series of doublets in the absence of $\{^1H\}$ decoupling,[134] although *exo*-substituted boron atoms will of course generate singlets. It should be noted that if the doublet structure is not well resolved owing to excessive linewidth, then the apparent splitting will be somewhat less than $J(^{11}B-{}^1H)$; a number of quoted *J* values in the literature may therefore be too small. Apical positions often exhibit couplings towards the higher end of the range quoted, and relationships have been found with calculated percent *s*-character in boron orbitals,[135] with "cage umbrella angle" (which will also be related to hybridization),[136] with calculated electron density on the boron atom,[3] with calculated boron-hydrogen *s*-overlaps,[42] etc.

Known *endo*-terminal couplings $^1J(^{11}B-{}^1H)(endo)$ fall typically into two categories. The first category comprises those in the range of, say, 100–130 Hz, i.e., generally somewhat lower than related *exo*-terminal couplings. These higher *endo* couplings are generally associated with more open *arachno*-type features and may involve essentially sp^3 hybridized boron. The second category contains those with magnitudes of up to several tens of hertz which are often not resolved in typical ^{11}B spectra. Because of this lower coupling constant, these latter are sometimes described as having "unique terminal" or "pseudobridging" character,[73,137,138] although often the small coupling magnitude may originate more from the low *s*-overlap involved in a more straightforward two-electron two-center *endo* B–H bond which has high boron *p*-character, rather than from true bridging character.

The "genuinely" bridging two-electron three-center B–H–B bonds themselves typically exhibit values of $^1J(^{11}B-{}^1H)(bridge)$ from effectively zero to a few tens of hertz. In asymmetric species the two such couplings associated with the B–H–B bridge are typically different, often markedly so; this can be diagnostic in particular systems, for example in *arachno* four-vertex compounds.[118,143,169,170] Similar values obtain for *endo*-M – H – B-bridging bonds which are a common feature of metallaboranes and metallacarboranes. Some M – H – B bridges occur via *exo*-B – H bonds(68,139–141,191); little coupling information is yet available for these, but the indications are that they are somewhat larger than for *endo*-M – H – B linkages.(139,191)

These overall coupling ranges, especially when taken together with corresponding shielding behavior (Section 3.5 below) are generally diagnostic of *exo*-terminal, *endo*-terminal and bridging proton types. The smaller couplings are generally not resolved in the ^{11}B spectra of intermediate and heavier polyhedral borane species without the aid of line-narrowing techniques.(142) Selective ^{11}B-{^{1}H(*bridge*)} spectroscopy is also of some use in this area,(117) as is deuterium labeling of bridging hydrogen sites,(142) which may often be effected quite easily.

A general point arising from the three-center two-electron nature of the B – H – B bridging bonds is that it is not clear what factor should be introduced to take account of this and other multicenter bonding modes in simple interpretations of the Fermi contact mechanism (Chapter 4). In this context also it should be noted that, although in nonpolyhedral chemistry the descriptors ^{n}J are generally used to describe coupling paths extending over n two-electron two-center σ bonds, in "electron deficient" polyhedral clusters (not only of boron) they are generally used to define the number of polyhedral edges on the shortest edge-path linking the two atoms of interest. In the general case this will *not* be associated with n two-electron two-center bonds.(95)

Longer-range boron–proton couplings across the polyhedral framework are generally not resolved in all but the smallest species,(152) for example ^{2}J(^{11}B–B–^{1}H) of (–) 5 Hz in [B_4H_{10}].(143) Couplings of up to a few hertz to protons in *exo*-substituents are often more apparent, for example ^{2}J(^{11}B–C–^{1}H) of ca. (–) 6 Hz in methylboranes,(144,145) and ^{3}J(^{11}B–S–C–^{1}H) of up to ca. 2 Hz in {SMe_2}-substituted species.(117)

3.3.3. Boron–Boron Coupling Constants

Intracluster *endo* couplings ^{1}J(^{11}B–^{11}B)(*endo*) commonly take values from zero to one or two tens of hertz, and are often resolvable in the ^{11}B-{^{1}H} spectra of the smaller and more mobile polyhedral boron-containing species.(97,152) The apical-to-basal coupling ^{1}J(^{11}B–^{11}B)(*endo*) in [B_5H_9] is +18 Hz, for example (see also Figure 5 above).(145,146) With increasing cluster size, however, the coupling information is generally not so readily observable, although there are notable exceptions: e.g., fine structure arising from ^{1}J[^{11}B(2)–^{11}B(6)] of up to ca. 20 Hz is readily apparent in the ^{11}B spectra of nido-[$B_{10}H_{14}$] and its derivatives.(22,118,147,148) The more typical obscuring of the coupling information in larger clusters arises not only from increased ^{11}B relaxation rates, but also because a ^{11}B nucleus will generally have five different couplings to nearest-neighbor ^{11}B nuclei, and, furthermore, the spectrum peak will be a superposition of the various subspectra arising from the positional permutations within the six isotopomer combinations {$^{11}B_n\,^{10}B_{6-n}$}, $n = 1$–6. Few empirical correlations have yet been made, but it is apparent from work on smaller boranes,(51a,143) and from ^{11}B–^{11}B correlation spectroscopy (Section 3.3.2) that B–H–B and "long" B–B linkages are often associated with smaller couplings ^{1}J(^{11}B–^{11}B), and also that when significant two-electron two-center character is present, then couplings may often be relatively large, for example ^{1}J[^{11}B(2)–^{11}B(6)] in [$B_{10}H_{14}$] as cited above.

Intercluster *exo–exo* couplings are much larger, typically taking values of many tens of hertz.[34,149,192] As with $^1J(^{11}\mathrm{B}-^1\mathrm{H})(exo)$, couplings $^1J(^{11}\mathrm{B}-^{11}\mathrm{B})(exo\text{–}exo)$ involving apical boron atoms exhibit the largest magnitudes. For example, the (1,1′), (1,2′), and (2,2′) isomers of $[(B_5H_8)_2]$ (schematic geometries **23–25**, respectively) take intercluster $^1J(^{11}\mathrm{B}-^{11}\mathrm{B})(exo)$ values of 149, 115, and 79 Hz, respectively. The relationship between these *exo–exo* couplings and the degree of *s* character, etc., of the intercluster boron bonding orbitals has been considered in some detail.[149]

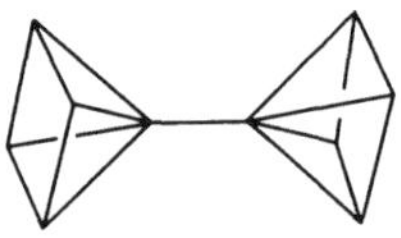

23

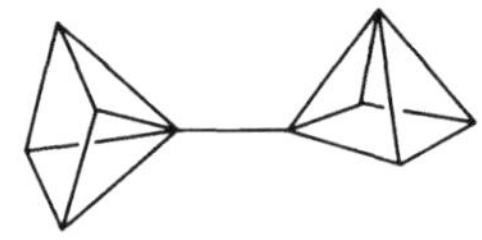

24

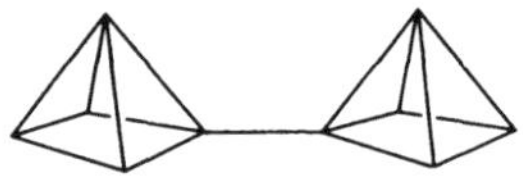

25

As with $^nJ(^{11}\mathrm{B}-^1\mathrm{H})$, no longer-range couplings $^nJ(^{11}\mathrm{B}-^{11}\mathrm{B})$ have been examined to any significant extent in the general case.

3.3.4. Couplings Between Boron and Other Nuclei

Principally because of ^{11}B-relaxation broadening in the spectra both of ^{11}B and of the other coupled nuclei,[150,151] these couplings have been as yet only little examined. Reasonable estimates of $S^2(0)$ for ^{11}B versus ^{13}C (Chapter 4, Section 4.1), together with the ratio $\gamma(^{11}\mathrm{B}):\gamma(^{13}\mathrm{C})$, suggest slightly larger values for $^1J(^{13}\mathrm{C}-^{11}\mathrm{B})$ versus $^1J(^{11}\mathrm{B}-^{11}\mathrm{B})$ for equivalent binding modes. The few examples reported,[97b] e.g., $^1J(^{13}\mathrm{C}-^{11}\mathrm{B})(exo)$ of (+) 72–75 Hz and 60–65 Hz, respectively, for 1-alkyl and 2-alkyl derivatives of $[B_5H_9]$, and of $^1J(^{13}\mathrm{C}-^{11}\mathrm{B})(endo)$ of ca. 18 Hz in $[C_2B_3H_5]$,[153] are not inconsistent with this. *Exo*-bound fluorine atoms in 2-FB_5H_8 and in *closo*-$[C_2B_{10}H_{11}F]$ isomers have $^1J(^{19}\mathrm{F}-^{11}\mathrm{B})$ of ca. 50 Hz,[111a,154a] and *exo*-bound triorganophosphine ligands typically exhibit values of $^1J(^{31}\mathrm{P}-^{11}\mathrm{B})(exo)$ in the range 150–200 Hz. Known *endo*-bound organophosphorus atoms in relatively open *arachno* configurations exhibit smaller couplings $^1J(^{31}\mathrm{P}-^{11}\mathrm{B})(endo)$ in the range 85–145 Hz,[52,74] although electronegative substitution can double these values.[53] Few coupling data are available for phosphorus-bound *endo* in more contiguous cluster positions.[154b]

Couplings $^1J(^{205}\mathrm{Tl}-^{11}\mathrm{B})(endo)$ of ca. 255 Hz and 175 Hz are observed in the anion *nido*-$[Me_2TlB_{10}H_{12}]^-$, together with couplings $^2J(^{205}\mathrm{Tl}-\mathrm{B}-^{11}\mathrm{B})$ of up to 85 Hz; all have the same sign, presumed to be positive.[95,155] Coupling constants $^1J(^{195}\mathrm{Pt}-^{11}\mathrm{B})$ for *endo*-bound platinum commonly take values of (+) 200 to (+) 300 Hz in platinaboranes.[29,32,73,101,117,118,124] Exceptions include $^1J(^{195}\mathrm{Pt}-^{11}\mathrm{B})(endo)$ of 565 Hz to B(3) in $[(PMe_2Ph)_2PtB_3H_7]$,[156] and $^1J(^{195}\mathrm{Pt}-^{11}\mathrm{B})(exo)$ of ca. 600 Hz in $[(PMe_2Ph)_2Pt\text{-}\eta^1,\eta^2\text{-}anti\text{-}B_{18}H_{20}]$,[157] together with smaller couplings of $<$ca. 100 Hz in 1-metalla *nido*-type six-vertex cluster species such as $[(PEt_3)_2PtBC_4H_2Me_3]$.[158] Some beryllium, yttrium, and zirconium couplings have also been reported.[64,159,160]

3.4. Relaxation Times

Longitudinal and transverse ^{11}B relaxation in the polyhedral species is generally quadrupolar dominated,[14,66,150,161–163] although there are probably often additional scalar contributions to T_2 arising from couplings $J(^{11}\mathrm{B}-^{11}\mathrm{B})$ with other rapidly relaxing ^{11}B nuclei (see Chapter 2, Section 2.5).[162] There has not been much extensive or systematic work in the area.

Boron quadrupolar coupling is not strong ($e^2qQ/\hbar$ of up to about 3 or 4 MHz),[66] and typical relaxation times T_1 and T_2 for molecular polyhedral species in solution (Table 6) vary from many tens of milliseconds for particular sites in small mobile species to somewhat less than 1 ms for sites in larger species such as polymetallaboranes which have a number of bulky ligands on each metal center.[150] For these latter a straightforward quadrupolar relaxation of the ^{11}B nucleus in question probably dominates the linewidth, but for the former significant contributions to linewidth can also derive from unresolved and/or partially collapsed couplings to other boron nuclei and also from scalar T_2 processes if the other coupled boron nuclei are relaxing rapidly; careful experimentation may be required to distinguish among, and hence make use of, these factors.[162] Linewidth and hence

Table 6. Reported $T_1(^{11}B)$ Values for Selected Polyhedral Borane Species[a,b]

Compound [number of atoms] (numbering scheme)	Conditions	$\delta(^{11}B)$	$T_1(^{11}B)$ (ms)	Position
nido-B_5H_9	−25°C, neat liquid	−51.0	ca. 200	1
[14 atoms]		−12.0	ca. 35	2, 3, 4, 5
(Figure 8.13)				
closo-2,4-$C_2B_5H_7$	+25°C, probably in $CDCl_3$	+12.5	38.6	3
[14 atoms]		+9.2	47.7	5, 6
(Figure 8.5)		−16.4	68.3	1, 7
closo-1-SB_9H_8-10-Br	+89°C, 1M in $CD_3C_6D_5$	+66.8	12,500	10
[19 atoms]		−6.5	18	2, 3, 4, 5
(structure 13; E = S,		−16.7	43	6, 7, 8, 9
Br on B)				
nido-$B_{10}H_{14}$	+21°C, 1M in $CD_3C_6D_5$	+12.3	17.3	1, 3
[24 atoms]		+10.2	4.8	6, 9
(Figure 8.13)		0.0	7.9	5, 7, 8, 10
		−36.0	28.1	2, 4
nido-$[Me_2TlB_{10}H_{12}]^{2-}$	+25°C, $[Me_2Tl]^+$ salt,	+8.2	2.9	8, 11
[29 atoms]	saturated solution in $(CD_3)_2CO$	+3.0	15.5	1
(structure as in $B_{11}H_{14}^-$,		−2.8	9.7	5
Figure 8.13, although H		−3.1	17.3	2, 3
atom disposition differs		−4.7	9.0	9, 10
from this)		−31.7	23.1	4, 6
6,6′-(*nido*-$B_{10}H_{13})_2$	+25°C, dilute in C_6D_6	+15.0	2.8	6
[46 atoms]		+13.5	15.0	1, 3
(numbering as in $B_{10}H_{14}$,		+11.7	4.0	9
Figure 8.13)		+2.7	8.2	5, 7, 8, 10
		+0.9	6.2	
		−33.9	20.0	2, 4
		−34.6	9.4	
arachno-4-$[(PMe_2Ph)_2PtB_8H_{12}]$	+21°C, saturated solution in $CDCl_3$	+21.6	3.2	1
[68 atoms]		+18.8	1.6	7
(numbering as in $B_9H_{14}^-$,		+1.7	1.9	5, 9
Figure 8.12, although H atom		−23.0	2.0	6, 8
disposition differs from this)		−31.6	5.6	2, 3

[a] Data from Refs. 73, 95, 164, 166b, and 168, and from other references cited therein.
[b] Note that $T_1(^{11}B)$ depends upon *solution conditions*, as well as electronic structure (see Chapter 5).

resolution may generally be improved by use of higher temperatures and/or less viscous solvents, which have the effect of reducing the molecular correlation time τ_c (Chapter 2, Section 2.5).[150]

Since both the quadrupolar coupling $e^2qQ/\hbar$ (Chapter 2, Section 2.5.4 and Chapter 5, Section 4.5) and the excited state nuclear deshielding subterm P_i (Chapter 3) are related to electronic symmetry about the nucleus, there are often parallels between $\sigma(^{11}B)$ and $T_1(^{11}B)$ or $T_2(^{11}B)$ within a particular molecule, so that a typical ^{11}B NMR spectrum of a species of larger molecular size is often characterized by broader resonance lines to the left and sharper ones to the right. There are many marked exceptions to this, however, since other electronic factors often predominate. A dramatic example of this latter is the low-field resonance at $\delta(^{11}B) = +65.6$ ppm in *closo*-[1,10$-SB_9H_8Br$], for which longitudinal relaxation times $T_1(^{10}B)$ and $T_1(^{11}B)$ can be as high as several tens of seconds under straightforward solution conditions.[19,20,156] In such instances of very long $T_1(^{11}B)$ values, contributions from the nonquadrupolar mechanisms (more typical for spin-1/2 nuclei) will become significant.

Different ^{11}B sites in a molecule will typically have different values for $T_1(^{11}B)$ under given solution conditions, and some empirical correlations have been made between $T_1(^{11}B)$ and structural position, for example in some selected small carboranes and metallacarboranes,[162] and in some *nido* 11-vertex heteroboranes[126]; these have had some (limited) diagnostic use in structural determination. A general use is inhibited because $T_1(^{11}B)$ depends upon the correlation time τ_c as well as upon the molecular electronic distribution (Chapter 2, Section 2.5.4, and Chapter 5, Section 2.5), and variations among different molecules are typically dominated more by τ_c variations than by $e^2qQ/\hbar$ variations.[126] This applies particularly to molecules of disparate polarity or size,[150] and with lighter boranes inertial isotope effects can also be significant.[164] In principle the use of molecular geometry plus dipolar-dominated proton relaxation $T_1(^1H)$ to determine τ_c under varying conditions,[66] permits, in conjunction with $T_1(^{11}B)$ measurements, the estimation of a standard longitudinal relaxation time $\theta_1(^{11}B)$ for a standard correlation time: this has potential for use as a "motion-independent" NMR parameter that depends only on electronic structure,[163] but this has not yet been applied in the general case. The converse use of relative T_1 rates to estimate relative size and/or motional differences in solute molecules of closely similar electronic structure has had some application.[101,165]

3.5. Polyhedral Species—Nuclei Other than Boron

This area is reviewed in more detail elsewhere,[9] but important points merit brief mention here. Some aspects of coupling constant behavior are in Section 3.3.4 above. $\{^{11}B\}$ decoupling and other multiple resonance experiments are of considerable use in this area (for example Refs. 29–32, 89, 109, 145, 151, 152, 156, 166, etc.), as is the phenomenon of "thermal decoupling."[150,151,167]

Proton spectroscopy is of obvious importance. It is found that *exo*-terminal proton shieldings parallel the shieldings of the boron nuclei to which they are bound: gradients $\delta(^1H){:}\delta(^{11}B)$ vary from 1:10 to 1:16 in the (limited) series of compounds that have been studied in detail (e.g., Figure 16).[28,32,83,101,117,122,168–170,189] Deviations from this correlation usually indicate particular structural features,[83,117,168] for example *endo*-terminal protons often resonate at $\Delta\delta(^1H)$ ca. -2 ppm from the plot, B–H–B bridging protons at $\Delta\delta(^1H)$ ca. -5 ppm, and M–H–B bridging protons at $\geqslant\Delta\delta(^1H)$

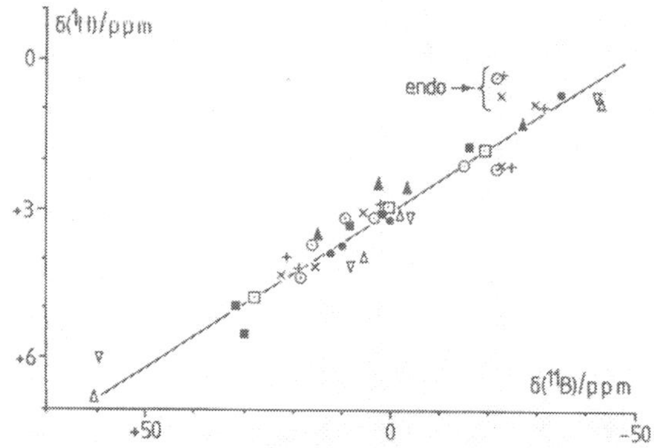

Figure 16. Proton–boron-11 nuclear shielding correlation plot (with bridging protons omitted) for the compounds *arachno*-4-$[(PMe_2Ph)_2PdB_8H_{12}]$ (×), *nido*-$[B_{10}H_{14}]$ (●), *arachno*-$[B_9H_{13}$-4-$(PMe_2Ph)]$ (○), *arachno*-6,9-$[(PMe_2Ph)_4PtPdB_8H_{10}]$ (■), *arachno*-6,9-$[(PMe_2Ph)_4Pt_2B_8H_{10}]$ (□), *nido*-7-$[(PMe_2Ph)_2PtB_{10}H_{12}]$ (▲), *arachno*-4-$[(PMe_2Ph)_2PtB_8H_{12}]$ (+), $[(PMe_2Ph)_2Pt_2B_{12}H_{18}]$ (△), and $[(PPh_3)_2Pt_2B_{12}H_{18}]$ (▽). The line drawn represents the ratio $\delta(^{11}B):\delta(^{1}H)$ 16:1 (Data from Ref. 101).

ca. −8 ppm. Solvent effects can be quite marked on $\delta(^{1}H)$ and have been studied in detail.(28,78,80,117,163,166) Interproton couplings $^{2}J(^{1}H-^{1}H)$ and $^{3}J(^{1}H-^{1}H)$ of up to one or two tens of hertz are generally present though often difficult to resolve.(18,117,144,152,155,166,168) It is apparent that they depend critically on the electronics of the intermediate boron (and other) atoms, but insufficient data are available for useful generalizations. These interproton couplings permit successful $^{1}H-^{1}H$ COSY experiments when these are carried out with simultaneous $\{^{11}B(\text{broad-band noise})\}$ decoupling;(104) it is likely that this technique will complement and often replace $^{11}B-^{11}B$ COSY experimentation as a predominant and powerful structural tool in polyhedral boron chemistry.(104,189) Similar magnitudes also occur for proton couplings to ^{31}P, both for boron-bound phosphines, and also for metal-bound phosphines and hydrides in metallaborane and metallacarborane species.(105,124,169,170) In this regard larger *transoid* (versus smaller *cisoid*) couplings across transition metals are often useful in assignments and in the determination of stereochemistry.(124,169,170) Proton couplings to ^{91}Zr, ^{103}Rh, ^{195}Pt, ^{199}Hg, and $^{203/205}Tl$ have been reported for metallaborane cluster species.(32,73,83,109,155,160,171,172) Some work on proton relaxation is given in Refs. 163 and 166.

^{13}C spectroscopy in carborane chemistry is not very extensive when considered in the light of the considerable amount of preparative and ^{11}B NMR work in the field The reader is referred to the review "Carbon-13 NMR Spectroscopy of Boron Compounds" for more detailed material.(173) There is a general parallel between $\delta(^{11}B)$ and $\delta(^{13}C)$ for atoms in structurally analogous polyhedral positions; for example, in a (limited) series of *closo* $\{C_2B_{10}\}$ cluster species the $\delta(^{11}B):\delta(^{13}C)$ correlation plot has gradient 0.74:1.(108) In the more general case $\delta(^{11}B):\delta(^{13}C)$ is ca. 0.4:1, with approximate intercept $\delta(^{11}B) = -40$, $\delta(^{13}C) = 0$ ppm.(70) There are also $\delta(^{11}B):\delta(^{13}C)$ shielding parallels when metallaboranes and organometallic compounds are compared, and the NMR parallels between borane clusters and isoelectronic "nonclassical" carbenium ions have been convincingly elucidated.(70)

Some ^{31}P data for phosphaboranes, i.e., species that contain an *endo*-bound phosphorus atom as part of the cluster, are in Refs. 112 and 154b. For pendant *exo*-bound tertiary phosphine ligands $\delta(^{31}P)$ values are similar to those for the same

Table 7. Coupling Constants $^1J(^{195}Pt-^{31}P)$ for Selected Compounds of General Formula *cis*-$[Pt(PMe_2Ph)_2XY]$[a]

Compound	X	Y	$^1J(^{195}Pt-^{31}P)$ (Hz) *trans* to X	$^1J(^{195}Pt-^{31}P)$ (Hz) *trans* to Y
$[PtCl_2(PMe_2Ph)_2]$	Cl	Cl	3548	3548
$[Pt(\mu-B_5H_8)Cl(PMe_2Ph)_2]$	μ-B_5H_8	Cl	2341	3714
$[PtClMe(PMe_2Ph)_2]$	Me	Cl	1730	4250
$[PtMe_2(PMe_2Ph)_2]$	Me	Me	1820	1820
$[Pt(\mu-B_5H_8)Me(PMe_2Ph)_2]$	μ-B_5H_8	Me	2510	1915
$[Pt(\mu-B_5H_8)Br(PMe_2Ph)_2]$	μ-B_5H_8	Br	2411	3658
$[Pt(\mu-B_5H_8)I(PMe_2Ph)_2]$	μ-B_5H_8	I	2405	3565
$[PtI_2(PMe_2Ph)_2]$	I	I	3364	3364
$[Pt(\eta^1,\eta^2-B_{18}H_{20})(PMe_2Ph)_2]$	μ-BB	σ-B	3335	1430
$[(Pt_3B_{12}H_{18})(PMe_2Ph)_4]$[b]	μ-BB	H	2601	1804
$[Pt(\eta^4-B_{10}H_{12})(PMe_2Ph)_2]$	η^4-BBBB		2534	2534
$[Pt(\eta^3-B_8H_{12})(PMe_2Ph)_2]$	η^3-BBB		2724	2724
$[Pt(\eta^3-B_3H_7)(PMe_2Ph)_2]$	η^3-BBB		2625	2625

[a] Data from Ref. 68, p. 636, and from Ref. 156.
[b] $[Pt_2(\eta^3-B_6H_9)(\eta^3-B_6H_8-\eta^2-PtH\{PMe_2Ph\}_2)(PMe_2Ph)_2]$ (Ref. 101).

ligands bound to transition metals such as platinum. There are considerable data for tertiary phosphine *exo*-ligands on metals incorporated in *endo* positions in metallaboranes and metallacarboranes (see Refs. 101, 124, 156, 174, 175, and other papers from the three schools cited), for which $^2J(^{31}P-M-^{11}B)$(*trans*) is commonly two or three tens of hertz though often unresolved.(117,151) The value of $^1J(^{195}Pt-^{31}P)$ *trans* to *endo* platinum–boron linkages is commonly in the range 2000–3000 Hz; variations in this coupling for platinaboranes have been used to compare the *trans* effects of metal-to-borane bonding modes with those of other ligands,(156,176,177) and the order $Cl > Br > I \gg \mu$-$BB > \sigma$-alkyl $\sim H > \sigma-B$ has been found (Table 7).(68) Some longer-range transcluster couplings such as $^4J(^{195}Pt-^{31}P)$ and $^5J(^{31}P-^{31}P)$ of up to several hertz occur in a variety of metallaborane systems.(101,124)

Although metallaborane and metallacarborane chemistry is extensive,(68,178) little information is available on metal nuclear shielding in these compounds, and is essentially limited to some nice work on ^{9}Be,(159) some isolated ^{195}Pt chemical shifts,(32,29,73,156,174) a $\delta(^{89}Y)$ datum for $[(thf)_2YB_3H_{12}]$,(64) and a $\delta(^{91}Zr)$ datum for $[ZrB_4H_{16}]$.(160) Very recent work has included ^{29}Si,(193–195) ^{45}Sc,(188) and ^{119}Sn.(195) Work in the area by direct observation is often severely hindered by boron coupling and quadrupolar broadening effects (e.g., Figure 17), and indirect measurement via $^1H-\{X\}$ spectroscopy can then be of use.(156)

3.6. Fluxionality

This is essentially an NMR phenomenon (see Chapter 3, Section 6), and non-dissociative dynamic behavior is common in polyhedral boron chemistry,(179) often resulting in simpler NMR spectra than required by the static structure.(100,180) Examples include simultaneous complete scrambling of all boron and all hydrogen positions (e.g., $[B_3H_8]^-$),(181) scrambling of all boron positions with retention of individual *exo*-B–H bonds (e.g., $[B_{11}H_{11}]^{2-}$),(182) exchange among only bridging protons on a

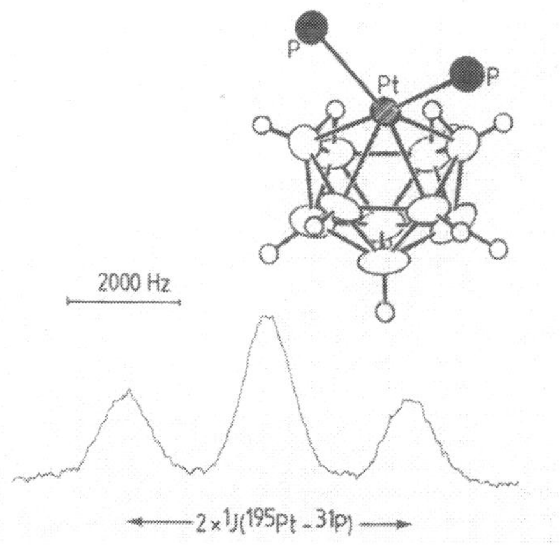

Figure 17. 77-MHz FT ^{195}Pt-{^{1}H(broad band noise)} NMR spectrum of $[(PMe_2Ph)_2 PtB_{10}H_{12}]$ (overnight accumulation, saturated solution in CD_2Cl_2 at 21°C). In solution the compound has effective mirror-plane symmetry and the two P atoms are equivalent. The triplet structure arises from $^{1}J(^{195}Pt–^{31}P)$ *ca.* 2530 Hz (Table 6) and the broadness arises from the (unresolved) couplings $^{1}J(^{195}Pt–^{11}B)$ of 200–300 Hz together (presumably) with longer-range couplings $^{n}J(^{195}Pt–^{11}B)$; spin systems from isotopomers containing ^{10}B will also contribute (Ref. 156).

static borane matrix (e.g., $[B_6H_{10}]$),[67] exchange between selected *exo*, *endo*, and bridging proton sites (e.g., metal complexes of $[B_3H_8]^-$),[68,183,197] etc. In metallaboranes and metallacarboranes there is in addition a large variety of fluxional modes that can be associated with the metal-to-borane bonding and/or with the metal-bound *exo*-polyhedral ligands (see, for example, Refs. 29, 68, 160, 174, 180, 183–186, 196, 197, and other references cited therein). It is of interest to note that the first fluxional compound discovered was in fact a metallaborane.[187]

REFERENCES

1. Dickinson, W. C. *Phys. Rev.* 1951, **81**, 717.
2. Schaeffer, R. *Prog. Boron Chem.* 1964, **1**, 417–462.
3. Eaton, G. R.; Lipscomb, W. N. *NMR Studies of Boron Hydrides and Related Compounds*; Benjamin: New York, 1969.
4. Williams, R. E. *Prog. Boron Chem.* 1970, **2**, 90–118.
5. Nöth, H.; Wrackmeyer, B. *Nuclear Magnetic Resonance Spectroscopy of Boron Compounds*; Springer: Berlin, 1978.
6. Todd, L. J., Siedle, A. R. *Prog. NMR Spectrosc.* 1979, **13**, 87–176.
7. Siedle, A. R. *Annu. Rep. NMR Spectrosc.* 1982, **12**, 177–261
8. Kidd, R. G. in *NMR of the Newly Accessible Nuclei*, Laszlo, P., Ed.; Academic: London, 1983, Vol. 2, Chap. 3, pp. 49–77.
9. Kennedy, J. D. *Annu. Rep. NMR Spectrosc.*, to be published.
10. *Spectroscopic Properties of Inorganic and Organometallic Compounds*; The Chemical Society: London, yearly, 1968–present.
11. *Gmelin Handbuch, New Supplement Series, Boron Compounds*; Springer: Berlin, 1977 ff; *Caveat*: note that in some sections of this work the $[BF_3(OEt_2)]$ = zero, high frequency positive, chemical shift convention is not universally applied.
12. Eaton, G. R. *J. Chem. Ed.* 1969, **46**, 547–556; Smith, W. L. *J. Chem. Ed.* 1977, **54**, 469–473.
13. Wrackmeyer, B.; Köster, R. in *Organoborverbindung III (Houben-Weyl: Methoden der Organischen Chemie, Volume XIII/3c)*, Köster, R., Ed.; Georg Thieme: Stuttgart and New York, 1984, Sec. D, pp. 377–611.

14. McFarlane, W. *J. Magn. Reson.* 1973, **10**, 98–99.
15. See, for example, Akitt, J. W. *J. Magn. Reson.* 1970, **3**, 411–414.
16. Mann, B. E. in *NMR and the Periodic Table*, Harris, R. K.; Mann, B. E., Eds.; Academic: London, 1978, Chap. 4C, pp. 91–93.
17. Kasler, F.; Tierney, M. *Mikrochim. Acta.* 1981, **2**, 301–312.
18. Boocock, S. K.; Greenwood, N. N.; Kennedy, J. D.; McDonald, W. S.; Staves, J. *J. Chem. Soc., Dalton Trans.* 1980, 790–796.
19. Smith, W. L.; Meneghelli, B. J.; McClure, N.; Rudolph, R. W. *J. Am. Chem. Soc.* 1976, **98**, 624–626.
20. Akitt, J. W.; McDonald, W. S. *J. Magn. Reson.* 1984, **58**, 401–412.
21. Beall, H.; Elvin, A. T.; Bushweller, C. H. *Inorg. Chem.* 1974, **13**, 2031–2033.
22. Clouse, A. O.; Moody, D. C.; Rietz, R. R.; Roseberry, T.; Schaeffer, R. *J. Am. Chem. Soc.* 1973, **95**, 2496–2501.
23. Akitt, J. W.; Savory, C. G. *J. Magn. Reson.* 1975, **17**, 122–124.
24. See, for example, Gaines, D. F.; Nelson, C. K.; Kunz, J. C.; Morris, J. H.; Reed, D. *Inorg. Chem.* 1984, **23**, 3252–3254.
25. Anderson, J. A.; Astheimer, R. J.; Odom, J. D.; Sneddon, L. G. *J. Am. Chem. Soc.* 1984, **106**, 2275–2283.
26. Allerhand, A.; Clouse, A. O.; Rietz, R. R.; Roseberry, T.; Schaeffer, R. *J. Am. Chem. Soc.* 1972, **94**, 2445–2448.
27. Rietz, R. R.; Schaeffer, R. *J. Am. Chem. Soc.* 1973, **95**, 4580–4582.
28. Kennedy, J. D.; Greenwood, N. N. *Inorg. Chim. Acta.* 1980, **38**, 93–96.
29. Boocock, S. K.; Greenwood, N. N.; Kennedy, J. D.; McDonald, W. S.; Staves, J. *J. Chem. Soc., Dalton Trans.* 1981, 2573–2584.
30. Onak, T. *Inorg. Chem.* 1968, **7**, 1043.
31. Tucker, P. M.; Onak, T. *J. Am. Chem. Soc.* 1969, **91**, 6869–6871.
32. Kennedy, J. D.; Wrackmeyer, B. *J. Magn. Reson.* 1980, **38**, 529–535.
33. Howarth, O. W.; Kennedy, J. D. 1982–1984, unpublished observations.
34. Boocock, S. K.; Cheek, Y. M.; Greenwood, N. N.; Kennedy, J. D. *J. Chem. Soc., Dalton Trans.* 1981, 1430–1437.
35. Greenwood, N. N.; Kennedy, J. D.; Spalding, T. R.; Taylorson, D. *J. Chem. Soc., Dalton Trans.* 1979, 840–846.
36. Webb, G. A. in *NMR of the Newly Accessible Nuclei*, Laszlo, P., Ed.; Academic: London, 1983, Vol. 1, Chap. 3, pp. 79–101.
37. Armstrong, D. R.; Perkins, P. G. *Chem. Commun.* 1965, 337–338.
38. Chakrabarty, W. R.; Thompson, C. C.; Brey, W. S. *Inorg. Chem.* 1967, **6**, 518–520.
39. Hensen, K.; Messer, K. P. *Theoret. Chim. Acta.* 1967, **9**, 17–25.
40. Davis, F. A.; Dewar, M. J. S.; Jones, R. *J. Am. Chem. Soc.* 1968, **90**, 706–708.
41. Kroner, J.; Nölle, D.; Nöth, H. *Z. Naturforsch., Teil B*, 1973, **28B**, 416–425.
42. Kroner, J.; Wrackmeyer, B. *J. Chem. Soc., Faraday Trans. II*, 1976, **72**, 2283–2290.
43. Cheremisin, A. A.; Schastnev, P. V. *J. Magn. Reson.* 1980, **40**, 459–468.
44. Spielvogel, B. F.; Purser, J. M. *J. Am. Chem. Soc.* 1977, **93**, 4418–4426.
45. Nöth, H.; Wrackmeyer, B. *Chem. Ber.* 1974, **107**, 3070–3088.
46. Spielvogel, B. F.; Nutt, W. R.; Izydore, R. A. *J. Am. Chem. Soc.* 1975, **97**, 1609–1610.
47. McFarlane, W.; Wrackmeyer, B.; Nöth, H. *Chem. Ber.* 1975, **108**, 3831–3841.
48. Good, C. D.; Ritter, D. M. *J. Am. Chem. Soc.* 1962, **84**, 1162–1166.
49. Nöth, H.; Wrackmeyer, B. *Chem. Ber.* 1974, **107**, 3089–3103.
50. Wrackmeyer, B. *Z. Naturforsch., Teil B* 1980, **35B**, 439–446.
51a. Hall, L. W.; Lowman, D. W.; Ellis, P. D.; Odom, J. D. *Inorg. Chem.* 1975, **14**, 580–583, and references cited therein.
51b. Negrebetskii, V. V.; Bogdanov, V. S.; Kessenikh, A. V.; Petrovskii,. P. V.; Bubriov, Y. N.; Mikhailov, B. M. *Zh. Obschch. Khim.* 1974, **44**, 1882–1888.
52. Bishop, V. L.; Kodama, G. *Inorg. Chem.* 1981, **20**, 2724–2727.
53. Odom, J. D.; Zozulin, A. J. *Inorg. Chem.* 1981, **20**, 3740–3745.
54a. Durig, J. R.; Riethmiller, S.; Kalasinsky, V. F.; Odom, J. D. *Inorg. Chem.* 1974, **13**, 2729–2735.
54b. Dietz, E. A.; Morse, K. W.; Parry, R. W. *Inorg. Chem.* 1976, **15**, 1–7; (c) Cowley, A. H.; Damasco, M. C. *J. Am. Chem. Soc.* 1971, **93**, 6815–6821.
55. Kennedy, J. D.; McFarlane, W.; Pyne, G. S.; Wrackmeyer, B. *J. Chem. Soc., Dalton Trans.* 1975, 386–390.

56. Kennedy, J.D.; McFarlane, W.; Wrackmeyer, B. *Inorg. Chem.* 1976, **15**, 1299–1302.
57. Fussstetter, H.; Nöth, H.; Wrackmeyer, B.; McFarlane, W. *Chem. Ber.* 1977, **110**, 3172–3182.
58. Wrackmeyer, B. *J. Magn. Reson.* 1983, **54**, 174–176.
59. Clippard, P. H.; Taylor, R. C. *Inorg. Chem.* 1969, **8**, 2802–2804.
60. McFarlane, H. C. E.; McFarlane, W.; Rycroft, D. S. *J. Chem. Soc., Faraday II* 1972, **68**, 1300–1305.
61. Whipple, E. B.; Brown, T. H.; Farrar, T. C.; Coyle, T. D. *J. Chem. Phys.* 1965, **43**, 1841–1842.
62. Fieldhouse, S. A.; Peat, I. R. *J. Phys. Chem.* 1969, **73**, 275.
63. Bird, P. H.; Wallbridge, M. G. H. *J. Chem. Soc.* 1965, 3923–3928.
64. Boiko, G. N.; Dravchenko, S. E.; Semenenko, K. N. *Izv. Akad. Nauk. S.S.S.R. Ser. Khim.* 1981, 1199–1207.
65a. Powell, P.; Nöth, H. *Chem. Commun.* 1966, 637–638.
65b. Lehman, D. D.; Schriver, D. F. *Inorg. Chem.* 1974, **13**, 2203–2207.
66. Boden, N.; Gutowsky, H. S.; Hansen, J. R.; Farrar, T. C. *J. Chem. Phys.* 1967, **14**, 2849–2850.
67. Shore, S. G. in *Boron Hydride Chemistry*, Muetterties, E. L., Ed.; Academic: New York, 1975, Chap. 3, pp. 79–174.
68. Kennedy, J. D. *Prog. Inorg. Chem.* 1984, **32**, 519–679, and 1986, **34**, 211–434.
69. Housecroft, C. E.; Fehlner, T. P. *Adv. Organomet. Chem.* 1982, **21**, 59–114.
70. Williams, R. E.; Field, L. D., in *Boron Chemistry-4*, Parry, R. W.; Kodama, G., Eds.; Pergamon: Oxford, 1980, pp. 131–150.
71. Dobson, J.; Schaeffer, R. *Inorg. Chem.* 1968, **7**, 402–408.
72. Ellis, P. D.; Chou, Y.-C.; Dobosh, P. A. *J. Magn. Reson.* 1980, **39**, 529–532.
73. Boocock, S. K.; Greenwood, N. N.; Hails, M. J.; Kennedy, J. D.; McDonald, W. S. *J. Chem. Soc., Dalton Trans.* 1981, 1415–1429.
74. See, for example, Beckett, M. A.; Kennedy, J. D. *J. Chem. Soc., Chem. Commun.* 1983, 575–576, and Fontaine, X. L. R.; Kennedy, J. D. *J. Chem. Soc., Dalton Trans.*, to be submitted.
75. Boer, F. P.; Hegstrom, R. A.; Newton, M. D.; Potenza, J. A.; Lipscomb, W. N. *J. Am. Chem. Soc.* 1966, **88**, 5340–5342.
76. Marynick, D. S.; Lipscomb, W. N. *J. Am. Chem. Soc.* 1972, **94**, 8692–8699.
77. Marynick, D. S.; Lipscomb, W. N. *J. Am. Chem. Soc.* 1972, **94**, 8699–8706.
78. Onak, T.; Inman, W.; Rosendo, H.; Distefano, E. W.; Nurse, J. *J. Am. Chem. Soc.* 1977, **99**, 6488–6492.
79. Dolanský, J.; Heřmánek, S.; Zahradník, R. *Coll. Czech. Chem. Commun.* 1981, **46**, 2479–2493.
80. Jarvis, W.; Inman, W.; Powell, B.; Distefano, E. W.; Onak, T. *J. Magn. Reson.* 1981, **43**, 302–315.
81. Streib, W. E.; Boer, F. P.; Lipscomb, W. N. *J. Am. Chem. Soc.*, 1963, **85**, 2331.
82. Onak, T.; Marynick, D.; Mattshei, P.; Dunks, G. *Inorg. Chem.* 1968, **7**, 1754–1757.
83. Beckett, M. A.; Crook, J. E.; Greenwood, N. N.; Kennedy, J. D. *J. Chem. Soc., Dalton Trans.* 1986, 1879–1893.
84. Sprecher, R. F.; Carter, J. C. *J. Am. Chem. Soc.* 1973, **95**, 2369–2370.
85. Aufderheide, B. E.; Sprecher, R. F. *Inorg. Chem.* 1974, **13**, 2286–2287.
86. Sprecher, R. F.; Aufderheide, B. E. *Inorg. Chem.* 1974, **13**, 2287–2289.
87. Sprecher, R. F.; Aufderheide, B. E., Luther, G. W.; Carter, J. C. *J. Am. Chem. Soc.* 1974, **96**, 4404–4410.
88. Swicker, F. S. *Diss. Abs. Int. B* 1972, **32B**, 4474.
89. Colquhoun, I. J.; McFarlane, W., results communicated to INTRABORON I, Strathclyde, May 1980 and INTRABORON III, Leeds, September 1982; also *J. Chem. Soc., Dalton Trans.* 1981, 2014–2016.
90. Venable, T. L.; Hutton, W. C.; Grimes, R. N.; *J. Am. Chem. Soc.* 1982, **104**, 4716–4717.
91. Reed, D. *J. Chem. Res. (S)* 1984, 198–199.
92. Venable, T. L.; Hutton, W. C.; Grimes, R. N. *J. Am. Chem. Soc.* 1984, **106**, 29–37
93. Jacobsen, G. B.; Meina, D. G.; Morris, J. H.; Thomson, C.; Andrews, S. J.; Reed, D.; Welch, A. J.; Gaines, D. F. *J. Chem. Soc., Dalton Trans.* 1985, 1645–1654.
94. Brewer, C. T.; Grimes, R. N. *J. Amer. Chem. Soc.* 1984, **106**, 2722–2723.
95. Beckett, M. A.; Howarth, O. W.; Kennedy, J. D. *J. Chem. Soc., Chem. Commun.* 1985, 855–856.
96a. Jasztal, M., results communicated to INTRABORON IV, Durham, September 1984 and Howarth, O. W.; Jasztal, M. J.; Taylor, J. G.; Wallbridge, M. G. H. *Polyhedron* 1985, **4**, 1461–1466.
96b. Meina, D., results communicated to INTRABORON IV, Durham, September 1984.
97a. Corcoran, E. W.; Sneddon, L. G. *J. Am. Chem. Soc.* 1985, **107**, 7446–7450.
97b. Astheimer, R. J.; Sneddon, L. G. *Inorg. Chem.* 1983, **22**, 1928–1934; and other references cited therein.
98. Wiersema, R. J.; Hawthorne, M. F. *J. Am. Chem. Soc.* 1974, **96**, 761–770.

99. Morris, J. H.; Reed, D. *J. Chem. Res.* 1980, 4522.
100. Heřmánek, S.; Plešek, J. *Z. Anorg. Allg. Chem.* 1974, **409**, 115–120.
101. Greenwood, N. N.; Hails, M. J.; Kennedy, J. D.; McDonald, W. S. *J. Chem. Soc., Dalton Trans.* 1985, 953–972.
102. See, for example, Salentine, C. G.; Hawthorne, M. F. *Inorg. Chem.* 1978, **17**, 1498–1501.
103. Bould, J. *Metallaboranes of the Platinum Group Metals*, thesis, University of Leeds, 1983.
104. Fontaine, X. L. R., results communicated to INTRABORON V, Killin, September 1985, and Fontaine, X. L. R.; Kennedy, J. D. *J. Chem. Soc., Chem. Commun.* 1986, 779–781.
105. Crook, J.; Elrington, M.; Greenwood, N. N.; Kennedy, J. D.; Thornton-Pett, M.; Woolins, J. D. *J. Chem. Soc., Dalton Trans.* 1985, 2407–2415.
106. Wong, K. S.; Scheidt, W. R.; Fehlner, T. P. *J. Am. Chem. Soc.* 1982, **104**, 1111–1113.
107. Beer, D. C.; Burke, A. R.; Engleman, T. R.; Storhoff, B. N.; Todd, L. J. *J. Chem. Soc., Chem. Commun.* 1971, 1611–1612.
108. Todd, L. J.; Siedle, A. R.; Bodner, G. M.; Kahl, S. B.; Hickey, J. P. *J. Magn. Reson.* 1976, **23**, 301–311.
109. Boocock, S. K.; Bould, J.; Greenwood, N. N.; Kennedy, J. D.; McDonald, W. S. *J. Chem. Soc., Dalton Trans.* 1982, 713–719.
110. Bould, J.; Crook, J. E.; Greenwood, N. N.; Kennedy, J. D. *J. Chem. Soc., Dalton Trans.* 1984, 1903–1910.
111a. Burg, A. B. *J. Am. Chem. Soc.* 1968, **90**, 1407–1411.
111b. Warren, R.; Paquin, D.; Onak, T.; Dunks, G.; Spielman, J. R. *Inorg. Chem.* 1970, **9**, 2285–2287.
112. Siedle, A. R.; Bodner, G. M.; Garber, A. R.; Beer, D. C.; Todd, L. J. *Inorg. Chem.* 1974, **13**, 2321–2324.
113. Heřmánek, S.; Plešek, J.; Gregor, V.; Štíbr, B. *J. Chem. Soc., Chem. Commun.* 1977, 561–563.
114. Heřmánek, S.; Gregor, V.; Plešek, J.; Janousek, Z.; Antonovich, V. A. *Coll. Czech. Chem. Commun.* 1976, **41**, 1492–1499.
115. Srebny, H. G.; Preetz, W.; Marsmann, H. C. *Z. Naturforsch., Teil B* 1984, **39B**, 6–13 and 189–196.
116. Tucker, P. M.; Onak, T.; Leach, J. B. *Inorg. Chem.* 1970, **9**, 1430–1441.
117. See, for example, Crook, J. E.; Greenwood, N. N.; Kennedy, J. D.; McDonald, W. S. *J. Chem. Soc., Dalton Trans.* 1984, 2487–2495.
118. Ahmed, R. *Reactions of Substituted Decaboranes and Nonaboranes with Platinium (II) Compounds*, thesis, University of Leeds, 1982.
119. Davan, T.; Morrison, J. A. *Inorg. Chem.* 1979, **18**, 3194–3197.
120. Saulys, D.; Morrison, J. A. *Inorg. Chem.* 1980, **19**, 3057–3059.
121. Gaines, D. F.; Iorns, T. V. *J. Am. Chem. Soc.* 1970, **92**, 4571–4574.
122. Beckett, M. A., results communicated to INTRABORON 4, Durham, September 1984. See also Beckett, M. A.; Greenwood, N. N.; Kennedy, J. D.; Thornton-Pett, M. *J. Chem. Soc., Dalton Trans.* 1986, 795–801.
123. Fontaine, X. L. R.; MacKinnon, P.; Salter, P., unpublished observations 1984–1987, University of Leeds.
124. Bould, J.; Greenwood, N. N.; Kennedy, J. D. *J. Chem. Soc., Dalton Trans.* 1984, 2477–2485.
125. Bown, M., unpublished observations 1984–1985.
126. Wright, W. F.; Garber, A. R.; Todd, L. J. *J. Magn. Reson.* 1978, **30**, 595–602.
127. Garret, P. M.; Ditta, G.S.; Hawthorne, M. F. *J. Am. Chem. Soc.* 1971, **93**, 1265–1266.
128. Štíbr, B.; Heřmánek, S.; Janousek, Z.; Plzák, Z.; Dolanský, J.; Plešek, J. *Polyhedron* 1982, **1**, 822–824.
129. Brice, V. T.; Johnson, H. D.; Shore, S. G. *J. Am. Chem. Soc.* 1973, **95**, 6629–6635.
130. Groszek, E.; Leach,. J. B.; Wong, G. T. F.; Ungermann, C.; Onak, T. *Inorg. Chem.* 1971, **10**, 2770–2775.
131. Spielman, J. R.; Dunks, G. B.; Warren, R. *Inorg. Chem.* 1969, **8**, 2172–2174.
132. Howard, J. W.; Grimes, R. N. *Inorg. Chem.* 1972, **11**, 263–267.
133. Miller, V. R.; Grimes, R. N. *Inorg. Chem.* 1972, **11**, 862–865.
134. Vickers, G. D.; Agahigian, H.; Pier, E. A.; Schroeder, H. *Inorg. Chem.* 1966, **5**, 693–696.
135. Onak, T.; Leach, J. B.; Anderson, S.; Frisch, M. J. *J. Magn. Reson.* 1976, **23**, 237–248.
136. Jarvis, W.; Abdon, Z. J.; Onak, T. *Polyhedron* 1983, **2**, 1067–1070.
137. Moody, D. C.; Schaeffer, R. *Inorg. Chem.* 1976, **15**, 233–236.
138. Jacobsen, G. B.; Morris, J. H.; Reed, D. *J. Chem. Soc., Dalton Trans.* 1984, 415–421.
139. Calabrese, J. C.; Fischer, M. P.; Gaines, D. F.; Lott, J. W. *J. Am. Chem. Soc.* 1974, **96**, 6318–6323.
140. Doi, J. A.; Teller, R. G.; Hawthorne, M. F. *J. Chem. Soc., Chem. Commun.* 1980, 80–82.
141. Elrington, M.; Greenwood, N. N.; Kennedy, J. D.; Thornton-Pett, M. *J. Chem. Soc., Chem. Commun.* 1984, 1398–1399.

142. See, for example, Heřmánek, S.; Fetter, K.; Plešek, J.; Todd, L. J.; Garber, A. R. *Inorg. Chem.* 1975, **14**, 2250–2253.
143. Hopkins, R. C.; Baldeschwieler, J. D.; Schaeffer, R.; Tebbe, F. N.; Norman, A. *J. Chem. Phys.* 1965, **43**, 975–983.
144. Leach, J. B.; Onak, T. *J. Magn. Reson.* 1971, **4**, 30–39.
145. Zozulin, A. J.; Jakobsen, H. J.; Moore, T. F.; Garber, A. R.; Odom, J. D. *J. Magn. Reson.* 1980, **41**, 458–466.
146. Lowman, D. W.; Ellis, P. D.; Odom, J. D. *Inorg. Chem.* 1973, **12**, 681–685.
147. Bruker promotional literature, ca. 1978, depicting $^{11}B-\{^1H\}$spectrum of $B_{10}H_{14}$.
148. Gaines, D. F.; Nelson, C. K.; Kunz, J. C.; Morris, J. H.; Reed, D. *Inorg. Chem.* 1984, **23**, 3252–3254.
149. Anderson, J. A.; Astheimer, R. J.; Odom, J. D.; Sneddon, L. G. *J. Am. Chem. Soc.* 1984, **106**, 2275–2283.
150. Beall, H.; Elvin, A. T.; Bushweller, C. H. *Inorg. Chem.* 1974, **13**, 2031–2033.
151. Kennedy, J. D.; Staves, J. *Z. Naturforsch., Teil B* 1979, **34B**, 808–813.
152. Nam, W.; Soltis, M.; Gordon, C.; Lee, S.; Onak, T. *J. Magn. Reson.* 1984, **59**, 399–405; and other references cited therein.
153. Onak, T.; Wan, E. *J. Chem. Soc., Dalton Trans.* 1974, 665–669; *J. Magn. Reson.* 1974, **14**, 66–74.
154a. Garber, A. R.; Bodner, G. M.; Todd, L. J.; Siedle, A. R. *J. Magn. Reson.* 1977, **28**, 383–390.
154b. Todd, L. J.; Little, J. L.; Silverstein, H. T. *Inorg. Chem.* 1969, **8**, 1698–1703.
155. Beckett, M. A.; Kennedy, J. D. *J. Chem. Soc., Dalton Trans.* to be submitted.
156. Kennedy, J. D.; McFarlane, W.; Reed, D.; Colquhoun, I. C.; Rogozinski, J., unpublished observations 1978–1984.
157. McInnes, Y. M.; Kennedy, J. D., unpublished observations 1982–1984.
158. Wrackmeyer, B., unpublished observations 1983–1984; personal communication.
159. Gaines, D. F.; Coleson, K. M.; Hillenbrand, D. F. *J. Magn. Reson.* 1981, **44**, 84–88.
160. Sayer, B. G.; Tompson, J. I. A.; Hao, N.; Birchall, T.; Eaton, D. R.; McGlinchey, M. J. *Inorg. Chem.* 1981, **20**, 3748–3750.
161. Stampf, E. J.; Garber, A. R.; Odom, J. D.; Ellis, P. D. *J. Am. Chem. Soc.* 1976, **98**, 6550–6554.
162. Weiss, R.; Grimes, R. N. *J.. Am. Chem. Soc.* 1978, **100**, 1401–1405.
163. Gibb, T. C.; Kennedy, J. D. *J. Chem. Soc., Faraday Trans. II* 1982, **78**, 525–536.
164. Allerhand, A.; Odom, J. D.; Moll, R. E. *J. Chem. Phys.* 1969, **50**, 5037–5039.
165. Boocock, S. K.; Greenwood, N. N.; Kennedy, J. D. *J. Chem. Research (S)* 1981, 50–51.
166. Bogdanov, V. S.; Kessenikh, A. V.; Negrebetsky, V. V. *J. Magn. Reson.* 1971, **5**, 145–150; Miller, V. R.; Grimes, R. N. *Inorg. Chem.* 1977, **16**, 15–20; Weiss, R.; Grimes, R. N. *J. Am. Chem. Soc.* 1977, **99**, 1036–1042.
167. Grace, M.; Beall, H.; Bushweller, C. H. *Chem. Commun.* 1970, 701.
168. Rogozinski, J. *A Nuclear Magnetic Resonance Study of Polyhedral Boranes and Metallaboranes*, thesis, University of Leeds, 1984.
169. Greenwood, N. N.; Kennedy, J. D.; Thornton-pett, M.; Woollins, J. D. *J. Chem. Soc., Dalton Trans.* 1985, 2397–2406.
170. Bould, J.; Greenwood, N. N.; Kennedy, J. D. *J. Organometal. Chem.* 1983, **249**, 11–21.
171. Hosmane, N. S.; Grimes, R. N. *Inorg. Chem.* 1979, **18**, 2886–2891.
172. Fowkes, H.; Kennedy, J. D.; Greenwood, N. N.; Thornton-Pett, M. *J. Chem. Soc., Dalton Trans.* 1986, 517–523.
173. Wrackmeyer, B. *Prog. NMR Spectroscopy* 1979, **12**, 227–259.
174. Green, M.; Spencer, J. L.; Stone, F. G. A. *J. Chem. Soc., Dalton Trans.* 1979, 1679–1686.
175. Baker, R. T.; Hawthorne, M. F.; Long, J. A.; Marder, T. B.; *et al. J. Am. Chem. Soc.* 1984, **106**, 2965–2978, 2979–2989, and 3004–3110.
176. Kennedy, J. D. *Proc. XIX Intern. Conf. Coord. Chem., Prague* 1978, **1**, 79–81.
177. Greenwood, N. N.; Kennedy, J. D.; Staves, J. *J. Chem. Soc., Dalton Trans.* 1978, 1146–1152.
178. Grimes, R. N. in *Comprehensive Organometallic Chemistry*, Wilkinson, G.; Stone, F. G. A.; Abel, E., Eds.; Pergamon: Elmsford, N. Y., 1982, Part I, Chap. 5.5, pp. 459–542.
179. Jesson, J. P.; Muetterties, E. L. in *Dynamic Nuclear Magnetic Resonance Spectroscopy*, Jackman, L. M.; Cotton, F. A., Eds.; Academic: New York, 1975, pp. 294–299.
180. Beckett, M. A.; Greenwood, N. N.; Kennedy, J. D.; Thornton-Pett, M. *J. Chem. Soc., Dalton Trans.* 1985, 1119–1129.
181. Bushweller, C. H.; Beall, H.; Grace, M.; Dewkett, H. J.; Bilofsky, H. S. *J. Am. Chem. Soc.* 1971, **93**, 2145–2149.

182. Middaugh, R. L.; Wiersema, R. J. *Inorg. Chem.* 1971, **10**, 423–424; Tolpin, E. I.; Lipscomb, W. N. *J. Am. Chem. Soc.* 1973, **95**, 2384–2386.
183. Gaines, D. F.; Hildebrandt, S. J. in *Metal Interactions with Boron Clusters*, Grimes, R. N., Ed.; Plenum: New York, 1982, Chap. 3, pp. 119–143.
184. Bould, J.; Greenwood, N. N.; Kennedy, J. D. *J. Chem. Soc., Dalton Trans.* 1982, 481–483.
185. Bould, J.; Crook, J. E.; Greenwood, N. N.; Kennedy, J. D.; McDonald, W. S. *J. Chem. Soc., Chem. Commun.* 1982, 346–348.
186. Hawthorne, M. F. in *Chemistry for the Future*, Grünewald, H., Ed.; Pergamon: Oxford, 1984, pp. 135–141, and references cited therein.
187. Ogg, R. A.; Ray, J. D. *Disc. Farad. Soc.* 1955, **19**, 239.
188. Mancini, M.; Bougeard, P.; Burns, R. C.; Mlekuz, M.; Sayer, B. G.; Thompson, J. I. A.; McGlinchy, M. J. *Inorg. Chem.* 1984, **23**, 1072–1078.
189. Beckett, M. A.; Bown, M.; Fontaine, X. L. R.; Greenwood, N. N.; Kennedy, J. D.; Thornton-Pett, M. *J. Chem. Soc., Dalton Trans.* 1987, submitted.
190. Brewer, C T.; Grimes, R. N. *J. Am. Chem. Soc.* 1984, **106**, 2722–2733.
191. Elrington, M.; Greenwood, N. N.; Kennedy, J. D.; Thornton-Pett, M. *J. Chem. Soc., Dalton Trans.* 1987, 451–456.
192. Barrett, S.A.; Greenwood, N. N.; Kennedy, J. D.; Thornton-Pett, M. *Polyhedron* 1985, **4**, 1981–1984.
193. Hosmane, N. S.; Sirmokadam, N. N. *Organometallics* 1984, **3**, 1119–1121.
194. Hosmane, N. S.; Dehghan, M.; Davies, S. *J. Am. Chem. Soc.* 1984, **106**, 6435–6436.
195. Hosmane, N. S.; Sirmokadam, N. N.; Herber, R. H. *Organometallics* 1984, **3**, 1665–1669.
196. Fontaine, X. L. R.; Fowkes, H.; Greenwood, N. N.; Kennedy, J. D.; Thornton-Pett, M. *J. Chem. Soc., Chem. Commun.* 1985, 1165–1166 and 1722–1723; *J. Chem. Soc., Dalton Trans.* 1986, 547–552, and 1987, in press.
197. Beckett, M. A.; Fontaine, X. L. R.; Thornton-Pett, M., *et al.*, to be submitted.
198. Bent, H. A. *Chem. Rev.* 1961, **61**, 275–311.
199. See, for example, Shoolery, J. N. *J. Chem. Phys.* 1959, **31**, 1427; and Grant, D. M.; Litchman, W. M. *J. Am. Chem. Soc.* 1965, **87**, 3994–3995.
200. Jameson, C. J.; Gutowsky, H. S. *J. Chem. Phys.* 1969, **51**, 2790–2803.
201. Extension and refinement of these rules have very recently been reported: Heřmánek, S.; Fusek, J.; Štíbr, B.; Plešik, J.; Jelínek, T. *Polyhedron* 1986, **5**, 1873–1879.

Chapter 9

ALUMINUM, GALLIUM, INDIUM, AND THALLIUM

J. W. AKITT

These four group III elements are all magnetically active and have been extensively studied. Thallium possesses two spin-1/2 nuclei but all the others are quadrupolar. It is therefore convenient to consider the first three elements as a group and then turn to the rather different properties of thallium NMR.

1. THE NUCLEAR PROPERTIES OF Al, Ga, AND In, THE QUADRUPOLAR NUCLEI

These are summarized in Table 1 to allow easy comparison between them. The width factor in the table shows how the linewidths of "identical" compounds of the elements should be related for each type of nucleus in the absence of chemical differences. (See the equation for the rate of quadrupole relaxation in Chapter 5.) The ratio of the receptivity to this figure then gives a number which indicates how the signal strengths of each type of nucleus might vary. Clearly, the ease of observation is in the order Al > Ga ≈ In, with ^{71}Ga (the gallium nucleus of lower natural abundance) and ^{115}In being the preferred gallium and indium nucleides to use. Table 1 also demonstrates how a high spin can offset the broadening influence of even a very high quadrupole moment.[1]

2. ALUMINUM

The element aluminum-27 occurs at 100% natural abundance and its nucleus has a good receptivity, though its quadrupole moment means that its resonances are broadened relative to those of spin-1/2 nuclei. However, this amounts to only a few hertz or tens of hertz in many cases, with the result that ^{27}Al has been considered worthy of study from the early days of NMR spectroscopy and has already been the subject of three reviews[2-4] and figures in a fourth,[5] and another is to appear.[259] Historically, the first observations were carried out in the derivative mode and with poor quality magnets, which were deemed sufficient for a quadrupolar nucleus; but as increasingly improved instrumentation became available, high-resolution absorption

J. W. AKITT • School of Chemistry, University of Leeds, Leeds LS2 9JT, England.

Table 1. Summary of the Properties of the Quadrupolar Nuclei ^{27}Al, ^{69}Ga, ^{71}Ga, ^{113}In, ^{115}In

Nucleus	^{27}Al	^{69}Ga	^{71}Ga	^{113}In	^{115}In
NMR frequency (MHz)	26.057	24.00	30.495	21.867	21.914
Quadrupole moment, $Q/10^{-28}$ m^2	0.149	0.178	0.112	1.14	1.16
$\frac{2I+3}{I^2(2I-1)}$	0.32	1.33	1.33	0.074	0.074
Width factor (Al = 1)[a]	1.0	5.93	2.34	13.5	14.0
Relative intensity[b]	0.206	0.007	0.024	0.0011	0.0237

[a] Width factor = $Q^2(2I+3)/I^2(2I-1)$ and is the nuclear contribution to quadrupole relaxation.
[b] Given by (receptivity/width factor).

spectra with locked fields started to show much more detail and multiple resonances. Linewidths are such, however, that much overlap occurs, and it was already apparent in 1971 that the then new superconducting magnets would prove of particular benefit in ^{27}Al spectroscopy.[6] Thirteen years later we are able to show some highly informative examples of spectra obtained with the most advanced instrumentation and which might well be described as high-resolution ^{27}Al spectra.

2.1. Operational Techniques

The minimum ^{27}Al linewidth so far observed is 2.0 Hz and the largest, in solution, some 60 kHz. The longest relaxation time to be encountered is then of the order of 160 ms so that the FT mode of data collection is preferred. Rapid rates of pulsing using 90° pulses are acceptable and the data accumulate extremely rapidly in memory, perhaps 20 s sufficing to provide 1000 transients. The FT method does, however, limit the width of the widest line that can be observed by a given equipment. In illustration we point out that if there is a receiver dead time of 20 μs then the apparent starting amplitude of a broad line decay will be reduced. A line of 15 kHz width will retain only a third of its real amplitude, and should this be mixed with transmitter breakthrough then a rolling base line may result while the detectable intensity in the next location will only be one-ninth. For many spectrometers the maximum observable linewidth seems to be near 5 kHz, though the newer, high-field spectrometers are better since they must by definition have much wider receiver bandwidths to be effective, and have larger maximum sweepwidths.

Many ^{27}Al spectra have now been reported which contain several chemically shifted resonances of a wide range of linewidths, and this also brings problems. The first is that it is not possible to be sure that all the resonances are narrow enough to observe on a given equipment. It is thus always wise to check, at least on some representative samples, that the FID initial intensity has the value to be expected for that concentration of aluminum, by comparison with, for example, an aqueous salt solution of known concentration. A useful aid to this end is an ability to obtain a numerical printout of the numbers in the first few memory locations.[7,8] A second danger is that despite the short relaxation times, it is possible to pulse too rapidly and partially saturate the narrower resonances, and this should not be ignored where quantitative data are required. A third problem arises with phase correction, which is particularly acute where some broad components are present, and/or base line roll. The latter may be reduced by a left shift of data, thus losing the offending data points, but this will probably lose most of the broad component also, and introduces a first-order phase error. Unfortunately, a first-order phase correction routine, where the correction varies continuously across the memory block, is not compatible with broad lines and introduces distortion of their shapes, thus confusing the phase adjustment.(9) Ideally, the *whole* of each line should receive the same phase adjustment and the usual procedure works well simply because most high-resolution lines can be regarded as delta functions. It is thus always best where the resonances to be recorded are broad, to set up the spectrometer with the delay between pulse and start of data collection optimized so as to minimize or avoid completely any need to make a first-order phase correction. The relevant theory, together with a method for reducing that perennial problem of base-line roll, has been discussed recently,(10) and suitable conditions can often be found empirically.

A note on sensitivity should also be given in practical terms. Both resonances in an aqueous solution of aluminum sulfate can be observed at an aluminum concentration of M/800 after 24 000 scans. Both lines are broadened owing to self hydrolysis and the weaker one corresponds to a concentration of 0.0001 *M*. The spectrometer used in this case was a Bruker HX 400 operating at 104.2 MHz.

2.2. Aluminum-27 NMR Parameters

By far the major effort in the field has gone into studies of chemical shift effects, and these will be studied first. Many of the older data have now been superceded, and where this is the case will not be referred to, an adequate cover being available in the earlier reviews.(2-4) Quite an impressive list of coupling constants has now been accumulated and these follow the shift/linewidth data with the few relaxation data last.

Aluminum will interact in solution strongly with any molecule which contains a donor atom so that the species present in a given system will depend on the solvent, the aluminum compound dissolved in it, and the nature of any other component codissolved, and indeed there are now reports showing that eight or more aluminum-containing species can be detected in some samples. The point has therefore been reached in aluminum NMR where it becomes convenient to present the data in a series of tables dealing with different solvents and ligands rather than to give a table ordered simply by chemical shift.

Broadly speaking, the chemical shift range observed can be separated into three regions: (a) alkylaluminum compounds 150 ppm and more to low field of the reference, $Al(H_2O)_6^{3+}$; (b) tetrahedrally coordinated aluminum with δ between 140 and 40 ppm; (c) octahedrally coordinated aluminum between 40 and -46 ppm. There are some exceptions to this generalization, notably AlI_4^-, which resonates at -26.7 ppm, but the shift is often a good indicator of coordination number. There are other recently noted exceptions and five-coordinated aluminum is said to resonate between the octahedral and tetrahedral regions; we will return to this point after we have considered the data.

The linewidths which are presented are a function of temperature, of concentration, and, in aqueous solution, of pH. The values given therefore are either a particular experimental value (and may not be particularly accurate, especially if estimated from publications—linewidth data are often not explicitly stated) or the extreme of a range, which may be quite wide. $Al(H_2O)_6^{3+}$, for instance, has a linewidth which may lie between 2 and at least 100 Hz, depending on experimental conditions. It is in fact likely that these changes may be a source of valuable information about the chemistry of these solutions, once we know the nature of the species involved. One source of broadening is likely to occur where a solution contains a component with a very broad, nonvisible line which is exchanging with the visible species. The width of the line observed will then depend in a complex way on the relevant equilibrium and the way this is affected by the solution parameters. The principal relaxation mechanism of any species will of course depend on the electric field gradient set up by the arrangement of ligands and the electron imbalance in the bonds to the aluminum. It will be observed, however, that complex stereochemistry is not a good indicator of when a given line may be broad or narrow, and some apparently nonsymmetrical species have remarkably narrow lines.(11)

The rather large charge-to-radius ratio of Al^{3+} means that its interaction with donors is exceptionally strong and ligand exchange is slow on the NMR time scale. A well-documented case is $Al(H_2O)_6^{3+}$, for which the rate of whole water molecule exchange is $0.6\ s^{-1}$ at 298 K though the protons independently exchange much more rapidly ($\sim 10^5\ s^{-1}$). This of course accounts for the many species visible together in these solutions and also means that spin–spin coupling, usually only over one bond, may be visible between aluminum and a ligand nucleus, provided the aluminum relaxation rate is small compared to $1/J$.

2.2.1. Aqueous Solutions

The Hexa-aquo Cation and its Hydrolysis. The relatively sharp resonance of $Al(H_2O)_6^{3+}$ is visible in all aqueous solutions of aluminum salts, is therefore easily accessible, and has been used as reference right from the start of aluminum-27 NMR. The only contender is the resonance of $Al(OH)_4^-$, but this line is a little broader and a suitable solution needs a degree of preparation. The aquo cation therefore remains the most suitable reference. Its use in this way, however, assumes that its chemical shift is independent of the way a salt solution is made up, or of which salt is used. While this certainly appears to be the case in practice, it is an assumption which, to the author's knowledge has never been tested. There is some evidence of small shifts in aluminum sulfate solutions,(12) but diagmagnetic susceptibility changes make it very difficult to measure the shifts in different salt solutions with sufficient precision. We have recently applied Becconsall's method to this problem at Leeds,(13) and find a small shift of 0.2 ppm between $AlCl_3$ and $Al_2(SO_4)_3$ solutions, the former being to high field of the latter. This small difference hardly exceeds two linewidths and so may be neglected for most purposes, but its existence may provide more information about interactions in these solutions.

The fact that the linewidth of $Al(H_2O)_6^{3+}$ varies with solution conditions has already been mentioned. It becomes broader if a concentrated solution is heated, but is narrowed if acid is added.(14) It is also broadened in the presence of the dimeric cation, especially if the solution is heated when it may not be detectable in older equipment.(15,16) It has now been discovered that in very dilute, pure salt solutions the line becomes very much broadened, a value of 150 Hz being obtained for 0.001 mol liter^{-1} $AlCl_3$*. High-field spectra at high temperatures show that the dimeric cation is very long lived, and the broadening has been shown to occur because of fast proton exchange between $Al(H_2O)_6^{3+}$ and $AlOH^{2+}$. A formation constant for the latter species is obtained in good agreement with earlier pH titration methods, and only the one hydrolysate is detected.[17] It is clear that the linewidth is determined by this autohydrolysis reaction even in quite concentrated and therefore acidic solutions. Earlier studies of the relaxation rate of ^{27}Al in $Al(H_2O)_6^{3+}$ propose a number of interpretations which unfortunately ignore these hydrolysis effects.[18-21] Even in dilute, acidic solution, we must expect the 2-Hz minimum linewidth observed to contain contributions from dipolar and quadrupolar relaxation; we should not neglect spin rotation relaxation; but above all we should be beware still of chemical exchange effects.

The study of the hydrolysis of Al(III) has in fact attracted interest for over a century, and while there was general agreement about the self-hydrolysis step discussed above, some two dozen species had been suggested to account for the changes seen when hydrolysis was forced by adding base. Aluminum-27 NMR has enabled significant progress to be made in this area. Early work[7,22,23] showed resonances which could be ascribed to a dimeric cation $(H_2O)_4Al(\mu\text{-}OH)_2Al(H_2O)_4^{4+}$ and a pseudospherical cation $AlO_4Al_{12}(OH)_{24}(H_2O)_{12}^{7+}$ which had both been characterized by x-ray crystallography. Fast proton exchange meant that they could not be detected by proton NMR.[24] Subsequent improvements in instrumentation have allowed it to be demonstrated that the result of a hydrolysis of aluminum is very dependent upon time and conditions of hydrolysis. A rapid reaction, or one carried out at low temperature, seems to proceed to "Al_{13}" via the dimer and involve no other species. Slow reaction at high temperature produces quite different products which seem to arise through decomposition and/or rearrangement of the Al_{13} first formed.[8,25-28] Similar studies have been reported on more dilute solutions,[29,30] though the method of hydrolysis differs from those used above and the low concentration means that any conclusions have to be based on a computer analysis of the intensities of the $Al(H_2O)_6^{3+}$ and AlO_4 (of the Al_{13}) resonances. This leads the authors to believe that no dimer is present though high field spectra have now demonstrated that the dimer, or at least the resonance ascribed to it, really does exist.[31] It has also been suggested that the (commercial?) chlorohydrate contains principally the Al_{13} polymer,[32,33] though comparison of the spectra with those reported above indicates that this is not the sole constituent. Hydrolysis also occurs when an aluminum salt is mixed with the salt of a weak acid, whose anion is protonated by the strong acid $Al(H_2O)_6^{3+}$. Thus a dimer is formed exclusively in the presence of sodium acetate[15,31] and is presumably the precursor of the di-hydroxotetraacetate which eventually precipitates from such solutions, a reaction which presumably explains some earlier results.[34] Hydrolysis can be reversed by adding acid. The way in which the tridecameric cation decomposes when acid is added has been studied by stopped flow ^{27}Al NMR spectroscopy.[35] The resonance of the tetrahedral Al in the cation was observed to disappear more rapidly than the aquo cation appeared and a two-stage decomposition is proposed.

Other Aqueous Systems. Several reports have appeared where bidentate ligands have been attached to aluminum either as *tris* complexes or as partially aquated complexes. Of particular interest is the use of ^{27}Al to study the macrocyclic

* See also Refs. 29, 30.

allumochromophores[36] which contain three hydroxamate ligands and give ^{27}Al resonances with a shift intermediate between the usual octahedral and tetrahedral ranges at 41.5 ppm. The *tris* acetylhydroxamate resonates close by at 36.55 ppm (at pH 7), so the shift appears to be a property of the hydroxamate ligand. The shift is attributed to the different electron densities on the two ligand atoms, i.e., to a distortion of the octahedral field. Decreasing the pH causes four resonances to be resolved which are assigned to complexes containing increasing numbers of water molecule ligands, and whose shifts lie progressively to high field. The *tris* oxalato complex has also been observed at 16.1 ppm, at much higher field than the hydroxamates. Replacement of oxalates by pairs of water molecules also causes the shifts to move high field, but their spread is now much less.[37] The resonance attributed to bidentate acetate complexes of the dihydroxodialuminum cation is observed at 38 ppm, in the same region as the hydroxamate complexes, and it is suggested that the small chelate rings may cause the low field shift.[15]

Complexes with phosphate, polyphosphates, and ADP/ATP have been studied.[38–40] The first paper reports two resonances at -2 to -4 and -5 to -7 ppm in most of the solutions observed.[38] The other two deal with the ATP/Al^{3+} system and use a multinuclear approach to determine the species present.[39,40] The cation binds to the β and γ phosphates of ATP and 2:2 and 2:1 complexes are formed, the latter with a ^{27}Al linewidth of typically 2400 Hz at ca. 0 ppm and the former with a 600-Hz-wide line at -10 ppm. There are clear differences between these two groups of workers.

Interactions with lactate, citrate and EDTA over a wide pH range give interesting spectral changes.[41] EDTA gives a complex with a broad line at 32 ppm whose linewidth increases with pH. In the presence of citrate the chemical shift varies with pH from 8 ppm at pH 2, 10 ppm at pH 3, to 12 ppm at pH 5–8. The spectra obtained in the presence of lactate are dependent upon the way the solutions are mixed and the chemical shifts of the resonances are pH dependent, as is also the number of resonances observed. Peaks are distinguished at about 14, 24, 53, 60, and 80 ppm, the latter at high pH.

The possibility of complexation of the hexa-aquo cation by chloride has been discussed,[20] and ^{27}Al relaxation studies suggest that close encounter complexes occur between hexa-aquo aluminum and hexa-aquo nickel cations in mixed solutions.[42] Complexation by fluoride has also received attention,[43] and it is shown that the chemical shifts of the fluoro species are small, measurements at high field giving 0.7 ppm for AlF^{2+} and 1.3 ppm for AlF_2^+.[17] The sulfate complex has also been extensively studied, combining ^{27}Al NMR with molar volume measurements. The anions and cations are closely associated in the second sphere, there being essentially no free Al^{3+}, and a small proportion of the anions enters the first sphere of the cations. The proportion so complexed increases with temperature, suggesting how alunite may be formed by heating at 250°C under pressure. The exchange processes present are studied in detail using the ^{27}Al.[12] Imidazolidine-2-one complexes Al^{3+} in aqueous solution and three resonances are observed at 0, -2, -3.6, and -5 ppm.[44] The ligand is presumably O bonded to give 1:1, 1:2, and 1:3 complexes, which undergo exchange among themselves and only slowly with the aquo cation.

In alkaline solution it was believed that only $Al(OH)_4^-$ is formed,[45,46] though some evidence has recently been obtained that other species predominate in certain highly alkaline aluminate solutions.[47] Interaction of $Al(OH)_4^-$ in alkaline TiO_2 suspension has been investigated.[48] Multiple resonances are seen in aluminosilicate solutions corresponding to different numbers of SiO units attached to the Al, the increasing substitution giving an increasing high-field shift.[49,50] It is claimed that $Al(acac)_3$ acts as a catalyst for silanol disproportionation via the formation of SiO–Al bonds, but no shifts are seen in this case so that the association may be less definate.[51]

The chemical shift data for these and other papers[52–56] will be found in Table 2, including some new data for heteropolyanions.

2.2.2. Alcoholic Solutions

Solutions of aluminum salts in alcohols have been widely studied using ^1H, ^{13}C, and ^{27}Al spectroscopy.[57–70] The early ^{27}Al results are consistent with the Al^{3+} cation being in an octahedral environment but the proton and carbon spectra suggest that the coordination number is less than six in the halide solutions and is six or a little less in the perchlorate solutions. The existence of ion pairing in $AlCl_3$ solution is thus postulated though there is a difference between the views of this taken by the Russian group, who seem to prefer solvent separated ion pairs and the other workers who

Table 2. Aluminum-27 Chemical Shifts in Aqueous Solutions

Compound	Chemical shift (ppm)	Linewidth (Hz)	Reference
ATP complexes	−10.1, −3.3, −0.9	700	38–40
Phosphate complexes	−7.7, −5, −3, −2	100	38, 52
$CH_2NHCONHCH_2$ complexes	−5, −3.6, −2		44
H_2SO_4 complex	−6.8	80	53
$Al(H_2O)_5(SO_4)^+$	−3.3	5	12, 53, 54
$Al(H_2O)_6^{3+}$ Reference	0.0	2	2, 17
$Al(OH)_6^{3-}$	0.0	Narrow	47
AlF^{2+}	0.7	620	17, 43, 55
AlF_2^+	1.3	800	17, 43, 55
$AlO_4Al_{12}(OH)_{24}(H_2O)_{12}^{7+}$ octahedra	12	Broad	31
$Al_2(\mu\text{-}OH)_2(OH)_n(H_2O)_{8-n}^{(4-n)+}$	4.2	500	31
Octahedral Al in polycations	6–12	Broad	27, 31
Citrate or lactate complexes	7–22		41
$(NH_4)_3\,AlO_6Mo_6O_{15}\cdot 10H_2O$	15.4	50	56
$Al(C_2O_4)_n(H_2O)_{2(3-n)}^{(3-2n)+}$	16.1, $n = 3$	150	
	13.1, $n = 2$	Broad	37
	6.3, $n = 1$	Broad	
Al(EDTA)	32		41
$Al(ONHCOCH_3)_3$	36.55	370	36
Acetate complexes of dimer	38	5000	15
Al(Alumichrome)	41.55	Broad	36
$AlO_4Al_{12}(OH)_{24}(H_2O)_{12}^{7+}$ central	62.5	20	8, 31
Tetrahedral Al in polycations	70–75	Broad	27, 31
$AlW_{12}O_{40}^{x-}$, central Al	71.2, 71.7	8, 2.5	56
Possible Al_2W_{18} anion	73.1	100	56
$Al(OH)_4^-$	80	25	45, 46
$AlO_{4-n}(OSi)_n$ in aluminosilicates	79.5, $n = 0$		
	74.3, $n = 1$		
	69.5, $n = 2$		49, 50
	64.2, $n = 3$		

favor contact pairs. More recent ^{27}Al spectra show a major resonance at about 10 ppm[67,71] which is slightly broadened and may be used to support either thesis. It could be due either to a solvated chloro complex or the hexasolvate. The problem could be resolved if we could detect a resonance near 0 ppm, even if weak. This will, of course, overlap with the main line at low fields, so we need recourse to a high-field instrument. A spectrum obtained at 104.2 MHz from a solution of $AlCl_3$ in anhydrous methanol is shown in Figure 1. This shows a resonance at about 0 ppm but it is broad.[68] It is concluded that the main species present is $AlCl_2(MeOH)_4^+$, and that the high-field line is broad because of exchange of hexasolvate with $AlCl^{2+}$. Diluting the solutions with $CDCl_3$ causes the high-field line to narrow appreciably, thus supporting this hypothesis.[70] The higher alcohols give broader lines in general and the results are less useful than with methanol, proton spectroscopy then being the preferred technique. Addition of water causes chloride to be lost and mixed solvates to be formed, *fac* $Al(H_2O)_3(ROH)_3^{3+}$ giving an obvious, narrow line in MeOH and EtOH.[68,70] Evidence is also obtained for complexing by the perchlorate anion (see Table 3).

2.2.3. *Solutions in Methyl Cyanide*

These solutions have also been extensively studied both anhydrous and with added water[72–85] or alcohol,[66] and while the systems now seem well characterized,

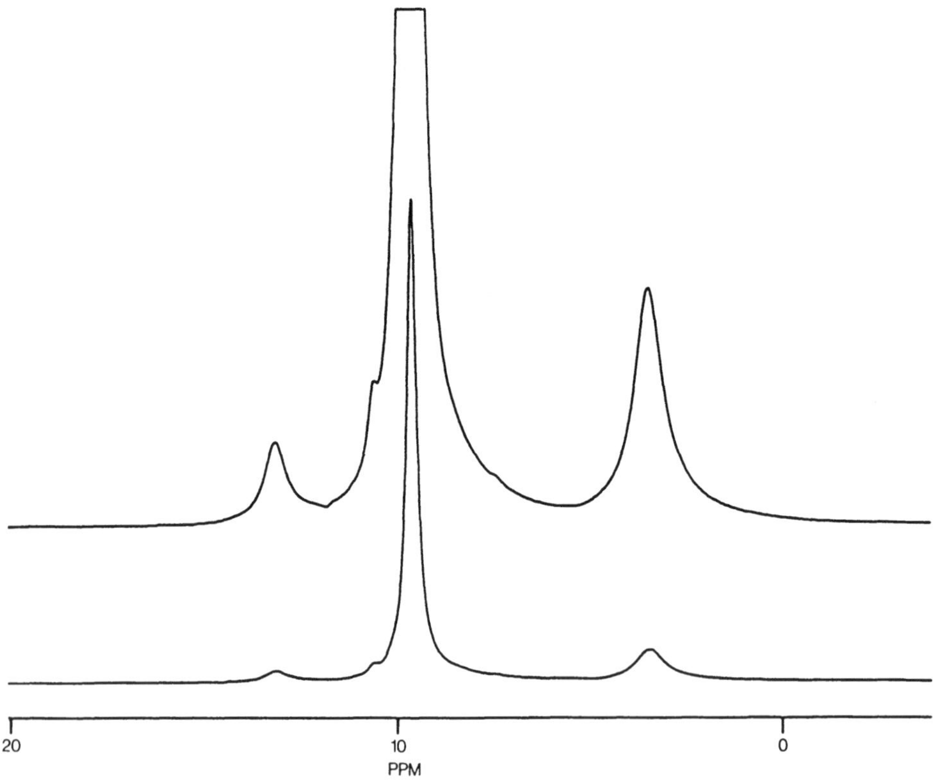

Figure 1. Aluminum-27 NMR spectrum at 104.2 MHz of a 0.5 M solution of $AlCl_3$ in anhydrous methanol, displayed at two magnifications. A solution of $Al(ClO_4)_3$ gives principally the highest field resonance, slightly chemically shifted. The resonances are believed to arise from $AlCl_3D_3$ (13.18 ppm) $AlCl_2D_4^+$ (9.66 ppm) and AlD_6^{3+} with some unspecified involvement of $AlClD_5^{2+}$ (2.6 ppm), (D = MeOH.)

Table 3. Aluminum-27 Chemical Shifts in Methanol and Ethanol[a]

System	Chemical shift (ppm)	Linewidth (Hz)	Reference
$AlCl_3/MeOH$	10	50	66
$AlCl_3(MeOH)_3$	13.15	60	68
$AlCl_2(MeOH)_4^+$	9.6	26	68
$Al(MeOH)_6^{3+}$ ($AlCl_3$ soln.)	3.4	161	68
($Al(ClO_4)_3$ soln.)	3.1	198	68
$AlCl_2(MeOH)_3(H_2O)^+$	8.4	ca. 190	68
$AlCl_2(MeOH)_2(H_2O)_2^+$	7.1	ca. 150	68
$Al(MeOH)_3(H_2O)_3^{3+}$	1.75	16 (*fac* isomer)	68
$Al(H_2O)_6^{3+}$ in MeOH	−0.8	13	68
$AlCl_2(MeOH)_4^+/CDCl_3$[b]	10.45	32	70
$Al(MeOH)_6^{3+}/CDCl_3$[b]	4.0	26	70
$AlCl_2(EtOH)_4^+$	11.3 and 8.01 (two isomers?)	260, 130	70
$AlCl_2(EtOH)_4^+/CDCl_3$[b]	9.6	365	70
$Al(EtOH)_6^{3+}/CDCl_3$[b]	2.85	13	70

[a] *n*-propanol and butanol give a main, broad resonance at 9.6 ppm.[70]
[b] 10% v/v alcohol in $CDCl_3$.

there remain some discrepancies between the results of the different groups.[78] Methyl cyanide is a weaker donor than are the alcohols or water so that chloride can compete effectively for the aluminum coordination sites, with the result that about 75% of the aluminum in trihalide solutions is present as the tetrahaloaluminate anion, and the remaining 25% is in the form of octahedral, mixed chloro solvates. The spectra thus contain a sharp, low field line due to the symmetrical anion, and a group of weak, broader lines to high field of the reference. These latter arise from the species AlN_6^{3+}, $AlClN_5^{2+}$, $AlCl_2N_4^+$, etc., where N = MeCN. The hexasolvate gives a sharp resonance at highest field, at −33 ppm, the chloride ligand giving a down field influence on shift, as in the alcoholic solutions. Some of the resonances are ascribed to geometrical isomers, e.g., *cis* and *trans* $AlCl_2N_4^{2+}$, though such isomers have not been observed in solutions of aluminum perchlorate. Here all the aluminum is in an octahedral environment and gives rise to a series of remarkably narrow resonances due to complexes containing one-, two-, three-, and four-coordinated perchlorate anions. The high-field spectrum of a mixed solution containing both chloride and perchlorate is shown in Figure 2.[81]

If water is added then octahedral species are formed with MeCN, H_2O, and Cl^- in various proportions. Notable is the recognition of a sharp resonance due to *fac* $AlN_3(H_2O)_3^{3+}$.[76] Alternatively, if pseudohalides are added then the octahedral species are suppressed and tetrahaloaluminate anions formed in which halide has been replaced by pseudohalide, such as $AlCl_3NCS^-$ or $AlCl_3NCO^-$. The resonances of these mixed complexes remain narrow and coupling is observed to ^{14}N indicating which atom of the pseudohalide is bonded to Al and how many are so bonded.[79,80] Their shifts are included in Table 4 and the coupling constants in Table 8.

The shifts given in Table 4 are generally the most recent values. Where there appears to be some discrepancy—for instance not all publications report a broad line in the tetrahedral region—these figures have been omitted and the original papers should be consulted. It seems likely that a full understanding of this system will only

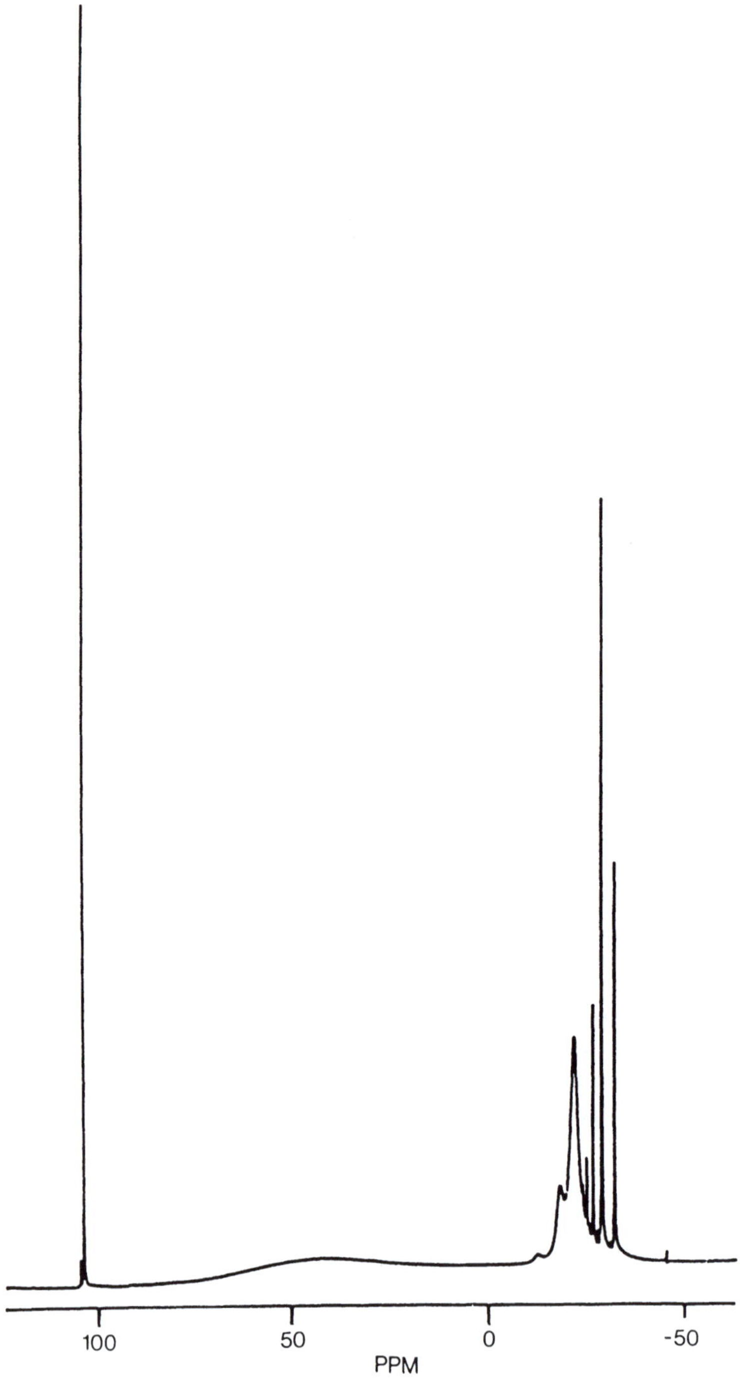

Figure 2. Aluminum-27 NMR spectrum at 104.2 MHz of a solution of composition $AlCl_2ClO_4$ in anhydrous methyl cyanide. Resonances are observed due to $AlCl_4^-$, $Al(MeCN)_n(ClO_4)_{6-n}^{(n-3)+}$, where $n = 0$, 1, 2, 3, due to $AlCl_n(MeCN)_{6-n}^{(3-n)+}$, where $n = 1$, 2, and possibly 3. The broad central resonance is due to aluminum in the solid material used in the probe construction. (Reprinted with permission from Akitt, *NMR and Chemistry*, 2nd Ed., Chapman and Hall, 1983.)

Table 4. Aluminum-27 Chemical Shifts in Methyl Cyanide and Phenyl Cyanide

Compound	Chemical shift (ppm)	Linewidth (Hz)	Reference
In MeCN (=N)			
$AlCl_4^-$	102–104	3 minimum	3, 74, 75, 78, 79, 83
AlN_6^{3+}	−33, −32.6, −34	20, 15	74–76, 78, 83
AlN_5Cl^{2+}	−23, −21.6, −24	150	74, 76, 78, 83
$AlN_4Cl_2^+$	−14.5, −12, −14	210 minimum	75, 78
$AlN_5(ClO_4)^{2+}$	−29.15	14	81
$AlN_4(ClO_4)_2^+$	−26.95	14	81
$AlN_3(ClO_4)_3$	−25.28	14	81
$AlN_2(ClO_4)_4^-$	−24.27	14	81
Al complex with $Cl^- + ClO_4^-$?	−18.4	14	81
AlN_5Br^{2+}	−30, −31	80	76, 77
$AlBr_4^-$	80	16	77, 79, 82
AlI_4^-	−26.7		76
$AlN_5(H_2O)^{3+}$	−27.4	900	76, 85
$AlN_4(H_2O)_2^{3+}$	−21.6	380	76, 85
$AlN_3(H_2O)_3^{3+}$	−15.4	23 (*fac* isomer)	76, 85
$Al/N/H_2O/Cl$ species	−24 to −4.7		76, 85
$AlCl_3N_3$	−6.3	1400 (*mer* isomer?)	85
$AlCl_3(NCS)^-$	88.7		79
$AlCl_2(NCS)_2^-$	73.3		79
$AlCl(NCS)_3^-$	56.0		79
$AlCl_3(NCO)^-$	89.8		79
$AlCl_2(NCO)_2^-$	78.7		79
$AlCl(NCO)_3^-$	66.7		79
$AlBr_3NCS^-$	76.7		79
In phenyl cyanide (=P)			
AlP_6^{3+}	−29	20	75
AlP_5Cl^{2+}	−18	1900	75

be obtained via a multinuclear approach. Relaxation measurements in this solvent are discussed in Section 2.2.8.[82,83]

2.2.4. *Other Solvents and Ligands*

If $AlCl_3$ is dissolved in a weak donor solvent then the spectra may resemble those obtained in methyl cyanide, or else the solvent may be incapable of causing the halide dimer to dissociate. Acetone is an example of the former type,[68] and $AlCl_4^-$ and octahedral species are formed, though this solvent is attacked by $AlCl_3$ and the composition of the octahedral species is uncertain. Benzene is an example of the latter type and Al_2Cl_6 gives a broad resonance in the tetrahedral region in this solvent. ^{27}Al chemical shifts are, however, observed in methyl benzenes and together with ^{13}C data are interpreted in terms of the presence of two complexes arene·$AlBr_3$ and arene·Al_2Br_6, the latter with a shift of up to 68 ppm low field of the free halide.*[86]

The weaker donor solvents are ideal media for the study of the interaction of aluminum with stronger ligands. Thus dimethylformamide in methyl nitrate gives

* The method of referencing used in this paper is not clear.

tetrahedral and octahedral ions with $AlCl_3$ and if dimethylsulfoxide is added seven mixed octahedral complexes are formed.[87–89] In tetrahydrofuran (THF) solution, depending upon conditions, several species are formed: $AlCl_4^-$ and $AlCl_2(THF)_4^+$, $AlCl_3 \cdot THF$, and $AlCl_3(THF)_2$.[90] The conclusions are supported by ir data and it is of particular interest to see the characterization of a five-coordinate complex, whose chemical shift lies between the usual octahedral and tetrahedral regions. Ether solutions of $AlCl_3$ have been examined at high field.[91] $AlCl_3$ ionizes in Et_2O to give $AlCl_4^-$, $AlCl_2(OEt_2)_2^+$, as well as the adduct. All species have similar chemical shifts and broad lines and would not be resolved at lower fields. THF behaves as described above, and exchange processes are also discussed. Up to six resonances are detected in monoglyme ($MeOCH_2CH_2OMe$) with $AlCl_4^-$ and an adduct seen in the tetrahedral region and resonances of the octahedral species $AlCl(glyme)_3^{2+}$ with both mono- and bidentate ligands, and a resonance possibly due to $Al(glyme)_3^{3+}$ to high field. $AlCl_3$ is more soluble in diglyme ($MeOCH_2CH_2OCH_2CH_2OMe$) than in monoglyme and different species are present: $AlCl_4^-$ and $AlCl_2(glyme)_2^+$. The resonance position of the latter changes upon dilution and a third resonance appears at 60 ppm. The linewidth changes observed with an increase in temperature suggest that two equilibria present.

The interaction of Al^{3+} with a variety of phosphate ligands has been reported extensively by the French group at Nancy,[92–96] and their displacement reactions with water studied. Spin–spin coupling with ^{31}P is observed in the most symmetrical complexes and this gives an independent determination of their coordination number, which is usually six but may be reduced to four for bulky ligands. Rates of ligand exchange were also studied.[96]

Methylene chloride is also a useful solvent for aluminum compounds and the mixed tetrahaloaluminate anions have been studied in a variety of mixtures.[97–98] Acetyl chloride gives a donor pair with $AlCl_3$ which exchanges very slowly with added Br^-.[98]

Aromatic solvents have been used in a number of cases. One or two molecules of triethylphosphine can be added to aluminum trihalides and the five-coordinate 2:1 adducts resonate in the intermediate shift range.[99] The spectra of aluminum *iso*-propoxide have been obtained, both of the pure tetramer[100] and of mixtures with ethylenediamine[101] or trimethylaluminum.[102] The tetramer gives two ^{27}Al resonances (Figure 3) a narrow one due to the central, octahedral Al, and a broad one due to the three tetrahedral Al atoms at 58.5 ppm.* This line narrows on heating and the spectrum can then be satisfactorily integrated. Addition of ethylenediamine produces principally a four-coordinate complex and no definite indication of a five-coordinate complex. Some resonances of intermediate shift were observed, though these were rather to high field for five-coordinate Al, unless of course the upfield shift caused by nitrogen ligands is influencing these results.[90,99] There is exchange between dimeric and tetrameric structures in the presence of Me_6Al_2.[102] A range of alkylaluminum compounds has recently been remeasured, together with some of their derivatives, and the linewidths reported are appreciably narrower than the early work had suggested.[103] The shifts depend only upon the coordination number of the Al, though the alkyl ligands cause a marked downfield shift relative to the all oxo-ligand complexes. Lanthanide reagents cause ^{27}Al shifts in aluminum *tris*-acetylacetonate.[104]

The presumed dimer $[Me_2Al \cdot \gamma \text{ picoline}]_2$ is red and gives a resonance at

* It has recently been pointed out that the original value had been multiplied 2×. R. H. Duncan, Thesis, Leeds 1979.

44.4 ppm, remarkably to high field of monomeric $Me_3AlNC_5H_4Me$, which is found at 174.7 ppm.[105] Results are reported in SO_2.[106] It is now believed that Et_2O solutions of Al_2I_6 contain only the adduct.[107] The anions $Al_2Cl_7^-$, $AlCl_4^-$, and mixed $AlCl_{4-n}I_n^-$ have been observed in low-temperature melts.[108–110]

These and other results are summarized in Tables 5 and 6.[111–127]

2.2.5. AlH_4^-, AlR_4^-, *and the Borohydrides*

AlH_4^- has been studied in a variety of systems. The Al–H coupling constant was found to be somewhat larger than early measurements had indicated, at about 170 Hz.[128,129] The form of its aluminum-27 spectrum depends strongly upon solvent, being a quintet if the anion is free but a singlet of variable breadth if ion pairing is possible, the former being the case in THF whereas diethyl ether permits ion pairing. The ^{27}Al shift is slightly dependent on the nature of the cation.[130] The anion interacts with alcohols and thiols,[131] and the species $AlH_{4-n}(OBu^t)_n^-$ have been detected for $n = 1$ and 2 but not 3.[132] In more recent studies[133,134] it is concluded that simple ion pairs are present at low concentrations and larger aggregates at high ones. Mixtures containing AlH_4 with AlD_4^- equilibrate to give all the mixed species $AlH_{4-n}D_n^-$, which in THF or diglyme give well-resolved coupling patterns.[84]

The coupling constants have been determined between aluminum and ^{13}C or ^1H in $AlMe_4^-$,[135,136] and in $AlEt_4^-$ [137,138] and ion pairing is detected here also.[139–142] The ^{27}Al shift and linewidth are affected by the presence of coordinating ligands in benzene solutions of $LiAl(SiMe_3)_4$.[143]

A few further studies have appeared of the aluminum borohydride system. Shifts and coupling constants are recorded for the mixed alkyl-hydride-borohydrides and their adducts.[144–147] The *tris* compounds have shifts in the region of 140 to 160 ppm and the adducts are somewhat to high field though all the resonances are broad. AlH_3 in THF appears to be tetracoordinated at ambient temperatures with a line 3300 Hz wide at 105 ppm. On cooling, this moves upfield and broadens so that the aluminum is probably becoming pentacoordinate. Addition of BH_3 gives the series of hydridoborohydrides, while BH_3 plus $LiAlH_4$ gives tetracoordinate species.[147] This set of results is summarized in Tables 7 and 8.[148–153]

2.2.6. *Solid State Aluminum-27 Spectroscopy—Magic Angle Spinning*

It is well known that solid state samples give broad, unresolved spectra due to dipole–dipole or quadrupole interactions or to chemical shift anisotropy of the nuclear environment studied. The quadrupolar interaction seems to be the most important one in the case of ^{27}Al, and with standard techniques the lines are very broad. Solid state spectroscopy has nevertheless received attention and the chemical shift range for aluminum is large enough that it is possible to distinguish octahedral and tetrahedral aluminum environments, and aluminum in symmetrical sites may give noticeably narrowed resonances. Thus the presence of $AlCl_4^-$ is confirmed in solid samples prepared from mixtures of $AlCl_3 + ICl + I_2$, the cationic components being I_3^+, I_5^+, I_2Cl^+, or ICl_2^+.[154] Changes in the signal intensity have been used to monitor changes in zeolites,[155] and six- and four-coordinate aluminum have been distinguished in a variety of solids with the AlF_6^{3-} in $(NH_4)_3AlF_6$ resonating at -10 ppm.[156] A few other recent references are also given,[157–161] the last being a wide-line study of zeolites.

Table 5. Aluminum-27 Chemical Shifts in Other Solvents

Compound/Solvent	Chemical shift (ppm)	Linewidth (Hz)	Reference
$Al(DMF)_6^{3+}/MeNO_2$	-2.6 ($AlCl_4^-$ ref.)	5.6	87, 113
$AlCl(DMF)_5^{2+}/MeNO_2$	2.4		87
$AlCl_2(THF)_4^+$	19		90, 91
$AlCl_3(THF)_2$	63, 64.4	*cis* form 500 *trans* form 240	90, 91
$AlCl_3(THF)$	94	102	90
$AlCl(glyme)_3^{2+}$/monoglyme	25.5		91
$Al(glyme)_3^{3+}$/monoglyme	15		91
$AlCl_2(glyme)_2^+$/diglyme	25		91
$[AlCl(S_2CNMe_2)_2]_2$	14		125
$Al(S_2CNMe_2)_3$	22	16	125
$MeNSi(Me)_2N(Me)Si(Me)_2N(Me)AlCl$	112		126
$Al(OP(OMe)_3)_6^{3+}/MeNO_2$	-20.5	3.2	96
$Al(OPMe(OMe)_2)_6^{3+}/MeNO_2$	-20.2		94
$Al(OPH(OMe)_2)_6^{3+}/MeNO_2$	-17.7		94
$Al(OP(OEt)_3)_6^{3+}/MeNO_2$	-21.6		96
$Al(OPEt(OEt)_2)_6^{3+}/MeNO_2$	-20.2		96
$Al(OP(NMe_2)_3)^{3+}/MeNO_2$	34.11	est. 8	94
$AlCl_3(OCClCH_3)/CH_2Cl_2$	92.4	ca. 600	98
$AlCl_3(Bu^tCOCH_3)$	92.8	296	98
$Al(ViCN)_6^{3+}$/ViCN	-33	83	111
$Al(POCl_3)_6^{3+}$	-21.2	55 to 77	112
$Al(EtNCS)_6^{3+}$	20	60	111
$AlCl_3(PhCOCl)$	77	2000	111
$Al[(\mu\text{-}OPr^i)_2Al(OPr^i)_2]_3$	-2.69 (central Al)	102	102
in toluene	58.5 (tetrahedral Al)	Broad	100
AlI_4^-/CH_3Cl_2	-27	24	97
$AlBrI_3^-/CH_2Cl_2$	8		97
$AlClI_3^-/CH_2Cl_2$ or melt	21.7	46	97, 109
$AlI_3 \cdot OEt_2/Et_2O$	39, 40	90, 81	111, 115
$AlBr_2I_2^-/CH_2Cl_2$	37		97
$AlBr_3(Et_3P)_2$/benzene	35.0		99
/toluene	34.7		
$AlClBrI_2^-/CH_2Cl_2$	47.7		97
$[Me_2AlNC_5H_4Me]_2/C_5H_{10}$	44.4	5080	105
$AlBr_3 \cdot OEt_2/Et_2O$	47	210	111
$AlCl_3 \cdot OEt_2/Et_2O$	50	160	111
$AlCl_3(Et_3P)_2$/benzene	55.9		99
/toluene	55.0		
$Al[OCH(CF_3)_2]_3 \cdot R_2PCH_2CH_2PR_2$	ca. 60		114
$AlCl_2I_2^-/CH_2Cl_2$ or melt	59.4, 58	57	97, 109
$AlBr_3I^-/CH_2Cl_2$	60.6		97
$AlClBr_2I^-/CH_2Cl_2$	69.3		97
Al_2Br_6/benzene	75	510	111
$AlCl_2BrI^-/CH_2Cl_2$	79		97
$Al_2Cl_4Bu_2^i$/neat	86	4000	115
$AlCl_3I^-/CH_2Cl_2$ or melt	86.2	42	97, 109
$AlClBr_3^-/CH_2Cl_2$	87.2		97

Table 5. (*Continued*)

Compound/Solvent	Chemical shift (ppm)	Linewidth (Hz)		Reference
$LiAl(SiMe_3)_4 \cdot DMF_2/C_6D_6$	91.6	50		116
$LiAl(SiMe_3)_4 \cdot (OEt_2)_3/C_6D_6$	91.1	120		116
$LiAl(SiMe_3)_4 \cdot (THF)_4/C_6D_6$	95.3	80		116
$LiAl(SiMe_3)_4/C_6D_6$	85.9	240		116
$AlCl_3 \cdot$ monoglyme/monoglyme	90			91
$Al_2(OMe)_2Cl_4$/neat	90			115
$Al_2Me_3Cl_3$	93			115
$AlBr_2Cl_2^-/CH_2Cl_2$	93, 94, 92.1	25		97, 98, 127
/acetone	94	43		
$AlBr_3 \cdot OEt_2/Et_2O$	95	104		111
$AlBr_3 \cdot PEt_3/CH_2Cl_2$	99			118
/benzene	99.2			99
/toluene	97.8			
$AlBr_3 \cdot PPh_3$	98			99
$AlCl_3Br^-/CH_2Cl_2$	99, 97.4	25		97, 98, 117, 127
/acetone	99	17		
Al_2Cl_6/toluene	91	300		115, 119
/chloroform	105	260		127
$AlBr_3 \cdot PMe_3/CH_2Cl_2$	100.8			118
$AlCl_3 \cdot OEt_2/Et_2O$	101, 102	110		115, 120, 91
$AlCl_4^-$/various	101–103	3.9 min.		97, 98, 109 111, 117
$AlCl_3 \cdot PMe_3/CH_2Cl_2$	108.2			121
$AlCl_3 \cdot PEt_3$/benzene	109.9			99
/toluene	109.1			
$Al_2Cl_7^-$/melt	115			108, 109
$AlCl_3 \cdot PR_2CH_2CH_2PR_2/CH_2Cl_2$	103.1 to 109.3			114
$AlCl_4^-$/ethers	104.2–103.4			91
$AlCl_2(OEt_2)_2^+/Et_2O$	98			91
$Et_2AlO(CH_2)_2NEt_3/d_8$ toluene	112	7200	Five-coordinate Al	103
$R_2AlO(CH_2)_2OMe/d_8$ toluene	121	1830 R = Me 5600 R = Et		103
$Et_2AlO(CH_2)_2OEt/d_8$ toluene	126	6000	Four-coordinate Al	103
Et_2AlOMe/d_8 toluene	146	3400		103
Et_2AlOEt/d_8 toluene	151	2450		103
Bu^n_3Al/d_8 toluene	152	7800		103
Et_3Al/d_8 toluene	154	2550		103
/benzene	142, 171	1000		122, 123
Me_3Al/d_8 toluene	157	850		103
/neat	156	450		115
Et_2AlR/d_8 toluene	157	4500 R = H		103
	160	1220 R = NEt_2		
	167	3300 R = Cl		
$AlMe_3OEt_2$/benzene	177	2000		124
Bu^i_3Al/d_8 toluene	255	6100	Three-coordinate Al	103
Pr^i_3Al/d_8 toluene	256	5900		103
Bu^i_3Al/d_8 toluene	276	6300		103
	220	6000		115

Table 6. Aluminum-27 Chemical Shifts in Mixed Octahedral Aquo Complexes $Al(H_2O)_{6-n}L_n^{3+}$ [a]

Ligand	Chemical shift (ppm) where *n* is 1	2	3	4	5	6
$OP(OMe)_3$	−3.7	−6.7	−10.0	−14.0	−17.5	−20.5
$OPMe(OMe)_2$	−3.5	−6.8	−10.1	−14.8	−17.5	−20.2
$OPH(OMe)_2$	−3.3	−6.6	−9.1	−14.0	−15.9	−17.7
$OP(OEt)_3$	−3.6	−7.0	−11.0	−14.7	−17.5	−21.6
$OPEt(OEt)_2$	−3.75	−7.5	−11.0	−14.8	−17.5	−20.2

[a] Solvent, $MeNO_2$; Refs. 94, 96.

The situation has, however, recently changed dramatically with the introduction of magic angle spinning into aluminum-27 spectroscopy.[162,163] This eliminates the dipolar, first-order quadrupole and chemical shift anisotropy effects on line broadening, provided the spinning rate is high enough. A second-order quadrupolar effect remains, but this is reduced at high magnetic fields and the technique is best associated with superconducting magnets. This second-order effect does have the consequence that the line position of an unsymmetrically sited atom may be shifted from its true position.[164] The lines are sufficiently narrow that not only may aluminum sites of different coordination number be distinguished, but like sites may have distinguishable chemical shifts, and linewidth differences are also significant. As in the

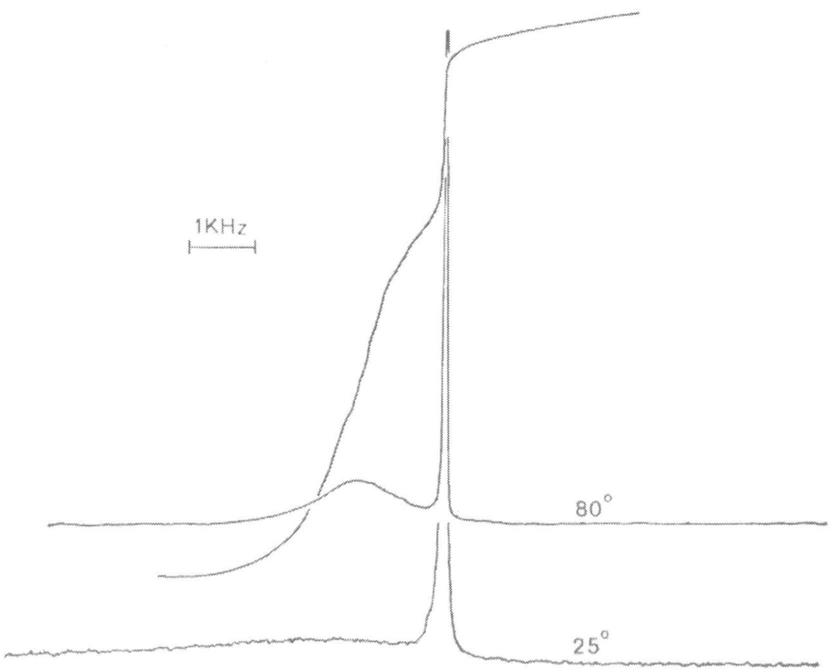

Figure 3. Aluminum-27 NMR spectrum at 23.45 MHz of the tetramer $[Al(OPr^i)_3]_4$ in toluene. The spectrum was run at two temperatures and demonstrates the well-known narrowing effect of higher temperatures on the spectra of quadrupolar nuclei. (Reprinted with permission from Akitt and Duncan, *J. Magn. Reson.*, 1974, **15**, 162.)

Table 7. Aluminum-27 Chemical Shifts of AlH_4^-, AlR_4^-, and Borohydride Derivatives[a]

Compound	Chemical shift (ppm)	Linewidth (Hz)	Reference
AlH_4^-	98–103	Variable	84, 128–133, 134
$AlH_2(OBu^t)_2^-$	ca. 100	ca. 1200	132
$AlH(OBu^t)_3^-$	ca. 73	ca. 700	132
$LiAl(SiMe_3)_4$	85.9	240	143
with donors	91.1 to 95.3	50–120	116
$AlH_3 \cdot THF$	105	3300	147
$Al(BH_4)_3/C_6H_6$	97.4	180	148
$Al(BH_4)_3 \cdot THF$	34.4, 37.8	60	147
$AlH(BH_4)_2 \cdot THF$	49, 45.6	550	147
$AlH_2BH_4 \cdot THF$	78, 74.5	1860	147
$Al(BH_4)_3 \cdot NMe_3$	71.1	9–20	148
$Al(BH_4)_3 \cdot PMe_3$	63.2	195	124
$AlH_2(BH_4)_2^-$	48.8		147
$AlH(BH_4)_3^-$	49.6		147
$Al(BH_4)_4^-$	34		147
$AlEt_2BH_4$	162	2400	145
$AlEt(BH_4)_2$	152	830	145
$AlEt_2BH_4OEt_2$	142	2900	145
$AlEt(BH_4)_2 \cdot OEt_2$	102	1700	145
$Al(BH_4)_3 \cdot NHMe_2$	67	180	124
$Al(BH_4)_3 \cdot OMe_2$	62	320	124
$Al(BH_4)_2Me \cdot NMe_3$	110	580	124
$Al(BH_4)_2Me \cdot PMe_3$	103	550	124
$Al(BH_4)Me_2 \cdot PMe_3$	148	1700	124

[a] Reference 124 contains data for a further 11 similar compounds.

liquid phase, not all the aluminum present in a given sample may contribute to the observed spectrum due to second-order quadrupole broadening.[258] Zeolites, because of their catalytic importance, are receiving much attention in conjunction with solid state ^{29}Si NMR. The area has been reviewed.[165] The preparation and dealumination of zeolites can be studied and aluminum within and without the lattice distinguished, and its quantitative loss measured.[166–174] Their structure and catalytic activity is also studied,[175–180] and for instance the four-line spectrum obtained for a dehydrated ZSM5 is notable.[176] Structural studies of other materials include that of the tridecameric cation $AlO_4Al_{12}(OH)_{24}(H_2O)_{12}^{7+}$, which gives a resonance only for the central AlO_4 unit, just as seen in the liquid.[181–184] In contrast, zunyite $(Al_{13}(OH, F)_{16}F_2Si_5O_{20}Cl)$, which has a structure related to that of the tridecamer, gives resonances from both octahedral and tetrahedral Al.[183] The differences are due to differences in the bridging between the octahedral units in the two compounds, and this also leads to a measurable chemical shift between the tetrahedral resonances, correlated with AlO_4 bond lengths.[47] The implication of this is that chemical shifts obtained from known structures in the solid may give a lead as to the structure in the liquid. The coordination number of Al has been investigated in a number of solid systems and the effects of reaction processes observed.[185–194] Thus during the hydration of monocalcium aluminate, a main constituent of high alumina cement, the Al coordination changes from four to six after an initial delay period.[188] When $Al(OH)_3$ is dehydrated to form alumina at 500°C, the cation is six coordinate at the

Table 8. Coupling Constants of Aluminum-27

Compound	J (Hz)	Remarks	Reference
To hydrogen			
AlH_4^-	170–173	Depends on solvent and counterion	129–133
$AlMe_4$	$^2J = 6.3, 7.1, 9.1$	Medium dependent	136, 142, 149
$AlEt_4$	$^2J = 7.3$ $^3J = 5.8$		150
$Al(BH_4)_3 \cdot NMe_3$	44–46		144, 148, 151
$Al(BH_4)_3 \cdot THF$	39	Nonet	147
To boron-11			
$Al(BH_4)_3 \cdot NMe_3$	ca 9		148
To carbon-13			
$AlMe_4^-$	71.5		135
Al_2Me_6	Terminal C = 110 Bridge C = 19	^{13}C relaxation	152
$Al_2Me_4X_2$ (X = Cl or Br)	105	^{13}C relaxation	152
$Al_2Et_4Cl_2$	105	^{13}C relaxation	152
Al_2Me_6/Et_2O	91	Adduct formed?	152
$AlEt_4$	$^1J = 73$ ^2J ca 1		137
$AlBu_4^{n-}$	$^1J = 71.6$ ^2J = ca 4		137
To nitrogen-14			
$Al(MeCN)_6^{3+}$	$^1J = 22$ Hz	^{27}Al scalar relaxation	83
$AlCl_3(NCS)$	$^1J = 40$	Al triplet	79
$AlCl_2(NCS)_2^-$	$^1J = 45$	Al quintet	79
$AlCl_3(NCO)$	$^1J = 40$	Al triplet	79
To phosphorus-31			
$AlCl_3 \cdot PMe_3$	263, 280		118, 153
$AlCl_3 \cdot PMe_2Ph$	290		153
$AlCl_3 \cdot PEt_3$	260, 263	Slightly medium dependent?	99
$AlBr_3 \cdot PMe_3$	248		118
$AlBr_3 \cdot PEt_3$	240		99, 118
$Al(BH_4)_3 \cdot PMe_3$	265		124
$Al(OP(OMe)_3)_6^{3+}$	19.5	Al septet	94
$Al(OPMe(OMe)_2)_6^{3+}$	15.0	Al septet	94
$Al(OPH(OMe)_2)_6^{3+}$	13.4	Al septet	94
$Al(OP(OEt)_3)_6^{3+}$	19.0	Al septet	96
$Al(OP(NMe_2)_3)_4^{3+}$	30	Al quintet	94
$Al((Me_2N)_2PON(Me)PO(NMe_2)_2)_3^{3+}$	19	By ^{31}P line-shape analysis	207
To chlorine-35			
$AlCl_4$	650	^{27}Al relaxation	82, 83
To bromine-81			
$AlBr_4^-$	750	^{27}Al relaxation	82, 83

start and at the completion of dehydration but four-coordinate aluminum is formed in the intermediate stages.[193] Clearly, we can expect to see a great deal of new information emerging as this technique becomes more generally applied. Glasses have also been examined.[195,196]

2.2.7. *Spin–Spin Coupling to Aluminum-27*

In order that spin–spin coupling is observed to ^{27}Al the nuclear relaxation time must be longer than the reciprocal of the coupling constant. This is a nontrivial constraint in the case of a quadrupolar nucleus and means that coupling effects are only likely to be observed if the aluminum environment is regular and symmetrical, though as we shall see there are some rather irregular species in which coupling is manifested. The coupling constants are listed in Table 8 in the order of the atomic number of the coupled nucleus.

Coupling has been observed directly to five other nuclei and indirectly via relaxation measurements to another three. For the type of compound where coupling can be measured we see that $^1J > {}^2J > {}^3J$.

2.2.8. *Aluminum-27 Relaxation Measurements*

Relatively few relaxation measurements have been made with this nucleus, though they should not be a very time-consuming operation, owing to the generally short relaxation times involved. Indeed, linewidth measurements should prove satisfactory in some cases. ^{13}C and ^{27}Al relaxation has recently been studied in detail for $AlEt_3$ and correlation times decrease in the order $Al > CH_2 > CH_3$.[197] Mixed scalar and quadrupole relaxation can be detected in symmetrical species and the scalar part can be used to estimate coupling constants between Al and other quadrupolar nuclei.[83,82] The linewidth of $AlCl_4^-$ has been estimated to be 3.9 Hz using a pulse sequence,[98] and is broadened by exchange in THF.[91] Other data will be found in Refs. 18 and 21.

2.3. Some Observations on the Parameters

It is clear that the ^{27}Al chemical shifts are determined by two main factors: coordination number and the nature of the ligands. The shifts move generally downfield as we pass from six-, through five-, to four-coordinate aluminum for the ionic or donor ligands. A similar progression is seen for bound alkyl ligands; five- > four- > three-coordinate Al, though these types of compound resonate appreciably to low field of the first type. There is overlap between shift regions for the latter so that ^{27}Al shifts can never be used on their own as diagnostic of coordination number and account must always be taken of other factors such as the type of ligand. The data are summarized on the chart of Figure 4. Of equal interest is an understanding of the influences which cause the small changes in shift among compounds with the same coordination number and with the same atoms directly bonded to the metal. In Table 2 we see that replacement of H_2O in $Al(H_2O)_6^{3+}$ by a charged ligand gives a small upfield shift, probably due to the extra negative charge on the bonding oxygen. In certain cases, however, the effect is the opposite one, but this is accompanied by broader lines, the alumichrome complex being a case in point.[27] It is known in many of these cases that the octahedra around Al are distorted, and this may well account

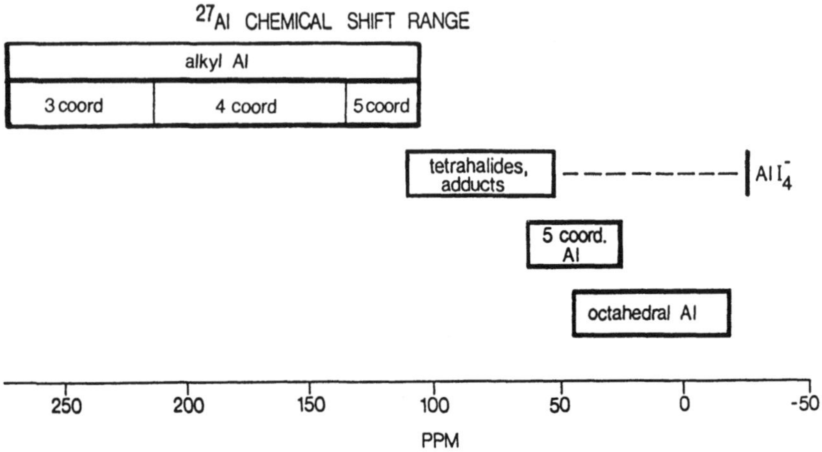

Figure 4. A chart showing the gross features of aluminum-27 chemical shift variations.

for the paramagnetic shift, due to electron imbalance in the bonds, an effect which should be particularly large for the alumichrome complex. The same tendency appears in the tetrahedral region, though it will be seen that there are some exceptions to this generalization. The species present in nonaqueous solvents are usually mononuclear so that steric distortion cannot be present though the ligands may be different. Tables 3–6 show that nitrogen ligands cause an upfield shift relative to oxygen or chloride ligands, in both four- and six-coordinate compounds. Organophosphates give a high-field shift, as indeed do the heavier halides, for reasons which are still far from clear. Shift progressions have been commented upon for series of compounds such as $AlX_nY_{4-n}^-$ or $AlL_nG_{6-n}^{p+}$, where X, Y are halides and L, G are ligands, possibly including halides. In some cases the shifts change nonlinearly with *n*. A good example is the tetrahaloaluminates, where the effect is ascribed to a pairwise additive effect of the ligands. This is also observed for the octahedral $Al(NCCH_3)_n(ClO_4)_{6-n}^{(3+n-6)+}$ species, and perhaps may arise because of the different ligand radii near the metal. This conclusion is reinforced by the observation that the shift progressions are much more nearly linear where the directly bonded groups are similar in size, such as $AlCl_n(NCS)_{4-n}^-$ or $Al(H_2O)_n(OP(OMe)_3)_{6-n}^{3+}$. There appears thus to be both a ligand–ligand effect and an electron donor effect. It is clear that coordination number does not determine the effect.

The linewidth of ^{27}Al resonances continues to provide some enigmas. The truly regular species have narrow lines as expected, and linewidths of about 3 Hz seem to be attainable when favorable conditions have been discovered. Solutions with broader linewidths than this must then contain the regular species in some modified environment. There appears to be much information available here, though to date it has received very little comment. Distorted environments give broad lines but mixing the ligands gives very variable results. Compare $AlF_n(H_2O)_{6-n}$ with $Al(SO_4)(H_2O)_5^+$ in Table 2, $AlCl(NCCH_3)_5^{2+}$ with $Al(ClO_4)(NCCH_3)_5^{2+}$ in Table 4, $AlCl_3(THF)$ with $Al(CH_3)_3\,OEt_2$ in Table 5 and wonder, as the writer does, at the very narrow line given by $Al(BH_4)_3\,NMe_3$. The explanation of these linewidth differences could form a very useful avenue of investigation and perhaps enable us to define more clearly what is meant by cubic symmetry in this context and thence discover more about the nature of the molecules that are being investigated. A start has of course been made in this direction.[11]

3. GALLIUM

Gallium has two magnetically active isotopes, ^{69}Ga and ^{71}Ga, whose properties are summarized in Table 1. The least abundant ^{71}Ga is the one preferred for NMR work because its higher receptivity and slightly less efficient quadrupolar relaxation means that the signal-to-noise ratio expected from this nucleus is about four times greater than for ^{69}Ga. The linewidths obtained are much greater than with ^{27}Al and indeed seem to be greater than suggested by the width factors quoted in Table 1. We thus expect to encounter relaxation times of ca. 4 ms and less. The broader lines which result of course mean that resolution of individual resonances will be more difficult, though this is aided by the increased chemical shift dispersion of the gallium nucleus. The prevalence of broad lines means also that the problem of nonobservable resonances is even more acute for gallium. The nuclei are in fact ideal for rapid FT spectroscopy, so that in practice the collection of data is equally easy for either isotope. Since these have different quadrupole moments, by obtaining spectra from both nuclei it should be possible to separate quadrupole relaxation effects from exchange effects. This could be quite a useful technique because exchange rates on gallium complexes tend to be rather faster than do those on aluminum.

3.1. Operational Techniques

The techniques outlined for aluminum spectroscopy apply also to gallium, though saturation effects are unlikely to be a problem at pulse repetition rates of less than $100\ s^{-1}$. Rapid pulsing is of course consistent with the wide sweep widths required. The rapid decay of the FIDs obtained allows a severe exponential multiplication to be used, and indeed there is no reason why data should be collected in large memory blocks. Small blocks of less than 1K will suffice and reduce the amount of noise that is collected while allowing the number of scans made in a given time to be proportionately increased.

3.2. Gallium-69 and Gallium-71 NMR Parameters

These have already been discussed in three previous reviews.[2,3,5] The number of reports is rather few, but a picture of the chemical shifts range is starting to emerge.

3.2.1. Aqueous Solutions

The quest for knowledge of the aqueous solution chemistry of gallium has received considerable impetus lately from our medical colleagues. The isotope ^{67}Ga has proved to be very successful as an antitumor scanning agent and the mode of its action is of interest. How, for instance, is it transported to the tumor cells, and can the efficiency of transport be increased? ^{71}Ga NMR has thus been used to study how various ligands may form complexes with the cation. Unfortunately the resonances of the complexes are so far too broad to observe and competition-type studies have to be made in which the intensity of the signal from a known complex with a narrow resonance is monitored as the ligand to be studied is added. Thus citrate forms a 1:1 complex with Ga^{3+} in $Ga(NO_3)_3$ solution but does not interact at high pH where the gallium is present as $Ga(OH)_4^-$.[198] ^1H and ^{13}C spectroscopy allow it to be inferred that small cationic gallium polymers are formed, and in this the system presumably will resemble the aluminum citrate system.[199] This work has been extended to other complexing agents such as EDTA or lactate and only $Ga(D_2O)_6^{3+}$ or $Ga(OD)_4^-$ are ever observed.[200] Ga^{3+} forms a series of complexes in the presence of phosphate: $GaH_3PO_4^{3+}$, $GaH_2PO_4^{2+}$, a complex with $H_5P_2O_8^-$ and a polymeric entity have been detected.[201] In this case the ^{71}Ga resonance is detectable, though the results rest principally upon ^{31}P spectroscopy, and follow closely the behavior of the corresponding aluminum/phosphoric acid system.[52] Pyridoxal 5′phosphate isonicotinyl hydrazone forms a 2:1 complex with gallium at high pH where the metal is in the form of $Ga(OD)_4^-$. The resulting complex may be six-coordinate, a question which could be answered easily if its gallium resonance could be seen.[202] Hydrolysis of Ga^{3+} with sodium carbonate causes the $Ga(H_2O)_6^{3+}$ resonance to decrease in intensity, linearly to zero at a hydrolysis ratio of 2.5 just as occurs with aluminum. Unfortunately the wealth of data obtained with this latter nucleus does not seem to be available for Ga, though perhaps a high-field investigation might prove fruitful.[203] ^{69}Ga and ^{71}Ga relaxation have been compared in $Ga(H_2O)_6^{3+}$, and in many samples the two rates are not in the required ratio.[204] As for Al relaxation, however, the no doubt marked influence of pH or hydrolysis on the gallium relaxation has been ignored, and the system is probably more complex than was initially envisaged. Recent preliminary investigations of solutions of $GaCl_3$ and varied amounts of HCl have confirmed earlier results[205] that $GaCl_4^-$ has a narrow line some 80 Hz in width but can exchange with other species if the HCl concentration is not high enough, giving a broadened line of variable chemical shift. Use of more modern apparatus allowed more dilute solutions to be studied, and it was found that if sufficiently dilute, $GaCl_3$ solutions contain only $Ga(H_2O)_6^{3+}$. If a solution giving the broadened resonance is heated then the equilibrium between the various species is moved in favor of $GaCl_4^-$. The resonance moves and narrows, and at high enough temperatures the gallium is present solely as this anion.[206] $Ga(H_2O)_6^{3+}$ is often present in such solutions at ambient temperatures and is not involved in the faster exchange processes though it is involved in slow equilibria. The exchange of chloride on $GaCl_4^-$ has been studied using chlorine NMR, though the ^{71}Ga resonance reported is broader than the minimum that can be obtained and there is a possibility that other species may be involved in the reactions observed.[207–209]

3.2.2. Nonaqueous Solvents

Solutions in or containing methanol have been studied with results comparable with those obtained for aluminum.[65,66] Trimethyl phosphate forms a hexacoordinate complex with Ga^{3+}.[210] Spin–spin coupling is observed and the exchange mechanism was investigated. $GaCl_3$ in methyl cyanide has a low solvation number, as determined by proton NMR, and the solutions contain $GaCl_4^-$, octahedral species, and probably Ga_2Cl_6 which gives a broad line near that of $GaCl_4^-$ and obvious in high-field spectra.[211,212] Addition of water gives mixed species, though only the ones of highest symmetry can be observed.[212] A number of tetrahalogallates have been characterized in MeCN solutions,[213,214] and mixed halide/pseudohalide anions exhibit gallium–nitrogen spin–spin coupling.[59,60] Exchange among the tetrahalogallates is slow in methylene chloride, the iodo anions showing the slowest rates. The formation of these latter is favored and the distribution of species is not statistical.[215] No dimeric species were observed. Gallium–hydrogen spin–spin coupling has been observed in the GaH_4^-

Table 9. Gallium-69 and Gallium-71 Chemical Shifts and Coupling Constants

Compound/solvent[a]	Chemical shift (ppm)	Other information[b]	Reference
$KGaH_4$/diglyme	110	$^1J(^{69}Ga–H) = 515$ Hz $^1J(^{71}Ga–H) = 650$ Hz	84, 128
$LiGaH_4/Et_2O$	682	Broad	205
$GaCl_4^-/H_2O$	240–275	Width = 80 Hz	83, 198, 205, 208, 217, 218
$GaCl_4^-$/MeCN	251.8	$T_1 = 15$ ms	211–214
$GaCl_4^-/CH_2Cl_2$	taken as 251 ppm for CH_2Cl_2 solutions	Width = 180 Hz	215
Ga_2Cl_6/MeCN	241.7	Broad	212
$GaCl_4^-$ in Ga_2Cl_4	223, 197	Solution and melt	205
$GaCl_3NCS^-$/MeCN	225	$^1J(^{71}Ga–^{14}N) = 95$ Hz	79
$Ga(OH)_4^-/H_2O$	220 (192)	Width 385 Hz minimum	198, 202, 205
$GaCl_3Br^-$/MeCN	212	$T_1 = 6.6$ ms width 200 Hz in CH_2Cl_2	214, 215
$GaCl_2(NCS)_2^-$/MeCN	194	$^1J(^{71}Ga–^{14}N) = 115$ Hz	79
$GaCl_2Br_2^-$/MeCN	168	$T_1 - 5.5$ ms width 215 Hz in CH_2Cl_2	214, 215
$GaCl(NCS)_3^-$/MeCN	157		79
$GaClBr_3^-$/MeCN	119	$T_1 = 6.1$ ms width 218 Hz in CH_2Cl_2	214, 215
$GaCl_3I^-/CH_2Cl_2$	123	Width 220 Hz	215
$GaI(NCS)_3^-$/MeCN	72		79
$GaCl_2BrI^-/CH_2Cl_2$	70	Width 255 Hz	215
$GaBr_4^-$/MeCN	68	$T_1 = 12$ ms	214
/CH_2Cl_2	64	Width 185 Hz	215
/H_2O	64, 69		205, 217
/Ga_2Br_4 melt	127	A solvated form?	205, 217
$Ga(H_2O)_6^{3+}/H_2O$	0	Reference minimum width 252 Hz	2, 198
Ga–phosphate complexes	−25.5	Width 1000/1500 Hz only 1/3 Ga visible	201
$Ga(DMF)_6^{3+}$	−25	Width 120 Hz	218
$Ga(MeCN)_3(H_2O)_3^{3+}$	−29.9	Relatively sharp line	212
$GaCl_2I_2^-/CH_2Cl_2$	−40		215
$GaBr_3I^-/CH_2Cl_2$	−49	Width 205 Hz	215
$GaI_2(NCS)_2^-$	−50		79
$Ga(MeCN)_6^{3+}$/MeCN	−72, −75, −63.2	Narrow line	210–213
$Ga(OP(OMe)_3)_6^{3+}/MeNO_2$	−77.5	$^2J(^{71}Ga–^{31}P) = 38$ Hz $T_1 = 3.6$ ms	210
$GaClBr_2I^-/CH_2Cl_2$	−13	Width 290 Hz	215
$GaClBrI_2^-/CH_2Cl_2$	−105	Width 340 Hz	215
$GaBr_2I_2^-/CH_2Cl_2$	−174	Width 220 Hz	215
$GaClI_3^-/CH_2Cl_2$	−235	Width 290 Hz	215
$GaI_3(NCS)^-$/MeCN	−234		79
$GaBrI_3/CH_2Cl_2$	−310	Width 210 Hz	79
GaI_4^-/CH_2Cl_2	−455		215
/MeCN	−500, −505		79, 213
/H_2O	−450, −479	Width 100 Hz	205, 218
Ga^+/dihalide melts	−413, −493	Ga_2Br_4, Ga_2Cl_4	205
/benzene solution	−652, −685		

[a] Some results for the tetrahalgallates have been obtained in both MeCN and CH_2Cl_2. Where the results are essentially the same, only the former is given.
[b] All linewidths are for ^{71}Ga.

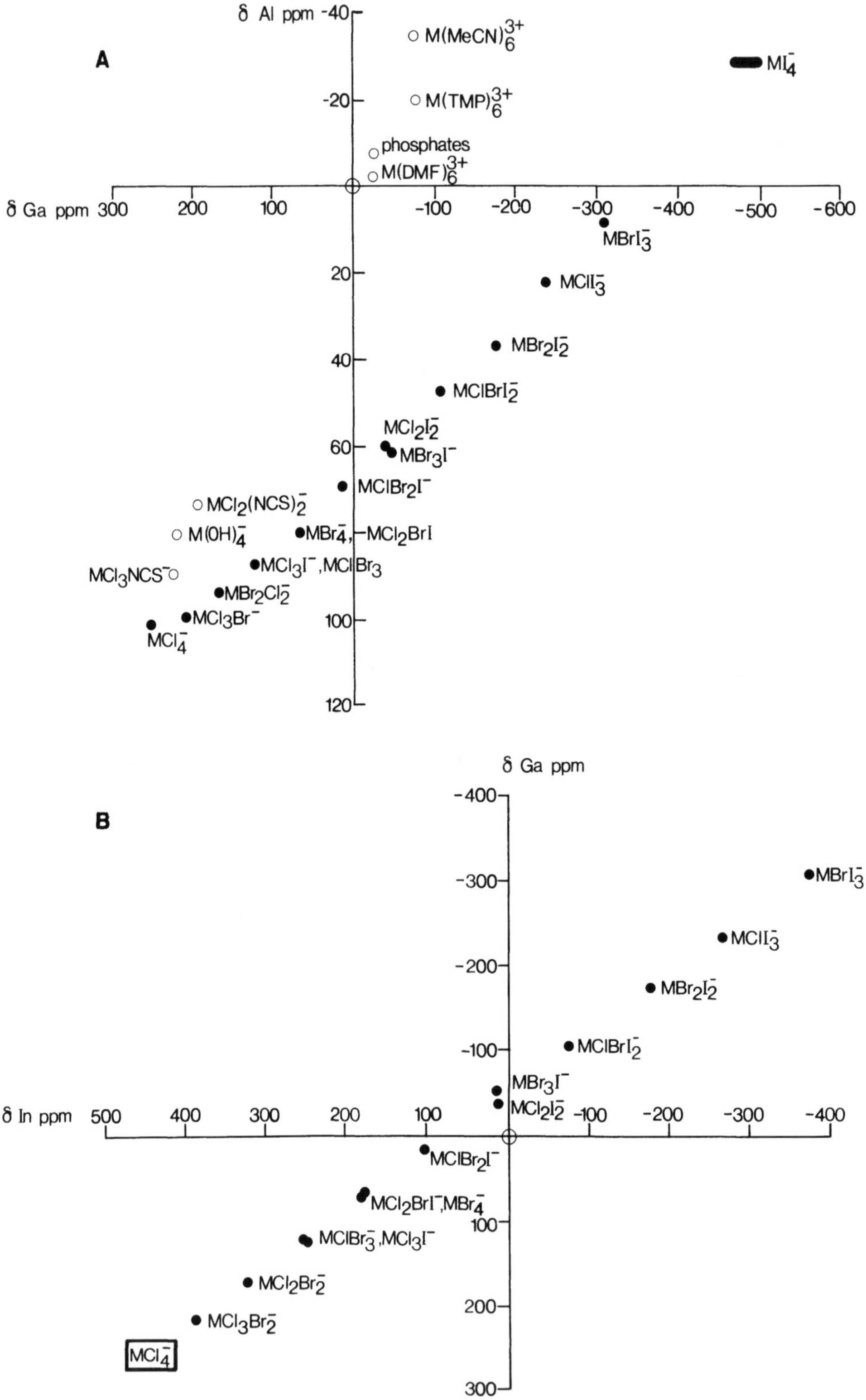

Figure 5. Correlations between the chemical shifts of pairs of nuclei. (A) ^{27}Al against ^{71}Ga. The octahedral species and tetrahalo anions appear to behave differently. (B) ^{71}Ga against ^{115}In.

anion for a solution of $KGaH_4$ in diglyme. The chemical shift is 110 ppm. In an earlier report a broad featureless line was obtained from a concentrated solution of $LiGaH_4$ in Et_2O with a chemical shift of 682 ppm. This is presumably a manifestation of ion pairing such as has been proposed for salts of AlH_4^-, though the difference in chemical shift between the forms is surprisingly large.[128] A chemical shift scale relative to the free Ga atom has been proposed.[216] The chemical shift and coupling data are summarized in Table 9. The chemical shift range, for like compounds, is about 5.7 times greater for Ga than for Al. The shifts for similar species of the two elements are correlated in Figure 5, where it is clear that the compounds studied fall into two groups, the tetrahalo anions and the octahedral species, with $M(OH)_4^-$ possibly belonging to the latter group. The influence of the ligands is thus different for the two elements in the two groups of compounds. One possible explanation is that in the octahedral complexes the smaller Al^{3+} brings about greater steric crowding of the ligand atoms whereas there is no such crowding in the tetrahedral complexes. However, the correlation involving the latter is distinctly nonlinear and this also indicates that the relative behavior of the two elements changes as the total ligand bulk changes.

4. INDIUM

Indium has two magnetically active isotopes, ^{113}In and ^{115}In, whose properties are summarized in Table 1. The quadrupole moment is large, and though this is offset by the large spin quantum number, the lines are expected to be broader than those of the other group III elements. Thus we are likely to be able to observe only the most symmetrical species. Collecting data should be very easy in the FT mode but observations will be limited by the recovery time of the equipment used. ^{115}In has the larger receptivity and is the preferred nucleus.

4.1. Indium-115 NMR Parameters

These have been summarized previously[3,5] and are relatively few. A selection of the earlier values are given together with those from some more recent publications in Table 10.[219–222] Some mixtures of molten InI_3–In have been examined. The In shifts are in the region expected for InI_4^- in the InI_3 rich region but move downfield as In is added. The relaxation behavior is believed to be determined by two mechanisms, quadrupole relaxation and an exchange mechanism involving the formation of short-lived paramagnetic In^{2+} centers.[223]

A correlation between the chemical shifts of similar ionic species of the three elements Al, Ga, and In is shown in Figure 5. The tetrahalo anions behave similarly for all three, though whereas the Ga/In correlation is linear, the Al/Ga one is noticeably curved. Approximate shift/constitution figures may be derived from these plots and show that In is 1.46 times more sensitive to change in chemical environment than is Ga, which in turn is 5.7 times more sensitive than is Al.

It is appropriate here to compare the linewidths of the three elements so far considered and see if these follow the sequence given in Table 1. It has already been noted that some gallium resonances seem broader than expected,[2] and the early work with indium gave rise to similar conclusions[3] that exchange processes may be important for these heavier elements or that the coordination may differ from that assumed and

Table 10. Indium-115 Chemical Shifts and Coupling Constants

Compound/solvent	Chemical shift (ppm)	Other information	Reference
$InCl_4^-$/various	415–480	Width 180 Hz in CH_2Cl_2 broader in other solvents	219–222
$In_2Cl_6^-/CH_2Cl_2$	439	Width 500 Hz	222
$InCl_3Br^-/CH_2Cl_2$	389		222
$InCl_5^{2-}/CH_2Cl_2$	326		222
$InCl_2Br_2/CH_2Cl_2$	324		222
$InClBr_3^-/CH_2Cl_2$	254		222
$InCl_3I^-/CH_2Cl_2$	251		222
$InCl_6^{3-}/CH_2Cl_2$	206		222
$InCl_2BrI^-/CH_2Cl_2$	183 est.		222
$In_2Br_6^-/CH_2Cl_2$	183	Width 330 Hz	222
$InBr_4^-/Et_2O$	180	Width 540, 375 Hz	219, 221
$/CH_2Cl_2$	177	Width 180 Hz	222
$InClBr_2I^-/CH_2Cl_2$	104		222
$InBr_3I^-/CH_2Cl_2$	12		222
$InCl_2I_2^-/CH_2Cl_2$	12		222
$In(H_2O)_6^{3+}$/aqueous	0–10	Width 375 min.	219–221
$In(SO_4)_2^-/H_2O$	−18	Width 300 Hz	221
$InClBrI_2^-/CH_2Cl_2$	−72		222
$InBr_2I_2^-/CH_2Cl_2$	−175		222
$InClI_3^-/CH_2Cl_2$	−265		222
$InBrI_3^-/CH_2Cl_2$	−371		222
In_2I_6/CH_2Cl_2	−548	Width 250 Hz	222
InI_4^-/aqueous	−440 to −620	Width 140 min.	219–221
InI_4^-/Et_2O	−583	Width 247 Hz	221
InI_4^-/CH_2Cl_2	−569	Width 150 Hz	221
In_2I_6/CH_2Cl_2	−590	Width 1354 Hz	222
InH_4^-		^1J(InH) est. 968 Hz	128
$In(OP(OMe)_3)_6^{3+}$		$^2J(^{115}In–^{31}P) = 48$ Hz $T_1 = 1.18$ ms	210

give quadrupolar effects. We need in fact to be able to compare like compounds of each element in the same solvents for meaningful correlations to be obtained. This has not so far been done systematically but it is possible to extract a few valid comparisons from the results printed here. Table 1 tells us that the linewidths should be in the ratio $^{27}Al:^{71}Ga:^{115}In = 1:2.34:14$. The results for four sets of molecular ions are

MCl_4^-: 4, 20, 180 Hz or 1:5:45 ratio

MBr_4^-: 16, 185, 180 Hz or 1:11.6:11.3 ratio

$M(H_2O)_6^{3+}$: 2, 25, 375 Hz or 1:12.5:187 ratio
Note that the value for Ga is based on a T_1 determination, whereas the linewidth is 252 Hz. For In, the value quoted is the minimum observed.

$M(OP(OMe)_3)_6^{3+}$: 3.2, 89, 270 Hz or 1:27.8:84.4 ratio

Clearly, it is not even possible to assert that the linewidths are always in the correct order, and their relative magnitudes deviate from those expected by almost ten times. It is important to try and understand the reason for these differences since they provide further and comparative evidence of interactions in solution.

5. THALLIUM

Thallium occurs as two magnetically active isotopes, thallium-203 and thallium-205. Their properties are summarized in Table located on pp. 625–629 (Appendix), where it will be seen that they are both sensitive spin-1/2 nuclei with ^{205}Tl preferred for spectroscopy because of its higher natural abundance. The lack of quadrupole moment, unique in group III, means that the relaxation times are much longer than found for the other elements discussed above. The operational techniques are therefore quite different and more like those appropriate to ^{13}C, though the relaxation times are not particularly long. Thus care is required if quantitative data are needed, but data collection is otherwise not particularly limited by the relaxation behavior. This is perhaps as well since the chemical shift range over which compounds of this element resonate is large so that pulse repetition rates can be quite high. The longer relaxation times of course mean that the linewidths are generally smaller than those of the other group III elements, so that Tl can in principle be observed in all possible chemical environments. The shift range is some 5500 ppm. The possession of spin 1/2 also means that spin–spin coupling is always possible and coupling constants form an appreciable part of the database. Coupling constants may be very large where thallium is involved and it may be difficult to observe all components of a multiplet unless very large sweepwidths are used to accommodate this. The shift and coupling data for thallium are summarized in Figures 6 and 7. It should be noted that the values for a given compound may vary widely with the conditions used for measurement such as temperature, solvent, concentration, and counterion. The data for thallium have been reviewed quite recently[3] and the literature covered up to 1976. Much of the detail used to produce the figures will be found in this source. We will limit ourselves here simply to a comment on recent data.

The species present in aqueous solutions of TlX_3 (X = Cl, Br, I) have been investigated using thallium NMR. Exchange is fast and a singlet is observed whose shift varies with solution composition and may be interpreted as being dependent on the nature of the ionic species whose formation constants are known from other

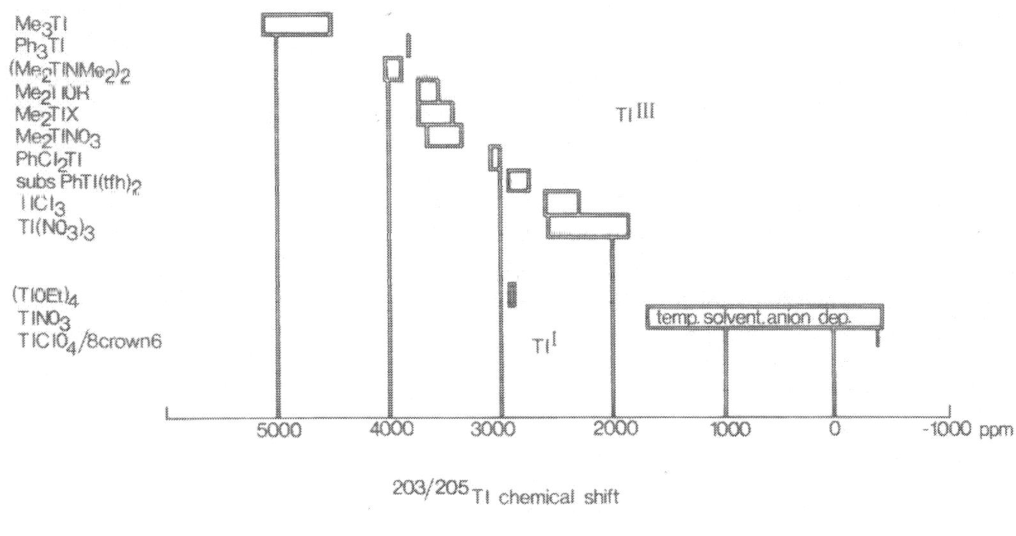

Figure 6. Chart of chemical shift ranges for thallium.

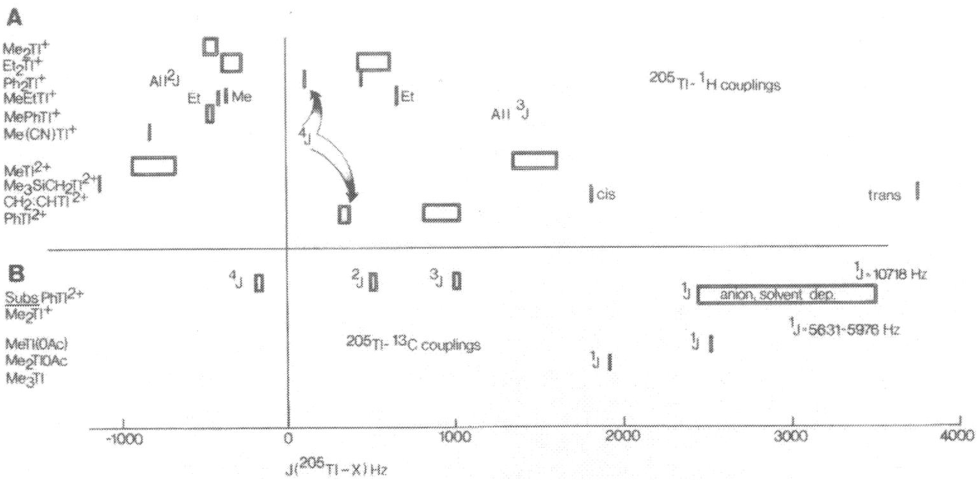

Figure 7. Chart of coupling constants between ^{205}Tl and (A) hydrogen and (B) ^{13}C.

studies. Thus $TlCl_3$ gives $TlCl_4^-$ ($\delta = 2630$ ppm), $TlCl_5^{2-}$, and $TlCl_6^{3-}$. $TlBr_3$ gives $TlBr_4^-$ ($\delta = 1318$ ppm) and a little $TlBr_5^{2-}$, and TlI_3 gives only TlI_4^- ($\delta = -1560$ ppm). The shifts given in brackets refer to $Tl(H_2O)_6^{3+}$ at about 2075 ppm.(224) Solutions of salts of the cation Me_2Tl^+ have a Tl chemical shift which is very strongly dependent on the sample parameters such as temperature, concentration, the anion type, and solvent.(225–230) Shifts have been observed in binary solvent mixtures(231–233) and have been analyzed in terms of the Covington model of preferential solvation.(233) The interaction of Tl^+ with a variety of antibiotics has been studied and profound shift and relaxation effects are observed.(234–238) The shifts span the range from 297.6 ppm to −700 ppm, indicating very different modes of bonding to the various ionophores. There is also an initial report of an investigation of the solid valinomycin-Tl complex which suggests that chemical shift anisotropy effects may be less for thallium than had been assumed from liquid state studies.(239) It is suggested that Tl^+ may be a good probe to be used to imitate the behavior of alkali metals in such systems. The interactions of Tl^+ has also been studied with macrocyclic ligands,(240–242) and with EDTA.(243) In liquid ammonia solution, the thallium shift varies in such a way as to suggest that the species present may be the free ion, fully solvated ions, ion pairs, and ion pair aggregates. At high concentrations of $TlNO_3$ the fully solvated contact ion pair $(NH_3)_3TlNO_3$ may be present.(244)

Spin–spin coupling between thallium and hydrogen, carbon, fluorine, and phosphorus is discussed in nine recent papers.(245–253) If the coupling constant is large enough, it may be possible to resolve coupling to both thallium isotopes in the spectrum of the coupled nucleus, but normally only a single value is observed, often described as an average value. ^{203}Tl–^{205}Tl coupling is observed in the Tl^I alkoxide tetramers, $(TlOR)_4$, and its magnitude varies considerably with changes in the nature of R, though no obvious pattern can be seen to explain this.(245) The one-bond coupling to ^{13}C changes markedly with n in the series of compounds $Me_nTl(OAc)_{3-n}$, where $n = 1, 2, 3$.(246) No correlation is found between $^1J(Tl–^{13}C)$ and C–Tl–C bond angle in a series of molecules containing Me_2Tl units.(247) Coupling for both thallium isotopes is reported for $PhTl(OOCCF_3)_2$.(248) Nine Tl–^{13}C coupling constants can be measured for the Tl complex of the antibiotic lasalocid and their magnitudes give a convincing picture of how the thallium is bonded.[249] Reduced temperatures are needed for the couplings to be resolved and at ambient, the spectrum is the same as that of the sodium salt. Tl substituted at the 2 position of norbornene derivatives gives resolvable coupling to all seven carbons varying from 5764 Hz to almost zero.[250] On the other hand ^{13}C spectra may be simplified and coupling to thallium removed by adding the nitroxide radical.[251] The fully coupled spectra may cover the whole spectral range. Coupling to ^{19}F has been observed in di-fluorophenol thallium bromides and nitrate. Approximate values of the couplings in $(C_6F_5)_2TlBr$ are $^3J = 800$ Hz, $^4J = 340$ Hz, and $^5J = 80$ Hz, but these values depend on the solvent and change if substituents are introduced into the ring.[252] In complexes of the type $(Ph_3P)(CO)(OAc)_2IrTl(OAc)_2$, the phosphines are *trans* and $J(Tl{-}^{31}P)$ lies in the range of 1202–1056 Hz depending on the nature of the phosphine and the ligands.[253]

In relaxation studies, it is reported that several processes affect the behavior of Tl^+,[254–255] and chemical shift anisotropy may make an important contribution. The two relaxation times T_1 and T_2 are now believed to differ markedly for Tl^+ in the presence of dissolved oxygen, and it is proposed that some scalar relaxation (i.e., electron delocalization) is present.[256] The relaxation rates of all the magnetically active nuclei in melts containing TlX (X = Cl, Br) and Tl have been measured and suggest that adding the metal produces paramagnetic centers which govern the observed behavior.[257]

REFERENCES

1. Akitt, J. W. *J. Magn. Reson.* 1970, **3**, 411.
2. Akitt, J. W. *Ann. Rep. NMR Spectrosc.* 1972, **5A**, 465.
3. Hinton, J. F.; Briggs, R. W. In *NMR and the Periodic Table*, Harris, R. K.; Mann, B. E., Eds.; Academic: New York, 1978.
4. Delpuech, J.-J., *NMR of the Newly Accessible Nuclei*, Vol. 2, Laszlo, P., Ed.; Academic: New York, 1983.
5. Wehrli, F. W. *Ann. Rep. NMR Spectrosc.* 1979, **9**, 126.
6. Reference 2, p. 469.
7. Akitt, J. W.; Greenwood, N. N.; Khandelwal, B. L.; Lester, G. D. *J. Chem. Soc. Dalton* 1972, 604.
8. Akitt, J. W.; Farthing, A. *J. Chem. Soc. Dalton Transl.* 1981, 1617.
9. Farthing, A. Thesis, University of Leeds, U.K., 1980.
10. Hoult, D. I.; Chen, C.-N.; Eden, H.; Eden, M. *J. Magn. Reson.* 1983, **51**, 110.
11. Akitt, J. W.; McDonald, W. S. *J. Magn. Reson.* 1984, **58**, 401.
12. Akitt, J. W.; Farnsworth, J. A.; Letelier, P. *J. Chem. Soc. Faraday Trans. I* 1985, **81**, 1930.
13. Akitt, J. W.; Elders, J. M. *J. Magn. Reson.* 1985, **63**, 587.
14. Akitt, J. W.; Greenwood, N. N.; Lester, G. D. *J. Chem. Soc. (A)* 1969, 803.
15. Akitt, J. W., Milic, N. B. *J. Chem. Soc. Dalton Trans.* 1984, 981.
16. Akitt, J. W.; Farthing, A. *J. Magn. Reson.* 1978, **32**, 345.
17. Akitt, J. W.; Elders, J. M. *J. Chem. Soc. Faraday Trans. I* 1985, **81**, 1923.
18. Hetrz, H. G.; Tutsch, R.; Versmold, H. *Ber. Bunsenges. Phys. Chem.* 1971, **75**, 1177.
19. Takahashi, A. *J. Phys. Soc. Jpn* 1977, **43**, 968 and 971.
20. Zaripov, U. M.; Nikiforov, E. A. *Russ. J. Phys. Chem.* 1979, **53**, 39.
21. Michaelov, V. I.; Yishik, V. I. *Vestnik. Leningrad Fiz. Khim.* 1982, **3**, 81.
22. Akitt, J. W.; Greenwood, N. N.; Lester, G. D. *J. Chem. Soc. Chem. Commun.* 1969, 988.
23. Akitt, J. W.; Greenwood, N. N.; Khandelwal, B. L. *Spectrosc. Lett.* 1971, **4**, 139.
24. Akitt, J. W. *J. Chem. Soc. Dalton* 1973, 1177.
25. Akitt, J. W.; Farthing, A. *J. Chem. Soc. Dalton Trans.* 1981, 1233.
26. Akitt, J. W.; Farthing, A. *J. Chem. Soc. Dalton Trans.* 1981, 1606.
27. Akitt, J. W.; Farthing, A. *J. Chem. Soc. Dalton Trans.* 1981, 1624.
28. Schoenherr, S.; Goerz, H.; Bertram, R.; Mueller, D.; Gessner, W. *Z. Anorg. Chem.* 1983, **502**, 113.

29. Bottero, J. Y.; Poirer, J. E.; Fiessinger, F. *Progr. Water Technol.* 1980, **12**, 601.
30. Bottero, J. Y.; Cases, J. M.; Fiessinger, F.; Poirier, J. E. *J. Phys. Chem.* 1980, **84**, 2933.
31. Akitt, J. W.; Mann, B. E. *J. Magn. Reson.* 1981, **44**, 584.
32. Teagarden, D. L.; Koslowski, J. F.; White, J. L.; Radavich, J. F.; Hem, S. L. *Trav. Com. Int. Etude Baux.* 1982, **12**, 267.
33. Teagarden, D. L.; Koslowski, J. F.; White, J. C.; Hem, S. L. *J. Pharm. Sci.* 1981, **70**, 758, 762, 808.
34. Toy, A. D.; Smith, T. D.; Pilbrow, J. R. *Aust. J. Chem.* 1973, **26**, 1889.
35. Akitt, J. W.; Farthing, A.; Howarth, O. W. *J. Chem. Soc. Dalton Trans.* 1981, 1609.
36. Llinas, M.; de Marco, A. *J. Am. Chem. Soc.* 1980, **102**, 2226.
37. Jaber, M.; Bertin, F.; Thomas-David, G. *Can. J. Chem.* 1977, **55**, 3689.
38. Karlik, S. J.; Elgavish, G. A.; Pillai, R. P.; Eichborn, G. L. *J. Magn. Reson.* 1982, **49**, 164.
39. Laussac, J. P.; Laurent, J. P. *C. R. Acad. Sci. Ser. C* 1980, **291**, 157.
40. Laussac, J. P.; Commenges, G. *Nouv. J. Chim.* 1983, **7**, 579.
41. Karlik, S. J.; Tarien, E.; Elgavish, G. A.; Eichhorn, G. L. *Inorg. Chem.* 1983, **22**, 525.
42. Holtz, M.; Friedman, H. L.; Tembe, B. L. *J. Magn. Reson.* 1982, **47**, 454.
43. Buslaev, Y. A.; Petrosyants, S. P. *Koord. Khim.* 1979, **5**, 163 (123).
44. Caminiti, R.; Lai, A.; Saba, G.; Crisponi, G. *Chem. Phys. Lett.* 1983, **97**, 180.
45. Dovbysh, N. G.; Volokhov, Y. A.; Mironov, V. E. *Tezisy Dokl. Vses. Soveshch "Termodin. Struct. Gidrosokompleksov Rastvorakh"* 2nd, 1974, 18.
46. Dovbysh, N. G.; Volokhov, Y. A.; Sizyakov, V. M.; Eremin, N. I. *Trav. Com. Int. Etude Baux.* 1982, **12**, 157.
47. Akitt, J. W.; Gessner, W. *J. Chem. Soc. Dalton Trans.* 1984, 147.
48. Burlamacchi, L.; Lai, A.; Moduzzi, M. *Chem. Phys. Lett.* 1983, **100**, 549.
49. May, L.; Rudzite, M.; Lagsdins, E. *Latv. PSR. Zinat. Acad. Vestis. Khim. Ser.* 1978, **4**, 428.
50. Mueller, D.; Hoebbel, D.; Gessner, W. *Chem. Phys. Lett.* 1981, **84**, 25.
51. Gaul, J. H.; Carr, T. M. *Spectrosc. Lett.* 1983, **16**, 651.
52. Akitt, J. W.; Greenwood, N. N.; Lester, G. D. *J. Chem. Soc. A* 1971, 2450.
53. Akitt, J. W.; Greenwoud, N. N.; Khandelwal, B. L. *J. Chem. Soc. Dalton Trans.* 1972, 1226.
54. Epperlein, B. W.; Lutz, O. *Z. Naturforsch.* 1968, **23a**, 1413.
55. Akitt, J. W. In *Nuclear Magnetic Resonance Spectroscopy of Nuclei Other Than Protons*, Axenrod, T.; Webb, G. A., Eds.; Whiley: New York, 1974; p. 339.
56. Akitt, J. W.; Farthing, A. *J. Chem. Soc. Dalton Trans.* 1981, 1615.
57. Nakamura, S.; Meiboom, S. *Xth Int. Conf. on Coord. Chem.*, Nikko, Japan, 1967, p. 42.
58. Grasdalen, H. *J. Magn. Reson.* 1971, **5**, 84.
59. Grasdalen, H. *J. Magn. Reson.* 1972, **6**, 336.
60. Haraguchi, H.; Fujiwara, S. *J. Phys. Chem.* 1969, **73**, 3467.
61. Martin, J. S.; Stockton, G. W. *J. Magn. Reson.* 1973, **12**, 218.
62. Stockton, G. W.; Martin, J. S. *Can. J. Chem.* 1974, **52**, 744.
63. Richardson, D.; Alger, T. D. *J. Phys. Chem.* 1975, **79**, 1733.
64. Wawer, I.; Kecki, Z. *Pol. J. Chem.* 1981, **55**, 219.
65. Buslaev, Y. A.; Tarasov, V. P.; Petrosyants, S. P.; Kirakosyan, G. A. *Dokl. Akad. Nauk. SSSR (Chem. Ser.)* 1978, **241**, 838(356).
66. Buslaev, Y. A.; Tarasov, V. P.; Petrosyants, S. P.; Kirkosyan, G. A. *Koord. Khim.* 1978, **4**, 1346(1019).
67. Lyashchenko, A. K. *Koord. Khim.* 1981, **7**, 1171.
68. Akitt, J. W.; Duncan, R. H.; Setchell, C. *J. Chem. Soc. Dalton Trans.* 1983, 2639.
69. Kon'shin, V. V.; Chernyshov, B. N. *Zh. Strukt. Khim.* 1983, **24**, 137.
70. Akitt, J. W.; Lelievre, J. *J. Chem. Soc. Dalton Trans.* 1985, 591.
71. Duncan, R. H. Thesis, University of Leeds, 1979.
72. Akitt, J. W.; Duncan, R. H. *J. Magn. Reson.* 1977, **25**, 391.
73. Akitt, J. W.; Duncan, R. H.; Beattie, I. R.; Jones, P. J. *J. Magn. Reson.* 1978, **34**, 435.
74. Beattie, I. R.; Jones, P. J.; Howard, J. A. K.; Suart, L. E.; Gilmore, C. J.; Akitt, J. W. *J. Chem. Soc. Dalton Trans.* 1979, 528.
75. Wehrli, F. W.; Hoerdt, R. *J. Magn. Reson.* 1981, **42**, 334.
76. Wehrli, F. W.; Wehrli, S. *J. Magn. Reson.* 1981, **44**, 197.
77. Dalibart, M.; Derouault, J.; Granger, P. *Inorg. Chem.* 1981, **20**, 3975.
78. Dalibart, M.; Derouault, J.; Granger, P.; Chapelle, S. *Inorg. Chem.* 1982, **21**, 1040.
79. Tarasov, V. P.; Petrosyants, S. P.; Kirakosyan, G. A.; Buslaev, Y. A. *Dokl. Acad. Nauk. SSSR* 1978, **242**, 156.

80. Tarasov, V. P.; Petrosyants, S. P.; Kirakosyan, G. A.; Buslaev, Y. A. *Koord. Khim.* 1980, **6**, 52(35).
81. Akitt, J. W.; Jones, P. J. unpublished results.
82. Tarasov, V.; Privalov, V. I.; Buslaev, Y. A. *Zhur. Strukt. Khim.* 1978, **19**, 1012(866).
83. Tarasov, V. P.; Privalov, V. I.; Kirakosyan, G. A.; Buslaev, Y. A. *Z. Naturforsch.* 1983, **38B**, 1173.
84. Tarasov, V. P.; Bakum, S. I.; Privalov, V. I.; Buslaev, Y. A.; Kusnetsov, A. M. *Koord. Khim.* 1983, **9**, 882.
85. Dalibart, M.; Derouault, J.; Granger, P. *Inorg. Chem.* 1982, **21**, 2241.
86. Erb, H. P.; Felbecker, H.; Bluhm, R. *Ber. Bunseges. Phys. Chem.* 1984, **88**, 122.
87. Schippert, E. *Adv. Mol. Relaxation Processes* 1976, **9**, 167.
88. Gudlin, D.; Schneider, H. *J. Magn. Reson.* 1974, **16**, 362.
89. Schneider, H. *Electrochem. Acta.* 1976, **21**, 711.
90. Derouault, J.; Granger, P.; Forel, M. T. *Inorg. Chem.* 1977, **16**, 3214.
91. Nöth, H.; Rurländer, R.; Wolfgardt, P. *Z. Naturforsch.* 1982, **37B**, 29.
92. Delpuech, J.-J.; Peguy, A.; Khaddar, M. R. *J. Magn. Reson.* 1972, **6**, 325.
93. Canet, D.; Delpuech, J.-J.; Khaddar, M. R.; Rubini, P. *J. Magn. Reson.* 1973, **9**, 329.
94. Delpuech, J.-J.; Khaddar, M. R.; Peguy, A.; Rubini, P. *J. Chem. Soc. Chem. Commun.* 1974, 154.
95. Canet, D.; Delpuech, J.-J.; Khaddar, M. R.; Rubini, P. *J. Magn. Reson.* 1974, **15**, 325.
96. Delpuexh, J.-J.; Khaddar, M. R.; Peguy, A.; Rubini, P. *J. Am. Chem. Soc.* 1975, **97**, 3373.
97. Kidd, R. G.; Truax, D. R. *J. Am. Chem. Soc.* 1968, **90**, 6867.
98. Wilinski, J.; Kurland, R. J. *J. Am. Chem. Soc.* 1978, **100**, 2233.
99. Vriezen, W. H. N.; Jellinek, F. *Rec. Trav. Chim.* 1970, **89**, 1306.
100. Akitt, J. W.; Duncan, R. H. *J. Magn. Reson.* 1974, **15**, 162.
101. Akitt, J. W.; Duncan, R. H. *J. Chem. Soc. Dalton Trans.* 1976, 1119.
102. Uetsuki, M.; Fujiwara, Y. *Bull. Chem. Soc. Jpn* 1977, **50**, 673.
103. Benn, R.; Rufinska, A.; Lehmkuhl, H.; Janssen, E.; Krueger, C. *Angew. Chem. Int. Edn.* 1983, **22**, 779.
104. Hirayama, M.; Kitami, K. *J. Chem. Soc. Chem. Commun.* 1980, 1030.
105. Dorogy, W. E.; Schramm, E. P. *Inorg. Chim. Acta.* 1983, **72**, 187.
106. Glavincevski, B.; Brownstein, S. K. *J. Org. Chem.* 1982, **47**, 1005.
107. Ogren, P. J.; Cannon, J. P.; Smith, C. F. *J. Phys. Chem.* 1971, **75**, 282.
108. Anders, O.; Plambeck, J. A. *J. Inorg. Nucl. Chem.* 1978, **40**, 387.
109. Gray, J. L.; Maciel, G. E. *J. Am. Chem. Soc.* 1981, **103**, 7147.
110. Wilkes, J. S.; Frye, J. S.; Reynolds, G. F. *Inorg. Chem.* 1983, **22**, 3870.
111. Haraguchi, H.; Fujiwara, S. *J. Phys. Chem.* 1969, **73**, 3467.
112. Kidd, R. G.; Truax, B. R. *J. Chem. Soc. Chem. Commun.* 1969, 160.
113. Movius, W. G.; Matwiyoff, N. A. *Inorg. Chem.* 1967, **6**, 847.
114. Laussac, J.-P. *Organic Magn. Reson.* 1979, **12**, 237.
115. O'Reilly, D. E. *J. Chem. Phys.* 1960, **32**, 1007.
116. Rosch, L.; Altnau, G. *Chem. Ber.* 1979, **112**, 3934.
117. Jones, D. E. H. *J. Chem. Soc. Dalton* 1972, 567.
118. Vriezen, N. H. N.; Jellinek, F. *Chem. Phys. Lett.* 1967, **1**, 284.
119. O'Reilly, D. E.; Poole, C. P.; Belt, R. F.; Scott, H. *J. Polymer Sci.* 1964, **A2**, 3257.
120. Ohtaki, H. *Bull. Chem. Soc. Jpn* 1970, **43**, 2463.
121. Cohen, B. M.; Cullingworth, A. R.; Smith, J. D. *J. Chem. Soc. (A)* 1969, 2193.
122. Swift, H. E.; Poole, C. P.; Itzel, J. F. *J. Phys. Chem.* 1964, **68**, 2509.
123. DiCarlo, E. N.; Swift, H. E. *J. Phys. Chem.* 1964, **68**, 551.
124. Oddy, P. R. Thesis, University of Warwick, 1975.
125. Nöth, H.; Konrad, P. *Chem. Ber.* 1983, **116**, 3552.
126. Wannagat, U.; Blumenthal, T.; Brauer, D. J.; Buerger, H. *J. Organomet. Chem.* 1983, **249**, 33.
127. Akitt, J. W.; Setchell, C. unpublished results.
128. Tarasov, V. P.; Bakum, S. I. *J. Magn. Reson.* 1975, **18**, 64.
129. Santini, R. E.; Grutzner, J. B. *J. Magn. Reson.* 1976, **22**, 155.
130. Heřmanek, S.; Kriz, O.; Plešek, J. P.; Hanslick, T. *Chem. Ind. (Lond.)* 1975, 42.
131. Huet, J.; Durand, J.; Infarnet, Y. *Org. Magn. Reson.* 1976, **8**, 382.
132. Horne, D. A. *J. Am. Chem. Soc.* 1980, **102**, 6011.
133. Nöth, H. *Z. Naturforsch.* 1980, **35B**, 119.
134. Nöth, H.; Rurlaender, R.; Wolfgardt, P. *Z. Naturforsch.* 1981, **36B**, 31.
135. Yamamoto, O. *Chem. Lett.* 1975, 511.
136. Ross, J. F.; Oliver, J. P. *J. Organomet. Chem.* 1970, **22**, 503.

137. Westmoreland, T.D.; Bhacca, N. S.; Wander, J. D.; Day, M. C. *J. Am. Chem. Soc.* 1973, **95**, 2019.
138. Ahmad, N.; Day, M. C. *J. Am. Chem. Soc.* 1977, **99**, 941.
139. Gray, J. L.; Maciel, G. E. *J. Phys. Chem.* **87**, 1983, 5290, 5418.
140. Day, M. C.; Medley, J. H.; Ahmad, N. *Can. J. Chem.* 1983, **61**, 1719.
141. Oliver, J. P.; Wilkie, C. A. *J. Am. Chem. Soc.* 1967, **89**, 163.
142. Gore, E. S.; Gutowsky, H. S. *J. Phys. Chem.* 1969, **73**, 2515.
143. Rosch, L.; Altnau, G. *Chem. Ber.* 1979, **112**, 3934.
144. Boiko, G. N.; Malov, Y. I.; Sensevenko, K. N. *Izv. Acad. Nauk.* 1973, **22**, 1143(1100).
145. Oddy, P. R.; Wallbridge, M. G. H. *J. Chem. Soc. Dalton Trans.* 1976, 2076.
146. Oddy, P. R.; Wallbridge, M. G. H. *J. Chem. Soc. Dalton Trans.* 1978, 572.
147. Nöth, H.; Rurländer, R. *Inorg. Chem.* 1981, **20**, 1062.
148. Lauterbur, P. C.; Hopkins, R. C.; King, R. W.; Ziebarth, U. V.; Heitsch, C. W. *Inorg. Chem.* 1968, **7**, 1025.
149. Weibel, A. T.; Oliver, J. P. *J. Organometal. Chem.* 1974, **74**, 155.
150. Westmoreland, T. D.; Bhacca, N. S.; Wander, J. D.; Day, M. C. *J. Organometal. Chem.* 1972, **38**, 1.
151. Bird, P. H.; Wallbridge, M. G. H. *J. Chem. Soc.* 1965, 3923.
152. Yamamoto, O. *J. Chem. Phys.* 1975, **63**, 2988.
153. Laussac,. J. P.; Laurent, J. P.; Commenges, G. *Org. Magn. Reson.* 1975, **7**, 72.
154. Merryman, D. J.; Corbett, J. D.; Edwards, P. E. *Inorg. Chem.* 1975, **14**, 428.
155. Resing, H. A.; Rubenstein, M. *J. Colloid. Interface Sci.* 1978, **64**, 48.
156. Mueller, D.; Gessner, W.; Grimmer, A. R. *Z. Chem.* 1977, **17**, 453.
157. Harly, S. F.; Williams, C. D.; Tipswold, R. F. *J. Chem. Phys.* 1976, **64**, 4815.
158. Sook Lee; Brodbeck, C. M.; Chi Chung Yang *Phys. Revs. B* 1977, **15**, 2469.
159. Okuda, T.; Ishihara, H.; Yamada, K.; Negita, H. *Bull. Chem. Soc. Jpn* 1978, **51**, 1273.
160. Proskho, Y. A.; Litovchenko, A. S.; Bagmut, N. N.; Matyash, I. V. *Ukr. Fiz. Zh.* 1978, **23**, 249.
161. Bosacek, V.; Freude, D.; Fröhlich, T.; Pfiefer, H.; Schmiedel, H. *J. Colloid. Interface Sci.* 1982, **85**, 502.
162. Muller, D.; Gessner, W.; Behrens, H.-J.; Scheler, G. *Chem. Phys. Lett.* 1981, **79**, 59.
163. Freude, D.; Behrens, H.-J. *Crystal Res. Tech.* 1981, **16**, K36.
164. Mueller, D. *Ann. Phys.* 1982, **39**, 451.
165. Thomas, J. M.; Klinowsky, J.; Ramadas, S.; Anderson, M. W.; Fyfe, C. A.; Gobbi, G. C. *Angew. Chem. Int. Ed.* 1983, **22**, 259.
166. Fyfe, C. A.; Gobbi, G. C.; Hartman, J. S.; Lenkinski, R. E.; O'Brien, J. H.; Beange, E. R.; Smith, M. A. R. *J. Magn. Reson.* 1982, **47**, 168.
167. Fyfe, C. A.; Gobbi, G. C.; Hartman, J. S.; Klinowski, J.; Thomas, J. M. *J. Phys. Chem.* 1982, **86**, 1247.
168. Klinowski, J.; Thomas, J. M.; Fyfe, C. A.; Gobbi, G. C. *Nature* 1982, **296**, 533.
169. Fyfe, C. A.; Gobbi, G. C.; Klinowski, J.; Thomas, J. M.; Ramdas, S. *Nature* 1982, **296**, 530.
170. Fyfe, C. A.; Gobbi, G. C.; Kennedy, G. J. *Chem. Lett.* 1983, 1551.
171. Thomas, J. M.; Klinowski, J.; Anderson, M. W. *Chem. Lett.* 1983, 1555.
172. Freude, D.; Frohlich, T.; Hunger, M.; Pfeifer, H.; Scheler, G. *Chem. Phys. Lett.* 1983, **98**, 541.
173. Englehardt, G.; Fahlke, B.; Magi, M.; Lippmaa, E. *Zeolites* 1983, **3**, 292.
174. Thomas, J. M.; Klinowski, J.; Wright, P. A.; Roy, R. *Angew. Chem. Int. Ed.* 1983, **22**, 614.
175. Fyfe, C. A.; Gobbi, G. C.; Klinowski, J.; Putnis, A.; Thomas, J. M. *J. Chem. Soc. Chem. Commun.* 1983, 556.
176. Kentgens, A. P. M.; Scholle, K. F. M. G. J.; Veeman, W. S. *J. Phys. Chem.* 1983, **87**, 4357.
177. Mastikhin, V. M.; Shubin, A. A.; Nekipelov, V. M.; Zamaraev, K. I. *Kinet. Katal.* 1983, **24**, 756.
178. Freude, D.; Froelich, T.; Pfeifer, H.; Scheler, G. *Zeolites* 1983, **3**, 171.
179. Anderson, M. W.; Thomas, J. M. *J. Chem. Soc. Chem. Commun.* 1983, 525.
180. Ione, K. G.; Echevskii, G. V.; Nosyreva, G. N. *J. Catal.* 1984, **85**, 287.
181. Schonherr, S.; Gorz, H.; Muller, D.; Gessner, W. *Z. Anorg. Chem.* 1981, **476**, 188.
182. Muller, D.; Gessner, W.; Schonherr, S.; Gorz, H. *Z. Anorg. Chem.* 1981, **483**, 153.
183. Lampe, F. v.; Muller, D.; Gessner, W.; Grimmer, A. R.; Scheler, G. *Z. Anorg. Chem.* 1982, **489**, 16.
184. Klinowski, J., personal communication.
185. Mueller, D.; Grunze, I.; Hallas, E.; Ladwig, G. *Z. Anorg. Chem.* 1983, **500**, 80.
186. Gessner, W.; Mueller, D. *Z. Anorg. Chem.* 1983, **505**, 195.
187. Mueller, D.; Gessner, W.; Scheler, G. *Polyhedron* 1983, **2**, 1195.
188. Mueller, D.; Rettel, A.; Gessner, W.; Scheler, G. *J. Magn. Reson.* 1984, **57**, 152.
189. Klinowski, J.; Mackay, A. L. *Struct. Perform. Cements* 1983, 1.

190. Gessner, W.; Muller, D.; Behrens, H.-J.; Scheler, G. *Z. Anorg. Chem.* 1982, **486**, 193.
191. Weinberger, M.; Mueller, D.; Gessner, W. *Z. Chem.* 1983, **23**, 420. A.; Roy, R. *Angew. Chem. Int. Ed.* 1983, **22**, 614.
192. Mastikhin, V. M.; Krivornchko, O. P.; Zolotovskii, B. P.; Buyanov, R. A. *React. Kinet. Catal. Lett.* 1981, **18**, 117.
193. John, C. S.; Alma, N. C.; Hays, G. R. *Appl. Catal.* 1983, **6**, 341.
194. Akna, B. C. M.; Hays, G. R.; Samoson, A. V.; Lippmaa, E. T. *Analyt. Chem.* 1984, **56**, 729.
195. de Jong, B. H. W. S.; Schramm, C. M.; Parziale, V. E. *Geochim. Cosmochim. Acta.* 1983, **47**, 1223.
196. Hallas, E.; Haubenreisser, U.; Haehnert, M.; Mueller, D. *Glastech. Ber.* 1983, **56**, 63.
197. Vestin, U.; Kowalewski, J.; Henriksson, U. *Org. Magn. Reson.* 1981, **16**, 119.
198. Glickson, J. D.; Pitner, T. P.; Webb, J.; Gams, R. A. *J. Am. Chem. Soc.* 1975, **97**, 1679.
199. Chang, C. H. F.; Pitner, T. P.; Lenkinski, R. E.; Glickson, J. D. *J. Am. Chem. Soc.* 1977, **99**, 5858.
200. Chang, C. H. F.; Pitner, T. P.; Lenkinski, R. E.; Glickson, J. D. *Bioinorg. Chem.* 1978, **8**, 11.
201. Lenkinski, R. E.; Chang, C. H. F.; Glickson, J. D. *J. Am. Chem. Soc.* 1978, **100**, 5383.
202. Haran, R.; Gairin, J.; Commenges, G. *Inorg. Chim. Acta.* 1980, **46**, 63.
203. Akitt, J. W.; Khandelwal, B. L. unpublished results.
204. Tarasov, V. P.; Buslaev, Y. A. *Mol. Phys.* 1972, **24**, 665.
205. Akitt, J. W.; Greenwood, N. N.; Storr, A. *J. Chem. Soc.* 1965, 4410.
206. Akitt, J. W.; Stilwell, C. unpublished results.
207. Lincoln, S. F.; Sandercock, A.; Stranks, D. R. *J. Chem. Soc. Chem. Commun.* 1972, 1069.
208. Lincoln, S. F.; Sandercock, A. C.; Stranks, D. R. *J. Chem. Soc. Dalton Trans.* 1975, 669.
209. Buslaev, Y. A.; Tarasov, V. P.; Petrosyants, S. P.; Melnikov, N. N. *Koord. Khim.* 1975, **1**, 1435(1195).
210. Rodehuser, L.; Rubini, P. L.; Delpuech, J.-J. *Inorg. Chem.* 1977, **16**, 2837.
211. Lincoln, S. F. *Australian J. Chem.* 1972, **25**, 2705.
212. Wehrli, F. W.; Wehrli, S. L. Bruker Report, No. 1, 1981.
213. Buslaev, Y. A.; Tarasov, V. P.; Petrosyants, S. P.; Melnikov, N. N. *Zh. Strukt. Khim.* 1974, **15**, 617 (525).
214. Tarasov, V. P.; Privalov, V. I.; Buslaev, Y. A.; Kuz'min, I. A. *Dokl. Akad. Nauk. SSSR.* 1977, **234**, 636.
215. McGarvey, B. R.; Taylor, M. J.; Tuck, D. J. *Inorg. Chem.* 1981, **20**, 2010.
216. Kodweiss, J.; Lutz, O.; Messner, W.; Mohn, K. R.; Nolle, A.; Stutz, B.; Zepf, D. *J. Magn. Reson.* 1981, **43**, 495.
217. Movius, W. G.; Matwiyoff, N. A. *Inorg. Chem.* 1969, **8**, 925.
218. Fratiello, A.; Lee, R. E.; Schuster, R. E. *Inorg. Chem.* 1970, **9**, 82.
219. Haraguchi, H.; Fuwa, K.; Fujiwara, S. *J. Phys. Chem.* 1973, **7**, 1497.
220. Fratiello, A.; Davis, D. D.; Peak, S.; Schuster, R. E. *Inorg. Chem.* 1971, **10**, 1627.
221. Cannon, T. H.; Richards, R. E. *Trans. Faraday Soc.* 1966, **62**, 1378.
222. McGarvey, B. R.; Trudell, C. O.; Tuck, D. G.; Victoriano, L. *Inorg. Chem.* 1980, **19**, 3432.
223. Ichikawa, K.; Warren, W. W. *Phys. Revs. B Cond. Matter* 1979, **20**, 900.
224. Glaser, J.; Henriksson, U. *J. Am. Chem. Soc.* 1981, **103**, 6642.
225. Burke, P. J.; Matthews, R. W.; Cresshull, I. D.; Gillies, D. G. *J. Chem. Soc. Dalton Trans.* 1981, 132.
226. Briggs, R. W.; Metz, K. R.; Hinton, J. F. *J. Soln. Chem.* 1979, **8**, 479.
227. Schramm, C.; Zink, J. I. *J. Magn. Reson.* 1977, **26**, 513.
228. Burke, P. J.; Matthews, R. W.; Gillies, D. G. *J. Organomet. Chem.* 1976, **118**, 129.
229. Zink, J. I.; Srivanavit, C.; Dechter, J. J. *Organomet. Polymer (Symp.)*, Carraher, C. E.; Sheats, J. E.; Pittam, C. U., Eds.; Academic: New York, 1978.
230. Hinton, J. F.; Briggs, R. W. *J. Magn. Reson.* 1977, **25**, 555.
231. Briggs, R. W.; Hinton, J. F. *J. Soln. Chem.* 1978, **7**, 1.
232. Briggs, R. W.; Hinton, J. F. *J. Soln. Chem.* 1977, **6**, 827.
233. Briggs, R. W.; Hinton, J. F. *J. Soln. Chem.* 1979, **8**, 579.
234. Briggs, R. W.; Etzkorn, F. A.; Hinton, J. F. *J. Magn. Reson.* 1980, **37**, 523.
235. Briggs, R. W.; Hinton, J. F. *J. Magn. Reson.* 1979, **33**, 363.
236. Briggs, R. W.; Hinton, J. F. *J. Magn. Reson.* 1978, **32**, 155.
237. Briggs, R. W.; Hinton, J. F. *Biochem.* 1978, **17**, 5576.
238. Hinton, J. F.; Turner, G. L.; Millett, F. S. *J. Magn. Reson.* 1981, **45**, 42.
239. Hinton, J. F.; Metz, K. R.; Millett, F. S. *J. Magn. Reson.* 1981, **44**, 217.
240. Gudlin, D.; Schneider, H. *Inorg. Chim. Acta.* 1979, **33**, 205.
241. Srivanavit, C.; Zink, J. I.; Dechter, J. J. *J. Am. Chem. Soc.* 1977, **99**, 5876.

242. Shamsipur, M.; Rounaghi, G.; Popov, A. I. *J. Soln. Chem.* 1980, **9**, 701.
243. Popov, K. I.; Vekshin, V. V.; Chuvaev, V. F.; Vechurova, N.I. *Izv. Akad. Nauk. SSSR, Ser. Khim.* 1978, 1195.
244. Hinton, J. F.; Metz, K. R. *J. Soln. Chem.* 1980, **9**, 197.
245. Burke, P. J.; Matthews, R. W.; Gillies, D. G. *J. Chem. Soc. Dalton Trans.* 1980, 1439.
246. Hoad, C. S.; Matthews, R. W.; Thakur, M. M.; Gillies, D. G. *J. Organomet. Chem.* 1977, **124**, C31.
247. Burke, P. J.; Gray, L. A.; Hayward, P. J. C.; Matthews, R. W.; McPartlin, M.; Gillies, D. G. *J. Organomet. Chem.* 1977, **136**, C7.
248. Lallemand, J.-Y.; Duteil, M. *Org. Magn. Reson.* 1976, **8**, 328.
249. Lallemand, J.-Y.; Michon, V. *J. Chem. Res. (S)* 1978, **162(M)**, 2081–2090.
250. Barron, P. F.; Doddrell, D.; Kitching, W. *J. Organomet. Chem.* 1977, **132**, 35.
251. Morishima, I.; Inubushi, T.; Uema, S.; Miyoshi, H. *J. Am. Chem. Soc.* 1978, **100**, 354.
252. Deacon, G. B.; Slade, R. M.; Vince, D. G. *J. Fluorine Chem.* 1978, **11**, 57.
253. Van Vliet, P. I.; Vrieze, K. *J. Organomet. Chem.* 1977, **139**, 337.
254. Hinton, J. F.; Ladner, K. H. *J. Magn. Reson.* 1978, **32**, 303.
255. Hinton, J. F.; Briggs, R. W. *J. Magn. Reson.* 1977, **25**, 329.
256. Baugerter, B. W. *J. Magn. Reson.* 1977, **28**, 141.
257. Dechmukh, V. G. L.; Seymour, E. F. W.; Styles, G. A. *Phys. Lett.* 1979, **74A**, 43.
258. Kunwar, A. C.; Thompson, A. R.; Gutowski, M. S.; Oldfield, E. *J. Magn. Reson.* 1984, **60**, 467.
259. Akitt, J. W. *Progr. NMR Spectrosc.*, to be published.

Chapter 10

CARBON

BRIAN E. MANN

A massive database of ^{13}C NMR chemical shifts of inorganic and organometallic compounds has now been built up: more than 1000 relevant papers were published in 1983 alone. There exists a comprehensive listing of ^{13}C NMR data published up to 1978 for organometallic compounds.(1) The Specialist Periodical Report "Spectroscopic Properties of Inorganic and Organometallic Compounds" provides the only comprehensive update of papers containing relevant data.(2) *Chemical Abstracts* is not a comprehensive source as synthetic papers containing ^{13}C NMR data are often not listed in the ^{13}C NMR section.

There are many compounds in the literature that are difficult to categorize and cannot therefore be included in a short chapter such as this. Only the "parent" ligands are considered here. Shift ranges are given in Figures 1–9, coupling constants in Table 1 and Figures 9–12.

Despite the extensive data, it is clear that the ranges are frequently limited by the compounds that have been reported, and are potentially greater. This is shown by the methyl group. In tungsten derivatives, reported chemical shifts cover the range δ-35 to δ90 because of the variety of compounds, whereas for methyl-chromium derivatives the choice is very limited and the range restricted. Therefore these range plots can only be considered to be a guide as to where to find a given type of carbon resonance, and cannot be used to determine structure in the same way as the corresponding plots for organic compounds. New groups of compounds may therefore produce ^{13}C chemical shifts which lie outside the ranges given in the plots. The plots are based on the comprehensive listing of data in *^{13}C NMR Data for Organometallic Compounds*(1) and the reader should refer to this book for details of the data and references.

Figure 1 shows the chemical shift ranges for the methyl group. Most resonances fall between δ-25 and δ25, with the ranges for the transition metals generally more substantial than for the main group metals. In the latter group the methyl chemical shift reflects both charge and steric effects, in the same way as has been observed for organic compounds. Thus Me_4Sn has δ ca. 9, but Me_3SnLi has δ-1.5 and the change is attributable to charge, while on going to Me_3SnBu^t, the methyl signal moves to δ-12.2. For the transition metals various correlations can be seen. Thus coligands of (η^5-C_5H_5) often show increased shielding, as in *cis*-[$PtMe_2(AsMe_3)_2$], with the platinum-methyl signal at δ-4.1, whereas [$PtMe(\eta^5$-$C_5H_5)(CO)$] has $\delta(^{13}CH_3)$ at δ-47.8. Tungsten provides an extreme example with [WMe_6] at δ83.6 and [$WMe(\eta^5$-$C_5H_5)(CO)_3$] at δ-35.

BRIAN E. MANN • Department of Chemistry, University of Sheffield, Sheffield S3 7HF, England.

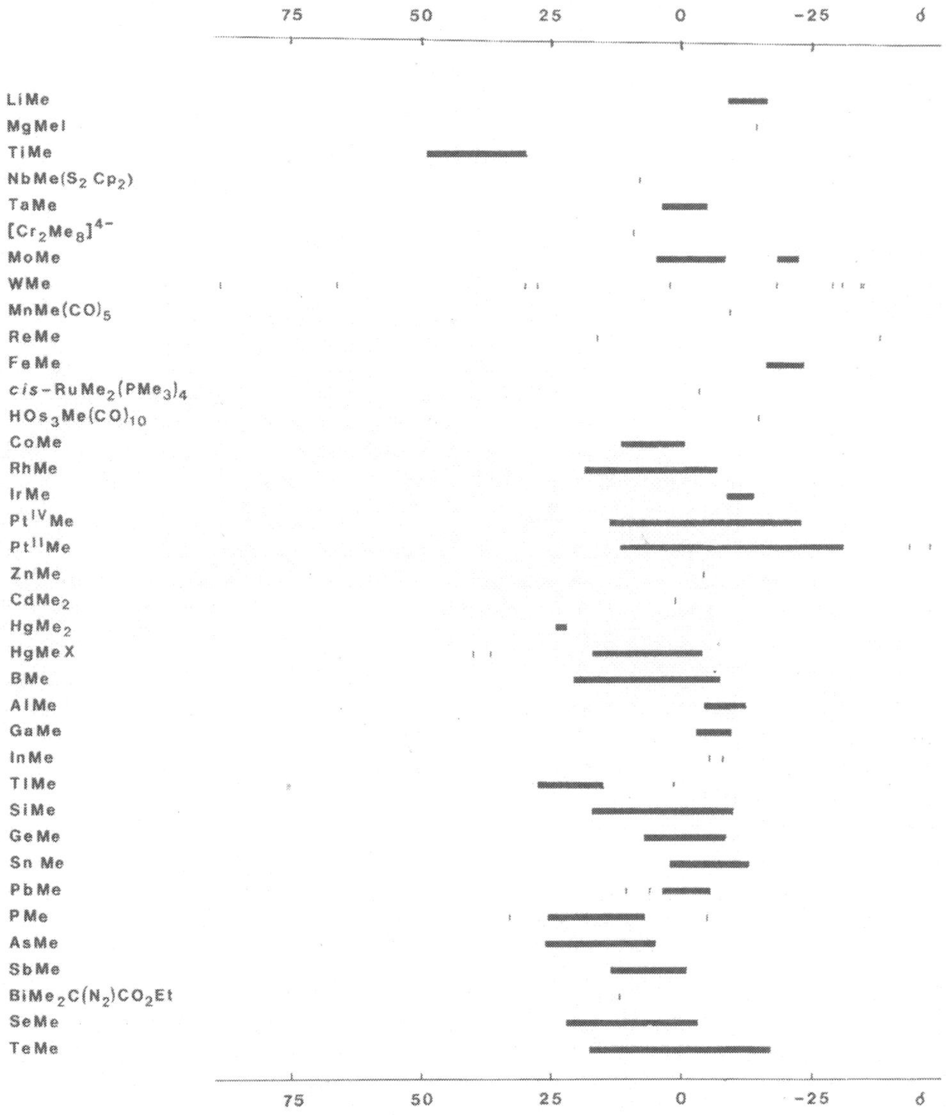

Figure 1. Chemical shift ranges for methyl groups attached to metals.

Figure 2 shows the chemical shift ranges for the phenyl group. At first sight the ranges for all but the phenyl–platinum and phenyl–mercury ipso carbons are narrow, but for the transition metals this arises from the lack of compounds examined. The chemical shifts of the *meta*-carbons are generally close to $\delta 128.5$, but there is some danger of this being a cyclic argument as the signal closest to $\delta 128.5$ is often assigned to the *meta*-carbon. It has been suggested that the chemical shifts of the *para*-, and, to a certain extent, the *ortho*-carbons are related to π-electron density and hence give a measure of metal-phenyl π-bonding.[3]

Figure 3 gives chemical shift ranges for different sorts of metal-carbenes, covering a very wide range, $\delta 170$–$\delta 430$. Very minor changes can cause very marked changes in chemical shift, e.g., $[(OC)_5Cr\mathit{C}Ph_2]$ has $\delta 399.4$, yet $[Cr(CO)_5\mathit{C}(2\text{-furanyl})_2]$ has

Table 1. Some Typical Coupling Constants of Metals to ^{13}C for Organometallic Compounds

Compound	Nucleus	Coupling constant (Hz)
$[LiCH_3]_4$	^7Li	+15
$[Y(CH_2SiMe_3)_4]^-$	^{89}Y	28
$[V(CO)_6]^-$	^{51}V	146
$Mo(CO)_6$	^{95}Mo	68
$W(CH_3)_6$	^{183}W	43.2
$W(CO)_6$	^{183}W	126
$Fe(CO)_5$	^{57}Fe	23.4
$[Co(CO)_4]^-$	^{59}Co	287
$Rh(CH_3)(dmgH)_2(OH_2)$	^{103}Rh	22.5
$[Rh(CO)_4]^-$	^{103}Rh	74.7
$[Pt(CH_3)_6]^{2-}$	^{195}Pt	434
cis-$Pt(CH_3)_2(cod)$	^{195}Pt	773
$Pt(CO)_2Cl_2$	^{195}Pt	1576
$Ag(CH_2PPh_3)_2Cl$	^{109}Ag	96.5
$Cd(CH_3)_2$	^{113}Cd	537.5
$Hg(CH_3)_2$	^{199}Hg	690
$[B(CH_3)_4]^-$	^{11}B	22
$Tl(CH_3)_3$	^{205}Tl	1930
$Si(CH_3)_4$	^{29}Si	−50
$Ge(CH_3)_4$	^{73}Ge	18.7
$Sn(CH_3)_4$	^{119}Sn	340
$Pb(CH_3)_4$	^{207}Pb	250
$Te(CH_3)_2$	^{125}Te	160

δ286.7. Smaller chemical shifts are encountered when electronegative groups such as oxygen or nitrogen are attached to the carbene carbon atom. As a consequence of the large ranges of chemical shifts and limitations imposed by the types of compounds examined, Figure 3 must be used with caution.

Little information is available for metal carbynes. This is collected in Figure 4, which can only be considered as a crude guide. Metal carbonyls have provided extensive ^{13}C NMR data, gathered in Figure 5, sufficient to give correlations. With charge in the metal down the Periodic Table, or from left to right, the signals move to low frequency. However, of more value is the observation that for a given series of compounds for a given metal, the ^{13}C signal moves to lower frequency with increase in the carbon–oxygen stretching frequency, or, better, force constant.[4] Hence all the arguments usually used to explain $\nu(CO)$ can be applied to $\delta(^{13}CO)$. As carbonyl coligands are replaced by weaker π acceptors $\delta(^{13}CO)$ moves to higher frequency.

When an olefin is attached to a metal the carbon changes from sp^2 towards sp^3 hybridization and the ^{13}C resonance moves to lower frequency, the more so the greater the interaction. Unfortunately as the chemical shift of even a methyl group is difficult to predict, these olefin coordination shifts can only be used in a qualitative sense. The limited data on ethylene coordination are collected in Figure 6; extensive data are available only for rhodium and platinum.

The two types of carbon atoms in the η^3-allyl group, the terminal C^1 and middle C^2, are generally well separated with C^1 at lower frequency; see Figure 7. The C^2 carbon atom retains much of its olefinic character while the C^1 carbon atom can move very strongly towards sp^3.

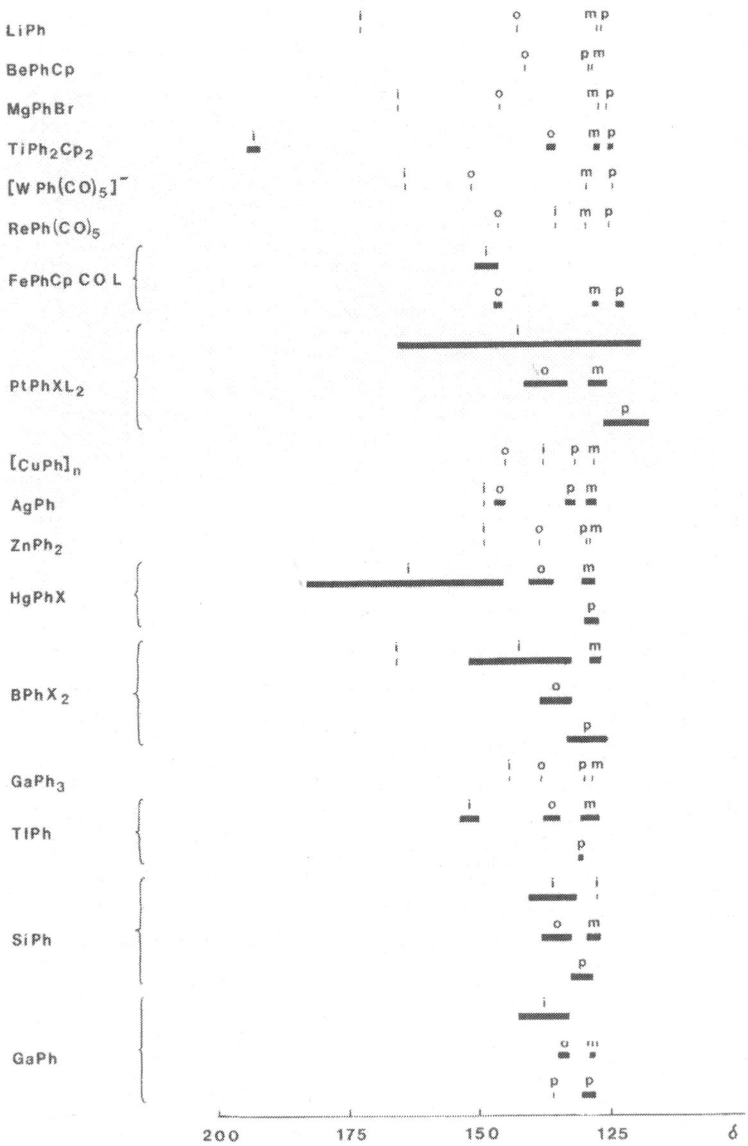

Figure 2. Chemical shift ranges for phenyl groups attached to metals.

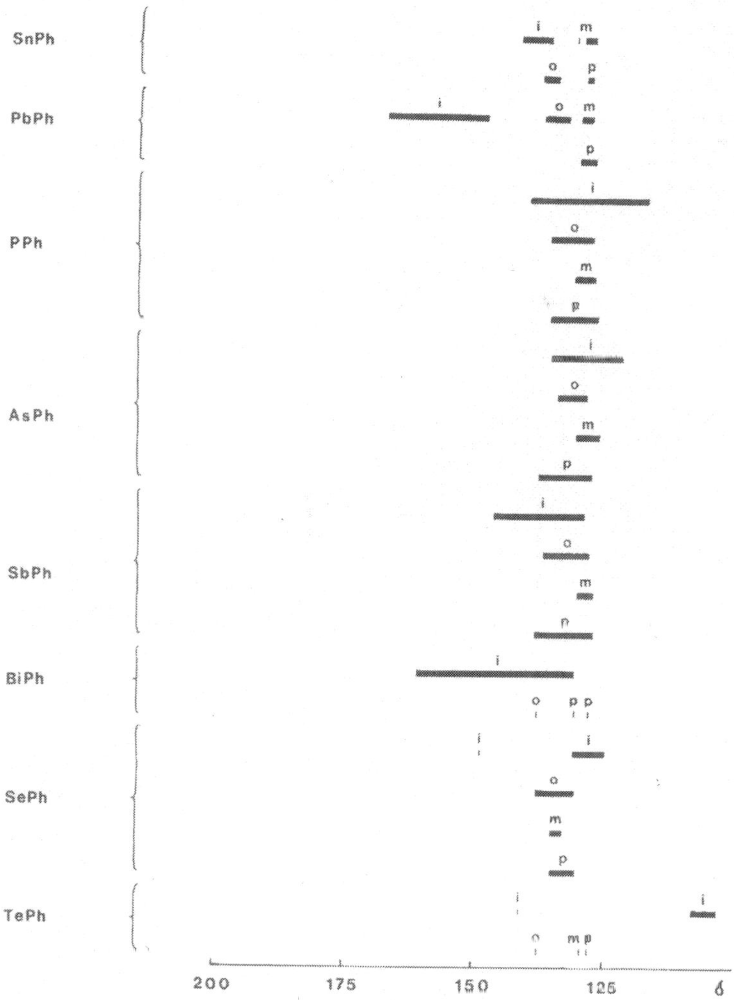

Figure 2. (*Continued*).

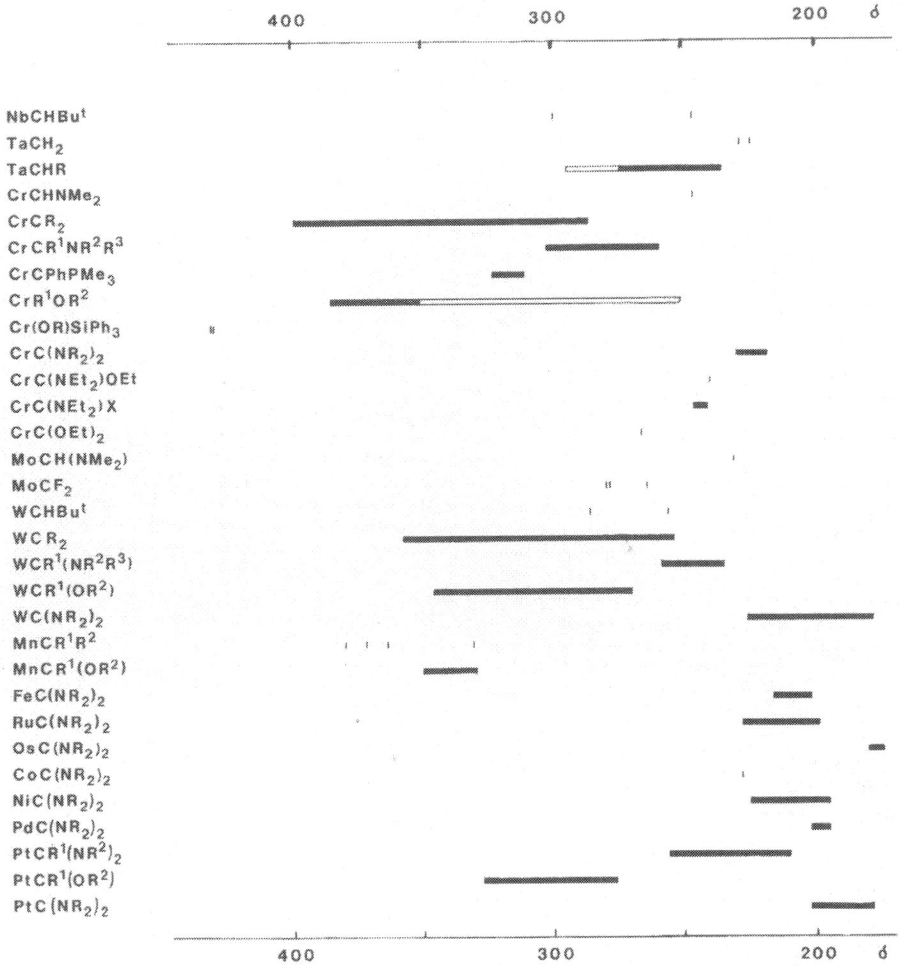

Figure 3. Chemical shift ranges for carbene groups attached to metals.

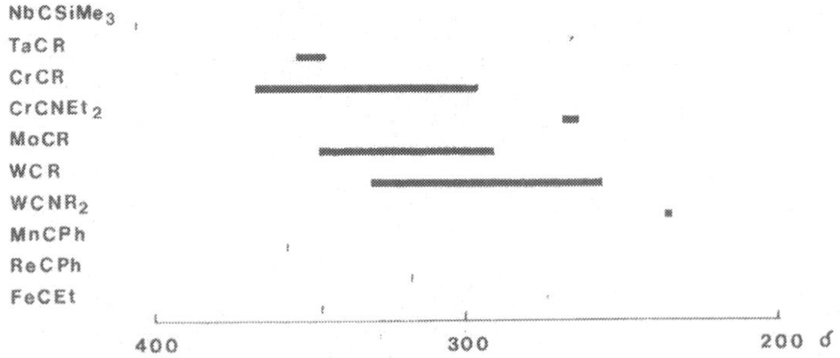

Figure 4. Chemical shift ranges for carbyne groups attached to metals.

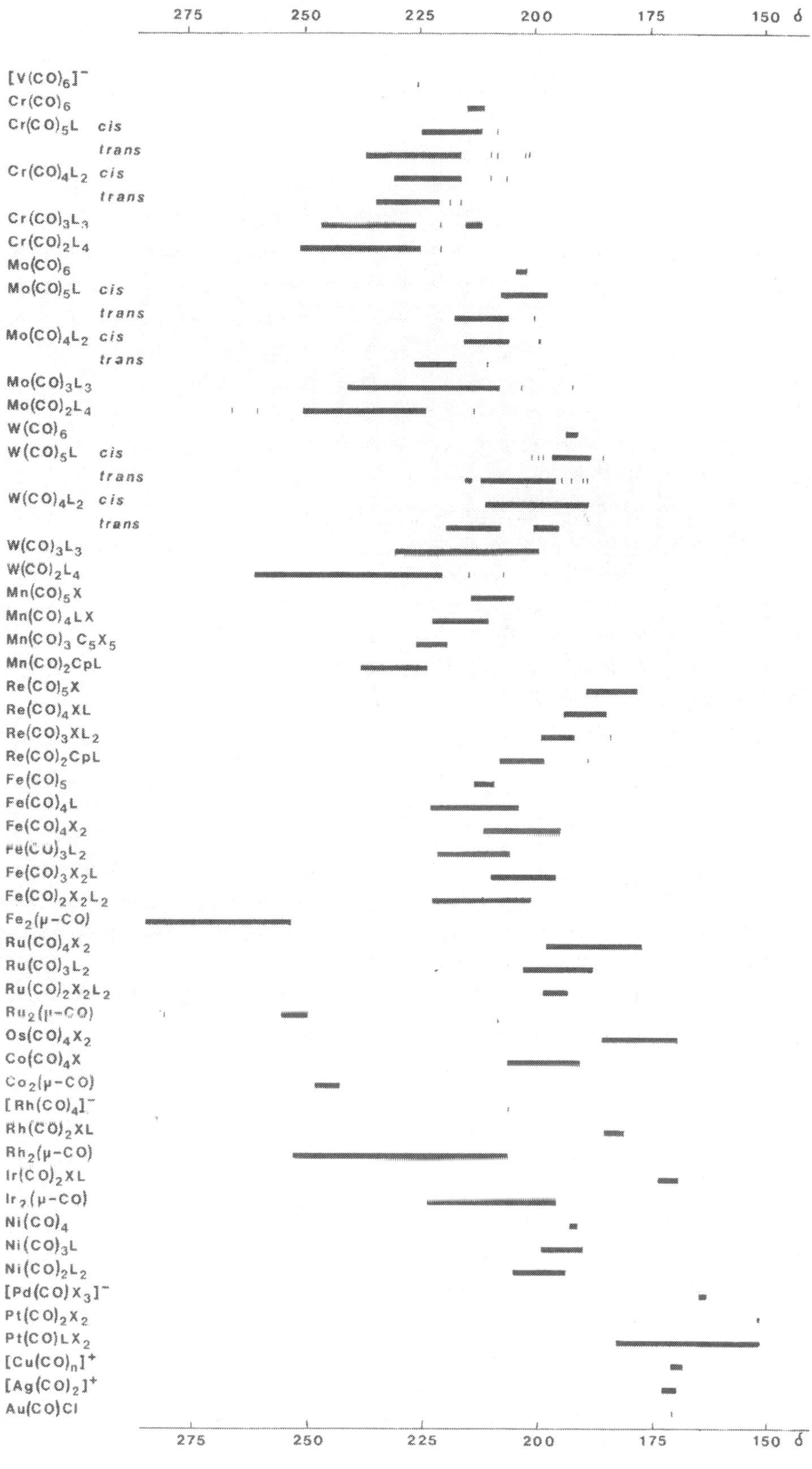

Figure 5. Chemical shift ranges for carbonyl groups attached to metals.

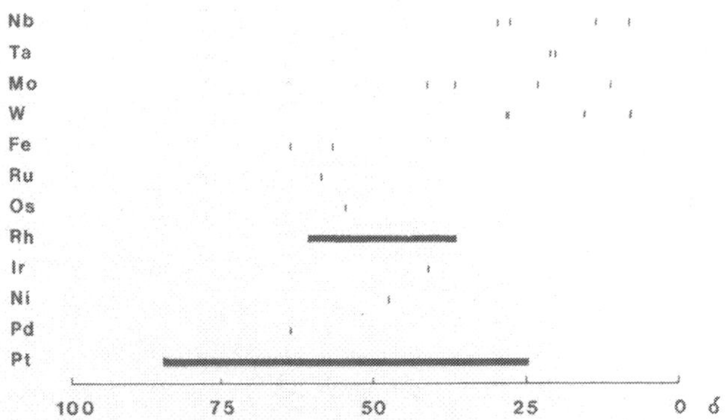

Figure 6. Chemical shift ranges for olefins attached to metals.

The chemical shift of the η^5-C_5H_5 group covers the range $\delta 123$–$\delta 68$; see Figure 8. Unlike most other organometallic metallic compounds, fairly narrow chemical shift ranges are found for a given type of compound. This is mainly because the shift is insensitive to other substituents on the metal, thus providing little structural information.

Apart from chromium very little information is available for η^6-benzene complexes. Hence Figure 9 should only be treated as a crude guide.

Metal–carbon coupling constants are dominated by the Fermi contact interaction and can generally be rationalized qualitatively using the simplified equation

$$J_{AB} \propto \frac{\gamma_A \gamma_B S_A^2(O)\, S_B^2(O)\, \alpha_A^2 \alpha_B^2}{{}^3\Delta E}$$

where γ_A, γ_B are the gyromagnetic ratios for nuclei A and B, $S_A^2(O)$, $S_B^2(O)$ are the valence *s*-electron densities at the nuclei A and B, α_A^2, α_B^2 are the fraction of *s*-electron usage by atoms A and B in forming the AB bond, and ${}^3\Delta E$ is a mean triplet excitation energy. The direct corollary is that changing the carbon hybridization from sp^3 to sp^2, as from CH_3 to CO, causes a marked increase in the coupling constant, e.g., ${}^1J({}^{183}W, {}^{13}C) = 43.2$ Hz in WMe_6 and 126 Hz in $W(CO)_6$. The lack of a precise relationship arises from ignoring changes in the other terms in the equation and from

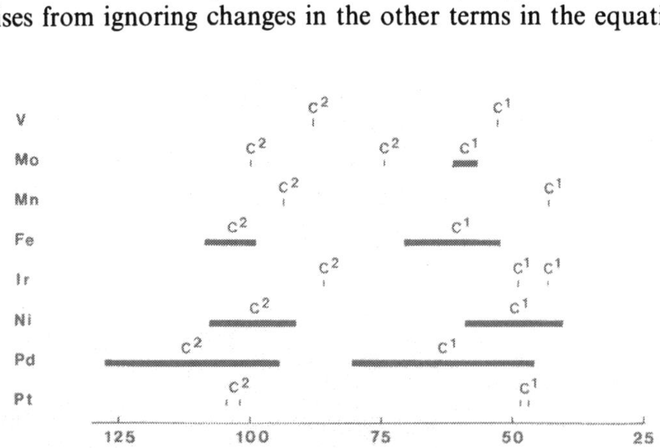

Figure 7. Chemical shift ranges for η^3-allyl groups attached to metals.

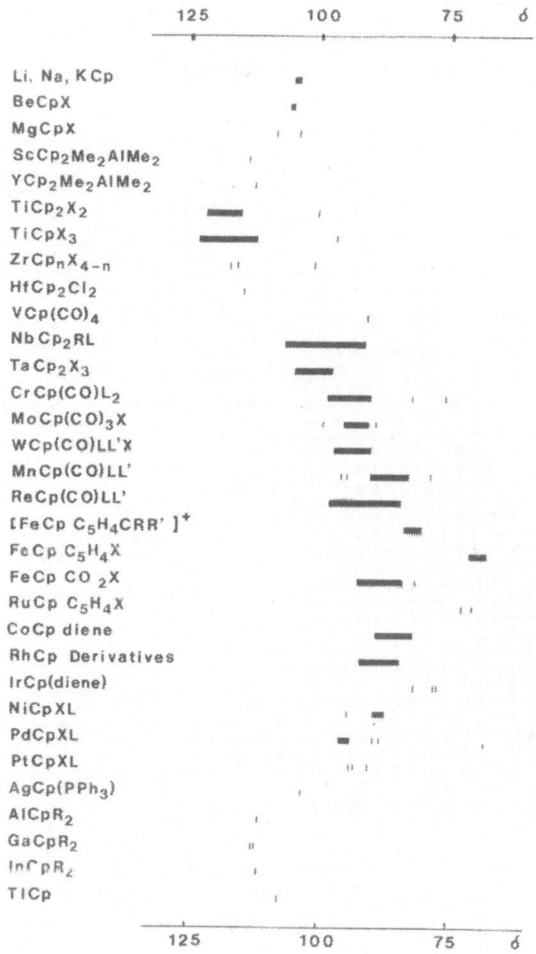

Figure 8. Chemical shift ranges for η^5-cyclopentadienyl groups attached to metals.

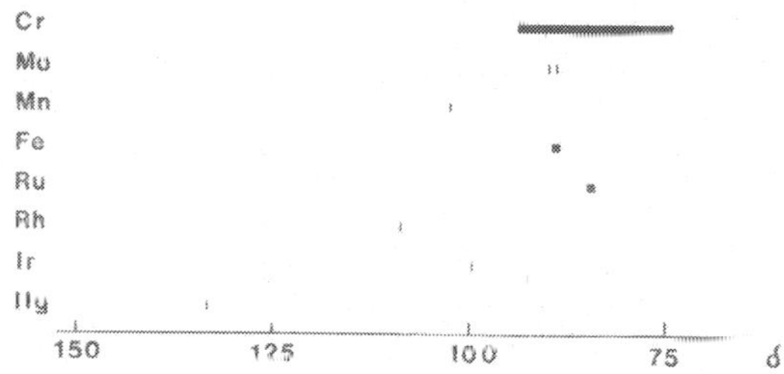

Figure 9. Chemical shift ranges for η^6-arene groups attached to metals.

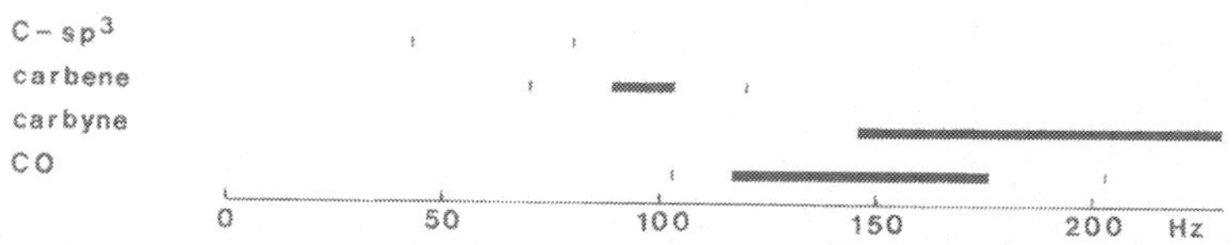

Figure 10. Ranges of coupling constants found for common ligands on ^{183}W.

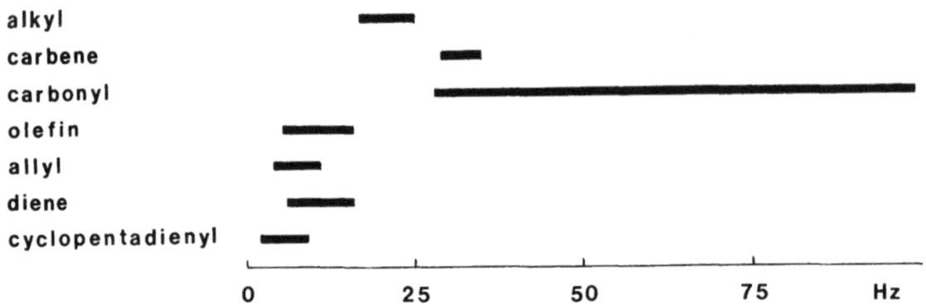

Figure 11. Ranges of coupling constants found for common ligands on ^{103}Rh.

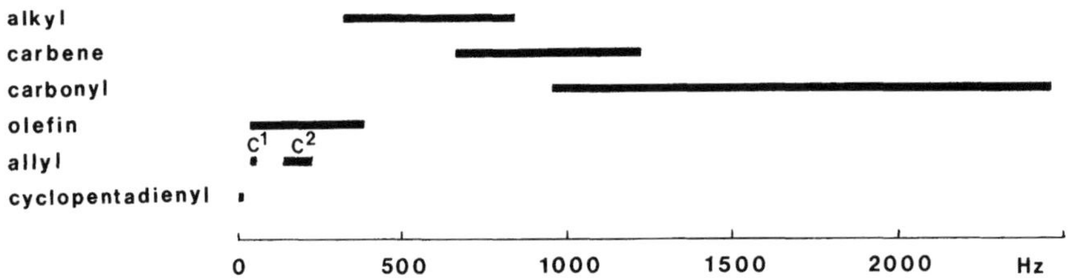

Figure 12. Ranges of coupling constants found for common ligands on ^{195}Pt.

the approximate nature of the equation. The magnitude of the coupling constant is quite dependent on the *trans*-ligand, e.g., $^1J(^{195}Pt, ^{13}C)$ is 673 Hz in *trans*-$Pt(CH_3)Cl(PMe_2Ph)_2$ but 495 Hz in *trans*-$Pt(CH_3)(CN)(PMe_2Ph)_2$. The coupling constant follows the *trans*-influence series. Polyhapto ligands have much smaller $^1J(M, ^{13}C)$ than σ-bonded ligands, reflecting the greater usage of carbon *p* orbitals in the π bonding.

Figures 10–12 show the ranges found for coupling constants for common ligands on ^{183}W, ^{103}Rh, and ^{195}Pt. Other metals that commonly show coupling to carbon are given in Table 1.

REFERENCES

1. Mann, B. E.; Taylor, B. F. *^{13}C NMR Data for Organometallic Compounds*; Academic: London, 1981.
2. Akitt, J. W. (Vols. 1–3), Blackborow, J. R.; Crosbie, K. D. (Vol. 4), and Mann, B. E. (Vols. 5–20), Specialist Periodical Report, "Spectroscopic Properties of Inorganic and Organometallic Compounds," Senior Reporters Greenwood, N. N. (Vols. 1–9), Ebsworth, E. A. V. (Vols. 10–20), and Adams, D. M. (Vols. 11–20), Chap. 1, The Royal Society of Chemistry, 1968–1987.
3. Nelson, G. L.; Levy, G. C.; Cargioli, J. D. *J. Am. Chem. Soc.*, 1972, **94**, 3089.
4. See, for example, Mann, B. E. *J. Chem. Soc., Dalton Trans.*, 1973, 2012.

ADDITIONAL REFERENCES

Books

Breitmaier, E.; Bauer, G. *^{13}C-NMR-Spektroskopie*; Georg Thieme Verlag: Stuttgart, 1977.
Breitmaier, E.; Voelter, W. *^{13}C-NMR Spectroscopy*, Verlag Chemie: Weinheim, 1974.
Clerc, J. T.; Pretsch, E.; Sternhell, S. *^{13}C-Kernresonanzspektroskopie*; Akademische Verlagsgesellschaft: Frankfurt am Main, 1973.

Levy, G. C.; Lichter, R. L.; Nelson, G. L. *Carbon-13 Nuclear Magnetic Resonance Spectroscopy*, 2nd ed.; Wiley-Interscience: New York, 1980.
Stothers, J. B. *Carbon-13 NMR Spectroscopy*; Academic: New York, 1972.
Wehrli, F. W.; Wirthlin, T. *Interpretation of Carbon-13 NMR Spectra*; Heyden: London, 1976.

Reviews—General

Chisholm, M. H.; Godleski, S. "Applications of C-13 NMR in Inorganic Chemistry," *Progr. Inorg. Chem.* 1975, **20**, 299–436.
Gansow, O. A.; Vernon, W. D. "Carbon-13 NMR Studies of Organometallic and Transition Metal Complex Compounds," Topics in C-13 NMR Spectroscopy 1976, **2**, 269–341.
Jolly, P. W.; Mynott, R. "The Application of ^{13}C NMR Spectroscopy to Organo–Transition Metal Complexes," *Adv. Organomet. Chem.* 1981, **19**, 257–304.
Mann, B. E. "^{13}C NMR Chemical Shifts and Coupling Constants of Organometallic Compounds," *Adv. Organomet. Chem.* 1974, **12**, 135–213.
Pregosin, P. S.; Kunz, R. W. *^{31}P and ^{13}C NMR of Transition Metal Phosphine Complexes, NMR Basic Principles Progr.* 1976, **16**, 65–74, 141–156.

Relaxation

Lyerla, J. R.; Levy, G. C. *Topics Carbon 13 NMR Spectrosc.* 1974, **1**, 79–148.
Wright, D. A.; Axelson, D. R.; Levy, G. C. *Topics Carbon-13 NMR Spectrosc.* 1979, **3**, Chap. 2.

Coupling Constants

Hansen, P. E. "Carbon–Carbon Coupling Constants: Discussion," *Ann. Rep. NMR Spectrosc.* 1981, **11A**, 66–98.
Wray, V.; Hansen, P. E. "Carbon–Carbon Coupling Constants: Data," *Ann. Rep. NMR Spectrosc.* 1981, **11A**, 99–81.

Substituent Effects etc.

Maciel, G. E. "Substituent Effects on ^{13}C Chemical Shifts," *Topics Carbon-13 NMR Spectrosc.* 1974, **1**, 53–77.
Nelson, G. L.; Williams, E. A. "Electronic Structure and ^{13}C NMR," *Progr. Phys. Org. Chem.* 1976, **12**, 229–342.
Craik, D. J. "Substituent Effects on Nuclear Shielding," *Ann. Rep. NMR Spectrosc.* 1983, **15**, 17–75.

Dynamic Processes

Aime, S.; Milone, L. *Progr. NMR Spectrosc.* 1977, **11**, 183–210.
Band, E.; Muetterties, E. L. *Chem. Rev.* 1978, **78**, 639–658.
Jackman, L. M.; Cotton, F. A. Eds., *Dynamic NMR Spectroscopy*; Academic: New York, 1975.
Mann, B. E. In *Comprehensive Organometallic Chemistry*; Wilkinson, G.; Stone, F. G. A.; Abel, E. W., Ed.; Pergamon: Oxford, 1982, Chap. 20.

Chapter 11

SILICON, GERMANIUM, TIN, AND LEAD

JOHN D. KENNEDY and W. McFARLANE

1. INTRODUCTION

Of the groups of the Periodic Table, IV is perhaps the most interesting to the NMR spectroscopist since with the exception of germanium each of its members has at least one spin-1/2 isotope of moderate or good receptivity. Notwithstanding this, previous reviews have tended to concentrate on only a single element, and relatively little attempt has been made to give comparative treatments. This is perhaps understandable in the case of carbon with its different chemical behavior and its importance in organic chemistry, but is harder to justify for the other elements, and the present chapter now aims at providing an account of the NMR of silicon, germanium, tin, and lead which emphasizes the similarities. In this connection germanium presents a problem owing to the quadrupole moment of ^{93}Ge, the only magnetic isotope, which has severely limited the range of molecules studied, while the immense number of compounds with silicon–oxygen bonds has no parallel among the other elements of the group. Previous reviews of the NMR of these elements are in Refs. 1–5 (silicon), 6 (germanium), 6–11a (tin), and 6 (lead), and up-to-date reports of chemical shifts and coupling constants are available in the Annual Specialist Reports on NMR.(12)

2. EXPERIMENTAL ASPECTS

The summary of the nuclear properties in Table 1 shows that only in the case of tin is any choice of isotope necessary. Usually ^{119}Sn is selected since its receptivity is greater than that of ^{117}Sn, although for certain heteronuclear double resonance experiments the particular resonance frequency of the latter has occasionally offered special advantages. It is difficult to envisage circumstances in which ^{115}Sn might be chosen, except perhaps when there is a requirement for very high spin dilution to avoid homonuclear interactions. The negative magnetogyric ratios of all of the nucleides except ^{207}Pb can lead to disadvantageous NOE behavior, which can be avoided(13) by inverse gated decoupling, or the use of a relaxation time additive such

JOHN D. KENNEDY • Department of Inorganic and Structural Chemistry, University of Leeds, Leeds LS2 9JT, England. W. McFARLANE • Department of Chemistry, Sir John Cass School of Science and Technology, City of London Polytechnic, London EC3N 2EY, England.

Table 1. NMR Properties of Magnetic Isotopes of Silicon, Germanium, Tin, and Lead

Nuclide	Spin	Natural abundance (%)	Receptivity[a]	Resonance frequency (MHz)[b]
^{29}Si	1/2[c]	4.7	2.09	19.867184
^{73}Ge	9/2[c,d]	7.76	0.617	3.488315
^{115}Sn	1/2[c]	0.35	0.069	32.718780
^{117}Sn	1/2[c]	7.61	19.5	35.632295
^{119}Sn	1/2[c]	8.56	25.2	37.290665
^{207}Pb	1/2	22.6[e]	11.7	20.920597

[a] Relative to ^{13}C.
[b] For the reference compound Me_4M in a magnetic field strength for which Me_4Si has a proton resonance of 100 MHz.
[c] γ is negative.
[d] Quadrupole moment = -2.0×10^{-29} m^2.
[e] Varies with source.

as [$Cr(acac)_3$]. Although much of the important earlier work on ^{119}Sn and some of that on ^{29}Si used indirect ^1H–{X} double resonance methods of detection,[14,15] the preferred method now in almost all cases is direct Fourier transform (FT) observation, a technique which can often be greatly improved by using special pulse sequences such as INEPT[16] and DEPT[17] to effect spin-polarization transfer from protons and hence achieve dramatic increases in signal-to-noise ratio with corresponding reductions of spectral accumulation time. This approach also eliminates the negative γ problem and difficulties with long relaxation times, but it does require appreciable X–^1H coupling of approximately known magnitude (just as in the case of the older indirect detection methods). It is thus particularly suitable for systems with $(CH_3)_nSi$ [$^2J(^{29}Si–^1H) \sim 6$ Hz], $(CH_3)_nSn$ [$^2J(^{119}Sn–^1H) \sim 40–100$ Hz], and $(C_6H_5)_nSn$ [$^3J(^{119}Sn–^1H) \sim 20–60$ Hz] groups,[18] and also in principle for corresponding lead compounds, although this potential use seems not to have been used much. Couplings to ^{73}Ge are not often quoted, but nonetheless INEPT experiments have been reported[19] which enhance the signals in some simple relatively symmetrical molecules, and indirect detection via protons has also been used for this nucleus.[6,20] In all cases the standard reference compound is the tetramethyl derivative, $(CH_3)_4M$, and the chemical shift range is such that it can often be used externally, although this is less important on modern spectrometers that permit the use of frequency standards.

In principle direct observation of ^{29}Si, ^{119}Sn, or ^{207}Pb will provide the best sensitivity for the measurement of couplings to carbon, but there will often be problems of dynamic range due to the low abundance of ^{13}C and it may be difficult to assign the splittings in molecules of any complexity; ^{13}C observation is therefore normally the preferred approach for such measurements.

Two-dimensional homonuclear *J*-resolved and COSY ^{29}Si spectra have been used[21] to study polymeric species in aqueous potassium silicate solutions enriched in ^{29}Si, and a two-dimensional chemical exchange ^{119}Sn spectrum has been reported for a five-coordinate ditin compound.[22] Much work has been done on ^{29}Si NMR in the solid state using magic angle spinning (MASS) and cross-polarization (CP) methods,[23–25] and this technique has potential[26] also for ^{119}Sn and ^{207}Pb.

3. CHEMICAL SHIFTS

3.1. Isotope Effects

Only in the case of tin is the question of a *primary* isotope effect (i.e., a dependence of the shielding upon which nucleide of tin is studied) apposite, and it has been established[27] that any such effect is less than 0.1 ppm and therefore irrelevant in most work. Secondary isotope effects arising from a change in a neighboring atom can be more important, especially when H is replaced by D. The one-bond D/H effect can be quite large, e.g., -1.62 ppm in Bu_3SnH for the ^{119}Sn resonance, but the effect is an order of magnitude smaller when two or more bonds intervene.[28,29] However, multiple substitution [e.g., $(CH_3)_4Sn \rightarrow (CD_3)_4Sn$] can additively produce much larger effects.[29] Interesting stereochemical dependences of these longer-range effects have been found, and clearly have diagnostic value.[28] Thus $\delta(^{119}Sn)_D - \delta(^{119}Sn)_H$ has values of $+0.14$ and 0.0 ppm for I(a) and (b), respectively.[28]

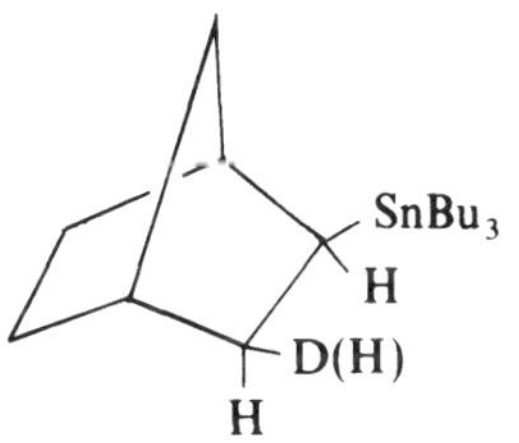

Ia

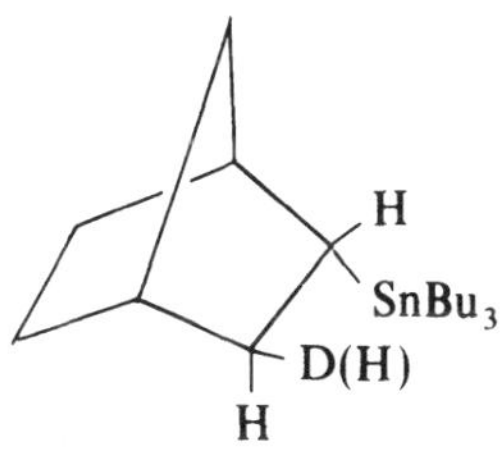

Ib

The corresponding effects upon other group IV resonances are less well studied but can be expected to be greater for ^{207}Pb, and less for ^{29}Si and ^{73}Ge. Other isotopic substitutions such as $^{13}C/^{12}C$ have much smaller effects[30] owing to the smaller relative-mass change, and can generally be ignored. The $^{16}O/^{18}O$ effect upon $\delta(^{29}Si)$ has potential for mechanistic studies.[31]

3.2. Solvent and Temperature Effects

In the absence of any specific chemical interaction leading perhaps to a change in coordination number of the atom of the observed nucleus, changes in solvent affect the group IV chemical shifts relatively slightly in comparison with their overall chemical shift ranges, i.e., a few ppm at most. The effects are greatest for lead and least for silicon. Similarly, changes in temperature are also relatively unimportant in the absence of concurrent chemical changes, although variations in conformational populations can have a significant effect in this regard.[32] By contrast, if there is a chemically significant change then very large shielding differences can occur, and can provide one of the most sensitive methods for studying such behavior.

3.3. Chemical Shift Patterns

Charts I and II, and Tables 3–5, exhibit the main features of silicon, tin, and lead nuclear shieldings in different types of compound, and Table 2 summarizes the position for germanium for which the available data are severely limited. The overall chemical shift ranges presently known are ca. 600, 3000, and 9000 ppm for silicon, tin,

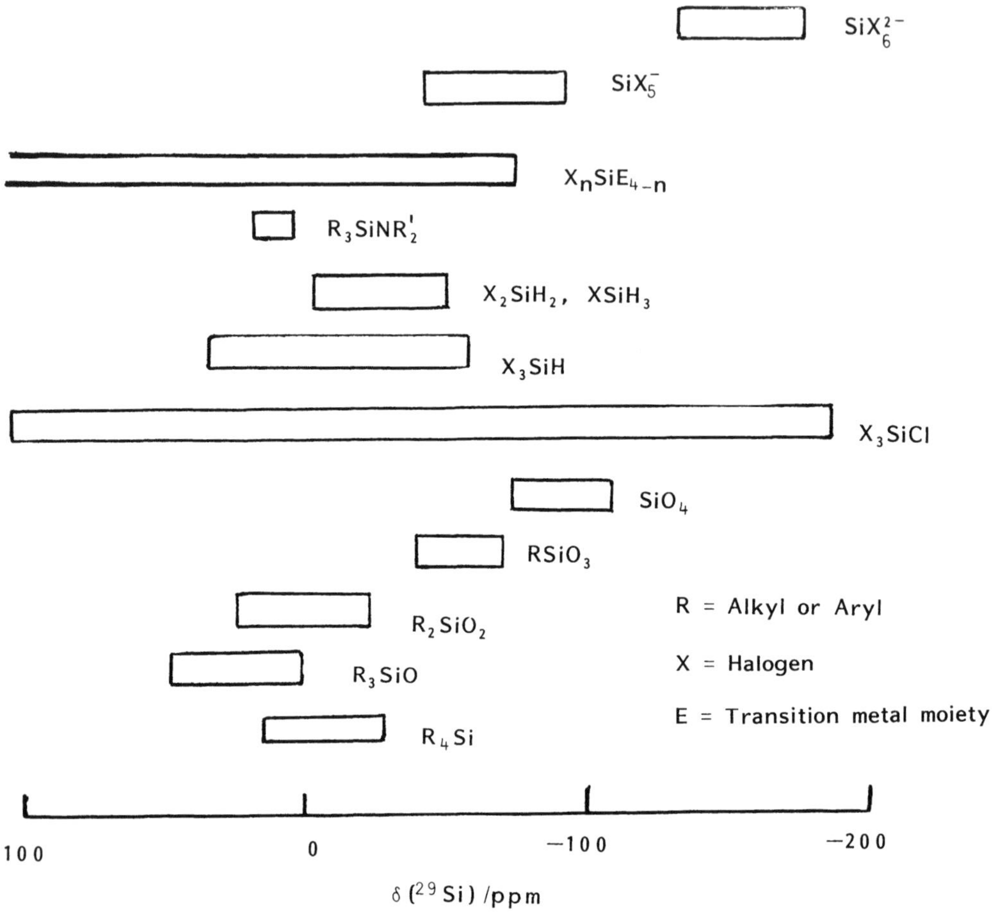

Chart 1. ^{29}Si Chemical Shift Patterns.

and lead, respectively, and while these to some extent reflect the types of compound that have actually been studied, they also parallel the trends in the radial expansion term $\langle r^{-3} \rangle_n$ of equation (10) in Chapter 3 with increasing atomic number. This kind of dependence upon $\langle r^{-3} \rangle_n$ was demonstrated[33] in 1977 (republished 1983[34]) by noting that a plot of $\delta(^{207}Pb)$ against $\delta(^{119}Sn)$ in similar compounds is a straight line of slope ca. 3.0. More-recently measured (and extreme) chemical shift values have also been found to confirm this; for example,[35] $\delta(^{119}Sn) = -1600$ ppm in the cluster $SnPb_8^{4-}$ compared with $\delta(^{207}Pb) = -4152$ for the corresponding lead atom in Pb_9^{4-}, and $\delta(^{207}Pb) = -2670$ in $PbSn_8^{4-}$ compared with $\delta(^{119}Sn) = -1250$ ppm for the corresponding tin atom in Sn_9^{4-}. Similarly the sensitivity of $\delta(^{119}Sn)$ to changes of R′ in $R_3SnOC(O)R'$ has been found[36] to be considerably greater than that of $\delta(^{29}Si)$ in $R_3SiOC(O)R'$, which is again partially attributable to the differences in $\langle r^{-3} \rangle_n$, a result that has been confirmed[34] by plotting $\delta(^{119}Sn)$ against $\delta(^{29}Si)$ as was done for the lead/tin pair.[33] A similar Ge/Sn relationship has also been found.[37] Thus while the correlation is not exact (and the relationship with $\langle r^{-3} \rangle_n$ even less so) a knowledge of the chemical shift of one group IV nucleus can provide a useful guide to others in analogous compounds.

One very striking feature of group IV chemical shifts is that while in EX_nY_{4-n} for X, Y = halogen there is an approximately linear relationship between δ(E) and n,[86]

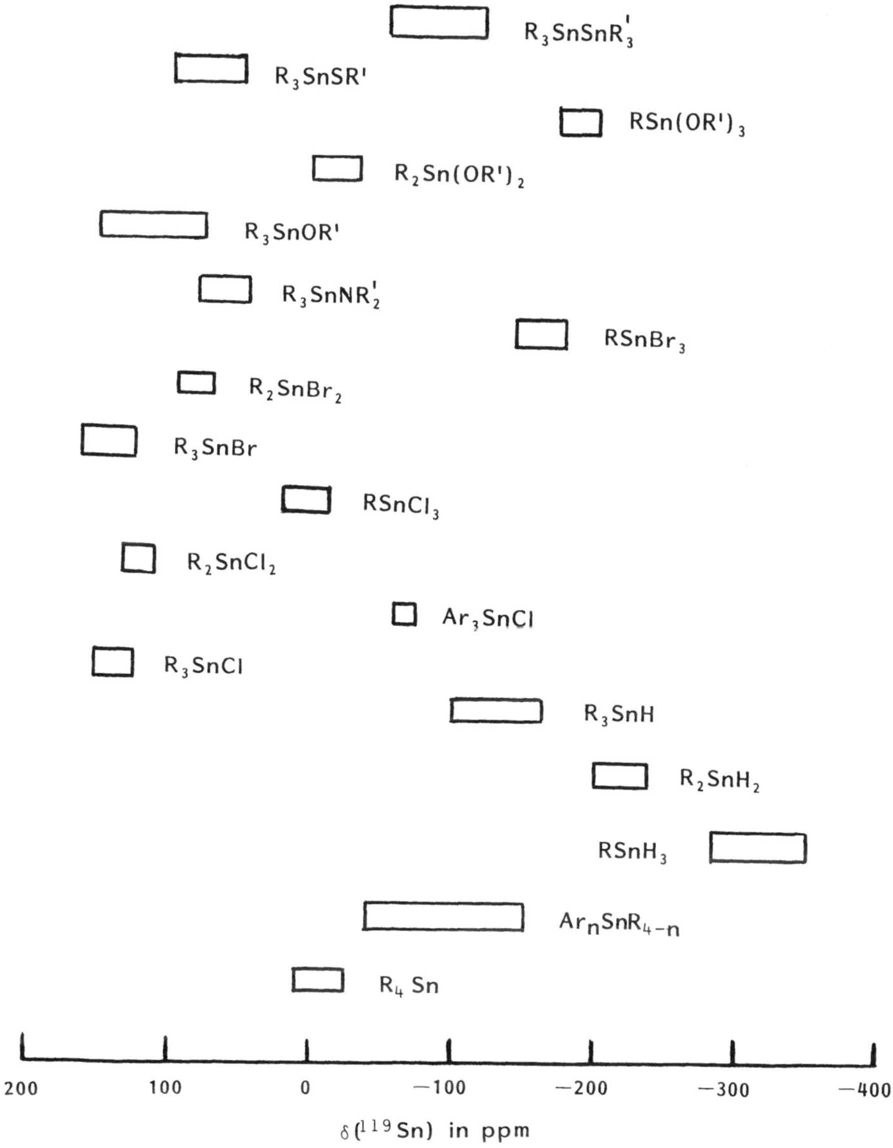

Chart 2. ^{119}Sn Chemical Shift Patterns in Organotin Compounds.

this is often not the case when either X or Y is not a halogen. Instead, plots of *shielding* σ (i.e., $-\delta$) against n have a characteristic "U"-shaped appearance with a degree of asymmetry which depends upon the values of $\delta(E)$ for $n=0$ and 4. This is illustrated in Figure 1 for $\delta(^{119}\text{Sn})$, and similar graphs are available for ^{29}Si and ^{207}Pb.[1-6] It is particularly to be noted that for $n=1$ or 2 $\delta(E)$ may exceed the values for $n=0$ and 4.

3.4. Factors Influencing Shielding

It is now generally agreed that variation of the "paramagnetic term," σ^p, [Chapter 3, equation (10)] is the dominating factor for all four elements, and discussion usually centers on the effects of changing the radial expansion terms, the electron

Table 2. ^{73}Ge Chemical Shifts (in ppm) relative to Me_4Ge

Compound	$\delta(^{73}Ge)$	Reference
Me_4Ge [a]	.0[a]	217
Et_4Ge [b]	17.3	217
n-Pr_4Ge	2.4	218
n-Bu_4Ge	6.0	218
Ph_4Ge [c]	−31.6	217
$(2\text{-furyl})_4Ge$ [d]	−112.8	217
GeH_4	−283.7	218
Me_2GeH_2	−127.6	218
Ge_2H_6	−300.5	218
$(MeO)_4Ge$	−36.0	220
$(MeS)_4Ge$	153	221
$GeCl_4$	30.9	220, 222
$GeBr_4$	−311.3	218
GeI_4	−1081	220
$GeCl_3Br$	−48	222
$GeCl_2Br_2$	−131	218
$GeClBr_3$	−219	222
$GeCl_2I_2$	−524	218
$GeClBrI_2$	−614	218
$MeOGe(OCH_2CH_2)_3N$	−60.6	20
$Me_3SiGe(OCH_2CH_2)_3N$	−73.8	20

[a] Linewidth = 1.3 Hz.
[b] Linewidth = 1.7 Hz.
[c] Linewidth = 6.0 Hz.
[d] Linewidth = 11.0 Hz.

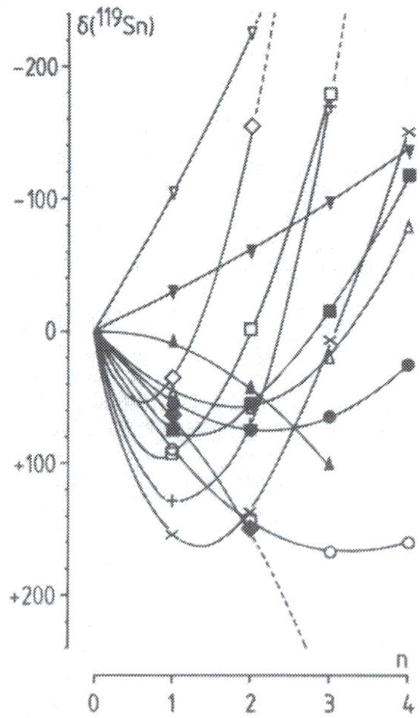

Figure 1. Plot of $\delta(^{119}Sn)$ (in ppm) against n for series of monomeric tin compounds $Me_{4-n}SnX_n$, where X = I (◇), Br (+), Cl (×), O^{t}Bu (□), SMe (○), S^{t}Bu (●), SeMe (△), NMe_2 (■), H (▽), Ph (▼), and $Mn(CO)_5$ (◆). Also included are data for the series $Me_3SnCH_{3-n}Br_n$ (▲).

imbalance terms, and the electronic excitation term, although the last of these is often ignored. Additionally there is the question of the relative importance of the *p* and *d* orbitals, and "for simplicity" the latter are also often ignored. It is, however, quite difficult to account for the effects of heavy atom substituents in this way.

3.4.1. *Effective Nuclear Charge*

Substituents which lead to electron withdrawal should increase the paramagnetic term σ^p and hence give resonances to higher frequency (similar behavior is also predicted if the diamagnetic term σ^d is considered to dominate), but simple consideration of species such as $(CH_3)_3SiX$ or $(CH_3)_3SnX$ yields very poor correlations of $\delta(^{29}Si)$ or $\delta(^{119}Sn)$ with the electronegativity of the atom X. If the site of variation is further removed from the observed nucleus so that inductive effects might be expected to predominate then good agreement is obtained, as in the case of the carboxylates mentioned above,[36] and in many substituted phenyl derivatives of silicon and tin. Thus while it is clear that electronegative substituents do lead to higher-frequency silicon and tin (and presumably germanium and lead) chemical shifts, the effects are often masked by larger ones due to other causes. In this connection it is worth noting that in anionic clusters the high tin or lead shielding[35] can be attributed either to high electron density on the metal or to large diamagnetic circulation in an extended molecular orbital system.

3.4.2. *Interbond Angles*

In molecules in which the metal atom is part of a ring system the silicon, tin, or lead chemical shifts are often found to be anomalous when compared with those of their acyclic counterparts. This behavior is particularly well documented for tin and lead for species such as II–VI[33,38,39] and it appears that a reduction of the cyclic inter-

II, $\delta(^{119}Sn) = +54$

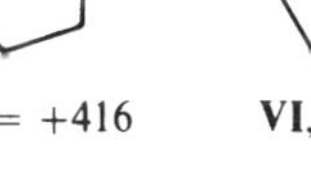

III, $\delta(^{119}Sn) = -43$

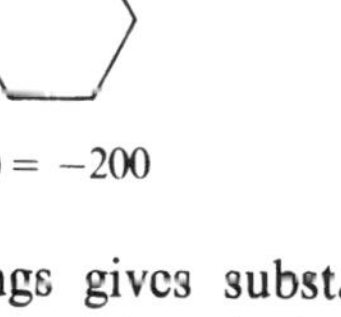

IV, $\delta(^{119}Sn) = +190$

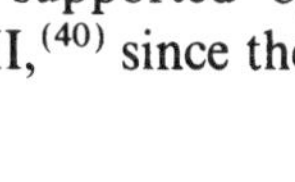

V, $\delta(^{207}Pb) = +416$

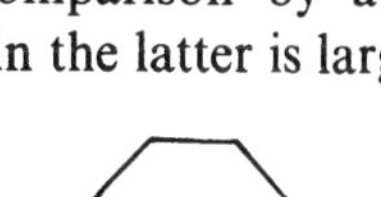

VI, $\delta(^{207}Pb) = -200$

bond angles at the metal in the five-membered rings gives substantially reduced shielding, while an enlargement of them in the six-membered rings increases the shielding. This view is supported by a comparison by a comparison of the tin shieldings in VII and VIII,[40] since the ring in the latter is large enough for there to be

VII

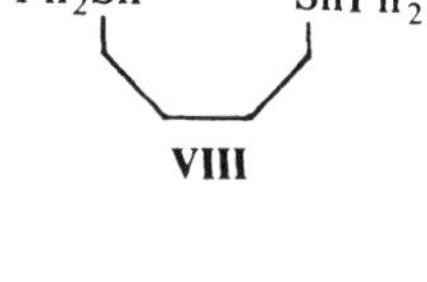

VIII

little or no angular deformation. It is tempting to attribute many of the apparently erratic variations in tin and lead chemical shifts among mixed tetraalkyl derivatives to similar deformations of interbond angles according to the bulk of substituents, but it is difficult to achieve a rational pattern in this way, and it should be remembered that quite small changes in vibrational and rotational energies may have a pronounced effect, as is demonstrated by multiple deuterium substitution in methyl groups.[29]

Incorporation of the silicon atom in a four- or five-membered ring decreases its nuclear shielding by 15–20 ppm and it has been suggested[41] that comparison of the behavior in four- and five-membered rings can provide an indication of the amount of angle strain in the former. In contrast with this, however, incorporation of silicon in a three-membered ring *increases* the shielding by some 40–60 ppm.[42,43]

3.4.3. Coordination Number

It is well established that an increase in coordination number from four to five or six will lead to significant increases in the shielding of tetravalent silicon, tin, and lead. The limited available evidence[20] indicates that this is also true for germanium in that the germatranes IX with five-coordinate germanium have $\delta(^{73}Ge)$ some 20–40 ppm

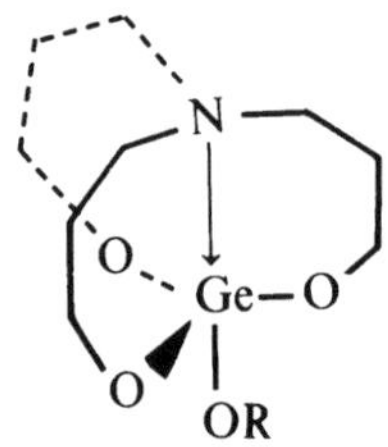

IX

lower in frequency than the four-coordinate $Ge(OMe)_4$. In cationic and neutral hexacoordinate silicon complexes of the types X and XI $\delta(^{29}Si)$ ranges from −198 to

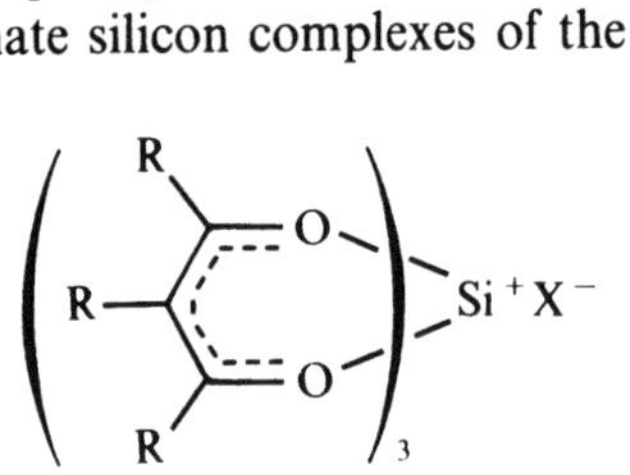

X

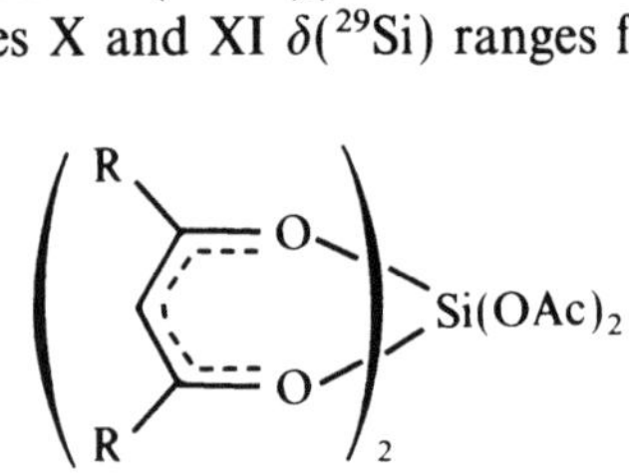

XI

−188 ppm, i.e., substantially to low frequency of four-coordinate species, and shows little variation as *R* is changed, or indeed upon the formal charge on silicon, although this will of course be substantially delocalized onto the substituents.[44] In the five-coordinate XII $\delta(^{29}Si)$ is −176 ppm, again well to low frequency of four-coordinate species.[44]

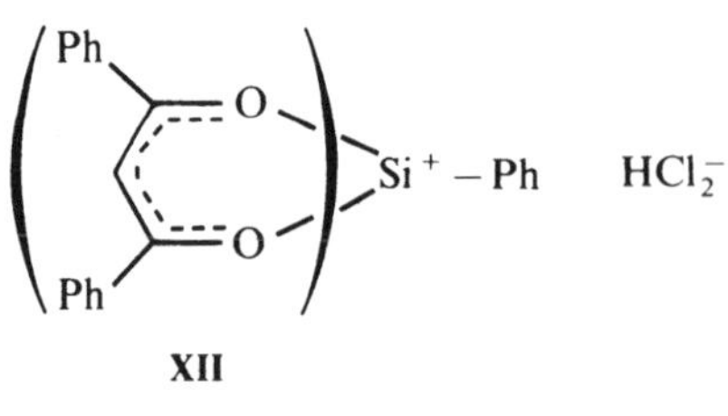

XII

This kind of behavior is best displayed and documented in the case of tin, although it should be emphasized that this is primarily a consequence of chemistry rather than an NMR feature. In general, increase in tin coordination number from four to five and then to six leads to increases in tin shielding of ca. 150 ppm per step,[14] with an increase of a further 100 ppm being associated with sevenfold coordination[45] although this last effect is less well established. This behavior has been variously attributed to π bonding, and to occupancy of the tin 5*d* orbitals, and it leads to an extreme variability of tin chemical shifts in donor solvents like DMSO.[46] Tin compounds which are capable of autoassociation also have temperature- and concentration-dependent chemical shifts for the same reason.[47] Lead compounds have been less well studied in this regard, but tend to exhibit the same behavior[39,48] with significantly larger changes in $\delta(^{207}Pb)$.

3.4.4. Other Effects

These are grouped together because their precise nature is less well established. Exceptionally, silicon can participate in $p\pi$–$p\pi$ bonding, and in the silene XIII the sp^2-hybridized silicon atom has $\delta(^{29}Si) = +41.4$ ppm.[49] Analogous tin and lead compounds are not known, but the tin chemical shift of $(\eta^5\text{-}C_5H_5)_2Sn$, a compound of tin (II), is very low frequency at -2199 ppm.[6,50] Silicon is often involved in $d\pi$–$p\pi$ bonding, but although this can be expected to affect $\delta(^{29}Si)$ it is not yet possible to use this diagnostically. Comparable behavior for tin and lead is much less common and again the shieldings cannot be used as a reliable criterion of this phenomenon.

Exceptionally high-frequency shifts are found for tin bound to one or more first row transition metals,[51,52] and these have been attributed[51] to a reduction of ΔE [equation (10), Chapter 3] resulting from $d\pi$–$d\pi$ overlap between tin and the transition metal. Such interaction is less favorable for silicon, and silicon chemical shifts apparently do not show such large effects,[53] although XIV provides an exception to this with $\delta(^{29}Si) = +173$ ppm.[54] Similarly, $d\pi$–$d\pi$ overlap is less efficient between tin and a second or third row transition metal, with the result that shifts to relatively low frequency are found.[51]

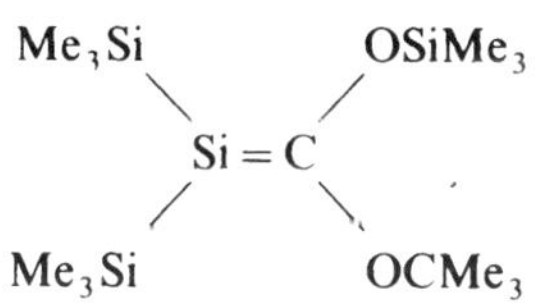

XIII

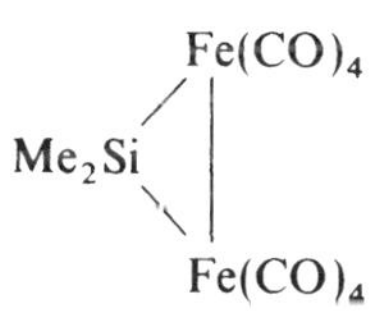

XIV

An additional contributing factor to this last behavior is probably the so-called "bulky substituent" effect. It is found that silicon, tin, and lead shieldings are substantially increased by directly bound heavy atoms such as iodine, bromine, and selenium[14,38,55–57] and some of the heavier transition metals themselves.[51] To some extent this effect is undoubtedly due to diamagnetic circulation on the polarizable large atom(s),[58,59] but calculations have indicated that other factors must also be involved.[60,60a]

3.5. Chemical Shifts in Specific Classes of Compound

It should be clear from the foregoing that while a great deal has been published on the influence of various factors upon the shielding of the group IV nuclei, our detailed understanding is far from complete, and it is not possible to transfer the general discussion in Chapter 3 to specific cases without detailed commentary. In most cases the chemical shift depends upon several competing factors, and for practical purposes it is usually necessary to rely upon one of the many tabulations that are available.[1–11a] Broad indications are given by charts 1–2, and Tables 2–5 are provided

Table 3. Selected ^{29}Si Chemical Shifts (Relative to Me_4Si)

Compound	$\delta(^{29}Si)$ (ppm)	Notes	Reference
SiH_4	−9.9		61, 62
Me_2SiH_2	−37.3		62
$Si(CH=CH_2)_4$	−30.3		63
$Me_3SiC\equiv CSiMe_3$	+19.3		64
$Me_3SiC^+Ph_2$	+5.4		65
$Si(C_3H_5)_4$	−18.9 to −49.6	Range of isomers	66
SiF_4	−109.0		67
$SiCl_4$	−20.0		61
Me_2SiCl_2	+32.2		68
$SiBr_4$	−93.6		61
SiI_4	−346.2		61
Me_3SiCN	−12.1		69
H_3SiNCO	−58.1		69
Me_3SiNCO	+7.4		70
$H_3SiNCSe$	−53.5		70
$(Me_3Si)_4Si$	−9.8(a), −135.5	(a) Me_3Si	71
$[(Me_3Si)_3Si]Li$	−5.2(a), −188.0	(a) Me_3Si	71
$Me_2Si[SiPh_3]_2$	−45.9(a), −17.0	(a) Me_2Si	72
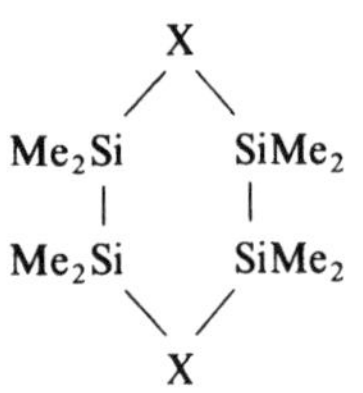	+2.7	X = O	73
	−5.4	X = S	73
	−9.8	X = NH	73
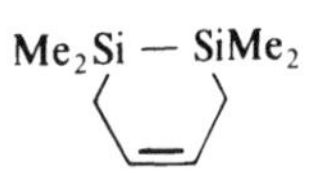	−22.3		74
$K_3Ni(SiPh_3)_3(thf)_3$	+29.5	In thf solution	75
$(\eta^5\text{-}C_5H_5)Fe(CO)_2SiMe_3$	+41.9	In CS_2	76
$(Me_3SiO)_4Si$	+7.5 to +8.8	Me_3Si, pH dependent	77
$(Me_3SiO)_4Si$	−104 to −105	$Si(O_{1/2})_4$, pH dependent	77
SiO_4^{4-}	−71	pH dependent	78
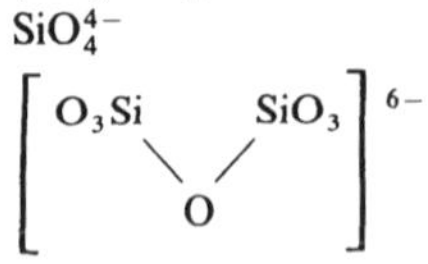	−79.6	pH dependent	78
$(Me_3C_6H_2)_2Si=Si(C_6H_2Me_3)_2$	−64	Calculated on basis that $\delta = -\frac{1}{3}(\sigma_{xx}+\sigma_{yy}+\sigma_{zz})$	78a

Table 4. Selected ^{119}Sn Chemical Shifts (Relative to Me_4Sn)

Compound	$\delta(^{119}Sn)$ (ppm)	Notes	Reference
$(Me_3Sn)_4C$	+49.8		79
$(Me_3Sn)_2CH_2$	+23.3		79
$(CD_3)_4Sn$	+2.6		29
Pr^i_4Sn	−43.9		80
$MeBu^t_3Sn$	−25.4		81
Me_3SnCCl_3	+85		40
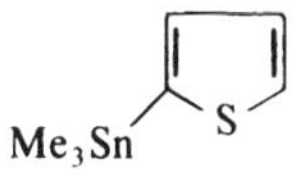	−27.5		82
$Sn(C{\equiv}CH)_4$	−279		83
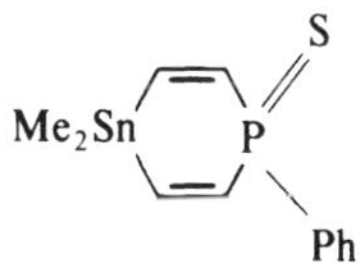	−178.9		84
$(C_3H_5)_4Sn$	−143 to −234	Mixture of 15 isomers	85
$SnCl_4$	−150		86, 87
SnI_4	−1701		86, 87
Me_2SnCl_2	+137	In nondonor solvents	40
$SnCl_6^{2-}$	−732	Bu_4N^+ salt, −30°C in CH_2Cl_2	87
$SnBr_6^{2-}$	−2064	Bu_4N^+ salt, −30°C in CH_2Cl_2	87
$[SnBr_4(PBu_3)_2]$	−935	−30°C in CH_2Cl_2	87
$SnCl_2$	−388.1	In 12 *M* HCl, tin (II) species	88
$(Me_3Sn)_2O$	+109.5	Solvent dependent	89
Me_3SnOMe	+129		90
$MeSn(OBu^t)_3$	−117		90
$Me_4Sn_2(OAc)_2$	−128	Sn–Sn bond, bridging OAc	147
$BuSn(oxinate)_3$	−561	Seven-coordinate	45
$Sn(OH)_6^{2-}$	−591.0	Na^+ salt, aq.	88
$(Me_3Sn)_2S$	+86.5		89
$(Me_3Sn)_2Se$	+44.5		89
$(MeS)_4Sn$	+160		92
$(MeSe)_4Sn$	−80.5		93
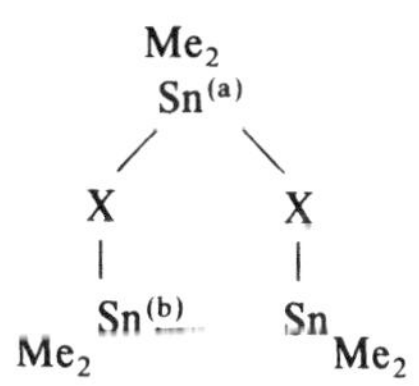		(a) (b)	
	X = S	+44 +176	
	X = Se	+21 +82	94
	X = Te	−38 −164	
$[Me_2SnS]_3$	+128	Cyclic trimer	14
$[Me_2SnSe]_3$	+42	Cyclic trimer	94
$[SnTe_4]^{4-}$	−1828	Naked cluster anion	95
Me_3SnNMe_2	+75		96
$(Me_2N)_4Sn$	−118		96
$(Me_3Sn)_3P$	+36.3		97
$(Me_3Sn)_3PW(CO)_5$	+63.4		97
$Me_3SnN(CMe_3)_2$	+40.7		71

Table continued

Table 4. (*Continued*)

Compound	$\delta(^{119}Sn)$ (ppm)	Notes	Reference
$(Me_3Sn)_2AsPh$	−1.7		71
$SnCl_4(PBu_3)_2$	−573		87
$(Me_3Sn)_4Si$	−34.0		71
$Me_3SnSiMe_3$	−126.7		98
Me_3SnLi	−183	Solvent dependent	99
$[(2,6\text{-}Et_2C_6H_3)_2Sn]_3$	−416.5	Cyclotristannane	99a
Sn_4^{4-}	−1895	Naked cluster anion	95
Sn_9^{4-}	−1230	Naked cluster anion	35
$[Sn_8Tl]^{5-}$	−1167	Naked cluster anion	95
$Me_3SnMo(CO)_2(\eta^5C_5H_5)$	+121		51
$MeSn[Co(CO)_4]_3$	+483		51
trans$[(Cl_3Sn)_2Pt(PEt_3)_2]$	−84.5		100
$[(Cl_3Sn)_5Pt]^{3-}$	−133.4		101
$[(Cl_3Sn)_nRhCl_{6-n}]^{3-}$	−991.6 to −100.5	$n = 1\text{–}5$	102

as a sample of individual values with some emphasis upon more recent measurements. In particular cases it should be borne in mind that quite small changes at a site remote from the monitored position sometimes can have remarkable effects, for example, in the various *cis* and *trans* isomers of $(C_3H_5)_4M$,[66,86] and in $SnCl_3$/Rh complexes.[102] A few classes of compound merit special mention.

3.5.1. *Tetraorgano Derivatives*

The sensitivity to quite minor conformational and configurational changes at sites relatively remote from the central atom (for example, in norbornyl compounds[107] and the various *cis* and *trans* propenyl compounds[66,85]) precludes simple correlations

Table 5. Selected ^{207}Pb Chemical Shifts Relative to Me_4Pb

Compound	$\delta(^{207}Pb)$ (ppm)	Comments	Reference
Et_4Pb	+73.3		33
Ph_4Pb	−178.0		103
$Pb(C_3H_5)_4$	−192 to −440	Mixture of *cis* and *trans* isomers	66
$(MeC\equiv C)_4Ph$	−685.5		104
Me_3PbCl	+432		33
Me_3PbI	+203.6		33
Ph_2PbI_2	−391.8		103
$(o\text{-}MeOC_6H_4)_4Pb$	−195.6		103
$(p\text{-}MeOC_6H_4)_4Pb$	−153		105
Ph_3PbOAc	−93		105
$Pb(OAc)_4$	−1869		105
Me_3PbSMe	+239		33
$Me_3PbSeMe$	+98		33
Me_3PbNEt_2	+242		33
$(Me_3Pb)_3N$	+294		71
Me_3PbPPh_2	+40		33
$Me_3PbPbMe_3$	−281		33
$Me_3PbSnMe_3$	−324		33
Pb_9^{4-}	−4152	Naked anion cluster	35

in many cases, although in substituted aromatic compounds there are correlations of shielding with Hammett σ constants that can be attributed to changes in effective electronegativity.[106]

3.5.2. Halogen Derivatives

The dominating effect here appears to be factors associated with the bulk and/or polarizability of the halogen,[60a] with iodides at lowest frequency, and the pattern in multiply substituted species indicating that effective electronegativity plays a relatively small part. This consideration applies also with other substituent atoms that have lone pairs (e.g., N, Se, etc.).

3.5.3. Oxygen Derivatives

^{29}Si NMR has proved to be invaluable for elucidating the structures of silicones[1] and of the multifarious species present in aqueous solutions of inorganic silicates.[78] A selection of results is presented below,[108,109] and it has been found possible to use the wide ^{29}Si chemical shift dispersion allied to empirically established trends to identify various isomeric species. Note that $M \equiv Me_3SiO_{1/2}$, $D \equiv Me_2Si(O_{1/2})_2$, $T \equiv MeSi(O_{1/2})_3$ and $Q \equiv Si(O_{1/2})_4$.

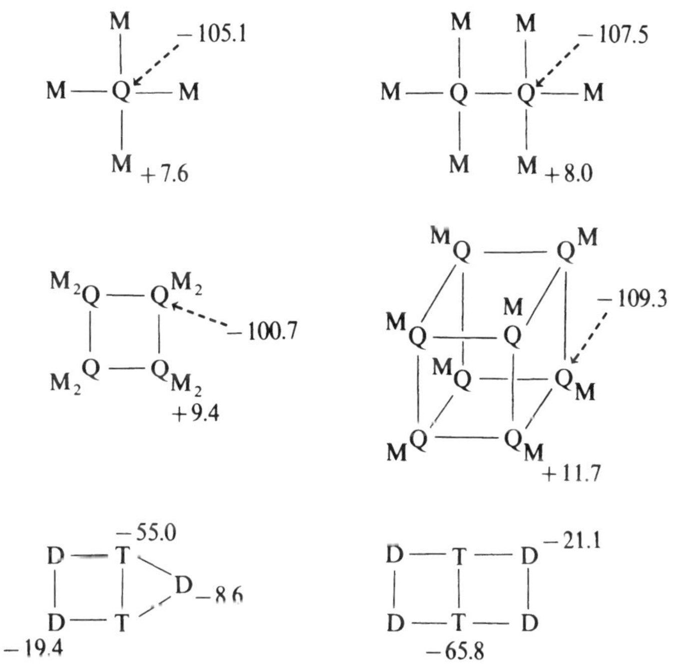

Tin and lead do not display the same tendency to form this type of compound, but oxygen does tend to coordinate to already four-coordinate tin (and lead) with a concomitant substantial increase in metal-atom nuclear shielding. Thus tin NMR has been invaluable for the study of the autoassociation and complexation of many different tin compounds, the large differences in $\delta(^{119}Sn)$ permitting the relatively precise determination of thermodynamic parameters.[47,90] For example, in n-BuSn(OR)$_3$ $\delta(^{119}Sn)$ ranges from −322 at 20°C to −205 ppm at 135°C for R = *sec*-Bu as the tin coordination number changes from five to four. For this reason, caution must always be exercised in interpreting chemical shifts measured in coordinating solvents such as DMSO and pyridine, and also for species capable of autoassociation.

3.5.4. Metal Derivatives

Many polysilanes have been studied, and $\delta(^{29}Si)$ has been correlated with α, β, and γ effects and the extent of chain branching[110,111]; such effects have been used to deduce structures.[112] Many di- and tri-tins have also been studied.[79] The very low-frequency tin and lead chemical shifts in anionic metal-atom clusters could be due to massive diamagnetic circulation in an extended molecular-orbital system, and an extended molecular orbital interpretation can also be advanced to account for the extreme sensitivity[102] of $\delta(^{119}Sn)$ to changes in remote substituents in $[(SnCl_3)_nRhCl_{6-n}]^{3-}$. Overall many ^{119}Sn shifts in transition metal complexes have now been determined,[11a,51,102,113–115] but the distribution of values is such that detailed discussion is not appropriate at present.

4. COUPLING CONSTANTS

Almost all considerations of spin–spin couplings involving group IV elements have assumed dominance by the Fermi contact mechanism (see Chapters 2 and 4), and this is the approach adopted here. Furthermore, provided that the atoms of *both* nuclei are reasonably electropositive and have no nonbonded valence shell electron pairs it appears permissible to make the mean electronic excitation energy approximation (cf. Chapter 4) which leads to direct dependence upon the hybridization (i.e., *s* character of the relevant atomic orbitals) and effective nuclear charge of the participating atoms. In addition, for couplings over more than one bond the characteristics of the intervening atom(s) will also play their part and in general a more empirical approach is needed for these. It should be borne in mind that for many couplings involving a group IV element $K(AB)$ and $J(AB)$ are of opposite sign owing to the negative magnetogyric ratios of ^{29}Si, ^{73}Ge, and $^{115/117/119}Sn$, although this is of no chemical significance.

Owing to its quadrupole moment, relatively few couplings involving ^{73}Ge have been reported. These are from highly symmetrical species and it seems likely that many others could be measured in similar molecules. It also seems likely that couplings to ^{73}Ge will closely parallel those for ^{29}Si in corresponding compounds.

4.1. One-Bond Couplings

The most satisfactory rationalizations have been obtained for these, probably because there is no need to take into account any properties of an intervening atom. Table 6 gives a *selection* of values of such couplings to many different elements, but it is emphasized that it is impossible to be truly representative in the space available here.

For couplings to the more electropositive elements where the mean electronic excitation energy approximation is appropriate there are generally found to be close parallels among the couplings with the *reduced* coupling constants increasing in the order $^1K(SiX) < {}^1K(SnX) < {}^1K(PbX)$. Indeed, these couplings often approximate to the ratios expected from the relative sizes of $\psi^2(O)$ for the valence *s* electrons.

However, for couplings to less electropositive and more polarizable elements this neat relationship breaks down, because it is necesary to use the fuller expression [Chapter 4, equation (21)] for the Fermi contact interaction which incorporates the mutual polarizability π_{xy} of the atoms whose nuclei are coupled. In these circumstances π_{xy} can vary substantially among silicon, tin, and lead, even in analogous molecules, and can therefore account for apparently erratic relationships between corresponding couplings.

4.1.1. $^1J(M–^1H)$

For the tetra-coordinate elements this coupling is "well behaved" and appears to show the expected correlations with the *s* character of the atomic orbitals of MH bonds and with effective nuclear charge of the central group IV metal atom. For $^1J(^{29}Si–^1H)$ numerous correlations have been established with electronic properties of substituents which are fairly consistent with the foregoing statement[1,148] and its behavior is not unlike that of $^1J(^{13}C–^1H)$ apart from its opposite sign which is due to the negative magnetogyric ratio of ^{29}Si. The more limited data available for ^{119}Sn indicate a similar pattern of behavior, although a notable exception[116] is provided by SnH_3^-, in which the electron lone pair on tin leads to very low *s* character for the hybrid orbitals it uses in the Sn–H bonds and hence to a very small magnitude for the coupling and probable breakdown of the mean excitation energy approximation.

4.1.2. $^1J(M–^{13}C)$

Good correlations are again found between the coupling constant and the *s* character and/or effective nuclear charge of the central atom except in special circumstances. For example, in Me_3SiX $^1J(^{29}Si–^{13}C)$ is linearly related[149] to the electronegativity of X, and similar correlations are found for the series $Me_nSi(OEt)_{4-n}$,[149] $(Me_3Si)_3COR$,[150] tetraorganosilanes,[151] and silyl acetylenes.[152] These last two examples also illustrate the effects of changing the carbon hybridization. Similar correlations are often found for tin–carbon couplings, with negative values [corresponding to positive $^1K(SnC)$] being the norm. Thus $^1J(^{119}Sn–^{13}C)$ is relatively small in species such as $Me_3SnSiMe_3$(−245.8 Hz)[139] and $[Me_3SnBH_3]^-$(−110 Hz)[118] which have electropositive atoms attached to tin, and is often large in species where the tin has one or more bulky and/or electron withdrawing substituents. The long-range effect of changing X in $Ph_3SnOC(O)C_6H_4X$ has also been studied.[153] In more extreme bonding situations these generalizations break down, as does the mean excitation energy approximation itself. For example, in $Me_3Sn^-Li^+$ $^1K(^{119}Sn–^{13}C)$ is negative because the tin atom has an electron lone pair that accommodates most of the *s* character.[121] Very small values of $^1J(^{119}Sn–^{13}C)$ are found in η^5-C_5H_5 tin(II) compounds,[50] which is attributed to low carbon *s* character in the penta-hapto link to tin. It is also interesting that in $(BrMe_2Sn)_2CHCH_2CH_2OH$ values of $^1J(^{119}Sn–^{13}C)$ differing by up to 20 Hz have been found for the inequivalent methyl carbon nuclei, indicating a significant stereochemical dependence of this coupling. Insofar as they have been investigated, the pattern[66,105,125] of lead–carbon couplings is very similar to that found for tin–carbon, although there is a greater tendency to abnormal values in unusual situations. Thus $^1K(PbC)$ may be negative[154] in $Et_3Pb–C{\equiv}CH$ whereas $^1K(SnC)$ is positive in the tin analog.

Table 6. Selected One-Bond Couplings in Hertz Involving ^{29}Si, ^{119}Sn, or ^{207}Pb (References in Parentheses)

Coupled nucleus X	$^{1}J(^{29}Si\text{–}X)$	$^{1}J(^{119}Sn\text{–}X)$	$^{1}J(^{207}Pb\text{–}X)$
^{1}H	Me_3SiH, −184 (113); F_3SiH, −381.7 (114) H_3SiCN, −238 (69); SiH_4, −202.5 (69)	Me_3SnH, −1744 (115); SnH_4, −1930 (113) SnH_3^-, ±109.4 (116); SnH_3^+, −2960 (117)	Me_2PbH_2, +2454.6 (113)
^{11}B	$[Me_3SiBH_3]^- Li^+$, −74.0 (118)	$[Me_3SnBH_3]^- Li^+$, −554 (118); $Me_3SnB(NMe_2)_2$, −953 (118)	
^{13}C	$(CH_3)_4Si$, −50.3 (115, 116); H_3SiCN, −62.5 (69); $(Me_3Si)_4C$, −30.9 (119); Me_2Si, 16.4 (119)	$(CH_3)_4Sn$, −340 (120); $MeSnBr_3$, −640 (120); $Me_3SnLi \cdot 3thf$, +155 (121); $(Me_3Sn)_4C$, −107 (79); $Sn(C\equiv CBu)_4$, −1160 (122)	$(CH_3)_4Pb$, +320 (123); $Ph_2Pb(OAc)_2$, 1203 (125); Ph_4Pb, +481 (124); $PhPb(OAc)_3$, 2100 (125); *cis* $Me_3PbCH=CHMe$, +268 (66)
^{15}N	$(H_3Si)_3N$, +6 (126); H_3SiNCO, +18.6 (69); $Me_3SiNHPh$, 15.7 (127); $(HNSiMe_2)_4$, 16.9 (127)	Me_3SnNEt_2, *ca* 0 (11a); $(Me_3Sn)_3N$, 79.6 (11a); $Me_2Sn(NMePh)_2$, +24 (128); $Sn(NMePh)_4$, +175 (128)	$Me_3PbNHPh$, +261 (129); $[Pb(cydta)]^{2-}$, 188 (130)

^{19}F	SiF_4, +172 (131); SiF_6^{2-}, +110 (131) $SiFH_2Me$, +279.8 (132); $SiFBr_3$, +369 (133)	$(PhCMe_2CH_2)_3SnF$, +2298 (134) SnF_6^{2-}, +1550 (135, 137) *fac*$[Sn(OH)_3F_3]^{2-}$, 1605 (136)	
^{29}Si	$(Me_3Si)_4Si$, +52.5 (71); $Ph_3SiSiMe_3$, −86.5 (138); $[(Me_3Si)_3Si]^- Li^+$, 19.0 (71)	$(Me_3Sn)_4Si$, 220 (71); $Me_3SnSiMe_3$, +656 (139)	
^{31}P	Me_3SiPH_2, +16.2 (140); $(Me_3Si)_3P$, 27.3 (141)	$(Me_3Sn)_3P$, +832 (97); $(Me_3Sn)_3PW(CO)_5$, +375 (97); $[SnBr_4(PBu_3)_2]$, 1960 (87)	Me_3PbPPh_2, +1335 (139)
^{77}Se	$(H_3Si)_2Se$, +110.6 (142)	$Me_3SnSeMe$, +1015 (143); $(Me_3Sn)_2Se$, −1060 (89); $Sn(SeMe)_4$, 1520 (143); $[Sn(SePh)_3]^-$, 710 (144)	$Me_3PbSeMe$, −1170 (129)
^{119}Sn	See earlier entry	$Me_3SnSnMe_3$, +4460 (145); $(Bu^t_3Sn)_2$, <60 (146); $Me_4Sn_2(OAc)_2$, 14980 (147); $(Me_3Sn)_3SnLi$, −5200 (93) Sn_4^{2-}, 1281 (95); Sn_9^{4-}, 266 (35)	$Me_3SnPbMe_3$, −3570 (129); $[Sn_{9-x}Pb_x]^{4-}$, *ca.* 560 (35)

4.1.3. *Other Group IV Couplings*

The best studied is $^1J(^{119}Sn–^{119}Sn)$, which is normally positive. [Note that K and J are of the same sign in this case, and that it is often easier to *measure* $^1J(^{119}Sn–^{117}Sn)$.] For small or only moderately bulky tin substituents there is a clear dependence upon tin hybridization with bulky groups generally believed to reduce the s character of the hybrid orbitals used to form the tin–tin bond.[93,129,145] This analysis has been confirmed by a considerable number of measurements of polyorgano di- and tri-tins.[146] However, for very bulky substituents the tin–tin s-overlap integral will become small as the bond weakens and the tin–tin mutual polarizability may change sign thus leading to very small or even negative $^1J(^{119}Sn–^{119}Sn)$ as in $Bu^t_6Sn_2$[146] or more extremely in $(Me_3Sn)_3Sn^-Li^+$, where one tin atom has an electron lone pair.[93] This kind of behavior is more strongly accentuated in $Me_3PbSnMe_3$, where the value of -3570 Hz for $^1J(^{207}Pb–^{119}Sn)$ (i.e., K positive) is of considerably smaller magnitude than would be expected simply from consideration of the likely tin and lead hybridizations and must therefore be due to reduced lead/tin s overlap.[129] The effect is even more extreme in the easily dissociated compound $Me_3PbPbMe_3$, where $^1J(^{207}Pb–^{207}Pb)$ is only $+290$ Hz.[155] In di- and polysilanes $^1J(^{29}Si–^{29}Si)$ (presumably positive) correlates well with the inductive effects of substituents[73,138] and with hybridization.[110,156]

4.1.4. $^1J(M–^{19}F)$

It may be that terms in addition to the Fermi contact one contribute significantly to these couplings, and in any case they represent an extreme breakdown of the mean excitation energy approximation to the extent that the reduced coupling has a large and negative multiplicative contribution from the mutual polarizability term. In practice no values for lead–fluorine coupling are known, while with the exception of $(PhCMe_2CH_2)_3SnF$ all those to tin are in substituted SnF_6^{2-} species and range from $+1500$ to $+3000$ Hz with marked dependences on stereochemistry.[135–137] Even in straightforward series of compounds the effects of substituents upon $^1J(^{29}Si–^{19}F)$ tend to be erratic and it is difficult to separate the influences of electronegativity and hybridization, and also contributions from non-Fermi-contact terms. The presently known range is from 110 to 488 Hz for this coupling.

4.1.5. $^1J(M–^{15}N)$ *and* $^1J(M–^{31}P)$

For these couplings variations in the s-overlap integral and hence in the mutual polarizability terms are often as important as changes in hybridization and substituent electronegativity.[128] The available data are too sparse for much useful comment, although trends in $^1J(^{119}Sn–^{15}N)$ in $Me_{4-n}Sn(NMePh)_n$ for $n = 1$–4 can be accounted for by effective electronegativity changes, and values for $^1J(^{31}P–^{29}Si)$ of 35.4, 43.0, and 50.0 Hz for $n = 4$, 5, and 6, respectively, in the cyclic phosphasilanes $MeP(SiMe_2)_n$ indicate a dependence upon interbond angles.[157] $^1K(SnP)$ is usually negative in four-coordinate tin compounds, but its behavior parallels that of $^1K(PC)$ in that there is a substantial positive increment accompanying the transformation $P^{III} \rightarrow P^{V}$ which can be achieved by coordination of the phosphorus to a metal atom,[97,158] even though the reduced coupling itself usually remains negative. Much larger values of $^1J(^{119}Sn–^{31}P)$ of presently unknown sign occur in phosphine complexes of hexa-coordinate tin.[87]

4.1.6. $^1J(M\text{–}^{77}Se)$ and $^1J(M\text{–}^{125}Te)$

Small *s*-overlap integrals are expected to lead to negative reduced coupling constants, and this has been confirmed in practice. Rather few results are available for silicon but $^1J(^{125}Te\text{–}^{29}Si)$ is ca. 280 Hz in $Me_3Si\text{–}Te$ derivatives and is presumably negative since both nuclei have negative magnetogyric ratios.[159] Trends in $^1J(^{119}Sn\text{–}^{77}Se)$ in the series $Me_{4-n}Sn(SeMe)_n$ can be accounted for in terms of changing effective nuclear charge at tin,[143] and this interpretation is generally confirmed by more recent measurements of other species.[94,160] $^1J(^{125}Te\text{–}^{119}Sn)$ is typically some 2–3 times larger than $^1J(^{119}Sn\text{–}^{77}Se)$, and is normally negative (i.e., K also negative).[89,94] In the cluster $[SnTe_4]^{4-}$ it is 2804 Hz,[95] and in $(Me_3Sn)_2Te$ it is 2770 Hz and not 1385 Hz as originally reported.[81,89]

4.1.7. Coupling to Transition Metals

The ease with which oxidative addition reactions can be used to prepare $SnCl_3^-$ (isoelectronic and isostructural with PR_3) complexes of transition metals has led to the report of many tin–metal couplings for these complexes. These are often very large, and some typical values are as follows:

$^1J(^{195}Pt\text{–}^{119}Sn)$[164–166]: *trans*-$[Pt(SnCl_3)_2(PEt_3)_2]$, 20,585 Hz[100] $[PtH(SnCl_3)(PEt_3)_2]$, 9067 Hz[161] $[PtCl_2(SnCl_3)_2]^{2-}$, 27,754 Hz[161] $[Cl_2(CO)PtSnCl_2Pt(CO)Cl_2]$, 15,264 Hz.[162] This coupling is negative.[163]

$^1J(^{119}Sn\text{–}^{103}Rh)$[164,170]: $[Rh(SnCl_3)Cl_5]^{3-}$, 864 Hz[102] $[Rh(SnCl_3)_5]^{2-}$, 806 Hz[102] $[Rh(SnCl_3)(NBD)(PPh_3)_2]$, 580 Hz.[169] Couplings from ^{119}Sn to ^{99}Ru,[167,168] ^{187}Os,[167] and ^{183}W,[52] have also been reported. Related couplings to ^{29}Si include $^1J(^{103}Rh\text{–}^{29}Si) = 25.4$ Hz in $[Rh(\eta^5\text{-}C_5Me_5)(CO)HSiR_3]$,[169a] and for lead values of $^1J(^{207}Pb\text{–}^{195}Pt)$ of 14,500 to 18,500 Hz are found[171] in *trans*-$[Pt(PR_3)_2(PbPh_3)_2]$ and related species. Owing to the proximity of the ^{195}Pt and ^{207}Pb resonance frequencies the spectra can display second order heteronuclear coupling features which has excited some surprise. In the cluster Pb_9Pt $^1J(^{207}Pb\text{–}^{195}Pt)$ is 4122 Hz.[172]

4.2. Two-Bond Couplings

These are found to be sensitive to the same factors as couplings between the nuclei of directly bound atoms, and in addition to depend upon properties of the intervening atom. In consequence they can be of considerable value for determining stereochemistry. It is not possible in the space available to give more than a fairly general survey.

4.2.1. $^2J(M\text{–}X\text{–}^1H)$

For X = C these couplings are well documented, especially for ^{119}Sn. Typically $^2J(^{29}Si\text{–}^1H)$ can be up to +10 Hz (K is negative) and therefore some care is needed if it is to be measured from routine spectra. There are fewer values in the literature than might be expected, although the advent of the INEPT pulse sequence is having some impact.[173] There is evidence that this coupling is dependent upon stereochemistry.[174]

$^2J(^{119}Sn\text{–}C\text{–}^1H)$ usually parallels the corresponding $^1J(^{119}Sn\text{–}^{13}C)$ particularly in substituted alkyl groups, but is of opposite sign (i.e., normally positive for sp^3

hybridized carbon) with magnitudes of ca. 40–150 Hz.[175] Exceptionally, the coupling can be negative, as in Me_3SnLi where the tin atom has an electron lone pair, and although a plot of $^1J(^{119}Sn–^{13}C)$ against $^2J(^{119}Sn–C–^1H)$ for methyl groups on tin is approximately linear and can accommodate this species, it does not pass through the origin (Figure 2). Thus it is unwise to use either coupling as a direct measure of tin hybridization unless a closely similar model compound is available. This problem is accentuated by the effects of changing the hybridization and/or substituents of the carbon atom. For example, in $Sn(CH=CH_2)_4$ $^2J(^{119}Sn–^1H)$ across the sp^2 hybridized carbon is -99.7 Hz,[176] and in Ph_3SnCH_2I its magnitude is only 23.4 Hz.[177] An increase in tin coordination number is usually accompanied by an increase in $^2J(^{119}SnH)$. The situation for lead is less well studied but can be expected to be similar. Normally $^2J(^{207}Pb–C–^1H)$ is negative with a magnitude somewhat greater than for the corresponding $^2J(^{119}Sn–C–^1H)$. Values of 188 and 194 Hz, respectively,[66] in *cis* and *trans* $Me_3PbCH=CHMe$ indicate the sensitivity to minor geometrical changes, and significant dependences upon solvent polarity have been reported for a number of species.

Much less is known about two-bond couplings to hydrogen when atoms other than carbon intervene, $^2J(^{29}Si–Si–^1H)$ varies between 1.4 and 5.6 Hz in polysilanes[157] and is probably negative, whereas in $Me_3SnPHPh$ $^2J(^{119}Sn–P–^1H)$ is probably positive.[178] Tin to proton couplings via a transition metal are of theoretical and diagnostic interest, typical values being 1740 Hz for the *trans* coupling in *trans*-$[PtH(SnMe_3)(PPh_3)_2]$[179] and 27 Hz for the *cis* coupling in $[PtH_2(SnPh_3)_2\,bipy]$.[180] The first of these is known to be negative.[163]

4.2.2. $^2J(M–X–^{13}C)$

The greatest body of data is available for tin with magnitudes ranging from close to zero to 235 Hz in $Sn(C\equiv CBu)_4$.[122] For tetracoordinate tin and sp^3 hybridized intervening carbon the coupling is positive [as with $^2J(^{119}Sn–C–^1H)$], but in other circumstances, negative values are known, e.g., -141.8 Hz in $Et_2Sn(C\equiv CH)_2$.[154] The dependence upon geometrical factors is illustrated by values of 16.5 and 12.8 Hz for C_x and C_y respectively in **XV**. Other typical values of $^2J(^{119}Sn–C–^{13}C)$ are as follows: in

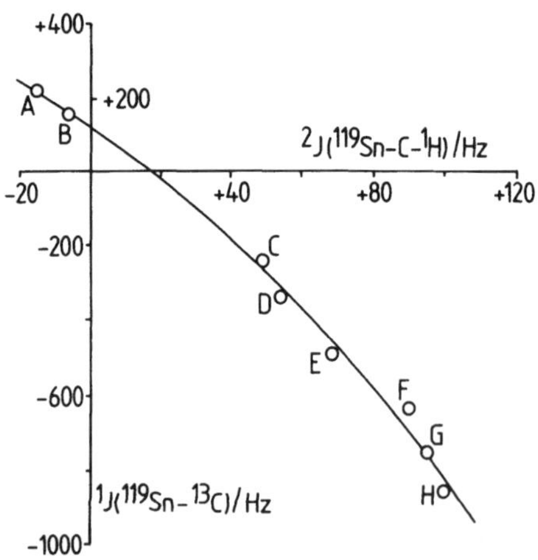

Figure 2. Plot of $^1J(^{119}Sn–^{13}C)$ against $^2J(^{119}Sn–C–^1H)$ in representative methyltin compounds. (A) Me_3SnLi in $(Me_2N)_3PO$; (B) Me_3SnLi in thf; (C) Me_6Sn_2; (D) Me_4Sn; (E) Me_3SnBr in H_2O; (F) $MeSnBr_3$ in C_6H_6; (G) Me_2SnCl_2 in wet Me_2CO; (H) Me_2SnCl_2 in H_2O. (Data from Refs. 121 and 145.)

Et_4Sn, +23.5 Hz;[181] in Ph_2SnCl_2, −63 Hz;[182,183] in $[Sn(C_5H_4Me)_2]$, 13.3 Hz;[50] and in XVI, 14.6 Hz.[184] In VII $^2J(^{119}Sn–C–^{13}C)$ is −19.6 Hz whereas in VIII it is

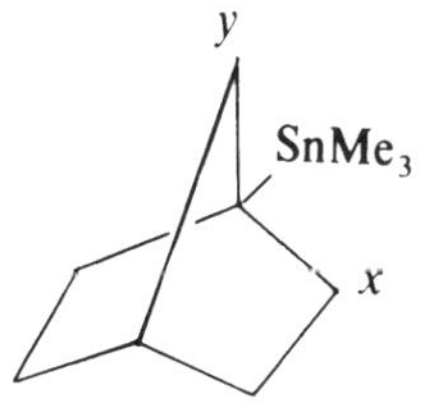

XV

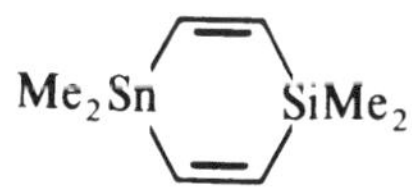

XVI

+19.4 Hz. This is because in the former compound two coupling routes are available so that the observed coupling is really the sum of ^2J and ^3J, which are of opposite sign, with $|^3J| \sim 2\,|^2J|$.[40,185] For other intermediate atoms the couplings usually lie in the above ranges, and in metal carbonyl complexes the tin to carbonyl couplings have geometrical dependence.[164,186]

4.2.3. *Other Two-Bond Couplings*

Many of these are known, with again the best studied being those involving tin. In transition metal complexes of $SnCl_3^-$ there are parallels among the behaviors of $^2J(^{119}Sn–M–^{31}P)$, $^2J(^{31}P–M–^{31}P)$ and $^2J(^{119}Sn–M–^{119}Sn)$, which apply not only to the magnitudes[187] [which can be very large when tin is a participant; see also $^3J(^{119}Sn–^{119}Sn)$ etc., Section 4.3.3 below] but also to the signs.[163] Typical magnitudes of $^2J(^{119}Sn–^{119}Sn)$ are as follows: in *cis*$[(SnCl_3)_2PtCl_2]^{2-}$, 2601 Hz;[188] in *trans*-$[(SnCl_3)_2Pt(PEt_3)_2]$, 36,286 Hz;[165] and in $[Pt(SnCl_3)_5]^{3-}$, 6520 Hz [189] (fluxionally averaged value).

Substantial though less spectacular variations are also found when the intervening atom is not a transition metal. Some typical two-bond couplings involving ^{19}F are 26.3 Hz (^{73}Ge) in $(CF_3)_4Ge$,[190] 531 Hz (^{119}Sn) in $Sn(CF_3)_4$,[191] 240 Hz (^{207}Pb) in Me_3PbCF_3,[191] and 379 Hz in $Me_2Pb(CF_3)_3$.[191] The last two are consistent with a marked dependence upon effective nuclear charge. Considerable numbers of $^{29}Si–^{19}F$ couplings have been listed.[1]

4.3. Three-Bond Couplings

These are often somewhat larger than two-bond couplings in the same molecule, a point that often surprises newcomers to the subject.

4.3.1. *Coupling to ^1H and ^{13}C*

For ^{29}Si and ^{73}Ge these couplings are only a few hertz[174] and are seldom reported; detailed discussion here is not warranted. The reduced couplings to tin and to lead are usually positive (i.e., ^3J is negative in the case of ^{119}Sn) and slow marked stereochemical dependence.[175,192,193] This probably takes the form of a Karplus type of relationship with dihedral angle (see Chapter 10), although difficulties associated with the analysis of complex proton spectra have made this difficult to establish with certainty for $^3J(^{119}Sn–^1H)$ and $^3J(^{207}Pb–^1H)$. In olefin derivatives a *cis*-coupling is normally smaller than a *trans*, e.g., in $(Me_3Sn)_2C{=}CH_2$ the values are 124 and

208 Hz, respectively.[194] In **XVII** a value of 79 Hz for the coupling between ^{119}Sn and the *N*-methyl protons is attributed to N → Sn transannular interaction which generates a ^3J pathway.[195]

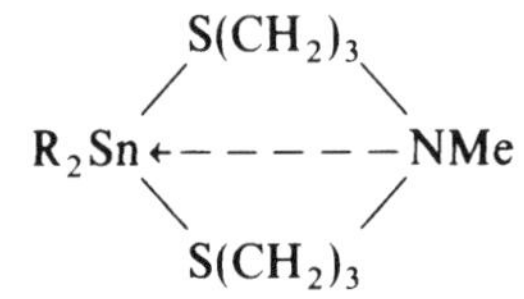

XVII

It has similarly been found that $^3J(^{119}Sn{-}^{13}C)$ is larger for a *trans* than a *cis* relationship,[85] and indeed a good relationship against dihedral angle[11a] has been established for this coupling in SnCCC systems, the values being ca. 34 Hz at 0° (i.e., eclipsed), ca. 10 Hz at 90°, and ca. 60 Hz at 180°.

4.3.2. Other Three-Bond and Longer-Range Couplings

Many are known, but the only one to be at all systematically studied is $^3J(^{119}Sn{-}^{119}Sn)$, which apparently behaves in a similar manner to $^3J(^{119}Sn{-}^{13}C)$, as values of 491 and 1012 Hz in *cis* and *trans* $Me_3SnCH{=}CHSnMe_3$, respectively, indicate.[11a] When several *additive* pathways are available the couplings are larger,[84] and the exceptionally large value of 25,430 Hz is found[196] in **XVIII**. In $Me_3PbC{\equiv}$

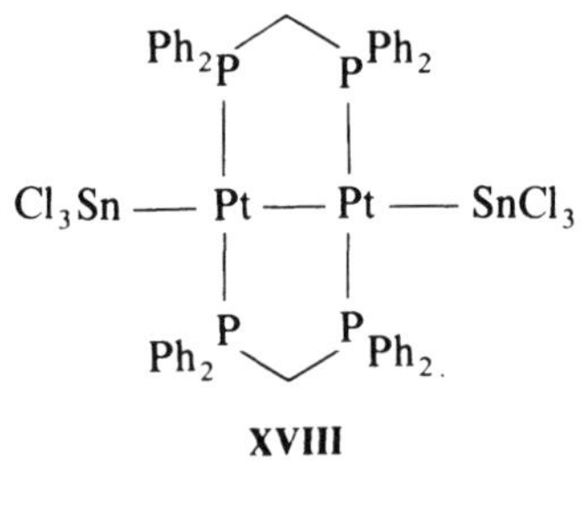

XVIII

$CSnMe_3$ $^3J(^{207}Pb{-}^{119}Sn)$ is 32 Hz.[154] Longer-range couplings involving a group IV element have been reported sporadically, but discussion would be inappropriate at this stage.[197–201]

5. RELAXATION BEHAVIOR

For ^{73}Ge the dominant relaxation mechanism is quadrupolar and T_1 is readily available from the linewidth; however, no specially significant conclusions have been reached. For ^{29}Si the (negative) proton NOE can often be significant and, when allied to T_1 measurements, can be used to assess segmental motion in silicone polymers.[202] Other work has tended to concentrate on small molecules and can be of diagnostic value for compound type.[64] Relaxation studies[203] of $SnCl_4$ have yielded a value of ca. 27 Hz for $^1J(^{35}Cl{-}^{29}Si)$ in this molecule.

Earlier suggestions that the SR mechanism dominates ^{119}Sn relaxation in small molecules have been largely borne out by more recent work.[204] This is also the case for ^{207}Pb at low field strengths,[205] although CSA is important in Me_3PbCl (in the presence of DMSO which may coordinate the lead atom) at 7.05 T. On occasion $^{119}Sn/^{1}H$ dipolar relaxation can be significant, and may lead to problems with the negative NOE.

6. MISCELLANEOUS AND SOLID STATE WORK

Many of the applications of group IV element NMR are relatively straightforward from a spectroscopic point of view and are therefore not discussed here. Worthy of special note, however, are ^{29}Si studies of aqueous silicates and silicones, molten silicates,[28] and silyl transition metal derivatives[207]; the use of $^{29}Si/^{1}H$ 2D correlation spectra to make assignments[208,209]; and the substantial and growing body of body of ^{29}Si MAS and/or CP work on solid materials such as zeolites,[210,211] other silicate minerals,[212] polysilanes,[213] and surface-immobilized materials.[214] The potential of solid-state NMR for ^{119}Sn and ^{207}Pb has yet to be fully realised,[26] although solid trimethyl stannyl hydroxide[215] and tributyl stannyl fluoride[216] have been studied.

Two comprehensive reviews discuss the applications of ^{29}Si (and other nuclide) NMR to zeolites,[217,218] and provide a comprehensive guide to the literature in this area for the 227-year period up to 1983. While early solid state work on stationary samples (SSSS)[219,220] demonstrated the potential of this approach, it was the pioneering efforts of Lippmaa and Engelhardt[221,222] that brought out the ability of MAS experiments to provide resolution down to a few ppm and hence make possible very fine levels of distinction. Thus for condensed silicates the following ^{29}Si chemical shift ranges are found: neosilicates (Q^0) -60 to -83; sorosilicates (Q^1) -68 to -84; sorosilicates with nearly linear Si–O–Si links -86 to -97; inosilicates (Q^2) -73 to -93; phyllosilicates (Q^3) -91 to -98; and silica polymorphs -106 to -115 ppm.[223] Replacement of some of the silicon in the framework by aluminum reduces the shielding of the remaining silicon atoms, the resulting chemical shifts again being characteristic of the environment of the silicon, and providing useful confirmation of crystal structures.[224] Indeed, in certain circumstances MAS ^{29}Si NMR can be superior to x-ray diffraction studies, as in an examination of the highly siliceous zeolites ZSM-5 and ZSM-11 which have very similar lattice structures, but whose uniqueness was demonstrated by high-field experiments.[211]

Figure 3 shows ^{29}Si MAS spectra from synthetic zeolites with different Si/Al ratios together with computer simulations and deconvolutions which determine the Si/Al ratio with the aid of the following equation

$$(\mathrm{Si/Al})_{\mathrm{NMR}} = \sum_{n=0}^{4} I_{\mathrm{Si}(n\mathrm{Al})} \Big/ \sum_{n=0}^{4} \frac{n}{4} I_{\mathrm{Si}(n\mathrm{Al})} \tag{1}$$

where $I_{\mathrm{Si}(n\mathrm{Al})}$ is the intensity of the (deconvoluted) signal due to Si(nAl) units.[225] This is now a standard method for determining Si/Al ratios,[226] and has been used to confirm that the "aluminum avoidance principle" is usually adhered to.[227] Furthermore, in some poorly ordered aluminosilicates there is more long-range order than had been previously supposed, this being a long-range manifestation of this rule.[228] In the FAU and LTA frameworks a correlation of $\delta(^{29}Si)$ with the number of aluminum atoms was

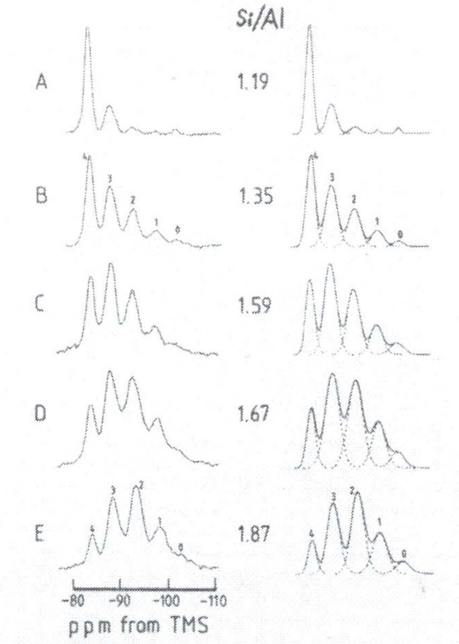

Figure 3. High-resolution ^{29}Si MAS NMR spectra of synthetic zeolites Na–X and Na–Y at 79.8 MHz. Experimental spectra are in the left column, and Si(nAl) signals are identified by the value of n placed above the peaks. Computer-simulated spectra using Gaussian line shapes are in the right column, and broken lines indicate the individual deconvoluted peaks. Taken from Ref. 225.

indicative of second-neighbor effects, $\Delta\delta(^{29}Si)$ being 6.28 and 0.70 ppm per added aluminum atom as first and second neighbor, respectively.[229]

^{29}Si MAS experiments have also been used to determine the effects of chemical modification of zeolites, for example by heat treatment, and quite substantial spectral changes can occur.[230–235] In this connection relaxation time measurements can be important,[214] and it may be noted that since the chemical shift anisotropy of ^{29}Si in silicates is small, problems of spinning sidebands do not arise in their study by MAS. The presence of Pr_4N^+ templates increases $T_1(^{29}Si)$ by an order of magnitude or more in ZSM-5 zeolites.[236] Finally, it should be realized that studies of other nuclei, notably ^{17}O and ^{27}Al, can complement ^{29}Si work on silicate minerals, and these are treated in Chapters 9 and 14, while the general background to solid state NMR is in Chapter 2, Section 6.1.

REFERENCES

1. Williams, E. A.; Cargıoli, J. D. *Ann. Rep. NMR Spectrosc.* 1979, **9**, 221–318.
2. Harris, R. K. *NMR and the Periodıc Table*, Harrıs, R. K., and Mann, B. E., Eds.; Academic: New York, 1978; pp. 309–340.
3. Schraml, J.; Bellama, J. M. *Det. Org. Struct. Phys. Meth.* 1979, **6**, 203–269.
4. Marsmann, H. *NMR, Basıc Princıples Progr.* 1981, **17**, 65–235.
5. Coleman, B. *NMR of the Newly Accessible Nuclei*, Laszlo, P., Ed.; Academıc· New York, 1983; Vol. 2, pp. 197–228.
6. Kennedy, J. D.; McFarlane, W. *NMR and the Perıodıc Table*, Harris, R. K., and Mann, B. E., Eds.; Academic: New York, 1978; pp. 340–371.
7. Smith, P. J.; Smith, L. *Inorg. Chım. Acta Rev.* 1973, **7**, 11.
8. Smıth, P. J.; Tupčiauskas, A. *Ann. Rep. NMR Spectrosc.* 1978, **8**, 291–370.

9. Kennedy, J. D.; McFarlane, W. *Rev. Silicon. Germanium, Tin Lead Compounds*, 1974, **1**, 235–298.
10. Hani, R.; Geanangel, R. A. *Coord. Chem. Rev.* 1982, **44**, 229–246.
11. Petrosyan, V. S. *Prog. NMR Spectrosc.* 1977, **11**, 115–148.
11a. Wrackmeyer, B. *Ann. Rep. NMR Spectrosc.* 1985, **16**, 73–186.
12. *Spec. Per. Rep. NMR*, Royal Society of Chemistry, Vols. 1–15, 1971–1986.
13. Blunden, S. J.; Frangou, A.; Gillies, D. G. *Org. Magn. Reson.* 1982, **20**, 170–174.
14. Davies, A. G.; Harrison, P. G.; Kennedy, J. D.; Puddephatt, R. J.; Mitchell, T. N.; McFarlane, W. *J. Chem. Soc. A* 1969, 1136–1141.
15. McFarlane, W. *Ann. Rep. NMR Spectroscop.* 1968, **1**, 135–163; 1972, **5A**, 353–393.
16. Morris, G. A.; Freeman, R. *J. Am. Chem. Soc.* 1979, **101**, 760–762.
17. Doddrell, D. M.; Pegg, D. T.; Brooks, W.; Bendall, M. R. *J. Am. Chem. Soc.* 1981, **103**, 727–728.
18. Jakobsen, H. J.; Ellis, P. D.; Inners, R. R.; Jensen, C. F. *J. Am. Chem. Soc.* 1971, **104**, 7442–7452.
19. Mackay, K. M.; Watkinson, P. J.; Wilkins, A. L. *J. Chem. Soc., Dalton Trans.* 1984, 133–139.
20. Pestunovich, V. A.; Tandura, S. N.; Shterenberg, B. Z.; Khromova, N. Y.; Gar, T. K.; Mironov, V. F.; Voronkov, M. G. *Izv. Akad. Nauk., SSSR Ser. Khim.* 1980, **4**, 959.
21. Harris, R. K.; O'Connor, M. J.; Curzon, E. H.; Howarth, O. W. *J. Magn. Reson.* 1984, **57**, 115–122.
22. Wynants, C.; van Binst, G.; Mägge, C.; Jurkschat, K.; Tzschach, A.; Pepermans, H.; Gielen, M.; Willem, R. *Organometallics* 1985, **4**, 1906–1909.
23. Thomas, J. M.; Ramdas, S.; Millward, G. R.; Klinowski, J.; Audier, M.; Gonzalez-Calbert, J.; Fyfe, C. A. *J. Solid State Chem.* 1982, **45**, 368–380.
24. Sindorf, D. W.; Maciel, G. E. *J. Am. Chem. Soc.* 1983, **105**, 3767–3776.
25. Engelhardt, G.; Jancke, H.; Lippmaa, E.; Samoson, A. *J. Organometall. Chem.* 1981, **210**, 295–301.
26. Davies, A. G.; Harris, R. K.; Sebald, A. Personal Communication, 1985.
27. McFarlane, H. C. E.; McFarlane, W.; Turner, C. J. *Molec. Phys.* 1979, **37**, 1639–1642.
28. Quintard, J. P.; Degueil-Castaing, M.; Dumartin, G.; Barbe, B.; Petraud, M. *J. Organometall. Chem.* 1980, **234**, 27–40.
29. Lassigne, C. R. O.; Wells, E. J. *J. Magn. Reson.* 1977, **26**, 55–69.
30. Kerschl, S.; Sebald, A.; Wrackmeyer, B. *Magn. Reson. Chem.*, in press.
31. McFarlane, W., unpublished observations.
32. Gasparis-Ebeling, T.; Nöth, H.; Wrackmeyer, B. *J. Chem. Soc. Dalton* 1983, 97–100.
33. Kennedy, J. D.; McFarlane, W.; Pyne, G. S. *J. Chem. Soc. Dalton* 1977, 2332–2336.
34. Mitchell, T. N. *J. Organometall. Chem.* 1983, **255**, 279–285.
35. Rudolph, R. W.; Wilson, W. L.; Parker, F.; Taylor, R. C.; Young, D. C. *J. Am. Chem. Soc.* 1978, **100**, 4629–4630.
36. McFarlane, W.; Wood, R. *J. Organometall. Chem.* 1972, **40**, C17–C20.
37. Watkinson, P. J.; Mackay, K. M. *J. Organometall. Chem.* 1984, **275**, 39–42.
38. Kennedy, J. D.; McFarlane, W.; Pyne, G. S. *Bull. Soc. Chem. Belge.* 1975, **84**, 289–298.
39. Wrackmeyer, B. *Rev. Silicon, Germanium, Tin, Lead Compound* 1982, **6**, 75.
40. Davies, A. G.; Tse, M. W.; Kennedy, J. D.; McFarlane, W.; Pyne, G. S.; Ladd, M. F. C.; Povey, D. C. *J. Chem. Soc., Perkin Trans. II* 1981, 369–375.
41. Krapivin, A. M.; Mägi, M.; Svergun, V. I.; Zaharjan, R. Z.; Babich, E. D.; Ushakov, N. V. *J. Organometall. Chem.* 1980, **190**, 9–33.
42. Seyferth, D.; Annarelli, D. C. *J. Am. Chem. Soc.* 1975, **97**, 2273–2275.
43. Lambert, R. L.; Seyferth, D. *J. Am. Chem. Soc.* 1972, **94**, 9246–9248.
44. Cella, J. A.; Cargioli, J. D.; Williams, E. A. *J. Organometall. Chem.* 1980, **186**, 13–17.
45. Otera, J.; Hinoishi, T.; Okawara, R. *J. Organometall. Chem.* 1980, **202**, C93–C94.
46. Holeček, J.; Nádvorník, M.; Handlír, K.; Lyčka, A. *J. Organometall. Chem.* 1983, **241**, 177–184.
47. Kennedy, J. D.; McFarlane, W.; Smith, P. J.; White, R. F. M.; Smith, L. *J. Chem. Soc., Perkin Trans. II* 1973, 1785–1788.
48. Harrison, P. G.; Healy, M. A.; Steel, A. T. *J. Chem. Soc., Dalton Trans.* 1983, 1845–1848.
49. Brook, A. G.; Nyburg, S. C.; Reynolds, W. F.; Poon, Y. C.; Chang, Y. M.; Lee, J.-S.; Picard, J.-P. *J. Am. Chem. Soc.* 1979, **101**, 6550–6552.
50. Bonny, A.; McMaster, A. D.; Stobart, S. R. *Inorg. Chem.* 1978, **17**, 935–988.
51. Harris, D. H.; Lappert, M. F.; Poland, J. S.; McFarlane, W. *J. Chem. Soc., Dalton Trans.* 1975, 311–316.
52. du Mont, W. W.; Kroth, H. J. *Z. Naturforsch. B* 1980, **35B**, 700–702.
53. Pannell, K. H.; Bassindale, A. R. *J. Organometall. Chem.* 1982, **229**, 1–9.
54. Bikovetz, A. L.; Kuzmin, O. V.; Vdovin, V. M.; Krapivin, A. M. *J. Organometall. Chem.* 1980, **194**, C33–C34.

55. Vongehr, M.; Marsmann, H. C. *Z. Naturforschung B* 1976, **318**, 1423–1424.
56. Phillips, L.; Wray, V. *J. Chem. Soc., Perkin Trans. II* 1972, 214–219.
57. Maciel, G. E.; Dallas, J. L. *J. Am. Chem. Soc.* 1973, **95**, 3039–3040.
58. Mason, J. *J. Chem. Soc. (A)* 1971, 1038–1047.
59. Mason, J. *Adv. Inorg. Chem. Radiochem.* 1976, **18**, 197–228; 1979, **22**, 199–237.
60. Kennedy, J. D.; McFarlane, W.; Pyne, G. S.; Clarke, P. L.; Wardell, D. L. *J. Chem. Soc., Perkin Trans. II* 1975, 1234–1239.
60a. Cheremisin, A. A.; Schastnev, P. V. *J. Magn. Reson.* 1980, **40**, 459–468.
61. Niemann, U.; Marsmann, H. C. *Z. Naturforsch. B* 1975, **30B**, 202–206.
62. Löwer, R.; Vongehr, M.; Marsmann, H. C. *Chem. Ztg.* 1975, **99**, 33.
63. Delmulle, L.; Van der Kelen, G. P. *J. Molec. Struct.* 1980, **66**, 309–314.
64. Pannell, K. H.; Bassindale, A. R.; Fitch, J. W. *J. Organometall. Chem.* 1981, **209**, C65–C68.
65. Olah, G. A.; Berrier, A. L.; Field, L. D.; Surya Prakash, G. K. *J. Am. Chem. Soc.* 1982, **104**, 1349–1355.
66. Mitchell, T. N.; Marsmann, H. C. *Org. Magn. Reson.* 1981, **15**, 263–267.
67. Marsmann, H. C.; Horn, H. G. *Chem. Ztg.* 1972, **96**, 456.
68. van den Berghe, E. V.; Van der Kelen, G. P. *J. Organometall. Chem.* 1973, **59**, 175–187.
69. Arnold, D. E. J.; Cradock, S.; Ebsworth, E. A. V.; Murdoch, J. D.; Rankin, D. W. H.; Skea, D. C. J.; Harris, R. K.; Kimber, B. J. *J. Chem. Soc., Dalton Trans.* 1981, 1349–1351.
70. Marsmann, H. C. *Chem. Ztg.* 1972, **96**, 288.
71. Biffar, W.; Gasparis-Ebeling, T.; Nöth, H.; Storch, W.; Wrackmeyer, B. *J. Magn. Reson.* 1981, **44**, 54–61.
72. Williams, E. A.; Cargioli, J. D.; Donahue, P. E. *J. Organometall. Chem.* 1980, **192**, 319–327.
73. Engelhardt, G.; Radeglia, R.; Kelling, H.; Stendel, R. *J. Organometall. Chem.* 1981, **212**, 51–58.
74. Marchand, A.; Gerval, P.; Barbe, B.; Petraud, M.; Caubit, A.; Joanny, M.; Mazerolles, P. *J. Organometall. Chem.* 1981, **217**, 19–33.
75. Uhlig, E.; Hipler, B.; Müller, P. *Z. Anorg. Allg. Chem.* 1978, **442**, 11–18.
76. Sakurai, H.; Kamiyama, Y.; Mikoda, A.; Kobayashi, T.; Sasaki, K.; Nakadaira, Y. *J. Organometall. Chem.* 1980, **201**, C14–C18.
77. Engelhardt, G.; Jancke, H.; Mägi, M.; Pehk, T.; Lippmaa, E. *J. Organometall. Chem.* 1971, **28**, 293–300.
78. Harris, R. K.; Knight, C. T. G.; Hull, W. E. *J. Am. Chem. Soc.* 1981, **103**, 1577–1578.
78a. Zilm, K. W.; Grant, D. M.; Michl, J.; Fink, M. J.; West, R. *Organometallics* 1983, **2**, 193–194.
79. Mitchell, T. N.; El-Behairy, M. *J. Organometall. Chem.* 1979, **72**, 293–298.
80. Hunter, B. K.; Reeves, L. W. *Can. J. Chem.* 1968, **46**, 1399–1414.
81. Kennedy, J. D.; McFarlane, W.; Pyne, G. S., unpublished work.
82. Wrackmeyer, B. *Z. Naturforsch. B* 1979, **34B**, 235–237.
83. McFarlane, W.; Maire, J. C.; Delmas, M. *J. Chem. Soc., Dalton Trans.* 1972, 1862–1865.
84. Berger, H.-O.; Nöth, H.; Wrackmeyer, B. *Chem. Ber.* 1979, **112**, 2866–2883.
85. Mitchell, T. N.; Kummetat, C. *J. Organometall. Chem.* 1978, **157**, 275–279.
86. Burke, J. J.; Lauterbur, P. C. *J. Am. Chem. Soc.* 1961, **83**, 326–331.
87. Colton, R.; Dakternieks, D.; Harvey, C.-A. *Inorg. Chim. Acta* 1982, **61**, 1–7.
88. Lassigne, C. R.; Wells, E. J. *Can. J. Chem.* 1977, **55**, 927–931.
89. Kennedy, J. D.; McFarlane, W. *J. Organometall. Chem.* 1975, **94**, 7–14.
90. Kennedy, J. D. *J. Chem. Soc., Perkin Trans. II* 1977, 242–248.
91. Mitchell, T. N.; Mathiasch, B. *J. Organometall. Chem.* 1980, **185**, 351–357.
92. Kennedy, J. D.; McFarlane, W. *J. Chem. Soc., Perkin Trans. II* 1974, 146–149.
93. Kennedy, J. D.; McFarlane, W. *J. Chem. Soc., Dalton Trans.* 1976, 1219–1223.
94. Blecher, A.; Mathiasch, B.; Mitchell, T. N. *J. Organometall. Chem.* 1980, **184**, 175–180.
95. Rudolph, R. W.; Wilson, W. L.; Taylor, R. C. *J. Am. Chem. Soc.* 1981, **103**, 2480–2481.
96. van den Berghe, E. V.; van der Kelen, G. P. *J. Organometall. Chem.* 1973, **61**, 197–205.
97. McFarlane, W.; Rycroft, D. S. *J. Chem. Soc., Dalton Trans.* 1974, 1977–1979.
98. Kennedy, J. D.; McFarlane, W.; Pyne, G. S.; Wrackmeyer, B. *J. Chem. Soc., Dalton Trans.* 1975, 386–390.
99. Harrison, P. G.; Zuckerman, J. J.; Ulrich, S. E. *J. Am. Chem. Soc.* 1971, **93**, 5398–5402.
99a. Masamune, S.; Sita, L. R.; Williams, D. L. *J. Am. Chem. Soc.* 1983, **105**, 630–632.
100. Koch, B. R.; Fazakerly, C. V.; Dijkstra, E. *Inorg. Chim. Acta* 1980, **45**, L51–L53.
101. Moriyama, H.; Aoki, T.; Shinoda, S.; Saito, Y. *J. Chem. Soc., Chem. Commun.* 1982, 500–502.
102. Moriyama, H.; Aoki, T.; Shinoda, S.; Saito, Y. *J. Chem. Soc., Dalton Trans.* 1981, 639–644.

103. van Beelen, D. C.; van der Kooi, H. O.; Wolters, H. *J. Organometall. Chem.* 1979, **179**, 37–41.
104. Cooper, M. J.; Holliday, A. K.; Makin, P. A.; Puddephatt, R. J.; Smith, P. J. *J. Organometall. Chem.* 1974, **65**, 377–382.
105. Mitchell, T. N.; Gmehling, J.; Huber, F. *J. Chem. Soc., Dalton Trans.* 1978, 960–964.
106. Ernst, C. R.; Spialter, L.; Buell, G. R.; Wilhite, D. L. *J. Am. Chem. Soc.* 1974, **96**, 5375–5381.
107. Kennedy, J. D. *J. Organometall. Chem.* 1976, **104**, 311–315.
108. Hoebbel, D.; Garzó, G.; Engelhardt, G.; Jancke, H.; Franke, P.; Wieker, W. *Z. Anorg. Allg. Chem.* 1976, **424**, 115–127.
109. Jancke, H.; Engelhardt, G.; Mägi, M.; Lippmaa, E. *Z. Chem.* 1973, **13**, 435–436.
110. Hahn, J. *Z. Naturforsch. B* 1980, **35B**, 282–296.
111. Ishikwaa, M.; Iyoda, J.; Ikeda, H.; Kotake, K.; Hashimoto, T.; Kumada, M. *J. Am. Chem. Soc.* 1981, **103**, 4845–4850.
112. Marsmann, H. C.; Raml, W.; Hengge, E. *Z. Naturforsch. B* 1980, **35B**, 35–37.
113. Schumann, C.; Dreeskamp, H. *J. Magn. Reson.* 1970, **3**, 204–217.
114. Ebsworth, E. A. V.; Turner, J. J. *J. Chem. Phys.* 1962, **36**, 2628–2634.
115. Flitcroft, N.; Kaesz, H. D. *J. Am. Chem. Soc.* 1963, **85**, 1377–1380.
116. Birchall, T.; Pereira, A. *Chem. Commun.* 1972, 1150–1151.
117. Webster, J. R.; Jolly, W. C. *Inorg. Chem.* 1971, **10**, 877–879.
118. Biffar, W.; Nöth, N.; Pommerening, H.; Schwerthöffer, R.; Storch, W.; Wrackmeyer, B. *Chem. Ber.* 1981, **114**, 49–60.
119. Wrackmeyer, B.; Biffar, W. *Z. Naturforsch. B* 1979, **34B**, 1270–1274.
120. McFarlane, W. *J. Chem. Soc. (A)* 1967, 528–530.
121. Kennedy, J. D.; McFarlane, W. *J. Chem. Soc. Chem. Commun.* 1974, 983–984.
122. Mitchell, T. N. *J. Organometall. Chem.* 1977, **141**, 289–293.
123. McFarlane, W. *Molec. Phys.* 1967, **13**, 587–588.
124. de Vos, D. *J. Organometall. Chem.* 1976, **104**, 193–197.
125. de Vos, D.; van Beelen, D. C.; Wolters, J. *J. Organometall. Chem.* 1979, **172**, 303–308.
126. Anderson, D. W. W.; Bentham, J. E.; Rankin, D. W. H. *J. Chem. Soc., Dalton Trans.* 1973, 1215–1217.
127. Kupče, E.; Liepiņš, E.; Pudova, O.; Lukevics, E. *J. Chem. Soc., Chem. Commun.* 1984, 581–582.
128. Kennedy, J. D.; McFarlane, W.; Pyne, G. S.; Wrackmeyer, B. *J. Organometall. Chem.* 1980, **195**, 285–290.
129. Kennedy, J. D.; McFarlane, W.; Wrackmeyer, B. *Inorg. Chem.* 1976, **15**, 1299–1302.
130. Alcock, N. W.; Herron, N.; Moore, P. *J. Chem. Soc., Dalton Trans.* 1979, 1488–1491
131. Brownstein, S. *Can. J. Chem.* 1980, **58**, 1407–1411.
132. Ebsworth, E. A. V.; Frankiss, S. G. *Trans. Farad. Soc.* 1963, **59**, 1518–1524.
133. Hengge, E. *Monatsh. Chem.* 1971, **102**, 734–740.
134. McFarlane, W.; Wood, R. J. *J. Organometall. Chem.* 1972, **40**, C17–C20.
135. Ragsdale, R. A.; Stewart, B. B. *Inorg. Chem.* 1965, **4**, 740–742.
136. Dean, P. A. W.; Evans, D. F. *J. Chem. Soc. A* 1968, 1154–1166.
137. Chernysov, B. N.; Didenko, N. A.; Ippolitov, E. G. *Koord. Khim.* 1983, **9**, 210.
138. Williams, E. A.; Cargioli, J. D.; Donahue, P. E. *J. Organometall. Chem.* 1980, **192**, 319–327.
139. Kennedy, J. D.; McFarlane, W.; Pyne, G. S.; Wrackmeyer, B. *J. Chem. Soc. (Dalton Trans)* 1975, 386–390.
140. Fritz, G.; Schäfer, H. *Z. Anorg. Allg. Chem.* 1974, **409**, 137–151.
141. van Linhthoudt, J. P.; van den Berghe, G. P.; van der Kelen, G. P. *Spectrochim. Acta* 1980, **A36**, 17–21.
142. Pfisterer, G.; Dreeskamp, H. *Ber. Bunsenges. Phys. Chem.* 1969, **73**, 654–661.
143. Kennedy, J. D.; McFarlane, W. *J. Chem. Soc., Dalton Trans.* 1973, 2134–2139.
144. Arsenault, J. J. I.; Dean, P. A. W. *Can. J. Chem.* 1983, **61**, 1516–1523.
145. McFarlane, W. *J. Chem. Soc. (A)* 1968, 1630–1634.
146. Mitchell, T. N.; Walter, G. *J. Chem. Soc., Perkin Trans. II* 1977, 1842–1847.
147. Mitchell, T. N.; Mathiasch, B. *J. Organometall. Chem.* 1980, **185**, 351–357.
148. Nagai, Y.; Matsumoto, A.; Nakano, T.; Watanabe, H. *Bull. Chem. Soc. Jpn* 1972, **45**, 2560–2565.
149. Harris, R. K.; Kimber, B. *J. Magn. Reson.* 1975, **17**, 174–188; *Org. Magn. Reson.* 1975, **7**, 460–464.
150. Benn, R.; Rufinska, A. *J. Organometall. Chem.* 1982, **238**, C27–C30.
151. Filleux-Blanchard, M. L.; An, N. D.; Manuel, G. *J. Organometall. Chem.* 1977, **137**, 11–22.
152. Kamienska-Trela, K.; Biedrzycka, Z.; Machinek, R.; Knieriem, B.; Lüttke, W. *Org. Magn. Reson.* 1984, **22**, 317–322.
153. Holeček, J.; Handlíř, K.; Nádvorník, M.; Lyčka, A. *J. Organometall. Chem.* 1983, **258**, 147–153.

154. Wrackmeyer, B. *J. Magn. Reson.* 1981, **42**, 287–297.
155. Kennedy, J. D.; McFarlane, W. *J. Organometall. Chem.* 1974, **80**, C47–C49.
156. Brook, A. G.; Abdesaken, F.; Gutekunst, G.; Plavac, N. *Organometallics* 1982, **1**, 994–998.
157. Newman, T. H.; West, R.; Oakley, R. T. *J. Organometall. Chem.* 1980, **197**, 159–168.
158. Schumann, H.; Feldt, W. *Z. Anorg. Allg. Chem.* 1979, **458**, 257–262.
159. Jones, C. H. W.; Sharma, R. D. *J. Organometall. Chem.* 1984, **268**, 113–118.
160. Dean, P. A. W. *Can. J. Chem.* 1982, **60**, 2921–2926.
161. Ostoja-Starzewski, K.-H. A.; Pregosin, P. S. *Ang. Chem.* 1980, **92**, 323–324.
162. Goodfellow,. R. J.; Herbert, I. R. *Inorg. Chem. Acta* 1982, **65**, L161–L162.
163. McFarlane, W.; Rees, N H., to be published.
164. Kretschmer, M.; Pregosin, P. S.; Rüegger, H. *J. Organometall. Chem.* 1983, **241**, 87–98.
165. Ostoja-Starzewski, K. H. A.; Pregosin, P. S.; Rüegger, H. *Helv. Chim. Acta* 1982, **65**, 785–797.
166. Holt, M. S.; McDougall, J. J.; Mathey, F.; Nelson, J. H. *Inorg. Chem.* 1984, **23**, 449–453.
167. Farrugia, L. J.; James, B. R.; Lassigne, C. R.; Wells, E. J. *Can. J. Chem.* 1982, **60**, 1304–1316.
168. Brevard, C.; Granger, P. *Inorg. Chem.* 1983, **22**, 532–535.
169. Garralda, M.; Garcia, V.; Kretschner, M.; Pregosin, P. S.; Rüegger, H. *Helv. Chem. Acta* 1981, **64**, 1150–1157.
169a. Aylett, B. J.; Beckett, M. A., to be published.
170. Kretschner, M.; Pregosin, P. S.; Garralda, M. *J. Organometall. Chem.* 1983, **244**, 175–181.
171. Carr, S.; Colton, R.; Dakternieks, D. *J. Organometall. Chem.* 1982, **240**, 143–151; *J. Magn. Reson.* 1982, **47**, 156–158.
172. Teixidor, F.; Leutkens, M. L.; Rudolph, R. W. *J. Am. Chem. Soc.* 1983, **105**, 149–150.
173. Helmer, B. J.; West, R. *Organometallics* 1972, **1**, 877–879.
174. Grignon-Dubois, M.; Laguerre, M.; Barbe, B.; Petraud, M. *Organometallics* 1984, **3**, 359–362.
175. Petrosyan, V. S. *Prog. NMR Spectrosc.* 1977, **11**, 115–148.
176. Lunazzi, L.; Taddei, F. *Spectrochem. Acta* 1969, **25A**, 611–617.
177. Harrison, P. G.; Molloy, K. *J. Organometall. Chem.* 1978, **152**, 53–62.
178. Harrison, P. G.; Ulrich, S.E.; Zuckerman, J. J. *Inorg. Nucl. Chem. Lett.* 1971, **7**, 865–868.
179. Ostoja-Starzewski, K. H. A.; Rüegger, H.; Pregosin, P. S. *Inorg. Chem. Acta* 1979, **36**, L445.
180. Almeida, J. F.; Azizian, H.; Eaborn, C.; Pidcock, A. *J. Organometall. Chem.* 1981, **210**, 121–133.
181. Weigert, F. J.; Roberts, J. D. *J. Am. Chem. Soc.* 1969, **91**, 4940–4941.
182. Mitchell, T. N. *J. Organometall. Chem.* 1973, **59**, 189–197.
183. McFarlane, W.; McHarg, I. R., unpublished results.
184. Berger, H.-O.; Nöth, H.; Wrackmeyer, B. *Chem. Ber.* 1979, **112**, 2866–2883.
185. Davies, A. G.; Tse, M.-W.; Kennedy, J. D.; McFarlane, W.; Pyne, G. S.; Ladd, M. F. C.; Povey, D. C. *J. Chem. Soc., Chem. Commun.* 1978, 791.
186. Vancea, L.; Pomeroy, R. K.; Graham, W. A. G. *J. Am. Chem. Soc.* 1976, **98**, 1407–1413.
187. Ostoja-Starzewski, K. H. A.; Pregosin, P. S. *Adv. Chem. Ser.* 1982, **196**, 23–41.
188. Alcock, N. W.; Nelson, J. H. *J. Chem. Soc., Dalton Trans.* 1981, 2415–2418.
189. Anderson, G. K.; Clark, H. C.; Davies, J. A. *Inorg. Chem.* 1983, **22**, 427–433.
190. Eujen, R.; Mellies, R. *J. Fluorine Chem.* 1983, **22**, 263–280.
191. Eujen, R.; Lagow, R. J. *J. Chem Soc., Dalton Trans.* 1978, 541–544.
192. Quintard, J.-P.; Degueil-Castaing, M.; Dumartin, G.; Barbe, B.; Petraud, M. *J. Organometall. Chem.* 1982, **234**, 27–40.
193. Matorasso-Tchiroukhine, E.; Cadiot, P. *Can. J. Chem.* 1983, **61**, 2712–2485.
194. Mitchell, T. N.; Amamria, A. *J. Organometall. Chem.* 1983, **252**, 47–56.
195. Tzschach, A.; Jurkschat, K.; Scheer, M. *Z. Anorg. Allg. Chem.* 1983, **507**, 196–198.
196. Grossel, M. C.; Moulding, R. P.; Seddon, K. R. *Inorg. Chim. Acta. Lett.* 1982, **64**, L275–L277.
197. Jousseaume, B.; Duboudin, J. G.; Petraud, M. *J. Organometall. Chem.* 1982, **238**, 171–175.
198. Adcock, W.; Kok, G. B.; Abeywickrema, A. N.; Kitching, W.; Drew, G. M.; Olszowy, H. A.; Schott, I. *J. Am. Chem. Soc.* 1983, **105**, 290–292.
199. Schaeffer, C. D.; Lefferts, J. L.; Zuckerman, J. J. *Org. Magn. Reson.* 1984, **22**, 125–126.
200. Wrackmeyer, B.; Bihlmayer, C.; Schilling, M. *Chem. Ber.* 1983, **116**, 3182–3191.
201. Mathiasch, B.; Kunze, U. *Inorg. Chim. Acta* 1983, **75**, 209–213.
202. Saluvere, T.; Puskar, J.; Past, J.; Lippmaa, E.; Engelhardt, G. *Magn. Reson. Relax. Phenom. Proc. Congr. Ampère 19th*, 1976, pp. 333–336.
203. Briguet, A.; Duplon, J. C.; Delmau, J. *J. Magn. Reson.* 1981, **42**, 141–146.
204. Lassigne, C. R.; Wells, E. J. *J. Magn. Reson.* 1977, **26**, 55–69; *Can. J. Chem.* 1977, **55**, 927–931.
205. Hays, G. R.; Gillies, D. G.; Blaauw, L. P.; Clague, A. D. H. *J. Magn. Reson.* 1981, **45**, 102–107.

206 Stebbins, J F, Murdoch, J B, Schneider, E, Carmichael, I S E, Pines, A *Nature* 1985, **314**, 250–252

207 Krentz, R, Pomeroy, R K *Inorg Chem* 1985, **24**, 2976–2980

208 Liepinsh, E, Sekacis, I, Lukevics, E *Magn Reson Chem* 1985, **23**, 10–11

209 Nardin, R, Vincendon, M *J Magn Reson* 1985, **61**, 338–342

210 Engelhardt, G, Fahlke, B, Maegi, M, Lippmaa, E *Z Phys Chem (Leipzig)* 1985, **266**, 239–245

211 Fyfe, C A, Kokotailo, G T, Kennedy, G J, De Schutter, C *J Chem Soc, Chem Commun* 1985, 306–308

212 Kinsey, R A, Kirkpatrick, R J, Hower, J, Smith, K A, Oldfield, E *Am Mineral* 1985, **70**, 537–548

213 Harris, R K, Metcalfe, K, Hengge, E *Polyhedron* 1985, **4**, 1319–1321

214 Fyfe, C A, Gobbi, G C, Kennedy, G J *J Phys Chem* 1985, **89**, 277–281

215 Harris, R K, Packer, K J, Reams, P *J Magn Reson* 1985, **61**, 564–566

216 Harris, R K, Packer, K J, Reams, P *Chem Phys Lett* 1985, **115**, 16–18

217 Klinowski, J *Progr NMR Spectrosc* 1984, **16**, 237–310

218 Fyfe, C A, Thomas, J M, Klinowski, J, Gobbi, G *Angew Chem Internat Edn* 1983, **95**, 257

219 Holzman, G R, Lauterbur, P C, Anderson, J H, Koth, W *J Chem Phys* 1956, **25**, 172

220 Grimmer, A-R, Peter, R, Fechner, E, Moldegney, G *Chem Phys Lett* 1981, **77**, 331

221 Lippmaa, E, Alla, M, Pehk, T J, Engelhardt, G *J Am Chem Soc* 1978, **100**, 1929

222 Lippmaa, E, Magi, M, Samoson, A, Engelhardt, G, Grimmer, A-R *J Am Chem Soc* 1980, **102**, 4889

223 Magi, M, Lippmaa, E, Samoson, A, Engelhardt, G, Grimmer, A-R *J Phys Chem* 1984, **88**, 1518

224 Lippmaa, E, Magi, M, Samoson, A, Tarmak, M, Engelhardt, G *J Am Chem Soc* 1981, **103**, 4992

225 Klinowski, J, Ramdas, S, Thomas, J M, Fyfe, C A, Hartman, J S *J Chem Soc Faraday 2* 1982, **78**, 1025

226 Blackwell, C S, Pluth, J J, Smith, J V *J Phys Chem* 1985, **89**, 4420–4422

227 Loewenstein, W *Am Mineral* 1954, **39**, 92–96

228 Wilson, M A, McCarthy, S A *Anal Chem* 1985, **57**, 2733–2735

229 Newsam, J M *J Phys Chem* 1985, **89**, 2002–2005

230 Klinowski, J, Thomas, J M, Fyfe, C A, Gobbi, G C *Nature* 1982, **296**, 533

231 Maxwell, I E, van Erp, W A, Hays, G R, Couperus, T, Huis, R, Clague, A D H *J Chem Soc Chem Commun* 1982, 523

232 Engelhardt, G, Lohse, U, Patzelova, V, Magi, M, Lippmaa, E *Zeolites* 1983, **3**, 239

233 Bosacek, V, Freude, D, Frolich, T, Pfeifer, H, Schmiedel, H *J Colloid Interface Sci* 1982, **85**, 502

234 Artaki, I, Sinha, S, Jonas, J *J Non-Cryst Solids* 1985, **72**, 391–402

235 Murdoch, J B, Stebbins, J F, Carmichael, I S E *Am Mineral* 1985, **70**, 332–343

236 Van de Ven, L J M, Post, J G, Van Hooff, J H, De Haan, J W *J Chem Soc Chem Commun* 1985, 214–216

Chapter 12

NITROGEN

JOAN MASON

1. NITROGEN NMR SPECTROSCOPY

Nitrogen NMR spectroscopy has been well served in the review literature,[1-11] as befits the importance of nitrogen chemistry and the potentialities and difficulties of NMR work with the two magnetic nuclei, ^{14}N and ^{15}N. Both have rather low magnetogyric ratios (γ), and so rather low NMR sensitivity—about one-tenth that of ^{13}C. ^{15}N has spin 1/2 but low natural abundance (0.365%) and negative NOE factors (γ being negative) and may be slow to relax. ^{14}N is nearly 100% abundant but is quadrupolar ($I = 1$), though with a relatively small quadrupole moment ($Q = 0.017\ 10^{-28}\ m^2$)[12] so that broadening may not be excessive. There is also the advantage that information is available from ^{14}N NQR spectroscopy. Until the 1970s most nitrogen NMR spectroscopy used the ^{14}N nucleus. Since then ^{15}N has become increasingly accessible to work in natural abundance, particularly with wide-bore higher-field spectrometers, and polarization-transfer techniques for sensitivity enhancement. Enrichment may not be very expensive if starting material such as $^{15}NH_4Cl$ or $^{15}NO_2^-$ can be used.

This chapter aims to provide a reasonably comprehensive survey of the source material in nitrogen NMR spectroscopy, with leading references to the growth areas.

1.1. Nitrogen Referencing

^{14}N and ^{15}N shifts are interchangeable, as primary isotope effects are very small.[13]

Nitrogen NMR spectroscopy has suffered from a proliferation of reference standards, as shown in Table 1, but a consensus is developing in favor of the use of neat liquid nitromethane as external reference.[1,2,4,6,7,14,15] This is accurately reproducible, it can be obtained enriched with ^{15}N or ^2D, the ^{15}N signal is little affected by proton decoupling, and the ^{14}N line is fairly sharp. Internal referencing is unsatisfactory because of the magnitude and variability of medium effects, as illustrated in Table 1

Difficulties have arisen because of the desire of some workers to minimize negative signs, as achieved for ^1H and ^{13}C with TMS. A major source of nitrogen NMR information[1] uses shielding constants (low frequency positive) instead of chemical shifts (and some workers defy the IUPAC sign convention).[16] Another

JOAN MASON • Department of Chemistry, The Open University, Milton Keynes MK7 6AA, England.

Table 1. Nitrogen Reference Standards and Shielding Scales

	δ^a (ppm)	Reference
NH_3(equil vap, 302 K)	-399.3^b	18
NH_3(liq, 303 K)	−382.1	18
	−381.9	4, 14
(liq, 300 K)	−380.4	15
(liq, 298 K)	−380.2	3, 17
$\mathit{NH_4}NO_3$(aq, 12.3 M)^c	−359.6	4, 14
(aq 2 M HNO_3, 5 *M*)	−361.2	16
NH_4Cl(aq, 5.64 *M*)^c	−359.0	19
	−352.9	4, 14
$(CH_3)_4NCl$(aq, 2 *M*)	−339.0	16
(aq, 6.03 *M*)^c	−336.7	14, 20
$(C_2H_5)_4NCl$(aq, 4.58 *M*)^c	−315.8	14, 20
KNCO(aq, 6.21 *M*)^c	−302.9	4, 14
dmf(liq)	−277.0	4, 14
	−276.4	3, 17
CH_3CN(liq)	−136.4	4, 14
KCN(aq, 8.5 *M*)^c	−102.5	4, 14
HNO_3(aq, 9 *M*)	−14.4	3, 17
(aq, 7 *M*)	−12.6	4, 14
(aq, 1 *M*)	−6.2	16
	−4.4	4, 14
$NH_4\mathit{NO_3}$(2 M aq HNO_3, 5 *M*)	−4.6	19
(aq, 12.3 *M*)^c	−4.0	4, 14
$NaNO_3$(aq, 7.93 *M*)	−3.7	4, 14
CH_3NO_2(CCl_4, 0.3 *M*)	−7.1	4, 14
($CHCl_3$, 0.3 *M*)	−3.8	4, 14
(liq, 18.4 *M*)	**0.0**^{d}	—
CD_3NO_2(liq)	0.1	22
CH_3NO_2(dmso, 0.3 *M*)	2.0	4, 14
CH_3NO_2(liq + 0.1 *M* [Cr(acac)$_3$])	2.75^e	22
$NaNO_2$(aq, 7.56 *M*)^c	228.9	4, 14

^{a} ^{14,15}N shifts at 303 ± 2 K (unless a temperature is given) uncorrected for effects of magnetic susceptibility, except for those from Ref 15 Those of Refs 4 and 13 are ^{14}N measurements in spherical containers to avoid susceptibility effects

^{b} Absolute shielding $\sigma = 263.5$ ppm [15]

^{c} Saturated solutions

^{d} $\Xi = 10.1367673$ MHz[61], absolute shielding $\sigma = -135.8$ ppm, obtained from the value for NH_3 (equil vap) combined with the chemical shift [15]

^{e} Shift proportional to concentration of paramagnetic reagent

source[3,17] recommends pure nitromethane as practical reference but then presents the shifts relative to anhydrous liquid ammonia at 25°C. This unfortunately is an impractical reference, as Table 1 demonstrates, with problems of temperature dependence,[15,18] reproducibility, and susceptibility corrections. Other reference standards in use are 1 *M* HNO_3,[16] 5 *M* NH_4NO_3 in 2 *M* HNO_3,[19] and $(CH_3)_4N^+$ and $(C_2H_5)_4N^+$ ions.[20] In the solid state, NH_4^+ is commonly used.[21]

Although [Cr(pd)$_3$]* has been called a "shiftless" reagent, the effects of this and other paramagnetic relaxation reagents are not negligible in precise work, particularly in comparisons of measurements made in superconducting magnets (in which the field is parallel to the NMR tube) with those in electromagnets (perpendicular field). The shift observed[22] in a superconducting magnet for CH_3NO_2 containing 1 *M*

* pd is the 2,4-pentanedionate ligand, fomerly acac.

$[Cr(pd)_3]$ as quoted in Table 1 agrees with the value calculated from the magnetic moment of Cr(III), 3.9 μ_B, correction being made for the diamagnetic susceptibility of the ligands. The susceptibility shift in an electromagnet is half the magnitude and opposite in sign. Larger shifts are likely if there are specific interactions between the nitrogen and the paramagnetic reagent ($[Cr(pd)_3]$ interacts weakly with acidic protons).[23]

Nitrogen shifts can be referred to an absolute scale[15,24] based on spin-rotation measurements[25] of NH_3; values for other compounds are obtained by subtraction of the chemical shift (Table 1). Absolute shieldings are of interest for comparison with the values given by quantum-mechanical calculations, or with estimated effects of substituents, as well as in comparisons involving different nuclei, as with ^{13}C shieldings in Section 4.

1.2. Medium Effects

Various effects of change in solvent, concentration, counterion, or pH are illustrated in Table 1. The sensitivity of nitrogen resonance to medium effects has made this a useful probe of intra- and intermolecular influences and equilibria.[1–3,25a] Shifts may be pronounced if hydrogen bonding or acid–base interactions occur. Direct hydrogen bonding to nitrogen usually decreases the shielding in saturated groups (NH_3, RNH_2, etc.)[26a] but increases the shielding if the nitrogen is part of a delocalized system, as in nitriles or aromatic azines. These shifts are in the same direction as protonation shifts (Section 4.1) though smaller in magnitude. Indirect hydrogen bonding, however, through hydrogen attached to nitrogen, increases the nitrogen shielding[26b]; thus maximal shielding in NH_4^+ is observed in liquid HF.[28] Many systems therefore show complex behavior, particularly when self-association or bifunctionality is present.[26b] ^{14}N Knight shifts have been used in studies of solvated electrons in alkali metal–ammonia[27a] and methylamine[27b] solutions.

Coupling constants to nitrogen are less sensitive than chemical shifts to medium effects: thus in anilines, increases of 3–5 Hz or less in the absolute values of $^1J(^{15}NH)$ or $^1J(^{13}C\,^{15}N)$ are observed with strong hydrogen bonding, as in dmso. Larger changes may be observed when equilibria are present in which coupling is averaged over different sites.[2]

1.3. Solid State Measurements in High Resolution

^{15}N spectra in the solid state are gradually becoming accessible (though usually with enrichment) by the "dilute spin" techniques as developed for ^{13}C of cross-polarization (CP) from protons, with high-power ^1H dipolar decoupling, magic angle spinning (MAS),[21,29] suppression of spinning side-bands to give the isotropic shift (TOSS),[30] extraction of shielding tensor components from the spinning sidebands,[31] etc. Tensor measurements are given in Table 2. Highest shielding (σ_{33}) is observed for the axial component ($\sigma_{\parallel}$) in linear molecules, or for the perpendicular component ($\sigma_{\perp}$) if the nitrogen is part of a planar π system, since the in-plane components (like the perpendicular components in linear molecules) mix $n(\sigma)$ and π orbitals, with lower excitation energies (cf. Section 4). The tensor elements thus contain useful information on bonding and structure, as well as on shielding anisotropy contributions to relaxation.

Because of the biological importance of nitrogen linkages, including hydrogen

Table 2. Examples of Nitrogen Shielding Tensors and Anisotropies[a]

Compound	Group	Method[b]	$\bar{\sigma}$	σ_{33}	σ_{22}	σ_{11}	$\Delta\sigma$	Reference
$^{14}NH_3$		sr, g	263.5	278.0	278.0	237.7	−40.3	25
$^{14,15}NH_4^+$			224[c]				Small	21
Various	$-{}^{14,15}NH_3^+$		Close to NH_4^+				ca. 10	21, 41
glyglyHCl · H_2O	$-CO{}^{15}NH-$	CP/MAS	129	184.2	173.6	28.6[d]	150.3	38b
$CH_3{}^{14}NC$		lc	130	370	10	10	360	51
Trytophan-26	$-{}^{15}NH=$	lc	126	185	116	70 (∥ NH)	88	36
$^{14}\mathit{N}{}^{14}NO$		sr, g	104	349	−18	−18	367	52
$^{15}\mathit{N}{}^{15}NO$		lc	104	349	−18	−18	367	53
$^{14}N{}^{14}\mathit{N}O$		sr, g	20	359	−149	−149	508	52
$^{15}N{}^{15}\mathit{N}O$		lc	25	364	−144	−144	508	53
Imidazole	^{15}NH		74					44
	$-{}^{15}N=$		2					
Histidine HCl · H_2O						(∥ NH)		41
Cationic	^{15}NH (π)		69	190	36	−19	181	
	^{15}NH (τ)		72	182	51	−17	165	
Neutral	$-{}^{15}N=$ (π)		2	222	−58	−158	330	
	^{15}NH (τ)		77	183	59	−11	159	
Anionic	$-{}^{15}NH=$ (π)		−4	216	−86	−142	330	
	^{15}NH (τ)		77	189	62	−20	168	
$CH_3C{}^{15}N$		CP	21	346	−142	−142	488	29
$HC{}^{14}N$		sr, g	−27	348	−215	−215	563	54
Pyridine-^{15}N			−53	395	−168	−387	672	55
$^{15}N_2$		sr, g	−101	335	−319	−319	654	56
		dd					603	57
$^{15}NH_4{}^{15}NO_3$		CP	−131	9	−188[e]	−214[e]	210	21
$Ph{}^{15}NO_2$			−134	30	−32	−399	398	58
$Ph{}^{14}NO_2$		relaxation	−134	30	−32	−399	398	59
$[Ru({}^{15}NO)_2(PPh_3)_2]$		CP/MAS						43
	$Ru\hat{N}O = 175°$		−173	48	−284	−284	332	
	$Ru\hat{N}O = 168°$		−176	35	−281	−281	316	
$[RuCl({}^{15}NO)_2(PPh_3)_2](BF_4)$		CP/MAS						43
	$Ru\hat{N}O = 178°$		−162	133	−309	−309	442	
	$Ru\hat{N}O = 138°$		−439	−94	−229	−994	832	
$P{}^{14}N$		sr, g	−349	350	−698	−698	1047	60

[a] All values are in ppm. The shielding terms follow the conventions of Chapter 2, namely, low frequency positive, $\sigma_{33} \geqslant \sigma_{22} \geqslant \sigma_{11}$, with the isotropic shielding $\bar{\sigma} = (\sigma_{33} + \sigma_{22} + \sigma_{11})/3$, the shielding anisotropy $\Delta\sigma = \sigma_{\parallel} - \sigma_{\perp}$ for axial, symmetry, or more generally, $\Delta\sigma = \sigma_{33} - (\sigma_{22} + \sigma_{11})/2$, and the asymmetry parameter $\eta = (\sigma_{22} - \sigma_{11})/(\sigma_{33} - \sigma)$. With the value $\sigma(MeNO_2) = -135.8$ ppm,[15] shieldings (σ) may be expressed as chemical shifts (δ) by use of the equation $\delta = -135.8 - \sigma$.

[b] Measurements are of solids unless otherwise described. g = gas, l = liquid, lc = liquid crystal, sr = spin-rotation, dd = dipole–dipole interaction.

[c] Assumed (cf. Section 1.1).

[d] The axis of lowest shielding lies in the plane of the peptide linkage and nearly orthogonal to the C–N bond; single crystal study.

[e] Axial symmetry perturbed by hydrogen bonding to NH_4^+.

bonding, high NMR technology is increasingly being applied to natural products, usually with bio-^{15}N enrichment. Double cross-polarization, the transfer of proton magnetization to either ^{15}N or ^{13}C, then from the one to the other, has been used to study protein turnover in soybean leaves grown on $^{15}NH_4{}^{15}NO_3$ with $^{13}CO_2$ pulse-labeling.[32] The metabolism of labeled ^{13}C–^{15}N bonds in static cell components can thus be observed *in vivo* or in the lyophilized powder in high resolution.[32,33] Nitrate metabolism has been studied by CP/MAS in mycelia of *Neurospora crassa.*[34]

Strategies have been outlined for biopolymers and applied to bacterial and viral DNA (and protein) to show hydrogen bonding and neighbor effects within and between strands.[35] The dipolar coupling may be used to establish N–H geometries.[36–38] Dilute ^{15}N spin exchange (NOESY) techniques were used to locate neighboring ^{15}N sites in α-helical bacteriophage coat proteins,[39] and tryptophan-^{15}N shielding tensor elements and dipolar couplings obtained for this protein *in vivo*, as it can be oriented parallel to the magnetic field.[36] CP/MAS techniques can characterize copolyamides and copolypeptides that cannot be sequenced by other methods (including ^1H and ^{13}C NMR), and show the tacticity and secondary structure.[40]

Table 2 includes tensor measurements of molecules of biological importance, such as histidine[41,42] and glycylglycine.[38] The tensor elements are particularly valuable because of their sensitivity to bond type and to the arrest of processes such as proton exchange, which are rapid in the liquid phase. Whereas deprotonation of the imidazole ring nitrogen in histidine decreases the isotropic shielding (σ) by 70 ppm, it decreases each of the in-plane tensor elements (which mix n_N or σ with π orbitals) by over 100 ppm, so that the anisotropy doubles. The anisotropy of the ring nitrogen (N–π) next to the side chain is significantly increased by hydrogen bonding to $-NH_3^+$.[41,42]

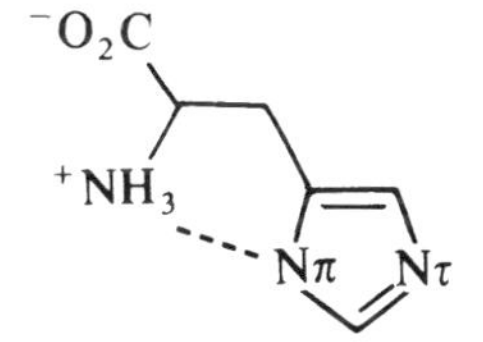

Neutral histidine

CP/MAS ^{15}N NMR spectroscopy (with and without TOSS) has been applied to nitrosyl complexes which are known from X-ray crystallography to have different MNO angles in the solid state.[43] For [Ru(NO)$_2$(PPh$_3$)$_2$], which has a distorted tetrahedral geometry and nonequivalent near-linear nitrosyls (with M$\hat{N}$O 175° and 168°) two isotropic shifts differing by 4 ppm were measured, but the difference in $\sigma_{\parallel}$ and in the anisotropy $\Delta\sigma$ were 13 and 16 ppm, respectively. The five-coordinate complex [RuCl(NO)$_2$(PPh$_3$)$_2$](BF$_4$) has a square pyramidal structure in the solid state with one linear and one bent apical nitrosyl. The solid state spectra show a very large anisotropy in the bent ligand (900 ppm), with very strong deshielding for the tensor element mixing n_N (nitrogen lone pair) and π^* orbitals (Table 2). This compound shows rapid and complex bent-linear fluxionality in solution, so the solid state measurements of the isotropic shifts are essential to the understanding of this process (Section 1.4).

^{14}N work can have certain advantages, not only from the higher natural abundance. The quadrupole couplings may increase resolution by spreading the spectra over a larger range, more than compensating for the line broadening, as in ^{13}C and ^{14}N measurements of the CCN geometry and ^{14}N quadrupole coupling tensor in single crystals of glycine.[44a] ^{14}N spectra of ammonium hydrogen oxalate show differences in the (small) electric field gradients at nonequivalent NH_4^+ ions, arising from different hydrogen-bond distances.[44b] Fundamental and overtone ^{14}N NMR has been used to derive peptide plane orientations from the quadrupole splitting, second order quadrupole shift, chemical shift, and NH dipolar splitting, in an oriented sample.[45]

Bonding to ^{14}N is well known to produce broadening or asymmetric doublet splitting of ^{13}C resonances in the solid state. Analysis of ^{13}C line shapes in CP–MAS spectra has been used to obtain information on the quadrupole coupling tensor of adjacent ^{14}N in amide, nitro, amino, and azole groupings[46a] and two-dimensional methods used to distinguish ^{14}N dipolar and quadrupolar effects.[46b]

Nitrogen NMR can be a powerful tool for adsorption studies, distinguishing different sites such as Lewis or Brønsted acids, cations, channels for physisorption, etc. A standard 90 MHz spectrometer was used to demonstrate strong SiOH ··· N bonding of pyridine to silica gel,[47] and interactions with zeolites of NH_3, Me_3N, MeCN, and pyridine.[48] CP–MAS ^{15}N spectra are reported for pyridine on γ-alumina and mordenite, etc.[49a] and NH_3 on γ-alumina[49b]; $^{15}N_2$ also has been studied on mordenites.[50]

1.4. Isotope Effects and Tracer Studies

Primary isotope effects on nitrogen shifts appear to be commensurate with experimental uncertainties,[13] but secondary (one-bond) effects are observable; normally a heavier isotopic substituent increases the shielding, as observed for the end nitrogen in $^{15}N_2O$ relative to $^{15}N\,^{14}NO$,[62] *cf.* also studies of semienriched species of N_2O_3, $N_2O_3^{2-}$, and related compounds.[63] Similarly, replacement of (each) ^{16}O by ^{18}O increases the ^{15}N shielding by 0.13–0.14 ppm in nitrite ion or 0.06 ppm in nitrate[64]; these isotope effects have been used to monitor acid-catalyzed oxygen exchange between nitrite and water[64a] and the production of nitrite by the nitrosomonas bacterium by oxidation of ammonia or hydroxylamine.[64b]

Various deuterium isotope effects on $\delta\ ^{14,15}N$ have been reported, including equilibrium isotope effects arising from perturbation of tautomeric equilibria[65a] or hydrogen bonding.[65b]

An $^{14,15}N$ equilibrium isotope effect has been observed, arising from fluxionality in the five-coordinate dinitrosyl ion $[RuCl(NO)_2(PPh)_3)_2]^+$, in which there is rapid interconversion of bent and linear nitrosyls. ^{15}NO is very slightly favoured in the linear geometry in the semienriched species, which gives an averaged ^{15}N shielding which is greater by 0.6 ppm at 300 K (1.6 ppm at 183 K) than in the $(^{15}NO)_2$ species.[66] The magnitude of this isotope effect arises from the large difference in zero-point energy (the stretching frequency is ca. 200 cm^{-1} greater for the linear than for the bent ligand) and in ^{15}N shift, 277 ppm in the solid state, as discussed in Section 4.1.[43]

As described in other sections, ^{15}N is important as a biological tracer in view of the central part played by nitrogen in many vital processes, including hydrogen bonding, and the absence of a useful radioisotope, since ^{13}N has a half-life of 10 min.

2. ^{15}N NMR SPECTROSCOPY

Because of the low natural abundance, low sensitivity, relaxation problems, and negative NOE (γ being rather small, and negative), working with ^{15}N presents various choices: how to optimize the relaxation and NOE, whether to use ^{15}N-enrichment or sensitivity enhancement, and so on.

2.1. ^{15}N Relaxation and NOE Factors

Processes contributing to ^{15}N relaxation are illustrated in Table 3 (cf. also Chapter 2, Table 1, and Chapter 5). In the absence of protons to relax ^{15}N by intra- or intermolecular dipole–dipole (dd) interactions, small molecules or ions such as N_2,[67,68] ND_3,[69] NO_3^-,[70] etc. can find relaxation in *spin-rotation* (sr) processes, as also can freely rotating groups such as nitrite in larger molecules.[71] This mechanism is more effective the higher the temperature (as, e.g., for liquid N_2, NH_3, or pyridine) since the rate is proportional to kT; in contrast to dipolar (dd) rates, which are disfavored by higher temperatures. *Shielding anisotropy* (sa) relaxation can be important for linear or planar nitrogen groups such as CN^- [72] or dinitrogen ligands in metal complexes,[73] nitro, nitrate, etc. The rate is proportional to $(\Delta\sigma)^2 B_0^2$, and the anisotropy can be quite large for nitrogen (Table 2). Advantages of higher-field working are evident for the dinitrogen complex in Table 3. The use of *paramagnetic relaxation reagents* has been researched in detail[3,23]; some electron–nuclear relaxation times are included in Table 3. [Cr(pd)$_3$], which is commonly used, can show weak outer-sphere interactions, as of the C=O groups with amine protons in pyrrole or ^{n}BuNH$_2$. [Gd(dpm)$_3$] and [Gd(pd)$_3$] interact more strongly with basic nitrogen sites, and can be used for spin labeling.[23]

Although relaxation progresses are qualitatively similar for ^{15}N and ^{13}C in comparable locations, as in CH_2 and NH_2, ^{15}N shows the greater diversity because of the variety of nitrogen bond types, and the lower efficiency of *intramolecular dipolar relaxation* by protons.[74] This is about four times faster for ^{13}C–^1H than for a corresponding ^{15}N–^1H, the γ^2 dependence of the relaxation rate giving a six-fold advantage to ^{13}C which is offset somewhat by the r^{-6} dependence, as bonds to nitrogen are somewhat shorter.[23] ^{15}N T_{1dd} measurements can be used to locate labile protons.[75] The contribution per proton at a distance of 1 Å is given by $1/T_{1dd} = 5\times10^9\,\tau_c$, where τ_c is an effective correlation time. For small to medium-sized molecules in normal solvents this corresponds to T_{1dd} values of 15–80 s for one attached proton, and to larger values in the absence of nearby protons. In practice faster relaxation is usually observed because of the presence of paramagnetic impurities, usually transition metal ions, to which basic nitrogen is very sensitive; this must be borne in mind in relaxation studies.[75a] Hydrogen bonding, also, can play an important role. (Effects of dissolved oxygen are relatively small.) *Intermolecular dipolar relaxation* by protons can be important for "exposed" nitrogen, as, e.g., in pyridine.[55] This rate has an R^{-1} dependence, where R is the distance of closest approach. Up to 40% of the relaxation rate of dinitrogen ligands measured at 2.11 T was due to dipolar interaction with protons in the solvent and phosphine coligands.[73]

If dipolar relaxation is inconveniently slow (giving rise to NOE factors which are detrimental because partial) it can be speeded up by lengthening the correlation time τ_c. This is proportional to η/kT, where η is the microviscosity, so cooling greatly increases τ_c through the η and T^{-1} terms (and increases intrinsic sensitivity). The rate can effectively be doubled by solvents such as dioxan or ethanol in place of chloroform or acetone, exploiting the η factor. More drastically, glycerol may be used as solvent for polar compounds, or sugar added to aqueous solutions, with additional advantages of hydrogen-bonding; toluene saturated with polystyrene may be used as organic solvent.[76] Very high macroviscosities can be used in practice without danger of line-broadening due to increase in T_2, since the microviscosity is smaller than the macroviscosity (by a factor of 6–7). Working with larger molecules, also, is advantageous, and Table 3 shows dramatic reductions of T_{1dd} in oligopeptides[77] or small proteins such as lysozyme.[78] Internal motions may reduce the effective τ_c, as for guanidine $-NH_2^+$ in lysozyme.

These advantages of increase in τ_c require that the "extreme narrowing condition"

$\omega^2\tau_c^2 \ll 1$ be still met. T_{1dd} is proportional to $\tau_c(1+\omega^2\tau_c^2)$, because of the dependence on fluctuations of local fields. With increasingly long correlation times T_{1dd} goes through a minimum near $\omega\tau_c = 1$, then begins to increase (with concomitant decrease in NOE). This minimum moves to shorter τ_c with increase in the resonance frequency ω, so that working at higher fields is now detrimental. T_{2dd}, however, has an additional τ_c dependence which is independent of ω, and continues to decrease with increase in τ_c so that uncertainty broadening may become a hazard. Thus amide protons were deuterated to improve ^{15}N resolution in work on unfolded hemoglobin(79) (this effect of the γ^2 dependence of the dipolar relaxation rate is illustrated by the comparison of NH_3 with ND_3 in Table 3).

Table 3. Examples of ^{15}N Spin-Lattice Relaxation Times at Ambient Temperatures

Compound	T_1 (s)	B_0 (T)	η_{obs}[a]	$10^{12}\tau_c$[b] (s)	Mechanism[c]	Reference
PhCN(liq)	ca. 420	1.41	−1.6	(7.5)	dd, sr, e	74
$PhNO_2$(liq)	450	1.41	−1.8	(10)	sr, dd	74
	170	7.42	−0.6	4	sa, sr	58
ND_3(liq)	413	1.41	−0.5	0.35	sr, dd	69
$Ca(NO_3)_2$(0.6:1 w/w in D_2O)	300	1.4, 0.94	0		sr, sa	70
(0.7:1 w/w in H_2O)	180		−2.0		sr, dd	70
NH_3(liq)	186	1.41	−3.7	0.25	dd, sr	69
(g)	0.3	6.35	0		sr	86
$NH_4\mathit{NO}_3$(1:1 w/w in H_2O)	140	1.41	−0.9		sr	74
$(SiHMe_2)_2NH$(1:1 in $CDCl_3$)	90	4.70	−4.93		dd	23,87a
CH_3CN(liq)	90	6.34				23
with 5 10^{-4} *M* [$Gd(dpm)_3$]	53				e	
Pyridine(liq)	85	6.34	−0.4			23
with 0.05 *M* [$Cr(pd)_3$]	3.7				e	
Pyridine(liq)	55	3.3, 7.4		1.8	sa, dd, sr	55
trans-PhN=NPh(2.7 *M* in $CDCl_3$)	56	2.11			sr, dd, sa	71
NH_4NO_3(1:1 w/w in H_2O)	37	1.41	−4.93		dd	70
(0.7 *M* in 0.5 *M* HCl)	46.4	4.70	−4.78	1.07	dd	87b
$K_3[Co(^{13}CN)_6]$(0.2 *M* in D_2O)	35	7.04		73		88
with Mg^{2+} (ion pair)	27			ca. 100		
$NaNO_2$(0.6:1 w/w in D_2O)	27	1.41			sr	70
(0.7:1 w/w in H_2O)	23		−0.4		sr	
mer-[$OsCl_2(N_\alpha N_\beta)(PMe_2Ph_3)$](thf)						73
N_α	23	2.11	−1.2	ca. 45	dd, sa	
N_β	31		−0.9			
N_α	5.7	9.34	−0.23		sa, dd	
N_β	4.0		−0.35			
HCONHMe(liq)	23	6.34	−4.93		dd	23
with 5 10^{-4} *M* [$Gd(dpm)_3$)]	11				e	
^nBuONO(liq)	24	2.11			sr(int)	71
KCN(1 *M* in 1:1 H_2O/D_2O, pH 12.5)	21	1.41	0		sa, sr	72
	14.5	4.23				
$HCONH_2$ (with 10% dmso-d^6)	21–25	4.23	−4.93		dd	89a
$MeCONH_2$ (5.3 *M* in H_2O)	14	1.41	−4.93	7.5	dd	90
$PhNH_2$(liq)	13	6.34	−4.93		dd, sr(int)	23
with 0.05 *M* [$Cr(pd)_3$]	3				e	
N_2(liq, 77 K)	15	0.70		0.2[d]	sr	67, 68
(liq, 126 K)	1.5	1.02		1[d]	sr	
$^{14}N\,^{15}N$(g, 294 K)	$T_2 = 0.77(1.1)$	8.46		0	sc	89b

Table 3. (*Continued*)

Compound		T_1 (s)	B_0 (T)	η_{obs}[a]	$10^{12}\tau_c$[b] (s)	Mechanism[c]	Reference
NH_4Cl(s)		4	4.70		14	dd	91
$C(NH_2)_2^+ NH(CH_2)_3 CH(N^+H_3) CO_2^-$ (arginine)							
	NH_3^+	4.6	4.23	−4.93	15	dd, sr(int)	78
	NH_2^+	2.8		−4.93	37	dd	
	NH	4.5		−4.93	43	dd	
TPP[e](s, 180 K)	NH	0.67		0.7		dd	84
	N	0.05		0.1[f]		dd	
$TPPH_2^{2+}$ (0.06 *M* in $CHCl_3$, with 0.08 *M* CF_3COOH)[e]		1.9	4.23	−4.93		dd	92
ZnTPP (0.006 *M* in $CHCl_3$)[e]		56	4.23	0		e, etc.	92
Lysozyme (9.4 mM in 0.1 *M* citrate, pH 5)							78
	amide	0.16	4.23	−0.5	5000	dd	
	NH_2^+	0.33		−3.5	3500	dd	
Oligopeptides (0.1 *M* in dmso-d^6)							77
	gly, phe	0.03–1.6	6.34		ca. 250	dd, pe	
Adenosine monophosphate (0.8 *M* in H_2O, pH 8.5)							3
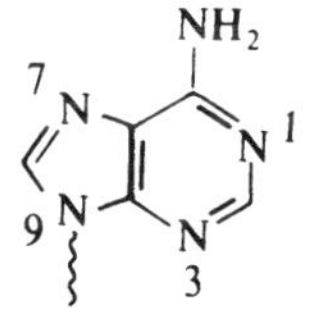	N-1	2.1		Small			
	N-3	3.8		Medium			
	NH_2	0.7		~Full		dd	
	N-7	1.7		Small			
	N-9	7.0		~Full		dd	

[a] Approximate NOE factor.
[b] Effective τ_c/ps. Some τ_q values are given in parentheses; cf. also the values in Table 4.
[c] Relaxation mechanisms in descending order of their contribution. dd = intra-, intermolecular dipole–dipole interaction, sr = spin-rotation, (int) = internal, sa = shielding anisotropy, pe = proton exchange, e = electron nuclear (values[77] normalized to a viscosity of 1 cP), sc = scalar coupling.
[d] τ_{sr}.
[e] TPP = *meso*-tetraphenylporphine.
[f] See text.

The maximal NOE factor for ^{15}N wholly relaxed by ^{15}N–H dipolar interactions is

$$\eta_{max} = \gamma(^1H)/2\gamma(^{15}N) = -4.93$$

where

$$\eta_{obs} = (I - I_0)/I_0 = -4.93\,(T_1)_{obs}/T_{1dd}$$

giving a maximal gain of −3.93 in signal intensity I on proton decoupling (due to cross-relaxation), in the "extreme narrowing" region. This compares favorably with the ^{13}C value, 2.99, but the negative sign means that the enhanced signal is inverted. The signal may therefore be annulled ($I = 0$, $\eta_{obs} = -1$) if only 20% of the relaxation is by the ^{15}N–^1H dipolar mechanism. This could arise in various ways (cf. Table 3): if insufficient of a relaxation agent were added (to reduce T_1 and eliminate an unfavorable NOE), with NOE averaging by fast chemical exchange, with partial proton decoupling, or else with increase in correlation time (as described above) so that the ^{15}N–H dipolar relaxation became less efficient.[80] Outside the "extreme narrowing regime" the NOE diminishes with increase in τ_c,[81] and becomes uniformly unfavorable, the sooner the higher the resonance frequency. In this respect also higher field working may be disadvantageous. The maximal NOE factor shrinks from −4.93 to −0.12, crossing the −1 mark for correlation times of ca. 6 ns at 2.1 T but 1 ns at 11.7 T. Useful graphs are available which show the diminution of the ^{15}N NOE with increasing τ_c as a function of field strength[2,82] (cf. Figure 4 on p. 143, for ^{13}C); also the variation of T_1 and T_2 with τ_c for different relaxation mechanisms.

Clearly the NOE can be useful for signal enhancement, for spectral assignment,[82a] to identify spin-label sites,[23] and as a probe of molecular dynamics,[82b] including segmental motions of large molecules, such as double-stranded DNA.[83] In *meso*-tetraphenylporphine the NOE and T_1 values characterize the HH migration between the central 4N: the unprotonated nitrogen relaxes mainly by the shielding anisotropy mechanism, also by dipolar interaction with the "distant" inner protons and by magnetization transfer by the migrating proton.[84] A detrimental NOE can be avoided[85] by gated decoupling, or by the provision of a more efficient relaxation mechanism, as by paramagnetic additives.

2.2. Sensitivity Enhancement

^{15}N sensitivity can be improved by the use of lower temperatures, higher fields, wide-bore spectrometers, proton decoupling, and the NOE, if favorable. More dramatic enhancements can be achieved by pulse sequences which transfer polarization (magnetization) from a more sensitive coupled nucleus, usually the proton.[7] The enhancement is a product of the $\gamma(^1H)/\gamma(^{15}N)$ ratio (=10, the sign of γ being immaterial) and a factor due to the ability to pulse at a rate determined by the fast proton relaxation. For slowly relaxing nitrogen the enhancement may be a hundredfold or more. In the solid state cross-polarization transfers magnetization via dipolar couplings[21]; in the liquid phase scalar (J) coupling may be used, as in J cross-polarization (JCP),[93] selective polarization transfer (SPT),[94] INEPT (insensitive nuclei enhanced by PT),[95] DEPT (distortionless EPT),[96] SINEPT,[97] etc. (cf. Chapter 2, Table 3). Thus JCP techniques applied to solutions of nylon polyamide gave ^{15}N shifts in natural abundance which are sensitive to chain length, sequencing, conformation, solvation, etc.[98] Extensions of the methods may be used to measure coupling constants[99,100] and their signs,[101,102] to assign connectivities,[103] for heteronuclear shift correlation,[97] for selective enhancement for nonprotonated nitrogen,[104] to speed up relaxation time measurements,[87] for new types of relaxation information,[77] etc. Such methods are increasingly used for natural abundance work[100,103]; some (e.g., INEPT, DEPT) are commercially available. $\{^{31}P\}$ ^{15}N INEPT also has been demonstrated.[100b]

Multiple quantum methods are being developed which involve a double transfer of magnetization $^1H \rightarrow {}^{15}N \rightarrow {}^1H$, greatly increasing the enhancement, and with the additional advantage of obtaining ^{15}N NMR information in two-dimensional spectra at proton frequencies.[105]

3. ^{14}N NMR SPECTROSCOPY

Although the ^{14}N quadrupole moment is relatively small ($Q = 0.017\ 10^{-28}\ m^2$),[12] ^{14}N NMR spectroscopy is dominated by quadrupolar relaxation. Thus in the paramagnetic complexes $K_3[M(CN)_6]$ (M = Cr, Mn, Fe) the nitrogen shifts were determined by the contact interaction, but the ^{14}N relaxation was wholly quadrupolar.(106) Even in solid NH_4Cl, a switch in the NH_4^+ orientation in the lattice changes the electric field gradient (efg), relaxing the nuclear electric quadrupole and therefore the nuclear spin, so that the dipolar mechanism takes over only below 220 K.(107)

As shown by Table 4, ^{14}N linewidths vary widely, depending on two factors, the nitrogen bond type and the molecular or ionic mobility. High-resolution work is possible if the local electronic symmetry, and the mobility, are both high; and ^{14}N is a useful probe nucleus for studies of molecular motions.

3.1. ^{14}N Quadrupolar Relaxation(108)

The ^{14}N linewidth ($W_{1/2}$) is related to the quadrupolar relaxation time T_q and to the structural and motional factors by

$$\pi W_{1/2} = 1/T_q = \tfrac{3}{8}\chi^2(1 + \eta^2/3)\,\tau_q$$

where $T_q = T_2 = T_1$, for mobile species with $\omega^2\tau^2 \ll 1$, and $\tau_q = \tau_c$, the (effective) rotational correlation time of the nuclear quadrupole. If the motions are anisotropic, τ_q is a weighted average of the values for the different axes of rotation. $\chi = e^2qQ/\hbar$ is the nuclear quadrupolar coupling constant (NQCC) in radians and eq is the efg at the nucleus (in fact eq_z, where $|q_z| \geqslant |q_y| \geqslant |q_x|$). $\eta = (q_x - q_y)/q_z$ is the asymmetry parameter ($0 \leqslant \eta \leqslant 1$), and $W_{1/2}$ is the true width at half-height of the (Lorentzian) line or component of a multiplet. If a derivative is measured $T_2\ (= T_q) = 1/\sqrt{3}\,\pi W$, where W is the peak-to-trough separation.

Thus linewidths for the different kinds of nitrogen in the N_2O molecule are found to be in the ratio of the respective χ^2 values.(109) NQCCs are obtained from microwave spectroscopy of gases and NQR spectroscopy of solids, also by NMR methods on solids and liquid crystals. Values in the liquid phase are usually close to those in the solid, and 10%–15% less than in the gas phase. Determining the NQCC from estimates of τ_q requires knowledge of the molecular motions. Anisotropic or additional internal motions may be monitored by relaxation measurements on two or more nuclei in the molecule, as of ^{14}N, ^{17}O, and ^{2}H in dimethyl formamide.(110)

Electric field gradients are small ($\chi < 1$ MHz, say) for linear two-coordinate, planar three-coordinate, or four-coordinate nitrogen, if the ligands are comparable in electronegativity. Examples in Table 4 are N*N*O (MeNC, N*N*N$^-$), NO_3^-, and R_4N^+: note the greater values for $MeNO_2$, or cobalt ammine complexes. The efg is greatly increased if nitrogen carries a lone pair, in which electronic charge lies closer to the nucleus (χ is very large for NF_3). Thus χ is much greater for MeCN than MeNC (and greater still for N_2), greater also for NO_2^- than NO_3^-, for R_3N than R_4N^+, and so on. As a corollary, nitrogen χ values are reduced on coordination or protonation, or by delocalization of the lone pair in conjugated systems, this tending to increase the bond angle for two-coordinate nitrogen, or flatten a pyramidal group, as in anilines or amides. Relatively small χ values are observed for a "lone pair" of π type, as in pyrrole. Hydrogen bonding to the lone pair slightly reduces the NQCC, but hydrogen bonding of hydrogen attached to nitrogen may have the opposite effect, as shown by pyrrole in pyridine.(111) Effects of bond angle distortions have been demonstrated in 3–7-membered rings containing nitrogen.(112) Linewidths are sensitive to changes in χ as they

Table 4. Examples of ^{14}N Relaxation Times and Linewidths at Ambient Temperatures

	$W_{1/2}$ (Hz)	T_q (ms)	NQCC (MHz) (η)	η'(cP)	$10^{12}\tau_q{}^a$ (s)	Reference
$NaNO_3$(single cryst.)		ca. 10^5	0.745			118a
$Me_3NR^+I^-$(0.05 *M* in D_2O, 35°C)						
R = Me		$\geqslant 10^4$	v. small	0.72		118b
Et		800				
Pr^i		1000				
Bu^t		1350				
Hex^n		350				
Me_4NBr(1 *M* aq, 43°C)		10^4				112
Bu^n_4NI(0.07 *M* aq, 62°C)	0.18	1800	0.04	0.45	20	112
(0.05 *M* in benzene, 62°C)	3.7	85	0.23	0.39	15	
NH_4Cl(1 *M* aq, 43°C)		1600		1.08	(5.9, H_2O)	112
(1 *M* in D_2O, 43°C)		1200			(7.2, D_2O)	
N_2O(15 atm. in C_6H_{14}) N*N*O		2960	0.238		20	109a
*N*NO		280	0.792			
MeNC(liq)	0.26	1220	0.27	0.35	30	119
Bu^tNC(liq)	1.6	180				120
NH_4Cl, ND_4Cl(s, 295 K)		700–800	0.016		20	91, 107
$C_6H^n_{13}NMe_3Br$(0.55 *M* in D_2O)		260	0.083		20	117
$C_{12}H^n_{25}NMe_3Cl$(49% w/w in D_2O) (C_{12}TAC)		$T_1 = 24$ $T_2 = 3.7$	0.083		370, fast 5×10^4, slow	115
$C_{16}H^n_{33}NMe_3Br$(0.4 *M* aq) (C_{16}TAB)		$T_1 = 10$ $T_2 = 1$	0.083		560, fast 2×10^5, slow	117
NF_4AsF_6(HF)	3					28
NO_2AsF_6(HF)	5					28
$NaNO_3$(1 *M* aq, 22°C)		85			1.43 ($\tau_\perp$)	120a
$EtONO_2$(liq)		45	0.9		2	121
Pd-$CNBu^t$ complexes ($CDCl_3$)	4–12	30–80				120
NH_3(liq)		35	4.08(g) 3.47(s)	0.14	0.18	113
ND_3(liq)		25			0.23	113
$MeNO_2$(liq)		22	1.7(0.424)	0.6	1.03	121, 122
NOF_2AsF_6(HF)	18	18				28
$^{14}N_2$(liq, 126 K)		18.5	5.39		0.15	67, 68
(liq, 77 K)		11.7		0.074	0.2	
$^{14}N\,^{15}N$(g, 294 K)		15.3				89b
HNCO(C_6H_{12})	20					123
MeNCO(liq)	35					123
Pyridine N-oxide(0.2 *M* in CCl_4)	28.3	11.2	1.1		3.4, 11.9, 2.9	124
$MeNH_2$(liq)	30		3.98	0.18	0.55	125
Me_2NH(liq)	125		5.05	0.17	1.0	125
Me_3N(liq)	100		5.2	0.16	0.8	125
Quinuclidine $CH(CH_2CH_2)_3N$ (0.5 *M* in $CHCl_3$, 30°C)			5.4	0.51	0.8 ($\tau_c \parallel$) 6.7 ($\tau_c \perp$)	125a

Table 4. (*Continued*)

	$W_{1/2}$ (Hz)	T_q (ms)	NQCC (MHz) (η)	η'(cP)	$10^{12}\tau_q{}^a$ (s)	Reference
$PhNO_2$(1:1 v/v in hexane)		6.2	1.426	1.83	3.6, 11, 4.4	59
(liq)		3.2			7.0, 21, 6.3	
NaN_3(1 *M* aq) N*N*N$^-$		29.3	1.03			109b
*N*N*N*$^-$		7.4	1.79			
(2 *M* aq, pH 6.4) *N*N*N*$^-$	50					126
(+0.5 *M* aq Hg^{2+}) *N*N*N*$^-$	60					
ClCN(liq)		6	3.219			127
MeCN(1% v/v in hexane)	53.5		3.74	0.29	0.82	128
(liq)	78			0.34	0.67	
EtONO(liq)		5.2	3.8		0.9	121
$C_5H'_9ONO$(liq)		1.6			3.0	121
Pyrrole (17 mol % in pyridine)		4	2.12	1.2	12	111
Pyrrole (liq)		2	2.06		8	111
(NO) AsF_6(HF)	95					28
$(CH_2)_6N_4$(0.1 *M* in hexane)	122	2.6	4.19	0.29	1.48	129
(0.1 *M* in CCl_4)	294		4.17	0.82	3.3	
(0.1 *M* in $CDCl_3$)	1052		4.31	0.5	11.1	
MeSCN(liq)		2.0	3.75	0.72	2.4	121
Me_2NCHO(liq)		2.0	3.6(0.06)			110
Pyridine(liq)		1.65	4.584(0.4)	0.89	1.95	130
CCl_3CN(liq)		1.5	4.1	0.75	2.7	131
NC−C≡C−CN(∞ diln. in hexane)	244		4.14	0.30	4.75	132
(liq)	394			0.60		
HC−C≡C−CN(5 *M* in tol-d_8)	189	1.7	4.14	0.45	2.36	133
$(NH_2F)(CF_3SO_3)$(HF) +10°C	300					28
−40°C	550					
NF_3(liq) −165°C		1	7.068		1.35	108
2-pyrrolidone(2 *M* aq)	380		3.9		>282, 1, 9	134
(2 *M* CCl_4)	690					
$K_3[Fe(CN)_6]$(0.2–1.0 *M* aq)	780–1040					135
$K_4[Fe(CN)_6]$(0.2–0.8 *M* aq)	550–650					
$[Co(NH_3)_6]Cl_3$(aq)	370		3.62		10.3	136
$[Co(NH_3)_6](ClO_4)_3$(0.24–0.5 *M* in dmso-d^6)		0.29	3.3		20	137
$K_3[Co(CN)_6]$(aq)	570		3.62		16.0	136
$[Co(en)_3]Cl_3$(aq)	600				22.1	136
$Na[Co(NH_3)_2(NO_2)_4]$(aq) ONO^-	700		4.25		14.25	136
NH_3	170		2.09			
$N(CH_2CH_2)_3N$(s, 415 K)	484	0.658	4.925		4.23	138
	2100	0.152			18.4	
PhN=N̲(O)Ph(liq, 11°C)	700		1.4		10	74
Choline Me_3N- head-group in aq. sphingomyelin (47°C)		$T_1 = 37$ $T_2 = 2.0$	0.135			114a

[a] Cf. the values in Table 3, obtained in ^{15}N resonance. If three values are given, these correspond to $\tau_{x,y,z}$: the *x* direction is that of the axial bond to nitrogen, and the *y* direction is in the molecular plane.

are proportional to χ^2, though less sensitive to changes in the asymmetry factor, as they are proportional to $(1+\eta^2/3)$, and $\eta < 1$ by definition. Thus ^{14}N linewidths may be helpful in assignments.

Motional factors determining the linewidths can be envisaged in terms of a frictional model (Debye–Stokes–Einstein theory) according to which

$$\tau_c = V\eta'/kT$$

where $V = 4\pi a^3/3$ is the volume of the molecule, assumed spherical, and η' the viscosity. Real molecules usually have shorter correlation times and a microviscosity factor of ca. 1/6 may be used. The relaxation of asymmetric molecules has been described in terms of rotational diffusion constants[75a]:

$$D_i = 1/6\tau_i\ (=kT/6V\eta')\ \text{about an axis}\ i$$

Increases in linewidth and in τ_q with molecular size, as from NH_3 or MeCN to ammine or cyano complexes, or from Me_4N^+ to long-chain surfactants, are evident in Table 4. A direct dependence of linewidth on viscosity is observed for a wide range of molecules. ^{14}N NMR of fluoronitrogen compounds is favored by their high mobility. Linewidths may be sharply reduced by warming the sample, since viscosity decreases exponentially with increase in temperature.

Departures from direct viscosity dependence may be observed for molecules which are small, or linear or flat, so that they can rotate or librate for a while in between neighboring molecules. Three correlation times τ_x, τ_y, τ_z are given for some such molecules in Table 4. Departures are observed also for high or low viscosities. Molecules such as ammonia which rotate rather freely in the liquid phase are better described by an inertial rotation model, according to which

$$\tau_c = \tfrac{1}{2}(\pi I/3kT)^{1/2}$$

where I is a moment of inertia, and the temperature dependence is less marked. In the conditional inertial rotation model, the rotation requires an expanded lattice or lattice defect.[113] At the other extreme, in Table 4, is the use of wideline ^{14}N NMR of a highly symmetric group such as Me_3N^+- to study oriented systems. Examples in Table 4 are the choline headgroup as a probe of orientational order in phospholipids, as in a phase transition in sphingomyelin from bovine brain membrane,[114a] cf. egg phosphatidylcholine studies in which the ^{14}N quadrupolar splittings are reduced by reduction in C–N ordering, with added metal ions or anaesthetics.[114b] In surfactants such as C_nTAC and C_nTAB spherical and rod-shaped micelles could be distinguished, also fast and slow motions, the faster motion being rotation within a micelle, the slower motion micellar diffusion.[115]

For ions in aqueous solution, quadrupolar relaxation is induced by the reorientation of solvent dipoles, and buffeting of neighboring ions, which may be direct in smaller ions, or transmitted by ligand vibrations. The relaxation is then a useful probe of processes such as ion-pairing or ligand exchange. In mobile solutions (with $\omega^2\tau^2 \ll 1$) an approximate formulation is[108]

$$\pi W_{1/2} = 1/T_q = 8\pi(\beta eqQ/\hbar)^2\left[\mu_s^2 c_s r_0^{-5}\tau_s + \sum_i z_i^2(3a_i^3)^{-1}c_i\tau_i\right]$$

The amplification of the efg is given by

$$\beta = P(1 - \gamma_\infty)$$

where $P = (2\varepsilon + 3)/5\varepsilon$ is the polarizability of the solvent with dielectric constant ε, and γ_∞ is the Sternheimer antishielding factor, which gives the distortion of the inner electron shells by the external efg's. Reported values for ^{14}N are 18 for $1 - \gamma_\infty$ in solid (disordered) NH_4Cl,[91] and -4 for γ_∞ in ammonium aluminum alum.[116] The first and second terms in the square brackets represent effects of solvent dipoles (s) and ionic charges (i), respectively: c_s, c_i are concentrations, r_0, a_i are distances of closest approach, μ_s the dipole moment, z_i ionic charges, and τ_s, τ_i the respective correlation times. Further terms are required to take account of exchange processes. As before, activation parameters may be derived from the temperature dependence. The part played by the water can be gauged by substitution of D_2O, in which the motions are slowed somewhat by the higher mass and stronger hydrogen bonds (cf. Table 4).[117]

4. PATTERNS OF NITROGEN SHIELDING[6]

The extensive literature on nitrogen shifts has been reviewed recurrently.[1–11,139] Shift ranges are given for functional groups in Figure 1, and for ligands in metal complexes in Figure 2, and the parallelisms are evident. Figure 3 compares nitrogen and carbon shielding for corresponding bond types, since the similarities and differences show up the fundamental determinants of nitrogen shielding. The carbon shifts are plotted on twice the scale of the nitrogen shifts as the radial term $\langle r^{-3} \rangle_{2p}$ for the free atoms is twice as large for nitrogen as for carbon—2.46 vs. 1.23 a.u.—reflecting the greater electronegativity of nitrogen (cf. Chapter 3, Section 3). Correlations of nitrogen shifts with the atomic charge on nitrogen in amines, nitroalkanes, isonitriles and azines follow similar trends to those for carbon and oxygen.[140]

The figures show that high shielding, as for singly bonded carbon, is observed for singly bonded nitrogen (N–H, B–N, C–N, N–N, Si–N, P–N, etc., also M–N as in metal ammines). Additive substituent parameters express nitrogen shifts in alkylamines and hydrochlorides[26] and are comparable to those for alkyl carbon, the C_α and C_β parameters reflecting inductive effects, the C_γ parameter steric and conformational factors.

Nitrogen shielding tends to decrease from saturated to linear groups such as $-NCO$ (i.e., MNCO, RNCO, NCO^-), also $-N_3$, $-NCS$, $-CNR$, $-NC$, $-NCR$, $-CN$, etc., then to aromatics and doubly bonded $C{=}N$ compounds. Strongest deshielding is observed for bent diazene ($XN{=}NX$) and nitrosyl or nitroso ($XN{=}O$) groups. Thus comparable shielding of nitrogen and carbon is seen in (isoelectronic) NH_4^+ and CH_4; N_2H_4 and C_2H_6; CN^-, NO^+ and CO; NO_3^- and CO_3^{2-}; and so on. The decrease in shielding from single to triple to double bond can be discerned in nitrogen as in carbon resonance, this being the order of increase in asymmetry of the local electronic charge, as shown also by the NQCC values in Table 3. The energy term, also, may decrease in this sequence: the *ab initio* orbital energy gaps (for charge circulation) are 29, 23, and 20 eV for C_2H_6, C_2H_2, and C_2H_4, respectively, and 25, 23, and 12 eV for N_2H_4, N_2, and N_2H_2,[140a] with a particularly small (n_N–π^*) gap for the diazene $HN{=}NH$.

The major difference between the nitrogen and carbon charts in Figure 3 is that

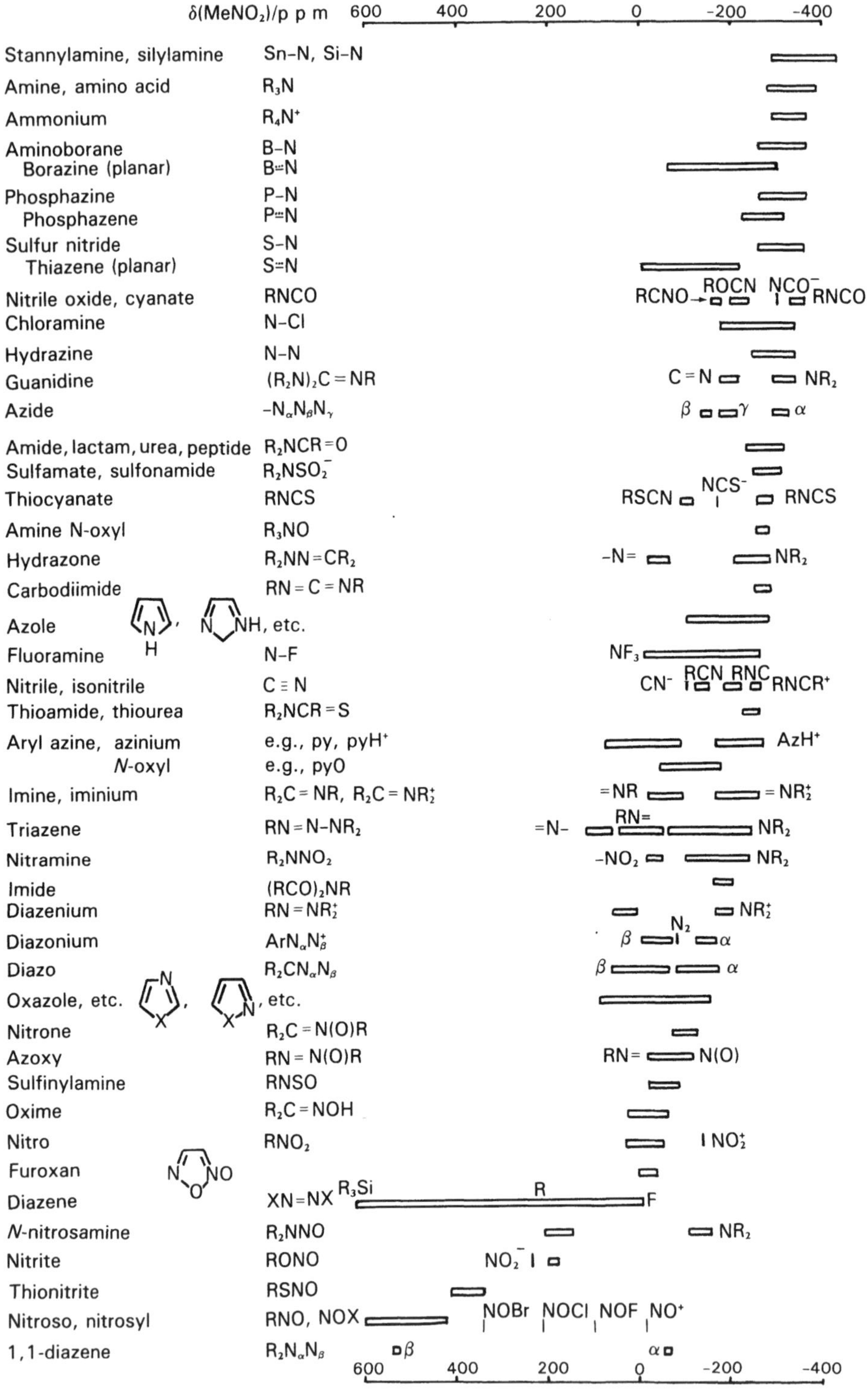

Figure 1. Ranges of nitrogen shifts in compounds with main-group elements.

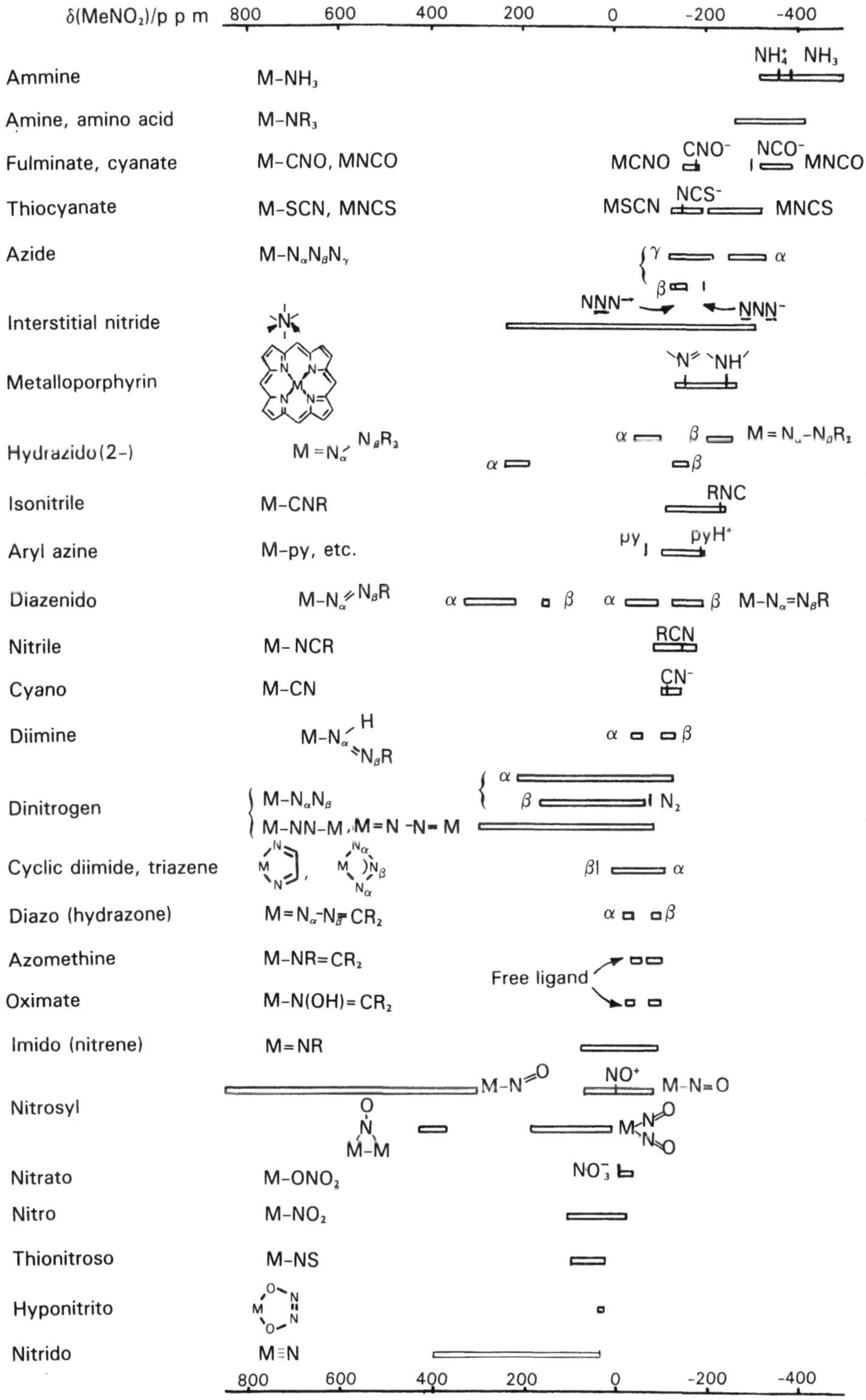

Figure 2. Ranges of nitrogen shifts in metal complexes.

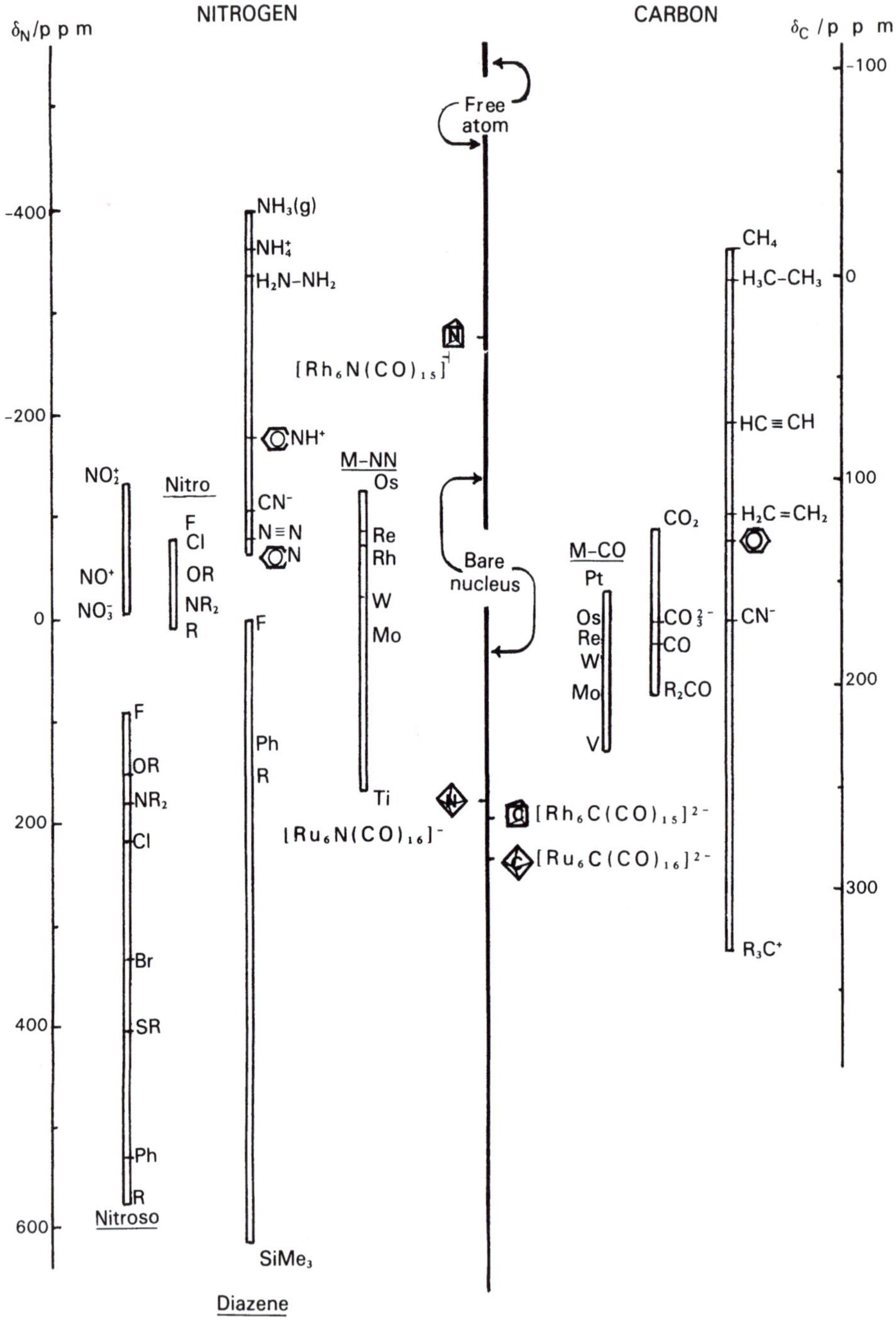

Figure 3. Comparison of nitrogen and carbon chemical shifts.

the nitrogen shifts cover three times the range of carbon shifts (in diamagnetic compounds). Individual ranges are more than twice as great for nitrogen as for (isoelectronic) carbon, e.g., for ligating nitrogen in dinitrogen complexes (M–*N*N) compared with M–^{13}CO. There is also an extension of the nitrogen range towards lower shielding for π-bonded nitrogen carrying a lone pair of electrons. The larger range for nitrogen is thus attributable to all three shielding factors (discussed in Chapter 3): the larger radial term $\langle r^{-3}\rangle_{2p}$ for nitrogen than for carbon, large asymmetries in the valence shell when nitrogen carries a lone pair, and some very low-lying $n_N \rightarrow \pi^*$ states, as in the nitrosyl group, these differences being related to the greater electronegativity of nitrogen than carbon.

Effects of the energy term are observed as δ/λ correlations, notably for NOX,[141] XN=NX,[142] and YNN[143] compounds. The nitrogen shielding increases as the $n_N \rightarrow \pi^*$ band moves to longer wavelength (λ) in a given series, from the fluoro compounds (NOX, N_2F_2), which are colorless, to Bu^{t}NO and $Me_3SiN{=}NSiMe_3$, which are blue; cf. also 1,1-diazenes, which are purple and show strong nitrogen deshielding.[144] (The shifts often correlate also with the inverse of the lone pair ionization energy.[142]) Comparable effects are observed for nitro compounds, even for dinitrogen complexes, those with the highest nitrogen shielding being colorless.[145] Comparing groups with $n_N \rightarrow \pi^*$ states, we find that δ(N) and λ both increase in the sequence (N=C) < (N=N) < (N=O) as increasing electronegativity of the partner in the double bond lowers the π^* relative to the n_N orbital. The nitrogen shift is further increased in this sequence by increase in the radial term, also by the presence of a lone pair on the partner, as this raises the n_N HOMO.

Important evidence of influences of the energy and charge asymmetry terms is becoming available from shielding tensors, as shown in Table 2. In linear systems, highest shielding (σ_{33}) is observed for the axial component ($\sigma_{\parallel}$) corresponding to a free diamagnetic circulation ($\Delta E = 0$, $\sigma_p = 0$), and it is interesting to see the near-constant values (335–370 ppm) of σ_{33} in CH_3NC, *N*NO, N*N*O, CH_3CN, HCN, N_2, and PN. In planar ($p\pi$) systems such as nitrate ion, or aromatics, the highest shielding component is perpendicular to the plane, since this axis mixes σ and σ^* orbitals. Strongest deshielding is observed for axes which mix n_N and π^* orbitals with small ΔE, notably $\sigma_{11} = -994$ for the bent nitrosyl ligand.[43]

Clearly, nitrogen shifts can be used as carbon shifts are. Thus there are substituent parameters for alkylamines,[26] amino acids and peptides,[40,146] ureas, amides,[2] and so on, although two-dimensional ^{15}N–^{1}H correlation spectra are increasingly being used to assign nitrogen resonances in complex molecules such as peptides or proteins.[147] Correlation analyses for nitrogen shifts against shifts of other nuclei, electron density, and Hammett-like substituent parameters (linear free energy relationships) have been surveyed.[2,3,148] But nitrogen NMR spectroscopy is more versatile than ^{13}C NMR, because of the greater range of nitrogen bond types and the importance of nitrogen as a center of reaction.

4.1. Nitrogen NMR Criteria of Structure

Because of the sensitivity of nitrogen shielding (and coupling constants) to the presence of a lone pair on nitrogen, and to π-delocalization, useful criteria of structure can be formulated:

(i) Groups which may be bent or linear at nitrogen, such as diazene or nitrosyl, show sizable deshielding as the angle at nitrogen decreases. In diazenido (M–N=NR) complexes of Rh or Ir, with comparable coligands, ligating nitrogen (N_α) may be deshielded by 200 ppm on bending of MNN.[149,150] Similarly for triazenes, as in the [Ts–NNN–Ts]$^-$ ion (Ts = tosyl),[151] the central nitrogen is deshielded by 286 ppm compared to the linear azide ion. In nitrosyl complexes deshieldings of 300–800 ppm are found for strongly bent (ca. 120°) M ^{15}NO[152,153] relative to comparable linear nitrosyls[154] (cf. the deshielding of 300 ppm for

Me_2NNO relative to NNO). Intermediate shielding is observed for intermediate MNO angles[155] (cf. also the deshielding of nitrosyl ^{15}N if an *N*-nitrosamine is distorted from planarity[156]). Bent nitrogen is less deshielded in aza-arenes and azoles, and quite highly shielded in azides.[157]

Thus nitrogen NMR can monitor reactions in which the shift of a *d*-electron pair onto nitrogen, as it bends, allows the metal to accept an incoming ligand (and the nitrogen to coordinate a proton or other Lewis acid). A comparable *d*-electron pair shift onto nitrogen is also possible in imido and nitrido ligands.[150,158]

(ii) The flattening of pyramidal (saturated) nitrogen, with π-delocalization of the lone pair, also involves deshielding of nitrogen. This is well known in substituted anilines[2,3]; an inorganic example is the hydrazido(2-) ligand $M{=}N{-}NH_2$,[158] in which the $-NH_2$ group is planar, so that the nitrogen shift (ca. -230) resembles that of an amide (e.g., -265 for formamide) rather than that of a hydrazine (ca. -320). Planarity allows of lower-energy $\pi \leftrightarrow \sigma$ paramagnetic circulations, as compared with $n_N \rightarrow \sigma^*$ in the saturated compound.

(iii) Protonation of π-bonded nitrogen, as of pyridine, is well known to correlate with increased shielding, by 100 ppm or more, as the $n_N \rightarrow \pi^*$ band disappears from the uv spectrum. Similarly, the protonation of a dinitrogen ligand to diazenido then hydrazido(2-)

$$W{-}N{\equiv}N \longrightarrow W{-}N{=}NH \longrightarrow W{=}N{-}NH_2$$

increases the shielding of N_β by 135 ppm then 55 ppm.[149,150,158] Protonation of a bent diazenido group attached to Pt(II), to give a hydrazido(1-) ligand,

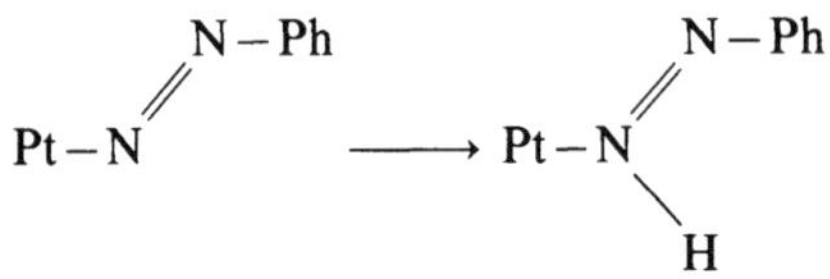

increases the N_α shielding by 340 ppm, showing the involvement of *d* orbitals in the deshielding of the ligating nitrogen.[150] Interestingly, nitrogen deshieldings of ca. 70 ppm are observed on protonation of oxygen in enaminones $RCOCH{=}CHNR_2$ (compared with a few ppm only in amides).[159] Negatively charged nitrogen in anions made by deprotonation of 2,4,6-trimethylpyridine is shielded by ca. 100 ppm relative to the parent compound.[160]

(iv) Protonation of singly bonded nitrogen, as in ammonia or alkylamines, usually leads to a deshielding of up to 20 ppm (despite the increase in local symmetry) probably with increase in the radial factor due to the positive charge. Small *in*creases in shielding are observed, however, for amines in which there is branching at the α-carbon. This is related to the cross-over in the lines which correlate nitrogen shifts in primary and secondary amines and their ammonium ions, with ^{13}C shifts for carbon similarly situated in the corresponding alkanes.[8]

(v) Hydrogen bonding to a lone pair on nitrogen leads to nitrogen shifts in the same direction as the protonation shifts, but smaller in magnitude; though nitrogen shifts with hydrogen bonding of attached hydrogen are usually in the opposite direction.[26b] The sensitivity of nitrogen shielding (and coupling constants) to tautomeric, acid–base, and related processes makes nitrogen NMR a sensitive probe of intra- and intermolecular interactions, particularly when information is available also from the NOE and relaxation.

(vi) Coordination shifts often resemble protonation shifts: thus coordination of BH_3 or BMe_3 to NH_3 or alkylamines deshields the nitrogen somewhat, whereas coordination to π-bonded nitrogen increases the shielding, e.g., by 74 ppm for py:BH_3. Coordination to transition metals may shield or deshield nitrogen, as discussed below.

(vii) Although few examples are known as yet it is likely that nitrogen in a given ligand is deshielded (as is carbon) when bridging two transition metals: thus bridging nitrosyls (Ru_2NO, Os_2NO)[161] give nitrogen shifts in the same region as bent nitrosyls. Bridging dinitrogen (MNNM) is strongly deshielded when coordinated to a π-acceptor metal (d^0, d^2, etc.).[162]

(viii) Basic nitrogen sites are readily probed by paramagnetic shift reagents.[3,23]

4.2. Inorganic Azines and Azenes

A large number of compounds in which nitrogen is (formally) singly or doubly bonded to a nonmetallic heteroelement have been studied in ^{14}N or ^{15}N resonance.[1-3] Topologic, electronic, and steric contributions to ^{15}N and ^{31}P shielding relationships in the P(III)–N bond have been deduced from aminophosphine studies,[163] and for amine–boranes, correlations observed of $\delta\,^{14,15}N$ and $\delta\,^{11}B$ with $\delta\,^{13}C$ in the corresponding alkanes,[164] also of $^1J(^{15}N\,^{11}B)$ with $^1J(^{31}P\,^{11}B)$ in the analogous phosphine–borane adducts.[164b]

For corresponding N–X compounds, the ranges of nitrogen shifts show periodic trends reflecting inductive effects on the $\langle r^{-3}\rangle_{2p}$ and ΔE terms. The nitrogen shielding decreases across the row of the heteroelement (from B–N to C–N to N–N to N–O to N–F, from Si–N similarly to N–Cl, and so on), with increase in electronegativity, and increases down the group of the neighbor (e.g., from N–F to N–Cl to N–Br) with decrease in electronegativity, increase in polarizability, and increased relativistic effects of the neighbor,[6] as discussed in Chapter 3, Section 3. σ-fluoro effects, therefore, are an extreme example: the nitrogen shielding decreases by 280 ppm from NH_4^+ to NF_4^+, showing the dominance of the radial term.[28] In contrast to this are the π-fluoro effects in nitrogen NMR spectroscopy[165,166] which resemble the "perfluoro effects" used in photoelectron and electronic spectroscopy to distinguish σ from π orbitals. Fluorination stabilizes the σ relative to the π manifold, in which the inductive and conjugative effects are opposed. The lone pair orbital being linked to the σ manifold, increased nitrogen shielding is observed on fluorination of azabenzenes and other planar species (cf. NOF, NO_2F, N_2F_2, etc.), despite the opposition of the radial term; there are smaller π-chloro effects in the same sense.[166]

As already described, nitrogen shielding decreases when n_N electrons delocalize into $p\pi$ or $d\pi$ orbitals on a neighbor, with opening out of the nitrogen bond angle(s), as in aminoboranes, borazines, silazanes, phosphazenes, thiazenes, etc.[6] The nitrogen shielding may then be sensitive to the substituents on the neighbor, through conjugative effects: thus the ranges of ^{15}N and ^{31}P shifts in cyclophosphazenes $(NPX_2)_n$ are similar, 80–90 ppm, although the substituents are attached to phosphorus. Interestingly, the nitrogen shielding in $(NPX_2)_3$ increases as $X = Br < Cl < SR < F \lesssim OR$, reflecting the influence of the energy term, whereas the phosphorus shielding increases as $X = SEt < Cl < OEt \lesssim F < Br$, showing mixed influences of the energy and radial terms. The nitrogen shielding decreases from trimer to tetramer, with increase in PNP angle, whereas the phosphorus shielding increases (with increase in NPN angle).[167] In silatranes the nitrogen shielding decreases with shortening of the transannular bond to silicon, from the gaseous to the crystalline state, or with increase in electronegativity of the exocyclic substituent on silicon.[168] In sulfur–nitrogen compounds, the wide range spanned by the nitrogen shifts, about 500 ppm from S_7NMe to aromatic or other planar thiazenes, gives them particular value in preparative or structural work.[169] The nitrogen shift as a criterion of aromaticity in phosphazenes, thiazenes, etc. has been discussed.[6] Chemical shifts are similar for nitrogen in rings and in corresponding acyclic compounds, as found for carbon in arenes and alkenes.

4.3. Coordination Shifts

Coordination shifts in metal complexes can be gauged from Figure 2, since shifts for the free ligand (or near relation) are included. In the higher shielding range coordination shifts are fairly small (as observed for ^{13}C, ^{19}F, ^{31}P, etc.). Coordination shifts are usually larger for transition than for nontransition metals.

Ambidentate ligands are well distinguished by nitrogen NMR: thus nitrogen is more shielded in M–NCS than in NCS^- or MSCN, following the stabilization of the nonbonding orbital when nitrogen coordinates.[170] Similarly in covalent azides ($X–N_\alpha N_\beta N_\gamma$) N_α is more shielded than N_β or N_γ; the N_α and N_γ (rather than N_β) resonances respond to changes in the X group, as explained by the nature of the frontier orbitals.[165] Coordination of aromatic azines moves the nitrogen resonance closer to that of the azinium ion. In (diamagnetic) metalloporphyrins the nitrogen line lies between those for the pyrrole-type (–NH–) and pyridine type (=N–) nitrogen, which can be differentiated at low temperature with slowing down of the tautomeric exchange.[84] A π-coordination shift analogous to those observed for ^{13}C in arene–metal complexes is shown by the increased shielding (by 47 ppm) of ^{14}N in the borazine on forming the complex $[Cr(CO)_3(h^6\text{-}B_3N_3Me_6)]$,[171] as also of η^2 (asymmetrically bridging) cyanide, by 82 ppm, relative to the terminal ligand.[171a]

Relatively low shielding is observed for ligating nitrogen in doubly bonded groups such as hydrazide(2-), $M{=}N{-}NH_2$,[158] and for π-acceptor ligands such as N_2, NO (linear or bent), NS,[172] etc. in which nitrogen is deshielded by circulations using low-lying *d* orbitals. The importance of the ligand field splittings is shown by the periodic dependence of the nitrogen shifts on the transition metal, by *trans*-effects, and so on.[145] Such influences may be amplified when nitrogen is close to the metal, as in nitrido complexes,[150,158] and when the HOMO–LUMO gap is particularly small, as in bent[152,153] or bridging[161] NO ligands.

Evidence is accumulating on the complex question of *trans* influences on NMR shifts. Again, useful parallels may be drawn with substituent effects in organic chemistry, such as the increase in shielding of ^{15}N in anilines and pyridines, and of ^{15}N and ^{13}C in benzonitriles and enzamides, with *para*-substituents which are π-donors (or deshielding with π-acceptors).[2,148] In linear NO and N_2R (diazenido) complexes of Mo(0), similarly, the N_α shielding is increased by *trans*-donors (e.g., OH) and decreased by *trans*-acceptors (CN, MeCN)[150]; and the N_α shielding in the N_2 ligand increases with σ-donor ability of the R group in a *trans*-benzonitrile (NCR) ligand.[173] In nitrobenzenes, however, *para* donors *de*shield ^{15}N and ^{17}O, and both nitrogens are similarly deshielded in arene–diazonium compounds; and N_2 ligand nitrogens in Mo(0) complexes are deshielded on replacement of *trans*-N_2 by the more electron-releasing benzonitrile ligand.[173] Also relevant is the strength of binding of the *trans* ligand, since the $d\sigma$ orbital splitting increases with softness of the ligands, for metals in low oxidation states.[174] Such correlations may be monitored by comparisons with other physical properties, such as ir stretching frequencies (which reflect back-bonding to ligands such as N_2), redox potentials, electronic absorption, orbital energies from PES, also spin–spin coupling constants, and chemical shifts of the metal or other ligand nuclei.

Particularly interesting examples of "multicoordination" shifts are those of interstitial atoms in metal (carbonyl) clusters, because of the large shift ranges, and unexpected differences in shift patterns for interstitial nitride and carbide when these are isoelectronic. In $[M_6N(CO)_{15}]^-$ trigonal prisms the nitride, unexpectedly, is relatively shielded, with $\delta\,^{15}N$ -184 ppm (Co_6) or -273 (Rh_6),[175] whereas carbide in the corresponding 2^- anions is *de*shielded, with $\delta\,^{13}C$ 330 and 265 ppm, respectively. Interestingly, the long-wavelength electronic absorption is at longer wavelengths for the carbides. Similarly in the butterfly $[Fe_4(CO)_{12}]^{n-}$ clusters, $\delta\,^{15}N$ is 216 compared with $\delta\,^{13}C$ 478 ppm.[161] The importance of the site symmetry is shown by the near equivalence (according to Figure 3) of the interstitial shifts in octahedral $[Ru_6(CO)_{16}]^{n-}$ clusters, with $\delta\,^{13}C$ 288 and $\delta\,^{15}N$ 179 ppm. The Ru_6N cluster, also, shows longer-wavelength visible absorption than the Rh_6N trigonal prism.[161] For a given cluster type the interstitial ^{15}N or ^{13}C shielding follows the normal pattern of increase down the group of the metal, and perhaps also across the row; in butterfly mixed-metal clusters the interstitial shift is determined by the wing-tip metal. ^{15}N NMR is particularly useful in this field for following transformations of nitrosyls, protonation reactions, and so on.[161]

5. NITROGEN SPIN–SPIN COUPLING

Coupling constants to ^{15}N and ^{14}N, and the reduced coupling constant K, for a given molecule, are related by

$$-{}^nJ(^{15}NX) = 1.403\ {}^nJ(^{14}NX) = {}^nK(NX) \times h\gamma_N\gamma_X/4\pi^2$$

the magnetogyric ratio γ being positive for ^{14}N and negative for ^{15}N. In this chapter if no sign is given for J no sign should be inferred, and signs are specified as relevant. In Figures 4–6 and the discussion the couplings are expressed as $J(^{15}NX)$, as ^{14}N lines are commonly too broad for resolution or accurate measurement of the splitting, which is relatively small because $\gamma(^{14}N)$ is small; although a number of $J(^{14}NX)$ values have been measured, particularly for ^1H and ^{19}F. The theoretical discussion uses K values for generality. For reasons of space, examples must be limited and reference made to the reviews.[1–4,7,9,11,176–182]

Nitrogen J couplings are peculiarly valuable in structure determination because of their sensitivity to the presence and orientation of a lone pair on the nitrogen. Thus they are usually diagnostic of coordination, chelation, protonation, or hydrogen bonding of the nitrogen, and reflect departures from planarity, conformational relationships, *syn–anti* or *cis–trans* stereochemistry, solvent effects and solution equilibria, and so on.

$^1J(NH)$ observation may require rigorous purification of the sample to avoid exchange [alternatively, exchange-averaged values of $^1J(NH)$ afford information on the dynamics]. In accordance with the Fermi contact (Fc) mechanism (cf. p. 9), $^1K(NH)$ is positive and tends to increase with s character in the N–H orbital, from sp^3

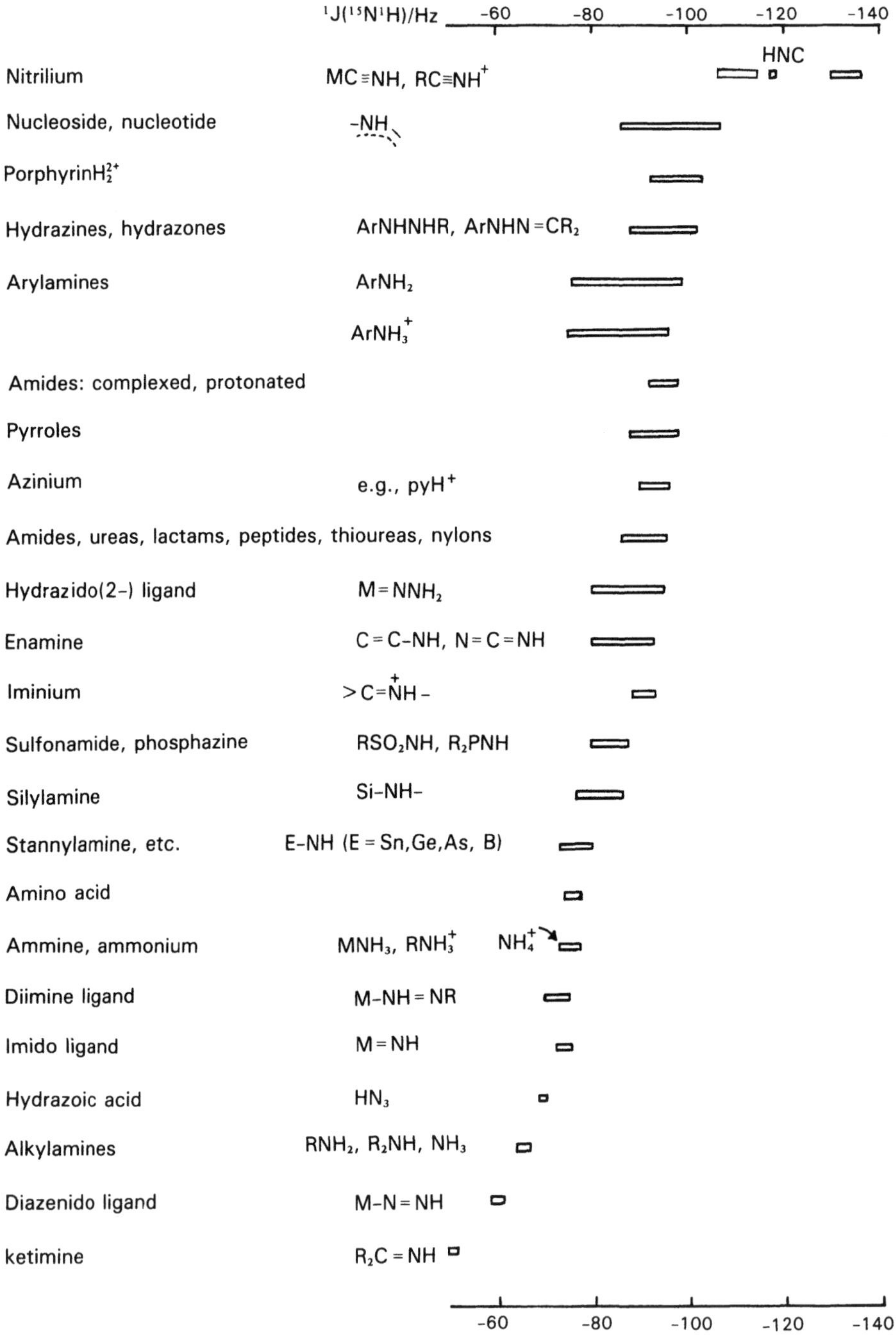

Figure 4. Ranges of ${}^1J({}^{15}N\,{}^1H)$ values.

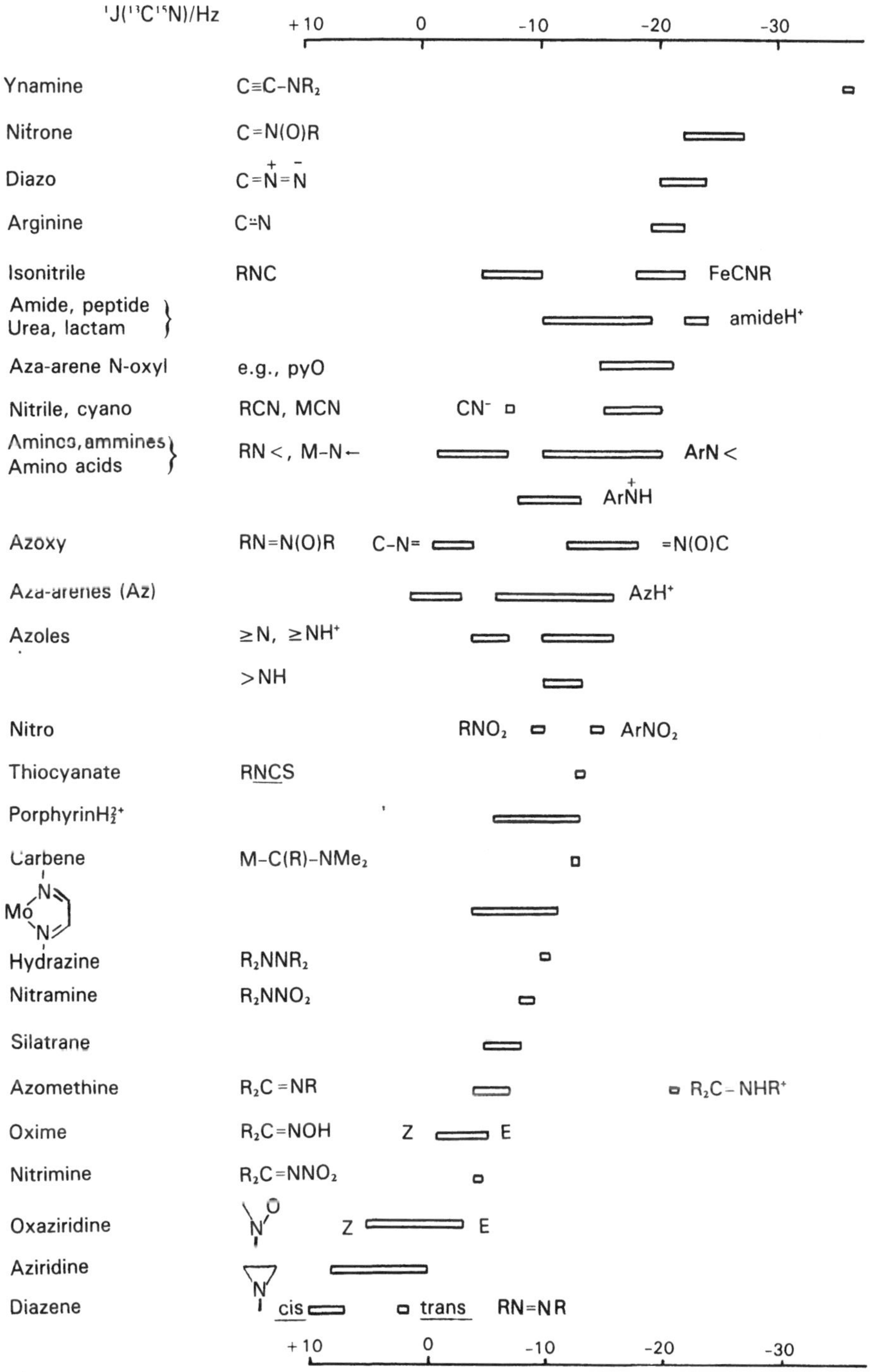

Figure 5. Ranges of $^1J(^{13}C\,^{15}N)$ values.

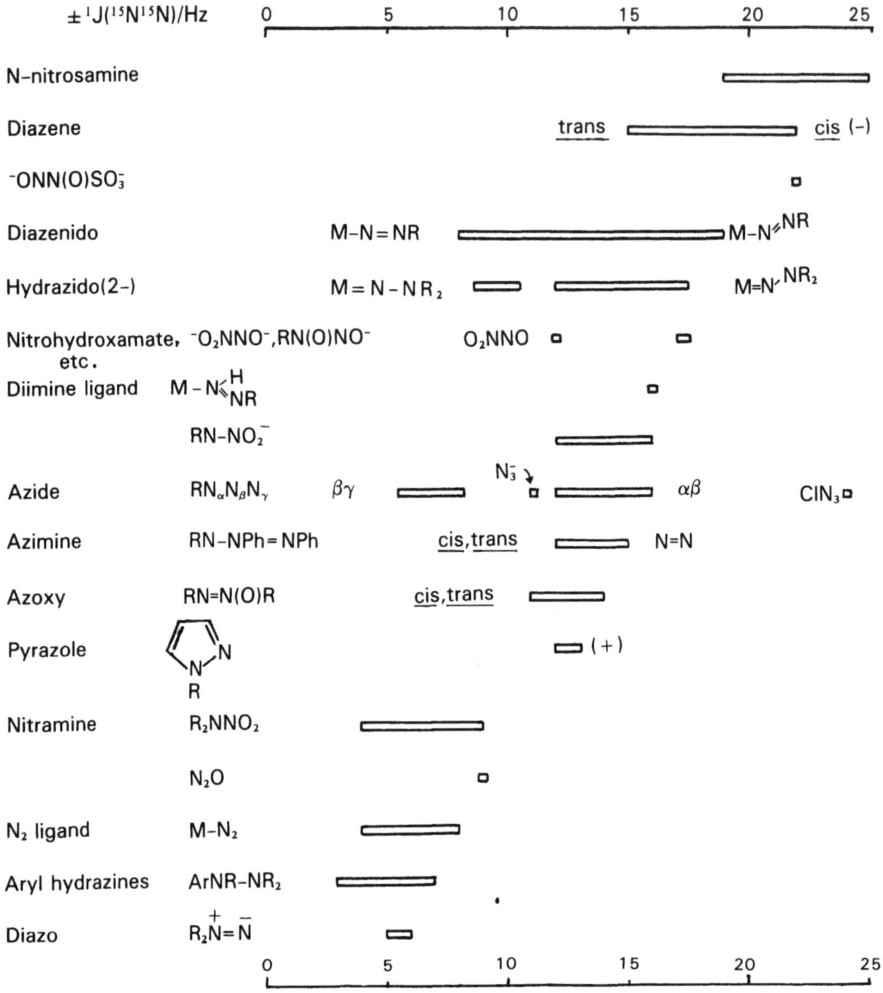

Figure 6. Ranges of $^1J(^{15}N\,^{15}N)$ values.

to sp^2 to sp hybridization of the nitrogen, also with loss of n_s (valence shell s-type lone pair) electron density on nitrogen, as by coordination, or to minimize steric repulsion. The small $^1J(NH)$ values in Figure 4 are for imino or imido $\ddot{N}$–H groups.

$^1K(NH)$ in amides (peptides, etc.) is larger if the proton is *trans* to oxygen, and decreases with departure from planarity. $^1K(NH)$ increases also with electronegativity of the substituents, e.g., in the sequence $Si < B \sim As < P < S$.[183] The largest $^1K(NH)$ values are those of nitrilium ions (Figure 4). Solvent and substituent effects have been reviewed.[179]

C–N and N–N couplings have been analyzed by coupled Hartree–Fock calculations,[184–185] and the conclusions can help to explain couplings to H, P, Si, metals etc. If neither of the coupled atoms bears a lone pair with s character, $^1K(NX)$ like $^1K(CC)$ is positive and increases with s character in the bond orbitals (e.g. $sp^3 < sp^2 < sp$[179]). The Fc term normally dominates, but the orbital and spin-dipolar terms increase with π bond order and become relatively large for triple bonds. These terms depend on $\langle r^{-3}\rangle_{2p}$, and so increase with effective nuclear charge of the nitrogen: the very large $^1J(CN) = (-)77$ Hz for PhCNO has large contributions from all three coupling mechanisms.[184]

If one of the coupled atoms bears a lone pair ^1K is decreased algebraically by an amount that increases with the *s* character of the lone pair. The *s* character increases from zero for pyrrole (in which the "lone pair" is $p\pi$), through $MeNH_2(sp^3)$ and pyridine (sp^2) to HCN (sp). This "one-bond lone pair effect"[184] can be traced to opposing signs of Fc contributions from lone pair and σ-bonding electrons: the lone pair terms may be large as the energy denominator is small. Thus the net Fc contribution to $^1K(CN)$ is calculated to be small for $MeNH_2$, smaller for pyridine, changing sign for RCN, in which the other mechanisms may dominate. As Figure 5 shows, $^1J(CN)$ is known to change sign for *cis*-diazenes, aziridines, pyridine, etc. but becomes more negative if the nitrogen is protonated. In oximes, $^1J(CN)$ is less negative for the *E* than for the *Z* form, in which the lone pair *s* character falls to minimize *cis*-repulsions.[184]

If both coupled atoms carry a lone pair, $^1K(NX)$ or $^1K(NN)$ is very sensitive to the dihedral angle ϕ between the $n(N)$ orbitals, since the HOMOs are bonding and antibonding combinations of these. Model calculations[185] on $^{15}N_2H_4$, in which ϕ is normally 90–95°, give $^1J(NN)$ values (now the same sign as ^1K) of -11 Hz for $\phi=0$, going to zero for $\phi=115°$ (also -8 Hz for $H_2NNH_3^+$ with only one lone pair; cf. -13 Hz reported for N-phenylpyrazole[101]). For $\phi=180°$ $^1K(NN)$ should be relatively large and positive for pyramidal *or* planar nitrogen (i.e., with *s or p* lone pairs). Similarly the Fc term is expected to be larger and negative for *cis*-diazenes $(\phi=0)$, larger and positive for *trans*-diazenes $(\phi=180°)$, but smaller and negative for diazo compounds, with only one lone pair. These relationships carry over to nitrogen ligands in metal complexes.

$^2K(NH)$ and $^2K(CN)$ values are commonly small and negative. They increase algebraically if the nitrogen carries a *cis* or proximate lone pair, maximally if the lone pair is coplanar with the coupling pathway. $^2J(NH)$ is -11 Hz in pyridine as compared with -3 Hz in pyH^+, and the coupling is diagnostic of O, N chelation in 8-quinolinol (oxine).[186] $^3K(NH)$ values are similarly increased by a *cis* or proximate lone pair, as in oximes or nitrites.[181] They often show a Karplus-type dependence on dihedral angle (which can be determined in this way in peptides) and so are useful in conformational analysis.[186a]

Clearly, certain $J(N)$ criteria of structure can be formulated, parallel to those described for $\delta(N)$ in Section 4.1, the regularities established for organic compounds applying correspondingly in organometallic and inorganic chemistry. As Figures 4 and 5 show, $^1J(NH)$ and $^1J(CN)$ are maximal (most negative) for linear systems. $^1J(CN)$ is diminished by lone pair *s* character on nitrogen *or* carbon: thus $^1J(C{\equiv}N)$ is enhanced on complexation of CN^-, RCN, or RNC. $^1J(NH)$ values in the hydrazido(2-) ligand, which is flat, resemble those in amides,[158] those for hydrazines (with pyramidal nitrogen) being less negative. Protonation of diazenido ligands with MNN linear, as in $[WBr(N_2H)(dppe)_2]$, or bent, as in $[PtCl(N_2Ph)(PEt_3)_2]$, enhances $J(N)$[150].

$^1J(^{15}NH)$ 60 87 $^1J(^{195}Pt\,^{15}N)$ 157 510 Hz

$^1J(NN)$ values are more difficult to rationalize as they may be the resultant of components differing in sign, either sign is possible (as shown by the N_2H_4 calculations,[185] and few signs have been determined. They are plotted regardless of sign in Figure 6, which may therefore be misleading in this respect. Certain regularities are, however, evident, such as the similar values in XNN compounds—azides, diazo compounds, and dinitrogen ligands.

6. DYNAMICS

Dynamic processes in a variety of systems with a range of timescales have been investigated by nitrogen NMR; examples are intramolecular motions (rotations, inversions, fluxions), tautomerism, ligand exchange, and chemical and biochemical reactions[3,6,7,187] (cf. also Sections 2.1 and 3.1). The dynamics, thermodynamics and quantum mechanics of isomerization processes involving N–X bonds have been reviewed comprehensively.[187c]

^{15}N CIDNP has been particularly valuable in the study of radical-mediated processes including nitration,[188] azo-coupling[189] and the decomposition of diazenes[190] and triazenes.[191a] For stable nitroxide radicals, however, ^{15}N NMR spectra have been observed and hyperfine constants measured.[191b]

7. BIOMOLECULES[1–3,7,33–42,77–79,139,192–197]

Nitrogen is closely concerned with many or most vital processes. With 95%–99% bioenrichment, which can be achieved with relatively cheap materials such as $^{15}NH_4^+$ salts, ^{15}N is several times more NMR sensitive than naturally-abundant ^{13}C, and can give useful labeling information. Sensitivity enhancement methods such as DEPT or INEPT are of great value, with or without ^{15}N-enrichment.[7,97,103–105] Also useful, particularly for *in vivo* work, is indirect detection of ^{15}N by double quantum proton NMR.[198] Applications of solid state techniques to biomolecules in ^{15}N or ^{14}N resonance were described in Section 1.3.

The use of ^{14}N in highly symmetrical locations, as in RNH_3^+, in which internal rotation is possible, should not be neglected. The metabolism of inorganic nitrogen in roots has been studied by ^{14}N NMR,[199] and *in vivo* studies of metabolism in algae, fungi, and higher plants by ^{13}C, ^{15}N and ^{14}N NMR have been reviewed.[197]

REFERENCES

1. Witanowski, M.; Stefaniak, L.; Webb, G. A. *Annu. Rep. NMR Spectrosc.* 1981, **11B**, 1–502; 1986, **18**, 1–761.
2. Martin, G. J.; Martin, M. L.; Gouesnard, J.-P. *^{15}N-NMR Spectroscopy*, NMR Basic Princ. Prog. 1981, **18**, 1–382.
3. Levy, G. C.; Lichter, R. L. *Nitrogen-15 NMR Spectroscopy*; Wiley-Interscience: New York, 1979, 1–221.
4. Witanowski, M.; Stefaniak, L.; Webb, G. A. *Annu. Rep. NMR Spectrosc.* 1977, **7**, 117–244.
5. Witanowski, M.; Webb, G. A., Eds.; *Nitrogen NMR*; Plenum: London, 1973.
6. Mason, J. *Chem. Rev.* 1981, **81**, 205–227. *Chem. Brit.* 1983, 654–658.
7. Von Philipsborn, W.; Müller, R. *Angew. Chem. Int. Ed. Engl.* 1986, **25**, 383–486.
8. Roberts, J. D. In *Bloch Festschrift*; Rice University Press: Houston, Texas, 1980.

9. Lichter, R. L. Chap. 10 in *The Multinuclear Approach to NMR Spectroscopy*, Lambert, J. B.; Riddell, F. G., Eds.; Reidel: Dordrecht, 1983, pp. 207–244.
10. Lichter, R. L. In *Determination of Organic Structures by Physical Methods*, Nachod, F. C.; Zuckerman, J. J., Eds.; Academic: New York, 1971, Vol. 4, pp. 195–231.
11. Randall, E. W.; Gillies, D. G. *Prog. NMR Spectrosc.* 1970, **6**, 119–174.
12. Jost, J. W.; O'Konski, C. T. *J. Mol. Struct.* 1983, **111**, 387–396.
13. Becker, E. D.; Bradley, R. B.; Axenrod, T. *J. Magn. Reson.* 1971, **4**, 136–141.
14. Witanowski, M.; Stefaniak, L.; Szymanski, S.; Januszewski, H. *J. Magn. Reson.* 1977, **28**, 217–226.
15. Jameson, C. J.; Jameson, A. K.; Oppusunggu, D.; Wille, S.; Burrell, P. M.; Mason, J. *J. Chem. Phys.* 1981, **74**, 81–88.
16. Duthaler, R. O.; Roberts, J. D. *J. Am. Chem. Soc.* 1978, **100**, 4969–4973.
17. Srinivasan, P. R.; Lichter, R. L. *J. Magn. Reson.* 1977, **28**, 227–234.
18. Litchman, W. M.; Alei, M.; Florin, A. E. *J. Chem. Phys.* 1969, **50**, 1031–1032; Alei, M.; Florin, A. E.; Litchman, W. M.; O'Brien, J. F. *J. Phys. Chem.* 1971, **75**, 932–938.
19. Briggs, J. M.; Randall, E. W. *Mol. Phys.* 1973, **26**, 699–703.
20. Becker, E. D. *J. Magn. Reson.* 1971, **4**, 142–147.
21. (a) Gibby, M. G.; Griffin, R. G.; Pines, A.; Waugh, J. S. *Chem. Phys. Lett.* 1972, **17**, 80–81. (b) Marino, R. A.; Bulusu, S. *J. Energetic Materials* 1985, **3**, 57–74.
22. Stenhouse, I. A., unpublished observations.
23. Levy, G. C.; Dechter, J. J.; Kowalewski, J. *J. Am. Chem. Soc.* 1978, **100**, 2308–2314.
24. Mason, J.; Vinter, J. G. *J. Chem. Soc., Dalton Trans.* 1975, 2522–2526.
25. Kukolich, S. G. *J. Am. Chem. Soc.* 1975, **97**, 5704–5707.
25a. Webb, G. A.; Witanowski, M. Chapter 6 in *Topics in Molecular Interactions*, Orville–Thomas, W. J.; Ratajczak, H.; Rao, C. N. R., Eds.; Elsevier, Amsterdam, 1985, pp. 241–290.
26. (a) Duthaler, R. O.; Roberts, J. D. *J. Am. Chem. Soc.* 1978, **100**, 3889–3895; (b) Duthaler, R. O.; Roberts, J. D. *J. Magn. Reson.* 1979, **34**, 129–139.
27. (a) Niibe, M.; Nakamura, Y. *J. Phys. Chem.* 1984, **88**, 5608–5614; Holton, D. M.; Edwards, P. P.; McFarlane, W.; Wood, B. *J. Am. Chem. Soc.* 1983, **105**, 2104–2108.
28. Mason, J.; Christe, K. O. *Inorg. Chem.* 1983, **22**, 1849–1853.
29. Kaplan, S.; Pines, A.; Griffin, R. G.; Waugh, J. S. *Chem. Phys. Lett.* 1974, **25**, 78.
30. Dixon, W. T. *J. Chem. Phys.* 1982, **77**, 1800–1809.
31. Herzfeld, J.; Berger, A. E. *J. Chem. Phys.* 1980, **73**, 6021–6030.
32. Schaefer, J.; Stejskal, E. O.; McKay, R. A. In *NMR Spectroscopy. New Methods and Applications*, Levy, G. C., Ed.; ACS Symposium Series, 1982, Vol. 191, pp. 187–198; *J. Magn. Reson.* 1984, **57**, 471–485.
33. Schaefer, J.; Stejskal, E. O.; Sefcik, M. D.; McKay, R. A. *Phil. Trans. R. Soc. London A* 1981, **299**, 593–608. Schaefer, J.; Stejskal, E. O.; Garbow, J. R., McKay, R. A. *J. Magn. Reson.* 1984, **59**, 150–156.
34. Jacob, G. S.; Schaefer, J.; Stefskal, E. O.; McKay, R. A. *Biochem. Biophys. Res. Commun.* 1980, **97**, 1176–1182.
35. Cross, T. A.; Diverdi, J. A.; Opella, S. J. *J. Am. Chem. Soc.* 1982, **10**, 1759–1761.
36. Cross, T. A.; Opella, S. J. *J. Am. Chem. Soc.* 1983, **105**, 306–308.
37. Diverdi, J. A.; Opella, S. J. *J. Am. Chem. Soc.* 1982, **104**, 1761–1762.
38. (a) Munowitz, M.; Aue, W. P.; Griffin, R. G. *J. Chem. Phys.* 1983, **77**, 1686–1689; (b) Harbison, G. S.; Jelinski, L. W.; Stark, R. E.; Torchia, D. A.; Herzfeld, J.; Griffin, R. G.; *J. Magn. Reson.* 1984, **60**, 79–82.
39. Cross, T. A.; Frey, M. H.; Opella, S. J. *J. Am. Chem. Soc.* 1983, **105**, 7471–7473.
40. Krichelhof, H. R. *Org. Magn. Reson.* 1981, **15**, 162–177; *Pure Appl. Chem.* 1982, **54**, 467–481.
41. Harbison, G.; Herzfeld, J.; Griffin, R. G. *J. Am. Chem. Soc.* 1981, **103**, 4752–4754.
42. Munowitz, M.; Bachovchin, W. W.; Herzfeld, J.; Dobson, C. M.; Griffin, R. G. *J. Am. Chem. Soc.* 1982, **104**, 1192–1196.
43. Mason, J.; Mingos, D. M. P.; Schaefer, J.; Sherman, D.; Stejskal, E. O. *J. Chem. Soc. Chem. Commun.* 1985, 444–446.
44. (a) Haberkorn, R. A.; Stark, R. E.; van Willigen, H.; Griffin, R. G. *J. Am. Chem. Soc.* 1981, **103**, 2534–2539; (b) Wolff, E. K.; Griffin, R. G.; Watson, C. *J. Chem. Phys.* 1977, **66**, 5433–5438.
45. Tycko, R.; Stewart, P. L.; Opella, S. J. *J. Am. Chem. Soc.* 1986, **108**, 5419–5424; Tycko, R.; Opella, S. J. *J. Am. Chem. Soc.* 1986, **108**, 3531–3532.
46. (a) Naito, A.; Ganapathy, S.; Akasaka, K.; McDowell, C. A. *J. Chem. Phys.* 1981, **74**, 3190–3197; Naito, A.; Ganapathy, S.; McDowell, C. A. *J. Magn. Reson.* 1982, **48**, 367–381. (b) Naito, A.; Barker, P. B.; McDowell, C. A. *J. Chem. Phys.* 1984, **81**, 1583–1591.

47. Michel, D.; Germanus, A.; Pfeifer, H. *J. Chem. Soc., Faraday Trans. 1* 1982, **78**, 237–254.
48. Bernstein, T.; Kitaev, L.; Michel, D.; Pfeifer, H.; Fink, P. *J. Chem. Soc., Faraday Trans. 1* 1982, **78**, 761–769; Juenger, I.; Meiler, W.; Pfeifer, H. *Zeolites* 1982, **2**, 310–312.
49. (a) Ripmeester, J. A. *J. Am. Chem. Soc.* 1983, **105**, 2925–2927. Maciel, G. E.; Haw, J. F.; Chuang, I.-S.; Hawkins. B. L.; Early, T. A.; McKay, D. R.; Petrakis, L. *J. Am. Chem. Soc.* 1983, **105**, 5529–5535. (b) Majors, P. D.; Raidy, T. E.; Ellis, P. D. *J. Am. Chem. Soc.* 1986, **108**, 8123–8129. Majors, P. D.; Ellis, P. D. *J. Am. Chem. Soc.* 1987, **109**, 1648–1653.
50. Hayhurst, D. T.; Sefcik, M. D. *ACS Symp. Ser.* 1983, **218** *(Intrazeolite Chem.)*, 333–344.
51. Yannoni, C. S. *J. Chem. Phys.* 1970, **52**, 2005–2010.
52. Casleton, K. H.; Kukolich, S. G. *J. Chem. Phys.* 1975, **62**, 2696–2699.
53. Bhattacharyya, P. K.; Dailey, B. P. *J. Chem. Phys.* 1973, **59**, 5820–5823.
54. Garvey, R. M.; De Lucia, F. C. *J. Mol. Spectrosc.* 1974, **50**, 38–44.
55. Schweitzer, D.; Spiess, H. W. *J. Magn. Reson.* 1974, **15**, 529–539.
56. Chan, S. I.; Baker, M. R.; Ramsey, N. F. *Phys. Rev. A*, 1964, **136**, 1224–1228; Baker, M. R.; Anderson, C. H.; Ramsey, N. F. *Phys. Rev. A*, 1964, **133**, 1533–1536.
57. Ishol, L. M.; Scott, T. A. *J. Magn. Reson.* 1977, **27**, 23–28.
58. Schweitzer, D.; Spiess, H. W. *J. Magn. Reson.* 1974, **16**, 243–251.
59. Stark, R. E.; Vold, R. L.; Vold, R. R. *Chem. Phys.* 1977, **20**, 337–345.
60. Raymonda, J.; Klemperer, W. *J. Chem. Phys.* 1971, **55**, 232–233.
61. Live, D. H.; Davis, D. G.; Agosta, W. C.; Cowburn, D. *J. Am. Chem. Soc.* 1984, **106**, 1939–1941.
62. Jameson, C. J.; Jameson, A. K.; Parker, H.; Cohen, S. M.; Lee, C. L. *J. Chem. Phys.* 1978, **68**, 2861–2865.
63. Schultheiss, H.; Fluck, E. *Z. Naturforsch.* 1977, **32b**, 257–264.
64. (a) van Etten, R. L.; Risley, J. M. *J. Am. Chem. Soc.* 1981, **103**, 5633–5636; (b) Andersson, K. K.; Philson, S. B.; Hooper, A. B. *Proc. Natl. Acad. Sci. U.S.A.* 1982, **79**, 5871; Andersson, K. K.; Hooper, A. B. *FEBS Lett.* 1983, **164**, 236–240.
65. (a) Lyčka, A.; Hansen, P. E. *Org. Magn. Reson.* 1984, **22**, 569–572; (b) Lyčka, A.; Hansen, P. E. *Magn. Reson. Chem.* 1985, **23**, 973–976, and references therein.
66. Mason, J.; Mingos, D. M. P.; Sherman, D.; Wardle, R. W. M. *J. Chem. Soc., Chem. Commun.* 1984, 1223–1225.
67. Ishol, L. M.; Scott, T. A.; Goldblatt, M. *J. Magn. Reson.* 1976, **23**, 313–320.
68. Krynicki, K.; Rahkamaa, E. J.; Powles, J. G. *Mol. Phys.* 1975, **29**, 539–548.
69. Litchman, W. M.; Alei, M. *J. Chem. Phys.* 1972, **56**, 5818–5822.
70. Saluvere, T.; Lippmaa, E. *Chem. Phys. Lett.* 1970, **7**, 545–548.
71. Lambert, J. B.; Netzel, D. A. *J. Magn. Reson.* 1977, **25**, 531–538.
72. Loftus, P.; Bearden, W. H.; Roberts, J. D. *Nouv. J. Chim.* 1977, **1**, 283–285.
73. Donovan-Mtunzi, S.; Hawkes, G. E.; MacDonald, C. J.; Mason, J.; Richards, R. L. *J. Chem. Soc. Dalton Trans.* 1985, 2473–2477.
74. Lippmaa, E.; Saluvere, T.; Laisaar, S. *Chem. Phys. Lett.* 1971, **11**, 120–123.
75. Jackman, L. M.; Trewella, J. C. *J. Am. Chem. Soc.* 1979, **101**, 6428–6429.
75a. Boeré, R. T.; Kidd, R. G. *Annu. Rep. NMR Spectrosc.* 1982, **13**, 319–385, Sec. VIB.
76. Bammel, B. P.; Evilia, R. F. *Anal. Chem.* 1982, **54**, 1318–1321.
77. Marion, D.; Garbay-Jaureguiberry, C.; Roques, B. P. *J. Am. Chem. Soc.* 1982, **104**, 5573–5576.
78. Gust, D.; Moon, R. B.; Roberts, J. D. *Proc. Natl. Acad. Sci. U.S.A.* 1975, **72**, 4696–4700.
79. Lapidot, A.; Irving, C. S. *J. Am. Chem. Soc.* 1977, **99**, 5488–5490.
80. Hawkes, G. E.; Litchman, W. M.; Randall, E. W. *J. Magn. Reson.* 1975, **19**, 255–258.
81. Doddrell, D.; Glushko, W.; Allerhand, A. *J. Chem. Phys.* 1972, **56**, 3683–3689.
82. Brévard, C.; Granger, P. *Handbook of High Resolution Multinuclear NMR*; Wiley: New York, 1981.
82a. Griffey, R. H.; Jarema, M. A.; Kunz, S.; Rosevear, P. R.; Redfield, A. G. *J. Am. Chem. Soc.* 1985, **107**, 711–712.
82b. Levy, G. C.; Godwin, A. D.; Hewitt, J. M.; Sutcliffe, C. *J. Magn. Reson.* 1978, **29**, 553–562.
83. James, T. L.; James, J. L.; Lapidot, A. *J. Am. Chem. Soc.* 1981, **103**, 6748–6750.
84. Hennig, J.; Limbach, H.-H. *J. Am. Chem. Soc.* 1984, **106**, 292–298. Limbach, H.-H.; Hennig, J.; Kendrick, R.; Yannoni, C. S. *J. Am. Chem. Soc.* 1984, **106**, 4059–4060; Schlabach, M.; Wehrle, B.; Limbach, H.-H.; Bunnenberg, E.; Knierzinger, A.; Shu, A. Y. L.; Tolf, B.-R.; Djerassi, C. *J. Am. Chem. Soc.* 1986, **108**, 3856–3858.
85. Martin, M. L.; Delpuech, J.-J.; Martin, G. J. *Practical NMR Spectroscopy*; Heyden: London, 1980.
86. Clague, A. D. H.; van Dongen, J. P. C. M., personal communication.

87. (a) Kowalewski, J.; Morris, G. A. *J. Magn. Reson.* 1982, **47**, 331–338; (b) Perrin, C. L.; Gipe, R. K. *J. Am. Chem. Soc.* 1986, **108**, 1088–1089.
88. Chacko, V. P.; Bryant, R. G. *J. Magn. Reson.* 1984, **57**, 79–84.
89. (a) Hawkes, G. E.; Randall, E. W.; Hull, W. E. *J. Chem. Soc., Perkin Trans. 2* 1977, 1268–1275; (b) Friedrich, J. O.; Wasylishen, R. E. *J. Chem. Phys.* 1985, **83**, 3707–3708.
90. Giannini, D. D.; Armitage, I. M.; Pearson, H.; Grant, D. M.; Roberts, J. D. *J. Am. Chem. Soc.* 1975, **97**, 3416–3419.
91. Hunter, B. K.; Brown, R. J. C. *J. Magn. Reson.* 1982, **46**, 227–232.
92. Gust, D.; Roberts, J. D. *J. Am. Chem. Soc.* 1977, **99**, 3637–3640.
93. Bertrand, R. D.; Moniz, W. B.; Garroway, A. N.; Chingas, G. C. *J. Magn. Reson.* 1978, **32**, 465–467.
94. Jakobsen, H. J.; Brey, W. S. *J. Am. Chem. Soc.* 1979, **101**, 774–775; *J. Chem. Soc., Chem. Commun.* 1979, 478–479.
95. Morris, G. A.; Freeman, R. *J. Am. Chem. Soc.* 1979, **101**, 760–762; Morris, G. A. *J. Am. Chem. Soc.* 1980, **102**, 428–429.
96. Doddrell, D. M.; Pegg, D. T.; Bendall, M. R. *J. Magn. Reson.* 1982, **48**, 323–327.
97. Jakobsen, H. J.; Bilsøe, H.; Dønstrup, S.; Sørensen, O. W. *J. Magn. Reson.* 1984, **57**, 324–330; Hansen, P. E.; Jakobsen, H. J. *Org. Magn. Reson.* 1984, **22**, 668–670.
98. Holmes, B. S.; Chingas, G. S.; Moniz, W. B.; Ferguson, R. C. *Macromolecules*, 1981, **14**, 1785–1787.
99. Städeli, W.; Bigler, P.; von Philipsborn, W. *Org. Magn. Reson.* 1981, **16**, 170–172.
100. (a) Chen, B. C.; von Philipsborn, W.; Nagarajan, K. *Helv. Chim. Acta* 1983, **66**, 1537–1555; Schenker, K. V.; von Philipsborn, W. *J. Magn. Reson.* 1985, **61**, 294–305; (b) Berners Price, S. J.; DiMartino, M. J.; Hill, D. T.; Kuroda, R.; Mazid, M. A.; Sadler, P. J. *Inorg. Chem.* 1985, **24**, 3425–3434.
101. Berkhoudt, T.; Jakobsen, H. J. *J. Magn. Reson.* 1982, **50**, 323–327.
102. Martineau, H.; Trierweiler, M.; Martin, M. L. *Org. Magn. Reson.* 1981, **17**, 182–185.
103. Davis, D. G.; Live, D. H.; Agosta, W. C.; Cowburn, D. *J. Magn. Reson.* 1983, **53**, 350–354.
104. Bax, A.; Niu, C. H.; Live, D. *J. Am. Chem. Soc.* 1984, **106**, 1150–1151.
105. Bodenhausen, G.; Ruben, D. J. *Chem. Phys. Lett.* 1980, **69**, 185–189; Mueller, L.; Schniksnis, R. A.; Opella, S. J. *J. Magn. Reson.* 1986, **66**, 379–384.
106. Shporer, M.; Ron, G.; Loewenstein, A.; Navon, G. *Inorg. Chem.* 1965, **4**, 358–361.
107. Cooke, D. F.; Jeffrey, K. R. *J. Magn. Reson.* 1975, **18**, 455–470.
108. Lehn, J.-M.; Kintzinger, J.-P. In Ref. 5, pp. 80–161.
109. (a) Loewenstein, A.; Brenman, M. *J. Magn. Reson.* 1979, **34**, 193–198; (b) Loewenstein, A. *J. Magn. Reson.* 1982, **49**, 332–334.
110. Wallach, D.; Huntress, W. T. *J. Chem. Phys.* 1969, **50**, 1219–1227.
111. Ratajczak, H. M.; Ladd, J. A.; Orville-Thomas, W. J.; Ratajczak, H. *Mol. Phys.* 1977, **34**, 1019–1025.
112. Kintzinger, J.-P.; Lehn, J.-M. *Helv. Chim. Acta* 1975, **58**, 905–917.
113. Atkins, P. W.; Loewenstein, A.; Margalit, Y. *Mol. Phys.* 1969, **17**, 329–340.
114. (a) Siminovitch, D. J.; Jeffrey, K. R. *Biochim. Biophys. Acta* 1981, **645**, 270–278; (b) Siminovitch, D. J.; Brown, M. F.; Jeffrey, K. R. *Biochemistry* 1984, **23**, 2412–2420.
115. Eriksson, P.-O.; Khan, A.; Lindblom, G. *J. Phys. Chem.* 1982, **86**, 387–393.
116. Bailey, W. C.; Story, H. S. *J. Chem. Phys.* 1974, **60**, 1952–1953.
117. Henriksson, U.; Ödberg, L.; Eriksson, J. C.; Westman, L. *J. Phys. Chem.* 1977, **81**, 76–82.
118a. Gourdji, M.; Guibé, L. *C.R. Acad. Sci. Paris* 1965, **260**, 1131–1133; Gourdji, M.; Guibé, L.; Peneau, A. *J. Phys.* 1974, **35**, 497–507.
118b. Larsen, D. W. *J. Phys. Chem.* 1971, **75**, 509–513.
119. Moniz, W. B.; Poranski, C. F. *J. Phys. Chem.* 1969, **73**, 4145–4147.
120. Yamamoto, Y.; Uzawa, J. *Chem. Lett.* 1978, 1213–1216.
120a. Nicholas, A. M. de P.; Wasylishen, R. E. *J. Phys. Chem.* 1985, **89**, 5446–5450.
121. Moniz, W. B.; Gutowsky, H. S. *J. Chem. Phys.* 1963, **38**, 1155–1162.
122. Suchanski, W.; Canepa, P. S. *J. Magn. Reson.* 1979, **33**, 389–399.
123. Chew, K. F.; Derbyshire, W.; Logan, N.; Norbury, A. H.; Sinha, A. I. P. *J. Chem. Soc., Chem. Commun.* 1970, 1708.
124. Pedersen, E. J.; Vold, R. L.; Vold, R. R. *Mol. Phys.* 1980, **41**, 811–822.
125. Loewenstein, A.; Waiman, R. *Mol. Phys.* 1973, **25**, 49–59.
125a. Maliniak, A.; Kowalewski, J.; Panas, I. *J. Phys. Chem.* 1984, **88**, 5628–5631.
126. Raj, T.; Bryant, T. *J. Magn. Reson.* 1979, **34**, 537–542.
127. Gillen, K. T.; Noggle, J. H. *J. Chem. Phys.* 1970, **52**, 4905–4906.
128. Tiffon, B.; Ancian, B. *J. Chem. Phys.* 1982, **76**, 1212–1216.

129. Ancian, B.; Tiffon, B. *J. Chem. Soc., Faraday Trans. 2* 1984, **80**, 1067–1076.
130. Kintzinger, J.-P.; Lehn, J.-M. *Mol. Phys.* 1971, **22**, 273–287.
131. Lindman, B.; Forsén, S. *NMR Basic Princ. and Prog.* 1976, **12**.
132. Vold, R. R.; Vold, R. L.; Szeverenyi, N. M. *J. Chem. Phys.* 1979, **70**, 5213–5217.
133. Szeverenyi, N. M.; Vold, R. R.; Vold, R. L. *Chem. Phys.* 1976, **18**, 23–30, 31–36.
134. Marchal, J.-P.; Canet, D. *J. Chem. Soc., Faraday Trans. 2* 1982, **78**, 435–445.
135. Shporer, M.; Ron, G.; Loewenstein, A.; Navon, G. *Inorg. Chem.* 1965, **4**, 361–364.
136. Hartmann, H.; Sillescu, H. *Theor. Chim. Acta* 1964, **2**, 371–385.
137. Jordan, R. B. *J. Magn. Reson.* 1980, **38**, 267–275.
138. Wasylishen, R. E.; Pettitt, B. A.; Dong, R. Y. *J. Chem. Soc., Faraday Trans. 2* 1980, **76**, 571–573.
139. *NMR Spec. Per. Rep. Chem. Soc.* 1972–1986, vols. 1–15, *et seq.*
140. Comeau, M.; Béraldin, M. T.; Vauthier, E. C.; Fliszar, S. *Can. J. Chem.* 1985, **63**, 3226–3232.
140a. Jørgensen, W. L.; Salem, L. *The Organic Chemist's Book of Orbitals*, Academic Press: New York, 1973.
141. Andersson, L.-O.; Mason, J.; van Bronswijk, W. *J. Chem. Soc. (A)* 1970, 296–299.
142. Kroner, J.; Schneid, W.; Wiberg, N.; Wrackmeyer, B.; Ziegleder, G. *J. Chem. Soc., Faraday Trans. 2* 1978, **74**, 1909–1919.
143. Morris, D. G.; Murray, A. M. *J. Chem. Soc., Perkin Trans. 2* 1976, 1579–1584.
144. Dervan, P. B.; Squillacote, M. E.; Lahti, P. M.; Sylwester, A. P.; Roberts, J. D. *J. Am. Chem. Soc.* 1981, **103**, 1120–1122.
145. Donovan-Mtunzi, S.; Richards, R. L.; Mason, J. *J. Chem. Soc., Dalton Trans.* 1984, 469–474.
146. Sogn, J. A.; Gibbons, W. A.; Randall, E. W. *Biochemistry* 1973, **12**, 2100–2105.
147. Gray, G. A. *Org. Magn. Reson.* 1983, **21**, 111–118; Hawkes, G. E.; Lian, L. Y.; Randall, E. W. *J. Magn. Reson.* 1984, **56**, 539–542.
148. Craik, D. J. *Annu. Rep. NMR Spectrosc.* 1983, **15**, 75–81.
149. Dilworth, J. R.; Kan, C.-T.; Richards, R. L.; Mason, J.; Stenhouse, I. A. *J. Organomet. Chem.* 1980, **201**, C24–C26.
150. Haymore, B. L.; Hughes, M.; Mason, J.; Richards, R. L. to be published.
151. Casewit, C.; Roberts, J. D. *J. Am. Chem. Soc.* 1980, **102**, 2364–2368.
152. Bell, L. K.; Mason, J.; Mingos, D. M. P.; Tew, D. G. *Inorg. Chem.* 1983, **22**, 3497–3503.
153. Bultitude, J.; Larkworthy, L. F.; Mason, J.; Povey, D. C.; Sandell, B. *Inorg. Chem.* 1984, **23**, 3629–3633.
154. Botto, R. E.; Kolthammer, B. W. S.; Legzdins, P.; Roberts, J. D. *Inorg. Chem.* 1979, **18**, 2049–2051.
155. Evans, D. H.; Mingos, D. M. P.; Mason, J.; Richards, A. *J. Organomet. Chem.* 1983, **249**, 293–298.
156. Gouesnard, J. P.; Martin, G. J. *Org. Magn. Reson.* 1979, **12**, 263–270.
157. Müller, R.; Wallis, J. D.; von Philipsborn, W. *Angew. Chem. Int. Ed.* 1985, **24**, 513–515.
158. Donovan-Mtunzi, S.; Richards, R. L.; Mason, J. *J. Chem. Soc., Dalton Trans.* 1984, 1329–1332.
159. Kozerski, L.; von Philipsborn, W. *Helv. Chim. Acta* 1982, **65**, 2077–2087.
160. Konishi, K.; Yoshino, A.; Katoh, M.; Matsumoto, H.; Takahashi, K.; Iwamura, H. *Chem. Lett.* 1982, 169–170.
161. Gladfelter, W. L. *Adv. Organometallic Chem.* 1985, **24**, 41–86.
162. Donovan-Mtunzi, S.; Richards, R. L.; Mason, J. *J. Chem. Soc., Dalton Trans.* 1984, 2429–2433.
163. Gouesnard, J. P.; Dorie, J.; Martin, G. J. *Can. J. Chem.* 1980, **58**, 1295–1304.
164. (a) Nöth, H.; Wrackmeyer, B. *Chem. Ber.* 1974, **107**, 3070–3088; (b) Wrackmeyer, B. *J. Magn. Reson.* 1986, **66**, 172–175.
165. Kanjia, D. M.; Mason, J.; Banks, R. E.; Venayak, N. D. *J. Chem. Soc., Perkin Trans. II* 1981, 975–979.
166. Mason, J. *J. Chem. Soc., Faraday Trans. 2* 1982, **78**, 1539–1549.
167. Thomas, B.; Grossmann, G.; Meyer, H. *Phosphorus and Sulphur* 1981, **10**, 375–382.
168. Pestunovich, V. A.; Shterenberg, B. Z.; Lippmaa, E. T., *et al. Dokl. Akad. Nauk. SSSR* 1981, **258**, 1410–1413.
169. Chivers, T.; Oakley, R. T.; Scherer, O. J.; Wolmershäuser, G. *Inorg. Chem.* 1981, **20**, 914–917; Chivers, T.; Cordes, A. W.; Oakley, R. T.; Pennington, W. T. *Inorg. Chem.* 1983, **22**, 2429–2435.
170. Howarth, O. W.; Richards, R. E.; Venanzi, L. M. *J. Chem. Soc.* 1964, 3335–3337; Yamaguchi, T.; Yamamoto, K.; Ohtaki, H. *Bull. Chem. Soc. Jpn.* 1985, **58**, 3235–3243.
171. Prinz, R.; Werner, H. *Angew. Chem., Int. Ed. Engl.* 1967, **6**, 91–92.
171a. Aspinall, H. C.; Deeming, A. J.; Donovan-Mtunzi, S. *J. Chem. Soc., Dalton Trans.* 1983, 2669–2671; Deeming, A. J.; Donovan-Mtunzi, S. D. *ibid.* 1985, 1609–1612.

172. Minelli, M.; Hubbard, J. L.; Lichtenberger, D. L.; Enemark, J. H. *Inorg. Chem.* 1984, **23**, 2721–2722.
173. Donovan-Mtunzi, S.; Mason, J.; Richards, R. L. *J. Chem. Soc., Dalton Trans.* 1984, 2729–2731.
174. Kanda, K.; Nakatsuji, H.; Yonezawa, T. *J. Am. Chem. Soc.* 1984, **106**, 5888–5892.
175. Martinengo, S.; Ciani, G.; Sironi, W.; Heaton, B. T.; Mason, J. *J. Am. Chem. Soc.* 1979, **101**, 7095–7097.
176. Kowalewski, J. *Annu. Rep. NMR Spectrosc.* 1982, **12**, 82–176.
177. Freyer, W. *Z. Chem.* 1981, **21**, 47–58.
178. Wasylishen, R. E. *Annu. Rep. NMR Spectrosc.* 1977, **7**, 262–279.
179. Axenrod, T. Chap. 6 in ref. 179, pp. 81–94.
180. Axenrod, T.; Webb, G. A., Eds.; *NMR Spectroscopy of Nuclei other than Protons*; Wiley: New York, 1974.
181. Wasylishen, R. E. Chap. 8 in Ref. 179, pp. 105–125.
182. Axenrod, T. Chap. 5 in Ref. 5, pp. 261–317.
183. Cowley, A. H.; Schweiger, J. R. *J. Am. Chem. Soc.* 1973, **95**, 4179–4185.
184. Schulman, J. M.; Venanzi, T. *J. Am. Chem. Soc.* 1976, **98**, 4701–4705, 6739–6741.
185. Schulman, J. M.; Ruggio, J.; Venanzi, T. J. *J. Am. Chem. Soc.* 1977, **99**, 2045–2048.
186. Jain, V. K.; Mason, J.; Mehrotra, R. C.; Saraswat, B. S. *Polyhedron* 1985, **4**, 2089–2096; *J. Organomet. Chem.* 1986, **309**, 45–54.
186a. Karplus, S.; Karplus, M. *Proc. Nat. Acad. Sci. U.S.A.* 1972, **69**, 3204–3206.
187. (a) Gouesnard, J. P.; Martin, G. J. *Org. Magn. Reson.* 1979, **12**, 263–270; (b) Chapter 8 in reference 2; (c) Martin, M. L.; Sun, X. Y.; Martin, G. J. *Annu. Rep. NMR Spectrosc.* 1985, **16**, 187–292.
188. Clemens, A. H.; Ridd, J. H.; Sandall, J. P. B. *J. Chem. Soc., Perkin Trans. II* 1984, 1659–1665, 1667–1672; 1985, 1217–1225, 1227–1231.
189. Bubnov, N. N.; Bilevitch, K. A.; Poljakova, L. A.; Okhlobystin, O. Yu. *J. Chem. Soc., Chem. Commun.* 1972, 1058.
190. Porter, N. A.; Dubay, G. R.; Green, J. G. *J. Am. Chem. Soc.* 1978, **100**, 920–925.
191. (a) Lippmaa, E.; Saluvere, T.; Pehk, T.; Olivson, A. *Org. Magn. Reson.* 1973, **5**, 429–436; (b) Grigor'ev, I. A.; Volodarsky, L. B.; Gogolev, A. Z.; Sagdeev, R. Z. *Chem. Phys. Lett.* 1985, **122**, 46–50.
192. Blomberg, F.; Rüterjans, H. *Biol. Magn. Reson.* 1983, **5**, 21–73.
193. Kanamori, K.; Roberts, J. D. *Acc. Chem. Res.* 1983, **16**, 35–41.
194. Kyogoku, Y. *Appl. Spectrosc. Rev.* 1981, **17**, 279–335.
195. Roberts, J. D. *Japanese J. Antibiotics* 1979, **32**, S112–S121.
196. Breitmaier, E. *Pharm. Unser. Zeit.* 1983, **12**, 161–180.
197. Martin, F. *Physiol. Veg.* 1985, **23**, 463–490.
198. Juretschke, H. P. *Arch. Int. Physiol. Biochim.* 1985, **93**, 141–150.
199. Belton, P. S.; Lee, R. B.; Ratcliffe, R. G. *J. Exp. Biol.* 1985, **36**, 190–210.

Chapter 13

PHOSPHORUS TO BISMUTH

KEITH R. DIXON

1. PHOSPHORUS-31

1.1. Introduction

Phosphorus-31 is spin 1/2, 100% abundant and has a receptivity relative to the proton of 0.0663. In consequence, this is an easy nucleus to observe, and even prior to 1970 a great deal of information had been obtained for simple compounds using continuous wave instruments. The advent of pulsed NMR techniques has greatly accelerated development of the field (especially in coordination chemistry and biochemistry), and ^{31}P is now one of the most common of all NMR experiments. With the additional constraints imposed by the dearth of recent comprehensive review literature, it is impossible to provide comprehensive coverage within a single chapter of limited length. I will aim at an outline of the main developments with sufficient lead references to enable individuals to trace details applicable to their own work; but I apologize in advance for the inevitable omissions and personal bias in the selection of material. Literature coverage in the present article extends to about the end of 1982 with minor additions during the prolonged development of the book.

The wealth of available information is both a blessing and an embarrassment to the worker in ^{31}P NMR; the latter because the mass of data has effectively deterred any attempts at comprehensive review since 1969. Indeed, a truly comprehensive compilation may now be almost impossible as the technique becomes routine and the information recedes into fine print in the experimental sections of papers! The early book by van Wazer *et al.*[1] is a comprehensive compilation of data to 1967 and an authoritative presentation of the available theory of phosphorus chemical shifts. The treatment remains a reasonably valid basis for most subsequent discussions of $\delta(P)$, but unfortunately the book has little to say about coupling constants. A second book[2] which extends coverage to 1969 contains considerably more information on coupling constants, and a selective review to mid 1982 has appeared.[2a] Phosphorus-phosphorus coupling constants have been reviewed to 1970[3] and sources of other coupling constants are compiled in Section 4. For coordination complexes containing phosphorus ligands the early work is covered to 1968 in an important review by Nixon and Pidcock,[4] and articles by Verkade[5] and Appleton *et al.*[6] also contain much valuable data. A 1979 book has appeared[7] outlining the main developments in

KEITH R. DIXON • Department of Chemistry, University of Victoria, Victoria, British Columbia, V8W 2Y2, Canada.

^{31}P NMR of transition metal phosphine complexes in the period 1970–1977, but unfortunately the tabulations of data are representative rather than comprehensive.

For recent comprehensive information the worker in ^{31}P NMR is dependent mainly on reviews of particular compound types, such as those appearing in *Topics in Phosphorus Chemistry*,[8] and on the excellent annual surveys provided by the Specialist Periodical Reports published by the Royal Society of Chemistry. The reports on NMR,[9] Organophosphorus Chemistry,[10] and Spectroscopic Properties of Inorganic and Organometallic Compounds,[11] are especially relevant. Biochemical applications of ^{31}P NMR have been reviewed,[12,12a] and will not be considered further in the present chapter. Recent biomedical work on ^{31}P is reviewed in Chapter 23.

^{31}P NMR needs little operational description since spectra are easily observed on modern instruments with readily available pretuned probes or suitable tuning of a wide band instrument. Decoupling of protons by broadband ("noise") irradiation of the sample is standard procedure and usually results in dramatic reductions in line widths and enhancements of signal:noise. Measurements of the nuclear Overhauser enhancement show a wide variation from essentially zero to the theoretical maximum of 124% (see Section 2 and Ref. 13). Most workers seem to prefer the combination of relatively rapid pulse repetition (~ 1 s) with short pulse duration ($\sim 30°$), but there is not yet sufficient information on relaxation times to enable the selection of optimum acquisition parameters for particular compound types. Values of the chemical shift anisotropy in phosphorus compounds can be quite large, ranging up to 1376 ppm in PN,[14] but a range from -224 ppm (P_4S_3) to $+326$ ppm (P_4O_{10}) is more typical.[14–16] This is not generally a problem for solution spectra in fields up to 2.35 T, but the author has encountered several examples of line broadening in coordination complexes at 5.87 T.

A problem exists in the choice of chemical shift reference since the reactivity of phosphorus compounds has precluded the use of internal standards. The most common choice is 85% H_3PO_4 as an external reference, and this has been adopted in this chapter. It suffers from several disadvantages, not the least of which is the central position of the resonance relative to the full range of phosphorus shifts. This can make it difficult to decide which sign convention an author is using; the older high field-positive or the more modern high frequency-positive as employed in this book. Great caution is necessary as even much recent work still employs the older convention. Also the 85% H_3PO_4 resonance is broad and there is little agreement on Ξ relative to $SiMe_4$ with values being quoted in the range from 40, 480, 720 Hz to 40, 480, 754 Hz.[17] Corrections for diamagnetic susceptibility are rarely made and there are also the possibilities of temperature, solvent, and concentration dependence for both reference standards and the compounds under study. Although these problems were pointed out by Mann in 1978,[17] the situation has changed little since then and unduly large variations (up to 5 ppm in some cases) in reported chemical shifts may be encountered. For this reason shift values in the present chapter have been rounded to the nearest integer.

Several other possible reference standards, including P_4O_6 (+113 ppm),[1] $P(OMe)_3$ (+141 ppm),[1] and $PO(OMe)_3$ (-2.4 ppm),[18] have been suggested and used by various authors but none has found general acceptance. $P(OMe)_3$ is a particularly interesting possibility since it is widely available and used as a resolution standard for ^{31}P NMR and also provides a sharp reference signal when proton decoupled. The principal problem in using this reference to date seems to have been the adoption of TMP as an abbreviation, since some authors have used this for trimethylphosphate! The best usage would seem to be TMP for trimethylphosphite and TMPO for the phosphate.

1.2. Spin-Lattice Relaxation (T_1)

In view of the ease of observation of ^{31}P and the wealth of available chemical shift and coupling constant information, there are surprisingly few published T_1 measurements for simple phosphorus compounds.* There appear to be no proposals of generalizations or correlations which would be useful to the experimentalist. Most authors have analyzed their results in terms of three contributions to the relaxation process:

$$T_{1obs}^{-1} = T_{1D}^{-1} + T_{1SR}^{-1} + T_{1SA}^{-1}$$

where D = dipolar, SR = spin rotation, and SA = shielding anisotropy. The dipolar term normally arises from interactions with protons within the molecule and its contribution can be assessed by measurement of the nuclear Overhauser enhancement (NOE) produced by broadband irradiation of the protons. This enhancement is given by $\eta - 1 = (\gamma_j/2\gamma_i)\, T_{1OBS}/T_{1D}$, where γ is the magnetogyric ratio of the nucleus in question. Thus for $j = {}^1H$ and $i = {}^{31}P$ the expression is $\eta - 1 = 1.24 T_{1OBS}/T_{1D}$, i.e., a maximum enhancement of 124% if relaxation is entirely dipolar.† T_{1D} and T_{1SA} increase with increase in temperature and T_{1SR} decreases and T_{1SA} also depends on the square of the applied field. Thus with a combination of temperature and field dependence studies and NOE measurements it should normally be possible to assess the relative importance of the three relaxation processes.

A recent study[19] of trimethylphosphine provides a typical example: T_1 is 3.7 s at 166 K, rises to a maximum of 14.0 s at 210 K, and then falls again to 2.8 s at 353 K. Over the same range, the NOE decreases steadily from 1.05 ($T_{1D} = 4.4$ s) at 166 K to zero ($T_{1D} > 150$ s) above 300 K. Thus the rising T_1 trend at low temperatures is ascribed to domination by the dipolar mechanism and the falling trend at high temperatures to the spin rotation mechanism. The dipolar relaxation is primarily intramolecular and T_{1SA} is considered negligible ($|\sigma_{\parallel} - \sigma_{\perp}| < 8$ ppm) at the field strength (2.3 T) employed.[19] The latter point has been confirmed by other workers, who find no field dependence.[20]

Selected T_1 measurements for simple phosphorus compounds are presented in Table 1. Unfortunately, the early studies did not have NOE measurements available to assess the dipolar contribution to the total relaxation, and their interpretations of relaxation mechanisms should be approached with caution. The first study to use NOE measurements was the work of Harris and McVicker,[21] who found relaxation in P_2Me_4 to be largely due to spin-rotation at 35°C (NOE = 0.05) but with a substantial dipolar contribution at −55°C (NOE = 0.85). The larger molecule $P_2Bu^t_4$ relaxed mainly by a dipolar mechanism even at ambient temperature (NOE = 1.12). In

* As noted in the Introduction, biochemical applications are not considered here.

† Literature practice in reporting NOE values varies, with some authors quoting the enhancement (maximum value, 1.24 or 124%) and others quoting η (maximum value, 2.24 or 224%). The former method will be used here.

general, most studies of three-coordinate compounds have given results compatible with those presented above for PMe_3. Thus relaxation of P_4O_6,(22–24) $P(OMe)_3$,(20) PH_3,(25,26) PH_2Ph,(27) PEt_3,(28) and $P(NMe_2)_3$(20) is dominated by spin-rotation contributions at normal temperatures with the possibility of dipolar contributions at low temperatures. The halides, PCl_3 and PBr_3,(22,29–32) and the phenyl derivatives, PCl_2Ph, $PClPh_2$,(20,33,34) and PPh_3(28,33–36) also relax primarily by spin-rotation, but opinions differ as to the other mechanisms involved. For PPh_3 even the measurements are problematical with ambient temperature T_1 values reported from 1.9 to 30.8 s (see Table 1)! Going from three to four coordination generally decreases the relative contribution of the spin-rotation mechanism,(21,28) although it remains important in surprisingly large molecules and is still significant in five-coordinate PCl_nPh_{5-n} ($n = 2$, 3, or 4) species(34) and in the PF_6^- ion.(37) $POPh_3$ is a rare example where authors agree on an important role for relaxation by chemical shift anisotropy,(28,34,36) apparently because the P=O bond disturbs the isotropic electron distribution around the phosphorus.(28) In general, the positive identification of relaxation by chemical shift anisotropy seems to have been a problem in many molecules, and reliable field-dependent studies are needed. Effects similar to those in $POPh_3$ might be expected in sulfides(22,34,36) and in phosphate esters, $PO(OR)_3$,(20,22,38,39) but there is less consistency between workers for these compounds.

Studies on phosphate solutions present an interesting cautionary tale, with early workers reporting complex oscillations of T_1 with variation of pH, with maxima corresponding to the ions PO_4^{3-}, HPO_4^{2-}, $H_2PO_4^-$, and H_3PO_4. However, a recent paper(40) demonstrates that these oscillations are due to traces of paramagnetic metal ions and that T_1 in very pure Na_3PO_4 solutions is almost independent of pH. At 25°C and pH 10.3 in 20% D_2O/H_2O the relaxation at 60.74 MHz has the following contributions:

$$T_{1OBS}^{-1} = 6.22 = \underset{3.88}{T_{1DD}^{-1}} + \underset{0.29}{T_{1SA}^{-1}} + \underset{2.05}{T_{1OTHER}^{-1}}$$

Table 1. Selected ^{31}P Spin-Lattice Relaxation Data (T_1) for Simple Phosphorus Compounds

Compound		Solvent[a]	Field[b] (T)	T_1 (s) ambient[c]	Temperature range (K)	T_1 (s) range[d]	References
P_4O_6		Solid	0.5		250–290	600–40	24
		N	0.5, 1.1	17.5	286–378	21–7.5	22, 24
PCl_3 [e]		N	0.5	2	140–370	1.5–30	30
PBr_3 [e]		N	1.0	4.5	190–450	1.5–4.5	31
$P(NMe_2)_3$		N	1.4	11.2			20
PCl_2Ph		N, C	1.4	11.5	200–400	3–13	20, 33, 34
$PClPh_2$		N, C	1.4	1.4	243–430	1.3–12	33, 34
PR_3	R = H	N	1.0	0.2	100–300	3–0.2	25, 26
	Me	T	2.3	4.7	166–353	14.0–2.8	19
	Et	C	2.3	10.4	254–331	15.3–3.4	28
	Pr^n	M	*f*	15.4			7
	Ph	C	1.4	1.9	243–343	6.8–1.4	34, 35
		C	2.3	30.8	302–323	30.8–25.8	28
		A	2.3	17.3	288–308	23.2–17.3	36
		N	1.4		360–430	10–18	33
	p-tol	M	*f*	26.7			7

Table 1. (*Continued*)

Compound	Solvent[a]	Field[b] (T)	T_1 (s) ambient[c]	Temperature range (K)	T_1 (s) range[d]	References
$Ph_2P(CH_2)_nPPh_2$	C	2.3	22	288–308	16.5–22.7	36
PH_2Ph	N	1.4	15	210–310	3.5–1.2	27
$P(OR)_3$ R = Me	N	1.1, 1.4	6.9	273–378	9.9–4.0	20, 22
Et	N	1.1	12.3	273–378	19.2–8.6	22
Pr^i	N	1.1	18.5	273–378	19.6–11.4	22
Bu	N	1.1	11.8	273–378	11.1–14.7	22
Ph	C	*f*	19.8			7
$[PR_2]_2$ R = Me	B	2.3	9.9			21
Bu^t	B	2.3	19.2			21
Ph	B	2.3	29.0			21
$POCl_3$	N	0.5	6	250–400	30–2	30
POF_3	N	0.5	1.5	230–320	5–1.5	30
$POPh_3$	C	1.4	13.8	243–343	7.8–13.8	34
	C	2.3	18.7	288–336	14.4–25.4	28, 36
$PSPh_3$	C	1.4	32.7	243–343	5.2–32.7	34, 36
$Ph_2P(X)(CH_2)_nP(Y)Ph_2$ (X, Y = O or S)	C	2.3	6.3–11.9	288–308	3.8–11.9	36
$PO(OR)_3$ R = Me	N, S[g]	1.1, 2.3	11.3	273–378	8.8–14.1	20, 22, 39
Et	N, S[g]	1.1, 2.3	18.6	273–378	18.9–13.2	22, 38
Bu^n	N	1.1	9.3	273–378	6.1–16.7	22
	A, S[g]	2.1, 2.3	28.9, 24.0			38, 39
Ph	S[g]	2.3	25.0			39
$PS(OMe)_3$	N	1.1	14.3	273–378	14.9–7.7	22
$PO_2(OR)_2^-$ [h]	H	2.3	17.4–19.7			39
$PO_3(OR)^{2-}$ [h]	H	2.3	14.0–21.7			39
PO_4^{3-}	HD	3.5	16	273–340	7–18	40
H_3PO_4	N	*f*	0.14			43
PR_4^+ $R_4 = Et_4$	C	2.3	9.0	302–331	9.0–11.0	28
Ph_4	C	1.4, 2.3	11.9	243–343	2.8–11.9	28, 34
$MePh_3$	C	2.3	12.6	288–308	9.8–12.6	36
R_4	D, MD	1.4, 2.1	2.1–6.5			20, 38
$[P(S)Me_2]_2$	C	2.3	17.1	283–321	14.9–17.3	21
$[P(S)R_1R_2]_2$	C	2.3	9.5–21.9			21
$S[P(S)R_2]_2$	C	2.3	14.6–23.6			42
$PPhCl_4$	C	1.4	7.0	243–343	8.9–5.1	34
PPh_2Cl_3	C	1.4	19.6	243–343	19.6–11.8	34
PPh_3Cl_2	C	1.4	18.7	243–343	18.7–12.1	34
PF_6^-	D	*f*	5	270–330	6–5	37
cis-$[PdCl_2(PR_3)_2]$	C	2.1	10.8–13.8			41
trans-$[PdCl_2(PR_3)_2]$	C	2.1	3.6–13.7			41
R = Pr^n	C	2.1	6.3	243–323	2.1–8.2	41

[a] N = neat liquid, C = $CDCl_3$, T = toluene-d^8, M = CD_2Cl_2, A = acetone-d^6, B = C_6D_6, S = CS_2, H = H_2O, D = D_2O, HD = 20% D_2O in H_2O at pH 10.3, MD = methanol-d_4

[b] 2.35 T gives ^{31}P resonances at 40.5 MHz (equivalent ^1H frequency, 100 MHz)

[c] Exact temperatures for these measurements vary but are generally in the range 298–308 K. Where disagreement exists between measurements made by different authors, the most recent values are generally quoted

[d] Since the temperature dependence of T_1 is not necessarily monotonic, the extreme T_1 values may not correspond to the temperature extremes. For details of the temperature dependence, the original literature should be consulted

[e] More recent studies on PCl_3/PBr_3 mixtures[22,29,32] have given results and interpretations significantly different from the measurements on pure compounds

[f] Not reported

[g] These compounds gave very significantly different T_1 values when dissolved in H_2O or D_2O[39]

[h] T_1 values for these anionic species varied significantly with change of cation. Values quoted are for Me_4N^+ salts

T_{1OTHER} is considered to arise from a "quasi-spin-rotation" involving hydrogen bond shifts around a PO_4 unit locked into the local solvent structure.[40]

Only a very few T_1 measurements have been reported on phosphorus coordination compounds.[41] The relaxation is probably dominated by dipolar processes since NOE measurements gave values of 0.68–1.00 for the palladium complexes listed in Table 1. Decrease in the relaxation time with decreasing temperature, decease with molecular weight in alkyl phosphine complexes, and longer T_1 for aryl phosphines relative to alkyl, are all consistent with the dipolar mechanism. The results also showed a marked solvent dependence—which may be important from the practical point of view—with T_1 up to 50% longer in CD_2Cl_2 than in $CDCl_3$.

1.3. Chemical Shifts

1.3.1. Noncoordination Compounds

1.3.1a. Introduction. The reported chemical shifts of phosphorus compounds cover a very wide range, from the almost incredibly highly deshielded phosphinidene bridge in $Bu^tP\{Cr(CO)_5\}_2$, +1362 ppm,[43a] to the highly shielded P_4 molecule, −461 ppm.[1] The extreme downfield portion of this range includes phosphide in a metal cluster[43b] and some unusual two-coordinate derivatives,[44,44a–c] but the most deshielded of "normal" phosphorus compounds is $PF_2(CH_3)$ at +245 ppm.[17] Figure 1 shows the typical chemical shift ranges found for various phosphorus coordination numbers from 1 to 6, and two major trends are immediately apparent. Firstly, the ranges of chemical shifts are very much larger for two- and three-coordinate compounds where the phosphorus atom has lone pairs of electrons than they are for four-, five-, and six-coordination where all the phosphorus electrons are involved in bonding. Secondly, there is a general tendency to increased shielding of the phosphorus as the coordination increases from 2 to 6. For monocoordinate phosphorus, only a very few examples are known, namely, the phospha-alkynes, $P{\equiv}CY$, where Y = F, H, Ph, Bu^t, or $SiMe_3$,[45] and it is impossible to say whether these general trends may be extended to this coordination number.

To a first approximation $\delta(P)$ is determined by the number and type of directly bound atoms and is relatively little affected by considerations of charge and structure. This is illustrated by the data collected in the first two rows of Table 2 for phosphorus tetracoordinated by oxygen. This table also shows the reduction in chemical shift range caused by a change from tricoordinate PR_3 to tetracoordinate PR_3H^+ derivatives and the slight increase in chemical shift range caused by introducing a polarizable atom such as sulfur, relative to corresponding oxygen-containing compounds.

On the basis of Figure 1, the phenomenon of increasing shielding with increasing coordination number should be approached with caution since the "most typical ranges" are inevitably liable to some subjective selection. However, Table 3 shows some examples where series can be constructed using identical ligands. For electronegative ligands the effect is clear cut, but for hydrogen the trend apparently reverses, although the data are very limited and the true coordination number for PH_2^- may be >2 in solution. Similar considerations apply to PPh_2^-, but the reversal of the trend from PPh_3 to PPh_4^+ is definite and in fact, all the simple PR_3 derivatives

Table 2. Effects of Charge and Structure on Phosphorus Chemical Shifts (ppm)

					References
$P(OH)_4^+$	$P(OMe)_4^+$	$P(OPh)_4^+$	$OP(OMe)_3$	$OP(OPh)_3$	1, 18
−2.7	+1.4	−18	−2.4	−18	46
$OP(OCH_2)_3CMe$	$OP(OMe)_2O^-$	$OP(OMe)\,O_2^{2-}$	OPO_3^{3-}	cyclic-$P_3O_9^{3-}$	1, 39
−8.0	+1.8	+4.8	+4.2	−21	
PH_3	PMe_3	PEt_3	PPh_3	PBu^t_3	47
−238	−62.2	−19.2	−5.4	+61.9	
PH_4^+	PMe_3H^+	PEt_3H^+	PPh_3H^+	$PBu^t_3H^+$	47
−101	−3.2	+22.5	+6.8	+58.3	
$PO(SMe)_3$	$POMe_3$	$POCl_3$	$POClBr_2$	$POBr_3$	1
+66	+36.2	+1.9	−64.8	−103	
$PS(SMe)_3$	$PSMe_3$	$PSCl_3$	$PSClBr_2$	$PSBr_3$	1
+98	+59.1	+28.8	−61.4	−112	

(R = Me, Et, Pr^n, Bu^n)[48] are more shielded than the corresponding PR_4^+ cations.[49,50] This admittedly fragmentary evidence may reflect a situation similar to that for the phosphorus coordination shifts induced by metals, namely, that deshielded ligands become more shielded on coordination, while shielded ligands become less shielded (see Section 1.3.2a).

1.3.1b. The Less Common Coordination Numbers (1, 2, 5, and 6). Figures 2 and 3 show the chemical shift ranges for the various coordination numbers in more detail. As mentioned above, there is insufficient data for monocoordination to justify a detailed plot.[45] For two-coordination,[44,44a–d,45] the overall shift range is very large,

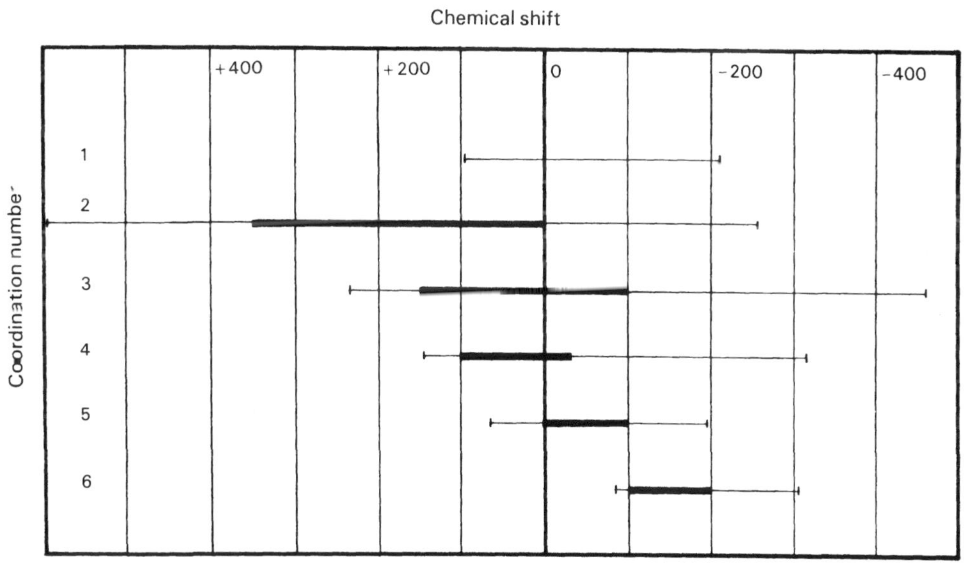

Figure 1. ^{31}P Chemical shift ranges for various coordination numbers. Shifts in ppm relative to 85% H_3PO_4 as zero. Solid bars indicate the most usual ranges and the lighter lines the limits of known values. NOTE: Subsequent to the preparation of these Figures, the shift range for two-coordinate compounds has been extended to +1362 ppm (see page 376).[43a,44b–c]

Table 3. Effect of Coordination Number on Phosphorus Chemical Shift[a]

Y	PY_2^-	PY_3	PY_4^+	PY_5	PY_6^-	References
Cl		+219	+87	−80	−295	2
F		+97		−80	−145	2
OEt		+139	−2.7	−71		1, 18
Ph	−25 → 0[b]	−6	+21	−89	−181[c]	1, 2
H	−255[b]	−238	−101			1, 47

[a] In ppm.
[b] The actual coordination number of these species in solution is open to question.
[c] $Y_6 = (o,o\text{-biphenyl})_3$.

from $Bu^tP\{Cr(CO)_5\}_2$ +1362 ppm,[43a] to PH_2^-, −255 ppm, but the negative portion of the range is occupied primarily by anionic species such as PH_2^-, PR_2^-, $P_2R_2^{2-}$, and $[RPPY_2]^-$. There is considerable doubt as to whether these species are truly two-coordinate in solution, although $P(CN)_2^-$ at −193 ppm may be an exception to this statement since it was studied as an $[Na(\text{crown ether})]^+$ salt. The only other examples showing negative shifts are $HP{=}CF_2$, −62 ppm, and a few compounds in which the phosphorus is bound to two phosphorus atoms, e.g., $Et_3P{=}P{-}P(O)(OEt)_2$, −217.5 ppm.[44] With these exceptions two-coordinate compounds have positive shifts. They are especially sensitive to vicinal substitution, and, for example, PhP=CXY derivatives range from $PhP{=}C(SiMe_3)_2$, +376 ppm, to $PhP{=}\overline{C(CH_2)_3N}Me$, +49 ppm. More generally, shifts for phospha-alkenes lie in the range from +300 to +100 ppm, with values >300 occurring only if the carbon has a silicon substituent and values <100 only if the carbon or phosphorus have a hydrogen substituent or are part of a ring.[45] Other highly deshielded two-coordinate species include phosphorus cations (phosphenium ions), +110 to +513 ppm,[44a,44b] compounds with double bonds to P, As, and Sb, +450 to +670 ppm,[44c] and RP groups bridging two metal atoms where there is no M–M bond, to +1362 ppm.[43a,44d]

Five-coordinated phosphorus compounds are typified by relatively high shielding of the phosphorus atom and the symmetrically substituted compounds which have been studied occupy a very narrow chemical shift range.

$P(OEt)_5$[1]	PCl_5[2]	PF_5[2]	$P(aryl)_5$[54]	PBr_5[1]
−71	−80	−80	−88 ± 1	−101 ppm

The presence of mixed substituents generally reduces the phosphorus shielding, for example, PR_nX_{5-n} derivatives (X = halogen) have been reported from +65 to −67 ppm[1]; and compounds more highly shielded than PBr_5 are very rare.[55] Most of the reported compounds contain two or more types of ligating atoms around the phosphorus and are often rather unusual and specific structures. This makes it difficult to designate typical ranges for various types of substituent. However, five-coordinate compounds containing fluorine have been well documented,[56] and a review of cyclic oxy-phosphoranes has also been published.[57] The oxy-phosphoranes are noteworthy in showing clear and consistent deshielding of the phosphorus when it is incorporated in five- and six-numbered rings.[58,59] This topic is discussed in more detail in Section 1.3.1e.

Six-coordinate phosphorus compounds are similar to the five-coordinate examples in showing very high shielding. The overall range is from a pyridine oxyphosphorane adduct at −82 ppm[61] through PF_6^- (−145 ppm)[2] and P(*o*, *o*-biphenyl)$_3^-$ (−181.2 ppm)[2] to various PCl_6^- salts ca. −300 ppm.[62] Other examples are rare.[63]

1.3.1c. The Common Coordination Numbers (3 and 4). The overall range of chemical shifts for three-coordinate phosphorus compounds extends from $PF_2(CH_3)$, +245 ppm,[17] to P_4, −461 ppm,[1] and for four-coordinate from $[PF(CH_3)_3]^+$, +143 ppm,[64] to $SPBrI_2$, −314.8 ppm.[65] However, these exceptional shifts apart, the ranges for both three and four coordinates are well defined by the symmetrically substituted PY_3, POY_3, and PSY_3 compounds shown in Table 4, and general ranges for other compounds are shown in Figures 2 and 3.

Correlations with established measures of electron withdrawing power such as electronegativity, or Hammett or Taft parameters are occasionally possible for restricted groups of compounds, for example, substituted phenyl phosphoric acids[68] or substituted triphenyl phosphines.[69] However, from the totally different sequences of shielding for the three types of compound in Table 4 and from our treatment of mixed substituent chemical shifts in Section 1.3.1d, it is obvious that neither correlations with electronegativity nor prediction of shifts using simple additive substituent constants will be generally effective. The latter approach has achieved some success only with restricted groups of compounds involving ranges of organic substituents. Early workers in this field defined substituent contributions (σ^P) relative to methyl derivatives ($\sigma^P = 0$) or in terms of the absolute group contributions (GC) to chemical shifts. Since PMe_3 resonates at −62 ppm relative to 85% H_3PO_4, the relationship between the two parameters is

$$\sum_{n=1}^{3} GC_n = \sum_{n=1}^{3} \sigma_n^P - 62$$

or approximately $GC = \sigma^P - 21$.[70] In the interests of uniformity, several equations originally quoted in terms of GC have been recast in σ^P in the following table. Also the equations have been modified to accord with the modern sign convention so that

Table 4. Typical Phosphorus Shifts (ppm) in Three- and Four-Coordinate Compounds[a]

Y −	Br	Cl	OMe	SMe	NMe_2	F	NCS	Bu^t	CF_3
PY_3	227	219	141	125	122	97	86	63[47]	−2.6
POY_3	−103	2	−2.4	66	23	−36	−62	−41[66]	2.3
PSY_3	−112	29	73	98	82		−10		

Y =	Ph	Me	C≡CH	C_6F_5[67]	CN	H	$SiMe_3$
PY_3	−6	−62	−91	−77	−136	−238	−330
POY_3	25	36	−56[66]	−8			
PSY_3	40	59		−9			

[a] Data from Ref. 1 unless otherwise noted.

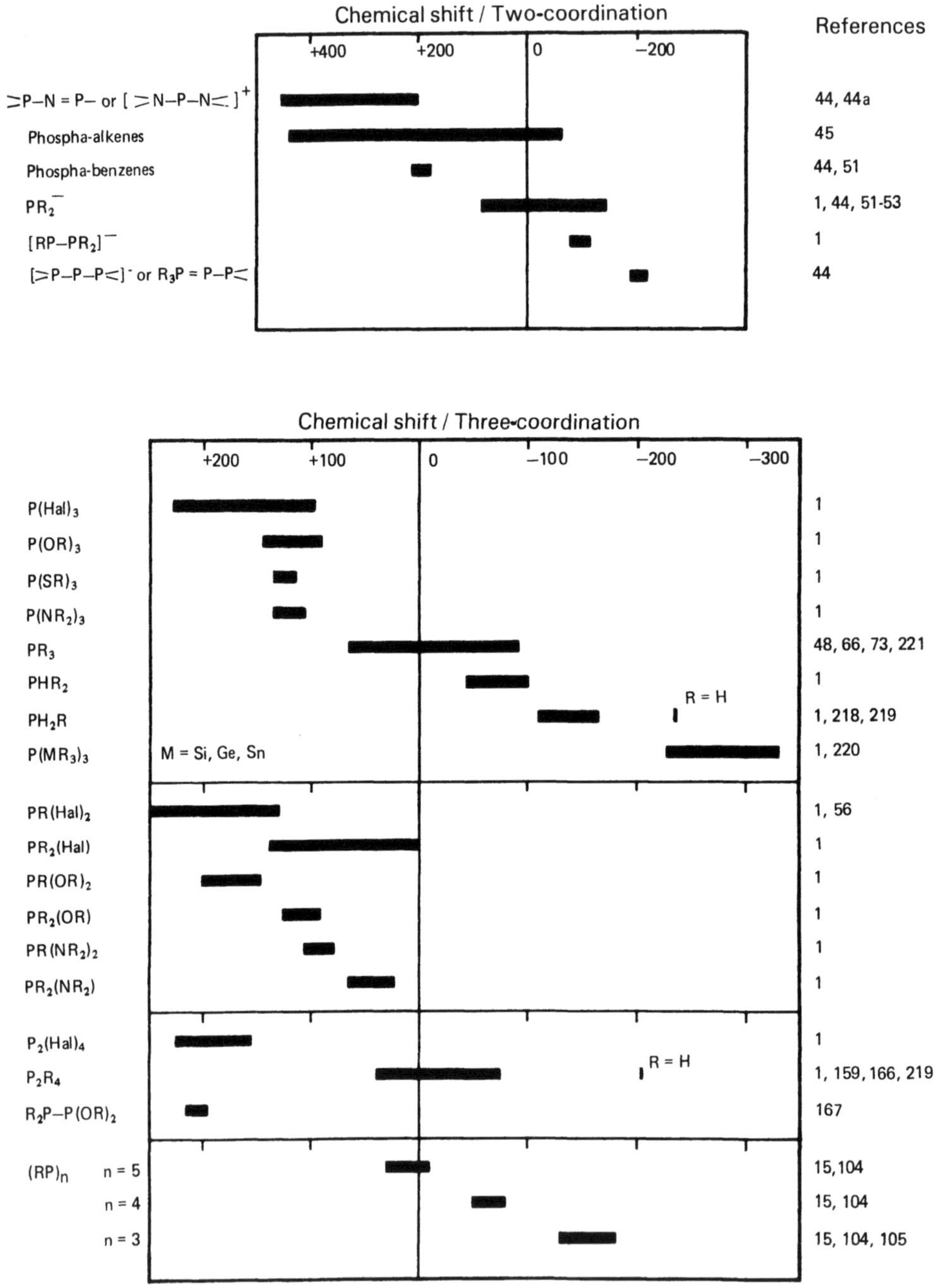

Figure 2. ^{31}P Chemical shift ranges for two- and three-coordinate phosphorus compounds. Shifts relative to 85% H_3PO_4 as zero. NOTE: Subsequent to the preparation of these Figures, the shift range for two-coordinate compounds has been extended to +1362 ppm (see page 376).[43a,44b–c]

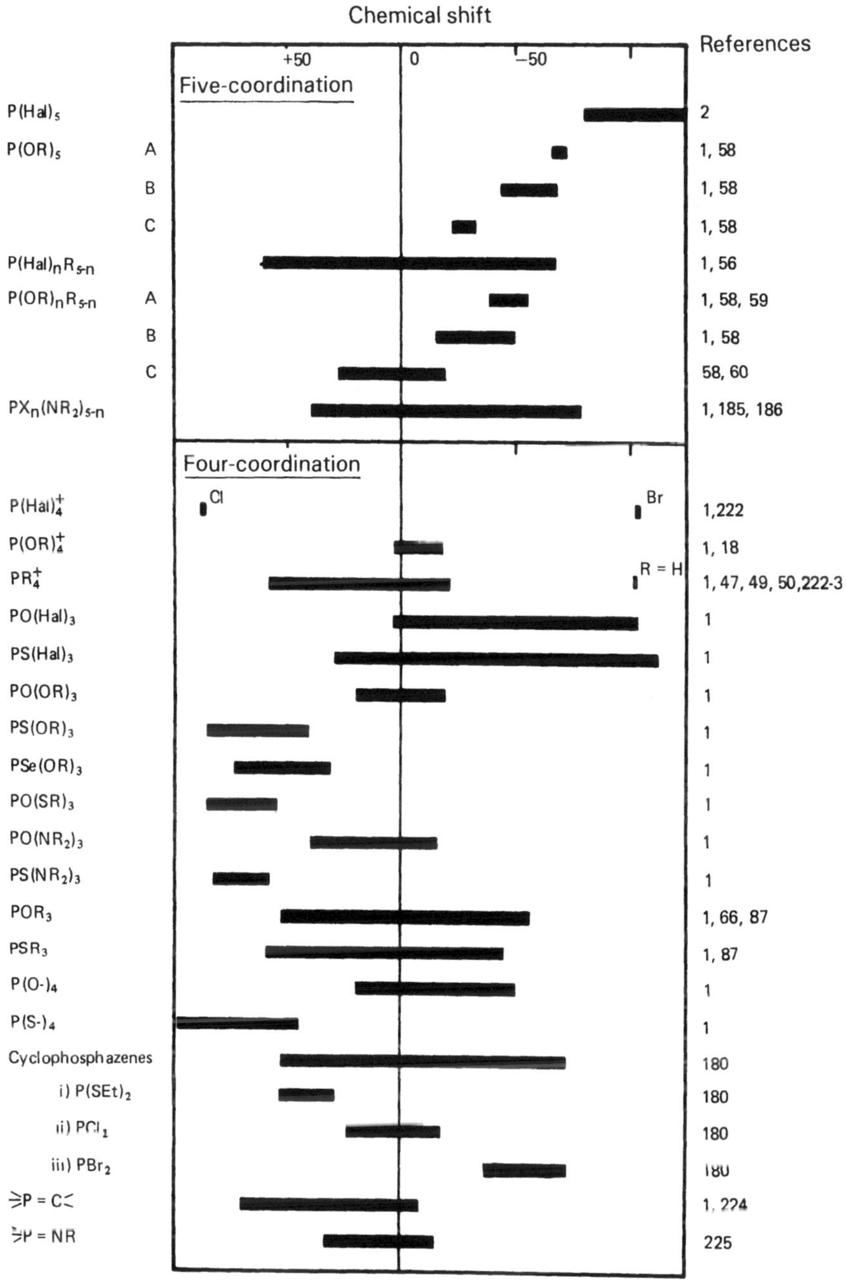

Figure 3. ^{31}P Chemical shift ranges for four- and five-coordinate phosphorus compounds. Shifts relative to 85% H_3PO_4 as zero.

positive values of σ^P (or GC) represent a deshielding contribution (downfield shift). With these changes, the following approximate equations can be obtained:

$$\delta(P) = A + B\sum_{1}^{n} \sigma_n^P$$

Compound type	n	A	B	References
RPH_2	1	−163.5	2.5	71
R_2PH	2	−99	1.5	71
R_2P^-	2	−130	2.3	53
R_3P	3	−62	1.0	47, 70
R_3PH^+	3	−3.2	0.56	47, 72
R_4P^+	4	25.3	0.26	49

For each of the ionic derivatives listed above, the original authors noted that slightly different equations are needed when phenyl groups are involved,[49,53,72] and for R_4P^+ compounds an extended equation making special allowance for allyl, benzyl, phenyl, and cyclohexyl groups was proposed.[49] This is a further limitation on the usefulness of the equations, which can in any case yield only rather approximate predictions of shifts.

Using values from the above references and more recent values for alkynyl[66] and alkenyl[73] compounds we obtain the following substituent constants:

R:	Ethynyl	Z-1-propenyl	Me	neopentyl	Bu^i	allyl	n-C_nH_{2n+1} $(n \geqslant 3)$
σ^P:	−10	−6	0	+3	+6	+9	+10 ppm

R:	Et	vinyl	benzyl	Ph	cyclopentyl	cyclohexyl	Bu^s	Pr^i	t-amyl	Bu^t
σ^P:	+14	+15	+17	+18	+21	+23	+24	27	42	+45 ppm

The sequence of σ_P contributions shown above is not readily interpreted in terms of electronegativity or steric bulk. A proposal that hyperconjugation of the α protons might be important led to the idea of subdividing σ^P for alkyl substituents into contributions depending on the numbers of β and γ carbons.[70] As is the case for ^{13}C and ^{15}N chemical shifts, it appears that the presence of β carbons has a deshielding effect whereas γ carbons are shielding. The relation $\delta(P) = \delta(\text{parent}) + m\beta + n\gamma$ has been proposed, where δ(parent) refers to the methyl derivative, m is the total number of β carbons in all R groups, and n is the number of γ carbons; i.e., $\sum \sigma^P = m\beta + n\gamma$. The best fit values of β and γ for RPH_2, R_2PH, R_3P, R_3PH^+, R_4P^+, and R_3PO are tabulated in Ref. 74, which is also a good source of references to data on these compound types.

The shielding effect of γ substitution appears to be general for the shifts of heavy nuclei (i.e., ^{13}C, ^{19}F, ^{31}P, etc.) and it has been proposed that it is primarily a bond angle distortion effect.[75] The significance of γ substitution arises from "coupling" of torsional angles to bond angles since the X–Y–Z angle is typically 2–6° larger in a *gauche* V–X–Y–Z fragment than in a corresponding *trans* fragment. Enlargement of the X–Y–Z angle causes increased screening of Y and in consequence γ substitution, conformational changes, and incorporation of the phosphorus into ring structures can all affect $\delta(P)$. Similar arguments can be applied to the conformational dependence of coupling constants.[75] Although present theory certainly predicts a bond angle dependence of $\delta(P)$ it is not adequate to account quantitatively for the observed changes, especially since it appears that very strained rings produce upfield shifts (see Section 1.3.1e).

Other examples of the shielding effect of γ substitution occur in alkenyl derivatives (compare σ^P for E-1-propenyl, +13 ppm, with Z-1-propenyl, −6 ppm)[73] and in *o*-tolyl phosphines[69] and phosphine sulfides[76] which are upfield of *m* and *p* analogs. Examples of conformational dependence of $\delta(P)$ are extremely numerous[10] and temperature changes may drastically affect $\delta(P)$ if equilibria of tautomers or conformers are involved.[77]

1.3.1d. Derivatives with Mixed Substitutents. As discussed in the previous section, it is often possible to use simple addition of substituent constants to predict chemical shifts of phosphorus compounds substituted by mixtures of different hydrocarbon radicals. Unfortunately, these relations fail immediately the range of compounds is widened to include other ligating atoms. Some selected examples are collected in Figure 4 and Table 5. The figure shows $\delta(P)$ for derivatives $PEt_{3-n}Y_n$ for a wide range of Y groups. The approximately linear relationships with other hydrocarbon groups are clearly evident, but with other substituents a "sagging curve" is typically obtained with the PEt_2Y and $PEtY_2$ derivatives less shielded than predicted by a straight line interpolation between PEt_3 and PY_3. The magnitude of the deshielding relative to the "straight line predictions" is roughly equal for PEt_2Y and $PEtY_2$, and it increases as the shifts of PEt_3 and PY_3 become more disparate.

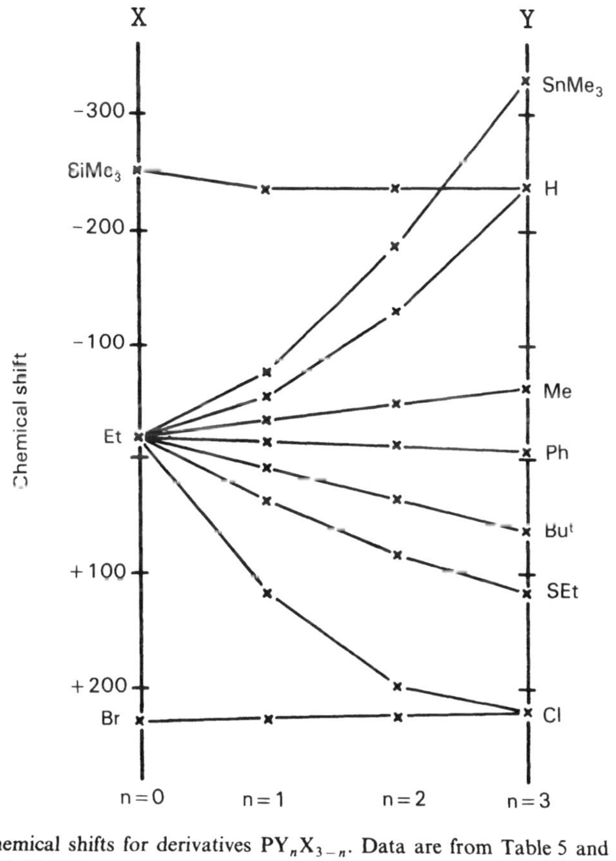

Figure 4. ^{31}P Chemical shifts for derivatives PY_nX_{3-n}. Data are from Table 5 and Refs. 85 and 220 and are relative to 85% H_3PO_4 as zero.

However, inspection of Table 5 will show that neither of these guidelines is a reliable, general predictor. It is generally true that derivatives $PX_{3-n}Y_n$ will have similar shifts if PX_3 and PY_3 are similar, and Figure 4 includes data for $PBr_{3-n}Cl_n$ and $P(SiMe_3)_{3-n}H_n$ to illustrate this point. At the other extreme the deshielding effect can be very large; for example, the deshielding relative to a straight line interpolation is

Table 5. Chemical Shifts of Derivatives with Mixed Substituents

Compounds		Chemical shifts (ppm)				References
		$n=0$	$n=1$	$n=2$	$n=3$	
(A) $PX_{3-n}Y_n$						
X = Cl	Y = Br	219	224	227	227	82
	OMe		181	169	141	1
	SMe		206	188	125	1
	NMe_2		166	160	122	1
	CF_3		131	51	−3	80
	Ph		162	81	−6	1
	Et		198	118	−19	83
	Me		192	97	−62	1
	CN		93	−34	−137	81
X = OMe	Y = NMe_2	141	148	138	122	84
	CF_3		136	96	−3	80
	Ph		159	116	−6	1
X = SMe	Y = CF_3	125	10	35	−3	80
X = NMe_2	Y = Ph	122	100	64	−6	1
	Et		100	64	−19	1
	Me		86	39	−62	1
X = Ph	Y = SPh	−6	39	100	130	85
	Et		−12	−17	−19	48
	Me		−28	−48	−62	48
	C≡CPh		−33	−61	−88	86
	CN		−36	−76	−137	87
	$SnPh_3$		−56	−163	−330	88
X = Et	Y = Bu^t	−19	8	35	+63	89
	Me		−34	−49	−62	1
	H		−55	−131	−238	1
	$SiMe_3$		−88	−167	−252	89
	$SnMe_3$		−76	−187	−329	89
(B) $OPX_{3-n}Y_n$						
X = Me	Y = NMe_2	36	44	38	23	1
	Cl		63	45	3	1
	OMe		52	32	−2	18, 90a
	F		66	27	−36	1
	Br		51	9	−103	1
X = Ph	Y = Cl	25	43	34	3	1
	F		40	11	−36	1
X = NMe_2	Y = Cl	23	30	17	3	1
	F		4	−17	−36	90b
X = Cl	Y = OMe	3	6	6	−2	1
	F		0	−15	−36	1
	Br		−30	−65	−103	1
X = Bu^t	Y = C≡CH	−41	48	7	−91	66

Table 5. (*Continued*)

Compounds		Chemical shifts (ppm)				References
		$n=0$	$n=1$	$n=2$	$n=3$	
(C) $SPX_{3-n}Y_n$						
X = OMe	Y = Cl	73	73	59	29	1
X = Me	Y = Cl	36	87	80	29	1
	Br		63	21	−112	1
X = Ph	Y = Cl	40	80	75	29	1
X = Cl	Y = Br	29	−15	−61	−112	1
	I	32	−112	−260	[−411][a]	65
X = Br	Y = I	−108	−215	−315	[−411][a]	65
(D) $PX_{4-n}Y_n^+$						
X = Ph	Y = Pr^i	23	24	27	30[b]	50
	Cl		67	95	106[c]	91
X = OMe	Y = Me	1	54	98	101[d]	18, 50, 90a
	NMe_2	1	17	29	38[e]	92

[a] Predicted value.
[b] $n=4$, $\delta(P)=32$.
[c] $n=4$, $\delta(P)=88$.
[d] $n=4$, $\delta(P)=25$.
[e] $n=4$, $\delta(P)=43$.

almost 200 ppm in the series PF_3, +97 ppm; PF_2Me, +245 ppm; $PFMe_2$, +186 ppm, PMe_3, −62 ppm.[1,2] Plots similar to Figure 4 can be constructed for four-coordinate phosphorus (see Table 5), and there is one almost complete series for five-coordination: PPh_5, −89, PPh_3Cl_2, −65, PPh_2Cl_3, −73, $PPhCl_4$, −39, PCl_5, −80.[78]

Although Table 5 shows most "mixed" compounds as deshielded relative to the straight line interpolation between the symmetrical derivatives, and this is indeed the most usual situation, there are a significant number of examples for which a shielding effect is observed. These all seem to involve chloride or bromide, and for example the series $PCl_{3-n}Y_n$, where Y = NEt_2,[79] NMe_2,[1] NCO,[1] CF_3,[80] or CN,[81] all involve more negative shifts for the mixed derivatives. However, when Y = Br,[82] OMe,[1] SMe,[1] Ph,[1] Et,[83] or Me[1] the shifts are more positive for the mixed derivatives. The difference between the two sets of substituents is not immediately obvious!

1.3.1e. Effects of Ring Strain. In general, the detailed evidence discussed below shows that in four- and five-coordinate compounds, those containing phosphorus in a five-membered ring are always significantly deshielded relative to corresponding acyclic derivatives. This effect is much smaller and possibly nonexistent in three-coordinate phosphorus compounds. For all three coordination numbers, acyclic compounds and those containing phosphorus in a six-membered ring have similar shifts. The six-membered rings usually contribute a small shielding effect but there are several examples of deshielding. Also for all three coordination numbers, very small rings (four and three member) produce large shielding contributions to $\delta(P)$. Thus it appears that moderate angular restraint produces deshielding whereas severe ring strain has the opposite effect. The chemical shifts of five-coordinate oxy-phosphoranes exhibit a particularly clear and consistent dependence on ring size as is shown by the data tabulated in Refs. 58 and 59. Relative to the acyclic compounds, six-membered rings contribute a small deshielding effect ($\sim$4 ppm per ring) and incorporation of the phosphorus into a five-membered ring causes significant deshielding ($\sim$17–23 ppm per ring).[58,59] There is, however, some evidence that inclusion of the phosphorus in a four-membered ring produces increased shielding relative to acyclic derivatives.[93] Deshielding by five-membered rings is also observed in fluorophosphoranes.[94]

For four-coordinate compounds, a small shielding effect of six-membered rings has been observed for PR_4^+,[94,95] $P(O)(OR)_3$,[94] $P(O)R(OR')_2$,[94] $P(O)R_2(OR')$,[94] and $P(O)Cl(OR)_2$,[94] but for $P(S)(OEt)(OR)_2$ compounds there is a small deshielding effect.[96] All of these compound types exhibit deshielding of the phosphorus when it is in a five-membered ring.[94–96] For five- and six-membered cyclic phosphate esters, ester anions, related acyclic compounds, and Li_3PO_4, it has been proposed that a general relation exists between $\delta(P)$ and the O–P–O bond angle. The relation is bell-shaped with maximum shielding at $\sim 107°$ and with both smaller (e.g., five-membered rings) and larger angles (e.g., acyclic diester, free acids, and Li_3PO_4) leading to deshielding.[97] Unfortunately it is not at all clear how this proposal relates to the more general trends enunciated at the beginning of this section, since an extension of the proposed bell-shaped relationship would predict a deshielding effect for small rings (3 and 4 member) whereas in fact these derivatives are highly shielded. An example for four-coordinate phosphorus is $OP(Bu^t)CH(Bu^t)CH(Bu^t)$ (-1.3 ppm), which is highly shielded relative to $OP(Bu^t)(CH_2Bu^t)_2$ ($+52.2$ ppm).[98]

Moving to three-coordinate phosphorus, the effect of rings on chemical shifts of tertiary phosphites is much less definite than those described above for oxy-phosphoranes. There is some evidence that both six-member and five-member rings produce shielding, with the former exhibiting the largest effect. Thus the series $(PhO)P(OMe)_2$, 135 ppm, $(PhO)PO(CH_2)_2O$, 124 ppm, $(PhO)POCH_2C(Me)_2CH_2O$, 115 ppm,[1] and $P(OEt)_3$, 137 ppm, $P(OCH_2)_3C(H)O$, 105 ppm, $P(OCH_2)_3CMe$, 91.5 ppm,[99] support this trend, but there is no clear distinction in the overall ranges observed: acyclic, 127–143 ppm, five-member rings, 124–138 ppm, six-member rings, 115–138 ppm.[1] For tertiary phosphines six-member rings tend to contribute a shielding effect[74,100] but five-member rings cause little change in saturated systems (e.g., $MeP(CH_2)_3CH_2$, -32 ppm, $PMeEt_2$, -34 ppm).[74,101] However, the unsaturated five-member ring compound, $MeP(CH)_3CH$, -8.7 ppm, is deshielded relative to divinylphosphines, ca. -21 ppm.[102] In three-membered rings, the phosphorus in the phosphirans, $RPCH_2CHR'$ (-234 to -341 ppm) is very highly shielded relative to acyclic analogs.[103]

For three-coordinate phosphorus, the best correlations of $\delta(P)$ with ring size concern the polyphosphorus compounds, $(RP)_n$. The phosphorus chemical shift is approximately linearly correlated with average endocyclic P–P–P bond angles[104] and definite ranges can be established for $n=5$, $+30$ to -10 ppm,[15,104] $n=4$, -50 to -80 ppm,[15,104] and $n=3$, -130 to -180 ppm.[15,104,105] The rapid increase in shielding with decrease in bond angle is consistent with the very high shielding in the P_4 molecule, −461 ppm.[1] Only one example is known for $n=6$, namely, $(PhP)_6$, −22 ppm, but this also lies on the shift/bond angle correlation.[104]

1.3.1f. Theory of Phosphorus Chemical Shifts. The lack of correlations of δ(P) with substituent electronegativity and the other interpretive problems outlined in the preceding sections make this a difficult area for theory. No fully satisfactory theory exists and the semiempirical treatment proposed by Letcher and Van Wazer[1] in 1967 remains the basis for most subsequent discussion. The approach follows previous authors in assuming that chemical shifts for heavy atoms arise primarily from variations in the paramagnetic contribution from electrons in valence orbitals and attempts to account for the shifts using three contributions: substituent electronegativity, bond angles, and the extent of π bonding.

1.3.2. Coordination Compounds

Study of coordination complexes is perhaps the most rapidly expanding area of ^{31}P NMR. Although several relatively recent reviews are available,[4–7,106,107] none are comprehensive and workers requiring details of particular compounds are again referred to the Specialist Periodical Reports published by the Royal Society of Chemistry.[9–11]

1.3.2a. Monodentate Ligands. Coordination of a phosphorus(III) compound to a metal (or other Lewis acid) results in a "coordination shift" from the original resonance position of the noncoordinated ligand. This shift may occasionally be very large (the current record is probably −291 ppm for $Cr(CO)_5(PI_3)$[108]), and there are also many cases of erratic variations between seemingly similar ligands. Despite these limitations and its admittedly empirical nature, the concept of coordination shifts ($\Delta\delta$) remains the only basis for discussion of δ(P) in metal complexes.[7] The wide variation of ligand types renders attempts to construct typical shift ranges for particular metals neither practical nor useful, and the present state of theory of chemical shifts is not helpful.

Early workers in the field studied a range of tertiary organo-phosphines with a variety of metals.[109] With this restricted ligand set approximate linear correlations were found between the coordination shift, $\Delta\delta$, and the chemical shift δ(P). The coordination shifts were mostly in the direction of decreased shielding relative to the free ligand and the shifts were larger (more positive) with highly shielded ligands. This work is summarized in Ref. 109 and more detailed accounts have been published for some individual sets of compounds, namely, $[M(CO)_5L]$ (M = Cr, Mo, W);[110] *cis* and *trans*-$[RuCl_2(CO)_2L_2]$;[111] *trans*-$[RhCl(CO)L_2]$;[112] *mer*-$[MCl_3L_3]$, (M = Rh, Ir);[113] *cis* and *trans*-$[MCl_2L_2]$ (M = Pd,[114] Pt[115]). The much greater body of information now available allows some of these series to be extended beyond the original restricted ligand set. The following paragraphs, while not intended to be comprehensive, discuss some illustrative examples and show that the above ideas require considerable modification. This is especially so with respect to the linearity of the plots. However, some useful guidelines do remain.

Table 6 collects together values of coordination shifts for two representative and well-studied series, namely, $[M(CO)_5L]$, where L = Cr, Mo, W, and *cis*-$[PtCl_2L_2]$, and these data are plotted in Figures 5 and 6. Figure 5 for the $[M(CO)_5L]$ systems

Table 6. Ligand Chemical Shifts (ppm) and Coordination Shifts (ppm)[a] in Cr, Mo, W, and Pt Complexes

Ligand	Ligand shift	$Cr(CO)_5L$ $\Delta\delta$	$Mo(CO)_5L$ $\Delta\delta$	$W(CO)_5L$ $\Delta\delta$	*cis*-$PtCl_2L_2$ $\Delta\delta$	References
PH_3	−239	109	74	51		108
$P(CH_2O)_3CBu$	−81		40			138
PMe_3	−62	69	45	26		139
PMe_2Ph	−48		63		32	115, 140
PBu^n_3	−33	63	45	26	34	110, 115
PPr^n_3	−33				33	115
$PMePh_2$	−28	63	43	24	27	110, 115
PPr^n_2Ph	−27				25	115
PBu^n_2Ph	−26	61	44	25	24	110, 115
PEt_3	−20		40		30	115, 139
PPr^nPh_2	−18		44		25	110, 115
PEt_2Ph	−17				21	115
PBu^nPh_2	−17	62	44	25	24	110, 115
$PEtPh_2$	−12	60	42	24	22	110, 115
PPh_3	−6	61	44	27		110
PPr^iPh_2	0	59	43	26		110
PBu^tPh_2	17	55	40	24		110
$PClPh_2$	80				−8	141
$P\{N(Me)CH_2\}_3CPen$	87			25		138
$P(OCH_2)_3CPen$	93			22		138
$P(OCH_2)_3CEt$	93		44			138
PF_3	97		50		−27	132, 140
$PF_2(OPh)$	110				−41	132
$Me_2Si\{(CH_2)_3OPPh_2\}_2$	111				−28	141
$Ph_2Si\{(CH_2)_3OPPh_2\}_2$	112				−30	141
$PPh_2(OMe)$	116		30			140
$P(NC_5H_{10})_3$	117				−61	142
$P(NMe_2)_3$	123		23			139
$PF(OPh)_2$	123				−57	132
$P(SMe)_3$	125		5			139
$P(OPh)_3$	128		26		−61	126, 140
P(O-o-tol)$_3$	130				−71	131
$P(OEt)_3$	138		18			140
$P(OMe)_3$	141	39	21	−4	−68	131, 139
$PF_2(NMe_2)$	144				−56	143
$PF(NMe_2)_2$	151				−65	143
$P(OEt)_2Ph$	159				−66	130
$P(OMe)_2Ph$	159		18			140
$[P(F)N(Bu^t)]_2$	164				−79	142
$PCl(OCH_2)_2$	167				−51	141
$PCl(OC_3H_2)_2$	168				−57	141
TPPB[b]	178	20	2	−22		144
PI_3	179	−291				108
PCl_2Me	191				−82	145
$[P(Cl)N(Bu^t)]_2$	211				−119	142
PCl_3	218	−31		−120	−102	108, 142
PBr_3	227	−132	−166	−231		108
(Mes)P=CPh$_2$	233				−105	146

[a] Coordination shift($\Delta\delta$) = δ(complex) − δ(ligand).
[b] 2,4,6-Triphenylphosphabenzene.

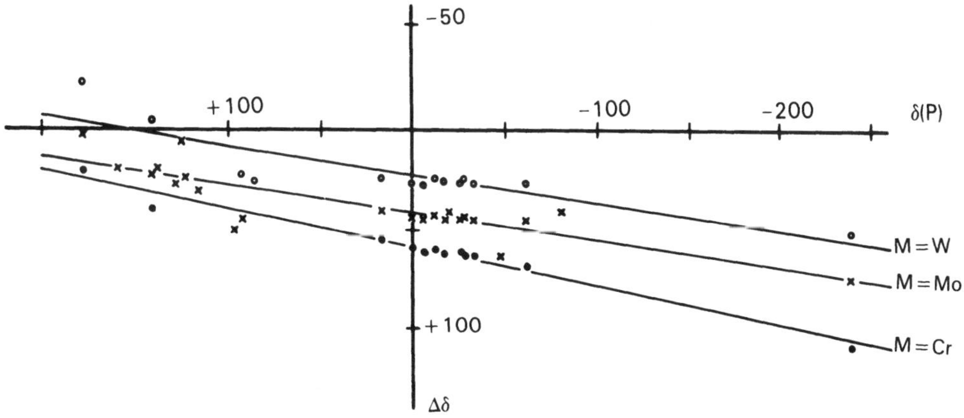

Figure 5. Coordination shifts ($\Delta\delta$) for complexes $M(CO)_5L$ plotted against the chemical shifts of the free ligand. Data are from Table 6 and are relative to 85% H_3PO_4 as zero. Note that data for $L = P(hal)_3$, hal = Cl, Br, I are not included in the plots and do not fit the correlations (see text).

shows that the linear correlations for the tertiary organophosphine ligands (grouped from $\delta(P) = -62$ to $+17$ ppm) applies only in the most approximate way when the series is extended to include tertiary phosphites and other highly deshielded ligands. The correlation fails totally for the halides PCl_3, PBr_3, and PI_3. In contrast, at the extreme shielding limit PH_3 lies approximately on the overall correlation. For any given ligand there is a consistent increase in shielding (more negative $\Delta\delta$) in the sequence Cr < Mo < W. For zero-valent complexes there is also a considerable body of data available for ML_n complexes, where $n = 2$, 3, or 4 and M = Ni,[116–123] Pd,[116,124,125] or Pt.[116–118,125–128] The author has constructed plots similar to Figure 5 and, for example, $\Delta\delta$ for PtL_4 complexes lies in the range from 0 to +20 ppm for tertiary phosphine ligands ($\delta(P)$, negative) and 0 to −30 ppm for tertiary phosphite

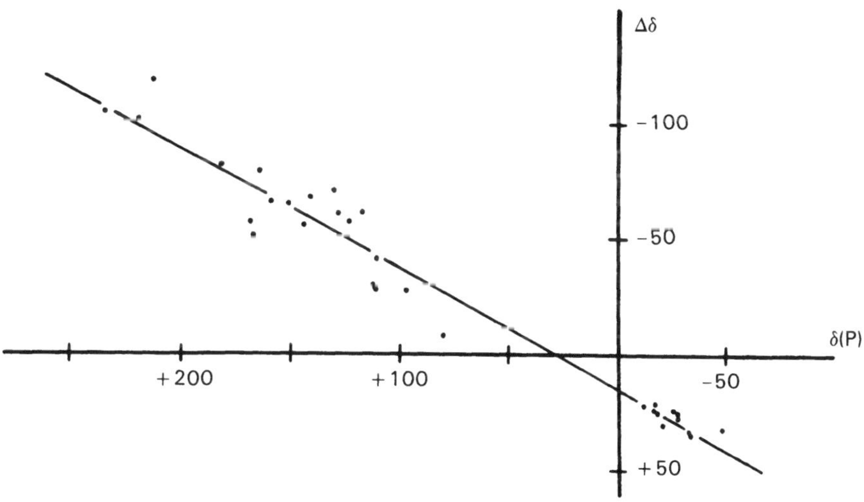

Figure 6. Coordination shifts ($\Delta\delta$) for complexes *cis*-$[PtCl_2L_2]$ plotted against the chemical shifts of the free ligand. Data are from Table 6 and are relative to 85% H_3PO_4 as zero. (The author thanks Drs. J. F. Nixon and A. Pidcock and their graduate students at the University of Sussex for assistance in compiling this figure and data.)

ligands (δ(P), positive), but within the two groups the points are scattered and there is no suggestion of a linear correlation. The situation is similar for NiL_4 derivatives except that in this case the majority of coordination shifts are positive (0 to +40 ppm) even for deshielded ligands such as phosphites. Data for PdL_4 are limited mainly to tertiary phosphine complexes, but they do confirm the sequence of increasing shielding, i.e., 1st row < 2nd row < 3rd row, observed for the zero-valent group VI metals. For example, $\Delta\delta$ values for $[M(PMe_3)_4]$ complexes are Ni, 40; Pd, 27; Pt, 9 ppm and for $[M(PEt_3)_4]$ complexes, Ni, 23; Pd, 19; Pt, 2 ppm.[119,124,127]

Turning to the data for *cis*-$[PtCl_2L_2]$ complexes presented in Table 6 and Figure 6, we see that some overall correlation exists between coordination shift, $\Delta\delta$, and ligand shift, δ(P), over a wide range of δ(P) values. This plot also makes it clear that the earlier observations of mainly deshielding (positive) $\Delta\delta$ values were due to the restricted ligand set and that in general more shielding values may be expected with highly deshielded ligands. Data for the related *cis*-$[PdCl_2L_2]$ complexes[114,129-132] produce a plot similar to Figure 6, albeit with a considerably greater scatter of points, and where identical ligands are involved the Pd complexes are less shielded than the Pt analogs. For example, values of $\Delta\delta$ in *cis*-$[PdCl_2L_2]$ complexes where L = PMe_2Ph (+54),[114] $PEtPh_2$ (+42),[114] PF_3 (−14),[132] and $P(OPh)_3$ (−45),[131] may be compared with corresponding values from Table 6 of +32, +22, −27, and −61 ppm for platinum complexes. Data for *trans*-$[PdCl_2L_2]$ complexes,[114,129,133] are similar to those for corresponding *cis* derivatives except that the *trans* complexes are generally 8–12 ppm more shielded than the *cis*. However, the dependence on stereochemistry cannot be generalized to other metals since in $[PtCl_2L_2]$ complexes *trans* derivatives[115] are 3–9 ppm less shielded than *cis* (Table 6) and the situation apparently reverses again in Pt^{IV} complexes.[7] It is also interesting that the slope of the plot in Figure 6 for Pt^{II} is much greater (ca. −0.52) than for the M^0 complexes discussed above (ca. −0.1 to −0.2). Moreover a plot for Rh^I complexes[112,134] has a slope ca. −0.3 and for Rh^{III} and Ir^{III} [113] ca. −0.7 to −0.8. The limited data on *trans*-$[Ru(CO)_2Cl_2L_2]$, slope ca. −0.3,[111] and $[PtCl_3L]^-$,[132,135,136] slope ca. −0.6, disturb the pattern somewhat, but there remains a fairly clear tendency for the slope of the $\Delta\delta$ against ligand chemical shift plot to increase in magnitude as the metal oxidation state increases.

In addition to the nature of the metal, its oxidation state and geometry, and the chemical shift of the noncoordinated ligand, one might expect $\Delta\delta$ to be affected by the steric bulk of the ligand[7] and the *cis* and *trans*-influences of other ligands.[6,7] Clearly the situation is very complex and the correlations discussed are not capable of accurate predictions of chemical shifts for individual complexes. The most serious errors occur for highly deshielded ligands (especially PX_3, X = Cl, Br, I) in low oxidation state compounds. Bearing in mind these limitations, the following principles are reasonably valid: (1) There is a general tendency for shielded ligands to become less shielded on coordination and for deshielded ligands to become more shielded. In favorable cases, such as sets of similar ligands, linear correlations may be obtained. (2) The sensitivity of the coordination shift to changes in ligand chemical shift increases with the oxidation state of the metal. (3) For similar complexes, shielding generally increases for heavier transition metals, i.e., 1st row < 2nd row < 3rd row.

Finally one may note that the ligands such as P(halogen)$_3$, for which the coordination shift correlations fail, are also the ligands which undergo the most dramatic chemical shift changes on protonation, alkylation, or oxidation. Indeed, for $[PHY_3]^+$,[108,137] $[PMeY_3]^+$,[1] $[PPhY_3]^+$,[1] and $[POY_3]^+$ [1] derivatives, one can easily construct $\Delta\delta$ plots which are broadly similar to those described above for metal complexes. This point does not seem to have been previously noted in the literature.

1.3.2b. Chelate Ligands and Other Ring Systems. The discussion in the previous section was concerned with the coordination shift for monodentate ligands. Corresponding compounds involving chelating phosphorus ligands show gross deviations from the linear relationships even for simple ditertiary phosphine ligands. For example, the series $[PtMe_2\{(CH_2)_n(PPh_2)_2\}]$, yields ^{31}P shifts of -40 ($n=1$), $+55$ ($n=2$), and $+3$ ($n=3$) ppm. These may be compared with the closest available monodentate analog, $[PtMe_2(PPh_2Me)_2]$ at $+6$ ppm. This effect has been interpreted in terms of a specific ring contribution, Δ_R, to the coordination shift, and a valuable article by Garrou[(107)] reviews the field and also serves as a comprehensive source of ^{31}P data for chelate complexes. The ring contribution effect is general for diphosphines, metallated organophosphines, and alkenylphosphines. For incorporation of the phosphorus in a four-membered ring Δ_R is strongly shielding, five-membered rings cause deshielding, and six-membered rings, shielding. For the $[PtMe_2\{(CH_2)_n(PPh_2)_2\}]$ complexes, Δ_R is -52, $+33$, and -14 ppm, respectively. Garrou emphasizes the differences between Δ_R and the effects of rings in simple organophosphorus compounds and concludes that Δ_R is not simply a carry-over of the same effects.[(107)] However, it is clear that the general trends developed in Section 1.3.1e for noncoordinated compounds (namely, small rings ($\leqslant 4$), shielding; five-member, deshielding; six-member, variable) are essentially the same as those for Δ_R in coordination complexes. An explanation based on bond angles seems the most obvious, but unfortunately there are conflicting data on this point. The proposal for phosphate esters (see Section 1.3.1e) is for a δ(P)-bond angle relationship having a *maximum* shielding at $\sim 107°$, with smaller and larger angles producing deshielding. In contrast the general pattern for oxy-phosphoranes and $(RP)_n$ rings and for Δ_R in metal chelates seems to require a relationship with a *minimum* shielding for five-membered rings and smaller rings (angles) giving higher shielding.

A further interesting chemical shift effect, possibly related to bond angle at the phosphorus or possibly specific to the presence of metal–metal bonds, is observed in complexes containing two metal atoms bridged by a phosphido ligand. The effect consists of a very large deshielding of the phosphorus in a three-membered ring containing a metal–metal bond. For example δ(P) for compound **I** is -127 ppm but $+205$ ppm for **II**,[(147)] the opposite effect to that observed in $(RP)_3$ rings (see Section 1.3.1e). The relation with M–P–M bond angle has been explored by Carty and co-workers in compounds of type **III** where M = Fe, Ru, Os,[(148)] and the importance of the M–M bond is illustrated by examples for M = Mo[(149a)] or W.[(149b)] The latter case is especially interesting since $[(CO)_4W(\mu\text{-}PPh_2)_2W(CO)_4]^{2-}$, with no W–W bond, has δ(P) -99 ppm, and the corresponding neutral complex of the same formula has a W–W bond and δ(P), $+180$ ppm.[(149b)] Several recent papers have reported similar effects, and a summary may be found in Ref. 149c.

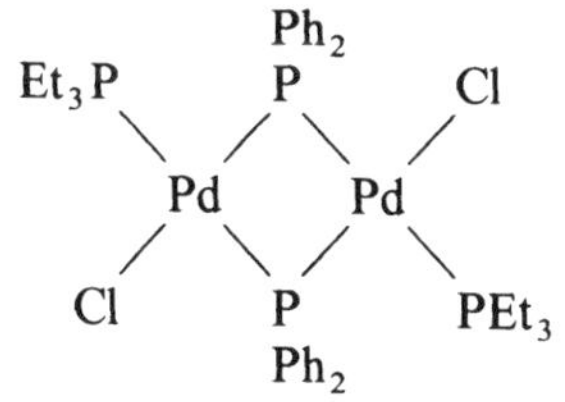

I

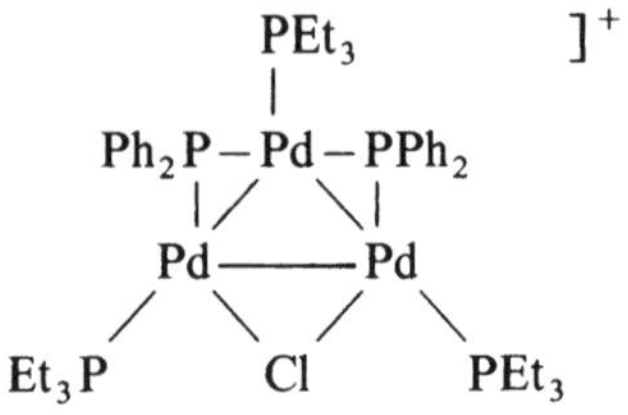

II

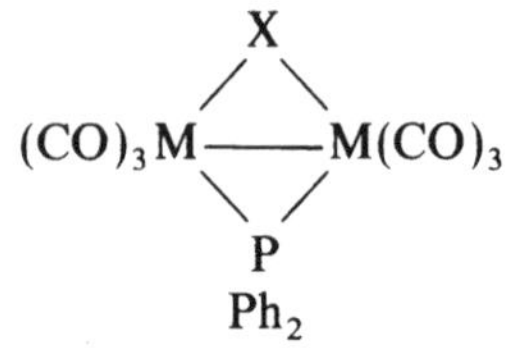

III

1.4. Coupling Constants

1.4.1. Introduction

Coupling constants involving ^{31}P have been measured for most of the magnetically active nuclei described in this book. Their magnitudes can be very large, ranging to 12 kHz in some Hg complexes.(7) For reduced coupling constants between directly bonded nuclei [i.e., $^{1}J(P\text{–}X)/\gamma_P\gamma_X$], the following pattern of signs has been established(150):

	H	B	C	N	O	F
three coordinate P:	+	+	−	−	−	−
four or five coordinate P:	+	+	+	±	−	−

Table 7. Representative Coupling Constants (Hz)[a] between Directly Bound Phosphorus Atoms

Compound	^1J(P–P)	References
(a) P^{III}–P^{III}		
P_2F_4	−229	2
$F_2PP(CF_3)_2$	234	158
F_2PPH_2	−211	158
P_2H_4	−108.2	3
$H(CF_3)PPH(CF_3)$	−115 → −204	162
P_2Me_4	−179.7	157
P_2Et_4	221	159
$Me_2PP(CF_3)_2$	−256	3
$P_2Pr^i_4$	302	159
$Me_2PPBu^t_2$	−318	157
$P_2Bu^t_4$	449	157
$R^1R^2PPR^3R^4$	167–449	3, 157, 159, 165, 166
$R^1_2PP(OR^2)_2$	208–260	167
$\overline{C(CF_3)=C(CF_3)P(CF_3)P}(CF_3)$	55	3
$(RP)_4CH_2$	−237 → −316	168
$(PhP)_6$	−106	169
$(RP)_5$	−214 → −343	163, 164
$(RP)_4$	100–149	15
$(RP)_3$	184–201	170
P_4Se_3	157	3

Table 7. (*Continued*)

Compound	^1J(P–P)	References
$P(PF_2)_3$	323	3
HPhPP(Ph) PHPh	197–226	171
$M_2(PPh)_{3\text{ or }4}$	$-150 \rightarrow -334$	172
(b) P^{III}–P^{V}		
$Me_2PP(S)Me_2$	−220	16
$Y^1Y^2PP(O)Y^3Y^4$ (Y = R or OR)	157–392	3, 158, 173, 194
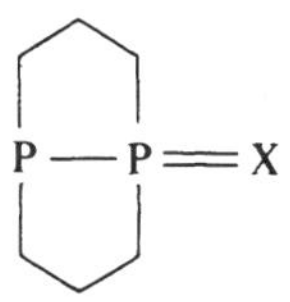 X = O	503	165
X = S	188	165
$R_2PP(R)_2CHPPh_2$	211–301	174
$CF_3P=PMe_3$	−436.4	175
$(NMe_2)_2P^+-P(NMe_2)_3$	340	175
$H_2P-PF_2\cdot BH_3$	−255	17
$(CF_3)_2P-PF_2\cdot BH_3$	316	17
(c) P^V–P^V		
$[O_2FP-PFO_2]^{2-}$	+766	156
$[O_2FP-PO_3]^{3-}$	+650	156
$[O_2HP-PO_3]^{3-}$	+465.5	3
Me_3P-PF_5	+715	158
$Me_2P(O)P(S)Me_2$	27	176
$R_2P(S)P(O)(OMe)_2$	150–175	177
$(EtO)_2P(S)P(O)(OR)_2$	583–590	3
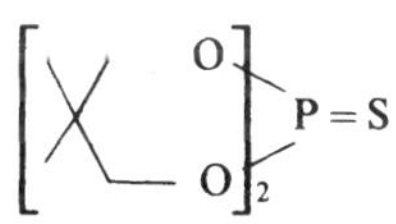	+475	158
$Me_2P(S)P(S)Me_2$	−18.7	157
$R^1R^2P(S)P(S)R^3R^4$	18.7–118	3, 157, 159
$Me_2P(S)P(Se)Me_2$	−40	157
$Me_2P(Se)P(Se)Me_2$	−67	157
$Bu^n_3P^+-P(S)(R)S^-$	108–118	3
$M-P(Me)_2P(Me)_2-M'$ (M = M(CO)$_n$)	132–158	178

[a] Signs are given where definite determinations have been made. The absence of a sign indicates it has not been determined.

The reduced one-bond coupling constants which have been measured between ^{31}P and the heavier main group elements ^{29}Si, ^{119}Sn, ^{77}Se, and ^{125}Te, are all negative,(17) whereas those for transition metals ^{103}Rh, ^{113}Cd, ^{183}W, ^{195}Pt, and ^{199}Hg are all positive.(7,17) The couplings are very sensitive to the phosphorus substituents, hybridization, and oxidation state and generally increase in magnitude with substituent electronegativity and show an algebraic increase in reduced coupling constant from P^{III} to P^V. An important exception to the last effect is ^1J(P–F) where P^{III} and P^V compounds share a common range from 800 to 1500 Hz.(17)

The remainder of the present chapter deals with coupling between phosphorus nuclei. Details of phosphorus coupling to the "less common" NMR nuclei will be found elsewhere in this book or its predecessor.[17] Space precludes detailed treatment of couplings to the "common" nuclei (^{1}H, ^{11}B, ^{13}C, ^{15}N, and ^{19}F) but these are well documented in the existing literature. Some important sources of coupling constant data are as follows: ^{1}H,[2,151] ^{11}B,[152] ^{13}C,[2,153] ^{15}N,[154] ^{19}F,[2,56,155] transition metals,[4–7,17,106] all nuclei.[17]

1.4.2. $^{1}J(P\text{–}P)$

Coupling constants between directly bonded phosphorus atoms cover a very wide range, from +766 Hz in $P_2F_2O_4^{2-}$ [156] to −451 Hz in $P_2Bu^t_4$,[157] and representative data are collected in Table 7. Early workers in the field[2,3,158] proposed (a) that P^V–P^V couplings are more positive than related P^{III}–P^{III} values; (b) that compounds with electronegative substituents have more positive coupling constants; and (c) that compounds with bulky substituents have more negative coupling constants. Points (a) and (c) appear to have stood the test of time fairly well, but point (b) requires careful qualification. It is true that the most positive coupling constants (e.g., $P_2F_2O_4^{2-}$, +766,[156] $Me_3P \rightarrow PF_5$, +715 Hz[158]) involve formal P^V–P^V bonds *and* electronegative substituents; but where P^{III}–P^{III} bonds are involved electronegative substituents do not produce more positive coupling constants. Thus P_2F_4, −230 Hz,[158] and $P_2(CH_3)_4$, −179.7 Hz,[157] have ^{1}J(PP) couplings of similar sign and magnitude.

In general P^{III}–P^{III} couplings vary from 55 Hz in the highly strained $\overline{C(CF_3{=}C(CF_3)-P(CF_3)-P}(CF_3)$,[2] or −108 Hz in P_2H_4[3] to −451 Hz in $P_2Bu^t_4$.[157] In agreement with point (a) above, all sign determinations have shown negative values.[2,3,157–159] The most intensively studied group of compounds are P_2R_4 and related asymmetric derivatives and in these the magnitude of ^{1}J(PP) increases steadily with substituent bulk (i.e., point (c) above). Since bulky substituents should increase the R–P–R angles, thereby increasing the *s* character in the P–R bonds and decreasing *s* character in the P–P bond, a simple rationalization in terms of *s* character and the Fermi contact term is precluded.[159] However, calculations for P_2H_4 have shown the Fermi contact term to be dominant[160] and it follows that the decreased *s* character must influence ^{1}J(PP) by producing a more negative value for the mutual polarizability of the valence *s* orbitals.[159] The effect of substituent bulk is also related to differences in rotamer populations but this is a smaller effect than the *s* character contribution.[157]

As Table 7 shows, those examples where the sign of a P^{III}–P^V coupling has been determined are all negative and where signs of P^V–P^V couplings are known they are often positive. However, ^{1}J(PP) in the $[P(S)R_2]_2$ series is negative and depends on substituent bulk in a manner similar to that described above for P_2R_4 compounds.[157,159]

Coupling constants between directly bound phosphorus atoms are also of interest in showing a large stereochemical dependence, related especially to rotation about the P–P bond when lone pairs are present. Ab initio SCF MO calculations on P_2H_4[160] predict that ^{1}J(PP) should change from −283 to +11 Hz as the dihedral angle between the phosphorus lone pairs changes from 0 to 180°. This gives a value of −166 Hz for the equilibrium conformation, in reasonable agreement with the observed −108 Hz.[160] In $[PR^1R^2]_2$ derivatives[157,161,162] the conformational dependence can be observed directly if the *meso* and *racemic* isomers have different rotamer populations. The effect can be very large and for example ^{1}J(PP) varies from −114.8 (−100°C) to −135.2 Hz (35°C) in *meso*-$[PH(CF_3)]_2$ and from −203.4 to −183.7 Hz in *racemic*-$[PH(CF_3)]_2$.[162] Conformational effects are also observed in $(PR)_n$ ring systems,[15,163,164] and for example the $^1J(PP)$ values in $(PCH_3)_5$ range from -248.6 to -310.3 Hz depending on the stereochemistry at each pair of coupled atoms.[163]

1.4.3. $^2J(P\text{–}E\text{–}P)$, *E = Nonmetal*

From an extension of the ideas presented above for $^1J(P\text{–}P)$, we may expect a coupling $^2J(P\text{–}E\text{–}P)$ to depend on the oxidation states of the phosphorus atoms, the electronegativity and steric bulk of their substituents, and the degree of rotation about the P–E bonds. Any π bonding between P and E and the interbond angles at E could also be important.[179] The most extensively studied case is where E = N.[3,180]

$^2J(P^{III}\text{–}N\text{–}P^{III})$: For the acyclic compounds $Y^1Y^2P\text{–}N(R)\text{–}PY^3Y^4$ values of $^2J(P\text{–}P)$ are usually large (>100 Hz) and positive, extending to +665 Hz for $Br_2P\text{–}N(Me)\text{–}PBrCl$. Compounds studied include examples where Y = halogens, alkyl or aryl, NMe_2, OMe, and SMe.[179] However, when steric constraints are placed on the possible conformations about the P–N bonds, the range of $^2J(P\text{–}P)$ is extended downward to −35 Hz. Particularly dramatic is the change for $\{Ph_2P\text{–}N(R)\}_2PPh_2$ from 280 Hz for R = Me to 25 Hz for R = Et.[181] The general trend is well illustrated by data for the series $Ph_2PN(R)PPhCl$: R = Me (+333.5 Hz); Et (+158 Hz); Pr^n (+145 Hz); Pr^i (−29 Hz); Bu^t (−35 Hz).[181] These changes are considered to result from the effect of the steric bulk of R on the relative populations of conformers A and B, with A being favored for small R groups.[179,181] Steric constraints also influence $^2J(P\text{–}P)$ in ring systems since in the cyclodiphosph(III)azanes (I; X, Y = F, Cl, OMe, NMe_2, Me) the values range from −10 to +50 Hz and in several instances *cis* and *trans* isomers have couplings of opposite sign.[182]

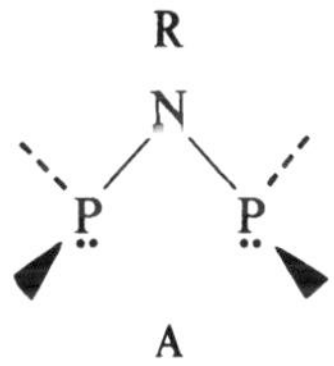

A

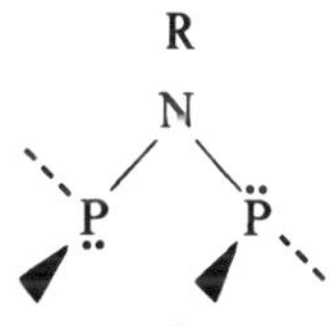

B

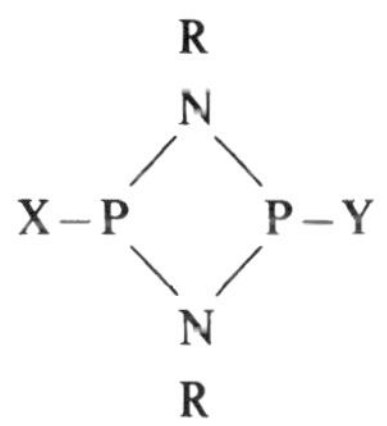

(I)

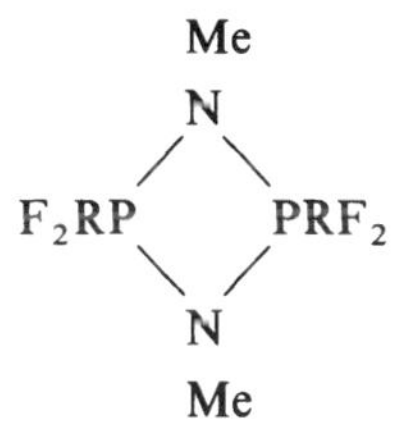

(II)

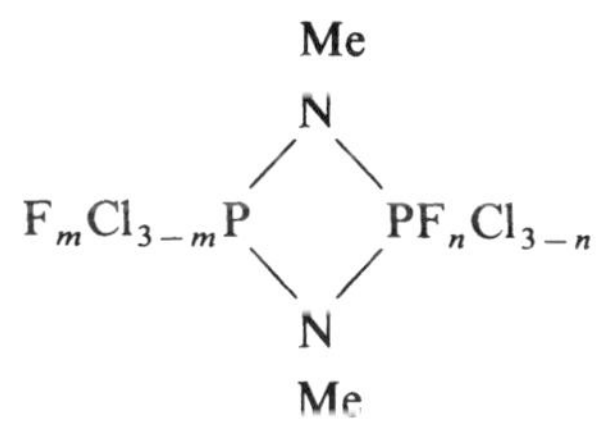

(III)

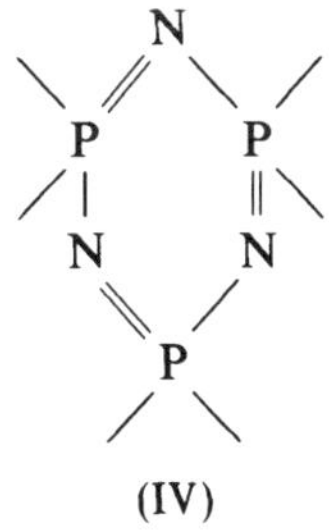

(IV)

$^2J(P^V\text{–}N\text{–}P^V)$: Oxidation of simple acyclic compounds usually results in a reduction of ^2J(P–P). For example, the range of 11–43 Hz found for the compounds $X^1X^2P(O)N(Me)P(O)X^3X^4$ (X = Cl, NMe_2) and their corresponding disulfides[183] may be compared with the values ~300 Hz found in corresponding P^{III}–N(Me)–P^{III} derivatives.[179] Similarly ^2J(P–P) is 437 Hz in $MeN(PF_2)_2$[3] and only 32 Hz in $MeN\{P(O)F_2\}_2$.[184] However, the reduction of ^2J(P–P) on oxidation can easily be obscured by the dramatic conformational effects mentioned above. This is seen by comparing the ranges for the related ring systems, I (−10 to +50 Hz),[182] II (22 to 55 Hz, *gauche* isomers, 82–92 Hz, *trans* isomers)[185] and III (+129 to +210 Hz).[186]

However, the most intensively studied group of compounds involving P^V–N–P^V linkages are of course the cyclophosphazenes, IV, and related compounds with different ring sizes.[3,180,187] Values of ^2J(P–P) are generally positive[188] and show a very wide variation; for example, the symmetrically substituted compounds $P_3N_3X_6$, X = F (199.2 Hz), Cl (42.8 Hz), Br (4.8 Hz) have been studied recently.[189] There is a rough overall correlation with substituent electronegativity[180] and apparently some effect of ring size with derivatives of $N_4P_4Cl_8$ generally showing much smaller ^2J(P–P) values than corresponding $N_3P_3Cl_6$ derivatives.[188]

$^2J(P^{III}\text{–}N\text{–}P^V)$: The typical range for this coupling in simple acyclic compounds (+50 to +150 Hz) apparently lies between that for corresponding P^{III}–P^{III} and P^V–P^V derivatives.[190] The effect of steric bulk of substituents on nitrogen is much less pronounced than in P^{III}–P^{III} systems but conformational effects do influence the $^2J(P^{III}\text{–}P^V)$ coupling, producing more negative values when the P^V atom is essentially *trans* to the P^{III} lone pair. This effect is apparent in the cage molecules, $P_4(NMe_2)_6S_n$,

Table 8. Typical Ranges for Phosphorus–Phosphorus Coupling in Metal Complexes

Geometry/metal	^2J(P–M–P) Hz *cis*	^2J(P–M–P) Hz *trans*	References
Octahedral			
$Cr^{(0)}$	(−)78 to +79	(−)34 to 0	5, 7
$Mo^{(0)}$	(−)55 to +121	(+)49 to (+)312	5, 7, 204
$W^{(0)}$	(−)38 to +155	(+)65 to (+)315	5, 7
Rh^{III}	−35 to +9	+370 to +567	132, 150, 199, 200, 204
Ir^{II}	−22 to −15	(+)320 to +539	150, 199, 200
Pt^{IV}	−13 to −12	+547 to +660	199, 200
Square planar			
Rh^{I}	−65 to −22	(+)306 to (+)538	7, 132
Pd^{II}	−25 to +80	(+)441 to +1269	5, 7, 132, 133, 200
Pt^{II}[a]	−44 to −2	(+)364 to +775	5, 7, 132, 200, 204, 205
Trigonal bipyramidal			
Ni^{II}	60 to 100		206
Co^{II}[b]	47 to 210		207
Tetrahedral			
$Ni^{(0)}$	0 to (+)38		5, 7, 118, 208
$Pd^{(0)}$	+22 to +101		118, 208
$Pt^{(0)}$	+81 to +103		118, 208

[a] The range for *cis* Pt^{II} couplings may extend to positive values since many signs have not been determined. An early value of 90 Hz in *trans*-$[PtCl_2(PEt_3)_2]$ from line-shape analysis is almost certainly in error.[5,7]

[b] These values for Co^{II} have been subdivided into axial–axial, axial–equatorial, and equatorial–equatorial couplings, and some data for square pyramidal complexes are also available.[207]

where ^2J(P–P) varies from -15 to -28 Hz[190] and in derivatives of the ring systems, I, in which one phosphorus has been oxidized[190] or quaternized.[191]

Other ^2J(P–E–P) Couplings: These generally exhibit effects of oxidation state, substituent electronegativity, and conformation similar to those described above for X = N, and values have been collected in Ref. 3.

E = C: P^V–P^V couplings via carbon are usually smaller than P^{III}–P^{III} as illustrated by recent data for the stereo isomers of MePhP(Se)CH_2P(Se)MePh (-20.9 {*meso*} and -16.5 {*rac*} Hz) and MePhPCH_2PMePh ($+90$ {*meso*} and $+112$ {*rac*} Hz).[192] The possibility of electron delocalization also seems important since deprotonation of the methylene group in $(EtO)_2P(O)CH_2P(O)(OEt)_2$ raises ^2J(P–P) from a very small value to >25 Hz[3] and in addition couplings up to $+190$ Hz have been observed in phosphorus ylids.[174,193]

E = P: Two bond couplings via phosphorus have been measured mainly in the cyclic $(RP)_n$ compounds[163,164] and related derivatives.[168] Values range from -8 to $+3$ Hz and exhibit a strong conformational dependence.

E = O, S, Se: Couplings via oxygen are usually of small magnitude (<35 Hz) in both P^{III} and P^V derivatives,[3,42,184] and even with very electronegative substituents,[184] although a few larger values (up to 111 Hz) have been reported.[194,195] Rather few signs have been determined.[3] In contrast P^{III}–P^{III} couplings via sulfur[195,196] and selenium[197] can be very large, ranging up to $+392.9$ Hz in the low-temperature conformer of F_2P–S–PF_2.[196] P^V–P^V couplings via sulfur are much smaller, for example in $R_2P(S)$–S–$P(S)R_2$ derivatives values are ~15 Hz and probably positive,[42] and in $F_2P(S)$–S–$P(S)F_2$, 25.1 Hz and probably negative.[184]

1.4.4. ^2J(P–M–P), M = Metal

The measurement of two bond phosphorus–phosphorus coupling constants where the intervening atom is a metal has been an area of special interest for its value in making stereochemical assignments. Illustrations of the utility of the method and tabulations of data have been well documented in the review literature.[3–5,7,106]

Many early values of ^2J(P–M–P) were derived from band shape analysis for the X region of $X_nAA'X'_n$ spin systems, where X = ^{1}H or ^{19}F and A = ^{31}P. In the limit of large $J(AA')$ the X spectrum can appear as a "deceptively simple" triplet, a circumstance which led to the concept of "virtual coupling" as a criterion for *trans* stereochemistry. However, it should be noted that a number of examples exist where *cis*-^2J(P–M–P) is sufficiently large to give virtually coupled spectra.[198] In general values of ^2J(P–M–P) derived from band shape analysis should be approached with extreme caution except possibly where X = ^{19}F and the analysis is facilitated by large ^1J(A–X) couplings. Fortunately a large body of reliable data is now available from more direct measurements and a summary of the more intensively studied metals and geometries is given in Table 8. The large variations in values may be usefully discussed under these headings: (1) the complex geometry; (2) the nature of the metal; and (3) the nature of the substituents on phosphorus. The last of these is perhaps the simplest with ^2J(P–M–P) generally increasing in magnitude as electronegativity of substituents increases.[5,7,133]

The effect of complex geometry is straightforward with couplings between phosphorus atoms in mutually *trans* relationships being large and positive whereas *cis* phosphorus leads to small and often negative values. The only major exception to this generalization occurs for $Cr^{(0)}$ complexes where, although *trans* couplings are more positive than *cis*, both are of relatively small magnitude. The effect of changing the metal is more complex. It has been suggested that there is a general tendency for ^2J(P–M–P) to become more positive as one descends a transition metal triad,(7) but unfortunately this trend is perfectly obeyed only for *cis* complexes of $Cr^{(0)}$, $Mo^{(0)}$, and $W^{(0)}$. For example, *cis*-[$M(CO)_4(PH_3)_2$]: −26.2, −18.9, −13.4; [$M(CO)_4\{EtN(PF_2)_2\}$]: +78.9, +121.0, +155.3 Hz. For *trans* complexes, there is a large algebraic increase from $Cr^{(0)}$ to $Mo^{(0)}$ but between $Mo^{(0)}$ and $W^{(0)}$ the trend is erratic; for example, *trans*-[$M(CO)_4(PBr_3^n)(PPh_3)$], 25, 50, 65 Hz; *trans*-[$M(CO)_4\{P(NMe_3)_3\}_2$], −17, +101, +81 Hz. In other triads data are limited but both *cis* (negative) and *trans* (positive) couplings via Ir^{III} seem slightly smaller in magnitude than via Rh^{III}.(113,199) There is a considerable body of data for Pd and Pt complexes, and couplings via Pd are usually more positive than those via Pt; e.g., *trans*-[$MCl_2(PMe_3)_2$], +610, +510; *cis*-[$MCl_2\{P(OMe)_3\}_2$], +78, −18 Hz.(200) Unfortunately the series has rarely been extended to Ni(118,201) but the limited data do indicate more negative couplings: e.g., *trans*-[$MI_2(PMe_2H)_2$], 355 (Ni), 447 (Pt);(5) M[$PF(OPh)_2]_4$, +21.7 (Ni), +102.5 (Pt);(118) [$MCl_2(Me_2PCH_2CH_2PPh_2)$], 76 (Ni), 10 (Pd), <2 (Pt).(201) The signs were not determined for the last example but to be consistent with all of the foregoing should be negative for Ni and positive for Pd. In general we may conclude that the most usual algebraic sequence of ^2J(P–M–P) values within a transition metal triad is 1st row ≪ 2nd row > 3rd row but that considerable variations exist between the second and third row.

Although the theoretical situation is not in general very satisfactory, some success in rationalizing the above trends has been achieved in terms of an extended Fermi contact interaction. Gross simplifying assumptions are necessary and the results have little predictive value.(7)

Another aspect of ^2J(P–M–P) values which has attracted some attention is the relative weighting of the contributions via the metal and via the ligand "backbone" in phosphorus chelate complexes. Authors working on Cr, Mo, and W complexes, [$M(CO)_4L_2$] where $L_2 = Ph_2P(CH_2)_nOPPh_2$(202) or $Ph_2P(CH_2)_nPR'R''$,(203) have suggested that the observed couplings can be divided into metal, ^2J(P–M–P), and backbone, ^nJ(PP), contributions where the latter is assumed to be similar to the coupling in the free ligand. The magnitudes of the couplings would then require the two contributions to be of opposite sign, for example, $Ph_2PCH_2PR'R''$ ligands have average J(PP) 114 Hz in the free ligand compared with 24, 15, and 21 Hz in Cr, Mo, and W complexes, respectively.(203) However, other workers have pointed out that a similar interpretation for MCl_2L_2 complexes, where M = Ni, Pd, or Pt and $L_2 = Me_2PCH_2CH_2PPh_2$ or $Ph_2PCH{=}C(R)PR'R''$, would lead to "metal" contributions quite outside the normal range for ^2J(P–M–P) couplings between monodentate ligands.(201) This suggests that in fact the "backbone" contribution is reduced on coordination and that the "metal" contribution dominates. We may also note that such a reduction of ^nJ(PP) "backbone" would be expected from our discussion (Sections 1.4.2 and 1.4.3) of the normal differences between P^{III}–P^{III} and P^V–P^V coupling.

1.4.5. Long-Range P–P Coupling

Phosphorus–phosphorus couplings over more than two bonds are often less than the spectrometer resolution.[3] However, a number of examples of large three-bond couplings are known ranging up to 212 Hz[209] and in naphthalene derivatives bearing two $CH_2P(O)(OEt)_2$ groups, detectable coupling has been observed over as many as nine bonds.[210] In derivatives of the cage compounds $[P(NMeNMe)_3P]$ the magnitude of $^3J(P\text{–}P)$ increases in the sequence $P^{III}\text{–}P^{III} < P^{III}\text{–}P^{V} < P^{V}\text{–}P^{V}$.[211]

2. ARSENIC, ANTIMONY, AND BISMUTH

None of these elements have been the subject of any extensive NMR study. This is somewhat surprising in that all three possess NMR active nuclei of high abundance and relatively high sensitivity. Thus ^{75}As is 100% abundant with a receptivity of 2.5×10^{-2} relative to the proton; ^{121}Sb is 57% abundant with receptivity 9.2×10^{-2}, and ^{209}Bi is 100% abundant with receptivity 0.14. Somewhat less favorable is ^{123}Sb with receptivity 1.9×10^{-2}. The high spin quantum numbers (^{75}As, $I = 3/2$; ^{121}Sb, $I = 5/2$; ^{123}Sb, $I = 7/2$; ^{209}Bi, $I = 9/2$) and the associated quadrupole moments lead to short relaxation times, broad NMR lines, and difficulties of observation with all but symmetrical molecules. Nevertheless, it does seem that these nuclei may repay further study.

The information available on arsenic, antimony, and bismuth NMR has been well documented in two reviews[17,212] and a summary of the main data is given in Table 9. There appears to be very little more recent work except for some further applications of ^{75}As in solid-state physics[213]; observations of phase transitions in solid AsF_6^- salts[214]; measurements of quadrupole coupling constants in aluminum halide adducts with antimony and bismuth halides[215]; application of the pairwise shift additivity model to assign configurations and ^{121}Sb shifts in the series, $SbBr_{6-n}Cl_n^-$ [216]; and measurements of ^{209}Bi shifts and relaxation times in $Bi/BiBr_3$ melts.[217]

Table 9. NMR Parameters for Compounds of ^{75}As, ^{121}Sb, and ^{209}Bi[a]

Molecule or ion	δ (ppm)	^1J (Hz)	T_1 (μs)	$W_{1/2}$ (Hz)
AsH_4^+	−291	555		115
AsF_6^-	0	930	57–8350	94
AsR_4^+	206–258			113–1300
AsO_4^{3-}	369			1125
$Me_3N \cdot AsF_5$		840 (axial) 1048 (equatorial)		
AsH_3		92.1	48–95	
$SbBr_6^-$	−2436			300
$SbCl_6^-$	0			300
SbF_6^-	88	1820–1950		300
$Sb(OH)_6^-$	296			1800
$SbCl_5$	509			8000
$Sb(CH_3)_4^+$	780			
SbS_4^{3-}	1032			550
BiF_6^-		2700		

[a] Data from Refs. 17 and 212.

REFERENCES

1. Crutchfield, M. M.; Dungan, C. H.; Van Wazer, J. R. *Top. Phosphorus Chem.* 1967, **5**, 1.
2. Mavel, G. *Annu. Rep. NMR Spectrosc.* 1973, **5B**, 1.
2a. Gorenstein, D. *Prog. NMR Spectrosc.* 1983, **16**, 1.
3. Finer, E. G.; Harris, R. K. *Prog. NMR Spectrosc.* 1971, **6**, 61.
4. Nixon, J. F.; Pidcock, A. *Annu. Rev. NMR Spectrosc.* 1969, **2**, 346.
5. Verkade, J. G. *Coord. Chem. Rev.* 1972, **9**, 1.
6. Appleton, T. G.; Clark, H. C.; Manzer, L. E. *Coord. Chem. Rev.* 1973, **10**, 378.
7. Pregosin, P. S.; Kunz, R. W. *^{31}P and ^{13}C NMR of Transition Metal Phosphorus complexes*, No. 16 in series *NMR Basic Principles and Progress*, Diehl, P.; Fluck, E.; Kosfeld, R., Eds.; Springer-Verlag: New York, 1979.
8. *Top. Phosphorus Chem.*, Grayson, M.; Griffith, E. J., Eds.; Interscience: New York, Vols. 1–.
9. *Spec. Per. Rep.: NMR*; R.S.C.: London, Vols. 1–.
10. *Spec. Per. Rep.: Organophosphorus Chemistry*; R.S.C.: London, Vols. 1–.
11. *Spec. Per. Rep.: Spectrosc. Prop. Inorg. Organomet. Compounds*; Royal Society of Chemistry: London, Vols. 1–.
12. O'Neill, I. K.; Richards, C. P. *Annu. Rep. NMR Spectrosc.* 1980, **10A**, 134. Gadian, D. G.; Radda, G. K.; Richards, R. E.; Seeley, P. J. In *Biological Applications of Magnetic Resonance*, Shulman, R. G., Ed.; Academic Press: New York, 1979; p. 463. Hollis, D. P. In *Biological Magnetic Resonance*, Berliner, L. J.; Rouben, J., Eds.; Plenum: New York, 1980; Vol. 2, p. 1.
12a. *Phosphorus NMR, Principles and Applications*, Gorenstein, D., Ed.; Academic Press: Orlando, Florida, 1984.
13. Yeagle, P. L.; Hutton, W. C.; Martin, R. B. *J. Am. Chem. Soc.* 1975, **97**, 7175.
14. Appleman, B. R.; Dailey, B. P. *Adv. Magn. Reson.* 1974, **7**, 279.
15. Albrand, J. P.; Cogne, A.; Robert, J. B. *J. Am. Chem. Soc.* 1978, **100**, 2600.
16. Kennedy, J. D.; McFarlane, W. *J. Chem. Soc., Chem. Commun.* 1976, 666. Wong, T. C.; Ashe, A. J. *J. Mol. Struct.* 1978, **48**, 219. Grimmer, A. R. *Z. Chem.* 1978, **18**, 233.
17. *NMR and the Periodic Table*, Harris, R. K.; Mann, B. E., Eds.; Academic Press: London, 1978.
18. Murray, M.; Schmutzler, R.; Grundemann, E.; Teichmann, H. *J. Chem. Soc. (B)* 1971, 1714.
19. Robert, J. B.; Taieb, M. C.; Tabony, J. *J. Magn. Reson.* 1980, **38**, 99.
20. Tabony, J. *Spectrochim. Acta, Part A* 1979, **35**, 217.
21. Harris, R. K.; McVicker, E. M. *J. Chem. Soc., Faraday Trans. 2* 1976, **72**, 2291.
22. Dale, S. W.; Hobbs, M. E. *J. Phys. Chem.* 1971, **75**, 3537.
23. Mowthorpe, D. J.; Chapman, A. C. *Mol. Phys.* 1968, **15**, 429.
24. Aksnes, D. W. *Acta. Chem. Scand.* 1969, **23**, 1078.
25. Sawyer, D. W.; Powles, J. G. *Mol. Phys.* 1971, **21**, 83.
26. Boden, N.; Folland, R. *Chem. Phys. Lett.* 1975, **32**, 127.
27. Seymour, S. J.; Jonas, J. *J. Magn. Reson.* 1972, **8**, 376.
28. Koole, N. J.; DeKoning, A. J.; De Bie, M. J. A. *J. Magn. Reson.* 1977, **25**, 375.
29. Jordan, A. D.; Cavell, R. G.; Jordan, R. B. *J. Chem. Phys.* 1972, **56**, 483.
30. Aksnes, D. W.; Rhodes, M.; Powles, J. G. *Mol. Phys.* 1968, **14**, 333.
31. Rhodes, M.; Aksnes, D. W.; Strange, J. H. *Mol. Phys.* 1968, **15**, 541.
32. Leipert, T. K.; Freeman, W. J.; Noggle, J. H. *J. Chem. Phys.* 1975, **63**, 4177.
33. Seymour, S. J.; Jonas, J. *J. Chem. Phys.* 1971, **54**, 487.
34. Wilkie, C. A. *J. Magn. Reson.* 1979, **33**, 127.
35. Wilkie, C. A. *J. Am. Chem. Soc.* 1978, **100**, 2288.
36. Ramarajan, K.; Herd, M. D.; Berlin, K. D. *Phosphorus Sulfur* 1981, **11**, 199.
37. Lui, N. I.; Jonas, J. *J. Chem. Phys.* 1971, **55**, 463.
38. Beierbeck, H.; Martino, R.; Saunders, J. K.; Benezra, C. *Can. J. Chem.* 1976, **54**, 1918.
39. Glonek, T.; Van Wazer, J. R. *J. Phys. Chem.* 1976, **80**, 639.
40. McCain, D. C.; Markley, J. L. *J. Am. Chem. Soc.* 1980, **102**, 5559.
41. Pregosin, P. S.; Bosch, W. *Helv. Chim. Acta* 1979, **62**, 838.
42. Harris, R. K.; McVicker, E. M.; Hagele, G. *J. Chem. Soc., Dalton Trans.* 1978, 9.
43. Morgan, W. E.; Van Wazer, J. R. *J. Am. Chem. Soc.* 1975, **97**, 6347.
43a. Huttner, G.; Borm, J.; Zsolnai, L. *J. Organomet. Chem.* 1984, **263**, C33.
43b. Chini, P.; Ciani, G.; Martinengo, S.; Sironi, A.; Longhetti, L.; Heaton, B. T. *J. Chem. Soc., Chem. Commun.* 1979, 188.
44. Fluck, E. *Top. Phosphorus Chem.* 1980, **10**, 193.

44a. Cowiey, A. H.; Cushner, M. C.; Lattman, M.; McKee, M. L.; Szobota, J. S.; Wilburn, J. C. *Pure Appl. Chem.* 1980, **52**, 789.
44b. Cowley, A. H.; Kemp, R. A. *Chem. Rev.* 1985, **85**, 367.
44c. Cowley, A. H.; Kilduff, J. E.; Lasch, J. G.; Mehrotra, S. K.; Norman, N. C.; Pakulski, M.; Whittlesey, B. R.; Atwood, J. L.; Hunter, W. E. *Inorg. Chem.* 1984, **23**, 2582.
44d. Lang, H.; Orama, O.; Huttner, G. *J. Organomet. Chem.* 1985, **291**, 293.
45. Appel, R.; Knoll, F.; Ruppert, I. *Agnew. Chem., Int. Ed. Engl.* 1981, **20**, 731.
46. Glonek, T.; Van Wazer, J. R. *J. Magn. Reson.* 1974, **13**, 390.
47. Olah, G. A.; McFarland, C. W. *J. Org. Chem.* 1969, **34**, 1832.
48. Mann, B. E. *J. Chem. Soc., Perkin Trans. 2* 1972, 30.
49. Grim, S. O.; Davidoff, E. F.; Marks, T. J. *Z. Naturforsch., Teil B* 1971, **26**, 184.
50. Schiemenz, G. P. *Phosphorus* 1973, **3**, 125.
51. Ashe, A. J.; Sharp, R. R.; Tolan, J. W. *J. Am. Chem. Soc.* 1976, **98**, 5451.
52. Quin, L. D.; Orton, W. L. *J. Chem. Soc., Chem. Commun.* 1979, 401. Bangerter, B. W.; Beatty, R. P.; Kouba, J. K.; Wreford, S. S. *J. Org. Chem.* 1977, **42**, 3247.
53. Grim, S. O.; Molenda, R. P. *Phosphorus* 1974, **4**, 189.
54. Schlosser, M.; Kardibelhan, T.; Steinhoff, G. *Annalen.* 1971, **743**, 25.
55. Burger, K.; Ottlinger, R.; Frank, A.; Schubert, U. *Agnew. Chem., Int. Ed. Engl.* 1978, **17**, 774. Michalski, J., Mikolajczak, J.; Pakulski, M.; Skowronska, A. *Phosphorus Sulfur* 1978, **4**, 233.
56. Schmutzler, R. *Adv. Fluorine Chem.* 1965, **5**, 31. Holmes, R. R. "Penta-Coordinated Phosphorus, Vol. 1," ACS Monograph No. 175: Washington, 1980.
57. Arbuzov, B. A.; Polezhaeva, N. A. *Russ. Chem. Rev.* 1974, **43**, 414.
58. Chang, B. C.; Conrod, W. E.; Denney, D. B.; Denney, D. Z.; Edelman, R.; Powell, R. L.; White, D. W. *J. Am. Chem. Soc.* 1971, **93**, 4004.
59. Denney, D. B.; Denney, D. Z.; Chang, B. C.; Marsi, K. L. *J. Am. Chem. Soc.* 1969, **91**, 5243.
60. Hall, C. D.; Bramblett, J. D.; Lin, F. F. S. *J. Am. Chem. Soc.* 1972, **94**, 9264.
61. Ramirez, F.; Prasad, V. A. V.; Maracek, J. F. *J. Am. Chem. Soc.* 1974, **96**, 7269.
62. Dillon, K. B.; Lynch, R. J.; Reeve, R. N.; Waddington, T. C. *J. Inorg. Nucl. Chem.* 1974, **36**, 815.
63. Sheldrick, W. S.; Hewson, M. J. C. *Z. Naturforsch., Teil B* 1978, **33**, 834. Skowronska, A.; Pakulski, M.; Michalski, J. *J. Am. Chem. Soc.* 1979, **101**, 7412.
64. Seel, F.; Bassler, H. J. *Z. Anorg. Allg. Chem.* 1975, **418**, 263.
65. Dillon, K. B.; Dillon, M. G. C.; Waddington, T. C. *Inorg. Nucl. Chem. Lett.* 1977, **13**, 349.
66. Rosenberg, D.; Drenth, W. *Tetrahedron* 1971, **27**, 3893.
67. Fild, M.; Glemser, O. *Fluorine Chem. Rev.* 1969, **3**, 129.
68. Mitsch, C. C.; Freedman, L. D.; Moreland, C. G. *J. Magn. Reson.* 1970, **3**, 446.
69. Grim, S. O.; Yankowsky, A. W. *Phosphorus Sulfur* 1977, **3**, 191.
70. Grim, S. O.; McFarlane, W.; Davidoff, E. F. *J. Org. Chem.* 1967, **32**, 781.
71. Maier, L. *Helv. Chim. Acta* 1966, **49**, 1718.
72. Grim, S. O.; McFarlane, W. *Can. J. Chem.* 1968, **46**, 2071.
73. Grim, S. O.; Molenda, R. P.; Mitchell, J. D. *J. Org. Chem.* 1980, **45**, 250.
74. Quin, L. D.; Breen, J. J. *Org. Magn. Reson.* 1973, **5**, 17.
75. Gorenstein, D. G. *J. Am. Chem. Soc.* 1977, **99**, 2254.
76. Pinnell, R.; Megerle, C. A.; Manatt, S. L.; Kroon, P. A. *J. Am. Chem. Soc.* 1973, **95**, 977.
77. Gordon, M. D.; Quin, L. D. *J. Am. Chem. Soc.* 1976, **98**, 15. Ali, A. A. M.; Bocelli, G.; Harris, R. K.; Fild, M. *J. Chem. Soc., Dalton Trans.* 1980, 638. Verkade, J. G. *Phosphorus Sulfur* 1976, **2**, 251.
78. Latscha, H. P. *Z. Naturforsch., Teil B* 1968, **23**, 139.
79. Van Linthoudt, J. P.; Van den Berghe, E. V.; Van der Kelen, G. P. *Spectrochim. Acta, Part A* 1979, **35**, 1307.
80. Burg, A. B. *Inorg. Nucl. Chem. Lett.* 1977, **13**, 199.
81. Dillon, K. B.; Dillon, M. G. C.; Waddington, T. D. *J. Inorg. Nucl. Chem.* 1976, **38**, 1149.
82. Dillon, K. B.; Waddington, T. C. *Inorg. Nucl. Chem. Lett.* 1974, **10**, 777.
83. Van Linthoudt, J. P.; Van den Berghe, E. V.; Van der Kelen, G. P. *Spectrochim. Acta, Part A* 1980, **36**, 315.
84. Burgada, R. *Bull. Soc. Chim. Fr.* 1971, 136.
85. Razumov, A. I.; Krasil'nikova, E. A.; Zykova, T. V.; Sinitsyna, N. I.; Salakhutdinov, R. A.; Bankovskaya, N. N. *J. Gen. Chem. USSR (Engl. Transl.)* 1972, **42**, 1245.
86. Chekunina, L. I.; Bokanov, A. I.; Stepanov, B. I. *J. Gen. Chem. USSR, (Engl. Transl.)* 1972, **42**, 985.
87. Fluck, E.; Seng, N. *Z. Anorg. Allg. Chem.* 1972, **393**, 126.
88. Schmidpeter, A.; Brecht, H. *Inorg. Nucl. Chem. Lett.* 1968, **4**, 563.

89. Van Linthoudt, J. P.; Van den Berghe, E. V.; Van der Kelen, G. P. *Spectrochim. Acta, Part A* 1980, **36**, 17.
90a. Radeglia, R.; Teichmann, H. *Z. Chem.* 1974, **14**, 282.
90b. Robinson, E. A.; Lavery, D. S. *Spectrochim. Acta, Part A* 1972, **28**, 1099.
91. Buder, W.; Schmidt, A. *Spectrochim. Acta, Part A* 1974, **30**, 1311.
92. Pressl, K. D.; Schmidt, A. *Z. Anorg. Allg. Chem.* 1977, **434**, 175.
93. Vedejs, E.; Snoble, K. A. J.; Fuchs, P. L. *J. Org. Chem.* 1973, **38**, 1178. Vedejs, E.; Snoble, K. A. J. *J. Am. Chem. Soc.* 1973, **95**, 5778.
94. Blackburn, G. M.; Cohen, J. S.; Weatherall, I. *Tetrahedron* 1971, **27**, 2903.
95. Allen, D. W.; Tebby, J. C. *J. Chem. Soc. (B)* 1970, 1527.
96. Makarova, N. A.; Mukmenev, E. T.; Arbuzov, B. A. *Bull. Acad. Sci. USSR, Div. Chem. Sci.* 1974, **23**, 1769.
97. Gorenstein, D. G. *J. Am. Chem. Soc.* 1975, **97**, 898.
98. Quast, H.; Heuschmann, M. *Angew. Chem., Int. Ed. Engl.* 1978, **17**, 867.
99. Bertrand, R. D.; Verkade, J. G.; White, D. W.; Gagnaire, D.; Robert, J. B.; Verrier, J. *J. Magn. Reson.* 1970, **3**, 494.
100. Tavs, P. *Angew. Chem., Int. Ed. Engl.* 1969, **8**, 751.
101. Breen, J. J.; Engel, J. F.; Myers, D. K.; Quin, L. D. *Phosphorus* 1972, **2**, 55.
102. Quin, L. D.; Bryson, J. G.; Moreland, C. G. *J. Am. Chem. Soc.* 1969, **91**, 3308.
103. Chan, S.; Goldwhite, H.; Keyzer, H.; Rowsell, D. G.; Tang, R. *Tetrahedron* 1969, **25**, 1097.
104. Smith, L. R.; Mills, J. L. *J. Am. Chem. Soc.* 1976, **98**, 3852.
105. Baudler, M.; Carlsohn, B.; Kloth, B.; Koch, D. *Z. Anorg. Allg. Chem.* 1977, **432**, 67.
106. Pidcock, A. In *Catalytic Aspects of Metal Phosphine Complexes*, No. 196 in series *Advances in Chemistry*, Alyea, E. C.; Meek, D. W., Eds.; American Chemical Society: Washington, 1982; p. 1. Stelzer, O. *Top. Phosphorus Chem.* 1977, **9**, 1. Tolman, C. A. *Chem. Rev.* 1977, **77**, 313. Meek, D. W.; Mazanec, T. J. *Acc. Chem. Res.* 1981, **14**, 266.
107. Garrou, P. E. *Chem. Rev.* 1981, 229.
108. Moser, E.; Fischer, E. O.; Bathelt, G.; Gretner, L.; Knauss, L.; Louis, E. *J. Organomet. Chem.* 1969, **19**, 377.
109. Mann, B. E.; Masters, C.; Shaw, B. L.; Slade, R. M.; Stainbank, R. E. *Inorg. Nucl. Chem. Lett.* 1971, **7**, 881, and references therein.
110. Grim, S. O.; Wheatland, D. A.; McFarlane, W. *J. Am. Chem. Soc.* 1967, **89**, 5573.
111. Gill, D. F.; Mann, B. E.; Shaw, B. L. *J. Chem. Soc., Dalton Trans.* 1973, 311.
112. Mann, B. E.; Masters, C.; Shaw, B. L. *J. Chem. Soc. (A)* 1971, 1104.
113. Mann, B. E.; Masters, C.; Shaw, B. L. *J. Chem. Soc., Dalton Trans.* 1972, 48 and 704.
114. Mann, B. E.; Shaw, B. L.; Slade, R. M. *J. Chem. Soc (A)* 1971, 2976.
115. Grim, S. O.; Keiter, R. L.; McFarlane, W. *Inorg. Chem.* 1967, **6**, 1133.
116. Tolman, C. A.; Siedel, W. C.; Gerlach, D. H. *J. Am. Chem. Soc.* 1972, **94**, 2669.
117. Gerlach, D. H.; Kane, A. R.; Parshall, G. W.; Jesson, J. P.; Muetterties, E. L. *J. Am. Chem. Soc.* 1971, **93**, 3543.
118. Crocker, C.; Goodfellow, R. J. *J. Chem. Soc., Dalton Trans.* 1977, 1687.
119. Tolman, C. A.; Siedel, W. C.; Gosser, L. W. *J. Am. Chem. Soc.* 1974, **96**, 53.
120. Reddy, G. S.; Schmutzler, R. *Inorg. Chem.* 1967, **6**, 823.
121. Micoud, M. H.; Savariault, J. M.; Cassoux, P. *Bull. Soc. Chim. Fr.* 1972, 3774.
122. Tolman, C. A. *J. Am. Chem. Soc.* 1970, **92**, 2956.
123. Nixon, J. F.; Sexton, M. D. *J. Chem. Soc. (A)* 1969, 1089.
124. Mann, B. E.; Musco, A. *J. Chem. Soc., Dalton Trans.* 1975, 1673.
125. Kruck, T.; Baur, K. *Z. Anorg. Allg. Chem.* 1969, **364**, 192.
126. Pregosin, P. S.; Sze, S. N. *Helv. Chim. Acta.* 1977, **60**, 1371.
127. Mann, B. E.; Musco, A. *J. Chem. Soc., Dalton Trans.* 1980, 776.
128. Al-Ohaly, A. R.; Nixon, J. F. *Inorg. Chim. Acta* 1980, **47**, 105.
129. Verstuyft, A. W.; Nelson, J. H.; Cary, L. W. *Inorg. Nucl. Chem. Lett.* 1976, **12**, 53.
130. Couch, D. A.; Robinson, S. D.; Wingfield, J. N. *J. Chem. Soc., Dalton Trans.* 1974, 1309.
131. Ahmad, N.; Ainscough, E. W.; James, T. A., Robinson, S. D. *J. Chem. Soc., Dalton Trans.* 1973, 1148.
132. Crocker, C.; Goggin, P. L. *J. Chem. Res. (M)* 1981, 701 and 742.
133. Verstuyft, A. W.; Redfield, D. A.; Gary, L. W.; Nelson, J. H. *Inorg. Chem.* 1976, **15**, 1128.
134. Garrou, P. E.; Hartwell, G. E. *Inorg. Chem.* 1976, **15**, 646.
135. Crocker, C.; Goggin, P. L.; Goodfellow, R. J. *J. Chem. Soc., Dalton Trans.* 1976, 2494.
136. Mather, G. G.; Pidcock, A.; Rapsey, G. J. N. *J. Chem. Soc., Dalton Trans.* 1973, 2095.

137. McFarlane, W.; White, R. F. M. *J. Chem. Soc., Chem. Commun.* 1969, 744. Grim, S. O.; McFarlane, W. *Can. J. Chem.* 1968, **46**, 2071. Vande Griend, L. J.; Verkade, J. G. *J. Am. Chem. Soc.* 1975, **97**, 5958.
138. Keiter, R. L.; Verkade, J. G. *Inorg. Chem.* 1969, **8**, 2115.
139. Mathieu, R.; Lenzi, M.; Poilblanc, R. *Inorg. Chem.* 1970, **9**, 2030.
140. Milbrath, D. S.; Verkade, J. G. *Inorg. Nucl. Chem. Lett.* 1976, **12**, 921.
141. Warrens, C., unpublished results.
142. Burckett St. Laurent, J. C.; D. Phil Thesis, University of Sussex, 1982.
143. Nixon, J. F. *Adv. Inorg. Rad. Chem.* 1970, **13**, 363. Nixon, J. F., unpublished observations.
144. Deberity, J.; Noth, H. *J. Organomet. Chem.* 1973, **49**, 453.
145. Johnstone, T. A.; Nixon, J. F.; unpublished results.
146. Kroto, H. W.; Nixon, J. F.; Taylor, M. J.; Frew, A.; Muir, K. W. *Polyhedron* 1982, **1**, 89.
147. Dixon, K. R.; Rattray, A. D. *Inorg. Chem.* 1978, **17**, 1099. Brandon, J. B.; Dixon, K. R. *Can. J. Chem.* 1981, **59**, 1188.
148. Mott, G. N.; Carty, A. J. *Inorg. Chem.* 1979, **18**, 2926. Carty, A. J. In *Catalytic Aspects of Metal Phosphine Complexes*, No. 196 in series *Advances in Chemistry*, Alyea, E. C.; Meek, D. W., Eds.; American Chemical Society: Washington, 1982; p. 164.
149a. Petersen, J. L.; Stewart, R. P. *Inorg. Chem.* 1980, **19**, 186.
149b. Keiter, R. L.; Madigan, M. J. *Organometallics* 1982, **1**, 409.
149c. Berry, D. E.; Bushnell, G. W.; Dixon, K. R.; Moroney, P. M.; Wan, C. *Inorg. Chem.* 1985, **24**, 2625.
150. McFarlane, H. C. E.; McFarlane, W. *J. Chem. Soc., Chem. Commun.* 1978, 531.
151. Brazier, J. F.; Houalla, D.; Loening, M.; Wolf, R. *Top. Phosphorus Chem.* 1976, **8**, 99.
152. Muylle, E.; Van der Kelen, G. P.; Claeys, E. G. *Spectrochim. Acta, Part A* 1976, **32**, 1149.
153. *Interpretation of Carbon-13 NMR Spectra*, Wehrli, F. W.; Wirthlin, T., Eds.; Heyden and Son: London, 1978, p. 60.
154. Witanowski, M.; Stefaniuk, G.; Webb, G. A. *Annu. Rev. NMR Spectrosc.* 1981, **11B**, 1. Mason, J. *Chem. Rev.* 1981, **81**, 205.
155. Emsley, J. W.; Phillips, L.; Wray, V. *Prog. NMR Spectrosc.* 1977, **10**, 645–701. Nixon, J. F., *Adv. Inorg. Rad. Chem.* 1970, **13**, 363. Wray, V. *Ann. Rep. NMR Spectrosc.* 1980, **10B**, 1, and earlier volumes in this series.
156. Falius, H.; Murray, M. *J. Magn. Reson.* 1973, **10**, 127.
157. McFarlane, H. C. E.; McFarlane, W.; Nash, J. A. *J. Chem. Soc., Dalton Trans.* 1980, 240.
158. Schultz, C. W.; Rudolph, R. W. *J. Am. Chem. Soc.* 1971, **93**, 1898.
159. Aime, S.; Harris, R. K.; McVicker, E. M.; Fild, M. *J. Chem. Soc., Dalton Trans.* 1976, 2144.
160. Albrand, J. P.; Faucher, H.; Gagnaire, D.; Robert, J. B. *Phys. Chem. Lett.* 1976, **38**, 521.
161. Albrand, J. P.; Gagnaire, D. *J. Am. Chem. Soc.* 1972, **94**, 8630.
162. Albrand, J. P.; Robert, J. B.; Goldwhite, H. *Tetrahedron Lett.* 1976, 949.
163. Albrand, J. P.; Gagnaire, D.; Robert, J. B. *J. Am. Chem. Soc.* 1973, **95**, 6498.
164. Albrand, J. P.; Robert, J. B. *J. Chem. Soc., Chem. Commun.* 1974, 644.
165. Issleib, K.; Thorausch, P.; Meyer, H. *Org. Magn. Reson.* 1977, **10**, 172.
166. Harris, R. K.; Norval, E. M.; Fild, M. *J. Chem. Soc., Dalton Trans.* 1979, 826.
167. Foss, U. L.; Veits, Y. A.; Kudinova, V. V.; Borisenko, A. A.; Lutsenko, I. F. *J. Gen. Chem. USSR (Engl. Transl.)* 1973, **43**, 994.
168. Hoffman, P. R.; Caulton, K. G. *Inorg. Chem.* 1975, **14**, 1997. Baudler, M.; Tolls, E., Clef, E.; Kloth, B.; Koch, D. *Z. Anorg. Allg. Chem.* 1977, **435**, 21.
169. Albrand, J. P.; Cogne, A.; Robert, J. B. *Chem. Phys. Lett.* 1976, **42**, 498.
170. Hodges, H. L.; Rudolph, R. W. *Inorg. Chem.* 1972, **11**, 2845.
171. Baudler, M.; Koch, D., Carlsohn, B. *Chem. Ber.* 1978, **111**, 1217.
172. Hoffman, P. R.; Caulton, K. G. *J. Am. Chem. Soc.* 1975, **97**, 6370.
173. Okruszek, A.; Stec, W. J.; Harris, R. K. *Org. Magn. Reson.* 1977, **9**, 497.
174. Karsch, H. H. *Z. Naturforsch., Teil B* 1979, **34**, 1171.
175. Cowley, A. H.; Cushner, M. C. *Inorg. Chem.* 1980, **19**, 515.
176. Appel, R.; Milker, R. *Chem. Ber.* 1977, **110**, 3201.
177. Abrahams, K. M.; Van Wazer, J. R. *Phosphorus* 1975, **6**, 23.
178. Beurich, H.; Vahrenkamp, H. *J. Chem. Res. (M)* 1977, 1066.
179. Colquhoun, I. J.; McFarlane, W. *J. Chem. Soc., Dalton Trans.* 1977, 1674.
180. Allcock, H. R. *Phosphorus-Nitrogen Compounds*; Academic Press: New York, 1972; pp. 73–85.
181. Cross, R. J.; Green, T. H.; Keat, R. *J. Chem. Soc., Dalton Trans.* 1976, 1424.
182. Keat, R.; Thompson, D. G. *J. Chem. Soc., Dalton Trans.* 1978, 634.

183. Hagele, G.; Harris, R. K.; Wazeer, M. I. M.; Keat, R. *J. Chem. Soc., Dalton Trans.* 1974, 1985.
184. Charlton, T. L.; Cavell, R. G. *Inorg. Chem.* 1972, **11**, 1583.
185. Harris, R. K.; Wazeer, M. I. M.; Schlak, O.; Schmutzler, R. *J. Chem. Soc., Dalton Trans.* 1976, 17.
186. Harris, R. K.; Wazeer, M. I. M. *J. Chem. Soc., Dalton Trans.* 1976, 302.
187. Krishnamurthy, S. S.; Sau, A. C.; Woods, M. *Adv. Inorg. Rad. Chem.* 1978, **21**, 41. Allcock, H. R. *Chem. Rev.* 1972, **72**, 315.
188. Keat, R.; Shaw, R. A.; Woods, M. *J. Chem. Soc., Dalton Trans.* 1976, 1582. Krishnamurthy, S. S.; Ramachandran, K.; Sau, A. C.; Sudheendra Rao, M. N.; Vasudeva Murthy, A. R.; Keat, R.; Shaw, R. A. *Phosphorus Sulfur* 1978, **5**, 117.
189. Thomas, B.; Grossmann, G. *J. Magn. Reson.* 1979, **36**, 333.
190. Bulloch, F.; Keat, R.; Rycroft, D. S.; Thompson, D. G. *Org. Magn. Reson.* 1979, **12**, 708.
191. Zeiss, W.; Feldt, C.; Weis, J.; Dunkel, G. *Chem. Ber.* 1978, **111**, 1180.
192. Colquhoun, I. J.; McFarlane, W. *J. Chem. Res. (S)* 1978, 368.
193. Harris, R. K.; Woplin, J. R.; Issleib, K.; Lindner, R. *J. Magn. Reson.* 1972, **7**, 291.
194. Foss, V. F.; Veits, Y. A.; Lutsenko, I. F. *Phosphorus Sulfur* 1977, **3**, 299.
195. Kang, D. K.; Servis, K. L.; Burg, A. B. *Org. Magn. Reson.* 1971, **3**, 101.
196. Rudolph, R. W.; Newmark, R. A. *J. Am. Chem. Soc.* 1970, **92**, 1195.
197. Arnold, D. E. J.; Dryburgh, J. S.; Ebsworth, E. A. V.; Rankin, D. W. H. *J.Chem. Soc., Dalton Trans.* 1972, 2518.
198. Bancroft, G. M.; Libbey, E. T. *Can. J. Chem.* 1973, **51**, 1482.
199. Goggin, P. L.; Goodfellow, R. J.; Knight, J. R.; Norton, M. G.; Taylor, B. F. *J. Chem. Soc., Dalton Trans.* 1973, 2220.
200. Goodfellow, R. J.; Taylor, B. F. *J. Chem. Soc., Dalton Trans.* 1974, 1676.
201. Carty, A. J.; Johnson, D. K.; Jacobsen, S. E. *J. Am. Chem. Soc.* 1979, **101**, 5612.
202. Grim, S. O.; Briggs, W. L.; Barth, R. C.; Tolman, C. A.; Jesson, J. P. *Inorg. Chem.* 1974, **13**, 1095.
203. Grim, S. O.; Barth, R. C.; Mitchell, J. D.; Del Gaudio, J. *Inorg. Chem.* 1977, **16**, 1776.
204. McFarlane, W.; Rycroft, D. S. *J. Chem. Soc., Faraday Trans. 2* 1974, **70**, 377.
205. Goodfellow, R. J.; Hardy, M. J.; Taylor, B. F. *J. Chem. Soc., Dalton Trans.* 1973, 2450.
206. Meier, P. F.; Merback, A. E.; Dartiguenave, M.; Dartiguenave, Y. *Inorg. Chim. Acta.* 1980, **39**, 19.
207. Du Bois, D. L.; Meek, D. W. *Inorg. Chem.* 1976, **15**, 3076.
208. Crocker, C.; Goodfellow, R. J. *J. Chem. Res. (S)* 1979, 378.
209. Bertrand, R. D.; Allison, D. A.; Verkade, J. G. *J. Am. Chem. Soc.* 1970, **92**, 71. Maier, L. *Phosphorus* 1973, **2**, 229. Dean, P. A. W. *Can. J. Chem.* 1979, **57**, 754. Hill, W. E.; Silva-Trivino, L. M. *Inorg. Chem.* 1979, **18**, 361. Grossman, G.; Walther, B.; Gastrock-Mey, U. *Phosphorus Sulfur* 1981, **11**, 259.
210. Ernst, L. *J. Chem. Soc., Chem. Commun.* 1977, 375. Ernst, L. *Org. Magn. Reson.* 1977, **9**, 35.
211. Kroshefsky, R. D.; Verkade, J. G. *Phosphorus Sulfur* 1979, **6**, 397.
212. Wehrli, F. W. *Annu. Rep. NMR Spectrosc.* 1979, **9**, 177.
213. Dumas, K. A.; Soest, J. F.; Sher, A.; Swiggard, E. M. *Phys. Rev. B* 1979, **20**, 4406.
214. Gabuda, S. P.; Maifat, M. A.; Zilberman, B. D.; Panich, A. M.; Ippolitov, E. G. *Koord. Khim.* 1979, **5**, 850.
215. Okuda, K.; Yamada, K.; Ishihara, H.; Negita, H. *Bull. Chem. Soc. Jpn.* 1977, **50**, 3136.
216. Kidd, R. G.; Spinney, H. G. *Can. J. Chem.* 1981, **59**, 2940.
217. Dupree, R.; Gardner, J. A. *J. Phys. Colloq. (Orsay, Fr.)* 1980, **C8**, 20.
218. Maier, L. *Phosphorus* 1974, **4**, 41.
219. Junkes, P.; Baudler, M.; Dobbers, J.; Rashwitz, D. *Z. Naturforsch., Teil B* 1972, **27**, 1451.
220. Schuman, H.; Kroth, H. J. *Z. Naturforsch., Teil B* 1977, **32**, 513.
221. De Ketelaere, R.; Muylle, E.; Vanerman, W.; Claeys, E.; Van der Kelen, G. P. *Bull. Soc. Chim. Belg.* 1969, **78**, 219.
222. Dillon, K. B.; Waddington, T. C. *Spectrochim. Acta, Part A* 1971, **27**, 1381.
223. Grim, S. O.; Yankowsky, A. W. *J. Org. Chem.* 1977, **42**, 1236.
224. Schmidbauer, H.; Buchner, W.; Scheutzow, D. *Chem. Ber.* 1973, **106**, 1251. Issleib, K.; Lischewski, M.; Zschunke, A. *Org. Magn. Reson.* 1973, **5**, 401. Schmidbauer, H.; Eberlein, J.; Richter, W. *Chem. Ber.* 1977, **110**, 677. Grim, S. O.; Ambrus, J. H. *J. Org. Chem.* 1968, **33**, 2993. Koster, R.; Simic, D.; Grassberger, M. A. *Annalen* 1970, **739**, 211.
225. Flindt, E. P.; Rose, H.; Marsmann, H. C. *Z. Anorg. Allg. Chem.* 1977, **430**, 155. Goldwhite, H.; Gysegem, P.; Schow, S.; Swyke, C. *J. Chem. Soc., Dalton Trans.* 1975, 12.

Chapter 14

OXYGEN

W. McFARLANE and H. C. E. McFARLANE

1. INTRODUCTION

With a natural abundance of only 0.037%, a receptivity that is 0.61 of that of ^{13}C, and a moderate quadrupole moment ($eQ = -2.6 \times 10^{-26}$ e cm^2), ^{17}O ($I = 5/2$), the only magnetic isotope of oxygen, is poorly suited for study by NMR. Nonetheless the chemical importance of the element oxygen has contributed to considerable interest in its NMR spectroscopy, and following the earliest pioneering studies[1-6] the advent of Fourier transform methods and especially high-field spectrometers have now established the technique on a routine basis.

Although linewidths commonly in excess of 100 Hz prevent the study of couplings to oxygen in the majority of cases, there is an additional parameter available in the form of the quadrupolar coupling constant which can convey considerable information on chemical bonding. Furthermore, as a nonintegral spin nucleus ^{17}O can yield sharp lines from the $+1/2 \rightarrow -1/2$ transition in spectra from solid samples under magic angle spinning conditions, and this is a very promising area. There have been several reviews of ^{17}O NMR,[7-12] and of course mention is made of the technique in several more general treatments of NMR.

Many experiments involving ^{17}O NMR use isotopically enriched samples, (which may also be enriched in ^{18}O), and it is worth mentioning that there can be significant $^{16/17/18}O$ isotope effects upon the shielding of other nuclei including ^{13}C and ^{31}P, for which many biochemical applications have been devised.[13-18]

2. EXPERIMENTAL ASPECTS

The problem of probe ring-down times in ^{17}O NMR appears to be decaying with the passage of time as practical resonance frequencies rise, and probe design improves. Even so, the use of special pulse sequences[19] to avoid rolling base lines can be efficacious, because the normal instrumental delays required after the pulse can lead to substantial signal attenuation when lines several hundred hertz wide are to be recorded. Whether or not isotopic enrichment is necessary is very much a function of instrument sensitivity, and of sample concentration and molecular weight; surprisingly good results can be obtained at natural abundance for small molecules. It is also

W. McFARLANE and H. C. E. McFARLANE • Department of Chemistry, Sir John Cass School of Science and Technology, City of London Polytechnic, London EC3N 2EY, England.

worth stressing the matter of sample viscosity, as quite substantial reductions in linewidth can be achieved by reducing this as much as possible, either by raising the temperature when this is feasible or using a suitable solvent. Perhaps the ultimate was to use a liquid close to its critical point when ^{17}O linewidths of only 1.9 Hz at half-height were obtained for CO_2 at 28°C.[20] Another potential source of broadening in ^{17}O NMR is relaxation as a result of spin coupling to another quadrupolar nucleus. This is the case in the $NbW_5O_{19}^{3-}$ and similar ions where linewidths were reduced by a factor of as much as 30 by broadband ^{93}Nb decoupling.[21] In the same work selective ^{51}V decoupling was used to aid assignments of the ^{17}O spectra of the $V_{10}O_{28}^{6-}$ and $(VO_3)_n^{n-}$ ions.[21] Experimental aspects of solid state ^{17}O NMR are dealt with in Section 7.

3. CHEMICAL SHIFTS

At 9.2 T a linewidth of even 500 Hz corresponds to 10 ppm, so it is entirely realistic to obtain chemical shifts to significantly better than ± 1 ppm in most routine work. With a chemical shift range of over 1000 ppm considerable spectral dispersion is therefore available. Isotope effects upon ^{17}O chemical shifts are of virtually no practical importance, but of theoretical significance is the observation[22] that $\delta(^{17}O)_{^{13}C} - \delta(^{17}O)_{^{12}C} = -0.11$ ppm in CO and -0.143 ppm in CO_2. These are about four times larger than the $^{17/16}O$ effect upon $\delta(^{13}C)$ in the same molecules and thus correspond with the chemical shift ranges of the two nuclei. The derived values of $\partial\sigma/\partial r$ correspond well with those from *ab initio* calculations done using the method of Jameson and Osten.[23] Isotope effects in the uranyl ion have also been observed.[23a]

The usual chemical shift reference is water because it is readily available in enriched form, but it has many disadvantages, such as dependence on temperature,[24] hydrogen bonding,[25,26] and other factors,[27,28] and nitromethane and acetone are also used, normally as external standards or by substitution, the conversions in ppm being

$$\delta(\text{rel} \cdot H_2O) = \delta(\text{rel} \cdot MeNO_2) + 605 = \delta(\text{rel} \cdot Me_2CO) + 569$$

Of course, on modern spectrometers the Ξ convention can be used, $\Xi(^{17}O)$ for acetone being 13.564269 MHz for a neat sample[29] at 301 K, but variations in $\delta(^{17}O)$ for this material of up to 6.9 ppm over the temperature range 181–325 K have been reported[30] and attributed[30] to a dimer/monomer equilibrium. Table 1 lists ^{17}O shieldings in a range of species. It should be borne in mind that the linewidths in particular can be very sensitive to the conditions of measurement and so they are omitted from this table. Many more oxygen chemical shifts of organic compounds are listed by Kintzinger,[10] and more recent studies are available of *p*-substituted nitrobenzenes,[61,62] benzohydroxamic acids,[63] cyclohexanols,[64] 1,3-dioxans,[64] cyclic ketones and esters,[65,82] alkyl sulfoxides and sulfones,[66] unsaturated and aromatic ethers,[67] amino acids,[68,69,80] adenine nucleotides,[70] monosaccharides,[71] organo-silicon compounds,[72,73] species with boron–oxygen bonds,[74] furans,[75] oxiranes,[76] analogs of ATP,[76] boratranes,[77] germatranes,[77] silatranes,[77] nitro-aromatics,[78] tertiary alcohols and ethers,[79] pyridine N-oxides,[81] substituted chalcones[83] and nitrostyrenes,[83] five-membered heterocycles,[84] polycyclic quinones and hydroquinones,[85] components of synthetic fuels,[86] and substituted benzoyl cations.[87]

Many semiempirical correlations have been established, mostly using the concepts discussed in Chapter 3, and in addition the position of oxygen as a first row element has made its shielding the subject of considerable theoretical attention. On the basis of the ^{17}O spin-rotation coupling constant for carbon monoxide(88) and a calculated value(89) of σ^d, it has been proposed that for this molecule $\sigma(^{17}O) = -42.3 \pm 17.2$ ppm on an absolute scale.(31) *Ab initio*(90,91) extended Hückel MO,(92) CNDO/2,(93,95) and INDO(93,94) calculations have been applied to oxygen shielding with no more than moderate success. More recently there have been calculations of shieldings(96,97) in H_2O, H_3O^+, OH^-, and O_3, and of shielding tensor components(96) in CO, NNO, and CO_2, with results that are difficult to compare with experiment. Calculations of $^{2/1}H$ isotope effects on oxygen shielding in water appear to be satisfactory,(98,99) although it is of course necessary to allow for the effects of hydrogen bonding.

The general pattern of ^{17}O shieldings is exhibited in Chart 1, and there are significant parallels with the behavior of selenium (and presumably sulfur, although the available data are at present too limited to be able to claim that this has been established). For example, in alcohols and ethers increased α-branching of alkyl substituents increases $\delta(^{17}O)$, and there is a good correlation between the ^{17}O shieldings in carbonyl groups and the ^{77}Se shieldings in corresponding selenocarbonyl

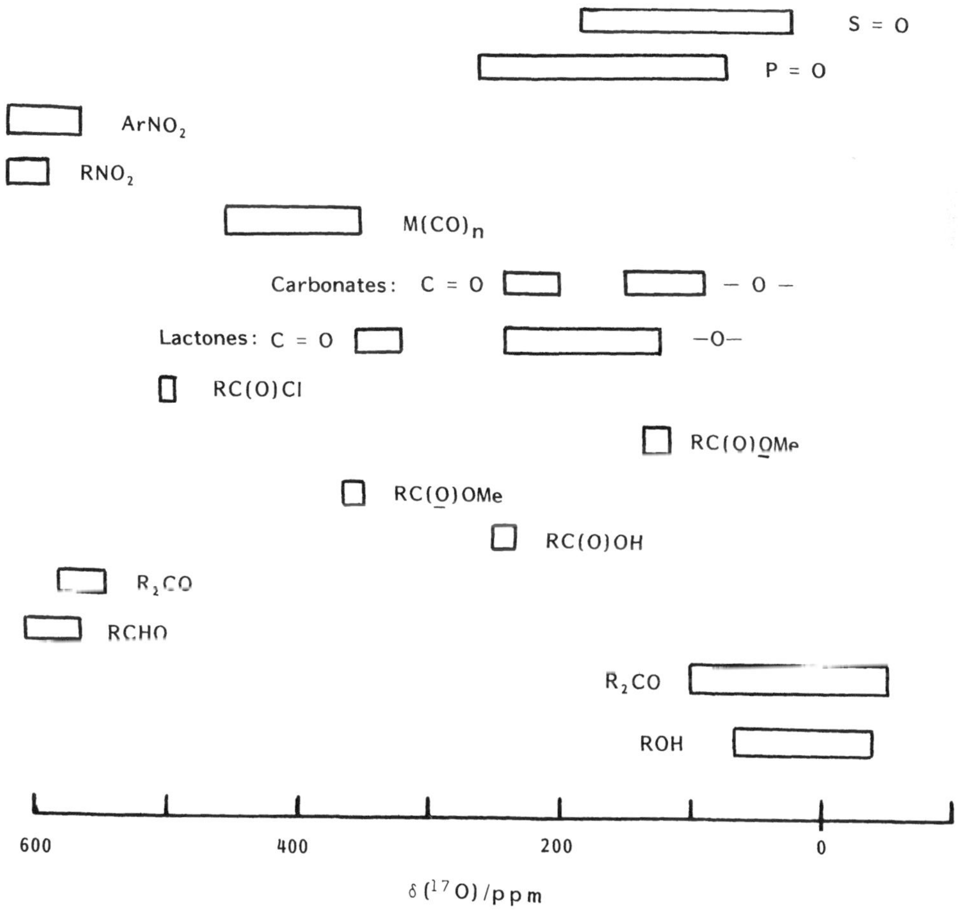

Chart 1. Pattern of ^{17}O Chemical Shifts.

compounds.[100] These latter are known to correlate with mean excitation energies,[101] and it is therefore not surprising that as long ago as 1962 $\delta(^{17}O)$ was found[6] to correlate with the wavelengths of the lowest energy transition in the electronic spectra of compounds with C=O and N=O groups. Similar correlations with the $n \rightarrow \pi^*$ transition frequency have been obtained for acetone under various conditions.[102,103] In other species with C=O and N=O bonds there are also correlations between $\delta(^{17}O)$ and $\delta(^{13}C)$[50] and $\delta(^{14}N)$[59], with slopes that reflect the relative values of the radial expansion terms. In species with carbon–oxygen single bonds there are correlations of $\delta(^{17}O)$ with $\delta(^{13}C)$ which have been interpreted in terms of relative charge densities on oxygen and carbon.[49,103–105]

The effects of changes in charge density at oxygen [increasing charge reduces $\delta(^{17}O)$] are brought out in studies of *p*-substituted nitrobenzenes[61] where there is resonance involving the substituent and the NO_2 group, and from which it is deduced

Table 1. Selected ^{17}O Chemical Shifts Relative to H_2O

Species	$\delta(^{17}O)$ (ppm)	Reference
H_2O liquid	0.0	31
H_2O gas	−36.1	31
CO_2 gas (7 atm)	64.5	31
N_2O gas (3 atm)	107.4	31
OCS gas (9 atm)	200.0	31
OCS liquid	204.9	31
CO gas (2 atm)	350.2	31
OF_2 gas (5 atm)	781	31
$POCl_3$ liquid	215	32
$POCl_3$ in HNO_3	187	32
$POBr_3$	260	33
POF_3	66	33
SO_2	513	6
SO_3	188	6
$SOCl_2$	292	6
$ClOF_2$	311	34
ClO_3F	302	34
$Ni(CO)_4$	362	35
$Fe(CO)_5$	388	35
$Mn_2(CO)_{12}$	355	35
$Co_4(CO)_{12}$ Bridging	501	36
$Co_4(CO)_{12}$ Terminal	366–382	36
$Te(OH)_6$	120	37
OsO_4	796	38
RuO_4	1106	38
$XeOF_4$	313	39
$Xe(OH)_6$	278	40
O_3 Terminal	1032	41
O_3 Central	1598	41
SO_4^{2-}	167	42
SeO_4^{2-}	204	6
ClO_3^-	290	43
BrO_3^-	297	6
IO_3^-	206	44
VO_4^{3-}	568	45

Table 1. (*Continued*)

Species	$\delta(^{17}O)$ (ppm)	Reference
CrO_4^{2-}	835	6
MoO_4^{2-}	530	6[a]
WO_4^{2-}	420	6
MnO_4^-	1230	47
TcO_4^-	749	6
ReO_4^-	569	6
$VO_3(OO)^{3-}$ (V=O)	568	48
$VO(OO)_3^{3-}$ (V=O)	990	48
MeOH	−37	5
EtOH	6	5
i-PrOH	38	5
t-BuOH	70	5
Me_2O	−53	49
MeOEt	−23	49
Et_2O	7	49
t-Bu_2O	76	49
PhOMe	44	10
Ph_2O	101	10
MeCHO	592	50
Me_2CO	569	50
CH_3COOH (C=O)	251	5
CH_3COCl	502	51
CH_3CONH_2	303	52
CH_3NO_2	605	53
$PhNO_2$	569	54
o-$ClC_6H_4NO_2$	610	54
m-$ClC_6H_4NO_2$	567	54
	565	55
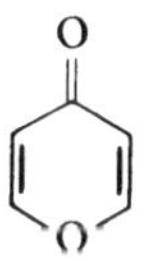 C=O	464	55
C−O−C	174	55
p-$ClC_6H_4(C(O)OMe)$ (C=O)	342	56
p-$ClC_6H_4(C(O)OMe)$ (OMe)	129	56
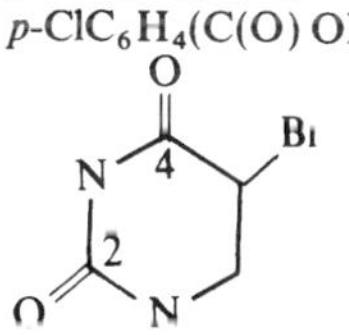 O_2	235	57
O_4	311	57
$(EtO)_4Si$	9	58
$Me\underline{O}NO$	420	59
$(MeO)_3P{=}O$ (P=O)	19	54
(−O−)	68	54
Me_2SO	20	60
Me_2SO_2	163	60
$(MeO)_2SO_2$ (S=O)	150	5
(−O−)	102	5

[a] The value of 831 ppm in Ref. 46 is probably[219] wrong.

the sensitivity of $\delta(^{17}O)$ to change in charge density is ca. 300 ppm/*e*. Similar conclusions can be drawn from studies of substituted benzyl alcohols,(106) benzoic acids,(56) and other species for which standard linear free energy relationships have been established.(56) Dissolution of $POCl_3$ in strong acids leads to protonation at oxygen which produces changes in $\delta(^{17}O)$ that can also be understood in terms of charge density.(32)

On a more empirical level still, additivity and other rules have been established for oxygen chemical shifts. Deformation of interbond angles by incorporation in one or more cyclic systems can affect $\delta(^{17}O)$ significantly as shown by results for bicyclic ethers (deshielded relative to corresponding acyclic derivatives),(105,107) small ring ketones (shielded),(50) lactones,(65) and sulfones and sulfoxides.(60) However, no overall rationale has yet emerged. Other evidence of steric effects upon $\delta(^{17}O)$ is to be found in studies of tetrahydropyrans(108) and the observation of separate resonances from diastereotopic oxygens in sulfones(109) and cyclic 2-deoxyadenosine-3′-5′-monophosphate.(110) In carbocyclic and heterocyclic ketones the carbonyl–oxygen shielding is sensitive to ring fusion as a result of changes in double bond character arising from resonance with the aromatic system.(55)

In the tetrahedral species MO_4^{n-} ($n = 0$–4) $\delta(^{17}O)$ decreases as the metal M becomes heavier in the same group. This is probably due to increased diamagnetic circulation about the heavier atom. There is also a correlation of $\delta(^{17}O)$ with ΔE^{-1} as measured from uv/visible spectra.(6) In polyoxometalates the oxygen chemical shift patterns are often reliable indicators of structural type.(111) For example, in the molybdenum(VI) series $\delta(^{17}O)$ is found to be in the nonoverlapping ranges 810–950, 350–710, 230–320, ca. 120, ca. 50, and ca. 20 ppm, for oxygen atoms bound to 1, 2, 3, 4, 5, and 6 molybdenum atoms respectively.(112–114) Similar behavior is displayed by polytungstates,(115) polyniobiates,(115) polytantalates,(115) polyvanadates,(116) and arsenomolybdates.(117) In peroxovanadates three chemical shift regions are found(48): 0–100, 555–650, and 1000–1250 ppm arising, respectively, from coordinated water, O^{2-} or OH^- coordinating tetrahedrally, and O^{2-} or OH^- coordinating octahedrally. Signals from O_2^{2-} are not observed in these, but they are in some molybdenum and platinum species.(118)

The carbonyl oxygens of transition metal carbonyls also give quite sharp lines, and various structural correlations have been identified which supplement those available from the ^{13}C resonances.(35,36,119–124) In some phosphine substituted molybdenum tetracarbonyls the correlation of $\delta(^{17}O)$ with $\delta(^{13}C)$ and $\delta(^{31}P)$ was good for carbonyl groups in positions *trans* to the phosphine ligands, but poor for the *cis*.(124)

4. SPIN COUPLING

With a few exceptions the quadrupolar broadening has limited the observation of spin coupling involving ^{17}O to the directly bonded interaction $^1J(X–^{17}O)$. This has seriously limited the number of experimentally determined signs of the couplings, and many are therefore based on theoretical conjecture. Note that γ for ^{17}O is negative so that K(X–O) and $J(X–^{17}O)$ are of opposite sign unless γ_x is also negative. Table 2 lists a substantial proportion of the one-bond coupling constants currently available. Examples of two-bond couplings to oxygen are 38, 7.5, and 10.5 Hz for the O–CH,

Table 2. One-Bond Couplings to Oxygen in Hz

Nucleus X	Species	$^1J(X\text{–}^{17}O)$	Reference
^1H	H_2O	(−)83	125, 137
	H_3O^+	(−)106	126
	CH_3OH	−85.5	127, 128
^{13}C	CO	16.4	22
	CO_2	16.1	22
	Me_2CO	22	129
	Me_2COH^+	74	126
^{19}F	OF_2	300	31
	F_2O_2	424	130
^{31}P	POF_3	(+)184	131
	$(MeO)_3P$	+153	132, 54
	$(MeO)_3PO$ (P=O)	+165	132, 54
	(O–)	+80	132, 54
	PO_4^{3-}	(+)91	133
^{35}Cl	ClO_3F	106	134
	ClO_4^-	85.5	43
^{51}V	VO_4^{3-}	62	45
	$V_2O_7^{4-}$	ca. 50	45
^{53}Cr	CrO_4^{2-}	ca. 10	6
^{55}Mn	MnO_4^-	29	135, 129
^{95}Mo	MoO_4^{2-}	40	46
^{99}Ru	RuO_4	23	38
^{99}Tc	TcO_4^-	131	136
^{129}Xe	$XeOF_4$	692	39
^{187}Re	ReO_4^-	ca. 50	138

O–CH_3, and HC=O interactions in methyl formate,(125,139) 11 Hz for $^2J(^{17}O\text{–}H)$ in Me_2NCHO,(140) 21 Hz for $^2J(^{17}O\text{–}H)$ in formic acid(140) (note that this is the mean of the couplings to the two oxygens), 39 Hz for $^2J(^{19}F\text{–}^{17}O)$ in CH_3COF,(141) and 31 Hz for $^2J(^{19}F\text{–}^{17}O)$ in POF_3.(131) Many others could no doubt be obtained by line-shape analysis of ^{17}O spectra, allied to decoupling experiments as appropriate.

Oxygen has the distinction that its spin coupling behavior has been better studied (though not necessarily understood) theoretically than experimentally. From its position in the Periodic Table the couplings should resemble those to fluorine,(142,143) and this is borne out by the positive and negative signs of $^1K(OH)$ and $^1K(PO)$, respectively.(129,132) On this basis it seems likely that most of the reduced one-bond couplings will be negative, but the determination of further signs should be given some priority. It has been noted that in the series VO_4^{3-}, CrO_4^{2-}, MnO_4^- ^1K decreases from left to right in contrast with expectation,(135) while $^1J(^{31}P{=}^{17}O)$ shows some correlation(144) with υ_{PO}, and its values have been rationalized using finite perturbation theory.(145)

5. RELAXATION BEHAVIOR

Except when ^{17}O is spin-coupled to a nucleus with a large quadrupole moment such as ^{93}Nb, its relaxation behavior is dominated by its own quadrupole moment in diamagnetic compounds. Thus in the absence of chemical exchange T_1 is equal to T_2, and can be obtained from the linewidth (with due allowance for any unresolved spin coupling[146]). Equation (1) relates T_1 to the quadrupolar coupling constant e^2qQ/h, the asymmetry parameter η, and the correlation time for isotropic reorientation τ_c.

$$\frac{1}{T_1}=\frac{4\pi^2}{125}\left(1+\frac{\eta^2}{3}\right)\left(\frac{e^2qQ}{h}\right)^2\tau_c \tag{1}$$

The relationship between η and e^2qQ/h depends upon the oxygen hybridization, while τ_c may be estimated, or can be determined from the relaxation behavior of another nucleus.[134] For example, in liquid SO_2 the linewidth of the ^{33}S resonance gave τ_c, and hence a value of 6.6 ± 0.7 MHz was deduced for e^2qQ/h of ^{17}O from the width of its resonance.[147] Table 3 gives a few values of e^2qQ/h and η for ^{17}O. The quadrupole coupling constant is found to be reduced by as much as 2.6 MHz by strong hydrogen bonding,[156] and also falls with reducing electronegativity of the atom directly bound to oxygen,[152,153] but is insensitive to longer-range effects.[153] Quadrupole coupling constants have also been derived from the splittings in the ^{17}O spectra of molecules partially oriented in an electric field[158] or by dissolution in a nematic liquid crystal.[157,159]

Conversely, T_1 (i.e., in practice linewidth) measurements allied to (estimated) values of the quadrupolar coupling constants have been used to determine τ_c and hence to study the motional behavior of various species. Thus τ_c increases with the hydrocarbon chain-length in aliphatic ethers,[160] and in acetone ^{17}O measurements can give information not available from ^{13}C, ^1H, or ^2H owing to rapid methyl group rotation.[161] Cyclic ketones,[162] cyclic ethers,[164] and crown ethers and their complexes[165] have also been studied, as have water reorientation in a lyotropic lamellar phase[157] and motions of the sulfate ion in aqueous solution.[166] Detailed studies of polynuclear osmium carbonyl dynamics have been undertaken, and allied to

Table 3. Nuclear Quadrupole Coupling Constants and Asymmetry Parameters for ^{17}O

Species	e^2qQ/h (MHz)	η (%)	Reference
O_2(g)	−8.42	0	148
CO(g)	4.43	0	149
H_2O(ice)	6.9	86.5	150
SO_2(l)	6.6	—	147
MoO_4^{2-}(l)	0.7	—	46
OsO_4(l)	0.97	0	38
HOC_2H_4OH (s)	8.9	97	151
$PhNO_2$ (s)	13.1	58	152
PhC(O)F (s)	8.5	35	153
HCOOH (C=O) (s)	7.8	7	154
(OH) (s)	−6.9	8	154
$(PhCH_2)_2\underline{O}$ (s)	10.9	92	155
$Na^+OO\overline{C}H$ (s)	7.3	46	153
Me_2CO (l)	11.5	—	157

^{13}C measurements have provided estimates of the ^{17}O quadrupolar coupling constant.[167,168] In molybdenum hexacarbonyl ^{95}Mo and ^{97}Mo linewidths gave the reorientational correlation time, which in turn made it possible to determine the quadrupole coupling constant from the ^{17}O linewidth.[169]

6. APPLICATIONS

The chemical importance of oxygen is such that despite the experimental difficulties ^{17}O NMR has been applied to the solution of a wide range of problems. Straightforward applications of chemical shifts were used in studies of the conformational analysis of methyl substituted thiane 1-oxide and 1,1-dioxide,[170] the protonation of urea,[171] mixed vanadium–tungsten polyoxo-complexes,[172] the protonation of molybdenum heteropolyanions,[173] *cis/trans* isomerism of *N*-acetyl-L-proline,[174] the interaction of lanthanide^{3+} cations with adenosine 5′-triphosphate,[175] proton transfer in methanesulfonic acid/water mixtures,[176] peptides,[177] rhodium(III) chloroaquo complexes,[178] reaction mechanisms in carbohydrate chemistry,[179] deviations from planarity of ortho-substituted nitrobenzenes[180] and acetophenones,[181] the hydration of 2′,3′-O-iso-propylideneuridene at the C_4O rather than the C_2O carbonyl group,[182] the absence of extensive aromaticity in 4-pyrone,[183] the mono-oxo-bridged dinuclear structure[184] of $Mo_2O_5[Me_2N(CH_2)_2NHCH_2CMe_2S]_2$, F^-/H_3PO_3 hydrogen bonding interactions,[185] molybdenum carbonyl complexes,[186] and the partial π-bond order of the C–O bond in protonated acetone.[126]

It was originally proposed that the value of 107 Hz for $^1J(^{17}O–H)$ in H_3O^+ was consistent with a planar geometry,[187] but this view has since been corrected[126] (the oxygen still retains an electron lone pair, and the ion is pyramidal) despite its ready acceptance.[12] The pattern of ^{99}Ru satellites (six lines) in the ^{17}O spectrum of RuO_4 demonstrated[38] that the former nucleus has spin 5/2.

^{17}O NMR has been used to study a range of inorganic reaction mechanisms,[188] including those between OF_2 and COF_2 to give CF_3OOOCF_3,[189] and between F_2O_2 and SO_2.[190] The mechanisms of many substitution reactions of transition metal complexes have also been studied by ^{17}O NMR, separate signals usually being obtained for free and complexed water. Examples include water exchange of $[Pd(H_2O)_4]^{2+}$, $[Al(H_2O)_6]^{3+}$, $[V(H_2O)_6]^{3+}$, and $[Pt(H_2O)_4]^{2+}$ under varying conditions of temperature and pressure,[191–194] the catalytic effect of nitrate ions on cobalt (III) substitution reactions,[195] $[Ru(H_2O)_6]^{2+}/[Ru(H_2O)_6]^{3+}$ self-exchange reactions,[196] and solvent exchange reactions of nickel complexes.[197] The binding of H_2O to various paramagnetic ions was studied by ^{17}O NMR at an early stage,[198–206] in many cases the paramagnetic shift being great enough for there to be a separate signal from the bound water even in the presence of quite rapid exchange. For exchange rates too rapid for this to be so, the change in ^{17}O linewidth can often be used to evaluate the exchange rate,[207–210] using the Swift–Connick method, and a considerable number of ligands binding to such ions as Mn^{2+}, Fe^{2+}, Co^{2+}, Ni^{2+}, Cu^{2+}, UO_2^+, NpO_2^+, and PuO_2^{2+} have been studied in this way.[211–212] The effects of lanthanide reagents in aqueous solution upon ^{17}O chemical shifts have been reported,[213] and their interpretation[214] disagrees with results from proton and ^{13}C work. The binding of water to paramagnetic metal sites in proteins has been examined[215,216] via $T_1(^{17}O)$, and biophysical applications of ^{17}O NMR have been reviewed.[217]

7. THE SOLID STATE

Under conditions of magic angle sample spinning (MASS) the $+1/2 \rightarrow -1/2$ transition of quadrupolar nuclei with nonintegral spin (i.e., $I = 3/2$, 5/2, 7/2, or 9/2) can give a sharp resonance.(218) This has now been exploited(219,220) in ^{17}O studies of a range of oxides and oxyanions highly enriched in ^{17}O at 11.7 T. From the results quadrupole coupling constants ranging from 0 to over 5 MHz were obtained, together with chemical shift anisotropies of up to 300 ppm. At the measuring frequency used (67 MHz) this latter corresponds to over 20 kHz and so the sample spinning rates used of 3–4.5 kHz produced substantial sidebands in many cases. The use of lower measuring frequencies to avoid this difficulty was inappropriate as the linewidths varied from 20 to 2700 Hz (0.3–40 ppm). Typical results are those obtained from Forsterite, Mg_2SiO_4, in which three crystallographically distinct oxygen sites were partially resolved, and it was possible to determine the chemical shifts (61, 62, and 47 ppm), the quadrupole coupling constants (2.35, 2.35, and 2.7 MHz), and the electric field gradient tensor asymmetry parameters (0.2, 1.0, and 0.3) associated with each. This is much more information than is available from MASS ^{29}Si NMR since the material has only one type of silicon site. In K_2WO_4 three types of oxygen with $\delta(^{17}O) = 437$, 429, and 422 ppm were clearly resolved, and from the pattern of the intensities of the spinning sidebands as a function of spinning speed individual components of the shielding tensors were obtained for each type.(219) From similar studies of other silicate minerals, α-Al_2O_3, B_2O_3, ZnO, H_2O, *p*-chlorophenol, tetrahydropyran, and other substances it was found(219) that the quadrupole coupling constant is related to *I*, the percent ionic character of the bonds to oxygen by equation (2) with a correlation coefficient of 0.96.

$$e^2qQ/h = -0.203I + 14.78 \qquad (2)$$

It was also pointed out that there is a straightforward set of relationships among the quadrupolar coupling constant, the bond length (energy), λ_{max}, the ionic radius, and $\delta(^{17}O)$, which may be of use for predicting the last of these. It has also been found that variable angle sample spinning (VASS) can be advantageous,(220) especially for removing resonances arising from transitions other than $+1/2 \rightarrow -1/2$. The method has been used to study(221) several molydenum carbonyl derivatives including $Mo(CO)_6$ itself, and $Mo(CO)_4$ bipy, and has great promise in many other areas, although the need for substantial isotopic enrichment is not likely to be eliminated in the near future.

REFERENCES

1. Alder, F.; Yu, F. C. *Phys. Rev.* 1951, **81**, 1067–1068.
2. Weaver, H. E.; Tolbert, B. M.; LaForce, R. C. *J. Chem. Phys.* 1955, **23**, 1956–1957.
3. Dharmatti, S. S.; Rao Sundora, J. K.; Vijayoraghan, R. *Nuovo Cimento* 1959, **11**, 656–659.
4. Christ. H. A. *Helv. Phys. Acta* 1960, **33**, 572–576.
5. Christ, H. A.; Diehl, P.; Schneider, H. R.; Dahn, H. *Helv. Chim. Acta* 1961, **44**, 865–880.
6. Figgis, B. N.; Kidd, R. S.; Nyholm, R. S. *Proc. R. Soc. London A* 1962, **269**, 469–480.
7. Silver, B. L.; Luz, Z. *Quart. Rev.* 1967, **21**, 458–473.
8. Rodger, C.; Shepherd, N. *NMR and the Periodic Table*, Harris, R. K.; Mann, B. E., Eds.; Academic: New York, 1978; pp. 383–400.
9. St. Amour, T. E.; Fiat, D. *Bull. Magn. Reson.* 1980, **1**, 118–129.
10. Kintzinger, J. P. *NMR Basic. Principles and Progress*, Diehl, P.; Fluck, E.; Kosfeld, R., Eds.; Springer: Heidelberg, 1981; Vol. 17, pp. 1–64.

11. Klemperer, W. G. *Angew. Chem. Int. Ed. Engl.* 1978, **17**, 246–254.
12. Kintzinger, J. P. *NMR of the Newly Accessible Nuclei*, Laszlo, P., Ed.; 1983; Vol. 2, pp. 79–104.
13. Sankawa, U.; Ebizuka, Y.; Noguchi, H.; Isikawa, Y.; Kitaghawa, S.; Yamamoto, Y.; Kotayashi, T.; Iztak, Y.; Seto, H. *Tetrahedron* 1983, **39**, 3583.
14. Gorst-Allman, C. P.; Steyn, P. S.; Vleggar, R. *J. Chem. Soc. Perkin Trans. 1* 1983, 1357–1359.
15. Holker, J. S. E.; O'Brien, E.; Moore, R. N.; Vedesas, J. C. *J. Chem. Soc., Chem. Commun.* 1983, 192.
16. De Brosse, C. W.; Villafranca, J. J. *Magn. Reson. Biol.* 1983, **2**, 1.
17. Tsai, M. D.; Bruzik, K. *Biol. Magn. Reson.* 1983, **5**, 129.
18. Joseph, A. P.; Bolton, P. H. *J. Am. Chem. Soc.* 1984, **106**, 437–439.
19. Patt, S. L. *J. Magn. Reson.* 1982, **49**, 161.
20. Robert, J. M.; Evilia, R. F. *J. Am. Chem. Soc.* 1985, **107**, 3733–3735.
21. Besecker, C. J.; Klemperer, W. G.; Maltkie, D.; Wright, D. A. *Inorg. Chem.* 1985, **24**, 1027–1032.
22. Wasylishen, R. E.; Friedrich, J. O.; Mooibroek, S.; Macdonald, J. B. *J. Chem. Phys.* 1985, **83**, 548–551.
23. Jameson, C. J.; Osten, H.-J. *J. Chem. Phys.* 1984, **81**, 4300.
23a. Jung, W. S.; Ikeda, Y.; Tomiyasu, H.; Fakutomi, H. *Bull. Soc. Chem. Jpn* 1984, **57**, 2317–2318.
24. Florin, A. E.; Alei, M. *J. Chem. Phys.* 1967, **47**, 4268–4269.
25. Alei, M.; Florin, A. E. *J. Phys. Chem.* 1969, **73**, 863–867.
26. Sardella, D. J.; Stothers, J. B. *Can. J. Chem.* 1969, **47**, 3089–3902.
27. Hertz, H. G.; Klute, R. *Z. Phys. Chem.* 1970, **69**, 101–107.
28. Lutz, O.; Oehler, H. *Z. Naturforsch.* 1977, **32a**, 131–133.
29. McFarlane, W. Private Communication.
30. Ancian, B.; Tiffon, B.; Dubois, J. E. *Chem. Phys.* 1983, **74**, 171.
31. Wasylishen, R. E.; Mooibroek, S.; MacDonald, J. B. *J. Chem. Phys.* 1984, **81**, 1057–1059.
32. Hibbert, R. C.; Logan, N. *J. Chem. Soc. (Dalton)* 1985, 865–866.
33. Grossmann, G.; Gruner, M.; Seifert, G. *Z. Chem.* 1976, **16**, 362.
34. Virlet, J.; Tantot, G. *Chem. Phys. Lett.* 1976, **44**, 296.
35. Bramley, R.; Figgis, B. N.; Nyholm, R. S. *Trans. Faraday Soc.* 1962, **58**, 1893.
36. Aime, S.; Osella, D.; Milone, L.; Hawkes, G. E.; Randall, E. W. *J. Am. Chem. Soc.* 1981, **103**, 5920–5922.
37. Luz, Z.; Pecht, I. *J. Am. Chem. Soc.* 1966, **88**, 1152–1154.
38. Brevard, C.; Granger, P. *J. Chem. Phys.* 1981, **75**, 4175–4177.
39. Shamir, J.; Selig, H.; Samuel, D.; Reuben, J. *J. Am. Chem. Soc.* 1965, **87**, 2359.
40. Reuben, J.; Samuel, D.; Selig, H.; Shamir, J. *Proc. Chem. Soc.* 1963, 270.
41. Solomon, T. J.; Kieth, J. N.; Kacmarek, A. J.; Ranay, J. K. *J. Am. Chem. Soc.* 1968, **90**, 5408–5411.
42. Lutz, O.; Nepple, W.; Nolle, A. *Z. Naturforsch.* 1976, **31a**, 978–980.
43. Alei, M. *J. Chem. Phys.* 1965, **43**, 2904–2906.
44. Dwek, R. A.; Luz, Z.; Peller, S.; Shporer, M. *J. Am. Chem. Soc.* 1971, **93**, 77–79.
45. Lutz, O.; Nepple, W.; Nolle, A. *Z. Naturforsch.* 1976, **31a**, 1046–1050.
46. Vold, R. R.; Vold, R. L. *J. Chem. Phys.* 1974, **61**, 4360–4361.
47. Jackson, J. A.; Taube, H. *J. Phys. Chem.* 1965, **69**, 1844–1849.
48. Harrison, A. T.; Howarth, O. W. *J. Chem. Soc. (Dalton)* 1985, 1173–1177.
49. Delseth, C.; Kintzinger, J. P. *Helv. Chim. Acta* 1978, **61**, 1327–1334.
50. Delseth, C.; Kintzinger, J. P. *Helv. Chim. Acta* 1976, **59**, 466–475; **59**, 1411.
51. Delseth, C.; Nguyên, T. T. T.; Kintzinger, J. P. *Helv. Chim. Acta* 1980, **63**, 498–503.
52. Burgar, M. I.; St. Amour, T. E.; Fiat, D. *J. Phys. Chem.* 1981, **85**, 502–510.
53. Lippmaa, E. *et al. Org. Magn. Reson.* 1972, **4**, 153–160, 197–202.
54. Christ, H. A.; Diehl, P. *Helv. Phys. Acta* 1963, **36**, 170–182.
55. Chandrasekaran, S.; Wilson, W. D.; Boykin, D. W. *Org. Magn. Reson.* 1984, **22**, 756–759.
56. Balakrishnan, P.; Baumstark, A. L.; Boykin, D. W. *Org. Magn. Reson.* 1984, **22**, 753–756.
57. Moore, P. *Pure Appl. Chem.* 1985, **57**, 347–354.
58. Harris, R. K.; Kimber, B. J. *Org. Magn. Reson.* 1975, **7**, 460–464.
59. Andersson, L. O.; Mason, J. *J. Chem. Soc. Dalton Trans.* 1974, 202–205.
60. Block, E.; Bazzi, A. A.; Lambert, J. B.; Wharry, S. M.; Anderson, K. K.; Dittmer, D. C.; Bhalchardra, H. P.; Smith, D. J. H. *J. Org. Chem.* 1980, **45**, 4807–4810.
61. Fraser, R. R.; Ragauskas, A. J.; Stothers, J. B. *J. Am. Chem. Soc.* 1982, **104**, 6475.
62. Lipkowitz, K. B. *J. Am. Chem. Soc.* 1982, **104**, 2647.
63. Lipczynska-Kochany, E.; Iwamura, H. *J. Org. Chem.* 1982, **47**, 5277.

64. Eliel, E. L.; Liu, K. T.; Chandrasekaran, S. *Org. Magn. Reson.* 1983, **21**, 179.
65. Delseth, C.; Kintzinger, J. P. *Helv. Chim. Acta* 1982, **65**, 2273.
66. Dyer, J. C.; Harris, D. L.; Evans, S. A. *J. Org. Chem.* 1982, **47**, 3660.
67. Kalabin, G. A.; Kushnarev, D. F.; Valeyev, R. B.; Trifomov, B. A.; Fedotov, M. A. *Org. Magn. Reson.* 1982, **18**, 1.
68. Hunston, R.; Gerothanassis, I. P.; Lauterwein, J. *Org. Magn. Reson.* 1982, **18**, 120.
69. Gerothanassis, I. P.; Hunston, R.; Lauterwein, J. *Helv. Chim. Acta* 1982, **65**, 1764.
70. Gerlt, J. A.; Demou, P. C.; Mehdi, S. *J. Am. Chem. Soc.* 1982, **104**, 2848.
71. Gerothanassis, I. P.; Lauterwein, J.; Sheppard, N. *J. Magn. Reson.* 1982, **48**, 431.
72. Kupce, E.; Liepin, E.; Lukevics, E. *J. Chem. Soc., Dalton Trans.* in press.
73. Maijs, L. *Latv. PSR Zinat. Akad. Vestis, Kim. Ser.* 1982, 646.
74. Wrackmeyer, B.; Koester, R. *Chem. Ber.* 1982, **115**, 2022.
75. Chimichi, S.; Nesi, R.; De Sio, F. *Org. Magn. Reson.* 1984, **22**, 55.
76. Gerlt, J. A.; Reynolds, M. A.; Demou, P. C.; Kenyon, G. L. *J. Am. Chem. Soc.* 1983, **105**, 6469.
77. Liepins, E.; Zicmane, I.; Zelcans, G.; Lukevics, E. *Zh. Obshch. Khim.* 1983, **53**, 245.
78. Balakrishnan, P.; Boykin, D. W. *J. Org. Chem.* 1985, **50**, 3661–3663.
79. Mansharan, M.; Eliel, E. L. *Magn. Reson. Chem.* 1985, **23**, 225–231.
80. Steinschneider, A.; Valentine, B.; Burgar, M. I.; Fiat, D. *Magn. Reson. Chem.* 1985, **23**, 104–110.
81. Boykin, D. W.; Baumstock, A. L.; Balakrishnan, P. *Magn. Reson. Chem.* 1985, **23**, 276–279.
82. Orsini, F.; Ricci, G. S. *Org. Magn. Reson.* 1984, **22**, 653–657.
83. Boykin, D. W.; Baumstock, A. L.; Balakrishnan, P.; Perjessy, A.; Honciar, P. *Spectrochimica Acta* 1984, **40A**, 887–891.
84. Chimichi, S.; Nesi, R.; De Sio, F. *Org. Magn. Reson.* 1984, **22**, 55–56.
85. Chandrasekaran, S.; Wilson, W.; Boykin, D. W. *Org. Magn. Reson.* 1984, **22**, 757–760.
86. Grandy, D. W.; Petrakis,L.; Young, D. C.; Gates, B. C. *Nature* 1984, **308**, 175–177.
87. Olah, G. A.; Iyer, P. S.; Prakash, G. K. S.; Krishnamurthy, V. V. *J. Org. Chem.* 1984, **49**, 4317–4319.
88. Frerking, M. A.; Larger, W. D. *J. Chem. Phys.* 1981, **74**, 6990.
89. Amos, R. D. *Chem. Phys. Lett.* 1979, **68**, 536.
90. Ditchfield, R.; Miller, D. P.; Pople, J. A. *J. Chem. Phys.* 1970, **53**, 613; 1971, **54**, 4186.
91. Ditchfield, R. *Mol. Phys.* 1974, **27**, 789.
92. Velenik, A.; Lynden-Bell, R. M. *Mol. Phys.* 1970, **19**, 371.
93. Ebraheem, K. A. K.; Webb, G. A. *J. Magn. Reson.* 1977, **25**, 399.
94. Moniz, W. B.; Paranski, Jr., C. F. *J. Magn. Reson.* 1973, **11**, 62.
95. Sadlej, J.; Sadlej, A. J. *J. Magn. Reson.* 1974, **14**, 97.
96. Schindler, M.; Kutzelnigg, W. *Mol. Phys.* 1983, **48**, 781.
97. Fukui, H.; Miura, K.; Tada, F. *J. Chem. Phys.* 1983, **79**, 6112.
98. Fowler, P. W. *Mol. Phys.* 1983, **48**, 153.
99. Lutz, O.; Oehler, H. *Z. Naturforsch., Teil A* 1977, **32**, 131.
100. Wong, T. C.; Guziec, Jr., F. S.; Moustakis, C. A. *J. Chem. Soc. Perkin Trans. 2* 1983, 1471.
101. Cullen, E. R.; Guziec, F. S.; Murphy, C. J.; Wong, T. C.; Anderson, K. K. *J. Am. Chem. Soc.* 1981, **103**, 7055–7057.
102. De Jeu, W. H. *Mol. Phys.* 1970, **18**, 31.
103. Sardella, D. J.; Stothers, J. B. *Can. J. Chem.* 1969, **47**, 3089–3092.
104. Eliel, E. L.; Pietrusiewicz, K. M.; Jewell, L. M.; Kenau, W. R. *Tetrahedron Lett.* 1979, 3649–3652.
105. Nguyên, T. T. T.; Delseth, C.; Kintzinger, J. P.; Carrupt, P. A.; Vogel, P. *Tetrahedron* 1980, **36**, 2793–2797.
106. Balakrishnan, P.; Baumstark, A. L.; Boykin, D. W. *Tetrahedron Lett.* 1984, **25**, 169–172.
107. Iwamura, H.; Sugawara, T.; Kawada, Y.; Tori, K.; Muneyuki, R.; Woyori, R. *Tetrahedron Lett.* 1979, 3449–3452.
108. McKelvey, R. D.; Kawada, Y.; Sugawara, T.; Iwamura, H. *J. Org. Chem.* 1981, **46**, 4948–4952.
109. Kobayashi, K.; Sugawara, T.; Iwamura, H. *J. Chem. Soc. Chem. Commun.* 1981, 479–480.
110. Coderre, J. A.; Mehdi, S.; Demou, P. C.; Weber, R.; Traficante, D. D.; Gerlt, J. A. *J. Am. Chem. Soc.* 1981, **103**, 1870–1872.
111. Filowitz, M.; Ho, R. K. C.; Klemperer, W. G.; Shum, W. *Inorg. Chem.* 1979, **18**, 93–103.
112. Day, V. W.; Fredrich, M. F.; Klemperer, W. G.; Shum, W. *J. Am. Chem. Soc.* 1977, **99**, 6146–6148.
113. Filowitz, M.; Klemperer, W. G.; Messerle, L.; Shum, W. *J. Am. Chem. Soc.* 1976, **98**, 2345.
114. Day, V. W.; Fredrich, M. F.; Klemperer, W. G.; Shum, W. *J. Am. Chem. Soc.* 1977, **99**, 952.
115. English, A. D.; Jesson, J. P.; Klemperer, W. G.; Mamouneas, T.; Messerle, L.; Shum, W.; Tramontano, A. *J. Am. Chem. Soc.* 1975, **97**, 4785–4786.

116. Klemperer, W. G.; Shum, W. *J. Am. Chem. Soc.* 1977, **99**, 3544.
117. Filowitz, M.; Klemperer, W. G. *J. Chem. Soc. Chem. Commun.* 1977, 201.
118. Postel, M.; Brevard, C.; Arzoumanian, H.; Reiss, J. G. *J. Am. Chem. Soc.* 1983, **105**, 4922.
119. Doyle, G.; Heaton, B. T.; Occhiello, E. *Organometallics* 1985, **4**, 1224–1225.
120. Horwitz, C. P.; Shriver, D. F. *Organometallics* 1984, **3**, 756–758.
121. Cozak, D.; Butler, I. S.; Hickey, J. P.; Todd, L. J. *J. Magn. Reson.* 1979, **33**, 149–157.
122. Kawada, Y.; Sugawara, T.; Iwamura, H. *J. Chem. Soc. Chem. Commun.* 1979, 291–292.
123. Kump, R. L.; Todd, L. J. *J. Chem. Soc., Chem. Commun.* 1980, 292–293.
124. Gray, G. M.; Redmill, K. A. *J. Organometall. Chem.* 1985, **280**, 105–114.
125. Canet, C.; Goulon-Ginet, C.; Marchal, J. P. *J. Magn. Reson.* 1976, **22**, 539; 1977, **25**, 397.
126. Olah, G. A.; Berrier, A. L.; Prakash, G. K. S. *J. Am. Chem. Soc.* 1982, **104**, 2373–2376.
127. Crandall, J. K.; Centcen, M. A. *J. Org. Chem.* 1979, **44**, 1183–1184.
128. Jenkins, W. B.; McFarlane, W. *J. Chem. Soc. Chem. Commun.* 1977, 531.
129. Broze, M.; Luz, Z. *J. Phys. Chem.* 1969, **73**, 1600.
130. Solomon, I. J.; Raney, R. K.; Kacmarek, A. J.; Maguire, R. G.; Noble, G. A. *J. Am. Chem. Soc.* 1967, **89**, 2015.
131. Gray, G. A.; Albright, T. A. *J. Am. Chem. Soc.* 1977, **99**, 3243.
132. McFarlane, H. C. E.; McFarlane, W. *J. Chem. Soc., Chem. Commun.* 1978, 531.
133. Gerlt, J. A.; Demou, P. C.; Mehdi, S. *J. Am. Chem. Soc.* 1982, **104**, 2848.
134. Virlet, J.; Tantot, G. *Chem. Phys. Lett.* 1976, **44**, 296–299.
135. Lutz, O.; Nepplc, W.; Nolle, A. *Z. Naturforsch.* 1976, **31a**, 1046.
136. Franklin, K. J.; Lock, C. J. L.; Sayer, B. G.; Schrobilgcn, G. J. *J. Am. Chem. Soc.* 1982, **104**, 5303.
137. Luz, Z.; Yagil, G. *J. Phys. Chem.* 1966, **70**, 554.
138. Dwek, R. A.; Luz, Z.; Shporer, M. *J. Phys. Chem.* 1970, **74**, 2232.
139. Earl, W. L.; Niederberger, W. *J. Magn. Reson.* 1977, **27**, 351.
140. Delseth, C.; Kintzinger, J. P.; Nguyên, T. T. T.; Neiderberger, W. *Org. Magn. Reson.* 1978, **11**, 38–39.
141. Reuben, J.; Brownstein, S. *J. Mol. Spectr.* 1967, **23**, 96.
142. Pople, J. A.; Santry, D. P. *Mol. Phys.* 1964, **8**, 1.
143. Jameson, C. J.; Gutowsky, H. S. *J. Chem. Phys.* 1969, **51**, 2790.
144. Grossman, G.; Gruney, M.; Seifert, M. *Z. Chem.* 1976, **16**, 362.
145. Gray, G. A.; Albright, T. A. *J. Am. Chem. Soc.* 1976, **98**, 3857; 1977, **99**, 3243.
146. Delseth, C.; Kintzinger, J. P.; Nguyên, T. T. T.; Niederberger, W. *Org. Magn. Reson.* 1978, **11**, 38–39.
147. Wasylishen, R. E.; Macdonald, J. B.; Friedrich, J. O. *Can. J. Chem.* 1984, **62**, 1181–1183.
148. Gerber, P. *Helv. Phys. Acta* 1972, **45**, 655–682.
149. Rosenblum, B.; Nethercot, A. H. *J. Phys. Chem.* 1957, **27**, 828–829.
150. Edmonds, D. T.; Goren, S. D.; Mackay, A. L.; White, A. A. L. *J. Magn. Reson.* 1976, **23**, 505–514.
151. Brosnan, S. G. P.; Edmonds, D. T. *J. Magn. Reson.* 1980, **38**, 47–63.
152. Cheng, C. P.; Brown, T. L. *J. Am. Chem. Soc.* 1980, **102**, 6418–6421.
153. Cheng, C. P.; Brown, T. L. *J. Am. Chem. Soc.* 1979, **101**, 2327–2334.
154. Brosnan, S. G. P.; Edmonds, D. T.; Poplett, I. J. F. *J. Magn. Reson.* 1981, **45**, 451–460.
155. Butler, L. G.; Cheng, C. P.; Brown, T. L. *J. Phys. Chem.* 1981, **85**, 2738–2740.
156. Butler, L. G.; Brown, T. L. *J. Am. Chem. Soc.* 1981, **103**, 6541–6549.
157. Chachaty, C.; Quaegebeur, J. P. *Molec. Phys.* 1984, **52**, 1081–1104.
158. Ruessink, B. H.; MacLean, C. *Molec. Phys.* 1984, **53**, 421–428.
159. Tricot, Y.; Niederberger, W. *Helv. Chim. Acta* 1984, **67**, 1033–1039.
160. Delseth, C.; Kintzinger, J. P. *Helv. Chim. Acta* 1978, **61**, 1327.
161. Serianni, A. S.; Barker, R. *J. Magn. Reson.* 1982, **49**, 335.
162. Ancian, B.; Tiffon, B.; Dubois, J. E. *Chem. Phys.* 1983, **74**, 171.
163. Gerothanassis, I. P.; Lauterwein, J.; Sheppard, N. *J. Magn. Reson.* 1982, **48**, 431.
164. Gerlt, J. A.; Demou, P. C.; Mehdi, S. *J. Am. Chem. Soc.* 1982, **104**, 2848.
165. Eliasson, B.; Larson, K. M.; Kowalewski, J. *J. Phys. Chem.* 1985, **89**, 258–261.
166. Masada, Y.; Sono, M.; Yamatera, H. *J. Chem. Soc. (Farad. I)* 1985, **81**, 127–136.
167. Hawkes, G. E.; Randall, E. W.; Aime, S.; Osella, D.; Elliott, J. E. *J. Chem. Soc. (Dalton)* 1984, 279–284.
168. Aime, S.; Gobetto, R.; Osella, D.; Hawkes, G. E.; Randall, E. W. *J. Chem. Soc. (Dalton)* 1984, 1863–1866.
169. Brownlee, R. T. C.; O'Connor, M. J.; Shehan, B. P.; Wedd, A. G. *J. Magn. Reson.* 1985, **61**, 22–27.
170. Barbarella, G.; Dewbech, P.; Tugnoli, V. *Org. Magn. Reson.* 1984, **22**, 402–407.

171. Valentine, B.; St. Amour, T. E.; Fiat, D. *Org. Magn. Reson.* 1984, **22**, 697–700.
172. Maksimovskaya, R. I.; Il'yasova, A. K.; Begalieva, D. U.; Takezhanova, D. F.; Akhmetora, A. K. *Izv. Akad. Nauk. SSSR, Ser. Khim.* 1984, 2169–2174.
173. Kotvanova, M. K.; Fedotov, M. A.; Kazanskii, L. P.; Torchenkova, E. A. *Koord. Khim.* 1984, **10**, 1062–1067.
174. Lauterwein, J.; Gerothanossis, I. P.; Hanston, R. N. *J. Chem. Soc. Chem. Commun.* 1984, 367–369.
175. Shyy, Y. J.; Tsai, T. C.; Tsai, M. D. *J. Am. Chem. Soc.* 1985, **107**, 3478–3484.
176. Khutsishvili, V. G.; Serebryanskoya, A. I.; Bogacher, Yu. S.; Kirilova, A. P.; Shapet'ko, N. N. *Zh. Fiz. Khim.* 1985, **59**, 790–791.
177. Valentine, B.; Steinschneider, A.; Dhawan, D.; Burgar, M. I.; St. Amour, T.; Fiat, D. *Int. J. Pept. Protein. Res.* 1985, **25**, 56–68.
178. Fedotov, M. A.; Belyaev, A. V. *Koord. Chim.* 1984, **10**, 1236–1242.
179. Dessinges, A.; Castillon, S.; Olesker, A.; Ton, T. T.; Lukacs, G. *J. Am. Chem. Soc.* 1984, **106**, 450–451.
180. Christ, H. A.; Diehl, P. *Helv. Phys. Acta* 1963, **36**, 170.
181. Sardella, D. J.; Stothers, J. B. *Can. J. Chem.* 1969, **47**, 3089.
182. Kriecheldorf, H. R. *Proc. IUPAC Macromol Symp.*, 28th (*C.A.*, 1983, **99**, 88821).
183. Williams, D. S.; Wong, T. C. *J. Mol. Struct.* 1983, **101**, 297.
184. Marabella, C P.; Enemark, J. H.; Miller, K. F.; Bruce, A. E.; Pariyadath, N.; Corbin, J. L.; Stiefel, E. I. *Inorg. Chem.* 1983, **22**, 3456.
185. Emsley, J.; Lucas, J.; Parker, R. T.; Overill, R. E. *Polyhedron* 1983, **2**, 19.
186. Gray, G. M.; Kraihanzel, C. S. *J. Organomet. Chem.* 1982, **238**, 209; 1983, **241**, 201.
187. Mateescu, G. D.; Benedikt, G. M. *J. Am. Chem. Soc.* 1979, **101**, 3959–3960.
188. Swift, T. J.; Connick, R. E. *J. Chem. Phys.* 1962, **37**, 307.
189. Solomon, I. J.; Kacmarek, A. J.; Sermida, W. K.; Roney, J. K. *Inorg. Chem.* 1972, **11**, 195.
190. Solomon, I. J.; Kacmarek, A. J.; Raney, J. *Inorg. Chem.* 1968, **7**, 1221.
191. Helm, L.; Elding, L. I.; Merbach, A. E. *Helv. Chim. Acta* 1984, **67**, 1453–1460.
192. Hugi-Cleary, D.; Helm, L.; Merbach, A. E. *Helv. Chim. Acta* 1985, **68**, 545–555.
193. Hugi, A. D.; Helm, L.; Merbach, A. E. *Helv. Chim. Acta* 1985, **68**, 508–521.
194. Helm, L.; Elding, L. I.; Merbach, A. E. *Inorg. Chem.* 1985, **24**, 1719–1721.
195. Aygen, S.; Hanssum, H.; Van Eldik, R. *Inorg. Chem.* 1985, **24**, 2853–2855.
196. Bernhard, P.; Helm, L.; Rapaport, I.; Ludi, A.; Merbach, A. E. *J. Chem. Soc. Chem. Commun.* 1984, 302–303.
197. Moore, P.; Sachinidis, J.; Willey, G. R. *J. Chem. Soc. (Dalton)* 1984, 1323–1327.
198. Fiat, D.; Chemlnick, A. M. *J. Am. Chem. Soc.* 1971, **93**, 2875.
199. Connick, R. E.; Fiat, D. *J. Chem. Phys.* 1966, **44**, 4103.
200. Neely, J. W.; Connick, R. E. *J. Am. Chem. Soc.* 1972, **94**, 3419.
201. Fiat, D.; Chemlnick, A. M. *Proc. Colloq. Ampère* 1967, **14**, 1157.
202. Chemlnick, A. M.; Fiat, D. *J. Chem. Phys.* 1967, **47**, 3986.
203. Fiat, D.; Luz, Z.; Silver, B. L. *J. Chem. Phys.* 1968, **49**, 1376.
204. Chemlnick, A. M.; Fiat, D. *J. Magn. Reson.* 1972, **8**, 325.
205. Reuben, J.; Fiat, D. *Inorg. Chem.* 1967, **6**, 579.
206. Chemlnick, A. M.; Fiat, D. *J. Chem. Phys.* 1969, **51**, 4238.
207. Swift, T. J.; Connick, R. E. *J. Chem. Phys.* 1964, **41**, 2553.
208. Desai, A. G.; Dodgen, H. W.; Hunt, J. P. *J. Am. Chem. Soc.* 1969, **91**, 5001.
209. Murray, R.; Dodgen, H. W.; Hunt, J. P. *Inorg. Chem.* 1964, **3**, 1576.
210. Rablen, D. P.; Dodgen, H. W.; Hunt, J. P. *J. Am. Chem. Soc.* 1972, **94**, 1771.
211. Stengle, R. T.; Langford, C. H. *Coord. Chem. Rev.* 1967, **2**, 349.
212. Hunt, J. P. *Coord. Chem. Rev.* 1971, **7**, 1.
213. Bryden, C. C.; Reilly, C. N.; Desreux, J. F. *Rare Earths Mod. Sci. Technol.* 1982, **3**, 53.
214. Sherry, A. D.; Singh, M. *J. Less-Common Met.* 1983, **94**, 343.
215. Meirovitch, E.; Kalb, A. J. *Biochem. Biophys. Acta* 1973, **303**, 258–263.
216. Bertini, F.; Luchinat, C.; Messori, L. *Biochem. Biophys. Res. Commun.* 1981, **101**, 577–583.
217. Fiat, D. *Bull. Magn. Reson.* 1984, **6**, 30–37.
218. Meadows, M. D.; Smith, K. A.; Kinsey, R. A.; Rothgelb, T. M.; Skarjune, R. P.; Oldfield, E. *Proc. Natl. Acad. Sci. U.S.A.* 1982, **79**, 1351.
219. Schramm, S.; Oldfield, E. *J. Am. Chem. Soc.* 1984, **106**, 2502–2506.
220. Schramm, S.; Kirkpatrick, R. J.; Oldfield, E. *J. Am. Chem. Soc.* 1983, **105**, 2483.
221. Oldfield, E.; Keniry, M. A.; Shinoda, S.; Schramin, S.; Brown, T. L.; Gutowsky, H. S. *J. Chem. Soc. Chem. Commun.* 1985, 791–793.

Chapter 15

SULFUR, SELENIUM, AND TELLURIUM

H. C. E. McFARLANE and W. McFARLANE

1. INTRODUCTION

The physical NMR properties of sulfur, selenium, and tellurium are given in Table 1, and as with other groups of the Periodic Table there are expected to be considerable parallels between the chemical aspects of the NMR behavior of these elements (and indeed with oxygen too). In practice the problems associated with ^{33}S NMR have restricted the types of compound studied, and it is not yet appropriate to combine the treatment of all three. However, the more recent work has served mainly to confirm the similarities of selenium and tellurium referred to previously,[1] and so these two elements are considered together. There is a brief review of ^{33}S NMR,[2] and there are several of ^{77}Se and ^{125}Te NMR.[2-5]

2. SULFUR

The only naturally occurring isotope of sulfur with a nuclear spin is ^{33}S with properties that are far from ideal for NMR work, but nonetheless some hundreds of compounds have now been examined successfully. These are usually ones with a fairly high degree of symmetry so as to reduce the electric field gradient at the sulfur nucleus and hence the quadrupolar broadening experienced.

Experimentally it is best to use a spectrometer whose electronics have solid state capabilities since it is necessary to have short 90° pulse lengths (typically 5–15 μs, which demands up to 1 kW rf power) and as short as possible a receiver dead-time.[6] Even so, rolling base lines will interfere seriously with lines that are more than 1 kHz wide when single 90° pulses are used, but special pulse sequences are now available[7-9] which can substantially eliminate many of these problems. Effects due to spin coupling are normally eliminated by quadrupolar relaxation, but in SF_6 $^1J(^{33}S-^{19}F)$ is 251.8 Hz, and this symmetrical species yields[10] a beautiful binomial septet for its ^{33}S spectrum, the central line having a T_1 of 10 s. In the sulfone I $^3J(^{33}S-H)$ is ca. 6 Hz and in the absence of proton decoupling the ^{33}S spectrum is a poorly resolved triplet.[6] As a general rule in ^{33}S NMR $T_1 = T_2$ and can be obtained

H. C. E. McFARLANE and W. McFARLANE • Department of Chemistry, Sir John Cass School of Science and Technology, City of London Polytechnic, London EC3N 2EY, England.

Table 1. NMR Properties of Magnetic Isotopes of Sulfur, Selenium, and Tellurium

Nuclide	Spin	Natural abundance (%)	Receptivity[a]	Resonance frequency[b]	Usual reference material
^{33}S	3/2[c]	0.76	0.097	7.670123	2 *M* Cs_2SO_4 in H_2O at pH 7.5
^{77}Se	1/2	7.58	2.98	19.071523	Me_2Se
^{123}Te	1/2[d,e]	0.87	0.89	26.169773	[Me_2Te]
^{125}Te	1/2[d]	6.99	12.5	31.549802	Me_2Te

[a] Relative to that of ^{13}C.
[b] For the reference compound in a magnetic field strength for which Me_4Si has a proton resonance frequency of 100 MHz.
[c] Quadrupole moment -6.4×10^{-30} m^2.
[d] Negative γ.
[e] Half life 2×10^{13} yr.

directly from the linewidth; values ranging from 10 s to ca. 100 μs have been reported, and it is normally possible to pulse quite rapidly.

Since linewidths commonly exceed 100 Hz it is clearly advantageous to use as high a magnetic field strength as possible so as to maximize the dispersivity, but even so, the very unpredictability of linewidths in particular compounds is liable to have an inhibiting effect upon the growth of ^{33}S NMR. It should further be noted that the linewidth can be very sensitive to solvent effects; for example, for sulfolane at 4.2 T it varies from 17 Hz at 1 *M* concentration in acetone, to 32 Hz in the neat liquid.[(6)]

I

The width of the signal from the symmetrical SO_4^{2-} ion is pH dependent, presumably owing to protonation at oxygen, and can be as great as 200 Hz. Its shielding is temperature dependent, but nevertheless SO_4^{2-} is the preferred reference, and in this chapter chemical shifts will be quoted relative to a 2 *M* Cs_2SO_4 solution at pH 7.5 when the line width is 6.5 Hz.[(6)] It has been suggested[(10)] that on an absolute shielding scale σ for the sulfate ion is ca. 250 ppm.

Table 2 gives a selection of ^{33}S chemical shifts from the literature, and Chart I depicts the main trends. It will be noticed that the overall range of just under 1,000 ppm is less than for oxygen and very much less than for ^{77}Se; however, this must surely be a reflection of the limited range of species for which data are available.

It appears[(13)] that sulfur chemical shifts closely parallel those of ^{77}Se, but are ca. 2.3 times less sensitive to structural and electronic changes. This is to be expected in view of the known correlation of ^{77}Se and ^{125}Te shieldings.[(1)] The relatively narrow lines obtained for sulfones have permitted the study of comparatively small electronic and structural effects,[(14)] and it is apparent that electron withdrawal from sulfur produces high-frequency shifts. Thus in a series of sodium alkyl and aryl sodium sulfonates, RSO_3Na, $\delta(^{33}S)$ varies from +4.2 (R = Et) to −16.5 (R = *m*-$NO_2C_6H_4$) ppm, and parallels the behavior of $\delta(^{13}C)$ in the corresponding sodium carboxylates RCOON.[(15)] The extreme low-frequency shift in $(NH_4)_2S_n$ can

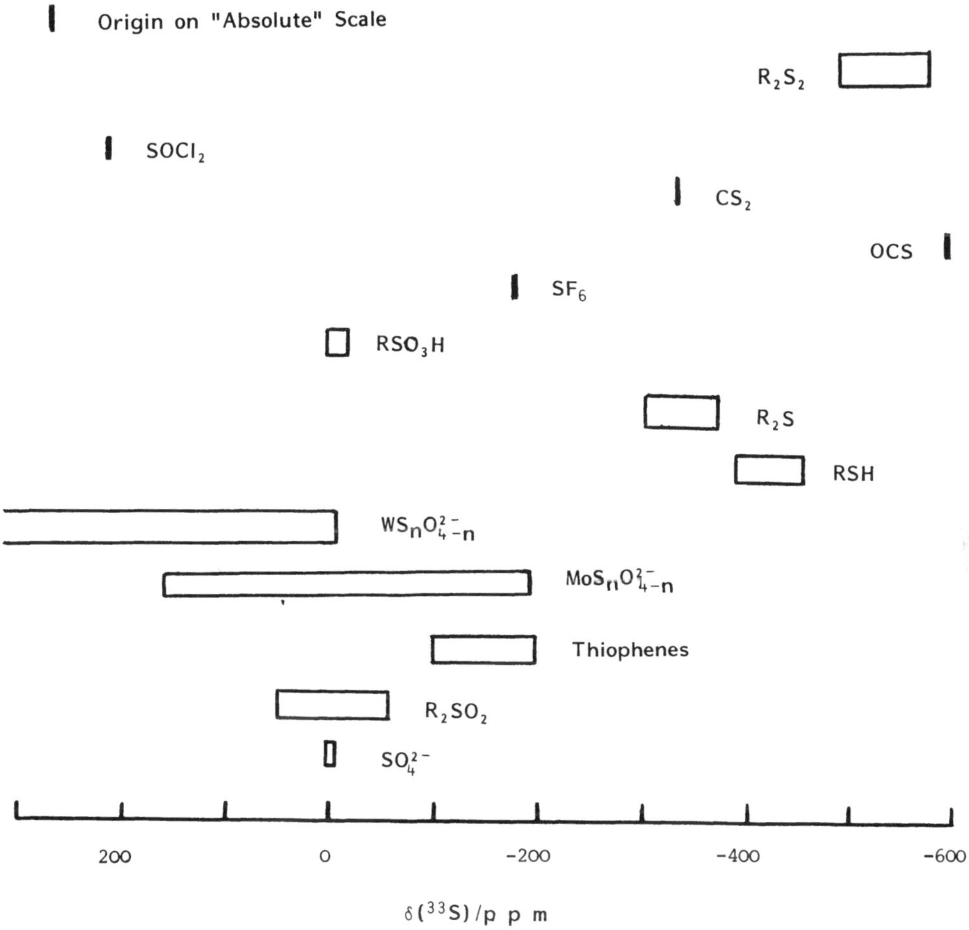

Chart 1. Pattern of ^{35}S Chemical Shifts.

be attributed to the negative charge on sulfur, and changes in electron density on sulfur can also account for the shifts in the series:[13]

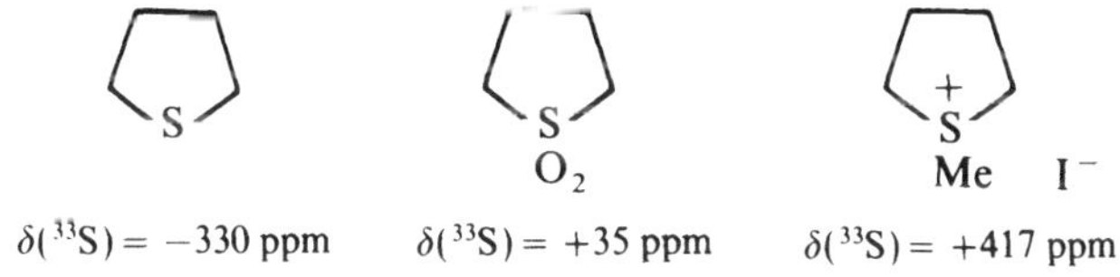

The results for cyclic sulfones show an interesting dependence upon ring size,[6,13] which can probably be attributed to distortion of the interbond angle at sulfur, since longer-range geometrical effects of substituents appear to be quite small. It has been suggested[6] that the results for $MoO_nS_{4-n}^{2-}$ and $WO_nS_{4-n}^{2-}$ can be rationalized by appealing to π overlap between the sulfur and metal atom d orbitals.

The data on linewidths are not yet susceptible of systematic analysis, although it may be noted that apart from SO_4^{2-}, the sulfones and sulfuric acid are the only classes of molecules that can be relied upon to give reasonably narrow lines, and even then increase of molecular weight can lead to problems. This characteristic was used to

Table 2. ^{33}S Chemical Shifts in ppm (Relative to SO_4^{2-}) and Linewidths in Hz for Selected Species

Species	$\delta(^{33}S)$	$W_{1/2}$ (Hz)	Comments	Reference
Cs_2SO_4	0.0	6.5	2 *M* in H_2O	6
H_2SO_4	−108	2200	Neat liquid	11
H_2SO_4	−14	160	5 *M* in water	11
$Na_2S_2O_3$	35	37	Central S atom	12
$HFSO_3$	−22			10
SF_6	−176.6	1	Liquid	10
H_2S	−503.0			10
OCS	−594	440	Liquid	10
CS_2	−333	350	Neat liquid	6
SO_2	375		Liquid	10
O_2SF_2	−40.8		Liquid at −50°	10
$SOCl_2$	224			10
WS_4^{2-}	159	13	In water	6
$WS_2O_2^{2-}$	−56	220	In water	6
MoS_4^{2-}	345	38	In water	6
$MoSO_3^{2-}$	−25	200	In water	6
$(NH_4)_2S_n$	−584	2200	In water	6
Et_2S_2	−500	5000	Neat liquid	11
	−330	5500		13
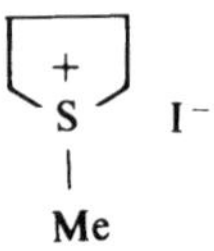	417	9000		13
Thiophene	−119	1450		13
Me_2SO	−20	4900	Neat liquid	6
MeSH	−458		Neat liquid	10
n-BuSH	−415	2100	Neat liquid	6
Me_2SO_2	−13.6	8.5	In acetone	6
Ph_2SO_2	−24.5	42	In acetone	6
$(CH_2)_3SO_2$ (cyclic)	−2.3	15	In $CHCl_3$	6
$(CH_2)_4SO_2$ (cyclic)	34.7	15	In $CHCl_3$	6
$(CH_2)_5SO_2$ (cyclic)	−11.4	44	In $CHCl_3$	6
PhMeS(O)NMe	−31	1200		13
$(MeO)_2SO_2$	−12.6	1400	Neat liquid	6

demonstrate[16] that a reaction product was **II** and not **III**, since an experimental linewidth of 283 Hz was obtained.

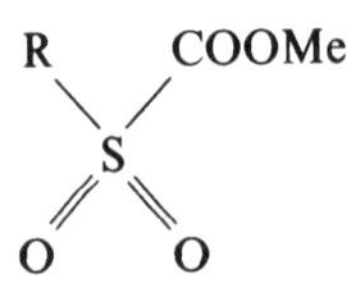

II

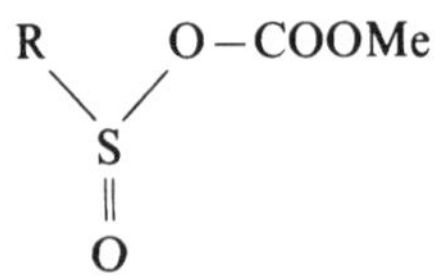

III

In summary, it can be said that ^{33}S NMR has some promise in a number of special situations, but these are likely to be limited by the excessive linewidths and their unpredictability. Indeed it must be borne in mind that quoted values of this parameter can vary by a factor of 2 or more, primarily as a result of different solvent viscosity. The likely parallel of sulfur and selenium shieldings should also be of diagnostic value.

3. SELENIUM AND TELLURIUM

^{77}Se and ^{125}Te are of moderate and good NMR receptivity respectively, but in practice rather more work has been done on the former, probably as a consequence of greater overall activity in selenium chemistry. Many of the earlier results for both of these nuclei were obtained by indirect double resonance methods,[1,17] but modern multinuclear F.t. instrumentation has now developed to the point where direct observation is usually to be preferred. Although ^{125}Te has a negative magnetogyric ratio, proton–tellurium distances are usually so great that the NOE is normally negligible, and standard spin-1/2 methods (i.e., proton decoupling, moderate pulse angles, pulse repetition rates of from 0.5 to a few seconds) can be used in the majority of cases for this nucleus and for ^{77}Se. Coupling constants to nuclei such as the proton, ^{13}C, and ^{31}P are normally obtained from the positions of the satellites in the spectra of these nuclei. ^{123}Te has little potential, except for the measurement of $J(^{125}\text{Te}–^{123}\text{Te})$ in symmetrical species such as the cluster **IV**.[18,19]

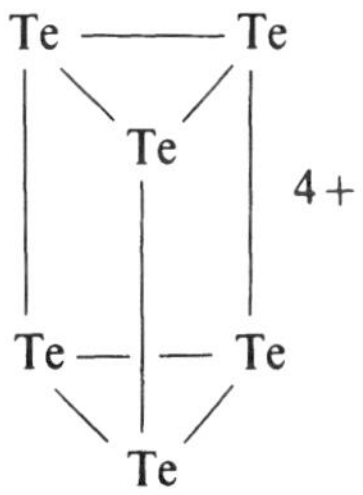

IV

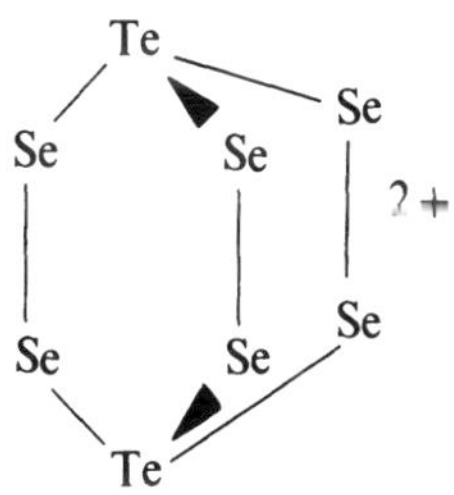

V

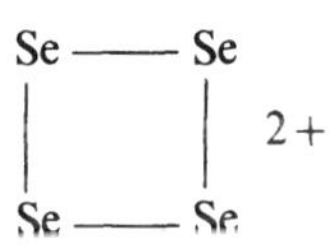

VI

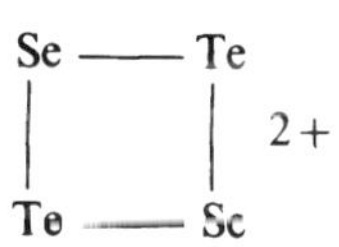

VII

3.1. Chemical Shifts

Tables 3 and 4 give ^{77}Se and ^{125}Te chemical shifts relative to Me_2Se and Me_2Te, respectively, in a selection of compounds chosen to exemplify the whole ranges as far as possible, and Charts II and III summarize the overall position. Other reference compounds have been used including $SeO_2 \cdot aq$, $SeOCl_2$, selenophene, and $Te(OH)_6$. In considering the data of Tables 3 and 4 it should be borne in mind that variations up to ca. 10 ppm with solvent and temperature can occur for ^{77}Se, and about twice this amount for ^{125}Te. Detailed studies of this behavior for Me_2Se and Me_2Te in 30 different solvents indicate a degree of correlation with ν_{max} for the solvent.(57)

An important feature is that tellurium and selenium shieldings run closely parallel in equivalent compounds,(1) a plot of $\delta(^{125}Te)$ against $\delta(^{77}Se)$ being linear with a slope of ca. 1.8. More recent measurements of more extreme values have tended to confirm this original observation,(1) although with somewhat greater deviations. Studies of sets of closely related compounds also support this, for example $\delta(^{125}Te) = 1.61\ \delta(^{77}Se)$ in R_1R_2E, where R_1 and/or R_2 is unsaturated.(52) The most extreme deviations from this general conclusion occur in inorganic derivatives of the two elements where the greatest differences in the type of bonding occur. To some extent the great sensitivity of $\delta(^{125}Te)$ to electronic changes may be attributed to a larger radial expansion term for tellurium, but as the ratio $\langle r_{5p}^{-3}\rangle(Te)/\langle r_{4p}^{-3}\rangle(Se) =$ ca. 1.25 it is clear that some other factor must also be important; this is possibly the greater polarizability of tellurium.

A number of systematic trends have been identified in the magnetic shielding of these nuclei, but their interpretation in the terms of Chapter 3 is often not

Table 3. Selected ^{77}Se Chemical Shifts in ppm Relative to Me_2Se

Species	$\delta(^{77}Se)$	Comments	Reference
$SeOF_2$	1378		20
$SeOCl_2$	1479	Liquid	21
SeF_4	1083	Liquid	20
SeF_6	610	Liquid	21
$ClSeF_5$	599		20
$K^+SeOF_5^-$	632		20
$F_2Se(OSeF_5)_2$	1027	Se(IV)	20
$F_2Se(OSeF_5)_2$	606	Se(VI)	20
H_2Se	−226	Liquid	21
$(NH_4)_2Se$	−511	0.5 *M* in D_2O	22
$SeCl_4$	1154	In DMF	21
Se_4^{2+} [VI]	1958	In SO_3/H_2SO_4	19
$Se_2Te_2^{2+}$ [VII]	1638	In SO_3/H_2SO_4	19
$Se_6Te_2^{2+}$ [V]	584	In SO_3/H_2SO_4 Structure II	23
CSe_2	299		24
K^+SeCN^-	−322	In EtOH	24
$Cd(Se_2CNBu_2)^-$	717	In $CDCl_3$	24
$K^+(Se_2CNBu_2)^-$	582	In D_2O	24
$Se(SiH_3)_2$	−666		25
$Se(GeH_3)_2$	−612		25
$Li^+(SeSiH_3)^-$	−736		26
$Se(SnMe_3)_2$	−547		27
Me_2Se	0	Liquid	17
MeSeH	−116	Liquid	17
$MeSe^-Na^+$	−332	In H_2O	17
$Me_3Se^+I^-$	253	In H_2O	17
Me_2Se_2	275	Liquid	17
Me_2SeCl_2	448		17
$MeSeCl_3$	890		17
Me_2SeO	812	In H_2O	17
Me_2SeO_2H	1216	In H_2O	17

Table 3. (*Continued*)

Species	$\delta(^{77}Se)$	Comments	Reference
$MeSeO_3^-K^+$	1045	In H_2O	17
MeSeCN	125		27
$MeSeCH_2SeMe$	117		28
$MeSeCH_2CH_2SeMe$	121		28
$MeSeCH_2CH_2CH_2SeMe$	74		28
cis-MeSeCH=CHSeMe	172		28
Et_2Se	233		17
$EtSe^-Na^+$	−150	In H_2O	17
$Et_3Se^+I^-$	377	In H_2O	17
EtSeC(O) OEt	428	In H_2O	17
EtSeC(S) OEt	543		29
EtSeC(O) SEt	537		29
Ph_2Se	402	Liquid	30
PhSeH	152	Liquid	17
Ph_2Se_2	460		30
Ph_2SeO	738		31
$PhSeCH_2SePh$	346		28
$PhCH_2CH_2CH_2SePh$	284		28
p-$NO_2C_6H_4SeMe$	233	Liquid	32
p-FC_6H_4SeMe	200	Liquid	32
o-MeC(O) C_6H_4SeCl	1087		33
o-HC(O) C_6H_4SeCl	1097		33
$(PhSe)_3 Sn^-$	164		34
$(PhSe)_3 Pb^-$	127		34
$(PhSe)_3 As$	266		34
$(PhSe)_3 Sb$	397		34
$(MeO)_2 SeO$	1339		17
$(MeO)_2 SeO_2$	1053		17
Selenophene	605	Proposed reference	27
2-Cl-Selenophene	564		31
3-Cl-Selenophene	591		31
$Ph_3P=NSeCl_3$	1144		35
$[(Ph_3P=N)_2 SeCl]^+ SbCl_6^-$	1334		35
p-$MeC_6H_4SO_2N=SePh_2$	652		35
$Bu^t_2C=Se$	2131		36
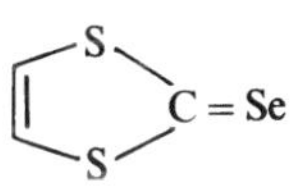	880		37
cis-$(Me_2Se)_2 PtCl_2$	120		17
trans-$(Me_2Se)_2 PtCl_2$	135		17
$(\eta^5\text{-}C_5H_5)_2 Ti(SePh)_2$	847		38
$(\eta^5\text{-}C_5H_5)_2 Zr(SePh)_2$	673		38
$(\eta^5\text{-}C_5H_5)_2 Hf(SePh)_2$	550		38
Me_2PSeMe	58		39
$Me_2P(S) SeMe$	196		39
Me_3PSe	−235		40
$(MeO)_3 PSe$	−396		40
Ph_2PSe	−275		41
cis-MeOP(Se)$(NBu^t)_2$ POMe	−173	Contains four-membered P_2N_2 ring	42
trans-MeOP(Se)$(NBu^t)_2$ POMe	−51		42
CF_3SeH	287		43

straightforward. It is reasonably well established that electron withdrawal from selenium or tellurium will lead to decreased shielding, as exemplified by series such as $MeSe^-$, Me_2Se, Me_3Se^+, and also by the low-frequency shifts of R_3PE, which can be attributed to predominance of the $R_3P^+–E^-$ canonical form (E = Se or Te).[40,58] In the series CF_3SeX (X = Cl, Br, CN, H, Ag) $\delta(^{77}Se)$ correlated well with the electronegativity of X,[43] and in various aromatic derivatives there are usually correlations of $\delta(^{77}Se)$ and $\delta(^{125}Te)$ with Hammett σ and Swain and Lupton substituent parameters.[17] Species exhibiting this behavior include 4-4′-disubstituted diphenyl diselenides,[59,60] selenophenes,[61] benzo[*b*]selenophenes,[62] selenoanisoles,[63,64] selenothiazoles,[65] selenoisoselenazoles,[65] selenenyl sulfides,[66] nitrobenzo[*b*]-selenophenes,[62] selenolo [2.3.-*b*] selenophenes,[67] nitrobenzo-[*b*] selenophenes,[68] diaryl tellurides[54,69] and ditellurides and diaryl dichlorides,[69] tellurophenes,[70] benzo[*b*]tellurophenes,[68,70] miscellaneous phenyl derivatives,[71] the compounds ArTeSeTe(O)Ar,[72] 2-substituted tellurophenes,[53] and the species **VIII**.[73]

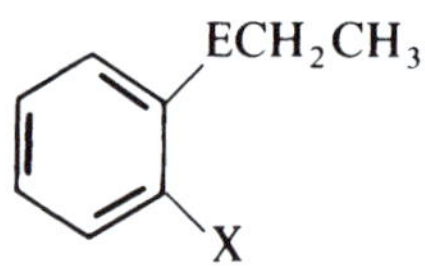

E = Se, Te

X = H, F, Cl, Br, I

VIII

Other evidence of the dependence of $\delta(E)$ upon the electron withdrawing ability of substituents comes from work on tellurium salts[74] and the species R_2TeX_2,[75] and the substantial increase in resonance frequency that occurs when R_2E forms a metal complex by lone pair donation.[1,17,76–78] In addition, estimates of charge densities at selenium based on CNDO calculations in **IX**, **X** and other heterocycles support this view.[79–82]

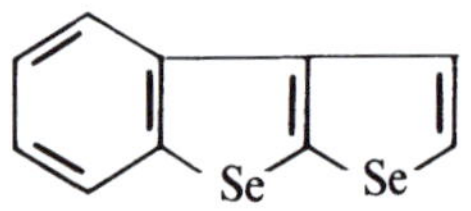

IX

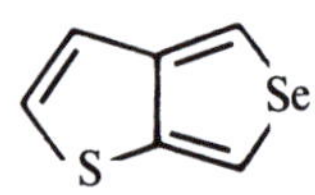

X

Table 4. Selected ^{125}Te Chemical Shifts in ppm Relative to Me_2Te

Species	$\delta(^{125}Te)$	Comments	Reference
$Te(OH)_6$	207	In H_2O	44
TeF_4	1317	In SO_2	48
$TeCl_4$	1237	In HCl/H_2O	45
$TeBr_4$	1282	In HBr/H_2O	45
TeI_4	845	In HI/H_2O	45
TeI_2	3219	In $DMSO/H_2O$	45
Te_4^{2+} (cyclic)	2811	H_2SO_4/SO_3	19
$Se_2Te_2^{2+}$ (cyclic)	3102	H_2SO_4/SO_3	19
Te_6^{4+} (cyclic)	152	H_2SO_4/SO_3	19
$Te_2Se_6^{4+}$ (cyclic)	1613	H_2SO_4/SO_3 (Structure II)	23
TeF_6	545	In BrF_5	47
$HOTeF_5$	601	MeCN	44

Table 4. (*Continued*)

Species	$\delta(^{125}Te)$	Comments	Reference
$XeOTeF_5$	576	In SbF_5	47
$F_2Te(OTeF_5)_2$	740	Te(IV)	20
	343	Te(VI)	20
$Te(OTeF_5)_4$	1105	In MeCN; Te(IV)	48
	553	In MeCN; Te(VI)	48
$Te(OTeF_5)_4$	1313	In SO_2ClF; Te(IV)	48
	552	In SO_2ClF; Te(VI)	48
TeF_3^+	1382	In SO_2	48
Me_2Te	0	Liquid	1
$Me_3Te^+I^-$	443	In DMSO	1
Me_2TeCl_2	749	In CH_2Cl_2	1
Me_2TeCl_2	1218	In MePh	45
Me_2Te_2	688	In CH_2Cl_2	1
$MeTeCl_3$	758	In C_6H_6	45
$MeTeBr_3$	647	In C_6H_6	45
Me_2TeBr_2	858	In PhMe	45
Et_2Te	380	In C_6H_6	1
$Et_3Te^+Br^-$	573	In D_2O	1
Ph_2Te	688	In CH_2Cl_2	1
Ph_2Te_2	422	In CH_2Cl_2	1
Ph_2TeCl_2	981	In CH_2Cl_2	49
$PhTeCl_3$	1238	In DMSO	50
$PhTeBr_3$	1193	In $DMSO/CH_2Cl_2$	50
$PhTeI_3$	1101	In $DMSO/CH_2Cl_2$	50
(p-$MeOC_6H_4$) $TeBr_3$	1204	In $DMSO/CH_2Cl_2$	50
(p-MeC_6H_4) $TeCl_3$	1234	In $DMSO/CH_2Cl_2$	50
(p-BrC_6H_4) $TeCl_3$	1218	In $DMSO/CH_2Cl_2$	50
$TeBr_6^{2-}$	1348	In $DMSO/CH_2Cl_2$	50
$TeCl_6^{2-}$	1531	In DMSO	50
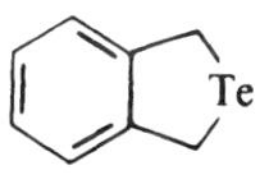	268	In $CDCl_3$	51
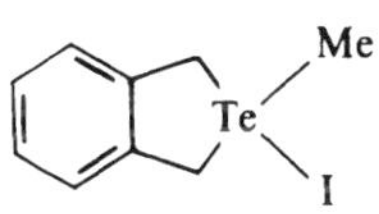	652	In $CDCl_3$	51
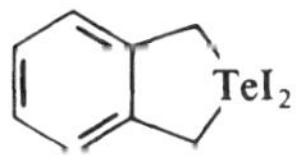	829	In $CDCl_3$	51
Tellurophene	782	Me_2CO	52
2-Br-Tellurophene	960	Me_2CO	53
2-$S(CH_3)$-Tellurophene	886	Me_2CO	53
$(CF_3)_2Te$	1368		43
$(CF_3)_2Te_2$	686		43
$(Me_3Si)_2Te$	−60	In C_6D_6	54
$(Me_3Sn)_2Te$	−1214	In CH_2Cl_2	1
Bu^t_3PTe	−80	In $CDCl_3$	54
$(EtO)_4Te$	1503	In $CDCl_3$	55
$Fe_3Te_2(CO)_9$	1123	In $CDCl_3$; *Nido* cluster	56
$Fe_2Te_2(CO)_6$	−733	In $CDCl_3$; *Arachno* cluster	56

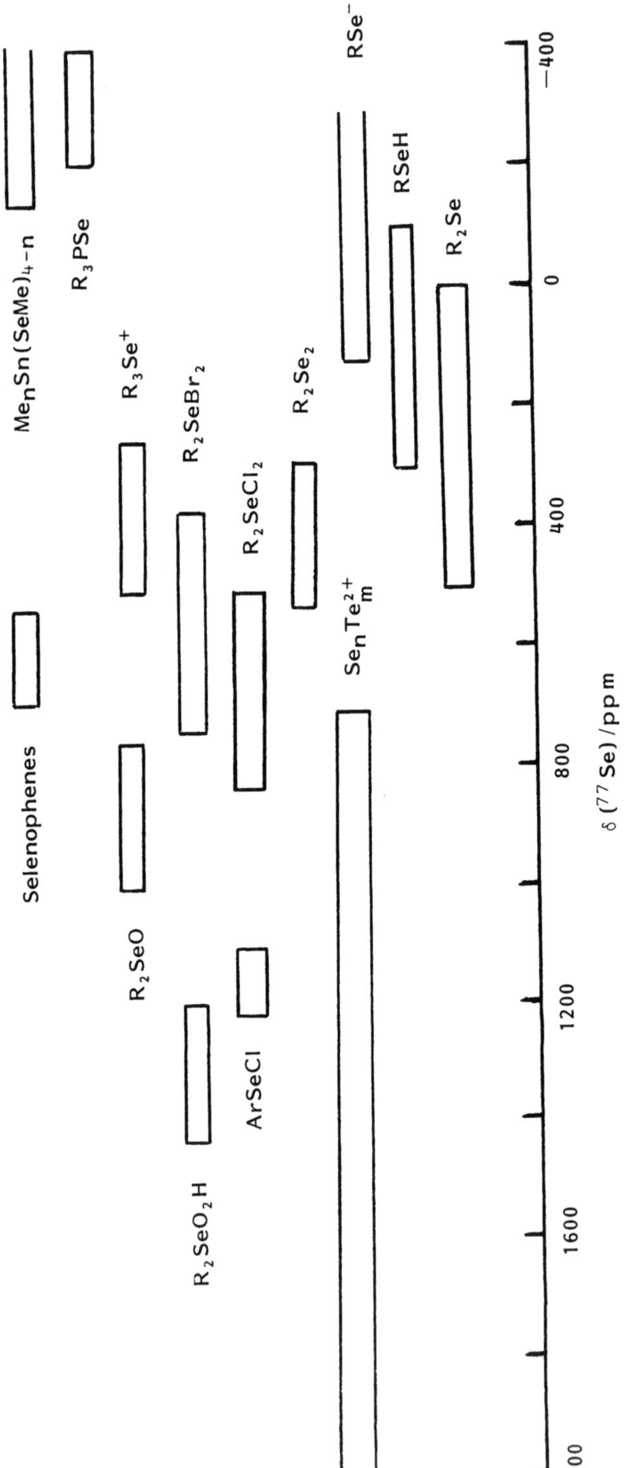

Chart 2. Pattern of ^{77}Se Chemical Shifts.

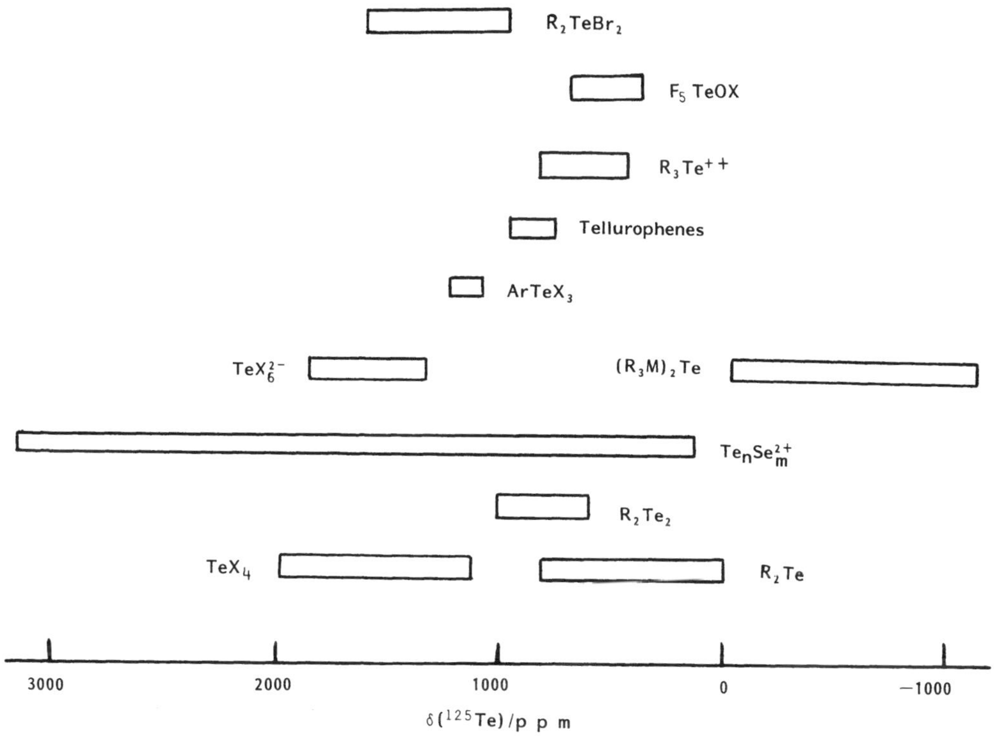

Chart 3. Pattern of ^{125}Te Chemical Shifts.

Nonetheless it must be emphasized that there are significant deviations from this behavior, notably in some species p-$XC_6H_4TeCl_3$ where greater electron release by the aryl group (as predicted by the characteristics of X) increases $\delta(^{125}Te)$,[50] in the patterns of chemical shifts in the series $H_3SeO_3^+$, H_2SeO_3, $HSeO_3^-$, and SeO_3^{2-} obtained from SeO_2 under various conditions[83] and in various mixed MeO/OH/F derivatives of $Te(OH)_6$ where the more electronegative substituent (fluorine) produces increased tellurium shielding.[44,84,85] Thus attempts to use $\delta(^{77}Se)$ or $\delta(^{125}Te)$ to derive substituent electronegativities may not always be soundly based,[86] although the pattern of shieldings in the mixed clusters $Te_{4-n}Se_n^{2+}$ and $Te_{6-n}Se_n^{2+}$ can be readily interpreted by invoking the greater electronegativity of selenium.[18,19]

The probable reality of an electron-density/shielding correlation still leaves open the question of whether it derives primarily from the diamagnetic or the paramagnetic term (Chapter 3). Evidence for the importance of the latter comes from correlations of shielding with electronic excitation energies in certain classes of compound which can be regarded as arising from the ΔE component in the shielding equation. For example, in **XI** and **XII** $\delta(^{77}Se)$ is found to correlate with the wavelength of the $n \rightarrow \pi^*$ transition for the C=Se group,[37] and similar results were earlier reported[36] for t-Bu_2C=Se, PhC(Se) OEt and other molecules. Additional evidence comes from the enhanced temperature dependences of $\delta(^{77}Se)$ and $\delta(^{125}Te)$ in diselenides and ditellurides, respectively.[30,1]

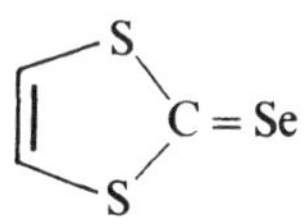

XI

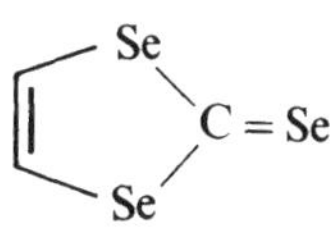

XII

Other correlations of shielding with molecular parameters are strictly empirical, but are none the less valuable for that, and indeed their range and variety serve to emphasize the impracticality at present of seriously using theoretical values in any but the simplest of molecules. In alkyl derivatives R_2E and R_3E^+ there is a clear dependence upon the amount of α-chain branching in R which leads to changes in $\delta(E)$ in the opposite direction to that predicted on electronegativity grounds.[1,17] For example $\delta(^{77}Se)$ in Pr^i_2Se is well to high frequency of that in Pr^n_2Se. Similar behavior is found[87] in RSeC(O)R′, and it may be noted that in this respect the behavior of selenium and tellurium parallels that of phosphorus in many tertiary phosphines R_3P.

The importance of stereochemistry in determining selenium (and presumably tellurium) chemical shifts is well established. $\delta(^{77}Se)$ differs by 2.1 ppm in the two diastereomers of the ester of 2-octanol and 2-phenylselenopropionic acid,[88] while the difference is as large as 17 ppm for the meso and DL forms of $[PhMeP(Se)]_2CH_2$.[89] The greater deshielding of Se^A in **XIII** compared with **XIV** is attributed to the NO_2 group being closer in the former,[90] and there are large differences between the *cis* and *trans* isomers of cyclophosphazene diselenides.[42] In the ligands $RSe(CH_2)_nSeR$ [R = Me, Ph; n = 1-3] $\delta(^{77}Se)$ is strongly dependent upon n in a nonsystematic way which may depend upon stereochemistry,[28] and in their platinum-(IV) complexes the selenium shielding depends upon ring size and conformation to the extent that it can be used to distinguish five- and six-membered chelate rings.[75]

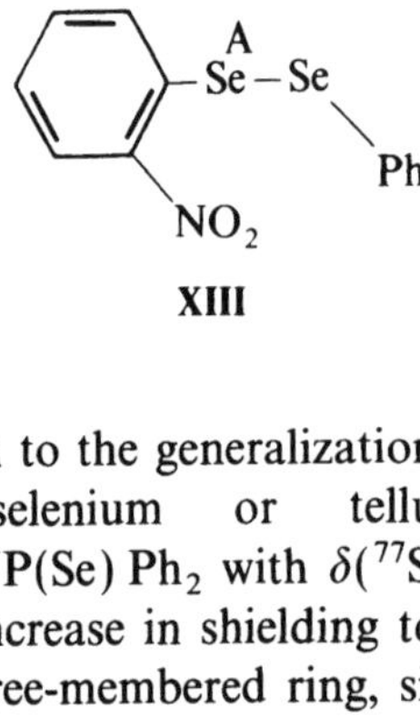

XIII

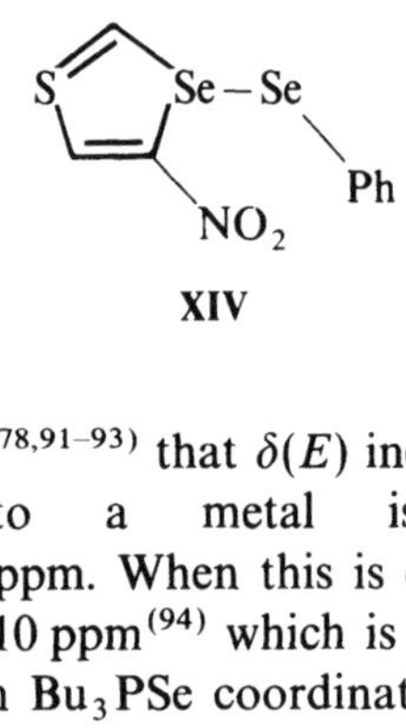

XIV

An exception to the generalization[1,17,24,75–78,91–93] that $\delta(E)$ increases upon coordination of selenium or tellurium to a metal is provided by η^5-$C_5H_5(CO)_3WP(Se)Ph_2$ with $\delta(^{77}Se) = 160$ ppm. When this is converted into **XV** there is a huge increase in shielding to $\delta = -910$ ppm[94] which is clearly due to formation of the three-membered ring, since when Bu_3PSe coordinates to mercury (II) via Se the shielding decreases by 147 ppm.[95]

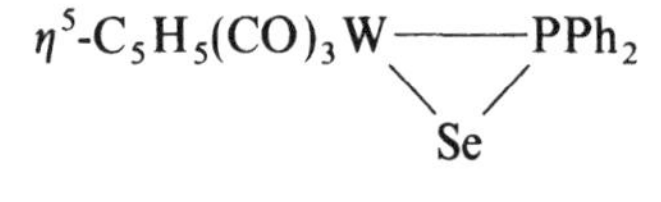

XV

Other examples of selenium bound to a metal are in $(\eta^5$-$C_5H_5)_2M(SePh)_2$, where a systematic decrease in $\delta(^{77}Se)$ in the series M = Ti, Zr, Hf can be attributed to greater diamagnetic circulation on the bulkier metal atoms.[38] This interpretation can also account for the relative values of $\delta(^{77}Se)$ in $Sn(SePh)_3^-$ and $Pb(SnPh)_3^-$ of 164 and 127 ppm, respectively, but not for those in $As(SePh)_3$ and $Sb(SePh)_3$, which are 266 and 397 ppm, respectively.[34]

In some clusters derived from $Fe_3(CO)_{12}$ and other polynuclear transition-metal carbonyls with tellurium atoms incorporated in the metal core the ^{125}Te chemical shift pattern is found to depend upon the number of metal-to-metal bonds subtended by the tethering tellurium groups and not upon the electron count of the cluster.[56] $\delta(^{125}Te)$ is thus determined by magnetic anisotropies associated with the metal–metal bonds.

3.2. Coupling Constants

$\gamma(^{125}Te)$ is -1.65 times $\gamma(^{77}Se)$, and together with the larger value of $\psi(0)^2_{nS}$ for tellurium this leads to coupling constants involving the ^{125}Te being some 2–3 times the magnitude of those involving ^{77}Se in corresponding situations, and of opposite sign. Of course exceptions to this generalization are known, but these usually occur in situations where the chemical bonding is abnormal. Owing to the only moderate abundances of ^{77}Se and ^{125}Te it is often most convenient to determine $J(^{125}TeX)$ or $J(^{77}SeX)$ from the position of the appropriate satellites in the spectrum of X; however, each case should be considered on its merits.

3.2.1. One-Bond Couplings

Table 5 gives a selection of these to a range of different nuclei. Since in most cases the selenium or tellurium atom retains at least one electron lone pair, breakdown of the mean excitation energy approximation is common (as with ^{31}P) and negative reduced couplings are often found.

3.2.1a. Coupling to ^1H. ^1K is positive, although rather few examples are known, and there is a general dependence upon the electron withdrawing ability of the other substituents. Replacement of ^1H by ^2H increases $^1K(SeH)$ by ca. 5%.[110]

3.2.1b. Coupling to ^{13}C. ^1K is negative in derivatives of coordination number 2, and remains so when the coordination of selenium or tellurium increases. However, there is a parallel with the behavior of phosphorus in that in $MeSeO_3^-K^+$, where exceptionally the selenium atom has lost all its electron lone pairs, $^1J(^{77}Se–^{13}C)$ is only -13 Hz and has thus become significantly less negative than in comparable two- and three-coordinate species. There are the usual dependences of $^1J(^{77}Se–^{13}C)$ and $^1J(^{125}Te–^{13}C)$ upon the hybridization[69] and/or effective nuclear charge of the participating nuclei as exemplified by selenoethers,[111] complex selenides,[112] 2-substituted tellurophenes,[53] and various heterocycles.[64] Correlations with Hammett σ-constants in substituted aromatic derivatives[63] and with the polarity of the Te–C bond[113] are also found.

3.2.1c. Coupling to ^{31}P. The reduced coupling is negative and has been best studied for the P=Se link, for which excellent correlations with the electron withdrawing ability of the phosphorus substituents have been found, the more electronegative substituents producing the larger magnitudes[40,104,112,114,115]; it seems likely that $^1J(^{125}Te–^{31}P)$ will behave similarly.[54] On formation of metal complexes by donation from selenium $^1J(^{77}Se–^{31}P)$ decreases by several hundred hertz,[89,92,116,117] and there is evidence for a stereochemical dependence of this coupling.[42,118,119] The single bond $^{77}Se–^{31}P$ and $^{125}Te–^{31}P$ couplings are less well studied and depend substantially upon the nature and orientation of adjacent substituents.[39,54,112,120]

Table 5. Selected One-Bond Couplings in Hz between ^{77}Se or ^{125}Te and Various Nuclei[a]

Nucleus X	$^1J(^{77}\text{Se–X})$	$^1J(^{125}\text{Te–X})$
^{1}H	H_2Se, 64.5 (96)	H_2Te, −59 (98)
	MeSeH, 44 (97)	
	PhSeH, 56 (97)	
^{13}C	Me_2Se, −62 (99)	Me_2Te, 162 (99)
	$EtSeO_3^-K^+$, −18.5 (99)	MeTeC≡CBu, 531 (69)
	$Bu^t_2C{=}Se$, −211 (36)	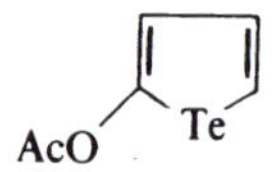, C_2: 274, C_5: 318 (53)
	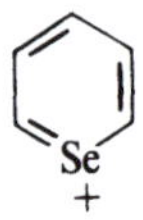, −155 (100)	
^{19}F	SeF_6, −1432 (101)	TeF_6, 3688 (102)
	SeF_5OCl, 1350(ax), 1453(eq) (101)	$HOTeF_5$, 3339 (44)
		$(HO)_5TeF$, 2754 (44)
	SeF_4, 284(ax), 1206(eq) (20)	$F_2Te(OTeF_5)_2$, 2812(Te^{IV}) (20)
		$F_2Te(OTeF_5)_2$, 3419(ax), 3647(eq) (20)
^{29}Si(γ, −ve)		$(Me_3Si)_2Te$; −278 (103)
^{31}P	Me_2PSeMe, −205 (39)	Bu^t_3PTe, 1600 (54)
	$Me_2P(S)SeMe$, −341 (39)	$[Bu^t_2P]_2Te$, 451 (54)
	Me_3PSe, −684 (40, 104)	$[Bu^t_2P]_2TeCr(CO)_5$, 324 (54)
	$(MeO)_3PSe$, −963 (40, 104)	
^{77}Se	MeSeSePh, 22 (105)	MeSeTeMe, −169 (106)
	$MeSeSeCF_3$, (107)	*cis*-$Te_2Se_2^{2+}$, ±470 (19)
	(90)	*trans*-$Te_2Se_2^{2+}$, ±550 (19)
^{119}Sn(γ, −ve)	$Me_3SnSeMe$, 1015 (108)	$(Me_3Sn)_2Te$, −2770 (109)
	$Sn(SeMe)_4$, 1520 (108)	
	$(Me_2SnSe)_3$, 1228 (109)	$(Me_2SnTe)_3$, −3090 (109)
	$(PhSe)_3Sn$, 710 (34)	

[a] References given in parentheses.

3.2.1d. Coupling to Metals. ^1K(SnSe) is negative, and its magnitude in $Me_{4-n}Sn(SeMe)_n$ ($n = 1$–4) correlates well with many of the other NMR parameters of these species,(108) and can also be described in the standard terms of changes in hybridization and effective nuclear charge. Where they have been studied the corresponding couplings involving ^{125}Te behave similarly, but are of ca. $2\frac{1}{2}$ times the magnitude.(109) In transition metal complexes the behavior is very dependent upon the other metal substituents, e.g., $^1J(^{195}\text{Pt–}^{77}\text{Se})$ has values of 259 and −68 Hz in $[PtCl_5SeMe_2]^-$ and $[PtI_5SeMe_2]^-$, respectively.(121) Typical values are available in Refs. 56, 75, 77, and 89 for several metals.

3.2.2. Longer-Range Couplings

Stereochemical dependences which are probably of Karplus-like form have been found for couplings to ^{1}H and ^{13}C for both nuclei,(64,122–125,129) but some typical values (in Hz) in species where there is essentially free rotation are as follows: (references in parentheses)

$^2J(^{77}\mathrm{Se}-^1\mathrm{H})$: Me_2Se, 10.5(97); Me_3Se^+, 9.3(99); $MeSeO_3^-$, −8.8(97); selenium$^+$, 45.3(100).
$^2J(^{125}\mathrm{Te}-^1\mathrm{H})$: Me_2Te, −20.7(99); Me_3Te^+, −20.9(99); tellurophene, −92(53).
$^3J(^{77}\mathrm{Se}-^1\mathrm{H})$: Et_2Se, 10.8(126); Et_3Se^+, 17.4(97); selenophene, 47.0(127).
$^3J(^{125}\mathrm{Te}-^1\mathrm{H})$: Et_2Te, −22.7(126); tellurophene, −12(53); *trans*-$MeOTeF_4OSnMe_3$, 120(128).
$^2J(^{77}\mathrm{Se}-^{13}\mathrm{C})$: Et_2Se, 8.5(126); $(MeO)_2SeO$, 0.9(97); $(MeO)_2SeO_2$, 7.1(97).

Many other longer-range couplings involving ^{77}Se and ^{125}Te are conveniently available in Ref. 130, but are not amenable to systematic treatment at present. Among the more striking are $^3J(^{77}\mathrm{Se}-^{77}\mathrm{Se}) = 12.0$ and 96.5 Hz in the *trans* and *cis* isomers of **XVI**, respectively[131]; $^2J(^{125}\mathrm{Te}-^{125}\mathrm{Te}) = 1400$ Hz in $F_2Te(OTeF_5)_2$[20]; $^3J(^{125}\mathrm{Te}-^{125}\mathrm{Te}) = 120$ Hz in **V**.[23]

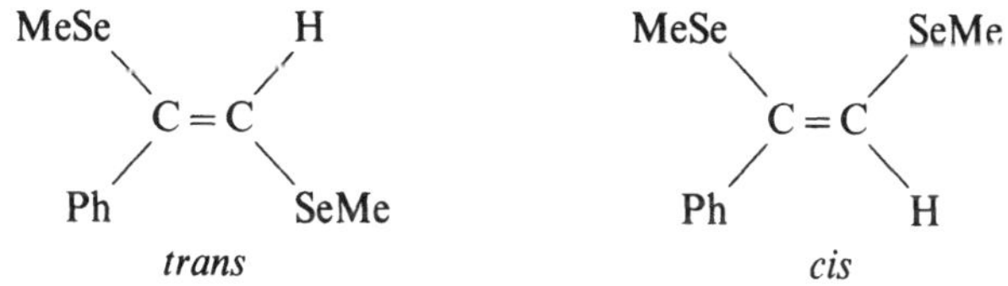

XVI

3.3. Relaxation Behavior

Values of T_1 ranging from under 1 s (H_2Se in D_2O,[22] Me_2Te[132]) to over 4 s (Me_2Se at −60°[22]) have been found for ^{77}Se and ^{125}Te in the rather small number of studies that have been undertaken.[22,24,55,75,110,121,132–138]

The dipole–dipole mechanism is seldom important,[136] which can be attributed to large E–H distances arising from the large covalent radii of selenium and tellurium. However, intermolecular NOEs (implying some dipole–dipole relaxation) have been reported for aqueous selenious acid.[139] For small freely tumbling molecules the spin rotation mechanism usually dominates, but in larger ones at high field strengths chemical shift anisotropy may be important.[136–138] This is especially true for species with P=Se and C=Se double bonds for which the selenium shielding anisotropy may be large.[137–138]

3.4. Applications

Most chemical applications of selenium and tellurium NMR depend upon standard methods of interpretation, and so are not discussed here. However, several studies do merit special mention. Tellurium forms a large number of perfluoroderivatives, often containing the powerful ligand $OTeF_5^-$, which have ^{19}F spectra that are highly second order and difficult to analyze. By contrast the ^{125}Te spectra are much simpler (although still having second-order features) and can be invaluable for the identification of the components of unseparated reaction mixtures.[20,48] Couplings between ^{129}Xe and ^{125}Te of several thousand Hertz have also been used in $OTeF_5$/Xe species.[47,82,140] Many polymeric mixed Se/Te cations have been identified from the pattern of ^{77}Se, ^{123}Te, and ^{125}Te satellites in the various spectra,[18,19,23] and this approach is also of value for transition metal clusters.[56] Changes in $^{1}J(^{77}Se–^{31}P)$ and $^{1}J(^{125}Te–^{31}P)$ were found to be indicative of complex formation by phosphine selenides and tellurides in liquid SO_2 solution,[41] and ^{125}Te measurements were used to detect[44] all possible octahedral species arising from the reaction of $Te(OH)_6$ with HF to give $Te(OH)_{6-n}F_n$ $[n=1–6]$. Mixed triselenides and tritellurides have been characterized by ^{77}Se and ^{125}Te NMR,[141] and new types of selenophene have been identified similarly. In liquid selenium at temperatures up to 1550°C $\delta(^{77}Se)$ and $T_1(^{77}Se)$ are sensitive to the degree of polymerization because they are determined by paramagnetic centers at the chain ends.[143]

REFERENCES

1. McFarlane, H. C. E.; McFarlane, W. *J. Chem. Soc., (Dalton)* 1973, 2416–2418.
2. McFarlane, H. C. E.; McFarlane, W. *NMR and the Periodic Table*; Harris, R. K.; Mann, B. E., Eds.; Academic Press, New York, 1978; pp. 402–419.
3. Lardon, M. *Organic Selenium Compounds: Their Chemistry and Biology*; Klayman, D. L.; Gunther, W. H. H., Eds.; Wiley: New York, 1973; pp. 933–939.
4. Iwamura, H.; Nakanishi, W. *Yuki Gosei Kagaku Kyokaishi* 1981, **39**, 795–804.
5. McFarlane, H. C. E.; McFarlane, W. *NMR of Newly Accessible Nuclei*; Laszlo, P., Ed.; Academic Press, New York, 1983; Vol. 2, pp. 275–299.
6. Belton, P. S.; Cox, I. J.; Harris, R. K. *J. Chem. Soc., Farad Trans. 2* 1985, **81**, 63–75.
7. Ellis, P. D. *NATO ASIC 103 Conference*, Stirling, North Britain, 1981.
8. Patt, S. L. *J. Magn. Reson.* 1982, **49**, 161–163.
9. Canet, D.; Brondeau, D.; Marchal, J. P.; Robin-Lherbier, B. *Org. Magn. Reson.* 1982, **20**, 51–53.
10. Wasylishen, R. E.; Connor, C.; Friedrich, J. O. *Can. J. Chem.* 1984, **62**, 981–985.
11. Retcofsky, H. L.; Friedel, R. A. *J. Am. Chem. Soc.* 1972, **94**, 6579–6584.
12. Hinton, J. F.; Buster, D. *J. Magn. Reson.* 1984, **58**, 324–327.
13. Annunziata, R.; Barbarella, G. *Org. Magn. Reson.* 1984, **22**, 250–254.
14. Harris, D. L.; Evans, S. A., Jr. *J. Org. Chem.* 1982, **47**, 3355–3358.
15. Cassidei, L.; Siracorelli, O. *J. Magn. Reson.* 1985, **62**, 529–532.
16. Farrar, T. C.; Trost, B. M.; Tong, S. L.; Springer-Wilson, S. E. *J. Am. Chem. Soc.* 1985, **107**, 262–264.
17. McFarlane, W.; Wood, R. J. *J. Chem. Soc. Dalton* 1972, 1397–1402.
18. Lassigne, C. R.; Wells, C. J. *J. Chem. Soc., Chem. Commun.* 1978, **21**, 956–957.
19. Schrobilgen, G. J.; Burns, R. C.; Granger, P. *J. Chem. Soc., Chem. Commun.* 1978, **21**, 957–960.
20. Damerius, R.; Huppmann, P.; Lentz, D.; Seppelt, K. *J.C.S. Dalton* 1984, 2821–2826.
21. Birchall, T.; Gillespie, R. J.; Vekris, S. L. *Can. J. Chem.* 1965, **43**, 1672–1679.
22. Odom, J. D.; Dawson, W. H.; Ellis, P. D. *J. Am. Chem. Soc.* 1979, **101**, 5815–5822.
23. Collins, M. J.; Gillespie, R. J. *Inorg. Chem.* 1984, **23**, 1975–1978.
24. Pan, W. H.; Fackler, J. P. *J. Am. Chem. Soc.* 1978, **100**, 5783–5789.
25. Arnold, D. E. J.; Dryburgh, J. S.; Ebsworth, E. A. V.; Rankin, D. W. H. *J. Chem. Soc. Dalton* 1972, 2518–2522.
26. Cradock, S.; Ebsworth, E. A. V.; Rankin, D. W. H.; Savage, W. J. *J. Chem. Soc. (Dalton)* 1976, 1661–1665.
27. Christiaens, L. *Org. Magn. Reson.* 1976, **8**, 354–356.
28. Gulliver, D. J.; Hope, E. G.; Levason, W.; Murray, S. G.; Potter, D. M.; Marshall, G. L. *J. Chem. Soc. Perkin II* 1984, 429–434.
29. Sturm, B.; Gattow, G. *Z. Anorg. Allg. Chem.* 1984, **513**, 183–188; **514**, 120–128.
30. Lardon, M. *J. Am. Chem. Soc.* 1970, **92**, 5063–5066.
31. Fredga, A.; Gronowitz, S.; Hörnfeldt, A. B. *Chem. Scr.* 1975, **8**, 15–20.

32. Kalabin, G. A.; Kushnarev, D. F.; Chmutova, G. A.; Koshuvnikova, L. V. *Zh. Org. Khim.* 1979, **15**, 24–31.
33. Llabrès, G.; Baiwir, M.; Piette, J. L.; Christiaens, L. *Org. Magn. Reson.* 1981, **15**, 152–154.
34. Arsenault, J. J. J.; Dean, P. A. W. *Can. J. Chem.* 1983, **61**, 1516–1523.
35. Roesky, H. W.; Weber, K.-L.; Seseke, V.; Pinkert, W.; Noltemeyer, M.; Clegg, W.; Sheldrick, G. M. *J. Chem. Soc. Dalton* 1985, 565–571.
36. Cullen, E. R.; Guziec, F. S.; Murphy, C. J.; Wong, T. C.; Anderson, K. K. *J. Am. Chem. Soc.* 1981, **103**, 7055–7057.
37. Poleschner, H.; Radeglia, R.; Meyer, H. *Org. Magn. Reson.* 1984, **22**, 480–485.
38. Granger, P.; Gautheron, B.; Tainturier, G.; Pouly, S. *Org. Magn. Reson.* 1984, **22**, 701–704.
39. McFarlane, W.; Nash, J. A. *Chem. Commun.* 1969, 913–914.
40. McFarlane, W., Rycroft, D. S. *J. Chem. Soc. (Dalton)* 1973, 2162–2166.
41. Dean, P. A. W. *Can. J. Chem.* 1979, **57**, 754–761.
42. Keat, R.; Rycroft, D. S.; Thompson, D. G. *Org. Magn. Reson.* 1979, **12**, 391–392.
43. Gombler, W. *Z. Naturforsch* 1981, **36B**, 535–543.
44. Tötsch, W.; Peringer, P.; Sladky, F. *J. Chem. Soc., Chem. Commun.* 1981, **16**, 841–842.
45. Schumann, H.; Magerstödt, M. *J. Organomet. Chem.* 1982, **232**, 147–150.
46. Fazerkerley, G. V.; Celotti, M. *J. Magn. Reson.* 1979, **33**, 219–220.
47. Keller, N.; Schrobilgen, G. J. *Inorg. Chem.* 1981, **20**, 2118–2129.
48. Collins, M. J.; Schrobilgen, G. J. *Inorg. Chem.* 1985, **24**, 2608–2614.
49. McFarlane, W.; Berry, F. J.; Smith, B. C. *J. Organometall. Chem.* 1976, **113**, 139–141.
50. Chadka, R. K.; Miller, J. M. *Can. J. Chem.* 1982, **60**, 2256–2258.
51. Al-Rubaie, A. Z.; McWhinnie, W. R.; Granger, P.; Chapelle, S. *J. Organomet. Chem.* 1982, **234**, 287–298.
52. Kalabin, G. A.; Valeev, R. B.; Kushnarev, D. F. *Zh. Org. Khim.* 1981, **17**, 947–953.
53. Martin, M. L.; Trierweiler, M.; Galasso, V.; Fringuelli, F.; Taticchi, A. *J. Magn. Reson.* 1982, **47**, 504–506.
54. DuMont, W. W.; Kroth, H. J. *Z. Naturforsch.* 1981, **36B**, 332–334.
55. Denney, D. B.; Denney, D. Z.; Hammond, P. J.; Hsu, Y. F. *J. Am. Chem. Soc.* 1981, **103**, 2340–2347.
56. Lesch, D. A.; Rauchfuss, T. B. *Inorg. Chem.* 1983, **22**, 1854–1857.
57. Taft, R. W.; Kamlet, M. J. *Inorg. Chem.* 1983, **22**, 250–254.
58. Wrackmeyer, B. *Z. Naturforsch.* 1984, **39B**, 533.
59. Gronowitz, S.; Konar, A.; Hörnfeldt, A. B. *Org. Magn. Reson.* 1977, **9**, 213–216.
60. Kalabin, G. A.; Kushnarev, D. F.; Kamaeva, L. M.; Kashurnikova, L. V.; Vinokurova, R. J. *Zh. Org. Khim.* 1978, **14**, 2478–2479.
61. Gronowitz, S.; Johnson, J.; Hörnfeldt, A. B. *Chem. Scr.* 1975, **8**, 8–12.
62. Baiwir, M.; Llabrès, G.; Christiaens, L.; Piette, J. L. *Org. Magn. Reson.* 1981, **16**, 14–16.
63. Kalabin, G. A.; Kushnarev, D. F.; Bzesovsky, V. M.; Tschmutova, G. A. *Org. Magn. Reson.* 1979, **12**, 598–603.
64. Kalabin, G. A.; Kushnarev, D. F.; Mannafov, T. G. *Zh. Org. Khim.* 1980, **16**, 505–511.
65. Onyamboko, N. V.; Renson, M.; Chappell, S.; Granger, P. *Org. Magn. Reson.* 1982, **19**, 74–77.
66. Luthra, N. P.; Dunlop, R. B.; Odom, J. D. *J. Magn. Reson.* 1982, **46**, 152–157.
67. Konar, A.; Litvinov, V. P. *Chem. Scr.* 1983, **22**, 22.
68. Baiwir, M.; Habir, G.; Christiaens, L.; Piette, J.-L. *Org. Magn. Reson.* 1982, **18**, 33–37
69. Kalabin, G. A.; Kushnarev, D. F.; Valeev, R. B. *Zh. Org. Khim.* 1981, **17**, 1139–1142.
70. Drakenberg, T.; Hörnfeldt, A. B.; Gronowitz, S.; Talbot, J. M.; Piette, J. L. *Chem. Scr.* 1979, **13**, 152–154.
71. Lohner, W.; Praefcke, K. *J. Org. Chem.* 1981, **208**, 39–42.
72. Lohner, W.; Praefcke, K. *J. Org. Chem.* 1981, **208**, 43–46.
73. Baiwir, M.; Llabrès, G.; Luxea, A.; Christiaens, L.; Piette, J. L. *Org. Magn. Reson.* 1984, **22**, 312–316.
74. Al-Rubaie, A. Z.; McWhinnie, W. R.; Granger, P.; Chapelle, S. *J. Organomet. Chem.* 1982, **234**, 287–298.
75. Zumbulyadis, N.; Gysling, H. J. *J. Org. Chem.* 1980, **192**, 183–188.
76. Abel, E. W.; Orrell, K. G.; Platt, A. W. G. *J. Chem. Soc. Dalton* 1983, 2345–2351.
77. Gysling, H. T.; Zumbulyadis, N.; Robertson, J. A. *J. Organ. Chem.* 1981, **209**, 41–44.
78. Baiwir, M.; Llabrès, G.; Piette, J. L.; Christiaens, L. *Org. Magn. Reson.* 1980, **14**, 293–295.
79. Abronin, I. A.; Djumanazarova, A. Z.; Litvinov, V. P. *Chem. Scr.* 1982, **19**, 75–77.
80. Gronowitz, S.; Konar, A.; Hörnfeldt, A. B. *Chem. Scr.* 1982, **19**, 5–12.

81 Talbot, J M, Piette, J L, Christiaens, L, Drakenberg, T, Gronowitz, S, Llabres, G, Baiwir, M *Chem Scr* 1981, **18**, 147–151
82 Birchall, T, Myers, R D, de Waard, H, Schrobilgen, G J *Inorg Chem* 1982, **21**, 1068–1073
83 Kolshorn, H, Meier, H *J Chem Res Synop* 1977, 338–339
84 Totsch, W, Sladky, F *J Chem Soc Chem Commun* 1980, **19**, 927–928
85 Totsch, W, Sladky, F *Z Naturforsch, Teil B* 1983, **38**, 1025
86 Collins, M J, Rao, U R K, Schrobilgen, G J *J Magn Reson* 1985, **61**, 137–140
87 Baiwir, M, Llabres, G, Piette, J L, Christiaens, L *Spectrochimica Acta A* 1982, **38A**, 575
88 Michelsen, P, Arnby, U, Gronowitz, S *Chem Scr* 1984, **24**, 251–252
89 Colquhoun, I J, McFarlane, W *J Chem Res Synop* 1978, 368–369
90 Granger, P, Chapelle, S, Paulmier, C *Org Magn Reson* 1980, **14**, 240–243
91 Anderson, S J, Goggin, P L, Goodfellow, R *J Chem Soc Trans* 1976, 1959–1964
92 Colton, R, Dakternieks, D *Aust J Chem* 1980, **33**, 1463–1470
93 Colquhoun, I J, McFarlane, W *J Chem Soc Dalton Trans* 1981, 658–660
94 Malisch, W, Maisch, R, Meyer, A, Greissinger, D, Gross, E, Colquhoun, I J, McFarlane, W *Phosphorus Sulphur* 1983, 299–302
95 Colquhoun, I J, McFarlane, W *J Chem Soc Dalton Trans* 1981, 658–660
96 Dreeskamp, H, Pfisterer, G *Mol Phys* 1968, **14**, 295–297
97 McFarlane, W, Rycroft, D S, Turner, C J *Bull Soc Chim Belge* 1977, **86**, 457–463
98 Glidewell, C, Rankin, D W A, Sheldrick, G M *Trans Farad Soc* 1969, **65**, 1409–1412
99 McFarlane, W *Mol Phys* 1967, **12**, 243–248
100 Sandor, P, Radics, L *Org Magn Reson* 1981, **16**, 148–155
101 Seppelt, K *Z Anorg Allg Chem* 1973, **399**, 65–69, 1975, **416**, 12–15
102 Fraser G W, Peacock, R D, McFarlane, W *Mol Phys* 1969, **17**, 291–297
103 Jones, C H W, Sharma, R D *J Organomet Chem* 1984, **268**, 113–118
104 McFarlane, W, Rycroft, D S *Chem Commun* 1972, 902–903
105 McFarlane, W *J Chem Soc A* 1969, 670–672
106 Pfisterer, G, Dreeskamp, H *Ber Bunsen-gesell Phys Chem* 1969, **73**, 654–661
107 Gombler, W *J Magn Reson* 1983, **53**, 69–75
108 Kennedy, J D, McFarlane, W *J Chem Soc Dalton Trans* 1973, 2134–2139
109 Blecker, A, Mathiasch, B, Mitchell, T N *J Organ Chem* 1980, **184**, 175–180
110 Jakobsen, H J, Zozulin, A J, Ellis, P D, Odom, J D *J Magn Reson* 1980, **38**, 219–227
111 Gulliver, D T, Hope, E G, Levason, W, Murray, S G, Potter, D M, Marshall, G L *J Chem Soc Perkin Trans* 1984, 429
112 Dean, P A W *Can J Chem* 1982, **60**, 2921–2926
113 Chadka, R K, Miller, J M *J Chem Soc Dalton Trans* 1982, 117–120
114 Colquhoun, I J, McFarlane, W *J Chem Soc Dalton Trans* 1982, 1915
115 Socol, S M, Verkade, J G *Inorg Chem* 1984, **23**, 3487
116 Grim, S O, Walton, E D, Satek, L C *Inorg Chim Acta* 1978, **27**, L115–117
117 Dean, P A W, Carson, G K *Can J Chem* 1983, **61**, 1800
118 Kroschefsky, R D, Verkade, J G *Inorg Chem* **18**, 469–472
119 Gombler, W, Kinas, R W, Stec, W T *Z Naturforsch, Teil B* 1983, **38**, 815
120 du Mont, W-M, Severengiz, T, Meyer, B *Angew Chem, Int Ed Engl* 1983, **22**, 983
121 Gansow, O A, Vernon, W D, Dechter, J J *J Magn Reson* 1978, **32**, 19–21
122 Olsson, K, Almquist, S O *Acta Chem Scand* 1969, **23**, 3271–3272
123 Reich, H J, Trend, J E *Chem Commun* 1976, 310–311
124 Dewan, J C, Jennings, W B, Silver, J, Tolley, M S *Org Magn Reson* 1978, **11**, 449–452
125 Abel, E W, Bhargava, S K, Bhatti, M M, Kite, K, Mazid, M A, Orrell, K G, Sik, V, Williams, B L, Hursthouse, M B, Malik, K M A *J Chem Soc Dalton Trans* 1982, 2065
126 Breuninger, V, Dreeskamp, H, Pfisterer, G *Ber Bunsengesell Phys Chem* 1966, **70**, 613–617
127 Simonin, M P, Pouet, M J, Cense, J M, Paulmier, F *Org Magn Reson* 1976, **8**, 508–515
128 Totsch, W, Sladky, F *J Fluorine Chem* 1982, **19**, 213
129 Nakanishi, W, Ikeda, Y *Bull Soc Chem Jpn* 1983, **56**, 1661
130 *Spec Per Rep NMR*, The Chemical Society, 1971–1985, Vols 1–14
131 Johannsen, I, Eggert, H *J Am Chem Soc* 1984, **106**, 1240–1243
132 Chapelle, S, Granger, P *Mol Phys* 1981, **44**, 459–467
133 Dawson, W H, Odom, J D *J Am Chem Soc* 1977, **99**, 8352–8354
134 Granger, P, Chapelle, S, Brevard, C *J Magn Reson* 1981, **42**, 203–207

135. Granger, P.; Chapelle, S. *J. Magn. Reson.* 1980, **39**, 329–334.
136. Abel, E. W.; Orrell, K. G.; Platt, A. W. G. *Org. Magn. Reson.* 1983, **21**, 196–198.
137. Wong, T. C.; Guziec, F. S.; Moustakis, C. A. *J. Chem. Soc., Perkin Trans. 2*, 1983, 1471.
138. Wong, T. C.; Ang, T. T.; Guziec, F. S.; Moustakis, C. A. *J. Magn. Reson.* 1984, **57**, 463.
139. Koch, W.; Lutz, O.; Nolle, A. *Z. Naturforsch. A* 1978, **33**, 1026–1030.
140. Jacob, E.; Lentz, D.; Seppelt, K.; Simon, A. *Z. Anorg. Allg. Chem.* 1981, **472**, 7–25.
141. Dereu, N. L. M.; Zingaro, R. A.; Meyers, E. A. *Organometallics* 1982, **1**, 111–115.
142. Konar, A.; Gronowitz, S. *Tetrahedron* 1980, **36**, 3317–3323.
143. Warren, W. W., Jr.; Dupree, R. *Phys. Rev. B* 1980, **22**, 2257–2275.

Chapter 16

FLUORINE

CYNTHIA J. JAMESON

1. ^{19}F NMR MEASUREMENTS

Because of the high inherent NMR sensitivity of ^{19}F nuclei, even the novice will have no difficulty observing ^{19}F NMR spectra. The liquid reference which has evolved as a standard is $CFCl_3$. This reference suffers from the same problems as most external reference liquids in that it has a sizable temperature dependence.[1] Use of $CFCl_3$ or C_6F_6 as an internal reference is not recommended; both have fairly sizable solvent shifts.[2,3] In general, ^{19}F solvent effects are greater than those of ^{1}H, so that a report of chemical shifts ideally should include the conditions under which the sample is observed, e.g., neat liquid, 298 K.

Bulk susceptibility contributions to the shielding are not negligible, although relatively less important than in ^{1}H work. This is of course due to the fact that all nuclei in a sample suffer the same absolute bulk susceptibility contribution to shielding; but because of the substantially ($\sim$40 times) larger shifts in ^{19}F compared to ^{1}H, the relative error in ignoring bulk susceptibility corrections is smaller in ^{19}F. However, other medium effects are proportionally larger for ^{19}F than ^{1}H. For example, gas-to-liquid shifts for ^{19}F in various molecules are typically 2–10 ppm (the liquid appears at higher resonance frequency than the gas), with an exceptionally large value, ca. 26 ppm, for ClF.[1] ^{19}F solvent shifts range from 2 to 16 ppm,[4] F^- ion in aqueous solution shifts by as much as 30.6 ppm, depending on the identity and the concentrations of the other ions present.[5] There are also medium effects on coupling constants involving F nuclei. Solvent shifts of coupling constants are typically of the order of 0.1%–10%,[6] occasionally much larger—for example, a change of nearly 9 Hz in 30 Hz for ^2J(FF) in $CF_2=CH_2$. $^1J(^{11}$BF) in BF_3 in the gas phase varies from 19 Hz at 240 K to 17 Hz at 330 K,[7] in the liquid phase from 24.4 Hz at 176 K to 18.5 Hz at 300 K.[8] $^1J(^{35}$ClF) in ClF $= 855 \pm 10$ Hz in the gas phase and 770 ± 10 Hz in the liquid phase. Coupling constants also depend on temperature, even when there are no exchange or internal rotation averaging processes involved. A review of the temperature coefficients of F spin–spin coupling shows that these coefficients can be of either sign and as large as 2.4 Hz/deg.[6]

CYNTHIA J. JAMESON • Department of Chemistry, University of Illinois, Chicago, Illinois 60680.

2. ^{19}F CHEMICAL SHIFTS

Tables of ^{19}F chemical shift data have been published, including literature values from 1951 to 1978.[2,9–14] In these compilations of chemical shift values, the liquid reference which has been used to reduce ^{19}F chemical shift data to a common scale is $CFCl_3$, although the signs are not uniform from one review to another so care must be exercised. The information in these compilations and in more recent papers up to 1982 has been summarized in Figure 1. Unlike nuclei in the other chapters, ^{19}F has been observed in bonds to nearly every element in the Periodic Table. Thus, Figure 1 is arranged according to the atomic number of the element to which F is attached rather than in the order of increasing or decreasing shielding. The range of ^{19}F shielding is large, the most shielded molecule ClF to the least shielded FOOF spanning 1313 ppm.

2.1. Absolute Shielding Scale

^{19}F is the only nucleus for which an absolute shielding scale includes a large number of free molecule values (i.e., extrapolated to zero density).[1,18] The molecular beam data for the primary reference molecule, on which the shielding scale is based, is of good precision,[19,20] and the diamagnetic term and vibrational corrections were calculated using wave functions of very good quality.[18,21] In addition, spin-rotation constants for several molecules have been measured, each one providing an indepen-

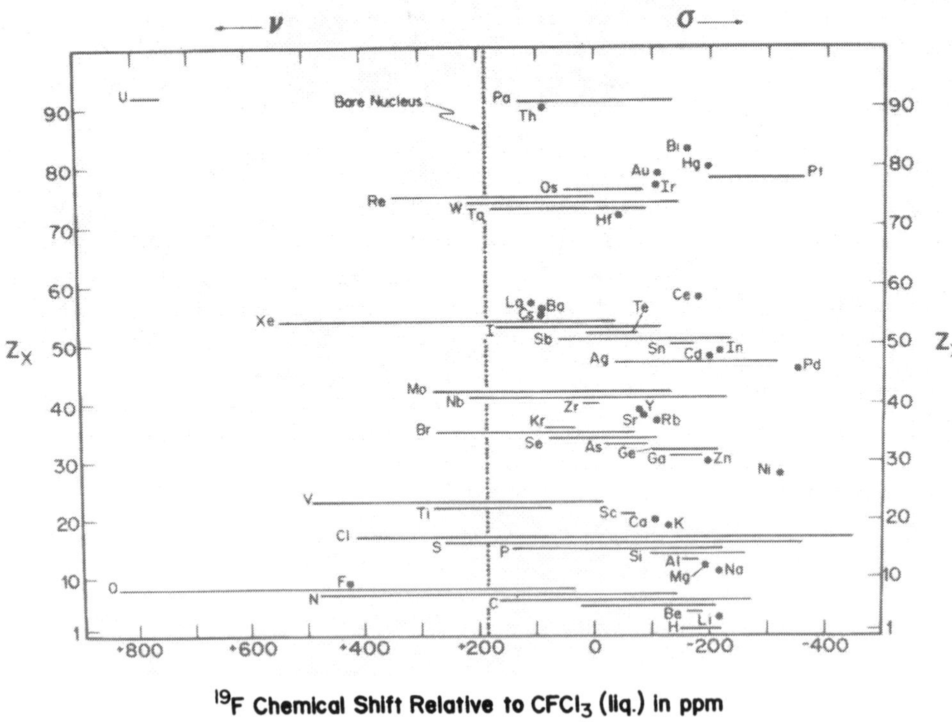

Figure 1. Fluorine chemical shifts relative to liquid $CFCl_3$, arranged according to the atomic number of the atom to which the fluorine is attached. Absolute shielding increases in going from right to left; note the dotted vertical line, which corresponds to the bare fluorine nucleus. Data taken from Refs. 2, 6, 10–17.

dent check on the absolute shielding scale. As we have seen in Chapter 3, components of the absolute shielding for the free molecule can be calculated from spin-rotation constants,[22] which are available for the following ^{19}F sites: CH_2CF_2,[23] DF,[24] ClF,[25,26] COF_2,[27] FC≡CH,[28] CH_2F_2,[29] HF, F_2,[20] CF_4, SiF_4, GeF_4,[30] SF_6,[31] CH_3F,[32] LiF, NaF, KF, RbF, CsF, InF, TlF,[33] BF_3.[29] It has been found that the isotropic shielding derived from these spin-rotation constants agree with the free-molecule NMR data where available, to within the precision of the spin-rotation measurement.[1] Actually, the free molecule NMR data[1,18] are easier to obtain than the molecular beam-derived data, so that the former serves as a check on the latter. Values of ^{19}F absolute shielding for the free molecules, based on Hindermann and Cornwell's analysis of HF as primary reference, are given in Table 1.

Table 1. ^{19}F Absolute Isotropic Shielding of the Rovibrationally Averaged Free Molecule at 300 K[1,34] Based on HF Molecule,[18] Compared with Absolute Shielding Derived from Spin-Rotation Constants

		From spin-rotation constants	
	σ (ppm)	σ (ppm)	Reference
ClF	637.1	630.7 ± 0.3	25, 17
HF	410.0		
SiF_4	363.2	371.1 ± 3.3/370	35, 17/32, 22
BF_3	327.2	327	32, 17
POF_3	288.2		
PF_3	228.3		
COF_2	221.6	228.0 ± 10	27
ClF_3	175.9, 50.0		
SO_2F_2	164.7		
SF_6	139.6	132 ± 26	31
NF_3	50.3		
ClO_2F	−136.5		
F_2	−232.8	−219/−216 ± 1.3	20, 22/36, 17
CH_3F	471.0	471 ± 9	32, 22
CF_2H_2	339.1	338	29
$CFHCl_2$	276.3		
CF_3H	274.1		
CF_2HCl	267.4		
CF_4	259.0	260.8 ± 10.8/258	35, 37/32, 17
CF_3CN	252.1		
CF_3Cl	224.4		
CF_3Br	213.8		
CF_2Cl_2	202.6		
CF_3I	199.6		
$CFCl_3$	195.6		
$CF_2{=}CFH$	297.5, 325.3, 404.2		
$CF_2{=}CFCl$	300.1, 316.7, 340.6		
$CF_2{=}CFBr$	295.4, 314.2, 341.4		
$CF_2{=}CFI$	286.8, 310.5, 346.3		
$CF_2{=}CF_2$	330.7		
$CF_2{=}CH_2$	279.0	272 ± 21	23
CF_2HCH_3	306.4		
CF_3CF_3	283.0		
CF_3CF_2Cl	281.9, 269.7		

Hindermann and Cornwell's absolute shielding for liquid $CFCl_3 = 188.7$ ppm[18] (cyl $\perp B_0$, 303 K) can be used to convert the scale in Figure 1 from chemical shifts to absolute shielding. We have drawn the +188.7 ppm vertical line on this chemical shift scale to indicate the ^{19}F bare nucleus. We note that some of the high-frequency ^{19}F resonances correspond to negative absolute shielding, that is, the paramagnetic circulations offset the diamagnetic circulations to lead to a greater local magnetic field at the nucleus than would be experienced by a bare ^{19}F nucleus. Examples of these chemical environments are F_2 ($\sigma = -232.8$),[1] FNO ($\sigma = -290$),[38] MoF_6 ($\sigma = -89$), XeF_4 ($\sigma = -251$), XeF_6 ($\sigma = -357$), ReF_7 ($\sigma = -158$), and UF_6 ($\sigma = -557$).[17] On this scale, ClF free molecule has isotropic $\sigma = 637.1$[1] and liquid FOOF has $\sigma = -676$ ppm.[39,40] It is quite remarkable that F bonded to Cl spans 860 ppm from ClF to $FClF_4$.

2.2. Empirical Patterns of ^{19}F Nuclear Shielding

^{19}F chemical shifts were found to correlate with electronegativity in 1951 by Gutowsky and Hoffmann,[41] with the shielding in the binary fluorides decreasing with increasing electronegativity of the other element (i.e., ^{19}F resonance frequencies increase with increasing electronegativity). This can be interpreted roughly by equation (10) of Chapter 3. The bonds to the less electronegative atoms are more ionic, the bonds to the more electronegative atoms more covalent. A larger paramagnetic contribution to σ comes with the latter due to a larger deviation from spherical (F^- ion) symmetry. Figure 1 shows a rather remarkable periodic pattern which was previously noted by Mason in the binary fluorides.[17] She found that the diamagnetic contribution to σ itself, calculated by the method of Gierke and Flygare,[22] varies periodically, as does the paramagnetic term derived from spin-rotation data.[17] The periodic patterns in the paramagnetic term may be interpreted

Table 2. Selected ^{19}F Absolute Shielding Tensors from Solid NMR Spectra[45] and from Spin-Rotation Constants[a]

	σ_{11}	σ_{22}	σ_{33}	σ_{iso}	$\Delta\sigma$	Methods[b]
$C_6F_{6(s)}$, 40 K	310.8	310.8	465.9	362.5	155.1	PW, MP
$PhF_{(s)}$, 77 K	247.5	312.5	356.6	305.5		PW, MP
$CHF_{3(s)}$, 4 K	242.5	242.5	347.5	277.5	105[c]	CT, CW
$CH_3F_{(s)}$, 1.3 K	427.5	493.5	493.5	471.5	−66	CT, CW
CH_3F, free	426	493	493	471	−67	SR, MB
$F_{2(s)}$, 29 K	−588.5	−588.5	461.5	−238.5	1050	SM, CW
F_2, free	−569	−569	480	−219	1049	SR, MB
ClF, free	488	806	806	700	−318	SR, MB
HF, free	393	393	480	421	87	SR, MB
MgF_2, RT	368.5	383.5	398.5	383.5		SC, MP
ZnF_2, RT	370.5	392.5	414.5	392.5		SC, MP
XeF_4	58.5	58.5	528.5	248.5	470	SM, CW
XeF_2	381.5	381.5	486.5	416.5	105	SM, CW

[a] From Refs. 44, 45, and 22 and references cited therein. All values in ppm, based on Hindermann and Cornwell's F shielding scale (see text).[18] Other tables of fluorine shielding tensors are given by Appleman and Dailey[46] and Haeberlen.[47]
[b] MB = molecular beam; PW = powder spectrum; SC = single crystal; CT = clathrate; MP = multiple pulse; SM = second moment; CW = wide line spectrum; SR = calculated from spin-rotation constants.
[c] Assumed axial symmetry.

on the basis of the simple "atom-in-a molecule" approach, equation (10) in Chapter 3, using other empirical information such as the tendency of the lowest molecular excitation energy of the binary fluorides to increase across a row of the Periodic Table and down a group. In addition, there is an associated variation in the XF bonds (in terms of the deviations from spherical symmetry of orbital populations) accompanying the periodicity in the electronegativity of X. While this simple model offers a picture drawn in fairly broad strokes, it is quite clear that circulations due to *d* electrons on the other atom should become important for the ^{19}F shielding in transition metal fluorides, as suggested by Buckingham and Stephens for the ^{1}H in transition metal hydrides.(42)

The chemical shift is sensitive to the total charge. ^{19}F shielding generally decreases from the neutral molecule to the anion, e.g., $\sigma(SiF_4) > \sigma(SiF_6^{2-})$, $GeF_4 > GeF_6^{2-}$, $PF_5 > PF_6^-$, $ZrF > ZrF_6^{2-}$, etc. ^{19}F shielding decreases from the neutral molecule to the cation as well, e.g., $\sigma(NF_3) > \sigma(NF_4^+)$, $ClF_5(ave) > ClF_6^+$. The more F's attached to the X atom, the less shielded is the ^{19}F nucleus in the XF_n molecule or ion. This is not unlike the shielding behavior of a typical central atom, for example $\sigma(^{31}P)$: $PCl_3 > PCl_4^+ > PCl_5 > PCl_6^-$.(43)

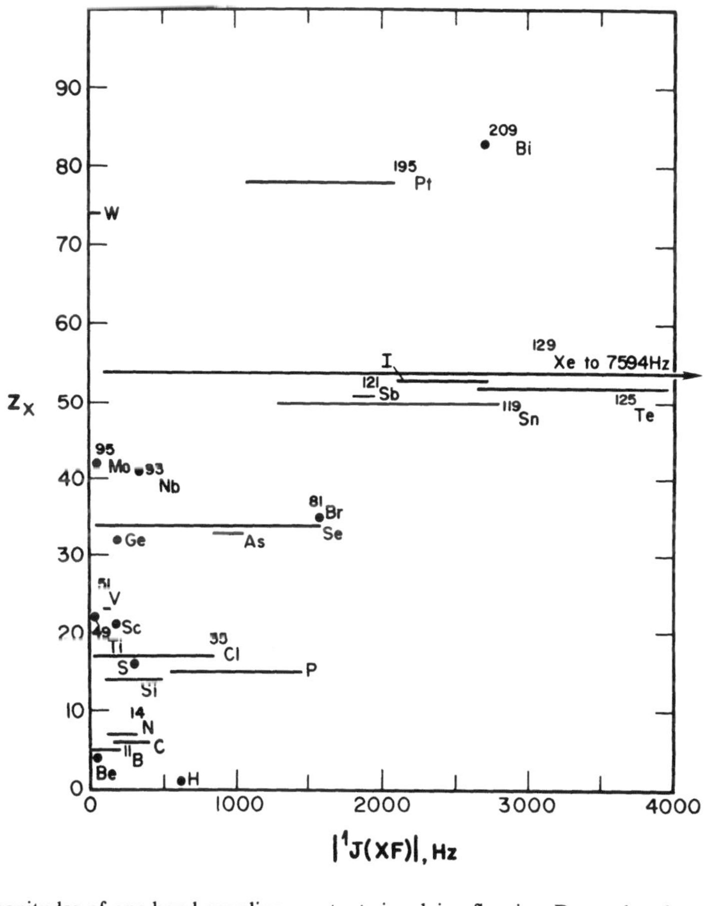

Figure 2. Magnitudes of one-bond coupling constants involving fluorine. Data taken from Refs. 6, 10–14, 48. (The corrected value of the ^{209}BiF coupling in BiF_6^- is 3823 ± 3 Hz.[71]) For a plot of the reduced coupling constants for the binary fluorides see Figure 4 of Chapter 4.

2.3. Anisotropy of the ^{19}F Shielding Tensor

^{19}F shielding tensors measured in solids have been reviewed by Mehring.[45] In solids the convention is to arrange the components in increasing order, so that $\sigma_{11} < \sigma_{22} < \sigma_{33}$, each one being referred to the isotropic value and the isotropic value referred to C_6F_6. In Table 2 we have converted these tensor components to absolute shielding by using $\sigma(C_6F_6$ liquid, room temperature$) = 355.5$ ppm[1] based on HF molecule. Spin-rotation-derived shielding components from Gierke and Flygare[22] are also shown in Table 2. For these molecules, the agreement between solid state and spin-rotation-derived free molecule components of the shielding tensor is quite good. For monofluorobenzene the spin-rotation constants (not shown) have large uncertainties which lead to σ components which do not agree with the solid state results.

3. SPIN–SPIN COUPLING INVOLVING ^{19}F

Tables of coupling constants involving ^{19}F have been compiled, including literature values over the same period as the chemical shifts.[6,10–14] The one-bond

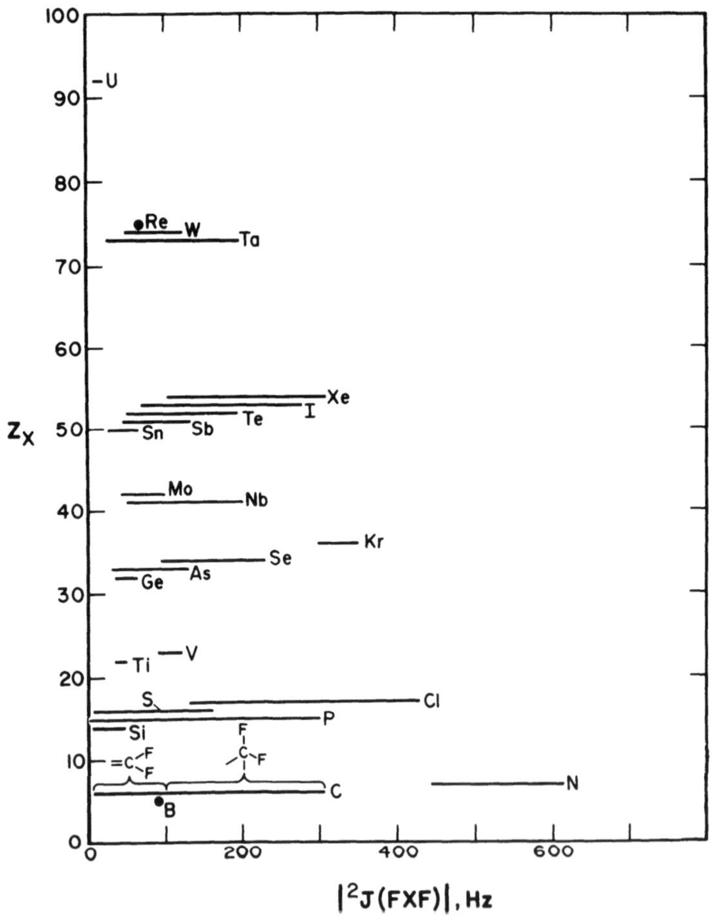

Figure 3. Magnitudes of FF two-bond coupling constants arranged according to the atomic number of the intervening atom. Data taken from Refs. 6, 10, 15, 16, 51–55.

couplings $^1J(XF)$ are shown in Figure 2. Since this chart is intended for practical reference, values plotted are in hertz and displayed according to atomic number for ease of reference.

We have already seen in Chapter 4 that the reduced one-bond coupling constants $^1K(XF)$ for the binary fluorides have a periodic behavior with atomic number of the X element, which follows along the periodicity of the density of the wave function at the nucleus. The general trends of $^1K(AF)$ in $AF_{3-n}L_n$, in tetrahedral $AF_{4-n}L_n$, and octahedral $AF_{6-n}L_n$ compounds have been summarized, A being main group element as well as transition metal.[49,50]

The two bond couplings $^2J(FXF)$ are shown in Figure 3. Some signs are known for a selected number of compounds. However, unlike ^1J, the signs of ^2J couplings are not characteristic of the coupled nuclei. Therefore, the few signs which are available do not allow generalization of this information to other values known only in absolute magnitude. The isotropic $^2J(FCF)$ has been found to be positive where the sign has been determined, and is very sensitive to structure and substituents, ranging from <5 to 109.9 Hz for sp^2 carbon, 105 to 343 Hz for saturated carbon, and can have a large anisotropy. For example, $^2J(FCF)$ in $F_2C=CH_2$ has components $J_{aa}=342\pm264$ Hz, $J_{bb}=-1215\pm132$, $J_{cc}=970\pm132$, where a is perpendicular to the plane of the molecule and b is along the C=C direction.[6]

$^3J(HCCF)$ in olefins have characteristic ranges: trans (10–100 Hz) > cis (−4 to +20 Hz). In olefins and other sp^2–sp^2 carbon paths $^3J(FCCF)$ is large negative (up to 143.4 Hz) for trans, and smaller (from −18.7 to +40) for cis. If one carbon in the

Table 3. Typical Values of Spin–Spin Coupling Constants between ^{19}F and Other Nuclei, $^nJ(XF)$, Hz.

Atomic number	X	$^2J(XF)^a$	$^3J(XF)$	$^nJ(XF)$ 0 up to
1	H	108(B), 11–91.7(C), 45–50(N), 43–96(Si)	1–58.1	$^4J=46.5$, $^5J=25.8$, $^6J=11.9$, $^7J=0.9$, $^8J=0.5$
5	^{11}B	29.6–33.6(C)		
6	C	8–102.5(C), 11.8–19.1(S), 15–23.4(As), 13.5–21(Sb)	1–11.5	$^4J=10.3$, $^5J=30$
7	^{15}N	13.7–73.8(C), 52–102(N), 2.5–24.6(P), 1.8–58(W), 77(Mo)	0–3.6	$^4J=7.0$, $^5J=1.8$
8	O	39(C)		
9	F	2.5(Si) to 615(N); see Figure 3	0–322 (see text)	$^4J=196.4$, $^5J=174$, $^6J=43.2$, $^7J=6.6$
14	Si	26.7–90.5(Si)	3.5–15.7	
15	P	195–240(B), 17.2–193.3(C), 700 P = CF_2, 15–184(P), 1.8(Si) 20–165(Rh), 7–9(Ta)	<0.2 to 209.2^b	$^4J=20$, $^5J=5.4$
34	Se	6–39.5(C), 165(P)		
45	Rh	6–14(C), 16–31.5(P)	1.6	
50	^{119}Sn	222–531(C)		
54	Xe		30–61	
74	W	27–32(P)		
78	Pt	160–753(C), 1371.6–1445.1(P)	25–472	$^4J=5.8$ to 101
80	Hg	480–820(C)	33–597	$^4J=26.5$ to 215
81	Tl		910–1500	$^4J=85.2$

^{a} Nucleus in parentheses is the intervening atom, e g, C in $^2J(HCF)$ Taken primarily from compilations,[6 10] updated with additional data[51 53 57 58]
^{b} All observed in PF_2NHPF_4

path is sp^3, ^3J(FCCF) is 0–40 Hz. Large substituent effects on ^3J(FCCF) dominate any possible dependence on dihedral angle so this coupling constant [unlike ^3J(HCCF)[6]] is of no use as an indication of conformational structure. ^3J(FF) for other coupling paths are 2.2–141 Hz for FCOF, 8–265 Hz for FCNF, 2–15 Hz for FCPF, 0–67 Hz for FCSF, ca. 5 Hz for FCSiF, 99–322 Hz for FNNF, 2.5–21.8 Hz for FNSF, 6.1–17.4 Hz for FSOF.[6,10]

^4J(FF) is often larger in magnitude than ^3J(FF), values range from 0 to 196.4 Hz, and it has a stereochemical dependence which is useful for structural determination. For example, in perfluoromethylcyclohexane, ^4J(ax, ax) = 26 Hz, ^4J(eq, eq) = −6 to −9 Hz, and ^4J(ax, eq) ~ 0.[6] ^nJ(FF) for $n > 4$ values can be atypically large, unlike long-range coupling between most nuclei. This has been attributed to the overlapping electron distributions for two F nuclei in close proximity, thereby allowing correlation of electron spins (and nuclear spins) by direct overlap in addition to correlations of electron-pair-bond to electron-pair-bond via the molecular frame. There are also some unusually large long-range FF couplings where the two nuclei are not in close spatial proximity, but are connected by an intervening pi network. Examples of FF long-range couplings: ^4J(FF) up to 196.4 Hz and ^5J(FF) up to 170 Hz have been observed, as well as smaller values for ^nJ(FF) for $n = 9$ to 14, e.g., 0.2 Hz for ^{14}J(FF).[56]

Table 4. Typical ^{19}F Spin-Lattice Relaxation Times

Phase	Conditions ρ, T	T_1	Reference
Gas phase			
CF_4	0.2–8.0 amag, RT	2.16 ms/amag	59
	45–405 amag, 273–373 K	0.079–1.920 s	60
SiF_4	0.02–6.0 amag, RT	18.9 ms/amag	59
F_2	4.15–6.0 amag, 187–333 K	9.7–180 μs/amag $T_1/\rho \propto T^{-3/2}$	61, 62
CH_3F	0.04–10 amag, RT	54.1 ms/amag	63
CHF_3	0.1–1.0 amag, RT	3.33 ms/amag	63
GeF_4	0.02–6.0 amag, RT	32.4 ms/amag	59
SF_6	0.02–10 amag, 238–350 K	2.34 ms/amag at 293 K, $T_1/\rho \propto T^{-1.53 \pm 0.05}$	59
ClF	13–28 amag, 295 K	0.0167–0.0379 s	64
BF_3	0.04–2.0 amag, RT	4.87 ms/amag	63, 65
$FClO_3$	6.13–10.4 bars, RT	0.6 ms/bar	66
Liquid phase			
ClF	125–171 K	3.97–4.11 s	64
$FClO_3$	130–368 K	0.1–2.05 s SR dominant	66
CF_4	88–143 K	2.08–3.64 s	67
$CFCl_3$	152–455 K	0.70–8.40 s	68
C_6D_5F	30–350°C, 1–3500 bar	0.64–24.8 s	69
1,3,5$C_6D_3F_3$	450–570 K	3.5–8 s	70
Solid phase			
ClF	112.8 K	3.25 s	64
Powder or single crystals e.g., CF_3COOAg, CaF_2		Different components of σ have different relaxation rates. T_1 varies with orientation w.r.t. field	45

[a] ρ, density; RT, room temperature; amag = amagat = density of an ideal gas at 1 atm, 273 K.

Typical values of $^2J(XF)$, $^3J(XF)$ and long-range couplings involving F and other nuclei are given in Table 3.

4. ^{19}F RELAXATION

In the gas phase, above the T_1 minimum, the ^{19}F spin-lattice relaxation time is linear with density ρ of the gas, and the slope (T_1/ρ) is found to be consistent with $(T_1/\rho) \propto T^{-3/2}$ in cases where temperature-dependent measurements of ^{19}F relaxation have been carried out. This implies that the relaxation is dominated by the spin-rotation mechanism in the gas phase. In the liquid phase, the results for several molecules indicate that spin-rotation mechanism is dominant, especially at higher temperatures; and intermolecular dipole–dipole mechanism contributes to the relaxation at lower temperatures, e.g., below 200 K in $FClO_3$. Intramolecular dipole–dipole mechanism, chemical shift anisotropy and scalar coupling contributions are usually found to be negligible in ^{19}F relaxation. Where intermolecular dipole–dipole and spin-rotation mechanisms have been used to interpret results using Arrhenius temperature dependence for both, the activation energies are usually not equal. Some typical values of ^{19}F spin-lattice relaxation times are shown in Table 4.

REFERENCES

1. Jameson, C. J.; Jameson, A. K.; Burrell, P. M. *J. Chem. Phys.* 1980, **73**, 6013–6020; Jameson, C. J.; Jameson, A. K.; Oppusunggu, D. *J. Chem. Phys.* 1984, **81**, 85–90.
2. Dungan, C. H.; Van Wazer, J. R. *Compilation of Reported ^{19}F NMR Chemical Shifts*; Wiley: New York, 1970.
3. Cyr, N.; Raza, M. A.; Reeves, L. W. *Mol. Phys.* 1972, **24**, 459–460.
4. Rummens, F. H. A. *NMR Basic Princ. Prog.* 1975, **10**, 1–118.
5. Tong, J. P. K.; Langford, C. H.; Stengle, T. R. *Can. J. Chem.* 1974, **52**, 1721–1731.
6. Emsley, J. W.; Phillips, L.; Wray, V. *Prog. NMR Spectrosc.* 1975, **10**, 83–756.
7. Jameson, A. K.; Moyer, J.; Jameson, C. J. *J. Chem. Phys.* 1978, **68**, 2873–2877.
8. Bacon, J.; Gillespie, R. J.; Hartman, J. S.; Rao, U. R. K. *Mol. Phys.* 1970, **18**, 561–570.
9. Emsley, J. W.; Phillips, L. *Prog. NMR Spectrosc.* 1971, **7**, 1–526.
10. Wray, V. *Annu. Rep. NMR Spectrosc.* 1980, **10B**, 1–507.
11. Jones, K.; Mooney, E. F. *Annu. Rep. NMR Spectrosc.* 1971, **4**, 391–495.
12. Fields, R. *Annu. Rep. NMR Spectrosc.* 1972, **5A**, 99–304.
13. Cavalli, L. *Annu. Rep. NMR Spectrosc.* 1976, **6B**, 43–222.
14. Fields, R. *Annu. Rep. NMR Spectrosc.* 1977, **7**, 1–117.
15. Holloway, J. H.; Schrobilgen, G. J. *Inorg. Chem.* 1981, **20**, 3363–3368.
16. Holloway, J. H.; Schrobilgen, G. J. *Inorg. Chem.* 1980, **19**, 2632–2640.
17. Mason, J. *J. Chem. Soc., Dalton Trans.* 1975, 1426–1431.
18. Hindermann, D. K.; Cornwell, C. D. *J. Chem. Phys.* 1968, **48**, 2017–2024
19. Weiss, R. *Phys. Rev.* 1963, **131**, 659–665.
20. Baker, M. R.; Anderson, C. H.; Ramsey, N. F. *Phys. Rev. A* 1964, **133**, 1533–1536.
21. Stevens, R. M.; Lipscomb, W. N. *J. Chem. Phys.* 1964, **41**, 184–194.
22. Gierke, T. D.; Flygare, W. H. *J. Am. Chem. Soc.* 1972, **94**, 7277–7283.
23. Casleton, K. H.; Gierke, T. D.; Wang, J. H. S.; Kukolich, S. G. *J. Chem. Phys.* 1976, **64**, 471–476.
24. Muenter, J. S.; Klemperer, W. *J. Chem. Phys.* 1970, **52**, 6033–6037.
25. Davis, R. E.; Muenter, J. S. *J. Chem. Phys.* 1967, **57**, 2836–2838.
26. Fabricant, B.; Muenter, J. S. *J. Chem. Phys.* 1977, **66**, 5274–5277.
27. Wang, J. H. S.; Kukolich, S. G. *J. Am. Chem. Soc.* 1973, **95**, 4138–4141.
28. Weiss, V. M.; Todd, H. D.; Lo, M. K.; Gutowsky, H. S.; Flygare, W. H. *J. Chem. Phys.* 1967, **47**, 4021–4029.

29. Flygare, W. H. *Chem. Rev.* 1974, 653–687.
30. Ozier, I.; Lee, S. S.; Ramsey, N. F. *J. Chem. Phys.* 1976, **65**, 3985–3993.
31. Ozier, I.; Yi, P. N.; Ramsey, N. F. *J. Chem. Phys.* 1977, **66**, 143–146.
32. Wofsy, S. C.; Muenter, J. S.; Klemperer, W. *J. Chem. Phys.* 1971, **55**, 2014–2019.
33. White, R. L. *Rev. Mod. Phys.* 1955, **27**, 276–288.
34. Jameson, C. J.; Jameson, A. K.; Honarbakhsh, J. *J. Chem. Phys.* 1984, **81**, 5266–5267.
35. Ozier, I.; Crapo, L. M.; Lee, S. S. *Phys. Rev. A* 1968, **172**, 63–82.
36. Landolt-Börnstein *Zahlenwerte und Funktionen aus Naturwissenschaften und Technik*, Neue Serie, Group II; Springer-Verlag: Berlin, 1974; Vol. 6, pp. 2–383 to 2–434.
37. Tigelaar, H. L.; Flygare, W. H. *Chem. Phys. Lett.* 1970, **7**, 254–256.
38. Holmes, J. R.; Stewart, B. B.; MacKenzie, J. S. *J. Chem. Phys.* 1962, **37**, 2728–2729.
39. Lawrence, N. J.; Ogden, J. S.; Turner, J. J. *J. Chem. Soc., Chem. Commun.* 1966, 102–103.
40. Nebgen, J. W.; Metz, F. I.; Rose, W. B. *J. Am. Chem. Soc.* 1967, **89**, 3118–3121.
41. Gutowsky, H. S.; Hoffman, C. J. *J. Chem. Phys.* 1951, **19**, 1259–1267.
42. Buckingham, A. D.; Stephens, P. J. *J. Chem. Soc. A* 1964, 2747–2759, 4583–4587.
43. Andrew, E. R. *Philos. Trans. R. Soc. London Ser. A* 1981, **299**, 505–520.
44. Garg, S. K.; Ripmeester, J. A.; Davidson, D. W. *J. Magn. Reson.* 1980, **39**, 317–323; *J. Chem. Phys.* 1982, **77**, 2847–2851.
45. Mehring, M. *NMR Basic Princ. Progr.* 1976, **11**, 1–243.
46. Appleman, B. R.; Dailey, B. P. *Adv. Magn. Reson.* 1974, **7**, 231–320.
47. Haeberlen, U. *Adv. Magn. Reson.* 1976, Suppl. 1, 1–186.
48. Lindman, B.; Forsen, S. *NMR Basic Princ. Prog.* 1976, **12**, 1–365.
49. Shustorovich, E.; Dobosh, P. A. *J. Magn. Reson.* 1980, **39**, 79–99.
50. Shustorovich, E.; Dobosh, P. A. *J. Magn. Reson.* 1980, **39**, 101–117.
51. Ewing, D. F. *NMR Spec. Per. Rep. R.S.C.* 1983, **12**, 68–95.
52. Ewing, D. F. *NMR Spec. Per. Rep. R.S.C.* 1982, **11**, 71–105.
53. Ewing, D. F. *NMR Spec. Per. Rep. R.S.C.* 1981, **10**, 74–111.
54. Ewing, D. F. *NMR Spec. Per. Rep. R.S.C.* 1980, **9**, 68–100.
55. Gillespie, R. J.; Netzer, A.; Schrobilgen, G. J. *Inorg. Chem.* 1974, **13**, 1455–1459.
56. Mitchell, P. J.; Phillips, L.; Roberts, S. J.; Wray, V. *Org. Magn. Reson.* 1974, **6**, 126–127.
57. Witanowski, M.; Stefaniak, L.; Webb, G. A. *Annu. Rep. NMR Spectrosc.* 1981, **11B**, 1–502.
58. Martin, G. J.; Martin, M. L.; Gouesnard, J. P. *NMR Basic Princ. Prog.* 1981, **18**, 1–382.
59. Courtney, J. A.; Armstrong, R. L. *Can. J. Phys.* 1972, **50**, 1252–1261.
60. Finney, R. J.; Wolfe, M.; Jonas, J. *J. Chem. Phys.* 1977, **67**, 4004–4011.
61. Courtney, J. A.; Armstrong, R. L. *J. Chem. Phys.* 1970, **52**, 2158–2159.
62. Dybowski, C. R.; Chien, M.; Wade, C. G. *J. Chem. Phys.* 1972, **56**, 4229–4231.
63. Armstrong, R. L.; Courtney, J. A. *Can. J. Phys.* 1972, **50**, 1262–1272.
64. Gillen, K. T.; Douglass, D. C.; Griffiths, J. E. *J. Chem. Phys.* 1978, **69**, 461–467.
65. Hinshaw, W. S.; Hubbard, P. S. *J. Chem. Phys.* 1971, **54**, 428–431.
66. Maryott, A. A.; Farrar, T. C.; Malmberg, M. S. *J. Chem. Phys.* 1971, **54**, 64–71.
67. Rugheimer, J. H.; Hubbard, P. S. *J. Chem. Phys.* 1963, **39**, 552–569.
68. Gillen, K. T.; Douglass, D. C.; Malmberg, M. S.; Maryott, A. A. *J. Chem. Phys.* 1972, **57**, 5170–5179.
69. DeZwaan, J.; Finney, R. J.; Jonas, J. *J. Chem. Phys.* **60**, 3223–3230.
70. Bull, T. E. *J. Chem. Phys.* 1975, **62**, 222–226.
71. Morgan, K.; Sayer, B. G.; Schrobilgen, G. J. *J. Magn. Reson.* 1983, **52**, 139–142.

Chapter 17

THE QUADRUPOLAR HALIDES

Chlorine, Bromine, and Iodine

J. W. AKITT

1. INTRODUCTION

The three heavy halogen elements all have nuclei with appreciable quadrupole moments and quadrupole relaxation effects are expected to dominate their NMR spectroscopy. When encountered as ions in salt solutions their behavior is complementary to that already described for the alkali and alkaline earth metals, the cations and anions often being studied in each other's presence as counterion. A full description of such solutions thus demands an understanding of the parameters of the anion, cation, and solvent resonances, and to this end the results of this chapter and Chapter 7 should be taken together. The halogens, though, do have a very well developed covalent chemistry in which they are univalent and so possess a marked electric field gradient at the nucleus, efficient relaxation mechanisms, and broad to very broad resonances. The chemical shifts of these compounds cannot therefore be determined with any accuracy, and such study has been relatively infrequent because of the poor signal-to-noise ratios of the broad lines. The advent of Fourier transform techniques has, however, made such studies much more accessible and a good data base is starting to accumulate. Relaxation studies of these covalent halides enable theories of molecular motion to be tested. The subject matter of this chapter has been particularly well reviewed by Lindman and Forsen[1,2] in what amounts to a review textbook of 340 pages and a supplementary review to update this material to 1978.[2] The inorganic, physical, and biological aspects are fully covered and the latter are also reviewed by Stevens and Bryant.[3] Here we will summarize the content of these earlier reviews and add comments on the new material which has been published up to approximately mid-1982.

The nuclear properties of these elements are outlined in Table 1, where the parameters for ^{39}K are also included to act as a link with Chapter 7. The width factors, $(2I+3)/I^2(2I-1)\cdot Q^2(1+\gamma_\infty)^2$, for the free ions as defined in equation (1) are given in the table. When these are multiplied by the infinite dilution values of T_1 for the aqueous solutions we find a series of very similar values which indicate that the solution chemistry of these anions must be very similar, the ion parameters being the principal determinants of the relaxation behavior, and the rather short relaxation

J. W. AKITT • School of Chemistry, University of Leeds, Leeds LS2 9JT, England.

Table 1. Nuclear Properties of the Quadrupolar Halides

Nucleus	Quadrupole moment Q (10^{-28} m^2)	$1+\gamma_{ion}$	Width factor $^{35}Cl = 1$ (WF)	T_1 (X^- dil. aquo.) (ms)	T_1 (WF)
^{35}Cl	-0.082	57.6	1.00	40.0	40.0
^{37}Cl	-0.0650	57.6	0.62	—	—
^{79}Br	0.33	100	51	—	—
^{81}Br	0.27	100	34	0.953	32
^{127}I	-0.79	180	230	0.19	44
^{39}K	5.5×10^{-2}	18.3	0.044	57	2.5

times observed being a result of the large Sternheimer antishielding effect of the more diffuse charge on the anions. A comparison of Cl^- with the isoelectronic K^+ shows that the cation relaxation time is at least ten times shorter than that of the anion, after all these parameters have been taken into account, so that the anion relaxation processes are much less effective, presumably reflecting more rapid solvent motion around them.

2. EXPERIMENTAL TECHNIQUES

The receptivities of this group of nuclei do not vary over a very wide range and any may be chosen for study, with possibly a tendency to choose ^{35}Cl before ^{37}Cl. The ions in aqueous solution are the easiest to study since here the lines are at their narrowest, and allow effects due to binding to sites where the relaxation times are very short to be observed via exchange with the free anions. The narrowest resonances likely to be seen in such work are ClO_4^- (1.2 Hz) or Cl^- (8 Hz) and rapid scan pulsed Fourier transform techniques are indicated as the best approach. The covalent chloro compounds have linewidths in the range of about 1000–22,000 Hz and wide bandwidth equipment is essential for their satisfactory observation, though provided the spectrometer window is wide enough, the data collection is straightforward. Covalent compounds of bromine or iodine have much broader resonances and are not observed, presumably because a spectral window in excess of 2 MHz is required to observe even the narrowest covalent bromine or iodine resonances.

3. NMR PARAMETERS

NMR parameters need to be dealt with under two headings: (a) relaxation and molecular motion, chemical shifts and spin–spin coupling in the covalent compounds, and (b) relaxation and chemical shifts in the ionic solutions.

3.1. Covalent Compounds

3.1.1. Nuclear Relaxation

The ^{35}Cl relaxation times of a wide range of compounds have been measured and are assembled in Table 2. The few results for bromine and iodine are in the following

Table 2. Chlorine-35 Relaxation Times and Derived Correlation Times for Covalent Halides

Compound	T_1 (μs)	T_2 (μs)	τ_c (ps)	References
CCl_4		18–25	1.7–2	6–11
$CHCl_3$, $CDCl_3$		16–31	1.3–2.7	6–8, 12–18
CH_2Cl_2		40–50	1.0–1.2	8, 11, 19, 20
CH_3Cl		100	0.55	8
$CFCl_3$	30.7, 38.3	38.6	1.3, 1.04	21, 22
CF_2Cl_2		40	1.1	23
$CHCl_2F$	38.4		1.03	22
$CHClF_2$	51		0.78	24
$MeCCl_3$		21.5	2.05	17
Me_3CCl		42.5	1.53	25
$(CN)CCl_3$		13.5	2.7	26
ClCN		29.8	1.22	27
C_6H_5Cl		12.9–13.2	4.0, 4.1	7, 8, 28
MeCOCl		66.8	1.15	8
$SiCl_4$	46	52–79	1.9–3.3	5, 6, 9, 29
$SiHCl_3$		112, 124	1.6, 1.4	30
$GeCl_4$		41	2.3	9
$SnCl_4$		40–19.2	2.7–5.6	6, 9, 31–33
$PbCl_4$		7.15, 5	17.2	32
$TiCl_4$	310	211–415	4.1–8.1	7, 9, 11, 29, 34
VCl_4		82–120	—	11, 34, 35
PCl_3	40, 45	33.9–60	1.6–2.7	9, 12, 36–39
$AsCl_3$		16.9, 23	4.4, 5.9	9, 36
BCl_3		108, 122	1.27, 1.11	26, 40
HCl	770, 765		0.072, 0.073	38, 41
Cl_2		14.1–40	0.52–1.5	8, 9, 42–44
$VOCl_3$		140–161	2.9–3.4	9, 26, 29
CrO_2Cl_2	80	92, 107	2.4–3.2	9, 29
ClO_3F	1300, 1420		0.53, 0.48	45–47
$POCl_3$		27	2.8	9
$SbCl_5$		16	4.8	9
Cl_3^- in aquo.		15(t), 3(c)	6.3, 6.4	48
ClO_3^- in aquo.		23.6, 15.7	3, 4.5	7, 8
ClO_4^- in aquo.	270×10^3	270×10^3	—	11
$GaCl_4^-$		8–15	16–30	49

table, Table 3. The relaxation times are short and are quoted in microseconds (a T_2 of 31.8 μs is equivalent to a linewidth of 10,000 Hz). They are very temperature dependent and no attempt has been made to reproduce this in the table, which simply gives the ranges of values observed. Because the quadrupole coupling constants of ^{35}Cl are easily determined by other methods it is possible to derive values of τ_c from the quadrupole relaxation equation describing the relaxation of covalently bound nuclei:

$$\frac{1}{T_1}=\frac{1}{T_2}=\frac{3\pi^2}{10}\frac{2I+3}{I^2(2I-1)}\left(\frac{eqQ}{h}\right)^2\left(1+\frac{\eta^2}{3}\right)\tau_c \tag{1}$$

where η is the asymmetry parameter $(V_{xx}-V_{yy})/V_{zz}$, and is zero for the axial symmetry that may be assumed here. The correlation times obtained are usually much shorter than is given by simple hydrodynamic theory. If the relaxation time of a spin-1/2 nucleus can also be studied in the same molecule as the halogen then it may

Table 3. Bromine-81 and Iodine-127 Relaxation Times and Derived Correlation Times for Covalent Halides[a]

Compound	T_1 (μs)	T_2 (μs)	τ_c (ps)	References
HBr	14, 17		0.07–0.09	41
PBr_3	0.33	0.32	5.8, 5.7	12, 50
$SnBr_4$		0.748	31	51
$ZnBr^+$ aqu.		0.07–1.0		52
$ZnBr_4^{2-}$ aqu.		0.25		52
SnI_4	0.15		3.67	31
IO_4^-		200		
IF_6^+ in HF	~1000			53
IF_7	57			54

[a] Indirect methods are used to obtain the very short relaxation times of compounds of these nuclei.

be possible to obtain the correlation time for changes in angular momentum, from spin rotation contributions to the spin-1/2 relaxation. This then allows theories of molecular diffusion in liquids to be tested.[1,5]

3.1.2. Chemical Shifts

These have only been determined for compounds of chlorine and are summarized in the charts of Figures 1–3, including recent data.[74,75] The shifts in the MCl_4 series may be partly explained by changes in the electronegativity of the central atom, though the large changes observed with change in substituent size in the chloromethane and -silane series shows that other important effects operate. Shape, and therefore medium, effects are likely to be important[1] and the number of halogens bonded to one atom also has a significant effect. The chlorine and fluorine chemical shifts correlate well when there are only one or two halogen substituents[1,74,75] but not for higher substitution. It is now possible to make an interesting comparison of linewidths and chemical shifts, and this is done in Figure 4. If the shift changes were due to halogen overlap or orbital type distortion brought about by intermolecular effects then relaxation of the chlorine should also be affected, shorter relaxation times and paramagnetic shifts being associated. This is clearly the case for the chloromethanes and -silanes, but unfortunately the entirely opposite effect is seen for metallo-chlorides (Table 4). A medium/overlap effect is reasonable for the former, where there are presumably negligible hybridization changes, but in the second series, the hybridization changes also give a more symmetrical electronic distribution, perhaps more *p* character and better developed lone pair orbitals? It is certainly remarkable that the longest correlation times in this series are associated with the longest relaxation times, so that any viscosity effects can be ruled out. Recent work reports ^{35}Cl shifts of 75 ppm and ^{81}Br shifts of 200 ppm throughout the solid/liquid region for CuX, though in the melt, the chloride seems to be ionic.[81,82] Substantial ^{127}I shifts are observed in solid MIO_4 salts.[83]

3.1.3. Spin–Spin Coupling

In situations where the halogen relaxation times are long spin–spin coupling can be measured directly, and the few examples of this will be found in Table 4 marked D.

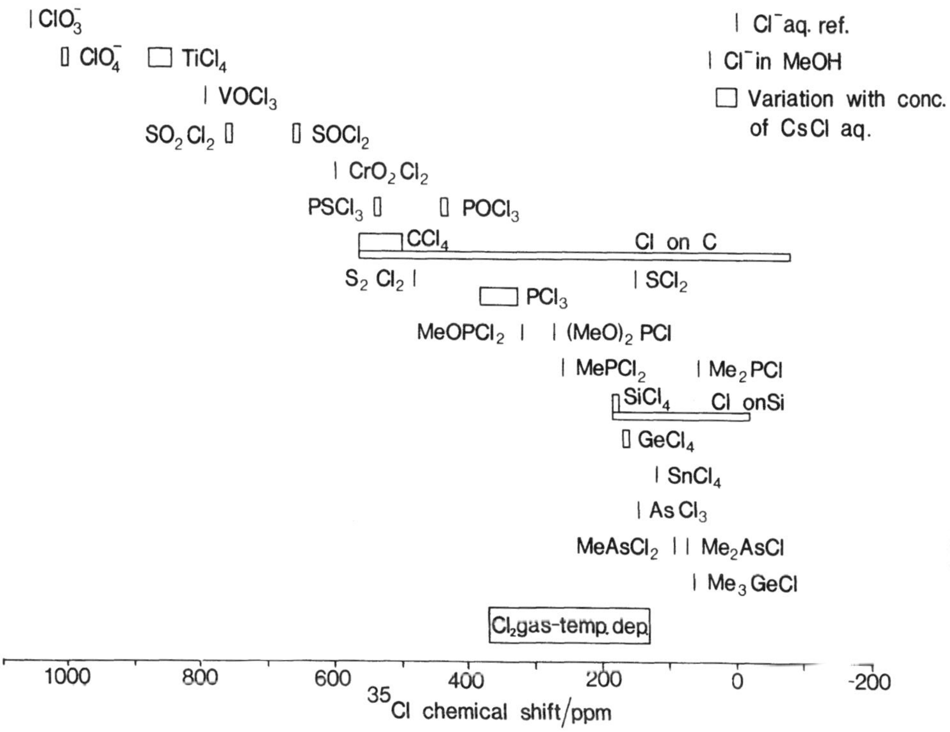

Figure 1. Chart of the ^{35}Cl chemical shifts of covalent chlorine compounds (Cl^- aqu. reference). The width of the bars reflects the spread of results and not the accuracy of the measurements, which depends on the linewidth and is typically ± 50 ppm to ± 100 ppm. The data are extracted from Refs. 8, 9, 11, 34, 35, 43, 72–75.

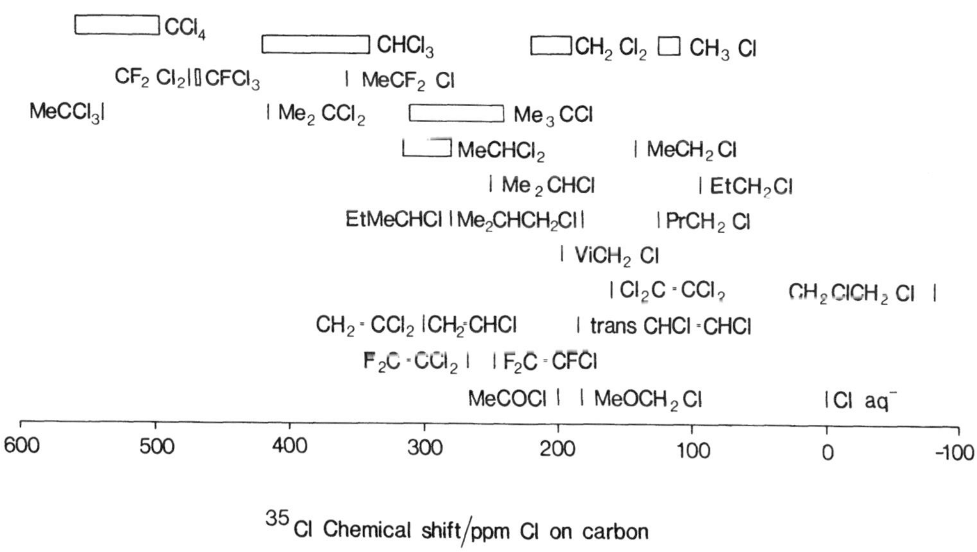

Figure 2. Chart of the ^{35}Cl chemical shifts of compounds with chlorine bonded to carbon. This is an expansion of part of Figure 1.

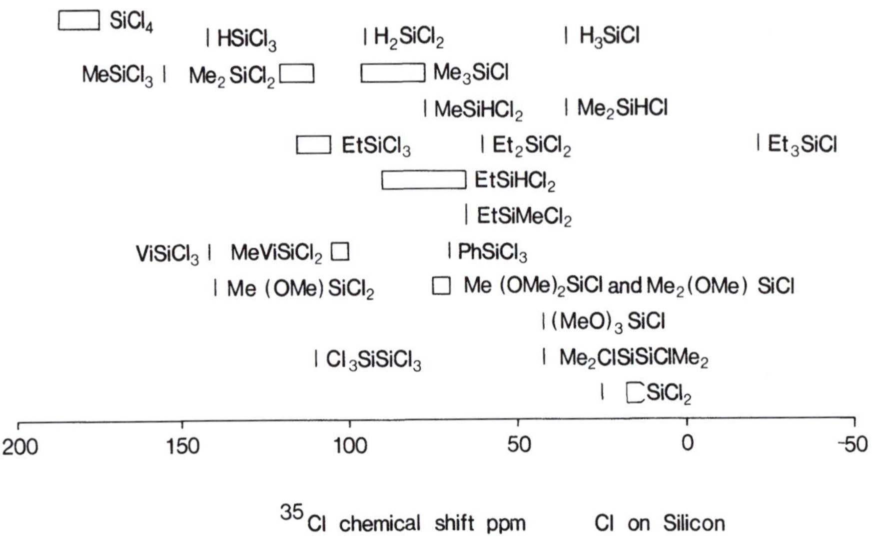

Figure 3. Chart of the ^{35}Cl chemical shifts of compounds with chlorine bonded to silicon. This is an expansion of part of Figure 1.

In all other cases the coupling has to be measured indirectly via the scalar relaxation of a coupled spin-1/2 nucleus.[1] It is of course necessary to separate the scalar contribution from other relaxation mechanisms of the spin-1/2 nucleus, but if this is done then the T_1 and T_2 values will be different according to equations (2) and (3):

$$\frac{1}{T_{1I}}=\frac{8}{3}\pi^2 J_{IS}^2 S(S+1)\frac{T_{2S}}{1+(\omega_I-\omega_S)^2\,T_{2S}^2} \tag{2}$$

$$\frac{1}{T_{2I}}=\frac{4}{3}\pi^2 J_{IS}^2 S(S+1)\left[T_{1S}+\frac{T_{2S}}{1+(\omega_I-\omega_S)^2\,T_{2S}^2}\right] \tag{3}$$

where I is the spin-1/2 nucleus and S is the quadrupolar nucleus. Alternatively, $T_{1\rho}$ measurements may be made.[1,84–86] 17.2 and 17.3 require large J and short T_{2S} to be effective, unless the nuclear frequencies (w) are similar. Interpretation of the relaxation results is complicated by the necessity usually to have to take into account the existence of more than one S isotope. Unfortunately this has sometimes been ignored so that some published results are in error. This is discussed in Ref. 1, and the results

Table 4. ^{35}Cl Chemical Shifts and Relaxation Times of Selected Chlorine Compounds

Compound	Chemical shift (ppm)	Typical T_1 or T_2 (μs)	τ_c (ps)
$AsCl_3$	150	20	4.4–5.9
$SnCl_4$	120	30	2.7–5.6
$GeCl_4$	165	41	2.3
$SiCl_4$	180	65	1.9–3.3
CrO_2Cl_2	600	99	2.4–3.2
$VOCl_3$	790	150	2.9–3.4
$TiCl_4$	860	300	4.8–8.1

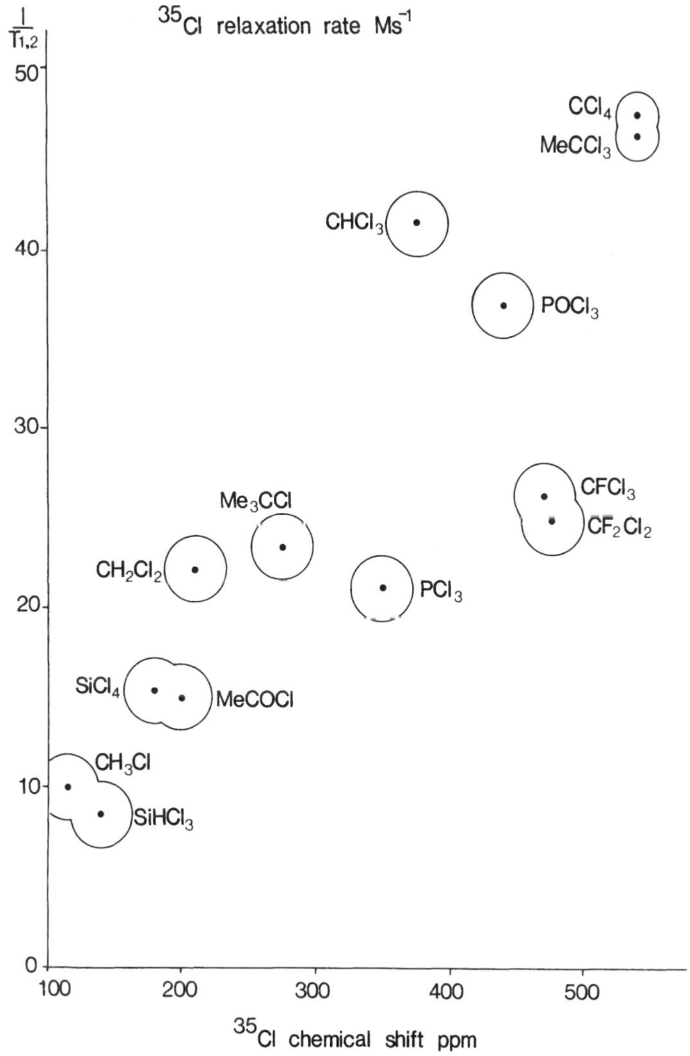

Figure 4. Correlation between the ^{35}Cl linewidths and chemical shifts of compounds containing Cl–C or Cl–Si bonds. The data are derived from Figure 1 and Table 2.

obtained in this way and quoted in Tables 5 and 6 (marked *R*) are the corrected values taken from this source. In addition, if there are several halides coupled to the spin-1/2 nucleus, there may be cross-correlation effects on the spin-1/2 relaxation which have to be allowed for. This reduces the calculated coupling from 50 to 30 Hz for $^{1}J(Si–^{35}Cl)$ in $SiCl_4$.[(67)] In another report[(58)] it is noted that the presence of two bromine isotopes causes the ^{13}C relaxation in bromomethanes to be nonexponential, and coupling constants are extracted. It has also been found that there is a marked change in coupling in ClF on going from the liquid to gaseous state.[(60)] The method has also been applied to the tetrahaloaluminate ions.[(66)] The ^{27}Al relaxation time in the symmetrical species is sufficiently long that the scalar component can be detected and *J* values calculated. This is apparently the only example involving a non-spin-1/2 nucleus as the *I* spin.

Table 5. One-Bond Spin–Spin Coupling between Halogens and Spin-1/2 Nuclei, Aluminum-27 and Oxygen-17

Molecule	J (Hz)	How determined	References
$^{1}H\,^{35}Cl$	Liquid 35, 41 Gas 47	R	41,55, 56, 2
$^{1}H\,^{81}Br$	62	R	41
$^{13}CH\,^{35}Cl_3$	23.3, 49	R	24, 57
$^{13}C\,^{79}Br_4$	151	R	58
$^{13}CD\,^{79}Br_3$	127	R	58
$^{13}CD_2\,^{79}Br_2$	95	R	58
$^{13}CH_3\,^{127}I$	<60	R	59
$^{19}F\,^{35}Cl$	Gas 855	R	60
	Liquid 770	R	60
$^{35}Cl\,^{19}F_3$	260 (only one value)	R	61
$^{35}Cl\,^{19}F_5$	axial 192 equat. $\leqslant 20$	R	61
$^{35}Cl\,^{19}F_6^+$	337	D	62, 63
$^{37}Cl\,^{19}F_6^+$	281	D	62, 63
$^{81}Br\,^{19}F_6^+$	1697	D	64
$^{79}Br\,^{19}F_6^+$	1575	D	64
$^{127}I\,^{19}F_6^+$	2730	D	53
$^{127}I\,^{19}F_7$	~ 2100	D	54, 65
$^{35}Cl\,^{19}FO_3$	278	D	45
$^{35}Cl\,^{19}F_3O$	195	R	61
$^{27}Al\,^{37}Cl_4^-$	650	R	66
$^{27}Al\,^{81}Br_4^-$	750	R	66
$^{29}Si\,^{35}Cl_4$	30	R	67
$^{29}Si\,^{35}Cl_3H$	25	R	67
$^{31}P\,^{35}Cl_3$	120, 127	R	12, 37–39
$^{31}P\,^{81}Br_3$	380, 330	R	12, 39
$^{35}ClF\,^{17}O_3$	108	D	78
$^{35}Cl\,^{17}O_4^-$	85.5, 81.4	D	78–80
$^{31}P\,^{79}Br_3$	350, 296	R	50
$^{119}Sn\,^{35}Cl_4$	375	R	31
$^{119}Sn\,^{81}Br_4$	920	R	51
$^{119}Sn\,^{127}I_4$	940	R	31
$^{119}Sn\,^{35}ClI_3$	421	R	33
$^{119}SnCl\,^{127}I_3$	1097	R	33
$^{119}Sn\,^{35}Cl_3I$	378	R	33
$^{119}SnCl_3\,^{127}I$	1638	R	33
$^{207}Pb\,^{35}Cl_4$	705	R	32

Table 6. Two-Bond Spin–Spin Coupling between Chlorine and Spin-1/2 Nuclei[a]

Molecule	J (Hz)	References
$C\,^{1}H\,^{35}Cl_3$	5.1–7.3	12, 13, 68–70
$CH\,^{19}F\,^{35}Cl_2$	27.6	22
$C\,^{19}F\,^{35}Cl_3$	13.7, 11.9	21, 22
$C\,^{19}F_2\,^{35}Cl_2$	28, 12	23, 24
$Si\,^{1}H\,^{35}Cl_3$	9.3	30

[a] All the above were derived from relaxation measurements.

3.2. Ionic Solutions

3.2.1. *Relaxation Studies*

The rates of relaxation of the halide anions in aqueous solution have been much studied[76,87–106] and some of the theoretical considerations discussed in Chapter 7 also apply here. For instance, the place of viscosity is considered.[102,104] The sequences in which cations have an increasing effect on anion relaxation differ between anions and also change if viscosity is taken into account. As a generalization it can be said that the cations near the centers of their respective groups (K^+, Ca^{2+}) have the least effect, though the spread between them is not great. The alkylammonium cations also have a marked effect on the relaxation rate of chloride, and this is believed to be due to unsymmetrical hydration of the anion in the aqueous cage around the hydrophobic part of the cation.[107–111]

In mixed solvents, the relaxation rate depends on the countercation and the concentration and the plots of rate of relaxation versus solvent composition are more or less curved, often with maxima which can only partly be explained by changes in viscosity.[100,112–118] For the uni–uni charged salts, the anion resonances give access to preferential solvation phenomena, though the results do not seem to be quite so amenable to analysis as the cation results described in Chapter 7. In the case of pure, nonaqueous solvents, or of solutions of highly charged cations in all solvents, the studies shade off into investigations of ligand exchange. Halide ions complex many cations and the kinetics and extent of such interaction has been extensively studied by halogen NMR,[11,49,52,119–132] and the linewidths observed may also be used to monitor whether the chlorine present in a system is in the simple ionic form or not.[133–136] Recent studies deal with halide exchange in the systems $CoCl_2$,[137] $MnCl_2$ or $MnBr_2$,[138] $NiCl_2$ or $NiBr_2$[139] and for a range of lanthanide chlorides in the presence of certain symmetrical ligands.[140] The reaction between iodine or chlorine and the iodide ion has also been studied,[44,48,141,142] and the results critically reviewed.[1]

The perchlorate anion is also potentially a source of information on interactions in solution since it has a very narrow ^{35}Cl resonance and small changes can be monitored accurately. In addition, the chlorine nucleus is well protected from its environment and a linewidth change must result from a restriction in the rotation of the ion, or a destruction of its tetrahedral symmetry, or both. Surprisingly then, such studies are much more recent than those of the chloride ion. The infinite dilution value of T_1 for ClO_4^- in water is 270 ms[106] and remains relatively long in all aqueous solutions since perchlorate competes only weakly with the solvent. Effects are nevertheless measurable.[143–147] The dissociation of perchloric acid is usefully studied in this way,[148,149] and paramagnetic cations are shown to complex ClO_4^-,[146,150] the ^{35}Cl relaxation being dominated by the electron spin interaction in the presence of Mn^{2+}. Complexing by perchlorate is much more likely in nonaqueous or mixed media, depending upon the donor power of the solvent. Thus DMSO or DMF permit no significant broadening, whereas alcohols, acetone, MeCN, or ethyl acetate do, particularly in the presence of Mg. The linewidths can either increase or decrease with concentration, and in the latter case it appears to be implied that more symmetrical complexes may be formed at high concentration.[151] Adding water reduces the linewidths.[152] Recent studies in a variety of solvents are reported[153,154] and Ca^{2+} has been shown to complex ClO_4^- particularly strongly in HMPTA, where Mg^{2+} shows no, and Ba^{2+} very little, interaction.[155] The anion coordinates Yb^{3+} strongly in aqueous acetone and the ^{35}Cl resonance is broadened, though here the conclusion is based in the first place on ^{1}H NMR.[156] The metal resonance may also be a useful indicator of such interaction and an example will be found in Chapter 9.

3.2.2. *Chemical Shifts*

The chemical shifts seen with the halogen resonances as a function of salt concentration are illustrated in Figure 5. The shifts move steadily down field with increasing concentration, except for LiCl and NaCl, and the larger cations tend to have the largest effect, though the order for the lighter ones differs from that of their weights, being Na < K < Li. Mutual shift effects are also caused by the anions and are large enough to measure easily.[72,73,76,77,157–160] There is also an isotope effect for the solvents which increases in the order Cl < Br < I.[11,73,161–163] The chemical shifts are also sensitive to solvent, though relatively few measurements have been made. It has been suggested that the shifts are related to charge-transfer-to-solvent effects,[164] which mechanism may also of course influence relaxation. Large shifts are seen in the presence of paramagnetic cations[139,140] and the perchlorate anion also has solvent-dependent shifts.[155]

3.2.3. *Macromolecular and Biological Systems*

The halide anions are in general absorbed weakly by large polyelectrolyes and exist in both bound and free forms which undergo rapid exchange. Study of the temperature or concentration dependence of the free anion linewidths can thus in principle give information on the bound state. This field has recently been particularly well reviewed[2,3,165] and it is necessary here merely to indicate the nature of the latest work.

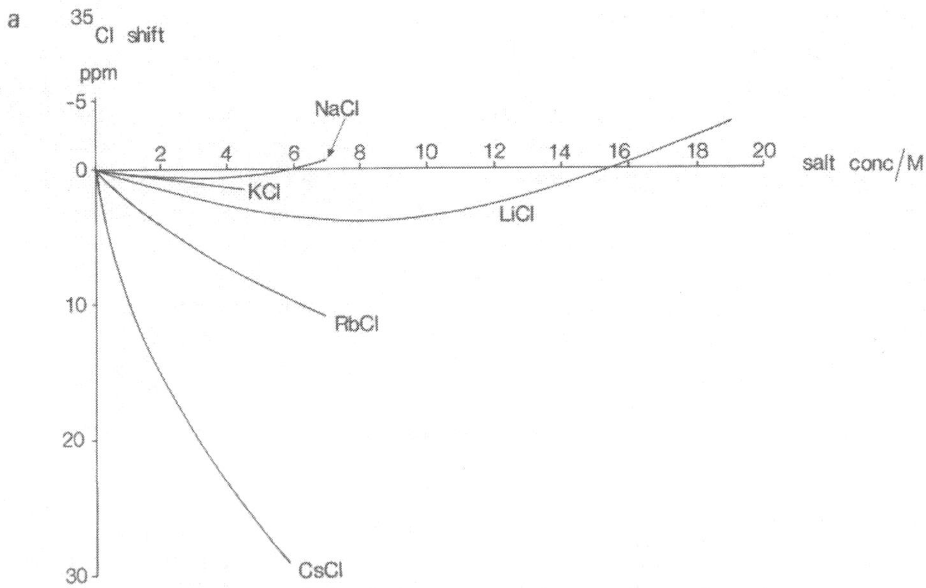

Figure 5a. Halogen chemical shifts in aqueous MX salt solutions as a function of concentration. (a) ^{35}Cl shifts in chlorides.

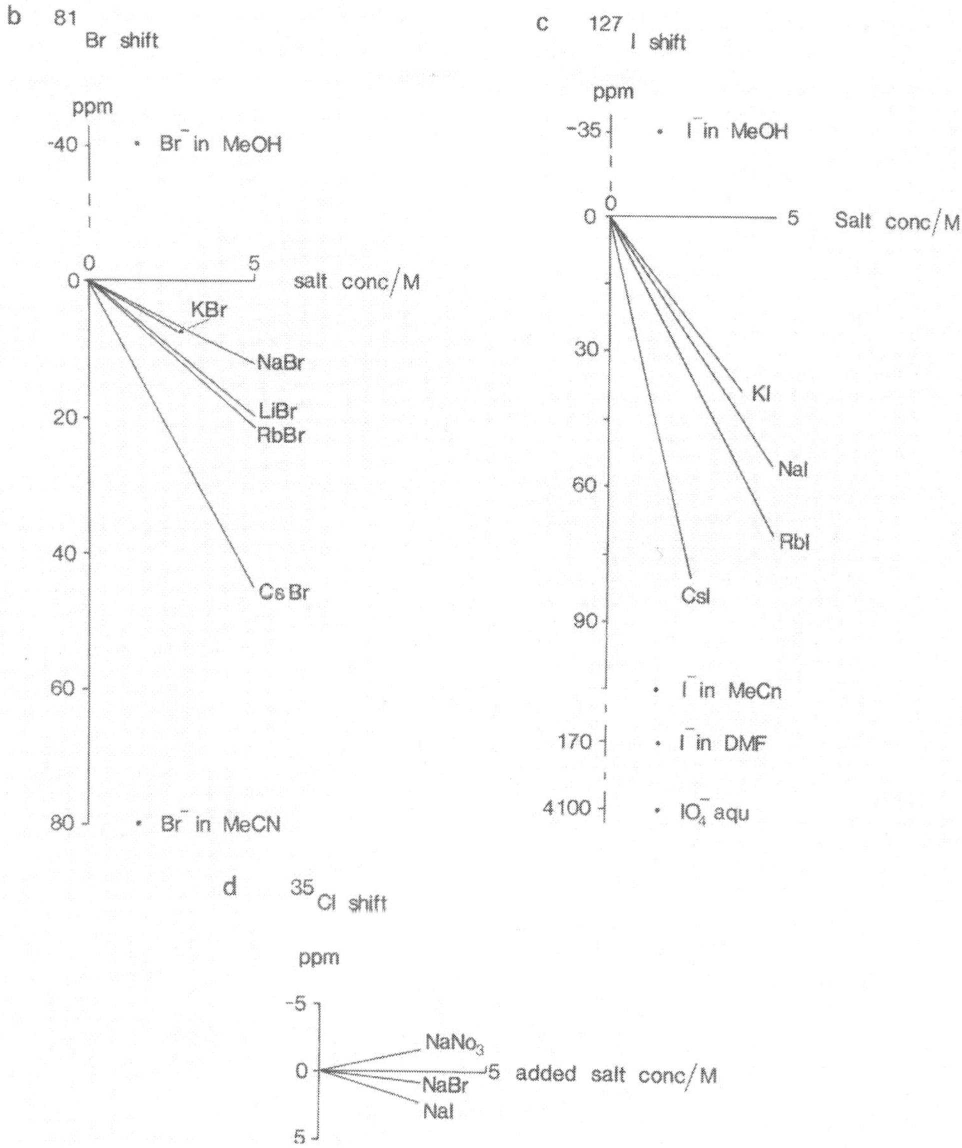

Figure 5b, c, d. (b) ^{81}Br shifts in bromides. Two values in nonaqueous media are also included in this figure. (c) ^{127}I shifts in iodides. The shift in IO_4^- aqu. and in three nonaqueous media are also indicated. (d) The effect of anions on the ^{35}Cl chemical shift in solutions containing 1 *M* NaCl and various concentrations of NaX, $X = I^-$, Br^- or NO_3^-. References 1, 11, 72, 76, 77.

Chloride is bound by the sols formed upon the extensive hydrolysis of Al(III) salt solutions.[166] Binding to liquid crystal systems is also studied[167] and chloride and bromide bind to the micelles formed by hexadecyltrimethylammonium salts. The relaxation rate changes when the micelles change from spheres to rods and there is a marked difference between the behavior of the two ions.[168,169] Many enzymes bind chloride and this may affect their activity. Thus horse liver alcohol dehydrogenase is stabilized by low concentrations of Cl^- which binds to two sites, neither of which is Zn. Phosphate ligands compete for the binding sites.[170,171] Interaction between Zn^{2+}, nucleotides, and Cl^- have been studied also.[172] NADH influences the binding of Cl^- to lactate dehydrogenase[173] and NaCl activates the reduction of dihydrofolate to tetrahydrofolate by the chicken liver dihydrofolate reductase system.[174] Binding to other species is also studied,[175–177] and the processes affecting relaxation are discussed.[178]

REFERENCES

1. Lindman, B.; Forsen, S. "*Chlorine, Bromine and Iodine NMR: Physicochemical and Biological Applications*," in *NMR, Basic Principles and Progress*; Diehl, P.; Fluck, E.; Kosfeld, R., Eds.; Springer: Berlin, 1976; Vol. 12.
2. Lindman, B.; Forsen, S. "*The Halogens*," in *NMR and the Periodic Table*; Harris, R. K.; Mann, B. E., Eds.; Academic: New York, 1978; Chap. 13.
3. Stevens, R. S.; Bryant, R. G. *Mol. Cell. Biochem.* 1976, **13**, 101.
4. Lindman, B. *J. Magn. Reson.* 1978, **32**, 39.
5. Briguet, A.; Duplan, J. C.; Delmau, J. *Mol. Phys.* 1980, **39**, 1541.
6. Diehl, P. *Helv. Phys. Acta.* 1956, **29**, 219.
7. O'Reilly, D. E.; Schacher, G. E. *J. Chem. Phys.* 1963, **39**, 1768.
8. Saito, Y. *Can. J. Chem.* 1965, **43**, 2530.
9. Johnson, K. D.; Hunt, J. P.; Dodgen, H. W. *J. Chem. Phys.* 1969, **51**, 4493.
10. Gillen, K. T.; Noggle, J. H.; Leipert, T. K. *Chem. Phys. Lett.* 1972, **17**, 505.
11. Results described in Refs. 1 and 2.
12. Winter, J. M. *C. R. Acad. Sci. Paris* 1959, **249**, 1346.
13. Ottavi, H. *C. R. Acad. Sci. Paris* 1961, **252**, 1439.
14. Huntress, W. T. *Adv. Magn. Reson.* 1970,. **4**, 1.
15. Huntress, W. T. *J. Phys. Chem.* 1969, **73**, 103.
16. Shoup, R. R.; Farrar, T. C. *J. Magn. Reson.* 1972, **7**, 48.
17. Hogenboom, D. L.; O'Reilly, D. E.; Peterson, E. M. *J. Chem. Phys.* 1970, **52**, 2793.
18. Brussau, R. G.; Sillescu, H. *Ber. Buns. Phys. Chem.* 1972, **76**, 31.
19. Fung, B. M.; Gerace, M. J.; Gerace, L. S. *J. Phys. Chem.* 1970, **74**, 83.
20. O'Reilly, D. E.; Peterson, E. M.; Yasaitis, E. L. *J. Chem. Phys.* 1972, **57**, 890.
21. Gillen, K. T.; Douglass, D. C.; Malmberg, M. S.; Maryott, A. A. *J. Chem. Phys.* 1972, **57**, 5170.
22. Sears, R. E. J. *J. Chem. Phys.* 1972, **56**, 983.
23. Blicharski, J. S.; Blicharska, B. *Acta. Phys. Pol.* 1970, **38A**, 289.
24. Sears, R. E. J.; Hahn, E. L. *J. Chem. Phys.* 1966, **45**, 2753.
25. O'Reilly, D. E.; Peterson, E. M.; Scheie, C. E.; Seyfarth, E. *J. Chem. Phys.* 1973, **59**, 3576.
26. Gillen, K. T.; Noggle, J. H. *J. Chem. Phys.* 1970, **53**, 801.
27. Gillen, K. T.; Noggle, J. H. *J. Chem. Phys.* 1970, **52**, 4905.
28. O'Reilly, D. E.; Peterson, E. M. *J. Chem. Phys.* 1971, **55**, 2155.
29. Masuda, Y. *J. Phys. Soc. Jpn* 1956, **11**, 670.
30. Briguet, A.; Duplan, J. C.; Graveron-Demilly, D.; Delmau, J. *Mol. Phys.* 1974, **28**, 177.
31. Sharp, R. R. *J. Chem. Phys.* 1972, **57**, 5321.
32. Hawk, R. M.; Sharp, R. R. *J. Chem. Phys.* 1974, **60**, 1009.
33. Sharp, R. R.; Tolan, J. W. *J. Chem. Phys.* 1976, **65**, 522.
34. Pratt, D. W. *U.S. Atomic Energy Comm.* 1967, UCRL-17406.
35. Pratt, D. W.; Myers, R. J. *J. Chem. Phys.* 1969, **50**, 555.
36. Diehl, P. *Helv. Phys. Acta.* 1956, **29**, 219.
37. Strange, J. H.; Morgan, R. E. *Magn. Reson. Radiofrequ. Spectrosc., Proc. XV Colloque Ampere*, Ed Averbuch, P., Grenoble, 1968, 260.
38. Strange, J. H.; Morgan, R. E. *J. Phys. C., Solid State Phys.* 1970, **3**, 1999.
39. Jordan, A. D.; Cavell, R. G.; Jordan, R. B. *J. Chem. Phys.* 1972, **56**, 483.
40. Allerhand, A. *J. Chem. Phys.* 1969, **17**, 397.
42. Obermeyer, R. T.; Jones, E. P. *J. Chem. Phys.* 1973, **58**, 1677.
43. Hall, C.; Kydon, D. W.; Richards, R. E.; Sharp, R. R. *Mol. Phys.* 1970, **18**, 711.
44. Dodgen, H. W.; Jordan, A. D.; Jordan, R. B. *J. Phys. Chem.* 1973, **77**, 2149.

45. Bacon, J.; Gillespie, R. J.; Quail, J. W. *Can. J. Chem.* 1963, **41**, 3063.
46. Maryott, A. A.; Farrar, T. C.; Malmberg, M. S. *J. Chem. Phys.* 1971, **54**, 64.
47. Farrar, T. C.; Maryott, A. A.; Malmberg, M. S. *Ber. Buns. Phys. Chem.* 1971, **75**, 246.
48. Hall, C.; Kydon, D. W.; Richards, R. E.; Sharp, R. R. *Proc. R. Soc. London* 1970, **A318**, 119.
49. Lincoln, S. F.; Sandercock, A. C.; Stranks, D. R. *J. Chem. Soc. Dalton Trans.* 1975, 669.
50. Rhodes, M.; Aksnes, D. W.; Strange, J. H. *Mol. Phys.* 1968, **15**, 541.
51. Sharp, R. R. *J. Chem. Phys.* 1974, **60**, 1149.
52. Hertz, H. G. *Z. Electrochem.* 1961, **65**, 36.
53. Brownstein, M.; Selig, H. *Inorg. Chem.* 1972, **11**, 656.
54. Gillespie, R. J.; Quail, J. W. *Can. J. Chem.* 1964, **42**, 2671.
55. Blicharski, J. S.; Krynicki, K. *Acta. Phys. Pol.* 1962, **22**, 409.
56. O'Reilly, D. E.; Peterson, E. M. *J. Chem. Phys.* 1968, **49**, 2872.
57. Freeman, R.; Ernst, R. R.; Anderson, W. A. *J. Chem. Phys.* 1967, **46**, 1125.
58. Yamamoto, D.; Yamagisawa, M. *J. Chem. Phys.* 1977, **67**, 3803.
59. Farrar, T. C.; Druck, S. J.; Shoup, R. R.; Becker, E. D. *J. Am. Chem. Soc.* 1972, **94**, 699.
60. Gillen, K. T.; Douglass, D. C.; Griffiths, J. E. *J. Chem. Phys.* 1978, **69**, 461.
61. Alexandre, M.; Rigny, P. *Can. J. Chem.* 1974, **52**, 3676.
62. Christe, K. O.; Hon, J. F.; Pilipovich, D. *Inorg. Chem.* 1973, **12**, 84.
63. Christe, K. O. *Inorg. Nucl. Chem. Lett.* 1972, **8**, 741.
64. Gillespie, R. J.; Schrobilgen, G. J. *J. Chem. Soc. Chem. Commun.* 1974, 90.
65. Muetterties, E. L.; Packer, K. J. *J. Am. Chem. Soc.* 1964, **86**, 293.
66. Tarasov, V.; Privalov, V. I.; Buslaev, Y. A. *Zhur. Strukt. Khim.* 1978, **19**, 1012(866).
67. Briguet, A.; Duplan, J. C.; Delmau, J. *J. Magn. Reson.* 1981, **42**, 141.
68. Kitchlew, A.; Kumar, A.; Nageswara Rao, B. D. *Chem. Phys. Lett.* 1968, **2**, 216.
69. Deitrich, W.; Kosfeld, R. *Z. Naturforsch.* 1969, **24A**, 1209.
70. Clementi, E. *J. Chem. Phys.* 1962, **36**, 33.
71. Kreshkov, A. P.; Andronov, V. F.; Drozdov, V. A. *Russian J. Phys. Chem.* 1972, **46**, 183.
72. Deverell, C.; Richards, R. E. *Mol. Phys.* 1969, **16**, 421.
73. Blazer, J.; Lutz, O.; Steinkilberg, W. *Z. Naturforsch.* 1972, **27A**, 72.
74. Barlos, K.; Kroner, J.; Noth, H.; Wrackmeyer, B. *Chem. Ber.* 1978, **111**, 1833.
75. Barlos, K.; Kroner, J.; Noth H.; Wrackmeyer, B. *Chem. Ber.* 1980, **113**, 3716.
76. Hertz, H. G. *Ber. Buns. Phys. Chem.* 1963, **67**, 311.
77. Stengle, T. R.; Pan, Y. C. E.; Langford, C. H. *J. Am. Chem. Soc.* 1972, **94**, 9037.
78. Virlet, J.; Tantot, G. *Chem. Phys. Lett.* 1976, **44**, 296.
79. Lutz, O.; Nepple, W.; Nolle, A. *Z. Naturforsch.* 1976, **31A**, 1046.
80. Alei, M. *J. Chem. Phys.* 1963, **43**, 2904.
81. Becker, K. D. *J. Chem. Phys.* 1978, **68**, 3785.
82. Boyce, J. B.; Mikkelsenm, J. C. *J. Phys. C* 1977, **10**, L41.
83. Segel, S. L.; Vyas, H. M. *J. Chem. Phys.* 1980, **72**, 1406.
84. Abragam, A. *Principles of Nuclear Magnetic Resonance*; Oxford University: London, 1961.
85. Morgan, R. E.; Strange, J. H. *Mol. Phys.* 1965, **9**, 565
86. Strange, J. H.; Morgan, R. E. *J. Phys. C* 1970, **3**, 1999.
87. Hertz, H. G. *Z. Elektrochem.* 1961, **65**, 20.
88. Hertz, H. G.; Stalidis, G.; Versmold, H. *J. Chim. Phys. Physicochim. Biol.* Oct. 1969, 177.
89. Hertz, H. G. *Ber. Buns. Phys. Chem.* 1973, **77**, 531.
90. Valiev, K. A. *Sov. Phys. JETP.* 1960, **10**, 77.
91. Valiev, K. A. *Sov. Phys. JETP.* 1960, **11**, 853
92. Valiev, K. A.; Khabibullin, B. M. *Russian J. Phys. Chem.* 1961, **35**, 1118.
93. Valiev, K. A. *J. Strukt. Chem.* 1962, **3**, 630.
94. Valiev, K. A. *J. Strukt. Chem.* 1964, **5**, 477.
95. Deverell, C. *Mol. Phys.* 1969, **16**, 491.
96. Deverell, C. *Progr. NMR Spectrosc.* 1969, **4**, 235.
97. Hertz, H. G. *Ber. Buns. Phys. Chem.* 1973, **77**, 688.
98. Hertz, H. G.; Holtz, M.; Klute, R.; Stalidis, G.; Versmold, H. *Ber. Buns. Phys. Chem.* 1974, **78**, 24.
99. Hertz, H. G.; Holtz, M.; Keller, G.; Versmold, H.; Yoon, C. *Ber. Buns. Phys. Chem.* 1974, **78**, 493.
100. Richards, R. E.; Yorke, B. A. *Mol. Phys.* 1963, **6**, 289.
101. Deverell, C.; Frost, D. J.; Richards, R. E. *Mol. Phys.* 1965, **9**, 565.
102. Richards, R. E. *Mol. Phys.* 1964, **7**, 500.

103. Hall, C.; Richards, R. E.; Schultz, G. N.; Sharp, R. R. *Mol. Phys.* 1969, **16**, 529.
104. Arnold, M. St. J.; Packer, K. J. *Mol. Phys.* 1968, **14**, 241.
105. Holtz, M.; Weingartner, H. *J. Magn. Reson.* 1977, **27**, 153.
106. Reimarsson, P.; Wennerstrom, H.; Engstrom, S.; Lindman, B. *J. Phys. Chem.* 1977, **81**, 789.
107. Wennerstrom, H.; Lindman, B.; Forsen, S. *J. Phys. Chem.* 1971, **75**, 2936.
108. Lindman, B.; Wennerstrom, H.; Forsen, S. *J. Phys. Chem.* 1970, **74**, 754.
109. Hertz, H. G.; Holtz, M. *J. Phys. Chem.* 1974, **78**, 1002.
110. Yudasaka, M.; Sugawara, T.; Iwamura, H.; Fujiyama, T. *Bull. Chem. Soc. Jpn* 1981, **54**, 1933.
111. Yudasaka, M.; Sugawara, T.; Iwamura, H.; Fujiyama, T. *Bull. Chem. Soc. Jpn* 1982, **55**, 311.
112. Hall, C.; Haller, G. L.; Richards, R. E. *Mol. Phys.* 1969, **16**, 377.
113. Melenderes, C.; Hertz, H. G. *J. Chem. Phys.* 1974, **61**, 4156.
114. Lindman, B. *Acta Chem. Scand.* 1975, **29A**, 935.
115. Weingartner, H.; Hertz, H. G. *Ber. Buns. Phys. Chem.* 1977, **81**, 1204.
116. Holtz, M.; Weingartner, H.; Hertz, H. G. *J. Solution Chem.* 1978, **7**, 705.
117. Holtz, M. *J. Chem. Soc. Faraday Trans. I* 1978, **74**, 644.
118. McCain, D. C. *J. Inorg. Nucl. Chem.* 1980, **42**, 1185.
119. Zeltmann, A. H.; Morgan, L. O. *Inorg. Chem.* 1971, **10**, 2739.
120. Wertz, J. E. *J. Chem. Phys.* 1956, **24**, 484.
121. Chesnut, D. B. *J. Chem. Phys.* 1960, **33**, 1234.
122. Barbalat-Rey, F.; Bene, G.; Peter, M. *Helv. Phys. Acta* 1967, **40**, 787.
123. Barbalat-Rey, F. *Helv. Phys. Acta* 1969, **42**, 515.
124. Connick, R. E.; Poppel, C. P. *J. Am. Chem. Soc.* 1959, **81**, 6389.
125. Zeltman, A. H.; Morgan, L. O. *J. Phys. Chem.* 1966, **70**, 2807.
126. Hertz, H. G. *Z. Elektrochem.* 1960, **64**, 53.
127. O'Reilly, D. E.; Schacher, G. E.; Schug, K. *J. Chem. Phys.* 1963, **39**, 1756.
128. Gentzler, R. E.; Stengle, T. R.; Langford, C. H. *J. Chem. Soc. Chem. Commun.* 1970, 1257.
129. Zeltmann, A. H.; Matwiyoff, N. A.; Morgan, L. O. *J. Phys. Chem.* 1969, **73**, 2689.
130. Lincoln, S. F.; Aprile, F.; Dodgen, H. W.; Hunt, J. P. *Inorg. Chem.* 1968, **7**, 929.
131. Lincoln, S. F.; Sandercock, A.; Stranks, D. R. *J. Chem. Soc. Chem. Commun.* 1972, 1069.
132. Lincoln, S. F.; Sandercock, A.; Stranks, D. R. *Aust. J. Chem.* 1975, **28**, 1901.
133. Dalibart, M.; Derouault, J.; Granger, P. *Inorg. Chem.* 1981, **20**, 3975.
134. Beattie, I. R.; Jones, P. J.; Howard, J. A. K.; Suart, L. E.; Gilmore, C. J.; Akitt, J. W. *J. Chem. Soc. Dalton Trans.* 1979, 528.
135. Deshmukh, V. G. I.; Seymour, E. F. W.; Styles, G. A. *Phys. Lett.* 1979, **74A**, 432.
136. Anders, U.; Plambeck, J. A. *J. Inorg. Nucl. Chem.* 1978, **40**, 387.
137. Gulya, A. P. *Khimiya Koordinats. Neorg. Organ. Soedin.* 1978, 10 (Chem. Abs. **90**, 31566).
138. Helm, L.; Hetrz, H. G. *Ber. Buns. Phys. Chem.* 1981, **85**, 158.
139. Weingartner, H.; Muller, C.; Hertz, H. G. *J. Chem. Soc. Faraday Trans. I* 1979, **75**, 2712.
140. Bryden, C. C.; Reilly, C. N.; Desreux, J. F. *Analyt. Chem.* 1981, **53**, 1418.
141. Myers, O. E. *J. Chem. Phys.* 1958, **28**, 1027.
142. Genser, E. E.; Connick, R. E. *J. Chem. Phys.* 1973, **58**, 990.
143. Cahen, Y. M.; Handy, P. R.; Roach, E. T.; Popov, A. I. *J. Phys. Chem.* 1975, **79**, 80.
144. Baum, R. G.; Popov, A. I. *J. Solution Chem.* 1975, **4**, 441.
145. Popov, A. I. *Pure Appl. Chem.* 1975, **41**, 275.
146. Reimarsson, P.; Lindman, B. *Inorg. Nucl. Chem. Lett.* 1977, **13**, 449.
147. Contreras, M.; Hertz, H. G. *Faraday Diss. Chem. Soc.* 1978, **64**, 33; Macklin, J. W., *et al. J. Am. Chem. Soc.* 1985, **107**, 1190, 1193.
148. Masuda, Y.; Kanda, T. *J. Phys. Soc. Jpn* 1954, **9**, 82.
149. Akitt, J. W.; Covington,. A. K.; Freeman, J. G.; Lilley, T. H. *Trans. Faraday Soc.* 1969, **65**, 2701.
150. Reimarsson, P. *J. Magn. Reson.* 1980, **38**, 245.
151. Berman, H. A.; Stengle, T. R. *J. Phys. Chem.* 1975, **79**, 1001.
152. Berman, H. A.; Yeh, H. J. C.; Stengle, T. R. *J. Phys. Chem.* 1975, **79**, 2551.
153. Greenberg, M. S.; Popov, A. I. *J. Solution Chem.* 1976, **5**, 653.
154. Grandjean, J.; Laszlo, P. *Helv. Chim. Acta* 1977, **60**, 259.
155. Echegoyen, L.; Nieves, I.; Thompson, J.; Rosa, F.; Concepcion, R. *J. Phys. Chem.* 1981, **85**, 3697.
156. Kavum, V. Y.; Chernyshow, B. N. *Koord. Khim.* 1979, **5**, 53(41).
157. Itoh, J.; Yamagata, Y. *J. Phys. Soc. Jpn* 1958, **13**, 1182.
158. Hindman, J. C. *J. Chem. Phys.* 1962, **36**, 1000.

159. Halliday, J. D.; Richards, R. E.; Sharp, R. R. *Proc. R. Soc. London* 1969, **A313**, 45.
160. Sazonov, A. M.; Porai-Koshits, A. B.; Olevskii, V. M.; Skobelev, V. N.; Shmuilovich, G. A. *J. Struct. Chem. USSR* 1973, **14**, 752.
161. Loewenstein, A.; Shporer, M.; Lauterbur, P. C.; Ramirez, J. E. *J. Chem. Soc. Chem. Commun.* 1968, 214.
162. Halliday, J.; Hill, H. D. W.; Richards, R. E. *J. Chem. Soc. Chem. Commun.* 1969, 219.
163. Gustavsson, H.; Lindman, B. *Proc. Int. Conf. Colloid Surface Science*; Wolfram, E., Ed.; Vol. 1, 625, Budapest, 1975.
164. Stengle, T. R.; Pan, Y. C. E.; Langford, C. H. *J. Am. Chem. Soc.* 1972, **94**, 9037.
165. Reference 1, pp. 249–325.
166. Fedotov, M. A.; Krivoruchko, O. P.; Byanov, R. A. *Zh. Neorg. Khim.* 1978, **23**, 2326(1282).
167. Gustavsson, H.; Lindman, B. *J. Am. Chem. Soc.* 1978, **100**, 4647.
168. Fabre, H.; Kamenka, N.; Khan, A.; Lindblom, G.; Lindman, B. *J. Phys. Chem.* 1980, **84**, 3428.
169. Ulmius, J.; Lindman, B.; Lindblom, G.; Drakenberg, T. *J. Colloid Interfac. Sci.* 1978, **65**, 88.
170. Andersson, I.; Katzberg, D.; Lindman, B.; Zeppezauer, M. *Dev. Halophilic Microorg.* 1978, **1**, 33.
171. Andersson, I.; Zeppezauer, M.; Bull, T.; Einarsson, R.; Norne, J. E.; Lindman, B. *Biochemistry* 1979, **18**, 3407.
172. Happe, J. A.; Ward, R. L. *J. Phys. Chem.* 1979, **83**, 3457.
173. Reimarsson, P.; Bull, T.; Norne, J. E.; Lindman, B. *Dev. Halophilic Microorg.* 1978, **1**, 41.
174. Subramanian, S.; Shindo, H.; Kaufman, B. T. *Biochemistry* 1981, **20**, 3226.
175. Ward, R. L. *Chem. Abs.* **92**, 2266.
176. Chiancone, E.; Kniper, H. A.; Antonini, E.; Norne, J. E.; Forsen, S. *Invertebr. Oxygen Binding Proteins* 1979; Lamy, J., Ed.; Dekker: New York, 1981, p. 647.
177. Reimarsson, P.; Lindman, B.; Werber, M. M. *Arch. Biochem. Biophys.* 1980, **202**, 664.
178. Leyte, J. C. *Chem. Phys. Lett.* 1979, **66**, 417.

Chapter 18

THE NOBLE GASES

CYNTHIA J. JAMESON

1. INTRODUCTION

Excluding ^{3}He, the rare-gas nuclei amenable to NMR investigation are ^{21}Ne, ^{83}Kr, ^{129}Xe, and ^{131}Xe. Of these only Kr and Xe form compounds that are stable under ordinary conditions, and only ^{129}Xe NMR spectra have been observed in chemical environments other than the pure element, although Kr compounds have been characterized by ^{19}F NMR spectroscopy.(1) Thus, we will divide this chapter into two parts: (1) ^{129}Xe NMR studies of bonding and structure of xenon compounds, and (2) ^{3}He, ^{21}Ne, ^{83}Kr, ^{129}Xe, and ^{131}Xe NMR studies of nonspecific intermolecular interactions.

An excellent comprehensive review of this subject up to 1978 is available in Schrobilgen's chapter in "NMR and the Periodic Table."(2) In addition to the compounds included in that chapter, the following have been studied by ^{129}Xe NMR: $FXeN(SO_2F)_2$,(3) $Xe[N(SO_2F)_2]_2$,(4) $[XeN(SO_2F)_2]_2F^+AsF_6^-$,(4) $XeOTeF_5^+$,(5) $XeOSO_2F^+$,(5) $FXeF\text{–}BrOF_2^+$,(5) $FXeFXe'OTeF_5$,(5,6) $Xe(OTeF_5)_4$,(6,7) $O{=}XeF_n(OTeF_5)_{4-n}$, $n = 0\text{–}3$,(7) and $FXeF \cdots (MOF_4)_n$ (M = Mo, W; $n = 1\text{–}4$).(8) Seppelt has reviewed developments in noble gas chemistry,(9,10) including crystallographic and spectroscopic evidence, in which ^{129}Xe NMR studies are included.

^{129}Xe relaxation times (T_1) in various species in solution range from 285 to 780 msec.(11,12) Reasonable pulse repetition rates and medium sensitivity (31.8 times that of natural abundance C-13), 1/2 spin and very large shifts (chemical shift range 7540 ppm) make ^{129}Xe NMR spectroscopy relatively easy with a modern FT spectrometer. However, the explosive nature of many of its compounds has limited the number of papers involving ^{129}Xe NMR spectroscopy.

2. ^{129}Xe NMR STUDIES OF BONDING AND STRUCTURE OF XENON COMPOUNDS

Xenon is known to form bonds in compounds isolatable under ordinary laboratory conditions with seven elements: F, O, Cl, N, Xe, B, and C. There are, of course, the very cold van der Waals molecules, and species which have been observed only in matrix isolation studies, which we are not including in this category.

CYNTHIA J. JAMESON • Department of Chemistry, University of Illinois, Chicago, Illinois 60680.

These bonds are the following types[10]:

1. Normal Xe–F bonds, e.g., in the binary fluorides XeF_2, XeF_4, XeF_6;
2. Bridging Xe–F bonds, e.g., $[(FXe)_2-F]^+ AsF_6^-$, $[(F_5Xe)_2-F]^+ AuF_6^-$;
3. Weaker Xe–O bonds, e.g., $Xe(OSeF_5)_2$;
4. Stronger Xe=O bonds, e.g., XeO_3, $O=XeF_4$;
5. Weak Xe–Cl bonds, e.g., $Cs_9(XeO_3Cl_2)_4Cl$;[13]
6. Xe–N bonds, e.g., $FXeN(SO_2F)_2$, $Xe[N(SO_2F)_2]_2$;
7. Xe–Xe bond in Xe_2^+ in solution, observed by EPR;[14]
8. Xe–B bond in $FXe-BF_2$;[15] and
9. Xe–C bond in $Xe(CF_3)_2$.[16]

In these compounds, xenon is in one of the following oxidation states: Xe(II) in XeA_2E_3 types (three lone pairs) in the terminology of Gillespie and Nyholm's VSEPR theory of molecular structure, Xe(IV) in XeA_4E_2 types, Xe(VI) in XeA_6E or $O=XeA_4$, and Xe(VIII) in XeO_6^{-4}. Several reviews of noble gas chemistry, spectroscopy, and structure determination discuss the early work soon after the first discovery of a xenon compound in 1962.[17–19] Two comprehensive reviews published in 1970 and 1973,[20,21] and the 1982 review by Seppelt and Lentz cover the later developments.[10]

Of the types of covalent compounds formed by Xe, only those in XeF, XeO, and XeN bonds have been observed by ^{129}Xe NMR spectroscopy. Although many $Xe(OL)_2$ and FXeOL compounds exist, all except the $OL=OTeF_5$, $OSeF_5$, and OSO_2F compounds must be considered as explosives, so it is not surprising that only these three types have been observed by ^{129}Xe NMR. In addition, the species $F-Xe-O-WF_5(WOF_4)_n$, $n=1, 2$ have been identified only by NMR measurements in solution.[8,22] These Xe(II) derivatives have linear F–Xe–O and O–Xe–O structures, as do XeF_2 and the F–Xe–F in $(FXe)_2F^+$

Table 1. ^{129}Xe Chemical Shifts, ppm Relative to Neat Liquid $O=XeF_4$, and $^1J(XeF)$[a]

δ (ppm)	Reference	Molecule	$^1J(XeF)$ (Hz)	Reference
Xe(0)				
ca. −5460	(see text)	Xe atom		
−5331	11	Xe in n-C_6F_{14}, 25°C		
Xe(II)				
−2327/−2447.4	6, 12	$Xe(OTeF_5)_2$		
−2289	12	$Xe(OTeF_5)(OSeF_5)$		
−2200	12	$Xe(OSeF_5)_2$		
−2051/−2067	6, 12	$FXeOTeF_5$	5670/5743	6, 12
−1952	12	$FXeOSeF_5$	5630	12
−2101	4	$Xe[N(SO_2F)_2]_2$		
−1997.3/−2053	3, 4, 22	$FXeN(SO_2F)_2$	5572/5624	3, 4, 22
−1933	4	$[(FSO_2)_2NXe]_2^+ AsF_6^-$		
−1592/−2009	6, 8, 11, 12, 22	XeF_2	5550/5665	6, 8, 11, 12, 22, 23
−1320/−1613	5, 22	$Xe(OSO_2F)_2$		
−1407/−1666	8, 22	$FXeOSO_2F$	5830/6051	8, 22, 23
−1472/−1608	5	$XeOTeF_5^+$		
−1278/−1306	5	$XeOSO_2F^+$		

Table 1. (*Continued*)

δ (ppm)	Reference	Molecule	^1J(XeF) (Hz)	Reference
−1342	8, 22	$FXeOS(=O)FOMoOF_4$	5853/5971	8, 22
−1335	8, 22	$FXeOS(=O)FOWOF_4$	5992/6131	8, 22
−1258	22	$[(FXeO)_2S(=O)F]^+$	6428	22
−1359	5	$FXeFBrOF_2^+$	5680(ave)	5
−1381/−1441	8, 22	$FXeF{-}MoOF_4$	6018/6140(term) 5000/5117(br)	8, 22
−1315/−1331	8, 22	$FXeF{-}WOF_4$	6058/6196(term) 5000/5051(br)	8, 22
−1338	8, 22	$FXeF{-}MoOF_4(MoOF_4)$	6159(term) 5036/5110(br)	8, 22
−1321	8, 22	$FXeF{-}MoOF_4(MoOF_4)_2$	6156(term) 5029(br)	8, 22
−1189	8, 22	$FXeF{-}WOF_4(WOF_4)$	6260/6268(term) 4964/5000(br)	8, 22
−1170	8, 22	$FXeF{-}WOF_4(WOF_4)_2$	6304(term) 4996(br)	22
−1146, 1633(Xe′)	5, 6	$FXeFXe'OTeF_5$	5747(ave)	5, 61
−1051/−1059	5, 6, 22	$(FXe)_2F^+$	6662/6740(term) 4828/4865(br)	5, 6, 22
−955	8, 22	$FXeO{-}WF_5(WOF_4)$	6315/6373	8, 22
−906	8, 22	$FXeO{-}WF_5(WOF_4)_2$	6330/6373	8, 22
−574/−911	6, 18, 22	XeF^+	6703/7594	6, 11, 22
Xe(IV)				
−637/−662.8	6, 7	$Xe(OTeF_5)_4$		
−436	22	$FXe(OTeF_5)_3$	3506	22
−243	22	$F_2Xe(OTeF_5)_2$ *cis*	3714	22
−216	22	$F_2Xe(OTeF_5)_2$ *trans*	3503	22
−26	22	$F'F_2Xe(OTeF_5)$	3552, 3733	22
+166.1/+259	6, 11, 22	XeF_4	3801/3900	6, 11, 22
+595	22	$XeF_2F'^+$	2384, 2609(F′)	22
Xe(VI)				
−204/−211.8	6, 7, 22	$O=Xe(OTeF_5)_4$		
−133/−157	7, 22	$O=XeF(OTeF_5)_3$	1056/1082	7, 22
−92/−106	7, 22	$O=XeF_2(OTeF_5)_2$(*trans*)	984/990	7, 22
−81/−118	7, 22	$O=XeF_2(OTeF_5)_2$(*cis*)	1074/1231	7, 22
−40/−66	7, 22	$O=XeF_3(OTeF_5)$	1127/1148, and 945	7, 22
0	7, 22	$O=XeF_4$	1115/1131	6, 7, 11, 22
−35/−45	11, 24	XeF_6	330	11
−39/−60.8	11, 22, 24	$(XeF_6)_4$	331.7	22
−23.9/+131.8	22	$XeF_4F'^+$	1377/1400 165/159(F′)	22
+131	22	$O_2Xe(OTeF_5)_2$		
+154	22	$O_2XeF(OTeF_5)$	1046	22
+17/73	22	XeO_2F_2	1217	22
+217	11, 22	XeO_3		
+238	22	$O=XeF_2F'^+$	1018, 434(F′)	22
+600	22	XeO_2F^+	95	22
Xe(VIII)				
+2077	22	XeO_6^{-4}		

[a] Range of values are indicated by extremes of range separated by slash. br = bridging, term = terminal.

Xe(IV) compounds are expected to be less stable than those of Xe(II), and the stable compounds XeF_4 and $Xe(OTeF_5)_4$ have been observed by Xe NMR spectroscopy.

Xe(VI) compounds with ligands $=O$, $-OTeF_5$, and $-F$ which have been observed are listed in Table 1. Because of the lone pair electrons, XeF_6 is not exactly octahedral and forms a tetramer, whose structure has been established from its ^{129}Xe and ^{19}F NMR spectra.[9,10]

2.1. ^{129}Xe Chemical Shifts

Table 1 lists the ^{129}Xe chemical shifts which have been measured, and Figure 1 is a chart indicating the range of values of these shifts in various chemical environments. There are some general trends worth noting:

(a) ^{129}Xe absolute shielding depends largely on oxidation number and is in the order

$$^{129}\mathrm{Xe}\ \sigma\ :\quad \mathrm{Xe(0)} > \mathrm{Xe(II)} > \mathrm{Xe(IV)} > \mathrm{Xe(VI)} > \mathrm{Xe(VIII)}$$

We know this from the range of ^{129}Xe chemical shifts relative to neat liquid $O{=}XeF_4$; $\delta = \sigma(\mathrm{OXeF_4\ liq}) - \sigma(\mathrm{sample}) = -5460$ to -5125 ppm for Xe(0), -2447 to -574 ppm for Xe(II), -663 to $+595$ ppm for Xe(IV), -212 to $+600$ ppm for Xe(VI), and $+2077$ ppm for Xe(VIII).

(b) The Xe shielding in the Xe(II) oxidation state is in the order

$$^{129}\mathrm{Xe}\ \sigma\ :\quad \mathrm{XeA_2} > \mathrm{FXeA} > \mathrm{(AXe)_2\,F^+} > \mathrm{XeA^+}$$

This is found to hold for $A = OTeF_5$, OSO_2F, $N(SO_2F)_2$ and F, without exception.

(c) For all oxidation states, the Xe nucleus of the cation with one less F is less shielded:

$$^{129}\mathrm{Xe}\ \sigma\ :\quad \mathrm{XeF}_n > \mathrm{XeF}_{n-1}^+$$

The chemical shifts of XeF_n and XeF_{n-1}^+ are, respectively, $n=2$: $\delta = -2009$ to -1592 ppm, -911 to -574 ppm; $n=4$: $\delta = +253$, $+595$ ppm; $n=6$: $\delta = -60.8$, -23.9 to $+131.8$. This is also typified by

$$^{129}\mathrm{Xe}\ \sigma\ :\quad (\mathrm{O}{=})_m\,\mathrm{XeF}_n > (\mathrm{O}{=})_m\,\mathrm{XeF}_{n-1}^+$$

Two known examples are (1) $m=2$, $n=2$: O_2XeF_2 and O_2XeF^+ have Xe chemical shifts of $+173.2$ and $+600$ ppm, respectively, and (2) $m=1$, $n=4$: $O{=}XeF_4$ and $O{=}XeF_3^+$ have Xe chemical shifts of 0 and $+238$ ppm, respectively.

(d) Physical effects on Xe chemical shifts are very large. Xe is very sensitive to solvent effects, and the sensitivity increases with increasing number of lone pairs on Xe. For example, the ^{129}Xe spectrum of XeF_2 shifts over a 400 ppm range depending on the solvent, concentration, and temperature, Xe in $Xe(OTeF_5)_2$ has a 200 ppm range, and Xe(0) has a range of chemical shifts over 300 ppm in the pure gas, liquid, and solid phase as well as in organic solvents and clathrates. These physical effects are so large that comparisons of Xe shielding in different compounds can only be made in the same solvent. For example, the Xe chemical shifts in XeF_2 and $Xe(OTeF_5)_2$ separately measured in different solvents were thought to support the notion that the

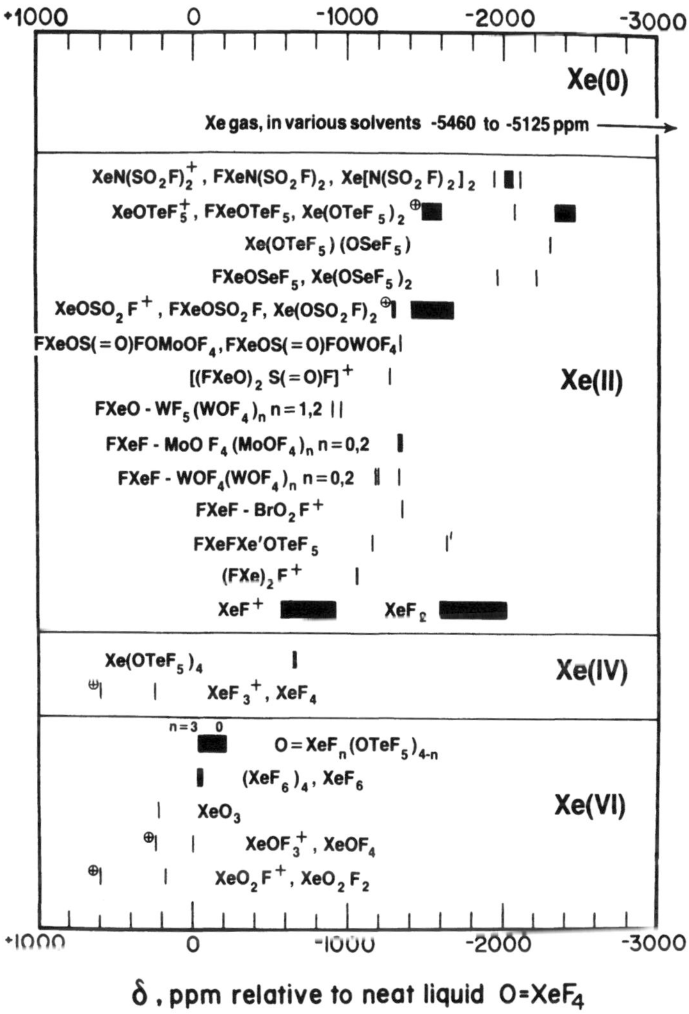

Figure 1. ^{129}Xe chemical shifts, in ppm relative to neat liquid $O{=}XeF_4$ at 24°C. The isolated ^{129}Xe atom corresponds to an absolute shielding of $\sigma = 7040$ ppm[25] from a relativistic calculation, and σ(neat liquid $O{=}XeF_4$) = 1580 ppm (see text). Thus, δ(Xe atom) $\simeq -5460$ ppm and δ(bare nucleus) $\simeq +1580$ ppm.

$OTeF_5$ moiety had an electronegativity greater than F itself. However, measured in the same solvent, $\delta(XeF^+) - \delta(XeOTeF_5^+) = 898$ ppm and $\delta(XeF_2) - \delta(Xe(OTeF_5)_2) = 438$ ppm, which shows that F is indeed more electronegative than the $OTeF_5$ group.[6]

We have not marked the bare ^{129}Xe nucleus in the chemical shift chart given in Figure 1. This requires a measurement of the chemical shift of xenon gas in the zero-pressure limit (Xe isolated atom) relative to the neat liquid reference $O{=}XeF_4$. This is not available, but we can estimate it as follows: Xe gas dissolved in liquid n-C_6F_{14} at 25°C has a chemical shift of -5331 ppm relative to neat liquid $O{=}XeF_4$.[11] Xe shifts of Xe(0) have been measured in various organic solvents: $\sigma(\text{Xe in solution}) = \sigma(\text{Xe free atom}) + \sigma_b + \sigma_m$, where σ_b is the bulk susceptibility contribution and σ_m has been found to be -148 to -335 ppm at 25°C.[26,27] These organic solvents do not include n-C_6F_{14}, thus we can only make an estimate of its σ_m. Based on the straight line correlation of σ_m with the function $[(n^2-1)/(2n^2+1)]^2$ we take σ_m to be -130 ppm for Xe in n-C_6F_{14}. With this, δ(isol. atom) is $-5331-130 \simeq -5460$ ppm relative to neat liquid $O{=}XeF_4$. The calculated value of σ for an isolated Xe atom is 7040 ppm.[25] Thus, the absolute ^{129}Xe shielding of neat liquid $OXeF_4$ must be ca. 1580 ppm. It appears that the ^{129}Xe nucleus in the perxenate ion, XeO_6^{-4} is somewhat deshielded compared to the bare nucleus, and all the other Xe environments are more shielded than the bare Xe nucleus, the isolated Xe atom being the most shielded of all.

Although Xe absolute shielding components in various oxidation states have not been measured, theoretical arguments[28] lead to predictions of rather large anisotropies: $(\sigma_{\parallel}-\sigma_{\perp}) = +5125$ to $+7185$ ppm (depending on the bonding model used) for XeF_2, $+725$ to $+3940$ ppm for XeF_4 and $+365$ to $+1500$ ppm for $O{=}XeF_4$, where $\sigma_{\parallel}$ and $\sigma_{\perp}$ are the Xe shielding components parallel and perpendicular to the symmetry axis of the molecule. There have been no measurements of the ^{129}Xe shielding anisotropy, but there is independent experimental evidence that such large Xe shielding anisotropies may exist. In Chapter 3 we have established the relationship between the anisotropy of the paramagnetic part of the nuclear shielding and the electric field gradient, eq. Measurements of the Xe quadrupole coupling constants $|e^2qQ|$ in various chemical environments by Mössbauer spectroscopy yield values of this electric field gradient. Such quantities are an indirect measure of nuclear shielding anisotropy and they also allow us to put in context those compounds such as $XeCl_2$ and $XeCl_4$, and Kr compounds which have not been observed by NMR spectroscopy.

The beta decay of ^{129}I in an iodine compound often leads to a xenon compound with the same ligands. The Mössbauer spectrum of the gamma radiation emitted from the ^{129}Xe nuclei so formed shows hyperfine structure. This is due to the nuclear electric quadrupole moment eQ of the excited ^{129}Xe nucleus, which results in a doublet absorption spectrum if the electric field gradient eq $(=\partial^2V/\partial z^2)$ does not vanish. The observed quadrupole splitting equals $|e^2qQ/2|$ if the asymmetry parameter of the electric field gradient is negligible. (See Chapter 2.)

One might roughly consider the total electric field gradient due to the nonspherical symmetry of the electron distribution around the nucleus as follows[29,30]:

$$q = q_{\text{ext}} + q_{\text{ind}} + q_{\text{ct}} + q_{\text{pol}} + q_{\text{ov}}$$

q_{ext} are contributions to the efg due to the external electrical charges. q_{ind} are contributions from the induced efg in the electron shells of the atom of interest due to the external charges; the extent of this effect is measured by $\gamma_\infty = -q_{\text{ind}}/q_{\text{ext}}$, the Sternheimer electric shielding factor, which is -138 in Xe[29] and -67.8 in Kr.[29] q_{ct} is the efg due to charge transfer out of the p or d orbitals of the atom in question; q_{pol} is the efg due to polarization of the remaining electrons, especially s electrons, by the hole left in the atom due to the charge transfer. The latter is an indirect contribution from the s orbitals, since they do not have a direct efg contribution. q_{ov} is the efg due to reorthogonalization of Xe $5p$ orbitals due to overlap of Xe orbitals with ligand orbitals.

We can take the simplest approach and assume that only q_{ct} is important. If we do this, then we can roughly describe the Xe(II) and Xe(IV) compounds by Xe valence configurations $[(5s5p_x5p_y)^6\,5p_z^{(2-h_p)}]$ and $[(5s5p_z)^4(5p_x5p_y)^{(4-h_p)}]$, respectively, with h_p being the fractional $5p$ hole left in the Xe valence shell due to the transfer of charge from the $5p$ shell of Xe to the two or four electronegative ligands. To find

h_p, all that is needed is a measure of the electric field gradient q_p in a xenon with a unit $5p$ hole. In this simple charge transfer model,

$$q_{\text{obsd}} = h_p q_p \text{ for Xe(II)} \qquad \text{and} \qquad -2h_p q_p \text{ for Xe(IV)}$$

since all contributions to q other than q_{ct} are ignored. By spectroscopic and atomic beam determinations of the quadrupolar splitting in the $(5p^5 6s)\,^3P_2$ state of a ^{131}Xe atom[31] and comparative ^{131}Xe and ^{129}Xe Mössbauer studies in XeF_4,[32] the e^2qQ obtained for ^{131}Xe has been translated to e^2qQ for ^{129}Xe: $|e^2qQ(^{129}\text{Xe})| = 1745$ MHz for the $(5p^5 6s)\,^3P_2$ state of the Xe atom. This may directly be used as a measure of q_p in our simple model. Thus, h_p can be obtained from the ratios $|e^2qQ|/1745$ MHz. Owing to the charge transfer from the Xe to the electronegative ligands, leaving a fractional hole, h_p, in the $5p$ shell of Xe, there will be a negative q for Xe(II) and Xe(0)$(5p^5 6s)$, a positive q for Xe(IV) and $q = 0$ for Xe(VIII). The electric field gradient in Xe(VI) compounds is largely due to the deviation from octahedral symmetry caused by the lone pair electrons and is expected to be smaller.

In Table 2 we have indicated h_p and the approximate charge transferred per bond. This is a very rough approximation since we attributed the entire observed q to q_{ct}. The same model can be applied to ^{83}Kr efg's (Table 3). Whereas relatively large fractions of electronic charge are involved in Xe compounds, KrF^+ and KrF_2 are more like the less stable $XeCl_2$ than XeF_2.

2.2. Spin–Spin Coupling to Xenon

The coupling constants involving ^{129}Xe are summarized in Table 4; the one-bond couplings to F are shown in Figure 2 and listed in Table 1. Some general trends which can be noted are as follows:

(1) ^1J(XeF) is sensitive to the charge on the Xe, with the largest values observed in XeF^+ and $(F\text{Xe})_2F^+$.

(2) The ^1J(XeF) values have nonoverlapping ranges for different oxidation states of Xe, ^1J(XeF): Xe(II) > Xe(IV) > Xe(VI), therefore the magnitude of ^1J(XeF) serves as an important diagnostic tool. In Table 1 we note the following ^1J(XeF) ranges: normal Xe(II)F: 5583–7594 Hz, bridging Xe(II)F: 4828–5117 Hz, Xe(IV)F: 2384–3823 Hz, Xe(VI)F: 95–1400 Hz. This is in agreement with general trends in the dependence of ^1J on oxidation number discussed in Chapter 4.

(3) Other one-bond couplings which have been reported are $^1J(\text{Xe}^{\text{VI}}{=}^{17}\text{O})$ 692 Hz and $^1J(\text{Xe}^{\text{II}}{-}^{15}\text{N})$ 307.4 Hz, which can be expressed in reduced form to compare directly with the Xe–F one-bond couplings: $^1K(\text{Xe}^{\text{VI}}{=}\text{O}) = 1537$ is large compared to $^1K(\text{Xe}^{\text{VI}}\text{F}) - 30.4$–448. The $(\text{Xe}^{\text{VI}}{-}\text{O})$ would undoubtedly be smaller than the $(\text{Xe}^{\text{VI}}{=}\text{O})$ coupling. On the other hand, $^1K(\text{Xe}^{\text{II}}\text{N}) = 913.2$ while $^1K(\text{Xe}^{\text{II}}\text{F}) =$ 1786–2430, all in units of 10^{19} N A^{-2} m^{-3} (see Chapter 4).

(4) Two-bond couplings to ^{125}Te also have nonoverlapping ranges of values depending on the Xe oxidation number,

$$^2J(\text{XeTe}): \quad \text{Xe(VI)} > \text{Xe(IV)} > \text{Xe(II)}$$

but in the opposite order compared to ^1J. $^2J(\text{XeO}\,^{125}\text{Te}) = 470$–540 Hz for Xe(II),[12] 988–1192 Hz for Xe(IV), and 1281–1856 Hz for Xe(VI).[6,7] The reduced two-bond

Table 2. ^{129}Xe Nuclear Quadrupole Coupling Constants[a] in Various Compounds

Compound	$\|e^2qQ\|$ (MHz)	q_{compd}/q_p [b]	h_p [c]	Electron transferred per bond[d]
Xe(II)				
$XeF^+Sb_2F_{11}^-$	2650	+1.520	1.520	0.760
$XeF^+BiF_6^-$	2645	+1.515	1.515	0.757
$XeF^+TaF_6^-$	2610	+1.496	1.496	0.748
$(FXe)_2F^+BiF_6^-$	2645	+1.515	1.515	0.757
$(FXe)_2F^+AsF_6^-$	2642	+1.514	1.514	0.757
$XeF_2 \cdot 2WOF_4$	2620	+1.500	1.500	0.750
$XeF_2 \cdot MoOF_4$	2580	+1.477	1.477	0.738
XeF_2	2540	+1.456	1.456	0.728
$(FXeO)_2S(=O)F^+AsF_6^-$	2590	+1.426	1.426	0.713
$FXeN(SO_2F)_2$	2415	+1.383	1.383	0.691
$Xe(OTeF_5)_2$	2450	+1.402	1.402	0.701
$XeOTeF_5^+AsF_6^-$	2390	+1.370	1.370	0.685
$XeCl_2$ [e]	1800	+1.032	1.032	0.516
Xe(IV)				
$XeF_3^+BiF_6^-$	2642	−1.514	3.028	0.756
$XeF_3^+Sb_2F_{11}^-$	2530	−1.45	2.90	0.725
XeF_4	2620	−1.50	3.00	0.750
$XeCl_4$ [e]	1640	−0.94	1.98	0.495
Xe(VI)				
$O{=}XeF_4$	1156	0.663		
$XeO_2F^+SbF_6^-$	868	0.498		
$XeF_5^+Sb_2F_{11}^-$	765	0.438		
F_5XeOSO_2F	638	0.366		
$(F_5Xe)_2F^+BiF_6^-$	580	0.332		
XeF_6	492	0.282		
Xe(VIII)				
Na_4XeO_6	0			
Xe(0)				
$Xe(^3P_2\text{: }5p^56s)$ [f]	1745	+1.000	1.000	
XeHCl [g]	4.9	0.012	0.005	

[a] All data taken from Ref. 33 unless otherwise indicated, and converted from mm/s to MHz by the conversion factor 31.93 MHz = 1 mm/s, given by Ref. 34.
[b] A measure of the p electron imbalance, with the $^3P_2$: $5p^56s$ state of Xe atom corresponding to a unit hole in the $5p$ shell.
[c] Approximate, assuming the bonding involves transfer to the ligands of $5p$ electrons only, leaving h_p holes in the $5p$ shell of Xe.
[d] Approximate, according to above model.
[e] Reference 34.
[f] Reference 31.
[g] ^{131}Xe Refs. 29, 35.

coupling to Se, $^2K(Xe^{II}OSe) = 205.2$, is less than that to Te, $^2K(Xe^{II}OTe) = 448$ to 514, both in units of $10^{19}\ N\ A^{-2}\ m^{-3}$.

(5) Three-bond couplings depend on Xe oxidation number but less consistently so; they also depend on the coupling path geometry. For example, $^3J(XeOTeF) =$ 0–4 Hz[7] for axial fluorines, and 50–55 Hz[6,7] for equatorial fluorines in $O{=}XeF_n(OTeF_5)_{4-n}$. The relative magnitudes differ because the coupling paths to the equatorial fluorines are planar (dihedral angles 0° or 180°), whereas the coupling

Table 3. ^{83}Kr Nuclear Quadrupole Coupling Constants

Compound	$\lvert e^2qQ \rvert$ (MHz)	h_p	Reference
KrF_2	978	1.072	36
$KrF^+AsF_6^-$	1073, 1067	1.17	36
$KrF^+SbF_6^-$	1090	1.19	36
Kr^+ ion $(4p^5)^a$	915	1.00	36
KrClF	13.90	0.015	37
KrHF	10.227	0.011[b]	38, 39

[a] Derived by Ref. 36 from atomic beam data on the $Kr(^3P_2)$ state.
[b] Negligible charge transfer according to Ref. 37.

paths to the axial fluorines involve a 90° dihedral angle. This is in agreement with the general trends found for ^3J in other systems (see Chapter 4).

(6) There are large solvent and temperature effects on Xe spin–spin couplings. In $FXeOSO_2F$, for example, the values in HSO_3F solvent range from 6051 Hz at −100°C to 5975 Hz at −84°C, and in BrF_5 solvent, 5530 Hz at −77°C to 5848 Hz at −40°C.[22] The largest solvent dependence of Xe spin–spin couplings was observed in XeF^+, from 6703 Hz in HSO_3F at −70°C[11] to 7594 Hz in SbF_5 at 25°C.[22]

Table 4. Other Coupling Constants Involving ^{129}Xe, in Hz

	Hz	Molecule	Reference
$^1J(Xe{=}^{17}O)$	692	$O{=}XeF_4$	40
$^1J(Xe\,^{15}N)$	307.4	$FXeN(SO_2F)_2$	3
$^2J(Xe^{II}O\,^{125}Te)$	470	$Xe(OTeF_5)_2$	12
	540	$FXeOTeF_5$	12
	480	$Xe(OTeF_5)(OSeF_5)$	12
$^2J(Xe^{IV}O\,^{125}Te)$	988/1107	$Xe(OTeF_5)_4$	6, 7
	1032(E), 1292(A)	$FXe(OTeF_5)_3$	22
	1166	*trans* $F_2Xe(OTeF_5)_2$	22
	1059	*cis* $F_2Xe(OTeF_5)_2$	22
	1192	$F_2Xe(OTeF_5)$	22
$^2J(Xe^{VI}O\,^{125}Te)$	1281/1351	$O{=}Xe(OTeF_5)_4$	6, 7, 22
	1535	$OXeF_2(OTeF_5)_2$	22
	1364	$OXeF_3(OTeF_5)$	22
	1684	$O_2Xe(OTeF_5)_2$	22
	1856	$O_2XeF(OTeF_5)$	22
$^2J(Xe^{II}O\,^{77}Se)$	130	$Xe(OSeF_5)_2$	12
$^3J(Xe^{II}OTeF)$	18.5	$XeOTeF_5^+$	5
	30	$FXeOTeF_5$	12
	31	$Xe(OTeF_5)_2$	12
$^3J(Xe^{II}OSeF)$	37	$FXeOSeF_5$, $Xe(OSeF_5)_2$	12
$^3J(Xe^{II}NSF)$	18/18.7	$FXeN(SO_2F)_2$	3, 4
$^3J(Xe^{IV}OTeF)$	63/67.7	$Xe(OTeF_5)_4$	6, 7
$^3J(Xe^{VI}OTeF)$	54/55, 4(ax)	$O{=}Xe(OTeF_5)_4$	6, 7
	52	$O{=}XeF(OTeF_5)_3$	7
	53	$O{=}XeF_2(OTeF_5)_2$ *trans*	7
	51	$O{=}XeF_2(OTeF_5)_2$ *cis*	7
	50	$O{=}XeF_3(OTeF_5)$	7

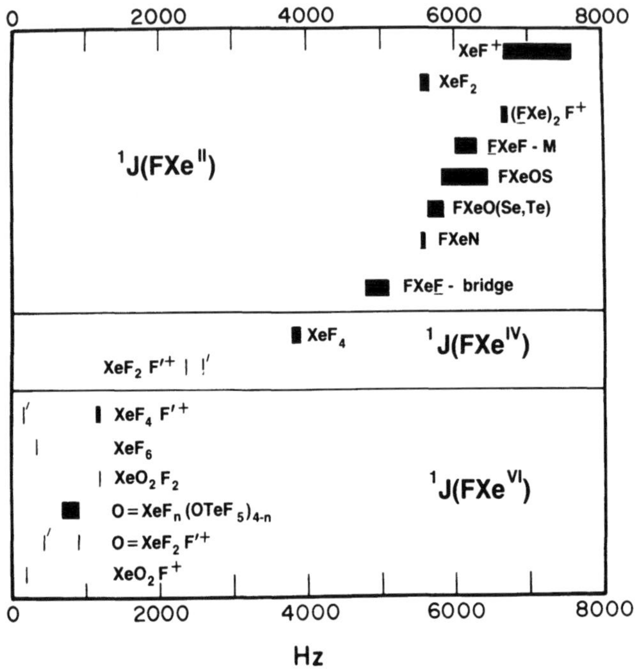

Figure 2. $^1J(^{129}XeF)$ values in various Xe environments. (See Table 1.)

There have been no determinations of absolute or relative signs of spin–spin couplings involving Xe.

The NMR evidence (^{129}Xe shifts, ^{19}F shifts and XeF coupling constants) and the related evidence of the electric field gradients provide an internally consistent picture of the electronic structure of the covalent xenon compounds: Schrobilgen shows that the empirical plot of Xe chemical shifts versus the F chemical shifts of the terminal fluorine in Xe(II) species containing F bridges is linear, as is the plot for species containing O bridges.(22) There is also a smooth curve correlating the ^1J(XeF) and F chemical shifts for all Xe oxidation numbers.(41) The covalency of the Xe–F bond is seen to increase with increasing Xe chemical shift, decreasing F chemical shift, and increasing XeF coupling constant. This is supported also by the decreasing bond length at least for the Xe(II) species.

The ionic character of the Xe–F bond has been used as a parameter in the early theoretical interpretation of the F and Xe chemical shifts as well as ^1J(XeF).(28) There have been no recent calculations of Xe nuclear shielding or spin–spin coupling constants. Owing to the high atomic number of Xe, there are significant relativistic corrections to these quantities. For example, the Xe diamagnetic shielding, which is the entire nuclear shielding in the free Xe atom, calculated with nonrelativistic wave functions by Malli and Froese,(42) is $\sigma(\text{Xe atom})_{\text{nonrel}} = 5642$ ppm. However, a relativistic calculation by Kolb and Johnson(25) gives $\sigma(\text{Xe atom})_{\text{rel}} = 7040$ ppm. The expectation value of r^{-3} for a xenon $5p$ electron which is important in the Xe nuclear shielding calculations in Xe compounds also should be evaluated relativistically, as should the electron spin density at the nucleus, which appears in the Fermi contact term of the spin–spin coupling. Pyykkö gives relativistic values for these quantities which are a factor of 1.4181 and 1.4242, respectively, as large as the nonrelativistic values.(43) Since relativistic corrections increase as the square of the atomic number, they are important for Xe, which has $Z=54$. The large number of electrons involved in the Xe species practically rules out any ab initio calculations of Xe nuclear shielding or J(XeF). There have been no theoretical attempts beyond the average energy approximation used in the early theory of Xe and F chemical shifts of XeF_2, XeF_4, XeF_6, and $OXeF_4$.[28]

3. PROBING NONSPECIFIC INTERMOLECULAR INTERACTIONS WITH NOBLE GAS NUCLEI

Relaxation times and medium shifts have been used to study intermolecular interactions between noble-gas atoms in the dilute gas all the way to the pure liquid and solid phases by ^{3}He, ^{21}Ne, ^{83}Kr, ^{129}Xe, and ^{131}Xe NMR studies. In the case of Xe there have been additional studies of intermolecular interactions with foreign gases,[44–48] organic solvents,[26,27] liquid crystal,[49] clathrate cages,[50,51] and zeolites.[51,52]

3.1. Medium Shifts

Even with 51 at. % ^{21}Ne (highly enriched compared to the natural 0.26 at. % abundance), attempts to determine shifts of the ^{21}Ne resonance in liquid and solid neon between 23 and 34 K were unsuccessful. The shifts were less than 1 ppm.[53] On the other hand, chemical shifts in the gas, liquid, and solid phases of krypton and xenon have been observed. The results from several studies are summarized in Ref. 54. The absolute shielding of Kr or Xe nuclei in the dilute gas can be written as a virial expansion in the powers of the density ρ, as shown in Chapter 3:

$$\sigma(T, \rho) = \sigma(\text{free atom}) + \sigma_1(T)\rho + \sigma_2(T)\rho^2 + \cdots$$

where $\sigma_1(T)$ is measured as $\lim_{\rho \to 0} -(1/v_0)(\partial v/\partial \rho)_T$. In condensed phases, it is no longer valid to use the virial expansion but it is still possible to measure the change $(\partial v/\partial \rho)$. For example, we can write

$$\sigma(T, \rho_{\text{liq}}) = \sigma(\text{free atom}) + \sigma_1^{\text{eff}} \rho_{\text{liq}}(T)$$

The same can be done for the solid phase. In Ref. 54 the results for the ^{83}Kr, ^{129}Xe, and ^{131}Xe chemical shifts in the pure gas, liquid, and solid have been summarized by listing the values of σ_1^{eff} for each phase. For solid krypton, the variation of the chemical shift with density is continuous upon melting, (see Figure 6 of Ref. 54) and the gas phase data pass over to the liquid phase data. The same is not true for xenon The xenon shifts are approximately 3.8 times as large as the krypton shifts. This is to be expected; the magnitudes of medium shifts and $\sigma_1(T)$ have been found to be related to the range of chemical shifts of the probe nucleus.

For an infinitely dilute solution of Xe in an organic solvent, $\sigma(T, C_A) = \sigma(\text{free Xe atom}) + \sigma_m(\text{Xe–A}) + \sigma_b(\text{A})$ where $\sigma_b(\text{A})$ is the bulk susceptibility contribution to Xe shielding. $\sigma_m(\text{Xe–A})$ depends on the solvent A, and its concentration C_A, which of course depends on the temperature, since a liquid expands with increasing temperature. The deshielding effect of the medium ($\sigma_m < 0$) is probably roughly additive in mixed solvents, although this experiment has not been carried out. σ_m has been found to vary from -148 (in CH_3OH) to -335 (in CH_2I_2) ppm, depending on the solvent, and is a roughly linear function of $[(n^2-1)/(2n^2+1)]^2$, where n is the refractive index of the pure liquid solvent.[26,27] Expressed in this way, σ_m(Xe–Xe) in pure liquid xenon is −165 ppm (at 244 K) to −230 ppm (at 161 K); σ_m(Xe–Xe) in pure solid xenon is −272 ppm (at 161 K) to −307 ppm (at ~0 K). Because the organic liquid solvent densities at room temperature are not directly comparable to each other, the medium effects expressed in this way are not reflective of the nature of the intermolecular interactions. It would make more sense if all the values were compared at the triple point or the critical point of the liquid solvents.

^{129}Xe NMR spectra have also been observed for xenon trapped in clathrates and in zeolites. In small and large clathrate deuteriohydrate cages, the ^{129}Xe resonances are distinct and are shifted +240 ppm and +150 ppm from the zero-pressure gas, that is, $\sigma_m = -240$ and −150 ppm, respectively.[50] By utilizing proton dipolar decoupling, ^{129}Xe–^{1}H cross-polarization, with and without magic angle spinning, Ripmeester has been able to observe σ_m for ^{129}Xe in these and other clathrates and in zeolites. He also measured the first ^{129}Xe shielding anisotropies imposed by the environment[51]: The cages are elongated along the symmetry axis, and the static ^{129}Xe line shape is characteristic of an axial shielding tensor with the component along the cage symmetry axis being greater than that perpendicular to the axis. With MAS the powder pattern collapses to a single line which gives σ_m. For Xe in β quinol clathrate, $\sigma_m = -222$ ppm and $\Delta\sigma = -160$ ppm. In β phenol clathrate there are one small cage and two large cage sites which are characterized by $\sigma_m = -229$, −250, −279 ppm, and shielding anisotropies $\Delta\sigma = -171$, −105, −53 ppm, respectively.[51] in a dehydrated Linde 13X zeolite sample at 160 K, separate signals are seen for solid and sorbed xenon. At 170 K (above the Xe melting point) the liquid and sorbed xenon lines are resolved. σ_m for the sorbed xenon was −250 ppm. From the integrated line intensities there are an average of 10–11 sorbed xenon atoms per zeolite supercage under the conditions of the experiment.[51] Experiments on other zeolites indicate 1–3 adsorbed xenon atoms per cage and shifts of +100 to +150 ppm from the dilute gas. ^{129}Xe is emerging as a very useful probe of metal-containing zeolites such as Pt- and Ni- NaY catalysts.[52]

3.2. Relaxation Times

Nuclear relaxation in the noble gases is different from polyatomic gases in that the relaxation mechanisms in the former have to be purely intermolecular rather than intramolecular. Thus, while T_1 for polyatomic gases increase with increasing gas density above the T_1 minimum ($T_1 = k\rho$), for the noble gases the opposite is true, i.e., $1/T_1 = k\rho$.

The nuclear spin-lattice relaxation times in the pure gas, liquid, and solid phases have been measured for ^{3}He,[55–57] ^{21}Ne (liquid and solid only[53]), ^{83}Kr,[54,58] ^{129}Xe,[59,60] and ^{131}Xe.[61] The spin–spin relaxation times (T_2) have also been measured for liquid and solid ^{129}Xe.[62] The relaxation of the quadrupolar nuclides are nearly entirely by means of the quadrupolar mechanism (although there is some dipolar contribution to ^{21}Ne near the melting point), the interaction of the nuclear quadrupole moment with the electric field gradients generated by deformations of the spherical atoms during collisions. The ^{129}Xe relaxation, on the other hand, is thought to be dominated by the spin-rotation mechanism in the transient diatomic molecule formed in collisions,[63,64] and ^{3}He relaxation is by dipolar interactions.[64]

The T_1 data for ^{21}Ne, ^{83}Kr, and ^{131}Xe have been reduced using the triple point data, that is, $T^* = T/T_{tr}$ and $T_1^* = T_1/T_1(T_{tr})$, and plotted in Figure 3. (The ^{21}Ne data

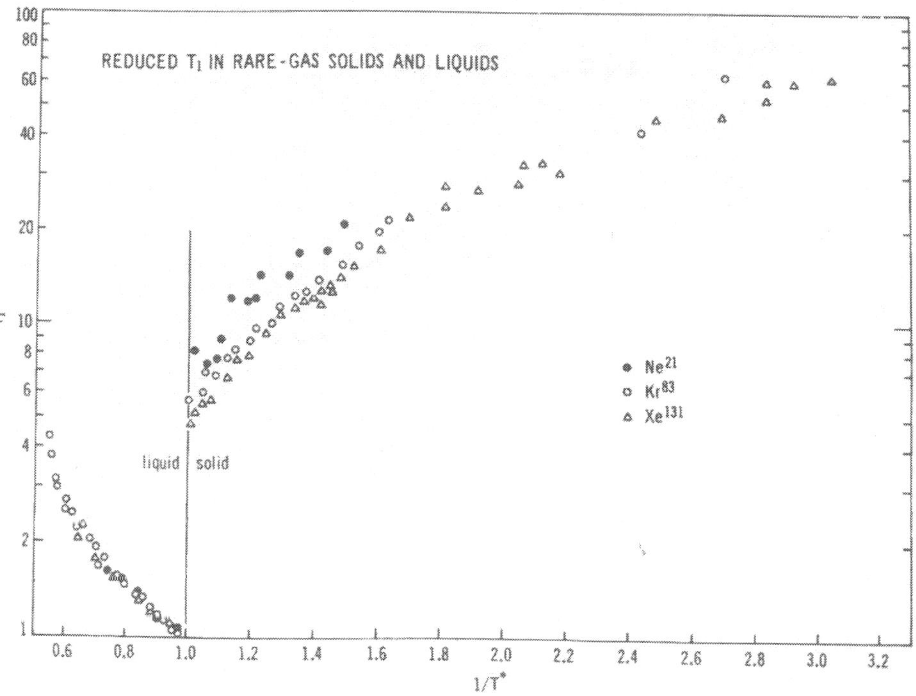

Figure 3. Relaxation times for ^{21}Ne, ^{83}Kr, ^{131}Xe liquid and solid phases, reduced to the triple point data. This figure reproduced from Ref. 54, with permission.

have had a dipolar contribution to T_1 removed from the solid near the melting point.) Although the observed relaxation times vary by nearly a factor of 10^3 among the three nuclides ($T_{1L} = 23.5$, 0.608, 0.041 for the liquid, $T_{1S} = 150$, 3.3, 0.190 seconds for the solid, for Ne, Kr and Xe, respectively), the ratio T_{1L}/T_{1S} is constant to ca. 30%. The density dependence of the liquid T_1 are also identical if expressed in reduced parameters: $T_1^* = T_1/T_1(T_{tr})$, $\rho^* = \rho/\rho(T_{tr})$. A linear equation $T_1^* = 6.2\text{–}5.2\rho^*$ is satisfied by all three nuclides. From the correspondence of the T_1 vs. $1/T$ curves and the T_1 vs. ρ curves, it may be concluded that the mechanism for nuclear quadrupole relaxation in the liquid and solid are the same for all three nuclides. In fact, the results can be intepreted by the same theory.[54]

REFERENCES

1. Gillespie, R. J.; Schrobilgen, G. J. *J. Chem. Soc., Chem. Commun.* 1974, 90–92.
2. Schrobilgen, G. J. In *NMR and the Periodic Table*; Harris, R. K., Mann, B. E., Ed.; Academic: London, 1978; pp. 439–454.
3. Sawyer, J. F.; Schrobilgen, G. J.; Sutherland, S. J. *J. Chem. Soc., Chem. Commun.* 1982, 210–211.
4. DesMarteau, D. D.; LeBlond, R. D.; Hossain, S. F.; Nothe, D. *J. Am. Chem. Soc.* 1981, **103**, 7734–7739.
5. Keller, N.; Schrobilgen, G. J. *Inorg. Chem.* 1981, **20**, 2118–2129.
6. Birchall, T.; Myers, R. D.; DeWaard, H.; Schrobilgen, G. J. *Inorg. Chem.* 1982, **21**, 1068–1073.
7. Jacob, E.; Lentz, D.; Seppelt, K.; Simon, A. *Z. Anorg. Allg. Chem.* 1981, **472**, 7–25.
8. Holloway, J. H.; Schrobilgen, G. J. *Inorg. Chem.* 1980, **19**, 2632–2640.
9. Seppelt, K. *Acc. Chem. Res.* 1979, **12**, 211–216.
10. Seppelt, K.; Lentz, D. *Prog. Inorg. Chem.* 1982, **29**, 167–202.

11. Seppelt, K.; Rupp, H. H. *Z. Anorg. Allg. Chem.* 1974, **409**, 331–337.
12. Seppelt, K.; Rupp, H. H. *Z. Anorg. Allg. Chem.* 1974, **409**, 338–342.
13. Willett, R. D.; Peterson, S. D.; Coyle, B. A. *J. Am. Chem. Soc.* 1977, **99**, 8202–8207.
14. Stein, L.; Norris, J. R.; Downs, A. J.; Minihan, A. R. *J. Chem. Soc., Chem. Commun.* 1978, 502–504.
15. Goetschel, C. T.; Loos, K. R. *J. Am. Chem. Soc.* 1972, **94**, 3018–3021.
16. Turbini, L. J.; Aikman, R. E.; Lagow, R. J. *J. Am. Chem. Soc.* 1979, **101**, 5833–5834.
17. *Noble-Gas Compounds*; Hyman, H. H., Ed.; University of Chicago Press: Chicago and London, 1963; 404 pp.
18. Claassen, H. H. *The Noble Gases*; D. C. Heath: Boston, 1966; 117 pp.
19. Holloway, J. H. *Noble-Gas Chemistry*; Methuen: London, 1968; 213 pp.
20. Horn, H. G. In *Gmelins Handbuch der anorganischen Chemie*, Erganzungswerk zur 8 Auflage; Verlag Chemie: Weinheim, 1970; Band 1, pp. 1–130.
21. Bartlett, N.; Sladky, F. O. In *Comprehensive Inorganic Chemistry*; Bailar, J. C., Ed.; Pergamon: New York, 1973; Vol. 1, pp. 213–330.
22. Schrobilgen, G. J.; Holloway, J. H.; Granger, P.; Brevard, C. *Inorg. Chem.* 1978, **17**, 980–987; Schumacher, G. A.; Schrobilgen, G. J. *Inorg. Chem.* 1984, **23**, 2923–2929.
23. Gillespie, R. J.; Netzer, A.; Schrobilgen, G. J. *Inorg. Chem.* 1974, **13**, 1455–1459.
24. Rupp, H. H.; Seppelt, K. *Angew. Chem. Int. Ed. Engl.* 1974, **13**, 612–613.
25. Kolb, D.; Johnson, W. R. *Phys. Rev. A* 1982, **26**, 19–31.
26. Stengle, T. R.; Reo, N. V.; Williamson, K. L. *J. Phys. Chem.* 1981, **85**, 3772–3775.
27. Miller, K. W.; Reo, N. V.; Uiterkamp, A. J. M. S.; Stengle, D. P.; Stengle, T. R.; Williamson, K. L. *Proc. Natl. Acad. Sci. USA* 1981, **78**, 4946–4949.
28. Jameson, C. J.; Gutowsky, H. S. *J. Chem. Phys.* 1964, **40**, 2285–2293.
29. Keenan, M. R.; Buxton, L. W.; Campbell, E. J.; Balle, T. J.; Flygare, W. H. *J. Chem. Phys.* 1980, **73**, 3523–3529.
30. Lucken, E. A. C. *Nuclear Quadropole Coupling Constants*; Academic: New York, 1968.
31. Faust, W. L.; McDermott, M. N. *Phys. Rev.* 1961, **123**, 198–204.
32. Perlow, G. J. *Phys. Rev. B* 1964, **135**, 1102–1105.
33. DeWaard, H.; Bukshpan, S.; Schrobilgen, G. J.; Holloway, J. H.; Martin, D. *J. Chem. Phys.* 1979, **70**, 3247–3253.
34. Perlow, G. J.; Perlow, M. R. *J. Chem. Phys.* 1968, **48**, 955–961.
35. Campbell, E. J.; Buxton, L. W.; Keenan, M. R.; Flygare, W. H. *Phys. Rev. A* 1981, **24**, 812–821.
36. Holloway, J. H.; Schrobilgen, G. J.; Bukshpan, S.; Hilbrants, W.; deWaard, H. *J. Chem. Phys.* 1977, **66**, 2627–2630.
37. Buxton, L. W.; Campbell, E. J.; Flygare, W. H. *Chem. Phys.* 1981, **59**, 55–59.
38. Campbell, E. J.; Keenan, M. R.; Buxton, L. W.; Balle, T. J.; Soper, P. D.; Legon, A. C.; Flygare, W. H. *Chem. Phys. Lett.* 1980, **70**, 420–424.
39. Buxton, L. W.; Campbell, E. J.; Keenan, M. R.; Balle, T. J.; Flygare, W. H. *Chem. Phys.* 1981, **54**, 173–181.
40. Shamir, J.; Selig, H.; Samuel, D.; Reuben, J. *J. Am. Chem. Soc.* 1965, **87**, 2359–2360.
41. Gillespie, R. J.; Schrobilgen, G. J. *Inorg. Chem.* 1974, **13**, 765–770.
42. Malli, G.; Froese, C. *Int. J. Quantum Chem., Suppl.*, 1967, **1**, 95–98.
43. Pyykkö, P.; Pajanne, E.; Inokuti, M. *Int. J. Quantum Chem.* 1973, **7**, 785–806.
44. Jameson, A. K.; Jameson, C. J.; Gutowsky, H. S. *J. Chem. Phys.* 1970, **53**, 2310–2321.
45. Jameson, C. J.; Jameson, A. K.; Cohen, S. M. *J. Chem. Phys.* 1975, **62**, 4224–4226.
46. Jameson, C. J.; Jameson, A. K.; Cohen, S. M. *J. Chem. Phys.* 1976, **65**, 3401–3406.
47. Jameson, C. J.; Jameson, A. K.; Cohen, S. M. *J. Chem. Phys.* 1977, **66**, 5226–5230.
48. Jameson, C. J.; Jameson, A. K.; Parker, H. *J. Chem. Phys.* 1978, **68**, 3943–3944.
49. Loewenstein, A.; Brenman, M. *Chem. Phys. Lett.* 1978, **58**, 435–436.
50. Ripmeester, J. A.; Davidson, D. W. *J. Mol. Struct.* 1981, **75**, 67–72.
51. Ripmeester, J. A. *J. Am. Chem. Soc.* 1982, **104**, 289–290.
52. Ito, T.; Fraissard, J. *J. Chem. Phys.* 1982, **76**, 5225–5229; DeMenorval, L.-C.; Fraissard, J. *J. Chem. Soc. Faraday Trans. 1* 1985, **81**, 2855–2865; Primet, M.; DeMenorval, L.-C.; Fraissard, J. *J. Chem. Soc. Faraday Trans. 1* 1985, **81**, 2867–2874; Scharpf, E. W.; Crecely, R. W.; Gates, B. C.; Dybowski, C. *J. Phys. Chem.* 1986, **90**, 9–11.
53. Henry, R. D.; Norberg, R. E. *Phys. Rev. B* 1972, **6**, 1645–1653.
54. Cowgill, D. F.; Norberg, R. E. *Phys. Rev. B* 1973, **8**, 4966–4974.
55. Volhardt, D.; Woelfle, P. *Physica B + C (Amsterdam)* 1981, **108**, 1055–1056.

56. Friedman, L. J.; Millet, P.; Richardson, R. C. *Physica B + C (Amsterdam)* 1981, **108**, 837–838.
57. Deville, G.; Bernier, M.; Delrieux, J. M. *Phys. Rev. B* 1979, **19**, 5666–5688.
58. Brinkmann, D.; Kuhn, D. *Phys. Rev. A* 1980, **21**, 163–167.
59. Streever, R. L.; Carr, H. Y. *Phys. Rev.* 1961, **121**, 20–25.
60. Hunt, E. R.; Carr, H. Y. *Phys. Rev.* 1963, **130**, 2302–2305.
61. Warren, W. W., Jr.; Norberg, R. E. *Phys. Rev.* 1966, **148**, 402–412.
62. Yen, W. M.; Norberg, R. E. *Phys. Rev.* 1963, **131**, 269–275.
63. Torrey, H. C. *Phys. Rev.* 1963, **130**, 2306–2312.
64. Shizgal, B. *Can. J. Phys.* 1976, **54**, 164–175.

Chapter 19

EARLY TRANSITION METALS, LANTHANIDES AND ACTINIDES

DIETER REHDER

1. INTRODUCTION

Among the early transition metals of the B groups III–VII, there are four spin-1/2 nuclei (^{89}Y, ^{169}Tm, ^{171}Yb, ^{183}W), but ^{89}Y and ^{183}W are the only ones for which a substantial amount of high-resolution NMR data has been reported. The other early transition metal nuclei have nuclear spins $>1/2$ (and, with the exception of Ce, there is at least one magnetic isotope for each metal). Relaxation for these nuclei is governed by the quadrupolar mechanism, causing comparatively broad resonance signals unless the nuclear quadrupole moment is small (^{51}V) and/or the local symmetry is such that the field gradient is very small (O_h, T_d, some cases of C_{3v}). A nuclear quadrupole has often been considered to seriously hamper NMR detection. In fact, there are a few nuclei with nuclear quadrupole moments so large as to cause severe problems (^{177}Hf, ^{179}Hf, ^{181}Ta, ^{185}Re, ^{187}Re, ^{235}U), but in general, the high relative receptivity for most of the nuclei, the large shielding sensitivity, and the fact that information on the electronic surroundings of the metal can be obtained by direct probing of the coordination center readily overcome disadvantages encountered with line broadening. Furthermore, the linewidth provides an additional parameter for the investigation of electronic interactions.

Shielding variations for group IIIb elements are limited by the tendency of the M^{3+} cations (which are comparable to the group IIIa metals in many aspects of their solution chemistry) to form rather stable and well-defined first coordination spheres occupied by solvent (H_2O) molecules. In contrast, the shielding ranges in groups Vb–VIIb encompass several thousand ppm. Since the shielding range for a specific nucleus more or less depends upon which systems have been studied, it is more convenient to compare relative intrinsic shielding sensitivities, which can be defined, e.g., as the shift differences $\Delta\delta(MO_4^q - M(CO)_6^q)$ or $\Delta\delta(MCl_6^q - MF_4^q)$, where q is a charge. These are displayed in Figure 1, which gives the total shielding range, and within this the range from MCl_6^q to MF_6^q and/or MO_4^q to $M(CO)_6^q$. As is shown for the ^{183}W shielding, the two sensitivity scales are about equivalent.

In the following treatment, emphasis will be laid upon solution NMR, i.e., measurements in isotropic media, in which the field gradient is modulated by the

---

DIETER REHDER • Institute of Inorganic and Applied Chemistry, University of Hamburg, D-2000 Hamburg 13, Federal Republic of Germany.

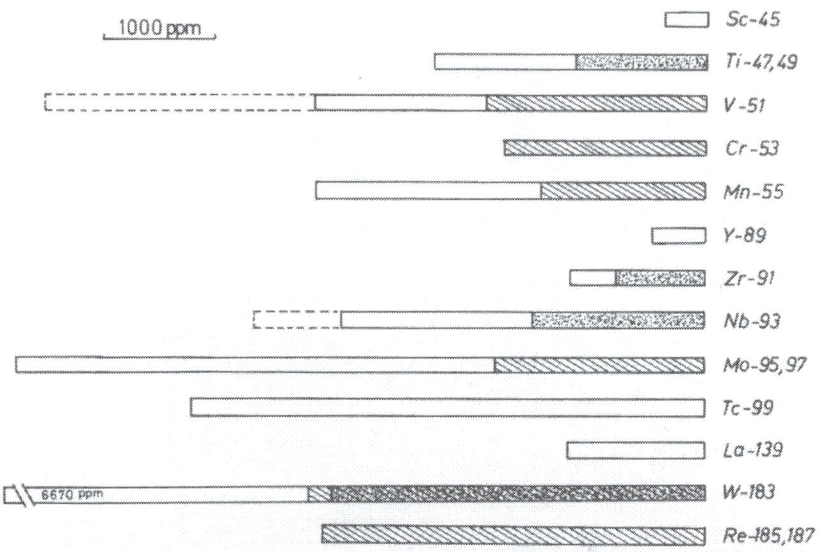

Figure 1. Shielding ranges and shielding sensitivities for the early transition metals. Shielding increases from left to right. Shaded areas: $[MO_4]^q/[M(CO)_6]^q$, punctuated areas: $[MCl_6]^q/[MF_6]^q$. Broken lines encompass measurements on solids (^{51}V, ^{93}Nb) or extrapolated data (^{93}Nb in $[NbTeBr_4]^-$).

Brownian motion and hence no quadrupole splitting is observed. A few key references will also be given of selected solid state studies (excluding intermetallics), where there is only a restricted number of other data, or where NMR parameters obtained under anisotropic conditions have been correlated with solution data (^{51}V, ^{55}Mn).

Transition metal NMR has been surveyed by Kidd and Goodfellow,[1] Kidd (nuclear shielding),[2] Rinaldi, Levy, and Choppin (selected topics only),[3] and Wehrli (quadrupolar nuclei).[4] Monographs have appeared on ^{51}V NMR primarily concerned with solution[5] and solid state measurements,[5a] the chromium triad,[213] and on the first transition series nuclei (Sc to Zn).[6] Some generalizations on nuclear shielding in carbonyl complexes of ^{51}V, ^{55}Mn, and ^{93}Nb have been reported.[7] A comprehensive account on the complete block of the *d*-metals has recently appeared.[214]

2. GROUP IIIb

2.1. Scandium

Although ^{45}Sc is one of the most sensitive nuclei (0.3 relative to ^1H) and has a sufficiently favorable quadrupole moment (-0.22×10^{-28} m^2), ^{45}Sc NMR studies are scarce. One of the reasons may be that well-defined Sc^{3+} complexes with covalent scandium-ligand bonds are less common; in aqueous solution, the coordination sphere is subject to rapid exchange processes, and the resonance observed usually corresponds to an average for all species present. On the other hand, since Sc^{3+} is a d^0 system and electronic absorption spectra are insufficiently helpful, NMR studies may well be considered an alternative source for information on coordination compounds of Sc. Selected shielding data are summarized in Table 1, referenced to dilute aqueous $Sc(ClO_4)_3$ or $ScCl_3$, which attain the same limiting ^{45}Sc shielding at infinite dilution.[8]

Table 1. ^{45}Sc Chemical Shifts Relative to Dilute Aqueous ScX_3 (X = Cl, ClO_4), i.e., $[Sc(OH)(H_2O)_5]^{2+}$

System	(Tentative) assignment	δ (ppm)	Reference
$ScCl_3$, MeCN	$ScCl_3 \cdot x$MeCN or $[\{(MeCN)_2ScCl_2\}_2(\mu\text{-}Cl)_2]$	+204	11
$ScCl_3$, MeCN, NEt_3	$ScCl_3 \cdot NEt_3$	+225	11
	$[ScCl_{3-n}(NEt_3)]\,Cl_n$	+158	11
$ScCl_3$, MeCN, NEt_3, HCl	$[ScCl_6]^{3-}$	+249	11
$ScCl_3$, MeCN, KSCN, or Me_4NSCN	$[ScCl_3(NCS)(MeCN)_2]^-$	+178	11
	$[ScCl_2(NCS)(MeCN)_3]$	+149	11
	$[ScCl_2(NCS)_2(MeCN)_2]^-$	+138	11
	$[ScCl_2(NCS)_3(MeCN)]^{2-}$	+121	11
	$[ScCl(NCS)_3(MeCN)_2]^-$	+92	11
	$[ScCl(NCS)_4(MeCN)]^{2-}$	+83	11
	$[Sc(NCS)_5(MeCN)]^{2-}$	+40	11
	$[Sc(NCS)_6]^{3-}$	+27	11
$[Sc(NCS)_6]^{3-}$, MeCN, Me_4NCl	$[ScCl_2(NCS)_4]^{3-}$	+102	11
	$[ScCl(NCS)_5]^{3-}$	+66	11
$ScCl_3$, $4H_2O$, THF		+193	8
ScX_3 (X = Cl, Br, NO_3), THF	Contact ion pairs	+202/273/10	8
$ScCl_3$, H_2O (1 *M*), HCl 12–10 *M*	$[ScCl_4(H_2O)_2]^{-}$ [a]	+144 to +124	8, 10
$ScCl_3$, H_2O (1 *M*), HCl 8–6 *M*	$[ScCl_2(H_2O)_4]^{+}$ [a]	+88 to +73	8, 10
$ScCl_3$, H_2O (1 *M*), HCl 3–1 *M*	$[ScCl(H_2O)_5]^{2+}$ [a]	+19 to +0.5	8, 10
$ScCl_3$ (0.005 molal), NaOH	$[Sc(OH)_4]^-$	−115.7(4)	9a
$Sc(ClO_4)_3$, H_2O, cryptand	Weak Sc-cryptate	+22	8
$Sc_2(SO_4)_3$, H_2O (0.02–2 *M*)	One to two anions in the first coordination sphere	+9.4 to +6.7	8
$Sc(NO_3)_3$, H_2O (0.03–2 *M*)		+5.0 to −0.1	8
ScI_3, H_2O (0.1–3 *M*)		+2.8 to −5.7	8
$ScBr_3$, H_2O (3 *M*)		−6.5	8
$Sc(ClO_4)_3$, H_2O (3 *M*)		−14.8	8
$Sc(acac)_3$, benzene		+85	16
$ScCl_3$, $OP(OEt)_3$, MeCN	$[ScCl_4\{OP(OEt)_3\}_2]^-$	+166	11a
	m/f-$[ScCl_3\{OP(OEt)_3\}_3]$	+134/+119	11a
	c/t-$[ScCl_2\{OP(OEt)_3\}_4]^+$	+82/+67	11a
	$[ScCl\{OP(OEt)_3\}_5]^{2+}$	+27	11a
	$[Sc\{OP(OEt)_3\}_6]^{3+}$	−26 5	11a
$ScCl_3$, DMSO	$[Sc(DMSO)_6]^{3+}$	+115.5	11b
$ScCl_3$, MeCN	$[Sc(NCMe)_6]^{3+}$	+200	11b
$ScCl_3$, THF	$[Sc(THF)_6]^{3+}$	+203	11b
$Sc(BH_4)_3$, benzene		+113.3	11c
$Cp_2Sc(BH_4)$, benzene		+67.5	11b
Cp_3Sc,[b] benzene		+13.5	11b

[a] According to a recent study by Lutz *et al*,[9a] rapid exchange equilibria between complexes of varying composition rather than defined species are present

[b] At 213 K and in toluene-d_8, a second signal (+53 7 ppm) arises which is due to the formation of Cp-bridged dimers with two inequivalent Sc centers

A solution of $ScCl_3$ (0.11 molal) in 0.05 molal HCl has $T_1 = 7.2(3)$ ms and $W_{1/2} =$ 29 Hz.[9a]

Lutz has reported the isotope shift of $ScCl_3$ in D_2O (−6.2[9] and −5.9[9a] ppm relative to $ScCl_3$ in H_2O). This large shift is indicative of solvent molecules in direct contact with Sc^{3+}. A small inverse, i.e. positive isotope shift of +1.0(4) ppm is observed for $ScCl_3$ (0.005 molal) in >10 molal NaOH; the species present presumably is $[Sc(OH)_4]^-$.[9a] Linewidths (and chemical shifts) reported for ScX_3 (X = Cl^-, Br^-,

I^-, NO_3^-, ClO_4^-, SO_4^{2-})[8,9a,9b] vary with the concentration and anion. In general, shielding increases with increasing concentration, and the linewidths are significantly greater in more concentrated solutions, which is attributed to complex or contact ion-pair formation. In contrast, $ScCl_3$ exhibits a deshielding as the concentration increases, more pronounced on addition of HCl, which may be due to increasing introduction of Cl^- into the first coordination sphere. A plot of $\delta(^{45}Sc)$ vs. molarity of HCl displays three concentration ranges for which the chemical shift remains fairly constant, and these might correspond with $[Sc(OH)(H_2O)_5]^{2+}/[ScCl(H_2O)_5]^{2+}$ (0–4 M, $\delta = 0.3$–14.3), $[ScCl_2(H_2O)_4]^+$ (6–8 M, $\delta = 72.8$–87.8) and $[ScCl_4(H_2O)_2]^-$ (10–12 M, $\delta = 124$–144 ppm).[8] Tarasov[10] assigned signals observed in the system $ScCl_3/H_2O/HCl$ to appropriate species, but an independent study by Lutz and co-workers did not show any steps in the δ/concentration curve attributable to defined specimens. Rather, a smooth increase of δ (and T_1 and $W_{1/2}$) was observed at $c(HCl) \geqslant 4$ molal, while there is a slight decrease in the concentration range 0–0.1.[9a]

Buslaev, Kirakosjan, and Tarasov[11] studied water-free systems ($ScCl_3/SCN^-$ and $[Sc(SCN)_6]^{3-}/[Me_4N]Cl/MeCN$,[11] and $ScCl_3/OP(OR)_3$ (R = Et, Bu) in MeCN[11a]), where there might be a reliable chance for the formation of Sc complexes. ^{45}Sc resonances have been assigned by these authors to a considerable number of distinct species of the general composition $[ScCl_{6-x-y}(SCN)_x(MeCN)_y]^{(3-y)-}$ and $[ScCl_{6-n}\{OP(OR)_3\}_n]^{(3-n)-}$ (Table 1).

Shielding increases in the order Cl (< MeCN) < NCS, paralleling observations for Nb(+V) complexes. Rather low shielding is also obtained in the system $ScCl_3/MeCN/NEt_3$ (+225 ppm); this resonance is shifted to +249 ppm on addition of HCl ($[ScCl_6]^{3-}$?). Further studies will have to elucidate the question of whether the deshielding of the ^{45}Sc nucleus for $ScCl_3$ in nonaqueous solvents (200–300 ppm) has to be accounted for in terms of contact ion-pair formation as suggested for $ScCl_3$/THF,[8] or complex formation as proposed for $ScCl_3$/MeCN[116] and $ScCl_3/NEt_3$.[11]

The following coupling data are available: $^1J(^{45}Sc–^{19}F)$ in $[ScF_6]^{3-}$, measured in ^{19}F (172 Hz[12]) and ^{45}Sc resonance (180(10) Hz[13,60a]), $^2J(^{45}Sc–^{31}P)$ in $[Sc\{OP(OMe)_3\}_6]^{3+}$ (35 Hz, from ^{31}P NMR[14]) and several $[ScCl_{6-n}\{OP(OR)_3\}_n]^{(3-n)-}$ (30–40 Hz, from ^{45}Sc NMR[11a]), $J(^{45}Sc–^{11}B)$ (15.5) and $J(^{45}Sc–^1H)$ (30 Hz) in $Sc(BH_4)_3$.[11c]

2.2. Yttrium

^{89}Y ($N = 100\%$) has spin 1/2 and should be a favorable nucleus. Its sensitivity is close to that of ^{13}C (0.668 relative to ^{13}C), but the small (and negative) magnetic moment causes problems as to its observability with common NMR equipment. Further, relaxation times T_1 are rather long (vide infra). On the other hand, dipolar relaxation plays an important role in the overall relaxation of ^{89}Y, and the potentially large NOEF values (the theoretical maximum for dipolar relaxation by ^1H is −10.2[17]) should make this nucleus an important probe for the study of metal interaction with large organic molecules or matrices.

The first ^{89}Y resonance [in $Y(NO_3)_3$] was detected in 1949.[18] With improved measuring techniques,[19] several ^{89}Y NMR studies have been carried out in the past few years. T_1 values for aqueous $Y(NO_3)_3$ are 60 to 85 (±15%) s in 3 M solutions,[17,20] and larger in 1 M solutions, ranging from 130 (338 K) to 270 (278 K) s; NOEF values are −0.7 and −10, respectively, with an absolute error of

Table 2. ^{89}Y Chemical Shifts[a]

Compound	Conc. (mole/liter)	δ (ppm)
K[Y(edta)]	1.0	+125.9
$K_3[Y(iminodiacetate)_3]$	1.0	+98.9
$K_3[Y(nitrilotriacetate)_2]$	1.0	+85.7
$Y(CH_3CO_2)_3$	1.0	+37.2
YCl_3	2.8	+8.9
YBr_3	2.0	+6 8
$Y(NO_3)_3$	0.75	−1 9
$Y(NO_3)_3$	3.0	−18.0
$(MeC_5H_4)_3Y(THF)$	0.90	−371
$(MeC_5H_4)_2YCl(THF)$	1.5	−103
$[(MeC_5H_4)_2YCl]_2$ [b]	1.5	−97
$[(MeC_5H_4)_2Y(\mu\text{-H})(THF)]_2$ [c]	0.28	−92
$\{[(C_5H_5)_2Y(\mu\text{-H})]_3(\mu_3\text{-H})\}^-$ [d]	0.41	−67
$[(MeC_5H_4)_2Y(\mu\text{-Me})]_2$ [b]	0.26	−15
$(MeC_5H_4)_2YMe(THF)$	0.72	+40

[a] Chemical shifts are taken from Ref 20 (80% H_2O/20% D_2O, pH 5 5, relative to 1 5 *M* aqueous $Y(ClO_4)_3$, signs adjusted to the discussion in Ref 20, and Ref 21a (organoyttrium compounds, THF solution, relative to YCl_3 in H_2O)
[b] In toluene
[c] Triplet $J(Y-(\mu\text{-H}))=27$ Hz
[d] Multiplet

less than 15%.[17] At 308 K, 25% of the relaxation is accounted for by dipolar interaction with ^1H of H_2O in the solvation sphere of Y^{3+}. Relaxation studies for mixtures of $Y(NO_3)_3$ and crown-ethers show a high NOEF (around −10) in the case of 12-crown-4. A slight high-frequency shift [+1.5 ppm relative to $Y(NO_3)_3$] is thought to support the formation of an yttrium crown-ether complex.[17]

The solvent isotope effect for Y^{3+} in D_2O is −4.3(5) ppm.[21] The concentration dependence of δ (maximum effects in parentheses) for aqueous solutions of YCl_3 (+2 ppm), $Y(NO_3)_3$ (−18 ppm), and $Y(ClO_4)_3$ (−5 ppm) at pH = 0 has been determined for concentrations up to 3 mole/liter; $\delta(^{89}Y)=0$ is the shift for infinite dilution.[21] Very few shielding data have been reported (Table 2). The shift scale is flanked by $[Y(edta)]^-$ (+126)[20] and organo-yttrium compounds such as $[(MeC_5H_4)_3Y(THF)]$ (−371 ppm).[21a]

$^2J(^{89}Y-^{31}P)=5.1$ Hz has been measured in the ^{31}P NMR of $[Y\{S_2P(OEt)_2\}_4]^-$ at 223 K, indicating that there is considerable covalency in the yttrium-ligand bond.[22] At room temperature, a broad signal due to rapid exchange between coordinated and noncoordinated dithiophosphinate is observed. In $\{[(C_5H_5)_2Y(\mu-H)]_3(\mu_3-H)\}^-$, the coupling constants are $^1J(Y-(\mu_2-H))=29.7$, $^1J(Y-(\mu_3-H))=17$, $J(Y-Y)=4$ Hz.[22a]

2.3. Lanthanum

There are two magnetically active nuclei: ^{139}La (N = 99.91%, $Q=0.21\times10^{-28}$ m^2, $I=7/2$, $s=0.059$) and ^{138}La (0.09%, $Q=0.45$,[23] $I=5$). The ^{138}La resonance has been observed by Lutz and Oehler[23,24] in a 0.6 *M* aqueous (D_2O) $LaCl_3$ solution. NMR studies on ^{139}La have recently been applied to bioinorganic problems: $1/T_1$ values were shown to increase linearly with the concentration of protein (bovine serum albumine,[25] prothrombine[26]) added to a solution of $LaCl_3$ in water. The molecular correlation time τ_c for large molecules of this type is around 10^{-8} s; i.e., T_1 becomes frequency-dependent. A general treatment is given by Reuben and Luz.[27]

Some earlier (and partially contradicting) studies dealing with the concentration and counterion dependence of δ and $W_{1/2}$ of lanthanum salts[24,28–30] have been supplemented by more recent reports (some still of preliminary character) on various La^{3+} species, including a number of lanthanum coordination compounds.[9b,31–34] Although several points still remain dubious, these investigations have led to a first set of shift data (Table 3).

Linewidths are broad except where presumably octahedral complexes are formed; the relatively narrow lines ($W_{1/2} = 200–600$ Hz) observed for $[LaX_6]^{3-}$ (X = Cl, Br, NO_2, NO_3) and also for $[La\{P(O)(NMe_2)_3\}_6]^{3+}$ are thought to be indicative of local O_h symmetry. Support for this assumption comes from the considerable broadening of the resonance signals in the systems $La(ClO_4)_3-CH_3CN-L$ (L = P(O)$(NMe_2)_3$, H_2O, NO_3^-, H_2CNMe_2), if the amount of L is kept beyond the level necessary for attaining the maximum coordination number.

The concentration behavior of $\delta(^{139}La)$ parallels that of $\delta(^{45}Sc)$ and $\delta(^{89}Y)$, i.e., in aqueous (H_2O, D_2O) solutions of $La(NO_3)_3$, shielding increases with increasing concentration, while there is a slight decrease for $LaCl_3$. Linewidths follow the same pattern and are 117 Hz in H_2O and 143 Hz in D_2O (infinite dilution). The solvent isotope effect (D_2O) amounts to -1.6 ppm and is the smallest observed in the scandium triad.[24]

In aqueous methanol with up to 66% v/v MeOH, there is no indication of chloride ions being in direct contact with La^{3+}, but the rapid decrease in shielding as the percentage of methanol is increased clearly shows that Cl^- moves from the outer to the inner sphere, until a limiting value of $+293$ ppm is reached in 100% MeOH. The rather large linewidths (1100 to 4600 Hz) indicate the formation of lower symmetry complexes due to H_2O/Cl^- exchange.[33]

Studies on ^{139}La have focused quite generally on the problem of outer- and inner-sphere complexes formed by various (potential) ligands, i.e., on equilibria of the kind

$$[La(H_2O)_n]^{3+} + xL^- \rightleftharpoons [La(H_2O)_{n-x}L_x]^{(3-x)+} + xH_2O$$

where $n = 8$ in concentrated and $\geqslant 9$ in dilute solutions. La^{3+} in 10 *M* HCl ($\delta = 60–100$[9b,34] and in 5 *M* NH_4SCN ($\delta = +36$ ppm)[34] also shows partial inner-sphere coordination of Cl^- and NCS^-. No shifts relative to La^{3+}(aq) were observed in the presence of trichloroacetate,[34] whereas δ values of $+60$[31,32] to $+100$ ppm[34] have been reported to be caused by the acetate ligand. Intermediate shifts are produced by mono- and dichloroacetate,[34] oxidodiacetate and glycolate[31,32] with only one carboxylato ligand in the first coordination sphere. As all nine water molecules are substituted for three oxidodiacetates, a shift of $+160$ ppm is found.[32] Only the carboxylato group is thought to induce variations in $\delta(^{139}La)$; the contributions of hydroxo-, ether, and nitrogen functions are assumed to be zero.

On the other hand, Evans and Missen[32] showed that, in $[La(en)_n]^{3+}$ (en = ethylenediamine or diethylenetriamine), there is a monotonic decrease of the ^{139}La shielding as the number of nitrogens coordinated to La^{3+} increases (ca. 50 ± 10 ppm per N). This shielding decrease is in accord with increased metal-ligand covalency on replacement of oxygen ligands (H_2O) by the less electronegative nitrogen.

Table 3. ^{139}La Chemical Shifts (Relative to $LaCl_3/H_2O$ at Infinite Dilution, i.e., $[La(H_2O)_n]^{3+}$, $n = 9$) and Linewidths

System	δ (ppm)	$W_{1/2}$ (Hz)	Reference
$[LaBr_6]^{3+}$	+1090	500	32
$[LaCl_6]^{3+}$	+851	200	32
$[La(S_2CNEt_2)_4]^-$	+710	2100	32
La^{3+}, edta	+570, 0[a]		34
$[La\{HN(CH_2CH_2NH_2)_2\}_3]^{3+}$	+450	2400	32
$\{LaHN(CH_2CH_2NH_2)_2\}^{3+}$	+122	7700	32
$[La(H_2NCH_2CH_2NH_2)_4]^{3+}$	+400	3800	32
$\{La(H_2NCH_2CH_2NH_2)\}^{3+}$	+40	3800	32
$[La(NO_2)_6]^{3-}$	+342	600	32
$LaCl_3$, MeOH	+293	4640	33
$LaCl_3$, $MeOH/H_2O$ 2/1 [b]	+16	1100	33
$[La\{O(CH_2CO_2)_2\}_3]^{3-}$	+160		31
$[La\{N(CH_2CO_2)_3\}_2]^{3-}$	+202	7800	32
$\{LaN(CH_2CO_2)_3\}$	+106	500	31
La^{3+}, $CH_3CO_2^-$	+100[c]		34
$\{La(O_2CCH_3)\}^{2+}$	+60[c]	350	31
La^{3+}, $CH_2ClCO_2^-$	+50[c]	ca. 1300	34
La^{3+}, $CHCl_2CO_2^-$	+22[c]		34
La^{3+}, $CCl_3CO_2^-$	0		34
$\{LaO(CH_2CO_2)_2\}^+$	+55[c]	2200	31
$\{LaO_2CCH_2OH\}^{2+}$	+33		31
$[La\{OS(CH_2)_3CH_2\}_8]^{3+}$	+93	2600	32
$[La(OSMe_2)_8]^{3+}$	+82	1800	32
$[La\{P(O)(NMe_2)_3\}_6]^{3+}$ [b]	+66	930	32
$[La(NCS)_7]^{4-}$ [b]	+196	4400	32
La^{3+}, NH_4SCN 5 *M*	+36	1000	34
$[LaCH(COCF_3)_{24}]^-$	+33	7000	32
La^{3+}, HCl 10 *M*	+100	1300	34
$LaCl_3$, H_2O (3 *M*)	+7		34
$La(ClO_4)_3$, H_2O (0.5 *M*)	0		34
$La(NO_3)_3$, H_2O (2.7 *M*)	−8.5		24
$La(NO_3)_3$, D_2O (2.7 *M*)	−13		24
$[La(NO_3)_6]^{3-}$ [b]	−60	260	32
$La(NO_3)_3$ (0.29 *M*)	0[d]	700[d]	9b[d]
$La(NO_3)_3$ (3 *M*)	−75		9b
$La(NO_3)_3$ (0.29 *M*), HNO_3 (10.5 *M*)	−75	1700	9b
$LaCl_3$ (0.4 *M*)	+3	250	9b
$LaCl_3$ (3 *M*)	+10		9b
$LaCl_3$ (0.4 *M*), HCl (10 *M*)	+60	1500	9b
$La(ClO_4)_3$ (0.36 *M*)	0	250	9b
$La(ClO_4)_3$ (3 *M*)	−25		9b
$La(ClO_4)_3$ (0.36 *M*), $HClO_4$ (7.5 *M*)	−25	2000	9b
$[La(ONC_5H_5)_8]^{3+}$	+4	2800	32
$[La(Me_2NCH_2)_8]^{3+}$ [b]	−39	1700	32
$[La(ONC_5H_3Me_2)_8]^{3+}$	−122	2700	32
$[La(NCMe)_n]^{3+}$	−129	380	32
LaB_6 solid	−128		24

[a] The signal at $\delta = 0$ refers to La^{3+}(aq)
[b] Data for complexes containing less than the indicated (maximum) number of ligands are also reported
[c] Limiting shifts for LaX^{2+}(aq)
[d] All data from Ref 9b are estimates from a graphical presentation

Corresponding trends are observed for S relative to O coordination ($[La(S_2CNEt_2)_4]^-$, $[La\{CH(COCF_3)_2\}_4]^-$), and for the pair $[LaBr_6]^{3-}$/$[LaCl_6]^{3-}$. In all these cases, $\delta(^{139}La)$ follows an "inverse order" typical of d^0 metals.

2.4. The Lanthanides

The nuclear properties are unfavorable for most of the lanthanides, with low transition frequencies and receptivities, or large quadrupole moments. Exceptions are ^{141}Pr ($s=0.258$, $I=5/2$), ^{151}Eu (0.175, 5/2), ^{159}Tb (0.03, 3/2) and ^{165}Ho (0.102, 7/2). The detectability of ^{151}Eu and ^{165}Ho is again hampered by Q values between 1 and $3\times10^{-28}\,m^2$, and that of ^{159}Tb by the low receptivity. There are two spin-1/2 nuclei: ^{169}Tm ($N=100\%$, $s=5.5\times10^{-4}$) and ^{171}Yb (14.31, 5.5×10^{-3}). Cerium has no stable magnetic nucleus. Yb^{2+} and Lu^{3+} are diamagnetic and these ions at least should have been probed by direct NMR, but no solution NMR on lanthanides has been reported. In contrast, solid state investigations of paramagnetic samples are quite common and a list of selected recent studies is given below:

^{141}Pr: $PrVO_4$;[35] PrN, PrP, PrAs, PrSb, PrS, PrSe;[36] $CsNaPrCl_6$;[37] $Pr(NO_3)_3\cdot6H_2O$, $Pr(ReO_4)_3\cdot4H_2O$, $Pr_2(SO_4)_3\cdot8H_2O$, $Pr_2(SeO_4)_3\cdot5H_2O$, PrF_3, $Pr(O_2CMe)_3\cdot2H_2O$;[38,39]
^{153}Eu: in $EuSe_{1-x}Te_x$ ($x=0.3$);[40,41]
^{159}Tb: Rb_2NaTbF_6;[42] $Cs_2NaTbCl_6$;[37]
^{165}Ho: Rb_2NaHoF_6;[42,43] $Cs_2NaHoCl_6$;[44] ^{165}Ho and ^{51}V in $HoVO_4$;[45,46]
^{169}Tm: $Tm(SO_4Et)_3$;[47] $Cs_2NaTmCl_6$;[37] $LiTmF_4$;[38,48,49] ^{169}Tm and ^{51}V: in $TmVO_4$;[50] ^{169}Tm and ^{31}P: in $TmPO_4$.[50a]

The ^{141}Pr resonance in $PrVO_4$ at 4.2 K, e.g., shows a nice five-line pattern due to first-order quadrupole splitting by the $I=5/2$ ^{141}Pr nucleus, if the magnetic field is applied parallel to the [110] plane of the crystal.[35]

The ^{19}F NMR of $(C_6F_5)_2Yb(THF)_4$ exhibits a satellite system due to ^{171}Yb–^{19}F coupling of 120 Hz.[51] The coupling constant is tentatively assigned to the fluorine atoms in ortho position. The more common use of lanthanide ions as shift reagents has recently been reviewed by Williams.[52]

2.5. Actinium and the Actinides

There is only one single direct NMR observation of an actinide nucleus, viz., ^{235}U in neat UF_6 at 380 K and 4 bar. The measured resonance frequency is 9.2098(40) MHz, the magnetogyric ratio for this spin-7/2 nucleus $|\gamma|=492.6(2)\,rad\,s^{-1}\,G^{-1}$, the linewidth $W_{1/2}$ 20 kHz.[52a] ^{235}U has a large Q of $4.1\times10^{-28}\,m^2$ which renders its direct observation extremely difficult. Indirect observation on ^{235}U has been extracted from ^1H, ^{11}B, ^{13}C, ^{14}N, ^{17}O, ^{19}F, ^{27}Al, ^{31}P, ^{63}Cu, ^{103}Rh, ^{119}Sn, and ^{127}I NMR and has recently been surveyed by Kannellakopulos.[53] Glebov[54] has analyzed ^1H and ^{17}O relaxation by NpO_2^+ and PuO_2^{2+} with respect to the electronic structure of these ions. Seppelt and Bartlett[55] observed an unstructured, broad ($W_{1/2}=550$ Hz) resonance for ^{19}F in UF_6 (96% ^{235}U enriched, $\nu_0=56.4$ MHz) between 273 and 323 K, while Ursu *et al.*[56,57] estimated the $^1J(^{235}U–^{19}F)$ from measurements of the longitudinal and transversal relaxation times in the ^{19}F NMR spectra of UF_6 at $\nu_0 = 60$ MHz. The data reported are $T_1(^{235}U) = 21.3 \times 10^{-9}$ s and $^1J_{UF} = 4240$ Hz at 338 K, $T_1 = 23.7 \times 10^{-9}$ s and $^1J_{UF} = 4566$ Hz at 356 K.

3. GROUP IVb: TITANIUM, ZIRCONIUM, AND HAFNIUM

Despite the importance of Ti and Zr in catalysis, there are only a few isolated reports on $^{47,49}Ti$ and ^{91}Zr NMR. The receptivities are similar to that of ^{13}C. ^{91}Zr has a medium Q $(-0.21 \times 10^{-28}\ m^2)$; the quadrupole moments for the two Ti isotopes are of the same order of magnitude [$Q(^{47}Ti) = 0.29 \times 10^{-28}\ m^2$, $Q(^{49}Ti) = 0.24 \times 10^{-28}\ m^2$]. This is not the case with the two hafnium isotopes ^{177}Hf and ^{179}Hf ($Q = 3 \times 10^{-28}\ m^2$), which appear not to have been studied as yet.

The earliest reports on Ti NMR deal with $TiCl_4$ and $H_2[TiF_6]$[58] and solid $Sr[TiO_3]$.[59] ^{47}Ti and ^{49}Ti are of comparable natural abundance (7.28% and 5.51%). The gyromagnetic ratios are remarkably close, and consequently the separation between the ^{47}Ti and ^{49}Ti resonances for each distinct species is small [271(3) ppm[60]]. The shifts of the titanium tetrahalides are +1162 ($TiCl_4$), +1645 ($TiBr_4$), and +2440 (TiI_4) relative to that of $[NH_4]_2[TiF_6]/H_2O$.[60a] This "inverse" dependence has been investigated systematically in various TiX_4 (X = Cl, Br, I, OiPr, NEt_2)[60a] and Cp_2TiX_2 complexes[60a,60b] (X = halogen, N_3, NCS). For the latter, a $\delta(^{47,49}Ti)/\sum \chi_X$ correlation (χ = electronegativity) very similar to that of VOX_3 compounds[5] was found. The inverse δ/χ dependence is commonly observed in systems with a d^0 coordination center (see also ^{45}Sc and ^{139}La in the previous sections) and will be discussed more comprehensively in Section 4.2.1.

The ^{49}Ti–^{19}F coupling (33 Hz) has been estimated from the linewidth of the unresolved ^{19}F resonance in $[TiF_6]^{2-}$,[61] confirmed recently by the direct observation of well-resolved ^{47}Ti and ^{49}Ti septets ($J = 34.0$ Hz).[60a] Tarasov and Buslaev[15] obser-

Table 4. ^{91}Zr Chemical Shifts (Relative to $[Zr(Cp)_2Br_2]$,[a] Scalar Coupling Constants, and Linewidths

Complex	δ (ppm)	$W_{1/2}$ (Hz)	Reference
$[Zr(NEt_2)_4]$	+875		60a
$H_2[ZrCl_6]$/conc. HCl	+601.2	245 ± 10	62
$[Zr(Cp)_2I_2]$	+126.0	134 ± 10	62
$[Zr(C_5Me_5)_2Cl_2]$	+82	90	61a
$[Zr(\eta^3\text{-}C_4H_7)(Cp)(\eta^4\text{-}C_4H_6)]$	+142	2500	61a
$[Zr(COT)(\eta^4\text{-}C_4H_6)]$	+148	220	61a
$[Zr(BH_4)_4]$[b]	+40.7[c]	184	63
$[Zr(Cp)_2Br_2]$	0	19 ± 2	62
$[Zr(Cp)_2(Br)Cl]$	−65.9	237 ± 10	62
$[Zr(Cp)_2Cl_2]$	−121.9	276 ± 10	62
	−113	250	61a
$[NH_4]_2[ZrF_6]/D_2O$	−191.0	50 ± 5	62
$[Zr(Cp)_2(\eta^4\text{-}C_4H_6)]$	−384	3100	61a
$[Zr(Cp)_2(\eta^4\text{-}C_5H_8)]$	−324	2200	61a

[a] In THF (Refs. 60a, 62, 63) and THF + 10% C_6D_6 (Ref. 61a), respectively.
[b] $(^1J + {}^3J)(^{91}Zr\text{–}^1H) = 28$, $^1J(^{91}Zr\text{–}^1H) = 37$ (estimated), $^2J(^{91}Zr\text{–}^{11}B) = 18$, $^2J(^{91}Zr\text{–}^{10}B) = 6$ Hz (rationalized); see also Ref. 11c.
[c] There is a slight high-frequency isotope shift for $\{Zr\,^{10}B\}$ relative to $\{Zr\,^{11}B\}$.

ved an eight-line pattern in the ^{19}F NMR of 71.5% ^{49}Ti enriched $[NH_4]_2[TiF_6]$ with enhanced flanking peaks characteristic of quadrupolar relaxation dominating over exchange broadening. NQCCs have been obtained for $[^{49}TiF_6]^{2-}$ (4.5 MHz),[15] $^{47}TiCl_4$ (2.76 MHz), and $^{49}TiCl_4$ (2.73 MHz) ($TiCl_4$ in 80% toluene-d_8 at 297 K).[60a]

Three reports on ^{91}Zr NMR[61a,62,63] provide a first insight into shielding and coupling conditions encountered with this nucleus (Table 4). An inverse halogen dependence is found. ^{91}Zr chemical shifts of organozirconium compounds with π ligands (η^8-COT, η^5-Cp, η^3-allyl, η^4-dienes) show that shielding is larger by several hundred ppm in 18- than in 16-electron systems.[61a] The multinuclear study on $[Zr(BH_4)_4]$[63] is a nice example of how a thorough NMR treatment can yield a great deal of coupling information.

4. GROUP Vb

4.1. Vanadium

There are two magnetic vanadium nuclei: ^{51}V ($N = 99.76\%$, $I = 7/2$) and ^{50}V (0.24%, $I = 6$), and from the natural abundances it is clear why practically all investigations have been carried out on ^{51}V. This is one of the most suitable nuclei for NMR experiments, considering its high receptivity (0.381) and low Q (-0.052×10^{-28} m^2) which, in general, gives rise to sufficiently narrow lines even in compounds where there is virtually no symmetry.

The ^{51}V nucleus has also attracted wide interest in solid state NMR, and the vanadium–oxygen systems (including vanadates) have been surveyed in a book.[5a] More recent examples are V_6O_{18} and other vanadium oxides,[63a] and $Ln[VO_4]$.[50,63b] Vanadium catalysts for the SO_2 oxidation have also been studied by ^{51}V NMR.[63c] High-resolution ^{51}V NMR spectra obtained by magic-angle or near-to-magic-angle spinning have been reported for NH_4VO_3, $NaVO_3$,[64,64a] and V_2O_5.[65] The solution NMR of ^{51}V and some aspects of ^{51}V in anisotropic systems have been reviewed (covering the literature to the summer of 1983).[5,6] A comprehensive view on electronic and steric factors influencing metal shielding in low-valent carbonyltransition metal complexes with special respect to ^{51}V is given in Ref. 7. Key references are contained in Figure 2, where the shift data are presented graphically and listed relative to neat $VOCl_3$, which is the standard adopted in solution NMR. The shift for the "metavanadate" ion (pH 8–9) relative to $VOCl_3$ is -572 to -574 ppm. At this pH, the species present is $[H_2V_2O_7]^{2-}$[75] and not (as thought earlier) $[H_2VO_4]^-$. Not included in the chart are $\delta(^{51}V)$ for thiovanadates (up to $+1392$ ppm[66]), for powder spectra of Cu(I) and Tl chalcogenovanadates (up to $+3885$ ppm[67]; cf. Section 4.2.1), and more recent results on [η^2-CORV(CO)$_3$(diphos)] (-931 to -1082),[67a] carbonylvanadium complexes containing a vanadium–metal bond, extending the δ scale to -2054 ppm $[V(Cp)(SnPh_3)(CO)_3]^-$,[102b] the halide complexes $[V(Cp)(X)(CO)_3]^-$ (with $\delta = +417$ for X = F at the low field margin),[102a] and cyclo-thiazeno complexes such as $[VCl_3(N_3S_2)]_2^{2-}$ ($\delta = +594$).[74g]

4.1.1. *Oxovanadium (V) and Vanadates (V)*

The ratio $eQ(^{50}V)/eQ(^{51}V)$ has been determined from ^{50}V and ^{51}V relaxation measurements on neat $VOCl_3$ [4.11(15)[68,68a]] and $Sm[VO_4]$ (4.019[68b]), giving a

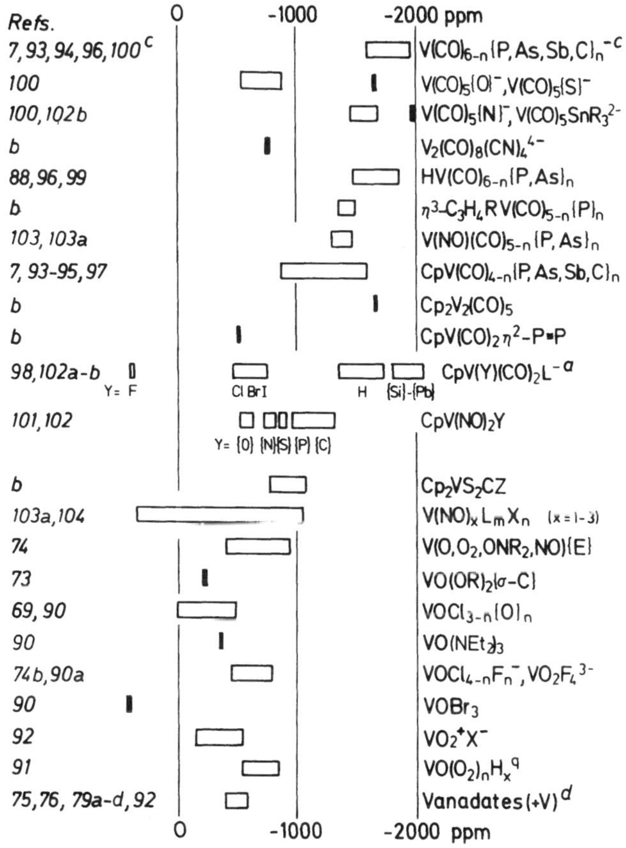

Figure 2. ^{51}V shielding relative to $VOCl_3$. Not included are thiovanadates (the range is flanked by $[VS_4]^{3-}$, $\delta = +1392$ ppm[(66)]). Abbreviations: Cp = η^5-C_5H_5, Tp = η^7-C_7H_7, P≡P = $Ph_2PCH{\equiv}CHPPh_2$, X = halogen or pseudohalogen, R = alkyl, H, or aryl, L = miscellaneous; {E} is a composite ligand coordinating via E. [a]L = CO except for Y = H (L = {P}). [b]From Refs. 5 and 6. [c]The shift for $[Et_4N][V(CO)_6]$ in THF at 295 K is −1952 ppm. Shift values reported for $[V(CO)_6]^-$ earlier (Nakano[(190)]; Rehder *et al.* prior to 1977, see Ref. 90) are in error. [d]Includes phospho-[(79)] and tungstovanadates.[(79d,79e)]

$Q(^{50}V) = 0.21 \times 10^{-28}\ m^2$. The NQCC of $VOCl_3$ has been redetermined for polycrystalline samples [5.7(2) MHz, $\eta = 0.09$] and in the liquid state (7.5 MHz),[(69)] employing a molecular correlation time $\tau_c = 2.41 \times 10^{-12}$ s. Agreement with the data published by Habayeb and Hileman[(70)] is achieved, if their τ_c (15×10^{-12} s, based on the inadequate assumption that, in liquid $VOCl_3$, the hydrodynamic model applies) is recalculated with the correction factors given by Gierer and Wirtz.[(71)] Various compounds of the general formula $VOCl_{3-n}(OR)_n$ have also been investigated in the solid (pseudopolycrystalline) state, in meso-phases, and in isotropic systems.[(69,72)] In liquid crystal matrices, a seven-line pattern due to first-order quadrupole perturbation is observed. The NQCC decreases in the order $VOCl_3 > VOCl(OR)_2 > VO(OR)_3$, paralleling an increase of the sum of the electronegativities of the groups attached to VO^{3+} and also correlating with increased nuclear shielding. Selected NQCCs are compiled in Table 5. Two recent reports deal with vanadyl complexes containing a vanadium-carbon-σ bond {e.g., $[Me_3SiCH_2VO(O\text{-}tBu)_2]$[(73)]} and with nitrosyl- and hydroxylamidovanadium(III and V),[(74)] respectively. For the latter, an increase of shielding is found in the sequence $O^{2-} < O_2^{2-} < Et_2NO^- < H(Me)NO^- < H_2NO^-$.

Table 5. ^{51}V NMR Data on Selected Compounds

Compound	NQCC [a] (MHz)	Reference	δ (ppm)[b]	$W_{1/2}$ [b] (Hz)	Reference
VOF_3	8.9(4)	80	−757		90
$VOCl_3$	5.7(2)	69	0	17.7	69
	7.5[c]	69			69
$VOCl_2(Oi\text{-}Pr)$	4.3[c]	69	−309	15.7	69
$VOCl(Oi\text{-}Pr)_2$	5.1(2)	69	−506	56.0	69
$VO(Oi\text{-}Pr)_3$	4.4(2)	69	−628	14.0	69, 90
$VO(On\text{-}Bu)_3$	5.0[d]	69	−548	23	69, 90
V_2O_5	0.805	81			
$Na[VO_3]$	3.65(6)	82	−574[e]	80[e]	91
$Mg[VO_3]_2$	6.79	83			
$Na_3[VO_4]$	1.05(3)[f]	84	−536	27	78
$La[VO_4]$	5.21(7)	85			
$Ni_2[V_2O_7]$	3.55(6)	86			
V_3Si	2.924(15)	87			
$[V(CO)_3Tp]$[g]	2.4(2)[c]	88			
$[VCp(CO)_4]$	2.79(1)	89	−1534	12	5, 7

[a] Nuclear quadrupole coupling constant (polycrystalline samples if not indicated otherwise); absolute errors in parentheses.
[b] Chemical shifts and linewidths are given for isotropic systems (neat liquids or solutions).
[c] From relaxation measurements in isotropic systems.
[d] From quadrupole splittings in nematic liquid crystals.
[e] Data for $[H_2V_2O_7]^{2-}$ ("metavanadate").
[f] For aqueous solutions, a NQCC of 0.57 MHz is reported.[123a]
[g] $Tp = \eta^7\text{-}C_7H_7^+$.

$^1J(^{51}V-^{14}N) = 92$–112 Hz have been observed in various $[(RO)_3V{=}NR']$ complexes (1:1:1 triplets; $\delta = -406$ to -751).[74a] A similar coupling (105 Hz) was reported for $VO(OiPr)_2NEt_2$ ($\delta = -593$ ppm).[74f] $[(TPPH_2)^{2+}(VOCl_3^-)_2]$, which appears to be the first VO^{3+} porphyrine complex (TPP = mesotetraphenylporphine), exhibits a singlet ($\delta = -481$ ppm) despite the fact that a vanadium–nitrogen bond exists.[74f] Apparently, the two quadrupolar nuclei are effectively relaxation decoupled. Renewed attention (by ^{19}F and ^{51}V NMR) has been paid to the system oxovanadium/H_2O/F^-.[74b] The signals observed at -770 to -777 ppm are assigned to the anion $[VO_2F_4]^{3-}$ [quintet, $^1J(^{51}V-^{19}F) = 117$ Hz]. The quintet of $J = 140$ Hz reported earlier by Howell and Moss[74c] is attributed to $[VOF_4]^-$, and it is assumed that this anion exists in water-free media only,[74b] which statement may be doubted on the grounds that a recent crystal structure of $Rb[VOF_4(H_2O)]$ clearly demonstrates the presence of the VOF_4^- structure unit in an environment where water plays an integral part.[74e] The oxo anions $[VOCl_4]^-$ and $[VO_2Cl_2]^-$ show δ values of $+50$ and -360 ppm, respectively[74g]; the anionic imino complex $[V(NtBu)Br_4]^-$ resonates at $+418$.[74a]

The isopolyvanadate systems with their various condensed species taking part in pH-dependent equilibria have again been studied by combined methods (^{51}V and ^{17}O,[75] ^{51}V and emf[76]), but some of the central questions still remain open. There is a general consensus on the existence and the chemical shifts of the mono-, di-, tetra-, hexa-, and decameric species. Trimeric and pentameric species have been assigned by Howarth[75] and by Tytko,[77] but have not been confirmed by Petterson *et al.*[76] $^1J(^{51}V-^{17}O)$ coupling ranges from 31 ($[V_2O_7]^{4-}$) to 62.6 Hz ($[VO_4]^{3-}$),[75] the latter being in accord with the coupling constants observed previously by Lutz *et al.*[78] The isotope effect in $[V(^{16}O)_{4-n}(^{18}O)_n]^{3-}$ amounts to -0.19 ppm per ^{18}O.[78a]

$[PV_{14}O_{42}]^{7-}$ has been shown to exhibit two ^{51}V resonances (−515 and −574 at pH 6.1; −523 and −575 at pH 3.8) in a 1:6 intensity ratio. The low-frequency shift of the signal of lower intensity upon acidification is due to protonation of the capping VO_5 units in the bicapped Keggin structure.(79) A well-resolved spectrum of the same compound obtained by Howarth shows three resonances at −529, −580, and −596 ppm (1:4:2).(66) Pope *et al.* have reported on the spectra of phosphotungstovanadates α-$[P_2W_{18-n}V_nO_{62}]^{(6+n)-}$ ($n=1-3$; $\delta=-503$ to −570 ppm, $W_{1/2}=33$ to 180 Hz; at pH 4.7).(79d) Correlations with ESR data of the reduced forms are also discussed. The structures of $[EVW_{11}O_{40}]^{n-}$ (E = P, $n=4$; E = Si, $n=5$; E = B, $n=6$) have been established by combined ^{183}W and ^{51}V NMR, including ^{51}V–^{51}V two-dimensional homonuclear correlation spectroscopy.(79e)

4.1.2. *Vanadium Carbonyl and Related Compounds*

In the field of low-valent vanadium complexes, studies have been carried out on derivatives of $[V(Cp)(CO)_4]$ and $[V(CO)_6]^-$ with ligands of the type o-$C_6H_4(EPh_2)E'Ph_2$ (E, E' = P, As, Sb, Bi),(93,96) $Ph_2P(CH_2)_{5,6}PPh_2$, and $R_2P(CH_2)_2PR'(CH_2)_2PR_2$ (R, R' = Ph, Cy),(94–95a) As_2Ph_4 and $Ph_2ECH_2EPh_2$ (E = P: dppm, As, Sb).(97) In the photoinduced reaction between $[V(Cp)(CO)_4]$ and dppm, which leads to *cis*-$[V(Cp)(CO)_2\,dppm]$ as the final product, an intermediate dinuclear complex $[\{V(Cp)(CO)_3\}_2\,\mu$-dppm$]$ is detected in which vanadium is deshielded relative to $[V(Cp)(CO)_3\,dppm]$. Former studies on phosphine derivatives of $[VH(CO)_6]^{(98)}$ have been supplemented.(99) In these systems, introduction of the H^- ligand deshields the metal nucleus, which contrasts with findings on $[HV(CO)_5]^{2-}$,(99b) $[V(Cp)(CO)_4]/[V(Cp)H(CO)_3]^-$ (98) and on Nb, Mo, and W complexes. fac-$[V(CO)_3$ hexaphos$]^-$ gives rise to a resolved quartet [$\delta=-1778$ ppm, $^1J(^{51}V–^{31}P)=224$ Hz; hexaphos = $(Ph_2PCH_2CH_2)_2P(CH_2)_2P(CH_2CH_2PPh_2)_2$] to low frequency of the unresolved signal for the meridional isomer (−1734 ppm).(99)

$[V(CO)_5L]^-$ (L = thf: −1387,(99a) py: −1464, NCMe: −1579, $S(O)Me_2$: −1659, CNCy: −1901 ppm,(100) $[V(CO)_5L]^{2-}$ (L = CN: −1864, $SnPh_3$: −1994(102b)) and a large number of $[VCp(NO)_2L]$ complexes(101,102) have been characterized by ^{51}V NMR. Shielding increases in the series {O} < amine < nitrile < {S} < PR_3(large cone angle) < CN^- and isonitrile < PR_3(small cone angle) < $P(OR)_3$ < CO and PF_3 < $SnPh_3$. This "normal" dependence of δ on the electronegativity and/or polarizability (Pearson basicity) of the donating function, and the increase of $|\delta|$ with the π acceptor power of the ligand, are discussed in some detail in Section 5.2.2. A normal dependence is also found in the series $[VCp(X)(CO)_3]$ for X = F (+417), Cl (−515), Br (−578), I (−742), and CN^- (−1394),(102a) and in the series $[VCp(E^{IV}Ph_3)(CO)_3]^-$ for E^{IV} = Si (−1860), Ge (−2026), Sn (−2054); for the weakly π interacting $PbPh_3^-$ ligand, a shift of −1789 ppm is found.(102b) η^2-acyl complexes $[RCOV(CO)_3\,diphos]$ display a substituent effect (R being a substituted phenyl)(67a) similar to that discussed in the molybdenum section.

A rather wide range of chemical shifts is covered by complexes derived from $[V(CO)_5NO]$.(103,103a) For the parent compound, $\delta=-1489$ ppm at 245 K; a short-lived signal observed at −2008 ppm may correspond to its precursor $[V(CO)_6NO]$.(103a) Substituting CO by a phosphine or arsine ligand deshields the ^{51}V nucleus with respect to the corresponding carbonylphosphinevanadate(−I). A wide shielding range if also observed for carbonyl-free nitrosyl complexes of vanadium:

$[VCl_2(thf)_2(NO)]$ −517, $[VCl_2(NO)_3]_n$ −605, $[VCl(NCMe)_2(NO)_2]_x$ −223, $[V(CNtBu)_4(NO)_2]^+$ −1040 ppm.(104) A normal halogen dependence is observed in $[VX(NO)_2(CNR)_3]$ (X = Cl, Br, I), while the invariance of $\delta(^{51}V)$ towards variation of X in dinitrosyl complexes containing other than the isonitrile ligand demonstrates that these complexes have to be formulated *cis*-$[V(NO)_2(thf)_{4-n}L_n]X$ ($n = 1$–4, L, e.g., $=E^VR_3$, SPhMe, $OCMe_2$, NCMe).(103a) In the case of phosphine ligands, complex coupling patterns arise.

The large shielding ranges which are found with many low-valent transition metal complexes and quite often readily overlap with the shielding ranges for the metals in their highest oxidation states indicate that it is misleading to correlate shielding data with formal oxidation states, thus implying substantial influence on the shielding arising from local diamagnetic contributions. Dia- and paramagnetic terms, calculated on the basis of MO results (Table 6) demonstrate that (i) the local diamagnetic term is constant within ca. 30 ppm for a given nucleus, (ii) ΔE and the paramagnetic contributions to the over-all shielding qualitatively agree with the observed shielding, and (iii) variations in the paramagnetic term are chiefly due to the main component of the symmetry allowed electronic *d–d* transition (see also Section 6.1.2).

Coupling constants involving the ^{51}V nucleus have been obtained for $[V(CO)_{6-n}(^{13}CO)_n]^-$ (99b,99c) $[^1J(^{13}C-^{51}V) = 116\ Hz]$, $[V(^{13}CO)_5]^{3-}$ $[^1J(^{13}C-^{51}V) = 139\ Hz]$, and $[VH(^{13}CO)_5]^{2-}$ $[^1J(^{13}C-^{51}V) = 124.5$; $^1J(^1H-^{51}V) = 27.6\ Hz]$,(92b) and $[V(Cp)H(CO)_3]^-$ $[^1J(^1H-^{51}V) = 20.6\ Hz]$(98) both from the ^{51}V and the ^{13}C or ^1H NMR spectra, and for $[VCp(SnCl_3)(CO)_3]^{2-}$ $[J(SnV) = 900\ Hz]$.(219) Of special interest is the binomial sextet in the ^{51}V NMR of $[V(^{13}CO)_5]^{3-}$, since it provides a

Table 6. Calculated (from SCF-MO Data) and Observed Shielding Parameters of ^{51}V and ^{55}Mn Complexes[a]

Complex	*d*-HOMO (eV)[b]	*d*-LUMO (eV)[b]	ΔE (eV)	σ(para)[c]	σ(dia) calc.	δ (*M*) (ppm)
$[VO_4]^{3-}$				−6595	1703	0
$[VCp(CO)_4]$	11.60	9.60	2.00		1712	−998
$[VCp(CO)_3PF_3]$	11.80	9.60	1.94		1712	−1031
$[VCp(CO)_3PH_3]$	11.17	9.23	1.94		1712	−844
$[V(CN)Cp(CO)_3]^-$	9.89	9.21	0.68		1712	−834
$[V(CO)_6]^-$	10.75	6.04	4.71	−4502	1730	−1416
	7.50[d]	4.10[d]	3.40[d]	−5993	1718	*e*
$[V(CO)_5PF_3]^-$	11.22	6.25	5.07	−4188	1730	−1425
$[V(CO)_5PH_3]^-$	10.74	6.68	4.06	−5215	1730	−1343
$[V(CN)(CO)_5]^{2-}$	10.48	6.60	3.88	−5467	1730	−1328
$[MnO_4]^-$				−7015	1900	0
$[Mn(CO)_6]^+$	9.77[d]	4.39[d]	5.38[d]	−6102	1915	−1445[e]
$[MnCl(CO)_5]$	10.40[f]		(10.414)	−2722	1897[g]	−1005
$[MnBr(CO)_5]$	10.15[f]		(10.475)	−2696	1898	−1160
$[MnI(CO)_5]$	9.95[f]		(10.465)	−2615	1899	−1485

[a] If not indicated otherwise, data were taken from Ref. 7 (^{51}V, partially recalculated) and Ref. 190 (^{55}Mn). Vanadium shifts are given relative to orthovanadate $[\delta(VOCl_3) = 536\ ppm]$.
[b] HOMO and LUMO with substantial *M*(*d*) character, i.e., 7*e* and $10a_1$ for C_{4v}, and $2t_{2g}$ and $3e_g$ for O_h.
[c] Calcd. with $\Delta E(e \rightarrow a_1^*)$ for C_{4v} complexes and $\Delta E(t_{2g} \rightarrow e_g^*)$ for O_h complexes of vanadium and $[Mn(CO)_6]^+$, and with $\sum \Delta E$ for $[MnX(CO)_5]$.
[d] From Beach, N. A.; Gray, H. B. *J. Am. Chem. Soc.* 1968, **90**, 5713.
[e] Shifts in Ref. 190 are in error.
[f] From Fenske, R. F.; DeKock, R. L. *Inorg. Chem.* 1970, **9**, 1053.
[g] Compare also the value calculated by Nakatsuji *et al.* (2267 ppm), Ref. 197.

direct proof for the existence of this superreduced species in liquid ammonia solution. The well-resolved patterns for the hexacarbonyl vanadates are reminiscent of the spectra reported for $[V(PF_3)_6]^-$ $[^1J(^{31}P-^{51}V) = 510$, $^2J(^{19}F-^{51}V) = 10.3$ Hz].[6,118] $^1J(^{31}P-^{51}V)$ in carbonylphosphine vanadium complexes usually range from ca. 130 (alkyl and aryl phosphines) to 330 (phosphites) and 500 Hz (PF_3).[5,6] In the system $[V(CO)_{6-n}](^{13}CO)_n]^-$, an isotope effect of -0.2 to -0.4 ppm per ^{13}CO substituting ^{12}CO has been noted.[99c]

4.2. Niobium

^{93}Nb is one of the most sensitive nuclei ($s = 0.482$). Its $Q = -0.22 \times 10^{-28}$ m^2, about four times that of ^{51}V, makes it rather more difficult to observe in compounds of low point symmetry. There is no generally adopted standard in ^{93}Nb NMR spectroscopy. $[NbF_6]^-$ (in 48% HF) was originally used as a reference and later replaced by $[NbCl_6]^-$ (or $NbCl_5$/MeCN), which is deshielded with respect to $[NbF_6]^-$/HF by ca. 1495 ppm (1550 ppm relative to $[NbF_6]^-$/MeCN). A solution of $NbCl_5$ in (wet ?) MeCN shows three resonances, which have been assigned to $[NbCl_6]^-$ (0 ppm, $W_{1/2} < 30$ Hz), $[NbCl_5 \cdot MeCN]$ (-49 ppm, $W_{1/2} = 200$ Hz), and $[NbCl_4(MeCN)_2]^+$ (-500 ppm, $W_{1/2} = 5$–7 kHz).[105] In dry MeCN, there is only one resonance at -49 ppm.[106] For comparison with ^{51}V shielding data, $NbOCl_3$ may be considered a more suitable standard, although, dissolved in MeCN, it produces an unfavorably broad line (700 Hz[106]). In the shielding chart (Figure 3), we have nonetheless quoted ^{93}Nb shifts relative to $NbOCl_3$. The shift of $NbOCl_3$ relative to that of $[NbCl_6]^-$ is not accurately known, which is mainly due to confusion as to the exact nature of the species present in MeCN $\{[NbCl_6]^-$ or $[NbCl_5 \cdot MeCN]$, $NbOCl_3$, $[NbOCl_4]^-$, $[NbOCl_3 \cdot MeCN]$, or $[NbOCl_4(MeCN)]^-\}$ and, correspondingly, to the assignment of the resonances and the zero point of the respective scales. In relating

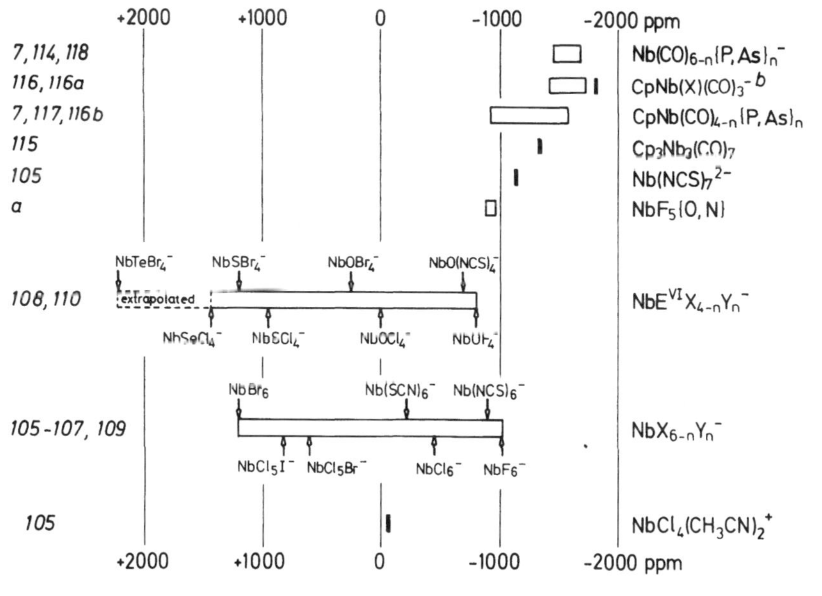

Figure 3. ^{93}Nb chemical shifts relative to $NbOCl_3$/MeCN. See Figure 2 for abbreviations. [a]Cited in Ref. 1. [b]$X = E^{IV}R_3$ (E^{IV} = Ge, Sn, Pb): high-frequency range; X = H, D: low-frequency bar.

the publishing data to $\delta(NbOCl_3) = 0$ ppm, we have employed $\delta[NbCl_6]^- = +460$ (and $\delta[NbF_6]^- = -1035$) ppm. $\Delta\delta = 460 \pm 10$ ppm is the shielding difference we have measured between saturated solutions of freshly prepared and resublimed $NbCl_5$ and $NbOCl_3$ in highly purified acetonitrile (free of O_2, H_2O, and amines).

4.2.1. Niobium (V) Halogen and Pseudohalogen Complexes

Compounds of the general formulas $[NbX_{6-n}Y_n]^-$ and $[NbEX_{4-n}Y_n]^-$ (X, Y = F, Cl, Br, NCS, SCN, NCO; E = O, S, Se, Te) have been studied extensively by Kidd and Spinney[106,107] and by Buslaev and Tarasov.[105,108–110] The chemical shift data of the chalcogeno complexes are collected in Figure 4.

The general trends are a decrease of ^{93}Nb shielding in the order F > NCS ≈ NCO ≫ Cl > SCN ≫ Br and O > S > Se (> Te; the data for the Te complexes have been extrapolated).

These trends (which, in the case of the halogens, have been designated "inverse"[2]) are commonly observed in all high-valent transition metal compounds such as TiX_4 and $[Zr(Cp)_2X_2]$ (X = Cl, Br, I), VOX_3 (X = F, Cl, Br), and $[VOX_{4-n}Y_n]^-$ (X, Y = F, Cl), WX_6 (X = F, Cl), $[LaX_6]^{3-}$ (X = Cl, Br), $[MoE_4]^{2-}$ (E = O, S, Se) (see the respective sections) and $M^I_3[M^VE_4]$ (M^I = Cu, Tl; M^V = V, Nb, Ta; E = S, Se, Te)[67].

$Cu_3[VS_4]$ (620)	$Cu_3[VSe_4]$ (1710)	$Cu_3[VTe_4]$ (3885)
$Cu_3[NbS_4]$ (840)	$Cu_3[NbSe_4]$ (1560)	$Cu_3[NbTe_4]$ (2970)
$Cu_3[TaS_4]$ (2740)	$Cu_3[TaSe_4]$ (3730)	$Cu_3[TaTe_4]$ (5720)

(^{51}V and ^{93}Nb relative to $MOCl_3$; ^{181}Ta relative to $K[TaO_3]$).

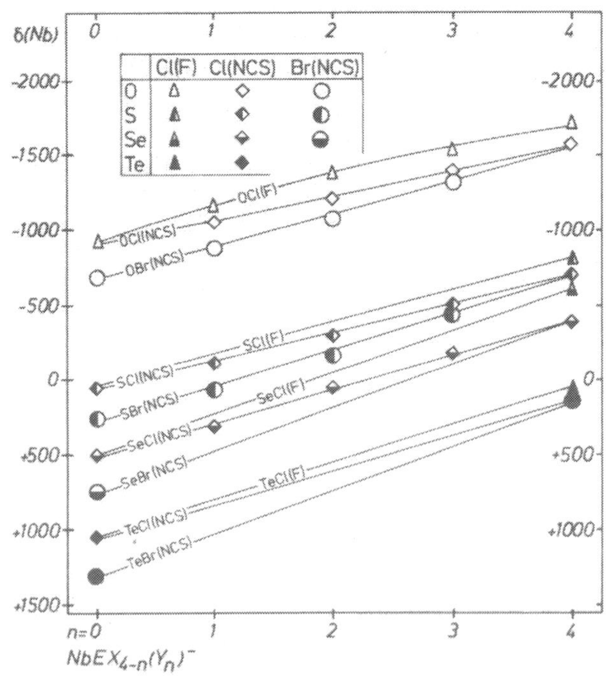

Figure 4. ^{93}Nb chemical shifts of $[Nb(E)X_{4-n}Y_n]^-$ (E = O, S, Se, Te; X = Cl, Br; Y = F, NCS) relative to $NbOCl_3$. Data from Refs. 108 and 110.

An explanation for these shielding patterns may lie in an increased E → M π donation and involvement of the metal *d* orbitals (*sd*^{3} hybrids in the case of tetrahedral complexes) in the M–E or M–X bond and hence in an increase of *d* character of the orbitals involved in electronic transitions.

Similar shielding is observed for niobyl cyanate and thiocyanate complexes, indicating that these ligands exclusively bond via nitrogen[108,110] (see, however, the discussion below). Investigations on $[NbEX_{4-n}Y_n]^-$ in MeCN show that MeCN, if coordinated to the niobyl, does not affect the ^{93}Nb shielding to a great extent, which is in accord with the rather long ONb–L distance (for L trans to O) in pseudo-octahedral niobium complexes.[110] The variations of δ for stepwise X/Y exchange follow a linear relationship. This has led to the incorrect assumption that varying non-local diamagnetic contributions, amounting to 20%–30%, are important for variations of the overall shielding.[105,109]

The data on $[NbCl_{6-n}Br_n]^-$ and $[NbCl_{6-n}(NCS\ or\ SCN)_n]^-$, which were published independently by Tarasov[105,109] and Kidd,[106,107] need some comment because they are, in part, contradictory, as demonstrated for selected shift values on thiocyanato complexes:

cis-$[NbCl_4(NCS)_2]^-$:	−25	*trans*-$[NbCl_4(NCS)_2]^-$:	−110	Ref. 105
cis-$[NbCl_4(NCS)_2]^-$:	+3	*trans*-$[NbCl_4(SCN)_2]^-$:	+240	Ref. 107

and

cis-$[NbCl_2(NCS)_4]^-$:	−325	*trans*-$[NbCl_2(NCS)_4]^-$:	−490	Ref. 105
cis-$[NbCl_2(NCS)_4]^-$:	−442	*trans*-$[NbCl_2(SCN)_4]^-$:	−15	Ref. 107

The clue for these deviations possibly lies in the sample preparation. Whereas Kidd and Spinney prepared the complexes and dissolved them in dry MeCN, Tarasov and Buslaev used mixtures of $NbCl_5/NbBr_5$ and $NbCl_5/NH_4SCN$, and their MeCN may have contained water and amines which (i) give solvent effects (which may account for variations around 20–30 ppm), (ii) initiate hydrolysis, which may give anionic species, and (iii) may steer the coordination of the thiocyanate in such a manner that only complexes containing N-bonded SCN^- are formed. In dry MeCN, resonances for the neutral complexes $[NbCl_{5-n}Br_n(MeCN)]$ are observed.[106]

Both groups assigned their signals on the basis of the pairwise additivity model introduced by Vladimiroff and Malinowski.[111] This model recognizes the mutual ligand influence via changes in the wave functions of a ligand set $(L_i)_n$ caused by a ligand L_j. δ is calculated from $\delta = \sum c_{ij} p_{ij}$, where summation is carried out for all adjacent (*cis*-standing) pairs L_iL_j as exemplified for the three complexes shown in Figure 5. The *p* parameters (in ppm) are *p*(Cl–Cl) = 0, *p*(Cl–Br) = +32.8, *p*(Br–Br) = +60.9, *p*(S–S) = −65.0, *p*(S–N) = −115.9. Signal assignments to cis/trans and mer/fac isomers have been supported by linewidth considerations (see below).

Relaxation studies have been carried out in the series $[NbOCl_{4-n}F_n]^-$ [110] and $[NbCl_{6-n}Br_n]^-$ [109] and used, in the latter case, to assign the ten resonances observed for the different species. Relaxation rates at 229 K, obtained from the half-widths of the resonance signals, vary from 0.04($[NbCl_6]^-$) to 3.3 units (*trans*-$[NbX_2Y_4]^-$) and are in good qualitative agreement with data calculated on the basis of the point charge model (see Refs. 109 and 110 for the explicit analytical expressions for the field

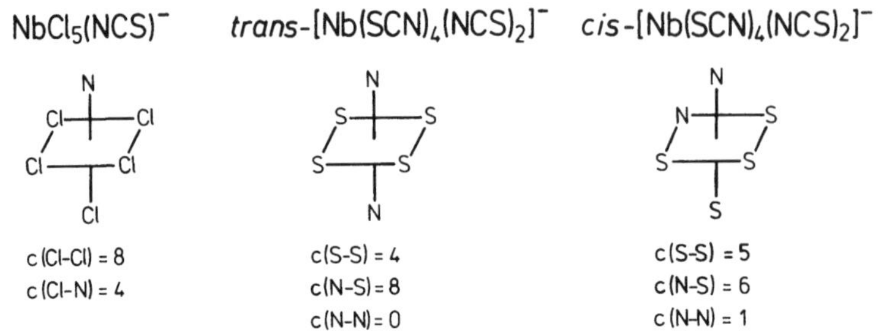

Figure 5. Examples for the calculation of ^{93}Nb chemical shifts (relative to $[NbCl_6]^-$) for $[NbX_{6-n}Y_n]^-$ complexes (X, Y = Cl, Br, NCS, SCN) on the basis of the pair-wise additivity model.

gradient invariant). $^1J(^{93}Nb-^{19}F)$ coupling constants have been obtained for $[NbOF_4]^-$ (410),[112] $[NbOCl_2F_2]^-$ (405),[110] $[NbF_6]^-$ and fac-$[NbCl_3F_3]^-$ (335 and 281 Hz at 263 K).[105]

A ^{93}Nb (and ^{23}Na) single crystal study of Na$[NbF_6]$[113] showed small tetragonal distortions (1 pm) of the NbF_6 octahedron, leading to a NQCC = 6.0 MHz below 143 K. Magic angle spinning of a polycrystalline sample of Na$[NbO_3]$ resulted in a well-resolved second-order powder pattern, from which the NQCC was determined as 19.5 MHz.[65]

4.2.2. Niobium Carbonyl and Related Compounds

The temperature dependences of $W_{1/2}$ and δ for $[Nb(CO)_6]^-$,[114] $[NbCp(CO)_4]$,[115] and $[NbCp(H, D)(CO)_3]^-$ [116] have been investigated. $W_{1/2}$ values for $[Et_4N][Nb(CO)_6]$ in THF vary from 9.8 (323 K) to 23.0 (203 K) Hz. The temperature gradients of δ are 0.18 ppm/degree for $[Nb(CO)_6]^-$ ($[V(CO)_6]^-$: 0.31[5]) and 0.40 for $[NbCp(CO)_4]$ ($[VCp(CO)_4]$: 0.61[5]). There is an almost linear increase of shielding on going to lower temperatures. Investigation of the temperature dependences of δ and $W_{1/2}$ for $[(\eta^5\text{-Cp})_3\,\eta^2(\mu_3\text{-C}, \mu_2\text{-O})(CO)_6Nb_3]$ have revealed a dynamic component contributing to the ^{93}Nb shielding above 303 K.[115] As observed in carbonyl cyclopentadienyl complexes of vanadium, molybdenum, and tungsten, shielding of the ^{93}Nb nucleus increases as a CO is replaced by H^- ($[NbCp(CO)_4]$: −1566 ppm, $[NbCp(H)(CO)_3]^-$: −1802 ppm).[116] Substituting H for D results in additional shielding (−1808 ppm). $\delta(^{93}Nb)$ values for $[NbCp(E^{IV}Ph_3)(CO)_3]^-$ and $[NbCp(AuPPh_3)(CO)_3]^-$ [116a] range from −1390 for the gold compound to −1700 ppm for the tin derivative. A lower limit of $^1J(^{93}Nb-^1H) = 36$ Hz is obtained from the ^{1}H NMR, which is ca. twice that of $^1J(^{51}V-^1H)$ in $[VCp(H)(CO)_3]$.[98] $^1J(^1H-^{93}Nb)$ in $[Nb(H)(CO)_5]^{2-}$ ($\delta = -1622$ ppm in liquid NH_3) is ca. 80 Hz.[99b] The ^{13}C NMR spectra of $[Nb(CO)_6]^-$ and $[NbCp(CO)_4]$ form a 10-line system; $^1J(^{93}Nb-^{13}C)$ are 236 and 221 Hz, respectively.[7]

Replacing CO by mono- or bidentate phosphine and arsine ligands gives high-frequency shifts of the resonance signals and considerable line broadening. In general, the electronic and steric effects observed in $[NbCp(CO)_{4-n}L_n]$[117] and $[Et_4N][Nb(CO)_{6-n}L_n]$[114] $\{n = 1, 2;\ L_n = PZ_3,\ Ph_2P(CH_2)_mPPh_2\ [m = 0\text{–}5],$ *cis*-$Ph_2PCH{=}CHPPh_2,\ Ph_2PCH_2CH_2AsPh_2,\ o\text{-}C_6H_4(PPh_2)AsPh_2\}$ and $[Nb(H)(CO)_3$ triphos$]$[114a] are comparable to those of the corresponding ^{51}V[7] and ^{95}Mo compounds (see Section 5.2.2 for discussion). Narrow lines are again observed for $[Nb(PF_3)_6]^-$ [118] and fac-$[Nb(CO)_3(PF_3)_3]^-$,[7] demonstrating that even in these covalently bound complexes application of the point-charge model (which predicts a vanishing electric field gradient in pseudo-octahedral C_{3v} complexes[215] has a certain justification. Similarly, it has been shown that in $[V(CO)_3$ hexaphos$]^-$ the facial isomer gives rise to a resolved quartet, while the meridional isomer is represented by a broad line.[99] The ^{93}Nb NMR spectrum of $[Nb(PF_3)_6]^-$ is a septet $[^1J(^{93}Nb-^{31}P) = 1050$ Hz$]$ with a fine structure due to $^2J(^{93}Nb-^{19}F) = 53$ Hz. The ^{31}P NMR spectrum of this complex is a superposition of ten equidistant binomial quartets ($^{31}P-^{19}F$ coupling) of approximately the same integrated intensity. $\delta(^{93}Nb)$ for the two PF_3 complexes is -1644 and -1617 ppm ($[Nb(CO)_6]^-$: -1670), i.e., a shielding difference between the two O_h complexes of 53 ppm. A relative deshielding of the metal in fully substituted PF_3 complexes is also observed for $[V(CO)_6]^-/[V(PF_3)_6]^-$ and $[VCp(CO)_4]/[VCp(PF_3)_4]$ and has been ascribed to PF_3 being a slightly weaker π acceptor than CO.[7]

A first-order quadrupole interaction pattern is observed in the partly ordered liquid crystal systems $[NbCp(CO)_4]$/nematic phase 4 (nine-line pattern). Polycrystalline $[NbCp(CO)_4]$ exhibits a comparable spectrum at 1.53 T; but a second-order pattern (split central component) arises at 0.27 T.[67]

4.3. Tantalum

The large Q of ^{181}Ta (3×10^{-28} m^2) and the relatively low receptivity (less than 1/10 that of ^{93}Nb or ^{51}V) have prevented systematic NMR studies on this nucleus. There is an early report on ^{181}Ta NMR in solution[119]: a 17% Ta solution in concentrated HF/HNO_3 1/1, where the presumed species is $[TaF_6]^-$, shows a ^{181}Ta resonance at 7.285 MHz ($W_{1/2} = 40.4$ kHz) at room temperature and 7.289 MHz ($W_{1/2} = 12.7$ kHz) at 353 K (applied magnetic field $B_0 = 1.43$ T), the deshielding with increasing temperature being in accord with theory. The broad line is caused by second-order quadrupole effects; the decrease of $W_{1/2}$ as the temperature is raised has been explained by motional narrowing (decreasing τ_c). Within the limits of error, the signal appears at the same position as the resonance obtained earlier for solid $K[TaO_3]$ (7.2859 MHz, extrapolated to the same B_0).[120] On the other hand, the ^{181}Ta nucleus is considerably deshielded in the chalcogeno complexes $Cu_3[TaE_4]$[67] (see the previous section). The shift difference between $[TaS_4]^{3-}$ and $[TaSe_4]^{3-}$ (1900 ppm) is almost exactly the same as that for the pair $[MoS_4]^{2-}/[MoSe_4]^{2-}$; the shielding sensitivity for these two nuclei should therefore be comparable. A first shift scale for ^{181}Ta solution NMR has been established, flanked by $[TaCl_6]^-$ at the low-field and $[Ta(CO)_6]^-$ at the high-field margin ($\Delta\delta = 3450$ ppm). For these octahedral complexes, $W_{1/2}$ are 4.3 and 6.7 KHz, respectively, in organic solvents.[216] $T_1(^{181}Ta)$ measurements have been reported for TaS_2 and its pyridine intercalation compound,[121] and also for 2H–$TaSe_2$[121a] and 2H–TaS_2.[121b] At 4.2 K, the latter exhibits two different Ta sites with $\eta = 0.00$ and 0.52(2), respectively.

5. GROUP VIb

5.1. Chromium

The only magnetic isotope, ^{53}Cr ($N = 9.54$%, $I = 3/2$) has a very low receptivity (8.62×10^{-5}) which may be the reason for the small number of ^{53}Cr NMR studies carried out to date, in contrast to the large amount of systematic data now available for molybdenum and tungsten.

Solution studies have been reported for the chromate ion[78,78a,122,123,123a] and for hexacarbonylchromium.[123] These are highly symmetric complexes with a vanishing field gradient at the nucleus. Narrow signals are therefore to be expected, and in fact, $[CrO_4]^{2-}$ and $[Cr(CO)_6]$ show $W_{1/2}$ values of 11 and 17 Hz, respectively.[123] This is still much more than the metal resonances in $[^{95}MoO_4]^{2-}$, $[^{95}Mo(CO)_6]$, or $[^{51}V(CO)_6]^-$, and in disagreement with the low quadrupole moments that have recently been reported for the ^{53}Cr nucleus (0.07, estimated as a lower limit from ^{53}Cr hyperfine interaction in a single crystal of $Rb_2[CrCl_4]$;[124] $-0.15(5) \times 10^{-28}$ m^2 by laser-rf double resonance[124a]), or the $Q = 0.038$ "rationalized" by Kidd from a comparative linewidth study on $[MO_4]^{n-}$ species.[125] The latter study may suffer from the fact that $W_{1/2}$ for the ^{99}Tc resonance in $[TcO_4]^-$ is not merely reflecting a (slightly disturbed) T_d environment (cf. Section 6.2). Based on a comparison of $W_{1/2}$ in hexacarbonyltransition metal complexes (Table 7), $Q(^{53}Cr)$ should be of medium magnitude.

The separation between $[CrO_4]^{2-}$ and $[Cr(CO)_6]$ (1795 ppm; as usual, the carbonyl is the more shielded species) is comparable to that of other $[MO_4]^{n-}/[M(CO)_6]^q$ pairs (M = W, Re excluded; Table 7). Phosphine derivatives of $[Cr(CO)_6]$ have not yet been studied by ^{53}Cr NMR; but $^1J(^{53}Cr–^{31}P) = 153$ Hz was obtained from the satellite system of the ^{31}P resonance of carbonylchromium complexes with bidentate phosphonites.[126] It is likely that satellite systems have been overlooked in ^{13}C and ^{31}P NMR spectra of chromium compounds.

In the chromate system, the $[CrO_4]^{2-}/[Cr_2O_7]^{2-}$ equilibrium,[122] concentration, and counterion effects[123] were studied. The influence of the bulk susceptibility was also investigated[123]: A cylindrical sample of $K_2[CrO_4]$ in D_2O is deshielded by ca. 2 ppm with respect to a spherical sample. $^1J(^{53}Cr–^{17}O) = 10 \pm 2$ Hz was found by analyzing the shoulders observed in the ^{53}Cr NMR spectra of $[Cr(^{16}O)_3{}^{17}O]^{2-}$ and $[Cr(^{16}O)_2(^{17}O)_2]^{2-}$.[78] There is an isotope shift of -0.44 ppm per ^{18}O in $[Cr(^{16}O)_{4-n}(^{18}O)_n]^{2-}$ [78a] The NQCC for the $[CrO_4]^{2-}$ anion in aqueous solution is 0.75 MHz.[123a]

Table 7. Linewidths, Electric Nuclear Quadrupole Moments and Shift Differences $\delta[MO_4]^{n-} - \delta[M(CO)_6]^q$ for Transition Metal Hexacarbonyls

Complex	$W_{1/2}$ ^{a} (Hz)	Q (10^{-28} m^2)	$\Delta\delta$ (ppm)	Reference
$[^{51}V(CO)_6]^-$	<1.3	−0.052	1416	5, 7
$[^{53}Cr(CO)_6]$	17	b	1795	123
$[^{55}Mn(CO)_6]^+$	<2	0.55	1445	7, 189
$[^{93}Nb(CO)_6]^-$	10	−0.2	1170^c	114
$[^{95}Mo(CO)_6]$	<0.7	0.12	1857	131
$[^{97}Mo(CO)_6]$	6	1.1	1857	131
$[^{181}Ta(CO)_6]^-$	6700	3	2120^c	216
$[^{183}W(CO)_6]$	—	—	3500	169
$[^{185}Re(CO)_6]^+$	2600	2.8	3400	189
$[^{187}Re(CO)_6]^+$	2250	2.6	3400	189

^{a} At room temperature.
^{b} A recent laser-rf double resonance experiment gave a Q value of $-0.15(5) \times 10^{-28}$ m^2.[124a]
^{c} Using an extrapolated $\delta[NbO_4]^{3-} = -500$ ppm (rel. $NbOCl_3$) and $\delta[TaO_4]^{3-} = -1330$ ppm (rel. $[TaCl_6]^-$).

Some of the more recent solid state studies include $Er[CrO_3]$,[127,127a] $Fe[Cr_2S_4]$,[128] $Cu[Cr_2S_{4-n}Se_n]$,[129] and $Y[CrO_3]$.[130]

5.2. Molybdenum

Molybdenum has two spin-5/2 nuclei with low relative receptivities (^{95}Mo: 5.07×10^{-4}, ^{97}Mo: 3.25×10^{-4}). The rather large Q of ^{97}Mo (1.1×10^{-28} m^2, ^{95}Mo: 0.12) and correspondingly large $W_{1/2}$ in all but highly symmetric compounds, make ^{97}Mo a difficult nucleus for NMR detection. The measured $W_{1/2}$ in $[^{97}Mo(CO)_6]$ is 6–7 Hz,[131,131a] indicating that reported widths in $[^{95}Mo(CO)_6]$ (3 Hz,[132] 4 Hz,[133] 0.7 Hz[134]) represent line broadening induced by field inhomogeneity. The same is true for $W_{1/2}$ values of $[^{95}MoO_4]^{2-}$ (0.5 Hz; $[^{97}MoO_4]^{2-}$: 35 Hz) and $[^{95}MoS_4]^{2-}$ (0.3 Hz; $[^{97}MoS_4]^{2-}$: 25 Hz).[135] Since molybdenum NMR spectroscopy was reviewed by Kidd in 1980,[2] the number of compounds studied by ^{95}Mo NMR has increased substantially, initiated also by the potential importance of ^{95}Mo NMR for biological systems. As seen in Figure 6, the chemical shift range spans almost 5500 ppm, with $[MoCp(CO)_3]^-$ exhibiting the highest shielding (−2120 ppm[137]). On the high-frequency side, the scale extends to +4026 ppm ($[Mo_2(O_2CCF_3)_4]$[149c]), approaching the Knight shift for ^{95}Mo in molybdenum powder (+5900 ppm[138]). An

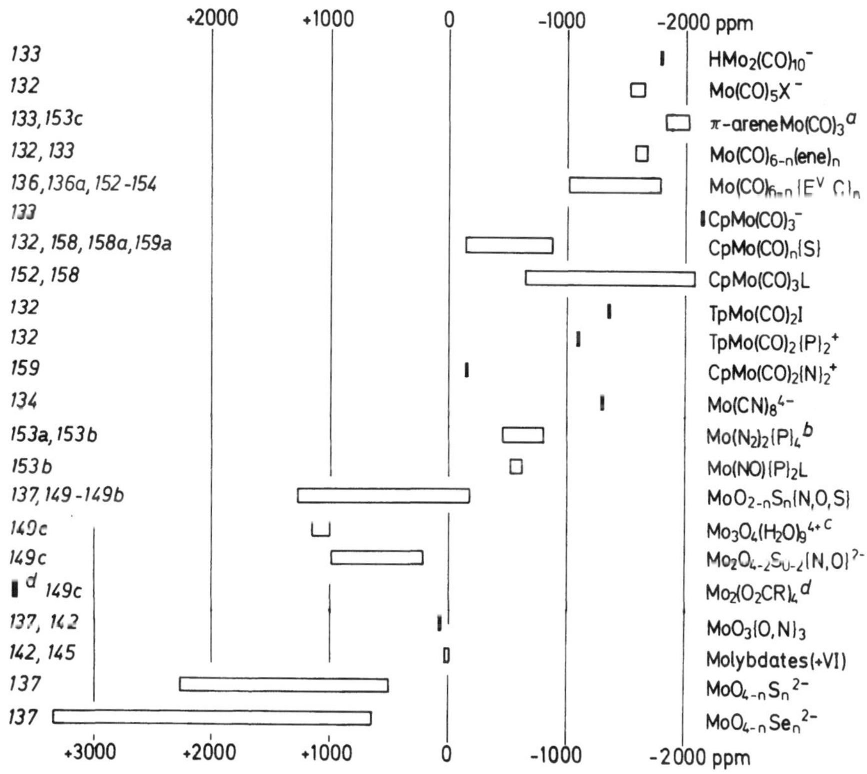

Figure 6. ^{95}Mo chemical shifts relative to $[MoO_4]^{2-}$ at pH 11 ± 1. For abbreviations cf. Figure 2. ^{a}Includes $[Mo_2(CO)_8P\{(p\text{-Tol})_2(\eta^6\text{-}p\text{-Tol})\}]$[153c]. ^{b}L = N_3, Br, MeCN. ^{c}And related complexes. ^{d}R = nPr (δ = +3730), R = CF_3 (δ = +4026 ppm).

overview of ^{95}Mo NMR, containing 14 selected references for the period 1975–1980, has appeared.[137a] The information available up to mid 1985 has recently been reviewed by Enemark *et al.*[213]

The reference generally applied is a 1–2 *M* aqueous $Na_2[MoO_4]$ solution at pH 11–12. The pH dependance of $\delta[MoO_4]^{2-}$ has been pointed out in several instances,[78,138–142] and an elaborate study by Lutz and co-workers[138] reveals that there is also a significant dependence upon concentration (ca. 2 ppm between 0 and 2.5 *M*). Consequently, chemical shifts reported by the Tübingen group refer to molybdate solutions at infinite dilution. Another factor limiting the accuracy or, rather, the reproducibility of shift data, is the temperature dependence of $\delta(^{95}Mo)$, which, between 295 and 311 K and for $[Mo(\pi\text{-arene})(CO)_3]$, amounts to ca. 0.2 ppm/degree (higher shielding at lower temperatures).[133] Further, solvent and counterion effects have to be taken into account.[136]

5.2.1. *Molybdenum (II, IV, V, and VI)*

Earlier work on molybdates concentrated on relaxation measurements,[135,138,141] isotope effects,[134,135,138,141,143,144] and medium effects.[134,138,141] The relaxation times for $[^{95}MoO_4]^{2-}$ and $[^{97}MoO_4]^{2-}$ are 6.5 and 840 ms, respectively (1 *M* aqueous solution, 3 *M* NaCl, pH 9–12)[141]; ^{97}Mo linewidths for $K_2[MoO_4]$ in H_2O vary from 35 Hz at low concentrations to 65 Hz in 2.6 *M* solutions.[138] With increasing H_3O^+ concentration, relaxation times decrease,[138,141] which has been ascribed by Vold and Vold[141] to the formation of a protonated $[MoO_3(OH)]^- \cdot xH_2O$ {or $[MoO_2(OH)_3(H_2O)]^-$} and (at pH <7.7) condensation to $[Mo_7O_{24}]^{6-}$. In a more recent report it has been shown that the two signals corresponding to $[MoO_4]^{2-}$ or $[MoO_3(OH)]^-$, and $[Mo_7O_{24}]^{6-}$ (+40 ppm) can be observed simultaneously at pH values around 6.[142] The ^{95,97}Mo relaxation measurements by Vold also allowed for a detailed analysis of the kinetics involved in the $[MoO_4]^{2-}/[MoO_3(OH)]^-$ exchange.

An initial survey of polyoxomolybdates (including some heteropolymolybdates) has appeared. The compounds resonate in the range of −18 to +140 ppm. Linewidths (16–716 Hz) mainly reflect site symmetries. In two species ($[Mo_7O_{24}]^{6-}$ and β-$[Mo_8O_{26}]^{4-}$), two signals corresponding to two different Mo sites are observed. For α-$[Mo_8O_{26}]^{4-}$, a single sharp resonance arises.[145]

The molybdenum shift in $M_2[MoO_4]$ depends upon the counterion M.[138] While there is an increase by ca. +2 ppm with increasing concentration for M = Na and K, ν_0 decreases for M = Li (−6.5 ppm) and NH_4 as in $[^{53}CrO_4]^{2-}$ and $[^{55}MnO_4]^-$ for which an overall decrease of ν_0 is observed independent of M. The solvent isotope effect (H_2O/D_2O) amounts to −1.0(0.2) ppm (D_2O) and is comparable to that of $[CrO_4]^{2-}$ and $[MnO_4]^-$.[138] An intramolecular isotope effect was observed for $[Mo(^{32}S)_3{}^{34}S]^{2-}$ (0.09(1) ppm[143]) and in the series $[M(^{16}O)_{4-n}{}^{18}O_n]^{2-}$ (n = 0–4) (0.25 ppm per ^{18}O).[144] The ^{95}Mo nucleus is shielded upon introducing the heavier nucleus, following the general pattern outlined by Lauterbur.[146] In the solid state ($Ag_2[MoO_4]$ powder), the ^{95}Mo and ^{97}Mo resonances are shifted to higher frequencies (+82 ppm) relative to $[MoO_4]^{2-}$ in solution.[138] T_1 values are 1 s and 40 ms, respectively; no quadrupole splitting was observed. A small NQCC (^{95}Mo) was found for MoF_6 [0.14(2) MHz, asymmetry parameter $\eta = 0.3$].[147]

The occurrence of Mo in enzymatically active metalloproteins, such as the FeMo-cofactor of the N_2-reducing nitrogenase system present in several procaryotes, has prompted investigations into the ^{95}Mo characteristics of sulfur-containing molybdenum compounds. The simplest class, $[MoO_{4-n}S_n]^{2-}$ ($n = 1$–4) was first studied by Lutz and his collaborators,[134,148] and later by Enemark,[149] and O'Connor and Wedd,[137] who also extended this series to the selenium homologues. In thiomolybdates, low-lying, empty molecular orbitals of π character[150] are available for low-energy electronic transitions, which in turn induce high paramagnetic deshielding contributions. These low-lying, π-accepting orbitals make the thiomolybdates excellent ligands for coordination to metal ions. Examples for species with highly deshielded ^{95}Mo nuclei are $[(CNM^I)(\mu\text{-}S)_2MoS_2]^{2-}$ (M^I = Cu: +1863; Ag: +1921)[137,137b] and $[(PhSCu)MoS_4]^{2-}$ (+1903).[137c] In this context a ^1H and ^{95}Mo ENDOR study on the nitrogenase FeMo-cofactor, yielding an upper limit for NQCC = 0.28 MHz, is of some interest.[151]

The high-frequency shift on replacing O by S is 500–660 ppm per sulfur atom in $[MoO_{4-n}S_n]^{2-}$ and cupromolybdates, and 650 to more than 1000 ppm per Se.[137,137b,137c] The resonance positions are solvent dependent,[137] indicating participation of solvent molecules in the coordination sphere of Mo. This high-frequency shift is also observed in octahedrally coordinated complexes. Examples are $[MoO_2(Et_2NO)_2]$ (−174 ppm), $[MoOS(Et_2NO)_2]$ (+545), and $[MoS_2(Et_2NO)_2]$ (+1235)[149b]; $[Mo_2O_4(edta)]^{2-}$ (+609) and $[Mo_2O_2S_2(edta)]^{2-}$ (+982)[149c]; $[MoO_2\{OO\}\{ON\}]$ (+34) and $[MoO_2\{OS\}\{ON\}]$ (+222)[149]; $[MoO_2\{ONNO\}]$ (+8) and $[MoO_2\{SNNS\}]/[MoO_2\{NS\}_2]$ (+452 to +666)[137,149,149b]; $[MoO_2(ONEt_2)_2]$ (−169), $[MoOS(ONEt_2)_2]$ (+544), $[MoOSe(ONEt_2)_2]$ (+865).[149d] The shift differences on exchanging O for N are only minor as exemplified by $[MoO_2\{OO\}\{ON\}]$ (+34) and $[MoO_2\{OO\}_2]$ (+58 to −45),[149] and in the series $[MoO_3L]$ with L a tetradentate ligand coordinating via N and O (+76 to +63 ppm).[137,141,149b] A slight substituent effect in MoO^{2+} Schiff base complexes has been reported.[149a]

Paralleling the inverse chalcogen dependence, an inverse halogen dependence of ^{95}Mo shielding has been noted in octahedral complexes of the composition $[MoO_2(L)X]$ [L = HB(3,5-Me_2-pyrazolyl)$_3^-$; X = Br^-, Cl^-, NCS^-, OMe^-; shielding increases in this order].[149f]

Linewidths have been reported for various complexes containing the MoO_2[149–149e] and MoO_3[141] group, but no apparent correlation with the complex geometry or the coordinating functions was observed. This is also true for a number of molybdenum cluster compounds containing Mo in the oxidation states II, IV, and V.[149c] The shift ranges for these species are as follows:

$Mo^{IV}(aq) = [Mo_3O_4(H_2O)_9]^{4+}$ and related complexes: +990 to +1162

$Mo^{V}(aq) = [Mo_2O_4(H_2O)_6]^{2+}$ and related complexes: +535 to +635
(for thio compounds see above)

$[Mo_2^{II}(O_2CR)_4]$ (R = Pr, CF_3): +3730 and +4027 ppm

In some cases, e.g., $[Mo_2O_4(edta)]^{2-}$, two different Mo sites can be detected, with two distinct ^{95}Mo NMR signals. The dinuclear, ligand-bridged complex $[Mo_2O_5L'_2]$ (L′ = 1 − n − 2 − α-hydroxibenzylimidazole) which, according to an x-ray analysis, contains five- and six-coordinate Mo, shows two signals of $\delta = +76$ and +47 ppm.[151a]

5.2.2. Molybdenum Carbonyl and Related Compounds

Shift values reported for $[Mo(CO)_6]$ are around -1857 ppm.[132,133,136,136a,152,153] The resonance position is slightly solvent dependent,[136,152] and the temperature gradient is 0.29 ppm/deg.[136a] Second-order quadrupole splitting was observed in the $[Mo(CO)_6]$ powder spectrum [NQCC = 0.021(0.0023) MHz, $\eta < 0.1$; $\delta = -1852$ ppm].[154] From relaxation studies on $[Mo(CO)_6]$ in $CDCl_3$, NQCC(^{95}Mo) = 0.133(15) and NQCC(^{97}Mo) = 1.43(16) MHz have been calculated. Quadrupole relaxation is exclusively caused by rotational reorientation of a permanent quadrupole (and not by momentarily induced electric field gradients arising from molecule collisions or vibrational imbalances).[131a] Practically the same shift values as for $[Mo(CO)_6]$ have been reported for the ^{95}Mo and ^{97}Mo resonances of some $[Mo(CO)_{6-n}(PF_3)_n]$ complexes.[131] Whereas well-resolved multiplets are observed in the ^{95}Mo spectra, the ^{97}Mo resonances are broadened to an extent where coupling is unresolved {estimated $^1J(^{97}Mo\text{–}^{31}P)$ in $[Mo(CO)_5PF_3]$: 254 Hz}. Other phosphine derivatives of $Mo(CO)_6$ show similar {$[Mo(CO)_5P(OMe)_3]$: -1864[136a,152–153a]} or distinctly lower (less negative) shift values {$[Mo(CO)_5PPh_3]$: -1721,[153] -1740[132]} than the parent carbonyl. The first systematic investigation into the family of $[Mo(CO)_{6-n}(PR_3)_n]$ complexes (R = OMe, Ph) was carried out by Jaitner and Wohlgenannt.[153] Steric and electronic effects, which have been discussed in detail for carbonyl–phosphinevanadium complexes,[96,99a] have also been observed in series of $[Mo(CO)_nPPh_2XR]$ complexes ($n = 5, 4$; X = O, S, NH; R = alkyl, aryl, $SiMe_3$).[153d–g] The large nephelauxetic effect of $SbPh_3$ induces a higher shielding ($[Mo(CO)_5SbPh_3]$: -1862[132]); the poor π-accepting ability of N-donors such as pyridine gives rise to substantial deshielding ($[Mo(CO)_5py]$: -1387 ppm).[132,136,153b] This low-field shift is also observed in $[Mo(L)(CO)_n]^-$ ($n = 5, 4$; $L = S_2P(OEt)_2$, S_2CNR_2, and S_2COEt).[153h] Still lower shielding is found in the related N_2 complexes *trans*-$[Mo(N_2)_2(R_2PCH_2CH_2PR_2)_2]$ (R = substituted phenyl; $\delta = -774$ to -798[153a,153b]) and *trans*-$[Mo(N_2)_2(PMe_2Ph)_4]$ ($\delta = -447$ ppm[153b]). As in vanadium complexes, the introduction of a nitrosyl ligand also leads to a deshielding of the metal nucleus. Examples are *trans*-$[Mo(NO)(NCMe)(dppe)_2]^+$ (-553 ppm)[153b] (dppe = $Ph_2PCH_2CH_2PPh_2$) and $[Mo(NO)_2(ox)_2]^{2-}$ ($+201$ ppm).[153i] The low metal shielding induced in low-valent metal complexes by ligands bonding via nitrogen appears to be quite a general aspect in transition metal NMR.

As in bis(phosphine) chelate complexes of other metals, there is a maximum shielding in five-ring structures and a minimum in chelate-4 rings.[136a] ^{95}Mo shielding decreases with increasing degree of substitution and is larger in *trans*- and meridional than in *cis*- and facial complexes.[136,136a,153–153b] While this latter effect contrasts with findings on $[^{51}V(CO)_{6-n}L_n]^-$, most of the shielding trends in $[Mo(CO)_{6-n}L_n]$ parallel those in carbonyl complexes of ^{51}V, ^{55}Mn, ^{93}Nb, and ^{183}W and may be interpreted in terms of variations of the paramagnetic deshielding contribution to the overall shielding.

Roughly, the three molecular parameters which can influence the paramagnetic term are related to molecular properties in the following manner (k' is the orbital reduction factor):

$$\Delta E \leftrightarrow \text{ligand strength, local symmetry}$$

$$\langle r^{-3} \rangle \leftrightarrow \text{nephelauxetic effect}$$

$$k' \text{ or } c(\mathrm{M}) \leftrightarrow \text{covalency of the metal–ligand bond}$$

One has to keep in mind, of course, that these quantities are interdependent. A strongly π-interacting phosphine will give rise to a larger ΔE(HOMO–LUMO) and small c(M) values. The factors act in the same direction and diminish the paramagnetic term. The importance of the $\delta/\Delta E$ dependence has been demonstrated by Gray and Kraihanzel[153e]: $\delta(^{95}Mo)$ for a series of $[Mo(CO)_{5/4}(PPh_2Z)_{1/2}]$ complexes correlates linearily with the lowest energy absorption band in the uv–visible spectrum. Small ΔE are expected for electronically weak phosphines. A bulky phosphine is a weakly interacting ligand due to nonideal bond lengths and often unfavorable bond angles. Deshielding effects similar to those for bulky phosphines have been observed in strained chelate-ring structures.[7,96] On the other hand, a weak ligand (be it primarily weak on electronic grounds or indirectly from its steric requirement) will induce increased π delocalization into π^*(CO) orbitals, thus increasing r and decreasing $\langle r^{-3} \rangle$ and the paramagnetic term. This effect seems to be a nonnegligible factor in $[Mn(CO)_{6-n}(PZ_3)_n]^+$ complexes. Where π interaction is weak, σ influences may also come in; this will be outlined in connection with neutral pentacarbonylmanganese complexes (Section 6.1.2). Soft ligands, i.e., those with a large nephelauxetic effect and a tendency to form typically covalent bonds (I, Se, SbR_3, SnR_3, and, to a lesser extent, Br, S, AsR_3, GeR_3) will influence the shielding mainly via $\langle r^{-3} \rangle$ c(M), and give rise to an increase of shielding on going down the groups, as normally observed in *low*-valent transition metal complexes ("normal" halogen dependence[2]). Finally, it should be noted that, since any ligand L is a weaker π acceptor than CO (except, perhaps, PF_3; Section 4.2.2), increasing replacement of CO will increasingly weaken the overall strength of the ligand field. Further, with successive lowering of the symmetry, the degeneracy of the metal–*nd* system is lifted and the number of allowed transitions increases; a net deshielding is achieved.

Coupling constants $^1J(^{95}Mo–^{31}P)$, obtained from ^{95}Mo[131,132,136,136a,152–153b] and ^{31}P NMR spectra,[139,140] vary between 284 $\{[Mo(CO)_5PF_3]\}$ and 114 Hz $\{$*trans*-$[Mo(CO)_4(PEtPh_2)_2]\}$ and correlate with the electronegativity EN of the substituents attached to phosphorus and the CO stretching frequency, thus with the π-acceptor strength of the phosphine. Molybdenum–phosphorus coupling constants of the same order of magnitude have also been obtained for a couple of $[XHgMo(Cp)(CO)_3P(OMe)_3]$ complexes.[158a] $^1J(^{95}Mo–^1H)$ in $[Mo_2H(CO)_{10}]^-$ ($\delta = -1937$ ppm) is 15 Hz,[133,153a] $^1J(^{95}Mo–^{13}C)$ in $[Mo(CO)_6]$ 68 Hz,[155] and $^1J(^{95}Mo–^{14}N)$ in $[Mo(Cp)(CO)_2NO]$ 46 Hz.[155a] The magnitudes of J are usually explained on the basis of the Fermi contact contribution. To account for the J/π-acceptor correlation, indirect influences upon the terms $|S(O)|^2$ (s-densities at the coupling nuclei) and $\sigma(s)$ (valence-s electron density in the bond) have to be considered. Large coupling constants in a given series such as $[(CO)_nM(PZ_3)]$ may then be caused by (i) enhanced $|S(O)|^2_M$ due to an increase of the effective charge on the metal nucleus through withdrawal of dπ electrons; (ii) enhanced $|S(O)|^2_P$ induced by charge delocalization from phosphorus to strongly electronegative Z; (iii) synergically increased phosphorus-to-metal donation [increase of $\sigma(s)$]. The synergic increase in s-character by s-electron delocalization into orbitals representing the bond to the more electropositive bonding partner Y in a molecule Y–E–Z has been referred to as Bent's rule[156] and is discussed in the context of carbonylphosphinemetal complexes [where Y is the electropositive $\{(CO)_nM\}$ moiety and E is phosphorus] in Ref. 157.

Masters, O'Connor *et al.*[133] have obtained narrow signals ($W_{1/2}$ around 6 Hz) for the complexes $[Mo(\pi\text{-arene})(CO)_3]$ (-1907 to -2034) and $[Mo(\eta^6\text{-}C_7H_8)(CO)_3]$ (-1684 ppm). The sharp lines are indicative of idealized C_{3v} symmetry where, in the narrowing of the point-charge model, quadrupolar relaxation becomes ineffective.[215] This work is a good demonstration of how even minor changes in the second coordination sphere (substituents at the phenyl ring) are easily detectable by metal NMR:

[(Mes) $Mo(CO)_3$] −1907	[(*m*-Xyl) $Mo(CO)_3$] −1971	[(*p*-Xyl) $Mo(CO)_3$] −1979
[(*o*-Xyl) $Mo(CO)_3$] −1988	[(Tol) $Mo(CO)_3$] −2034	

It appears that this is also an ordering of increasing Mo-arene bond strength. Enhanced steric hindrance in the mesitylene compound leads to a decrease of Mo-arene interaction, hence enhanced paramagnetic deshielding. The high shielding of the ^{95}Mo nucleus in these complexes as a *class* may well be due to favorable π-electron delocalization into accepting π-type orbitals associated with the arene, increasing ΔE and producing a similar (but more pronounced) effect as that responsible for high nuclear shielding in phosphine complexes containing strongly π-accepting phosphines [PF_3, $P(OR)_3$]. A substituent effect is also apparent in [$Mo(CO)_5PPh_2XR$] (X = O, NH; R = 1–4 carbon alkyl)[153d] and in [$Mo(N_2)_2\{R_2P(CH_2)_2PR_2\}$], where R is C_6H_4R' and R′ = H, 4-Me, 3-Me, 4-Cl, 4-CF_3, 4-OMe.[153b] Alyea *et al.*[153c] have shown that $\delta(^{95}Mo)$ can be used to distinguish between P-bonded and η^6-arene bonded $P(aryl)_3$ in $Mo(CO)_nPR_3$ complexes. The shifts are −1739 (R = *m*-Tol) and −1746 (*p*-Tol), and −1823 (Mes) and −1999 (*p*-Tol), respectively.

Carbonylcyclopentadienylmolybdenum and related complexes have been studied by several workers.[132,152,153i,155a,158–158b,159–159b] These complexes cover the astonishing range of ca. 2000 ppm, thus exhibiting similarities with [$VCp(CO)_4$] and {$WCp(CO)_3$} derivatives. Molybdenum–mercury complexes of the general formula [$XHgMo(Cp)(CO)_2L$] [X = Cl, Br, I, S_2COEt, S_2CNEt_2, $S_2P(OEt)_2$; L = CO, $P(OMe)_3$] with a strongly covalent Mo–Hg bond (^{199}Hg evidence) fall within this range (−1706 to −1939 ppm).[158a] Here, sulfur-containing complexes have the lowest ^{95}Mo-shielding, but there is a "normal" halogen dependence in the series [$Mo(Cp)X'(CO)_3$] (X′ = Cl: −863, Br: −956, I: −1248) and [$Mo(Cp)(NO)_2X'$] (X′ = Cl: −852, Br: −883, I: −985).[153i] On going from [$Mo\{(3,5\text{-}Me_2\text{-}pyrazolyl)_3BH\}(CO)_2NO$] to the corresponding thionitrosyl complex, shielding substantially decreases (−743 and −192 ppm, respectively).[158b] Also, in [$(XHg)Mo(Cp)(CO)_3$], the halogen attached to Hg induces a (slight) inverse effect upon $\delta(^{95}Mo)$.[152,158] A normal halogen dependence was also found for [$Mo(X)(CO)_5$]$^-$ (X = Cl: −1513 to −1534, depending on the counterion; X = Br: −1540, −1551; X = I: −1660 ppm).[132,153g] The complex [$MoCp(NN)(CO)_2$]$^+$ [{NN} = 3-RC_5NH_4, R = CH=N−CH(Ph)(Me)] was investigated by Enemark, Brunner *et al.*[159] The compound contains both a chiral metal and carbon center, and the two diastereomers are clearly distinguished by two sharp resonances at −154 and −168 ppm. More generally, in related complexes, $\Delta\delta$ between diastereomers are between 6 and 32 ppm.[159a]

For the benzyl derivatives [$MoCp(CO)_3CH_2C_6H_4R$], there is again a substituent effect on ^{95}Mo shielding. Shielding increases in the series R = CF_3 < Cl < F < Me, OMe, 2-Me, 2-OMe < 3-Me, 3-OMe < 4-Me, 4-OMe < 2,4,6-Me_3, flanked by R = 4-CF_3 (−1551) and the mesityl substituent (−1599 ppm). The $\delta(^{95}Mo)$ correlate linearly with the Hammett substituent constants σ^-.[159b]

5.3. Tungsten

^{183}W ($N = 14.27\,\%$) is a spin-1/2 nucleus of very low receptivity (1.04×10^{-5}), and until recently, most of the data collected for ^{183}W had been obtained by $^{19}F\{^{183}W\}$, $^{1}H\{^{183}W\}$, $^{31}P\{^{183}W\}$, and $^{1}H\{^{183}W, ^{31}P\}$ double and triple resonance techniques,[160,161] exceptions being direct ^{183}W measurements carried out on solid WO_3[162] and liquid samples of WCl_6, WF_6, and $[WO_4]^{2-}/H_2O$.[163] In the last few years, however, a rapidly increasing number of direct observations have been reported on polytungstates,[79d,79e,137b,164–168j] including a ^{183}W NMR study in paramagnetic heteropolytungstates.[168f] Direct observations of $[WCp(PR_2)(CO)_3]$,[169] $[W(CO)_6]$, and $[W(CO)_5PR_3]$ (the latter with the use of $[Cr(acac)_3]$ as a relaxation reagent),[169a] a $^{183}W\{^{1}H\}$ and $^{183}W\{^{31}P\}$ INEPT study on *cis*-$[W(CO)_4(PPhMe_2)_2]$[170] and a ^{183}W two-dimensional INADEQUATE on polyoxotungstates[79e] have also appeared. Apart from these new developments, McFarlane's early investigations of carbonyltungsten[160] (supplemented very recently by new data[136a,159c,159d]) and fluorotungsten complexes[161] remain an abundant and valuable source for shielding data and their interpretation.

On the other hand, reports of coupling constants involving the ^{183}W nucleus, mainly obtained from the satellite systems of ^{1}H, ^{13}C, and ^{31}P resonances, have increased to the extent that a comprehensive treatment would clearly be beyond the slope of this chapter. As far as carbonyltungsten complexes are concerned, $^{1}J(^{183}W–^{13}C)$ are listed and discussed, e.g., in Refs. 171–172a, and $^{1}J(^{183}W–^{31}P)$ in Refs. 173, 173a. Examples of one-bond coupling constants in other than carbonyl complexes are $[WOCl_2(PR_3)_2]$, $J(WP) = 418–442$[174]; $[WX(NNH_2)py(PMe_2Ph)_3]Y$, $J(WP) = 288–347$[175]; $[W\{P(OMe)_3\}_6]$, $J(WP) = 449$[176]; $[WCl_2H_2(PMe_2Ph)_4]$, $J(WP) = 166$ and 190[176a]; $[W(CN)_7]^{5-}$ and $[W(CN)_7H]^{4-}$, $J(WC) = 57.3$ and 62.9, $J(WH) = 34.0$[177]; $[WH_2\{P(OMe)_3\}_5]$ and $[WH_4\{P(OMe)_3\}_4]$ $J(WH) = 18.6$ and 29.9[175]; $[W_2(\mu\text{-}H)(OiPr)_7]_2$, $J(WH) = 96$ Hz.[178]

Chemical shifts, collected in Figure 7, encompass a range of ca. 6200 ppm. For better comparison with ^{53}Cr and ^{95}Mo shift data, $\delta(^{183}W)$ are given relative to $[WO_4]^{2-}$, although this standard has not commonly been used. The shift of WF_6 (the reference usually employed until ca. 1980) relative to $[WO_4]^{2-}$ is -1120 ppm; the

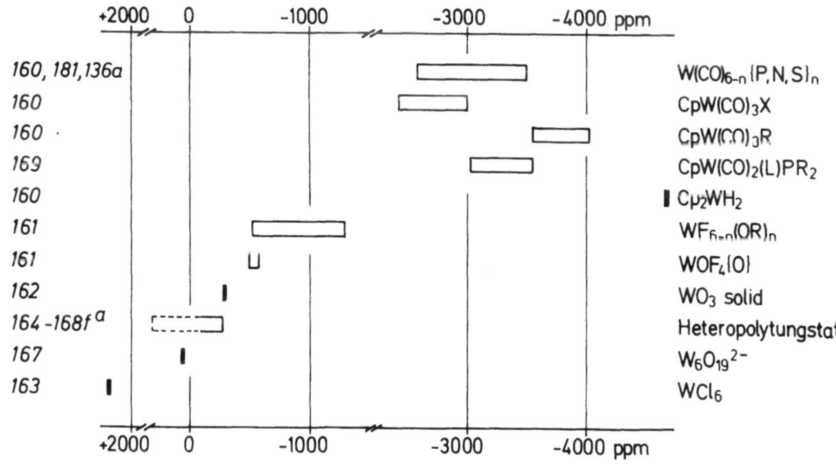

Figure 7. ^{183}W chemical shifts relative to $[WO_4]^{2-}/H_2O$. Note that there are two gaps in the shift scale. [a]The dashed range for heteropolytungstates encompasses paramagnetic species.[168f]

shift of $[W(CO)_6]$ relative to $[WO_4]^{2-}$ is -3505 ppm.[136a] The standard which has been adopted by the groups working on polytungstates is a ca. 1 *M* $Na_2[WO_4]$ solution in D_2O at pD 9, and this reference is recommended for future studies.

The earliest direct solution study on ^{183}W was undertaken by Banck and Schwenk,[163] who investigated WCl_6/CS_2, WF_6 (neat), and $K_2[WO_4]/H_2O$. Relaxation times for WF_6 are $T_1 = 4.2(0.6)$ s and $T_2 = 2.1(0.4)$ s; $^1J(^{183}W-^{19}F) = 40.7(1.5)$ Hz. The $^{183}W-^{17}O$ coupling in $[W(^{16}O)_3{}^{17}O]^{2-}$ is not resolved, but an upper limit of 10 Hz is estimated from a fit between the observed and calculated spectrum. The resonances for $[WO_4]^{2-}$ and WCl_6 show a temperature dependence of 0.16 and 0.34 ppm/degree. Substituting the oxo by the sulfido ligand gives rise to a deshielding effect of ca. 850 ppm per S^{2-}.[137b]

WF_6 is included in a comprehensive study on fluoro and oxofluoro complexes $[WF_{6-n}(OR)_n]$ ($R = Me$, Ph; $n = 5$) and $WOF_4 \cdot L$ [$L = OMe_2$, dmso, $OP(OMe)_2Me$].[161] ^{183}W shielding decreases as F replaces the less electronegative OR, the more so in $[WF_5(OPh)]$ than in $[WF_5(OMe)]$, paralleling similar trends observed in $[M(CO)_{6-n}(PZ_3)_n]$ complexes, where Z = F, OMe, OPh.

Isopolytungstates studied to date are $[W_6O_{19}]^{2-}$ ($\delta = +58.9$),[167] $[W_{10}O_{22}]^{4-}$ (-43.1), and $[H_2W_{12}O_{42}]^{10-}$ (-95.7 and -128.1 ppm),[168g] but there is a considerable body of data on heteropolytungstates.[164–168j] The spectra show extremely sharp signals, tungstophosphates showing coupling [$J(POW) \approx 1$ Hz];[167,168] cf. also $J(WF) = 32$ Hz, $J(WW) = 5$ and 22 Hz in α-$[H_2FW_{12}O_{39}]^{5-}$; $J(WW) = 8$ and 20 Hz in β-$[SiW_{12}H_4O_{40}]$.[165] More generally, $^2J(WOW)$ are around 9 Hz for edge coupling and 20 Hz for corner coupling.[168d] For vanadotungstates, the corresponding $J(WOV)$ are 9–15.5 and 20–30 Hz.[79e] The direct connectivities between tungsten sites as furnished by two-dimensional ^{183}W NMR have been used for an unequivocal assignment of the tungsten sites in $[PW_{11}O_{39}]^{7-}$, $[SiW_{11}O_{39}]^{8-}$, and $[PPbW_{11}O_{39}]^{5-}$.[168h] δ and J have widely been employed for structure assignments (α/β form of the Keggin structure) and in the discussion of intramolecular interconversion mechanisms.[164–167] In $[MW_{12}O_{40}]^{n-}$ the highly symmetric (T_d) α-structure shows one single ^{183}W resonance, whereas the β-form (C_{3v}) should exhibit three peaks in the intensity ratio 1:2:1. In $[(PhPO_3)_2W_5O_{15}]^{4-}$, the ^{183}W resonance (-155.9 ppm) shows coupling to two equivalent ^{31}P nuclei ($J = 1.95$ Hz). The equivalence of the W atoms is achieved by ring pseudorotation.[166] Another interesting feature of the $[MW_{12}O_{40}]^{n-}$ tungstates is the increase of shielding in the order $M = P/n = -3$ (-98.8 [-86.7]), $M = Si/n = -4$ (-103.8 [-92.1]), and $M = B/n = -5$ (-130.4) (in D_2O at pD 1.5 from Ref. 168 and [square brackets] in CD_3CN/DMF from Ref. 167), attributed to increasing overall electron density on the tungsten ions.[167] ^{183}W spectra may show up to nine resonances (α_2-$[P_2W_{17}O_{61}]^{10-}$ [164] and $[P_2W_{15}Zn_2(H_2O)O_{56}]_2^{16-}$ [168c]), corresponding to structurally distinct ($8 \times 2 + 1$) W atoms, spaced by as much as 114 ppm, very impressively showing the sensitivity of ^{183}W shielding to even minor environmental changes. In α-$[TiCpPW_{11}O_{39}]^{4-}$, all six resonances were assigned to six specific tungsten sites[167] on the basis of a charge distribution scheme, but this method has been criticized. An alternative method, based on the analysis of the coupling pattern produced by the ^{183}W isotopomers, has given a convincing assignment for $[TiW_{11}PO_{40}]^{5-}$ [168d] Some of the earlier assignments were criticized also by Acerete *et al.*,[168b,168j] who have developed a theoretical interpretation of variations in $\delta(^{183}W)$. For Keggin-type tungstates, a linear $\Delta E/\delta$ correlation is established.

Comparison of the diamagnetic $Ce^{4+}[PW_{11}O_{39}]_2^{10-}$ ($\delta = -175.7$ to -105.8) and the paramagnetic $Ce^{3+}[PW_{11}O_{39}]_2^{11-}$ (−114.8 to +349 ppm) reveals a substantial paramagnetic high-frequency shift of the ^{183}W resonances, indicating that electron density is transferred from the central Ce^{3+} to W via oxygen bridges.(168f)

Three classes of low-valent carbonyltungsten complexes have been investigated: $[W(CO)_{6-n}L_n]$ (where L usually is a phosphine),(136a,160) $[WCp(CO)_3R]$ (R = halogen, H, D, alkyl, $SnMe_3$, $WCp(CO)_3$),(160) and $[WCp(PR_2)(CO)_3]$ plus derivatives.(169) Shielding decreases as CO is successively replaced by a weaker π-accepting ligand (Section 5.2.2; compare, e.g., $[W(CO)_6]$: −3505 and $[W(CO)_5PMePh_2]$: −3324 ppm(136a)), and the ^{183}W nucleus is more shielded in *trans*- than in *cis*-$[W(CO)_4L_2]$, as in $[Mo(CO)_4L_2]$ and $[Mn(CO)_4L_2]^+$, but in contrast to $[V(CO)_4L_2]^-$. For $[W(CO)_5(PZ_3)]$, shielding steadily decreases as the electronegativity of Z decreases and the overall bulk of PZ_3 increases (Section 5.2.2). Further, a chelate-ring effect for $\delta(^{183}W)$ in *cis*-$[W(CO)_4$ diphos] (diphos = dppm: −2978, dppe: −3313, dppp: −3179, dppb: −3138 ppm)(159c) comparable to that in *cis*-$[M(CO)_4$ diphos$]^-$ (M = V,(96) Nb(114)) is observed. There is also a close parallel between W and Mo shielding. The slope of a plot of $\delta(^{183}W)/\delta(^{95}Mo)$ for $[M(CO)_5PZ_3]$ is 1.7,(136a) reflecting a greater intrinsic shielding sensitivity of the nucleus ^{183}W and/or reduced π back-bonding in tungsten complexes, leading to an increase of $\langle r^{-3}\rangle_{nd}$.

The ratio $^1J(^{183}W-^{31}P)/^1J(^{95}Mo-^{31}P)$ in corresponding complexes is 1.76/1 and can be attributed to the larger $|S(O)|^2(^{183}W)$ of the Fermi contact term.(136a) A similar ratio, for the δ values, (1.67(3)) has been noted for $[MO_{4-n}S_n]^{2-}$ and $[(CNM^I)(\mu\text{-}S)_2MS_2]$ (M = Mo, W; M^I = Cu, Ag).(137b) There is a relation between the electronegativities of Z and $J(WP)$,(179,180) and $J(WP)$ have been correlated with CO stretching frequencies and interpreted on the basis of qualitative MO considerations.(180a) In *trans*-$[W(CO)_4(PR_3)_2]$, $^1J(WP)$ is larger by a factor of approximately 1.2 than in the *cis*-isomer(180b); similar results have been reported for carbonylphosphinevanadium complexes.(180c) For discussion, see Section 5.2.2. $^1J(W-PF_3) = 485$ Hz(157b) is the largest scalar W–P interaction constant known.

In cyclopentadienylcarbonyltungsten complexes of the general type $[WCp(CO)_3R]$, ^{183}W shielding is very sensitive to the nature of R. The high shielding imposed as a hydride ligand enters the coordination sphere is in accord with similar observations for $[VCp(CO)_4]/[VCp(H)(CO)_3]^-$,(98) $[NbCp(CO)_4]/[NbCp(H)(CO)_3]^-$,(116) and $[MoCp(CO)_3]/[MoCp(H)(CO)_3]$.(158) The very highest shielding has been detected for the dihydrido compound $[WCp_2H_2]$.(160) The isotope shift on going from $[WCp(H)(CO)_3]$ to $[WCp(D)(CO)_3]$ is −10 ppm(160); cf. −6 ppm for $[NbCp(H)(CO)_3]^-/[NbCp(D)(CO)_3]^-$.(116) For R = Cl, Br, I, a "normal" halogen dependence (Sections 5.2.2 and 6.1.2) is observed. $^1J(^{183}W-^{119}Sn)$ in $[WCp(CO)_3SnMe_3]$ is −150 Hz.(160)

6. GROUP VIIb

6.1. Manganese

The relatively high receptivity of ^{55}Mn (0.175) and its medium size quadrupole moment (0.55×10^{-28}) make this nucleus a favorable one for NMR detection. There is, however, only a restricted number of diamagnetic complexes: except for the permanganate ion and a series of $[Mn(NO)_3L]$(181a) and isonitrile complexes,(181b) the only compounds which have been investigated are carbonylmanganese (+I, O, and −I) complexes. Shift data, relative to aqueous K[MnO_4], are represented in Figure 8. Not included in Figure 8 are recent data on fac-[$MnX(CO)_3LL$], where LL is a phosphinothioformamide or -thioformimido ester ($\delta = +70$ to -730 ppm),[190b] on [$\overline{MnSC(NMe)P}Ph_2(CO)_{4-n}(PR_3)_n$] ($\delta = -1090$ to -1530 ppm),[196] and on [$Mn(CNR)_6$]$^+$ (-1292 to -1380 ppm).[181b]

6.1.1. Permanganate

This ion has been subject to several studies focusing on isotope effects,[144,182] concentration and counterion effects,[182] solvent effects,[183] relaxation phenomena,[123a,144,184,185,185a] and solid state investigations.[186,187] There is the usual low-frequency shift of the metal resonance as the concentration of an aqueous solution is increased. This shielding effect depends upon the counterion (Li < Mg < K ≪ Ba) and is -0.8 ppm for 0.08 *M* K[MnO_4] relative to K[MnO_4] at infinite dilution.[182] The temperature dependence is +0.3 Hz/degree, and the solvent isotope effect (D_2O) is -0.76 ppm,[182] which is rather a large effect considering that the Mn^{+7} ion is isolated from the water molecules by its tetrahedral O^{2-} sphere. An intramolecular isotope effect was observed on substituting ^{16}O for ^{18}O (-0.59 ppm per ^{18}O).[144] Relaxation times for [$Mn(^{16}O)_{4-n}(^{18}O)_n$]$^-$ are around 12 s,[144] which compares to $T_1 = 19$ s for [$Mn^{17}O_4$]$^-$ [184] and the relaxation times measured by Britt and Yen.[185] T_1 in aqueous [MnO_4]$^-$ is practically independent of concentration, counterion, and the solvent (H_2O or D_2O); $(T_1)_0$ is 83(4) ms.[123a] The significant relaxation rate indicates that [MnO_4]$^-$ undergoes some intramolecular deformation which introduces an electric field gradient at the nucleus. This might be related to ion pair formation or to distortions caused by collisions with other molecules (see

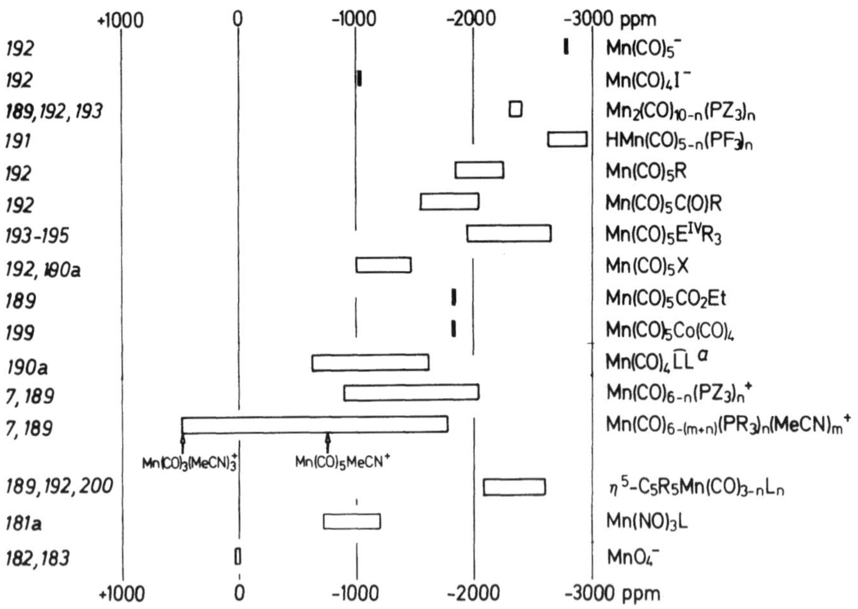

Figure 8. ^{55}Mn chemical shifts relative to [MnO_4]$^-$/H_2O. For abbreviations see Figure 2. ^{a}LL = $R_2PC(S)NHR$, $R_2PC(NR)S^-$, $R_2P(S)C(NR)S^-$, ZCS_2^- (Z = O, OR, SR, NR_2, $P(S)R_2$, $AsPh_2$, Ph).

also the discussion in the Tc section). The coupling constant $^1J(^{55}Mn-^{17}O)$ is 30(3) Hz.[78,144,184]

Rapid sample rotation of polycrystalline K[MnO_4] at 54°44′ removes broadening arising from dipole–dipole interaction, shielding anisotropy and scalar coupling, leading to a considerably sharpened central line ($-1/2 \rightarrow +1/2$ transition). From the second-order quadrupole splitting which, under these conditions is easily observed, an NQCC value of 1.60(2) MHz was calculated,[186] and this again demonstrates the non-ideal T_d symmetry at the ^{55}Mn nucleus. In aqueous solution, NQCC is 1.29 MHz.[123a]

6.1.2. Manganese Carbonyl and Related Compounds

These complexes cover the whole ^{55}Mn shift range with the hydrido complexes as usual at the low-frequency margin. The first shift value for (solid) $[Mn(CO)_6]^+$ was reported by Spiess and Sheline (−1500 ppm)[188] and agrees much better with that reported recently for $[Mn(CO)_6]Cl \cdot HCl$ dissolved in THF (−1445)[189] than the apparently erroneous $\delta(^{55}Mn)$ for $[Mn(CO)_6]^+$/acetone suggested by Nakano (−935).[190] Whereas in the hexacarbonyls of V, Nb, Mo, and W, replacement of CO results in a decrease of metal shielding (unless the substituent is PF_3), derivatives of $[Mn(CO)_6]^+$ may exhibit both a substantial increase and decrease of ^{55}Mn shielding.[7,189] Introduction of MeCN always produces a shift to high frequency with fac-$[Mn(CO)_3(MeCN)_3]^+$ (+490 ppm)[7] being the least shielded diamagnetic manganese compound.

In $[Mn(CO)_4LL]$ and $[MnBr(CO)_3LL]$ complexes, where LL is a heterosubstituted dithioformato, thioformamido or thioformamide ligand coordinating via **S, S′** or **S, P**, the range of the ^{55}Mn shift has been found to be a valuable tool for the classification of the complexes. Shielding decreases in the sequences $[Mn(CO)_4LL] > [MnX(CO)_3LL]$ (S, P coordination, chelate four-rings); $[Mn(CO)_4S, P] > [Mn(CO)_4S, S]$ (chelate four-rings); $[Mn(CO)_4S, S]$ (five-ring) > $[Mn(CO)_4S, S]$ > (four-ring).[190a,190b] A normal halogen dependence (Cl < Br < I) is observed for $[MnX(CO)_3LL]$,[190b] and a substituent effect has been noted for the complexes $[Mn(CO)_4S_2CZ]$, where Z is NR_2, OR, Ph, $AsPh_2$, PR_2, and SR.[190a] The complexes $[Mn(CO)_{4-n}(PR_3)_nS, P]$ [chelate four-rings; R = OMe, OPh, Me, Et, Ph; $n = 1, 2$ (R = OMe only)] exhibit deshielding of the ^{55}Mn nucleus with an increasing number of phosphine ligands coordinated to Mn.[196] In a few cases,[7,189,196] $^1J(^{55}Mn-^{31}P)$ coupling has been observed in $[Mn(CO)_5L]^+$ [L = $PPh(OPh)_2$: 270, PPh_2H: 170] and *cis*-$[Mn(CO)_4\{P(OMe)_3\}_2]^+$ (330 Hz), fac-$[Mn(P, S)(CO)_3P(OPh)_3]$ (400 and 140 Hz), *cis*-$[Mn(P, S)(CO)_2\{P(OMe)_3\}_2]$ [triplet of doublets: $J(Mn-P(OMe)_3) = 405$, $J(Mn-P, S) = 165$ Hz at 333 K].[190c] $^1J(^{55}Mn-^{13}C)$ coupling constants in $[Mn(CNR)_6]^+$, obtained from the six peak multiplets in the ^{13}C NMR, are 111–121 Hz. The ^{13}C isotope effect in the ^{55}Mn spectra amounts to −0.9 ppm; the temperature gradient for $\delta(^{55}Mn)$ between 238 and 298 K is 0.67 ppm/deg.[181b]

Phosphines, owing to their empty 3*d* orbitals, are efficient π acceptors and influence the paramagnetic deshielding term via the π-interacting *e* level (d_{xz}, d_{yz}) of the Mn–3*d* set under C_{4v} symmetry.[189] If the ligand is comparable to CO in its π-acceptor ability, as is PF_3, or if the ligand–metal interaction does not allow for much π contribution (as is the case for the halogens and $E^{IV}R_3$, E^{IV} = Si, Ge, Sn, Pb), σ influences upon the HOMO-LUMO separation (vide infra) may become predominant for variations of δ; examples are $[MnH(CO)_{5-n}(PF_3)_n]$,[191] [Mn(Hal)

Table 8. ^{55}Mn NQR and NMR Data on Selected Manganese Complexes

Complex	|NQCC| ^{a} (MHz)	Reference	$W_{1/2}$ ^{b} (Hz)	δ ^{b} (ppm)	Reference
$[Mn_2(CO)_{10}]$	3.05	201	83	−2227 to −2331	189/192/193
$[MnCp(CO)_3]$	64.28	188/202	10,040	−2225 to −2280	7/192
$Na[Mn(CO)_5]$	65.0	204	10,585	−2780	192
$[MnH(CO)_5]$	45.66 [0.0134]	203	4,340	−2630/−2578	192/191
$[MnCl(CO)_5]$	13.86 [0.0433]	188	331	−1005	192
	16.2^c	195			
$[MnBr(CO)_5]$	17.46 [0.0421]	188	688	−1160	192
	17.5^c	195			
$[MnI(CO)_5]$	19.85 [0.0242]	188	1,013	−1485	192
	16.8^c	195			
$[Mn(GePh_3)(CO)_5]$	24.64 [0.2026]	205		−2430	195
	23.47 [0.3739]				
$[Mn(SnPh_3)(CO)_5]$	18.33	205	3000/2610	−2510/−2610	193/195
$[Mn(PbPh_3)(CO)_5]$	12.04 [0.1155]	205		−2350	195
	10.6 [0.11]				
$[Mn(SnCl_3(CO)_5]$	7.3	205	330	−1910 to −2024	193–195
$[Mn(CO)_5Co(CO)_4]$	19.94	199		−1840	199
$[Mn(CO)_5Re(CO)_5]$	8.67	206			

^{a} Nuclear quadrupole coupling constants and asymmetry parameters [in brackets] are from single crystal studies. NQCC(^{55}Mn) values have also been predicted from partial quadrupole splitting parameters for $[Mn(CO)_5L]$ (L = CN, CNR, PPh_3, NCS, N_3, $SbPh_3$,[198] and H[195]), and calculated from $W_{1/2}$ of the NMR resonance for $[MnH(CO)_{5-n}(PF_3)_n]$ (n = 0–5).[191]
^{b} Linewidths and chemical shifts are given for solutions of the respective complexes.
^{c} Calculated from partial quadrupole splitting parameters.

$(CO)_5]$,[192] $[MnE^{IV}R_3(CO)_5]$.[193–195] Attempts have been made to explain the σ influence phenomenologically from linear correlations between $\delta(^{55}Mn)$ and the NQCC.[188,191,195] Although there are some interesting parallels between these two parameters (Table 8), one should keep in mind that (i) the NQCC is a ground state variable whereas δ depends on electronic transitions between ground and excited states, (ii) the NQCC depends on σ *and* π interactions (via the electric field gradient tensor), and (iii) the NQCC cannot wholly represent $W_{1/2}$, the latter being also very sensitive to solvent/solute interaction as quantified by the molecular τ_c (or rotational τ_q) correlation time. An alternative approach has been initiated by Onaka *et al.*,[193] based on the assumption that, where π interaction is minimal, the b_1 and a_1 MOs $[\sigma^*(d_{x^2-y^2}), \sigma^*(d_{z^2})]$ are those influenced by the ligand, and further that it is the metal-d contribution to these levels rather than ΔE^{-1} and $\langle r^{-3}\rangle$ factors which is responsible for variations in the paramagnetic term. Thus, $c(M)$ decreases (covalency increases) in the sequence Cl, Br, I (the increase in covalency is also reflected in the NQCC) and so does the overall shielding in $[Mn(Hal)(CO)_5]$. Consequently, there is a correlation between $\delta(^{55}Mn)$ and $\nu(CO)$ in that a decrease of $\nu(CO)$ (corresponding to an increase of ligand-to-Mn σ donation) is accompanied by an increase in shielding.[193,195] A linear relation also exists between $\delta(^{55}Mn)$ and the ^{119}Sn Mössbauer isomer shift of $[Mn(SnR_{3-n}X_n)(CO)_5]$ and is thought to support the σ thesis.[194]

Nakano[190] and Nakatsuji *et al.*[197] have calculated dia- and paramagnetic terms for $[MnX(CO)_5]$ (X = Hal, H, CN, Me), $[Mn(CO)_6]^+$, and $[MnO_4]^-$. A qualitative agreement is found between the calculated paramagnetic σ(para) term and the observed shielding. As expected, the local diamagnetic terms are practically constant (for X = Hal compare Table 6 in Section 4.1.2). According to these calculations, the main contribution to σ(para) arises from the *d–d* excitations. Overall shielding increases with increasing π-donating ability and, via d_σ, with increasing hardness, the two partly counteracting effects giving rise to an ordering H < CN < Me < Cl < Br < I.

Linewidths usually vary in a random manner and correlations between $W_{1/2}$ and δ are accidental except where τ_q is approximately constant for molecules of comparable size and shape. This appears to be the case for [MnX(CO)$_5$] (X = H, Me, Hal),(192) for which a rotational correlation time of 1.3 ps was calculated from the slope of a $W_{1/2}$/NQCC graph (including [Mn$_2$(CO)$_{10}$] and [MnCp(CO)$_3$]).(198) In pursuing former work by Bancroft,(195) Ireland and co-workers(198) estimated partial quadrupole splittings (pqs) for CO and L in [Mn(CO)$_5$L] applying the Mössbauer pqs scale for the isoelectronic [Fe(CO)$_5$L]$^+$ series. Linewidths have been claimed to reflect σ influences only, of the ligand L,(188) π-influences only,(198) or σ plus π influences,(195) and this again illustrates the present state of uncertainty in understanding this parameter.

Quite interestingly, in [Mn(NO)$_3$L] [L – P(*t* Bu)$_{3-n}$(E^{IV}Me$_3$)$_n$, CO, THF], the $\delta(^{55}Mn)$ values correlate linearily with the lowest *d–d* transition in the electronic absorption spectra.(181a)

6.2. Technetium

The isotope used in studying technetium is ^{99}Tc (half-life 5×10^5 yr). The nuclear properties ($I = 9/2$, $s = 0.375$, $Q = -0.19 \times 10^{-28}$ m^2) are quite favorable for NMR detection. Several reports on ^{99}Tc NMR, dealing with aqueous solutions of the pertechnate anion, have appeared. Kidd(125) obtained a ^{99}Tc resonance in 0.63 *M* Na[TcO$_4$]. A NQCC value of 7.2 MHz was calculated from $W_{1/2} = 29$ Hz and $\tau_c = 8.1$ ps. This gives a nuclear electric field gradient $eq = 0.33 \times 10^{15}$ esu/cm^3, which was employed to calculate "rationalized" nuclear quadrupole moments Q for ^{51}V, ^{53}Cr, ^{97}Mo, ^{185}Re, ^{189}Os (and ^{55}Mn) in T_d environments. An interesting feature is that the *eq* in the tetraoxoanions are remarkably uniform, although they are probably not absolute values (see below). Thornback,(207) Tarasov,(208) Schrobilgen,(209) and co-workers have studied aqueous [NH$_4$][TcO$_4$] solutions and found that $W_{1/2}$ for the ^{99}Tc resonances are ca. 3 Hz between 280 and 298 K (10 Hz at 350 K). This is considerably less than in Na$^+$[TcO$_4$]$^-$ and suggests ion-pairing phenomena (leading to a reduction of the T_d symmetry) as a contribution to the linewidth in aqueous sodium pertechnate. It has been shown by NQR measurements on perrhenates that the distortional effect induced by polarization by the cation is more effective for Na$^+$ than for the large NH$_4^+$ by a factor of 2.55.(210)

The natural abundance ^{17}O NMR gave the expected ten-line spectrum with $^1J(^{17}O\ ^{99}Tc) - 131.6$ Hz,(207) confirmed also by ^{17}O and ^{99}Tc NMR measurements in ^{17}O and ^{18}O-enriched H$_2$O (133.3(5)(208) and 131.4 Hz(209)). For the species [Tc^{17}O(^{16}O)$_n$(^{18}O)$_{3-n}$]$^-$, a ^{99}Tc isotope shift of –0.43(1) ppm per ^{18}O(78a,208) and –0.22 ppm per unit mass(209) was found. The effective field gradient at the ^{99}Tc nucleus in [TcO$_4$]$^-$ calculated from $W_{1/2} = 2.7$ Hz, $\tau_c = 8.1$ ps, and $Q = -0.19(5) \times 10^{-28}$ m^2 is $-0.15 \cdot 10^{15}$ esu,(209) which is about half of that estimated by Kidd from $W_{1/2} = 29$ Hz and $Q = 0.3 \times 10^{-28}$ m^2 (125).

Schrobilgen and co-workers have also investigated several oxofluoro technetium compounds ([TcO$_3$F]: $\delta = +43.7$ ppm relative to [TcO$_4$]$^-$, $W_{1/2} = 23$ Hz, $^1J(^{17}O–^{99}Tc) = 139.8$ Hz; [{TcO$_2$F$_2$}$_2\mu$–O]: $\delta = +245.9$, $W_{1/2} = 135$, $^1J(^{17}O–^{99}Tc) =$

87(5), $^1J(^{19}F\text{–}^{99}Tc) = 259$; $[TcO_2F_3]$: $\delta = +396.3$, $W_{1/2} = 375$; the latter two being tentative assignments) and $[TcO_3][AsF_6]$ ($\delta = +160.7$, $W_{1/2} = 670$), and further extended the ^{99}Tc shift scale to *trans*-$[TcO_2(CN)_4]^{3-}$ ($\delta = +806.0$, $W_{1/2} = 642$) and $[TcH_9]^{2-}$ [$\delta = -3672$, $W_{1/2} = 22$, $^1J(^1H\text{–}^{99}Tc) = 24$].[209] The decrease of shielding as O^{2-} is replaced by F^- contrasts with what one observes in $[WF_{6-n}(OR)_n]$[161] and $[VOF_3]/[VO(OR)_3]$ complexes[90] and appears to be one of the rare examples where a "normal" shielding/electronegativity relation is found in a d^0 system.

Tarasov, Buslaev, and co-workers observed a nine-line pattern for polycrystalline $K[TcO_4]$, arising from first-order quadrupole perturbation. The NQCC, calculated for $\eta = 0$ from the spacing between two symmetrically disposed satellites (217 kHz) is 5.2(2) MHz,[212] and so exceeds the values for $K[MnO_4]$ (1.6) and $Na_3[VO_4]$ (1.05 MHz).

6.3. Rhenium

The two rhenium isotopes ^{185}Re (N = 37.07%) and ^{187}Re (62.93%) have medium receptivities (0.049 and 0.086), a nuclear spin of 5/2, and large Q values (2.8 and 2.6×10^{-28} m^2), and thus impose great problems as to their NMR detection. $[ReO_4]^-$,[211] $[ReS_4]^-$ [217] and $[Re(CO)_6]^+$ [189] have been investigated. The half-height widths for ^{185}Re (which resonates at lower frequency) and ^{187}Re are 14.2 and 13.5 (0.8 *M* $Na[ReO_4]$), and 2.6 and 2.25 kHz {0.3 *M* $[Re(CO)_6]Cl \cdot HCl$ in MeCN}, respectively. In the hexacarbonyl, the rhenium nuclei are substantially more shielded ($\delta = -3400$ ppm) than in $[ReO_4]^-$ ($\delta = 0$) and $[ReS_4]^-$ ($\delta = +3500$ ppm; the δ value reported in Ref. 218 is in error); the shift difference is virtually the same as observed for $[^{183}WO_4]^{2-}$ vs. $[^{183}W(CO)_6]$.[136a] $W_{1/2}$ of aqueous perrhenate solutions is independent of the concentration (0.25–1.2 *M*) and does not change on addition of NaCl. Ion-pairing is excluded as a line-broadening effect for this reason, and the large $W_{1/2}$ are thought to be caused by the formation of short-lived collision complexes, with instantaneous distortion of the tetrahedron. On the other hand, line widths decrease by an order of magnitude in organic solvents and with $[Ph_4P]^+$ as the counter-ion.[217]

REFERENCES

1. Kidd, R. G.; Goodfellow, R. J. *NMR and the Periodic Table*; Harris, R. K.; Mann, B. E., Eds.; Academic: London, 1978; Chap. 8.
2. Kidd, R. G. *Annual Reports on NMR Spectroscopy*; Webb, G. A., Ed.; Academic: London, 1980; Vol. 10A, pp. 1–79.
3. Rinaldi, P. L.; Levy, G. C.; Choppin, G. R. *Rev. Inorg. Chem.* 1980, **2**, 53.
4. Wehrli, F. W. *Annual Reports on NMR Spectroscopy*; Webb, G. A., Ed.; Academic: London, 1979; Vol. 9, pp. 125–219.
5. Rehder, D. *Bull. Magn. Reson.* 1982, **4**, 33.
5a. Pletnev, R. N.; Gubanov, V. A.; Fotiev, A. A. *YaMR v Oksidnykh Soyedineniyakh Vanadiya*; Nauka: Moscow, 1979.
6. Rehder, D. *Magn. Reson. Rev.* 1984, **9**, 125.
7. Rehder, D.; Bechthold, H.-Ch.; Keçeci, A.; Schmidt, H.; Siewing, M. *Z. Naturforsch.* 1982, **37b**, 631.
8. Melson, G. A.; Olzanski, D. J.; Rahimi, A. Z. *Spectrochim. Acta* 1977, **33A**, 301.
9. Lutz, O. *Phys. Lett.* 1969, **29A**, 58.
9a. Haid, E.; Köhnlein, D.; Kössler, G.; Lutz, O.; Messner, W.; Mohn, K. R.; Nothaft, G.; van Rickelen, B.; Schich, W.; Steinhauser, N. *Z. Naturforsch.* 1983, **A38**, 317.

9b. Tarasov, V. P.; Kirakosyan, G. A.; Trots, S. V.; Buslaev, Yu. A.; Panyushkin, V. T. *Koord. Khim.* 1983, **9**, 205.
10. Kirakosyan, G. A.; Tarasov, V. P. *Koord. Khim.* 1982, **8**, 261.
11. Buslaev, Yu. A.; Kirakosyan, G. A.; Tarasov, V. P. *Koord. Khim.* 1980, **6**, 361.
11a. Buslaev, Yu. A.; Kirakosyan, G. A.; Tarasov, V. P. *Dokl. Akad. Nauk SSSR* 1982, **264**, 1405.
11b. Bougeard, P.; Mancini, M.; Sayer, B. G.; McGlinchey, M. J. *Inorg. Chem.* 1985, **24**, 93.
11c. Mancini, M.; Bougeart, P.; Burns, R. C.; Mlekuz, M.; Sayer, B. G.; Thompson, J. I. A.; McGlinchey, M. J. *Inorg. Chem.* 1984, **23**, 1072.
12. Pfadenhauer, E. H.; McCain, D. C. *J. Phys. Chem.* 1970, **74**, 3291.
13. Buslaev, Yu. A.; Petrosyans, S. P.; Tarasov, V. P.; Chagin, V. I. *Russ. J. Inorg. Chem.* 1974, **19**, 1790.
14. Pisaniello, D. L.; Lincoln, S. F.; Williams, E. H. *Chem. Commun.* 1978, 1047.
15. Tarasov, V. P.; Buslaev, Yu. A. *J. Magn. Reson.* 1977, **25**, 197.
16. Huttig, G. F.; Danschen, W. *Monatsh. Chem.* 1951, **82**, 796.
17. Levy, G. C.; Rinaldi, P. L.; Bailey, J. T. *J. Magn. Reson.* 1980, **40**, 167.
18. Brun, E.; Oeser, J.; Staub, H. H.; Telschow, C. G. *Phys. Rev.* 1949, **76**, 1528.
19. Kronenbitter, J.; Schwenk, A. *J. Magn. Reson.* 1977, **25**, 147.
20. Adam, R. M.; Fazakerley, G. V.; Reid, D. G. *J. Magn. Reson.* 1979, **33**, 655.
21. Haßler, C.; Kronenbitter, J.; Schwenk, A. *Z. Phys. A* 1977, 117.
21a. Evans, W. J.; Meadows, J. H.; Kostka, A. G.; Closs, G. L. *Organometallics* 1985, **4**, 324.
22. Pinkerton, A. A.; Earl, W. L. *J. Chem. Soc. Dalton Trans.* 1979, 1347.
22a. Evans, W. J.; Meadows, J. H.; Hanussa, T. P. *J. Am. Chem. Soc.* 1984, **106**, 4454.
23. Lutz, O.; Oehler, H. *J. Magn. Reson.* 1980, **37**, 261.
24. Krüger, H.; Lutz, O.; Oehler, H. *Phys. Lett.* 1977, **62A**, 131.
25. Reuben, J. *J. Am. Chem. Soc.* 1975, **97**, 3822.
26. Scott, M. E.; Marsh, H. C.; Harris, D. L.; Hiskey, R. G.; Koehler, K. A., cited in Ref. 4.
27. Reuben, J.; Luz, Z. *J. Phys. Chem.* 1976, **80**, 1357.
28. Reuben, J. *J. Phys. Chem.* 1975, **79**, 2154.
29. Nakamura, K.; Kawamura, K. *Bull. Chem. Soc. Jpn.* 1971, **44**, 330.
30. Smith, Jr., L. S.; McCain, D. C.; Wertz, D. L. *J. Am. Chem. Soc.* 1976, **98**, 5125.
31. Vijverberg, C. A. M.; Peters, J. A.; Kieboom, A. P. G.; van Bekkum, H. *Recl. Trav. Chim. Pays-Bas* 1980, **99**, 287. Vijverberg, C. A. M.; Peters, J. A.; Kieboom, A. P. G.; van Bekkum, H. "Abstract of Papers," 22nd. Int. Conf. Coord. Chem., Budapest, Hungary, Aug. 1982; p. 154.
32. Evans, D. F.; Missen, P. H. *J. Chem. Soc. Dalton Trans.* 1982, 1929.
33. McCain, D. C. *J. Inorg. Nucl. Chem.* 1980, **42**, 1185.
34. Rinaldi, P. L.; Khan, S. A.; Choppin, G. R.; Levy, G. C. *J. Am. Chem. Soc.* 1979, **101**, 1350.
35. Bleaney, B.; Harley, R. T.; Ryan, J. F.; Wells, M. R.; Wiltshire, M. C. K. *J. Phys.* 1978, **11C**, 3059.
36. Schirber, J.; Weaver, H. T.; Ginley, D. S. *J. Appl. Phys.* 1978, **49**, 1469.
37. Bleaney, B.; Stephen, A. G.; Walker, P. J.; Wells, M. R. *Proc. R. Soc. Lond.* 1980, **A370**, 131.
38. Konov, I. S.; Teplov, M. A. *Paramagn. Rezon.* 1976, **12**, 77.
39. Bol'shakov, I. G.; Teplov, M. A. *Fiz. Tverd. Tela* 1979, **21**, 2142.
40. Kojima, K.; Tadamiki, T.; Kamigaichi, T. *J. Phys. Soc. Jpn.* 1980, **49**, 2419.
41. Hihara, T.; Kojima, K.; Kamigaichi, T. *J. Phys. Soc. Jpn.* 1981, **50**, 1499.
42. Tagirov, M. S.; Teplov, M. A.; Livanova, L. D. *Fiz. Tverd. Tela* 1979, **21**, 3488.
43. Egorov, V. A.; Livanova, L. D.; Tagirov, M. S.; Teplov, M. A. *Fiz. Tverd. Tela* 1980, **22**, 2836.
44. Bleaney, B.; Stephen, A. G.; Walker, P. J.; Wells, M. R. *Proc. R. Soc. London* 1981, **A376**, 235.
45. Bleaney, B.; Robinson, F. N. H.; Wells, M. R. *Proc. R. Soc. London* 1978, **A362**, 179.
46. Bleaney, B.; Loftus, K. V.; Rosenberg, H. M. *Proc. R. Soc. London* 1980, **A372**, 9.
47. Aminov, L. K.; Tagirov, M. S.; Teplov, M. A. *Zh. Eksp. Teor. Fiz.* 1982, **82**, 224.
48. Egorov, A. V.; Eremin, M. V.; Tagirov, M. S.; Teplov, M. A. *Zh. Eksp. Teor. Fiz.* 1979, **77**, 2375.
49. Aminov, L. K.; Tagirov, M. S.; Teplov, M. A. *Zh. Eksp. Teor. Fiz.* 1980, **79**, 1322.
50. Bleaney, B.; Wells, M. R. *Proc. R. Soc. London* 1980, **A370**, 131.
50a. Bleaney, B.; Pasman, J. H. T.; Wells, M. R. *Proc. R. Soc. London* 1983, **A387**, 75.
51. Deacon, G. B.; Ravety, W. D.; Vince, D. G. *J. Organomet. Chem.* 1977, **135**, 103.
52. Williams, R. J. P. *Struct. Bond.* 1982, **50**, 79.
52a. Le Bail, H.; Chachaty, C.; Rigny, P.; Bougon, R. *J. Phys. (Paris)* 1983, **44**, L 1017.
53. Kanellakopulos, B. *Gmelin Handbuch der Anorganischen Chemie*; Uran Suppl. Vol. A5, Springer-Verlag: Berlin, 1982; p. 251.
54. Glebov, V. A. *Radiokhimiya* 1979, **21**, 793.

55. Seppelt, K.; Barlett, N. *Z. Anorg. Allgem. Chem.* 1977, **436**, 122.
56. Ursu, I.; Demco, D. E.; Simplaceanu, V.; Valcu, N.; Ilie, N. *Proc. Congr. Ampere* 18th, 1974, 533.
57. Ursu, I.; Demco, D. E.; Simplaceanu, V.; Valcu, N. *Rev. Roum. Phys.* 1977, **22**, 705.
58. Jeffries, C. D. *Phys. Rev.* 1953, **92**, 1262
59. Frisch, R. C.; Forman, R. A. *J. Chem. Phys.* 1968, **48**, 5187.
60. Kidd, R. G.; Matthews, R. W.; Spinney, H. G. *J. Am. Chem. Soc.* 1972, **94**, 6686.
60a. Hao, N.; Sayer, B. G.; Denes, G.; Bickley, D. G.; Detellier, C.; McGlinchey, M. J. *J. Magn. Reson.* 1982, **50**, 50.
60b. Dormond, A.; Fauconet, M.; Leblanc, J. C.; Moise, C. *Polyhedron* 1984, **3**, 897.
61. Dean, P. A.; Evans, D. F. *J. Chem. Soc. A* 1967, 698.
61a. Benn, R.; Rufinska, A. *J. Organomet. Chem.* 1984, **273**, C 51.
62. Sayer, B. G.; Hao, N.; Denes, G.; Bickley, D. G.; McGlinchey, M. J. *Inorg. Chim. Acta* 1981, **48**, 53.
63. Sayer, B. G.; Thompson, J. I. A.; Hao, N.; Birchall, T.; Eaton, D. R.; McGlinchey, M. J. *Inorg. Chem.* 1981, **20**, 3748.
63a. Yasuoka, H.; *et al. J. Magn. Magn. Mater.* 1983, **31–34**, 287, 343, 345.
63b. Bleaney, B.; de Oliveira, A. C.; Wells, M. R. *J. Magn. Magn. Mater.* 1983, **31–34**, 31, 713; *J. Phys.* 1982, **C15**, 5305.
63c. Mastikhin, V. M.; Lapina, O. B.; Lyakhova, V. F.; Simonova, L. G. *React. Kinet. Catal. Lett.* 1981, **17**, 109.
64. Oldfield, E.; Kinsey, R. A.; Montez, B.; Ray, T.; Smith, K. A. *Chem. Commun.* 1982, 254.
64a. Ganapathy, S.; Schramm, S.; Oldfield, E. *J. Chem. Phys.* 1982, **77**, 4360.
65. Meadows, M. D.; Smith, K. A.; Kinsey, R. A.; Rothgeb, T. M.; Skarjune, R. P.; Oldfield, E. *Proc. Natl. Acad. Sci. U.S.A.* 1982, **79**, 1351.
66. Harrison, A. T.; Howarth, O. W. *J. Chem. Soc. Dalton Trans.* 1986, 1405. Hayden, Y. T.; Edwards, J. O. *Inorg. Chim. Acta* 1986, **114**, 63.
67. Becker, K. D.; Berlage, U. *J. Magn. Reson.* 1983, **54**, 272.
67a. Schiemann, J.; Weiss, E. *J. Organomet. Chem.* 1983, **255**, 179.
68. Tarasov, V. P.; Privalov, V. I.; Buslaev, Yu. A. *Dokl. Akad. Nauk. SSSR* 1981, **257**, 678; *Bruker Report* 1981, 10.
68a. Lutz, O.; Messner, W.; Mohn, K. R.; Kroneck, P. *Z. Phys.* 1981, **A300**, 111.
68b. Bleaney, B.; Gregg, J. F.; Wells, M. R. *J. Phys.* 1982, **15**, L 349.
69. Paulsen, K.; Rehder, D. *Z. Naturforsch.* 1982, **37a**, 139. Priebsch, W.; Rehder, D. *Inorg. Chem.* 1985, **24**, 3058.
70. Habayeb, M. A.; Hileman, Jr., O. E. *Can. J. Chem.* 1980, **58**, 2115.
71. Gierer, A.; Wirtz, K. *Z. Naturforsch.* 1953, **8a**, 532.
72. Basler, W.; Lechert, H.; Paulsen, K.; Rehder, D. *J. Magn. Reson.* 1981, **45**, 170.
73. Preuss, F.; Ogger, L. *Z. Naturforsch.* 1982, **37b**, 957.
74. Rehder, D.; Wieghardt, K. *Z. Naturforsch.* 1981, **36b**, 1251.
74a. Preuss, F.; Towae, W.; Kruppa, V.; Fuchslocher, E. *Z. Naturforsch.* 1984, **B39**, 1510.
74b. Gillespie, R. J.; Rao, U. R. K. *J. Chem. Soc. Chem. Commun.* 1983, 422.
74c. Howell, J. A. S.; Moss, K. C. *J. Chem. Soc. A*, 1971, 270.
74d. Konovalova, A. A.; Bainova, S. V.; Kopanev, V. D.; Buslaev, Yu. A. *Koord. Khim.* 1982, **8**, 1211, 1364.
74e. Schabert, M.; Pausewang, G. *Z. Anorg. Allgem. Chem.* 1983, **506**, 169.
74f. Hoch, M.; Rehder, D.; Weidemann, C. *Inorg. Chim. Acta* 1984, **92**, L5.
74g. Hanich, J.; Krestel, M.; Müller, U.; Dehnicke, K.; Rehder, D. *Z. Naturforsch.* 1984, **B39**, 1686.
75. Heath, E.; Howarth, O. W. *J. Chem. Soc. Dalton Trans.* 1981, 1105.
76. Petterson, L.; Hedman, B.; Andersson, I.; Ingri, N. *Chem. Script.* 1983, **22**, 254. Petterson, L.; Hedman, B.; Nenner, A. M.; Andersson, I. *Acta Chem. Scand.* 1985, **A39**, 499.
77. Tytko, K.-H.; Baethe, G.; Hirschfeld, E. R.; Mehmke, K.; Stellhorn, D. *Z. Anorg. Allgem. Chem.* 1983, **503**, 43.
78. Lutz, O.; Nepple, W.; Nolle, A. *Z. Naturforsch.* 1976, **31a**, 1046.
78a. Tarasov, V. P.; Privalov, V. I.; Buslaev, Yu. A. *Z. Naturforsch.* 1984, **B39**, 1230.
79. Kato, R.; Kobayashi, A.; Sasaki, Y. *Inorg. Chem.* 1982, **21**, 240.
79a. Howarth, O. W.; Jarrold, M. *J. Chem. Soc. Dalton Trans.* 1978, 503.
79b. Kazanskii, L. P.; Fedotov, M. A.; Ptushkina, M. M.; Spitsyn, V. I. *Dokl. Akad. Nauk. SSR* 1978, **240**, 117.
79c. Maksimovskaya, R. I.; Fedotov, M. A.; Mastikhin, V. M.; Kuznetsova, L. I.; Matseev, K. I. *Dokl. Akad. Nauk. SSR* 1978, **240**, 422.

79d. Hamalker, S. P.; Leparuldo, M. A.; Pope, M. T. *J. Am. Chem. Soc.* 1983, **105**, 4286.
79e. Domaille, P. J. *J. Am. Chem. Soc.* 1984, **106**, 7677.
80. Allerhand, A. *J. Chem. Phys.* 1970, **52**, 2162.
81. France, P. W. *J. Magn. Reson.* 1979, **34**, 585.
82. Baugher, J. E.; Taylor, P. C.; Oja, T.; Bray, P. J. *J. Chem. Phys.* 1969, **50**, 4914.
83. Gubanov, V. A.; Luzukova, N. I.; Pletnev, R. N. *Zh. Neorg. Khim.* 1978, **23**, 655.
84. Pletnev, R. N.; Lazukova, N. I.; Gubanov, V. A. *Zh. Fiz. Khim.* 1977, **51**, 2359.
85. Basuev, G. V.; Pletnev, R. N.; Slebuchin, W. K.; Lisson, V. N.; Shveykin, G. P. *Izvest. Neorg. Mater.* 1978, **14**, 711.
86. Pletnev, R. N.; Lisson, W. N.; Belyukov, Yu. N.; Miller, J. J. *Inst. Khim. Akad. Nauk. SSR Ural'skij Filial Trudy.* (Sverdlovsk) 1975, **32**, 32.
87. Gossard, A. C. *Phys. Rev.* 1966, **149**, 246.
88. Burkert, P. K.; Eckel, M. F.; Fritz, H. P. *Proc. Congr. Ampere* 17th., 1972, 263.
89. Rehder, D.; Paulsen, K.; Basler, W. *J. Magn. Reson.* 1983, **53**, 500.
90. Rehder, D. *Z. Naturforsch.* 1977, **32b**, 771.
90a. Buslaev, Yu. A.; Kopanev, V. D.; Konovalova, A. A.; Bainova, S. V.; Tarasov, V. P. *Dokl. Akad. Nauk. SSR* 1978, **243**, 583.
91. Howarth, O. W.; Hunt, J. R. *J. Chem. Soc. Dalton Trans.* 1979, 1388.
92. O'Donnell, S. E.; Pope, M. T. *J. Chem. Soc. Dalton Trans.* 1976, 2290.
93. Talay, R.; Rehder, D. *Z. Naturforsch.* 1981, **36b**, 451.
94. Mühlbach, G.; Rausch, B.; Rehder, D. *J. Organomet. Chem.* 1981, **205**, 343.
95. Von Deuten, K.; Rehder, D.; Roose, W. *J. Organomet. Chem.* 1981, **214**, 71.
95a. Keçeci, A.; Rehder, D.; Roose, W.; Talay, R. *Chem. Ber.* 1982, **115**, 3257.
96. Rehder, D. *J. Magn. Reson.* 1980, **38**, 419.
97. Borowski, R.; Rehder, D.; von Deuten, K. *J. Organomet. Chem.* 1981, **220**, 45.
98. Puttfarcken, U.; Rehder, D. *J. Organomet. Chem.* 1980, **185**, 219.
99. Rehder, D.; Puttfarcken, U. *Z. Naturforsch.* 1982, **37b**, 348.
99a. Ihmels, K.; Rehder, D. *Chem. Ber.* 1985, **118**, 895; *Organometallics*, 1985, **4**, 1334 and 1340.
99b. Warnock, G. F. P.; Philson, S. B. P.; Ellis, J. E. *J. Chem. Soc., Chem. Commun.* 1984, 893. Warnock, G. F. P.; Ellis, J. E. *J. Am. Chem. Soc.* 1984, **106**, 5016.
99c. Ihmels, K.; Rehder, D.; Pank, V. *Inorg. Chim. Acta* 1985, **96**, L69. Hoch, M.; Rehder, D. *Inorg. Chim. Acta* 1986, **111**, L13.
100. Ihmels, K.; Rehder, D. *J. Organomet. Chem.* 1982, **232**, 151.
101. Näumann, F.; Rehder, D. *J. Organomet. Chem.* 1981, **204**, 411.
102. Herberhold, M.; Trampisch, H. *Z. Naturforsch.* 1982, **37b**, 614.
102a. Talay, R.; Rehder, D. *Inorg. Chim. Acta* 1983, **77**, L 175.
102b. Talay, R.; Rehder, D. *J. Organomet. Chem.* 1984, **262**, 25.
103. Schiemann, J.; Weiss, E.; Näumann, F.; Rehder, D. *J. Organomet. Chem.* 1982, **232**, 219.
103a. Näumann, F.; Rehder, D.; Pank, V. *Inorg. Chim. Acta* 1984, **84**, 117; *Z. Naturforsch.* 1984, **B39**, 1647, 1654.
104. Herberhold, M.; Trampisch, H. *Inorg. Chim. Acta* 1983, **70**, 143.
105. Il'in, E. G.; Tarasov, V. P.; Ershova, M. M.; Ermakov, V. A.; Glushkova, M. A.; Buslaev, Yu. A. *Koord. Khim.* 1978, **4**, 1370.
106. Kidd, R. G.; Spinney, H. G. *Inorg. Chem.* 1973, **12**, 1967.
107. Kidd, R. G.; Spinney, H. G. *J. Am. Chem. Soc.* 1981, **103**, 4759.
108. Tarasov, V. P.; Sinitsyna, S. M.; Kopanev, V. D.; Khlebodarov, V. G.; Buslaev, Yu. A. *Koord. Khim.* 1980, **6**, 1568.
109. Tarasov, V. P., Privalov, V. I.; Buslaev, Yu. A. *Zh. Strukt. Khim.* 1979, **20**, 274; Buslaev, Yu. A.; Kopanev, V. D.; Tarasov, V. P, *Chem. Commun.* 1971, 1175; Tarasov, V. P.; Privalov, V. I.; Buslaev, Yu. A. *Molec. Phys.* 1978, **35**, 1047.
110. Buslaev, Yu. A.; Tarasov, V. P.; Sinitsyna, S. M.; Khlebodarov, V. G.; Kopanev, V. D. *Koord. Khim.* 1979, **5**, 189.
111. Vladimiroff, T.; Malinowski, E. *J. Chem. Phys.* 1967, **46**, 1830.
112. Buslaev, Yu. A.; Il'in, E. G.; Kopanev, V. D.; Tarasov, V. P. *Zh. Strukt. Khim.* 1972, **13**, 930.
113. Gabuda, S. P.; Zil'berman, B. D.; Goncharuk, V. K. *Zh. Strukt. Khim.* 1978, **19**, 431.
113a. Devreux, F. *J. Phys.* 1982, **43**, 1489.
114. Bechthold, H.-Ch.; Rehder, D. *J. Organomet. Chem.* 1982, **233**, 215.
114a. Bachmann, K.; Rehder, D. *J. Organomet. Chem.* 1984, **276**, 117.

115. Herrmann, W. A.; Biersack, H.; Ziegler, M. L.; Weidenhammer, K.; Siegel, R.; Rehder, D. *J. Am. Chem. Soc.* 1981, **103**, 1692.

116. Näumann, F.; Rehder, D.; Pank, V. *J. Organomet. Chem.* 1982, **240**, 362.

116a. Pforr, I.; Näumann, F.; Rehder, D. *J. Organomet. Chem.* 1983, **258**, 189.

116b. Hoch, M.; Rehder, D. *Z. Naturforsch.* 1983, **B38**, 446.

117. Bechthold, H.-Ch.; Rehder, D. *J. Organomet. Chem.* 1981, **206**, 305.

118. Rehder, D.; Bechthold, H.-Ch.; Paulsen, K. *J. Magn. Reson.* 1980, **40**, 305.

119. Erich, L. C.; Gossard, A. C.; Hartless, R. L. *J. Chem. Phys.* 1973, **59**, 3911.

120. Bennett, L. H.; Budnick, J. I. *Bull. Amer. Phys. Soc.* (Ser. 2) 1959, **4**, 417.

121. Wada, S.; Nakamura, S.; Aoki, R.; Molinie, P. *J. Phys. Soc. Japn.* 1980, **48**, 786.

121a. Nishihara, H.; Scholz, G. A.; Naito, M.; Frindt, R. F.; Tanaka, S. *J. Magn. Magn. Mater.* 1983, **31–34**, 717.

121b. Nishihara, H.; Scholz, G. A.; Frindt, R. F. *Solid State Commun.* 1982, **44**, 507.

122. Egozy, Y.; Loewenstein, A. *J. Magn. Reson.* 1969, **1**, 494.

123. Epperlein, B. W.; Krüger, H.; Lutz, O.; Nolle, A.; Mayr, W. *Z. Naturforsch.* 1975, **30a**, 1237.

123a. Haid, E.; Köhnlein, D.; Kössler, G.; Lutz, O.; Schich, W. *J. Magn. Reson.* 1983, **55**, 145.

124. Le Dang, K.; Veillet, P.; Walker, P. J. *J. Phys. C* 1977, **10**, 4593.

124a. Ertmer, W.; Johann, U.; Mosmann, R. *Z. Phys.* 1982, **A309**, 1.

125. Kidd, R. G. *J. Magn. Reson.* 1981, **45**, 88.

126. Schmidt, H.; Ph.D. thesis; Hamburg, 1982.

127. Karnachev, A. S.; Klochan, V. A.; Kovtun, N. M.; Sanina, V. A.; Solov'ev, E. E.; Troitskii, G. A.; Khmara, V. M. *Fiz. Nizk. Temp.* 1978, **4**, 90.

127a. Karnachov, A. S.; Kovtun, N. M.; Kurkin, M. I.; Solov'ev, E. E.; Tkachenko, A. A. *Zh. Exsp. Teor. Fiz.* 1983, **85**, 224.

128. Kovtun, N. M.; Prokopenko, V. K.; Prokhorenko, Yu. I.; Shemyakov, A. A. *Fiz. Nizk. Temp.* 1978, **4**, 1477.

129. Fiegiel, H.; Kukucz, J.; Bombik, A. *J. Magn. Magn. Mater.* 1980, **15–18**, 701.

130. Nidolski, S.; Jedryka, E.; Piotrowski, K.; Wanklyn, B. *Phys. Status Solidi* 1978, **49a**, K193.

131. Bailey, J. T.; Clark, R. J.; Levy, G. C. *Inorg. Chem.* 1982, **21**, 2085.

131a. Brownlee, R. T.; O'Connor, M. J.; Shenan, P. P.; Wedd, A. G. *J. Magn. Reson.* 1985, **61**, 22.

132. Dysart, S.; Georgii, I.; Mann, B. E. *J. Organomet. Chem.* 1981, **213**, C10.

133. Masters, A. F.; Brownlee, R. T. C.; O'Connor, M. J.; Wedd, A. G. *Inorg. Chem.* 1981, **20**, 4183.

134. Lutz, O.; Nolle, A.; Kroneck, P. *Z. Naturforsch.* 1976, **31a**, 454.

135. Kroneck, P.; Lutz, O.; Nolle, A. *Z. Naturforsch.* 1980, **35a**, 226.

136. Alyea, E. C.; Lenkinski, R. E.; Somogyvari, A. *Polyhedron* 1982, **1**, 130; Alyea, E. C.; Somogyvari, A. *Chemistry and Uses of Molybdenum, Proc. Int. Conf. 4th*, Colden, Colorado 1982.

136a. Andrews, G. T.; Colquhoun, I. J.; McFarlane, W.; Grim, S. O. *J. Chem. Soc. Dalton Trans.* 1982, 2353.

137. Gheller, S. F.; Gazzana, P. A.; Masters, A. F.; Brownlee, R. T. C.; O'Connor, M. J.; Wedd, A. G. *Inorg. Chim. Acta* 1981, **54**, L131.

137a. Enemark, J. H. *Nitrogen Fixation*; Müller, A.; Newton, W. E., Eds.; Plenum Press: New York, 1983, pp. 329–339.

137b. Gheller, S. F.; Hambley, T. W.; Rodgers, J. R.; Brownlee, R. T. C.; O'Connor, M. J.; Snow, M. R.; Wedd, A. G. *Inorg. Chem.* 1984, **23**, 2519.

137c. Minelli, M.; Enemark, J. H.; Nicholson, J. R.; Garner, C. D. *Inorg. Chem.* 1984, **23**, 4384.

138. Kantt, W. D.; Krüger, H.; Lutz, O.; Maier, H.; Nolle, A. *Z. Naturforsch.* 1976, **31a**, 351.

139. Milbrath, D. S.; Verkade, J. G.; Clark, R. J. *Inorg. Nucl. Chem. Lett.* 1976, **12**, 921.

140. Andrews, G. T.; McFarlane, W. *Inorg. Nucl. Chem. Lett.* 1978, **14**, 215.

141. Vold, R. R.; Vold, R. L. *J. Magn. Reson.* 1975, **19**, 365.

142. Freeman, M. A.; Schultz, F. A.; Reilley, C. N. *Inorg. Chem.* 1982, **21**, 567.

143. Lutz, O.; Nolle, A. *Z. Phys.* 1977, **A282**, 157.

144. Buckler, K. U.; Hasse, A. R.; Lutz, O.; Miller, M.; Nolle, A. *Z. Naturforsch.* 1977, **32a**, 126.

145. Sheller, S. F.; Sidney, M.; Masters, A. F.; Brownlee, R. T. C.; O'Connor, M. J.; Wedd, A. G. *Aust. J. Chem.* 1984, **37**, 1825.

146. Lauterbur, P. C. *J. Chem. Phys.* 1965, **42**, 799.

147. Kind, R.; Virlet, J. *Proc. Congr. Ampere 19th*, Heidelberg, 1976, 427.

148. Lutz, O.; Nolle, A.; Kroneck, P. *Z. Naturforsch.* 1977, **32a**, 505.

149. Christensen, K. A.; Miller, P. E.; Minelli, M.; Rockway, T. W.; Enemark, J. H. *Inorg. Chim. Acta* 1981, **56**, L27.

149a. Alyea, E. C.; Topich, J. *Inorg. Chim. Acta* 1982, **65**, L95.
149b. Gheller, S. F.; Hambley, T. W.; Traill, P. R.; Brownlee, R. T. C.; O'Connor,. M. J.; Snow, M. R.; Wedd, A. G. *Aust. J. Chem.* 1982, **35**, 2183.
149c. Gheller, S. F.; Hambley, T. W.; Brownlee, R. T. C.; O'Connor, M. J.; Snow, M. R.; Wedd, A. G. *J. Am. Chem. Soc.* 1983, **105**, 1527.
149d. Minelli, M.; Enemark, J. H.; Wieghardt, K.; Hahn, M. *Inorg. Chem.* 1983, **22**, 3952.
149e. Capparelli, M. V.; Piggott, B.; Thorpe, S. D.; Wong, S. F.; Sheppard, R. N. *Inorg. Chim. Acta* 1985, **106**, 19.
149f. Minelli, M.; Yamanouchi, K.; Enemark, J. H.; Subramanian, P.; Kaul, B. B.; Spence, J. T. *Inorg. Chem.* 1984, **23**, 2554.
150. Müller, A.; Diemann, E.; Jostes, R.; Bögge, H. *Angew. Chem.* 1981, **93**, 957; *Intern. Ed. Engl.* 1981, **20**, 934.
151. Hoffmann, B. M.; Roberts, J. E. *J. Am. Chem. Soc.* 1982, **104**, 860.
151a. Piggott, B.; Sheppard, R. N.; Williams, D. J. *Inorg. Chim. Acta* 1984, **86**, L65.
152. Masters, A. F.; Brownlee, R. T. C.; O'Connor, M. J.; Wedd, A. G.; Cotton, J. D. *J. Organomet. Chem.* 1980, **195**, C17.
153. Jaitner, P.; Wohlgenannt, W. *Mh. Chem.* 1982, **113**, 699.
153a. Masters, A. F.; Bossard, G. W.; George, T. A.; Brownlee, R. T. C.; O'Connor, M. J.; Wedd, A. G. *Inorg. Chem.* 1983, **22**, 908.
153b. Donovan-Mtunzi, S.; Hughes, M.; Leigh, G. J.; Ali, H. M.; Richards, R. L.; Mason, J. *J. Organomet. Chem.* 1983, **246**, C1.
153c. Alyea, E. C.; Ferguson, G.; Somogyvari, A. *Organometallics* 1983, **2**, 668.
153d. Gray, G. M.; Gray, R. J. *Organometallics* 1983, **2**, 1026.
153e. Gray, G. M.; Kraihanzel, C. S. *Inorg. Chem.* 1983, **22**, 2959
153f. Gray, G. M. *Inorg. Chim. Acta* 1984, **81**, 157.
153g. Gray, G. M.; Gray, R. J.; Berndt, D. C. *J. Magn. Reson.* 1984, **57**, 347.
153h. Alyea, E. C.; Somogyvari, A. *Inorg. Chim. Acta* 1984, **83**, L49.
153i. Minelli, M.; Hubbard, J. L.; Enemark, J. H. *Inorg. Chem.* 1984, **23**, 970.
154. Nolle, A. *Z. Phys.* 1977, **A280**, 231.
155. Mann, B. E. *J. Chem. Soc. Dalton Trans.* 1973, 2012.
155a. Minelli, M.; Hubbard, J. L.; Christensen, K. A.; Enemark, J. H. *Inorg. Chem.* 1983, **22**, 2652.
156. Bent, H. A. *Chem. Rev.* 1961, **61**, 275.
157. Keiter, R. L.; Verkade, J. G. *Inorg. Chem.* 1969, **8**, 2115.
158. Le Gall, J. Y.; Kubicki, M. M.; Petillon, F. Y. *J. Organomet. Chem.* 1981, **221**, 287.
158a. Kubicki, M. A.; Kergoat, R.; Le Gall, J. Y.; Guerchais, J. E.; Douglade, J.; Mercier, R. *Aust. J. Chem.* 1982, **35**, 1543.
158b. Minelli, M.; Hubbard, J. L.; Lichtenberger, D. L.; Enemark, J. E. *Inorg. Chem.* 1984, **23**, 2721.
159. Minelli, M.; Rockway, T. W.; Enemark, J. H.; Brunner, H.; Muschiol, M. *J. Organomet. Chem.* 1981, **217**, C34.
159a. Brunner, H.; Beier, P.; Frauendorfer, E.; Muschiol, M.; Rastogi, D. K.; Wachter, J.; Minelli, M.; Enemark, J. H. *Inorg. Chim. Acta* 1985, **96**, L5.
159b. Brownlee, R. T. C.; Masters, A. F.; O'Connor, M. J.; Wedd, A. G.; Kimlin, H. A.; Cotton, J. D. *Org. Magn. Reson.* 1982, **20**, 73.
159c. Andrews, G. T.; Colquhoun, I. J.; McFarlane, W. *Polyhedron* 1983, **2**, 783.
159d. Colquhoun, I. J.; McFarlane, W.; Keiter, R. L. *J. Chem. Soc. Dalton Trans.* 1984, 455.
160. McFarlane, H. C. E.; McFarlane, W.; Rycroft, D. S. *J. Chem. Soc. Dalton Trans.* 1976, 1616.
161. McFarlane, W.; Noble, A. M.; Winfield, J. M. *J. Chem. Soc. A* 1971, 948.
162. Narath, A.; Wallace, D. C. *Phys. Rev.* 1962, **127**, 724.
163. Banck, J.; Schwenck, A. *Z. Phys.* 1975, **B20**, 75.
164. Acerate, R.; Harmalker, S.; Hammer, C. F.; Pope, M. T.; Baker, L. C. W. *Chem. Commun.* 1979, 777.
165. Lefebvre, J.; Chauveau, F.; Dappelt, P.; Brevard, C. *J. Am. Chem. Soc.* 1981, **103**, 4589.
166. Sethuraman, P. R.; Leparulo, M. A.; Pope, M. T.; Zonnevijlle, F. *J. Am. Chem. Soc.* 1981, **103**, 7665.
167. Gansow, O. A.; Ho, R. K. C.; Klemperer, W. G. *J. Organomet. Chem.* 1980, **187**, C27.
168. Acerete, R.; Hammer, C. F.; Baker, L. C. W. *J. Am. Chem. Soc.* 1979, **101**, 267.
168a. Pope, M. T. *Heteropoly and Isopoly Oxometalates (Inorganic Chemistry Concepts)*; Lippard, S., Ed.; Springer-Verlag: Berlin, 1983; Vol. 8.
168b. Acerete, R.; Hammer, C. F.; Baker, L. C. W. *J. Am. Chem. Soc.* 1982, **104**, 5384.
168c. Domaille, P. J.; Knoth, W. H. *Inorg. Chem.* 1983, **22**, 818.
168d. Knoth, W. H.; Domaille, P. J.; Roe, D. C. *Inorg. Chem.* 1983, **22**, 198.

168e. Finke, R. G.; Droege, M. W. *Inorg. Chem.* 1983, **22**, 1006.
168f. Kazansky, L. P.; Fedotov, M. A. *J. Chem. Soc. Chem. Commun.* 1983, 417.
168g. Fedotov, M. A.; Kazanskii, L. P.; Spitsyn, V. I. *Dokl. Akad. Nauk. SSSR* 1983, **272**, 1179.
168h. Brevard, C.; Schimpf, R.; Tourne, G.; Tourne, C. M. *J. Am. Chem. Soc.* 1983, **105**, 7059.
168i. Finke, R. G.; Droege, M. W. *J. Am. Chem. Soc.* 1984, **106**, 7274.
168j. Acerete, R.; Hammer, C. F.; Baker, L. C. W. *Inorg. Chem.* 1984, **23**, 1478.
169. Malisch, W.; Maisch, R.; Colquhoun, I. J.; McFarlane, W. *J. Organomet. Chem.* 1981, **220**, C1.
169a. Keiter, R. L.; van der Velde, D. G. *J. Organomet. Chem.* 1983, **258**, C34.
170. Brevard, C.; Schimpf, R. *J. Magn. Reson.* 1982, **47**, 528.
171. Todd, L. J.; Wilkinson, J. R. *J. Organomet. Chem.* 1974, **77**, 1.
172. Mann, B. E.; Taylor, B. F. *NMR Data for Organometallic Compounds*; Academic: London, 1981.
172a. Köhler, T. H.; Kalder, H. J.; Fischer, E. O. *J. Organomet. Chem.* 1976, **113**, 11.
173. Pregosin, P. S.; Kunz, R. W. *NMR—Basic Principles and Progress*; Diehl, P.; Fluck, E.; Kosfeld, R., Eds.; Springer-Verlag: Berlin, 1979; Vol. 16.
173a. Schenk, W.; Büchner, W. *Inorg. Chim. Acta* 1983, **70**, 189.
174. Mather, G. G.; Pidcock, A.; Rapsey, G. J. N. *J. Chem. Soc. Dalton Trans.* 1973, 2095.
175. Chatt, J.; Pearman, A. J.; Richards, R. L. *J. Chem. Soc. Dalton Trans.* 1978, 1766.
176. Choi, H. W.; Gavin, R. M.; Muetterties, E. L. *Chem. Commun.* 1979, 1085.
176a. Britt, A. D.; Yen, W. M. *J. Am. Chem. Soc.* 1961, **83**, 4516; Fakley, M. E.; Richards, R. L. *Transition Met. Chem.* 1982, **7**, 1.
177. Soares, A.-M.; Griffith, W. P. *J. Chem. Soc. Dalton Trans.* 1981, 1886.
178. Akiyama, M.; Chisholm, M. H.; Cotton, F. A.; Extine, M. W.; Haitko, D. A.; Leonelli, J.; Little, D. *J. Am. Chem. Soc.* 1981, **103**, 779.
179. Fischer, E. O.; Knauss, L.; Keiter, R. L.; Verkade, J. G. *J. Organomet. Chem.* 1972, **37**, C7.
180. Grim, S. O.; Singer, R. M.; Johnson, A. W.; Randall, F. J. *J. Coord. Chem.* 1978, **8**, 121; Grim, S. O.; McAllister, P. R.; Singer, R. M. *Chem. Commun.* 1969, 38.
180a. Abd El-Mottaleb, M. S. A. *J. Molec. Struct.* 1976, **32**, 203.
180b. Grim, S. O.; Wheatland, D. A. *Inorg. Chem.* 1969, **8**, 1716.
180c. Rehder, D.; Dorn, W. L.; Schmidt, J. *Trans. Metal. Chem.* 1976, **1**, 233.
181. Green, P. J.; Brown, T. H. *Inorg. Chem.* 1971, **10**, 206.
181a. Schumann, H.; Meissner, M. *Z. Naturforsch.* 1980, **35b**, 342.
181b. Nielson, R. M.; Wherland, S. *Inorg. Chem.* 1985, **24**, 3458.
182. Lutz, O.; Steinkilberg, W. *Z. Naturforsch.* 1974, **29a**, 1467.
183. Gudlin, D.; Schneider, H. *J. Magn. Reson.* 1975, **17**, 268.
184. Broze, M.; Luz, Z. *J. Phys. Chem.* 1969, **73**, 1600.
185. Britt, A. D.; Yen, W. M. *J. Am. Chem. Soc.* 1961, **83**, 4516.
186. Burton, D. J.; Harris, R. K. *J. Chem. Soc., Chem. Commun.* 1982, 256.
187. Nolle, A. *Z. Phys.* 1977, **A282**, 21.
188. Spiess, H. W.; Sheline, R. K. *J. Chem. Phys.* 1971, **54**, 1099.
189. Keçeci, A.; Rehder, D. *Z. Naturforsch.* 1981, **36b**, 20.
190. Nakano, T. *Bull. Chem. Soc. Jpn.* 1977, **50**, 661.
190a. Rehder, D.; Kramolowsky, R.; Steinhäuser, K. G.; Kunze, U.; Antoniadis, A. *Inorg. Chim. Acta* 1983, **73**, 243.
190b. Kunze, U.; Bruns, A.; Rehder, D. *J. Organomet. Chem.* 1984, **268**, 213.
191. Miles, Jr., W. J.; Garnett, B. B.; Clark, R. J. *Inorg. Chem.* 1969, **8**, 2817.
192. Calderazzo, F.; Lucken, E. A. C.; Williams, D. F. *J. Chem. Soc.* 1967, 154.
193. Onaka, S.; Miyamoto, T.; Sasaki, Y. *Bull. Chem. Soc. Jpn.* 1971, **44**, 1851.
194. Onaka, S.; Sasaki, Y.; Sano, H. *Bull. Chem. Soc. Jpn.* 1971, **44**, 726.
195. Bancroft, G. M.; Clark, H. C.; Kidd, R. G.; Rake, A. T.; Spinney, H. G. *Inorg. Chem.* 1973, **12**, 728.
196. Rehder, D.; Pank, V.; Kunze, U.; Bruns, A. *Inorg. Chim. Acta* 1985, **99**, L35.
197. Kanda, K.; Nakatsuji, H.; Yonezawa, T. *J. Am. Chem. Soc.* 1984, **106**, 5888.
198. Ireland, P. S.; Deckert, C. A.; Brown, T. L. *J. Magn. Reson.* 1976, **23**, 485.
199. Mooberry, E. S.; Sheline, R. K. *J. Chem. Phys.* 1972, **56**, 1852.
200. Williams, J. P.; Wojcicki, A. *Inorg. Chim. Acta* 1975, **15**, L19.
201. Lucken, E. A. C.; Williams, D. F. *Proc. Colloq. Amp.* 1966, **14**, 1121.
202. Anderson, W. P.; Brill, T. B.; Schönberg, A. R.; Stanger, Jr., C. W. *J. Organomet. Chem.* 1972, **44**, 161.
203. Ireland, P. S.; Brown, T. L.; Olson, L. W. *J. Am. Chem. Soc.* 1975, **97**, 3548.
204. Pribula, C. D.; Brown, T. L.; Minck, E. *J. Am. Chem. Soc.* 1974, **96**, 4149.

205. Slater, J. L.; Pupp, M.; Sheline, R. K. *J. Chem. Phys.* 1972, **57**, 2105.
206. Mooberry, E. S.; Spiess, H. W.; Sheline, R. K. *J. Chem. Phys.* 1972, **57**, 813.
207. Buckingham, M. J.; Hawkes, G. E.; Thornback, J. R.*I. Inorg. Chim. Acta* 1981, **56**, L41.
208. Tarasov, V. P.; Privalov, V. I.; Kirakosyan, G. A.; Gorbick, A. A.; Buslaev, Yu. A. *Dokl. Akad. Nauk SSSR* 1982, **263**, 1416.
209. Franklin, K. J.; Lock, C. J. L.; Sayer, B. G.; Schrobilgen, G. J. *J. Am. Chem. Soc.* 1982, **104**, 5303.
210. Burkert, B. K.; Eckel, M. F. *Z. Naturforsch.* 1973, **28b**, 5.
211. Dwek, R. E.; Luz, Z.; Shporer, M. *J. Phys. Chem.* 1970, **74**, 2232.
212. Tarasov, V.; Privalov, V. I.; Petrushin, S. A.; Kirakosjan, G. A.; Kryuchkov, S. V.; German, K. E.; Buslaev, Yu. A. *Dokl. Akad. Nauk. SSSR* 1983, **272**, 919.
213. Minelli, M.; Enemark, J. H.; Brownlee, R. T. C.; O'Connor, M. J.; Wedd, A. G. *Coord. Chem. Rev.* 1985, **68**, 169.
214. Rehder, D. *Chimia* 1986, **40**, 186.
215. Akitt, J. W.; McDonald, W. S. *J. Magn. Reson.* 1984, **58**, 401.
216. Rehder, D.; Basler, W. *J. Magn. Reson.* 1986, **68**, 157.
217. Müller, A.; Krickelmeyer, E.; Bögge, H.; Penk, M.; Rehder, D. *Chimia* 1986, **40**, 50.
218. Do, Y.; Simhon, E. D.; Holm, R. H. *Inorg. Chem.* 1985, **24**, 4635.
219. Hoch, M.; Rehder, D. *J. Organometal. Chem.* 1985, **288**, C25.

Chapter 20

GROUP VIII TRANSITION METALS

R. J. GOODFELLOW

1. INTRODUCTION

There is considerable chemical similarity between the group VIII transition metals despite the increments in valence electrons between the iron, cobalt, and nickel triads. Conversely, there is considerable divergence in the nuclear properties of their naturally occurring spin-active nuclei, which accounts for the differential development of the study of their NMR. Thus ^{59}Co was among the earliest nuclei to be studied, and the consequent recognition of the phenomenon of chemical shift must represent the starting point of the chemists' interest in NMR. By contrast, observations for nuclei of three of the metals (Ru, Rh, and Pd) as complexes in solution have only been made since the general survey of Harris and Mann,[1] and such measurements are still lacking for iridium. Recent applications in coordination chemistry[2] and possible biological uses[3] have been reviewed. Further articles have considered ^{195}Pt[4] and ^{103}Rh[5] NMR. As magnetic resonance in metals and alloys is of peripheral chemical interest and has been the subject of a recent volume,[6] it will not be considered further. The chemical shifts of transition metal nuclei have been discussed by Kidd.[7]

The most commonly measured NMR parameters, chemical shifts and indirect (J) coupling constants, largely relate to chemical properties of a molecule and are comparable between the nuclei of different metals provided some simple adjustments for nuclear parameters (mainly γ) are made. We can take advantage of this to apply the knowledge acquired from the extensive studies on ^{59}Co and ^{195}Pt to the interpretation of results for the other nuclei. To facilitate this process, we will approach the group as a whole rather than separate the discussions for each element.

2. OBSERVATION

The table of NMR Properties of the Elements in the Appendix gives some parameters relevant to the observation of the NMR active nuclei of the group VIII

The following abbreviations are used in this chapter: acac, pentane-2,4-dionato; bipy, 2,2′-bipyridyl; cod, cycloocta-1,5-diene; Cp, cyclopentadienyl; diars, o-phenylenebis(dimethylarsine); dmg, dimethylglyoximato; edta, ethylenediaminetetraacetate; en, ethylenediamine; gly, glycinato; phen, 1,10-phenanthroline; py, pyridine; together with the general symbols M for a metal, L and X for monodentate ligands, typically neutral and anionic respectively and R for alkyl or aryl groups.

R. J. GOODFELLOW • Department of Inorganic Chemistry, University of Bristol, Bristol BS8 1TS, England.

elements, while Brevard and Granger[8] give guidance on spectrometer settings. The values of receptivity should be considered in relation to quaternary carbon atoms as nuclear Overhauser enhancements are likely to be small (see below). Instrument problems (e.g., "acoustic ringing") attendant on the low observation frequencies of nuclei such as ^{99}Ru, ^{103}Rh, or ^{187}Os can reduce the achievable sensitivity of these nuclei. In principle, the sensitivities of iron, nickel, and osmium could be greatly improved by isotopic enrichment, but this seems to have been done only in the case of double resonance measurements on ^{57}Fe.[9,10] Since, in practice, observations are highly dependent on relaxation characteristics, these will be considered before discussing the observational strategies that have been employed.

2.1. Relaxation Behavior

Generally, efficient spin-lattice relaxation (short T_1) will assist observation, whereas rapid spin–spin relaxation (short T_2) and related phenomena which broaden resonances will be deleterious. Currently there are relatively few data on relaxation times of these nuclei and only in a very few cases have the contributions of different mechanisms been evaluated (other than the reasonable assumption of relaxation by electric quadrupole interactions for nuclei with spin $>1/2$).

Relaxation of a spin 1/2 nucleus can occur by spin-rotation, dipole–dipole, shielding anisotropy, and scalar mechanisms. All but the first are proportional to γ^2 and are thus very inefficient for ^{57}Fe, ^{103}Rh, and ^{187}Os. For these nuclei a significant contribution is therefore likely to be from spin-rotation interactions.

In the following sections the extreme narrowing condition is assumed, but this is not necessarily satisfied, especially at greater fields (see next chapter, Section 2.4.1).

2.1.1. Spin-Rotation Relaxation

Quite efficient relaxation is possible by the spin-rotation mechanism for small, freely rotating molecules. For example, little else is likely to contribute to the relaxation of ^{187}Os in liquid OsO_4 ($1 < T_1 < 26$ s, $0.7 < T_2 < 12$ s).[11] This mechanism decreases, however, with decrease in temperature, increase in viscosity, or increase in molecular size, so that relaxation times for ^{57}Fe, ^{103}Rh, and ^{187}Os are liable to be long. Thus for ^{103}Rh, values of T_1 greater than 60 s have been measured.[12,13] Note that similar values of T_2 imply very narrow lines (<0.01 Hz) in the absence of other line broadening effects.

For $[PtCl_4]^{2-}$ in aqueous solution, T_1 decreases with increasing temperature, indicating dominance of spin-rotation relaxation at the field used (1.1 T).[14] However, at higher fields, shielding anisotropy interactions are likely to become more important, especially for the less symmetrical molecular sites (see Section 2.1.3).

2.1.2. Dipolar Relaxation

Pesek and Mason[14] have shown that for Na_2PtCl_6 and Na_2PtCl_4 replacement of H_2O as solvent by D_2O has little effect on the relaxation rates, implying that the intermolecular dipole–dipole mechanism is not important. Measurements on the intramolecular dipole–dipole contribution seem lacking, but for the vast majority of complexes there are no protons close to the metal nucleus, suggesting that this mechanism, which depends on $(r_{M-H})^6$, will not often make a large contribution. Thus nuclear Overhauser enhancement will be minimal.

2.1.3. Shielding Anisotropy Relaxation

In view of the very large chemical shift range displayed by these metals, the anisotropy of the nuclear shielding, $\Delta\sigma$, may be large for less symmetric complexes, thus providing another relaxation path. Relaxation by shielding anisotropy depends on $(\Delta\sigma)^2$, but information on $\Delta\sigma$ is rather scarce. All the existing data derive from measurements on solids, although they should be applicable to complexes in solution since intermolecular contributions to the shifts of these nuclei are usually insignificant.

The shielding anisotropy should be relatively small for octahedral complexes, although even here values of ca. 1000 ppm have been obtained for $[Me_3Pt(acac)]_2$ and $(Me_3Pt)_2SO_4 \cdot 4H_2O$.[15] Results for $\Delta\sigma$ of square planar (d^8) complexes vary from 10,500 to 2,200 ppm (for $K_2[PtCl_4]$ and $K_2[Pt(CN)_4] \cdot 3H_2O$, respectively)[16] and are in general agreement with theoretical predictions that $\Delta\sigma$ should be $1\frac{1}{2}$ times the paramagnetic contribution to the chemical shift (see Section 3.3). This predicts a value of 10,000 to 15,000 for $\Delta\sigma$ of $[Pt(OH_2)_4]^{2+}$, which is much lower than that suggested by Gröning *et al.*,[17] but the rotational correlation time, τ_c could be much longer than they suggest in view of the likelihood of hydrogen bonding interactions between this complex ion and solvent water molecules. Recalculation of the result for $[Pt_2Cl_4(PBu^n_3)_2]$,[18] putting $\tau_c = 10^{-12}$, leads to a value of ca. 8000 ppm for $\Delta\sigma$,[15] which would be reasonable for the relationship between $\Delta\sigma$ and σ^p.

To judge from linear mercury systems (see next chapter), $\Delta\sigma$ could be large for linear complexes, e.g., $[Pt^0(PR_3)_2]$, but there is no definite information on less symmetrical situations. For some five-coordinate complexes, T_1 (^{103}Rh) is ca. 0.5 s compared to >30 s for $[Rh(CO)_2(acac)]$ at the same field (9.4 T).[19] If τ_c is estimated as 10^{-10} then $\Delta\sigma$ for the five-coordinate complexes would be ca. 5000 ppm to produce this relaxation rate.

The typical values of $\Delta\sigma$ make it impracticable to obtain a single narrow resonance by spinning solid samples at the "magic angle." Instead there results a series of lines spread over the range of $\Delta\sigma$ (see, for example, Ref. 15).

The shielding anisotropy relaxation contribution is proportional to τ_c and thus will be greater for larger molecules, more viscous solvents, and lower temperatures (doubly so because of the additional effect of temperature on viscosity). The contribution depends on ν^2 and is thus very much greater for ^{195}Pt than for the other spin-1/2 nuclei, for given $\Delta\sigma$ and field. Obviously it increases as B_0^2. This mechanism gives a spin–spin relaxation rate of 7/6 times that of the spin-lattice relaxation in the motional narrowing limit so its contribution to linewidth is unlikely to cause problems except for ^{195}Pt. Thus the relatively large anisotropy of $[Pt(OH_2)_4]^{2+}$ leads to linewidths of ca. 80 Hz at 8.5 T.[17] Such widths are small compared to the range of ^{195}Pt chemical shifts and coupling constants such as $^1J(PtP)$, and in practice shielding anisotropy relaxation is much more of an advantage (via T_1) than a disadvantage to ^{195}Pt observation.

Important, however, is the contribution, $[\pi T_1(^{195}Pt)]^{-1}$, made to the width of ^{195}Pt "satellites" in the spectra of other nuclei. Although this is less than the broadening of the ^{195}Pt spectrum, it is proportionately greater when compared to the normal linewidths of nuclei such as ^{1}H or ^{13}C. Thus the satellites appear broader (and lower) than expected and may disappear completely at higher fields.[20]

2.1.4. Relaxation by Scalar Coupling

Relaxation of a nucleus I may occur via spin–spin coupling to another nucleus (S, usually quadrupolar) which is rapidly relaxing. Since the contributions to T_1 and T_2 are proportional to J^2 and J is proportional to γ_I, effects will be very much greater for ^{195}Pt than the other spin-1/2 nuclei (for which they will rarely be significant). Only when the relaxation rate of the other nucleus, $(T^S)^{-1}$ is comparable with the difference between the resonance frequencies of the two nuclei, $(\omega_I - \omega_S)$, is there a significant contribution to $(T_1)^{-1}$. It reaches a maximum when $(\omega_I - \omega_S)\, T^S = 1$ and the maximum values for ^{195}Pt bonded to some likely donor atoms with quadrupolar nuclei are given in Table 1 taking $J(PtS)$ as 1000 Hz and B_0 as 2.35 T. The relaxation times of the quadrupolar nuclei are unlikely to be as short as those to give maximum ^{195}Pt relaxation. If they are ten times greater, then $T_1(Pt)^{-1}$ is one-fifth and is reduced by further factors of 10 for every further factor of 10 increase in T^S. If, as is likely, $T^S > (\omega_I - \omega_S)^{-1}$, then the contribution to $T_1(Pt)^{-1}$ will decrease with increasing field (as B_0^2 when distant from the maximum contribution).

The effects on T_1 are most noticeable in the spectra of nuclei such as ^{1}H or ^{13}C coupled to ^{195}Pt. Thus the satellites of the methyl protons of $[PtCl_3(AsMe_3)]^-$ are significantly broader than the "main" resonance. The effect is much more marked for $SbMe_3$ derivatives where the broadening of the satellites may cause them to be overlooked completely.[21] No effects seem to be observed for the other nuclei in Table 1. This is surprising for ^{127}I where the potential contribution is large and the line broadening factor suggests rapid quadrupolar relaxation. One concludes that $^{1}J(PtI)$ is much less than $^{1}J(PtAs)$ or $^{1}J(PtSb)$ (which are likely to be greater than 1000 Hz).[22] The other halides and ^{14}N are unlikely to make a noticeable contribution to $T_1(Pt)^{-1}$.

Much more important for observation of the metal nuclei is the scalar coupling contribution to T_2^{-1}, which decreases steadily with decreasing T_1^S of the quadrupolar nucleus. Table 1 indicates the linewidths caused when T_1^S is 10^{-6} s and J is 1000 Hz

Table 1. Relaxation Characteristics for ^{195}Pt Coupled to Some Quadrupolar Donor Atoms (S)

Quadrupolar nucleus (S)	Line-broadening factor[a]	$T_1^S(\max)$[b] (ns)	$T_1(Pt)^{-1}$[c] (s^{-1})	$W_{1/2}(Pt)$[d] (Hz)
^{14}N	13	11	0.3	8.4
^{75}As	1200	37	1.8	15.7
^{121}Sb	800	64	7.3	36.7
^{123}Sb	670	19	3.9	66.0
^{35}Cl	830	14	0.7	15.7
^{37}Cl	510	12	0.6	15.7
^{79}Br	1450	44	2.2	15.7
^{80}Br	1050	28	1.4	15.7
^{127}I	1500	114	13.1	36.7

[a] $(2I+3)Q^2/[I^2(2I-1)]/10^{-60}\ m^4$.
[b] For maximum scalar spin-lattice relaxation of ^{195}Pt at 2.35 T.
[c] Resulting value of $T_1(Pt)^{-1}$ when $J(PtS) = 1000$ Hz.
[d] Values of $W_{1/2}(Pt) = [\pi T_2(Pt)]^{-1}$ for $J(PtS) = 1000$ Hz, $T_1^S = 1\ \mu s$.

(the same widths result when T_1^S is 10^{-4} s and J is 100 Hz and so on). The effect of ^{75}As is usually noticeable and may cause difficulties, e.g., linewidths of 80–400 Hz for *cis*, bis(arsine) complexes but ca. half this for the *trans* isomers, for which J(PtAs) will be less.[23] Linewidths resulting from coupling to the antimony isotopes can be greater, e.g., 850 Hz for $[PtBr_3(SbMe_3)]^-$ (at 2.1 T)[22] sometimes causing nonobservance of the ^{195}Pt resonance.[24] However, for $[Pt_2Cl_2(CO)_2(SbPh_3)_2]$ at 220 K the linewidth is only 36 Hz[25] as a result of a lower ^1J(PtSb) opposite a Pt–Pt bond and faster antimony relaxation.

Although the coupling of ^{195}Pt to ^{14}N (ca. 400 Hz) is less than to As and Sb, the much slower relaxation of ^{14}N [cf. line-broadening factor] commonly leads to very significant broadening and loss of sensitivity. Sometimes it is possible to reduce ^{14}N relaxation by using low viscosity solvents and higher temperatures sufficiently to resolve the ^{14}N coupling pattern.[20] Alternatively one might be able to increase ^{14}N relaxation by lowering the temperature and increasing the viscosity enough to give reasonably narrow singlet platinum resonances. A more expensive approach is to use ^{15}N enriched ligands.[26–28] Broadening may also result from coupling to ^2H present in complexes of ND_3, etc., formed by slow deuteration of amines in D_2O.[20] No significant contribution to the linewidth seems to be observed for coupling to quadrupolar halide nuclei suggesting that the coupling constant is relatively small and/or relaxation of the halide is very fast.

2.1.5. *Electric Quadrupole Relaxation*

For group VIII nuclei of spin $>1/2$, relaxation should be dominated by the electric quadrupole mechanism. The rate depends on the correlation time, the square of the electric field gradient (efg) and the line broadening factor, $(2I+3)Q^2/[I^2(2I-1)]$ (see Appendix).

Although the efg should be zero for symmetrical, octahedral ML_6 and tetrahedral ML_4 complexes, significant electric quadrupole relaxation can still be present because of distortions caused by outer sphere interactions or vibrational motion.[29,30]

There have been several discussions of the efg in less symmetrical complexes, usually regarding each ligand as a point charge or dipole (i.e., covalency of the M–L bond is not taken into account).[29,31,32] The most significant conclusion is that *fac*-MA_3B_3 should have zero efg so such complexes should give rise to relatively narrow resonances. There is some disagreement[31,32] on other MA_nB_{6-n} species, but in the present author's hands the equations[29] give ε for MAB_5 and *cis*-MA_2B_4, 3ε for *mer*-MA_3B_3 and 4ε for *trans*-MA_2B_4 where $\varepsilon = 9(e_A r_A^{-3} - e_B r_B^{-3})$ regarding each ligand A as a point charge of e_A at a distance r_A etc. In particular, the predicted line broadening can provide a means of identifying the isomers of such complexes.[32]

Finally, as the relaxation rate is proportional to τ_c, low viscosity solvents and higher temperatures give narrower lines.

2.1.6. *Other Contributions to Linewidths*

The observed linewidths are often greater than those corresponding to the spin–spin relaxation processes described above, i.e., correspond to $(\pi T_2^*)^{-1}$.

Field inhomogeneity is rarely significant compared to the chemical shift range or the digital resolution employed.

The temperature dependence of the chemical shift can be large for these nuclei (see Section 3.5.1) so that temperature inhomogeneity during variable temperature measurements or ^{1}H decoupling can make a significant contribution to linewidths especially at higher fields and with large diameter sample tubes.[33]

The metal chemical shift is also affected by the isotopic identity of the donor atoms (see Section 3.5.5); if the contributions of the different isotopomers (e.g., $[Pt\,^{35}Cl_n\,^{37}Cl_{6-n}]^{2-}$ [34]) are not resolved, this adds to the linewidth.

When exchange between two chemically inequivalent sites (A and B) is rapid, i.e., much faster than that causing coalescence, the residual line-broadening is proportional to $(\nu_A - \nu_B)^2$. Because of the enormous chemical shift range for the metals, $(\nu_A - \nu_B)$ may well be large, requiring very fast exchange to narrow the resonances. Thus considerable broadening may be observed on the metal resonances even when the exchange rate is sufficient to give narrow resonances for ^{1}H or ^{13}C nuclei associated with the same species.

2.2. Methods of Observation

2.2.1. Direct Observation

Many of the data on ^{59}Co come from CW "wide-line" measurements and this technique may still have advantage over pulsed FT measurements in which the rapidly decaying FIDs of broad resonances can cause problems. The receptivities (see Appendix) of the other nuclei dictate the use of FT techniques, although some early measurements on ^{195}Pt were made with cw spectrometers.[35–37]

2.2.1a. Pulsed FT Measurements. The relatively high resonance frequencies of ^{59}Co and ^{195}Pt make it impracticable to cover the whole of their very large chemical shift ranges in a single frequency width. In fact, the span of resonance frequencies from a sample containing several metal environments often exceeds the maximum frequency width available, so more than one acquisition is necessary using different observation frequencies. Thus "folded-back" resonances and the missing of resonances in unobserved regions are hazards especially at higher fields. Clearly a wide frequency width (fast analog-to-digital converter) is very desirable. The rapid relaxation of the quadrupolar nuclei (and ^{195}Pt in less symmetric environments) allows one to take advantage of the short acquisition times (corresponding to wide frequency widths) by using a fast pulsing rate.

By contrast, long pulse intervals are frequently necessary because of the long relaxation times of ^{57}Fe, ^{103}Rh, and ^{187}Os. A paramagnetic relaxation agent such as $[Cr(acac)_3]$[38] or hydrated chromium trichloride[39] may be added to reduce T_1. However, as this relaxation process is dependent on γ^2, it is not very efficient for these nuclei (e.g., for ^{103}Rh it will be less than one-sixtieth of that for ^{13}C). Since shielding anisotropy can provide a significant relaxation mechanism for these nuclei, higher fields increase the relaxation rate.[19] As the increased resonance frequencies should also reduce instrumental problems associated with the observation of low frequencies, high field instruments are particularly desirable for measurements on these nuclei. Since the magnitude of $\Delta\sigma$ varies very greatly between species (and is not always predictable, see Section 2.1.3) observed intensities are very variable and resonances can remain undetected, especially for highly symmetric species.

2.2.1b. Steady State Measurements. An alternative to waiting for the spins to relax is to maintain a steady state free precession of the spins with frequent 90° pulses. The "Quadriga" technique developed by Schwenk[40] involves measurements at four frequencies followed by special FT processing of the data. This gives a sensitivity enhancement over free induction decay measurements of $(T_2/T_2^*)^{1/2}$ (where T_2^* includes effects of field inhomogeneity, broadening, etc.). This method provided the first measurements on ^{57}Fe and ^{187}Os in diamagnetic samples.[41,42] Complications arise when there is coupling between two chemically inequivalent nuclei (e.g., ^{103}Rh–^{103}Rh) against which special precautions are necessary.[13] Variations on this technique give (T_1/T_2) and $(T_1 + T_2)$,[43] which provide the best means of evaluating T_1 and T_2 when T_1 is long.

2.2.1c. Polarization Transfer. In contrast to the small nuclear Overhauser enhancements expected for these nuclei, an enhancement of (γ_S/γ_I) can, in principle, be obtained by polarization transfer from spin S (e.g., ^{1}H) to spin I via the coupling J(IS) using pulse sequences such as INEPT[44] or DEPT.[45] Further, the repetition rate depends on the relaxation of S rather than I which gave a time saving factor of 900 for ^{103}Rh.[46] The approximate value of J(IS) is required to adjust the pulse sequence and complications may arise if species with widely different values of J(IS) are present together.

Because of the low value of γ_I, the coupling J(IS) may be vanishingly small. Brevard and Schimpf have shown that INEPT experiments transferring polarization from ^{31}P to the metal nucleus via the much larger coupling ^1J(MP) are possible.[47] When the ^{31}P relaxation time, which can be quite long, was reduced with [Cr(acac)$_3$] an overall time saving factor of 36 was achieved for ^{103}Rh compared to direct measurement with natural polarization. Since ^1J(MP) is less variable than J(MH) it can be estimated with sufficient accuracy to set up an INEPT experiment. Thus it is not necessary to obtain ^1J(MP) from the ^{31}P spectrum (the low abundance of ^{57}Fe or ^{187}Os makes identification of the relevant satellites difficult).

2.2.2. Indirect Observation

When an abundant, easily observed nucleus (S) such as ^{1}H is coupled to a low γ nucleus I, the resonance position of I may be established by S–{I} double resonance experiments. The sensitivity and relaxation time are that of S so that the sensitivity advantage is $(\gamma_S/\gamma_I)^3$ compared to direct measurement. Using the INDOR method with low power levels of ν_I, spin–spin splitting patterns and linewidths of the I resonances may be observed. The presence of I–S coupling on the I resonance is usually a serious disadvantage compared to direct measurements where S can be decoupled. Further, resonances not coupled to S cannot be observed and separate experiments are necessary for each S (and associated I) resonance. (This does, however, give valuable correlation information.) Until the recent development of multinuclear FT spectrometers, double resonance using cw spectrometers was the common means of observing spin-1/2 metal nuclei. Double resonance experiments including "pseudo-INDOR"[48] are possible with pulsed FT instruments if the facility to irradiate the metal nucleus exists.

Frequently, coupling to ^{13}C or ^{31}P exists when coupling to ^{1}H does not and ^{13}C–{I} and ^{31}P–{I} experiments provide valuable alternative approaches. Since there may be ^{13}C–H coupling or ^{31}P–H coupling these experiments are often done with ^{1}H broadband irradiation. In fact, one can take advantage of the P–H coupling to provide an indirect link from ^{1}H to ^{31}P and thence to the metal nucleus. This link is activated by irradiating a ^{31}P resonance, e.g., an ^{1}H–{^{31}P, ^{103}Rh} triple resonance experiment.(49)

2.2.3. *Observation and Relaxation of Individual Nuclei*

The following comments should be taken with the general discussion in the sections above.

2.2.3a. ^{57}*Fe.* This difficult nucleus has been observed by the conventional FT technique (at 5.9 T) but up to a day was required for ca. 1 *M* concentrations.(50) The majority of measurements have been by steady state(51) or ^{13}C–{^{57}Fe} double resonance (using ^{57}Fe enriched samples.)(10) Polarization transfer would appear to be a useful technique for phosphine complexes.(47) Although Schwenk(40) states that T_1 and T_2 are >1 s for $[Fe(CO)_5]$ there do not seem to be any actual values for ^{57}Fe relaxation times.

2.2.3b. ^{59}*Co.* The high receptivity of ^{59}Co is some compensation for the line-broadening factor (Appendix), which results in broad lines for all but the most symmetric species. The majority of measurements concern octahedral complexes of Co^{III} but there are some results for low valent carbonyl or nitrosyl derivatives. Linewidths range from 7 Hz for $K_3[Co(CN)_6]$ in water(52) to 40 kHz for $[HgCo_2(CO)_8]$.(53) (Values of $W_{1/2}$ are included in the extensive tables of Ref. 1.)

The symmetric ions $[Co(CN)_6]^{3-}$ or $[Co(am)_6]^{3+}$ (where am = NH_3 or half a chelating diamine) should have very small efg and have been the subjects of several relaxation studies. For the amine complexes(54) {and $[Co(NO_2)_6]^{3-}$ (55)} there are significant contributions to T_2^{-1} from scalar coupling to ^{14}N. The temperature dependence of T_1(56,57) for $[Co(NH_3)_6]^{3+}$ suggests that spin-rotation may compete with quadrupolar relaxation which is otherwise dominant. The necessary efg may arise from vibrational motion,(30) H bonding with the solvent,(52) or ion pairing,(56,58) although the outside ion is not thought to give sufficient direct field.(59) The effect of $HClO_4$ concentration on T_1 for $[Co(OH_2)_6]^{3+}$ suggests that exchange with hydroxy species may be involved,(60) while the field dependence of the linewidth of $[Co(NH_3)_5CN]^{2+}$ may indicate a shielding anisotropy contribution.(61)

2.2.3c. ^{61}*Ni.* The line-broadening factor suggests similar relaxation rates to ^{59}Co but broad resonances will be much more difficult to detect because of the much lower receptivity. Thus ^{61}Ni studies are currently restricted to tetrahedral, zerovalent complexes.(62,63) For NiL_4, linewidths vary from 4 to 90 Hz.(63)

2.2.3d. ^{99}*Ru and* ^{101}*Ru.* Notwithstanding the low resonant frequencies and low receptivities of ^{99}Ru and ^{101}Ru it is rather surprising that successful measurements on compounds in solution were not reported until 1981.(64,65) The spin-lattice relaxation rate of ^{101}Ru in RuO_4 is 30 times that of ^{99}Ru,(64) in agreement with line-broadening factors. Thus the much narrower lines expected for ^{99}Ru more than compensate for the receptivity advantage of ^{101}Ru so future ruthenium NMR studies will concern ^{99}Ru, with the possible exception of relaxation studies of highly symmetric species. Dykstra and Harrison(66) give linewidths and comment that resonances more than 150 Hz wide would require concentrations of more than 0.5 *M* at 1.9 T. They failed to observe resonances for several species of lower symmetry, although *mer*-$[RuI_3(CO)_3]^-$ was observed with a linewidth less than that of *fac*-$[RuCl_3(CO)_3]^-$. Brevard and Granger[67] observed some lines with widths in excess of 500 Hz but were unable to observe resonances for the less symmetric and larger, $[Ru_2(CO)_6Cl_4]$.

In summary, ^{99}Ru looks a promising nucleus for chemical studies but it is still not clear how far measurements can be extended to less symmetric complexes. Finally, it should be noted that the ^{39}K resonance of potassium salts falls in the same region as ^{99}Ru resonances with a shift on the ^{99}Ru scale of ca. 13,300 ppm to high frequency of $[Ru(CN)_6]^{4-}$.[65,67]

2.2.3e. ^{103}Rh. The direct observation of ^{103}Rh resonances for complexes in solution was first reported in 1979.[38] In this and several subsequent studies, a paramagnetic relaxation agent was used. At higher fields, however, relaxation due to shielding anisotropy makes measurements much easier,[19] at least for low symmetry species. Direct ^{103}Rh measurements have also been made by steady state techniques[12,13] and ^{103}Rh–{^{1}H} INEPT[46] and ^{103}Rh–{^{31}P, ^{1}H} INEPT[47] polarization transfer methods.

The 100% natural abundance of ^{103}Rh greatly facilities indirect measurements of rhodium resonances, and these double resonance experiments provided an extensive set of data prior to direct measurements.[1] Since coupling to ^{1}H can be vanishingly small in many cases, ^{1}H–{^{31}P, ^{103}Rh} INDOR,[49] ^{31}P–{^{103}Rh},[68] and ^{13}C–{^{103}Rh}[69,70] have also been used. The last has been of particular value in the study of rhodium carbonyl clusters and their fluxional behavior, e.g., Ref. 71. Although direct ^{103}Rh measurements can be straightforward, at the current state of the art they cannot be relied upon to reveal all the ^{103}Rh resonances present. (This is not really surprising as it can happen even for the much more receptive ^{195}Pt). Thus Gansow *et al.*[72] only observe two ^{103}Rh signals for $[Rh_{12}(CO)_{30}]^{2-}$, whereas ^{13}C–{^{103}Rh}experiments[73] confirm the three sites expected for the established structure. Again, one of the three resonances of $[Rh_{13}(CO)_{24}H_3]^-$ is easily found by ^{1}H–{^{103}Rh}[74] and ^{13}C–{^{103}Rh} measurements at 2.35 T but is not visible in the directly measured spectrum at 8.5 T.[75] In this case the cause is exchange which has not reached the fast limit and results in a linewidth of ca. 1 kHz at the higher field. Thus indirect methods with their very considerable sensitivity advantage would still seem to have a place in ^{103}Rh studies.

At the relatively low field of 2.1 T, T_1 of $[Rh(acac)_3]$ (82 s) is not much greater than for the much less symmetric [RhCp(cod)] (60 s).[13] For these compounds the values of T_2 are less but still comparable with those of T_1, but for [RhCp(1,3-cycloheptadiene)] the ratio (T_1/T_2) is 82 no doubt because of the contribution of the rapid inversion of the ring system to T_2^{-1}.[13]

2.2.3f. ^{105}Pd. The only observation of ^{105}Pd in solution is for $K_2[PdCl_6]$.[8] The linewidth of 25 kHz is very large even in view of the high line-broadening factor. The anticipated broadness, combined with the low receptivity and resonant frequency will make ^{105}Pd observations very difficult. In the event, there are relatively few highly symmetric palladium complexes and tetrahedral Pd^0 derivatives would incur further broadening if low temperatures had to be used to slow down fluxional processes. Thus ^{105}Pd is not a promising nucleus for chemical studies.

2.2.3g. ^{187}Os and ^{189}Os. From the high line-broadening factor, broad lines are to be expected for ^{189}Os, and this is confirmed by the width (>1 kHz) observed for molten OsO_4,(76) the only compound for which an ^{189}Os resonance has been reported. Few other osmium compounds are likely to give narrow enough lines to make this nucleus of use to the chemist. At natural abundance, ^{187}Os has the lowest receptivity of any spin-1/2 nucleus. Again molten OsO_4 provides the only reported observations (including relaxation information) achieved by steady state techniques.(11)

While direct ^{187}Os measurements are clearly possible, results will be very difficult to obtain. Polarization transfer, especially $^{187}Os-\{^{31}P, ^{1}H\}$, should offer more promise,(47) and isotopic enrichment would help. The latter would greatly assist indirect double resonance measurements which have not been reported.

2.2.3h. ^{191}Ir and ^{193}Ir. The very large line-broadening factors for both iridium isotopes are similar to those for ^{185}Re and ^{187}Re, suggesting very broad lines indeed (cf. $[ReO_4]^- > 10$ kHz(77)). When combined with the low receptivity and resonance frequency, this is quite sufficient to explain the lack of any report of an iridium resonance from a compound in solution and removes the possibility of chemically useful iridium NMR results.

2.2.3i. ^{195}Pt. After some early direct ^{195}Pt measurements, double resonance measurements (usually $^{1}H-\{^{195}Pt\}$ using CW techniques) provided a wide ranging set of ^{195}Pt chemical shifts.(1) Direct FT observation is now relatively easy (although liable to the difficulties discussed above). Coupling of ^{195}Pt to other nuclei, although giving valuable structural information also reduces the intensities of the individual lines.

The magnitude of $^{1}J(PtH)$ and $\delta(^{1}H)$ of hydrides can result in incomplete

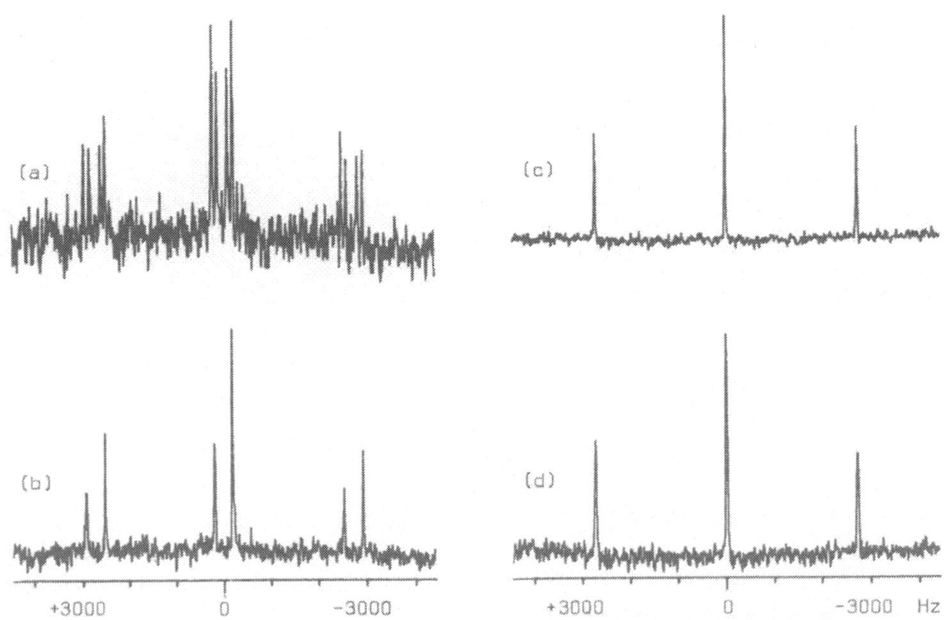

Figure 1. 19.1 MHz ^{195}Pt spectra of *trans*-$[PtHCl(PEt_3)_2]$ with ^{1}H irradiation, (a) broadband centered at $\delta(H) = 4.5$ ppm, (b) continuous wave at $\delta(H) = 4.5$ ppm, (c) broadband centered on hydride resonance, and (d) continuous wave at hydride resonance frequency (60 000 pulses each).

decoupling if the broadband irradiation is left centered on the usual region (e.g., at 4.5 δ). Figure 1 illustrates the undesirable consequences which are avoided if ν_H is set in the region of the hydride resonance. Broadband irradiation is still desirable to remove the small couplings to protons on the other ligands. An alternative is cw irradiation in the usual region (4.5 δ) to give an "off-resonance" partial decoupling of the hydride spins.

Despite the number of ^{195}Pt observations, there is a shortage of relaxation data. Values of T_1 for a range of salts are typically 0.1–1 s.[14,78] However, the linewidths reported by Pesek and Mason, despite corresponding to their values of T_2,[14] are much greater than those subsequently reported[34] for some of the same complexes. Thus it may be that field inhomogeneity and/or drift mars this set of results which were obtained without lock using an electromagnet at reduced field. Other values of T_1 and T_2 illustrate the contribution of shielding anisotropy to ^{195}Pt relaxation.[18,15]

3. CHEMICAL SHIFTS

There are no results to consider for palladium, osmium, and iridium and few for iron, nickel, and ruthenium. Salient features of the results for the last three metals are compared with those for related compounds of cobalt, rhodium, and platinum in Figure 2. The overall behavior is closely similar for all six metals, although there are variations between them just as there are in their chemical behavior. Thus we shall give a general discussion using the most suitable results from among those for the six metals.

Figure 2 illustrates a number of features of the chemical shifts of these nuclei. First there is the large range of the values, over 18,000 ppm for ^{59}Co. Only for ^{195}Pt can the "shielding sensitivity" be evaluated on the same basis as in Chapter 19. For $[PtF_6]^{2-}$ vs. $[PtCl_6]^{2-}$, $\Delta\delta = 7\,326$ ppm, over twice that for ^{183}W, which has the largest sensitivity of the early transition metals. If the value for $[Ru_3(CO)_{12}]$ can be approximated to that of $[M(CO)_6]$ then ^{99}Ru has a similar sensitivity to ^{183}W on the basis of $\Delta\delta\{MO_4 - [M(CO)_6]\}$. However, from Figure 2 it seems that ^{99}Ru lies between ^{103}Rh and ^{59}Co in sensitivity and that ^{59}Co has 1.5–2 times the sensitivity of ^{195}Pt. These large shielding sensitivities mean that chemical shifts are responsive to the molecules' environment (solvent, temperature, and pressure) and isotopic constitution.

The following trends are apparent from Figure 2:

i. Species with the same donor atoms are grouped together.
ii. Complexes of strong field donors resonate at low frequency.
iii. Donor atoms from lower rows of the Periodic Table cause lower frequency shifts.
iv. Lower oxidation states give lower frequency shifts, but there is considerable overlap between the ranges covered by different oxidation states.

3.1. Evaluation of Chemical Shifts

Unfortunately, not all the chemical shift values in the literature have been obtained in the same manner and conversions are required to make them compatible.

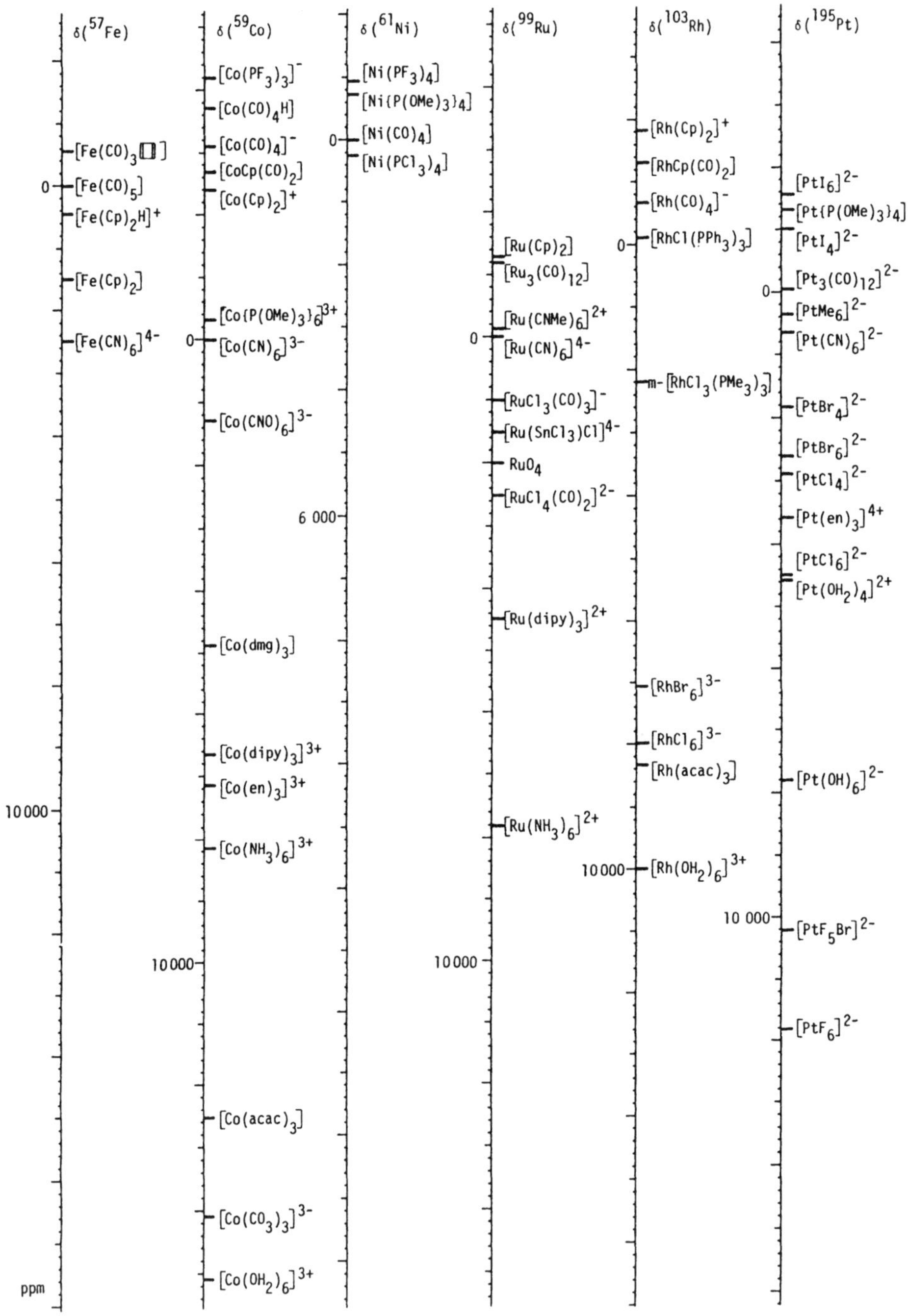

Figure 2. Representative chemical shifts for group VIII transition metal nuclei. (Scales are aligned using analogous complexes of each metal – $[M(CN)_6]^{x-}$ where possible.)

3.1.1. Scale Conversions

We shall take the standard definition of chemical shift as $\delta = 10^6 \cdot (\nu_s - \nu_r)/\nu_r$ (ppm), where s and r refer to the sample and reference resonances, respectively. However, many of the earlier measurements were in field sweep mode and calculated as $\delta' = (B_r - B_s)/B_r$ (or with the reverse sign). This does not give an identical result but may be converted to the frequency definition by dividing by $(1 - \delta' \times 10^{-6})$. Such a correction removes many of the apparent inconsistencies in the results for ^{59}Co,[1] e.g., the shift of $[Co(NH_3)_6]Cl_3$ of 8,100 ppm from field sweep measurements[31,79] converts to 8,166 ppm and is then in good agreement with the FT (frequency scale) result of 8,173 ppm.[80]

Even for frequency scale measurements, the ν_r^{-1} term requires due care; e.g., the shift of $[Co(NH_3)_6]Cl_3$ is 8,173 when $K_3[Co(CN)_6]$ is the reference but the shift of $K_3[Co(CN)_6]$ is −8,107 ppm when $[Co(NH_3)_6]Cl_3$ is taken as the reference. Because of the magnitudes of shifts for these nuclei, the true reference frequency should be used as denominator, but this may not always have been done. Further, the automatic data reduction on some multinuclear FT spectrometers uses the observing frequency not the reference frequency. In such cases, the safe rule is to reset the frequency to the reference frequency before data reduction.

The main problem arises with the conversion of results reported to a different reference. Here the equation $\delta = \delta(\text{ref}') + \delta' \cdot \nu_{r'}/\nu_r$ should be used [primes refer to the old scale with reference, ref′, which has a shift of $\delta(\text{ref}')$ on the new scale]. The term $\nu_{r'}/\nu_r$ can, of course, be replaced by $1 + \delta(\text{ref}') \times 10^{-6}$.

3.1.2. References

Nowadays, the shift of a metal nucleus is almost never measured with the reference substance in the same tube but rather it is a separate sample measured separately. Either the field is assumed to be the same for both measurements or more commonly the ^2H frequency of the solvent is used to ensure that they are (subject to differences in solvent which can be easily corrected for). In the latter case, the ^1H frequency of TMS may be measured or calculated so that Ξ(metal nucleus) can be determined. Because of the sensitivity of transition metal shifts to conditions, especially concentration and temperature, considerable variations are likely in frequencies measured for the metal references on different occasions. The ^2H frequencies of solvents and (particularly) the ^1H frequency of TMS are much less susceptible to variation and furthermore avoid diamagnetic corrections as they are internal. Thus rather than measure the reference substance on many occasions and in each laboratory, more consistent results will be obtained from a single well-determined value of Ξ for the reference, from which may be calculated the reference frequency at a particular field (e.g., given ^2H frequency and solvent). The consequence of this approach is that it is not necessary for the reference substance to be readily available or easily measured but merely for someone to have obtained a reliable value.

In fact, the value of Ξ need not refer to a real compound, which obviates the temptation to make a more accurate redetermination. Thus the round values $\Xi(^{195}Pt) = 21.4$ and $\Xi(^{103}Rh) = 3.16$ MHz, which fall towards the low frequency end of the range of shifts for these two metals, have been proposed as references and are adopted here. For the other nuclei, it would be preferable to use the simple Ξ values given in Table 2 as these should make all shifts positive. However, for compatibility

Table 2. Chemical Shift References for Group VIII Metals

Compound	Shift (ppm)	Ξ (MHz)	Reference
^{57}Fe			
$[Fe(CO)_5]$(neat)	0	3.237 798	51
Ferrocene (CS_2)	1560	3.242 849	51
$K_4[Fe(CN)_6]$ (H_2O pH 12.6)	2497	3.245 883	51
Ξ value	−2408	3.23	
^{59}Co			
$K_3[Co(CN)_6]$ (D_2O, 0.1 *M*)	0	23.727 118	22
$[Co(NH_3)_6]Cl_3(H_2O)$	8173	23.921 040	80
$[Co(en)_3]Cl_3(H_2O)$	7144	23.896 625	30
Ξ value	−5358	23.6	
^{61}Ni			
$[Ni(CO)_4]$	0	8.936 050	63
Ξ value	−677	8.93	
^{99}Ru			
$K_4[Ru(CN)_6]$ (D_2O)	0	4.605 127[a]	66
Ξ value	−1113	4.6	
^{103}Rh			
Ξ value	0	3.16	1
mer-$[RhCl_3(SMe_2)_3]$ (CH_2Cl_2)	3896	3.172 310	81
Rh-metal	−1370	3.155 671	68, 38
Ξ value (π)	−5825	3.141 593	82
^{195}Pt			
Ξ value	0	21.4	1
cis-$[PtCl_2(SMe_2)_2]$ ($CDCl_3$)	980	21.420 980	83
$Na_2[PtCl_6]$ (D_2O)	4522	21.496 770	1
$K_2[PtCl_4]$ (D_2O)	2888	21.461 795	78
trans-$[PtCl_2(pepy)_2]$[b] ($CHCl_3$)	2578	21.455 174	1
$K_2[Pt(CN)_6]$ (D_2O)	667	21.414 274	84

[a] $\Xi(^{101}Ru) = 5.161\ 344$ MHz.
[b] pepy = 4-*n*-pentyl-pyridine.

with the literature we give results in terms of the commonly used chemical references. Chemical comparability of the references for the group VIII metal (see Figure 2) would suggest the hexacyano-metallate anions as currently adopted for ^{99,101}Ru and ^{59}Co. We include these among the possible references whose details are given in Table 2.

3.2. Results

It is not practicable to reproduce here all the shifts reported for complexes of group VIII metals. The primary aim will be to direct the reader to papers covering particular types of complexes although chemical shift ranges will be given with individual values for the more important species. Where necessary, values have been converted as indicated in Section 3.1.1.

Table 3. Summary of ^{57}Fe Chemical Shifts

Complex	$\delta(^{57}Fe)$ (ppm)[a]
Fe^{II}	
$[Fe(Cp)_2]$ (benzene)(51)	1532
with substituents on rings(51,50,85)	2070 to 1530 [28]
except $[(Fc)_2CH]^{+\,b}$ (51)	2275
$[FcCHR]^{+\,b}$, R = H, Me(85)	1036, 1340 [2]
$[Fe(Cp)_2H]^+$ (86)	460
$[FeCp(\eta\text{-}C_6H_6)]^+$ (10)	1593
Allyl, dienyl etc. derivatives with CO and/or Cp(51,10)	2296 to −77 [6]
$K_4[Fe(CN)_6]$(51)	2497
Fe^0	
$[Fe(CO)_5]$ (neat)(51)	0
derivatives with olefins etc.(51,87)	1562 to −582 [20]
$[Fe(CO)_2(PMe_3)_2CS_2]$(47)	1220

[a] The number of reported shifts is given in square brackets.
[b] Fc = ferrocenyl.

3.2.1. ^{57}Fe Chemical Shifts

Results, largely for derivatives of ferrocene and for a wide range of alkene derivatives of $Fe(CO)_5$, are summarized in Table 3.

3.2.2. ^{59}Co Chemical Shifts

Although the vast majority of ^{59}Co results concern cobalt(III) there are some for low valent complexes with CO etc. The latter are summarized in Table 4 with cyclopentadiene derivatives. Results for the more symmetric complexes CoL_6 and $Co(L\text{–}L)_3$

Table 4. ^{59}Co Chemical Shifts of Some Carbonyl, Nitrosyl, Cyclopentadienyl, and Related Derivatives of Cobalt

Complex	$\delta(^{59}Co)$ (ppm)[a]
$[CoL_4]^-$ L = CO, PF_3(53)	−3090, −4200
$[CoHL_4]$ L = CO, PF_3(53)	−3700, −3900
$[Co_2(CO)_8]$(53,88,89)	−2100
$[Co(CO)_4]$—derivatives with metalloids etc.(53,90–92)	−1800 to −3270 [31]
$[Co_4(CO)_{12}]$(93,88,53)	−300, −1500
$[CoCp(CO)_2]$(53)	−2670
$[Co(CO)_3(NO)]$(94,53)	−1322
$[Co(CO)_2(NO)P(AMe_3)_3]$, A = C, Si, Ge, Sn(94)	−844 to −1144 [10]
$[CoX(NO)_2]_2$, X = Cl, Br or I(95)	3640 to 2790 [3]
$[CoCl(NO)_2PZ_3]$ } PZ_3 = phosphine, phosphite, etc.(95)	2040 to 1610 [9]
$[CoBr(NO)_2PZ_3]$ } PZ_3 = phosphine, phosphite, etc.(95)	1920 to 1380 [28]
$[CoI(NO)_2PZ_3]$ } PZ_3 = phosphine, phosphite, etc.(95)	1100 to 760 [9]
$[CoCp_2]^+$ (53)	−2400

[a] To high frequency of $K_3[Co(CN)_6]$; the number of reported shifts is given in square brackets.

where L–L is a symmetric chelating ligand are given in Table 5, while those for less symmetric species are summarized in Table 6. In view of the large linewidths encountered for many resonances, some variation in shift values is to be expected but those in Refs. 102 and 31 frequently differ considerably from other reported values.

Shielding tensors have been determined from single-crystal measurements in a few cases. For the low valent compounds $[Co_2(CO)_8]$[89] and $[Co(CO)_4X]$, where X is $Mn(CO)_5$,[90] GeI_3, $SnBr_3$, or $SnPh_3$[91] the anisotropy is small as it is in $[Co(NH_3)_5(CN)]Cl_2$, $[Co(NH_3)_4(CO_3)]Br$, and *trans*-$[Co(en)_2(NO_2)_2]NO_3$ but is much larger in *trans*-$[Co(en)_2Cl_2]Cl\cdot H_2O$ and $[Co(Cp)_2]NO_3\cdot H_2O$.[130]

3.2.3. ^{61}Ni Chemical Shifts

Results are restricted to Ni^0. The shifts of some complexes $[NiL_4]$ with $L = PF_3$, $P(OMe)_3$, $P(OEt)_3$, and PCl_3 range from −929 to 267 ppm.[63] The series $[Ni(CO)_3PR_3]$ where $PR_3 = PBu^t_2AMe_3$ (A = C, Si, Ge, or Sn) and $PBu^t(AMe_3)_2$ (A = Si or Ge) covers the narrow interval 57–93 ppm[62] showing similar behavior to the isoelectronic series $[Co(CO)_2(NO)PR_3]$.[94]

3.2.4. ^{99}Ru Chemical Shifts

There is a disturbing disagreement between the values of the three compounds which are repeated in the two papers reporting ^{99}Ru shifts.[66,67] Figure 2 depicts the

Table 5. ^{59}Co Chemical Shifts of Higher Symmetry Cobalt(III) Complexes

Complex	$\delta(^{59}Co)$ (ppm)	Reference
$[Co(OH_2)_6](ClO_4)_3$	15 110	60
$K_3[Co(CO_3)_3]$	14 100	31, 96–98
$K_3[Co(C_2O_4)_3]$	13 010	98–101, 31, 96[a]
$[Co(acac)_3]$	12 550	80, 30, 101, 31, 102, 96[b]
$[Co(S_2CS)_3]$	6 390[c]	103
$[Co\{S_2P(OEt)_2\}_3]$	8 980	101, 104
$[Co\{Se_2CNMe_2\}_3]$	6 760	101
$[Co(NH_3)_6]Cl_3$	8 173	80, 30, 100, 105, 98, 101, 31, 96[b]
$[Co(en)_3]Cl_3$	7 144[d]	30, 98–102, 106, 96[b]
$[Co\{NH_2(CH_2)_3NH_2\}_3]Cl_3$	8 340	107, 100, 30[e]
$[Co\{NH_2(CH_2)_4NH_2\}_3]^{3+}$	8 670	103
$[Co(bipy)_3][ClO_4]_3$	6 670	102
$[Co(phen)_3]Cl_3$	7 130	31
$[Co(NH_2OH)_6]Cl_3$	6 624	30, 31
$[Co(dmg)_3]$	4 880	103
$Na_3[Co(NO_2)_6]$	7 458	30, 100, 55, 101, 31, 88, 105, 96[b]
$[Co\{P(OMe)_3\}_6][BF_4]_3$	−305[f]	108, 109
$[Co(diars)_3]^{3+}$	−100[g]	103
$[Co(CNO)_6]^{3-}$	1 300	103
$K_3[Co(CN)_6]$	0	[b]

[a] Value in Ref 79 is clearly inapplicable
[b] References for the effect of conditions and isotopes may be found in Section 3 5
[c] Values for $[Co(S_2CX)_3]$ range from 5830 to 7230 ppm with $X = NH_2$, NR_2, SEt, Et, Pr^i, OR, CH_2Ph [103 110 101 31 104]
[d] Reduced by 140 ppm on loss of a proton [111]
[e] The ligand described as propylene diamine is clearly propane-1,3-diamine
[f] Other $[Co(phosphite)_6]^{3+}$ complexes cover the range −307 to ca −2000 ppm [108]
[g] Another supposedly $CoAs_6$ species has very different shift(s) [101]

Table 6. Summary of ^{59}Co Chemical Shift Data for Cobalt(III) Complexes with Asymmetric or Mixed Ligands

Complex[a]	$\delta(^{59}Co)$ (ppm)[b]
$[Co(NH_3)_{6-2n}(en)_n]^{3+}$, $n = 0$–3 [31, 98]	8 173 to 7 144 [5]
$[Co(amine)_6]^{3+}$ [96, 107, 112, 113]	7 214 to 8 273 [13]
$[Co(NH_3)_5F]^{2+}$ [106, 114]	9 530 or 8 790
$[Co(NH_3)_{6-n}Cl_n]^{3-n}$, $n = 1, 2$ [61, 98, 102c, 106, 114, 115]	8 890, 9 850
$[Co(amine)_{6-n}Cl_n]^{3-n}$, $n = 1, 2$ [96c, 98, 116]	7 180 to 9 350 [23]
$[Co(NH_3)_5Br]^{2+}$ [98, 106, 114]	8 840
$[Co(amine)_{6-n}Br_n]^{3-n}$, $n = 1, 2$ [98]	8 060 to 9 340 [5]
$[Co(NH_3)_5I]^{2+}$ [106, 114]	8 770 or 9 020
$[Co(NH_3)_{6-n}(OH\ or\ OH_2)_n]^{x}$, $n = 1, 2, 3$ [31c, 61, 98, 102c, 106, 114, 115]	9 110 to 10 830 [5]
$[Co(amine)_{6-n}(OH\ or\ OH_2)_n]^{x}$, $n = 1, 2, 3$ [98, 117]	8 280 to 10 910 [16]
$[Co(amine)_4O_2]^{4+}$ [61]	7 450 to 7 700 [12]
$[Co(NH_3)_{6-2n}(CO_3)_n]^{3-2n}$, $n = \frac{1}{2}, 1, 2$ [31, 96–98, 102c, 114, 115]	9 080 to 11 760 [3]
$[Co(amine)_{6-2n}(CO_3)_n]^{3-2n}$, $n = 1, 2$ [31c, 88c, 96, 98]	8 910 to 11 150 [3]
$[Co(amine)_{6-2n}(C_2O_4)_n]^{3-2n}$, $n = \frac{1}{2}, 1, 2$ [31, 79, 98]	8 640 to 10 980 [5]
$[Co(amine)_{6-2n}(amac)_n]^{3-n}$, $n = 1, 2$ [99, 100, 118, 119]	7 820 to 9 140 [7]
$[Co(NH_3)_{6-n}(oxy\text{-}anion)_n]^{x}$, $n = 1, 2$ [105, 114]	8 780 to 9 800 [6]
$[Co(NH_3)_{6-n}(N_3)_n]^{3-n}$, $n = 1, 2, 3$ [31c, 61, 98, 120]	8 670 to 10 010 [5]
$[Co(amine)_{6-n}(N_3)_n]^{3-n}$, $n = 1, 2$ [98]	7 920 to 8 848 [4]
$[Co(NH_3)_{6-n}(NO_2)_n]^{3-n}$, $n = 1$–5 [31, 61, 96, 98, 100, 102, 105, 106, 114, 115]	7 580 to 6 140 [7][d]
$[Co(amine)_{6-n}(NO_2)_n]^{3-n}$, $n = 1, 2$ [88, 96, 98, 100, 115, 121]	7 250 to 6 280 [11][d]
$[Co(amine)_n(amac)_m(NO_2)_{6-2m-n}]^{m+n-3}$, $n = 1, 2$ [119, 121]	9 010 to 7 270 [22]
$[Co(NH_3)_{6-n}(CN)_n]^{3-n}$, $n = 1$–3 [61, 100, 115, 122]	6 640 to 3 310 [5]
Other NH_3 complexes [79, 97, 105, 114, 115, 122, 124]	[55]
Other amine complexes [61]	[2]
$K[Co(edta)]$ [31]	10 400
$[Co(amac)_3]$ [99, 100, 119, 121]	1 410 to 2 550 [5]
$K[Co(gly)_2(CO_3)]$ isomers [119]	10 956, 10 979
$K_n[Co(gly)_{3-n}(C_2O_4)_n]$, $n = 1, 2$ [99, 100, 119]	10 610 to 11 880 [4]
$[Co(amac)_{3-n}(NO_2)_{2n}]^{n-}$, $n = 1, 2$ [119, 121]	9 076 to 7 588 [23]
$[Co(oxime)_2(NH_3)_2]^{x+}$, oxime = dmg, $\frac{1}{2}$bmip[e] [125]	5 120, 5 510
$[Co(dmg)_2Me(L)]$ [125, 126]	4 100 to 1 550 [15]
$[Co(dmg)_2(CNC_6H_{11})X]$, X = Cl, Br, I, NO_2, CNC_6H_{11} [127]	3 700 to 3 410 [5]
$[Co(dmg)_2R(L)]$, R = Bu, $SnPh_3$, $CHCl_2$ [125, 126]	3 090 to 3 580 [4]
$[Co(dmg)_2X(L)]$, X = Cl, I [126, 127]	5 170 to 3 590 [5]
$[Co(enimin)(NH_3)_2]Cl$, enimin = acacen, salen[f] [125]	8 300, 8 060
$[Co(enimin)Me(L)]$ [125, 126, 128]	7 440 to 6 800 [9]
Other salen complexes [61, 128]	[3]
$[Co(inaa)_3]$[g] [129]	6 670
$[Co(\beta\text{-diketonate})_3]$ [31, 113]	12 394 to 12 910 [21]
$[Co(CN)_5X]^{3-}$, X = Br, I [106]	1 210, 780
$[Co(CN)_5Y]^{x-}$, Y = OH^-, OH_2, NO_2^-[h] [31, 106]	1 830 to 1 400 [3]

[a] The term amine includes $\frac{1}{2}\times$en and NH_3 and mixtures but excludes complexes more specifically described in a higher entry This also applies to the terms amac (= amino-acido anion) and oxy-anion

[b] To high frequency of $K_3[Co(CN)_6]$, where applicable, the values correspond to the limits of n etc in the same order, the number of reported shifts is given in square brackets

[c] Values in this reference are not consistent with those in other references

[d] There are several significant discrepancies between the values in different references

[e] bmipH = 1,3-bis(biacetylmono-oximeimino)propane

[f] acacen^{2-} = NN'-ethylene bis(acetylacetoneiminato), salen^{2-} = NN'-ethylenebis(salicylaldiminato)

[g] inaaH = isonitrosoacetylacetone

[h] Value assigned to $K_3[Co(CN)_5NO_2]$ in reference 122 is clearly inapplicable

majority of the results but additional shifts are 5,040 and 5,003 for *mer*- and *fac*-tris[1-(2-pyridyl)-3,5-dimethylpyrazole]ruthenium(II),[67] 356 and 56 for *mer*- and *fac*-$[RuI_3(CO)_3]$,[66] and 800–1,200 ppm for $[Ru_2Cl_4(CO)_6]$ in coordinating media.[67]

3.2.5. ^{103}*Rh Chemical Shifts*

Recent years have seen a very considerable increase in the data on ^{103}Rh chemical shifts. Results for cluster compounds, an area of special interest, are collected in Table 7. Shifts for a representative set of simpler rhodium(III) complexes are given in Table 8, while other measurements are summarized in Table 9. There is an excellent tabulation of results for organometallic compounds by Maurer *et al.*[13]

3.2.6. ^{195}*Pt Chemical Shifts*

Even by 1977, the quantity of ^{195}Pt data was too great for a detailed listing to be given,[1] and ^{195}Pt work has increased considerably in subsequent years. Results are given for the higher symmetry platinum(IV) and platinum(II) complexes in Tables 10 and 12, respectively. Since the anions $[PtX_3L]^-$, X = Cl, Br, and I provide the most extensive series in which to compare the effect of neutral ligands (L) on the metal chemical shift, details are given in Table 13. Data for other Pt^{IV} and Pt^{II} complexes are summarized in Tables 11 and 14 where the complexes are arranged in order of increasing number of neutral ligands except that those of alkyls and similar σ-bonded organic ligands are collected together at the end of each table. Results for other, less

Table 7. Summary of ^{103}Rh Chemical Shifts in Metal Clusters

Complex		$\delta(^{103}Rh)$ (ppm)^a
$[Rh_3H_3(\eta\text{-}C_5Me_5)_3O]^+$ [131]		−136
$[Rh_4(CO)_{12-n}\{P(OPh)_3\}_n]$, $n = 1\text{–}4$ [132]		−351 to −792 [13]
$[Rh_4H_4(\eta\text{-}C_5Me_5)_4]^{2+}$ [133]		585
$[Rh_5(CO)_{15}]^{-\,b}$, eq, ax [134]		−149, −949
$[Rh_4Fe(CO)_{15}]^-$, eq, eq, ax [135]		59, −242, −1171
$[Rh_{6-n}Fe_n(CO)_{16}]^{n-}$, $n = 1, 2$ [135]		−401 to −502 [5]
$[Rh_6H(CO)_{15}]^-$ [136]		−374 to −538 [4]
$[Rh_6C(CO)_{13}]^{2-\,b}$ [71]		−615, −1003
$[Rh_6E(CO)_{15}]^{n-}$, E = N, C [71, 72]		−242, −313
$[Rh_6C(CO)_{13}H]^-$ [136]		−153, −416
$[Rh_{7-n}Ni_n(CO)_{16}]^{n-3}$, $n = 0, 1$ [137, 138]		+692 to −376 [5]
$[Rh_9P(CO)_{21}]^{2-}$ [72]		−1052 to −1146 [3]
$[Rh_{10}E(CO)_{22}]^{n-}$, E = As, S, P [139]		−739 to −1281 [6]
$[Rh_{12}(CO)_{30}]^{2-}$ [73]		168 to −560 [3]
$[Rh_{13}H_{5-n}(CO)_{24}]^{n-}$, $n = 1\text{–}4$ [74, 75]	int^c	6370 to 2917 [4]
	ext	−340 to −600 [8]
$[Rh_{14}H_{4-n}(CO)_{25}]^{n-}$, $n = 4$ [140]	int^c	5936
$n = 3, 4$	ext	320 to −1001 [8]
$[Rh_{17}S_2(CO)_{32}]^{3-}$ [72]	int^c	879
	ext	−639, −1458
Rh metal [141]		−1370

^{a} To high frequency of $\Xi(^{103}Rh) = 3.16$ MHz, the number of reported shifts is given in square brackets

^{b} Dynamic behavior probably increases the apparent symmetry

^{c} "int" indicates rhodium atom inside the cluster while "ext" are the rest

Table 8. ^{103}Rh Chemical Shifts in Some Complexes $[RhX_{6-n}L_n]^{n-3}$

Complex	$\delta(^{103}Rh)$ (ppm)[a]			Reference
	X = Cl	Br	I	
$[RhX_5(SOMe_2)]^{2-}$	5728			142
cis-$[RhX_4(PF_3)_2]^-$	1934	1243		143
cis-$[RhX_4(SbMe_3)_2]^-$	3755		1645	144
cis-$[RhX_4(SMe_2)_2]^-$	4882	4339	3070	145
cis-$[RhX_4(SOMe_2)_2]^-$	4145	3637		142
trans-$[RhX_4(NMe_3)_2]^-$	7942			144
trans-$[RhX_4(PMe_3)_2]^-$	4378	3617	1841	144
trans-$[RhX_4(AsMe_3)_2]^-$	4735	3950	2044	144
trans-$[RhX_4(SbMe_3)_2]^-$			1959	144
trans-$[RhX_4(SMe_2)_2]^-$	5226	4532	2958	145
trans-$[RhX_4(SOMe_2)_2]^-$	4506	3836		142
fac-$[RhX_3(PMe_3)_3]$[b]	1437	1229		144
fac-$[RhX_3(AsMe_3)_3]$	2125	1798		144
fac-$[RhX_3(SbMe_3)_3]$	1722	1297	457	144
fac-$[RhX_3(SMePh)_3]$[c]	3901	3529	2754	144, 81
mer-$[RhX_3(PMe_3)_3]$[d]	2202	1746	809	144
mer-$[RhX_3(AsMe_3)_3]$	2806	2276	1191	144
mer-$[RhX_3(SbMe_3)_3]$	2648	1996	721	144
mer-$[RhX_3(SMe_2)_3]$[e,f]	3897	3437	2448	145, 81
mer-$[RhX_3(SeMe_2)_3]$[e,f]	3904	3385		81
mer-$[RhX_3(TeMe_2)_3]$[e]	3179	2567	1352	145
mer-$[RhX_3(SOMe_2)_3]$	3476	3061		142
cis-$[RhX_2(PMe_3)_4]^+$	991	825		144
cis-$[RhX_2(AsMe_3)_4]^+$	1457	1222		144
cis-$[RhX_2(SbMe_3)_4]^+$	811	475		144
trans-$[RhX_2(PMe_3)_4]^+$	1564	1154	398	144
trans-$[RhX_2(AsMe_3)_4]^+$	1946	1488	658	144
trans-$[RhX_2(SMe_2)_4]^+$[g]	3081	2756		145

[a] To high frequency of $\Xi(^{103}Rh) = 3.16$ MHz.
[b] *fac*-$[RhX_3\{PF(OPh)_2\}_3]$, 989 X = Cl, 623 ppm X = Br.(143)
[c] *fac*-$[RhCl_3(SR_2)_3]$, 3658 R = Me, 3599 ppm R = Et.(81)
[d] *mer*-$[RhX_3(PBu^n_3)_3]$, 2770 X = Cl, 2334 ppm X = Br.(68)
[e] Also values for mixed halide complexes.
[f] Also values for SMePh derivatives.
[g] Isomers of $[RhCl_2\{MeS(CH_2)_2SMe\}_2]^+$ cover 2288 to 2586 ppm.(81)

common, oxidation states are covered in Table 15. The shifts reported for a mixed valence compound with thiourea do not seem likely for the structure proposed.(250) Some detailed listings of $\delta(^{195}Pt)$ values may be found in Ref. 4.

In addition to the results on the anisotropy of ^{195}Pt shielding discussed above (Section 2.1.3), measurements have been made on the one-dimensional conductor $K_2Pt(CN)_4Br_{0.3} \cdot 3H_2O$.(251) The adsorption of molecules onto platinum metal dispersed on alumina has been investigated by ^{195}Pt NMR and a PtO_6 environment was positively identified from its chemical shift.(252)

3.3. Theoretical Approaches

The range of shielding observed for symmetric cobalt(III) complexes led Griffith and Orgel to their expression for σ^p (the paramagnetic contribution to σ) based on the "atom-in-a-molecule" approach [see Chapter 3, equation (13)]. The predicted relationship of σ^p (and hence δ) to the reciprocal of the $A_{1g} \rightarrow T_{1g}$ spin allowed transition is exemplified by the results for strong field donors (Figure 2). Subsequent studies revealed a difference in behavior between complexes with donors from the second row of the Periodic Table(96) and those with donors from lower rows.(104,31,101,109,108) This is an example of the effect of "heavy atoms" which is most apparent with halide ligands, e.g., $[PtX_6]^{2-}$ and $[PtX_4]^{2-}$ in Figure 2.

Heavy donor atoms may make a relativistic contribution to the shielding of the metal nucleus (Chapter 3, Section 3.6.3). There do not seem to be reports of the likely magnitude for these metals, but we note that repeated halide substitutions (e.g., in $[PtX_{6-n}Y_n]^{2-}$ are nearly linearly additive(158,159,36) in contrast to the calculated spin–orbit contribution to the shielding of CX_4.(253) Increased covalency of the M–L bond is expected for donor atoms from lower rows of the Periodic Table. Within the formalism of Griffith and Orgel's expression, this could reduce σ^p (i.e., an increase in the overall shielding) by reducing the terms $\langle r^{-3} \rangle$ and/or $\langle 0 | \mathbf{L}^2 | 0 \rangle$. Contributions close to the nucleus dominate $\langle r^{-3} \rangle$ which is only halved between M(O) and M(VI) oxidation states.(254) This must represent the upper limit for variation in a given

Table 9. Summary of ^{103}Rh Chemical Shift Data for Monomeric and Dimeric Rhodium Complexes[a]

Complex	$\delta(^{103}Rh)$ (ppm)[b]	
Rh^{V}		
$[RhH_2(\eta\text{-}C_5Me_5)(SiEt_3)_2]$(146)	−1714	
Rh^{III}		
$[RhCl_{6-n}Br_n]^{3-}$, $n = 0\text{–}6$(147)	7985 to 7077	[10]
$[Rh(acac)_3]$(12)	8358	
$[RhCl_n(OH_2)_{6-n}]^{3-n}$, $n = 0\text{–}5$(39)	9992 to 8233	[10]
$[RhX_{6-n-m}(OSMe_2)_n(SOMe_2)_m]^{n+m-3}$, X = Cl, Br(142)	6853 to 3739	[12]
mer,trans-$[RhX_3(CO)(PR_3)_2]$, X = Cl, Br(148,149)	1588 to 983	[4]
cis,trans-$[RhX_2(R)(CO)(PMe_2Ph)_2]$, R = CF_3, CH_3(148)	707 to 52	[8]
trans,trans-$[RhCl_2H(CO)(PMe_2Ph)_2]$(148)	240	
trans-$[RhMe(CNBu^t)_4 L]^{2+}$ (1)	526 to 39	[11]
trans-$[RhX(R)(CNBu^t)_4]^{+}$ (1)	526 to 107	[7]
$[Rh(\eta\text{-}C_5Me_5)Cl_2]_2$(38)	2303	
$[Rh(\eta\text{-}C_5Me_5)(\text{oxy-anion})_2]$(38)	2644, 2365	
$[Rh(\eta\text{-}C_5Me_5)Cl(PMe_2Ph)_2]^{+}$ (38)	420	
$[Rh(Cp)_2]Cl$(13)	−1839	
$[Rh_2H_2Cl_4\{Bu^t_2P(CH_2)_5PBu^t_2\}_2]$[c](150)	2736	
$[RhHCl\{Bu^t_2P(CH_2)_2C(R)(CH_2)_2PBu^t_2\}]$[c], R = H, Me(150)	1011, 1029	
$[RhH(CO)\{Bu^t_2P(CH_2)_2CH(CH_2)_2PBu^t_2\}]$[c](151)	−716	
Rh^{II}		
$[Rh_2(\mu\text{-}O_2CCH_3)_4\{P(OMe)_3\}_2]$(152)	6694	
$[(\text{indenyl})Rh(\mu\text{-}CHCH_2)(\mu\text{-}CHMeCMe)Rh(\text{indenyl})]$(153)	−560, −574	
Rh^{I}		
$[Rh_2(\mu\text{-}X)_2(PF_3)_4]$, X = Cl, Br(143)	−353, −493	
cis-$[RhX_2(PF_3)_2]^{-}$, X = Cl, Br(143)	−382, −491	
$[RhX(PPh_3)_3]$, X = Cl, Br, I(68)	−81 to −268	[3]

Table 9. (*Continued*)

Complex	$\delta(^{103}Rh)$ (ppm)[a,b]
trans-$[RhF(CO)(PPh_3)_2]$[(82)]	−148
trans-$[Rh(oxy\text{-}anion)(CO)(PPh_3)_2]$[(82)]	−261 to −300 [3]
trans-$[RhX(CO)(PR_3)_2]$, X = Cl, NO_2, Br[(68,148,149,38,82)]	−368 to −448 [7]
trans-$[RhX(CO)(PR_3)_2]$, X = NCS, I[(148,82)]	−517 to −532 [3]
$[RhCl(Bu^t_2P{\sim}C{=}C{\sim}PBu^t_2)]$[(150)]	983 to 194 [5]
$[Rh(CO)\{Bu^t_2P(CH_2)_2CH{=}CH(CH_2)_2PBu^t_2\}]^+$[(150)]	−736
$[Rh_2L_2(CNMe)_2\{Bu^t_2P(CH_2)_5PBu^t_2\}_2]$, L = Cl, CNMe[(151)]	−854, ca −200
$[Rh_2(\mu\text{-}X)_2(CO)_4]$, X = Cl, Br[(69)]	+151, −1
cis-$[RhX_2(CO)_2]^-$, X = Cl, Br[(69)]	−17, +47
$[Rh(acac)(CO)_2]$[(69,19)]	290
$[Rh(acac)(diene)]$[(13)]	1760 to 1184 [5]
$[Rh(N\text{-}donor)_2(CO)_2]$[(46,69)]	46, −54
$[Rh(Bpz_4)(diene)]$[d (19,154)]	(914 to 2423)[e] [3]
$[RhCl(C_2H_4)(Me_2NCH_2CMe_2CH{=}CH_2)]$[(155)]	940
$[RhCl(N\text{-}donor)(Me_2NCH_2CMe_2CH{=}CH_2)]$[(155)]	2383 to 1948 [4]
$[Rh(CO)_2(\mu\text{-}polyene)]$[f (13)]	−710, −739
$[Rh(diene)(\mu\text{-}polyene)]$[f (13)]	131 to −133 [6]
$[RhCp(CO)_2]$[(13,156)]	−1322
$[RhCp(cyclo\text{-}C_4H_4)]$[(13)]	−2057
$[RhCp(diene)]$[(13)]	−348 to −1272 [18]
$[RhCp(\mu\text{-}polyene)]$[f (13)]	−676 to −1802 [8]
Rh^{-I}	
$[Rh(CO)_4]^-$[(69)]	−644

[a] There are additional values for partially characterised dimeric Rh^{III} complexes with PBu^n_3,[(68)] SMe_2,[(145)] and $SOMe_2$[(142)]

[b] To high frequency of $\Xi(^{103}Rh) = 3.16$ MHz Where applicable the values correspond to the limits of *n* etc, in the same order, the number of reported shifts is given in square brackets

[c] Five coordinate Rh^{III}

[d] $Bpz_4 = B(1\text{-pyrazole})_4^-$

[e] There is a lack of consistency between the two reports for these complexes

[f] Assignment of oxidation state is somewhat arbitrary for these dimeric complexes

Table 10. ^{195}Pt Chemical Shifts of Some Platinum(IV) Complexes, PtL_6

Complex	$\delta(^{195}Pt)$ (ppm)[a]	Reference
$[PtF_6]^{2-}$	11 847 (CH_2Cl_2)[b]	1
$[Pt(OH)_6]^{2-}$	7 823 (H_2O)	157[c]
$[PtCl_6]^{2-}$	4 522 (H_2O), 4 894 (acetone)	1, 78, 14, 158, 36
$[Pt(en)_3]^{4+}$	3 572 (H_2O)	14
$[PtBr_6]^{2-}$	2 619 (H_2O), 3 006 (acetone)	1, 14, 159,[d] 158,[d] 36[d]
$[Pt(S_5)_3]^{2-}$	2 067, 1 970 (MeOH)	160
$[Pt(SCN)_6]^{2-}$	1 034 (H_2O)	14
$[Pt(CN)_6]^{2-}$	667 (D_2O), 796 (acetone)	84,[e] 1, 14
$[Pt(Me)_6]^{2-}$	358 (Et_2O)	161
$[PtI_6]^{2-}$	−1 545 (H_2O)	14, 36

[a] To high frequency of $\Xi(^{195}Pt) = 21.4$ MHz

[b] $[PtF_5Br]^{2-}$, 10 235[(162)], *fac*-$[PtF_3Cl_3]^{2-}$, 7605[(1)]

[c] Including data on $[Pt(OH)_{6-n}Cl_n]^{2-}$ and effect of pH (see also Ref 163)

[d] Including data on $[PtCl_{6-n}Br_n]^{2-}$

[e] Including data on mixed cyano-halide anions

Table 11. Summary of ^{195}Pt Chemical Shift Data for Platinum(IV) Complexes

Complex	$\delta(^{195}Pt)$ (ppm)[a]	
$[PtX_{6-n}(NO_2)_n]^{2-}$, X = Cl, Br, n = 1–5(164)[b]	6363 to 3382	[16]
$[PtCl_5L]^-$, L = NMe_3, SMe_2, $SOMe_2$, CNMe, $AsMe_3$,[c] PMe_3(1 165 166)	4698 to 3060	[6]
$[PtBr_5L]^-$, L = NMe_3, SMe_2, $SOMe_2$, $AsMe_3$, CNMe, PMe_3(1 165 166)	3325 to 1551	[6]
$[PtI_5(PMe_3)]^-$ (165)	−1709	
$[PtCl_2(OH)_2(NH_2Pr^i)_2]$(20)	5406	
$[PtCl_4L_2]$, L = NMe_3, $SOMe_2$, SMe_2, CNMe, $AsMe_3$, PR_3(1 165 167 143)	4551 to 1822	[12]
$[PtBr_4L_2]$, L = NMe_3, $SOMe_2$, SMe_2, CNMe, $AsMe_3$, PR_3(1 165 167 143)	3606 to 546	[11]
trans-$[PtI_4(PMe_3)_2]$(165)	−1927	
trans-$[PtX_3H(PEt_3)_2]$, X = Cl, Br, I(167)	1167 to −1029	[6]
cis,trans-$[Pt(SiH_2Y)HI_2(PEt_3)_2]$, Y = I, CCH, $CCCF_3$(167)	−668 to −794	[3]
cis,trans-$[PtXYH_2(PEt_3)_2]$, X, Y = Cl, Br, I,(167) SeH,(168) CN(169)	514 to −772	[14]
mer-$[PtCl_3L_3]^+$, L = $AsMe_3$, PMe_3(1 165)	1449, 1348	
Alkyl and related derivatives		
$[PtMe_3(\mu\text{-}OH)]_4$(170)	3062	
$[PtMe_3(\mu\text{-}X)]_4$, X = SR, N_3, Cl, Br(170)	2552 to 2280	[6]
$[PtMe_3(\mu\text{-}I)]_4$(170)	1757	
fac-$[PtMe_3Y_3]^{x-}$, Y = OH_2, py, NO_2(171)	2534 to 2275	[5]
fac-$[PtMe_3(CN)_3]^{2-}$ (171)	256	
$[PtMe_2R(X)L_2]$, X = Cl, Br, I, L = amine, PR'_3, $AsMe_3$(172 173)	2152 to −146	[17]
$[PtR_2X_2Y_2]$, X = Cl, Br, Y = py,(172 174) MeOH(171)	3174 to 2282	[7]
$[PtR_2X_2Y_2]$, X = I, Y = py, X = Cl, Br, I, Y = PR'_3(172 175)	1423 to 143	[8]
$[PtMe_2BrY_3]^{x-}$, Y = OH, μ-OH, MeOH(171)	3654 to 3129	[3]
$[PtMe_2Y_4]^{x-}$, Y = OH, μ-OH, OH_2,(171) glycinato(176)	4427 to 3820	[3]
$[PtRX_3(PR'_3)_2]$, X = Cl, Br, R = C_6F_5,(175) Me(172)	2164 to 799	[3]

[a] To high frequency of $\Xi(^{195}Pt) = 21.4$ MHz, the number of reported shifts is given in square brackets
[b] Including data on mixed halide complexes
[c] The value for $[PtCl_5(AsMe_3)]^-$ is not as reported in Ref 1 but 3271 ppm

Table 12. ^{195}Pt Chemical Shifts of Some Platinum(II) Complexes PtL_4

Complex	$\delta(^{195}Pt)$ (ppm)[a]	Reference
$[Pt(OH_2)_4]^{2+}$	4553 (H_2O)	17, 163[b]
$[Pt(acac)_2]$	4093 ($CDCl_3$)[c]	177
$[PtCl_4]^{2-}$	2896 (H_2O), 3054 (CH_2Cl_2)	78, 164,[d] 17, 14, 36
$[Pt(NO_2)_4]^{2-}$	2346 (aqueous)	28,[e] 164, 14
$[PtBr_4]^{2-}$	1846 (H_2O), 2065 (CH_2Cl_2)	78, 164,[d] 14, 36
$[Pt(NH_3)_4]^{2+}$	1931 (H_2O)	26, 14
$[Pt(en)_2]^{2+}$	1486 (H_2O)	14
$[Pt(SCN)_4]^{2-}$	618 (H_2O), 543 (CH_2Cl_2)	37, 14, 178[f]
$[Pt(CN)_4]^{2-}$	−213 (D_2O), −174 [$(CD_3)_2CO$]	84, 14, 36
$[Pt(CNMe)_4]^{2+}$	−312 (CD_3CN)	166
$[Pt(PMe_3)_4]^{2+}$	−358 (H_2O)	179
$[Pt\{P(OMe)_3\}_4]^{2+}$ [g]	−571 (CH_2Cl_2)	179, 83
$[PtI_4]^{2-}$	−1000 (H_2O)	14

[a] To high frequency of $\Xi(^{195}Pt) = 21.4$ MHz
[b] Including the effect of pH
[c] Other bis-diketonates have shifts 4578–4155 ppm (177)
[d] Including data on mixed halide complexes
[e] Including data on hydrolysis products and $[Pt(NO_2)_2(C_2O_4)]^{2-}$
[f] Including data on $[Pt(SCN)_{4-n}(NCS)_n]^{2-}$
[g] $Pt[P]_4$ where [P] = $P(OH)(OMe)_2$, $P(OH)_3$, or $\frac{1}{2}(\mu\text{-}P_2O_5H_2)$ have shifts −436 to −640 (180)

Table 13. ^{195}Pt Chemical Shifts in Some Anionic Complexes, $[PtX_3L]^-$ [a]

L	$\delta(^{195}Pt)$ (ppm)		
	X = Cl	Br	I
$H_2C{=}CH_2(CDCl_3)$ (179,[b] 181)[c]	1748	1060	
$C{\equiv}NMe$ (166)	1487	806	−730
$C{\equiv}O$ (182)[b]	1265	564	−946
NH_2Me (183)	2635	2029	364
$NHMe_2$ (183)	2666	2107	528
NMe_3 (179)	2818	2282	
$N(Me){=}CMe_2(CDCl_3)$ (183)	2812	2197	525
py(4-Et) (22)	2772	2181	522
$N{\equiv}CMe$ (22)	2503	1827	
PMe_3 (179)[b]	1033	414	−973
$P(OMe)_3$ [d](179)[b]	1037	371	−1060
PF_3 [d](184)	907	183	−1370
$AsMe_3$ (179)[b]	1360	664	−913
$SbMe_3$ (179)[b]	1390	605	−1109
$OH_2(H_2O)$ (78)	3330		
SMe_2 (179)[b]	1776	1118	440
$TeMe_2$ (179)[b]	1474	707	−995
$SOMe_2$ (179,[b] 185,17)[e]	1535	892	
X (78)	3055	2066	

[a] As $[NPr_4]^+$ or $[NBu_4]^+$ salts in CH_2Cl_2 except where otherwise stated; shifts to high frequency of $\Xi(^{195}Pt) = 21.4$ MHz.
[b] Including data on mixed halide complexes.
[c] $[PtCl_3(CH_2{=}C{=}CH_2)]^-$, 2037.(186)
[d] Shifts for $[PtX_3\{PY_{3-n}(OR)_n\}]^-$, Y = F, Cl, Br cover ranges 1179–946 (X = Cl), 584–371 (X = Br), and −1023 to −1299 (X = I).(184,143)
[e] Including data on hydrolysis product.

oxidation state and it has been suggested that $\langle r^{-3}\rangle$ is largely insensitive to the ligands.(96,101,255) The shifts of $[PtI_6]^{2-}$ and $[PtF_6]^{2-}$ (Table 10) suggest that σ^p of the former is no more than one-sixth of that of the latter. This would require reduction of $\langle 0|\, \mathbf{L}^2\, |0\rangle$ as well as $\langle r^{-3}\rangle$, and several authors have made allowance by including orbital angular momentum reduction coefficients.(101,130,256–258) It is difficult to evaluate such terms(257–259) or to separate their variation from that of $\langle r^{-3}\rangle$,(108) but Juranić suggests that the former is dominant.(103) Although the effects of covalency are most noticeable with comparisons of donors from different rows of the Periodic Table, they also occur among donors from the same row.(110,256) The relationship of ^{59}Co shielding to the electronegativity of the ligand,(31,103) Co–L bond ionicity, and the nephelauxetic ratio β_{35}(110,103) has been noted with similar observations for $\delta(^{195}Pt)$.(37,216,83)

Modified forms of the Griffith and Orgel expression have been derived for Co^{III} complexes of lower symmetry.(99,258,260,53) In such cases, the appropriate average of all the electronic transitions related to the $A_{1g} \rightarrow T_{1g}$ transition in O_h symmetry should be used.(99) Using the expression for σ^p of symmetric square planar d^8 complexes(216,37) as a basis, reasonable agreement was achieved between the observed excitation energies and $\delta(^{195}Pt)$ of the anions, $[PtX_3L]^-$ provided different orbital coefficients were used for the halides (X) and neutral ligands (L) of each period.(179)

The expressions referred to above are derived from the ligand–field model, which is inappropriate for the highly covalent complexes of metals in low oxidation states and can lead to incorrect results.(91) Thus, MO approaches, tailored to the particular

Table 14. Summary of ^{195}Pt Chemical Shift Data for Platinum(II) Complexes

Complex	$\delta(^{195}Pt)$ (ppm)[a]	
$[PtX_{4-n}(NO_2)_n]^{2-}$, X = Cl, Br; n = 1–3 (164)[b]	2843 to 2209	[6]
cis-$[PtCl_2(SnCl_3)_2]^{2-}$ (187)	−200	
$[Pt(SnCl_3)_5]^{3-}$ [c] (187)	−1922	
$[Pt_2X_4(\mu\text{-}X)_2]^{2-}$, X = Cl, Br, I (1)	3325 to −600	[3]
$[Pt_2X_4(\mu\text{-}X)(\mu\text{-}SMe_2)]^-$, X = Cl, Br (179)	2336, 1583	
$[\{PtCl_2(CO)\}_2SnCl_2]^{2-}$ (188)	368	
$[Pt_2Cl_2(\mu\text{-}Cl)_2(H_2C{=}CHMe)_2]$ (189)	2058	
$[Pt_2Cl_2(\mu\text{-}Cl)_2L_2]$, L = $AsMe_3$, PR_3 (179,172,37,190)	1499 to 1117	[5]
$[Pt_2Br_2(\mu\text{-}Br)_2L_2]$, L = PMe_3, $AsMe_3$, $SbMe_3$ (179)	832 to 548	[3]
$[Pt_2I_2(\mu\text{-}I)_2L_2]$, L = SMe_2, PMe_3, $AsMe_3$, $SbMe_3$, (179) $PF(OPh)_2$ (191)	−262 to −934	[7]
$[Pt(NCS)_3L]^-$, L = NMe_3, PMe_3 (192)	2445, 1014	
$[Pt(SCN)_3L]^-$,[d] L = SMe_2, $SeMe_2$, AsR_3, PR_3, $TeMe_2$ (192)	466 to 92	[8]
$[Pt_2Cl_2(\mu\text{-}SR)_2(PMe_2Ph)_2]$, R = Me, Et, Ph (172)[e]	727 to 512	[3]
trans-$[Pt_2(\mu\text{-}H)_2\{Si(OEt)_3\}_2(PMeBu^t_2)_2]$ (193)	−44	
$[PtX_3(\mu\text{-}ER_2)$+], X = Cl, Br; ER_2 = SMe_2, (194,179) $SeMe_2$ (195)	1776 to 1146	[4]
$[PtClX(cod)]$, X = Cl, $SnCl_3$ (172,196)	1171, 747	
$[Pt(SnCl_3)_3L_2]^-$,[c] L_2 = diene, $(CO)_2$ (196)	−904 to −1191	[3]
$[PtCl_2(olefin)L]$, L = NH_2R, NHR_2, (197,198) py, (199,190) py-O, pip[f] (181)	1740 to 1421	[30]
$[PtCl(olefin)(prolinato)]$ (200)	2168 to 2001	[12]
cis-$[PtCl_2(olefin)PR_3]$ (186,201)	900 to 715	[4]
$[PtX_2(CO)PR_3]$, R = Et, (202) Me (149); X = Cl, Br, I	691 to −395	[8]
$[PtX_2(CNMe)_2]$, (166) X = Cl, Br, I	665 to −916	[3]
$[PtCl_2(NR_3)_2]$, NR_3 = NH_3, (203,26,204) NH_2R, (198) NHR_2, py, (37) NMe_3 (179)	2647 to 2297	[7]
$[PtX_2(NR_3)_2]$, X = Br, (37,179) NCO, NCS, Cl (205); NR_3 = py, NMe_3	2545 to 2309	[6]
$[PtI_2(4\text{-pentylpyridine})_2]$ (37)	1409	
$[PtX_2(NH_2R)_2]$, X = oxy-anion etc.; NH_2R = NH_3, (26,206,207) $\frac{1}{2}$en (208)	3417 to 2599	[12]
$[PtX_2(amine)L]$, X = Cl, Br; L = $SOMe_2$, (203) PR_3, AsR_3 (37,198,27)	1462 to 451	[17]
$[Pt(\overline{OC_6H_4CHN(CH_2)_2O})L]$, L = CNR, amine, PR_3, AsR_3, $SOMe_2$ (24)	2983 to 1863	[10]
Pt-glycinate complexes (203)	2913 to 1398	[4]
trans-$[PtCl_2L_2]$, L = AsR_3, PR_3 (37,167,83,179,209,190,210)	807 to 543	[16]
cis-$[PtCl_2L_2]$, L = AsR_3, PR_3 (37,167,83,179,172,209,190,23,143,210)	242 to −58	[44]
cis-$[PtCl_2(SbMe_3)_2]$ (179)	−79	
trans-$[PtBr_2L_2]$, L = AsR_3, PR_3 (37,167,179[b])	190 to 36	[8]
cis-$[PtBr_2L_2]$, L = AsR_3, PR_3 (37,179,[b] 172,210,[b] 143)	−41 to −312	[16]
cis-$[PtBr_2(SbMe_3)_2]$ (179)[b]	−486	
trans-$[PtI_2L_2]$, L = AsR_3, PR_3 (37,167,179,[b] 172,210[g])	−985 to −1144	[12]
cis-$[PtI_2L_2]$, L = AsR_3, PR_3 (37,179,[b] 172,143,210[g])	−521 to −727	[12]
trans- and *cis*-$[PtI_2(SbMe_3)_2]$ (179[b])	−1256, −1283	
trans-$[PtY_2L_2]$, Y = NCO, NCS; L = AsR_3, PR_3 (205,178)	819 to 623	[6]
cis-$[PtY_2L_2]$, Y = NCO, NCS, L = AsR_3, PR_3 (205)	416 to 99	[6]
$[PtY_2L_2]$, Y = SH, SeH, (168) SPh, SePh, (172) SCN (205,178); L = PR_3, AsR_3, $SbMe_3$	39 to −481	[12]
$[PtXYL_2]$, X = Y = SH, SeH, (168) SCN, NCS, (205,178) Cl, Br, I	777 to −653	[16]
trans-$[Pt(CN)_2L_2]$, L = PEt_3, (169) $AsPr^n_3$ (37)	−399, −448	
trans-$[PtY_2L_2]$, Y = $SnCl_3$, (209,211) $Pb(aryl)_3$ (212); L = PR_3, AsR_3	−380 to −807	[7]
$[Pt(SnCl_3)_3L_2]^-$ [c] (211,213)	−621 to −902	[5]
trans-$[Pt(CN)X(PEt_3)_2]$, X = Cl, Br, I, (169,167) SeH (168)	7 to −389	[4]
trans-$[PtX(EH_{3-n}Y_n)(PEt_3)_2]$, X = Cl, Br; E = Si, Ge (167,214)	−358 to −601	[18]
trans-$[PtI(EH_{3-n}Y_n)(PEt_3)_2]$, E = Si, Ge (167,214,215)	−615 to −822	[24]
$[PtCl(SnCl_3)L_2]$, L = PR_3, AsR_3 (211,209,213)	−221 to −357	[9]
trans-$[PtHX(PR_3)_2]$, X = RCO_2, NO_3, NO_2, Cl, Br, I, SCN, (216,167,83) SH, SeH, (168) $SSiH_3$, $SeSiH_3$, (214) $SnCl_3$ (213)	−212 to −824	[34]

Table 14. (*Continued*)

Complex	$\delta(^{195}Pt)$ (ppm)[a]	
cis-$[PtCl_2(PR_3)(SMe_2)]$, $PR_3 = PBu^n_3$,[37] PPh_3, PCl_3[83]	742 to 428	[3]
$[(PPh_3)_2Pt(\mu\text{-}H)_2PtH(PPh_3)_2]^+$,[217]	−605, −256[c]	
$[Pt_2(\mu\text{-}Cl)_2(PPr^n_3)_2]^{2+}$, $[Pt_3(\mu^3\text{-}S)_2(PMe_3)_2]^{2+}$[179]	278, 208	
trans-$[PtX_2(OH_2)_2]$, X = Cl,[17][h] Br[163]	3889, 3352	
$[PtCl_2L_2]$, L = SR_2, SeR_2, $SOMe_2$,[83 179 185] μ-SMe_2,[194] μ-$SeMe_2$[195]	1152 to 732	[30]
cis-$[PtCl_2(TeMe_2)_2]$[179]	164	
$[PtBr_2L_2]$, L = SR_2, SeR_2, $SOMe_2$[83 b 179 b]	654 to 492	[10]
$[PtI_2L_2]$, L = SR_2, SeR_2[83 b 179 b]	−153 to −531	[6]
trans-$[PtI_2(TeMe_2)_2]$[179]	−1131	
$[PtX_2L_2]$, X = NCS, SCN, Cl, L = SMe_2, $SeMe_2$[205]	1335 to 170	[10]
$[PtCl_2(PhCH=CH_2)(Bu^tN(CH)_2NBu^t)]$[c (27)]	2534	
trans-$[PtX(CO)(PR_3)_2]^+$, X = Cl, Br, $PR_3 = PMe_3$,[149] PPh_3[218]	62 to −61	[3]
$[PtX(CNMe)_3]^+$, X = Cl, Br, I[166]	203 to −334	[3]
cis-$[PtCl(NH_3)_2(amine)]^+$[203 204]	2210 to 2113	[8]
$[PtCl(amine)_2L]^+$, amine = NH_3, ½en, L = $OC(Me)NH_2$,[208] $SOMe_2$[203]	2894 to 1361	[4]
$[PtX(PR_3)_3]^+$, X = Cl, Br, I[179 83]	−137 to −400	[4]
$[PtX(AsMe_3)_3]^+$, X = Cl, Br, I[179]	−211 to −698	[3]
$[PtX(SbMe_3)_3]^+$, X = Cl, Br, I[179]	−646 to −1301	[3]
$[PtX_2(phosphole)_3]$,[c] X = Cl, Br, I[210][g]	−244 to −491	[13]
trans-$[PtH(PEt_3)_2PR_3]^+$, PR_3 = PEt_3, $PHMe_2$, PH_2Me, PH_3[219]	−571 to −745	[4]
$[PtX(SMe_2)_3]^+$, X = Cl, Br, I[179]	523 to 73	[3]
$[PtX(SeMe_2)_3]^+$, X = Cl, Br, I[179]	241 to −307	[3]
$[Pt(amine)_4]^{2+}$, amine = ½bipy, NH_3 etc[107 204]	2060 to 1775	[6]
$[Pt(NH_3)_3(SOMe_2)]^{2+}$,[203]	1283	
$[Pt(SnCl_3)L_4]^+$,[c] (L_4 = tetradentate P/As ligand)[211]	−802, −983	
Alkyl and related derivatives		
$[PtCl_2(R)L]^-$, $[Pt_2(\mu\text{-}Cl)_2R_2L_2]$ L = PR'_3,[202 220] CO[221]	965 to 642	[8]
$[PtR_2L_2]$, R = Me, Ph,[172 222 223] C_6F_5,[179] $(CH_2)_4$, L = PR_3, AsR_3	99 to −146	[8]
except $L_2 = (Ph_2P)_2CH_2$,[222] $Ph_2PC_6H_4CH=CH_2$,[224] cod[225 172] or $R_2 = \{C(CF_3)\}_6$[226]	1275 to 478	[15]
cis- and *trans*-$[Pt(C\equiv CR)_2(PR'_3)_2]$[227]	−185 to −280	[15]
$[Pt(B-B)(PMe_2Ph)_2]$, B−B = ½(B_8H_{10}), B_8H_{12}, $B_{10}H_{12}$[228]	−764 to −1085	[3]
$[PtX(R)L_2]$, R = Me,[172] CH_2Cl,[229] C_6F_5,[175] B_5H_8[230], X = Cl, ONO_2[224], L = PR_3, AsR_3	92 to −14	[10]
$[PtBr(R)L_2]$, R = Me,[172] C_6F_5,[175] L = PR_3, AsR_3	125 to −151	[5]
$[PtX(R)L_2]$, R = CF_3, Me,[172 167] C_6H_5[175], X = I, SePh, $SnCl_3$[213]	−99 to −435	[9]
cis-$[Pt(aryl)\{Pb(aryl)_3\}(PPh_3)_2]$[212]	−47 to −109	[8]
$[PtCl(R)(PR_3)L]$, R = Me,[224] Ph[202], L = olefin, CO, amine[220]	464 to 185	[7]
$[Pt_2(\mu\text{-}SMe)_2Me_2(PMe_2Ph)_2]$[172]	361	
$[PtCl(C\sim C\sim E)L]$, E = NR_2,[220 231] PR_2[222], L = NR_3, PR_3, $SOMe_2$	1643 to 635	[5]
$[Pt(C(O)C_6H_4O)(PR_3)_2]$[185] $[Pt(CH_2COCHCOCF_3)L_2]$, L = PR_3, $AsPh_3$[232]	287 to 112	[4]

[a] To high frequency of $\Xi(^{195}Pt) = 21.4$ MHz, the number of reported values is given in square brackets
[b] Including data on mixed halide complexes
[c] Five coordinate species
[d] Complexes with *N* and *S* bonded thiocyanate have intermediate shifts
[e] Reference 190 includes values for some incorrectly formulated but probably related species
[f] py-O = pyridine-*N*-oxide, pip = piperidine
[g] The values of Ref 210 for $[PtI_2(PR_3)_2]$ are not comparable with those of other authors, suggesting a systematic error
[h] Including the effect of pH
[i] Reference 233 gives some information on L = $NHMe_2$

Table 15. Summary of ^{195}Pt Chemical Shift Data for Complexes in Less Common Oxidation States

Complex	$\delta(^{195}Pt)$ (ppm)[a]	
Pt^{III}		
$[Pt_2(SO_4)_4XY]^{n-}$, $[Pt_2(PO_4H)_4XY]^{n-}$, X, Y = Cl, Br, OH_2; n = 2–4[234]	6504 to 5968	[14]
$[Pt_2(P_2O_5H_2)_4XY]^{4-}$, X, Y = Cl, Br, I, CH_3[235]	4236 to 5227	[5]
Pt^{I}		
$[Pt_2X_4(CO)_2]^{2-}$, X = Cl, Br, I[182][b]	373 to −176	[3]
$[Pt_2(CNMe)_6]^{2+}$[195]	−35	[1]
[codPt(μ-L−L') Ptcod], L−L' = $(CF_3)_2CO$, $(CF_3)_2NH$,[195] $MeCCCO_2Me$,[225] $CF_3CCHCF_3^+$[236]	824 to 190	[8]
$[Pt_2(\mu\text{-}R_2PS)_2\{P(OR')_3\}_2]$, R = R' = Me[195] R = Ph, C_6H_{11}, R' = Ph[237]	−606 to −677	[3]
Pt^{0}		
$[codPt(MeCCCO_2Me)]$,[225] $[(codPt)_2\ \mu\text{-}Me_3SiCCSiMe_2Bu^t]$[238]	756, 619	[2]
$[Pt(PR_3)_2R'C{\equiv}CR'']$[239,172,226,195]	53 to −208	[18]
$[Pt(PR_3)_2L]$, L = PhNSO, PhNSNPh,[240] olefin, R_2CO, R_2CSO,[226,172,179,241] $C_2B_7H_9$[242]	318 to −747	[11]
$[Pt(PR_3)_4]$, PR_3 = PMe_2Ph, $P(OMe)_3$,[179] $PF(OR)_2$[243]	−195 to −1297	[4]
$[\{Pt_3(CO)_6\}_n]^{2-}$, n = 1–5[244]	+122 to −558[c]	[9]
$[Pt_3(CO)_3(PR_3)_3]$[245,246]	110 to −28	[5]
$[Pt_3(CO)_3(PR_3)_4]$[246]	747 to 438	[10]
PtRh clusters[247]	989 to −6	[7]
Adducts with metal carbynes[248,249]	1651 to −477	[20]

[a] To high frequency of $\Xi(^{195}Pt)$ = 21.4 MHz; the number of reported values is given in square brackets.
[b] Data on mixed halide complexes are included.
[c] Higher values belong to the outermost triangles.

structural type, have been used in discussions of the shielding in cobalt(0)[53,95] and platinum(0)[239] complexes.

Plots of δ vs. ΔE^{-1} for Co^{III} complexes have been extrapolated to find the chemical shift corresponding to a zero σ^p contribution as given by the expression of Griffith and Orgel.[7,96,101,257,258,108] The most recent estimate, based on measurements at different temperatures, is $-6{,}500 \pm 1{,}000$ ppm {relative to $[Co(CN)_6]^{3-}$}.[110] This is between two earlier estimates, −5,000 and −7,000 ppm, based on the shifts in solid CoSi and $CoSi_2$[256] and the shielding tensors in solid $[CoCp_2]NO_3 \cdot H_2O$,[130] respectively. Results for $[PtX_3L]^-$[179] suggested that zero σ^p corresponds to −5,900 ppm [with respect to $\Xi(^{195}Pt)$ = 21.4 MHz], whereas −2,900 ppm was obtained[261] on the dubious basis that $d\sigma/dT$ should be proportional to σ^p (see Section 3.5.1). Further estimates can be made from shielding anisotropy measurements on square planar Pt^{II} complexes since $\Delta\sigma^p$ is predicted to be 1.5 times σ^p.[262] (This becomes ca. 1.7 σ^p if allowance is made for Δ_E^{-1} being ca. $0.85\Delta E_A^{-1}$.[179]) Thus results for solid salts of $[PtCl_4]^{2-}$, $[Pt(C_2O_4)_2]^{2-}$, $[Pt(NH_3)_4]^{2+}$, and $[Pt(CN)_4]^{2-}$[16] give values ranging from −1,700 to −3,300 with mean of −2,500 ppm. These estimates for the zero σ^p point are 3,000–6,000 ppm to low frequency of $[Pt(CN)_6]^{2-}$. Comparison with the estimates for ^{59}Co and between the shielding ranges in Figure 2 suggests that the equivalent points for ^{57}Fe, ^{99}Ru, and ^{103}Rh are likely to lie within a similar bracket with respect to $[M(CN)_6]^{n-}$.

3.4. Empirical Correlations

We have seen that the chemical shift of a metal nucleus in a complex depends on both the spectrochemical and nephelauxetic ranking of the ligands. As these, often conflicting, contributions are likely to vary with identity and oxidation state of the metal a single order for the shielding effect of ligands is not to be expected. Indications of the typical ranking of ligands can be obtained from Tables 3 to 15, especially the shifts of $Co^{III}L_6$ (Table 5), $Pt^{IV}L_6$ (Table 10) and $Pt^{II}L_4$ (Table 12), while the values for $[PtX_3L]^-$ (Table 13) provide a comparison of effects of a wide range of neutral ligands bonded to platinum(II).

3.4.1. Mixed Ligand Complexes

There have been several attempts to analyze the shifts of mixed ligand complexes in terms of contributions from each ligand. Each replacement of $(NH_3)_2$ in $[Co(NH_3)_6]^{3+}$ by ethylenediamine results in a similar change in $\delta(^{59}Co)$, but this is not so for replacement of NH_3 by anions,(98) which largely invalidates the ligand contributions derived by Nakashima *et al.*(263) The changes in $\delta(^{195}Pt)$ for successive replacements of one halide by another in $[PtX_6]^{2-}$ and $[PtX_4]^{2-}$ are almost constant.(158,159,36) (The small increase in the effect of the heavier halide as their number increases could be due to decrease in ΔE or a relativistic contribution). In other Pt^{II} complexes, the effect of different halides depends in a consistent way on the *trans* and *cis* ligands.(179) By contrast, successive replacements of Cl in $[PtCl_4]^{2-}$ by PMe_3 cause rapidly diminishing changes in $\delta(^{195}Pt)$.(179)

This variation in behavior may be understood qualitatively by applying the theoretical discussion of Section 3.3 to two extreme cases. If the replacement causes no change in ΔE but significant change in covalency of the M–L bond and if these changes are the same for successive replacements, the changes in σ^p should be constant as found with halide-for-halide replacements. However, if the replacement causes no change in covalency but a significant change in ΔE and the latter are the same for successive replacements, the change in σ^p, which depends on ΔE^{-1}, will not be constant. Instead, the changes will decrease as the value of ΔE increases as found for replacements of Cl by PMe_3. Bonding and steric interactions between ligands as well as the consequences of lower symmetry further complicate the picture. Thus $\delta(^{59}Co)$ for most of $[Co(NH_3)_{n-6}(NO_2)_n]$ lie outside the interval defined by the shifts of $[Co(NH_3)_6]^{3+}$ and $[Co(NO_2)_6]^{3-}$.(100)

Simple additivity rules predict the same shift for isomers, which is not found in practice. For a series of complexes $[CoX_4Y_2]$, the *cis* isomer had the greater shift (i.e., to high frequency),(98) while for $[PtX_2Y_2]$, McFarlane(83) found the reverse to be true $\{[PtI_2(PMe_3)_2]$ is an exception(179)$\}$.

Clearly, such generalizations are unlikely to be valid in view of the possible differences in ΔE and M–L covalency between isomers. Nonetheless, metal resonances provide a valuable means of distinguishing isomers.(117,98,179) This is still true when the isomerism does not involve different arrangements of the donor atoms, although the shift differences are less and probably depend mainly on steric effects.(112,113,264,67,83) There is almost 100 ppm between the ^{195}Pt resonances of the enantiomers of the purely inorganic $[Pt(S_5)_3]^{2-}$.(160)

3.4.2. *Effect of Substituents on the Donor Atom*

The ^{59}Co chemical shifts of Co^{III} complexes are clearly grouped together according to the identity of the donor atoms, i.e., into those with coordination spheres CoO_6 etc.[1,7] This no doubt results from similar nephelauxetic effects and not too dissimilar spectrochemical rankings for ligands with the same donor atom.

A like situation applies to ^{103}Rh and ^{195}Pt (e.g., SMe_2 vs. $SOMe_2$[145,142,179]) and should be true of the other metals. Thus metal chemical shifts can be a useful aid in the investigation of the coordination mode adopted by ambidentate ligands such as SCN^- and Me_2SO.[205,192,142] Interestingly, the shifts of $[Co_2(NH_3)_6(\mu\text{-}OH_2)(\mu\text{-}RCO_2)]$ do not vary with R although there are large differences in linewidth as the basicity of RCO_2^- affects the electric field gradient.[124]

The shifts of $[PtCl_3(PR_3)]^-$ and *cis*-$[PtCl_2(PR_3)_2]$ do not vary greatly with R and the ranges are not much extended when phosphite and fluorophosphite complexes are included[179,143] so that reasonable estimates of the shifts of complexes of other phosphines may be made. This also seems true for the Rh^I complexes *trans*-$[RhCl(CO)(PR_3)_2]$,[148,149,82] but the range seems greater for Rh^{III} (e.g., *mer*-$[RhX_3(PR_3)_3]$, Table 8), which may result from the greater steric interactions in octahedral complexes.

Replacement of the methyl group in $[PtCl_2(Me)CO]^-$ by larger alkyls causes little change in $\delta(^{195}Pt)$ in contrast to the behavior of $\delta(^{199}Hg)$ for HgR_2 and $[HgX(R)]$, X = Cl, Br, or I.[221] Further, the changes for $[PtCl_2(R)CO]^-$ and $[RhI(R)(CNBu^t)_4]^+$[1] are in the opposite direction to the mercury compounds.

An important example of the effect of substituents on the donor atom is the conversion of OH_2 to OH^-. Generally, this seems to produce only a small difference in shifts, e.g., $[Co(NH_3)_5OH_2]^{3+}$ +80,[265] $[PtCl_2(OH_2)_2]$ −45[17] $[Pt(OH_2)_4]^{2+}$ −110,[163] and *cis*-$[Pt(NH_3)_2(OH_2)_2]$ ca. 0 ppm.[26,206] However, for $[PtCl_n(OH_2)_{6-n}]$ much larger increases in shift occur.[157]

The formation of a chelating ring can be regarded as a substituent effect. For $[Co\{NH_2(CH_2)_nNH_2\}_3]^{3+}$ $\delta(^{59}Co)$ increases by more than 1,500 ppm as n goes from 2 to 4,[103] but for $[PtMe_2\{PPh_2(CH_2)_nPPh_2\}]$ the values of $\delta(^{195}Pt)$ for $n = 2$, 3, and 4 are all close to that for $[PtMe_2(PMePh_2)_2]$.[222] However, the shift of $[PtMe_2(PPh_2CH_2PPh_2)]$ is some 700 ppm higher, no doubt as a result of the strain in the four-membered ring. Similarly, increasing the size of the rings in $[RhClBu^t_2P(CH_2)_nCH{=}CH(CH_2)_mPBu^t_2]$, n, $m = 1$ or 2 reduces the value of $\delta(^{103}Rh)$.[150] Dimeric platinum complexes frequently involve four-membered rings and show ^{195}Pt chemical shifts several hundred ppm to high frequency of their monomeric analogs, e.g., $[Pt_2X_6]^{2-}$,[1] $[Pt_2X_4L_2]$,[179] $[PtMe_2X_2(\mu\text{-}OH)_2]$,[171] and $[Pt_2(\mu\text{-}OH)_2(NH_3)_4]^{2+}$.[26,206]

3.5. Effect of the Molecular Environment and Isotopes

3.5.1. *Temperature*

The large temperature dependence of ^{59}Co chemical shifts was noticed in 1951,[266] and subsequent investigation showed it to be associated with the variation of ΔE.[96] Benedek *et al.* ascribed this to the changes in the mean vibrational energies of the ground and excited electronic states (which form part of ΔE).[267] However, with the more recent estimates of the magnitude of σ^p, the temperature dependence of

$[Co(CN)_6]^{3-}$ exceeds the maximum possible for an octahedral complex with 15 normal coordinates. Alternatively, owing to the anharmonicity of vibrations, increasing the temperature will increase the mean M–L distance and so ΔE^{-1} (see Chapter 3, Section 6.2).

Such approaches cannot account for the significant difference in the temperature coefficients observed for $[Co(acac)_3]$ in toluene and chloroform (2.3 and 3.0 ppm K^{-1}, respectively).[96] The observed shift may be the weighted average of the shifts of several species if they are in sufficiently rapid equilibrium. In general, temperature will alter the equilibrium position and hence the average value. Weak solute–solvent and solute–solute complexes are possible contributors, and the low-frequency shift of $[Co(acac)_3]$ in chloroform[96] suggests that complex formation is increasing the temperature coefficient of $\delta(^{59}Co)$ in this solvent. Such a mechanism may, of course, decrease the temperature coefficient to give very small or even negative values.[25]

Brown and Cohen[261] state that the temperature coefficient is proportional to σ^p. However, the equation of Benedek *et al.*[267] shows it to be proportional to $K(\Delta E^{-1})^2$ and hence $K(\sigma^p)^2$, where K is not expected to be constant between complexes [but is not proportional to $(\sigma^p)^{-1}$].

When comparing results at different temperatures it is desirable to make some estimate of the change in δ due to the temperature difference. An assessment of the likely magnitude may be made from the temperature coefficients given in Table 16.

The resonances of ^{59}Co have been proposed as NMR thermometers.[33]

3.5.2. Pressure

Benedek *et al.* found that the resonances of $[Co(CN)_6]^{3-}$, $[Co(NO_2)_6]^{3-}$, and $[Co(acac)_3]$ moved 22.3, 46.4, and 15.1 ppm, respectively, to low frequency when a pressure of 1000 kg cm^{-2} was applied. The behavior is not strictly linear, $d\sigma/dP$ decreasing by ca. 25% by 10,000 kg cm^{-2}. They interpret the effect of pressure as reducing the metal–ligand distance and hence ΔE^{-1} and so σ^p.[267]

Table 16. Temperature Coefficients of Some Transition Metal Chemical Shifts

Complex	$d\delta(M)/dT$[a]	Complex	$d\delta(M)/dT$[a]
$[Fe(CO)_5]$	0.55[41]		
$[Co(OH_2)_6]^{3+}$	ca. 2.5[60]	$[Co(acac)_3]$	2.3[96]
$[Co(NH_3)_6]^{3+}$	1.42[267]	$[Co(NO_2)_6]^{3-}$	2.85[267]
$[Co(CN)_6]^{3-}$	1.38[96]		
$[RuO_4]$	0.13[67]	$[Ru(CN)_6]^{4-}$	>1[67]
mer-$[RhCl_3(SMe_2)_3]$	1.0[81]	$[Rh(acac)_3]$	1.6[12]
trans-$[RhCl(CO)(PMe_2Ph)_2]$	0.26[148]	$[RhH(CO)(PPh_3)_3]$	0.44[268]
$[PtCl_6]^{2-}$	0.84[158]	*cis*-$[PtCl_4(PPr^n_3)_2]$	0.84[269]
trans-$[PtCl_4(SMe_2)_2]$	0.25[165]	$[PtCl_3(PMe_3)]^-$	0.5[179]
trans-$[PtCl_2(PBu^n_3)_2]$	0.34[269]	*trans*-$[PtBr_2(PPr^n_3)_2]$	0.33[269]
cis-$[PtCl_2(PEtPh_2)_2]$	0.54[269]	*trans*-$[PtHCl(PPr^n_3)_2]$	ca. 0.1[269]
trans-$[PtHBr(PBu^n_3)_2]$	0.12[269]	*trans*-$[PtBr_2(SMe_2)_2]$	0.25[179]
cis-$[PtCl_2(SMePh)_2]$	0.4[83]	*trans*-$[PtCl_2(SMePh)_2]$	0.4[83]
trans-$[PtEtBr(PBu^n_3)_2]$	ca. 0.5[269]		

[a] ppm K^{-1}.

3.5.3. Medium

There is a good correlation between $\delta(^{195}Pt)$ of $[Pt(NCC{\equiv}CCN)(PPh_3)_2]$ and ΔE^{-1} in different solvents(239) and it is reasonable to assume that differences in metal chemical shifts between solvents correspond to small changes in ΔE caused by interactions with the metal or the ligands. Thus Gonzalez *et al.* have demonstrated the (inverse) relationship between $\delta(^{59}Co)$ of $[Co(en)_3](ClO_4)_3$ and the donor number of the solvent: greater donation by the solvent to the amine protons increases the amine interaction with the metal and hence ΔE.(270) In general, we may expect the sensitivity of the shift to solvent to be greater for complexes with greater σ^p—e.g., the range for $[PtCl_6]^{2-}$ is almost twice that of $[PtCl_4]^{2-}$ (14,78)—and for metals with larger "shielding sensitivities" (especially ^{59}Co). Although there are not very many studies of complexes in a representative range of solvents {e.g., $[Co(acac)_3]$,(96,271) *trans*-$[PtCl_2(AsBu^n_3)_2]$,(37) and $[PtN(Ph)NC(Ph)NNC_6H_4O(NH_3)]$(272) in addition to those above} it is clear that there is no common ranking of solvents for different complexes as is only to be expected in view of the different types of interaction that are possible.

The sensitivity of metal shifts to weak interactions with solvent and other species present in solution provides a convenient means of studying these processes. In this way the preferential solvation of $[Co(acac)_3]$,(271) hydrogen bonding between $[Co(CN)_6]^{3-}$ and acidic solvents,(52) the formation of outer sphere complexes between $[Co(en)_3]^{3+}$ and PO_4^{3-} and SO_4^{2-},(273,58) and ion pairing of tetraalkyl ammonium salts of $[Co(CN)_6]^{3-}$ (274) have been investigated. Lanthanide shift reagents can decrease $\delta(^{59}Co)$ of cobalt diketonates by more than 400 ppm.(275)

3.5.4. Magnetic Field

Although the theoretical analysis of Doddrell *et al.*(276) showed that even for ^{59}Co the field dependence of the chemical shift would be insignificant at currently available fields, they rashly made a comparison of $\delta(^{59}Co)$ at 0.34 and 2.10 T.(80) For $[Co(NH_3)_6]^{3+}$ and $[Co(acac)_3]$ {but not $[Co(CN)_6]^{3-}$} they found a significant difference (ca. 0.3 ppm), which has the opposite sense as well as much greater magnitude compared to their predictions.

3.5.5. Isotopes

Approximately equal decrements of 0.91 ppm in $\delta(^{59}Co)$ of $[Co(CN)_6]^{3-}$ occur on each replacement of ^{12}C by ^{13}C. This was interpreted by Lauterbur(277) in terms of the change in mean vibrational energy contributions to ΔE (cf. the temperature dependence above), whereas Loewenstein and Shporer(278) invoke the decrease in Co–C distance (R) and hence increase in ΔE which corresponds to Jameson's discussion of isotope effects in terms of anharmonicity.(279) According to this, the shift change should relate to $d\sigma/dR$ and is thus expected to be larger for nuclei with larger "shielding sensitivities." Other examples of chemical shift change with donor atom isotope are $^1H/^2H$ *trans*-$[PtHX(PEt_3)_2]$, $X = Cl$ -12.4, $X = CN$ -8.7(84); $^{12}C/^{13}C$ $[Ru(CN)_6]^{4-}$ -0.37(67) $[\{Rh(C_5H_5)\}_2C_7H_7]^+$ -0.41,(87) $[PtCl(CN)_5]^{2-}$ -0.77, $[Pt(CN)_4]^{2-}$ -0.28,(84) $[Pt_2Br_4(CO)_2]^{2-}$ -0.8(182); $^{16}O/^{17}O$ RuO_4 -0.02(67); $^{16}O/^{18}O$ $[Pt(OH_2)_2]^{2+}$ -1.00, *trans*-$[PtCl_2(OH_2)_2]$ -0.81, *trans*-$[PtCl_2(OH)_2]^{2-}$

−0.63[17]; $^{35}Cl/^{37}Cl$ $[PtCl_6]^{2-}$ −0.17, $[PtCl_4]^{2-}$ −0.16; $^{79}Br/^{81}Br$ $[PtBr_6]^{2-}$ −0.03 ppm[34] (all per single isotopic substitution). The effect of replacing ^{14}N by ^{15}N in $[Co(CN)_6]^{3-}$ is small (−0.03 ppm),[277] but each $^1H/^2H$ exchange in $[Co(en)_3]^{3+}$ or $[Co(NH_3)_6]^{3+}$ moves $\delta(^{59}Co)$ ca. 5 ppm to low frequency, which may be interpreted in terms of the greater basicity of the deuterated amine.[280,281]

4. SPIN–SPIN COUPLING

In contrast to ^{103}Rh and ^{195}Pt, there are few examples of spin–spin coupling to the other group VIII nuclei. In general, coupling constants to ^{57}Fe, ^{61}Ni, ^{99}Ru, and ^{187}Os are expected to be small following the values of γ. Satellites due to coupling of an observed nucleus to ^{57}Fe or ^{187}Os at their low natural abundance are difficult to locate. A highly symmetric environment is required to resolve coupling to a quadrupolar nucleus, and no examples have been reported for iridium or palladium. Despite the difficulties of observation, coupling of ^1H or ^{13}C to ^{187}Os or ^{57}Fe has provided valuable information on the structure and dynamic behavior of complexes.[282–284,9]

Table 17 gives typical ranges for some coupling constants to group VIII nuclei. If the bonding terms in the usual equation for one-bond coupling constants are the same, then the coupling constants to two different metals should be in the ratio of their values of $\gamma\ |\psi_s(0)|^2$. This suggests that there should be an approximately constant ratio between the values of coupling constants to two different metals and that this ratio could be used to estimate likely values of coupling constants from known values of those to another metal nucleus (especially ^{195}Pt or ^{103}Rh). One may obtain such guidelines for the relative magnitude of coupling constants by comparison of the coupling constants in similar compounds of different metals {e.g., $^1J(M\,^{13}C)$ in $[Co(CN)_6]^{3-}$ and $[Pt(CN)_6]^{2-}$}. Guideline ratios for couplings of metal nuclei relative to ^{103}Rh are given in Table 18. For ^{195}Pt, where there are several examples, $J(^{195}PtX)$ is 16–22 times $J(^{103}RhX)$ and we have used the round figure of 20. Table 18 shows that there is, indeed, a fair correlation between these ratios and the relative values of $\gamma\ |\psi_s(0)|^2$.

Several reviews have considered possible relationships between $^1J(^{195}Pt\,^{31}P)$ and structural features of the complexes.[296,310,300] Although there are reasonable correlationships with bond lengths, oxidation state, and *trans* influence, too detailed quantitative analysis is hazardous as illustrated by the results for $[PtR(PR')_3]^+$.[311] Koie *et al.* have demonstrated the correlation between $^1J(PtP)$ in some Pt^0 and Pt^{II} complexes and the mutual polarizability from extended Hückel MO calculations.[239,312] Relationships between structural features and $^1J(^{195}Pt\,^1H)$,[292,300,313] $^1J(^{195}Pt\,^{13}C)$ in CO,[314] CH_3, and cod[315] derivatives, $^2J(^{195}Pt\,^1H)$ in CH_3 derivatives, $^2J(^{195}Pt\,^{19}F)$ in CF_3 derivatives,[300] or $^1J(^{195}Pt\,^{119}Sn)$ in $SnCl_3$ derivatives[307] have also been considered. There is a good correlation between $^1J(^{195}Pt\,^{15}N)$ in *trans*-$[PtCl_2L(NH_2R)]$ and $^1J(^{195}Pt\,^{31}P)$ in *trans*-$[PtCl_2L(PBu_3]$.[198] The very large variation in values of $^1J(^{195}Pt\,^{19}F)$, $^1J(^{195}Pt\,^{77}Se)$, and $^1J(^{195}Pt\,^{125}Te)$ (see Table 17) is probably related to the presence of a lone pair on the donor atom.

An initial collection of values of $J(^{195}Pt\,^{195}Pt)$ displayed a remarkably wide range with little relationship to the structure of the complex.[195] Subsequent results for complexes of Pt^0,[245,246,316,317] Pt^I,[182,225,236,237,318] Pt^{II},[217,171,190,191,307,188,201,319] and

Pt^{III} [234] do not substantially alter the picture and it is clear that values of $^{1}J(^{195}Pt\,^{195}Pt)$ may be less than those of $^{2}J(^{195}Pt\,^{195}Pt)$.

Although there may be difficulties in quantitative interpretation, spin–spin coupling to metal nuclei especially from ^{31}P [320,296] and ^{15}N (using ^{15}N enriched ligands) [27,28,164,178,203,231] is a valuable tool for structural investigation. In cluster compounds, complex spin systems may arise because of the coupling between metal nuclei, e.g., $[AB_2]$, $A = B = {}^{195}Pt$ [317], $[AX]_n$, $n = 3, 4$, $A = {}^{103}Rh$, $X = {}^{1}H$, [131,321] whose successful analysis give useful structural information.

The variation of $^{1}J(^{195}Pt\,^{31}P)$ of *cis*- and *trans*-$[PtCl_2(PEt_3)_2]$ with solvent and temperature have been investigated.[322] The values increase with temperature as do those of $^{1}J(^{195}Pt\,^{13}C)$ in one isomer of $[PtCl(Ph)CO(PR_3)]$. However, the reverse is true for the other isomer, suggesting that hybridization changes with solvation are not a complete answer.[314]

4.1. Sign Determinations

The positive sign established for $^{1}J(^{195}Pt\,^{31}P)$ in *cis*- and *trans*-$[PtCl_2(PEt_3)_2]$[323] is the most common basis for the determination of the signs of other couplings in complexes of these metals, since, in view of its magnitude, it is unlikely to reverse in any other complex. Other one-bond couplings shown to be positive are $^{1}J(^{125}Te\,^{103}Rh)$ ($TeMe_2$ complexes),[145] $^{1}J(^{195}Pt\,^{1}H)$,[324] $^{1}J(^{195}Pt\,^{13}C)$ (^{13}CO complexes),[202] (C_2H_4 complex),[325] and $^{1}J(^{195}Pt\,^{195}Pt)$,[195] while $^{1}J(^{103}Rh\,^{1}H)$,[148] $^{1}J(^{103}Rh\,^{19}F)$,[82] $^{1}J(^{103}Rh\,^{31}P)$,[148,293,143] $^{1}J(^{195}Pt\,^{29}Si)$ (SiH_2Cl complex),[167] and $^{1}J(^{195}Pt\,^{125}Te)$ ($TeMe_2$ complexes)[304] are negative. All these follow the signs of the magnetogyric

Table 17. Ranges of Some Spin–Spin Coupling Constants with Group VIII Metals

J	Complex type/ligands	Range (Hz)
$^{1}J(^{57}Fe\,^{13}C)$	$[Fe(CO)_5]$ etc.[51,284–286]	23–32
	$[Fe(C_5H_5)_2]$ etc.[9,86,85]	2–6
$^{2}J(^{57}Fe\,^{13}C)$	Fe-dienyl[10]	0–1.4
$^{1}J(^{57}Fe\,^{15}N)$	Fe porphyrin derivatives[287]	7–8
$^{1}J(^{57}Fe\,^{31}P)$	$[Fe(CO)_4(PR_3)]$ etc.[288,47]	26–31
$^{1}J(^{59}Co\,^{13}C)$	$[Co(CO)_4]^-$,[53] $K_3[Co(CN)_6]$[289]	287, 126
$^{1}J(^{59}Co\,^{15}N)$	$[Co(NH_3)_6]^{3+}$ etc.[290]	63
$^{1}J(^{59}Co\,^{31}P)$	$[Co(PF_3)_4]^-$,[53] $[Co\{P(OR)_3\}_6]^{3+}$[109]	1222, 414
$^{1}J(^{61}Ni\,^{31}P)$	$[Ni(CO)_3\,PBu^t_3]$,[62] $[Ni\{P(OR)_3\}_4]$, $[Ni(PF_3)_4]$[63]	232–482
$^{1}J(^{99}Ru\,^{13}C)$	$K_4[Ru(CN)_6]$[67]	44.8
$^{1}J(^{99}Ru\,^{17}O)$	$[RuO_4]$[67]	23.4
$^{1}J(^{119}Sn\,^{99}Ru)$	$[Ru(SnCl_3)_5\,Cl]^{4-}$[291]	846
$^{1}J(^{103}Rh\,^{1}H)$	$[RhHL_4]$[292]	0–8
	Rh^{III} hydrides[292]	15–30
$^{2}J(^{103}Rh\,^{1}H)$	Methyl derivatives[292,148]	ca. 2
$^{3}J(^{103}Rh\,^{1}H)$	PMe_3 etc.[148,293]	0.3–1.5
$^{1}J(^{103}Rh\,^{13}C)$	Olefin, allyl etc.[294]	5–16
	CO(terminal)[294]	55–85
	Alkyl, aryl[294]	11–43
$^{1}J(^{103}Rh\,^{15}N)$	Amine[295]	13–18
$^{2}J(^{103}Rh\,^{17}O)$	$[RhCp(CO)_2]$[156]	7.2
$^{1}J(^{103}Rh\,^{19}F)$	*trans*-$[RhF(CO)(PPh_3)_2]$[82]	52.5

Table 17. (*Continued*)

J	Complex type/ligands	Range (Hz)
$^{2}J(^{103}Rh\,^{19}F)$	CF_3,[148] fluorophosphine[143]	12–32
$^{1}J(^{103}Rh\,^{31}P)$	phosphine[296]	74–194^a
$^{1}J(^{103}Rh\,^{103}Rh)$	(297, 131, 153)	4–17^b
$^{1}J(^{119}Sn\,^{103}Rh)$	$SnCl_3$ etc.[196,298]	452–986
$^{1}J(^{125}Te\,^{103}Rh)$	*mer*-$[RhX_3(TeMe_2)_3]$[145]	66–94
$^{1}J(^{195}Pt\,^{103}Rh)$	Mixed metal clusters[247]	15–73
$^{1}J(^{187}Os\,^{1}H)$	$[OsH_4(PR_3)_3]$,[299] CO clusters[282]	16–38
$^{1}J(^{187}Os\,^{13}C)$	$[Os_3(CO)_{12}]$ etc.[283]	84–121
$^{1}J(^{187}Os\,^{119}Sn)$	$[Os(SnCl_3)_5Cl]^{4-}$ [291]	1290
$^{1}J(^{195}Pt\,^{1}H)$	Pt^{II},[292,300] Pt^{IV} hydrides[167]	700–1370
$^{2}J(^{195}Pt\,^{1}H)$	Methyl[300]	40–90
$^{3}J(^{195}Pt\,^{1}H)$	PMe_3 etc.[293]	15–57
$^{1}J(^{195}Pt\,^{11}B)$	Borane[301]	200–400
$^{1}J(^{195}Pt\,^{13}C)$	Olefin, allyl etc.[294]	38–429
	CO(terminal), CNR[294]	884–2459
	Alkyl, aryl,[294] CN,[84] CCR[227]	38–429
$^{2}J(^{195}Pt\,^{13}C)$	Alkyl, aryl, vinyl,[294] CCR[227]	0–362
$^{3}J(^{195}Pt\,^{13}C)$	CCR,[227] alkyl, aryl,[294]	11–111
$^{1}J(^{195}Pt\,^{15}N)$	Amine[198,203,206]	138–410
	NCO, NCS,[205,192] NO_2[164]	280–757^c
$^{2}J(^{195}Pt\,^{15}N)$	CNR[166]	70–170^c
$^{3}J(^{195}Pt\,^{15}N)$	SCN[178]	10–20
$^{1}J(^{195}Pt\,^{19}F)$	$[PtF(PR_3)_3]^+$,[302] $[PtF_6]^{2-}$, etc.[303]	200–2000
$^{2}J(^{195}Pt\,^{19}F)$	CF_3,[300] fluorophosphine[143]	236–1045
$^{1}J(^{195}Pt\,^{29}Si)$	*trans*-$[PtCl(SiH_2Cl)(PEt_3)_2]$[167]	ca. 1600
$^{1}J(^{195}Pt\,^{31}P)$	Phosphine[296]	1000–4000^a
$^{2}J(^{195}Pt\,^{31}P)$	PO_4H^{2-} [234]	41–93
$^{1}J(^{195}Pt\,^{77}Se)$	$SeMe_2$[304,305]	−68–670
$^{1}J(^{195}Pt\,^{109}Ag)$	$[L_3PtN{\sim}NAgBr]$[306]	ca. 170
$^{1}J(^{195}Pt\,^{119}Sn)$	$SnCl_3$,[203,307] μ-$SnCl_2$[188]	5780–35 309
$^{1}J(^{195}Pt\,^{125}Te)$	$TeMe_2$(terminal),[308] μ-$TeMe_2$[304]	400–5923
$^{1}J(^{195}Pt\,^{183}W)$	$Cp(CO)_2W{=}C{-}$ [248]	78–177
$^{1}J(^{195}Pt\,^{195}Pt)$	(195)	188–9007
$^{2}J(^{195}Pt\,^{195}Pt)$	$Pt(\mu\text{-}Y)_2Pt$, $Pt(\mu\text{-}Y)Pt$[195,188]	0–2097
$^{1}J(^{207}Pb\,^{195}Pt)$	PbR_3[212]	14 200–18 515

^{a} Values for phosphite and fluorophosphite complexes[143] are larger (up to twice). Coupling to Rh^{III} and Pt^{IV} is ca. 2/3 of that to Rh^{I} and Pt^{II}, respectively.
^{b} Some much larger values are likely to be artifacts of incorrect analysis of the spin systems.[309]
^{c} Some values calculated from $^{1}J(^{195}Pt\,^{14}N)$.

Table 18. Relative Magnitudes of Spin–Spin Coupling Constants

	Ratio relative to ^{103}Rh			Ratio relative to ^{103}Rh	
Nucleus	Guideline^a	$\gamma\,\lvert\psi_s(0)\rvert^{2\,b}$	Nucleus	Guideline^a	$\gamma\,\lvert\psi_s(0)\rvert^{2\,b}$
^{57}Fe	0.15	0.56	^{99}Ru	1	−1.44
^{59}Co	3	4.38	^{187}Os	1.5	1.51
^{61}Ni	1.5	−1.74	^{195}Pt	20	15.9

^{a} See text for derivation.
^{b} $\lvert\psi_s(0)\rvert^2$ values for metal with zero charge[296]; sign is the sign of γ (negative for ^{103}Rh).

ratios, which is unusual for ^{19}F. According to oxidation state and the other substituents, $^{1}J(^{195}Pt\,^{77}Se)$ in $SeMe_2$ complexes may have either sign.(304,305)

Positive signs have also been found for $^{2}J(^{103}Rh\,^{1}H)$, CH_3,(148); $^{3}J(^{195}Pt\,^{1}H)$, CH_2CH_3,(221) $CH_2{=}C{=}CH_2$,(186) $P(CH_3)_3$ etc.,(323,293,326) SCH_3,(327) $Se(CH_3)_2$(305); $^{2}J(^{195}Pt\,^{14}N)$, CNMe(166,195); $^{2}J(^{195}Pt\,^{19}F)$, CF_3,(328) C_2F_4,(327) SiH_2F,(167) PF_3 etc.,(143) and $^{3}J(^{195}Pt\,^{19}F)$, $CF_3C{=}CCF_3$,(327) $(CF_3)_2C{=}O$(195) while $^{3}J(^{103}Rh\,^{1}H)$, $P(CH_3)_3$ etc.(293,148,149); $^{2}J(^{103}Rh\,^{19}F)$, CF_3,(148) PF_3 etc.(143); $^{4}J(^{195}Pt\,^{1}H)$, $P(OCH_3)_3$,(184) $CNCH_3$(166); $^{3}J(^{195}Pt\,^{31}P)$, $(\mu\text{-SMe})_2PtP$,(327) $(\mu\text{-X})_2PtP$(195,191) were negative. Although $^{2}J(^{195}Pt\,^{1}H)$ is positive in alkyl(224,167,329,172,221,327,173) or π-olefin(330,186) derivatives, it has the opposite sign for silyl and germyl compounds.(167,214) $^{2}J(^{195}Pt\,^{31}P)$ is negative when $Pt-\widehat{Pt}-P$ is ca. 90° but positive (and of greater magnitude) when the angle is 180° or nearly so.(195,237,312) Opposite signs have been found for $^{2}J(^{195}Pt\,^{195}Pt)$ in the singly bridged anion $[PtCl_3(\mu\text{-SeMe}_2)PtCl_2SeMe_2]^-$(195) and the doubly bridged $[PtMe(PMe_2Ph)(\mu\text{-SMe})_2PtMe(PMe_2Ph)]$.(327)

REFERENCES

1. Kidd, R. G.; Goodfellow, R. J. In *NMR and the Periodic Table*; Harris, R. K.; Mann, B. E., Eds.; Academic; London, 1978; Chap. 8.
2. Rinaldi, P. L.; Levy, G. C.; Choppin, G. R. *Rev. Inorg. Chem.* 1980, **2**, 53–92.
3. Forsén, S.; Lindman, B. *Methods Biochem. Anal.* 1981, **27**, 289–486
4. Pregosin, P. S. *Coord. Chem. Rev.* 1982, **44**, 247–291.
5. Mann, B. E. In *NMR of Newly Accessible Nuclei, Volume 2*; Laszlo, P., Ed.; Academic; New York, 1983; Chap. II.
6. Carte, G. C.; Bennet, L. H.; Kahan, D. J. *Prog. Materials Sci.* 1976, **20**.
7. Kidd, R. G. *Annu. Rep. NMR Spectrosc.* 1980, **10A**, 1–79.
8. Brevard, C.; Granger, P. *Handbook of High Resolution Multinuclear NMR*; Wiley-Interscience: New York, 1981.
9. Koridze, A. A.; Petrovskii, P. V.; Gubin, S. P.; Fedin, E. I. *J. Organomet. Chem.* 1975, **93**, C26–C30.
10. Koridze, A. A.; Petrovskii, P. V.; Astrakhova, N. M.; Vol'kenau, N. A.; Petrakova, V. A.; Nesmeyanov, A. N. *Dokl. Akad. Nauk. SSSR* 1980, **255**, 117–120; *Dokl. Chem. (Engl. Transl.)* 1980, 500–502.
11. Schwenck, A. *Z. Phys.* 1968, **213**, 482–502.
12. Grüninger, K.-D.; Schwenk, A.; Mann, B. E. *J. Magn. Reson.* 1980, **41**, 354–357.
13. Maurer, E.; Rieker, S.; Schollbach, M.; Schwenk, A.; Egolf, T.; von Philipsborn, W. *Helv. Chim. Acta* 1982, **65**, 26–45.
14. Pesek, J. J.; Mason, W. R. *J. Magn. Reson.* 1977, **25**, 519–529.
15. Doddrell, D. M.; Barron, P. F.; Clegg, D. E.; Bowie, C. *J. Chem. Soc., Chem. Commun.* 1982, 575–576.
16. Keller, H. J.; Rupp, H. H. *Z. Naturforsch. Teil A* 1971, **26**, 785–786; Rupp, H. H. *Z. Naturforsch. Teil A* 1971, **26**, 1937–1939.
17. Gröning, Ö.; Drakenberg, T.; Elding, L. I. *Inorg. Chem.* 1982, **21**, 1820–1824.
18. Lallemand, J. Y.; Soulié, J.; Chottard, J. C. *J. Chem. Soc., Chem. Commun.* 1980, 436–438.
19. Cocivera, M.; Ferguson, G.; Lenkinski, R. E.; Szczecinski, P. *J. Magn. Reson.* 1982, **46**, 168–171.
20. Ismail, I. M.; Kerrison, S. J. S.; Sadler, P. J. *Polyhedron* 1982, **1**, 57–59.
21. Fritz, H. P.; Schwarzhans, K. E. *J. Organomet. Chem.* 1966, **5**, 103–105.
22. Goodfellow, R. J. unpublished results.
23. Balimann, G.; Venanzi, L. M.; Bachechi, F.; Zambonelli, L. *Helv. Chim. Acta* 1980, **63**, 420–432.
24. Motschi, H.; Pregosin, P. S. *Inorg. Chim. Acta* 1980, **40**, 141–146.
25. Herbert, I. R. Ph.D. thesis, University of Bristol, 1984.
26. Boreham, C. J.; Bromhead, J. A.; Fairlie, D. P. *Aust. J. Chem.* 1981, **34**, 659–664.
27. van der Poel, H.; van Koten, G.; Grove, D. M.; Pregosin, P. S.; Ostoja Starzewski, K. A. *Helv. Chim. Acta* 1981, **64**, 1174–1182.
28. Wood, F. E.; Balch, A. L. *Inorg. Chim. Acta* 1983, **76**, L63–L64.

29. Valiyev, K. A.; Zripov, M. M. *Zh. Strukt. Khim.* 1966, **7**, 494–504; *J. Struct. Chem. (Engl. Transl.)* 1966, **7**, 470–478.
30. Doddrell, D. M.; Bendall, M. R.; Healy, P. C.; Smith, G.; Kennard, C. H. L.; Raston, C. L.; White, A. H. *Aust. J. Chem.* 1979, **32**, 1219–1230.
31. Yamasaki, A.; Yajimi, F.; Fujiwara, S. *Inorg. Chim. Acta* 1968, **2**, 39–42.
32. Wehrli, F. W.; Wehrli, S. *J. Magn. Reson.* 1981, **44**, 197–207. Akitt, J. W.; McDonald, W. S. *J. Magn. Reson.* 1984, **58**, 401–412.
33. Levy, G. C.; Bailey, J. T.; Wright, D. A. *J. Magn. Reson.* 1980, **37**, 353–356.
34. Ismail, I. M.; Kerrison, S. J. S.; Sadler, P. J. *J. Chem. Soc., Chem. Commun.* 1980, 1175–1176.
35. Drain, L. E. *J. Phys. Chem. Solids* 1963, **24**, 379–382.
36. von Zelewsky, A. *Helv. Chim. Acta* 1968, **51**, 803–807.
37. Pidcock, A.; Richards, R. E.; Venanzi, L. M. *J. Chem. Soc. (A)* 1968, 1970–1973.
38. Gill, D. S.; Gansow, O. A.; Bennis, F. J.; Ott, K. C. *J. Magn. Reson.* 1979, **35**, 459–461.
39. Mann, B. E.; Spencer, C. *Inorg. Chim. Acta* 1982, **65**, L57–L58.
40. Schwenk, A. *J. Magn. Reson.* 1971, **5**, 376–389.
41. Schwenk, A. *Phys. Lett. Sect. A* 1970, **31**, 513–514.
42. Kaufmann, J.; Schwenk, A. *Phys. Lett. Sect. A* 1967, **24**, 115–116.
43. Kronenbitter, J.; Schwenk, A. *J. Magn. Reson.* 1977, **25**, 147–165.
44. Morris, G. A.; Freeman, R. *J. Am. Chem. Soc.* 1979, **101**, 760–762.
45. Doddrell, D. M.; Pegg, D. T.; Bendall, M. R. *J. Magn. Reson.* 1982, **48**, 323–327.
46. Brevard, C.; van Stein, G. C.; van Koten, G. *J. Am. Chem. Soc.* 1981, **103**, 6746–6748.
47. Brevard, C.; Schimpf, R. *J. Magn. Reson.* 1982, **47**, 528–534.
48. Feeney, J.; Partington, P. *J. Chem. Soc., Chem. Commun.* 1973, 611–612.
49. Crocker, C.; Errington, R. J.; McDonald, W. S.; Odell, K. J.; Shaw, B. L.; Goodfellow, R. J. *J. Chem. Soc., Chem. Commun.* 1979, 498–499.
50. Haslinger, E.; Robien, W.; Schlögl, K.; Weissensteiner, W. *J. Organomet. Chem.* 1981, **218**, C11–C14.
51. Jenny, T.; von Philipsborn, W.; Kronenbitter, J.; Schwenk, A. *J. Organomet. Chem.* 1981, **205**, 211–222.
52. Eaton, D. R.; Rogerson, C. V.; Sandercock, A. C. *J. Phys. Chem.* 1982, **86**, 1365–1371.
53. Lucken, E. A. C.; Noack, K.; Williams, D. F. *J. Chem. Soc. (A)* 1967, 148–154.
54. Rose, K. D.; Bryant, R. G. *Inorg. Chem.* 1979, **18**, 1332–1335.
55. Rose, K. D.; Bryant, R. G. *Inorg. Chem.* 1979, **18**, 2130–2131.
56. Ader, R.; Loewenstein, A. *J. Magn. Reson.* 1971, **5**, 248–261.
57. Jordan, R. B. *J. Magn. Reson.* 1980, **38**, 267–275
58. Craighead, K. L.; Jones, P.; Bryant, R. G. *J. Phys. Chem.* 1975, **79**, 1868–1874.
59. Craighead, K. L.; Bryant, R. G. *J. Phys. Chem.* 1975, **79**, 1602–1603; Rose, K.; Bryant, R. G. *J. Magn. Reson.* 1979, **35**, 223–226.
60. Navon, G. *J. Phys. Chem.* 1981, **85**, 3547–3549.
61. Au-Yeung, S. C. F.; Eaton, D. R. *Inorg. Chim. Acta* 1983, **76**, L141–L144.
62. Schumann, H.; Meissner, M.; Kroth, H. J. *Z. Naturforsch., Teil B* 1980, **35**, 639–641.
63. Hao, N.; McGlinchey, M. J.; Sayer, B. G.; Schrobilgen, G. J. *J. Magn. Reson.* 1982, **46**, 158–162.
64. Brevard, C.; Granger, P. *J. Chem. Phys.* 1981, **75**, 4175–4177.
65. Dykstra, R. W.; Harrison, A. M. *J. Magn. Reson.* 1981, **45**, 108–117.
66. Dykstra, R. W.; Harrison, A. M. *J. Magn. Reson.* 1982, **46**, 338–342.
67. Brevard, C.; Granger, P. *Inorg. Chem.* 1983, **22**, 532–535.
68. Brown, T. H.; Green, P. J. *J. Am. Chem. Soc.* 1970, **92**, 2359–2362.
69. Brown, C.; Heaton, B. T.; Longhetti, L.; Povey, W. T.; Smith, D. O. *J. Organomet. Chem.* 1980, **192**, 93–99.
70. Heaton, B. T.; Strona, L.; Jonas, J.; Eguchi, T.; Hoffmann, G. A. *J. Chem. Soc., Dalton Trans.* 1982, 1159–1164.
71. Heaton, B. T.; Strona, L.; Martinengo, S. *J. Organomet. Chem.* 1981, **215**, 415–422.
72. Gansow, O. A.; Gill, D. S.; Bennis, F. J.; Hutchison, J. R.; Vidal, J. L.; Schoening, R. C. *J. Am. Chem. Soc.* 1980, **102**, 2449–2450.
73. Heaton, B. T.; Strona, L.; Martinengo, S.; Chini, P. *J. Organomet. Chem.* 1980, **194**, C29–C31.
74. Martinengo, S.; Heaton, B. T.; Goodfellow, R. J.; Chini, P. *J. Chem. Soc., Chem. Commun.* 1977, 39–40
75. Allevi, C.; Heaton, B. T.; Seregni, C.; Strona, L.; Goodfellow, R. J.; Chini, P.; Martinengo, S. *J. Chem. Soc., Dalton Trans.* 1986, 1375–1381.

76. Schwenk, A.; Zimmermann, G. *Phys. Lett. Sect. A* 1968, **26**, 258–259.
77. Dwek, R. A.; Luz, Z.; Shporer, M. *J. Phys. Chem.* 1970, **74**, 2232–2233.
78. Freeman, W.; Pregosin, P. S.; Sze, S. N.; Venanzi, L. M. *J. Magn. Reson.* 1976, **22**, 473–478.
79. Biradar, N. S.; Pujar, M. A. *Z. Anorg. Allg. Chem.* 1970, **379**, 88–94.
80. Bendall, M. R.; Doddrell, D. M. *J. Magn. Reson.* 1979, **33**, 659–663.
81. McFarlane, H. C. E.; McFarlane, W.; Wood, R. J. *Bull. Soc. Chim. Belg.* 1976, **85**, 864–871 (e to g in Table I should be read with notes d to h, respectively).
82. Colquhoun, I. J.; McFarlane, W. *J. Magn. Reson.* 1982, **46**, 525–528.
83. McFarlane, W. *J. Chem. Soc., Dalton Trans.* 1974, 324–328.
84. Brown, C.; Heaton, B. T.; Sabounchei, J. *J. Organomet. Chem.* 1977, **142**, 413–421.
85. Koridze, A. A.; Astakhova, N. M.; Petrovskii, P. V.; Lutsenko, A. I. *Dokl. Akad. Nauk. SSSR* 1978, **242**, 117–120; *Dokl. Chem. (Engl. Transl.)* 1978, **242**, 416–418.
86. Koridze, A. A.; Petrovskii, P. V.; Gubin, S. P.; Fedin, É. I.; Lutsenko, A. I.; Amiton, I. P.; Okulevich, P. O. *Izv. Akad. Nauk. SSSR, Ser. Khim.* 1975, 1675; *Bull. Acad. Sci. USSR, Div. Chem. Sci.* 1975, **24**, 1567.
87. Salzer, A.; Egolf, T.; von Philipsborn, W. *Helv. Chim. Acta* 1982, **65**, 1145–1157.
88. Haas, H.; Sheline, R. K. *J. Inorg. Nucl. Chem.* 1967, **29**, 693–698.
89. Mooberry, E. S.; Pupp, M.; Slater, J. L.; Sheline, R. K. *J. Chem. Phys.* 1971, **55**, 3655–3660.
90. Mooberry, E. S.; Sheline, R. K. *J. Chem. Phys.* 1972, **56**, 1852–1855.
91. Spiess, H. W.; Sheline, R. K. *J. Chem. Phys.* 1970, **53**, 3036–3041.
92. Wuyts, L. F.; van der Kelen, G. P. *J. Mol. Struct.* 1974, **23**, 73–79.
93. Cohen, M. A.; Kidd, D. R.; Brown, T. L. *J. Am. Chem. Soc.* 1975, **97**, 4408–4409.
94. Schumann, H.; Meissner, M.; Kroth, H. J. *Z. Naturforsch., Teil B* 1978, **33**, 1489–1490.
95. Rehder, D.; Schmidt, J. *Z. Naturforsch., Teil B* 1972, **27**, 625–630.
96. Freeman, R.; Murray, G. R.; Richards, R. E. *Proc. R. Soc. London, Ser. A* 1957, **242**, 455–466.
97. Biradar, N. S.; Pujar, M. A. *Z. Anorg. Allg. Chem.* 1972, **391**, 54–59.
98. Yajima, F.; Koike, Y.; Yamasaki, A.; Fujiwara, S. *Bull. Chem. Soc. Jpn.* 1974, **47**, 1442–1446.
99. Juranić, N.; Ćelap, M. B.; Vučelić, D.; Malinar, M. J.; Radivojša, P. N. *J. Magn. Reson.* 1979, **35**, 319–327.
100. Juranić, N.; Ćelap, M. B.; Vučelić, D.; Malinar, M. J.; Radivojša, P. N. *Spectrochim. Acta Part A* 1979, **35**, 997–1002.
101. Martin, R. L.; White, A. H. *Nature (London)* 1969, **223**, 394–396.
102. Dharmatti, S. S.; Kanekar, C. R. *J. Chem. Phys.* 1959, **31**, 1436–1437.
103. Juranić, N. *Inorg. Chem.* 1983, **22**, 521–525.
104. Kanekar, C. R.; Dhingra, M. M.; Marathe, V. R.; Nagarajan, R. *J. Chem. Phys.* 1967, **46**, 2009–2010.
105. Pujar, M. A.; Biradar, N. S. *J. Indian Chem. Soc.* 1980, **57**, 783–786.
106. Matwiyoff, N. A.; Wageman, W. E. *Inorg. Chim. Acta* 1970, **4**, 460–462.
107. Hawkins, C. J.; Holm, R. H.; Palmer, J. A.; Traficante, D. D. *Aust. J. Chem.* 1982, **35**, 1815–1828.
108. Weiss, R.; Verkade, J. G. *Inorg. Chem.* 1979, **18**, 529–530; Verkade, J. G. *Pure Appl. Chem.* 1980, **52**, 1131–1139.
109. Yamasaki, A.; Aoyama, T.; Fujiwara, S.; Nakamura, K. *Bull. Chem. Soc. Jpn.* 1978, **51**, 643–644.
110. Juranić, N. *J. Chem. Phys.* 1981, **74**, 3690–3693.
111. Wilinski, J.; Kurland, R. J. *Inorg. Chem.* 1973, **12**, 2202–2203.
112. Koike, Y.; Yajima, F.; Yamasaki, A.; Fujiwara, S. *Chem. Lett.* 1974, 177–180; Hawkins, C. J.; Holm, R. H.; Palmer, J. A.; Traficante, D. D. *Aust. J. Chem.* 1982, **35**, 1815–1828.
113. Johnson, A.; Everett, G. W. *Inorg. Chem.* 1973, **12**, 2801–2805.
114. Kanekar, C. R.; Biradar, N. S. *Curr. Sci.* 1966, **35**, 37–38.
115. Hartmann, H.; Sillescu, H. *Theoret. Chim. Acta (Berlin)* 1964, **2**, 371–385.
116. Dodwad, S. S.; Datar, M. G. *Indian J. Chem., Sect. A* 1979, **17**, 100–101; Patankar, V. P.; Dodwad, S. S.; Datar, M. G. *Curr. Sci.* 1981, **50**, 719–720.
117. Yajima, F.; Koike, Y.; Sakai, T.; Fujiwara, S. *Inorg. Chem.* 1972, **11**, 2054–2057.
118. Juranić, N.; Ćelap, M. B.; Vučelić, D.; Malinar, M. J.; Radivojša, P. N. *Inorg. Chem.* 1980, **19**, 802–805.
119. Juranić, N.; Ćelap, M. B.; Vučelić, D.; Malinar, M. J.; Radivojša, P. N. *Inorg. Chim. Acta* 1977, **25**, 229–232.
120. Pujar, M. A. *Indian J. Chem., Sect. A* 1980, **19**, 384–385.
121. Juranić, N.; Ćelap, M. B.; Vučelić, D., Malinar, M. J.; Radivojša, P. N. *J. Coord. Chem.* 1979, **9**, 117–123.
122. Biradar, N. S.; Pujar, M. A. *Inorg. Nucl. Chem. Lett.* 1971, **7**, 269–273.
123. Biradar, N. S.; Pujar, M. A. *Curr. Sci.* 1966, **35**, 385–386; Hyde, M. R.; Sykes, A. G. *J. Chem. Soc.*,

Dalton Trans. 1974, 1583–1590; Hackbusch, W.; Rupp, H. H.; Weighardt, K. *J. Chem. Soc., Dalton Trans.* 1975, 1015–1018.
124. Hackbusch, W.; Rupp, H. H.; Weighardt, K. *J. Chem. Soc., Dalton Trans.* 1975, 2364–2367.
125. Reichert, B. E.; West, B. O. *J. Chem. Soc., Chem. Commun.* 1974, 177–178.
126. LaRossa, R. A.; Brown, T. L. *J. Am. Chem. Soc.* 1974, **96**, 2072–2081.
127. Gulya, A. P.; Starysh, M. P.; Batyr, D. G. *Koord. Khim.* 1981, **7**, 108–111; *Sov. J. Coord. Chem. (Engl. Transl.)* 1981, **7**, 60–62.
128. Floriani, C.; Puppis, M.; Calderazzo, F. *J. Organomet. Chem.* 1968, **12**, 209–223.
129. Patel, N. J.; Haldar, B. C. *J. Inorg. Nucl. Chem.* 1967, **29**, 1037–1045.
130. Spiess, H. W.; Haas, H.; Hartmann, H. *J. Chem. Phys.* 1969, **50**, 3057–3064.
131. Nutton, A.; Bailey, P. M.; Maitlis, P. M. *J. Organomet. Chem.* 1981, **213**, 313–332.
132. Heaton, B. T.; Strona, L.; Della Pergola, R.; Garlaschelli, L.; Sartorelli, U.; Sadler, I. H. *J. Chem. Soc., Dalton Trans.* 1983, 173–175.
133. Espinet, P.; Bailey, P. M.; Piraino, P.; Maitlis, P. M. *Inorg. Chem.* 1979, **18**, 2706–2710.
134. Fumagalli, A.; Koetzle, T. F.; Takusagawa, F.; Chini, P.; Martinengo, S.; Heaton, B. T. *J. Am. Chem. Soc.* 1980, **102**, 1740–1742.
135. Ceriotti, A.; Longoni, G.; Manassero, M.; Sansoni, M.; Della Pergola, R.; Heaton, B. T.; Smith, D. O. *J. Chem. Soc., Chem. Commun.* 1982, 886–887.
136. Heaton, B. T.; Strona, L.; Martinengo, S.; Strumolo, D.; Goodfellow, R. J.; Sadler, I. H. *J. Chem. Soc., Dalton Trans.* 1982, 1499–1502.
137. Heaton, B. T.; Della Pergola, R.; Strona, L.; Smith, D. O.; Fumagalli, A. *J. Chem. Soc., Dalton Trans.* 1982, 2553–2555.
138. Brown, C.; Heaton, B. T.; Longhetti, L.; Smith, D. O.; Chini, P.; Martinengo, S. *J. Organomet. Chem.* 1979, **169**, 309–314.
139. Garlaschelli, L.; Fumagalli, A.; Martinengo, S.; Heaton, B. T.; Smith, D. O.; Strona, L. *J. Chem. Soc., Dalton Trans.* 1982, 2265–2267.
140. Heaton, B. T.; Brown, C.; Smith, D. O.; Strona, L.; Goodfellow, R. J.; Chini, P.; Martinengo, S. *J. Am. Chem. Soc.* 1980, **102**, 6175–6177.
141. Sogo, P. B.; Jeffries, C. D. *Phys. Rev.* 1955, **98**, 1316–1317.
142. Barnes, J. R.; Goggin, P. L.; Goodfellow, R. J. *J. Chem. Res.* 1979 (*S*) 118–119, (M) 1610–1656.
143. Crocker, C.; Goodfellow, R. J. *J. Chem. Res.* 1981, (*S*) 38–39, (M) 0742–0775.
144. Barnes, J. R. Ph.D. thesis, University of Bristol, 1979.
145. Anderson, S. J.; Barnes, J. R.; Goggin, P. L.; Goodfellow, R. J. *J. Chem. Res.* 1978, (*S*) 286–287, (M) 3601–3628.
146. Fernandez, M. J.; Maitlis, P. M. *J. Chem. Soc., Chem. Commun.* 1982, 310–311.
147. Mann, B. E.; Spencer, C. M. *Inorg. Chim. Acta* 1983, **76**, L65–L66.
148. Hyde, E. M.; Kennedy, J. D.; Shaw, B. L.; McFarlane, W. *J. Chem. Soc., Dalton Trans.* 1977, 1571–1576.
149. Browning, J.; Goggin, P. L.; Goodfellow, R. J.; Norton, M. G.; Rattray, A. J. M.; Taylor, B. F.; Mink, J. *J. Chem. Soc., Dalton Trans.* 1977, 2061–2067.
150. Crocker, C.; Errington, R. J.; Markham, R.; Moulton, C. J.; Odell, K. J.; Shaw, B. L. *J. Am. Chem. Soc.* 1980, **102**, 4373–4379.
151. Crocker, C.; Errington, R. J.; Markham, R.; Moulton, C. J.; Shaw, B. L. *J. Chem. Soc., Dalton Trans.* 1982, 387–395.
152. Boyar, E. B.; Robinson, S. D. *Inorg. Chim. Acta* 1982, **64**, L193–L194
153. Caddy, P., Green, M.; Smart, L. E.; White, N. *J. Chem. Soc., Chem. Commun.* 1978, 839–840.
154. Cocivera, M.; Ferguson, G.; Lalor, F. J.; Szczecinski, P. *Organometallics* 1982, **1**, 1139–1142.
155. Cocivera, M.; McAlees, A. J.; McCrindle, R.; Szczecinski, P. *J. Organomet. Chem.* 1982, **235**, 97–104.
156. Dykstra, R. W.; Harrison, A. M.; Dombek, B. D. *Rev. Sci. Instrum.* 1981, **52**, 1690–1696.
157. Carr, C.; Goggin, P. L.; Goodfellow, R. J. *Inorg. Chim. Acta* 1984, **81**, L25–L26.
158. Kerrison, S. J. S.; Sadler, P. J. *J. Magn. Reson.* 1978, **31**, 321–325.
159. Pregosin, P. S.; Kretschmer, M.; Preetz, W.; Rimkus, G. *Z. Naturforsch., Teil B* 1982, **37**, 1422–1424.
160. Riddell, F. G.; Gillard, R. D.; Wimmer, F. L. *J. Chem. Soc., Chem. Commun.* 1982, 332–333.
161. Creaser, C. S.; Creighton, J. A. *J. Organomet. Chem.* 1978, **157**, 243–245.
162. Shipachev, V. A.; Zemskov, S. V.; Tkachev, S. V.; Al't, L. Ya. *Koord. Khim.* 1980, **6**, 1877–1878; *Sov. J. Coord. Chem. (Engl. Trans.)* 1980, **6**, 959–960.
163. Shipachev, V. A.; Zemskov, S. V.; Al't, L. Ya. *Koord. Chim.* 1980, **6**, 932–935; *Sov. J. Coord. Chem. (Engl. Trans.)* 1980, **6**, 472–475.
164. Kerrison, S. J. S.; Sadler, P. J. *J. Chem. Soc., Dalton Trans.* 1982, 2363–2369.

165 Taylor, B F Ph D thesıs, Unıversıty of Brıstol, 1973
166 Browning, J , Goggın, P L , Goodfellow, R J *J Chem Res* 1978, (*S*) 328–329, (M) 4201–4247
167 Anderson, D W W , Ebsworth, E A V , Rankın, D W H *J Chem Soc*, *Dalton Trans* 1973, 2370–2373
168 Blacklaws, I M , Ebsworth, E A V , Rankın, D W H , Robertson, H E *J Chem Soc*, *Dalton Trans* 1978, 753–758
169 Blacklaws, I M , Brown, L C , Ebsworth, E A V , Reed, F J S *J Chem Soc*, *Dalton Trans* 1978, 877–879
170 Appleton, T G , Hall, J R *Aust J Chem* 1980, **33**, 2387–2403
171 Agnew, N H , Appleton, T G , Hall, J R *Aust J Chem* 1982, **35**, 881–894
172 Kennedy, J D , McFarlane, W , Puddephatt, R J , Thompson, P J *J Chem Soc*, *Dalton Trans* 1976, 874–879
173 Bramley, R , Hall, J R , Swıle, G A , Tomkıns, I B *Aust J Chem* 1974, **27**, 2491–2493
174 Mıyashıta, A , Takahashı, M , Takaya, H *J Am Chem Soc* 1981, **103**, 6257–6259
175 Crocker, C , Goodfellow, R J , Gımeno, J , Uson, R *J Chem Soc*, *Dalton Trans* 1977, 1448–1452
176 Agnew, N H , Appleton, T G , Hall, J R *Inorg Chım Acta* 1981, **50**, 137–140
177 Zharkova, G I , Igumenov, I K , Tkachev, S V , Zemskov, S V *Koord Khım* 1982, **8**, 74–81, *Sov J Coord Chem (Engl Transl)* 1982, **8**, 41–48
178 Pregosın, P S , Streıt, H , Venanzı, L M *Inorg Chım Acta* 1980, **38**, 237–242
179 Goggın, P L , Goodfellow, R J , Haddock, S R , Taylor, B F , Marshall, I R H *J Chem Soc*, *Dalton Trans* 1976, 459–467
180 Dos Remedıos Pınto, M A F , Sadler, P J , Neıdle, S , Sanderson, M R , Subbıah, A , Kuroda, R *J Chem Soc*, *Chem Commun* 1980, 13–15, Dıckson, M K , Fordyce, W A , Appel, D M , Alexander, K , Steın, P , Roundhıll, D M *Inorg Chem* 1982, **21**, 3857–3858
181 Pregosın, P S , Venanzı, L M *Helv Chım Acta* 1975, **58**, 1548–1551
182 Boag, N M , Goggın, P L , Goodfellow, R J , Herbert, I R *J Chem Soc*, *Dalton Trans* 1983, 1101–1107
183 Charnock, J M Ph D thesıs, Unıversıty of Brıstol, 1981
184 Crocker, C , Goggın, P L , Goodfellow, R J *J Chem Soc*, *Dalton Trans* 1976, 2494–2500
185 Motschı, H , Pregosın, P S , Ruegger, H *J Organomet Chem* 1980, **193**, 397–405
186 Brıggs, J R , Crocker, C , McDonald, W S , Shaw, B L *J Chem Soc*, *Dalton Trans* 1981, 121–125
187 Nelson, J H , Cooper, V , Rudolph, R W *Inorg Nucl Chem Lett* 1980, **16**, 263–265
188 Goodfellow, R J , Herbert, I R *Inorg Chım Acta* 1982, **65**, L161–L162
189 Benkeser, R A , Kang, J *J Organomet Chem* 1980, **185**, C9–C12
190 Soulıe, J , Lallemand, J Y , Rao, R C *Org Magn Reson* 1979, **12**, 67–70
191 Crocker, C , Goggın, P L *J Chem Res* 1978, (S) 93, (*M*) 1274–1286
192 Anderson, S J , Goodfellow, R J *J Chem Soc*, *Dalton Trans* 1977, 1683–1686
193 Cırıano, M , Green, M , Howard, J A K , Proud, J , Spencer, J L , Stone, F G A , Tsıpıs, C A *J Chem Soc*, *Dalton Trans* 1978, 801–808
194 Goggın, P L , Goodfellow, R J , Reed, F J S *J Chem Soc*, *Dalton Trans* 1974, 576–585
195 Boag, N M , Browning, J , Crocker, C , Goggın, P L , Goodfellow, R J , Murray, M , Spencer, J L *J Chem Res* 1978, (*S*) 228–229, (*M*) 2962–2983
196 Kretschmer, M , Pregosın, P S , Ruegger, H *J Organomet Chem* 1983, **241**, 87–98
197 Pregosın, P S , Sze, S N , Salvadorı, P , Lazzaronı, R *Helv Chım Acta* 1977, **60**, 2514–2521
198 Motschı, H , Pregosın, P S , Venanzı, L M *Helv Chım Acta* 1979, **62**, 667–677
199 Motschı, H , Sze, S N , Pregosın, P S *Helv Chım Acta* 1979, **62**, 2086–2090
200 Shınoda, S , Yamaguchı, Y , Saıto, Y *Inorg Chem* 1979, **18**, 673–678
201 Brıggs, J R , Crocker, C , McDonald, W S , Shaw, B L *J Chem Soc*, *Dalton Trans* 1982, 457–463
202 Anderson, G K , Cross, R J , Rycroft, D S *J Chem Res (S)* 1980, 240–241
203 Kerrıson, S J S , Sadler, P J *J Chem Soc*, *Chem Commun* 1977, 861–863
204 Clore, G M , Gronenborn, A M *J Am Chem Soc* 1982, **104**, 1369–1375
205 Anderson, S J , Goggın, P L , Goodfellow, R J *J Chem Soc*, *Dalton Trans* 1976, 1959–1964
206 Chıkuma, M , Pollock, R J *J Magn Reson* 1982, **47**, 324–327
207 Neıdle, S , Ismaıl, I M , Sadler, P J *J Inorg Bıochem* 1980, **13**, 205–212, Wood, F E , Hunt, C T , Balch, A L *Inorg Chım Acta* 1982, **67**, L19–L20
208 Kerrıson, S J S , Sadler, P J *J Chem Soc*, *Chem Commun* 1981, 61–62
209 Pregosın, P S , Sze, S N *Helv Chım Acta* 1978, **61**, 1848–1855
210 MacDougall, J J , Nelson, J H , Mathey, F *Inorg Chem* 1982, **21**, 2145–2153
211 Ostoja Starzewskı, K H A , Pregosın, P S , Ruegger, H *Helv Chım Acta* 1982, **65**, 785–797
212 Carr, S , Colton, R , Dakternıeks, D *J Organomet Chem* 1982, **240**, 143–151

213. Ostoja Starzewski, K. H. A.; Pregosin, P. S. *Angew. Chem. Int. Ed. Engl.* 1980, **19**, 316–317.
214. Ebsworth, E. A. V.; Edward, J. M.; Rankin, D. W. H. *J. Chem. Soc., Dalton Trans.* 1976, 1667–1672.
215. Ebsworth, E. A. V.; Edward, J. M.; Rankin, D. W. H. *J. Chem. Soc., Dalton Trans.* 1976, 1673–1677.
216. Dean, R. R.; Green, J. C. *J. Chem. Soc. (A)* 1968, 3047–3050.
217. Bracher, G.; Grove, D. M.; Pregosin, P. S.; Venanzi, L. M. *Angew. Chem. Int. Ed. Engl.* 1979, **18**, 155–156.
218. Anderson, G. K.; Clark, H. C.; Davies, J. A. *Inorg. Chem.* 1983, **22**, 427–433.
219. Ebsworth, E. A. V.; Edward, J. M.; Reed, F. J. S.; Whitelock, J. D. *J. Chem. Soc., Dalton Trans.* 1978, 1161–1170.
220. Briggs, J. R.; Crocker, C.; McDonald, W. S.; Shaw, B. L. *J. Chem. Soc., Dalton Trans.* 1981, 575–580.
221. Browning, J.; Goggin, P. L.; Goodfellow, R. J.; Hurst, N. W.; Mallinson, L. G.; Murray, M. *J. Chem. Soc., Dalton Trans.* 1978, 872–876.
222. Hietkamp, S.; Stufkens, D. J.; Vrieze, K. *J. Organomet. Chem.* 1979, **169**, 107–113.
223. Puddephatt, R. J.; Thomson, M. A.; Manojlović-Muir, L.; Muir, K. W.; Frew, A. A.; Brown, M. P. *J. Chem. Soc., Chem. Commun.* 1981, 805–806.
224. Bennett, M. A.; Bramley, R.; Tomkins, I. B. *J. Chem. Soc., Dalton Trans.* 1973, 166–168.
225. Boag, N. M.; Dias, G. H. M.; Green, M.; Spencer, J. L.; Stone, F. G. A.; Vicente, J. *J. Chem. Soc., Dalton Trans.* 1981, 1981–1987.
226. Browning, J.; Green, M.; Spencer, J. L.; Stone, F. G. A. *J. Chem. Soc., Dalton Trans.* 1974, 97–101.
227. Sebald, A., Wrackmeyer, B.; Beck, W. *Z. Naturforsch., Teil B* 1983, **38**, 45–56.
228. Boocock, S. K.; Greenwood, N. N.; Hails, M. J.; Kennedy, J. D.; McDonald, W. S. *J. Chem. Soc., Dalton Trans.* 1981, 1415–1429; Boocock, S. K.; Greenwood, N. N.; Kennedy, J. D.; McDonald, W. S.; Staves, J. *J. Chem. Soc., Dalton Trans.* 1981, 2573–2584.
229. Scherer, O. J.; Jungmann, H. *J. Organomet. Chem.* 1981, **208**, 153–159.
230. Greenwood, N. N.; Kennedy, J. D.; Staves, J. *J. Chem. Soc., Dalton Trans.* 1978, 1146–1152.
231. Al-Najjar I. M.; Green, M.; Kerrison, S. J. S.; Sadler, P. J. *J. Chem. Soc., Chem. Commun.* 1979, 311–312; Al-Najjar, I. M.; Green, M.; Sarhan, J. K. K.; Ismail, I. M.; Sadler, P. J. *Inorg. Chim. Acta* 1980, **44**, L187–L188.
232. Okeya, S.; Nakamura, Y.; Kawaguchi, S.; Hinomoto, T. *Inorg. Chem.* 1981, **20**, 1576–1580.
233. Al-Najjar, I. M.; Green, M.; Kerrison, S. J. S.; Sadler, P. J. *J. Chem. Res. (S)* 1979, 206–207.
234. Appleton, T. G.; Hall, J. R.; Neale, D. W.; Ralph, S. F. *Inorg. Chim. Acta* 1983, **77**, L149–L151.
235. Che, C. M.; Schaefer, W. P.; Gray, H. B.; Dickson, M. K.; Stein, P. B.; Roundhill, D. M. *J. Am. Chem. Soc.* 1982, **104**, 4253–4255.
236. Boag, N. M.; Green, M.; Stone, F. G. A. *J. Chem. Soc., Chem. Commun.* 1980, 1281–1282.
237. Messbauer, B.; Meyer, H.; Walther, B.; Heeg, M. J.; Maqsudur Rahman, A. F. M.; Oliver, J. P. *Inorg. Chem.* 1983, **22**, 272–277.
238. Boag, N. M.; Green, M.; Howard, J. A. K.; Stone, F. G. A.; Wadepohl, H. *J. Chem. Soc., Dalton Trans.* 1981, 862–872.
239. Koie, Y.; Shinoda, S.; Saito, Y. *J. Chem. Soc., Dalton Trans.* 1981, 1082–1088.
240. Meij, R.; Stufkens, D. J.; Vrieze, K.; Roosendaal, E.; Schenk, H. *J. Organomet. Chem.* 1978, **155**, 323–336.
241. Götzfried, F., Beck, W. *J. Organomet. Chem.* 1980, **191**, 329–338.
242. Green, M.; Spencer, J. L.; Stone, F. G. A.; Welch, A. J. *J. Chem. Soc., Chem. Commun.* 1974, 571–572.
243. Crocker, C.; Goodfellow, R. J. *J. Chem. Soc., Dalton Trans.* 1977, 1687–1689.
244. Brown, C.; Heaton, B. T.; Towl, A. D. C.; Chini, P.; Fumagalli, A.; Longoni, G. *J. Organomet. Chem.* 1979, **181**, 233–254.
245. Moor, A.; Pregosin, P. S.; Venanzi, L. M. *Inorg. Chim. Acta* 1981, **48**, 153–157.
246. Moor, A.; Pregosin, P. S.; Venanzi, L. M. *Inorg. Chim. Acta* 1982, **61**, 135–140.
247. Fumagalli, A.; Martinengo, S.; Chini, P.; Albinati, A.; Bruckner, S.; Heaton, B. T. *J. Chem. Soc., Chem. Commun.* 1978, 195–196; Green, M.; Howard, J. A. K.; Mills, R. M.; Pain, G. N.; Stone, F. G. A.; Woodward, P. *J. Chem. Soc., Chem. Commun.* 1981, 869–871; Green, M.; Mills, R. M.; Pain, G. N.; Stone, F. G. A.; Woodward, P. *J. Chem. Soc., Dalton Trans.* 1982, 1309–1319.
248. Ashworth, T. V.; Chetcuti, M. J.; Howard, J. A. K.; Stone, F. G. A.; Wisbey, S. J.; Woodward, P. *J. Chem. Soc., Dalton Trans.* 1981, 763–770; Chetcuti, M. J.; Marsden, K.; Moore, I.; Stone, F. G. A.; Woodward, P. *J. Chem. Soc., Dalton Trans.* 1982, 1749–1755.
249. Jeffery, J. C.; Sambale, C.; Schmidt, M. F.; Stone, F. G. A. *Organometallics* 1982, **1**, 1597–1604; Jeffery, J. C.; Razay, H.; Stone, F. G. A. *J. Chem. Soc., Dalton Trans.* 1982, 1733–1739; Chetcuti, M. J.; Howard, J. A. K.; Mills, R. M.; Stone, F. G. A.; Woodward, P. *J. Chem. Soc., Dalton Trans.* 1982, 1757–1764.
250. Arrizabalaga, P.; Castan, P.; Laurent, J. P.; Salesse, A. *Chem. Phys. Lett.* 1980, **76**, 548–552.

251. Niedoba, H.; Launois, H.; Brinkmann, D.; Keller, H. U. *J. Phys. (Paris) Lett.* 1974, **35**, L251–L253; Mali, M.; Kanert, O.; Mehring, M.; Brinkmann, D. *Phys. Rev. B* 1979, **20**, 4442–4446.
252. Rhodes, H. E.; Wang, P. K.; Stokes, H. T.; Slichter, C. P.; Sinfelt, J. H. *Phys. Rev. B* 1982, **26**, 3559–3568.
253. Cheremisin, A. A.; Schastnev, P. V. *J. Magn. Reson.* 1980, **40**, 459–468.
254. Dunn, T. M. *Trans. Farad. Soc.* 1961, **57**, 1441–1444.
255. Owen, J.; Thornley, J. H. M. *Rep. Prog. Phys.* 1966, **29**, 675–728.
256. Walstedt, R. E.; Wernick, J. H.; Jaccarino, V. *Phys. Rev.* 1967, **162**, 301–311.
257. Betteridge, G. P.; Golding, R. M. *J. Chem. Phys.* 1969, **51**, 2497–2500.
258. Fujiwara, S.; Yajima, F.; Yamasaki, A. *J. Magn. Reson.* 1969, **1**, 203–210.
259. Salzmann, J. J.; Schmidke, H. H. *Inorg. Chim. Acta* 1969, **3**, 207–219.
260. Juranić, N. *Spectrochim. Acta, Part A* 1980, **36**, 249–252.
261. Brown, T. H.; Cohen, S. M. *J. Chem. Phys.* 1973, **58**, 395–396.
262. Jameson, C. J.; Gutowsky, H. S. *J. Chem. Phys.* 1964, **40**, 1714–1724.
263. Nakashima, Y.; Sakaguchi, U.; Yoneda, H. *Bull. Chem. Soc. Jpn.* 1975, **48**, 762–765.
264. Craighead, K. L. *J. Am. Chem. Soc.* 1973, **95**, 4434–4435.
265. Yajima, F.; Yamasaki, A.; Fujiwara, S. *Inorg. Chem.* 1971, **10**, 2350–2352.
266. Proctor, W. G.; Yu, F. C. *Phys. Rev.* 1951, **81**, 20–30.
267. Benedek, G. B.; Englman, R.; Armstrong, J. A. *J. Chem. Phys.* 1963, **39**, 3349–3363.
268. Brown, T. H.; Green, P. J. *Phys. Lett. Sect. A* 1970, **31**, 148–149.
269. Cohen, S. M.; Brown, T. H. *J. Chem. Phys.* 1974, **61**, 2985–2986.
270. Gonzalez, G.; Mayer, U.; Gutmann, V. *Inorg. Nucl. Chem. Lett.* 1979, **15**, 155–159.
271. Frankel, L. S.; Langford, C. H.; Stengle, T. R. *J. Phys. Chem.* 1970, **74**, 1376–1381.
272. Balt, S.; Meuldijk, J. *Inorg. Chim. Acta* 1981, **47**, 217–223.
273. Martin, T. H.; Fung, B. M. *J. Phys. Chem.* 1973, **77**, 637–640; Fung, B. M.; Wei, S. C.; Martin, T. H.; Wei, I. *Inorg. Chem.* 1973, **12**, 1203–1205.
274. Delville, A.; Laszlo, P.; Stockis, A. *J. Am. Chem. Soc.* 1981, **103**, 5991–5998.
275. Hirayama, M.; Kawamata, Y.; Fujii, Y.; Nakano, Y. *Bull. Chem. Soc. Jpn.* 1982, **55**, 1798–1802.
276. Doddrell, D. M.; Pegg, D. T.; Bendall, M. R. *Aust. J. Chem.* 1979, **32**, 1–10.
277. Lauterbur, P. C. *J. Chem. Phys.* 1965, **42**, 799–800.
278. Loewenstein, A.; Shporer, M. *Mol. Phys.* 1965, **9**, 293–294.
279. Jameson, C. J. *J. Chem. Phys.* 1977, **66**, 4983–4988.
280. Bendall, M. R.; Doddrell, D. M. *Aust. J. Chem.* 1978, **31**, 1141–1143.
281. Sudmeier, J. L.; Blackmer, G. L.; Bradley, C. H.; Anet, F. A. L. *J. Am. Chem. Soc.* 1972, **94**, 757–761.
282. Constable, E. C.; Johnson, B. F. G.; Lewis, J.; Pain, G. N.; Taylor, M. J. *J. Chem. Soc., Chem. Commun.* 1982, 754–756.
283. Koridze, A. A.; Kizas, O. A.; Astakhova, N. M.; Petrovskii, P. V.; Grishin, Y. K. *J. Chem. Soc., Chem. Commun.* 1981, 853–855.
284. Aime, S.; Osella, D. *J. Organomet. Chem.* 1981, **214**, C27–C30.
285. La Mar, G. N.; Viscio, D. B.; Budd, D. L.; Gersonde, K. *Biochem. Biophys. Res. Commun.* 1978, **82**, 19–23.
286. Morishima, I.; Hayashi, T.; Inubishi, T.; Yonezawa, T.; Uermura, S. *J. Chem. Soc., Chem. Commun.* 1979, 483–485.
287. Morishima, I.; Inubushi, T. *J. Chem. Soc., Chem. Commun.* 1978, 106–107.
288. Mann, B. E. *J. Chem. Soc., Chem. Commun.* 1971, 1173–1174.
289. Dykes, D.; Huckerby, T. N.; Oldham, C. *Inorg. Nucl. Chem. Lett.* 1977, **13**, 63–64.
290. Yamasaki, A.; Miyakoshi, Y.; Fujita, M.; Yoshikawa, Y.; Yamatera, H. *J. Inorg. Nucl. Chem.* 1979, **41**, 473–475.
291. Farrugia, L. J.; James, B. R.; Lassigne, C. R.; Wells, E. J. *Can. J. Chem.* 1982, **60**, 1304–1316.
292. Jesson, J. P. In *Transition Metal Hydrides*; Muetterties, E. L., Ed.; Marcel Dekker: New York, 1971; Chap. 4.
293. Goggin, P. L.; Goodfellow, R. J.; Knight, J. R.; Norton, M. G.; Taylor, B. F. *J. Chem. Soc., Dalton Trans.* 1973, 2220–2226.
294. Mann, B. E.; Taylor, B. F. *^{13}C NMR Data for Organometallic Compounds*; Academic: London, 1981; Pregosin, P. S. *Annu. Rep. NMR Spectrosc.* 1981, **11A**, 227–271.
295. Bose, K. S.; Abbott, E. H. *Inorg. Chem.* 1977, **16**, 3190–3193.
296. Pregosin, P. S.; Kunz, R. W. *^{31}P and ^{13}C NMR of Transition Metal Phosphine Complexes*; Springer-Verlag: Berlin, 1979.

297. Lawson, R. J.; Shapley, J. R. *Inorg. Chem.* 1978, **17**, 2963–2965.
298. Moriyama, H.; Aoki, T.; Shinoda, S.; Saito, Y. *J. Chem. Soc., Dalton Trans.* 1981, 639–644; Garralda, M.; Garcia, V.; Kretschmer, M.; Pregosin, P. S.; Rüegger, H. *Helv. Chim. Acta* 1981, **64**, 1150–1157; Kretschmer, M.; Pregosin, P. S.; Garralda, M. *J. Organomet. Chem.* 1983, **244**, 175–181.
299. Mann, B. E.; Masters, C.; Shaw, B. L. *J. Chem. Soc., Chem. Commun.* 1970, 1041.
300. Appleton, T. G.; Clark, H. C.; Manzer, L. E. *Coord. Chem. Rev.* 1973, **10**, 335–422.
301. Kennedy, J. D.; Wrackmeyer, B. *J. Magn. Reson.* 1980, **38**, 529–535.
302. Cairns, M. A.; Dixon, K. R.; McFarland, J. J. *J. Chem. Soc., Dalton Trans.* 1975, 1159–1164.
303. Evans, D. F.; Turner, G. K. *J. Chem. Soc., Dalton Trans.* 1975, 1238–1243.
304. Goggin, P. L.; Goodfellow, R. J.; Haddock, S. R. *J. Chem. Soc., Chem. Commun.* 1975, 176–177.
305. McFarlane, W. *Chem. Commun.* 1968, 755–756.
306. van der Ploeg, A. F. M. J.; van Koten, G.; Brevard, C. *Inorg. Chem.* 1982, **21**, 2878–2881.
307. Albinati, A.; Naegeli, R.; Ostoja Starzewski, K. H. A.; Pregosin, P. S.; Rüegger, H. *Inorg. Chem. Acta* 1983, **76**, L231–L232.
308. Gysling, H. J.; Zumbulyadis, N.; Robertson, J. A. *J. Organomet. Chem.* 1981, **209**, C41–C44.
309. Mague, J. T.; Sanger, A. R. *Inorg. Chem.* 1979, **18**, 2060–2066.
310. Pidcock, A. In *Catalytic Aspects of Metal Phosphine Complexes*, Alyea, E. C.; Meek, D. W., Eds.; ACS Advances in Chemistry Series, 1982, **196**, 1–22; Nixon, J. F.; Pidcock, A. *Annu. Rev. NMR Spectrosc.* 1969, **2**, 345–422.
311. Pregosin, P. S.; Favaz, R.; Roulet, R.; Boschi, T.; Michelin, R. A.; Ros, R. *Inorg. Chim. Acta* 1980, **45**, L7–L9.
312. Koie, Y.; Shinoda, S.; Saito, Y. *Inorg. Chem.* 1981, **20**, 4408–4413.
313. Miyamoto, T. *J. Organomet. Chem.* 1977, **134**, 335–362.
314. Anderson, G. K.; Cross, R. J.; Rycroft, D. S. *J. Chem. Res.* 1979 (*S*) 120, (M) 1601 1609.
315. Chisholm, M. H.; Clark, H. C.; Manzer, L. E.; Stothers, J. B.; Ward, J. E. H. *J. Am. Chem. Soc.* 1975, **97**, 721–727 and references therein.
316. Bushnell, G. W.; Eadie, D. T.; Pidcock, A.; Sam, A. R.; Holmes-Smith, R. D. *J. Am. Chem. Soc.* 1982, **104**, 5837–5839.
317. Jeffery, J. C.; Moore, I.; Murray, M.; Stone, F. G. A. *J. Chem. Soc., Dalton Trans.* 1982, 1741–1747.
318. Brown, M. P.; Franklin, S. J.; Puddephatt, R. J.; Thomson, M. A.; Seddon, K. R. *J. Organomet. Chem.* 1979, **178**, 281–290; Grossel, M. C.; Moulding, R. P.; Seddon, K. *Inorg. Chim. Acta Lett.* 1982, **64**, L275–L277; Auburn, M.; Ciriano, M.; Howard, J. A. K.; Murray, M.; Pugh, N. J.; Spencer, J. L.; Stone, F. G. A.; Woodward, P. *J. Chem. Soc., Dalton Trans.* 1980, 659–666. Arnold, D. P.; Bennett, M. A.; Bilton, M. S.; Robertson, G. B. *J. Chem. Soc., Chem. Commun.* 1982, 115–116.
319. Brown, M. P.; Cooper, S. J.; Puddephatt, R. J.; Thomson, M. A.; Seddon, K. R. *J. Chem. Soc., Chem. Commun.* 1979, 1117–1119; Bracher, G.; Grove, D. M.; Venanzi, L. M.; Bachechi, F.; Mura, P.; Zambonelli, L. *Angew. Chem. Int. Ed. Engl.* 1978, **17**, 778–779; Tulip, T. H.; Yamagata, T.; Yoshida, T.; Wilson, R. D.; Ibers, J. A.; Otsuka, S. *Inorg. Chem.* 1979, **18**, 2239–2250; Brown, M. P.; Cooper, S. J.; Frew, A. A.; Manojlović-Muir, L.; Muir, K.; Puddephatt, R. J.; Thomson, M. A. *J. Chem. Soc., Dalton Trans.* 1982, 299–305; Fakley, M. E.; Pidcock, A. *J. Chem. Soc., Dalton Trans.* 1977, 1444–1448; Eaborn, C.; Odell, K. J.; Pidcock, A. *J. Chem. Soc., Dalton Trans.* 1978, 1288–1294; Anderson, G. K.; Clark, H. C.; Davies, J. A. *Inorg. Chem.* 1983, **22**, 434–438; Farrugia, L. J.; Howard, J. A. K.; Mitrprachachon, P.; Stone, F. G. A.; Woodward, P. *J. Chem. Soc., Dalton Trans.* 1981, 162–170.
320. Meek, D. W.; Mazanec, T. J. *Acc. Chem. Res.* 1981, **14**, 266–274.
321. Ricci, J. S.; Koetzle, T. F.; Goodfellow, R. J.; Espinet, P.; Maitlis, P. M. *Inorg. Chem.* 1984, **23**, 1828–1831.
322. Dixon, K. R.; Fakley, M.; Pidcock, A. *Can. J. Chem.* 1976, **54**, 2733 2738.
323. McFarlane, W. *J. Chem. Soc. (A)* 1967, 1922–1923.
324. McFarlane, W. *Chem. Commun.* 1967, 772 773.
325. Iwayanagi, T.; Saito, Y. *Chem. Lett.* 1976, 1193–1196.
326. Goggin, P. L.; Goodfellow, R. J.; Haddock, S. R.; Knight, J. R.; Reed, F. J. S.; Taylor, B. F. *J. Chem. Soc., Dalton Trans.* 1974, 523–533.
327. Kennedy, J. D.; Colquhoun, I. J.; McFarlane, W.; Puddephatt, R. J. *J. Organomet. Chem.* 1979, **172**, 479–489.
328. Kennedy, J. D.; McFarlane, W.; Puddephatt, R. J. *J. Chem. Soc., Dalton Trans.* 1976, 745–748.
329. Goodfellow, R. J.; Hardy, M. J.; Taylor, B. F. *J. Chem. Soc., Dalton Trans.* 1973, 2450–2453.
330. McMillin, D. R.; Drago, R. S. *Inorg. Chem.* 1974, **13**, 546–549.

Chapter 21

POST-TRANSITION METALS, COPPER TO MERCURY

R. J. GOODFELLOW

1. INTRODUCTION

Although compounds of group IIB and many of group IB have formally filled *d* shells, both groups are frequently considered along with the true transition metals, with which they show undoubted similarites. With the exception of gold, there is a significant chemistry involving the diamagnetic d^{10} cations. This and the lability on the NMR time scale of many complexes are in contrast to NMR studies of group VIII.

Of the six metals, only the resonance of ^{197}Au had not been observed in a chemical environment at the time of the survey by Harris and Mann,[1] and this situation has not altered. Cadmium and mercury were among the first demonstrations of the potential of "broadband" observation systems, and the general availability of such systems has greatly increased the interest in these nuclei. Brief details, particularly on observation, are given by Brevard and Granger.[2] Applications in the field of coordination chemistry have been discussed by Rinaldi *et al.*[3] Resonances of these metals (especially ^{113}Cd) in biological environments have received particular attention and have been reviewed by Forsén and Lindman.[4] Ellis *et al.*[5] give a listing of papers on this topic to which may be added Refs. 6–12. The studies by Andrew *et al.* on solid copper halides[13] and cadmium metal[14] were among the first applications of "magic angle spinning" and this technique is now showing promise for the investigation of the environment of ^{113}Cd in solids. Information on the resonances in metals and alloys may be found in a recent volume.[15]

2. OBSERVATION

Inspection of the table of NMR Properties of the Elements in the Appendix shows that for each magnetic nucleus of groups IB and IIB there is a group VIII nucleus with

The following abbreviations are used in this chapter: acac, pentane-2,4-dionato; bipy, 2,2′-bipyridyl; Cp, cyclopentadienyl; Cy, cyclohexyl; DMSO, dimethylsulphoxide; edta, ethylenediaminetetraacetate; en, ethylenediamine; phen, 1,10-phenanthroline; py, pyridine; THF, tetrahydrofuran; TPP, meso-tetraphenylporphyrin; together with the general symbols M for a metal, L and X for monodentate ligands, typically neutral and anionic respectively and R for alkyl or aryl groups.

R. J. GOODFELLOW • Department of Inorganic Chemistry, University of Bristol, Bristol BS8 1TS, England.

similar parameters and hence observational characteristics, viz., ^{63}Cu and ^{65}Cu can be compared with ^{59}Co; ^{67}Zn with ^{99}Ru; ^{107}Ag and ^{109}Ag with ^{103}Rh; ^{111}Cd, ^{113}Cd, and ^{199}Hg with ^{195}Pt; ^{197}Au and ^{201}Hg with ^{105}Pd. We may thereby use the detailed discussion of Chapter 20, Section 2 as a basis for the discussion of the observation and relaxation of group IB and IIB nuclei.

2.1. ^{63}Cu and ^{65}Cu

Unfortunately, the good receptivity of the copper isotopes can only be applied to diamagnetic compounds and the chemistry of Cu^I is limited. The quadrupolar relaxation corresponding to a quite large linewidth factor further reduces the scope. Thus only tetrahedral CuL_4 give sufficiently narrow lines which have been observed for $L = CN^-$,[16,17] NCMe, pyridine, phosphite, or phosphine.[18,19] No resonance could be found for $[Cu(bipy)_2]^+$ or $[Cu(phen)_2]^+$,[18] and in general CuL_3L' are not observable. The ^{31}P linewidth of $[Cu\{P(OMe)_3\}_3(NCMe)]^+$ implies a ^{63}Cu relaxation rate some 200 times that of $[Cu\{P(OMe)_3\}_4]^+$ and ^{63}Cu linewidths of 20–160 kHZ.[20] Rapid equilibrium with other, lower-symmetry species accounts for the broadening of CuL_4 resonances, e.g., $[Cu(NCMe)_4]^+$ with H_2O or pyridine[21,22] or dissociation of $[Cu(PR_3)_4]^+$ as the size of the phosphine increases[19] or of $[Cu(CN)_4]^{3-}$ in water.[16] McConnell and Weaver found $T_2 = 2 \times 10^{-5}$ s for 1 *M* CuCl in concentrated hydrochloric acid and from the effect of Cu^{2+} ions determined the rate for the Cu^+/Cu^{2+} electron exchange.[23] The relatively long T_1 (ca. 0.2 ms) of ^{63}Cu in molten CuCl shows that it is ionic rather than molecular.[24]

As linewidths for ^{63}Cu are only marginally greater than those of ^{65}Cu,[25] ^{63}Cu is the preferred isotope on the ground of receptivity.

Measurements on solids in which copper is in a symmetric environment supplement the solution information.[26,27] The inert complex ion $[Cu\{P(OMe)_3\}_4]^+$ has also been studied by ^{1}H–{^{31}P, ^{63}Cu} triple resonance methods.[28]

2.2. ^{67}Zn

From its receptivity, ^{67}Zn is clearly not an easy nucleus to observe. Earlier studies of zinc compounds[29,30] used concentrations greater than 0.2 *M*, but more recently satisfactory results have been obtained from 0.05 *M* solutions[31] at natural abundance or 2.6 mM for ca. 80% ^{67}Zn enrichment.[32] Of course, linewidths will critically affect the observability of this quadrupolar nucleus with a linewidth factor about one-tenth of that of ^{63}Cu. Linewidths down to 12 Hz are obtained for very dilute aqueous solutions of zinc salts,[32] but these increase with concentration,[33,29] e.g., 50, 65, and 83 Hz for 2 *M* solutions of $Zn(ClO_4)_2$, $ZnSO_4$, and $Zn(NO_3)_2$, respectively, presumably because of weak association with the anions. The widths for $[Zn(SR)_4]^{2-}$ increase with the asymmetry of the thiolate ligands from 55 Hz for SMe to 1200 for SPr^i or SCH_2CH_2S[34] paralleling ^{63}Cu resonances of pyridine[18] or phosphine complexes.[19] The inability to observe a zinc signal from $ZnCl_2$ or $ZnBr_2$ in organic solvents or from edta or polyaza complexes[29,33] implies very broad lines from low symmetry species.

It is thus surprising that a resonance is observed[30] for the whole of the range from $Zn(ClO_4)_2$ to $Zn(ClO_4)_2 + 6\,Cl^-$ when this average signal contains large contributions from lower symmetry species {e.g., *trans*-$[ZnCl_2(OH_2)_4]$} expected to have large relaxation rates. A possible explanation may lie in the exchange rates between species. If exchange is slow compared with molecular rotation (but still fast compared to chemical shift differences), the observed relaxation rate is the weighted average of the individual relaxation *rates*.[35] Since the less symmetric species have very much faster relaxation rates, these predominate in the average and the linewidth increases greatly with the presence of low symmetry species as found for copper (Section 2.1). At the other extreme of exchange rates faster than molecular rotation, the relaxation *time* is the weighted average of the individual relaxation *times*. In this case, the long relaxation times of the highly symmetric species predominate. Thus if the rate of exchange between zinc halide species in water (but not organic solvents) is fast enough, it will reduce the effective contribution from the low symmetry species and give an observably narrow signal.

Shimiza *et al.* have used ^{67}Zn linewidths to investigate the interaction of zinc with apothermolysin[32] and with imidazoles and carboxylate anions.[31] While giving interesting qualitative results, there are considerable quantitative uncertainties as the rates of exchange, correlation times and asymmetries of the zinc sites are unknown.

2.3. ^{107}Ag and ^{109}Ag

The nuclear resonance properties of the two silver isotopes are very similar, but the modest receptivity advantage of ^{109}Ag has made it the usual choice for study. The difficulties of low receptivity and low resonant frequency are greatly increased by particularly long spin-lattice relaxation times. Further, γ and, therefore, the potential NOE are negative. Brevard *et al.*[36] have demonstrated how normal ^{1}H decoupling of $AgNO_3$ in H_2O causes the ^{109}Ag signal to vanish. This can be avoided by gated decoupling, but at the expense of time. The relaxation times may be reduced by the addition of paramagnetic materials such as Ni^{2+} salts[37] or 2,2,6,6-tetramethyl-4-hydroxypiperidine-1-oxyl (TANOL), but these may induce shifts, e.g., 10 ppm for a 50:1 Ag^+:TANOL solution.[38]

Polarization transfer from ^{1}H should give an enhancement factor of more than 20 and a further time saving because of the dependence on T_1 of ^{1}H.[36] However, coupling between silver and protons is not commonly observed because of the lability of silver complexes and the low value of $\gamma(^{109}Ag)$. In principle, such systems could be studied by ^{1}H–{^{109}Ag} double resonance with a further sensitivity advantage. For phosphine complexes, polarization transfer from ^{31}P should be possible.[39] Conversely, Colquhoun and McFarlane have combined ^{31}P–{^{109}Ag, ^{1}H} *pseudo* INDOR FT measurements with a two-dimensional display which separates the $J(^{109}Ag\,^{31}P)$ multiplets belonging to each silver-phosphine species.[40]

Nuclei such as $^{107/109}$Ag are ideal candidates for the steady state techniques of Schwenk and co workers. Thus "Quadriga" observation of the resonance from 0.02 *M* $AgNO_3$[41] betters the sensitivity of normal FT measurements with paramagnetic doping.[38]

Steady state techniques are the only practical approach to the measurement of the long relaxation times of silver.[41] The amazingly long T_1 of ^{109}Ag in $AgNO_3/D_2O$ (1115 s) is not greatly reduced when the solvent is H_2O showing that the intermolecular dipole–dipole contribution is small. Kronenbitter *et al.* propose shielding anisotropy as the main mechanism, but there are no temperature or field dependence studies to confirm this. Spin–spin relaxation times are much shorter (30–160 s), probably because of rapid exchange in the secondary coordination shell. The slower exchange of the primary coordination shell was blamed for the nonobservance of resonances from solutions expected to contain $[Ag(OH_2)(NH_2Et)]$.[41]

2.4. ^{111}Cd and ^{113}Cd

Direct observations are relatively easy for both isotopes of cadmium with ^{113}Cd having a small advantage on receptivity (and sensitivity when enriched samples are used). By contrast, the scope for double resonance is limited by the lability of cadmium compounds. Accordingly, activity in cadmium NMR has increased greatly with the availability of multinuclear observation systems. Recent measurements on inorganic complexes have employed ca. 0.1 *M* concentrations with $[Cr(acac)_3]$ to reduce ^{113}Cd relaxation times[42] or lower concentrations of enriched material and a 3.6 s cycle time.[43] Spectra have been obtained from 1 mM ^{113}Cd in presence of serum albumin using a specially built spectrometer with transverse geometry in a 6 T field.[8] One complication that can arise from the use of highly enriched material is spin–spin splitting between ^{113}Cd nuclei in adjacent coordination sites in a macromolecule,[44] although homonuclear decoupling experiments can then lead to structural assignments.[45,10]

Despite their distance from the metal, protons of the ligands (including H_2O) can cause dipole–dipole relaxation, resulting in some of the potential -2.25 NOE enhancement. To avoid the undesirable consequences (e.g., see Ref. 46) gated decoupling may be necessary at lower fields. At higher fields, increased shielding anisotropy relaxation outweighs the dipole–dipole contribution[44,47] (and greatly increases the sensitivity).[9] With biological measurements it is often impractical to wait for full relaxation between pulses so that intensities are not representative of concentrations.[48] Sometimes, however, structural determination may depend on true intensities from fully relaxed (and gated decoupled) spectra.[45] A further hazard is broadening (and consequent nonobservance) of signals because of intermediate rates of exchange between chemically shifted environments in biological[6,49] or inorganic complexes.[50]

Measurements of ^{113}Cd in solids[51,52] are greatly facilitated by cross-polarization with ^{1}H spins (if present) and magic angle spinning (MAS).[53–57] The shielding anisotropy may not be very much less than the spinning rate so that several spinning sidebands remain but the sensitivity gain is still important.[56] DuBois Murphy *et al.* have discussed how static, MAS and off-angle spinning measurements may be combined to determine the values of $\Delta\sigma$ and σ_{iso} and both these parameters assist the making of assignments between different ^{113}Cd sites.[56,55] The isotropic chemical shifts from MAS measurements, referring as they do to a definite coordination arrangement, are particularly valuable since resonances from solutions frequently represent the rapid exchange average of several environments. Sometimes lowering the temperature can slow exchange sufficiently to give signals from each environment, as in mixed halide complexes.[58,59] However, this is less applicable in the case of aqueous cadmium chemistry. Nonetheless, Jakobsen and Ellis have succeeded in obtaining separate resonances (complete with ^{113}Cd–^{15}N splitting) from cadmium glycine complexes in aqueous emulsions supercooled to $-40°C$.[60]

2.4.1. ^{113}Cd Relaxation

Bailey *et al.* have emphasized the importance of relaxation time measurements, not least for optimizing data acquisition parameters. However, as they estimate that a week of instrument time may be required to determine T_1 and η for a biological system,[61] such studies are not lightly undertaken. In fact there is a body of data for ^{113}Cd relaxation on which to base estimates of T_1 etc. For $[Cd(edta)]^{2-}$ with $\tau_c = 1.1 \times 10^{-10}$ s, the extreme narrowing condition is not satisfied at larger fields. In particular, this reduces the maximum NOE enhancement η_{max} which is of considerable consequence when estimating the dipole–dipole relaxation from η_{obs}/η_{max}.[47] Further, the effect of temperature on the relaxation rates may be misleading if the resonance represents the average of species whose relative concentrations and rates of interconversion are temperature dependent.

Spin rotation makes a major contribution to the relaxation of aqueous $Cd(ClO_4)_2$[62] and $CdMe_2$,[46] but for larger molecules it is likely to be much less significant.[47]

For $Cd(ClO_4)_2$ in H_2O ca. 50% of the relaxation is via dipole–dipole interactions with consequent NOE enhancement of -1.2 and much longer T_1 in D_2O (>80 s).[46,62] Dipole–dipole relaxation involving the protons of H_2O occurs for $[Cd(edta)]^{2-}$ dissolved in H_2O showing that species such as $[Cd(edta)(OH_2)]^{2-}$ are part of a rapid equilibrium.[47] There is no such contribution for the cyclohexyl derivative although the acetate CH_2 groups were shown to be the origin of a significant NOE.[63] Surprisingly, dipole–dipole relaxation seems to be negligible for $CdMe_2$[46] and $CdBu^n_2$.[64]

Relaxation by shielding anisotropy is important for ^{113}Cd especially at higher fields. The value of $\Delta\sigma$ for $CdMe_2$ (3400 ppm) is about half that of $HgMe_2$,[65,66] suggesting that shielding anisotropy relaxation for $CdMe_2$ will be about a quarter of that of $HgMe_2$.[67] This would account for half the observed ^{113}Cd rate.[46] Shielding anisotropy is the only significant mechanism for the pyridine adduct of cadmium *meso*-tetraphenylporphyrin ($\Delta\sigma$ 105 ppm)[9] and dominates at higher field for $[Cd(edta)]^{2-}$ (pH 5.7) where $\Delta\sigma$ was estimated as 131 ppm.[47]

For $Cd(ClO_4)_2$ in $^{17}OD_2$, there is a contribution from scalar coupling of the first kind, i.e., $J(^{113}Cd\,^{17}O)$ is modulated by the exchange of $^{17}OD_2$ ligands rather than the much slower ^{17}O relaxation.[62] Since ^{113}Cd and ^{195}Pt have similar resonant frequencies, the conditions for a contribution from scalar coupling of the second kind to T_1 are similar (see Chapter 20, Section 2.1.4). However, as couplings to ^{113}Cd are about half those to ^{195}Pt such contributions are unlikely to be significant. Although the contribution to T_2 will also be less, ^{14}N can cause significant broadening and loss of sensitivity.[9]

Jensen *et al.* have identified an unusual relaxation behavior for $[Cd(edta)]^{2-}$ at high pH which they ascribe to relaxation controlled by the rate of chemical exchange to a rapidly relaxing species.[47]

Values of T_1 for ^{113}Cd complexed to quite large molecules (e.g., proteins)[61,49,68,48] are usually more than a second even though correlation times are likely to be long.

2.5. ^{197}Au

Despite 100% natural abundance, the nuclear properties of ^{197}Au suggest that it will be very difficult to observe. The chemical restriction is the almost total absence of complexes with sufficiently high symmetry to avoid excessive broadening by quadrupolar relaxation (consequent upon the large linewidth factor). Accordingly, the only reported observations are for the metal and alloys.[69]

2.6. ^{199}Hg and ^{201}Hg

It is not surprising that the resonance of the quadrupolar ^{201}Hg with parameters similar to ^{105}Pd has only been observed in the metal.[70] Its significance to the chemist is further reduced by the relative ease of observing the spin-1/2 nucleus ^{199}Hg. Thus in recent years direct measurements of ^{199}Hg using multinuclear FT systems have produced a wealth of data to add to that already obtained by INDOR.[1] Typically, concentrations of 0.1–0.5 *M* have been used in studies of "inorganic" mercury complexes. Some workers have added $[Cr(acac)_3]$,[71] but relaxation times are unlikely to be inconveniently long except for very symmetric complexes, e.g., tetrahedral $[HgX_4]^{2-}$.[72] Sudmeier and Perkins[12] observed ^{199}Hg resonances in the presence of human carbonic anhydrase B with a concentration of 7 mM 83% ^{199}Hg (equivalent to 35 mM at natural abundance). As this was at 2.1 T, greater sensitivity might be obtained at higher fields. Roberts *et al.* have described two pulsed FT methods of determining the ^{199}Hg resonance from ^{1}H observation which are applicable to lower concentrations (e.g., 4 mM), providing, of course, Hg–H coupling is resolved.[73,74]

The most likely difficulty with ^{199}Hg observation is excessive linewidth; even "normal" lines are not particularly narrow, e.g., 20–50 Hz for X–Hg–Mo derivatives.[75] Sebald and Wrackmeyer[76] found ^{199}Hg resonances at 4.7 T severely broadened by shielding anisotropy relaxation. However, more common causes are scalar coupling (see next section) and exchange which is not rapid enough compared to the difference in frequency between sites.[12]

2.6.1. ^{199}Hg Relaxation

Despite the number of ^{199}Hg studies, information on relaxation behavior is relatively scarce. The similarity in nuclear properties suggests that ^{113}Cd may be taken as a guide when ^{199}Hg data are absent, e.g., relaxation processes for aqueous Hg^{2+} which has T_1 0.5 s at 2.1 T.[77]

Relaxation by spin-rotation is important for $HgMe_2$ at low fields[78,79] but negligible for the slower moving $HgPh_2$ which has been studied in detail by Gillies *et al.*[67] They find a small dipole–dipole contribution which they believe to be intermolecular but the relaxation is dominated by the shielding anisotropy interaction. For linear HgX_2 complexes, $\Delta\sigma$ is large (3000–7500 ppm[66,80,67,72]) leading to efficient relaxation. This will dominate at higher fields and results in short relaxation times (<0.1 s).[72,81] The consequences are most noticeable in the spectra of other nuclei (e.g., ^{1}H or ^{13}C) coupled to ^{199}Hg which show broad satellites which may almost disappear for bulky compounds.[72,82] In the absence of significant shielding anisotropy relaxation in $[HgX_4]^{2-}$ (X = Cl or CN), T_1 is much longer (ca. 7 s),[72] although Bain[77] reports the much shorter value of 0.4 s for a more dilute solution of $[Hg(CN)_4]^{2-}$.

The frequencies of ^{199}Hg and ^{75}As are very close (<0.8 MHz at 2.35 T), suggesting that a significant contribution to T_1 from scalar coupling of the second kind is possible. This is believed to be the cause of broad ^{199}Hg satellites in the ^{1}H spectra of $AsMe_3$ complexes.[83] Relaxation via ^{199}Hg–^{127}I coupling has also been identified (see Chapter 5, Section 4.2).

A scalar coupling contribution to T_2 and hence line broadening is possible for several donor atoms with quadrupolar nuclei (see Chapter 20, Section 2.1.4, and

Table 1). It has been reported for ^{14}N[84] but does not seem to be significant for ^{127}I[85] and there are no data for arsine complexes.

Very considerable linewidths (up to 900 Hz) have been reported[86] for Hg_2^{2+} in aqueous solutions. A likely explanation is to be found in the rapid exchange between Hg_2^{2+} and Hg^{2+} in aqueous solution observed by Peringer, e.g., 0.5 M $Hg_2(ClO_4)_2$ + 0.01 M $Hg(ClO_4)_2$ has $W_{1/2} > 100$ Hz at 1.9 T.[87] Similarly, the kinetics of the exchange between $[Hg(C_6H_5)]^+$ and Hg^{2+} have been investigated by ^{199}Hg linewidth studies.[88]

The longer $T_1(^{199}Hg)$ of $HgMe_2$ when the solvent is benzene rather than cyclohexane (0.96 vs. 0.56 s) has been rationalized in terms of the effect of complex formation on spin-rotation and shielding anisotropy relaxation,[89] but this ignores the difference in viscosities.

3. CHEMICAL SHIFTS

There are no chemical shift data for gold and few for copper, zinc, and silver in contrast to the situation for cadmium and mercury. For all these metals (including ^{197}Au) the resonant frequency for the free atom has been measured by atomic beam or optical pumping techniques. The results are not accurate enough to be used as the references for chemical shift calibrations but provide a valuable basis on which to compare the shielding in complexes of the different metals. The aqueous metal cation is accepted as the reference for zinc and silver and is chosen here for cadmium and mercury (rather than $CdMe_2$ and $HgMe_2$). Results for Cu^+ are not available so the reference $[Cu(NCMe)_4]^+$[18] is used. Details for these references and the free atoms are given in Table 1.

Ideally, the values for the aqueous ions should be related to a salt of a noncoordinating anion extrapolated to infinite dilution. Concentration effects are negligible for $Zn(ClO_4)_2$[29] and $Cd(ClO_4)_2 < 0.5$ M,[90,95] but small corrections are needed in other cases. Thus the shift of 2 M $AgNO_3$ is taken as 24 ppm.[92] For 4.5 M $Cd(NO_3)_2$, the reference used by Colton and co-workers, Maciel and Borzo[64] report a shift of -49.4 ppm, which is at variance with the shift extrapolated from the results of Kostelnik and Bothner-By,[98] namely, -73 ppm. Our own measurements suggest that the latter value is more likely to be correct and so has been used to make conversions. The shift of $Hg(ClO_4)_2$ in water does not extrapolate to a constant value on dilution (e.g., 0.1 M, 8.0; 0.04 M, 15.2; 0.02 M, 24.1 ppm[93]), presumably because of ionization of $[Hg(OH_2)_n]^{2+}$ to $[Hg(OH)(OH_2)_{n-1}]^+$. Therefore we have chosen as reference the resonance of 0.1 M $Hg(ClO_4)_2$ in 0.1 M $HClO_4$ at ca. 20°C. By cross-referencing, the shifts of the references used by other authors have been assigned values as follows: 2 mmol HgO/ml 60% $HClO_4$ (Peringer and co-workers), -134; [Hg(OAc) Ph] 1 M in DMSO (Colton and co-workers), 816[78]; $Hg(NO_3)_2$[99,100] -210 ppm.

3.1. Results

In many cases the tabulation and interpretation of chemical shift measurements for these metals is frustrated by uncertainties about the identity of the species to which they refer. For most of the metals, more than one coordination geometry is commonly shown for a given oxidation state so that structures cannot be confidently predicted. Thus for the sequence $[ZnCl_n(OH_2)_m]^{2-n}$, there is a change from octahedral to

Table 1. Chemical Shift Reference Standards for Post-Transition Metals

	Shift (ppm)	Ξ (MHz)	Reference
^{63}Cu			
$[Cu(NCMe)_4]\,BF_4$ (0.1 *M* in CH_3CN)	0	26.515 473(30)[a]	25
Free atom	−1820(80)	26.467 3(21)	25
^{67}Zn			
$Zn(ClO_4)_2$ (inf. dil.)	0	6.256 782(9)	29
Free atom	−690(10)	6.252 44(6)	29, 90
^{109}Ag			
Ag^+ (inf. dil.)	0	4.653 623(8)[b]	91
Free atom	−940(17)	4.649 2(8)	92
^{113}Cd			
$Cd(ClO_4)_2$ (0.1 *M* in H_2O)	0	22.178 946[c]	93
$CdMe_2$ (neat)	641	22.193 173	94
Free atom	−1105(4)	22.154 44(5)	95
^{197}Au			
Free atom		1.712 69(11)	96
^{199}Hg			
$Hg(ClO_4)_2$ (0.1 *M* in 0.1 *M* $HClO_4$)	0	17.870 535	93
$HgMe_2$	2255	17.910 841	97
Free atom	−2428	17.827 146(30)	86

[a] $\Xi(^{65}Cu) = 28.403\,659(30)$.[25]
[b] $\Xi(^{107}Ag) = 4.047\,898(6)$.[91]
[c] $\Xi(^{111}Cd) = 21.201\,877$.[94]

tetrahedral, possibly via trigonal bipyramidal as *n* increases, but precise details are lacking.[30]

Many complexes have low coordination numbers, as in linear Ag^I and Hg^{II} species, and are liable to undergo significant "secondary" coordination of additional ligands (including the solvent). Although these additional ligands are often distinguished by their lability, it may be a matter of dispute whether or not they count towards the coordination sphere. Certainly, they can make quite large contributions to the metal chemical shift so its value varies significantly with the conditions. Thus, on addition of two equivalents of pyridine to $AgClO_4$ in acetonitrile {probably $[Ag(NCMe)_2]^{2+}$ further solvated by CH_3CN}, $\delta(^{109}Ag)$ decreases from 529 to ca. 485 ppm {$[Ag(py)_2]^{2+}$ solvated by CH_3CN}, but further addition of pyridine increases the shift again to 556 ppm {$[Ag(py)_2]^{2+}$ solvated by pyridine}.[101] Such effects have been long known and studied for organo-mercurials, but there is still considerable uncertainty over the nature of the interaction (see Section 3.2.1).

Ligand exchange reactions tend to be more rapid than for true transition metals so that the observed resonance may refer to the weighted average of the shifts of several species in rapid equilibrium. Rapid ligand exchange can be discounted on the observation of spin–spin coupling between the metal and a ligand nucleus, or of separate resonances for species that would be related by the exchange, e.g., in a solution containing complexes of closely related ligands. Although the situation may be easily established for some ligands (e.g., phosphines), considerable efforts may be required to establish fully the coordination sphere. This is illustrated by the investigations into the behavior of anions (X) in $[CdX_2(PR_3)_n]$ systems[42] and the coordination of the carboxylate groups in $[Cd(edta)]^{2-}$.[47] Frequently such information is lacking and the nature of the species studied must be regarded as uncertain.

Of course, if only one species is present, lability is no problem. However, the possibility of significant contributions to the observed shift from other species is difficult to exclude on the basis of other types of measurement, especially when the latter refer to different experimental conditions. Thus Heinrichs *et al.* could not account for the behavior of $\delta(^{109}Ag)$ for the silver–thiourea system unless they invoked a series of oligomeric species that had not been shown to be present by other means of investigation.[102]

In principle, the shifts of the individual species in rapid equilibrium may be extracted by the application of known equilibrium constants to the results from solutions covering a range of compositions. This has been done for zinc and cadmium halide systems. Such treatments assume that the shift of an individual species is independent of the concentrations of the other species present. However, results for the nonlabile complexes of group VIII metals suggest that there will be some dependency but it is difficult to estimate the possible magnitude for metals such as cadmium. The accuracy of a determined chemical shift is poor if the species does not have a large concentration in at least one solution measured. Finally, considerable errors may be involved if the NMR measurements are made under different conditions (e.g., higher concentration) from those for which the equilibrium constants were evaluated.

It is possible to evaluate the equilibrium constants as well as the chemical shifts but the errors are likely to be greater especially if several species are involved or their nature is not well established. In a useful discussion of the method, Bach *et al.* conclude that for weakly formed complexes, the equilibrium constant is quite well determined but the chemical shift of the complex is not.[103] Of course, the determination of the equilibrium constant depends on the chemical shift difference between species, and this may be insufficient to give proper discrimination, e.g., the small difference between CdX^+ and CdX_2 species.[50,104]

Since the coordination arrangement may be firmly established for crystals, measurements on solids can be particularly valuable for these metals providing reasonably narrow resonances can be observed, e.g., because of high symmetry or by means of MAS.

The application of these various approaches is illustrated by the studies of cadmium-halide species (see Section 3.1.4).

3.1.1. ^{63}Cu Chemical Shifts

The very limited amount of $\delta(^{63}Cu)$ data is summarized in Figures 1 and 2. The shifts of ten complexes, $[Cu(PR_3)_4]ClO_4$, R = alkyl or alkoxyl span 73 to 247 ppm with the higher values corresponding to more alkyl groups on the phosphorus ligand.[19] The shift of $[Cu(py)_4]^+$ (110 ppm) falls in the same region[18] while the highest values (ca. 800 ppm) are for $[Cu(CN)_4]^{3-}$ in various systems.[16,105]

The effect of counterion and solvent on $[Cu(NCMe)_4]^+$ and of temperature on this and $[Cu\{P(OMe)_3\}_4]^+$ have been studied, and it is concluded that rapid exchange with species of lower symmetry is involved.[22] Becker has measured $\delta(^{63}Cu)$ of solid CuCl, CuBr, and CuI over a wide range of temperature.[26]

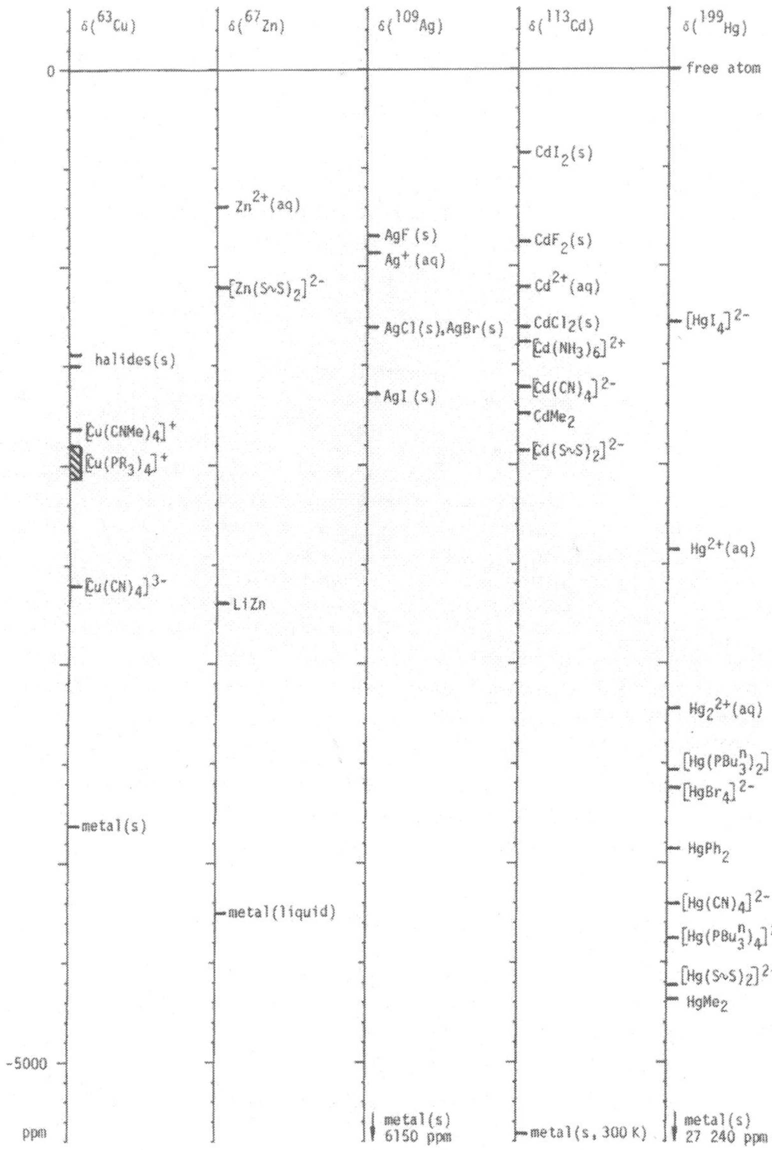

Figure 1. Comparison of the shieldings of ^{63}Cu, ^{67}Zn, ^{109}Ag, ^{113}Cd, and ^{199}Hg relative to the free atoms. [s = solid, S ~ S = $S(CH_2)_2S^{2-}$.]

3.1.2. ^{67}Zn Chemical Shifts

Most $\delta(^{67}Zn)$ measurements have been on aqueous solutions of simple salts and thus relate to the average of several species in rapid equilibrium.[29,30] The variation of the observed shift with concentration, temperature, and deuteriation of the solvent[29] no doubt follows changes in the equilibrium position. Using established equilibrium constants, $\delta(^{67}Zn)$ for $[ZnX_n(OH_2)_m]^{2-n}$, X = Cl and Br, $n = 1$–4 have been evaluated. The irregular sequences of values with maxima for $n = 2$ probably reflect changes in the coordination number between complexes.[30] The shift of $Zn(NO_3)_2$

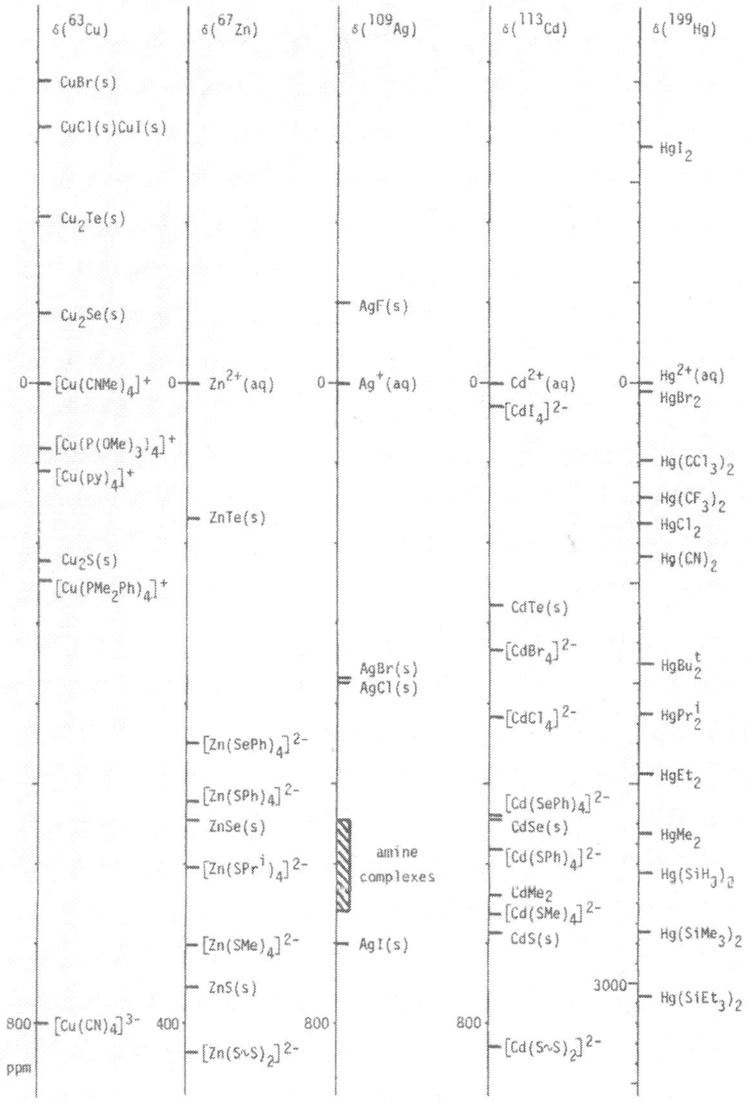

Figure 2. Some representative chemical shifts for copper, zinc, silver, cadmium, and mercury. [s = solid, $S \sim S = S(CH_2)_2S^{2-}$; HgX_2 are neat or as solutions in THF.]

goes to low frequency when the solvent is dimethylformamide or methanol instead of water.[33] The results of Carson and Dean[34] are believed to refer to the single anionic species $[Zn(SR)_4]^{2-}$, SR = SPh, SPr^i, SMe and $\frac{1}{2}S(CH_2)_2S$ (261–418 ppm) and $[Zn(SePh)_4]^{2-}$ (224 ppm) in D_2O solution and are comparable with the shifts in solid zinc chalcogenides[106] (see Figure 2).

3.1.3. ^{109}Ag Chemical Shifts

Most ^{109}Ag data refer to labile species with the uncertainties discussed above. There have been several studies of $AgClO_4$ or $AgNO_3$ in water or organic solvents (or liquid ligands).[92,107,37,101] Even the shift of $AgClO_4$ in water shows a marked concentration dependence (ca. 10 ppm mol^{-1} liter).[92] The effect of some S- and N-donors on $\delta(^{109}Ag)$ of $AgNO_3$ in DMSO [and on $\delta(^{13}C)$ of the donors] has been investigated. For the S-donors, the behavior is not clear cut and the presence of oligomers has been postulated, but with amines the results are consistent with the formation of $[Ag(amine)_2]^+$.[102,108] Similar species are likely to be dominant in other studies of $AgClO_4$ or $AgNO_3$ with amines in which a range of 232–556 ppm has been observed.[107,101,41,38] However, the shifts for aqueous silver halide/ethylamine systems depend strongly on the halide.[107,41] Lighter halides give lower shift values as do aqueous solutions of AgX + NaX[38] and solid AgX[109] (Figure 2).

There are some results for demonstrably nonlabile complexes.[36,110,111] Over 20 ppm separate the resonances of the diastereomers of $[Ag\{(S)(6\text{-Me-}2C_5H_3N)CH{=}NCHPhMe\}_2]^+$ (636 and 660 ppm). These values and that of the related cation $[Ag_2\{\mu\text{-}1,2\text{-}[(2\text{-}C_5H_5N)CH{=}N]_2\ \text{cyclohexane}\}]$ (604 ppm) are close to those of $[AgBrNRCHN(p\text{-tol})Pt\{2,6\text{-}(Me_2NCH_2)_2C_6H_3\}]$ (544–584 ppm) and, notably, $[Ag(py)_2]^+$ (556 ppm).[101] The shifts of $[Ag\{P(OEt)_3\}_n]NO_3$ increase by over 900 ppm as n goes from 1 to 4, whereas the increase is only 350 ppm when the anion is the better coordinator, SCN^-.[40]

3.1.4. Cadmium Chemical Shifts

Since there is no detectable primary isotope effect between ^{111}Cd and ^{113}Cd shielding,[112] we shall describe all shifts as $\delta(^{113}Cd)$, although in a few cases ^{111}Cd was in fact measured.

Several of the earlier studies concern solutions of simple salts in water[64,14,98,46,113] or organic solvents[98,113,114] and the $\delta(^{113}Cd)$ values cannot be related to properly defined species.[56] This is frequently true of results for such solutions after the addition of typical ligands,[98,113–116] arenes,[114] glucose,[116] or polyelectrolytes.[117] However, these measurements do give the direction and some indication of the magnitude of the change to be expected on complex formation with different ligands, e.g., N-donors to high frequency, O-donors to low frequency.[98] Studies of cadmium involving large biological molecules are discussed in a subsequent chapter (see also the beginning of this chapter for references), so here we shall concentrate on results for ^{113}Cd in well-defined environments made up of smaller ligands.

Studies of cadmium halide systems illustrate the various approaches that have been employed to obtain such information. For water or DMSO solutions of Cd^{2+} with halide ions, only a single peak is observed. Its shift depends on the halide concentration, and two groups have analyzed the data using established equilibrium constants to obtain values for each solvated CdX_n^{2-n} species (X = Cl, Br, I, or SCN).[50,104] The shifts for the chloro- and bromo-complexes in water show the largest step between $n = 2$ and $n = 3$, which probably corresponds to a change of geometry from octahedral to tetrahedral. The broad lines observed for some DMSO solutions indicate that exchange is slower in DMSO than in water.[50] The observation of separate signals for mixed halide complexes shows that there is no exchange between $[CdX_4]^{2-}$ and other species in solutions of $[NBu_4]_2[CdX_4]$ in CH_2Cl_2 at 208 K.[59] Finally, measurements on solid $[NR_4]_2[CdX_4]$ relate to the tetrahedral $[CdX_4]^{2-}$ anion as demonstrated by x-ray structure determinations.[56] The results for $[CdX_4]^{2-}$ from the three methods are in good agreement except for the shift of $[CdCl_4]^{2-}$ derived

Table 2. Summary of ^{113}Cd Chemical Shifts of Complexes in Solution

Complex[a]	$\delta(^{113}Cd)$ (ppm)[b]
$Cd(alkyl)_2$ (118,46,64)	641–434 [7]
$CdPh_2$ (46)	329
$[CdMe(OR)]$ (94)	382–258 [15]
$[CdMe(SR)]$ (94)	610, 597
$[Cd(CN)_4]^{2-}$ (46)	510
$[Cd(NH_3)_6]Cl_2$ (46)	274
$[Cd(TPP)L]$ (119,79)	454–413 [9]
$[Cd\{2,6\text{-}(RN=CMe)_2C_5H_3N\}_2]^{2+}$ (120)	289, 233
$[CdCl_2\{2,6\text{-}(RN=CMe)_2C_5H_3N\}]$ (120)	351–349 [3]
$[CdX_2\{2,6\text{-}(RN=CMe)_2C_6H_3N\}]$, X = NCS, I (120)	295, 270
$[CdX_2\{2,6\text{-}(RN=CMe)_2C_6H_3N\}]$, X = OAc, NO_3 (120)	69, 27
$[Cd(glycine)(OH_2)_n]^+$ (60)	54
$[Cd(glycine)_2(OH_2)_n]$ (60)	154
$[Cd(glycine)_3(OH_2)_n]^-$ (60)	263
$[Cd(edta)OH_2]$ (pH 5–11) (47,113)	85
$[CdCl_2(6\text{-mercaptopurine})OH_2]_2$ (121)	554
$[CdX_2(PR_3)]$, X = Cl, Br, dtp, dtc, I; R = Bu^n, Cy (42,122)c	452–207 [15]
$[CdX_2(PR_3)]$, X = OAc, NO_3; R = Cy, Bu^n (42)c	162–57 [4]
$[CdX_2(PR_3)_2]$, X = Cl, Br, I, dtp; R = Bu^n, Cy (42,58,122)c	602–417 [9]
$[CdX_2(PR_3)_2]$, X = OAc, NO_3, R = Bu^n, Cy (42)c	347–252 [4]
$[Cd(PR_3)_2](ClO_4)_2$, R = Cy, Bu^n (42)	287, 267
$[CdX_2(PBu^n_3)_3]$, X = Cl, Br, I, NO_3 (42)c	507–352 [4]
$[Cd(PBu^n_3)_3](ClO_4)_2$ (42)	412
$[Cd(\beta\text{-diketonate})_2(OH_2)_2]$ (123)	−24, −27
$[Cd(\beta\text{-diketonate})_2(amine)_2]$ (123)	43–39 [4]
$[Cd(thio\text{-}\beta\text{-diketonate})_2]$ (123)	237–191 [3]
$[Cd(OPCy_3)_4]^{2+}$ (124)	94
$[Cd\{(OPPh_2)_2(CH_2)_n\}_3]$, n = 1 or 2 etc. (124)	18 to −27 [3]
$[Cd(APR_3)_2]^{2+}$, A = S, Se, R = o-tolyl (124)	177, 130
$[Cd(APR_3)_3]^{2+}$, A = S, Se; R = Cy, o-tolyl (124)	503–440 [4]
$[Cd(APR_3)_4]^{2+}$, A = S, Se; R = o-tolyl, Ph, Cy (124)	618–522 [5]
$[Cd(SR)_4]^{2-}$, R = alkyl or aryl (43)d	666–622 [15]
$[Cd(SCH_2CHRS)_2]^{2-}$, R = H, Me etc. (43)d	829–778 [7]
$[Cd_{10}(SCH_2CH_2OH)_{16}](ClO_4)_4$ (113)	515, 382
$[Cd(SePh)_n(SPh)_{4-n}]^{2-}$, n = 0–4 (125)	590–541 [5]
$[NBu_4]_2[CdX_4]$, X = Cl, Br (59)e	417, 333
$[NBu_4]_2[CdI_4]$ (59)e	26
$[Cd\{Mn(CO)_5\}_2]$ (46)	552

[a] dtp = o,o-dialkyldithiophosphate, dtc = dialkyldithiocarbamate
[b] The number of reported values is given in square brackets
[c] Including data on mixed anion species,(58,42) $[Cd_2X_4(PBu_3)_3]$ and $[CdX_2HgX_2(PBu_3)_2]$ (58)
[d] Including studies on mixed ligand species at 193 K
[e] Including data on mixed halide species at 208 K

from equilibrium studies which may be affected by the presence of species with more than four chloride ions per cadmium.

Table 2 summarises results for species of reasonably certain composition in solution. Cadmium salts of alkyldithiophosphates or dialkyl dithiocarbamates are not included as they are rapidly exchanging even at 250 K.[122] This is likely to be true of the complexes with xanthate and $S_2CC(CN)_2^{2-}$, although the latter gives rise to a very deshielded ^{113}Cd environment, 813 ppm in DMSO solution.[113] A detailed listing of results for cadmium alkyls is given in Ref. 1.

A summary of $\delta(^{113}Cd)$ for solids with known Cd environment is given in Table 3 in order of increasing coordination number. There is a general trend of decreasing chemical shift with increasing coordination number (for the same donor). This has led Rodesiler and Amma to propose that when the resonance of ^{113}Cd surrounded by O-donors falls in the range -30 to -100 ppm, a coordination number of greater than six is indicated.[126] However, the shifts of $3CdSO_4 \cdot 8H_2O$ which contains two O_6 environments, fall in this region,[54,56] showing that such simple classifications are unlikely to be successful. Indeed, the bond lengths may differ considerably both within a structure and between structures, which is likely to significantly affect the chemical shift. The two cadmium environments in $[Cd_2(bipy)_4(NO_3)_2(OH_2)]$ differ only in the replacement of ONO_2^- in one by OH_2 in the other, yet this results in a 71 ppm difference in $\delta(^{113}Cd)$[57] showing that quite large differences in shift can occur for very similar coordination arrangements.

Table 3. Summary of ^{113}Cd Chemical Shifts in Known Solid Environments

Coordination[a]	Compound	$\delta(^{113}Cd)$ (ppm)[b]
N_4(Pl)	Cd(TPP)[5]	399[c]
S_4(Tet)	CdS[51]	687[c]
	$[Cd_{10}(SCH_2CH_2OH)_{16}](ClO_4)_4$[55]	673[c]
	$Cd(xanthate)_2$[56]	445, 414[c]
Se_4(Tet)	CdSe[51]	545[c]
Te_4(Tet)	CdTe[51]	278
Cl_4(Tet)	$[+]_2[CdCl_4]$[56]	483, 460[c]
Br_4(Tet)	$[+]_2[CdBr_4]$[56]	396 to 374[c]
I_4(Tet)	$[+]_2[CdI_4]$[56]	92, 79[c]
N_5(SBP)	[Cd(TPP)(py)][5]	432[c]
OS_4(TBP)	$[Cd_{10}(SCH_2CH_2OH)_{16}](ClO_4)_4$[55]	502[c]
S_5(SBP)	$Cd(S_2CNEt_2)_2$[56]	377
Cl_5(TBP)	$[Co(NH_3)_6][CdCl_5]$[56,104]	134
N_6(Oct)	$[Cd(en)_3]\ Cl_2 \cdot H_2O$[56]	380
	$[Cd(imidazole)_6][-]_2$[56]	272, 238
N_4O_2(Oct)	$[Cd_2(bipy)_4(NO_3)_2(OH_2)]$[57]	122, 51
N_2O_4(Oct)	$Na_2[Cd(edta)]$[56]	117, 107
O_6(Oct)	CdO[51,54]	414[c]
	$Cd(OH)_2$[56,53]	158[c]
	$[Cd(OH_2)_6](ClO_4)_2$[56]	0[c]
	$Cd(oxy\text{-}anion)_2$ etc.[56,126,53,54,51]	61 to -56[c]
O_3S_3(Oct)	$[Cd_{10}(SCH_2CH_2OH)_{16}](ClO_4)_4$[55]	409[c]
O_2Cl_4(Oct)	$CdCl_2 \cdot 2\frac{1}{2}H_2O$[56,53]	187
OCl_5(Oct)	$CdCl_2 \cdot 2\frac{1}{2}H_2O$[56,53]	213
Cl_6(Oct)	$CdCl_2$[51]	206[c]
	$[Co(en)_3]_2[CdCl_6]\ Cl_2 \cdot 2H_2O$[104]	171
I_6(Oct)	CdI_2[51]	-677
N_3O_4(PBP)	Cd(*o*-hydroxybenzoate)$_2 \cdot 3py$[126]	61
N_2O_5(PBP)	Cd(*p*-aminobenzoate)$_2 \cdot 3H_2O$[126]	34
O_7	$Cd(oxy\text{-}anion)_2 \cdot 2H_2O$[126,56]	-31 to -58
O_8	$Cd(NO_3)_2 \cdot 4H_2O$[56,53,54]	-100[c]
F_8	CdF_2[51]	-228

[a] The abbreviations Pl = planar, Tet = tetrahedral, SBP = square-based pyramid, TBP = trigonal bipyramid, Oct = octahedral, PBP = pentagonal bipyramid indicate the approximate coordination arrangement.

[b] To high frequency of 0.1 *M* $Cd(ClO_4)_2$. Where required to make conversions, a value of -100 ppm was used for solid $Cd(NO_3)_2 \cdot 4H_2O$.[56]

[c] Data on $\Delta\sigma$ and/or shielding tensor elements.

Comparison of the three shifts observed for solid $[Cd_{10}(SCH_2CH_2OH)_{16}](ClO_4)_4$ (Table 3) with the two found for solutions (Table 2) suggests that the S_4 sites become solvated to give OS_4 sites. This contrasts with $[Cd(SCH_2CH_2OH)_2]$ in DMSO where the shift of 655 ppm(113) suggests S_4 coordination.

Several of the studies on solids include information on $\Delta\sigma$ or shielding tensor elements as indicated in Table 3. Dalton *et al.* have evaluated $\Delta\sigma$ for $CdMe_2$ dissolved in a liquid crystal.(65) Of particular interest is the effect of the introduction of an axial pyridine ligand into Cd(TPP). The large change in $\sigma_\perp$ as well as $\sigma_\parallel$ implies that Cd is pulled out of the TPP plane.(9)

3.1.5. ^{119}Hg Chemical Shifts

The ^{199}Hg chemical shifts measured for aqueous solutions of simple mercury salts(127,86,128) will refer to poorly defined species. The dependence of the shift of $Hg(ClO_4)_2$ on its concentration (see above and Ref. 86) emphasizes the problem. The mercury(I) salts that have been measured have shifts in the region of 800 ppm.(86,129)

Table 4 sets out the ^{199}Hg chemical shifts of some complexes HgX_2 which hopefully have this linear formulation in solution [this may not be true for $Hg(SCN)_2$ in pyridine, the only solvent for which measurements are reported]. The shifts of HgX_2, X = Cl, Br, and I, in a wide range of "solvents" have been measured.(85) Studies on mixtures show that the halides are labile,(130,85) so the exact nature of the species present is uncertain, but for completeness the values for THF solutions are included in Table 4. There are too many reports for mixed complexes [HgXY] to give more than literature references. However, it has been noted that for many such complexes, the chemical shift lies quite close to the mean of the shifts of HgX_2 and HgY_2.(130,76) The largest discrepancy was with bromides and iodides and follows from the greater *trans* influence of Y compared with halide. Much better predications of the shifts of [HgBrY] and [HgIY] can be obtained if the values 400 and −100 ppm are used, respectively, in place of the observed shifts of $HgBr_2$ and HgI_2. Additional mixed species include the interesting series $[CH_{4-n}(HgX)_n]$, $n = 1$–4 (X = Cl, Br, I, or CN)(158) and derivatives with Cp,(132) olefins,(134) and CH_2SiMe_3.(76) Unfortunately the shift is unknown for $[HgMe(NO_3)]$ in methanol—the reference used for a study of adducts of pyridines and 2,2′-bipyridyls with $[HgMe(NO_3)]$.(172)

The shifts of complexes that are likely to be 3 or 4 coordinate in solution are summarized in Table 5. The anion $(aryl{-}N{=}N{-}N{-}aryl)^-$ is taken to be bidentate-towards mercury in view of the crystal structure of $[(2\text{-}ClC_6H_4)_2\text{-}N_3HgMn(CO)_5]$.(182) Coupling to ^{31}P confirms the number of phosphorus ligands coordinated to mercury, but there is little definite evidence for other ligands. The observation of separate resonances for mixed halide complexes of $[HgX_4]^{2-}$ (59) and $[HgX_2(PR_3)_2]$, R = Bu^n or 4-$MeOC_6H_4$(71,179) at low temperatures establishes that the halides are coordinated. Separate resonances could not be obtained for $[Hg(SR)_{4-n}(SR')_n]^{2-}$ in contrast to the cadmium analogs.(34)

The response of $\delta(^{199}Hg)$ to the concentrations of components has been analyzed to obtain formation constants for $HgCl_2 + Cl^-$,(183) [Hg(Me)(OAc)] with MeSSMe, PhSSPh, $P(OEt)_3$, and SBu^n_2,(103) methyl mercury halides with halide,(184,97) [HgMe(OH)] with H^+,(185) and $HgMe_2$ with benzene(89) as well as the self-association constants of $Hg(CH_2COCH_2R)_2$, R = H or CO_2Me.(144)

Despite the chemical similarity of mercury to cadmium, there has been very little interest in ^{199}Hg in biological environments. The effect of halide or cyanide on the

Table 4. ^{199}Hg Chemical Shifts of Linear Mercury(II) Compounds, HgX_2

X[a]	Representative $\delta(^{199}Hg)$ (ppm)	Reports on [HgXY][a b]
CH_3 (66 78 97 89)	2255 (neat)	[16] (130–132 76 103 97 133 66 78 128 127 134)
CH_2CH_3 (135 78 136 133)	1964 (neat)	[5] (76 137 133 138 78 134)
CH_2CH_2R [2] (133 135 136 127)	2049 (R = Me, neat)	[12] (76 129 133 138 139 134)
CH_2CHRR' [1] (133)	2131 (R = R' = Me, neat)	[4] (133 139 134)
$CH_2CRR'R''$ [1] (66)	2102 (R = R' = R'' = Me, neat)	[12] (76 128 140 139 134)
CH_2(aryl) [7] (135 141)	1554 (aryl = Ph, CH_2Cl_2)	[13] (141 143)
CH_2COR [3] (135 144 145)	1485 (R = OMe, CH_2Cl_2)	
$CH(CH_3)_2$ (133 135 136)	1660 (neat)	[2] (133 138 134)
$CHCH_3CH_2Me$ (133)	1745 (neat)	[1] (133)
Cyclohexyl etc [6] (146 147)	1756 (X = Cy, CH_2Cl_2/py)	[13] (146 78 147 76)
$C(CH_3)_3$ (133 138 100 66)	1399 ($CDCl_3$)	[1] (133 138 134)
CF_3 (130)	577 (THF)	[7] (130)
CCl_3 (130)	388 (THF)	[4] (130 78)
C_6H_5 (148 149 78 135)	1504 ($CHCl_3$)	[31] (146 130 78 135 150 76 151 149 152 153 128 129 154 142)
Aryl [30] (155 154 148 128)		[14] (142 154 78)
C_6F_5 (149)	1333 (acetone)	[2] (135 149)
$CH=CH_2$ (78 135 156)	1612 (neat)	[1] (76)
$CF=CF_2$ (131)	1296 (90%, C_6F_6)	[1] (131)
$CCl=CCl_2$ (150)	1082 ($CHCl_3$)	[13] (150 135 157)
$C\equiv CH$ (76)	1275 ($CDCl_3$)	[7] (76)
$C\equiv CR$ [8] (76)	1380 (R = Me, $CDCl_3$)	[10] (76)
CN (130 151 127)	866 (THF)	[20] (130 158 151 153 159 77)
SiH_3 (160)	2452 (C_6H_6)	[1] (160)
$SiR_2(CH_3)$ [5] (160 99 100 161)	2738 (R = Me, C_6H_6)	[9] (99 160)
$Si(C_2H_5)_3$ (99)	3068 (neat)	
GeH_3 (160)	2108 (C_6H_6)	[1] (160)
$Ge(CH_3)_3$ (162 100)	2370 (toluene)	[3] (99)
$Ge(C_2H_5)_3$ (162 163)	2764 (C_6H_6)	
$Ge(CF_3)_3$ (162)	1582 (toluene)	
$Ge(C_6F_5)_3$ (162)	1550 (toluene)	
$Sn(CH_2AMe_3)_3$, A = C, Si (100 164)	2414 (A = Si, toluene)	
$Sn(C_6F_6)_3$ (164)	1609 (toluene)	
$N(SiMe_3)_2$ (165)	1194 (neat)	[2] (165)
$P(CMe_3)_2$ (165)	1994 (THF)	[1] (165)
μ-PCy_2 (166) [c]	1397 (CH_2Cl_2)	
PR_3^+ [11] (130 167 168)	1093 (R = Me, D_2O)	[8] (130)
$P(O)R_2$ [3] (153 152 169 170)	927 (R = Et, MeOH)	[32] (153 152 165)
SR [3] (171 159)	1268 (R = Et, $CDCl_3$)	[9] (103 137 150–152 171)
SCN (159)	1258 (py)	[4] (152 150 151 153)
S_2CNR_2 [16] (171 159)	993 (R = Et, CH_2Cl_2)	[3] (75)
$S_2P(OR)_2$ [3] (171)	1221 (R = Bu, CH_2Cl_2)	
$Se(C_6H_5)$ (159)	849 (py)	[1] (159)
Cl (130 85 127)	703 (0 2 *M*, THF)	
Br (130 85)	41 (0 2 *M*, THF)	
I (130 85)	−1182 (0 2 *M*, THF)	
$Cr(CO)_3Cp$ (81)	2175 (DMSO)	[1] (81)
$Mo(CO)_3Cp$ (81 75)	2371 (DMSO)	[19] (81 75 171)
$W(CO)_3Cp$ (81)	1907 (DMSO)	[4] (81)

[a] The number of species is given in square brackets
[b] Including minor modifications of X, references follow Y in the same order as X
[c] $[Hg(PCy_2)]_5(O_3SCF_3)_5$

Table 5. Summary of ^{199}Hg Chemical Shifts of Mercury(II) Complexes with Coordination Number Greater Than 2

Complex	$\delta(^{199}Hg)$ (ppm)[a]
$[Hg\{(CH_2PMe_2)_2CH\}_2]$ (173)	2187
$[Hg(dat)_2]$[b] (84,174)	586–440 [16]
$[Hg(dat)Y]$, Y = Ph (84,174)	1093–1004 [11]
$[Hg(dat)Y]$, Y = $PO(OEt)_2$ (175)	969–918 [7]
$[Hg(dat)Y]$, Y = SCN, CN, $CClCCl_2$ (159,174,150)	929–862 [11]
$[Hg(dat)\{PO(OEt)_2\}(PPh_3)]$ (175)	1628, 1567
$[Hg(dat)PR_2]_n$, $n = 3, 4$; R = Bu^t, Cy (176)	2014–1820 [3]
$[Hg(OAc)PCy_2]$ (166)	1500
$[Hg\{PO(OEt)_2\}_n]^{2-n}$, $n = 3, 4$ (170)	1822, 1796
$[HgY\{PO(OEt)_2\}PR_3]$, Y = S_2CNEt_2, $S_2P(OEt)_2$, OAc, O_3SCF_3 (177)	1573–1007 [6]
$[HgY\{PO(OEt)_2\}(PR_3)_2]$, Y = Cl, Br, I, $PO(OEt)_2$, OAc, pyd[c] (177)	1965–1667 [11]
$[Hg\{PO(OEt)_2\}(PR_3)_3]O_3SCF_3$, R = Bu^n, Ph (177)	1972, 1872
$[HgX_2(PR_3)]$, X = SCN,[d] CN (168)	1767–1632 [3]
$[HgX_2(PR_3)]$, X = $S_2P(OEt)_2$, S_2CNEt_2 (122)	1537–1381 [14]
$[HgCl_2(PR_3)]_2$, R = Bu^n, 4-$MeOC_6H_4$ (71,178,179)	1507, 1462
$[HgBr_2(PR_3)]_2$, R = Bu^n, 4-$MeOC_6H_4$ (71,178,179)	1276, 1156
$[HgX_2(PR_3)]$, X = OAc, O_2CCF_3, NO_3 (168)	1151–846 [10]
$[HgI_2(PR_3)]_2$, R = Bu^n, 4-$MeOC_6H_4$ (71,178,179)	606, 341
$[HgX(PBu^n_3)_2]ClO_4$, X = Br, SCN, I (168)	1627–1567 [3]
$[HgX_2(PR_3)_2]$, X = CN, SCN, Cl (178,71,179,168,167)	2005–1757 [10]
$[HgX_2(PR_3)_2]$, X = Br, $S_2P(OEt)_2$, S_2CNEt_2 (178,71,179,122,167,58)	1812–1647 [14]
$[HgI_2(PR_3)_2]$, R_3 = Me_2Et, Bu^n_3, $(4\text{-}MeOC_6H_4)_3$ (178,71,179,167,58)	1699–1181 [3]
$[HgX_2(PR_3)_2]$, X = OAc, O_2CCF_3 (168)	1532–1186 [13]
$[HgX_2(PR_3)_3]$, X = SCN, Cl, Br, I, OAc, NO_3, O_2CCF_3 (168)	2082–1707 [10]
$[Hg(PBu^n_3)_n](ClO_4)_2$, $n = 3, 4$ (168)	1897, 1947
$[Hg\{P(aryl)_3\}_n](ClO_4)_2$, $n = 3, 4$ (168)	1712–1587 [9]
$[Hg(SR)_4]^{2-}$ (34)	2234–1668 [9]
$[Hg(SePh)_4]^{2-}$ (34)	1275
$[Hg(SCN)_2(APBu^n_3)_2]$, A = S, Se (180)	1743, 1526
$[HgX_2(SePBu^n_3)_2]$, X = Cl, Br, I (180)[e]	1280–86 [3]
$[HgCl_3]^-$ [d] (181)	1058
$[HgX_4]^{2-}$, X = CN, Cl (181,59)	1775, 1056
$[HgX_4]^{2-}$, X = Br, I (59,181)	1195, −1146

[a] The number of complexes is given in square brackets.
[b] dat = [(aryl) *N*=*N*−*N*(aryl)]$^-$.
[c] pyd = 2,5 pyrolidinedionato − *N*.
[d] Dimeric species.
[e] Somewhat different shifts are reported in Ref. 179, which includes data for $[HgX_2(SePBu_3)]$.

resonance of ^{199}Hg bound to human carbonic anhydrase B has been reported(12) and the response of $\delta(^{199}Hg)$ of ethyl mercury phosphate to various amino acids has been measured.(74)

The values of $\Delta\sigma$ for linear mercury complexes(66,80,67,72) are ca. twice those for the analogous cadmium species. Thus it will be more difficult to produce single, narrow lines from solids by "magic angle" spinning, and no examples seem to be reported.

3.2. Discussion

The shielding of these metal nuclei has been discussed by Kidd.(186) The observed range, relative to the free atom, is illustrated in Figure 1. The shielding sensitivity cannot be evaluated on the same basis as preceding chapters because of insufficient data, but it is clear from a comparison of Figure 1 with Figure 2 in Chapter 20 that the shielding range is much smaller than for group VIII metals. Most complexes of these post-transition metals are colorless, indicating a large value of ΔE and hence relatively small paramagnetic contribution to the shielding. Further, they have filled *d* shells of electrons, and the paramagnetic contributions from partially filled *p*-valence shells are expected to be less than from partially filled *d* shells.[187] Bonding in mercury complexes may involve significant proportions of *d*-orbital functions and this may contribute to the greater range (ca. $\times 2$) of shielding of ^{199}Hg compared to the other metals.[188]

The chemical shifts for copper, zinc, and silver complexes and some solid compounds are depicted in Figure 2 alongside representative results for cadmium and mercury (note the different scale factors consequent upon the different shielding ranges). While for ^{113}Cd and ^{199}Hg, increasing the atomic number of a bound halide has the usual result of increasing the shielding, ^{109}Ag quite clearly displays the reverse behavior. An intermediate position applies to copper and also to ^{67}Zn to judge from the results evaluated by means of equilibrium constants.[30] Even for ^{199}Hg the differences in the shifts of complexes involving different halides are very variable and become small for phosphine complexes, which may be indicative of relatively weak metal–halide bonding.[177,167]

The pattern of shieldings shown in Figures 1 and 2 is markedly different from that for group VIII metals (Figure 2 of Chapter 20). Complexes of the latter metals with "strong" ligands such as alkyl or cyanide show relatively high metal shieldings related to large ΔE values, but with the post-transition metals, these ligands give some of the lowest shieldings. The pure d^{10} ions (Cu^+, Ag^+, Zn^{2+}, Cd^{2+}, and Hg^{2+}) would have no paramagnetic shielding contribution. Deshielding in complexes may result from electrons, additional to the d^{10} shell, that formally originate from the ligands.[189] Thus ligands that more effectively donate electrons to the metal are expected to result in lower shielding (although variation in ΔE and other terms will modify the outcome). In agreement with this, methyl derivatives of cadmium and mercury are less shielded than phenyl analogs, whereas the reverse is true for ^{195}Pt in $[PtCl_2(CO)R]^-$.[138] However, such a relationship is only partially borne out by the results for phosphine complexes, where parallel electrochemical studies give an indication of electron donation to the metal.[71,179,168,190] Indeed, deshielding is also produced by removal of *d*-electron density from the metal, which is most likely for Cu^+, Ag^+, and Hg^{2+}.[188]

A more detailed consideration of the effect of variation of electron population in the valence orbitals is possible with the expression of Jameson and Gutowsky[187] for σ^p, the paramagnetic contribution to σ. If *d* orbitals could be neglected, this would take the relatively simple form $\sigma^p = -A\,\Delta E^{-1}\langle r^{-3}\rangle\, P_u$, where A is a constant and P_u involves the valence *p*-electron populations. For linear complexes of these metals, P_u has a maximum value of 1 (i.e., when the metal p_z orbital has equal share of an electron pair with the ligands). For bonding in two dimensions (e.g., trigonal planar coordination) or three dimensions (e.g., tetrahedral coordination) the maximum value is 3/2. Accordingly, if there are no great changes in ΔE we expect three- and four-coordinate complexes to have similar shifts, but two-coordinate complexes to be much more shielded. This pattern of shielding is shown by $[Hg\{PO(OEt)_2\}_n]^{2-n}$ with $\delta(^{199}Hg)$ 927 ($n=2$), 1822 ($n=3$), and 1796 ppm ($n=4$)[170] and also by phosphine complexes.[168,177] However, the change on going from two- to three-coordinate is rather small compared to the likely paramagnetic shielding contribution, which may be a consequence of a greater *d*-orbital contribution in the two-coordinate species.[188] Decreases in shielding also occur on addition of halide to $[Hg(PR_3)_2]^{2+}$,[167,71,168] HgX_2 (X = Cl, Br, I or CN)[1] and [HgX(R)].[184,97]

3.2.1. Solvent Effects on HgX_2

The sensitivity of $\delta(^{199}Hg)$ of HgX_2 to solvent has frequently been ascribed to coordination. According to the discussion above, coordination of additional ligands is expected to decrease the shielding, so solvents with the greatest donor ability are expected to give the least shielded resonances. Such behavior is shown by HgX_2, X = Cl, Br, I, or CN,[85,151] and in view of the readiness of these species to increase their coordination number can be confidently assigned to solvent coordination. However, for HgR_2, R = Me, Et, or aryl,[78,148] which show very little tendency to accept further ligands, the reverse order of shielding vs. solvent is observed. Between these two extremes {e.g., $Hg(C{\equiv}CR)_2$[76] and [HgX(R)][97,103]} mixed behavior occurs—the very strong donors such as pyridine decrease the shielding but other donors increase the shielding compared to nonpolar solvents. For $Hg(CCl{=}CCl_2)_2$, pyridine gives a shift similar to nonpolar solvents suggesting that both effects are operating simultaneously and largely canceling each other.[150] Since it seems that (at least) two types of interaction are possible, it is not surprising that the order of solvents according to $\delta(^{199}Hg)$ value differs with the mercury species and does not fully conform with any physical parameter of the solvent.[76]

The physical basis of the interaction between donor solvents and poor mercury acceptors is not clear other than that it seems distinct from the inner sphere coordination shown by good acceptors. For $HgMe_2$ dissolved in $C_6H_{6-n}F_n$, the shielding decreases as n increases, which also corresponds to increasing ionization potential of the benzenes. Thus formation of charge transfer complexes has been proposed as the cause of the changes in ^{199}Hg shielding.[89] Although intramolecular interaction is sup posed to occur between mercury and the aromatic ring of γ-phenylmercury chlorides, in this case decreasing the electron density on the ring causes shielding.[139] In fact, for both $HgMe_2$ in substituted benzenes[89] and γ-phenylmercury chlorides,[139] methyl substituents on the aromatic ring decrease the shielding, but for the former it was claimed that steric factors reduce the interaction with mercury while for the latter it was proposed that the greater electron density increased the interaction.

Regardless of whether mercury is a good or poor acceptor, increasing the temperature moves the ^{199}Hg resonance towards that in inert solvents[76,144,85] as expected for any sort of intermolecular association.

3.2.2. Substituent Effects in Organometallic Derivatives

Linear organomercurials provide a convenient basis for the investigation of the effect of substitution in the organic moiety on the metal chemical shift. In view of significant solvent effects, it is desirable to measure any such series of compounds in more than one solvent.[148] Increases in the shielding of ca. 150 ppm are observed for each substitution of Me for an α-H in Hg–CH_3 derivatives, while smaller decreases in shielding (ca. 30 ppm) occur on substitution of β-H's.[133,134] Cadmium alkyls seem to show a similar behavior,[1] but for silyl mercury compounds, replacements at both α and β positions result in deshielding.[160,99]

The first ionization potential of [HgCl(R)] which relates predominately to the

σ-HgC bond shows a correlation with $\delta(^{199}Hg)$.[133] However, increasing the number of β-carbon atoms, which decreases the ionization potential, should increase the electron density on mercury and decrease ΔE, both of which would cause deshielding, not shielding. Indeed, Table 4 shows that electron-withdrawing substituents usually increase the shielding. When the substituent effect is transmitted by $-C{\equiv}C-$ in $Hg(C{\equiv}CR)_2$, the shielding is roughly in agreement with inductive and mesomeric effects, the electron rich groups such as *tert*-butyl giving the most deshielded resonances, as expected.[76] This is also true for substituents on aromatic rings in $Hg(aryl)_2$, [HgCl(aryl)], and benzyl derivatives.[148,155,128,154,142,143,141]

4. SPIN–SPIN COUPLING

Of these six post-transition metals, ^{199}Hg has been the subject of the majority of spin–spin coupling reports. Lability of complexes is the main impediment to observation of coupling to the spin-1/2 nuclei of silver and cadmium. Coupling between ^{31}P and isotopes of copper in $[Cu(PR_3)_4]^+$ is the only example for the quadrupolar nuclei. To judge from the values of $\gamma\,|\psi_s(0)|^2$ (see Table 6) couplings to ^{67}Zn and ^{197}Au should be rather small, making their observation particularly difficult. Values of one-bond coupling constants (and to a lesser extent more indirect couplings) are predicted to be proportional to the term $\gamma\,|\psi_s(0)|^2$, values of which are given in Table 6 relative to ^{199}Hg. Also given are the guideline ratios derived from observed coupling constants (especially to ^{31}P) in closely related complexes of different metals. Coupling constants involving silver or cadmium nuclei are expected to have the opposite sign to those to ^{199}Hg because of the signs of the magnetogyric ratios.

Table 7 indicates the typical ranges for some coupling constants involving these metals. A recent review gives data on coupling to ^{13}C in organometallic compounds.[200] In alkyl mercury compounds, $^3J(^{199}Hg\,^{13}C)$ shows a Karplus type of dependency on the dihedral angle.[208] Metal–phosphorus coupling has been surveyed including theoretical considerations of its variation with structure.[191,209,210] The sequences $[Ag(PBu^n_3)_n]^+$,[191,210] $[Hg(PBu^n_3)_n]^{2+}$,[168] and $[Hg\{PO(OEt)_2\}_n]^{2-n}$,[170] $n = 2$, 3, and 4 show that $^1J(MP)$ follows closely the *s*-orbital coefficients as they alter with hybridization.

Extended Hückel MO calculations for $[Hg(Cl)\,CH_2CH(OMe)\,Me]$ give mutual

Table 6. Relative Magnitudes of Spin–Spin Coupling Constants

	Ratio relative to ^{199}Hg			Ratio relative to ^{199}Hg	
Nucleus	$\gamma\,|\psi_s(0)|^2$ [a]	Guideline[b]	Nucleus	$\gamma\,|\psi_s(0)|^2$ [a]	Guideline[b]
^{63}Cu[c]	0.37	0.3	^{67}Zn	0.13	—
^{109}Ag[d]	−0.105	0.1	^{113}Cd[e]	−0.71	0.6
^{197}Au	0.091		^{199}Hg	1.0[f]	1.0[g]

[a] $|\psi_s(0)|^2$ values for metal with zero charge.[196]
[b] See text for derivation.
[c] $J(^{65}CuX) = 1.0712J(^{63}CuX)$.
[d] $J(^{107}AgX) = 0.8698J(^{109}AgX)$.
[e] $J(^{111}CdX) = 0.9559J(^{113}CdX)$.
[f] 14.5 relative to ^{103}Rh (Chapter 20, Table 18).
[g] ca. 12 relative to ^{103}Rh (Chapter 20, Table 18).

Table 7. Ranges of Some Spin–Spin Coupling Constants with Post-Transition Metals

J^a	Complex type/ligands	Range (Hz)
$^1J(^{63}Cu\,^{31}P)$	$[Cu\{P(OR)_3\}_4]^+$ (19)	1109–1458
$^nJ(^{109}Ag\,^1H)$	Organometallic or N-donors(192)	0.3–14
$^1J(^{109}Ag\,^7Li)$	$[Ag_2Li_2(C_6H_4CH_2NMe_2\text{-}2)_4]$(193)	3.9
$^1J(^{109}Ag\,^{13}C)$	Organometallic(194,195)	96–139
$^nJ(^{109}Ag\,^{19}F)$	Fluoroalkyl(196)	8–37
$^1J(^{109}Ag\,^{31}P)$	PR_3, $P(OR)_3$(191,197,40,198,199)	253–1217
$^2J(^{109}Ag\,^{31}P)$	$[Ag(CHRPPh_3)X]$(195)	9–10
$^1J(^{195}Pt\,^{109}Ag)$	$[L_3PtN \sim NAgBr]$(111)	ca. 170
$^2J(^{113}Cd\,^1H)$	Alkyl(94,118)	49–90
$^3J(^{113}Cd\,^1H)$	O-alkyl, alkyl(94,118)	6–70
$^1J(^{113}Cd\,^{13}C)$	Alkyl,(200) CN(68)	271–1060
$^nJ(^{113}Cd\,^{13}C)$	Alkyl(200,46)	17–45
$^1J(^{113}Cd\,^{15}N)$	TPP,(119,9) amine(201)	75–150
$^1J(^{113}Cd\,^{17}O)$	$[Cd(OD_2)_6]^{2+}$ (62)	248
$^1J(^{113}Cd\,^{31}P)$	PR_3(42,202,122)	1142–2840
$^2J(^{113}Cd\,^{31}P)$	APR_3, A = O, S, Se(203,124)	2–80
$^1J(^{113}Cd\,^{77}Se)$	$[Cd(SePh)_m(SPh)_{4-m}]^{2-}$ (125)	46–126
$^2J(^{113}Cd\,^{113}Cd)$	Metallothionein(44)	30–50
$^1J(^{125}Te\,^{113}Cd)$	Solid CdTe(52)	ca. 645
$^2J(^{199}Hg\,^1H)$	Alkyls(204,205)	94–265
$^3J(^{199}Hg\,^1H)$	Organomercurials(206,204,149,135) PR_3(167)	38–311
$^1J(^{199}Hg\,^{13}C)$	Alkyl, aryl,(200) $C\equiv CR$,(76) CN(130)	264–3875
$^2J(^{199}Hg\,^{13}C)$	Alkyl, aryl,(200) $C\equiv CR$(76)	24–920
$^3J(^{199}Hg\,^{13}C)$	Alkyl, aryl(200)	93–205
$^1J(^{199}Hg\,^{15}N)$	Chelating amines(201)	318–395
$^2J(^{199}Hg\,^{19}F)$	Vinyl,(131) alkyl(207,130)	338–1911
$^nJ(^{199}Hg\,^{19}F)$	Alkyl,(207) vinyl,(131) aryl(149)	14–597
$^1J(^{199}Hg\,^{29}Si)$	$SiMe_3$(99)	957–1367
$^1J(^{199}Hg\,^{31}P)$	PR_3, $PO(OR)_2^-$ (191,168,167)	1980–12970
$^2J(^{199}Hg\,^{31}P)$	$SePBu^n_3$,(180) $(CH_2PMe_2)_2CH^-$ (173)	106–160
$^1J(^{199}Hg\,^{77}Se)$	$SePBu^n_3$ (180)	751–962
$^1J(^{199}Hg\,^{119}Sn)$	$Hg(SnR_3)_2$, R = CH_2SiMe_3, C_6F_5(164)	5920, 17550
$^1J(^{199}Hg\,^{183}W)$	$W(CO)_3Cp$(81)	151–706

^{a} $n \neq 1$ or any specifically mentioned value.

polarizabilities with signs in agreement with the signs of the two- and three-bond coupling constants. For a series of $[Hg(L)CH_2C(OMe)Me_2]$ the variations in ^1J(HgC) and ^2J(HgH) closely follow the calculated mutual polarizabilities and it is suggested that changes in the 6*s* contribution to the Hg–C bond are the main cause of changes in the magnitude of ^1J(HgC).[211] Similarly, it was concluded on the basis of relativistic calculations that the variation in ^1J(HgC) is due to changes in hybridization rather than charge. Compared to nonrelativistic calculations, larger values of ^1J(HgC) are derived in better agreement with the observed ones. However, the calculated anisotropy in ^1J(MC) is significantly less than that observed for $HgMe_2$ or $CdMe_2$ in nematic phases although the lesser anisotropy of ^1J(CdC) is reproduced.[212] The anisotropy of ^1J(TeCd) in solid CdTe has also been reported.[52]

The increase in the magnitude of ^2J(HgH) and ^1J(HgC) in organomercurials with the donor ability of the solvent is generally interpreted in terms of coordination.[204] The same order is observed for both poor and intermediate mercury acceptors, and so for the latter does not correspond to the behavior of $\delta(^{199}Hg)$.[97,76] Neither ^2J(HgC) nor ^3J(HgH) show any variation with solvent despite ^3J(HgH) being larger than ^2J(HgH).[204] In bis-fluoroalkyl mercurys, ^2J(HgF) decreases as the donor ability of the solvent increases, but for $[HgX(CF_3)]$[204] it increases except that for $Hg(CF_3)_2$ it shows an eratic response.[213,130] In contrast to ^3J(HgH), the coupling ^3J(HgF) shows considerable increases in the presence of donors.[214] Taken together, these features suggest that "coordination" is too simple a term to describe the interactions causing changes in spin–spin coupling to ^{199}Hg.

4.1. Sign Determinations

In $CdMe_2$, $^2J(^{113}Cd\,^1H)$ is positive,[156] but $^3J(^{113}Cd\,^1H)$ in $CdPr^n_2$ has the opposite sign.[118] Following the signs of the γ's, $^1J(^{113}Cd\,^{15}N)$ in [Cd(TPP) py] is positive[9] while $^1J(^{113}Cd\,^{13}C)$[156] and $^1J(^{113}Cd\,^{31}P)$[1] are negative.

Although $^2J(^{199}Hg\,^1H)$ is negative in alkyl derivatives of mercury,[206,130] it is positive in vinyl derivatives; the three-bond couplings are positive in both types of compound.[206,156] Couplings to ^{19}F in Hg–CF_3 complexes have the opposite sign to those to ^{1}H in the analogous Hg–CH_3 system,[130] but in aryl[149] and vinyl[131] compounds couplings of both ^{19}F and ^{1}H to ^{199}Hg are positive. Double resonance studies of CH_3,[215,206] CF_3, and PR_3 (R = Me or Et)[130,167] derivatives showed that ^1J(HgC) and ^1J(HgP) are positive, whereas ^2J(HgP) and ^1J(HgSe) in $SePBu^n_3$ derivatives are negative.[180]

REFERENCES

1 Kidd, R G, Goodfellow, R J In *NMR and the Periodic Table*, Harris, R K, Mann, B E, Eds, Academic London, 1978, Chap 8
2 Brevard, C, Granger, P *Handbook of High Resolution Multinuclear NMR*, Wiley-Interscience New York, 1981
3 Rinaldi, P L, Levy, G C, Choppin, G R *Rev Inorg Chem* 1980, **2**, 53–92
4 Forsen, S, Lindman, B *Methods Biochem Anal* 1981, **27**, 289–486
5 Ellis, P D, Inners, R R, Jakobsen, H J *J Phys Chem* 1982, **86**, 1506–1508
6 Armitage, I M, Schoot Uiterkamp, J A M, Chlebowski, J F, Coleman, J E *J Magn Reson* 1978, **29**, 375–392
7 Rodesiler, P F, Griffith, E H, Ellis, P D, Amma, E L *J Chem Soc, Chem Commun* 1980, 492–493
8 Martins, E O, Drakenberg, T *Inorg Chim Acta* 1982, **67**, 71–74
9 Jakobsen, H J, Ellis, P D, Inners, R R, Jensen, C F *J Am Chem Soc* 1982, **104**, 7442–7452
10 Otvos, J D, Olafson, R W, Armitage, I M *J Biol Chem* 1982, **257**, 2427–2431
11 Cave, A, Parello, J, Drakenberg, T, Thulin, E, Lindman, B *FEBS Lett* 1979, **100**, 148–152, Otvos, J D, Armitage, I M *Experientia, Suppl* 1979, **34** (Metallothionein), 249–257, Andersson, T, Drakenberg, T, Forsen, S, Thulin, E *Eur J Biochem* 1982, **126**, 501–505, Boulanger, Y, Armitage, I M *J Inorg Biochem* 1982, **17**, 147–153, Boulanger, Y, Armitage, I M, Miklossy, K A, Winge, D R *J Biol Chem* 1982, **257**, 13717–13719, Briggs, R W, Armitage, I M *J Biol Chem* 1982, **257**, 1259–1262, Evelhoch, J L, Bocian, D F, Sudmeier, J L *Biochemistry* 1981, **20**, 4951–4954
12 Sudmeier, J L, Perkins, T G *J Am Chem Soc* 1977, **99**, 7732–7733
13 Andrew, E R, Carolan, J L, Randall, P J *Chem Phys Lett* 1971, **11**, 298–299
14 Andrew, E R, Hinshaw, W S, Tiffen, R S *J Magn Reson* 1974, **15**, 191–195
15 Carter, G C, Bennet, L H, Kahan, D J *Prog Materials Sci* 1976, **20**
16 Yamamoto, T, Haraguchi, H, Fujiwara, S *J Phys Chem* 1970, **74**, 4369–4373
17 Ellis, P D, Walsh, H C, Peters, C S *J Magn Reson* 1973, **11**, 426–430
18 Lutz, O, Oehler, H, Kroneck, P *Z Naturforsch, Teil A* 1978, **33**, 1021–1024

19. Kroneck, P.; Lutz, O.; Nolle, A.; Oehler, H. *Z. Naturforsch., Teil A* 1980, **35**, 221–225.
20. Marker, A.; Gunter, M. J. *J. Magn. Reson.* 1982, **47**, 118–132.
21. Ochsenbein, U.; Schläpfer, C. W. *Helv. Chim. Acta* 1980, **63**, 1926–1931.
22. Kroneck, P.; Kodweiss, J.; Lutz, O.; Nolle, A.; Zepf, D. *Z. Naturforsch., Teil A* 1982, **37**, 186–190.
23. McConnell, H. M.; Weaver, H. E. *J. Chem. Phys.* 1956, **25**, 307–311.
24. Boyce, J. B.; Mikkelsen, J. C. *J. Phys. C.* 1977, **10**, L41–L43.
25. Lutz, O.; Oehler, H.; Kroneck. *Z. Physik A* 1978, **288**, 17–21.
26. Becker, K. D. *J. Chem. Phys.* 1978, **68**, 3785–3793.
27. Becker, K. D.; Schäfgen, H. P. *Solid State Commun.* 1979, **32**, 1107–1109.
28. McFarlane, W.; Rycroft, D. S. *J. Magn. Reson.* 1976, **24**, 95–101.
29. Epperlein, B. W.; Krüger, H.; Lutz, O.; Schwenk, A. *Z. Naturforsch., Teil A* 1974, **29**, 1553–1557.
30. Maciel, G. E.; Simeral, L.; Ackerman, J. J. H. *J. Phys. Chem.* 1977, **81**, 263–267.
31. Shimizu, T.; Kodaka, M.; Hatano, M. *Biochem. Biophys. Res. Commun.* 1982, **106**, 988–993.
32. Shimizu, T.; Hatano, M. *Biochem. Biophys. Res. Commun.* 1982, **104**, 1356–1362.
33. Li, Z.; Popov, A. I. *J. Solution Chem.* 1982, **11**, 17–26.
34. Carson, G. K.; Dean, P. W. *Inorg. Chim. Acta* 1982, **66**, 157–161.
35. Laszlo, P. *Prog. Nucl. Magn. Reson. Spectrosc.* 1979, **13**, 257–270.
36. Brevard, C.; van Stein, G. C.; van Koten, G. *J. Am. Chem. Soc.* 1981, **103**, 6746–6748.
37. Rahimi, A. K.; Popov, A. I. *Inorg. Nucl. Chem. Lett.* 1976, **12**, 703–707.
38. Endo, K.; Matsushita, K.; Deguchi, K.; Yamamota, K.; Suzuki, S.; Futaki, K. *Chem. Lett.* 1982, 1497–1500.
39. Brevard, C.; Schimpf, R. *J. Magn. Reson.* 1982, **47**, 528–534.
40. Colquhoun, I. J.; Mcfarlane, W. *J. Chem. Soc., Chem. Commun.* 1980, 145–147.
41. Kronenbitter, J.; Schweizer, U.; Schwenk, A. *Z. Naturforsch., Teil A* 1980, **35**, 319–328.
42. Dakternieks, D. *Aust. J. Chem.* 1982, **35**, 469–481.
43. Carson, G. K.; Dean, P. A. W.; Stillman, M. J. *Inorg. Chim. Acta* 1981, **56**, 59–71.
44. Otvos, J. D.; Armitage, I. M. *J. Am. Chem. Soc.* 1979, **101**, 7734–7736.
45. Otvos, J. D.; Armitage, I. M. *Proc. Natl. Acad. Sci. USA* 1980, **77**, 7094–7098.
46. Cardin, A. D.; Ellis, P. D.; Odom, J. D.; Howard, J. W. *J. Am. Chem. Soc.* 1975, **97**, 1672–1679.
47. Jensen, C. F.; Deshmukh, S.; Jakobsen, H. J.; Inners, R. R.; Ellis, P. D. *J. Am. Chem. Soc.* 1981, **103**, 3659–3666.
48. Forsén, S.; Thulin, E.; Lilja, H. *FEBS Lett.* 1979, **104**, 123–126.
49. Forsén, S.; Thulin, E.; Drakenberg, T.; Krebs, J.; Seamon, K. *FEBS Lett.* 1980, **117**, 189–194.
50. Drakenberg, T.; Björk, N. O., Portanova, R. *J. Phys. Chem.* 1978, **82**, 2423–2426.
51. Nolle, A. *Z. Naturforsch., Teil A* 1978, **33**, 666–671.
52. Nolle, A. *Z. Physik, Teil B* 1979, **34**, 175–182.
53. Cheung, T. T. P.; Worthington, L. E.; DuBois Murphy, P.; Gerstein, B. C. *J. Magn. Reson.* 1980, **41**, 158–168.
54. DuBois Murphy, P.; Gerstein, B. C. *J. Am. Chem. Soc.* 1981, **103**, 3282–3286.
55. DuBois Murphy, P.; Stevens, W. C.; Cheung, T. T. P.; Lacelle, S.; Gerstein, B. C.; Kurtz, D. M. *J. Am. Chem. Soc.* 1981, **103**, 4400–4405.
56. Mennitt, P. G.; Shatlock, M. P.; Bartuska, V. J.; Maciel, G. E. *J. Phys. Chem.* 1981, **85**, 2087–2091.
57. Turner, R. W.; Rodesiler, P. F.; Amma, E. L. *Inorg. Chim. Acta* 1982, **66**, L13–L15.
58. Colton, R.; Dakternieks, D. *Aust. J. Chem.* 1980, **33**, 1677–1684.
59. Colton, R.; Dakternieks, D. *Aust. J. Chem.* 1980, **33**, 2405–2409.
60. Jakobsen, H. J.; Ellis, P. D. *J. Phys. Chem.* 1981, **85**, 3367–3369.
61. Bailey, D. B.; Ellis, P. D.; Fee, J. A. *Biochemistry* 1980, **19**, 591–596.
62. Holz, M.; Jordan, R. B.; Zeidler, M. D. *J. Magn. Reson.* 1976, **22**, 47–52.
63. Bryden, C. C., Reilley, C. N. *J. Am. Chem. Soc.* 1982, **104**, 2697–2699.
64. Maciel, G. E.; Borzo, M. *J. Chem. Soc. Chem. Commun.* 1973, 394.
65. Dalton, J.; Kennedy, J. D.; McFarlane, W.; Wedge, J. R. *Mol. Phys.* 1977, **34**, 215–222.
66. Kennedy, J. D.; McFarlane, W. *J. Chem. Soc. Faraday Trans. 2* 1976, **72**, 1653–1660.
67. Gillies, D. B.; Blaauw, L. P.; Hays, G. R.; Huis, R.; Clague, A. D. H. *J. Magn. Reson.* 1981, **42**, 420–428.
68. Sudmeier, J. L.; Bell, S. J. *J. Am. Chem. Soc.* 1977, **99**, 4499–4500.
69. Narath, A. *Phys. Rev.* 1967, **163**, 232–237; 1968, **175**, 696.
70. Blumberg, W. D.; Eisinger, J.; Shulman, R. G. *J. Phys. Chem. Solids* 1965, **26**, 1187–1194.
71. Colton, R.; Dakternieks, D. *Aust. J. Chem.* 1980, **33**, 955–963.

72. Wasylishen, R. E.; Lenkinski, R. E.; Rodger, C. *Can. J. Chem.* 1982, **60**, 2113–2117.
73. Roberts, M. F.; Vidusek, D. A.; Bodenhausen, G. *FEBS Lett.* 1980, **117**, 311–314.
74. Vidusek, D. A.; Roberts, M. F.; Bodenhausen, G. *J. Am. Chem. Soc.* 1982, **104**, 5452–5456.
75. Kubicki, M. M.; Kergoat, R.; Le Gall, J. Y.; Guerchais, J. E.; Douglade, J.; Mercier, R. *Aust. J. Chem.* 1982, **35**, 1543–1554.
76. Sebald, A.; Wrackmeyer, B. *Spectrochim. Acta, Part A* 1982, **38**, 163–173.
77. Bain, R. L. *J. Inorg. Nucl. Chem.* 1981, **43**, 2481–2483.
78. Sens, M. A.; Wilson, N. K.; Ellis, P. D.; Odom, J. D. *J. Magn. Reson.* 1975, **19**, 323–326.
79. Lassigne, C. R.; Wells, E. J. *Can. J. Chem.* 1977, **55**, 1303–1313.
80. Jokisaari, J.; Diehl, P. *Org. Magn. Reson.* 1980, **5**, 359–362.
81. Albright, M. J.; Oliver, J. P. *J. Organomet. Chem.* 1979, **172**, 99–107.
82. Benn, R.; Günther, H.; Maercker, A.; Menger, V.; Schmitt, P. *Angew. Chem. Int. Ed. Engl.* 1982, **21**, 295–296.
83. Goggin, P. L.; Goodfellow, R. J.; Haddock, S. R.; Eary, J. G. *J. Chem. Soc., Dalton Trans.* 1972, 647–653.
84. Peringer, P. *Inorg. Chim. Acta* 1980, **42**, 129–132.
85. Peringer, P. *Inorg. Chim. Acta* 1980, **39**, 67–70.
86. Krüger, H.; Lutz, O.; Nolle, A.; Schwenk, A. *Z. Physik, Teil A* 1975, **273**, 325–330.
87. Peringer, P. *J. Chem. Res.* 1980, (*S*) 194, (*M*) 2757–2768.
88. Peringer, P.; Winkler, P. P. *J. Organomet. Chem.* 1980, **195**, 249–252.
89. Strelenko, Yu.; Sokolikova, O. K.; Griskin, Yu. K.; Ustynyuk, Yu. A. *Dokl. Akad. Nauk. SSSR* 1980, **252**, 924–927; *Dokl. Phys. Chem. (Engl. Transl.)* 1980, **252**, 444–447.
90. Spence, P. W.; McDermott, M. N. *Phys. Lett. Sec. A* 1967, **24**, 430–431.
91. Sahm, W.; Schwenk, A. *Z. Naturforsch., Teil A* 1974, **29**, 1763–1766.
92. Burges, C. W.; Koschmieder, R.; Sahm, W.; Schwenk, A. *Z. Naturforsch., Teil A* 1973, **28**, 1753–1758.
93. Goodfellow, R. J. unpublished observations.
94. Kennedy, J. D.; McFarlane, W. *J. Chem. Soc., Perkin Trans. 2* 1977, 1187–1191.
95. Krüger, H.; Lutz, O.; Schwenk, A.; Stricker, G. *Z. Phys.* 1974, **266**, 233–237.
96. Dahmen, H.; Penselin, S. *Z. Phys.* 1967, **200**, 456–466.
97. Goggin, P. L.; Goodfellow, R. J.; Hurst, N. W. *J. Chem. Soc., Dalton Trans.* 1978, 561–566.
98. Kostelnik, R. J.; Bothner-By, A. A. *J. Magn. Reson.* 1974, **14**, 141–151.
99. Mitchell, T. N.; Marsmann, H. C. *J. Organomet. Chem.* 1978, **150**, 171–177.
100. Larin, M. F.; Gendin, D. V.; Pestunovich, V. A.; Kruglaya, O. A.; Vyazankin, N. S. *Izv. Akad. Nauk. SSSR Ser. Khim.* 1979, 697–698; *Bull. Acad. Sci. USSR, Div. Chem. Sci.* 1979, **28**, 654.
101. Rahimi, A. K.; Popov, A. I. *J. Magn. Reson.* 1979, **36**, 351–358.
102. Henrichs, P. M.; Ackerman, J. J. H.; Maciel, G. E. *J. Am. Chem. Soc.* 1977, **99**, 2544–2548.
103. Bach, R. D.; Rajan, S. J.; Vardhan, H. B.; Land, T. J.; Albrecht, N. G. *J. Am. Chem. Soc.* 1981, **103**, 7727–7734.
104. Ackerman, J. J. H.; Orr, T. V.; Bartuska, V. J.; Maciel, G. E. *J. Am. Chem. Soc.* 1979, **101**, 341–347.
105. Mebs, R. W.; Carter, G. C.; Evans, B. J.; Bennett, L. H. *Solid State Commun.* 1972, **10**, 769–774.
106. Haller, M.; Hertler, W. E.; Lutz, O.; Nolle, A. *Solid State Commun.* 1980, **33**, 1051–1053.
107. Jucker, K.; Sahm, W.; Schwenk, A. *Z. Naturforsch., Teil A* 1976, **31**, 1532–1538.
108. Henrichs, P. M.; Sheard, S.; Ackerman, J. J. H.; Maciel, G. E. *J. Am. Chem. Soc.* 1979, **101**, 3222–3228.
109. Becker, K. D.; von Goldammer, E. *Chem. Phys.* 1980, **48**, 193–201.
110. van Stein, G. C.; van Koten, G.; Brevard, C. *J. Organomet. Chem.* 1982, **226**, C27–C30.
111. van der Ploeg, A. F. M. J.; van Koten, G.; Brevard, C. *Inorg. Chem.* 1982, **21**, 2878–2881.
112. Krüger, H.; Lutz, O.; Nolle, A.; Schwenk, A.; Stricker, G. *Z. Naturforsch., Teil A* 1973, **28**, 484–486.
113. Haberkorn, R. A.; Que, L.; Gillum, W. O.; Holm, R. H.; Liu, C. S.; Lord, R. C. *Inorg. Chem.* 1976, **15**, 2408–2414.
114. Damude, L. C.; Dean, P. A. W. *J. Organomet. Chem.* 1979, **168**, 123–138.
115. Armitage, I. M.; Pajer, R. T.; Schoot Uiterkamp, A. J. M.; Chlebowski, J. F.; Coleman, J. E. *J. Am. Chem. Soc.* 1976, **98**, 5710–5712; Birgersson, B.; Carter, R. E.; Drakenberg, T. *J. Magn. Reson.* 1977, **28**, 299–302.
116. Bain, A. D.; Eaton, D. R.; Hux, R. A.; Tong, J. P. K. *Carbohydr. Res.* 1980, **84**, 1–12.
117. Linse, P.; Gustavsson, H.; Lindman, B., Drakenberg, T. *J. Magn. Reson.* 1981, **45**, 133–141.
118. Turner, C. J.; White, R. F. M. *J. Magn. Reson.* 1977, **26**, 1–5.

119. Dominguez, D. D.; King, M. M.; Yeh, H. J. C. *J. Magn. Reson.* 1978, **32**, 161–165.
120. Alyea, E. C. *Inorg. Chim. Acta* 1983, **76**, L239–L240.
121. Griffith, E. A. H.; Amma, E. L. *J. Chem. Soc., Chem. Commun.* 1979, 1013–1014.
122. Bond, A. M.; Colton, R.; Dakternieks, D.; Dillon, M. L.; Hauenstein, J.; Moir, J. E. *Aust. J. Chem.* 1981, **34**, 1393–1400.
123. Maitani, T.; Suzuki, K. T. *Inorg. Nucl. Chem. Lett.* 1979, **15**, 213–217.
124. Dean, P. A. W. *Can. J. Chem.* 1981, **59**, 3221–3225.
125. Carson, G. K.; Dean, P. A. W. *Inorg. Chim. Acta* 1982, **66**, 37–39.
126. Rodesiler, P. F.; Amma, E. L. *J. Chem. Soc., Chem. Commun.* 1982, 182–184.
127. Maciel, G. E.; Borzo, M. *J. Magn. Reson.* 1973, **10**, 388–390.
128. Borzo, M.; Maciel, G. E. *J. Magn. Reson.* 1975, **19**, 279–282.
129. Peringer, P. *J. Inorg. Nucl. Chem.* 1980, **42**, 1501–1502.
130. Goggin, P. L.; Goodfellow, R. J.; McEwan, D. M.; Griffiths, A. J.; Kessler, K. *J. Chem. Res.* 1979, (*S*) 194–195, (*M*) 2315–2343.
131. Johannesen, R. B.; Duerst, R. W. *J. Magn. Reson.* 1971, **5**, 355–366.
132. Goodfellow, R. J.; Stobart, S. R. *J. Magn. Reson.* 1977, **27**, 143–145.
133. Baidin, V. N.; Chizhov, Yu. Y.; Timoshenko, M. M.; Sokolikova, O. K.; Grishin, Yu. K.; Ustynyuk, Yu. A. *Zh. Strukt. Khim.* 1981, **22**(4), 164–167; *J. Struct. Chem. (Engl. Transl.)* 1981, **22**, 616–618.
134. Olah, G. A.; Garcia-Luna, A. *Proc. Natl. Acad. Sci. USA* 1980, **77**, 5036–5039.
135. Tupčiauskas, A. P.; Sergeyev, N. M.; Ustynyuk, Yu. A.; Kashin, A. N. *J. Magn. Reson.* 1972, **7**, 124–126.
136. Dessy, R. E.; Flautt, T. J.; Jaffé, H. H.; Reynolds, G. F. *J. Chem. Phys.* 1959, **30**, 1422–1425.
137. Seyferth, D.; Song, L. C.; Henderson, R. S. *J. Am. Chem. Soc.* 1981, **103**, 5103–5107.
138. Browning, J.; Goggin, P. L.; Goodfellow, R. J.; Hurst, N. W.; Mallinson, L. G.; Murray, M. *J. Chem. Soc., Dalton Trans.* 1978, 872–876.
139. Kitching, W.; Drew, G. M.; Alberts, V. *Organometallics* 1982, **1**, 331–335.
140. Ibusuki, T.; Saito, Y. *Chem. Lett.* 1974, 311–314.
141. Kitching, W.; Drew, G.; Adcock, W.; Abeywickrema, A. N. *J. Org. Chem.* 1981, **46**, 2252–2260.
142. Michel, E.; Perie, J.; Lattes, A. *J. Organomet. Chem.* 1981, **204**, 1–12.
143. Strelenko, Yu. A.; Bundel, Yu. G.; Kasumov, F. H.; Rosenberg, V. I.; Reutov, O. A.; Ustynyuk, Yu. A. *J. Organomet. Chem.* 1978, **159**, 131–135.
144. Strelenko, Yu. A.; Grishin, Yu. K.; Kazankova, M. A.; Ustynyuk, Yu. A. *J. Organomet. Chem.* 1980, **192**, 297–303.
145. Meyer, R.; Gorrichon-Guigon, L.; Maroni, P. *J. Organomet. Chem.* 1980, **188**, 11–24.
146. Barron, P. F.; Doddrell, E.; Kitching, W. *J. Organomet. Chem.* 1977, **139**, 361–383.
147. Kitching, W.; Drew, G. M. *J. Org. Chem.* 1981, **46**, 2695–2706.
148. Grishin, Yu. K.; Strelenko, Yu. A.; Margulis, L. A.; Ustynyuk, Yu. A.; Golovchenko, L. S.; Peregudov, A. S.; Kravtsov, E. N. *Dokl. Akad. Nauk. SSSR* 1979, **249**, 892–896; *Dokl. Phys. Chem. (Engl. Transl.)* 1979, **249**, 1008–1010.
149. McFarlane, W. *J. Chem. Soc. (A)* 1968, 2280–2285.
150. Peringer, P.; Winkler, P. P. *Inorg. Chim. Acta* 1981, **52**, 257–260.
151. Peringer, P. *Inorg. Nucl. Chem. Lett.* 1980, **16**, 205–209.
152. Eichbichler, J.; Peringer, P. *Inorg. Chim. Acta* 1980, **43**, 121–124.
153. Peringer, P.; Eichbichler, J. *Inorg. Nucl. Chem.* 1981, **43**, 2033–2036.
154. Wells, P. R.; Hawker, D. W. *Org. Magn. Reson.* 1981, **17**, 26–27.
155. Grishin, Yu. K.; Roznyatovskii, V. A.; Vanchikov, A. N.; Lisichkina, I. N.; Tolstaya, T. P.; Ustynyuk, Yu. A. *Dokl. Akad. Nauk. SSSR* 1981, **257**, 919–922; *Dokl. Phys. Chem. (Engl. Transl.)* 1981, **257**, 280–292.
156. Hildenbrand, K.; Dreeskamp, H. *Z. Phys. Chem.* 1970, **69**, 171–182.
157. Wells, P. R. In *Determinations of Organic Structures by Physical Methods*, Vol. 4, Nachod, F. C.; Zuckerman, J. J., Eds.; Academic Press: New York, 1971; Chap. 5.
158. Breitinger, D. K.; Kress, W.; Sendelbech, R.; Ishiwada, K. *J. Organomet. Chem.* 1983, **243**, 245–251.
159. Peringer, P. *Monatsh. Chem.* 1980, **111**, 1245–1252.
160. Cradock, S.; Ebsworth, E. A. V.; Hosmane, N. S.; Mackay, K. M. *Angew. Chem. Int. Ed. Engl.* 1975, **14**, 167–168.
161. Gendin, D. V.; Larin, M. F.; Kruglaya, O. A.; Pestunovich, V. A.; Vyazankin, N. S. *Izv. Akad. Nauk. SSSR, Ser. Khim.* 1980, 2189.

162. Bochkarev, M. N.; Ermolaev, N. L.; Razuvaev, G. A.; Grishin, Yu. K.; Ustynyuk, Yu. A. *J. Organomet. Chem.* 1982, **229**, C1–C4.
163. Strelenko, Yu. A.; Grishin, Yu. K.; Bychkov, V. T.; Ustynyuk, Y. A. *Zh. Obshch. Khim.* 1979, **49**, 1172; *J. Gen. Chem. USSR (Engl. Transl.)* 1979, **49**, 1025.
164. Grishin, Yu. K.; Roznyatovskii, V. A.; Ustynyuk, Yu. A.; Bochkarev, M. N.; Kalinina, G. S.; Razuvaev, G. A. *Izv. Akad. Nauk. SSSR, Ser. Khim.* 1980, 2190.
165. Eichbichler, J.; Peringer, P. *Transition Met. Chem.* 1981, **6**, 313–314.
166. Peringer, P.; Eichbichler, J. *J. Organomet. Chem.* 1983, **241**, 281–287.
167. Bell, N. A.; Dee, T. D.; Goggin, P. L.; Goldstein, M.; Goodfellow, R. J.; Jones, T.; Kessler, K.; McEwan, D. M.; Nowell, I. M. *J. Chem. Res.* 1981, (*S*) 2–3, (*M*) 201–242.
168. Colton, R.; Dakternieks, D. *Aust. J. Chem.* 1981, **34**, 323–333.
169. Eichbichler, J.; Peringer, P. *J. Organomet. Chem.* 1982, **231**, 95–96.
170. Winkler, P. P.; Peringer, P. *Inorg. Chim. Acta* 1983, **76**, L59–L61.
171. Kubicki, M. M.; Kergoat, R.; Guerchais, J. E.; Bkouche-Waksman, I.; Bois, C.; L'Haridon, P. *J. Organomet. Chem.* 1981, **219**, 329–343.
172. Canty, A. J.; Barron, P.; Healy, P. C. *J. Organomet. Chem.* 1979, **179**, 447–458; Canty, A. J.; Marker, A.; Barron, P.; Healy, P. C. *J. Organomet. Chem.* 1978, **144**, 371–379.
173. Schmidbaur, H.; Gasser, O.; Fraser, T. E.; Ebsworth, E. A. V. *J. Chem. Soc., Chem. Commun.* 1977, 334–335.
174. Peringer, P. *Inorg. Nucl. Chem. Lett.* 1980, **16**, 461–467.
175. Eichbichler, J.; Peringer, P. *Inorg. Chim. Acta* 1981, **54**, L213–L215.
176. Eichbichler, J.; Peringer, P. *J. Chem. Soc., Chem. Commun.* 1982, 193–194.
177. Eichbichler, J.; Peringer, P. *Inorg. Nucl. Chem. Lett.* 1981, **17**, 305–309.
178. Goggin, P. L.; Goodfellow, R. J.; McEwan, D. M.; Kessler, K. *Inorg. Chim. Acta* 1980, **44**, L111–L112.
179. Colton, R.; Dakternieks, D. *Aust. J. Chem.* 1980, **33**, 1463–1470.
180. Colquhoun, I. J.; McFarlane, W. *J. Chem. Soc., Dalton Trans.* 1981, 658–660.
181. McEwan, D. M. Ph.D. thesis, University of Bristol, 1980.
182. Jaitner, P. E.; Peringer, P.; Huttner, G.; Zsolnai, L. *Transition Met. Chem.* 1981, **6**, 86–90.
183. Godfrey, P. D.; Heffernan, M. L.; Kerr, D. F. *Aust. J. Chem.* 1964, **17**, 701–704.
184. Lucchini, V.; Wells, P. R. *J. Organomet. Chem.* 1975, **92**, 283–290.
185. Sudmeier, J. L.; Birge, R. R.; Perkins, T. G. *J. Magn. Reson.* 1978, **30**, 491–496.
186. Kidd, R. G. *Annu. Rep. NMR Spectrosc.* 1980, **10A**, 1–79.
187. Jameson, C. J.; Gutowsky, H. S. *J. Chem. Phys.* 1964, **40**, 1714–1724.
188. Orgel, L. E. *J. Chem. Soc.* 1958, 4186–4190; *Mol. Phys.* 1958, **1**, 322–325.
189. Schneider, W. G.; Buckingham, A. D. *Discuss. Faraday Soc.* 1962, **34**, 147–155.
190. Bond, A. M.; Colton, R.; Dakternieks, D.; Hanck, K. W. *Inorg. Chem.* 1982, **21**, 117–122; Bond, A. M.; Colton, R.; Dakternieks, D.; Hanck, K. W.; Švestka, M. *Inorg. Chem.* 1983, **22**, 236–241.
191. Pregosin, P. S.; Kunz, R. W. *^{31}P and ^{13}C NMR of Transition Metal Phosphine Complexes*, Springer-Verlag: Berlin, 1979.
192. van Stein, G. C.; van der Poel, H.; van Koten, G.; Spek, A. L.; Duisenberg, J. M.; Pregosin, P. S. *J. Chem. Soc., Chem. Commun.* 1980, 1016–1018.
193. van Koten, G.; Jastrzebski, J. T. B. H.; Stam, C. H.; Niemann, N. C. *J. Am. Chem. Soc.* 1984, **106**, 1880–1881.
194. van Koten, G.; Noltes, J. G. *J. Organomet. Chem.* 1974, **82**, C53–C55.
195. Yamamoto, Y.; Schmidbaur, H. *J. Organomet. Chem.* 1975, **96**, 133–138; 1975, **97**, 479–486.
196. Dyatkin, B. L.; Martynov, B. I.; Martynova, L. G.; Kizim, N. G.; Sterlin, S. R.; Stumbrevichute, Z. A.; Federov, L. A. *J. Organomet. Chem.* 1973, **57**, 423–433.
197. Alyea, E. C.; Dias, S. A.; Stevens, S. *Inorg. Chim. Acta* 1980, **44**, L203–L204.
198. Sanghani, D. V.; Smith, P. J.; Allen, D. W.; Taylor, B. F. *Inorg. Chim. Acta* 1982, **59**, 203–206.
199. van der Ploeg, A. F. M. J.; van Koten, G.; Spek, A. L. *Inorg. Chem.* 1979, **18**, 1052–1060.
200. Mann, B. E.; Taylor, B. F. *^{13}C NMR Data for Organometallic Compounds*; Academic: London, 1981.
201. Curzon, E. H.; Herron, N.; Moore, P. *J. Chem. Soc., Dalton Trans.* 1980, 721–725.
202. Goel, R. G.; Henry, W. P.; Srivastava, R. C. *Inorg. Chem.* 1981, **20**, 1727–1731.
203. Dean, P. A. W.; Hughes, M. K. *Can. J. Chem.* 1980, **58**, 180–190; Dean, P. .A. .W.; Polensek, L. *Can. J. Chem.* 1980, **58**, 1627–1632; Otvos, J. D.; Alger, J. R.; Coleman, J. E.; Armitage, I. M. *J. Biol. Chem.* 1979, **254**, 1778–1780.
204. Petrosyan, V. S.; Reutov, O. A. *Pure Appl. Chem.* 1974, **37**, 147–159; *J. Organomet. Chem.* 1974, **76**, 123–169.

205. Federov, L. A. *Zh. Strukt. Khim.* 1976, **17**, 236–256; *J. Struct. Chem. (Engl. Transl.)* 1976, **17**, 207–224.
206. Anet, F. A. L.; Sudmeier, J. L. *J. Magn. Reson.* 1969, **1**, 124–138.
207. Federov, L. A.; Stumbrevichyute, Z. A.; Fedin, E. I. *Zh. Strukt. Khim.* 1975, **16**, 976–984; *J. Struct. Chem. (Engl. Transl.)* 1975, **16**, 899–906.
208. Anet, F. A. L.; Krane, J.; Kitching, W.; Doddrell, D.; Praeger, D. *Tetrahedron Lett.* 1974, 3255–3258; 1975, 759–762.
209. Buergi, H. B.; Fischer, E.; Kunz, R. W.; Parvez, M.; Pregosin, P. S. *Inorg. Chem.* 1982, **21**, 1246–1256.
210. Pidcock, A. In *Catalytic Aspects of Metal Phosphine Complexes*, Alyea, E. C.; Meek, D. W., Eds.; ACS Advances in Chemistry Series, 1982, **196**, 1–22; Nixon, J. F.; Pidcock, A. *Annu. Rev. NMR Spectrosc.* 1969, **2**, 345–422.
211. Iwayanagi, T.; Ibusuki, T.; Saito, Y. *J. Organomet. Chem.* 1977, **128**, 145–153.
212. Jokisaari, J.; Räisänen, K.; Lajunen, L.; Passoja, A.; Pyykkö, P. *J. Magn. Reson.* 1978, **31**, 121–132; Jokisaari, J.; Räisänen, K.; Kuonanoja, J.; Pyykkö, P.; Lajunen, L. *Mol. Phys.* 1980, **39**, 715–723.
213. Downs, A. J. *J. Chem. Soc.* 1963, 5273–5278.
214. Federov, L. A.; Stumbrevichyute, Z. A.; Dyatkin, B. L.; Martynov, B. I.; Sterlin, S. R. *Dokl. Akad. Nauk. SSSR* 1972, **204**, 1135–1138; *Dokl. Chem. (Engl. Transl.)* 1972, **204**, 520–522; Federov, L. A.; Stumbrevichyute, Z. A.; Martynov, B. I; Dyatkin, B. L. *Zh. Org. Khim.* 1975, **11**, 489–496; *J. Org. Chem. USSR (Engl. Transl.)* 1975, **11**, 487–493.
215. McLauchlan, K. A.; Whiffen, D. H.; Reeves, L. W. *Mol. Phys.* 1966, **10**, 131–140.

Chapter 22

NMR SPECTROSCOPY IN BIOINORGANIC CHEMISTRY

HENRY W. E. RATTLE

1. INTRODUCTION

Because of the sensitivity problems posed by other nuclei, probably 60% of current papers in biological NMR are taken up by proton studies, although the use of ^{13}C, with its much larger range of chemical shifts, is increasing. This chapter, however, is mainly concerned with biological NMR studies using nuclei other than ^{1}H and ^{13}C, ending with a list of key references. For a general introduction to "NMR in Molecular Biology," the reader is referred to the excellent volume of that title by G. C. K. Roberts and O. Jardetzky published by Academic Press in 1982. For detailed investigation, the subject index of Chemical Abstracts lists under "Nuclear Magnetic Resonance" abstracts of work done using specific nuclei.

Preceding the list of reviews is a short collection of examples of biological applications of a number of less commonly used nuclei.

2. SOME EXAMPLES OF BIOLOGICAL APPLICATIONS: ISOTOPE SHIFTS IN ^{31}P NMR

As Figure 1 shows, the substitution of ^{18}O for ^{16}O bonded to a phosphorus atom causes a shift of 0.0206 ppm in the ^{31}P resonance frequency, or 3 Hz at a spectrometer frequency of 145.7 MHz, and the effect is additive.[1] The isotope shift may therefore be used to label individual phosphorus atoms and to investigate their fate in biochemical reactions. Thus in the exchange reaction between the β phosphate of ADP and inorganic phosphate, which is catalyzed by polynucleotide phosphorylase, the use of $P_i(^{18}O_4)$ and $ADP(^{16}O_7)$ as starting materials, gives ADP containing $\alpha P(^{16}O_3{}^{18}O)$ and $\beta P(^{18}O_4)$. This shows that bond cleavage occurs between the α phosphorus and the bridge oxygen between α and β phosphorus atoms. With differential labeling of each phosphate of ATP the rates of ^{18}O transfer from nonphosphate to phosphate compounds and rates of scrambling of αP nonbridge oxygens in ATPase reactions can be measured.

Figure 2 illustrates the use of this method to determine the stereochemical course

HENRY W. E. RATTLE • Biophysics Laboratories, Portsmouth Polytechnic, Portsmouth PO1 2DT, England.

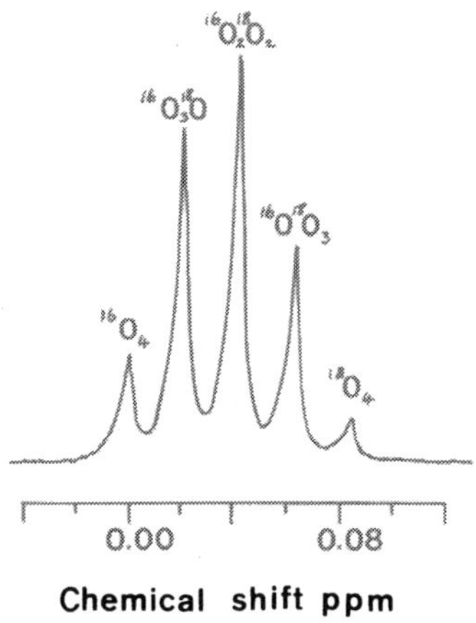

Figure 1. ^{31}P NMR spectrum of a randomized sample of $P_i(^{16}O, {}^{18}O)$. The solution contained 50 mM P_i (~44% ^{18}O) and 0.5 mM potassium EDTA in 2 ml of 50% D_2O, pH 8.2. NMR parameters: 145.7 MHz; 900 scans, acquisition time 8.2; 45° pulse angle. (From Ref. 1 with kind permission.)

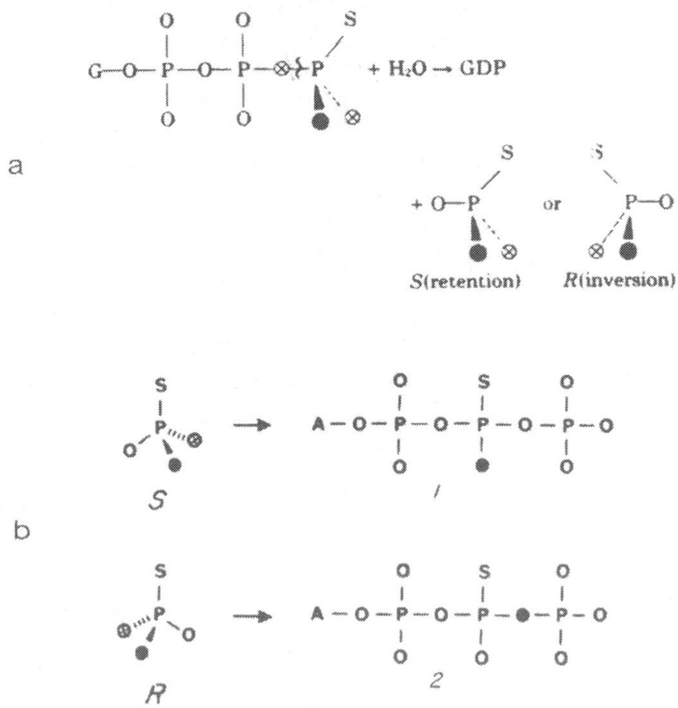

Figure 2. (a) Possible stereochemical courses of the GTP hydrolysis catalyzed by elongation factor G from *E. coli*. (b) The two ATPβS species which may be distinguished by ^{31}P NMR. ^{18}O is represented by ● and ^{17}O by ⊗. (From Ref. 2 by kind permission.)

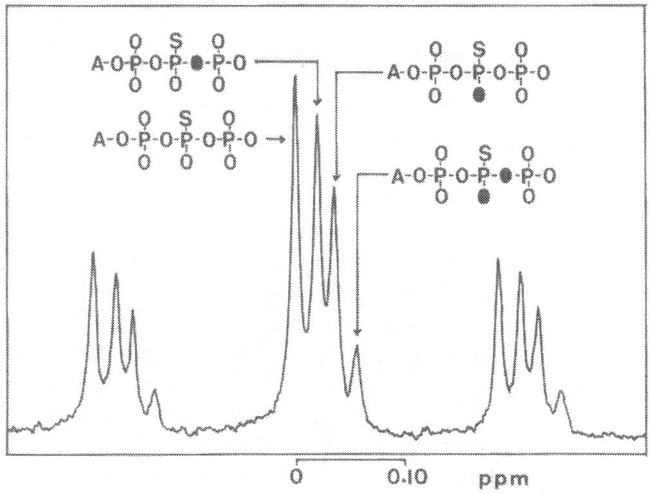

Figure 3. ^{31}P NMR spectrum of ATPβS, showing resonances from different isotopic arrangements. ^{18}O is represented by ●. (From Ref. 2 by kind permission.)

of the ribosome-dependent GTPase reaction of elongation factor *G* from *E. coli.* This enzyme is responsible for the hydrolysis of two molecules of GTP per amino acid addition in prokaryote protein synthesis. Guanosine 5′ (γ-thio) triphosphate was labeled stereospecifically with ^{17}O and ^{18}O on the γ-phosphorus atom, and hydrolyzed with unlabeled water. The resulting inorganic thiophosphate contains an ^{16}O atom, its enantiomeric form depending on the stereochemistry, and was then incorporated at the β position of adenosine 5′ (β-thio) triphosphate. Phosphates which retain the ^{17}O label do not give rise to observable ^{31}P resonances due to rapid quadrupole relaxation, but in molecules in which the ^{17}O is lost the S enantiomer will give ATP with a nonbridging ^{18}O label, while the R enantiomer gives a bridging ^{18}O atom (Figure 2), with different chemical shifts for the β-^{31}P (Figure 3). The spectrum was almost exactly that expected for 100% inversion of the phosphate configuration, and it is therefore likely that the original hydrolysis occurs in a single step, by direct in-line transfer of the phosphorus from GDP to a water oxygen without a phosphoenzyme intermediate. This in turn suggests that the GTP hydrolysis step is not an intrinsic part of the translocation of the ribosome on the DNA; the cleavage of GTP may be necessary only for the dissociation of the elongation factor from the ribosome to allow the elongation cycle to recommence.

3. SODIUM TRANSPORT THROUGH MEMBRANES USING ^{23}Na RESONANCE

The specific transport, active and passive, of metal ions across membranes is an exceedingly important biological process. It is, however, difficult to monitor in a "live" situation.(3) Very large lanthanide ions which remain on one side of a membrane cause paramagnetic shifts in the resonances of cations on that side only. One such shift reagent is triethanol-ammonium dysprosium nitrilotriacetate $[HN(CH_2CH_2OH)_3]_3Dy[N(CN_2CO_2)_3]_2$, which has been used to study ^{23}Na transport through large unilamellar vesicles of egg lecithin.[4] Sodium ions inside and outside the vesicles were readily distinguished in the NMR spectrum (Figure 4), and the time course of sodium transport following addition of gramicidin, which forms ion-transport channels, could be followed with ease. The transport occurred in stages: firstly a rapid transport of sodium from the vesicles in direct exchange for Li^+ ions from the medium outside, and subsequently a slower step in which a higher concentration of Li^+ was built up inside the vesicle as Na^+ was transported out, coupled with an equally slow process in which Na^+ transport was accompanied by passive nonfacilitated transport of $^{35}Cl^-$ ions out of the vesicles. The method of monitoring cation flow will clearly be of value in studying inherent and stimulated leakiness of unilamellar vesicles, but may also be extended to the study of transport in live systems; Balschi *et al.*[5] have monitored Na^+ efflux from yeast cells using the same shift reagent. Yeast cells normally accumulate K^+ in preference to Na^+ ions, but if incubated in an Na^+-rich medium which lacks K^+ they will take up Na^+. The separation of 364 Hz induced between

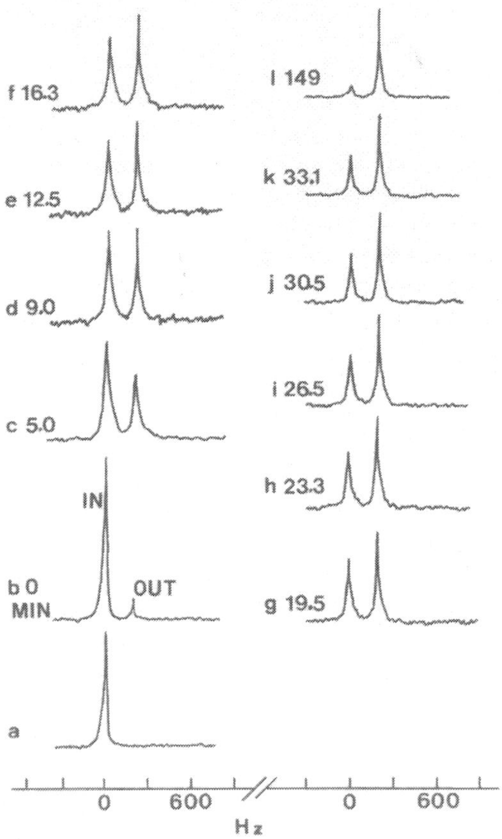

Figure 4. ^{23}Na NMR spectra (132.3 MHz, 11.74 T) of a dispersion of large unilamellar vesicles in ^2H_2O. (a) In the NMR tube, the final concentrations were egg lecithin, 6.0 mM; $NaCl_{in}$, 60 mM; $NaCl_{out}$, 0.27 mM; and $LiCl_{out}$, 35 nM; (b) concentrations were as in (a) except that the outside aqueous space was made 6.3 mM in $[HN(CH_2CH_2OH)_3]_3Dy[N(CH_2CO_2)_3]_2$. For a and b, 512 free-induction decays (FID) were accumulated in 208 s. (c-1) Spectra are labeled with the minutes elapsed after the solution was made 0.16 μM in gramicidin. The times recorded are those of the midpoints of the data accumulation periods, which were 156 FID in 64 s (c), 128 FID in 52 s (d–j), and 512 FID in 208 s (k and l). The temperature was ca. 297 K. (From Ref. 4 by kind permission.)

Na^+_{in} and Na^+_{out} by the shift reagent then permitted monitoring of the Na^+ efflux into a medium low in sodium, giving two first-order rate constants, in agreement with earlier results obtained by far more difficult biochemical methods. The mechanisms of cation transport through membranes must certainly become more accessible through the use of methods such as these.(6–8)

4. ACTIVE SITE INTERACTIONS IN FLUORINE-LABELED α-CHYMOTRYPSIN

The sidechain of methionine-192 plays an important role in α-chymotrypsin, being implicated in both the activation of the enzyme from its zymogen and the binding and orientation of α-chymotrypsin substrates during catalysis. X-ray diffraction studies(9) reveal that the sidechain may act as a "lid" over a hydrophobic sidechain binding pocket known as the "tosyl hole." Modification of the sidechain of Met-192 therefore has kinetic consequences; thus the addition of an aromatic moiety allows it to bind in the tosyl pocket and to compete for this site with substrate or inhibitor. The behavior of such an aromatic moiety may conveniently be followed via ^{19}F labeling(10–13) as in (trifluoromethyl) bromoacetanilides used as Met-192 sulfonium derivatives on α-chymotrypsin. The behavior of the single ^{19}F resonance was monitored as a function of concentration of indole, a competitive inhibitor of the enzyme. As Figure 5 shows, the CF_3 label is displaced from the tosyl pocket into a more solventlike environment. Kinetic studies(13) when correlated with the NMR work(12) suggest that the aromatic moiety is partitioned between this site and another, possibly the S_2 subsite of the peptide binding site. The ability to observe such interactions dynamically in molecules in solution makes ^{19}F NMR an invaluable adjunct to x-ray crystallographic work.

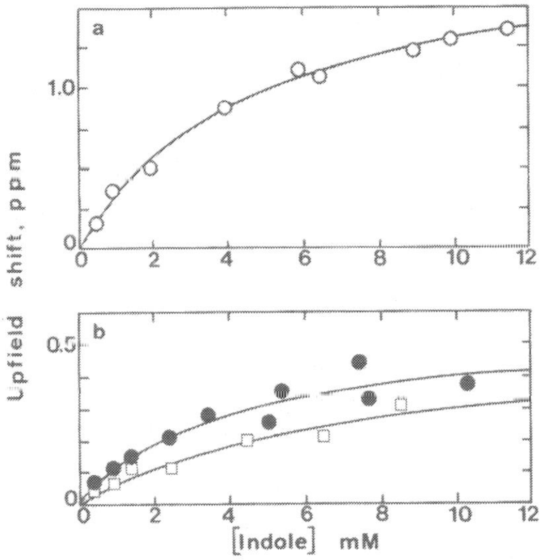

Figure 5. Upfield shift for derivatized chymotrypsins with increasing indole concentration. The solid line represents the theoretical fit to a single dissociation constant. (a) *p*-A-chymotrypsin, ○; (b) *m*-A-chymotrypsin, ●; *o*-A-chymotrypsin, □. (From Ref. 12 by kind permission.)

5. ^{113}Cd STUDIES OF ALKALINE PHOSPHATASE

The alkaline phosphatase of *E. coli* is a dimeric metalloenzyme which only attains full activity and stability when four zinc ions and two magnesium ions are bound to three distinct pairs of sites. The enzyme remains active when cadmium is substituted for zinc, though with a reaction rate some 10^2 times slower than normal. Enrichment of the cadmium with ^{113}Cd thus permits investigation of the metal binding sites by NMR.[14] In the absence of phosphate, the first two ^{113}Cd ions added to the dimeric apoenzyme give rise to a single resonance, implying that they bind to a pair of identical (symmetrical) sites, the A sites. After this ^{113}Cd ions occupy the B and C sites, but the resonance disappears, a phenomenon attributed to severe exchange broadening. Covalent phosphorylation of the active-site serine residues of the protein prevents this effect, so that three separate ^{113}Cd resonances are observable. Further, ^{31}P resonance experiments[15] reveal that enzyme containing only two equivalents of metal ion can bind only a single mole of phosphate per mole of enzyme dimer (half-of-sites reactivity) while enzyme containing its full complement of metal ions can bind two moles of phosphate tightly to indistinguishable sites on the dimer. The structural basis of this anticooperativity is discussed in Ref. 14; by using enzyme phosphorylated at the active site serine residues so that the three metal-ion sites are observed in the ^{113}Cd spectrum, it is possible to conclude that half-sites phosphorylation of the Cd_2^{2+} enzyme is accompanied by slow migration of half the Cd^{2+} originally located at A sites to the B sites on the phosphorylated subunits. Figure 6 shows ^{113}Cd spectra of $^{113}Cd^{2+}$ alkaline phosphatase; with two ^{113}Cd ions per dimer, a single (A site) resonance is observed. Addition of inorganic phosphate, which will bind to only one subunit in the presence of only two metal ions per dimer, causes the single resonance to become two, with the total integrated area divided equally between them, indicating that two different sites are now occupied. Assuming

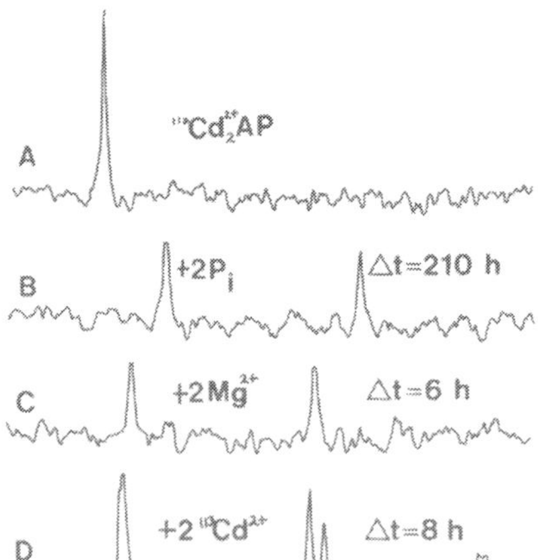

Figure 6. ^{113}Cd NMR spectra at 19.96 MHz of phosphoryl $^{113}Cd^{2+}$ alkaline phosphatase at pH 6.3. Each spectrum was acquired using a pulse angle of 90°, recycle time of 8.0 s, and 10000 transients (22.2 h); (A) 2.8 mM apophosphatase reconstituted with 2.2 equiv of $^{113}Cd^{2+}$; (B) A plus 2.2 equiv of P_{19}. (C) B plus 2.2 equiv of Mg^{2+}; (D) C plus 2.2 equiv of $^{113}Cd^{2+}$. The Δt values refer to the time interval between metal or P_i addition and the midpoint of spectral acquisition. (From Ref. 14 by kind permission.)

one of these to be the A site, even this has a different shift from the original spectrum, implying a structural change on phosphorylation. Addition of two Mg^{2+} ions causes a small shift in both resonances, and further addition of two $^{113}Cd^{2+}$ ions causes a doubling in the area of the resonances (but no shift), and a doubling of the amount of phosphate bound. All of this may be interpreted as demonstrating that a stable complex in a single subunit is formed only when phosphorylation is accompanied by metal-ion occupation of both A and B sites. Migration of metal ions from the A site of one subunit to the B site of the other occupies several days, and this is reflected in the rate of phosphorylation observed in the presence of insufficient metal ions. It is thus possible to explain the anticooperativity of the enzyme solely in terms of metal-ion occupancy of sites. The further implications of the NMR work, that the A and B sites are close together in the structure, are borne out by a recent x-ray study[16] which shows them to be some 5 Å apart.

6. ^{31}P NMR IN LIVING TISSUE

In recent years a vast area of useful application for NMR has been opened up as methods have developed for the rapid "noninvasive" observation of signals from whole organisms. The field is vast, extending into NMR tomography in which images are formed of sections through a limb, skull, or thorax, and considerable interest is being generated by topical magnetic resonance.[17,18] In this work a radiofrequency coil 2.5 cm in diameter was placed on the surface of a subject's arm and used to obtain ^{31}P spectra from the flexor digitorum superficialis muscle. As may be seen from Figure 7, resonances were observed from the α, β, and γ phosphates of ATP, from phosphocreatine, and from inorganic phosphate. In normal muscle at rest, the ratio of phosphocreatine to inorganic phosphate ($PCr:P_i$) is about 9, and this falls to about 1 on exercise, recovering >90% of its resting level after three minutes of relaxation. In two patients who were unable to tolerate exercise because of mitochondrial abnormalities, however, the level of phosphocreatine was abnormally low at the beginning of exercise, fell much more rapidly, and recovered far more slowly (Figure 8) than in

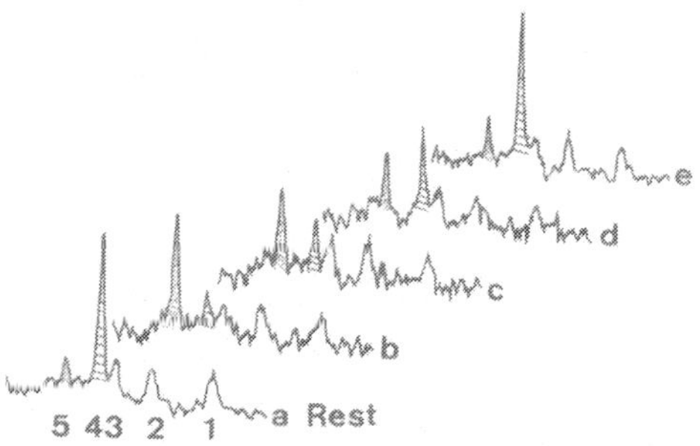

Figure 7. ^{31}P NMR spectra of one of the patients (R.O.) taken at rest, during exercise, and recovery. The signals are assigned as follows: 1, 2, and 3, the α, β, and γ phosphates of ATP; 4, phosphocreatine; 5, inorganic phosphate. The inorganic phosphate and phosphocreatine signals are shaded for clarity. All spectra were accumulated at 32.5 MHz using a train of radiofrequency pulses applied at intervals of 2 s; the number of pulses was 128 for spectrum a, 32 for spectra b–d, and 64 for spectrum e. Spectrum a was recorded at rest immediately before exercise. Spectrum b was recorded during the last minute of aerobic exercise, and spectra c–e were recorded during the recovery period at 5, 9.5, and 37 min, respectively, after the end of exercise. (From Ref. 17 by kind permission.)

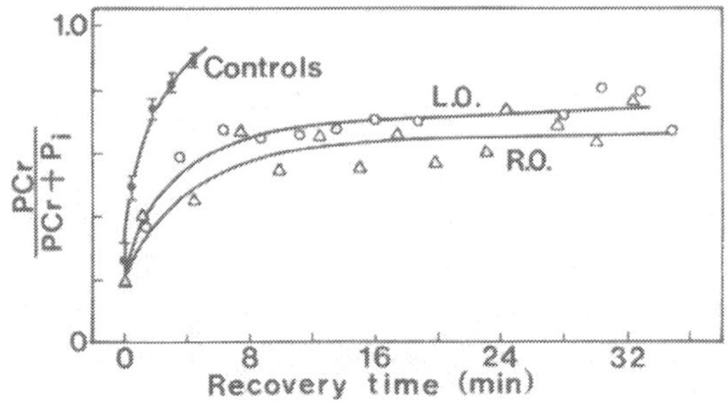

Figure 8. Recovery of PCr/(PCr + P_i) following exercise. Exercise was carried out by opening and closing the first every 2 s for 5 min (R.O. and L.O.) or squeezing a rubber bulb to a pressure of 100–300 mm Hg every 2 s for 7.5–21 min. Resting values of PCr/(PCr + P_i) were 0.83 (L.O.), 0.84 (R.O.), and 0.90 ± 0.01 (controls ± s.e.m.). Control values were obtained from four female and four male subjects aged 22–37 yr. Values are ± s.e.m. (From Ref. 17 by kind permission.)

normal subjects. The ability thus to monitor metabolite levels directly and continuously provides a vast improvement over biopsy methods, and further permits full assessment of treatments intended to alleviate the patients' intolerance of exercise.

7. ION BINDING TO CYTOCHROME *c* STUDIED BY NUCLEAR MAGNETIC QUADRUPOLE RELAXATION

Cytochrome *c* is one of the group of cytochromes in which the iron porphyrin is covalently linked to the protein. It forms part of the respiratory chain, relaying electrons from cytochrome c_1 to cytochrome oxidase. In a study of Cl^- binding to cytochrome *c*, use was made of the quadrupole moment of $^{35}Cl^-$. In aqueous solution, where the electronic environment of the ^{35}Cl nucleus is relatively symmetrical, quadrupole relaxation is inefficient and an NMR signal may be observed. On binding to an asymmetric site, however, the relaxation becomes extremely fast, to the extent that the resonance may become unobservable. In the case where there is a fast or intermediate exchange between solvent and the binding site, the transverse relaxation rate will be observed as being faster than that of ions in plain solvent, and the excess relaxation rate interpreted in terms of the fraction of ions bound.(19) As a function of NaCl concentration, pH, temperature, and the concentrations of various competing ions, the results indicate the presence of one class of high affinity sites and one of low-affinity sites for Cl^- on the protein. The effect of pH on the $^{35}Cl^-$ relaxation rate for both oxidation states is shown in Figure 9a; the most conspicuous feature is the hump at about pH 9.5 in the curve for the oxidized form of the protein. This pH corresponds to a conformational change in the protein, and is not found in experiments in which the cytochrome is carboxymethylated at methionine-80, the sixth heme ligand (Figure 9b). The p*K* values revealed by the relaxation curves at about pH 5.5 correspond exactly to the titration of histidine-33, indicating that this residue is involved in chloride ion binding in both oxidation states of the cytochrome, and that at about pH 10 may be associated with lysine-79. Taken in conjunction with competition experiments, these data may be interpreted as meaning that chloride is bound at the "front side" of the molecule and near the exposed heme edge, and that the surface

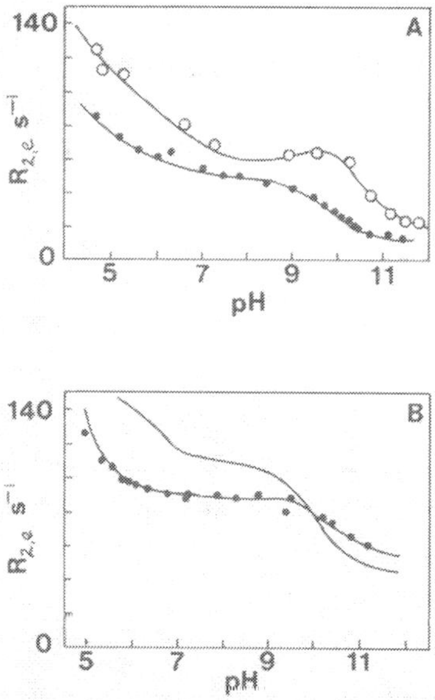

Figure 9. The pH dependence of the $^{35}Cl^-$ excess transverse relaxation rate for a 0.3 *M* NaCl solution containing oxidized (○) and reduced (●) cytochrome *c*(A) and carboxymethylated cytochrome *c*(B). The data are corrected to 3.0 mM protein concentration. Lines are theoretical curves. (From Ref. 19 by kind permission.)

structure or dynamics of this region are different in the two oxidation states. Further results and their interpretation are needed, and will undoubtedly cast considerable light on the redox mechanism of cytochrome *c*.

8. DEUTERIUM LABEL STUDIES OF MEMBRANES

Although much useful information can be obtained by labeling biological macromolecules with "foreign" elements such as ^{113}Cd or ^{19}F, it is better to avoid perturbation of the molecular structure and function by the use of "native" nuclei. Of course, if there are too many of them, problems of resonance overlap and assignment arise; proton spectra provide an extreme example of this difficulty. One solution is selective enrichment, say by ^{13}C or deuterium. One of the most important applications of deuterium labeling has been in the study of order and disorder in the lipid constituents of model and native membranes, and of the ordering effect of membrane proteins. Perhaps surprisingly, membrane proteins appear to have relatively small effects on lipid order, in contrast to the very large effects of cholesterol.[(20)] Presumably the small effect shown by the proteins is a function of their own very mobile outer surfaces. Biosynthetic deuterium enrichment with [γ-$^2H_6$] valine in bacteriorhodopsin, the photosynthetic protein of the purple membrane of *Halobacterium halobium*, gave dynamic information on the valine sidechain in the intact membrane. Figure 10 shows a set of inversion-recovery spectra taken at 37°C, a plot of peak heights against time for various temperatures, and the resultant Arrhenius plot, which reveals an activation energy for methyl group rotation of $\sim$2.4 kcal mol^{-1}. Overall the valine side-chain is a rather rigid structure in the membrane environment, with motion about the C_α–C_β

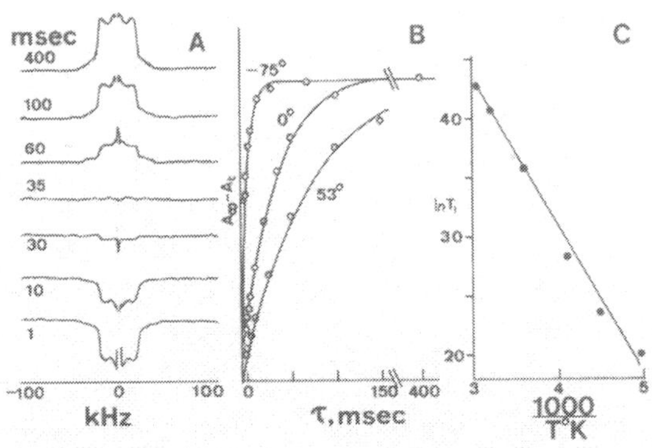

Figure 10. 55 MHz deuterium NMR spin-lattice relaxation results for [γ-$^2H_6$] valine-labeled bacteriorhodopsin in *H. halobium* purple membranes. A, partially relaxed Fourier transform NMR spectra obtained using a (180°-τ_3-90°-τ_1-90°-τ_2-Echo-T) sequence at 37°C for membranes dispersed in excess deuterium-depleted water, 2000 scans, 500-ms recycle time, $\tau_1 = \tau_2 = 55$ μs, 4 μs 90°C pulse widths, 2 MHz digitization rate, 2048 data points, line broadening = 400 Hz. The τ_3 values used are indicated on the figure. B, T_1 recovery curves, $(A_t - A_x)$ versus time, for [γ-$^2H_6$] bacterioehodopsin at −75°C, 0°C, and 52°C. The recycle time varied from 50 to 500 ms. C, Arrhenius plot of [γ-$^2H_6$] valine-labeled purple membrane bacteriorhodopsin spin-lattice relaxation time as a function of temperature. The activation energy is 2.4 ± 0.2 kcal mol^{-1}. (From Ref. 12 by kind permission.)

bond less than 10^5 s^{-1} at growth temperature. With further labeling and relaxation experiments, it should be possible to build up a detailed picture of the dynamics of all the amino acid sidechains in many cell membrane systems, investigating the effects of lipids, steroids, and other ligands on protein motions, and eventually to achieve new insights into the basis of biological energy transduction.

9. DIRECT DETERMINATION OF CORRELATION TIMES IN ENZYME COMPLEXES INVOLVING MONOVALENT CATIONS AND PARAMAGNETIC CENTERS

NMR methods are often used in the determination of the distance between a ligand bound to a macromolecule and a paramagnetic center, commonly an ion such as manganese, Mn^{2+}, substituted for the metal center in a metalloenzyme. The paramagnetic relaxation depends on the distance between them (Table 1, Chapter 2), and on the correlation times τ_c for the electron-nuclear interaction. Methods so far employed to determine τ_c have involved the use of two spectrometers working at different frequencies, but this poses obvious problems and the subsequent data analysis can be extremely complex. Alternatively another magnetic nucleus can be used, as in the relaxation enhancement[22] of lithium ions bound in a pyruvate kinase–Mn^{2+}–phosphoenolpyruvate–Li^+ complex. Earlier determinations of the Li^+–Mn^{2+} distance had varied from 4.8 to 5.8 Å according to the τ_c value chosen.[23,24] Relaxation measurements were carried out for both $^6Li^+$ and $^7Li^+$, and since many of the constants and the Li^+–Mn^{2+} distance could be assumed to be unchanged for the two isotopes, an unambiguous determination of τ_c was made, giving the Li^+–Mn^{2+} distance as 5.7 Å. The use of this method as suggested[22] for $^{15}NH_4^+$/$^{14}NH_4^+$ and $^{85}Rb^+$/$^{87}Rb^+$ monovalent cation pairs could point the way to clearer internuclear distance determinations in a number of enzymes which are activated by both monovalent and divalent cations.

INDEX OF REVIEWS

Deuterium Resonance

R1 Dynamic structure of membranes by deuterium NMR, Smith, R L, Oldfield, E *Science* 1984, **225**, 280–287 (48 references)

R2 The description of membrane lipid conformation, order and dynamics by deuterium NMR, Davis, J H *Biochim Biophys Acta* 1983, **737**, 117–171 (246 references)

R3 The use of isotopic hydrogen and nuclear magnetic resonance spectroscopic techniques for the analysis of biosynthetic pathways, Hutchinson, C R *J Nat Prod* 1982, **45**, 27–37 (45 references)

Nitrogen-15 Resonance

R4 Nitrogen-15 NMR in biological systems, Blomberg, F, Rueterjans, H *Biol Magn Res* 1983, **5**, 21–73 (108 references)

R5 Application of nitrogen-15 spectroscopy to studies of the intermolecular interaction of biomolecules, Kyogoku, Y *Appl Spect Rev* 1982, **17**, 279–334 (132 references)

R6 Application of nitrogen-15 NMR spectroscopy to the study of natural compounds and biomolecules, Martin, G J *Pharm Weekbl* 1984, **119**, 5–12 (23 references)

R7 Nitrogen-15 NMR studies of biological systems, Kanamori, K, Roberts, J D *Acc Chem Res* 1983, **16**, 35–41 Review with 48 references covering applications to proteins, nucleic acids, and in vivo metabolism and cellular dynamics

Oxygen-17 Resonance

R8 Oxygen-17 NMR and its biological applications, Fiat, D, St Amour, T E, Burgar, M I, Steinschneider, A, Valentine, B, Dhawan, D *Bull Mag Res* 1980, **1**, 18

Fluorine-19 Resonance

R9 Applications of fluorine NMR in biochemistry, Gerig, J T In *Biomedical Aspects of Fluorine Chemistry* Filler, R, Kobayashi, Y, Eds, Kodansha Tokyo, 1982, pp 163–189

R10 Fatty acyl chain structure, orientational order, and the lipid phase transition in *Acholeplasma laidlawii* membranes A review of recent fluorine-19 NMR studies, MacDonald, P M, Sykes, B D, McElhaney, R N *Can J Biochem Cell Biol* 1984, **62**, 1134–1150 (47 references)

Sodium-23 Resonance

R11 Sodium-23 NMR studies of cation binding by natural and synthetic ionophores, Laszlo, P *ACS Symp Ser* 1982, **191**, 63–95 Review with 117 references, covering thermodynamic and kinetic results

R12 NMR analysis of transporting epithelia, Civan, M M *Prog Clin Biol Res* 1983, **126**, 45–72 (79 references)

Phosphorus-31 Resonance

R13 A number of useful papers on biological applications of phosphorus resonance, including work on phospholipids and nucleic acids, will be found in *Phosphorus-31 NMR*, Gorenstein, D G Ed, Academic Orlando, 1984

R14 Phosphorus-31 NMR investigations of enzyme systems, Rao, B D Nageswara *Biol Magn Reson* 1983, **5**, 75–128 (110 references)

R15 Phosphorus-31 NMR chemical shifts of phosphate covalently bound to proteins, Matheis, G, Whitaker, J R *Int J Biochem* 1984, **16**, 867–873 (55 references)

R16 Phoshorus-31 resonance in living bodies, Hanley, P *Chem Br* 1981, **17**, 374–376 A short introduction (22 references)

R17. Fatigue in retrospect and prospect: phosphorus-31 NMR studies of exercise performance; Chance, B.; Sapega, A.; Sokolow, D.; Eleff, S.; Leigh, J. S.; Graham, T. Armstrong; Warnell, J. R. *Int. Ser. Sport Sci.* 1983, **13**, 895–908 (10 references).
R18. Phosphorus NMR studies of muscle metabolism; Gadian, D. E. *Spec. Pub. R. Soc. Chem.* 1984, **47**, 58–70 (32 references). Includes coverage of potential uses in meat chemistry.
R19. Study of *in vivo* cellular metabolism by phosphorus-31 NMR; Bernard, M.; Canioni, P.; Cozzone, P. J. *Biochimie* 1983, **65**, 449–470 (131 references).
R20. Phosphorus metabolites and the control of glycolysis studied by nuclear magnetic resonance; Dawson, M. J. *Int. Ser. Sport Sci.* 1983, **13**, 116–125 (19 references).
R21. Noninvasive pHi measurements of human tissue using phosphorus-31 NMR; Bore, P. J.; Chan, L.; Gadian, D. G.; Radda, G. K.; Ross, B. D.; Styles, P.; Taylor, D. J. *Kroc Found. Ser.* 1982, **15**, 527–535 (11 references).
R22. Phosphorus nuclear magnetic resonance of cardiac and skeletal muscles; Ingwal, J. S. *Am. J. Physiol.* 1982, **242**, H729–H744. A review with 82 references on metabolism of phosphorus compounds in muscle.
R23. Nucleotide conformational analysis by phosphorus-31 nuclear magnetic resonance spectroscopy; Gorenstein, D. G. *Ann. Rev. Biophys. Bioeng.* 1981, **10**, 355–386 (130 references).
R24. Phosphorus-31 studies of DNA conformation and dynamics; Cohen, J. S.; Chen, C. W. *ACS Symp. Ser.* 1982, **191**, 249–267 (50 references).
R25. High resolution phosphorus NMR spectroscopy of transfer nucleic acids; Gorenstein, D. G.; Goldfield, E. M. *Mol. Cell. Biochem.* 1982, **46**, 97–120. A review with 64 references, especially concerned with the analysis of conformational transitions in t-RNA.
R26. Applications of phosphorus-31 nuclear magnetic resonance to studies of protein structure and function; Sykes, B. D. *Can. J. Biochem. Cell Biol.* 1983, **61**, 155–164. A review with 32 references, emphasizing enzyme mechanistic studies and including alkaline phosphatase, glycogen phosphorylase, actin, and myosin.
R27. Phosphorus-31 NMR techniques and membrane lipid structure: Implications for mechanisms of membrane fusion; Cullis, P. R.; Farren, S. B.; Hope, M. J. *Can. J. Spect.* 1981, **26**, 89–95 (26 references).
R28. Techniques to study metabolic changes at the cellular and organ level; DeFuria, R. R.; Dygert, M. K. *Int. Rev. Cytol.* 1983, **83**, 27–62 (105 references). Review of ATP metabolism, with emphasis on phosphorus resonance.
R29. Isotope effects on phosphorus chemical shifts: Applications to enzyme mechanisms; DeBrosse, C. W.; Villafranca, J. J. *Magn. Res. Biol.* 1983, **2**, 1–52 (88 references).

Cadmium-113 Resonance

R30. Principles and applications of cadmium-113 NMR to biological systems; Armitage, I. M.; Otvos, J. D. *Biol. Magn. Reson.* 1982, **4**, 79–144 (93 references).
R31. Elucidation of metallothionein structure by cadmium-113 NMR; Otvos, J. D.; Armitage, I. M. in *Biochemical Structure Determination by NMR*; Bothner-By, A. A.; Glickson, J. D.; Sykes, B. D.; Eds.; Dekker: New York, 1982; pp. 65–96, 43 references. Covers structural characterization of metal-binding sites of metallothioneins from crab hepatopancreas and rabbit liver.

Reviews Covering More Than One Nucleus

R32. *In vivo* topical magnetic resonance; Shaw, D. *Org. Mag. Res.* 1983, **21**, 225–237 (43 references). Covers instrumental techniques, reviews applications, and future developments.
R33. *Topical Magnetic Resonance*; Gordon, R. E.; Hanley, P. E.; Shaw, D. *Prog. Nucl. Mag. Reson. Spect.* 1982, **15**, 1–47. A valuable introduction to this important medical and biochemical technique with 61 references, mainly on carbon and phosphorus resonance; includes a theoretical introduction.
R34. *Nuclear Magnetic Resonance Imaging in Medicine*; Kaufman, L.; Crooks, L. E.; Margulis, A. R., Eds.; Igaku-Shoin: New York, 1981.
R35. *NMR Imaging in Biomedicine*; Mansfield, P.; Morris, P. G.; Academic: New York, 1982.
R36. *NMR Data Handbook for Biomedical Applications*; Beall, P. T.; Amtey, S. R.; Kasturi, S. R.; Pergamon: Oxford, 1984.
R37. Deuterium and phosphorus NMR of microbial membranes; Smith, I. C. P.; Jarrell, H. C. *Acc. Chem. Res.* 1983, **16**, 266–272 (60 references).

R38. Phosphorus-31 and sodium-23 studies of diamagnetic metal ion binding to phosphatidylserine vesicles; Kurland, R. J. *NATO Adv. Stud. Inst. Ser. C* 1980, **61**, 655.

R39. Calcium and magnesium resonance in chemistry and biology; Forsen, S.; Lindman, B. *Ann. Rep. NMR Spectroscopy* 1981, **11A**, 183–226. 108 references. A good introduction to applications in biology.

R40. Magnesium-25, calcium-43, and cadmium-113 studies of regulatory calcium binding proteins; Forsen, F.; Andersson, A.; Drakenberg, T.; Teleman, O.; Thulin, E.; Vogel, H. J. *Dev. Biochem.* 1983, **25**, 121–131 (44 references).

R41. Chlorine-35 and bromine-81 NMR studies of carbonic anhydrase; Ward, R. L. In *Biophysics and Physiology of Carbon Dioxide*; Bauer, C.; Gros, G.; Bartels, H., Eds.; Springer: Berlin, 1980; pp. 262–273.

R42. The multinuclear approach to peptides: structure, conformations and dynamics; Deslauriers, R.; Smith, I. C. P. In *Biological Magnetic Resonance*; Berliner, L. J.; Reuben, J., Eds.; Plenum: New York, 1980; Vol. 2, pp. 243–344. Review mentions a number of nuclei in connection with peptide structure, but examples are mainly taken from carbon and proton spectroscopy.

R43. Multinuclear NMR approaches to the solution structures of alkaline phosphatase: carbon-13, fluorine-19, phosphorus-31, and cadmium-113 NMR; Coleman, J. E.; Armitage, I. M.; Chlebowski, J. F.; Otvos, J. D.; Schoot Uiterkamp, A. J. M. In *Biological Applications of Magnetic Resonance*; Shulman, R. G., Ed.; Academic: New York, 1979; pp. 345–395.

R44. Use of the nuclear magnetic resonance of nuclei other than hydrogen in tumor studies; Granger, P. *J. Biophys. Med. Nucl.* 1981, **5**, 137–140. A short review with 17 references, including oxygen, sodium, phosphorus, and potassium resonance.

R45. Use of phosphorus-31 (oxygen-18), phosphorus-31 (oxygen-17), and oxygen-17 NMR methods to study enzyme mecanisms involving phosphorus; Tsai, M-D. *Methods in Enzymology* 1982, **87**, 235–279 (92 references).

R46. Oxygen-18 and oxygen-17 effects on phosphorus-31 NMR as probes of enzymic reactions of phosphate compounds; Cohn, M. *Ann. Rev. Biophys. Bioeng.* 1982, **11**, 23–42 (61 references).

R47. NMR studies of metabolites in living tissue; Iles, R. A.; Stevens, A. N.; Griffiths, J. R. *Prog. NMR Spect.* 1982, **15**, 49–200. Excellent long introduction. Mainly focuses on phosphorus resonance, but covers other nuclei. 317 references. Some general introduction to NMR theory and coil design.

R48. NMR of membranes; Jacobs, R. E.; Oldfield, E. *Prog. NMR Spect.* 1981, **14**, 113–136. Mainly focuses on phosphorus and deuterium spectra. 172 references. Includes introductory theory.

R49. NMR and the biochemist; Moore, G. R.; Ratcliffe, R. G.; Williams, R. J. P. *Essays Biochem.* 1983, **19**, 142–195. Excellent general introduction with 110 references, covering phosphorus and cadmium as well as proton and carbon resonances.

R50. Application of carbon-13 and phosphorus-31 NMR to the study of hepatic metabolism; Cohen, S. M. *Fed. Proc., Fed. Am. Soc. Exp. Biol.* 1984, **43**, 2657–2662 (11 references).

R51. Use of phosphorus-31 and carbon-13 NMR to study enzyme mechanisms; Villafranca, J. J. *Fed. Proc., Fed. Am. Soc. Exp. Biol.* 1984, **11**, 2640–2647 (30 references). Chiefly concerned with glutamine synthetase and carbamyl phosphate synthase.

REFERENCES

1. Cohn, M.; Hu, A. *Proc. Natl. Acad. Sci. USA* 1978, **75**, 200–203.
2. Webb, M. R.; Eccleston, J. F. *J. Biol. Chem.* 1981, **256**, 7734–7737.
3. Pike, M. M.; Springer, C. S. *J. Magn. Res.* 1982, **46**, 348–353.
4. Pike, M. M., Simon, S. R.; Balschi, J. A.; Springer, C. S. *Proc. Natl. Acad. Sci. USA* 1982, **79**, 810–814
5. Balschi, J. A.; Cirillo, V. P.; Springer, C. S. *Biophys. J* 1982, **38**, 323–326.
6. Gupta, R. K.; Gupta, P. *J. Magn. Res.* 1982, **47**, 344–350.
7. Ting, D. Z.; Hagan, P. S.; Chan, S. I.; Doll, J. D.; Springer, C. S. *Biophys. J.* 1981, **34**, 189–216.
8. Civan, M. M.; Shporer, M. In *Biological Magnetic Resonance*; Berliner, L. J.; Reuben, J., Eds.; Plenum: New York, 1978; Vol. I, pp. 1–32.
9. Steitz, T. A.; Henderson, R.; Blow, D. M. *J. Mol. Biol.* 1969, **46**, 337–348.
10. Bittner, E. W.; Gerig, J. T. *J. Am. Chem. Soc.* 1970, **92**, 2114–2118.
11. Bittner, E. W.; Gerig, J. T. *J. Am. Chem. Soc.* 1970, **92**, 5001–5003.
12. Landis, B. H.; Berliner, L. J. *J. Am. Chem. Soc.* 1980, **102**, 5350–5354.
13. Landis, B. H.; Berliner, L. J. *J. Am. Chem. Soc.* 1980, **102**, 5354–5358.

14 Otvos, J D, Armitage, I M *Biochemistry* 1980, **19**, 4031–4043
15 Otvos, J D, Armitage, I M, Chlebowski, J F, Coleman, J E *J Biol Chem* 1979, **254**, 4707–4713
16 Sowadski, J M, Foster, B A, Wyckoff, H W *J Mol Biol* 1981, **150**, 245–272
17 Radda, G K, Bore, P J, Gadian, D G, Ross, B D, Styles, P, Taylor, D J, Morgan-Hughes, J *Nature* 1982, **295**, 608–609
18 Gadian, D G, Radda, G K *Ann Rev Biochem* 1981, **50**, 69–83
19 Andersson, T, Thulin, E, Forsen, S *Biochemistry* 1979, **18**, 2487–2493
20 Oldfield, E In *Membranes and Proteins*, Martonosi, A, Ed, Academic New York, 1981
21 Kinsey, R A, Kintanar, A, Tsai Ming-Daw, Smith, R L, Janes, N, Oldfield, E *J Biol Chem* 1981, **256**, 4146–4149
22 Raushel, F M, Villafranca, J J *J Am Chem Soc* 1980, **102**, 6618–6619
23 Hutton, W C, Stephens, E M, Grisham, C M *Arch Biochem Biophys* 1977, **184**, 166–171
24 Ash, D E, Kayne, F J, Reed, G H *Arch Biochem Biophys* 1978, **190**, 571–577

Chapter 23

BIOMEDICAL NMR

JOAN MASON

1. BIOMEDICAL NMR

Some of the most dramatic and far-reaching NMR developments are nowadays occurring in the biological and medical fields. The current literature is large and only key references can be given[1–22]; comprehensive annual reviews are available.[1c] Advances in instrumentation, notably in superconducting and resistive magnets, radiofrequency (rf) electronics and computing, now allow *in vivo* imaging and localized spectroscopy in species ranging in size from a whole adult human to the fruit fly Drosophila (i.e., NMR microscopy). (The field has been renamed in some medical circles as Magnetic Resonance (MR) and even Medical Resonance, since the word "nuclear" has some unwanted associations.)

Distinction must be made between NMR *imaging* as such (also called spin imaging or mapping, or zeugmatography),[23] and *localized* NMR *spectroscopy*. In NMR imaging, as in computed *tomography* (CT) in general, the objective is a picture of a particular section or slice (Greek *tomos*), perhaps 5 mm thick, of an intact organism or organ. This image is comparable in principle to those obtained by x-ray CT, ultrasound, or other scanning methods, and differs from the traditional x-ray picture, which is a shadowgraph, in which overlapping structures are not differentiated (although projection images can be produced also by NMR). The image can map the density, relaxation time T_1 or T_2, a chemical shift, or flow rate of the resonant nucleus.

In localized NMR spectroscopy a high-resolution spectrum in ^1H or ^{31}P (etc.) resonance is produced from a particular region of the organism, selected by techniques comparable to those used in imaging. Under favorable circumstances, metabolic processes may then be studied by standard NMR methods. Localization requires sufficiently large-volume rf coils, but superficial tissues may be studied with flat or shaped surface coils for better rf penetration, although the spatial response is then more complex. Such coils have been developed for different joints or limbs, the neck, head, spine, breasts, etc.

NMR examination *in vivo* is called noninvasive as it avoids the use of surgery, ionizing radiations, or ingested material. In practice, safe material may be ingested to improve the contrast, e.g., by promoting relaxation, or for isotopic enrichment. No adverse effects of the magnetic fields have so far been identified. Patients with metal implants such as pacemakers should not of course approach the magnet, and the field

JOAN MASON • Department of Chemistry, The Open University, Milton Keynes MK7 6AA, England.

switching and rf radiation levels must be carefully controlled.[24] On the other hand, it has been suggested that rf heating might be used to destroy tumors, with monitoring by NMR.

2. NMR IMAGING

NMR sensitivity is of prime importance, particularly since the use of higher magnetic fields carries certain disadvantages. Imaging, therefore, mainly uses the proton resonance of water and fat, either together or separately. Water forms some 55% of total body weight, from 3% in tooth enamel to 80% in muscle, and upwards of 90% in body fluids. The contrast is supplied by variations in density and even more in mobility, both fluid motions (as in the vascular system) and tumbling motions that determine the relaxation rates. Relatively immobile protons as in bone, proteins, DNA, and membrane lipids do not normally contribute, because of short T_2 and (usually) long T_1 values, or (as in bone) lower concentration. Since the signals are obtained from soft tissue, NMR imaging complements the x-ray CT scan, and gives valuable added information on the brain, and on the pelvic cavity, which is poorly penetrated by x-rays. Bones may be evident as dark regions enclosing a strong signal from the fat in bone marrow.

The mapping of the sample is produced by orthogonal *gradients* of the static magnetic field B_0 so that this becomes a function of the Cartesian coordinates of the sample volume. If a linear field gradient g_x is applied in the x direction the Larmor frequency ω_i $(=2\pi\nu)$ at a point i with coordinate x_i is given by

$$\omega_i = \gamma[B_0 - B_0\sigma + g_x x_i]$$

and similarly in the other dimensions, where σ is the nuclear magnetic shielding, opposite in sign to the chemical shift δ. $B_0\sigma_i$ is usually small in comparison with the field gradient term, with gradients of say 5 μT per mm. The B_0 dependence of the resonance frequency thus becomes a spatial dependence. The linear dependence was noted in 1951 by Gabillard[25] and applied in two dimensions 20 years later by Lauterbur[23] and Mansfield,[14] who with Ernst and others[26] developed a range of rf pulse and two-dimensional and three-dimensional FT techniques. The first reasonably good images made their appearance as recently as 1980, and progress since then has been exponential. Two-dimensional imaging commonly uses frequency-selective pulses with interleaved multisection acquisition to save time; three-dimensional imaging uses a broad-banded pulse to excite the whole volume of interest with (e.g.) phase-encoding gradients to differentiate the volume elements (voxels), and the computer then generates two-dimensional images.[14] Compared with x-ray CT scanning, NMR has the advantages of being able to image a section of any orientation, avoiding movement of equipment or patient, as well as the better differentiation of soft tissue. The main disadvantages of NMR imaging are the longer scanning times and the greater cost.

Some contrast in the NMR image arises from variations in proton density in biotissue, by a factor of 0.2 or so; thus grey matter contains 15% more water than white matter in the brain. Much greater is the variation in relaxation times, by a factor of 4 for T_1, and 10 or more for T_2, so that particularly high sensitivity may be attained by T_2 mapping. At typical field strengths, below 0.5 T or so, the most favorable T_1 and T_2 values, a few hundred milliseconds (rather more for T_1 than for T_2) are given by the rotational motions of protons in fats (CH_2) and water. Pure water has $T_1 = T_2 \simeq 3$ s, and the reduction in biotissues is due to the slowing down of the tumbling motion by hydrogen bonding to macromolecules such as proteins, with fast exchange between free and bound water. It was discovered early on that many tumors[27] and other kinds of diseased (e.g., inflamed) tissues are readily observable as they generally have increased T_1 and T_2 values. These may be due to the release of bound water, or even to accelerated cell division, or perhaps to free radicals. The properties of water in tissues are complex and little understood.

In contrast to mobile fluids, in which $T_1 \simeq T_2$, biotissues usually have $T_2 \ll T_1$ because of near-static dipolar interactions. The image can be made to depend on contrast in T_1 or T_2 by suitable pulse sequences and timing. Thus in saturation-recovery spin-echo images of the brain a shorter pulse repetition time reduces the signal from more slowly relaxing protons in more mobile phases (cerebrospinal fluid, oedema, neoplasia) as these protons with longer T_1 values tend to saturate sooner than protons in fat. On the other hand, lengthening the echo delay increases the signal from these more mobile protons because of their longer T_2 values relative to fat, muscle, etc. Paramagnetic relaxation reagents are often used *in vivo* to increase contrast, as shown in Figure 1.

Flow imaging has potential for monitoring vascular disease. Commonly blood vessels appear dark as excited nuclei flow out of the slice, and the signal may increase in the presence of thrombosis or atherosclerotic (fatty) plaque. Alternatively, blood flow may be imaged by increased signals from inflowing protons, by appropriate pulse sequences.

Four characteristics therefore determine the localized signal. This will normally be enhanced by increase in proton density, decrease in T_1, increase in T_2, and decrease in flow rate, and images may be needed at several T_1 and T_2 settings to resolve ambiguities or provide additional information.

Chemical shift imaging (ideally, chemical species imaging) is possible at sufficiently high field, say 1.5 T or so, depending on the gradient. The difference in resonance frequency between fat and water protons, corresponding to a shift difference of 3.7 ppm, may produce artifacts of displacement of a water–fat interface (which can be removed by appropriate techniques). Alternatively the shift difference can be used to produce separate water and lipid images, containing complementary information, by means of a selective pulse before imaging. A hybrid imaging technique collects fat and water images concurrently and combines them, so halving the overall acquisition time.

Other highly sensitive nuclei are usable also, ^{31}P for example. Added ^{19}F can be imaged, as in oxygen-carrying fluorocarbon emulsions used as blood substitutes (in which the T_1 is proportional to the oxygen concentration),[28] or inert fluorocarbon gases used to study lung ventilation. Imaging of sodium is of interest as ^{23}Na is the second most abundant magnetic nucleus in biotissue and is highly NMR sensitive, although quadrupolar. Contrast is good as there are large variations in concentration, and these can reflect tissue damage. The intracellular concentration is about one-tenth of that in the intercellular space, but breakdown of the sodium–potassium pump, or injury (oedema, neoplasia, etc.) expanding the intracellular space greatly increases the total concentration in the tissue. Much of the sodium is not observed, however, because of very short T_2 values arising from relaxation of the nuclear quadrupole in electric field gradients near macromolecules and membranes; a preacquisition delay of a few milliseconds is required to reduce distortions due to eddy currents produced by

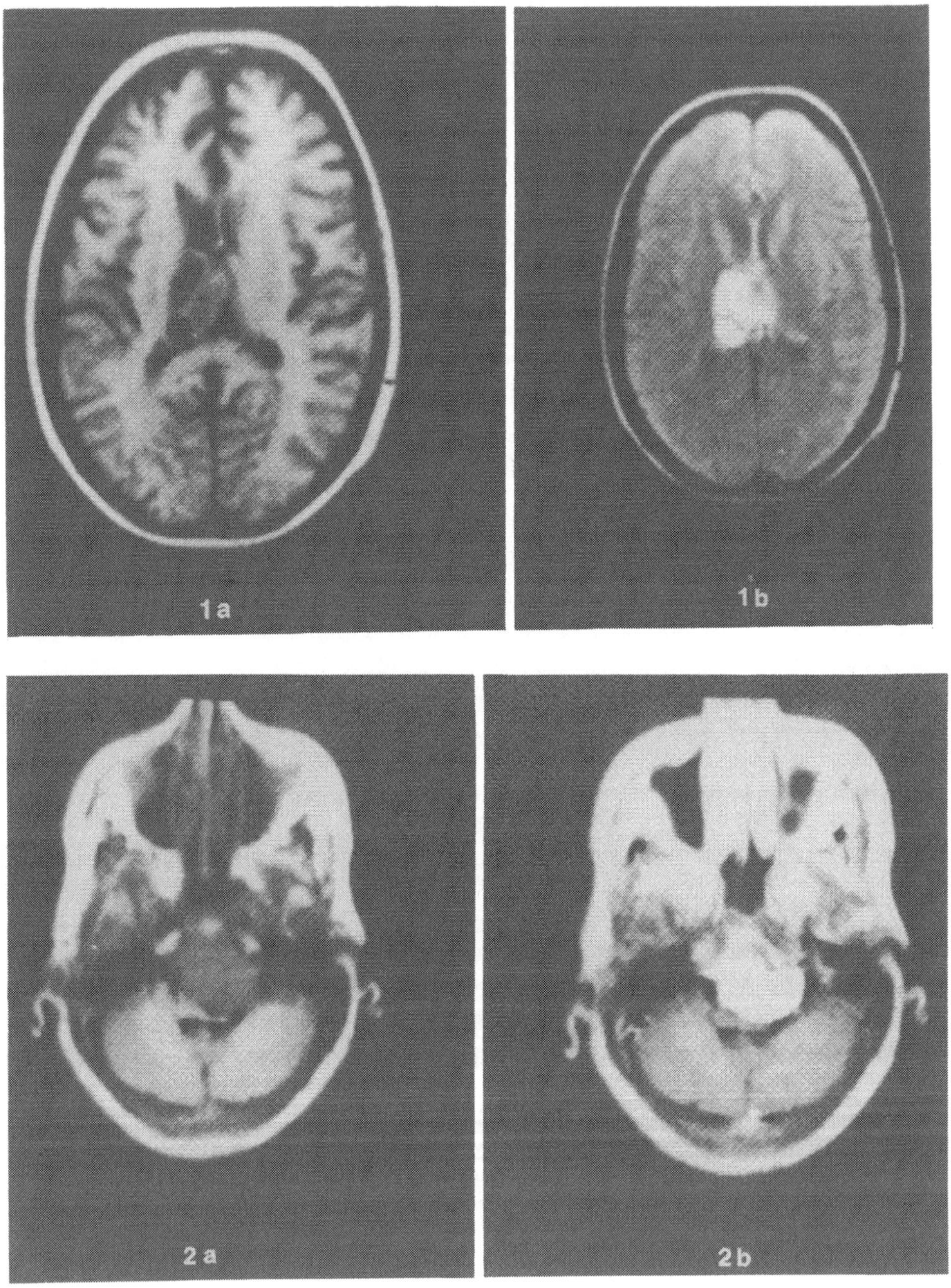

Figure 1. *In vivo* proton images measured with a cryogenic system at 0.15 T at the NMR Unit, Royal Postgraduate Medical School, Hammersmith Hospital, London, U.K. (1) Brain with a malignant tumor (astrocytoma grade III): (a) inversion recovery (T_1 weighted), with tumor dark, and (b) spin echo (T_2 weighted), with tumor light. (2) Brain with a benign tumor (meningioma): inversion recovery images (a) before and (b) after intravenous administration of a paramagnetic relaxation reagent gadolinium dtpa (diethylenetriaminepentaacetate) which highlights the tumor.

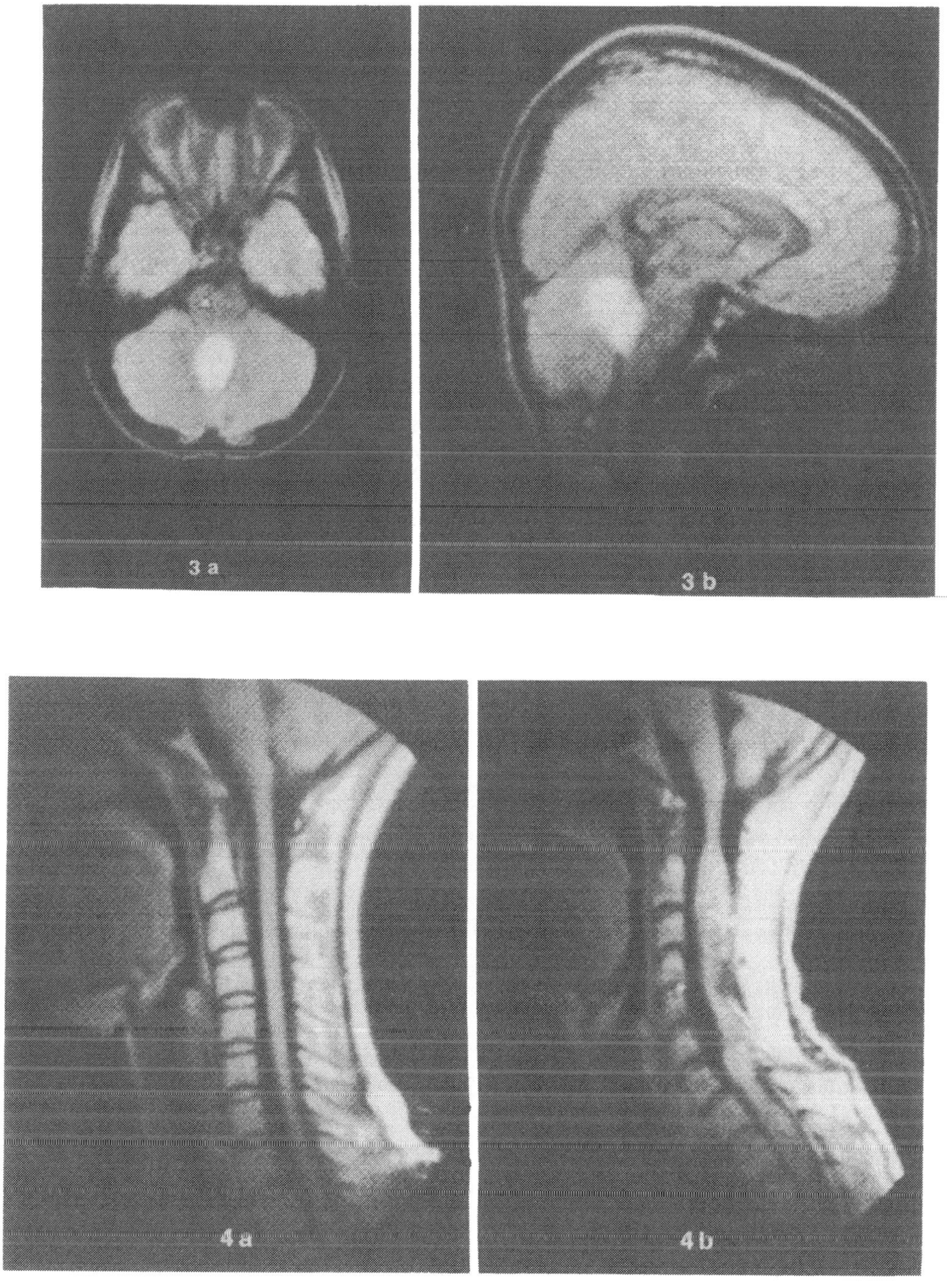

Figure 1. (*Continued*) (3) Brain with a malignant tumor of childhood (medulloblastoma) spin echo images in (a) transverse and (b) sagittal planes The tumor is much lighter than brain (4) Spinal cord, spin echo images (a) normal spine, and (b) cord expanded and irregular due to the presence of an intramedullary tumor

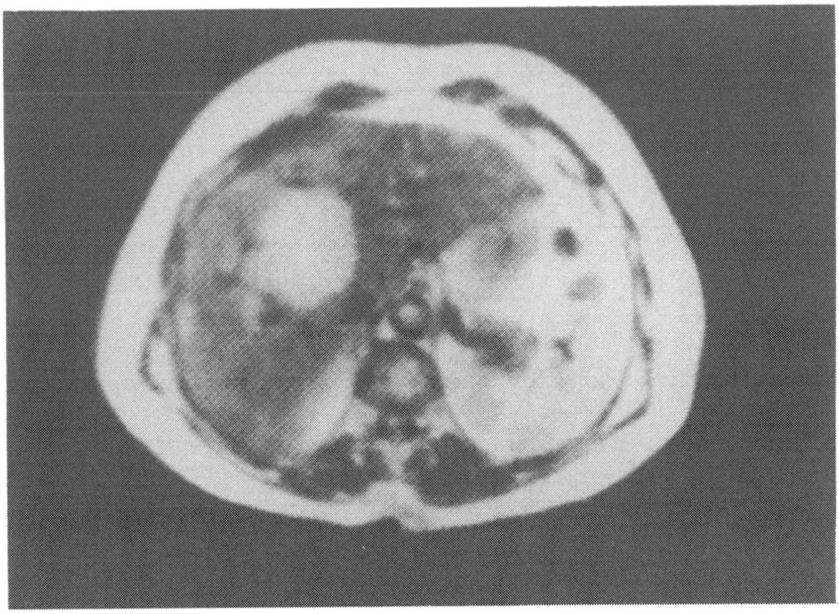

Figure 1. (*Continued*). (5) Liver (dark) with metastatic tumor (light): spin echo image.

the externally applied gradients. Lithium has important clinical uses, and can be imaged also.

The expected gain in sensitivity with increase in field strength is offset by several disadvantages for NMR imaging. Increases in T_1 values may require longer scan times to achieve the same contrast:noise ratio, although T_2 is less field dependent (cf. Figure 4, p. 143). Field homogeneity problems and chemical shift artifacts increase with the operating frequency. They may be reduced by higher field gradients, at the expense of the contrast. Higher rf frequencies increase the eddy currents in conducting phases, increasing the associated magnetic fields which reduce the homogeneity, and also increase the heat dissipation in the sample (as ω^2). A remedy is to use a circularly rather than a linearly polarized rf field, as this halves the power. Not least are the greater problems of expense and safety at higher fields.

For all these reasons small to medium fields are commonly used for whole-body clinical examination, say 0.07–2 T for imaging and 1.5–2 or more T for spectroscopy, reflecting technical and economic restraints. For whole-body work, with horizontal bore up to 1 m, air-cored resistive magnets cover the low-field range up to about 0.2 T, and superconducting magnets up to 2 T or so. For research purposes, and for smaller volumes (head, limbs, small creatures) higher fields, to 8 T or more, may be used for imaging and spectroscopy. Greater flexibility is achieved with magnets in which the field can be varied.

3. LOCALIZED NMR SPECTROSCOPY

For reasonable resolution and sensitivity spectroscopy needs higher fields and better homogeneity than imaging does, and may be based on a sufficiently wide-bore conventional high resolution NMR spectrometer. The technical problems are far greater than for imaging; at higher fields, also, lines may be broadened by inhomogeneity in magnetic susceptibility of the tissue, and by shielding anisotropy, shortening T_2. Thus in the short term a major use of the spectroscopy may be as an aid to the interpretation of NMR images, although "localized biochemistry" is developing rapidly, with specimens that are reasonably amenable to NMR study. Small organisms and plants[29] are increasingly being examined *in vivo*.

A notable success has been in spectroscopic studies of phosphorus metabolism (cf. also Chapter 22, p. 597ff). Phosphorus in DNA and most phospholipids is not readily observed *in vivo*, but the more nimble and abundant phosphates of energy metabolism, ATP, phosphocreatine, sugar phosphate, inorganic phosphate, etc., can be monitored quantitatively, as shown in Figure 2. These have a range of 25 ppm or more in chemical shift. The information content of a ^{31}P spectrum is potentially large.[30] The inorganic phosphate shift gives the local pH, and in favorable circumstances, splitting of the line reflects intra- and extracellular values, so that pH gradients across membranes can be studied.[31] Changes in health and disease, in exercise and at rest, can now be followed biochemically in living muscle,[32] as can brain development in infants,[4] in a way that would have astonished physiologists and clinicians of the 1970s.

The use of other nuclei is steadily increasing. Proton spectroscopy may be highly informative, although there is a problem of dynamic range in view of the small chemical shift span. In aqueous phases metabolites of interest are millimolar or less in concentration whereas water is about 100 molar, a disparity of 10^5 or so. Various devices are used, singly or in combination, to suppress the water signal, whether in biotissue or in body fluids (urine, plasma, saliva, cerebrospinal fluid, etc.) which are increasingly being studied by NMR methods. Selective saturation, paramagnetic ions which interact more with the water than with the organic material, fast exchange with ammonium ion, are some of the methods used. The proton signal from fat can be suppressed by T_2 effects. Deuterium resonance is of use in tracer and quadrupolar relaxation studies.

Natural abundance ^{13}C NMR spectroscopy is possible *in vivo* on compounds such as fats and glycogen which are present in high concentration. The potential for enrichment-tracer studies is obvious. Proton decoupling to enhance sensitivity is possible, but must avoid rf overheating of the tissue; alternatively, carbon may be monitored in proton resonance via the J coupling. ^{15}N is of value for organisms which can be enriched by a diet of (say) $^{15}NH_4Cl$. Biotissue is readily enriched with ^{17}O by the use of $H_2{}^{17}O$; since the ^{17}O relaxation is faster than transport processes in general, it affords another method for studying tissue water. ^{19}F is used to monitor fluoro anaesthetics or tranquillizers in the brain, antitumor agents, etc.; shielding anisotropy broadening may be a problem with large molecules. Quadrupolar relaxation studies with ^{23}Na yield information on muscular action, oedema, etc. Although less sensitive, ^{39}K can be used to follow potassium transport, and the linewidth again reflects the degree of mobility. Biomedical applications of all of these nuclei have been detailed[1,3,7,10,13–21] and Chapter 22 describes work with ^{23}Na, ^{113}Cd, and ^{2}H.

4. FURTHER APPLICATIONS

Many applications are evident in the description so far. Clearly, NMR now offers a method for the direct study of body water, or fats. As shown in Figure 1, imaging

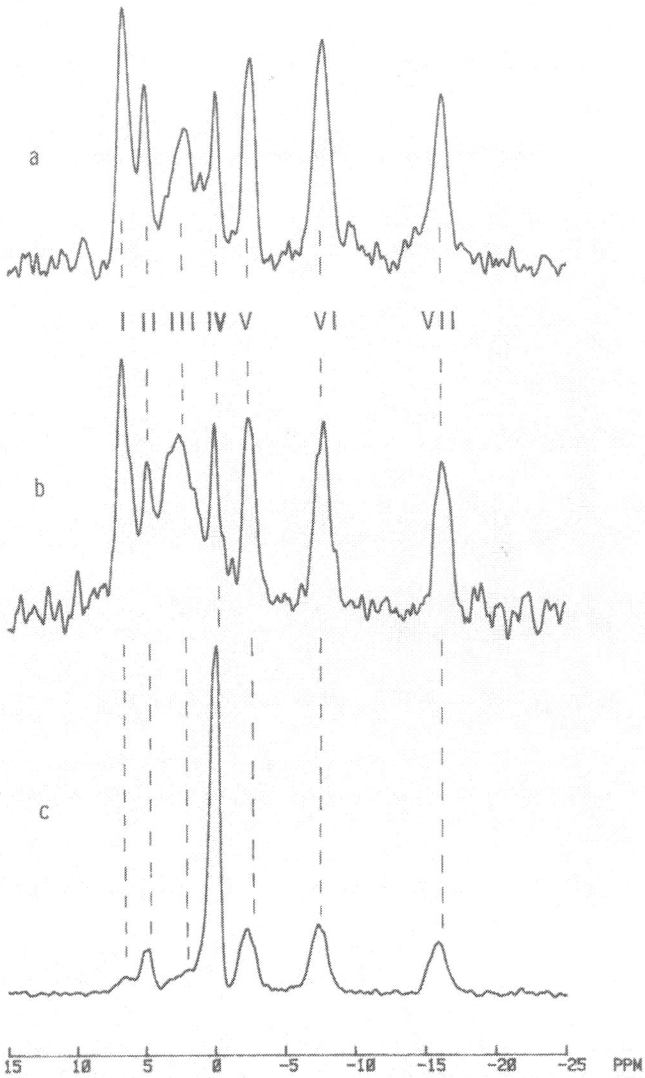

Figure 2. ^{31}P spectra from human tissues *in vivo*, measured by E. B. Cady and colleagues in the Department of Medical Physics and Bio-Engineering in the School of Medicine, University College London, U.K.; obtained with single turn surface coils, with a 90° flip angle at the coil center, in the acquisition times given (cf. Chapter 7 in Ref. 4). (a) From a malignant tumor (a rhabdomyosarcoma) on the back of the hand, 4 cm planar coil, 17 min, 512 pulses. (b) From a normal brain of a new-born baby, 7.4 cm planar coil positioned at the side of the cranium just above the ear, 38.5 min, 1024 pulses. (c) From normal resting forearm muscle, 6 cm contoured coil, 19.2 min, 512 pulses. Likely assignments are: I, In (a) and (b), phosphomonoesters; phosphoethanolamine has been identified as a metabolite in neuroblastoma, and is present together with phosphorylcholine in the brain. In (c), mainly phosphates. II, Inorganic phosphate. III, Phosphodiesters (probably several). IV, Phosphocreatine. V, Mainly ATP (γ-adenosine triphosphate), often with some ADP (adenosine diphosphate). VI, Mainly α-ATP, with some ADP, also NAD and NADH (nicotinamide adenine dinucleotides). VII, Probably mainly β-ATP. Peaks V, VI, and VII in (a) and (b) may contain contributions from other nucleotide triphosphates. In (c) they are mainly due to ATP with minimal amounts of other nucleotide triphosphates, but small amounts of ADP and NAD + NADH are present also. The ATP in all these spectra is probably complexed to Mg^{2+}.

The phospholipid signal is about 40–50 ppm broad, owing to the low mobility (short T_2), and has been removed to facilitate the measurement of peak areas and positions. It is of course more prominent in the brain than in the other tissues.

can locate tumors, including some unrevealed by x-ray CT, greatly aiding surgical or other therapy. In favorable cases, benign and malignant can be differentiated; on the other hand, tumors involving calcification are better imaged by x-radiation. NMR imaging can also distinguish other pathological conditions, from inflammation to degenerative diseases. In multiple sclerosis, for example, the major nerves of the brain and spinal cord lose their protective myelin (fatty) sheath, with increase in water content, so that T_1 and T_2 increase. In muscular dystrophy the fatty infiltration of muscle may be recognized. The beating heart can be examined with gating by the patient's ECG to give an image at any phase of the cycle,[3,5] and breathing synchronization methods are being developed for thorax studies. Alternatively the motion artifacts (ghosts) may be reduced by processing of the collected data; but irregular motions in the lungs or the gastrointestinal tract do degrade the image. Liver imaging can show the progress of hepatitis, cirrhosis, fatty infiltration, tumors (as in Figure 1), cysts, or abscesses: one kind of image can be subtracted from another (e.g., inversion recovery from proton density) to increase contrast.[3] As costs come down and sensitivity improves NMR will increasingly replace the invasive methods of biopsy, angiography, myelography, and the like. NMR methods may be used to monitor the metabolism of kidneys (etc.) for transplantation, and the result of the transplanting. NMR microscopy is under development.[14] The study of body tissues or fluids can monitor drug metabolism in treatment or overdose. ^{31}P spectroscopy is being extended to the brain and other organs, as shown in Figure 2.

Nonmedical applications of NMR imaging and spectroscopy include the study of interstitial or bound water in food products (as used to increase the return on delicatessen), oil in seeds, etc. There are parallel applications for abiological composite materials: fluids in porous (e.g., oil-bearing) rocks, plasticized resins, or rubber, in which the skeletal protons can be imaged, as in fats. But the field is developing so fast it is not easy to predict the new directions of advance (particularly with the advent of new superconductors that will not need liquid helium temperatures).

REFERENCES

1. (a) Foster, M. A.; Dodd, N. J. F.; Hutchison, J. M. S.; Smith, F. W. *Magnetic Resonance in Medicine and Biology*; Pergamon: Oxford, 1984. (b) Axenrod, T.; Ceccarelli, G., Eds., *NMR in Living Systems, NATO ASI Ser., Ser. C*, Vol. 164; Reidel; Dordrecht, 1986. (c) Morris, P. G. *NMR of Living Systems*, annual chapter in *NMR Spec. Per. Rep. Roy. Soc. Chem.*, 1984–1986, Vols. 13–15, *et seq.* (d) Bradbury, E. M.; Nicolini, C., Eds., *NMR in the Life Sciences, NATO ASI Ser. A*, Vol. 107, Plenum: New York, 1986.
2. Govil, G.; Khetrapal, C. L.; Saran, A., Eds. *Magnetic Resonance in Biology and Medicine*; Tata McGraw-Hill: Delhi, 1985.
3. Steiner, R. E.; Radda, G. K., Eds. *Nuclear Magnetic Resonance and its Clinical Applications; Br. Med. Bull.* 1984, **40**, 113–206; Roth, K. *NMR Tomography and Spectroscopy in Medicine. An Introduction*; Springer-Verlag: Berlin, 1984, 1–128.
4. Lerski, R. A., Ed. *Physical Principles and Clinical Applications of NMR*; Hospital Physicists' Association: London, 1985.
5. James, T. L.; Margulis, A. R., Eds. *Biomedical Magnetic Resonance*; Radiology Research and Education Foundation: San Francisco, 1984.
6. Morris, P. G. *NMR Imaging in Medicine and Biology*; Clarendon: Oxford, 1986.
7. Gadian, D. G. *NMR and its Application to Living Systems*; Clarendon: Oxford, 1982; Williams, S. R.; Gadian, D. G. "Tissue Metabolism Studied *in vivo* by NMR," *Quart. J. Exp. Physiol.* 1986, **71**, 335–360.
8. Gordon, R. E. *Magnets, Molecules and Medicine; Phys. Med. Biol.* 1985, **30**, 741–770.

9. Young, S. W. *NMR Imaging: Basic Principles*; Raven: New York, 1984.
10. Williams, R. J. P.; Andrew, E. R.; Radda, G. K., Eds. "NMR of Intact Biological Systems," *Phil. Trans. R. Soc. Lond., Ser. B* 1980, **289**, 379–553.
11. Newton, T. H.; Potts, D. G., Eds. *Advanced Imaging Techniques*; Vol. 2 in Modern Neuroradiology Ser. Clavadel: San Anselmo, California, 1983.
12. Gordon, R. E.; Hanley, P. E.; Shaw, D. "Topical Magnetic Resonance," *Prog. NMR Spectrosc.* 1982, **15**, 1–47.
13. Iles, R. A.; Stevens, A. N.; Griffiths, J. R. "NMR Studies of Metabolites in Living Tissue," *Prog. NMR Spectrosc.* 1982, **15**, 49–200.
14. Mansfield, P.; Morris, P. G. *NMR Imaging in Biomedicine*; Academic: New York, 1982.
15. Bailes, D. R.; Bryant, D. J. "NMR Imaging," *Contemp. Phys.* 1984, **25**, 441–475.
16. Kaufman, L.; Crooks, L. E.; Margulis, A. R., Eds. *NMR Imaging in Medicine*; Igaku-Shoin: New York-Tokyo, 1981.
17. Cohen, J. S., Ed. *Medical Imaging by NMR*; Wiley-Interscience: New York, Vol. 1, 1980 *et seq.*
18. Partain, C. L.; James, A. E.; Rollo, F. D.; Price, R. R., Eds. *NMR Imaging*; Saunders: Philadelphia, 1983.
19. Andrew, E. R. "NMR Imaging," *Acc. Chem. Res.* 1983, **16**, 114–122.
20. Bottomley, P. A. *Rev. Sci. Instrum.* 1982, **53**, 1319–1337.
21. Shulman, R. G., Ed. *Biological Applications of Magnetic Resonance*; Academic: New York, 1979; "NMR Spectroscopy of Living Cells," *Sci. Am.* 1983, **248**, 76–83.
22. Pykett, I. L. "NMR Imaging in Medicine," *Sci. Am.* 1982, **246**, 54–64.
23. Lauterbur, P. C. *Nature* 1973, **242**, 190–191.
24. "Guidelines for Evaluating Electromagnetic Exposure Risk for Trials of Clinical NMR Systems," U.S. Bureau of Radiological Health: Washington DC, 1982; "Exposure to NMR Clinical Imaging," National Radiological Protection Board: Harwell, Didcot, UK, 1982, "Advice on acceptable limits of exposure to NMR clinical imaging," *Radiography*, 1984, **50**, 220.
25. Gabillard, C. R. *C.R. Acad. Sci. (Paris)* 1951, **232**, 1551–1553; *Phys. Rev.* 1952, **85**, 694–695.
26. Kumar, A.; Welti, D.; Ernst, R. R. *J. Magn. Reson.* 1975, **18**, 69–83.
27. Damadian, R. *Science* 1971, **171**, 1151–1153.
28. Clark, L. C.; Ackerman, J. L.; Thomas, S. R.; Millard, R. W.; Hoffman, R. E.; Pratt, R. G.; Ragle-Cole, H.; Kinsey, R. A.; Janakiraman, R. *Adv. Exp. Med. Med. Biol.* 1984, **180**, 835–845.
29. Roberts, G. K. M. "Study of Plant Metabolism *in vivo* using NMR Spectroscopy," *Appl. Spectrosc. Rev.* 1984, **20**, 375.
30. e.g. Radda, G. K. *Biochem. Soc. Trans.* 1986, **14**, 517–525; Burt, C. T.; Roberts, M. F. "^{31}P NMR Observation of Less-Expected Phosphorus Metabolites," Chap. 15 in Ref. 5; Gorenstein, D. G., Ed. *Phosphorus-*31 *NMR*; *Principles and Applications*; Academic Press: Orlando, 1984, 1–604.
31. Feeney, J. "NMR Methods for Measuring Cation Concentrations in Living Systems, pp. 279–289 in Ref. 1(b).
32. Radda, G. K.; Taylor, D. J. *Int. Rev. Exptl. Path.* 1985, **27**, 1–58.

Many journals and review series carry articles on *in vivo* NMR: *Magnetic Resonance in Medicine* (Academic Press), *Magnetic Resonance Imaging* (Pergamon), the *Journal of Magnetic Resonance* (Pergamon), *NMR Imaging in Medicine* (Igaku-Shoin), *Reviews of Magnetic Resonance in Medicine* (Pergamon), *Magnetic Resonance in Biology* (Wiley), *Biological Magnetic Resonance* (Plenum), *Proceedings of the U.S. National Academy of Science*, etc.; as well as medical journals, particularly in the fields of computerized tomography, radiology, surgery, cancer and neurology.

SYMBOLS AND ABBREVIATIONS

ROMAN SYMBOLS

a_N	Electron (hyperfine) interaction constant with nucleus N
A_N	Natural abundance of nucleus N/%
$\mathbf{B}$	Magnetic field strength (flux density) in tesla: $1\mathrm{T} = 10^4$ gauss
$\mathbf{B}_0$	Static magnetic field strength of NMR spectrometer
$\mathbf{B}_1, \mathbf{B}_2$	Magnetic field strength of radiofrequencies applied to the sample; $\mathbf{B}_1$, observing field; $\mathbf{B}_2$, perturbing field
$\mathbf{C}_N$	Spin-rotation tensor of nucleus N
$C_\parallel, C_\perp$	Spin-rotation components for axial symmetry
$\mathbf{D}$	Direct dipolar coupling tensor
D_{IS}	(Mutual) translational diffusion constant for spins I and S
e	Elementary charge
e^2qQ/h	Nuclear quadrupole coupling constant (NQCC) in frequency units
g_e	Electronic g factor (Landé splitting factor)
g_N	Nuclear g factor $(=\mu_N/I_N)$
$G(\tau)$	$=\overline{f(t)\,f(t+\tau)}=$ autocorrelation function of $f(t)$
h	Planck's constant
$\hbar$	$h/2\pi$
$\hat{\mathscr{H}}$	Hamiltonian operator
i	$\sqrt{-1}$
$\mathbf{I}$	Nuclear spin angular momentum
I_N	Spin quantum number of nucleus N
I	Moment of inertia
$\mathbf{J}$	Spin spin (scalar, indirect) coupling tensor
$^nJ_{\mathrm{AB}}$, $^nJ(\mathrm{AB})$	Spin–spin coupling constant over n bonds between nuclei A and B; $^3J(\mathrm{ABCD})$ shows a coupling path
J^{Fc}	Fermi contact contribution to J
J^{sd}	Spin-dipolar contribution to J
J^o	Orbital contribution to J
$J(\omega)$	Spectral density at angular frequency ω
$^nK_{\mathrm{AB}}$	Reduced spin–spin coupling constant (as above) $$^nK_{\mathrm{AB}}=\frac{^nJ_{\mathrm{AB}}}{h}\left(\frac{2\pi}{\gamma_{\mathrm{A}}}\right)\left(\frac{2\pi}{\gamma_{\mathrm{B}}}\right)\Big/ N\,\mathrm{A}^{-2}\,\mathrm{m}^{-3}$$
L	Line-broadening factor for quadrupolar nucleus N $L=f(I_N)\,Q_N^2$, where $f(I_N)=(2I_N+3)/I_N^2(2I_N-1)$
m_e	Electron mass
m_p	Proton mass
M_I	Magnetic quantum number
$\mathbf{M}$	Bulk magnetization of nuclear spins

q or eq	Electric field gradient (efg)
Q or eQ	Nuclear electric quadrupole moment
S	1. Another spin interacting with I 2. Order parameter in partially oriented systems
T	Temperature
T	Tesla: $1\ \mathrm{T} = 10^4$ gauss
t_p	Pulse width (length, duration)
T_1	Longitudinal (spin-lattice) relaxation time
$T_{1\rho}$	T_1 in frame of reference rotating with B_1
T_2	Transverse (spin–spin) relaxation time
T_2^*	Observed value of T_2 $(=1/\pi W_{1/2})$, including effects of magnetic field inhomogeneity
T_q	Quadrupolar relaxation time
T_p	Pulse interval
u	In-phase (dispersion mode) signal
v	Out-of-phase (absorption mode) signal
$W_{1/2}$	Linewidth at half-height
W_I	Transition probability of nucleus I

GREEK SYMBOLS

α_0	Pulse angle (flip angle)
α, β	Spin states $(\pm 1/2)$
γ	Magnetogyric ratio of nucleus N/rad $\mathrm{T}^{-1}\ \mathrm{s}^{-1}$
$\bar{\gamma}$	$\gamma/2\pi$
γ_∞	Sternheimer antishielding factor
δ	Chemical shift/ppm
Δ	Difference (e.g., $\Delta\delta$)
$\Delta\sigma$	Shielding anisotropy $= \sigma_\parallel - \sigma_\perp$ for axial symmetry
$\Delta\chi$	Magnetic susceptibility anisotropy $= \chi_\parallel - \chi_\perp$ for axial symmetry
ε_0	Permittivity of a vacuum
ζ	Spin orbit coupling constant
η	1. NOE factor 2. Asymmetry of shielding tensor $= (\sigma_{22} - \sigma_{11})/(\sigma_{33} - \sigma_{av})$ 3. Asymmetry of efg tensor $= (q_{xx} - q_{yy})/q_{zz}$, where $\lvert q_{zz}\rvert \geqslant \lvert q_{yy}\rvert \geqslant \lvert q_{xx}\rvert$ 4. Viscosity
$\boldsymbol{\mu}$	Magnetic moment
μ_0	Permeability of a vacuum
μ_B	Bohr magneton
μ_N	Nuclear magneton
ν_0	1. Operating frequency of spectrometer 2. Larmor frequency/Hz
ν_1	Frequency of observing rf field B_1 in Hz
ν_2	Frequency of perturbing rf field B_2 in Hz
Ξ_X	Resonance frequency of standard substance for nucleus X at a magnetic field giving a resonance frequency of precisely 100 MHz for TMS protons under standard conditions
$\boldsymbol{\sigma}$	Nuclear magnetic shielding tensor
$\sigma_\parallel$, $\sigma_\perp$	Shielding tensor components for axial symmetry, otherwise σ_{11}, σ_{22}, σ_{33}, where $\sigma_{33} \geqslant \sigma_{22} \geqslant \sigma_{11}$

τ	Evolution time between rf pulses
τ_c	Correlation time for molecular tumbling
τ_{sr}	Correlation time for spin-rotation relaxation
χ	1. Magnetic susceptibility
	2. Nuclear quadrupole coupling constant $= e^2qQ/h$ (for χ in Hz)
χ_v	Volume magnetic susceptibility
ω_0	Larmor frequency/rad $s^{-1} = 2\pi v_0$
ω_1, ω_2	Angular frequencies of B_1, B_2 rf fields in rad s^{-1}

ABBREVIATIONS

$A - \{M, X\}$	Observation of nucleus A with irradiation also of nuclei M, X
ASIS	Aromatic solvent-induced shifts
CIDNP	Chemically induced dynamic nuclear polarisation
COSY	Correlated two-dimensional (2D) NMR spectroscopy
CSA	Chemical shift anisotropy – SA
CP	Cross-polarization
CW	Continuous wave
DD, dd	Direct dipolar (or dipole–dipole) interaction
DEFT	Driven equilibrium FT
DEPT	Distortionless enhancement by polarization transfer
ENDOR	Electron–nuclear double resonance
efg	Electric field gradient
ESR	Electron spin resonance
FID	Free induction decay
FT	Fourier transform
HOMO	High-lying occupied molecular orbital
INDOR	Internuclear double resonance
INEPT	Insensitive nucleus enhancement by polarization transfer
JCP	J cross-polarization
LUMO	Low-lying unoccupied molecular orbital
MAS	Magic angle spinning
MO	Molecular orbital
NMR	Nuclear magnetic resonance
NOE	Nuclear Overhauser effect
NOESY	Two-dimensional NOE spectroscopy
NQCC	Nuclear quadrupole coupling constant
NQR	Nuclear quadrupole resonance
ppm	Parts per million
ppt	Parts per thousand
rf	Radiofrequency
SA, sa	Shielding anisotropy
SECSY	Spin echo correlated 2D NMR spectroscopy
S:N	Signal-to-noise ratio
SPI	Selective population inversion
SPT	Selective population transfer
SR, sr	Spin-rotation interaction
TMS	Tetramethylsilane
WAHUHA	Pulse sequence of Waugh, Huber, and Haeberlen

SI UNITS AND FUNDAMENTAL CONSTANTS

These tables use the recommended SI (Système International) units, with some indication of their relation to other units in the literature, including electromagnetic units (e.m.u.) and the less common electrostatic units (e.s.u.), both in the centimeter gram second system (CGS).[1,2] The relationships are complex: four systems of equations have been used, and each of these has been written in nonrationalized and now rationalized forms.[3] With rationalization, explicit values and dimensions are given to the permittivity ε_0 of a vacuum, which was taken as unity and dimensionless in the e.m.u. system, and to the permeability μ_0, taken as unity and dimensionless in the e.s.u. system. (These two systems are mutually inconsistent: on the Maxwell theory the product $\varepsilon_0\mu_0$ is equal to c^{-2}, where c is the speed of light, approximately $3 \cdot 10^8$ m s^{-1}.) The physical relationships of the different systems of electrical and magnetic units have been ably described.[4]

The SI uses the unit of current (ampere) as fundamental unit (whereas for e.m.u. it is a unit magnetic pole, with $\mu_0 = 1$, and for e.s.u. a unit charge, with $\varepsilon_0 = 1$). It is a pity that what NMR spectroscopists call the magnetic field, B/tesla, is officially named the magnetic flux density, whereas the magnetic force (H/A m^{-1}) has been named the magnetic field intensity. The two are related by $H = B/\mu$, where μ/NA^{-2} is the permeability of the medium.

Although internationally agreed upon in 1960, the SI and the rationalizations are still not fully implemented; new editions or reprints of standard works in NMR, and even some current work, still use the old systems. Crooks's conclusion[4] is particularly relevant (now that we have the tesla as well!): "Many authors, especially in the older literature, skip lightly from e.m.u. to e.s.u., from gauss to oersted, with an agility born of long experience and deep understanding. Students who attempt to follow their reasoning must do so with care, or they will stumble into confusion." Such caution is needed with some of the "other units" given in the Tables. Far better is consistent use of the SI.

Unfortunately there is confusion also as to units and symbols that may be used with the SI. Although base units such as the kilogram and the meter have been used in the tables, all the decimal multiples (g; cm, mm, nm, pm, etc.) are also part of the SI, and may be used as convenient. Units recognized for continued use with the SI include the electron volt eV, a degree of angle, the minute, hour, and day. Names which are still in use for decimal multiples of SI units, although their continued use is discouraged, include the angstrom, barn, dyne, bar, erg, and poise. Definitions of these quantities in SI units are included in the tables.

The Fundamental Constants are the revised values of 1986.[5]

UNITS

Physical quantity	Symbol	SI Unit[a]	Symbol/definition	Other units	
Length, distance	l, r	meter	m	1 Å = 10 pm	
Time	t	second	s		
Plane angle	α, θ, ϕ	degree radian	° rad = 360°/2π	 1/2π cycles	
Frequency	ν, f	hertz	Hz = s^{-1}	1 cps or c/s (cycle per second)	
Angular frequency	$\omega = 2\pi\nu$	radian	rad		
Mass	m	kilogram	kg		
Reduced mass	μ	kg	$\mu = m_1 m_2/(m_1 + m_2)$		
Momentum	$\mathbf{p} = m\mathbf{u}$	kg m s^{-1} = N s			
Velocity	$\mathbf{u}$	m s^{-1}			
Angular momentum	$\mathbf{L} = \mathbf{r} \times \mathbf{p}$	kg m^2 s^{-1} = J s			
Wavelength	$\lambda = c/\nu$	meter	m		
Period, relaxation time, correlation time	$T, \tau = 1/\nu$	second	s		
Energy	U	joule	J = kg m^2 s^{-2}	1 erg = 10^{-7} J 1 NBS cal = 4.184 J 1 Int. cal = 4.186 J 1 eV = 1.602 19 10^{-19} J	
Entropy	S	J kg^{-1} K^{-1}		1 e.u. = cal g^{-1} $°C^{-1}$	
Temperature	T	kelvin	K	deg or °C (Celsius) 0 °C = 273.15 K	
Force	F	newton	N = kg m s^{-2} = J m^{-1}	1 dyne = 10^{-5} N	
Pressure	P, p	pascal	Pa = kg m^{-1} s^{-2} = N m^{-2} = J m^{-3}	1 atm = 101 325 Pa 1 torr = 1 atm/760 ≃ 1 mmHg 1 bar = 10^5 Pa	
Viscosity	η	poise	p = 0.1 Pa s		
				e.m.u./SI	*e.s.u./e.m.u.*
Electric current	I	ampere	A = C s^{-1}	10	1/c
Electric charge	Q, q	coulomb	C = A s	10	1/c
Electric potential difference	V	volt	V = kg m^2 s^{-3} A^{-1} = J A^{-1} s^{-1} = J C^{-1}	10^{-8}	c
Power	W	watt	W = kg m^2 s^{-3} = J s^{-1}	(1 W = 10^7 erg s^{-1})	
Electric field strength	E		V m^{-1}	10^{-6}	c
Magnetic flux	Φ	weber	Wb = V s	10^{-8} maxwell	c
Magnetic flux density (induction field)	B	tesla	T = V s m^{-2} = kg A^{-1} s^{-2}	1 G = 10^{-4}T gauss	c
Magnetic field intensity	H	ampere m^{-1}	A m^{-1}	1 Oe = $10^3/4\pi$ oersted	1/c
Self inductance	L	henry	H = Wb A^{-1}	10^{-9}	c^2
Magnetization	$M = (B/\mu_0) - H$		A m^{-1}	10^3	
Magnetic dipole moment	μ	ampere m^2	A m^2	10^{-3}	
Magnetogyric ratio	$\gamma = \mu/I\hbar$		rad T^{-1} s^{-1}	10^{-4} rad G^{-1} s^{-1}	
Nuclear electric quadrupole moment	Q	(barn)	10^{-28} m^2		
Reduced spin-spin coupling constant	$K_{NN'} = 4\pi^2 J_{NN'}/h\gamma_N\gamma_{N'}$		N A^{-2} m^{-3}	10 cm^{-3}	

[a] SI-prefixed decimal multiples and fractions are acceptable also (despite widespread belief to the contrary).

SI PREFIXES AND THEIR SYMBOLS

Multiple	Prefix	Symbol	Multiple	Prefix	Symbol
10^{-1}	deci	d	10	deca	da
10^{-2}	centi	c	10^{2}	hecto	h
10^{-3}	milli	m	10^{3}	kilo	k
10^{-6}	micro	μ	10^{6}	mega	M
10^{-9}	nano	n	10^{9}	giga	G
10^{-12}	pico	p	10^{12}	tera	T
10^{-15}	femto	f			
10^{-18}	atto	a			

$\pi = 3.141\ 5926$ $\quad$ $e = 2.718\ 282$ $\quad$ $\ln x = 2.302\ 585 \log x$

FUNDAMENTAL CONSTANTS

Constant	Symbol	Value	Multiplier and units: SI	Multiplier and units: CGS
Vacuum speed of light	c	2.997 925	10^{8} m s^{-1}	10^{10} cm s^{-1}
Vacuum permeability	μ_0	4π	10^{-7} J s^2 C^{-2} m^{-1}	
		1		e.m.u.
Vacuum permittivity	$\varepsilon_0 = 10^7/4\pi c^2$	8.854 188	10^{-12} J^{-1} C^2 m^{-1}	
		1		e.s.u.
Nuclear magneton	$\mu_N = e\hbar/2m_p$	5.050 787	10^{-27} J T^{-1}	10^{-24} erg G^{-1} (e.m.u.)
Bohr magneton	$\mu_B = e\hbar/2m_e$	9.274 015	10^{-24} J T^{-1}	10^{-21} erg G^{-1}
Magnetogyric ratio of free proton (proton in H_2O, with correction for diamagnetism)	$\gamma_p = \mu_p/I\hbar$	2.675 221	10^{8} rad s^{-1} T^{-1}	10^{4} rad s^{-1} G^{-1}
Free electron Landé factor	g_e	2.002 319		
Elementary charge	e	1.602 177	10^{-19} C	10^{-20} e.m.u.
		4.803 207		10^{-10} e.s.u.
Avogadro constant	N_A	6.022 137	10^{23} mol^{-1}	
Proton mass	m_p	1.672 623	10^{-27} kg	10^{-24} g
Electron mass	m_e	9.109 390	10^{-31} kg	10^{-28} g
Bohr radius	a_0	5.291 772	10^{-11} m	10^{-9} cm
Boltzmann constant	k	1.380 658	10^{-23} J K^{-1}	10^{-16} erg deg^{-1}
Gas constant	$R = N_A k$	8.314 51	J K^{-1} mol^{-1}	10^{7} erg deg^{-1} mol^{-1}
Planck constant	h	6.626 075	10^{-34} J s	10^{-27} erg s
	$\hbar = h/2\pi$	1.054 573	10^{-34} J s	10^{-27} erg s

REFERENCES

1. *Quantities, Units, and Symbols*; The Royal Society; London, 1975 (with 1981 addenda). A new edition will contain the revised values of the constants in Ref. 5.
2. McGlashan, M. L. *Physicochemical Quantities and Units*; *The Grammar and Spelling of Physical Chemistry*; 2nd ed., RIC; London, 1971.
3. Chapter 12 in Ref. 2.
4. Crooks, J. E. *The Spectrum in Chemistry*, Academic Press; London, 1978, pp. 280–283.
5. Taylor, B. N.; Cohen, E. R. *Recommended Values of the Fundamental Physical Constants*, 1986, CODATA Bulletin no. 64.

Appendix

NMR PROPERTIES OF THE ELEMENTS

The table includes some radioactive nuclei of reasonably long half-life ($t_{1/2}$) so as to cover as many of the elements as possible. Mass numbers (A) in parentheses are those of nuclei which are less NMR favorable than another of the same element.

The values of the magnetogyric ratio γ are based in the main on the nuclear magnetic moments μ_{max} ($=\hbar\gamma I/\mu_N$, where μ_N is the nuclear magneton) given by Lederer and Shirley,[1] with some from more recent literature. The ($\mu_N/\hbar$) conversion factor is given by $\gamma I = 4.789\,418 \times 10^7\,\mu$ for γ in rad s^{-1} T^{-1} and μ in nuclear magnetons, as tabulated[1,2] (cf. $\gamma I = 4.789\,415 \times 10^7\,\mu$ with the new values of the fundamental constants given in the Table, p. 621). The μ values[1] differ systematically from the ones given by Fuller,[2] used in other compilations,[3,4] in the introduction of a new value for the proton moment[5,6] (on which many other values are based), and also of relativistic values for the diamagnetic correction factors which give the magnetic moment for the bare nucleus.[1] The differences in the values of γ (and those derived for the receptivity R and the standard frequency Ξ) increase, therefore, with atomic mass. Values of μ and γ are negative when the magnetic moment is antiparallel to the nuclear spin I; a consequence is that NOE factors are negative.

The receptivity of a nucleus affords an estimate of the strength of signal to be expected. The receptivity R at constant field and for equal numbers of nuclei is the product of the natural abundance (N.A./%) and the NMR sensitivity, taken as $|\gamma^3|\, I(I+1)$ for that nucleus. The table gives values of R^C, the receptivity relative to that of ^{13}C. The low receptivity of low-frequency nuclei is well known; the increase in receptivity for nuclei with large I values (e.g., by a factor of 33, relative to $I = 1/2$ nuclei, for $I = 9/2$ nuclei such as ^{73}Ge or ^{93}Nb) is advantageous, if quadrupolar broadening is not excessive.

Precise values are given, if available, for the resonance frequency (Ξ) of a specified standard substance in a magnetic field (2.3488 T) such that the TMS protons resonate at exactly 100 MHz. (These and related values may be referenced in the appropriate chapter.) Otherwise the Ξ values given to three decimal places have been obtained from the γ value relative to that of the proton, and so refer to the bare nucleus. (Experimental values of Ξ to three decimal places, and linewidths, are given for samples of all observable nuclei in Brévard and Granger's handbook.[4]) Values of Ξ marked with an asterisk are rounded values as recommended by Goodfellow in Chapters 20 and 21, in view of referencing problems for nuclei with large shift ranges (cf. p. 533). This procedure could usefully be adopted more widely.

Standard solutions are not specified precisely if the Ξ value is not an experimental one. In practice, ionic standard solutions are commonly used in 1 M concentration in deuterated water, preferably with a counterion that has little tendency to associate

(ClO_4^-, NO_3^-, Cs^+, etc.). For more precise work extrapolation to infinite dilution (∞_{aq}) is often used.

Values of the nuclear electric quadrupole moment Q were obtained from Lederer and Shirley's Table,[1] and may be averages from parallel determinations; reference was made also to the Fuller Tables,[2] to Büttgenbach[7] for nuclei from the second and third transition series, and to the recent literature. The quadrupole moment is negative if the nuclear charge distribution is oblate. Q values may be subject to sizable uncertainties, by a factor of 50% or more, because of the difficulty of estimating the electric field gradient at the nucleus due to the electrons and other nuclei in the environment, and the effects of polarization of the core electrons. The latter effects, called Sternheimer shielding, increase sharply and periodically with atomic number.[8] Many of the Q values have been corrected for Sternheimer or other polarization, uncorrected values being marked with a superscript u; measurements of direct nuclear interactions give the true quadrupole moment. The compilations and the original literature should be consulted for further information.

The width factor (WF) affords an estimate of the susceptibility of the nucleus to line-broadening due to too-fast quadrupolar relaxation. The relaxation rate and the linewidth being proportional to $Q^2(2I+3)/I^2(2I-1)$ for a given nucleus (cf. Chapter 2), this function is variously called the nuclear width function (WF) or IQ. Clearly it increases sharply with Q, but increase in I is again advantageous, the width function decreasing from $5Q^2$ for $I=1$ to $0.027Q^2$ for $I=7$. Other factors, additionally, influence quadrupolar linewidths. With covalent bonding, they depend on the nuclear electric field gradient and the molecular rotation correlation time. For ions such as the monatomic ions of groups 1, 2, and 7 they depend on the Sternheimer antishielding, and in Chapters 7 and 17 the width factor is defined to include a Sternheimer factor.

REFERENCES

1. Lederer, C. M.; Shirley, V. S. *Table of Isotopes*, 7th edn., Wiley: New York, 1978; Appendix VII.
2. Fuller, G. H. *J. Phys. Chem. Ref. Data* 1976, **5**, 835–864.
3. Harris, R. K.; Mann, B. E., Eds.: *NMR and the Periodic Table*; Academic: London, 1978, 459 pp.; Harris, R. K. *NMR Spectroscopy: A Physicochemical View*; Pitman: London, 1983; 250 pp.
4. Brévard, C.; Granger, P. *Handbook of High Resolution Multinuclear NMR*; Wiley: New York, 1981; 229 pp.
5. Phillips, W. D.; Cooke, W. E.; Kleppner, D. *Phys. Rev. Lett.* 1975, **35**, 1619–1622; **36**, 1473 (erratum).
6. Phillips, W. D.; Cooke, W. E.; Kleppner, D. in *Atomic Masses and Fundamental Constants 5*, Sanders, J. H.; Wapstra, A. H., Eds., Plenum: New York, 1976; p. 492; Taylor, B. N.; Cohen, E. R., *Recommended Values of the Fundamental Physical Constants*, CODATA Bulletin no. 64, 1986.
7. Buttgenbach, S. Hyperfine Structure in Łd- *and 5d-Shell Atoms*; Springer-Verlag: Berlin, 1982; pp. 81–84.
8. Lucken, E. A. C. *Nuclear Quadrupole Coupling Constants*; Academic: London, 1969, Chap. 5.

NMR Properties of the Elements[a,b]

	A	I	N.A. (%) [$t_{1/2}$]	R^C	Q (10^{-28} m^2)	γ (10^7 rad s^{-1} T^{-1})	Ξ (MHz)	Reference standard	WF (10^{-59} m^4)
Hydrogen	1	1/2	99.985	5.67×10^3	—	26.7522	100.000 000	SiMe$_4$(1) (TMS)	—
	2	1	0.015	8.2×10^{-3}	2.87×10^{-3}	4.1066	15.351		0.04
	3	1/2	[12 y]	—	—	28.5350	106.664		—
Helium	3	1/2	0.00014	3.51×10^{-3}	—	−20.3801	76.181	He	—
Lithium	6	1	7.42	3.58	-6.4×10^{-4}	3.9371	14.717	Li$^+$ aq	2.0×10^{-3}
	7	3/2	92.58	1.54×10^3	-3.7×10^{-2}	10.3976	38.866		1.8
Beryllium	9	3/2	100	78.8	5.3×10^{-2}	−3.7606	14.057	Be^{2+} aq	3.7
Boron	(10)	3	19.58	22.1	8.5×10^{-2}	2.8747	10.743 657	BF$_3$ · Et$_2$O, 15%	1.4
	11	3/2	80.42	7.54×10^2	$4.1 \times \times 10^{-2}$	8.5847	32.083 971	v/v in CDCl$_3$	2.2
Carbon	13	1/2	1.108	1.00	—	6.7283	25.145 004	SiMe$_4$(1)	—
Nitrogen	14	1	99.63	5.70	1.67×10^{-2}	1.9338	7.226 455	MeNO$_2$(1)	1.4
	15	1/2	0.37	2.19×10^{-2}	—	−2.7126	10.136 767		—
Oxygen	17	5/2	0.037	6.11×10^{-2}	-2.6×10^{-2}	−3.6280	13.562	H$_2$O(1)	0.22
							13.564 269, δ 569	acetone(1) 301K	
Fluorine	19	1/2	100	4.73×10^3	—	25.1815	94.094 003	CFCl$_3$(1)	—
Neon	21	3/2	0.257	3.59×10^{-2}	9×10^{-2}	−2.1130	7.898		11
Sodium	23	3/2	100	5.24×10^2	0.10	7.0704	26.429	Na$^+$ aq	13
Magnesium	25	5/2	10.13	1.54	0.22	−1.6389	6.126	Mg^{2+} aq	15
Aluminum	27	5/2	100	1.17×10^3	0.14	6.9762	26.077	Al^{3+} aq	6.3
Silicon	29	1/2	4.70	2.10	—	−5.3190	19.867 184	SiMe$_4$(1)	—
Phosphorus	31	1/2	100	3.77×10^2	—	10.8394	40.480 747	H$_3$PO$_4$, 85% aq	—
Sulfur	33	3/2	0.76	9.78×10^{-2}	-6.4×10^{-2}	2.0557	7.670 123	Cs$_2$SO$_4$ aq, 2M	5.5
Chlorine	35	3/2	75.53	20.2	-8.2×10^{-2}	2.6242	9.809	Cl$^-$ aq	9.0
	37	3/2	24.47	3.78	-6.5×10^{-2}	2.1844	8.165		5.6
Argon	39	7/2	[269 y]		( )	−1.78	6.6		
Potassium	39	3/2	93.1	2.69	5.5×10^{-2}	1.2499	4.672	K$^+$ aq	4.0
	(40)	4	0.012	3.56×10^{-3}	-6.3×10^{-2}	−1.5544	5.810		0.45
	(41)	3/2	6.88	3.29×10^{-2}	6.0×10^{-2}	0.6861	2.565		4.8

(continued)

NMR Properties of the Elements (*Continued*)

	A	I	N.A. (%) $[t_{1/2}]$	R^C	Q $(10^{-28}\ m^2)$	γ $(10^7\ rad\ s^{-1}\ T^{-1})$	Ξ (MHz)	Reference standard	WF $(10^{-59}\ m^4)$
Calcium	(41)	7/2	$[10^5$ y]	—	( )	−2.1823	8.157	Ca^{2+} aq	
	43	7/2	0.145	5.29×10^{-2}	-5×10^{-2}	−1.8028	6.739		0.34
Scandium	45	7/2	100	1.72×10^{3}	−0.22	6.5088	24.330	Sc^{3+} aq	6.6
Titanium	(47)	5/2	7.28	0.867	0.29	−1.5106	5.64276	$TiCl_4$(1)	27
	49	7/2	5.51	1.18	0.24	−1.5110	5.64426		7.8
Vanadium	(50)	6	0.24	0.760	0.21	2.6721	9.988	$VOCl_3$(1)	0.19
	51	7/2	99.76	2.17×10^{3}	-5.2×10^{-2u}	7.0492	26.350		0.37
Chromium	53	3/2	9.55	0.485	−0.15	−1.5077	5.636	CrO_4^{2-} aq	30
Manganese	55	5/2	100	1.014×10^{3}	0.40	6.6453	24.840	MnO_4^- aq	51
Iron	57	1/2	2.19	4.25×10^{-3}	—	0.8687	3.237 798	$[Fe(CO)_5]$(1)	—
Cobalt	59	7/2	100	1.56×10^{3}	0.42	6.3015	23.727 118	$K_3[Co(CN)_6]$, 0.1 *M* in D_2O	24
Nickel	61	3/2	1.19	0.242	0.16	−2.3948	23.6*, δ–5358 ppm 8.936 050 8.93*, δ–677 ppm	$[Ni(CO)_4]$(1)	34
Copper	63	3/2	69.09	3.68×10^{2}	−0.22	7.1088	26.515 473	$[Cu(NCMe)_4]\ BF_4$, 0.1 *M* in MeCN	65
	65	3/2	30.91	2.02×10^{2}	−0.195	7.6104	28.403 659		51
Zinc	67	5/2	4.11	0.671	0.15	1.6778	6.256 782	$Zn(ClO_4)_2$ ∞aq	7.2
Gallium	(69)	3/2	60.4	2.39×10^{2}	0.17	6.4389	24.069	Ga^{3+} aq	39
	71	3/2	39.6	3.22×10^{2}	0.11	8.1812	30.581		16
Germanium	73	9/2	7.76	0.622	−0.17	−0.9360	3.488 315	$GeMe_4$(1)	2.1
Arsenic	75	3/2	100	1.44×10^{2}	0.29^u	4.5961	17.180	AsF_6^-(MeCN)	112
Selenium	77	1/2	7.58	3.02	—	5.1214	19.071 523	$SeMe_2$(1)	—
Bromine	(79)	3/2	50.54	2.28×10^{2}	0.33	6.7256	25.140	Br^- aq	145
	81	3/2	49.46	2.79×10^{2}	0.27	7.2498	27.100		97
Krypton	83	9/2	11.55	1.24	0.27	−1.0331	3.862		5.4
Rubidium	85	5/2	72.15	43.4	0.27	2.5923	9.690	Rb^+ aq	23
	87	3/2	27.85	2.80×10^{2}	0.13	8.7851	32.839		23
Strontium	87	9/2	7.02	1.08	0.16	−1.1635	4.349	Sr^{2+} aq	1.9
Yttrium	89	1/2	100	0.676	—	−1.3163	4.920	Y^{3+} aq	—

Zirconium	91	5/2	11.23	6.05	-0.21^u	−2.4975	9.336	$[Zr(\eta\text{-}Cp)_2Br_2]$ (thf)	14
Niobium	93	9/2	100	2.77×10^3	−0.32	6.5674	24.549	$NbOCl_3(MeCN)$	7.6
Molybdenum	95	5/2	15.72	2.92	−0.015	−1.7514	6.547	MoO_4^{2-} aq, pH 11	0.15
	(97)	5/2	9.46	1.87	0.17	−1.7884	6.685		3.8
Technetium	99	9/2	[2×10^5 y]		-0.13^u	6.0503	22.508 311	TcO_4^- aq	1.25
Ruthenium	99	5/2	12.72	0.815	7.6×10^{-2u}	−1.2286	4.605 127	$K_4[Ru(CN)_6](D_2O)$	1.8
							4.6*, δ–1113 ppm	—	
	101	5/2	17.07	1.54	0.44^u	−1.3773	5.161 344	$K_4[Ru(CN)_6](D_2O)$	62
Rhodium	103	1/2	100	0.180	—	−0.8468	3.16*	—	—
							3.155 671, δ–1370 ppm	Rh(s)	
Palladium	105	5/2	22.23	1.43	0.65^u	−1.2305	4.576	$K_2[PdCl_6]$ aq	135
Silver	(107)	1/2	51.82	0.198	—	−1.0878	4.047 898	Ag^+ ∞ aq	—
	109	1/2	48.18	0.280	—	−1.2519	4.653 623		—
Cadmium	(111)	1/2	12.75	7.01	—	−5.7046	21.201 877	$Cd(ClO_4)_2$, 0.1 *M* aq	—
							22.193 173, δ 641 ppm	$CdMe_2(1)$	—
	113	1/2	12.26	7.69	—	−5.9609	22.178 946		—
Indium	(113)	9/2	4.28	85.3	0.78^u	5.8845	21.996	In^{3+} aq	45
	115	9/2	95.72	1.92×10^3	0.86	5.8971	22.043		55
Tin	(115)	1/2	0.35	0.707	—	−8.8014	32.718 780	$SnMe_4(1)$	—
	(117)	1/2	7.61	19.9	—	−9.589	35.632 295		—
	119	1/2	8.58	25.7	—	−10.0318	37.290 665		—
Antimony	121	5/2	57.25	5.30×10^2	-0.33^u	6.4442	24.088	$SbCl_6^-(MeCN)$	35
	(123)	7/2	42.75	1.13×10^2	−0.68	3.4904	13.047		63
Tellurium	(123)	1/2	0.87 [2×10^{13} y]	0.906	—	−7.0576	26.169 773	$TeMe_2(1)$	—
	125	1/2	6.99	12.8	—	−8.5087	31.549 802		—
Iodine	127	5/2	100	5.41×10^2	−0.79	5.3896	20.146	I^- aq	200
Xenon	129	1/2	26.44	32.4	—	−7.4521	27.856	$XeOF_4(1)$	—
	(131)	3/2	21.18	3.38	-0.12^u	2.2091	8.257		19
Cesium	133	7/2	100	2.75×10^2	-3×10^{-3}	3.5339	13.210	Cs^+ aq	1.2×10^{-3}

(continued)

NMR Properties of the Elements (*Continued*)

	A	I	N.A. (%) [$t_{1/2}$]	R^C	Q (10^{-28} m^2)	γ (10^7 rad s^{-1} T^{-1})	Ξ (MHz)	Reference standard	WF (10^{-59} m^4)
Barium	(135)	3/2	6.59	1.87	0.18	2.6755	10.001	Ba^{2+} aq	43
	137	3/2	11.32	4.50	0.28[u]	2.9930	11.188		105
Lanthanum	138	5	0.09	0.481	0.51	3.5575	13.296	La^{3+} aq	15
	139	7/2	99.91	3.43×10^2	0.22	3.8085	14.236		6.6
Cerium	139	3/2	[137 d]	—	( )	(±)2.906	10.862		
Praseodymium	141	5/2	100	1.62×10^3	-4.1×10^{-2}	7.765	29.025		0.54
Neodymium	143	7/2	12.17	2.43	−0.56	−1.474	5.510		43
	145	7/2	8.3	0.393	−0.25	−0.913	3.413		8.5
Promethium	147	7/2	[2.6 y]	—	0.67[u]	3.613	13.51		61
Samarium	147	7/2	14.97	1.28	−0.18[u]	−1.1124	4.158		4.4
	149	7/2	13.83	0.665	5.6×10^{-2}	−0.9175	3.430		0.43
Europium	151	5/2	47.82	4.64×10^2	1.14	6.5477	24.475		416
	153	5/2	52.18	45.7	2.9	2.9371	10.979		2690
Gadolinium	155	3/2	14.73	0.124	1.6[u]	−0.8273	3.092		3410
	157	3/2	15.68	0.292	1.92	−1.0792	4.034		4910
Terbium	159	3/2	100	3.94×10^2	1.34	6.4306	24.038		1610
Dysprosium	161	5/2	18.88	0.509	2.38	−0.9206	3.441		1810
	163	5/2	24.97	1.79	2.51	1.2750	4.766		2020
Holmium	165	7/2	100	1.16×10^3	3.49	5.710	21.344		1660
Erbium	167	7/2	22.94	0.665	2.83	−0.7752	2.898		1090
Thulium	169	1/2	100	2.89	—	−2.1376	7.990		—
Ytterbium	171	1/2	14.31	4.50	—	4.7348	17.699		—
	173	5/2	16.08	1.23	2.8	−1.3025	4.869		2510
Lutetium	175	7/2	97.41	1.73×10^2	3.46	3.0589	11.434		1630
	176	7	2.59	5.83	8.1	2.1662	8.097		1750
Hafnium	177	7/2	18.50	1.47	3.3	1.086	4.059		1480
	179	9/2	13.75	0.427	3.7	−0.6821	2.550		1000
Tantalum	181	7/2	99.988	2.13×10^2	3.3	3.2445	12.128	$TaCl_6^-$ (MeCN)	1480
Tungsten	183	1/2	14.28	6.08×10^{-2}	—	1.1283	4.218	WO_4^{2-} aq	—

Rhenium	(185)	5/2	37.07	2.92×10^2	2.4^u	6.1057	22.823	$[PPh_4][ReO_4]$ (acetone)	1550
	187	5/2	62.93	5.11×10^2	2.2^u	6.1682	23.057		1590
Osmium	187	1/2	1.64	1.15×10^{-3}	—	0.6193	2.282 343	$OsO_4(CDCl_3)$	—
	(189)	3/2	16.1	2.23	0.91	2.1072	7.765 789		1100
Iridium	(191)	3/2	37.3	5.61×10^{-2}	0.86	0.4665	1.744		986
	193	3/2	62.7	0.122	0.78	0.5080	1.899		810
Platinum	195	1/2	33.8	19.9	—	5.8383	21.4*		—
							21.496 770, δ 4522 ppm	$Na_2[PtCl_6](D_2O)$	
Gold	197	3/2	100	0.153	0.55	0.4692	1.754		403
Mercury	199	1/2	16.84	5.68	—	4.8458	17.870 535	$Hg(ClO_4)_2$ (0.1 *M* in 0.1 *M* $HClO_4$)	
	(201)	3/2	13.22	1.12	0.42	−1.7888	6.612		235
Thallium	(203)	1/2	29.50	3.28×10^2	—	15.5394	57.072 641	Tl^+ aq	—
	205	1/2	70.50	8.07×10^2	—	15.6922	57.633 833		—
Lead	207	1/2	22.6	11.9	—	5.6264	20.920 597	$PbMe_4(1)$	—
Bismuth	209	9/2	100	8.19×10^2	−0.37	4.3752	16.355	$Bi(NO_3)_3$ (conc. HNO_3)	10
Polonium	209	1/2	[102 y]	—	—	7.4	28		—
Actinium	227	3/2	[21.8 y]	—	1.7	3.5	13.1		3850
Thorium	229	5/2	[7300 y]	—	4.4	0.40	1.5		6200
Protactinium	231	3/2	[3.3×10^4 y]	—	( )	3.21	12.0		( )
Uranium	235	7/2	0.72 [7×10^8 y]	5.4×10^{-3}	4.55	−0.4926	1.841	$UF_6(1)$, 380 K	2820
Neptunium	237	5/2	[2.1×10^6 y]	—	4.2	3.1	11.25		5600
Plutonium	239	1/2	[2.4×10^4 y]	—	—	0.972	3.63		—
Americium	241	5/2	[433 y]	—	4.9	1.54	5.76		7700
	243	5/2	[7370 y]	—	4.9	1.54	5.76		7700
Curium	247	9/2	[1.6×10^7 y]	—	( )	0.20	0.75		( )

[a] *A* is the mass number *I* is the nuclear spin quantum number R^C is the receptivity relative to that of ^{13}C *Q* is the nuclear electric quadrupole moment, with superscript *u*, it is uncorrected for Sternheimer or other polarization γ is the magnetogyric ratio Ξ is the resonance frequency of the standard substance in a magnetic field in which $SiMe_4$ (TMS) protons resonate at precisely 100 MHz δ is the chemical shift, the reference standard is that for $\delta = 0$, unless another value for δ is given WF is a width factor for quadrupolar nuclei the larger is WF, the broader the line, according to the values of *I* and *Q* The origin and derivation of the values of R^C, *Q*, γ, Ξ, and WF are explained in the Notes on pp 623–624

[b] *, rounded values (cf pp 533, 623)

INDEX

Because of the systematic organization of this book and its comprehensive nature, this index does not in general contain information that may be obtained from the table of contents. Thus a reader looking for general information on spin–spin coupling should refer to Chapter 4. Coupling constants involving particular nuclei, and their discussion in terms of molecular structure and theory may be found in the appropriate section(s) of the chapter(s) concerned with those nuclei. Similarly, many other topics are treated in some depth in the introductory Chapters 2–5, and examples and further discussion recur in Chapters 6–21, which follow the order of the Periodic Table.

Symbols and abbreviations are given on pp. 615–617, units and physical constants on pp. 619–621, and nuclear properties on pp. 623–629.

Kolibri 360 GmbH
Ferdinand-Jühlke-Straße 7
99095 Erfurt, Deutschland
produktsicherheit@kolibri360.de